Mechanics of Fluids

Providing a modern mathematical approach to classical fluid mechanics, this textbook presents an accessible and rigorous introduction to the field, with a strong emphasis on mathematical exposition.

It includes:

- A consistent treatment of a broad range of fluid mechanics topics, including vortical, potential, compressible, viscous, unstable, and turbulent flows
- Enhanced coverage of geometry, coordinate transformations, kinematics, thermodynamics, heat transfer, and nonlinear dynamics, to round out student understanding
- Robust emphasis on theoretical fundamentals and rigorous mathematical exposition, enabling students to gain confidence and develop a solid framework for further study
- 180 end-of-chapter problems, with full solutions and sample course syllabi available for instructors

With sufficient coverage for a one- or two-semester sequence, this textbook provides an ideal flexible teaching pathway for graduate students in aerospace, mechanical, chemical, and civil engineering, and applied mathematics.

Joseph M. Powers is a professor of Aerospace and Mechanical Engineering at the University of Notre Dame. His research uses computational science to elucidate the dynamics of high-speed reactive fluids as it applies to verification and validation of multiscale systems. He is the Editor-in-Chief of the *Journal of Propulsion and Power* and has previously published *Mathematical Methods in Engineering* (2015) and *Combustion Thermodynamics and Dynamics* (2016).

"An excellent first-level graduate textbook on fundamentals of fluid mechanics. By starting the book from the very basic vector notation, Professor Powers has made the book accessible to a large number of students who need to strengthen their mathematical background as well. This book will take the students all the way through rigorous understanding of hydrodynamic instabilities and turbulence. This is an excellent comprehensive book."

Bala Balachandar, *University of Florida*

"A rigorous mathematical treatise on the mechanics of fluids, in the spirit of Batchelor and Truesdell, something rarely seen today, and an exceptional counterpart to the many ad hoc books on this subject. For well-prepared students, this is a deeply technical introduction to this centrally important subject of physics and engineering."

Werner J. A. Dahm, *Arizona State University*

"A beautiful book on fluid mechanics: clear, insightful, comprehensive, rigorous, and detailed, with no stones unturned in derivations. This book will be a classic, one that I will often refer to when I need clarity and precision."

Tom Shih, *Purdue University*

"An enlightening 21st-century textbook in fluid mechanics which captures all the essence from the fundamentals of mechanics to the application of fluid dynamics. It comprehensively describes the intricate relationship between mathematics of statistical mechanics and physical observations of Newtonian fluids. This is a unique book which seamlessly relates 20th-century analytical mathematics-based fluid dynamics to 21st-century physics - and CFD-based fluid dynamics. I strongly recommend this book for an advanced undergraduate, or an introductory graduate-level, fluid dynamics course."

Chelakara S. Subramanian, *Florida Institute of Technology*

"This book achieves a rare combination of accessibility and mathematical rigor. It can provide a point of entry into contemporary fluid dynamics for the beginning graduate student while revealing fresh aspects of the subject to the seasoned researcher. I look forward to teaching from it."

William Eric Uspal, *University of Hawai'i at Mānoa*

Mechanics of Fluids

Joseph M. Powers
University of Notre Dame, Indiana

Shaftesbury Road, Cambridge CB2 8EA, United Kingdom

One Liberty Plaza, 20th Floor, New York, NY 10006, USA

477 Williamstown Road, Port Melbourne, VIC 3207, Australia

314–321, 3rd Floor, Plot 3, Splendor Forum, Jasola District Centre, New Delhi – 110025, India

103 Penang Road, #05-06/07, Visioncrest Commercial, Singapore 238467

Cambridge University Press is part of Cambridge University Press & Assessment, a department of the University of Cambridge.

We share the University's mission to contribute to society through the pursuit of education, learning and research at the highest international levels of excellence.

www.cambridge.org
Information on this title: www.cambridge.org/highereducation/isbn/9781316515693

DOI: 10.1017/9781009026307

© Joseph M. Powers 2024

This publication is in copyright. Subject to statutory exception and to the provisions of relevant collective licensing agreements, no reproduction of any part may take place without the written permission of Cambridge University Press & Assessment.

First published 2024

A catalogue record for this publication is available from the British Library.

A Cataloging-in-Publication data record for this book is available from the Library of Congress.

ISBN 978-1-316-51569-3 Hardback

Additional resources for this publication at www.cambridge.org/powers.

Cambridge University Press & Assessment has no responsibility for the persistence or accuracy of URLs for external or third-party internet websites referred to in this publication and does not guarantee that any content on such websites is, or will remain, accurate or appropriate.

Contents

Preface	page xv

Part I Continuum Equations of Fluid Mechanics

1 Introduction — 3
- 1.1 Mechanics — 4
- 1.2 Rational Continuum Mechanics — 5
  - 1.2.1 Notions from Newtonian Mechanics — 6
  - 1.2.2 Continuum Fields — 7
  - 1.2.3 Scalars, Vectors, and Tensors — 8
- 1.3 Molecular Limits of Continuum Theory — 9
- Summary — 13
- Problems — 13
- Further Reading — 14
- References — 14

2 Geometry — 16
- 2.1 Scalars, Vectors, and Tensors — 17
  - 2.1.1 Gibbs and Cartesian Index Notation — 17
  - 2.1.2 Rotation of Axes — 18
  - 2.1.3 Scalars — 24
  - 2.1.4 Vectors — 24
  - 2.1.5 Tensors — 27
- 2.2 Solution of Linear Algebraic Equations — 36
- 2.3 Eigenvalues, Eigenvectors, and Tensor Invariants — 38
- 2.4 Grad, Div, Curl, etc. — 46
  - 2.4.1 Gradient — 47
  - 2.4.2 Divergence — 48
  - 2.4.3 Curl — 48
  - 2.4.4 Laplacian — 49
  - 2.4.5 Biharmonic Operator — 49
  - 2.4.6 Time Derivative — 49
  - 2.4.7 Relevant Theorems — 50
- 2.5 General Coordinate Transformations — 53
- 2.6 Cylindrical Coordinates — 61

		2.6.1 Centripetal and Coriolis Accelerations	62
		2.6.2 Grad and Div	64
	2.7	Spherical Coordinates	70
	2.8	Quadratic Forms	72
Summary			75
Problems			76
Further Reading			83
References			83
3	**Kinematics**		**84**
	3.1	Motivating Problem	84
	3.2	Lagrangian Description	86
	3.3	Eulerian Description	87
		3.3.1 One-Dimensional	87
		3.3.2 Multi-Dimensional	88
	3.4	Material Derivative	89
		3.4.1 Simple Approach	89
		3.4.2 Coordinate Transformation Approach	90
	3.5	Streamlines	92
	3.6	Pathlines	93
	3.7	Streaklines	94
	3.8	Kinematic Decomposition of Motion	95
		3.8.1 Translation	97
		3.8.2 Rigid Body Rotation and Straining	97
		3.8.3 Principal Axes of Strain Rate	100
		3.8.4 Extensional Strain Rate Quadric	102
		3.8.5 Singular Value Decomposition	104
		3.8.6 Polar Decompositions	107
		3.8.7 Decomposition of a Material Line Element	110
	3.9	Convected Derivative	112
	3.10	Expansion Rate	115
		3.10.1 Material Volume	115
		3.10.2 Arbitrary Volume	116
	3.11	Invariants of the Strain Rate Tensor	118
	3.12	Two-Dimensional Kinematics	118
		3.12.1 General Two-Dimensional Flows	118
		3.12.2 Relative Motion Along 1 and 2 Axes	120
		3.12.3 Uniform Flow	122
		3.12.4 Pure Rigid Body Rotation	123
		3.12.5 Pure Extensional Motion (a Compressible Flow)	124
		3.12.6 Pure Shear Straining	124
		3.12.7 Ideal Corner Flow	125
		3.12.8 Couette Flow: Shear + Rotation	126

		3.12.9 Ideal Irrotational Vortex: Extension + Shear	127
	3.13	Three-Dimensional Kinematics: Summary	128
	3.14	Kinematics as a Dynamical System	129
	Summary		142
	Problems		143
	Further Reading		146
	References		146

4 Evolution Axioms 148

4.1	Mass		149
4.2	Linear Momenta		153
	4.2.1	Preliminary Vector Form	154
	4.2.2	Surface Forces	154
	4.2.3	Final Tensor Form	159
4.3	Angular Momenta		164
	4.3.1	General Relation for Polar Fluids	164
	4.3.2	Symmetry of the Stress Tensor for Nonpolar Fluids	166
	4.3.3	Cauchy's Stress Quadric	167
	4.3.4	Lamé Stress Ellipsoid	169
4.4	Energy		171
	4.4.1	Total Energy	171
	4.4.2	Work	172
	4.4.3	Heat Transfer	172
	4.4.4	Conservative Form	173
	4.4.5	Secondary Forms	174
4.5	Entropy Inequality		179
4.6	Integral Forms		182
	4.6.1	Mass	182
	4.6.2	Linear Momenta	184
	4.6.3	Energy	185
	4.6.4	General Expression	185
4.7	Summary of Axioms		185
	4.7.1	Conservative Form	186
	4.7.2	Nonconservative Form	187
	4.7.3	Physical Interpretations	188
4.8	Incompleteness of the Axioms		189
	Summary		189
	Problems		190
	References		192

5 Constitutive Equations 194

5.1	Frame Indifference	196

Contents

- 5.2 Second Law Restrictions and Onsager Relations — 197
  - 5.2.1 Weak Form of the Clausius–Duhem Inequality — 197
  - 5.2.2 Strong Form of the Clausius–Duhem Inequality — 200
- 5.3 Fourier's Law — 201
- 5.4 Stress–Strain Rate Relation for a Newtonian Fluid — 206
  - 5.4.1 Motivating Experiments — 207
  - 5.4.2 Analysis for an Isotropic Newtonian Fluid — 208
  - 5.4.3 Stokes' Assumption — 218
  - 5.4.4 Second Law Restrictions — 219
- 5.5 Irreversibility Production Rate — 223
- 5.6 Thermodynamic Equations of State — 223
- Summary — 226
- Problems — 226
- References — 228

6 Governing Equations: Summary and Special Cases — 229

- 6.1 Boundary and Interface Conditions — 229
- 6.2 Compressible Navier–Stokes Equations — 230
  - 6.2.1 Conservative Form — 230
  - 6.2.2 Nonconservative Form — 231
- 6.3 Fluid Statics — 234
- 6.4 Incompressible Navier–Stokes Equations — 236
  - 6.4.1 Mass — 236
  - 6.4.2 Linear Momenta — 237
  - 6.4.3 Energy — 237
  - 6.4.4 Poisson Equation for Pressure — 238
  - 6.4.5 Summary — 239
  - 6.4.6 Limits for One-Dimensional Diffusion — 240
- 6.5 Boussinesq Approximation — 240
  - 6.5.1 Cartesian Index Form — 243
  - 6.5.2 Gibbs Form — 243
- 6.6 Euler Equations — 243
  - 6.6.1 Conservative Form — 244
  - 6.6.2 Nonconservative Form — 244
  - 6.6.3 Alternate Forms of the Energy Equation — 245
- 6.7 Dimensionless Compressible Navier–Stokes Equations — 246
  - 6.7.1 Mass — 248
  - 6.7.2 Linear Momenta — 248
  - 6.7.3 Energy — 249
  - 6.7.4 Thermal State Equation — 250
  - 6.7.5 Caloric State Equation — 250
  - 6.7.6 Upstream Conditions — 250
  - 6.7.7 Reduction in Parameters — 250

		6.7.8 Alternate Scaling	251
	6.8	First Integrals of Linear Momenta	251
		6.8.1 Bernoulli Equation	251
		6.8.2 Crocco's Equation	255
	Summary		259
	Problems		259
	Further Reading		260
	References		260

Part II Solutions in Various Flow Regimes

7 Vortical Flow — 265

7.1	Streamlines and Vortex Lines	266
7.2	Incompressible Navier–Stokes Equations in Cylindrical Coordinates	268
7.3	Ideal Rotational Vortex	269
7.4	Ideal Irrotational Vortex	272
7.5	Helmholtz Vorticity Transport Equation	274
	7.5.1 General Development	274
	7.5.2 Bending and Stretching of Vortex Tubes	276
	7.5.3 Baroclinic (Non-Barotropic) Effects	277
	7.5.4 Incompressible, Conservative Body Force Limit	277
7.6	Kelvin's Circulation Theorem	281
7.7	Two-Dimensional Potential Flow of Ideal Point Vortices	282
	7.7.1 Two interacting Ideal Vortices	283
	7.7.2 Image Vortex	283
	7.7.3 Vortex Sheets	284
	7.7.4 Potential of an Ideal Irrotational Vortex	285
	7.7.5 Interaction of Multiple Vortices	286
	7.7.6 Pressure Field	289
Summary		291
Problems		292
Further Reading		294
References		295

8 Potential Flow — 296

8.1	Stream Functions and Velocity Potentials	297
8.2	Mathematics of Complex Variables	299
	8.2.1 Euler's Formula	300
	8.2.2 Polar and Cartesian Representations	300
	8.2.3 Cauchy–Riemann Equations	303
8.3	Elementary Complex Potentials	306
	8.3.1 Uniform Flow	306
	8.3.2 Sources and Sinks	307

		8.3.3 Point Vortices	309
		8.3.4 Superposition of Sources	309
		8.3.5 Flow in Corners	311
		8.3.6 Doublets	318
		8.3.7 Quadrupoles	320
		8.3.8 Rankine Half Body	320
		8.3.9 Flow over a Cylinder	321
	8.4	Forces Induced by Potential Flow	325
		8.4.1 Contour Integrals	325
		8.4.2 Laurent Series	327
		8.4.3 Pressure Distribution for Steady Flow	328
		8.4.4 Blasius Force Theorem	328
		8.4.5 Kutta–Zhukovsky Lift Theorem	331
	Summary		334
	Problems		334
	Further Reading		337
	References		337
9	**One-Dimensional Compressible Flow**		**338**
	9.1	Thermodynamics of Compressible Fluids	339
		9.1.1 Maxwell Relation	340
		9.1.2 Internal Energy from Thermal Equation of State	340
		9.1.3 Sound Speed	346
	9.2	Generalized One-Dimensional Flow	349
		9.2.1 Mass	350
		9.2.2 Linear Momentum	351
		9.2.3 Energy	353
		9.2.4 Summary of Equations	356
		9.2.5 Dynamical System Form	359
	9.3	Isentropic Flow with Area Change	361
		9.3.1 Isentropic Mach Number Relations	361
		9.3.2 Sonic Properties	365
		9.3.3 Effect of Area Change	365
		9.3.4 Choking	367
	9.4	Fanno Flow	369
	9.5	Rayleigh Flow	372
	9.6	Normal Shock Waves	376
		9.6.1 Analysis for a General Flux-Conservative System	377
		9.6.2 Rankine–Hugoniot Equations	379
		9.6.3 Rayleigh Line	380
		9.6.4 Hugoniot Curve	380
		9.6.5 Solution Procedure for General Equations of State	381
		9.6.6 Calorically Perfect Ideal Gas Solutions	381

		9.6.7 Weak Shock Limit	387	

- 9.7 Contact Discontinuities — 389
- 9.8 Throttling Device — 390
- 9.9 Flow with Area Change and Normal Shocks — 392
  - 9.9.1 Converging Nozzle — 392
  - 9.9.2 Converging–Diverging Nozzle — 393
- 9.10 Acoustics — 394
- 9.11 Method of Characteristics — 401
  - 9.11.1 Inviscid One-Dimensional Equations — 403
  - 9.11.2 Homeoentropic Flow of a Calorically Perfect Ideal Gas — 407
  - 9.11.3 Simple Waves — 414
  - 9.11.4 Centered Rarefaction — 417
  - 9.11.5 Unsteady Inviscid Bateman–Burgers Shock Formation — 419
- 9.12 Viscous Shock Waves — 425
  - 9.12.1 Unsteady Viscous Bateman–Burgers Shock Formation — 425
  - 9.12.2 Steady Viscous Bateman–Burgers Shocks — 427
  - 9.12.3 Steady Navier–Stokes Shocks — 431
- 9.13 Discontinuous Rarefactions for Nonideal Gases — 437
- 9.14 Taylor–Sedov Blast Waves — 441
  - 9.14.1 Governing Equations — 442
  - 9.14.2 Similarity Transformation — 443
  - 9.14.3 Transformed Equations — 445
  - 9.14.4 Dimensionless Equations — 447
  - 9.14.5 Reduction to Nonautonomous Form — 448
  - 9.14.6 Numerical Solution — 449
  - 9.14.7 Contrast with Acoustic Limit — 453
- Summary — 456
- Problems — 456
- Further Reading — 458
- References — 459

10 One-Dimensional Viscous Flow — 461

- 10.1 Flow with No Effects of Inertia — 462
  - 10.1.1 Incompressible Poiseuille Flow in a Slot — 462
  - 10.1.2 Incompressible Couette Flow — 470
  - 10.1.3 Incompressible Couette Flow with Pressure Gradient — 473
  - 10.1.4 Compressible Couette Flow — 476
- 10.2 Flow with Effects of Inertia — 480
  - 10.2.1 Stokes' First Problem — 480
  - 10.2.2 General Solution — 494
  - 10.2.3 Momentum Pulse — 494
  - 10.2.4 Unsteady Couette Flow — 495
  - 10.2.5 Stokes' Second Problem — 499

		10.2.6 Decay of an Ideal Vortex	501
	10.3	Non-Newtonian Flow	505
		10.3.1 Strain Rate Dependent Viscosity	505
		10.3.2 Viscoelastic Flow	510
	Summary		516
	Problems		516
	Further Reading		518
	References		519
11	**Multi-Dimensional Viscous Flow**		520
	11.1	Flow with No Effects of Inertia	521
		11.1.1 Stokes Equations	521
		11.1.2 Generalized Poiseuille Flow	533
		11.1.3 Lubrication Theory	539
	11.2	Flow with Effects of Inertia	542
		11.2.1 Blasius Boundary Layer	542
		11.2.2 Falkner–Skan Flow	557
		11.2.3 Jets	561
		11.2.4 Shear Layers	564
		11.2.5 von Kármán's Viscous Pump	565
		11.2.6 Natural Convection	569
		11.2.7 Compressible Boundary Layer	573
	Summary		576
	Problems		576
	Further Reading		579
	References		579
12	**Linearly Unstable Flow**		580
	12.1	Motivating Problem from Dynamics	581
	12.2	Planar Two-Dimensional Inviscid Instabilities	584
		12.2.1 Stationary Fluids with No Surface Tension: Rayleigh–Taylor Instability	591
		12.2.2 Stationary Fluids with Surface Tension	592
		12.2.3 Surface Waves	593
		12.2.4 Inviscid Shear Layer: Kelvin–Helmholtz Instability	594
		12.2.5 Kelvin–Helmholtz Instability with Surface Tension	594
	12.3	Parallel Viscous Flow Instability	595
	12.4	Thermal Convection: Rayleigh–Bénard Instability	600
	Summary		610
	Problems		610
	Further Reading		611
	References		611

13 Nonlinear Dynamics for Fluid Flow — 613

- 13.1 Traditional Fourier Series Expansion — 615
- 13.2 Galerkin Projection to a Low-order Dynamical System: Bateman–Burgers — 616
- 13.3 Landau Equation — 622
- 13.4 Lorenz Equations — 624
  - 13.4.1 Derivation from Boussinesq Approximation — 624
  - 13.4.2 Equilibrium — 627
  - 13.4.3 Linear Stability — 629
  - 13.4.4 Transition from Order to Chaos to Order — 631
- Summary — 639
- Problems — 639
- Further Reading — 640
- References — 641

14 Turbulent Flow — 642

- 14.1 Scaling Analysis — 644
  - 14.1.1 Mechanical Energy Evolution — 644
  - 14.1.2 Approximations at the Small Scale — 646
  - 14.1.3 Kolmogorov Microscales — 647
  - 14.1.4 Connection to the Mean Free Path Scale — 648
  - 14.1.5 Turbulent Kinetic Energy Cascade — 649
- 14.2 Reynolds-Averaged Navier–Stokes Equations — 650
  - 14.2.1 Time-Averaging — 650
  - 14.2.2 Averaged Incompressible Equations — 652
  - 14.2.3 Closure Problem — 654
- 14.3 Large Eddy Simulation — 655
- 14.4 Direct Numerical Simulation of a Rayleigh–Bénard Flow — 656
- Summary — 668
- Problems — 668
- Further Reading — 669
- References — 670

Bibliography — 672
Index — 685

Preface

This book considers the mechanics of fluids with a focus on topics seen in entry level graduate courses in aerospace and mechanical engineering. As reflected in its title and structure, it is first a presentation of mechanics, followed by application to fluids. It provides a rigorous presentation of standard topics with a more complete exposition of underlying mathematical details than is typically found. In an era in which students perhaps too hastily turn to experimental or computational methods to understand fluid behavior, this book provides a detailed presentation of the underlying theoretical foundations in a fashion designed to provide missing links in analysis that students often need to gain confidence in their understanding. An additional reason to include details is so that results may be reproduced by others, an important feature that should distinguish deterministic science from less quantifiable disciplines. The bulk of each chapter has a standard presentation; however, many chapters are augmented with material not always found in common texts. Some topics that receive enhanced emphasis include geometry, coordinate transformations, kinematics, thermodynamics, heat transfer, and nonlinear dynamics.

The book is built on lecture notes for two courses developed over three decades in the Department of Aerospace and Mechanical Engineering of the University of Notre Dame. The first is AME 60635, Intermediate Fluid Mechanics, and the second is AME 70731, Viscous Flow Theory. Additional topics have been included, and so there should be more than enough material available for a two-course sequence in introductory graduate fluid mechanics. Many of the later chapters are self-contained, and some can be omitted to suit instructor needs. The notes from which this book was drawn were themselves initially guided by the fine text of Panton (2013);[1] the reader will notice some similarities in choice of notation and ordering of a few topics. The influence of many other expositions on this subject, listed in an extensive bibliography, is evident as well.

The book is directed towards beginning graduate students and advanced engineering undergraduates. They have typically completed at least one undergraduate fluids course as well as courses in thermodynamics, linear algebra, vector calculus, and differential equations. Additionally, they have experience with basic numerical methods and modern software tools for solving such problems as root-finding, matrix inversion, determination of eigenvalues and eigenvectors, integration of nonlinear systems of ordinary differential equations, and some partial differential equations. A basic knowledge of such material is both necessary and sufficient preparation for the topics presented here.

---

[1] Panton, R. L. (2013). *Incompressible Flow*, 4th ed. New York: John Wiley.

Most of these students will later specialize in either experimental or computational fluid mechanics and will take additional specialized course work to these ends. All can benefit from a thorough preparation in theoretical foundations. As such, this book goes farther than most to bolster the connection between basic mathematical concepts relevant to the mechanics of fluids in an engineering context. While the material is relevant to experimental and computational fluid mechanics, it has minimal discussion either of underlying experiments or computations. Both of these vast subjects are well treated in other texts. That said, most of the topics developed here have clear relevance to problems in nature; however, in a few portions of the text, some unusual limits are considered to allow a more digestible exposition. These may involve specially prescribed flow fields, limits in which some physics are neglected, or limits in which competing mechanisms are unusually scaled so as to be in balance. This is often manifested by posing problems in which fluid properties such as density or viscosity are either zero or unity. A special emphasis is placed on problems that bolster the students' often tenuous confidence in using vector calculus, which is the most efficient language to describe the mechanics of fluids.

The book provides a survey of continuum fluid mechanics. Part I gives an extensive development of the compressible Navier–Stokes equations. It includes a review and exposition of the essential mathematics of differential geometry, kinematics, evolution axioms, constitutive equations, and a delineation of many special limits of the governing equations. Part II focuses on their solution in various limits: vortical, potential, compressible, viscous, unstable, chaotic, and turbulent flows. Most chapters contain example problems; some are mathematically motivated, and others are focused on quantitative problems involving fluid physics. A course that is more oriented towards development of the equations of continuum fluid mechanics may concentrate on Part I. Alternatively, one could give the briefest of introductions to governing equations and move straight to any of the chapters of Part II, which generally are self-supporting.

The expansiveness of these topics is such that many of them are only lightly treated; the reader should turn to more specialized books for additional detail. The emphasis here is on fluid physics and the mathematics necessary to efficiently describe the physics. Each chapter is concluded with exercises appropriate for homework. A detailed solution manual is available for instructors. Some of the problems require numerical methods that are routine for modern and widely available software tools; background for such tools and methods may be found in other sources. Specific hallmarks found throughout the book include (1) consideration of complete thermo-fluid systems so as to enable determination of velocity and temperature fields in compressible, viscous flows, (2) attention to the formalities of coordinate and similarity transformations, and (3) presentation of much classical material as well as a few novel or neglected topics selected to illustrate analysis of fluid physics. Some important topics are considered only briefly, for example (1) non-Newtonian flow, (2) stability, and (3) turbulence. The bulk of the book is devoted to fluids problems that are well described by a set of deterministic model equations with solutions that are well-behaved and amenable to causal inference. In a few instances, these nonlinear deterministic equations are pushed into regimes where the solutions acquire a more chaotic and random behavior. But we do not explicitly model the stochastic nature of fluid behavior or the effect of stochastic uncertainties in constitutive models; nor do we consider non-axiomatic frameworks, such as given by some data-driven modeling approaches that employ machine learning and artificial neural networks. While some computational results are presented, there is no treatment of the methods of computational fluid dynamics. The book provides the foundation for later courses that address these and several more advanced topics that are not typically considered in introductory graduate fluids courses.

The book is intended to be used as an instructional tool for students and advanced professionals who need to understand the fundamental mechanics of fluids. While in places, it gives a modern treatment of old subjects drawing upon recent scholarship, most of its topic matter is correctly described as "classical." As such, the extensive bibliography focuses on scholarly books, both historical and modern; for selected specialized topics, the underlying journal literature is drawn upon.

The author's gratitude is due to many, including former teachers, colleagues who commented on various drafts, inspiring family members, as well as the University of Notre Dame for providing support to bring this work to completion. Of those many, I am especially grateful to my talented colleague at Notre Dame, Prof. Jonathan F. MacArt, who carefully reviewed and contributed to Chapter 14 and skillfully performed the numerical simulations of laminar and turbulent Rayleigh–Bénard convective flow that both illuminate the cover and close the book. That said, my deepest appreciation is given to those who motivated me to prepare this book: the dozens of undergraduate and graduate students of AME 60635 and 70731 who have pushed me for excellence as I have pushed them. Many have gone on to careers of distinction in fluid mechanics, and for what small part I played in that, I am proud. This book is written with the hope that it can reach others who will be similarly motivated to learn and apply this beautiful science.

Part I

Continuum Equations of Fluid Mechanics

1 Introduction

This book considers the mechanics of a *fluid*, defined as a material that continuously deforms under the influence of an applied shear stress, as depicted in Fig. 1.1. Here the fluid, initially at rest, lies between a stationary wall and a moving plate. Nearly all common fluids stick to solid surfaces. Thus, at the bottom, the fluid remains at rest; at the top, it moves with the velocity of the plate. The vectors indicate the fluid displacement, a distance that grows with time. At early time, the displacement profile varies nonlinearly with distance from the stationary surface. At later times, the displacement profile becomes linear. For nearly all common fluids, it is observed that a nonzero shear stress is required to maintain this motion. As configured, this fluid will never come to rest. Such a definition allows both liquids and gases to be considered fluids. In contrast, a solid will deform but relax to an equilibrium configuration when subjected to an applied shear stress.

Motion in response to *transverse* shear forces is fundamental to fluids and induces such behavior as fluid rotation in a long persisting *vortex* such as seen in weather patterns and aerodynamic applications. In such a swirling environment, fluid particles often veer far from neighboring fluid particles. In contrast, solid particles almost always retain the same particles as neighbors; except for rotation as a rigid body, there is no clear counterpart to a vortex in typical solids. Motion in response to *longitudinal* normal forces is also fundamental to fluids and can result in volumetric compression and expansion as well as acceleration in the direction of the net normal force. Solids respond to longitudinal normal forces in a similar manner; they may induce weaker volumetric compression and expansion and certainly acceleration in the direction of the net normal force.

We present an approach to *fluid mechanics* founded on the general principles of *rational continuum mechanics*. These general principles apply to all continuous materials: solids, liquids, and gases. There are many paths to understanding fluid mechanics, and good arguments can be

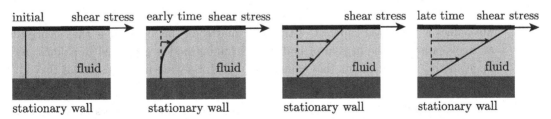

Figure 1.1 Diagram demonstrating the defining feature of a fluid: continuous deformation in response to an applied shear stress: snapshots of the fluid displacement profile at various times.

made for each. A typical first undergraduate class will combine a mix of basic equations, coupled with strong physical motivations, and allow the student to develop a knowledge that is of practical value, often driven by intuition. Such an approach works well within the confines of the intuition we develop in everyday life. It often fails when the engineer moves into unfamiliar territory. For example, lack of fundamental understanding of high Mach[1] number flows led to many aircraft and rocket failures in the 1950s. In such cases, a return to the formalism of a careful theory, one that clearly exposes the strengths and weaknesses of all assumptions, is invaluable in both understanding the true fluid physics, and applying that knowledge to engineering design.

Probably the most formal of approaches is that of the school of thought promoted by Truesdell,[2] who forcefully advocated for rational continuum mechanics. Truesdell developed a broadly based theory that encompassed all materials that could be regarded as continua, including solids, liquids, and gases, in the limit when the smallest volumes considered were sufficiently large so that the micro- and nanoscopic structure of these materials was unimportant. For fluids (both liquid and gas), such length scales are often at or below the order of microns, while for solids, the scales may be smaller, depending on the type of molecular structure. The difficulty of the Truesdellian approach is that it is burdened with a difficult notation and tends to become embroiled in proofs and philosophy, which while ultimately useful, can preclude learning basic fluid mechanics in the time scale of the typical student. It is possible, however, to give a discussion that respects the approach of Truesdell while also providing both technical rigor and accessibility. For example, Thorne and Blandford (2017) give a nuanced, detailed, and useful exposition of fluid mechanics in the context of physics, geometry, and experiment that complements well the formalism of rational continuum mechanics.

Here, we will attempt to steer between the fallible pragmatism of undergraduate fluid mechanics and the harsh formalism of the Truesdellian school. The presentation will pay due homage to rational continuum mechanics and will be geared towards a basic understanding of fluid behavior. We shall first spend some time carefully developing the governing equations for a compressible viscous fluid. We shall then study representative solutions of these equations in a wide variety of physically motivated limits in order to understand how the basic evolution principles of mass, linear momenta, angular momenta, and energy, coupled with constitutive relations, influence the behavior of fluids. In the end, it is hoped the reader will have an enhanced appreciation of the abilities and limitations of deterministic continuum mathematical physics to predict basic fluid behavior.

1.1 Mechanics

Mechanics is the broad superset of the topic matter of this book. It is the science that seeks an explanation for the motion of bodies based upon models grounded in axioms. Axioms, as in geometry, are statements that cannot be proved; they are useful insofar as they give rise to

---

[1] Ernst Mach, 1838–1926, Viennese physicist and philosopher who worked in optics, mechanics, and wave dynamics, and developed fundamental ideas of inertia.
[2] Clifford Ambrose Truesdell III, 1919–2000, American continuum mechanician and natural philosopher.

results that are consistent with empirical observation. A hallmark of science has been the struggle to identify the smallest set of axioms that are sufficient to describe our universe. When we find an axiom to be inconsistent with observation, it must be modified or eliminated. A familiar example of this is the Michelson[3]–Morley[4] experiment, which motivated Einstein[5] to modify the Newtonian[6] axioms of mass and mechanical energy into an axiom for mass-energy. There are many subsets of mechanics, for example statistical mechanics, relativistic mechanics, quantum mechanics, continuum mechanics, fluid mechanics, or solid mechanics. Each has its own set of axioms; often in certain physical limits, the axioms of one framework relax to those of another framework. For example, as the deviation of the velocity from the speed of light increases, the more robust axioms of Einstein's relativistic mechanics relax to those of Newton's mechanics.

1.2 Rational Continuum Mechanics

Newton introduces the concept of *rational* mechanics in the preface to the original 1687 edition of his *Principia* to distinguish it from both geometry and other types of mechanics that were common in his era. There Newton (1999, p. 382, originally presented 1687) states:

> ...*geometry* is commonly used in reference to magnitude, and *mechanics* in reference to motion. In this sense *rational mechanics* will be the science, expressed in exact proportions and demonstrations, of the motions that result from any forces whatever and of the forces that are required for any motions whatever.

Early mechanicians, such as Newton, dealt primarily with point masses and finite collections of distinct particles. Such systems are the easiest to study, and it makes more sense to grasp the simple before the complex. The discipline that considers such systems is often referred to as *classical mechanics*. Mathematically, such systems are generally characterized by a finite number of ordinary differential equations, and the properties of each particle (e.g. position, velocity) are taken to be functions of time only.

Continuum mechanics, generally attributed to Euler,[7] considers instead an infinite number of particles, which is in fact easier to analyze than a large finite number of particles. In continuum mechanics, every physical property (e.g. velocity, density, pressure) is taken to be a function of both time and space. Infinitesimal property variation from point to point in space is permitted. While variations are generally continuous, finite numbers of surfaces of discontinuous property variation are allowed. This models, for example, the contact between one continuous body and another or a shock wave within a inviscid fluid. Point discontinuities are not allowed, however. Finite valued material properties are required. Mathematically, such systems are characterized

---

[3] Albert Abraham Michelson, 1852–1931, Prussian-born American physicist.
[4] Edward Williams Morley, 1838–1923, American physical chemist.
[5] Albert Einstein, 1879–1955, German and later American physicist who developed the theory of relativity and made fundamental contributions to quantum mechanics and Brownian motion in fluid mechanics.
[6] Sir Isaac Newton, 1642–1727, English physicist, mathematician, and chief figure of the scientific revolution. Developed calculus, theories of gravitation, motion of bodies, and optics.
[7] Leonhard Euler, 1707–1783, Swiss-born mathematician and physicist.

by a finite number of partial differential equations in which the properties of the continuum material are functions of both space and time. It is possible to show that a partial differential equation can be thought of as an infinite number of ordinary differential equations, so this is consistent with our model of a continuum as an infinite number of particles.

In Truesdell's exposition on continuum mechanics, he suggests the following hierarchy:

- *bodies* exist,
- bodies are assigned to *place*,
- *geometry* is the theory of place,
- change of place in *time* is the *motion* of the body,
- a description of the motion of a body is *kinematics*,
- motion is the consequence of *forces*,
- the study of forces on a body is *dynamics*, sometimes called *kinetics*.

We will adapt this hierarchy in our exposition.

The modifier "rational" was applied by Truesdell to continuum mechanics to distinguish the formal approach advocated by his school, from less formal, though mainly not irrational, approaches to continuum mechanics. Rational continuum mechanics is developed with tools similar to those that Euclid[8] used for his geometry and Newton for his physics: formal definitions, axioms, and theorems, all accompanied by careful language and proofs. One of the hallmarks of rational continuum mechanics is a distinction between material-independent axioms, such as mass, momenta, and energy principles, from material-specific relations such as the ideal gas law. This book will recognize such distinctions, all the while following a less formal, albeit still rigorous, approach. The extensive literature associated with rational continuum mechanics considers a broad range of topics, and nuances of some aspects of its axioms, not relevant for this book, are not universally accepted; see for example Woods (1982) or Müller (2007) and references within.

1.2.1 Notions from Newtonian Mechanics

The following are useful notions from Newtonian mechanics. Here we use Newtonian to distinguish our mechanics from Einsteinian relativistic mechanics. Newton himself did not study continuum mechanics; however, notions from his studies of the mechanics of discrete sets of point masses extend to the mechanics of continua.

Space is three-dimensional and independent of time. An *inertial frame* is a reference frame in which the laws of physics are invariant; further, a body in an inertial frame with zero net force acting upon it does not accelerate. A *Galilean*[9] *transformation* specifies how to transform from one inertial frame to another inertial frame moving at constant velocity relative to the original frame. If a second inertial frame has constant velocity $\mathbf{v}_o = u_o\mathbf{i} + v_o\mathbf{j} + w_o\mathbf{k}$ relative to the original inertial frame, the Galilean transformation $(x, y, z, t) \to (x', y', z', t')$ is as follows:

---

[8] Euclid, Greek geometer of profound influence.
[9] Galileo Galilei, 1564–1642, Pisa-born Italian astronomer, physicist, and developer of experimental methods, first employed a pendulum to keep time, builder and user of telescopes used to validate the Copernican view of the universe, developer of the principle of inertia and relative motion.

$$x' = x - u_o t, \quad y' = y - v_o t, \quad z' = z - w_o t, \quad t' = t. \tag{1.1}$$

This must be accompanied with a transformation of the velocities:

$$u' = u - u_o, \quad v' = v - v_o, \quad w' = w - w_o. \tag{1.2}$$

1.2.2 Continuum Fields

In contrast to a single particle or finite set of particles for which time is the only independent variable, a fluid is modeled as a *continuum field* with properties that depend on both time and space. The notion of a continuum is rooted in Newtonian calculus; an example of a continuum from mathematics is the set of real numbers. A real number $x \in \mathbb{R}^1$ is a member of the set of real scalars, $\mathbb{R}^1$. It is often considered to reside on the x axis. The x axis may be finely partitioned. At the edges of a particular partition, x may have the respective values of x_n and x_{n+1}. The essence of the continuum assumption is that no matter how fine the partition, there exists an intermediate value $\tilde{x}$ with $x_n \leq \tilde{x} \leq x_{n+1}$. Here x is a geometrical property. Ordered pairs of real numbers $(x, y)^T \in \mathbb{R}^2$ reside in a plane, and ordered triples of real numbers $(x, y, z)^T \in \mathbb{R}^3$, reside in a volume. Here T is the transpose operator, employed so that the ordered pairs and triples are column vectors, by standard convention. We can extend these notions to fluid properties.

Density, ρ, is a material property of a fluid describing the concentration of its mass m within a volume V. Consider a sequence of shrinking volumes $V_1 > V_2 > V_3, \ldots, V_n$, each containing a sequence of shrinking masses, $m_1 > m_2 > m_3, \ldots, m_n$, sketched in Fig. 1.2. We expect the mass to be a function of the volume. Define the average density $\overline{\rho}_n$ as

$$\overline{\rho}_n = \frac{m_n}{V_n}. \tag{1.3}$$

In a true continuum, as we let $n \to \infty$, we could define the density at P to be

$$\rho(P) = \lim_{n \to \infty} \frac{m_n}{V_n}, \tag{1.4}$$

and importantly, expect the limiting value $\rho(P)$ to be finite and smoothly approaching its limiting value as $n \to \infty$. In this limit, we expect V_n to approach the infinitesimal volume dV surrounding the point P and m_n to approach the infinitesimal value dm. This gives

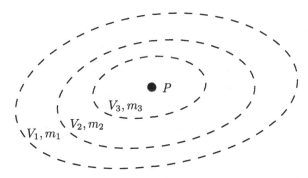

Figure 1.2 Diagram of sequence of volumes, each enclosing a respective mass, with the volume shrinking to a point P.

1 Introduction

$$\rho = \frac{dm}{dV}. \tag{1.5}$$

Every point P in the continuum would possess a set of spatial coordinates and a local value of ρ at each point. One can also allow for time variation so $\rho = \rho(x, y, z, t)$. Then with $dV = dx\, dy\, dz$, one could integrate Eq. (1.5) over a finite volume to get

$$m(t) = \int_{z_1}^{z_2} \int_{y_1}^{y_2} \int_{x_1}^{x_2} \rho(x, y, z, t)\, dx\, dy\, dz. \tag{1.6}$$

In this book, properties such as $\rho(x, y, z, t)$ will be assumed to exist and contain all the features of a mathematical continuum, except at jumps associated with material interfaces or shock waves. This approach has proven to work well as long as the volumes being considered are sufficiently large to contain many fluid molecules.

Density is not used in classical Newtonian mechanics, as that discipline only considers point masses. Continuum mechanics will treat macroscopic effects only and ignore individual molecular effects. For example, molecules bouncing off a wall exchange momentum with the wall and induce pressure. We could use Newtonian mechanics for each particle collision to calculate the net wall force. Instead our approach amounts to considering the *average* over space and time of the net effect of millions of collisions on a wall.

1.2.3 Scalars, Vectors, and Tensors

We briefly introduce here the notion of fields composed of what are known as *scalars*, *vectors*, and *tensors*. This important topic will be considered in more detail in Section 2.1. A diagram depicting the nature of scalars, vectors, and tensors is given in Fig. 1.3.

Density is an example of a scalar property. A scalar property associates a single number with each point in time and space. *Scalars possess magnitude but not direction.* We can think of this by writing the usual notation

$$\rho = \rho(x, y, z, t), \tag{1.7}$$

indicating that ρ has functional variation with position and time. Other properties are not scalar, but are vector properties. For example, the velocity vector

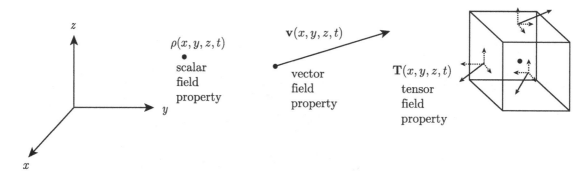

Figure 1.3 Diagram depicting the nature of scalars, vectors, and tensors.

$$\mathbf{v}(x,y,z,t) = u(x,y,z,t)\mathbf{i} + v(x,y,z,t)\mathbf{j} + w(x,y,z,t)\mathbf{k} = \begin{pmatrix} u(x,y,z,t) \\ v(x,y,z,t) \\ w(x,y,z,t) \end{pmatrix}, \quad (1.8)$$

associates three scalars u, v, w with each point in space and time. Here $\mathbf{i}$, $\mathbf{j}$, and $\mathbf{k}$ are the familiar set of orthonormal basis vectors associated with a Cartesian coordinate system. We will see that a vector can be characterized as a scalar associated with a particular direction in space; that is, *vectors possess both magnitude and direction*. Here we use a boldfaced notation for a vector. This is known as Gibbs[10] notation. We will study in Section 2.1 an alternate notation, developed by Einstein, known as Cartesian[11] index notation.

Other properties are not scalar or vector, but are what is known as tensors. The best known example is the *stress tensor*, whose physics and mathematics will be fully described in Section 4.2.2. One can think of a tensor as a quantity that associates a vector with a plane inclined at a selected angle passing through a given point in space. *Tensors possess magnitude, direction, and orientation relative to a plane*. For an infinitesimal cube surrounding a point, each of the faces of the cube can be associated with a unique vector. This is shown in Fig. 1.3, in which a distinct vector is associated with three orthogonal surfaces. Each vector on each surface is itself the sum of three orthogonal components. An example is the stress tensor $\mathbf{T}$. It can be thought of as associating three vectors (or nine scalars) with each spatial point. It is best expressed as a matrix:

$$\mathbf{T}(x,y,z,t) = \begin{pmatrix} T_{xx}(x,y,z,t) & T_{xy}(x,y,z,t) & T_{xz}(x,y,z,t) \\ T_{yx}(x,y,z,t) & T_{yy}(x,y,z,t) & T_{yz}(x,y,z,t) \\ T_{zx}(x,y,z,t) & T_{zy}(x,y,z,t) & T_{zz}(x,y,z,t) \end{pmatrix}. \quad (1.9)$$

The stress tensor will be important in the mechanics of fluids. It will describe, among other things, pressure forces normal to a surface and frictional forces tangential to a surface. It will be considered fully in Chapters 4 and 5. The set of spatial coordinates x, y, and z form a vector we call $\mathbf{x}$:

$$\mathbf{x} = x\mathbf{i} + y\mathbf{j} + z\mathbf{k} = \begin{pmatrix} x \\ y \\ z \end{pmatrix}. \quad (1.10)$$

Our scalar, vector, and tensor fields are typically functions of the vector $\mathbf{x}$ and time t, and may be compactly written in the form $\rho(\mathbf{x},t)$, $\mathbf{v}(\mathbf{x},t)$, $\mathbf{T}(\mathbf{x},t)$.

1.3 Molecular Limits of Continuum Theory

Continuum theory fails in scenarios in which the length and time scales are of comparable magnitude to molecular scales. Important applications for which the continuum assumption is inappropriate include rarefied gas dynamics (relevant for low Earth orbit vehicles), and

---

[10] Josiah Willard Gibbs, 1839–1903, American mechanical engineer who made fundamental contributions to vector analysis, statistical mechanics, thermodynamics, and chemistry.

[11] René Descartes, 1596–1650, French mathematician and philosopher who developed analytic geometry.

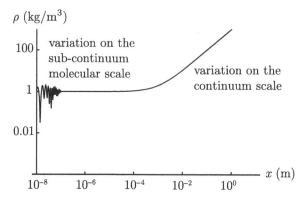

Figure 1.4 Diagram of possible density variation of a gas near atmospheric pressure as a function of length scale.

nanoscale heat transfer (relevant in cooling of computer chips). In commonly encountered physical scenarios, we expect the density to vary with distance on a macroscale, approach a limiting value at the microscale, and become ill-defined below a cutoff scale below which molecular effects are important. That is to say, when V_n from Fig. 1.2 becomes too small, such that only a few molecules are contained within it, we expect wild oscillations in ρ, and a unique value of ρ in the limit as $V_n \to 0$ formally does not exist.

To get some idea of the scales involved, we note that for air at atmospheric pressure and temperature, the time and distance between molecular collisions provide the limits of the continuum. Under these conditions, we observe for air that lengths $> 0.1\,\mu\text{m}$, and times $> 0.1\,\text{ns}$ will be sufficient to admit the continuum assumption. For denser gases, these cutoff scales are smaller. For lighter gases, these cutoff scales are larger. A depiction of a possible density variation in a gas near atmospheric pressure as a function of length scale is given in Fig. 1.4.

Details of collision theory can be found in texts such as that of Vincenti and Kruger (1965, pp. 12–26). The simplest model treats gases as composed of elastic spheres of diameter d moving within a volume that is mainly a vacuum. The model is valid in the limit in which the mean free path length λ between collisions is large relative to d. They show for air that λ is well modeled by:

$$\lambda = \frac{\mathcal{M}}{\sqrt{2}\pi\mathcal{N}\rho d^2}. \tag{1.11}$$

Here $\mathcal{M}$ is the molecular mass, $\mathcal{N}$ is Avogadro's number, and d is the molecular diameter, sometimes known as the *Lennard-Jones*[12] *diameter*.

Example 1.1 Find the variation of mean free path with density for air.

Solution
For a typical air mixture, the mixture molecular mass is taken as $\mathcal{M} = 28.97\,\text{kg/kmole}$. We turn to Vincenti and Kruger for other numerical parameter values: $\mathcal{N} = 6.022\,52 \times 10^{23}$ molecule/mole,

---

[12] John Edward Lennard-Jones, 1894–1954, British mathematician and physicist.

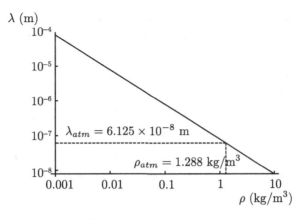

Figure 1.5 Mean free path length, λ, as a function of density, ρ, for air.

$d = 3.7 \times 10^{-10}$ m. One recognizes that these are mixture-averaged values for air, which is actually a mixture of mainly N_2 and O_2, each with its own molecular mass and molecular diameter. Thus,

$$\lambda = \frac{\left(28.97 \frac{\text{kg}}{\text{kmole}}\right)\left(1 \frac{\text{kmole}}{1000 \text{ mole}}\right)}{\sqrt{2}\pi \left(6.02252 \times 10^{23} \frac{\text{molecule}}{\text{mole}}\right) \rho \left(3.7 \times 10^{-10} \text{ m}\right)^2}, \tag{1.12}$$

$$= \frac{7.90864 \times 10^{-8} \frac{\text{kg}}{\text{molecule m}^2}}{\rho}. \tag{1.13}$$

The unit "molecule" is not really a dimension, but really is literally a "unit," that may well be thought of as dimensionless. We thus could report Avogadro's number as 6.02252×10^{23} mole^{-1}. Thus, we can safely say

$$\lambda = \frac{7.90864 \times 10^{-8} \frac{\text{kg}}{\text{m}^2}}{\rho}. \tag{1.14}$$

A plot of the variation of mean free path λ as a function of ρ is given in Fig. 1.5. Vincenti and Kruger consider an atmosphere with density of $\rho = 1.288\,\text{kg/m}^3$. For this density

$$\lambda = \frac{7.90864 \times 10^{-8} \frac{\text{kg}}{\text{m}^2}}{1.288 \frac{\text{kg}}{\text{m}^3}} = 6.14025 \times 10^{-8} \text{ m} = 6.14025 \times 10^{-2} \text{ μm}. \tag{1.15}$$

Vincenti and Kruger also show the mean molecular speed under these conditions is roughly $c = 500\,\text{m/s}$, so the mean time between collisions, t_c, is

$$t_c \sim \frac{\lambda}{c} = \frac{6.14025 \times 10^{-8} \text{ m}}{500 \frac{\text{m}}{\text{s}}} = 1.228 \times 10^{-10} \text{ s}. \tag{1.16}$$

Here and in other places in the book, we report more precision for many numbers than is justified by either the theory or measurements. Indeed there are good arguments against this practice. However, especially for those developing computational models, having several reported significant digits can be a useful aid in *solution verification*; see Roache (2009) or Oberkampf and Roy (2010) for a discussion of the nuances of *verification*, which connotes fidelity of a computational solution to a benchmark solution, and *validation*, which connotes fidelity of a computational solution to a relevant experimental measurement.

Diffusive transport properties, such as *viscosity* μ and *thermal conductivity* k, are continuum manifestations of molecular collisions; their accurate estimation requires a detailed consideration of statistical mechanics or careful experiment. Sometimes μ is described as the *dynamic viscosity*. Vincenti and Kruger give several simple estimates. Two provided can be inferred for an ideal gas to be

$$\mu = \sqrt{\frac{\Re T}{3\mathcal{M}}} \rho \lambda, \qquad k = c_v \sqrt{\frac{\Re T}{3\mathcal{M}}} \rho \lambda. \tag{1.17}$$

Here $\Re = 8.314\,472$ kJ/kmole/K is the *universal gas constant*, and c_v is the *specific heat at constant volume*. With this, the engineering gas constant R, often called the gas constant, for our air is

$$R = \frac{\Re}{\mathcal{M}} = \frac{8.314\,472\,\frac{\text{kJ}}{\text{kmole K}}}{28.97\,\frac{\text{kg}}{\text{kmole}}} = 0.287\,003\,\frac{\text{kJ}}{\text{kg K}}. \tag{1.18}$$

The viscosity and thermal conductivity are proportional to the mean free path and to the square root of the absolute temperature. Because the variation with temperature is weak, in applications for which the variation of temperature is mild, both μ and k are treated as constants. If we incorporate the mean free path estimate of Eq. (1.11), we find the viscosity and thermal conductivity do not depend on the density:

$$\mu = \mathcal{M} \frac{\sqrt{\frac{\Re}{\mathcal{M}} T}}{\sqrt{6}\pi d^2 \mathcal{N}}, \qquad k = c_v \mathcal{M} \frac{\sqrt{\frac{\Re}{\mathcal{M}} T}}{\sqrt{6}\pi d^2 \mathcal{N}}. \tag{1.19}$$

This yields an estimate for the *Prandtl number*,[13] Pr, generally defined as

$$Pr = \frac{\mu c_p}{k} = \frac{\mu/\rho}{k/(\rho c_p)} = \frac{\nu}{\alpha}. \tag{1.20}$$

Here c_p is the *specific heat at constant pressure*. We also introduce ν, the so-called *momentum diffusivity* and α the *thermal diffusivity*

$$\nu = \frac{\mu}{\rho}, \tag{1.21}$$

$$\alpha = \frac{k}{\rho c_p}. \tag{1.22}$$

It is more common to describe the momentum diffusivity as the *kinematic viscosity*. For the special case we consider, we find $Pr = c_p/c_v = \gamma$, where γ is the *ratio of specific heats*. For air, this crude estimate gives $Pr = 1.4$, while in fact it is closer to 0.7. Material property values such as μ, and k are often estimated by direct comparison to experimental results; when so done, they always carry with them experimental uncertainty. Such uncertainties can be modeled stochastically; however, we will not introduce such effects.

---

[13] Ludwig Prandtl, 1875–1953, German mechanician and father of aerodynamics, discoverer of the boundary layer, pioneer of dirigibles, and advocate of monoplanes.

Example 1.2 For the parameters used for air in the previous example, estimate the viscosity μ and thermal conductivity k of air at $T = 300\,\text{K}$.

Solution

We evaluate Eq. (1.19), neglecting the unit "molecule" in the same way as for the mean free path calculation, and find

$$\mu = \left(\frac{28.97\,\frac{\text{kg}}{\text{kmole}}}{\sqrt{6}\pi}\right) \frac{\sqrt{\left(\frac{8.314\,472\,\frac{\text{kJ}}{\text{kmole K}}}{28.97\,\frac{\text{kg}}{\text{kmole}}}\right)(300\,\text{K})\left(\frac{1000\,\frac{\text{m}^2}{\text{s}^2}}{\frac{\text{kJ}}{\text{kg}}}\right)}}{(3.7 \times 10^{-10}\,\text{m})^2\,(6.022\,52 \times 10^{23}\,\text{mole}^{-1})\left(1000\,\frac{\text{mole}}{\text{kmole}}\right)}, \quad (1.23)$$

$$= 13.40 \times 10^{-6}\,\frac{\text{kg}}{\text{m s}}. \quad (1.24)$$

This simple estimate gives the correct order of magnitude and compares well with the measured value of $\mu \sim 18.46 \times 10^{-6}$ kg/m/s, Incropera and DeWitt (1981, p. 775). We will often take values for physical constants from this source, as it is one of the more complete.

For thermal conductivity, we have

$$k = c_v \mu = \left(719.286\,\frac{\text{J}}{\text{kg K}}\right)\left(13.38 \times 10^{-6}\,\frac{\text{kg}}{\text{m s}}\right) = 9.64 \times 10^{-3}\,\frac{\text{W}}{\text{m K}}. \quad (1.25)$$

Again, the simple estimate gives the correct order of magnitude and can be compared to the measured value of 26.3×10^{-3} W/m/K, Incropera and DeWitt (1981, p. 775).

Molecules of fluid can be considered as distinct particles. They have a discrete size and distance between particles. Studies in this regime are in the realm of *molecular mechanics*, which is not the emphasis here. In restricting our attention to continuum fluid mechanics, we really are considering a fluid particle to be infinitesimally small. From here on, unless otherwise stated, we adopt the common continuum notion of a fluid particle as mass of fluid contained within an infinitesimally small volume. In this context, "fluid particle" in no way connotes a molecule or finite droplet; it is simply an idealization in which macro-properties are modeled as extending to an infinitely small length scale.

SUMMARY

This chapter reviewed basic concepts that form the foundation of the continuum mechanics of fluids. The overarching concept of mechanics was defined to describe bodies in motion and forces that induce the motion. The notion of a continuum was used to link Newtonian calculus to the description of fluids. The limits of the continuum were briefly considered by examining length scales at which molecular effects become dominant.

PROBLEMS

1.1 Present an argument for whether or not ordinary glass at room temperature is a fluid. Support your argument with appeals to the scientific literature.

1.2 A point mass m obeys Newton's second law of motion:
$$m\frac{d^2x}{dt^2} = \sum F_x.$$
With known $X(t)$, transform Newton's second law of motion to the reference frame given by $x' = x - X(t)$, $t' = t$. Find the most general conditions for $X(t)$ for which the transformation is invariant.

1.3 Estimate the viscosity μ of helium at $T = 100\,\text{K}$ using the molecular-based approach of this chapter. Compare to experiment. Search the literature for necessary data.

1.4 Examine the literature and discuss what happens to the viscosity of the isotope helium-4 when cooled to below $2.17\,\text{K}$.

1.5 Feynman[14] et al. (1963, Volume 2, Chapters 40, 41) gives a compact and lucid description of fluid mechanics, focusing on the underlying physics with carefully chosen mathematical models. It is a good example of a mathematically rigorous approach that is more intuitive than the approach of rational continuum mechanics. Read these chapters and write a one-page essay on some aspect of "The Flow of Dry Water" and "The Flow of Wet Water," explaining the difference between the two.

FURTHER READING

Aris, R. (1962). *Vectors, Tensors, and the Basic Equations of Fluid Mechanics*. New York: Dover.

Eringen, A. C. (1989). *Mechanics of Continua*, 2nd ed. Malabar, Florida: Krieger.

Gurtin, M. E., Fried, E., and Anand, L. (2013). *The Mechanics and Thermodynamics of Continua*, New York: Cambridge University Press.

Hirschfelder, J. O., Curtiss, C. F., and Bird, R. B. (1954). *Molecular Theory of Gases and Liquids*. New York: John Wiley.

Paolucci, S. (2016). *Continuum Mechanics and Thermodynamics of Matter*. New York: Cambridge University Press.

Truesdell, C. A. (1991). *A First Course in Rational Continuum Mechanics*, Vol. 1, 2nd ed. Boston, Massachusetts: Academic Press.

Truesdell, C., and Noll, W. (2004). *The Non-Linear Field Theories of Mechanics*, 3rd ed. Antman, S. S., ed. Berlin: Springer.

REFERENCES

Feynman, R. P., Leighton, R. B., and Sands, M. (1963). *The Feynman Lectures on Physics*, 3 vols., Reading, Massachusetts: Addison-Wesley.

Incropera, F. P., and DeWitt, D. P. (1981). *Fundamentals of Heat Transfer*. New York: John Wiley.

Müller, I. (2007). *A History of Thermodynamics: The Doctrine of Energy and Entropy*. Berlin: Springer.

---

[14] Richard Phillips Feynman, 1918–1988, American physicist and Nobel laureate.

References

Newton, I. (1999). Used with permission of University of California Press - Books, from *The Principia: Mathematical Principles of Natural Philosophy*. Cohen, I. B. and Whitman, A., trans. Berkeley, California: University of California Press; permission conveyed through Copyright Clearance Center, Inc.

Oberkampf, W. L., and Roy, C. J. (2010). *Verification and Validation in Scientific Computing*. Cambridge, UK: Cambridge University Press.

Roache, P. J. (2009). *Fundamentals of Verification and Validation*. Albuquerque, New Mexico: Hermosa.

Thorne, K. S., and Blandford, R. D. (2017). *Modern Classical Physics: Optics, Fluids, Plasmas, Elasticity, and Statistical Physics*. Princeton, New Jersey: Princeton University Press.

Vincenti, W. G., and Kruger, C. H. (1965). *Introduction to Physical Gas Dynamics*. New York: John Wiley.

Woods, L. C. (1982). Thermodynamic inequalities in continuum mechanics, *IMA Journal of Applied Mathematics*, **29**(3), 221–246.

2 Geometry

Here we give an introduction to topics in geometry that will be relevant to the mechanics of fluids. More specifically, we will consider elementary aspects of *differential geometry*. Geometry can be defined as the study of shape, and differential geometry connotes that methods of calculus will be used to study shape. It is well known that fluids in motion may transform location and shape, such as shown in Fig. 2.1. Here we have included a coordinate system with the coordinate axes labeled x_1, x_2, and x_3. The axes are oriented so that they obey the so-called *right-hand rule* in that the unit vectors associated with each axis $\mathbf{e}_1$, $\mathbf{e}_2$, and $\mathbf{e}_3$ are such that $\mathbf{e}_3 = \mathbf{e}_1 \times \mathbf{e}_2$. Here $\times$ is the well-known vector cross product. As will be discussed in Chapter 3, this motion can be decomposed locally into translation, rotation, and deformation of a fluid element's geometry.

The broad subject of differential geometry provides a language and set of tools to describe the location and shape of fluids; it gives particular emphasis to the *transformation* of one geometric form into another. This chapter especially and many parts of the book will focus on the notion of transformation. Much of this is an extension of concepts covered in an undergraduate class in the calculus of many variables. This is the language best suited to describe the mechanics of fluids. The bulk of this chapter will be confined to stationary spatial geometry; however, relevant time derivatives will be briefly considered in

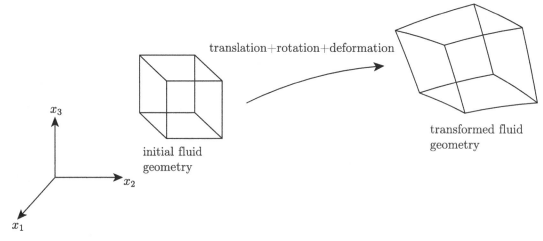

Figure 2.1 Diagram of fluid particle with initially cubical shape undergoing translation, rotation, and deformation of its geometry due to fluid motion.

Sections 2.4.6 and 2.4.7. Many of our choices for notation are guided by Panton (2013) who, along with Karamcheti (1980) and Kundu et al. (2015), gives a good exposition linking geometry and fluid mechanics.

2.1 Scalars, Vectors, and Tensors

In this section, we expand on ideas introduced in Section 1.2.3. Underlying our discussion is the idea that scalars, vectors, and tensors are concepts to describe physical entities that exist within three-dimensional geometries and whose character is independent of the particular coordinate system chosen to describe them. Scalars are entities whose value is the same regardless of the coordinate system. Vectors and tensors are intrinsic entities that have different representations in different coordinate systems. As long as the coordinate systems satisfy geometry-based transformation rules, the various representations will all possess the same values of coordinate-free features, such as the magnitude of a vector.

2.1.1 Gibbs and Cartesian Index Notation

Gibbs notation for vectors and tensors is familiar from undergraduate courses. It typically uses boldface, arrows, underscores, or over-bars to denote a vector or a tensor. Unfortunately, it also hides some of the structures that are actually present in the equations. Einstein realized this in developing the theory of general relativity and developed a useful alternate: index notation. In this book we often employ what is known as Cartesian index notation, which is restricted to Cartesian coordinate systems. Einstein also developed a more general index system for non-Cartesian systems. We will briefly touch on this later in Section 2.5 and Chapter 6 but refer the reader to books such as that of Aris (1962) for a full exposition. While it can seem difficult at the outset, use of index notation actually simplifies many common notions in fluid mechanics. Moreover, its use in the archival literature is widespread, so to be conversant in fluid mechanics, one must know index notation. Table 2.1 summarizes the correspondences between Gibbs, Cartesian index, and matrix notation. Here we adopt a convention for the Gibbs notation in which lower case italics font (a) indicates a scalar, lower case bold font (**a**) indicates a vector, and upper case bold font (**A**) indicates a second-order tensor. Also, following longstanding fluid mechanics tradition, we will break these conventions for the viscous stress tensor, $\boldsymbol{\tau}$, and for the scalar temperature, T. We will not use Gibbs notation for tensors whose order is greater than two. In Cartesian index notation, there is no need to use anything except italics, as all terms are thought of as scalar components of a more expansive structure, with the structure indicated by the presence of subscripts.

The essence of the Cartesian index notation is as follows. We can represent a three-dimensional vector **a** as a linear combination of scalars and orthonormal basis vectors found in a Cartesian coordinate system:

$$\mathbf{a} = a_x \mathbf{i} + a_y \mathbf{j} + a_z \mathbf{k} = \begin{pmatrix} a_x \\ a_y \\ a_z \end{pmatrix}. \tag{2.1}$$

Table 2.1 Scalar, vector, and tensor notation conventions

Quantity	Common parlance	Gibbs	Cartesian index	Matrix
Zeroth-order tensor	Scalar	a	a	(a)
First-order tensor	Vector	$\mathbf{a}$	a_i	$\begin{pmatrix} a_1 \\ a_2 \\ \vdots \\ a_n \end{pmatrix}$
Second-order tensor	Tensor	$\mathbf{A}$	a_{ij}	$\begin{pmatrix} a_{11} & a_{12} & \ldots & a_{1n} \\ a_{21} & a_{22} & \ldots & a_{2n} \\ \vdots & \vdots & \vdots & \vdots \\ a_{n1} & a_{n2} & \ldots & a_{nn} \end{pmatrix}$
Third-order tensor	Tensor	-	a_{ijk}	-
Fourth-order tensor	Tensor	-	a_{ijkl}	-
$\vdots$	$\vdots$	$\vdots$	$\vdots$	-

We choose now to associate the subscript 1 with the x direction, the subscript 2 with the y direction, and the subscript 3 with the z direction. Further, we replace the orthonormal basis vectors $\mathbf{i}$, $\mathbf{j}$, and $\mathbf{k}$, by $\mathbf{e}_1$, $\mathbf{e}_2$, and $\mathbf{e}_3$. Then the vector $\mathbf{a}$ is represented by

$$\mathbf{a} = a_1 \mathbf{e}_1 + a_2 \mathbf{e}_2 + a_3 \mathbf{e}_3 = \sum_{i=1}^{3} a_i \mathbf{e}_i = a_i \mathbf{e}_i = a_i = \begin{pmatrix} a_1 \\ a_2 \\ a_3 \end{pmatrix}. \tag{2.2}$$

Following Einstein, we have adopted the convention that a summation is understood to exist when two indices, known as dummy indices, are repeated, and have further left the explicit representation of basis vectors out of our final version of the notation. While we have indicated $\mathbf{a} = a_i$, it is perhaps more appropriate to interpret a_i as the ith scalar component of $\mathbf{a}$. We have also included a representation of $\mathbf{a}$ as a 3×1 column vector. We adopt the common standard that all vectors can be thought of as column vectors. Often in matrix operations, such as the *vector dot product*, we will need row vectors. They will be formed by taking the transpose, indicated by a superscript T, of a column vector. So $\mathbf{a}^T = (a_1, a_2, a_3)$. In the interest of clarity, full consistency with notions from matrix algebra, as well as transparent translation to the conventions of necessarily meticulous (as well as popular) software tools, we will scrupulously use the transpose notation. This comes at the expense of a more cluttered set of equations at times. We also note that most authors do not *explicitly* use the transpose notation, but its use is implicit.

2.1.2 Rotation of Axes

Cartesian index notation is developed to be valid under transformations from one Cartesian coordinate system to another Cartesian coordinate system. It is not applicable to either more general orthogonal systems (such as cylindrical or spherical) or nonorthogonal systems. It is straightforward, but complicated, to develop a more general system to handle generalized coordinate transformations, and Einstein did that as well. For our purposes, however, the simpler Cartesian index notation will suffice.

We will first consider a coordinate transformation that is a simple rotation of axes. This transformation preserves all angles; hence, right angles in the original Cartesian system will be right angles in the rotated, but still Cartesian system. It also preserves lengths of geometric features, with no stretching. We will require, ultimately, that whatever theory we develop must generate results in which physically relevant scalar quantities such as temperature, pressure, density, and velocity magnitude, are independent of the particular set of coordinates with which we choose to describe the system. To motivate this, let us consider a two-dimensional rotation from an unprimed system to a primed system. So, we seek a transformation that maps $(x_1, x_2)^T \to (x_1', x_2')^T$. We will rotate the unprimed system counterclockwise through an angle α to achieve the primed system. This is an example of a so-called *alias transformation*. In such a transformation, the coordinate axes transform, but the underlying object remains unchanged. So a vector may be considered to be invariant, but its representation in different coordinate systems may be different. Alias transformations are common in continuum mechanics. In contrast, an *alibi transformation* is one in which the coordinate axes remain fixed, but the object transforms. This mode of thought is common in fields such as robotics. In short, alias rotates the axes, but not the body; alibi rotates the body, but not the axes.

The rotation is depicted in Fig. 2.2. It is easy to show that the angle $\beta = \pi/2 - \alpha$. Here a point P is identified by a particular set of coordinates $(\tilde{x}_1, \tilde{x}_2)$. One of the keys to all of continuum mechanics is realizing that while the location (or velocity, or stress, ...) of P may be represented differently in various coordinate systems, ultimately it must represent the same physical reality. Straightforward geometry shows the following relation between the primed and unprimed coordinate systems for x_1':

$$\tilde{x}_1' = \tilde{x}_1 \cos\alpha + \tilde{x}_2 \cos\beta. \tag{2.3}$$

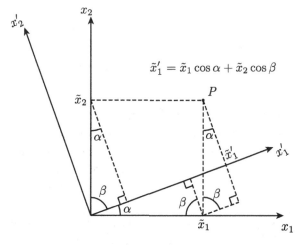

Figure 2.2 Diagram of coordinate transformation that is a rotation of axes.

More generally, we can say for an arbitrary point that $x'_1 = x_1 \cos \alpha + x_2 \cos \beta$. We adopt the following notation:

- (x_1, x'_1) denotes the angle between the x_1 and x'_1 axes,
- (x_2, x'_2) denotes the angle between the x_2 and x'_2 axes,
- (x_3, x'_3) denotes the angle between the x_3 and x'_3 axes,
- (x_1, x'_2) denotes the angle between the x_1 and x'_2 axes,
- $\vdots$

Thus, in two dimensions, we have

$$x'_1 = x_1 \cos(x_1, x'_1) + x_2 \cos(x_2, x'_1). \tag{2.4}$$

In three dimensions, this extends to

$$x'_1 = x_1 \cos(x_1, x'_1) + x_2 \cos(x_2, x'_1) + x_3 \cos(x_3, x'_1). \tag{2.5}$$

Extending this analysis to calculate x'_2 and x'_3 gives

$$x'_2 = x_1 \cos(x_1, x'_2) + x_2 \cos(x_2, x'_2) + x_3 \cos(x_3, x'_2), \tag{2.6}$$

$$x'_3 = x_1 \cos(x_1, x'_3) + x_2 \cos(x_2, x'_3) + x_3 \cos(x_3, x'_3). \tag{2.7}$$

These can be written in matrix form as

$$\begin{pmatrix} x'_1 & x'_2 & x'_3 \end{pmatrix} = \begin{pmatrix} x_1 & x_2 & x_3 \end{pmatrix} \begin{pmatrix} \cos(x_1, x'_1) & \cos(x_1, x'_2) & \cos(x_1, x'_3) \\ \cos(x_2, x'_1) & \cos(x_2, x'_2) & \cos(x_2, x'_3) \\ \cos(x_3, x'_1) & \cos(x_3, x'_2) & \cos(x_3, x'_3) \end{pmatrix}. \tag{2.8}$$

With shorthand notation, for example, $\ell_{11} = \cos(x_1, x'_1)$, we have

$$\underbrace{\begin{pmatrix} x'_1 & x'_2 & x'_3 \end{pmatrix}}_{\mathbf{x}'^T} = \underbrace{\begin{pmatrix} x_1 & x_2 & x_3 \end{pmatrix}}_{\mathbf{x}^T} \underbrace{\begin{pmatrix} \ell_{11} & \ell_{12} & \ell_{13} \\ \ell_{21} & \ell_{22} & \ell_{23} \\ \ell_{31} & \ell_{32} & \ell_{33} \end{pmatrix}}_{\mathbf{Q}}. \tag{2.9}$$

In Gibbs notation, defining the matrix of ℓ's to be $\mathbf{Q}$, and recalling that all vectors are taken to be column vectors, we can say

$$\mathbf{x}'^T = \mathbf{x}^T \cdot \mathbf{Q}. \tag{2.10}$$

The more commonly used alternate convention of not explicitly using the transpose notation for vectors would instead have our $\mathbf{x}'^T = \mathbf{x}^T \cdot \mathbf{Q}$ written as $\mathbf{x}' = \mathbf{x} \cdot \mathbf{Q}$. In fact, our use of the transpose notation is strictly viable only for Cartesian coordinate systems. It is common to allow Gibbs notation to represent vectors in non-Cartesian coordinates, for which the transpose operation is ill-suited. However, realizing that we will primarily focus on Cartesian systems, and that operations relying on the transpose are useful notions from linear algebra, it will be employed in a liberal fashion.

Taking the transpose of both sides of Eq. (2.10) and recalling the useful identities that $(\mathbf{A} \cdot \mathbf{b})^T = \mathbf{b}^T \cdot \mathbf{A}^T$ and $(\mathbf{A}^T)^T = \mathbf{A}$, we can also say

$$\mathbf{x}' = \mathbf{Q}^T \cdot \mathbf{x}. \tag{2.11}$$

We call $\mathbf{Q} = \ell_{ij}$ the matrix of *direction cosines* and $\mathbf{Q}^T = \ell_{ji}$ the rotation matrix. It can be shown that coordinate systems that satisfy the right-hand rule require that

$$\det \mathbf{Q} = 1. \tag{2.12}$$

Here "det" denotes the *determinant*. Matrices $\mathbf{Q}$ that have $|\det \mathbf{Q}| = 1$ are associated with volume-preserving transformations. Matrices $\mathbf{Q}$ that have $\det \mathbf{Q} = 1$ are volume- and orientation-preserving, and can be thought of a rotations. A matrix that had determinant -1 would be volume-preserving but not orientation-preserving. It could be considered as a reflection. A matrix $\mathbf{Q}$ composed of orthonormal column vectors, with $|\det \mathbf{Q}| = 1$ (thus either rotation or reflection matrices) is commonly known as *orthogonal*, though perhaps "orthonormal" would have been a more descriptive nomenclature. While $\mathbf{Q}$ may represent a rotation or reflection, except in a few cases, we will be concerned with rotation. A more general matrix $\mathbf{A}$ with $|\det \mathbf{A}| \neq 1$ is not volume-preserving. It is orientation-preserving if $\det \mathbf{A} > 0$, and orientation-reversing if $\det \mathbf{A} < 0$. If $\det \mathbf{A} \neq 0$, it has rank three, is nonsingular, and maps points in a volume uniquely to points in a transformed volume. If $\det \mathbf{A} = 0$, the matrix is singular; rank two matrices map points in a volume to a plane, rank one matrices map points in a volume to a line, and rank zero matrices map points in a volume to a point. In each case, there is no unique inverse mapping.

Another way to think of the matrix of direction cosines $\ell_{ij} = \mathbf{Q}$ is as a matrix with orthonormal basis vectors in its columns:

$$\ell_{ij} = \mathbf{Q} = \begin{pmatrix} \vdots & \vdots & \vdots \\ \boldsymbol{\alpha}^{(1)} & \boldsymbol{\alpha}^{(2)} & \boldsymbol{\alpha}^{(3)} \\ \vdots & \vdots & \vdots \end{pmatrix}. \tag{2.13}$$

Here $\boldsymbol{\alpha}^{(i)}$, $i = 1, 2, 3$, is a set of orthonormal basis vectors. In an important result, it can be shown that the *transpose of an orthogonal matrix is its inverse*:

$$\mathbf{Q}^T = \mathbf{Q}^{-1}. \tag{2.14}$$

Thus, we have

$$\mathbf{Q} \cdot \mathbf{Q}^T = \mathbf{Q}^T \cdot \mathbf{Q} = \mathbf{I}. \tag{2.15}$$

The equation $\mathbf{x}'^T = \mathbf{x}^T \cdot \mathbf{Q}$ is a set of three linear equations; for example the first is

$$x_1' = x_1 \ell_{11} + x_2 \ell_{21} + x_3 \ell_{31}. \tag{2.16}$$

More generally, we could say that

$$x_j' = x_1 \ell_{1j} + x_2 \ell_{2j} + x_3 \ell_{3j}. \tag{2.17}$$

Here j is a so-called *free index*, which for three-dimensional space takes on values $j = 1, 2, 3$. Some rules of thumb for free indices are (1) a free index can appear only once in each additive term, and (2) one free index (e.g. k) may replace another (e.g. j) as long as it is replaced in each additive term. We can simplify Eq. (2.17) further by writing

$$x'_j = \sum_{i=1}^{3} x_i \ell_{ij} = x_i \ell_{ij}. \tag{2.18}$$

We again note that it is to be understood that whenever an index is repeated, as has the index i here, that a summation from $i = 1$ to $i = 3$ is to be performed and that i is the *dummy index*. Some rules of thumb for dummy indices are (1) dummy indices can appear *only twice* in a given additive term, (2) a pair of dummy indices, say i, i, can be exchanged for another, say j, j, in a given additive term with no need to change dummy indices in other additive terms. We define the *Kronecker*[1] delta, δ_{ij} as

$$\delta_{ij} = \mathbf{I} = \begin{pmatrix} 1 & 0 & 0 \\ 0 & 1 & 0 \\ 0 & 0 & 1 \end{pmatrix} = \begin{cases} 0, & i \neq j, \\ 1, & i = j. \end{cases} \tag{2.19}$$

Direct substitution proves that what is effectively the law of cosines can be written as $\ell_{ij}\ell_{kj} = \delta_{ik}$. This is also equivalent to Eq. (2.15), $\mathbf{Q} \cdot \mathbf{Q}^T = \mathbf{I}$.

Example 2.1 Show for the two-dimensional system of Fig. 2.2 that $\ell_{ij}\ell_{kj} = \delta_{ik}$ holds.

Solution

Expanding for the two-dimensional system, we get

$$\ell_{i1}\ell_{k1} + \ell_{i2}\ell_{k2} = \delta_{ik}. \tag{2.20}$$

First, take $i = 1, k = 1$. We then get

$$\ell_{11}\ell_{11} + \ell_{12}\ell_{12} = \delta_{11} = 1, \tag{2.21}$$
$$\cos \alpha \cos \alpha + \cos(\alpha + \pi/2)\cos(\alpha + \pi/2) = 1, \tag{2.22}$$
$$\cos \alpha \cos \alpha + (-\sin(\alpha))(-\sin(\alpha)) = 1, \tag{2.23}$$
$$\cos^2 \alpha + \sin^2 \alpha = 1. \tag{2.24}$$

This is obviously true. Next, take $i = 1, k = 2$. We get then

$$\ell_{11}\ell_{21} + \ell_{12}\ell_{22} = \delta_{12} = 0, \tag{2.25}$$
$$\cos \alpha \cos(\pi/2 - \alpha) + \cos(\alpha + \pi/2)\cos(\alpha) = 0, \tag{2.26}$$
$$\cos \alpha \sin \alpha - \sin \alpha \cos \alpha = 0. \tag{2.27}$$

This is obviously true. Next, take $i = 2, k = 1$. We get then

$$\ell_{21}\ell_{11} + \ell_{22}\ell_{12} = \delta_{21} = 0, \tag{2.28}$$
$$\cos(\pi/2 - \alpha)\cos \alpha + \cos \alpha \cos(\pi/2 + \alpha) = 0, \tag{2.29}$$
$$\sin \alpha \cos \alpha + \cos \alpha(-\sin \alpha) = 0. \tag{2.30}$$

This is obviously true. Next, take $i = 2, k = 2$. We get then

---

[1] Leopold Kronecker, 1823–1891, German mathematician, critic of set theory, who stated "God made the integers; all else is the work of man."

$$\ell_{21}\ell_{21} + \ell_{22}\ell_{22} = \delta_{22} = 1, \tag{2.31}$$
$$\cos(\pi/2 - \alpha)\cos(\pi/2 - \alpha) + \cos\alpha\cos\alpha = 1, \tag{2.32}$$
$$\sin\alpha\sin\alpha + \cos\alpha\cos\alpha = 1. \tag{2.33}$$

Again, this is obviously true.

Using this, we can find the inverse transformation back to the unprimed coordinates via the following operations:

$$\ell_{kj}x'_j = \ell_{kj}x_i\ell_{ij} = \ell_{ij}\ell_{kj}x_i = \delta_{ik}x_i = x_k, \quad \text{so} \quad \ell_{ij}x'_j = x_i. \tag{2.34}$$

The Kronecker delta is also known as the substitution tensor as it has the property that application of it to a vector simply substitutes one index for another:

$$x_k = \delta_{ki}x_i. \tag{2.35}$$

Example 2.2 With

$$\mathbf{A} = \begin{pmatrix} 2 & 0 & 0 \\ 0 & 1/2 & 0 \\ 0 & 0 & 1 \end{pmatrix}, \tag{2.36}$$

show the squeezing transformation $\mathbf{x}' = \mathbf{A}^T \cdot \mathbf{x}$ is volume- and orientation-preserving. Show that orthogonal planes aligned with coordinate axes in the $\mathbf{x}$ system remain orthogonal in the $\mathbf{x}'$ system. Show that the transformation is not orthogonal. Show that the representation of a unit cube in $\mathbf{x}$ coordinates is a rectangular parallelepiped in $\mathbf{x}'$ coordinates. Show that the representation of a unit sphere in $\mathbf{x}$ coordinates is a triaxial ellipsoid in $\mathbf{x}'$ coordinates.

Solution

We see $\det \mathbf{A} = 1$, so this transformation is volume- and orientation-preserving. This is not true for all transformations. Consider two orthogonal planes aligned with the coordinate axes: $x_1 = a$, $x_2 = b$. In the transformed system, these planes are represented by orthogonal planes $x'_1 = 2a$ and $x'_2 = b/2$. Restricting attention to $x_3 = 0$, a unit square in the original (x_1, x_2) system is mapped to a rectangle in the (x'_1, x'_2) system; extending to three dimensions, we can easily envision the relevant rectangular parallelepiped. Orthogonality is maintained for the interior angles, but the geometry is deformed. The fact that the transformation is a deformation is verified when one sees that $\mathbf{A}$ is not an orthogonal matrix:

$$\mathbf{A}^T = \begin{pmatrix} 2 & 0 & 0 \\ 0 & 1/2 & 0 \\ 0 & 0 & 1 \end{pmatrix}, \quad \mathbf{A}^{-1} = \begin{pmatrix} 1/2 & 0 & 0 \\ 0 & 2 & 0 \\ 0 & 0 & 1 \end{pmatrix}, \quad \mathbf{A}^T \neq \mathbf{A}^{-1}, \quad \mathbf{A}^T \cdot \mathbf{A} \neq \mathbf{I}. \tag{2.37}$$

A unit circle in the $x_3 = 0$ plane is mapped into an ellipse. The unit sphere in $\mathbf{x}$ space is

$$x_1^2 + x_2^2 + x_3^2 = 1. \tag{2.38}$$

In the transformed space, this transforms to the triaxial ellipsoid

$$\left(\frac{x'_1}{2}\right)^2 + \left(\frac{x'_2}{1/2}\right)^2 + x_3'^2 = 1. \tag{2.39}$$

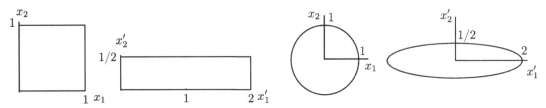

Figure 2.3 Unit square and unit circle in the (x_1, x_2) plane for which $x_3 = 0$ and stretched representations in the (x_1', x_2') plane for $x_1' = 2x_1$, $x_2' = x_2/2$, $x_3' = x_3 = 0$.

For this simple transformation, the deformation displayed by the ellipsoid is such that its principal axes remain aligned with the original coordinate axes. A more general transformation matrix **A** would map the unit sphere to an ellipsoid whose principal axes were rotated relative to the initial axes. Relevant plots are given in Fig. 2.3. The two-dimensional area-preserving version of this transformation in the (x_1, x_2) plane is known as a *squeeze mapping*. It is composed of a stretch along one coordinate axis and a reciprocal stretch on another.

2.1.3 Scalars

Scalars are quantities that do not change when coordinate axes are rotated. They possess magnitude, but not direction. A scalar ρ has identical representations

$$\rho' = \rho. \tag{2.40}$$

2.1.4 Vectors

Three scalar quantities v_i with $i = 1, 2, 3$ are scalar components of a vector if they transform in the same way

$$v_j' = v_i \ell_{ij}, \tag{2.41}$$

as ordinary distance vectors under a rotation of axes characterized by direction cosines ℓ_{ij}; see Eq. (2.18). In Gibbs notation, we would say

$$\mathbf{v}'^T = \mathbf{v}^T \cdot \mathbf{Q}, \quad \text{or} \quad \mathbf{v}' = \mathbf{Q}^T \cdot \mathbf{v}. \tag{2.42}$$

At a given point in space, a vector possesses a magnitude and direction. The magnitude and direction of the vector are independent of the coordinate system. The representation of the vector depends on the chosen coordinate system. We can also say that a vector associates a scalar with a chosen direction in space by an expression that is linear in the direction cosines of the chosen direction.

Example 2.3 A vector **v** has a representation in a Cartesian coordinate system of $\mathbf{v} = (1, 0, 0)^T$. A coordinate system rotated through an angle of $\pi/2$ about the x_3 axis has

$$\mathbf{Q} = \begin{pmatrix} 0 & 1 & 0 \\ -1 & 0 & 0 \\ 0 & 0 & 1 \end{pmatrix}. \tag{2.43}$$

Find the representation of **v** in the rotated coordinate system.

Solution
It is easily verified that $\mathbf{Q}^T = \mathbf{Q}^{-1}$. With this and because $\det \mathbf{Q} = 1$, the transformation is a rotation. Properties of the vector require that

$$\mathbf{v}' = \mathbf{Q}^T \cdot \mathbf{v} = \begin{pmatrix} 0 & -1 & 0 \\ 1 & 0 & 0 \\ 0 & 0 & 1 \end{pmatrix} \begin{pmatrix} 1 \\ 0 \\ 0 \end{pmatrix} = \begin{pmatrix} 0 \\ 1 \\ 0 \end{pmatrix}. \tag{2.44}$$

The physical entity of the vector has not changed; however, its representation in the new coordinate system is different than that of the original coordinate system. Clearly, the axes have rotated by $\pi/2$ as the representation changed from $(1, 0, 0)^T$ to $(0, 1, 0)^T$.

Example 2.4 Consider the set of scalars that describe the velocity in a two-dimensional Cartesian system:

$$v_i = \begin{pmatrix} v_x \\ v_y \end{pmatrix}, \tag{2.45}$$

where we return to the typical (x, y) coordinate system. Determine if v_i is a vector.

Solution
In a rotated coordinate system, using the same notation of Fig. 2.2, we find that

$$v'_x = v_x \cos \alpha + v_y \cos(\pi/2 - \alpha) = v_x \cos \alpha + v_y \sin \alpha, \tag{2.46}$$
$$v'_y = v_x \cos(\pi/2 + \alpha) + v_y \cos \alpha = -v_x \sin \alpha + v_y \cos \alpha. \tag{2.47}$$

This is linear in the direction cosines, and satisfies the definition for a vector. In matrix form, we have

$$\underbrace{\begin{pmatrix} v'_x \\ v'_y \end{pmatrix}}_{\mathbf{v}'} = \underbrace{\begin{pmatrix} \cos \alpha & \sin \alpha \\ -\sin \alpha & \cos \alpha \end{pmatrix}}_{\mathbf{Q}^T} \underbrace{\begin{pmatrix} v_x \\ v_y \end{pmatrix}}_{\mathbf{v}}, \quad \mathbf{v}' = \mathbf{Q}^T \cdot \mathbf{v}. \tag{2.48}$$

We thus easily show that

$$\mathbf{Q} = \begin{pmatrix} \cos \alpha & -\sin \alpha \\ \sin \alpha & \cos \alpha \end{pmatrix}, \quad \mathbf{Q}^T = \mathbf{Q}^{-1} = \begin{pmatrix} \cos \alpha & \sin \alpha \\ -\sin \alpha & \cos \alpha \end{pmatrix}. \tag{2.49}$$

Because $\mathbf{Q}^T = \mathbf{Q}^{-1}$ and $\det \mathbf{Q} = \cos^2 \alpha + \sin^2 \alpha = 1$, the transformation is a rotation. Recognize that most sets of scalars do not form a vector. For example, the set of thermodynamic scalars, pressure, density, and temperature, could be ordered $(p, \rho, T)^T$. However, subjecting this set of scalars to a coordinate rotation has no physical meaning. So it is not a vector.

We have the following vector algebra. Addition may be described as

$$w_i = u_i + v_i, \quad \mathbf{w} = \mathbf{u} + \mathbf{v}. \tag{2.50}$$

The dot product, also called the *vector inner product*, is

$$u_i v_i = b, \quad \mathbf{u}^T \cdot \mathbf{v} = b \quad \langle \mathbf{u}, \mathbf{v} \rangle = b. \tag{2.51}$$

Each notation requires $u_1 v_1 + u_2 v_2 + u_3 v_3 = b$. While u_i and v_i have scalar components that change under a rotation of axes, their dot product is a true scalar and is invariant under a rotation of axes. The dot product commutes: $u_i v_i = v_i u_i = b$. Here we have in the Gibbs notation explicitly noted that the transpose is part of the dot product. Most authors in fact assume the dot product of two vectors implies the transpose and do not write it explicitly, writing the dot product simply as $\mathbf{u} \cdot \mathbf{v} \equiv \mathbf{u}^T \cdot \mathbf{v}$.

Example 2.5 Demonstrate invariance of $\mathbf{u}^T \cdot \mathbf{v} = b$ by subjecting coordinates used to represent vectors $\mathbf{u}$ and $\mathbf{v}$ to transformation by an orthogonal matrix $\mathbf{Q}$.

Solution
Under the action of $\mathbf{Q}$, our representations transform as $\mathbf{u}' = \mathbf{Q}^T \cdot \mathbf{u}$, $\mathbf{v}' = \mathbf{Q}^T \cdot \mathbf{v}$. Thus, $\mathbf{Q} \cdot \mathbf{u}' = \mathbf{Q} \cdot \mathbf{Q}^T \cdot \mathbf{u} = \mathbf{u}$. and $\mathbf{Q} \cdot \mathbf{v}' = \mathbf{Q} \cdot \mathbf{Q}^T \cdot \mathbf{v} = \mathbf{v}$. Then consider the dot product

$$\mathbf{u}^T \cdot \mathbf{v} = b, \quad \text{so} \quad \left(\mathbf{Q} \cdot \mathbf{u}'\right)^T \cdot \left(\mathbf{Q} \cdot \mathbf{v}'\right) = b, \tag{2.52}$$

$$\mathbf{u}'^T \cdot \underbrace{\mathbf{Q}^T \cdot \mathbf{Q}}_{=\mathbf{I}} \cdot \mathbf{v}' = b, \quad \text{so} \quad \mathbf{u}'^T \cdot \mathbf{I} \cdot \mathbf{v}' = b, \quad \text{so} \quad \mathbf{u}'^T \cdot \mathbf{v}' = b. \tag{2.53}$$

The dot product is invariant under the transformation by the orthogonal matrix $\mathbf{Q}$. This invariance holds under either rotation or reflection.

Example 2.6 Given a vector $\mathbf{v}$, show that application of an orthogonal matrix $\mathbf{Q}$, either rotation or reflection, preserves the magnitude of $\mathbf{v}$.

Solution
We are given $\mathbf{v}$. We take its magnitude to be the scalar ℓ, and its square is given by

$$\ell^2 = \mathbf{v}^T \cdot \mathbf{v}. \tag{2.54}$$

We then transform via $\mathbf{v}' = \mathbf{Q}^T \cdot \mathbf{v}$. We also have $\mathbf{Q} \cdot \mathbf{v}' = \mathbf{v}$. Then we get

$$\ell^2 = \left(\mathbf{Q} \cdot \mathbf{v}'\right)^T \cdot \mathbf{Q} \cdot \mathbf{v}' = \mathbf{v}'^T \cdot \underbrace{\mathbf{Q}^T \cdot \mathbf{Q}}_{\mathbf{I}} \cdot \mathbf{v}' = \mathbf{v}'^T \cdot \mathbf{I} \cdot \mathbf{v}' = \mathbf{v}'^T \cdot \mathbf{v}'. \tag{2.55}$$

The magnitude of the vector is preserved by either rotation or reflection.

Generalizing from the previous example, we will occasionally use the concept of the *norm*; see Powers and Sen (2015, Chapter 6). We will restrict attention to so-called Euclidean norms, which for vectors in three-dimensional space is the magnitude or length ℓ of the vector. The standard notation for the Euclidean norm of a vector $\mathbf{v}$ is

$$||\mathbf{v}|| = \sqrt{\mathbf{v}^T \cdot \mathbf{v}} = \sqrt{v_1^2 + v_2^2 + v_3^2} = \ell. \tag{2.56}$$

It is sometimes called a *2-norm*, because of the 2 in $v_1^2 + v_2^2 + v_3^2 = \ell^2$. One can generalize to a *p-norm*, $p \geq 1$, with $|v_1|^p + |v_2|^p + |v_3|^p = \ell^p$, so that $||\mathbf{v}||_p = (|v_1|^p + |v_2|^p + |v_3|^p)^{1/p}$. These more general norms are useful aids in assessing the error of computational fluid dynamics (CFD) predictions. Vectors residing in a so-called *Hilbert space*[2] possess a 2-norm, while those residing in the more general *Banach space*[3] may possess a *p*-norm. The norm is a positive

---

[2] David Hilbert, 1862–1943, German mathematician.
[3] Stefan Banach, 1892–1944, Polish mathematician.

semi-definite scalar. It has value zero iff $\mathbf{v} = \mathbf{0}$. If a is a scalar $||a\mathbf{v}|| = |a|||\mathbf{v}||$. And it must satisfy the triangle inequality: $||\mathbf{u} + \mathbf{v}|| \leq ||\mathbf{u}|| + ||\mathbf{v}||$.

2.1.5 Tensors

A *second-order tensor*, or a rank two tensor, is a set of nine scalar components that under a rotation of axes transforms according to the following rule:

$$T'_{ij} = \ell_{ki}\ell_{lj}T_{kl}. \tag{2.57}$$

We could also write this as either

$$T'_{ij} = \sum_{k=1}^{3}\sum_{l=1}^{3}\ell_{ki}\ell_{lj}T_{kl}, \qquad \mathbf{T}' = \mathbf{Q}^T \cdot \mathbf{T} \cdot \mathbf{Q}. \tag{2.58}$$

In these expressions, i and j are both free indices, while k and l are dummy indices.

A tensor should be considered as an entity independent of a chosen coordinate system. At a given point in space, one can select a plane of arbitrary orientation. At this point and for this plane there exists a vector with magnitude and direction. As one continuously re-orients the selected plane, the associated vector continuously changes its magnitude and direction. This is a tensor. The representation of the tensor depends on the chosen coordinate system. Analogously to a vector, we say that a tensor associates a vector with each direction in space by an expression that is linear in the direction cosines of the chosen direction. For a given tensor T_{ij}, the first subscript is associated with the face of a unit cube (hence the mnemonic device, *first-face*); the second subscript is associated with the vector components for the vector on that face.

Tensors can also be expressed as matrices. All second-order tensors can be represented by two-dimensional matrices, but not all matrices are second-order tensors, as they do not necessarily satisfy the transformation rules. We can say

$$T_{ij} = \begin{pmatrix} T_{11} & T_{12} & T_{13} \\ T_{21} & T_{22} & T_{23} \\ T_{31} & T_{32} & T_{33} \end{pmatrix}. \tag{2.59}$$

The first row vector, $\begin{pmatrix} T_{11} & T_{12} & T_{13} \end{pmatrix}$, is the vector associated with the 1 face. The second row vector, $\begin{pmatrix} T_{21} & T_{22} & T_{23} \end{pmatrix}$, is the vector associated with the 2 face. The third row vector, $\begin{pmatrix} T_{31} & T_{32} & T_{33} \end{pmatrix}$, is the vector associated with the 3 face.

Example 2.7 A tensor $\mathbf{T}$ has a representation in a Cartesian coordinate system of

$$\mathbf{T} = \begin{pmatrix} 1 & 1 & 0 \\ 0 & 1 & 0 \\ 0 & 0 & 1 \end{pmatrix}. \tag{2.60}$$

A coordinate system rotated $\pi/2$ about the x_3 axis has

$$\mathbf{Q} = \begin{pmatrix} 0 & 1 & 0 \\ -1 & 0 & 0 \\ 0 & 0 & 1 \end{pmatrix}. \tag{2.61}$$

Find the representation of $\mathbf{T}$ in the rotated coordinate system.

Solution
It is easily shown that $\mathbf{Q}^T = \mathbf{Q}^{-1}$. Because of this and as $\det \mathbf{Q} = 1$, the transformation is a rotation. Note that $\mathbf{Q}^T \cdot (1,0,0)^T = (0,1,0)^T$ and $\mathbf{Q}^T \cdot (0,1,0)^T = (-1,0,0)^T$, as expected for a counter-clockwise rotation of $\pi/2$. Properties of the tensor require that

$$\mathbf{T}' = \mathbf{Q}^T \cdot \mathbf{T} \cdot \mathbf{Q} = \begin{pmatrix} 0 & -1 & 0 \\ 1 & 0 & 0 \\ 0 & 0 & 1 \end{pmatrix} \begin{pmatrix} 1 & 1 & 0 \\ 0 & 1 & 0 \\ 0 & 0 & 1 \end{pmatrix} \begin{pmatrix} 0 & 1 & 0 \\ -1 & 0 & 0 \\ 0 & 0 & 1 \end{pmatrix} = \begin{pmatrix} 1 & 0 & 0 \\ -1 & 1 & 0 \\ 0 & 0 & 1 \end{pmatrix}. \tag{2.62}$$

The physical entity of the tensor has not changed; however, its representation in the rotated coordinate system is different than that of the original coordinate system.

Levi-Civita Symbol

The *Levi-Civita*[4] *symbol*, ϵ_{ijk}, will soon be seen to be useful when we introduce the vector cross product and curl operator. It is defined as follows:

$$\epsilon_{ijk} = \begin{cases} 1 & \text{if } ijk = 123, 231, \text{ or } 312, \\ 0 & \text{if any two indices identical}, \\ -1 & \text{if } ijk = 321, 213, \text{ or } 132. \end{cases} \tag{2.63}$$

Another way to remember this is to start with the sequence 123, and associate that with $+1$. A sequential permutation, say from 123 to 231, retains the positive nature. A trade, say from 123 to 213, gives a negative value. The Levi-Civita symbol is also known as the *permutation symbol* or the *alternating symbol*. An identity that will be used extensively

$$\epsilon_{ijk}\epsilon_{ilm} = \delta_{jl}\delta_{km} - \delta_{jm}\delta_{kl}, \tag{2.64}$$

can be proved a number of ways, including direct substitution for all values of i, j, k, l, m.

Isotropic Tensors

We can define an *isotropic tensor* as a tensor whose components are invariant following an arbitrary rotation. All scalars are isotropic. There exist no vectors that are isotropic. The Kronecker delta is an isotropic tensor; the Levi-Civita symbol is isotropic, but, for reasons discussed in the following example, is often described as a *pseudo-tensor*.

---

[4] Tullio Levi-Civita, 1873–1941, Italian mathematician.

Example 2.8 Examine how the Kronecker delta and Levi-Civita symbol behave under rotation transformations, and verify both are isotropic.

Solution

First, let us consider the Kronecker delta. With Cartesian index notation, we have

$$\delta'_{pq} = \sum_{i=1}^{3}\sum_{j=1}^{3} \ell_{ip}\ell_{jq}\delta_{ij} = \sum_{i=1}^{3}\sum_{j=1}^{3} \ell_{ip}\ell_{iq} = \delta_{pq}. \tag{2.65}$$

The components of the Kronecker delta are unchanged by rotation, so it is an isotropic tensor. Gibbs notation is more easily understood, so we take $\delta_{ij} = \mathbf{I}$, the identity matrix.

$$\mathbf{I}' = \mathbf{Q}^T \cdot \mathbf{I} \cdot \mathbf{Q} = \mathbf{Q}^T \cdot \mathbf{Q} = \mathbf{Q}^{-1} \cdot \mathbf{Q} = \mathbf{I}. \tag{2.66}$$

The representation is identical in either coordinates. For our rotation matrix, we must have $\det \mathbf{Q} = 1$. For a reflection transformation with determinant of -1, we would also find $\mathbf{I}' = \mathbf{I}$. The behavior of the Levi-Civita symbol under transformation is better evaluated with index notation. Employing the formal summations, we have

$$\epsilon'_{pqr} = \sum_{i=1}^{3}\sum_{j=1}^{3}\sum_{k=1}^{3} \ell_{ip}\ell_{jq}\ell_{kr}\epsilon_{ijk}. \tag{2.67}$$

Detailed calculation for arbitrary rotation matrices verifies that the components of the transformed quantity are identical to those of the untransformed. So for rotations, one could say $\epsilon'_{ijk} = \epsilon_{ijk}$, rendering it isotropic under rotation. However, for arbitrary reflection matrices, the transformed components are additive inverses of the untransformed components; so for reflections, one has $\epsilon'_{ijk} = -\epsilon_{ijk}$. Depending on how precisely one defines a tensor, one finds various interpretations in the literature as to whether or not these quantities are tensors. A detailed discussion is given by Riley et al. (2006), who describe both δ_{ij} and ϵ_{ijk} as isotropic tensors in the context of rotations. Under reflection, they describe ϵ_{ijk} as a pseudo-tensor, in that its components are not invariant under a reflection transformation. Yih (1977) provides additional discussion of pseudo-tensors for Cartesian and non-Cartesian systems.

Some Secondary Definitions

Transpose The transpose of a second-order tensor, denoted by a superscript T, is found by exchanging elements about the diagonal. In Cartesian index notation, this is simply

$$T_{ij}^T = T_{ji}. \tag{2.68}$$

Written in full, if

$$T_{ij} = \begin{pmatrix} T_{11} & T_{12} & T_{13} \\ T_{21} & T_{22} & T_{23} \\ T_{31} & T_{32} & T_{33} \end{pmatrix}, \quad \text{then} \quad T_{ij}^T = T_{ji} = \begin{pmatrix} T_{11} & T_{21} & T_{31} \\ T_{12} & T_{22} & T_{32} \\ T_{13} & T_{23} & T_{33} \end{pmatrix}. \tag{2.69}$$

Symmetric A tensor D_{ij} is *symmetric* iff

$$D_{ij} = D_{ji}, \quad \mathbf{D} = \mathbf{D}^T. \tag{2.70}$$

A symmetric tensor has only six independent scalars. We will reserve $\mathbf{D}$ for tensors that are symmetric, and specifically use it to describe the deformation of a fluid element.

Anti-symmetric A tensor R_{ij} is *anti-symmetric* iff

$$R_{ij} = -R_{ji}, \qquad \mathbf{R} = -\mathbf{R}^T. \tag{2.71}$$

An anti-symmetric tensor must have zeros on its diagonal and only three independent scalars on off-diagonal elements. We will reserve $\mathbf{R}$ for tensors that are anti-symmetric and use it to characterize rotation of a fluid element. But $\mathbf{R}$ is *not* a rotation matrix in the sense that $\mathbf{Q}$ is. We will call $\mathbf{R}$ a *rotation tensor*.

Decomposition An arbitrary tensor T_{ij} can be separated into a symmetric and anti-symmetric pair of tensors:

$$T_{ij} = \frac{1}{2}T_{ij} + \frac{1}{2}T_{ij} + \frac{1}{2}T_{ji} - \frac{1}{2}T_{ji} = \underbrace{\frac{1}{2}\left(T_{ij} + T_{ji}\right)}_{\text{symmetric}} + \underbrace{\frac{1}{2}\left(T_{ij} - T_{ji}\right)}_{\text{anti-symmetric}}. \tag{2.72}$$

The first term must be symmetric, and the second term must be anti-symmetric. This can be seen by considering applying this to any matrix of actual numbers. If we define the symmetric part of the matrix T_{ij} by the following notation:

$$T_{(ij)} = \frac{1}{2}\left(T_{ij} + T_{ji}\right), \tag{2.73}$$

and the anti-symmetric part of the same matrix by the following notation:

$$T_{[ij]} = \frac{1}{2}\left(T_{ij} - T_{ji}\right), \tag{2.74}$$

we then have

$$T_{ij} = T_{(ij)} + T_{[ij]}. \tag{2.75}$$

Tensor Product

The *tensor product* between two arbitrary tensors yields a third tensor. It is the equivalent of matrix multiplication. For second-order tensors, we have the tensor product as

$$S_{ij}T_{jk} = A_{ik}, \qquad \mathbf{S} \cdot \mathbf{T} = \mathbf{A}. \tag{2.76}$$

Note that j is a dummy index, i and k are free indices, and that the free indices in each additive term are the same. In that sense they behave much as dimensional units, which must be the same for each term.

An important property of tensors is that, in general, the tensor product *does not commute*, $\mathbf{S} \cdot \mathbf{T} \neq \mathbf{T} \cdot \mathbf{S}$. In a very formal manifestation of Cartesian index notation, one would not commute the elements, and the dummy indices would appear next to another in adjacent terms as shown. However, it is of no great consequence to change the order of terms so that we can write $S_{ij}T_{jk} = T_{jk}S_{ij}$. That is, in Cartesian index notation, elements do commute. *But*, in Cartesian index notation, the order of the indices is important, and it is this order that does not commute: $S_{ij}T_{jk} \neq S_{ji}T_{jk}$ in general. The version presented for $S_{ij}T_{jk}$ in Eq. (2.76), in which the dummy index j is juxtaposed between each term, is slightly preferable as it maintains the order found in the Gibbs notation.

Example 2.9 For two general 2×2 tensors, **S** and **T**, find the tensor product.

Solution
The tensor product is

$$\mathbf{S} \cdot \mathbf{T} = \begin{pmatrix} S_{11} & S_{12} \\ S_{21} & S_{22} \end{pmatrix} \begin{pmatrix} T_{11} & T_{12} \\ T_{21} & T_{22} \end{pmatrix} = \begin{pmatrix} S_{11}T_{11} + S_{12}T_{21} & S_{11}T_{12} + S_{12}T_{22} \\ S_{21}T_{11} + S_{22}T_{21} & S_{21}T_{12} + S_{22}T_{22} \end{pmatrix}. \quad (2.77)$$

Compare with the commutation:

$$\mathbf{T} \cdot \mathbf{S} = \begin{pmatrix} T_{11} & T_{12} \\ T_{21} & T_{22} \end{pmatrix} \begin{pmatrix} S_{11} & S_{12} \\ S_{21} & S_{22} \end{pmatrix} = \begin{pmatrix} S_{11}T_{11} + S_{21}T_{12} & S_{12}T_{11} + S_{22}T_{12} \\ S_{11}T_{21} + S_{21}T_{22} & S_{12}T_{21} + S_{22}T_{22} \end{pmatrix}. \quad (2.78)$$

Clearly $\mathbf{S} \cdot \mathbf{T} \neq \mathbf{T} \cdot \mathbf{S}$. It can be shown that if both **S** and **T** are square and symmetric, that $\mathbf{S} \cdot \mathbf{T} = (\mathbf{T} \cdot \mathbf{S})^T$.

Tensor Inner Product

The *tensor inner product* of two tensors T_{ij} and S_{ji} is defined as follows:

$$T_{ij}S_{ji} = a, \qquad \mathbf{T} : \mathbf{S} = a, \quad (2.79)$$

where a is a scalar. The tensor inner product commutes as

$$\mathbf{T} : \mathbf{S} = \mathbf{S} : \mathbf{T} = \text{tr}(\mathbf{T} \cdot \mathbf{S}) = \text{tr}(\mathbf{S} \cdot \mathbf{T}) = a. \quad (2.80)$$

In forming an algorithm to calculate the tensor inner product, there is less ambiguity if one uses the formulations $T_{ij}S_{ji} = a$ or $\text{tr}(\mathbf{T} \cdot \mathbf{S}) = a$ to guide algorithm construction. It can be shown, and will be important in upcoming derivations, that the tensor inner product of any symmetric tensor **D** with any anti-symmetric tensor **R** is the scalar zero:

$$D_{ij}R_{ji} = 0, \qquad \mathbf{D} : \mathbf{R} = 0. \quad (2.81)$$

In contrast to the tensor product, which has one pair of dummy indices and one dot, the tensor inner product has two pairs of dummy indices and two dots.

Example 2.10 For all 2×2 matrices, prove the tensor inner product of general symmetric and anti-symmetric tensors is zero.

Solution
Take

$$\mathbf{D} = \begin{pmatrix} a & b \\ b & c \end{pmatrix}, \qquad \mathbf{R} = \begin{pmatrix} 0 & d \\ -d & 0 \end{pmatrix}. \quad (2.82)$$

By definition then

$$\mathbf{D} : \mathbf{R} = D_{ij}R_{ji} = D_{11}R_{11} + D_{12}R_{21} + D_{21}R_{12} + D_{22}R_{22}, \quad (2.83)$$

$$= a(0) + b(-d) + bd + c(0) = 0. \quad \text{QED}. \quad (2.84)$$

The theorem is proved. The proof can be extended to arbitrary square matrices. The common abbreviation QED at the end of the proof stands for the Latin *quod erat demonstrandum*, "that which was to be demonstrated."

Dual Vector of a Tensor

We define the *dual vector*, d_i, of a tensor T_{jk} as follows:

$$d_i = \frac{1}{2}\epsilon_{ijk}T_{jk} = \frac{1}{2}\underbrace{\epsilon_{ijk}T_{(jk)}}_{=0} + \frac{1}{2}\epsilon_{ijk}T_{[jk]}. \tag{2.85}$$

There is a lack of uniformity in the literature in this topic. For example, this definition differs from that given by Panton (2013, p. 39) by a factor of $1/2$. It is equivalent to that given by Aris (1962, p. 25). The term ϵ_{ijk} is anti-symmetric for any fixed i; for example for $i = 1$, we have

$$\epsilon_{1jk} = \begin{pmatrix} \epsilon_{111} & \epsilon_{112} & \epsilon_{113} \\ \epsilon_{121} & \epsilon_{122} & \epsilon_{123} \\ \epsilon_{131} & \epsilon_{132} & \epsilon_{133} \end{pmatrix} = \begin{pmatrix} 0 & 0 & 0 \\ 0 & 0 & 1 \\ 0 & -1 & 0 \end{pmatrix}. \tag{2.86}$$

Thus, when its tensor inner product is taken with the symmetric $T_{(jk)}$, the result must be the scalar zero. Hence, we also have

$$d_i = \frac{1}{2}\epsilon_{ijk}T_{[jk]}. \tag{2.87}$$

We can adapt the Gibbs notation of Aris (1962, p. 103), employing our convention of lower case for vectors, to say $\mathbf{t}_\times \equiv \epsilon_{ijk}T_{[jk]}$ that gives

$$\mathbf{d} = \frac{1}{2}\mathbf{t}_\times. \tag{2.88}$$

Let us find the inverse relation for d_i. Starting with Eq. (2.85), we take the dot product of d_i with ϵ_{ilm} to get

$$\epsilon_{ilm}d_i = \frac{1}{2}\epsilon_{ilm}\epsilon_{ijk}T_{jk}. \tag{2.89}$$

Employing Eq. (2.64) to eliminate the ϵ's in favor of δ's, we get

$$\epsilon_{ilm}d_i = \frac{1}{2}\left(\delta_{lj}\delta_{mk} - \delta_{lk}\delta_{mj}\right)T_{jk} = \frac{1}{2}(T_{lm} - T_{ml}) = T_{[lm]}. \tag{2.90}$$

Note that

$$T_{[lm]} = \epsilon_{1lm}d_1 + \epsilon_{2lm}d_2 + \epsilon_{3lm}d_3 = \begin{pmatrix} 0 & d_3 & -d_2 \\ -d_3 & 0 & d_1 \\ d_2 & -d_1 & 0 \end{pmatrix}. \tag{2.91}$$

And we can write the decomposition of an arbitrary tensor as the sum of its symmetric part and a factor related to the dual vector associated with its anti-symmetric part:

$$\underbrace{T_{ij}}_{\text{arbitrary tensor}} = \underbrace{T_{(ij)}}_{\text{symmetric part}} + \underbrace{\epsilon_{kij}d_k}_{\text{anti-symmetric part}}. \tag{2.92}$$

Vector-Tensor Product

The product of a vector and tensor, again that does not in general commute, comes in two flavors, pre-multiplication and post-multiplication, both important, and given in Cartesian index and Gibbs notation next.

Pre-multiplication

$$u_j = v_i T_{ij} = T_{ij} v_i, \qquad \mathbf{u}^T = \mathbf{v}^T \cdot \mathbf{T} \neq \mathbf{T} \cdot \mathbf{v}. \tag{2.93}$$

In Cartesian index notation, the first form is preferred as it has a correspondence with the Gibbs notation; both are correct representations given our summation convention.

Post-multiplication

$$w_i = T_{ij} v_j = v_j T_{ij}, \qquad \mathbf{w} = \mathbf{T} \cdot \mathbf{v} \neq \mathbf{v}^T \cdot \mathbf{T}. \tag{2.94}$$

Dyadic Product

As opposed to the dot product between two vectors, which yields a scalar, we also have the *dyadic product*, which yields a tensor. In Cartesian index and Gibbs notation, we have

$$T_{ij} = u_i v_j = v_j u_i, \qquad \mathbf{T} = \mathbf{u} \mathbf{v}^T \neq \mathbf{v} \mathbf{u}^T. \tag{2.95}$$

Notice there is no dot in the dyadic product; the dot is reserved for the inner product.

Example 2.11 Find the dyadic product between two general two-dimensional vectors. Show the dyadic product does not commute in general, and find the condition under which it does commute.

Solution
Take

$$\mathbf{u} = \begin{pmatrix} u_1 \\ u_2 \end{pmatrix}, \qquad \mathbf{v} = \begin{pmatrix} v_1 \\ v_2 \end{pmatrix}. \tag{2.96}$$

Then

$$\mathbf{u}\mathbf{v}^T = u_i v_j = \begin{pmatrix} u_1 \\ u_2 \end{pmatrix} \begin{pmatrix} v_1 & v_2 \end{pmatrix} = \begin{pmatrix} u_1 v_1 & u_1 v_2 \\ u_2 v_1 & u_2 v_2 \end{pmatrix}. \tag{2.97}$$

Compare this to the commuted operation, $\mathbf{v}\mathbf{u}^T$:

$$\mathbf{v}\mathbf{u}^T = v_i u_j = \begin{pmatrix} v_1 \\ v_2 \end{pmatrix} \begin{pmatrix} u_1 & u_2 \end{pmatrix} = \begin{pmatrix} v_1 u_1 & v_1 u_2 \\ v_2 u_1 & v_2 u_2 \end{pmatrix}. \tag{2.98}$$

By inspection, we see the operations in general do not commute. They do commute if $v_2/v_1 = u_2/u_1$. So in order for the dyadic product to commute, $\mathbf{u}$ and $\mathbf{v}$ must be parallel.
It can be seen that the dyadic product $\mathbf{v}\mathbf{v}^T$ is a symmetric tensor. For the two-dimensional system, we would have

$$\mathbf{v}\mathbf{v}^T = v_i v_j = \begin{pmatrix} v_1 \\ v_2 \end{pmatrix} \begin{pmatrix} v_1 & v_2 \end{pmatrix} = \begin{pmatrix} v_1 v_1 & v_1 v_2 \\ v_2 v_1 & v_2 v_2 \end{pmatrix}. \tag{2.99}$$

If we decompose a tensor into its symmetric and anti-symmetric parts, $T_{ij} = T_{(ij)} + T_{[ij]}$ and take $T_{(ij)} = D_{ij} = \mathbf{D}$ and $T_{[ij]} = R_{ij} = \mathbf{R}$, so that $\mathbf{T} = \mathbf{D} + \mathbf{R}$, the following common term can be expressed as a tensor inner product with a dyadic product:

$$x_i T_{ij} x_j = \mathbf{x}^T \cdot \mathbf{T} \cdot \mathbf{x}, \tag{2.100}$$

$$x_i \left(T_{(ij)} + T_{[ij]}\right) x_j = \mathbf{x}^T \cdot (\mathbf{D} + \mathbf{R}) \cdot \mathbf{x}, \tag{2.101}$$

$$x_i T_{(ij)} x_j = \mathbf{x}^T \cdot \mathbf{D} \cdot \mathbf{x}, \quad \text{or} \quad T_{(ij)} x_i x_j = \mathbf{D} : \mathbf{x}\mathbf{x}^T. \tag{2.102}$$

Contraction

We contract a general tensor, which has all of its subscripts different, by setting one of its subscripts to be the same as another. A single contraction will reduce the order of a tensor by two. For example, the contraction of the second-order tensor T_{ij} is T_{ii}, which indicates a sum is to be performed:

$$T_{ii} = T_{11} + T_{22} + T_{33}, \quad \text{tr } \mathbf{T} = T_{11} + T_{22} + T_{33}. \tag{2.103}$$

So, in this case the contraction yields a scalar. In matrix algebra, this particular contraction is the trace of the matrix.

Vector Cross Product

The *vector cross product* is defined in Cartesian index and Gibbs notation as

$$w_i = \epsilon_{ijk} u_j v_k, \quad \mathbf{w} = \mathbf{u} \times \mathbf{v}. \tag{2.104}$$

Expanding for $i = 1, 2, 3$ gives

$$w_1 = \epsilon_{123} u_2 v_3 + \epsilon_{132} u_3 v_2 = u_2 v_3 - u_3 v_2, \tag{2.105}$$

$$w_2 = \epsilon_{231} u_3 v_1 + \epsilon_{213} u_1 v_3 = u_3 v_1 - u_1 v_3, \tag{2.106}$$

$$w_3 = \epsilon_{312} u_1 v_2 + \epsilon_{321} u_2 v_1 = u_1 v_2 - u_2 v_1. \tag{2.107}$$

We could also say

$$\mathbf{w} = \mathbf{u} \times \mathbf{v} = \begin{vmatrix} \mathbf{e}_1 & \mathbf{e}_2 & \mathbf{e}_3 \\ u_1 & u_2 & u_3 \\ v_1 & v_2 & v_3 \end{vmatrix}. \tag{2.108}$$

The cross product does not commute, as $\mathbf{u} \times \mathbf{v} = -\mathbf{v} \times \mathbf{u}$. It can be shown that under an orthogonal transformation $\mathbf{Q}$ that $\mathbf{w}' = (\det \mathbf{Q})(\mathbf{u}' \times \mathbf{v}')$; thus, the cross product is invariant under rotation, for which $\det \mathbf{Q} = 1$. But it is not invariant under reflection, for which $\det \mathbf{Q} = -1$. Because of its failure to obey all vector transformation rules, while retaining many properties of ordinary vectors, terms such as $\mathbf{w}$ that arise from a vector cross product are often called *pseudo-vectors*.

Example 2.12 Using the determinant form of the cross product, show for our Cartesian coordinate system that obeys the right-hand rule that $\mathbf{e}_3 = \mathbf{e}_1 \times \mathbf{e}_2$.

Solution
We specialize Eq. (2.108) to get

$$\mathbf{e}_1 \times \mathbf{e}_2 = \begin{vmatrix} \mathbf{e}_1 & \mathbf{e}_2 & \mathbf{e}_3 \\ 1 & 0 & 0 \\ 0 & 1 & 0 \end{vmatrix} = 0\mathbf{e}_1 - 0\mathbf{e}_2 + (1)\mathbf{e}_3 = \mathbf{e}_3. \quad (2.109)$$

It is easy to show that $\mathbf{e}_2 \times \mathbf{e}_1 = -\mathbf{e}_3$. However, had we chosen a left-handed coordinate system, we would have required $\mathbf{e}_2 \times \mathbf{e}_1 = \mathbf{e}_3$.

Finally, as an aside, we note that given any vector $\mathbf{v}$ and any unit vector $\boldsymbol{\alpha}$, that $\mathbf{v}$ can be decomposed into a vector tangent to $\boldsymbol{\alpha}$, $\mathbf{v}_t = (\boldsymbol{\alpha}^T \cdot \mathbf{v})\boldsymbol{\alpha}$, and a vector normal to $\boldsymbol{\alpha}$, $\mathbf{v}_n = \boldsymbol{\alpha} \times (\mathbf{v} \times \boldsymbol{\alpha})$:

$$\mathbf{v} = \mathbf{v}_t + \mathbf{v}_n = (\boldsymbol{\alpha}^T \cdot \mathbf{v})\boldsymbol{\alpha} + \boldsymbol{\alpha} \times (\mathbf{v} \times \boldsymbol{\alpha}). \quad (2.110)$$

Vector Associated with a Plane

We often have to select a vector that is associated with a particular direction. Now for any direction we choose, there exists an associated unit vector and normal plane. Recall that our notation has been defined so that the first index is associated with a face or direction, and the second index corresponds to the components of the vector associated with that face. If we take n_i to be a unit normal vector associated with a given direction and normal plane, and we have been given a tensor T_{ij}, the vector t_j associated with that plane is given in Cartesian index and Gibbs notation by

$$t_j = n_i T_{ij}, \qquad \mathbf{t}^T = \mathbf{n}^T \cdot \mathbf{T}, \qquad \mathbf{t} = \mathbf{T}^T \cdot \mathbf{n}. \quad (2.111)$$

Here $\mathbf{t}$ is *not* a tangent vector; nor does have any relation to time t. This notation n_i or $\mathbf{n}$ is common for a unit normal vector; we will also employ $\boldsymbol{\alpha}_n = \mathbf{n}$ in portions of the text. Similarly, we will mainly use $\boldsymbol{\alpha}_t$ for a unit tangent vector. A diagram of a Cartesian element with the tensor components shown on the proper face is shown in Fig. 2.4.

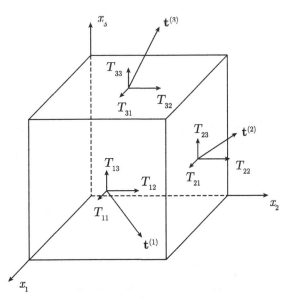

Figure 2.4 Sample Cartesian element that is aligned with coordinate axes, along with tensor components and vectors associated with each face.

Example 2.13 Find the vector associated with the 1 face, $\mathbf{t}^{(1)}$, as shown in Fig. 2.4,

Solution
We first choose the unit normal associated with the x_1 face, that is the vector $n_i = (1,0,0)^T$. The associated vector is found by doing the actual summation:

$$t_j = n_i T_{ij} = n_1 T_{1j} + n_2 T_{2j} + n_3 T_{3j}. \tag{2.112}$$

Now $n_1 = 1$, $n_2 = 0$, and $n_3 = 0$, so for this problem, we have

$$t_j^{(1)} = T_{1j}. \tag{2.113}$$

2.2 Solution of Linear Algebraic Equations

Because standard notions from linear algebra are often required to analyze fluid mechanics, we briefly discuss the solution of linear algebraic equations of the form

$$A_{ij} x_j = b_i, \qquad \mathbf{A} \cdot \mathbf{x} = \mathbf{b}. \tag{2.114}$$

Full details can be found in any text addressing linear algebra, for example Strang (2006) or Powers and Sen (2015, Chapter 7). Here, we are using standard notation from linear algebra. Of course, we could simply replace $\mathbf{b}$ by $\mathbf{x}'$ and interpret this as a coordinate mapping: $\mathbf{x}' = \mathbf{A} \cdot \mathbf{x}$. Let us presume that $\mathbf{A}$ is a known square matrix of dimension $N \times N$, $\mathbf{x}$ is an unknown column vector of dimension $N \times 1$, and $\mathbf{b}$ is a known column vector of dimension $N \times 1$. The following can be proved: (1) A unique solution for $\mathbf{x}$ exists iff $\det \mathbf{A} \neq 0$. (2) If $\det \mathbf{A} = 0$, solutions for $\mathbf{x}$ may or may not exist; if they exist, they are not unique. (3) *Cramer's*[5] *rule*, a method involving the ratio of determinants discussed in linear algebra texts, can be used to find $\mathbf{x}$; other methods exist, such as Gaussian[6] elimination. Let us consider a few examples for $N = 2$.

Example 2.14 Use Cramer's rule to solve a general linear algebra problem with $N = 2$.

Solution
Consider then

$$\begin{pmatrix} a_{11} & a_{12} \\ a_{21} & a_{22} \end{pmatrix} \begin{pmatrix} x_1 \\ x_2 \end{pmatrix} = \begin{pmatrix} b_1 \\ b_2 \end{pmatrix}. \tag{2.115}$$

The solution from Cramer's rule involves the ratio of determinants. We get

$$x_1 = \frac{\begin{vmatrix} b_1 & a_{12} \\ b_2 & a_{22} \end{vmatrix}}{\begin{vmatrix} a_{11} & a_{12} \\ a_{21} & a_{22} \end{vmatrix}} = \frac{b_1 a_{22} - b_2 a_{12}}{a_{11} a_{22} - a_{12} a_{21}}, \qquad x_2 = \frac{\begin{vmatrix} a_{11} & b_1 \\ a_{21} & b_2 \end{vmatrix}}{\begin{vmatrix} a_{11} & a_{12} \\ a_{21} & a_{22} \end{vmatrix}} = \frac{b_2 a_{11} - b_1 a_{21}}{a_{11} a_{22} - a_{12} a_{21}}. \tag{2.116}$$

---

[5] Gabriel Cramer, 1704–1752, Swiss mathematician at University of Geneva.
[6] Carl Friedrich Gauss, 1777–1855, Brunswick-born German mathematician who is considered the founder of modern mathematics. Worked in astronomy, physics, crystallography, optics, biostatistics, and mechanics.

If $b_1, b_2 \neq 0$ and $\det \mathbf{A} = a_{11}a_{22} - a_{12}a_{21} \neq 0$, there is a unique nontrivial solution for $\mathbf{x}$. If $b_1 = b_2 = 0$ and $\det \mathbf{A} = a_{11}a_{22} - a_{12}a_{21} \neq 0$, we must have $x_1 = x_2 = 0$. Obviously, if $\det \mathbf{A} = a_{11}a_{22} - a_{12}a_{21} = 0$, we cannot use Cramer's rule to compute as it involves division by zero. But we can salvage a nonunique solution if we also have $b_1 = b_2 = 0$, as we shall see.

Example 2.15 Find any and all solutions for

$$\begin{pmatrix} 1 & 2 \\ 2 & 4 \end{pmatrix} \begin{pmatrix} x_1 \\ x_2 \end{pmatrix} = \begin{pmatrix} 0 \\ 0 \end{pmatrix}. \tag{2.117}$$

Solution
Certainly $(x_1, x_2)^T = (0,0)^T$ is a solution. But there are more. Cramer's rule gives

$$x_1 = \frac{\begin{vmatrix} 0 & 2 \\ 0 & 4 \end{vmatrix}}{\begin{vmatrix} 1 & 2 \\ 2 & 4 \end{vmatrix}} = \frac{0}{0}, \qquad x_2 = \frac{\begin{vmatrix} 1 & 0 \\ 2 & 0 \end{vmatrix}}{\begin{vmatrix} 1 & 2 \\ 2 & 4 \end{vmatrix}} = \frac{0}{0}. \tag{2.118}$$

This is indeterminate! But the more robust Gaussian elimination process allows us to use row operations (multiply the top row by -2 and add to the bottom row) to rewrite the original equation as

$$\begin{pmatrix} 1 & 2 \\ 0 & 0 \end{pmatrix} \begin{pmatrix} x_1 \\ x_2 \end{pmatrix} = \begin{pmatrix} 0 \\ 0 \end{pmatrix}. \tag{2.119}$$

By inspection, we get an infinite number of solutions, given by the one-parameter family

$$x_1 = -2s, \qquad x_2 = s, \qquad s \in \mathbb{R}^1. \tag{2.120}$$

We could also eliminate s and say that $x_1 = -2x_2$. The solutions are linearly dependent. In terms of the language of vectors, we find the solution to be a vector of fixed direction, with arbitrary magnitude. In terms of a unit vector, we could write the solution as

$$\mathbf{x} = s \begin{pmatrix} -\frac{2}{\sqrt{5}} \\ \frac{1}{\sqrt{5}} \end{pmatrix}, \qquad s \in \mathbb{R}^1. \tag{2.121}$$

Example 2.16 Find any and all solutions that best satisfy

$$\begin{pmatrix} 1 & 2 \\ 2 & 4 \end{pmatrix} \begin{pmatrix} x_1 \\ x_2 \end{pmatrix} = \begin{pmatrix} 1 \\ 0 \end{pmatrix}. \tag{2.122}$$

Solution
Cramer's rule gives

$$x_1 = \frac{\begin{vmatrix} 1 & 2 \\ 0 & 4 \end{vmatrix}}{\begin{vmatrix} 1 & 2 \\ 2 & 4 \end{vmatrix}} = \frac{4}{0}, \qquad x_2 = \frac{\begin{vmatrix} 1 & 1 \\ 2 & 0 \end{vmatrix}}{\begin{vmatrix} 1 & 2 \\ 2 & 4 \end{vmatrix}} = \frac{-2}{0}. \tag{2.123}$$

There is no finite value of $\mathbf{x}$ that satisfies $\mathbf{A} \cdot \mathbf{x} = \mathbf{b}$!

There is, however, in some sense a *best* solution, that is, an $\mathbf{x}$ of minimum length that also minimizes $||\mathbf{A} \cdot \mathbf{x} - \mathbf{b}||$. Using the pseudo-inverse procedure described in Powers and Sen (2015, Section 7.15), we find there exists a nonunique set of $\mathbf{x} = (1/25 - 2s, 2/25 + s)^T$, $s \in \mathbb{R}^1$, for which the so-called error norm, e, takes on the same minimum value, $e = ||\mathbf{A} \cdot \mathbf{x} - \mathbf{b}|| = 2/\sqrt{5}$, for all values of s. For $s = 0$, we then get the "best" $\mathbf{x} = (1/25, 2/25)^T$ in that this $\mathbf{x}$ minimizes the error and is itself of minimum length.

More generally, the matrix $\mathbf{A}$ considered here is not useful as a coordinate mapping because it is singular with $\det \mathbf{A} = 0$. It maps general two-dimensional vectors $\mathbf{x}$ into a one-dimensional space; that is, it takes a plane into a line.

We have confined attention to square matrices of dimension $N \times N$ as those are the most common type that arise in fluid mechanics. The analysis may be extended for non-square matrices of dimension $N \times M$; such matrices may be physically important, for example in the stoichiometry of chemical systems; see Powers (2016), or robotics.

Example 2.17 Consider how $\mathbf{A} \cdot \mathbf{x} = \mathbf{b}$ transforms under rotation or reflection.

Solution
Using
$$\mathbf{A}' = \mathbf{Q}^T \cdot \mathbf{A} \cdot \mathbf{Q}, \qquad \mathbf{x}' = \mathbf{Q}^T \cdot \mathbf{x}, \qquad \mathbf{b}' = \mathbf{Q}^T \cdot \mathbf{b}, \tag{2.124}$$
we see that by pre-multiplying all equations by $\mathbf{Q}$, and post-multiplying the tensor equation by $\mathbf{Q}^T$ that
$$\mathbf{A} = \mathbf{Q} \cdot \mathbf{A}' \cdot \mathbf{Q}^T, \qquad \mathbf{x} = \mathbf{Q} \cdot \mathbf{x}', \qquad \mathbf{b} = \mathbf{Q} \cdot \mathbf{b}', \tag{2.125}$$
transforming $\mathbf{A} \cdot \mathbf{x} = \mathbf{b}$ to
$$\underbrace{\mathbf{Q} \cdot \mathbf{A}' \cdot \mathbf{Q}^T}_{\mathbf{A}} \cdot \underbrace{\mathbf{Q} \cdot \mathbf{x}'}_{\mathbf{x}} = \underbrace{\mathbf{Q} \cdot \mathbf{b}'}_{\mathbf{b}}, \quad \text{so} \quad \mathbf{Q} \cdot \mathbf{A}' \cdot \mathbf{x}' = \mathbf{Q} \cdot \mathbf{b}', \tag{2.126}$$
$$\underbrace{\mathbf{Q}^T \cdot \mathbf{Q}}_{\mathbf{I}} \cdot \mathbf{A}' \cdot \mathbf{x}' = \underbrace{\mathbf{Q}^T \cdot \mathbf{Q}}_{\mathbf{I}} \cdot \mathbf{b}', \quad \text{so} \quad \mathbf{A}' \cdot \mathbf{x}' = \mathbf{b}'. \tag{2.127}$$

Obviously, the form is invariant under rotation or reflection. The result holds if $\det \mathbf{Q} = \pm 1$.

2.3 Eigenvalues, Eigenvectors, and Tensor Invariants

For a given T_{ij}, it may be possible to select a plane for which the vector $t_j = n_i T_{ij}$ associated with that plane is aligned with the normal associated with the plane. For a three-dimensional element, it may be possible to choose three planes for which the vector associated with the given planes is aligned with the unit normal associated with those planes. We can think of this as finding a rotation as depicted in Fig. 2.5.

Mathematically, we can enforce this condition by requiring that
$$\underbrace{n_i T_{ij}}_{\text{vector associated with chosen direction}} = \underbrace{\lambda n_j}_{\text{scalar multiple of chosen direction}}. \tag{2.128}$$

2.3 Eigenvalues, Eigenvectors, and Tensor Invariants

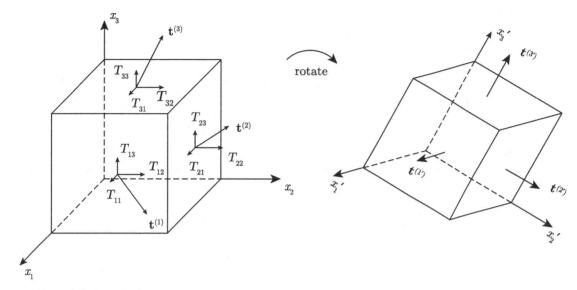

Figure 2.5 Sample Cartesian element that is rotated so that its faces have vectors that are aligned with the unit normals associated with the faces of the element.

Here λ is an as-of-yet unknown scalar. The vector n_i could be a unit vector, but does not have to be. We can rewrite this as

$$n_i T_{ij} = \lambda n_i \delta_{ij}, \qquad \mathbf{n}^T \cdot \mathbf{T} = \lambda \mathbf{n}^T \cdot \mathbf{I}. \tag{2.129}$$

The directions n_i that are found are aligned with the *principal axes* of the tensor. We might call λ the principal value of the tensor, but it is usually called the *eigenvalue*. In mathematics, this is known as a left eigenvalue problem. Solutions n_i that are nontrivial are known as *eigenvectors*. More specifically, they are left eigenvectors. We can also formulate this as a right eigenvalue problem by taking the transpose of both sides to obtain $\mathbf{T}^T \cdot \mathbf{n} = \lambda \mathbf{I} \cdot \mathbf{n}$. Here we have used the fact that $\mathbf{I}^T = \mathbf{I}$. The left eigenvectors of $\mathbf{T}$ are the right eigenvectors of $\mathbf{T}^T$. Eigenvalue problems are quite general and arise whenever an operator operates on a vector to generate a vector that leaves the original unchanged except in magnitude.

We can rearrange to form

$$n_i (T_{ij} - \lambda \delta_{ij}) = 0. \tag{2.130}$$

In matrix notation, this can be written as

$$\begin{pmatrix} n_1 & n_2 & n_3 \end{pmatrix} \begin{pmatrix} T_{11} - \lambda & T_{12} & T_{13} \\ T_{21} & T_{22} - \lambda & T_{23} \\ T_{31} & T_{32} & T_{33} - \lambda \end{pmatrix} = \begin{pmatrix} 0 & 0 & 0 \end{pmatrix}. \tag{2.131}$$

A trivial solution to this equation is $(n_1, n_2, n_3) = (0, 0, 0)$. But this is not interesting. As suggested by our understanding of Cramer's rule, we can get a nonunique, nontrivial solution if we enforce the condition that the determinant of the coefficient matrix be zero. As we have an unknown parameter λ, we have sufficient degrees of freedom to accomplish this. So, we require

$$\begin{vmatrix} T_{11} - \lambda & T_{12} & T_{13} \\ T_{21} & T_{22} - \lambda & T_{23} \\ T_{31} & T_{32} & T_{33} - \lambda \end{vmatrix} = 0. \tag{2.132}$$

We know from linear algebra that such an equation for a third-order matrix gives rise to a characteristic polynomial for λ of the form

$$\lambda^3 - I_T^{(1)}\lambda^2 + I_T^{(2)}\lambda - I_T^{(3)} = 0, \tag{2.133}$$

where $I_T^{(1)}, I_T^{(2)}, I_T^{(3)}$ are scalars that are functions of all the scalars T_{ij}. The I_T's are known as the *invariants* of the tensor T_{ij}. They can be shown, following a detailed analysis, to be given by

$$I_T^{(1)} = T_{ii} = \text{tr } \mathbf{T}, \tag{2.134}$$

$$I_T^{(2)} = \frac{1}{2}\left(T_{ii}T_{jj} - T_{ij}T_{ji}\right) = \frac{1}{2}\left((\text{tr } \mathbf{T})^2 - \text{tr}(\mathbf{T} \cdot \mathbf{T})\right) = (\det \mathbf{T})\left(\text{tr } \mathbf{T}^{-1}\right), \tag{2.135}$$

$$= \frac{1}{2}\left(T_{(ii)}T_{(jj)} + T_{[ij]}T_{[ij]} - T_{(ij)}T_{(ij)}\right), \tag{2.136}$$

$$I_T^{(3)} = \epsilon_{ijk}T_{1i}T_{2j}T_{3k} = \det \mathbf{T}. \tag{2.137}$$

It can also be shown that if $\lambda^{(1)}, \lambda^{(2)}, \lambda^{(3)}$ are the three eigenvalues, then the invariants can also be expressed as

$$I_T^{(1)} = \lambda^{(1)} + \lambda^{(2)} + \lambda^{(3)}, \tag{2.138}$$

$$I_T^{(2)} = \lambda^{(1)}\lambda^{(2)} + \lambda^{(2)}\lambda^{(3)} + \lambda^{(3)}\lambda^{(1)}, \tag{2.139}$$

$$I_T^{(3)} = \lambda^{(1)}\lambda^{(2)}\lambda^{(3)}. \tag{2.140}$$

In general these eigenvalues, and consequently, the eigenvectors are complex. Additionally, in general the eigenvectors are nonorthogonal. If, however, the matrix we are considering is symmetric, which is often the case in fluid mechanics, it can be formally proven that all the eigenvalues are real and all the eigenvectors are real and orthogonal. If, for instance, our tensor is the stress tensor, we will show in Section 4.3 that it is symmetric in the absence of external couples. The eigenvectors of a symmetric stress tensor can form the basis for an intrinsic coordinate system that has its axes aligned with the principal axes of stress on a fluid element. The eigenvalues themselves give the value of the principal stress. This is actually a generalization of the familiar Mohr's[7] circle from solid mechanics.

For strictly two-dimensional systems, we have

$$\begin{vmatrix} T_{11} - \lambda & T_{12} \\ T_{21} & T_{22} - \lambda \end{vmatrix} = 0, \tag{2.141}$$

which leads us to

$$\lambda^2 - \underbrace{(T_{11} + T_{22})}_{I_T^{(1)}}\lambda + \underbrace{(T_{11}T_{22} - T_{12}T_{21})}_{I_T^{(3)}} = 0. \tag{2.142}$$

---

[7] Christian Otto Mohr, 1835–1918, Holstein-born German civil engineer, railroad and bridge designer.

Here there are only two principal invariants:

$$I_T^{(1)} = \lambda^{(1)} + \lambda^{(2)} = \text{tr } \mathbf{T}, \tag{2.143}$$

$$I_T^{(3)} = \lambda^{(1)}\lambda^{(2)} = \det \mathbf{T}. \tag{2.144}$$

To be consistent with the common notation for three-dimensional systems, we maintain the numerical labels for I_T. For two-dimensional matrices, our defined $I_T^{(2)}$ is redundant with $I_T^{(1)}$; it can be proved for all 2×2 matrices that $\text{tr } \mathbf{T} = (\det \mathbf{T})(\text{tr } \mathbf{T}^{-1})$.

Just as for vectors, we can define the Euclidean norm of a tensor. Full details are given by Powers and Sen (2015, Chapter 6). The Euclidean norm of a tensor $\mathbf{T}$ is defined as

$$||\mathbf{T}|| = \sup ||\mathbf{T} \cdot \mathbf{n}||, \quad \text{for all} \quad \mathbf{n}^T \cdot \mathbf{n} = 1. \tag{2.145}$$

Here, sup is the supremum operator, formally the least upper bound of the norm of the vector $\mathbf{T} \cdot \mathbf{n}$ for all unit vectors $\mathbf{n}$. It can be shown that

$$||\mathbf{T}|| = \sqrt{\lambda_{max}}, \tag{2.146}$$

where λ_{max} is the largest eigenvalue of $\mathbf{T}^T \cdot \mathbf{T}$. As will be discussed in Section 3.8.5, this represents the largest so-called *singular value* of $\mathbf{T}$. *The product of a matrix and its transpose, for example* $\mathbf{T}^T \cdot \mathbf{T}$, *will be seen to be useful in a remarkable variety of applications.*

Example 2.18 Find the principal axes and eigenvalues if the tensor is

$$T_{ij} = \begin{pmatrix} 1 & 0 & 0 \\ 0 & 1 & 2 \\ 0 & 2 & 1 \end{pmatrix}. \tag{2.147}$$

Solution
A diagram of the element is shown in Fig. 2.6(a). We take the eigenvalue problem

$$n_i T_{ij} = \lambda n_j = \lambda n_i \delta_{ij}, \quad \text{so} \quad n_i (T_{ij} - \lambda \delta_{ij}) = 0. \tag{2.148}$$

This becomes for our problem

$$\begin{pmatrix} n_1 & n_2 & n_3 \end{pmatrix} \begin{pmatrix} 1-\lambda & 0 & 0 \\ 0 & 1-\lambda & 2 \\ 0 & 2 & 1-\lambda \end{pmatrix} = \begin{pmatrix} 0 & 0 & 0 \end{pmatrix}. \tag{2.149}$$

For a nontrivial solution for n_i, we must have

$$\begin{vmatrix} 1-\lambda & 0 & 0 \\ 0 & 1-\lambda & 2 \\ 0 & 2 & 1-\lambda \end{vmatrix} = 0. \tag{2.150}$$

This gives rise to the polynomial equation

$$(1-\lambda)\left((1-\lambda)(1-\lambda) - 4\right) = 0. \tag{2.151}$$

This has three solutions

$$\lambda = 1, \quad \lambda = -1, \quad \lambda = 3. \tag{2.152}$$

Notice all eigenvalues are real; this must be because the tensor is symmetric.

Now let us find the eigenvectors (aligned with the principal axes) for this problem. First, it can be shown that the vector product of a vector with a tensor commutes when the tensor is symmetric. Although this is not a crucial step, we will use it to write the eigenvalue problem in a slightly more familiar notation:

$$n_i \left(T_{ij} - \lambda \delta_{ij} \right) = 0 \implies \left(T_{ij} - \lambda \delta_{ij} \right) n_i = 0, \qquad \text{because scalar components commute.} \tag{2.153}$$

Because of symmetry, we can now commute the indices to get

$$\left(T_{ji} - \lambda \delta_{ji} \right) n_i = 0, \qquad \text{because indices commute if symmetric.} \tag{2.154}$$

Expanding into matrix notation, we get

$$\begin{pmatrix} T_{11} - \lambda & T_{12} & T_{13} \\ T_{21} & T_{22} - \lambda & T_{23} \\ T_{31} & T_{32} & T_{33} - \lambda \end{pmatrix} \begin{pmatrix} n_1 \\ n_2 \\ n_3 \end{pmatrix} = \begin{pmatrix} 0 \\ 0 \\ 0 \end{pmatrix}. \tag{2.155}$$

Substituting for T_{ji} and considering the eigenvalue $\lambda = 1$, we get

$$\begin{pmatrix} 0 & 0 & 0 \\ 0 & 0 & 2 \\ 0 & 2 & 0 \end{pmatrix} \begin{pmatrix} n_1 \\ n_2 \\ n_3 \end{pmatrix} = \begin{pmatrix} 0 \\ 0 \\ 0 \end{pmatrix}. \tag{2.156}$$

We get two equations $2n_2 = 0$, and $2n_3 = 0$; thus, $n_2 = n_3 = 0$. We can satisfy all equations with an arbitrary value of n_1. It is always the case that an eigenvector will have an arbitrary magnitude and a well-defined direction. Here we will choose to normalize our eigenvector and take $n_1 = 1$, so that the eigenvector is

$$n_j = \begin{pmatrix} 1 \\ 0 \\ 0 \end{pmatrix} \qquad \text{for} \qquad \lambda = 1. \tag{2.157}$$

Geometrically, this means that the original 1 face already has an associated vector that is aligned with its normal vector. Consider the eigenvector associated with the eigenvalue $\lambda = -1$. Again substituting into the original equation, we get

$$\begin{pmatrix} 2 & 0 & 0 \\ 0 & 2 & 2 \\ 0 & 2 & 2 \end{pmatrix} \begin{pmatrix} n_1 \\ n_2 \\ n_3 \end{pmatrix} = \begin{pmatrix} 0 \\ 0 \\ 0 \end{pmatrix}. \tag{2.158}$$

This is simply the system of equations $2n_1 = 0$, $2n_2 + 2n_3 = 0$, and $2n_2 + 2n_3 = 0$. Clearly $n_1 = 0$. We could take $n_2 = 1$ and $n_3 = -1$ for a nontrivial solution. Alternatively, let us normalize and take

$$n_j = \begin{pmatrix} 0 \\ \frac{\sqrt{2}}{2} \\ -\frac{\sqrt{2}}{2} \end{pmatrix}. \tag{2.159}$$

Finally consider the eigenvector associated with the eigenvalue $\lambda = 3$. Again substituting into the original equation, we get

$$\begin{pmatrix} -2 & 0 & 0 \\ 0 & -2 & 2 \\ 0 & 2 & -2 \end{pmatrix} \begin{pmatrix} n_1 \\ n_2 \\ n_3 \end{pmatrix} = \begin{pmatrix} 0 \\ 0 \\ 0 \end{pmatrix}. \tag{2.160}$$

This is the system of equations $-2n_1 = 0$, $-2n_2 + 2n_3 = 0$, and $2n_2 - 2n_3 = 0$. Clearly again $n_1 = 0$. We could take $n_2 = 1$ and $n_3 = 1$ for a nontrivial solution. Once again, let us normalize and take

$$n_j = \begin{pmatrix} 0 \\ \frac{\sqrt{2}}{2} \\ \frac{\sqrt{2}}{2} \end{pmatrix}. \tag{2.161}$$

In summary, the three eigenvectors and associated eigenvalues are

$$n_j^{(1)} = \begin{pmatrix} 1 \\ 0 \\ 0 \end{pmatrix}, \lambda^{(1)} = 1; \, n_j^{(2)} = \begin{pmatrix} 0 \\ \frac{\sqrt{2}}{2} \\ \frac{-\sqrt{2}}{2} \end{pmatrix}, \lambda^{(2)} = -1; \, n_j^{(3)} = \begin{pmatrix} 0 \\ \frac{\sqrt{2}}{2} \\ \frac{\sqrt{2}}{2} \end{pmatrix}, \lambda^{(3)} = 3. \tag{2.162}$$

The eigenvectors are mutually orthogonal, as well as normal. We say they form an orthonormal set of vectors. Their orthogonality, as well as the fact that all the eigenvalues are real can be shown to be a direct consequence of the symmetry of the original tensor. A diagram of the eigenvalues on the element rotated so that it is aligned with the principal axes is shown on the fluid element in Fig. 2.6(b). The three orthonormal eigenvectors, when cast into a matrix, form a rotation matrix $\mathbf{Q}$ with $\det \mathbf{Q} = 1$:

$$\mathbf{Q} = \begin{pmatrix} \vdots & \vdots & \vdots \\ \mathbf{n}^{(1)} & \mathbf{n}^{(2)} & \mathbf{n}^{(3)} \\ \vdots & \vdots & \vdots \end{pmatrix} = \begin{pmatrix} 1 & 0 & 0 \\ 0 & \frac{\sqrt{2}}{2} & \frac{\sqrt{2}}{2} \\ 0 & \frac{-\sqrt{2}}{2} & \frac{\sqrt{2}}{2} \end{pmatrix}. \tag{2.163}$$

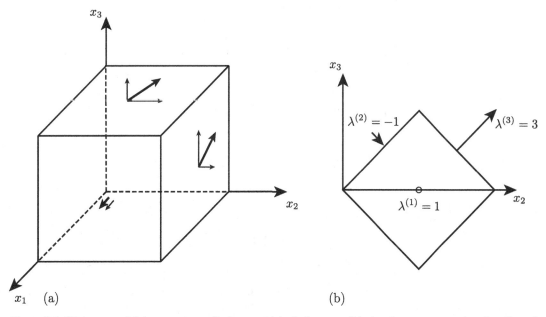

Figure 2.6 Diagrams of (a) stresses applied to a cubical element, (b) the element rotated to be aligned with its principal axes, along with the associated eigenvalues. The 1 face projects out of the page.

Example 2.19 For a given tensor, which we will take to be symmetric though the theory applies to asymmetric tensors as well,

$$T_{ij} = \mathbf{T} = \begin{pmatrix} 1 & 2 & 4 \\ 2 & 3 & -1 \\ 4 & -1 & 1 \end{pmatrix}, \tag{2.164}$$

find the three basic tensor invariants $I_T^{(1)}$, $I_T^{(2)}$, and $I_T^{(3)}$, and show they are truly invariant when the coordinate system is subjected to a rotation with direction cosine matrix of

$$\ell_{ij} = \mathbf{Q} = \begin{pmatrix} \frac{1}{\sqrt{6}} & \sqrt{\frac{2}{3}} & \frac{1}{\sqrt{6}} \\ \frac{1}{\sqrt{3}} & -\frac{1}{\sqrt{3}} & \frac{1}{\sqrt{3}} \\ \frac{1}{\sqrt{2}} & 0 & -\frac{1}{\sqrt{2}} \end{pmatrix}. \tag{2.165}$$

Find $||\mathbf{T}||$ and $||\mathbf{T}'||$.

Solution

Calculation reveals that $\det \mathbf{Q} = 1$, and that $\mathbf{Q} \cdot \mathbf{Q}^T = \mathbf{I}$, so that $\mathbf{Q}^T$ is a rotation matrix. The eigenvalues of $\mathbf{T}$ are calculated to be

$$\lambda^{(1)} = 5.28675, \qquad \lambda^{(2)} = -3.67956, \qquad \lambda^{(3)} = 3.39281. \tag{2.166}$$

The three invariants of T_{ij} are

$$I_T^{(1)} = \operatorname{tr} \mathbf{T} = \operatorname{tr} \begin{pmatrix} 1 & 2 & 4 \\ 2 & 3 & -1 \\ 4 & -1 & 1 \end{pmatrix} = 1 + 3 + 1 = 5, \tag{2.167}$$

$$I_T^{(2)} = \frac{1}{2} \left((\operatorname{tr} \mathbf{T})^2 - \operatorname{tr} (\mathbf{T} \cdot \mathbf{T}) \right), \tag{2.168}$$

$$= \frac{1}{2} \left(\left(\operatorname{tr} \begin{pmatrix} 1 & 2 & 4 \\ 2 & 3 & -1 \\ 4 & -1 & 1 \end{pmatrix} \right)^2 - \operatorname{tr} \left(\begin{pmatrix} 1 & 2 & 4 \\ 2 & 3 & -1 \\ 4 & -1 & 1 \end{pmatrix} \begin{pmatrix} 1 & 2 & 4 \\ 2 & 3 & -1 \\ 4 & -1 & 1 \end{pmatrix} \right) \right), \tag{2.169}$$

$$= \frac{1}{2} \left(5^2 - \operatorname{tr} \begin{pmatrix} 21 & 4 & 6 \\ 4 & 14 & 4 \\ 6 & 4 & 18 \end{pmatrix} \right) = \frac{1}{2}(25 - 21 - 14 - 18) = -14, \tag{2.170}$$

$$I_T^{(3)} = \det \mathbf{T} = \det \begin{pmatrix} 1 & 2 & 4 \\ 2 & 3 & -1 \\ 4 & -1 & 1 \end{pmatrix} = -66. \tag{2.171}$$

For $||\mathbf{T}||$, we prepare

$$\mathbf{T}^T \cdot \mathbf{T} = \begin{pmatrix} 1 & 2 & 4 \\ 2 & 3 & -1 \\ 4 & -1 & 1 \end{pmatrix} \begin{pmatrix} 1 & 2 & 4 \\ 2 & 3 & -1 \\ 4 & -1 & 1 \end{pmatrix} = \begin{pmatrix} 21 & 4 & 6 \\ 4 & 14 & 4 \\ 6 & 4 & 18 \end{pmatrix}. \tag{2.172}$$

Its eigenvalues are 27.9497, 13.5391, 11.5112, and the norm is the square root of the largest, so

$$||\mathbf{T}|| = \sqrt{27.9497} = 5.28675. \tag{2.173}$$

2.3 Eigenvalues, Eigenvectors, and Tensor Invariants

Now when we rotate the coordinate system, we get a transformed representation of $\mathbf{T}$ given by

$$\mathbf{T}' = \mathbf{Q}^T \cdot \mathbf{T} \cdot \mathbf{Q} = \begin{pmatrix} \frac{1}{\sqrt{6}} & \frac{1}{\sqrt{3}} & \frac{1}{\sqrt{2}} \\ \sqrt{\frac{2}{3}} & -\frac{1}{\sqrt{3}} & 0 \\ \frac{1}{\sqrt{6}} & \frac{1}{\sqrt{3}} & -\frac{1}{\sqrt{2}} \end{pmatrix} \begin{pmatrix} 1 & 2 & 4 \\ 2 & 3 & -1 \\ 4 & -1 & 1 \end{pmatrix} \begin{pmatrix} \frac{1}{\sqrt{6}} & \sqrt{\frac{2}{3}} & \frac{1}{\sqrt{6}} \\ \frac{1}{\sqrt{3}} & -\frac{1}{\sqrt{3}} & \frac{1}{\sqrt{3}} \\ \frac{1}{\sqrt{2}} & 0 & -\frac{1}{\sqrt{2}} \end{pmatrix}, \quad (2.174)$$

$$= \begin{pmatrix} 4.102\,38 & 2.522\,39 & 1.609\,48 \\ 2.522\,39 & -0.218\,951 & -2.912\,91 \\ 1.609\,48 & -2.912\,91 & 1.116\,57 \end{pmatrix}. \quad (2.175)$$

We then seek the tensor invariants of $\mathbf{T}'$. Leaving out some of the details, which are the same as those for calculating the invariants of $\mathbf{T}$, we find the invariants indeed are invariant:

$$I_{T'}^{(1)} = 4.102\,38 - 0.218\,951 + 1.116\,57 = 5, \quad (2.176)$$

$$I_{T'}^{(2)} = \frac{1}{2}(5^2 - 53) = -14, \quad (2.177)$$

$$I_{T'}^{(3)} = -66. \quad (2.178)$$

Finally, we confirm that the invariants are indeed related to the eigenvalues of the tensor:

$$I_T^{(1)} = \lambda^{(1)} + \lambda^{(2)} + \lambda^{(3)} = 5.286\,75 - 3.679\,56 + 3.392\,81 = 5, \quad (2.179)$$

$$I_T^{(2)} = \lambda^{(1)}\lambda^{(2)} + \lambda^{(2)}\lambda^{(3)} + \lambda^{(3)}\lambda^{(1)}, \quad (2.180)$$

$$= (5.286\,75)(-3.679\,56) + (-3.679\,56)(3.392\,81) + (3.392\,81)(5.286\,75) = -14, \quad (2.181)$$

$$I_T^{(3)} = \lambda^{(1)}\lambda^{(2)}\lambda^{(3)} = (5.286\,75)(-3.679\,56)(3.392\,81) = -66. \quad (2.182)$$

It is easily verified that

$$||\mathbf{T}'|| = ||\mathbf{T}|| = \sqrt{27.9497} = 5.286\,75. \quad (2.183)$$

If we instead subjected $\mathbf{T}$ to reflection, say by exchanging two rows of $\mathbf{Q}$, we would find the invariants to be unchanged.

Example 2.20 Given a two-dimensional tensor, which here we take to be asymmetric,

$$T_{ij} = \mathbf{T} = \begin{pmatrix} 2 & 1 \\ 2 & 2 \end{pmatrix}, \quad (2.184)$$

find the two basic tensor invariants $I_T^{(1)}$ and $I_T^{(3)}$ and show they are truly invariant when the tensor is subjected to a rotation with direction cosine matrix of

$$\ell_{ij} = \mathbf{Q} = \begin{pmatrix} \frac{1}{\sqrt{2}} & \frac{1}{\sqrt{2}} \\ -\frac{1}{\sqrt{2}} & \frac{1}{\sqrt{2}} \end{pmatrix}. \quad (2.185)$$

Find $||\mathbf{T}||$ and $||\mathbf{T}'||$.

Solution
Calculation reveals that $\det \mathbf{Q} = 1$ and that $\mathbf{Q} \cdot \mathbf{Q}^T = \mathbf{I}$, so that $\mathbf{Q}^T$ is a rotation matrix. The eigenvalue problem induces the condition

$$\begin{vmatrix} T_{11} - \lambda & T_{12} \\ T_{21} & T_{22} - \lambda \end{vmatrix} = 0. \quad (2.186)$$

This gives the characteristic polynomial

$$(T_{11} - \lambda)(T_{22} - \lambda) - T_{12}T_{21} = 0, \tag{2.187}$$

$$\lambda^2 - (T_{11} + T_{22})\lambda + (T_{11}T_{22} - T_{12}T_{21}) = 0, \tag{2.188}$$

$$\lambda^2 - I_T^{(1)}\lambda + I_T^{(3)} = 0. \tag{2.189}$$

Here we have for the two-dimensional system, the two invariants

$$I_T^{(1)} = T_{11} + T_{22} = \lambda^{(1)} + \lambda^{(2)} = \text{tr } \mathbf{T}, \tag{2.190}$$

$$I_T^{(3)} = T_{11}T_{22} - T_{12}T_{21} = \lambda^{(1)}\lambda^{(2)} = \det \mathbf{T}. \tag{2.191}$$

For this system, the eigenvalues of $\mathbf{T}$ are calculated to be $\lambda^{(1,2)} = 2 \pm \sqrt{2}$. The two invariants of $\mathbf{T}$ are

$$I_T^{(1)} = \text{tr } \mathbf{T} = \text{tr} \begin{pmatrix} 2 & 1 \\ 2 & 2 \end{pmatrix} = 2 + 2 = 4, \tag{2.192}$$

$$I_T^{(3)} = \det \mathbf{T} = \det \begin{pmatrix} 2 & 1 \\ 2 & 2 \end{pmatrix} = 2(2) - 1(2) = 2. \tag{2.193}$$

For $||\mathbf{T}||$, we prepare

$$\mathbf{T}^T \cdot \mathbf{T} = \begin{pmatrix} 2 & 2 \\ 1 & 2 \end{pmatrix} \begin{pmatrix} 2 & 1 \\ 2 & 2 \end{pmatrix} = \begin{pmatrix} 8 & 6 \\ 6 & 5 \end{pmatrix}. \tag{2.194}$$

Its eigenvalues are $(13 \pm 3\sqrt{17})/2$, and the norm is the square root of the largest, so $||\mathbf{T}|| = \sqrt{(13 + 3\sqrt{17})/2} = 3.56155$. Now when we rotate, we get

$$\mathbf{T}' = \mathbf{Q}^T \cdot \mathbf{T} \cdot \mathbf{Q} = \begin{pmatrix} \frac{1}{\sqrt{2}} & -\frac{1}{\sqrt{2}} \\ \frac{1}{\sqrt{2}} & \frac{1}{\sqrt{2}} \end{pmatrix} \begin{pmatrix} 2 & 1 \\ 2 & 2 \end{pmatrix} \begin{pmatrix} \frac{1}{\sqrt{2}} & \frac{1}{\sqrt{2}} \\ -\frac{1}{\sqrt{2}} & \frac{1}{\sqrt{2}} \end{pmatrix} = \begin{pmatrix} \frac{1}{2} & -\frac{1}{2} \\ \frac{1}{2} & \frac{7}{2} \end{pmatrix}. \tag{2.195}$$

By inspection, we see the invariants of $\mathbf{T}'$ are indeed the same as those of $\mathbf{T}$:

$$I_T^{(1)} = \frac{1}{2} + \frac{7}{2} = 4, \quad I_T^{(3)} = \left(\frac{1}{2}\right)\left(\frac{7}{2}\right) - \left(\frac{-1}{2}\right)\left(\frac{1}{2}\right) = 2. \tag{2.196}$$

Finally, we also see

$$I_T^{(1)} = \lambda^{(1)} + \lambda^{(2)} = (2 + \sqrt{2}) + (2 - \sqrt{2}) = 4, \tag{2.197}$$

$$I_T^{(3)} = \lambda^{(1)}\lambda^{(2)} = \left(2 + \sqrt{2}\right)\left(2 - \sqrt{2}\right) = 2. \tag{2.198}$$

It is easily verified that $||\mathbf{T}'|| = ||\mathbf{T}|| = \sqrt{(13 + 3\sqrt{17})/2} = 3.56155$.

2.4 Grad, Div, Curl, etc.

Thus far, we have considered the algebra of vectors and tensors. Now let us consider the calculus. For now, let us consider variables that are a function of the spatial vector x_i. We shall soon allow variation with time t also. We will typically encounter quantities such as $\phi(x_i)$, a scalar function of the position vector; $v_j(x_i)$, a vector function of the position vector; or $T_{jk}(x_i)$, a tensor function of the position vector.

2.4.1 Gradient

The *gradient* operator, sometimes denoted by "grad," is motivated as follows. Consider $\phi(x_i)$, which when written in full is

$$\phi(x_i) = \phi(x_1, x_2, x_3). \tag{2.199}$$

Forming the total differential, see for example Kaplan (2003, Chapter 2), gives

$$d\phi = \frac{\partial \phi}{\partial x_1} dx_1 + \frac{\partial \phi}{\partial x_2} dx_2 + \frac{\partial \phi}{\partial x_3} dx_3. \tag{2.200}$$

We adopt a nontraditional, but useful notation, ∂_i, for the partial derivative:

$$\partial_i \equiv \frac{\partial}{\partial x_i} = \mathbf{e}_1 \frac{\partial}{\partial x_1} + \mathbf{e}_2 \frac{\partial}{\partial x_2} + \mathbf{e}_3 \frac{\partial}{\partial x_3} = \nabla = \begin{pmatrix} \frac{\partial}{\partial x_1} \\ \frac{\partial}{\partial x_2} \\ \frac{\partial}{\partial x_3} \end{pmatrix} = \begin{pmatrix} \partial_1 \\ \partial_2 \\ \partial_3 \end{pmatrix}. \tag{2.201}$$

So the total differential is actually

$$d\phi = \partial_1 \phi \, dx_1 + \partial_2 \phi \, dx_2 + \partial_3 \phi \, dx_3 = \partial_i \phi \, dx_i. \tag{2.202}$$

After commuting so as to juxtapose the i subscripts, we have

$$d\phi = dx_i \, \partial_i \phi, \qquad d\phi = d\mathbf{x}^T \cdot \nabla \phi = d\mathbf{x}^T \cdot \text{grad } \phi. \tag{2.203}$$

We can also take the transpose of both sides, recalling that the transpose of a scalar is the scalar itself, to obtain

$$(d\phi)^T = \left(d\mathbf{x}^T \cdot \nabla \phi\right)^T = (\nabla \phi)^T \cdot d\mathbf{x} = \nabla^T \phi \cdot d\mathbf{x} = \begin{pmatrix} \partial_1 \phi & \partial_2 \phi & \partial_3 \phi \end{pmatrix} \begin{pmatrix} dx_1 \\ dx_2 \\ dx_3 \end{pmatrix}. \tag{2.204}$$

Note for the scalar $d\phi^T = d\phi$. Here we expand ∇^T as $\nabla^T = (\partial_1, \partial_2, \partial_3)$. When ∂_i or ∇ operates on a scalar, it is known as the gradient operator. The gradient operator operating on a scalar function gives rise to a vector function.

We next describe the gradient operator operating on a vector. For vectors in Cartesian index and Gibbs notation, we have, following a similar analysis,

$$dv_j = dx_i \, \partial_i v_j = \partial_i v_j \, dx_i, \tag{2.205}$$

$$d\mathbf{v}^T = d\mathbf{x}^T \cdot \nabla \mathbf{v}^T, \quad \text{so} \quad d\mathbf{v} = (\nabla \mathbf{v}^T)^T \cdot d\mathbf{x} = (\text{grad } \mathbf{v})^T \cdot d\mathbf{x}. \tag{2.206}$$

A more common approach, not using the transpose notation, would be to say here for the Gibbs notation that $d\mathbf{v} = d\mathbf{x} \cdot \nabla \mathbf{v}$. However, this is only works if we consider $d\mathbf{v}$ to be a row vector, as $d\mathbf{x} \cdot \nabla \mathbf{v}$ is a row vector. All in all, while at times clumsy, the transpose notation allows enhanced clarity and consistency with matrix algebra.

In Eq. (2.205), the quantity $\partial_i v_j$ is the gradient of a vector, that is a tensor. So the gradient operator operating on a vector raises its order by one. The Gibbs notation with transposes suggests properly that the gradient of a vector can be expanded as

$$\nabla \mathbf{v}^T = \begin{pmatrix} \partial_1 \\ \partial_2 \\ \partial_3 \end{pmatrix} \begin{pmatrix} v_1 & v_2 & v_3 \end{pmatrix} = \begin{pmatrix} \partial_1 v_1 & \partial_1 v_2 & \partial_1 v_3 \\ \partial_2 v_1 & \partial_2 v_2 & \partial_2 v_3 \\ \partial_3 v_1 & \partial_3 v_2 & \partial_3 v_3 \end{pmatrix} = \partial_i v_j. \tag{2.207}$$

Our choice of how to structure the matrix representation of the gradient of the vector is consistent with standard matrix notation for which elements of the first row are numbered 11, 12, 13, with other rows similarly numbered. One may find other sources that build schemes based on the transpose of our choice. The proper way to interpret our unusual notation $\nabla \mathbf{v}^T$ is an analog of a dyadic product (see Section 2.1.5), with the column operator ∇ operating on the row vector $\mathbf{v}^T$. It is *not* the transpose of $\nabla \mathbf{v}$, which in our notation scheme is ill-defined. There is another way this could be considered, recognizing $(d\mathbf{x}^T \cdot \nabla)$ as a scalar operator. Then we have

$$d\mathbf{v}^T = (d\mathbf{x}^T \cdot \nabla)\mathbf{v}^T, \quad d\mathbf{v} = (d\mathbf{x}^T \cdot \nabla)\mathbf{v}. \tag{2.208}$$

Lastly we consider the gradient operator operating on a tensor. For tensors in Cartesian index notation, we have, following a similar analysis

$$dT_{ij} = dx_k \ \partial_k T_{ij} = \partial_k T_{ij} \ dx_k. \tag{2.209}$$

Here the quantity $\partial_k T_{ij}$ is a third-order tensor. So the gradient operator operating on a tensor raises its order by one as well. The Gibbs notation is not straightforward as it can involve something akin to the transpose of a three-dimensional matrix.

2.4.2 Divergence

The contraction of the gradient operator on either a vector or a tensor is known as the *divergence* operator, sometimes denoted by "div." For the divergence of a vector, we have

$$\partial_i v_i = \partial_1 v_1 + \partial_2 v_2 + \partial_3 v_3 = \nabla^T \cdot \mathbf{v} = \text{div } \mathbf{v}. \tag{2.210}$$

The divergence of a vector is a scalar. A vector field that is divergence-free, $\nabla^T \cdot \mathbf{v} = 0$, is defined as *solenoidal*. For the divergence of a second-order tensor, we have

$$\partial_i T_{ij} = \partial_1 T_{1j} + \partial_2 T_{2j} + \partial_3 T_{3j} = \nabla^T \cdot \mathbf{T} = \text{div } \mathbf{T}. \tag{2.211}$$

The divergence operator, $\nabla^T \cdot$, operating on a tensor gives rise to a row vector. We will sometimes have to transpose this row vector in order to arrive at a column vector, for example we will have need for the column vector $\left(\nabla^T \cdot \mathbf{T}\right)^T$. As with the dot product, most texts assume the transpose operation is understood and write the divergence of a vector or tensor simply as $\nabla \cdot \mathbf{v}$ or $\nabla \cdot \mathbf{T}$.

2.4.3 Curl

The *curl* operator is the derivative analog to the cross product. We write it as

$$\omega_i = \epsilon_{ijk} \partial_j v_k, \quad \boldsymbol{\omega} = \nabla \times \mathbf{v} = \begin{vmatrix} \mathbf{e}_1 & \mathbf{e}_2 & \mathbf{e}_3 \\ \partial_1 & \partial_2 & \partial_3 \\ v_1 & v_2 & v_3 \end{vmatrix} = \text{curl } \mathbf{v}. \tag{2.212}$$

Expanding for $i = 1, 2, 3$ gives

$$\omega_1 = \epsilon_{123}\partial_2 v_3 + \epsilon_{132}\partial_3 v_2 = \partial_2 v_3 - \partial_3 v_2, \qquad (2.213)$$

$$\omega_2 = \epsilon_{231}\partial_3 v_1 + \epsilon_{213}\partial_1 v_3 = \partial_3 v_1 - \partial_1 v_3, \qquad (2.214)$$

$$\omega_3 = \epsilon_{312}\partial_1 v_2 + \epsilon_{321}\partial_2 v_1 = \partial_1 v_2 - \partial_2 v_1. \qquad (2.215)$$

2.4.4 Laplacian

The *Laplacian*[8] operator can operate on a scalar, vector, or tensor. It is a combination of first the gradient followed by the divergence. It yields a function of the same order as that on which it operates. For its operation on a scalar, it is denoted as follows:

$$\partial_i \partial_i \phi = \nabla^T \cdot \nabla \phi = \nabla^2 \phi = \text{div grad } \phi. \qquad (2.216)$$

Associated with the Laplacian operator is Laplace's equation:

$$\nabla^T \cdot \nabla \phi = 0. \qquad (2.217)$$

Functions $\phi(\mathbf{x})$ that satisfy Laplace's equation are known as *harmonic*. In viscous fluid flow, we will have occasion to have the Laplacian operate on vector:

$$\partial_i \partial_i v_j = \nabla^2 \mathbf{v} = \text{div grad } \mathbf{v}. \qquad (2.218)$$

Related to the Laplacian operator is the operator E^2. For many coordinate systems in which two axes are confined to a plane (e.g. Cartesian or cylindrical), $E^2 = \nabla^2$. For other coordinate systems, such as spherical, $E^2 \neq \nabla^2$. Details will not be considered here; however, this distinction will be necessary to properly describe some important problems, such as potential flow over a sphere, as will be considered in Problems 8.8, 8.9.

2.4.5 Biharmonic Operator

We will have occasional need for the *biharmonic operator* to operate on a scalar function. It arises from applying the Laplacian operator twice. It operates on a scalar as

$$\partial_i \partial_i \partial_j \partial_j \phi = (\nabla^T \cdot \nabla)(\nabla^T \cdot \nabla)\phi = \nabla^2 \nabla^2 \phi = \nabla^4 \phi = (\text{div grad})(\text{div grad})\phi. \qquad (2.219)$$

Related to the biharmonic operator is the operator $E^4 = E^2 E^2$. Similar to E^2, for many coordinate systems in which two axes are confined to a plane (e.g. Cartesian or cylindrical), $E^4 = \nabla^4$. For other coordinate systems, such as spherical, $E^4 \neq \nabla^4$. This distinction will be necessary to properly describe some important problems, such as slow viscous flow over a sphere, as will be seen in Section 11.1.1.

2.4.6 Time Derivative

We employ a compact, atypical notation for the partial derivative with respect to time:

$$\partial_o \equiv \frac{\partial}{\partial t}. \qquad (2.220)$$

---

[8] Pierre-Simon Laplace, 1749–1827, Normandy-born French astronomer.

2.4.7 Relevant Theorems

We will use several theorems that are developed in vector calculus. Here we give the simplest of motivations, and simply present them. The reader should consult a standard mathematics text for detailed derivations.

Fundamental Theorem of Calculus

The *fundamental theorem of calculus* is as follows:

$$\int_{x=a}^{x=b} f(x) \, dx = \int_{x=a}^{x=b} \left(\frac{d\phi}{dx}\right) dx = \phi(b) - \phi(a). \tag{2.221}$$

It effectively says that to find the integral of a function $f(x)$ – that is, the area under the curve – it suffices to find a function ϕ, whose derivative is f, and evaluate ϕ at each endpoint, and take the difference to find the area under the curve.

Gauss's Theorem

Gauss's theorem is the analog of the fundamental theorem of calculus extended to volume integrals. It applies to tensor functions of arbitrary order and is as follows:

$$\int_V \partial_i \left(T_{ijk...}\right) \, dV = \int_A n_i T_{ijk...} \, dA. \tag{2.222}$$

Here V is an arbitrary volume, dV is the element of volume, A is the surface that bounds V, n_i is the outward unit normal to A, and $T_{ijk...}$ is an arbitrary tensor function. The surface integral is analogous to evaluating the function at the end points in the fundamental theorem of calculus. In Gibbs notation, it is

$$\int_V \nabla^T \cdot \mathbf{T} \, dV = \int_A \mathbf{n}^T \cdot \mathbf{T} \, dA. \tag{2.223}$$

Note that both the left- and right-hand sides involve row vectors. If we want column vectors, we can transpose both sides to get

$$\int_V (\nabla^T \cdot \mathbf{T})^T \, dV = \int_A (\mathbf{n}^T \cdot \mathbf{T})^T \, dA. \tag{2.224}$$

If we take $T_{ijk...}$ to be the scalar of unity (whose derivative must be zero), Gauss's theorem reduces to

$$\int_A n_i \, dA = 0. \tag{2.225}$$

That is, the unit normal to the surface integrated over the surface cancels to zero when the entire surface is included. We will use Gauss's theorem extensively. It allows us to convert sometimes difficult volume integrals into easier interpreted surface integrals. It is useful to use this theorem as a means to toggle from one form to another.

Example 2.21 Demonstrate the validity of Gauss's theorem for the tensor field

$$\mathbf{T} = \begin{pmatrix} x_1 & x_2 & x_1 \\ x_2 & x_2 & x_3 \\ x_3 & x_3 & x_2 \end{pmatrix}, \qquad (2.226)$$

volume where the volume under consideration is the unit cube defined on the domain $x_1 \in [0,1]$, $x_2 \in [0,1]$, $x_3 \in [0,1]$.

Solution
We first note that $\mathbf{T}$ is asymmetric. We see that

$$\nabla^T \cdot \mathbf{T} = \begin{pmatrix} \partial_1 & \partial_2 & \partial_3 \end{pmatrix} \begin{pmatrix} x_1 & x_2 & x_1 \\ x_2 & x_2 & x_3 \\ x_3 & x_3 & x_2 \end{pmatrix} = \begin{pmatrix} 3 & 2 & 1 \end{pmatrix}. \qquad (2.227)$$

Integrating the constant row vector over the unit cube, we find

$$\int_V \nabla^T \cdot \mathbf{T} \, dV = \begin{pmatrix} 3 & 2 & 1 \end{pmatrix}. \qquad (2.228)$$

Then, we can evaluate the surface integral on each of the six faces and perform a set of six surface integrals. Leaving out the details, we do so, and find

$$\int_A \mathbf{n}^T \cdot \mathbf{T} \, dA = \begin{pmatrix} 3 & 2 & 1 \end{pmatrix}. \qquad (2.229)$$

This verifies Gauss's theorem for this case. For asymmetric tensors such as our $\mathbf{T}$, we need to be careful about commuting operators. For example, for this problem

$$\int_A \mathbf{T} \cdot \mathbf{n} \, dA = \begin{pmatrix} 2 \\ 2 \\ 0 \end{pmatrix} \neq \begin{pmatrix} 3 & 2 & 1 \end{pmatrix}. \qquad (2.230)$$

Stokes' Theorem

Stokes'[9] *theorem* is as follows:

$$\int_A n_i \epsilon_{ijk} \partial_j v_k \, dA = \oint_C v_i \alpha_{ti} \, ds, \qquad \int_A (\nabla \times \mathbf{v})^T \cdot \mathbf{n} \, dA = \oint_C \mathbf{v}^T \cdot \boldsymbol{\alpha}_t \, ds. \qquad (2.231)$$

Once again A is a bounding surface and $\mathbf{n}$ is its outward unit normal. The integral with the circle through it denotes a closed contour integral with respect to arc length s, and $\boldsymbol{\alpha}_t$ is the unit tangent vector to the bounding curve C.

A Useful Identity

It can be shown that a useful identity involving ∇, $\mathbf{v}$, and $\boldsymbol{\omega} = \nabla \times \mathbf{v}$ is as follows:

$$v_j \partial_j v_i = \partial_i \left(\frac{1}{2} v_j v_j \right) - \epsilon_{ijk} v_j \omega_k, \quad \left(\mathbf{v}^T \cdot \nabla \right) \mathbf{v} = \nabla \left(\frac{1}{2} \mathbf{v}^T \cdot \mathbf{v} \right) - \mathbf{v} \times \boldsymbol{\omega}. \qquad (2.232)$$

---

[9] Sir George Gabriel Stokes, 1819–1903, Irish-born British physicist and mathematician who developed, simultaneously with Navier, the governing equations of fluid motion, in a form that was more robust than that of Navier.

This can be proved by considering the right-hand side of Eq. (2.232), expanding, and using Eqs. (2.212) and then (2.64):

$$\partial_i \left(\frac{1}{2} v_j v_j\right) - \epsilon_{ijk} v_j \omega_k = v_j \partial_i v_j - \epsilon_{ijk} v_j \underbrace{\epsilon_{klm} \partial_l v_m}_{=\omega_k}, \tag{2.233}$$

$$= v_j \partial_i v_j - \epsilon_{kij} \epsilon_{klm} v_j \partial_l v_m, \tag{2.234}$$

$$= v_j \partial_i v_j - (\delta_{il}\delta_{jm} - \delta_{im}\delta_{jl}) v_j \partial_l v_m, \tag{2.235}$$

$$= \underbrace{v_j \partial_i v_j - v_j \partial_i v_j}_{=0} + v_j \partial_j v_i = v_j \partial_j v_i. \tag{2.236}$$

As an aside, we have what is sometimes defined as a *Lamb*[10] *vector*:

$$\mathbf{l} = \mathbf{v} \times (\nabla \times \mathbf{v}) = \mathbf{v} \times \boldsymbol{\omega}. \tag{2.237}$$

In the unusual case in which the Lamb vector is zero, the curl of the vector $\mathbf{v}$ is parallel to $\mathbf{v}$. Such a field is known as a *Beltrami*[11] *field*. For such fields, the vector $\mathbf{v}$ is everywhere a scalar multiple of its curl; $\boldsymbol{\omega} = \alpha \mathbf{v}$, with α as a scalar field variable.

Leibniz's Rule: General Transport Theorem for Arbitrary Volumes

Leibniz's[12] rule relates time derivatives of integral quantities to a form that distinguishes changes that are happening within the boundaries to changes due to fluxes through boundaries. This is the foundation of the so-called *control volume* approach. Using the nomenclature of Whitaker (1968, p. 92), we also call Leibniz's rule the *general transport theorem*. Leibniz's rule applied to an arbitrary tensorial function is as follows:

$$\frac{d}{dt} \int_{V_a(t)} T_{jk...}(x_i, t) \, dV = \int_{V_a(t)} \partial_o T_{jk...} \, dV + \int_{A_a(t)} n_l w_l T_{jk...} \, dA. \tag{2.238}$$

Here we have $V_a(t)$ as an arbitrary time-dependent volume, $A_a(t)$ as a bounding surface of the arbitrary moving volume, w_l as the velocity vector of points on the moving surface, and n_l as the unit normal to the moving surface. Say we have the special case in which $T_{jk...} = 1$; then Leibniz's rule reduces to

$$\frac{d}{dt} \int_{V_a(t)} dV = \int_{V_a(t)} \partial_o(1) \, dV + \int_{A_a(t)} n_k w_k(1) \, dA, \tag{2.239}$$

$$\frac{dV_a}{dt} = \int_{A_a(t)} n_k w_k \, dA. \tag{2.240}$$

This simply says the total volume, which we call V_a, changes in response to net motion of the bounding surface.

---

[10] Sir Horace Lamb, 1849–1934, English fluid mechanician.
[11] Eugenio Beltrami, 1835–1900, Italian mathematician with a specialty in differential geometry.
[12] Gottfried Wilhelm von Leibniz, 1646–1716, Leipzig-born German philosopher and mathematician. Invented calculus independent of Newton and employed a superior notation to that of Newton.

Material Volume: Reynolds Transport Theorem In the special case for which the volume contains the same fluid particles, the velocity of the boundary is the fluid particle velocity, $w_l = v_l$, and our general transport theorem becomes, again using the nomenclature of Whitaker (1968, p. 92), the *Reynolds*[13] *transport theorem*:

$$\frac{d}{dt}\int_{V_m(t)} T_{jk\ldots}(x_i,t)\, dV = \int_{V_m(t)} \partial_o T_{jk\ldots}\, dV + \int_{A_m(t)} n_l v_l T_{jk\ldots}\, dA. \quad (2.241)$$

The terms $V_m(t)$ and $A_m(t)$ denote the time-dependent *material volume* and *material surface area* to denote that the geometry in question always contains the same material particles.

Fixed Volume In the special case in which the volume is fixed in time, the velocity of the boundary is zero, $w_l = 0$, and our general transport theorem becomes

$$\frac{d}{dt}\int_V T_{jk\ldots}(x_i,t)\, dV = \int_V \partial_o T_{jk\ldots}\, dV. \quad (2.242)$$

In this case there is no time-dependency of the fixed volume V.

Scalar Function In the special case in which $T_{jk\ldots}$ is a scalar function f, Leibniz's rule reduces to

$$\frac{d}{dt}\int_{V_a(t)} f(x_i,t)\, dV = \int_{V_a(t)} \partial_o f(x_i,t)\, dV + \int_{A_a(t)} n_l w_l f(x_i,t)\, dA. \quad (2.243)$$

Further, considering one-dimensional cases only, we can then say

$$\frac{d}{dt}\int_{x=a(t)}^{x=b(t)} f(x,t)\, dx = \int_{x=a(t)}^{x=b(t)} \partial_o f\, dx + \frac{db}{dt} f(b(t),t) - \frac{da}{dt} f(a(t),t). \quad (2.244)$$

As in the fundamental theorem of calculus, for the one-dimensional case, we do not have to evaluate a surface integral; instead, we simply must consider the function at its endpoints. Here db/dt and da/dt are the velocities of the bounding surface and analogous to w_k. The terms $f(b(t),t)$ and $f(a(t),t)$ are equivalent to evaluating $T_{jk\ldots}$ on $A_a(t)$.

2.5 General Coordinate Transformations

Here we introduce – following Aris (1962), Kaplan (2003, Chapter 3.9), Powers and Sen (2015, Section 1.6), and many others – some standard notation from tensor analysis for general coordinate transformations. Such analysis has intrinsic value to better understand geometry; additionally, notions developed here can be applied to the generation of grids used in CFD. See Anderson et al. (2021) or Liseikin (2017). The presentation here extends our analysis of Section 2.1.2 that was confined to simple rotation transformations. In this notation, both sub- and superscripts are needed to distinguish between what are known as *covariant* and *contravariant* vectors, which are really different mathematical representations of the same quantity, just projected onto different basis vectors. The basis vectors may or may not be orthonormal. They

---

[13] Osborne Reynolds, 1842–1912, Belfast-born British engineer and physicist who did fundamental experimental work in fluid mechanics and heat transfer.

may not even be orthogonal. All they need be is linearly independent. In brief, we start with a general transformation from the non-Cartesian coordinate, defined here as x^i, to the Cartesian coordinate, defined here as ξ^k:

$$\xi^1 = \xi^1(x^1, x^2, x^3), \qquad \xi^2 = \xi^2(x^1, x^2, x^3), \qquad \xi^3 = \xi^3(x^1, x^2, x^3). \tag{2.245}$$

As an example, this form includes the transformation from a non-Cartesian cylindrical coordinate system to a Cartesian system; this will be taken up in detail in Section 2.6. We could also say

$$\xi^k = \xi^k(x^i), \qquad \boldsymbol{\xi} = \boldsymbol{\xi}(\mathbf{x}). \tag{2.246}$$

Local linearization of the transformation gives

$$\begin{pmatrix} d\xi^1 \\ d\xi^2 \\ d\xi^3 \end{pmatrix} = \underbrace{\begin{pmatrix} \frac{\partial \xi^1}{\partial x^1} & \frac{\partial \xi^1}{\partial x^2} & \frac{\partial \xi^1}{\partial x^3} \\ \frac{\partial \xi^2}{\partial x^1} & \frac{\partial \xi^2}{\partial x^2} & \frac{\partial \xi^2}{\partial x^3} \\ \frac{\partial \xi^3}{\partial x^1} & \frac{\partial \xi^3}{\partial x^2} & \frac{\partial \xi^3}{\partial x^3} \end{pmatrix}}_{\mathbf{J}} \begin{pmatrix} dx^1 \\ dx^2 \\ dx^3 \end{pmatrix}, \qquad d\boldsymbol{\xi} = \mathbf{J} \cdot d\mathbf{x}. \tag{2.247}$$

Here we define the local *Jacobian matrix*, $\mathbf{J}$, as

$$\mathbf{J} = \begin{pmatrix} \frac{\partial \xi^1}{\partial x^1} & \frac{\partial \xi^1}{\partial x^2} & \frac{\partial \xi^1}{\partial x^3} \\ \frac{\partial \xi^2}{\partial x^1} & \frac{\partial \xi^2}{\partial x^2} & \frac{\partial \xi^2}{\partial x^3} \\ \frac{\partial \xi^3}{\partial x^1} & \frac{\partial \xi^3}{\partial x^2} & \frac{\partial \xi^3}{\partial x^3} \end{pmatrix} = \frac{\partial \xi^k}{\partial x^i}. \tag{2.248}$$

Small notational differences may be found for $\mathbf{J}$ in the literature. Some sources may identify the Jacobian matrix as the transpose of ours. Others may define the Jacobian matrix in terms of the inverse function, yielding $\partial x^i / \partial \xi^k$.

For our notation, the transformation between a Cartesian representation $v^i = \mathbf{v}$ and a non-Cartesian representation $u^i = \mathbf{u}$ is given by

$$v^i = \frac{\partial \xi^i}{\partial x^j} u^j, \qquad \mathbf{v} = \mathbf{J} \cdot \mathbf{u}. \tag{2.249}$$

The rule for partial differentiation can be used to represent the gradient as

$$\underbrace{\begin{pmatrix} \frac{\partial}{\partial x^1} \\ \frac{\partial}{\partial x^2} \\ \frac{\partial}{\partial x^3} \end{pmatrix}}_{\nabla_{\mathbf{x}}} = \underbrace{\begin{pmatrix} \frac{\partial \xi^1}{\partial x^1} & \frac{\partial \xi^2}{\partial x^1} & \frac{\partial \xi^3}{\partial x^1} \\ \frac{\partial \xi^1}{\partial x^2} & \frac{\partial \xi^2}{\partial x^2} & \frac{\partial \xi^3}{\partial x^2} \\ \frac{\partial \xi^1}{\partial x^3} & \frac{\partial \xi^2}{\partial x^3} & \frac{\partial \xi^3}{\partial x^3} \end{pmatrix}}_{\mathbf{J}^T} \underbrace{\begin{pmatrix} \frac{\partial}{\partial \xi^1} \\ \frac{\partial}{\partial \xi^2} \\ \frac{\partial}{\partial \xi^3} \end{pmatrix}}_{\nabla_{\boldsymbol{\xi}}}, \qquad \nabla_{\mathbf{x}} = \mathbf{J}^T \cdot \nabla_{\boldsymbol{\xi}}. \tag{2.250}$$

Inverting, we find

$$\nabla_{\boldsymbol{\xi}} = \left(\mathbf{J}^T\right)^{-1} \cdot \nabla_{\mathbf{x}}. \tag{2.251}$$

This can be directly compared with Eq. (2.247). In the special case for which the transformation is a rotation, we have $\mathbf{J} = \mathbf{Q}$ and thus $\left(\mathbf{J}^T\right)^{-1} = \left(\mathbf{Q}^T\right)^{-1} = \mathbf{Q}$. In this case, we recover the simpler $d\boldsymbol{\xi} = \mathbf{Q} \cdot d\mathbf{x}$ and $\nabla_{\boldsymbol{\xi}} = \mathbf{Q} \cdot \nabla_{\mathbf{x}}$.

For Cartesian systems, we must have the classical formula for differential distance ds:

$$(ds)^2 = \left(d\xi^1\right)^2 + \left(d\xi^2\right)^2 + \left(d\xi^3\right)^2 = d\boldsymbol{\xi}^T \cdot d\boldsymbol{\xi}. \tag{2.252}$$

The differential distance must be an invariant in either coordinate representation, so

$$(ds)^2 = d\boldsymbol{\xi}^T \cdot d\boldsymbol{\xi} = (\mathbf{J} \cdot d\mathbf{x})^T \cdot (\mathbf{J} \cdot d\mathbf{x}) = d\mathbf{x}^T \cdot \underbrace{\mathbf{J}^T \cdot \mathbf{J}}_{\mathbf{G}} \cdot \mathbf{x} = d\mathbf{x}^T \cdot \mathbf{G} \cdot d\mathbf{x}. \tag{2.253}$$

Importantly, we see that

$$\left(d\xi^1\right)^2 + \left(d\xi^2\right)^2 + \left(d\xi^3\right)^2 \neq \left(dx^1\right)^2 + \left(dx^2\right)^2 + \left(dx^3\right)^2. \tag{2.254}$$

A modulation by the matrix $\mathbf{G}$ is required to maintain equality. So a naïve expression for distance in the transformed space as an unmodulated square root of a sum of squares is inconsistent with distance in the original space. However, in the special case that $\mathbf{J}$ is orthogonal, $\mathbf{J} = \mathbf{Q}$, distances are preserved without modulation as $\mathbf{G} = \mathbf{I}$. Noting again the importance of the product of a matrix with its transpose, we have defined the *metric tensor*

$$\mathbf{G} = \mathbf{J}^T \cdot \mathbf{J}, \qquad g_{ij} = \frac{\partial \xi^k}{\partial x^i} \frac{\partial \xi^k}{\partial x^j}. \tag{2.255}$$

We also define g as

$$g = \det \mathbf{G} = \det g_{ij}. \tag{2.256}$$

One can also show that

$$g^{ij} = \frac{1}{2} \epsilon^{imn} \epsilon^{jpq} g_{mp} g_{nq}, \qquad g_{ik} g^{kj} = \delta_i^j. \tag{2.257}$$

Here we have adopted common notations for the extension of the Levi-Civita symbol and the Kronecker delta to general coordinate systems. This requires the use of both sub- and superscripted indices. Details are given by Aris (1962) and many others. We also have the *Jacobian determinant*, J, defined as

$$J = \sqrt{g} = \det \frac{\partial \xi^k}{\partial x^i} = \frac{\partial \left(\xi^1, \xi^2, \xi^3\right)}{\partial \left(x^1, x^2, x^3\right)} = \det \mathbf{J}. \tag{2.258}$$

As will be discussed in Section 3.8.5, the singular values of $\mathbf{J}$ are $\sqrt{\lambda_i}$, where λ_i are the eigenvalues, guaranteed positive semi-definite real, of $\mathbf{G} = \mathbf{J}^T \cdot \mathbf{J}$. A vector's contravariant representation is given by v^i. Its covariant representation is given by v_i. The metric tensor links one representation to the other via

$$v_j = v^i g_{ij}, \qquad v^i = g^{ij} v_j. \tag{2.259}$$

At this point and for the remainder of this subsection, we present a slight modification of text first presented by Powers and Sen (2015, Section 1.6.5) to better understand the nature of vectors in terms of linear combinations of covariant and contravariant basis vectors. The only requirement we have for the basis vectors is linear independence: they must point in different directions. The basis vectors need not be unit vectors, and their lengths may differ from one another. Consider the nonorthogonal basis vectors $\mathbf{e}_1, \mathbf{e}_2$, aligned with the x^1 and x^2 directions shown in Fig. 2.7(a). The vector $\mathbf{v}$ can be written as a linear combination of the basis vectors:

$$\mathbf{v} = v^1 \mathbf{e}_1 + v^2 \mathbf{e}_2. \tag{2.260}$$

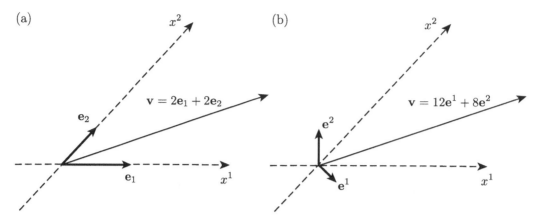

Figure 2.7 Schematic of a vector **v** represented in a nonorthogonal coordinate system in terms of its (a) basis vectors, and (b) reciprocal basis vectors.

We recognize v^1 and v^2 as the contravariant components of **v**. The vectors $\mathbf{e}_1$ and $\mathbf{e}_2$ are the contravariant basis vectors, despite being subscripted. The entity **v** can be considered as either an entity unto itself or perhaps as a column vector whose components are Cartesian. In matrix form, we have **v** as

$$\mathbf{v} = v^1 \begin{pmatrix} \vdots \\ \mathbf{e}_1 \\ \vdots \end{pmatrix} + v^2 \begin{pmatrix} \vdots \\ \mathbf{e}_2 \\ \vdots \end{pmatrix} = \underbrace{\begin{pmatrix} \vdots & \vdots \\ \mathbf{e}_1 & \mathbf{e}_2 \\ \vdots & \vdots \end{pmatrix}}_{\mathbf{J}} \begin{pmatrix} v^1 \\ v^2 \end{pmatrix}. \qquad (2.261)$$

The matrix of basis vectors can be considered to be a local Jacobian matrix, $\mathbf{J}$. It relates the Cartesian and nonorthogonal representations of **v**.

Vectors that comprise a *reciprocal* or *dual* basis have two important features: they are orthogonal to the original basis vectors with different indices, and the dot product of each reciprocal vector with respect to the original vector of the same index must be unity. The covariant basis vectors $\mathbf{e}^1, \mathbf{e}^2$ are reciprocal to $\mathbf{e}_1, \mathbf{e}_2$, as shown in Fig. 2.7(b). Specifically, we have $\mathbf{e}^{1T} \cdot \mathbf{e}_2 = 0$, $\mathbf{e}^{2T} \cdot \mathbf{e}_1 = 0$, $\mathbf{e}^{1T} \cdot \mathbf{e}_1 = 1$, and $\mathbf{e}^{2T} \cdot \mathbf{e}_2 = 1$. In matrix form, this is

$$\underbrace{\begin{pmatrix} \cdots & \mathbf{e}^{1T} & \cdots \\ \cdots & \mathbf{e}^{2T} & \cdots \end{pmatrix}}_{\mathbf{J}^{-1}} \underbrace{\begin{pmatrix} \vdots & \vdots \\ \mathbf{e}_1 & \mathbf{e}_2 \\ \vdots & \vdots \end{pmatrix}}_{\mathbf{J}} = \begin{pmatrix} 1 & 0 \\ 0 & 1 \end{pmatrix} = \mathbf{I}. \qquad (2.262)$$

Obviously, the matrix of reciprocal vectors can be formed by inverting the matrix of the original basis vectors. We can also represent **v** as

$$\mathbf{v} = v_1 \mathbf{e}^1 + v_2 \mathbf{e}^2. \qquad (2.263)$$

2.5 General Coordinate Transformations

In matrix form, we can think of $\mathbf{v}$ as

$$\mathbf{v} = v_1 \begin{pmatrix} \vdots \\ \mathbf{e}^1 \\ \vdots \end{pmatrix} + v_2 \begin{pmatrix} \vdots \\ \mathbf{e}^2 \\ \vdots \end{pmatrix} = \underbrace{\begin{pmatrix} \vdots & \vdots \\ \mathbf{e}^1 & \mathbf{e}^2 \\ \vdots & \vdots \end{pmatrix}}_{(\mathbf{J}^{-1})^T} \begin{pmatrix} v_1 \\ v_2 \end{pmatrix}. \tag{2.264}$$

We might also say

$$\mathbf{v}^T = \begin{pmatrix} v_1 & v_2 \end{pmatrix} \underbrace{\begin{pmatrix} \cdots & \mathbf{e}^{1T} & \cdots \\ \cdots & \mathbf{e}^{2T} & \cdots \end{pmatrix}}_{\mathbf{J}^{-1}}. \tag{2.265}$$

Because the magnitude of $\mathbf{v}$ is independent of its coordinate system, we can say

$$\mathbf{v}^T \cdot \mathbf{v} = \begin{pmatrix} v_1 & v_2 \end{pmatrix} \underbrace{\underbrace{\begin{pmatrix} \cdots & \mathbf{e}^{1T} & \cdots \\ \cdots & \mathbf{e}^{2T} & \cdots \end{pmatrix}}_{\mathbf{J}^{-1}} \underbrace{\begin{pmatrix} \vdots & \vdots \\ \mathbf{e}_1 & \mathbf{e}_2 \\ \vdots & \vdots \end{pmatrix}}_{\mathbf{J}}}_{\mathbf{I}} \begin{pmatrix} v^1 \\ v^2 \end{pmatrix}, \tag{2.266}$$

$$= \begin{pmatrix} v_1 & v_2 \end{pmatrix} \begin{pmatrix} v^1 \\ v^2 \end{pmatrix} = v_i v^i. \tag{2.267}$$

Now we can also transpose Eq. (2.261) to obtain

$$\mathbf{v}^T = \begin{pmatrix} v^1 & v^2 \end{pmatrix} \underbrace{\begin{pmatrix} \cdots & \mathbf{e}_1^T & \cdots \\ \cdots & \mathbf{e}_2^T & \cdots \end{pmatrix}}_{\mathbf{J}^T}. \tag{2.268}$$

Combining this with Eq. (2.261) to form $\mathbf{v}^T \cdot \mathbf{v}$, we also see

$$\mathbf{v}^T \cdot \mathbf{v} = \begin{pmatrix} v^1 & v^2 \end{pmatrix} \underbrace{\begin{pmatrix} \cdots & \mathbf{e}_1^T & \cdots \\ \cdots & \mathbf{e}_2^T & \cdots \end{pmatrix}}_{\mathbf{J}^T} \underbrace{\begin{pmatrix} \vdots & \vdots \\ \mathbf{e}_1 & \mathbf{e}_2 \\ \vdots & \vdots \end{pmatrix}}_{\mathbf{J}} \begin{pmatrix} v^1 \\ v^2 \end{pmatrix}. \tag{2.269}$$

Now with the metric tensor $g_{ij} = \mathbf{G} = \mathbf{J}^T \cdot \mathbf{J}$, we have

$$\mathbf{v}^T \cdot \mathbf{v} = \begin{pmatrix} v^1 & v^2 \end{pmatrix} \cdot \mathbf{G} \cdot \begin{pmatrix} v^1 \\ v^2 \end{pmatrix}. \tag{2.270}$$

One can now compare Eq. (2.270) with Eq. (2.267) to infer the covariant components v_1 and v_2. For the same vector $\mathbf{v}$, the covariant components are different than the contravariant components. Thus, for example,

$$v_i = \begin{pmatrix} v^1 & v^2 \end{pmatrix} \cdot \mathbf{G} = v^j g_{ij} = g_{ij} v^j. \tag{2.271}$$

Deducing from Eq. (2.257) that $g^{ij} = \mathbf{G}^{-1}$, we also see $g^{ij} v_i = v^j$.

In Cartesian coordinates, a basis and its reciprocal are the same, and so also are the contravariant and covariant components of a vector. For this reason Cartesian vectors and tensors are usually written with only subscripts.

Example 2.22 Consider the vector $\mathbf{v}$ that has the Cartesian representation of

$$\mathbf{v} = \begin{pmatrix} 6 \\ 2 \end{pmatrix}. \tag{2.272}$$

We are also given a pair of nonorthogonal basis vectors:

$$\mathbf{e}_1 = \begin{pmatrix} 2 \\ 0 \end{pmatrix}, \qquad \mathbf{e}_2 = \begin{pmatrix} 1 \\ 1 \end{pmatrix}. \tag{2.273}$$

Deduce the contravariant and covariant components, v^i and v_i, of the vector $\mathbf{v}$, which may be considered to be the covariant and contravariant representations of $\mathbf{v}$.

Solution
The Jacobian matrix has the basis vectors in its columns:

$$\mathbf{J} = \begin{pmatrix} \vdots & \vdots \\ \mathbf{e}_1 & \mathbf{e}_2 \\ \vdots & \vdots \end{pmatrix} = \begin{pmatrix} 2 & 1 \\ 0 & 1 \end{pmatrix}. \tag{2.274}$$

We find for J that

$$J = \det \mathbf{J} = (2)(1) - 1(0) = 2. \tag{2.275}$$

The contravariant components are v^i and can be found by solving

$$\mathbf{v} = \mathbf{J} \cdot \begin{pmatrix} v^1 \\ v^2 \end{pmatrix}, \qquad \begin{pmatrix} 6 \\ 2 \end{pmatrix} = \begin{pmatrix} 2 & 1 \\ 0 & 1 \end{pmatrix} \begin{pmatrix} v^1 \\ v^2 \end{pmatrix}. \tag{2.276}$$

We invert and find

$$\begin{pmatrix} v^1 \\ v^2 \end{pmatrix} = \begin{pmatrix} 2 \\ 2 \end{pmatrix}. \tag{2.277}$$

Thus, we can represent $\mathbf{v}$ in terms of the nonorthogonal basis as

$$\mathbf{v} = v^1 \mathbf{e}_1 + v^2 \mathbf{e}_2 = 2\mathbf{e}_1 + 2\mathbf{e}_2, \qquad \begin{pmatrix} 6 \\ 2 \end{pmatrix} = 2 \begin{pmatrix} 2 \\ 0 \end{pmatrix} + 2 \begin{pmatrix} 1 \\ 1 \end{pmatrix}. \tag{2.278}$$

The covariant basis vectors are found by inverting the Jacobian matrix:

$$\mathbf{J}^{-1} = \begin{pmatrix} \cdots & \mathbf{e}^{1T} & \cdots \\ \cdots & \mathbf{e}^{2T} & \cdots \end{pmatrix} = \begin{pmatrix} \tfrac{1}{2} & -\tfrac{1}{2} \\ 0 & 1 \end{pmatrix}. \tag{2.279}$$

Performing the transpose, we get

$$\mathbf{e}^1 = \begin{pmatrix} \tfrac{1}{2} \\ -\tfrac{1}{2} \end{pmatrix}, \qquad \mathbf{e}^2 = \begin{pmatrix} 0 \\ 1 \end{pmatrix}. \tag{2.280}$$

The metric tensor is easily found:

$$g_{ij} = \mathbf{G} = \mathbf{J}^T \cdot \mathbf{J} = \begin{pmatrix} 2 & 0 \\ 1 & 1 \end{pmatrix} \begin{pmatrix} 2 & 1 \\ 0 & 1 \end{pmatrix} = \begin{pmatrix} 4 & 2 \\ 2 & 2 \end{pmatrix}. \tag{2.281}$$

We also have

$$g = \det \mathbf{G} = (4)(2) - (2)(2) = 4. \tag{2.282}$$

We see that $\sqrt{g} = J$. The covariant components are easily found:

$$v_i = g_{ij} v^j = \begin{pmatrix} 4 & 2 \\ 2 & 2 \end{pmatrix} \begin{pmatrix} 2 \\ 2 \end{pmatrix} = \begin{pmatrix} 12 \\ 8 \end{pmatrix}. \tag{2.283}$$

Thus, the covariant representation of $\mathbf{v}$ is

$$\mathbf{v} = v_1 \mathbf{e}^1 + v_2 \mathbf{e}^2 = 12 \mathbf{e}^1 + 8 \mathbf{e}^2, \quad \begin{pmatrix} 6 \\ 2 \end{pmatrix} = 12 \begin{pmatrix} \frac{1}{2} \\ -\frac{1}{2} \end{pmatrix} + 8 \begin{pmatrix} 0 \\ 1 \end{pmatrix}. \tag{2.284}$$

In Cartesian coordinates we have the dot product of $\mathbf{v}$ with itself as

$$\mathbf{v}^T \cdot \mathbf{v} = \begin{pmatrix} 6 & 2 \end{pmatrix} \begin{pmatrix} 6 \\ 2 \end{pmatrix} = 40. \tag{2.285}$$

We see the dot product is invariant under coordinate transformation as in our nonorthogonal coordinate system, we find

$$v_i v^i = \begin{pmatrix} 12 & 8 \end{pmatrix} \begin{pmatrix} 2 \\ 2 \end{pmatrix} = 40. \tag{2.286}$$

The vectors represented in Fig. 2.7 are those of this example.

Example 2.23 Consider two-dimensional linear transformations with constant Jacobian $\mathbf{J}$. Determine if the following are area- and orientation-preserving, and if they preserve the unmodulated length of a vector when acted on by $\mathbf{J}$. Find the metric tensor $\mathbf{G}$ for each

$$\mathbf{J}_1 = \begin{pmatrix} \frac{\sqrt{2}}{2} & -\frac{\sqrt{2}}{2} \\ \frac{\sqrt{2}}{2} & \frac{\sqrt{2}}{2} \end{pmatrix}, \quad \mathbf{J}_2 = \begin{pmatrix} \frac{\sqrt{2}}{2} & \frac{\sqrt{2}}{2} \\ \frac{\sqrt{2}}{2} & -\frac{\sqrt{2}}{2} \end{pmatrix}, \quad \mathbf{J}_3 = \begin{pmatrix} 1 & 0 \\ 1 & 1 \end{pmatrix}, \quad \mathbf{J}_4 = \begin{pmatrix} -1 & 0 \\ 1 & 1 \end{pmatrix}. \tag{2.287}$$

Solution

We see $J = \det \mathbf{J}_1 = 1$, so it is area- and orientation-preserving. And $\mathbf{J}_1^T \cdot \mathbf{J}_1 = \mathbf{I}$, so $\mathbf{J}_1$ is a rotation matrix. Thus via Eq. (2.55), it preserves the length of vectors on which it acts. The metric tensor is $\mathbf{G}_1 = \mathbf{J}_1^T \cdot \mathbf{J}_1 = \mathbf{I}$.

We see $J = \det \mathbf{J}_2 = -1$, so it is area-preserving and orientation-reversing. And $\mathbf{J}_2^T \cdot \mathbf{J}_2 = \mathbf{I}$, so $\mathbf{J}_2$ is a reflection matrix. Via Eq. (2.55), it preserves the length of vectors on which it acts. The metric tensor is $\mathbf{G}_2 = \mathbf{J}_2^T \cdot \mathbf{J}_2 = \mathbf{I}$. The transformation associated with $\mathbf{J}_3$ is known as a *shear mapping*. We see $J = \det \mathbf{J}_3 = 1$, so it is area- and orientation-preserving. But $\mathbf{J}_3^T \cdot \mathbf{J}_3 \neq \mathbf{I}$, so $\mathbf{J}_3$ is not orthogonal. It does not preserve the unmodulated length of vectors; for example if $\mathbf{x} = \mathbf{i}$, then

$$\boldsymbol{\xi} = \mathbf{J}_3 \cdot \mathbf{x} = \mathbf{J}_3 \cdot \mathbf{i} = \begin{pmatrix} 1 & 0 \\ 1 & 1 \end{pmatrix} \begin{pmatrix} 1 \\ 0 \end{pmatrix} = \begin{pmatrix} 1 \\ 1 \end{pmatrix}. \tag{2.288}$$

The length of $\boldsymbol{\xi}$ is $\sqrt{2}$, while the unmodulated length of $\mathbf{x}$ is unity. We find
$$\mathbf{G}_3 = \mathbf{J}_3^T \cdot \mathbf{J}_3 = \begin{pmatrix} 2 & 1 \\ 1 & 1 \end{pmatrix}. \tag{2.289}$$

With modulation, lengths are preserved. For $\mathbf{x} = \mathbf{i}$, we have the modulated length as
$$\sqrt{\mathbf{x}^T \cdot \mathbf{G}_3 \cdot \mathbf{x}} = \sqrt{\begin{pmatrix} 1 & 0 \end{pmatrix} \begin{pmatrix} 2 & 1 \\ 1 & 1 \end{pmatrix} \begin{pmatrix} 1 \\ 0 \end{pmatrix}} = \sqrt{2}. \tag{2.290}$$

We see $J = \det \mathbf{J}_4 = -1$, so it is area-preserving and orientation-reversing. But $\mathbf{J}_4^T \cdot \mathbf{J}_4 \neq \mathbf{I}$, so $\mathbf{J}_4$ is not orthogonal. It does not preserve the unmodulated length of vectors, for example if $\mathbf{x} = \mathbf{i}$, then
$$\boldsymbol{\xi} = \mathbf{J}_4 \cdot \mathbf{x} = \mathbf{J}_4 \cdot \mathbf{i} = \begin{pmatrix} -1 & 0 \\ 1 & 1 \end{pmatrix} \begin{pmatrix} 1 \\ 0 \end{pmatrix} = \begin{pmatrix} -1 \\ 1 \end{pmatrix}. \tag{2.291}$$

The length of $\boldsymbol{\xi}$ is $\sqrt{2}$, while the unmodulated length of $\mathbf{x}$ is unity. We find
$$\mathbf{G}_4 = \mathbf{J}_4^T \cdot \mathbf{J}_4 = \begin{pmatrix} 2 & 1 \\ 1 & 1 \end{pmatrix}. \tag{2.292}$$

With modulation, lengths are preserved. For $\mathbf{x} = \mathbf{i}$, we have the modulated length as
$$\sqrt{\mathbf{x}^T \cdot \mathbf{G}_4 \cdot \mathbf{x}} = \sqrt{\begin{pmatrix} 1 & 0 \end{pmatrix} \begin{pmatrix} 2 & 1 \\ 1 & 1 \end{pmatrix} \begin{pmatrix} 1 \\ 0 \end{pmatrix}} = \sqrt{2}. \tag{2.293}$$

Example 2.24 Consider three-dimensional linear transformations with constant non-singular Jacobian matrix $\mathbf{J}$ of the form $\boldsymbol{\xi} = \mathbf{J} \cdot \mathbf{x}$. Show that a unit sphere represented in the Cartesian system $\boldsymbol{\xi}$ transforms to an ellipsoid when represented in the transformed $\mathbf{x}$ space.

Solution
We can understand this via the following analysis:
$$\mathbf{J} \cdot \mathbf{x} = \boldsymbol{\xi}, \tag{2.294}$$
$$\mathbf{J}^T \cdot \mathbf{J} \cdot \mathbf{x} = \mathbf{J}^T \cdot \boldsymbol{\xi}. \tag{2.295}$$

Now, $\mathbf{J}^T \cdot \mathbf{J}$ is symmetric, so it has the decomposition
$$\mathbf{J}^T \cdot \mathbf{J} = \mathbf{Q} \cdot \boldsymbol{\Lambda} \cdot \mathbf{Q}^T. \tag{2.296}$$

It is always possible to construct $\mathbf{Q}$ as a rotation, and we will consider it to be so done. Thus, our mapping can be written as
$$\mathbf{Q} \cdot \boldsymbol{\Lambda} \cdot \mathbf{Q}^T \cdot \mathbf{x} = \mathbf{J}^T \cdot \boldsymbol{\xi}, \tag{2.297}$$
$$\mathbf{x}^T \cdot \mathbf{Q} \cdot \boldsymbol{\Lambda} \cdot \mathbf{Q}^T \cdot \mathbf{x} = \mathbf{x}^T \cdot \mathbf{J}^T \cdot \boldsymbol{\xi}, \tag{2.298}$$
$$\mathbf{x}^T \cdot \mathbf{Q} \cdot \boldsymbol{\Lambda} \cdot \mathbf{Q}^T \cdot \mathbf{x} = (\mathbf{J} \cdot \mathbf{x})^T \cdot \boldsymbol{\xi}, \tag{2.299}$$
$$\left(\mathbf{Q}^T \cdot \mathbf{x} \right)^T \cdot \boldsymbol{\Lambda} \cdot (\mathbf{Q}^T \cdot \mathbf{x}) = \boldsymbol{\xi}^T \cdot \boldsymbol{\xi}. \tag{2.300}$$

Now on our unit sphere which defines $\boldsymbol{\xi}$, we have $\boldsymbol{\xi}^T \cdot \boldsymbol{\xi} = 1$, so the ellipsoid is defined by
$$\left(\mathbf{Q}^T \cdot \mathbf{x} \right)^T \cdot \boldsymbol{\Lambda} \cdot (\mathbf{Q}^T \cdot \mathbf{x}) = 1. \tag{2.301}$$

Here **Q** effects a rotation, and **Λ** effects stretching along principal axes. So the unit sphere given by $(\xi^1)^2 + (\xi^2)^2 + (\xi^3)^2 = 1$ transforms to the ellipsoid

$$\lambda_1 (x^{1'})^2 + \lambda_2 (x^{2'})^2 + \lambda_3 (x^{3'})^2 = \left(\frac{x^{1'}}{1/\sqrt{\lambda_1}}\right)^2 + \left(\frac{x^{2'}}{1/\sqrt{\lambda_2}}\right)^2 + \left(\frac{x^{3'}}{1/\sqrt{\lambda_3}}\right)^2 = 1. \quad (2.302)$$

Here $\mathbf{x}' = \mathbf{Q}^T \cdot \mathbf{x}$ is a rotated system, and $\lambda_1, \lambda_2, \lambda_3$ are the eigenvalues of $\mathbf{J}^T \cdot \mathbf{J}$. They are positive definite because $\mathbf{J}$ is non-singular. The lengths of the principal axes of the ellipsoid are proportional to the singular values of $\mathbf{J}^{-1}$. Moreover, *the action of a general linear alibi transformation matrix* $\mathbf{J}$ *on a vector is a composite of rotation and stretching.*

The *Christoffel*[14] *symbols* are given by

$$\Gamma^i_{jl} = \frac{1}{2} g^{ik} \left(\frac{\partial g_{lk}}{\partial x^j} + \frac{\partial g_{kj}}{\partial x^l} - \frac{\partial g_{jl}}{\partial x^k} \right) = \frac{\partial^2 \xi^p}{\partial x^j \partial x^l} \frac{\partial x^i}{\partial \xi^p}. \quad (2.303)$$

As discussed in detail by Powers and Sen (2015, Section 1.6), such notions as covariant derivatives of contravariant vectors can be defined and can be shown to take the form

$$\nabla_j v^i = \frac{\partial v^i}{\partial x^j} + \Gamma^i_{jl} v^l. \quad (2.304)$$

Here the notation ∇_j is that for a generalized covariant derivative. It is easy to show that any linear transformation, including rotation, reflection, and linear stretching, has $\Gamma^i_{jl} = 0$. However, more general transformations, such as Cartesian to cylindrical coordinates, have nontrivial Christoffel symbols. Their physical manifestation are terms such as *centripetal* and *Coriolis*[15] *accelerations*, as will be demonstrated in detail in Section 2.6.1. These terms are not based on the derivative of the vector, but are related to the vector itself. More generally, the concept of fluid velocity and its time rate of change, acceleration, will be considered more formally in Chapter 3; however, as it is such a familiar and important concept, a discussion of it in the context of coordinate systems is warranted.

2.6 Cylindrical Coordinates

Many problems are best described with a set of cylindrical coordinates. The transformation and inverse transformation to and from cylindrical $(r, \theta, \hat{z})$ to Cartesian (x, y, z) coordinates is given by the familiar

$$x = r \cos\theta, \qquad y = r \sin\theta, \qquad z = \hat{z}, \quad (2.305)$$

$$r = +\sqrt{x^2 + y^2}, \qquad \theta = \tan^{-1}\left(\frac{y}{x}\right), \qquad \hat{z} = z. \quad (2.306)$$

Because $z = \hat{z}$, we will occasionally simply use z in the cylindrical coordinate system in this chapter. Note that we have replaced (x_1, x_2, x_3) by the more traditional (x, y, z). Similarly,

---

[14] Elwin Bruno Christoffel, 1829–1900, German mathematician and physicist.
[15] Gaspard-Gustave de Coriolis, 1792–1843, Paris-born mathematician who taught with Navier, introduced the terms "work" and "kinetic energy" with modern scientific meaning and wrote on the mathematical theory of billiards.

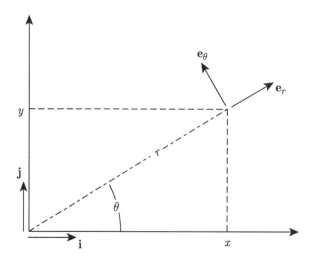

Figure 2.8 Representation of a point in Cartesian and polar coordinates along with unit vectors for both systems.

for this discussion, we will replace the Cartesian unit basis vectors e_2, e_2, e_3 with the more traditional **i**, **j**, **k**.

Most of the basic distinctions between the two systems can be understood by considering two-dimensional geometries. Often a pure two-dimensional representation is called "polar," while "cylindrical" is reserved for the three-dimensional extension. The representation of an arbitrary point in both two-dimensional (x, y) Cartesian and two-dimensional (r, θ) polar coordinate systems along with the unit basis vectors for both systems, **i**, **j**, and $\mathbf{e}_r$, $\mathbf{e}_\theta$, is depicted in Fig. 2.8.

2.6.1 Centripetal and Coriolis Accelerations

The fact that a point in motion is accompanied by changes in the basis vectors with respect to time in the cylindrical representation, but not for Cartesian basis vectors, accounts for the most striking differences in the formulations of the governing equations, namely the appearance of so-called centripetal and Coriolis accelerations in the cylindrical representation. Allusion was made to such accelerations in Section 2.5. In this section, for Cartesian systems, we will return to the traditional $u = v_1$, $v = v_2$, $w = v_3$ so that

$$\mathbf{v} = \begin{pmatrix} v_1 \\ v_2 \\ v_3 \end{pmatrix} = \begin{pmatrix} u \\ v \\ w \end{pmatrix}. \tag{2.307}$$

Consider two-dimensional representations of the velocity vector **v** in both coordinate systems:

$$\mathbf{v} = u\mathbf{i} + v\mathbf{j}, \quad \text{or} \quad \mathbf{v} = v_r \mathbf{e}_r + v_\theta \mathbf{e}_\theta. \tag{2.308}$$

Now the unsteady part of the acceleration vector of a particle is simply the partial derivative of the velocity vector with respect to time. Let us focus on that here. Later, in Chapter 3, we will see that acceleration also has a spatial dependency for spatially nonuniform velocity fields.

2.6 Cylindrical Coordinates

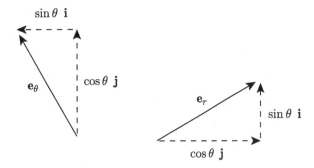

Figure 2.9 Geometrical representation of polar unit vectors in terms of Cartesian unit vectors.

Now formally, we must allow for variations of the unit basis vectors as well as the components themselves so that

$$\frac{\partial \mathbf{v}}{\partial t} = \frac{\partial u}{\partial t}\mathbf{i} + u\underbrace{\frac{\partial \mathbf{i}}{\partial t}}_{=0} + \frac{\partial v}{\partial t}\mathbf{j} + v\underbrace{\frac{\partial \mathbf{j}}{\partial t}}_{=0}, \qquad (2.309)$$

$$\frac{\partial \mathbf{v}}{\partial t} = \frac{\partial v_r}{\partial t}\mathbf{e}_r + v_r\frac{\partial \mathbf{e}_r}{\partial t} + \frac{\partial v_\theta}{\partial t}\mathbf{e}_\theta + v_\theta\frac{\partial \mathbf{e}_\theta}{\partial t}. \qquad (2.310)$$

Now the time derivatives of the Cartesian basis vectors are zero, as they are defined not to change with the position of the particle. Hence for a Cartesian representation, we have for the unsteady component of acceleration the familiar

$$\frac{\partial \mathbf{v}}{\partial t} = \frac{\partial u}{\partial t}\mathbf{i} + \frac{\partial v}{\partial t}\mathbf{j}. \qquad (2.311)$$

However, the time derivative of the polar basis vectors does change with time for particles in motion! To see this, let us first relate $\mathbf{e}_r$ and $\mathbf{e}_\theta$ to $\mathbf{i}$ and $\mathbf{j}$. Such a relationship is easy to find for this system, but requires consideration of additional nuances for more general systems; see Aris (1962, pp. 149–159) for discussion of unit basis vectors in orthogonal and nonorthogonal coordinate systems. From the diagram of Fig. 2.9, it is clear that

$$\mathbf{e}_r = \cos\theta\,\mathbf{i} + \sin\theta\,\mathbf{j}, \qquad \mathbf{e}_\theta = -\sin\theta\,\mathbf{i} + \cos\theta\,\mathbf{j}. \qquad (2.312)$$

This is a linear system of equations. We can use Cramer's rule, Section 2.2, to invert to find

$$\mathbf{i} = \cos\theta\,\mathbf{e}_r - \sin\theta\,\mathbf{e}_\theta, \qquad \mathbf{j} = \sin\theta\,\mathbf{e}_r + \cos\theta\,\mathbf{e}_\theta. \qquad (2.313)$$

Now, examining time derivatives of the unit vectors, we see that

$$\frac{\partial \mathbf{e}_r}{\partial t} = -\sin\theta\frac{\partial \theta}{\partial t}\mathbf{i} + \cos\theta\frac{\partial \theta}{\partial t}\mathbf{j} = \frac{\partial \theta}{\partial t}\mathbf{e}_\theta, \qquad (2.314)$$

$$\frac{\partial \mathbf{e}_\theta}{\partial t} = -\cos\theta\frac{\partial \theta}{\partial t}\mathbf{i} - \sin\theta\frac{\partial \theta}{\partial t}\mathbf{j} = -\frac{\partial \theta}{\partial t}\mathbf{e}_r, \qquad (2.315)$$

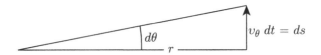

Figure 2.10 Diagram of relation of differential distance ds to velocity in angular direction v_θ.

so there is a formal variation of the unit vectors with respect to time as long as the angular velocity $\partial \theta/\partial t \neq 0$. So the acceleration vector is

$$\frac{\partial \mathbf{v}}{\partial t} = \frac{\partial v_r}{\partial t}\mathbf{e}_r + v_r\frac{\partial \theta}{\partial t}\mathbf{e}_\theta + \frac{\partial v_\theta}{\partial t}\mathbf{e}_\theta - v_\theta\frac{\partial \theta}{\partial t}\mathbf{e}_r, \qquad (2.316)$$

$$= \left(\frac{\partial v_r}{\partial t} - v_\theta\frac{\partial \theta}{\partial t}\right)\mathbf{e}_r + \left(\frac{\partial v_\theta}{\partial t} + v_r\frac{\partial \theta}{\partial t}\right)\mathbf{e}_\theta. \qquad (2.317)$$

Now from basic geometry, as depicted in Fig. 2.10, we have

$$ds = r\,d\theta, \quad \text{so} \quad v_\theta\,dt = r\,d\theta, \quad \text{so} \quad \frac{v_\theta}{r} = \frac{\partial \theta}{\partial t}. \qquad (2.318)$$

Consequently, we can write the unsteady component of acceleration as

$$\frac{\partial \mathbf{v}}{\partial t} = \left(\frac{\partial v_r}{\partial t} - \underbrace{\frac{v_\theta^2}{r}}_{\text{centripetal}}\right)\mathbf{e}_r + \left(\frac{\partial v_\theta}{\partial t} + \underbrace{\frac{v_r v_\theta}{r}}_{\text{Coriolis}}\right)\mathbf{e}_\theta. \qquad (2.319)$$

Two, apparently *new*, accelerations have appeared as a consequence of the transformation: centripetal acceleration, v_θ^2/r, directed towards the center, and Coriolis acceleration, $v_r v_\theta/r$, directed in the direction of increasing θ. These terms do not have explicit dependency on time derivatives of velocity. And yet when the equations are constructed in this coordinate system, they represent real accelerations, and are consequences of forces. As can be seen by considering the general theory of nonorthogonal coordinate transformations, terms like the centripetal and Coriolis acceleration are associated with the Christoffel symbols of the transformation.

2.6.2 Grad and Div

We can use the transformation rule for partial derivatives to develop expressions for grad and div in cylindrical coordinate systems. Consider the Cartesian

$$\nabla = \mathbf{i}\frac{\partial}{\partial x} + \mathbf{j}\frac{\partial}{\partial y} + \mathbf{k}\frac{\partial}{\partial z}. \qquad (2.320)$$

The transformation rule gives us

$$\frac{\partial}{\partial x} = \frac{\partial r}{\partial x}\frac{\partial}{\partial r} + \frac{\partial \theta}{\partial x}\frac{\partial}{\partial \theta} + \frac{\partial \hat{z}}{\partial x}\frac{\partial}{\partial \hat{z}}, \qquad (2.321)$$

$$\frac{\partial}{\partial y} = \frac{\partial r}{\partial y}\frac{\partial}{\partial r} + \frac{\partial \theta}{\partial y}\frac{\partial}{\partial \theta} + \frac{\partial \hat{z}}{\partial y}\frac{\partial}{\partial \hat{z}}, \qquad (2.322)$$

$$\frac{\partial}{\partial z} = \frac{\partial r}{\partial z}\frac{\partial}{\partial r} + \frac{\partial \theta}{\partial z}\frac{\partial}{\partial \theta} + \frac{\partial \hat{z}}{\partial z}\frac{\partial}{\partial \hat{z}}. \qquad (2.323)$$

Now, we have for derivatives of r that

$$\frac{\partial r}{\partial x} = \frac{2x}{2\sqrt{x^2+y^2}} = \frac{x}{r} = \cos\theta, \quad \frac{\partial r}{\partial y} = \frac{2y}{2\sqrt{x^2+y^2}} = \frac{y}{r} = \sin\theta, \quad \frac{\partial r}{\partial z} = 0. \quad (2.324)$$

For derivatives of θ, we have
$$\frac{\partial \theta}{\partial x} = -\frac{y}{x^2+y^2} = -\frac{\sin\theta}{r}, \quad \frac{\partial \theta}{\partial y} = \frac{x}{x^2+y^2} = \frac{\cos\theta}{r}, \quad \frac{\partial \theta}{\partial z} = 0. \tag{2.325}$$

For derivatives of $\hat{z}$, we have
$$\frac{\partial \hat{z}}{\partial x} = 0, \quad \frac{\partial \hat{z}}{\partial y} = 0, \quad \frac{\partial \hat{z}}{\partial z} = 1. \tag{2.326}$$

So the derivatives with respect to x, y, and z are
$$\frac{\partial}{\partial x} = \cos\theta \frac{\partial}{\partial r} - \frac{\sin\theta}{r}\frac{\partial}{\partial \theta}, \quad \frac{\partial}{\partial y} = \sin\theta \frac{\partial}{\partial r} + \frac{\cos\theta}{r}\frac{\partial}{\partial \theta}, \quad \frac{\partial}{\partial z} = \frac{\partial}{\partial \hat{z}}. \tag{2.327}$$

Grad

So now we are prepared to write an explicit form for ∇ in cylindrical coordinates:

$$\nabla = \underbrace{(\cos\theta \mathbf{e}_r - \sin\theta \mathbf{e}_\theta)}_{\mathbf{i}} \underbrace{\left(\cos\theta \frac{\partial}{\partial r} - \frac{\sin\theta}{r}\frac{\partial}{\partial \theta}\right)}_{\frac{\partial}{\partial x}}$$
$$+ \underbrace{(\sin\theta \mathbf{e}_r + \cos\theta \mathbf{e}_\theta)}_{\mathbf{j}} \underbrace{\left(\sin\theta \frac{\partial}{\partial r} + \frac{\cos\theta}{r}\frac{\partial}{\partial \theta}\right)}_{\frac{\partial}{\partial y}} + \mathbf{e}_{\hat{z}} \frac{\partial}{\partial \hat{z}}, \tag{2.328}$$

$$= \mathbf{e}_r \left((\cos^2\theta + \sin^2\theta)\frac{\partial}{\partial r} + \left(-\frac{\sin\theta\cos\theta}{r} + \frac{\sin\theta\cos\theta}{r}\right)\frac{\partial}{\partial \theta}\right)$$
$$+ \mathbf{e}_\theta \left((-\sin\theta\cos\theta + \sin\theta\cos\theta)\frac{\partial}{\partial r} + \left(\frac{\sin^2\theta}{r} + \frac{\cos^2\theta}{r}\right)\frac{\partial}{\partial \theta}\right)$$
$$+ \mathbf{e}_{\hat{z}} \frac{\partial}{\partial \hat{z}}, \tag{2.329}$$
$$= \mathbf{e}_r \frac{\partial}{\partial r} + \mathbf{e}_\theta \frac{1}{r}\frac{\partial}{\partial \theta} + \mathbf{e}_{\hat{z}} \frac{\partial}{\partial \hat{z}}. \tag{2.330}$$

We can now write a simple expression for the advective component, $\mathbf{v}^T \cdot \nabla$, of the acceleration vector:
$$\mathbf{v}^T \cdot \nabla = v_r \frac{\partial}{\partial r} + \frac{v_\theta}{r}\frac{\partial}{\partial \theta} + v_{\hat{z}}\frac{\partial}{\partial \hat{z}}. \tag{2.331}$$

Div

The divergence is straightforward. In Cartesian coordinates we have
$$\nabla^T \cdot \mathbf{v} = \frac{\partial u}{\partial x} + \frac{\partial v}{\partial y} + \frac{\partial w}{\partial z}. \tag{2.332}$$

In cylindrical coordinates, we replace derivatives with respect to x, y, z with those with respect to $r, \theta, \hat{z}$, so
$$\nabla^T \cdot \mathbf{v} = \cos\theta \frac{\partial u}{\partial r} - \frac{\sin\theta}{r}\frac{\partial u}{\partial \theta} + \sin\theta \frac{\partial v}{\partial r} + \frac{\cos\theta}{r}\frac{\partial v}{\partial \theta} + \frac{\partial w}{\partial \hat{z}}. \tag{2.333}$$

Now u, v, and w transform in the same way as $\mathbf{i}$, $\mathbf{j}$, and $\mathbf{k}$, so

$$u = v_r \cos\theta - v_\theta \sin\theta, \qquad v = v_r \sin\theta + v_\theta \cos\theta, \qquad w = v_{\hat{z}}. \tag{2.334}$$

Substituting and taking partials, we find that

$$\nabla^T \cdot \mathbf{v} = \cos\theta \left(\cos\theta \underbrace{\frac{\partial v_r}{\partial r} - \sin\theta \frac{\partial v_\theta}{\partial r}}_{A} \right) - \frac{\sin\theta}{r} \left(\underbrace{\cos\theta \frac{\partial v_r}{\partial \theta} - \sin\theta v_r}_{B} - \sin\theta \frac{\partial v_\theta}{\partial \theta} - \underbrace{\cos\theta v_\theta}_{C} \right)$$

$$+ \sin\theta \left(\underbrace{\sin\theta \frac{\partial v_r}{\partial r} + \cos\theta \frac{\partial v_\theta}{\partial r}}_{A} \right) + \frac{\cos\theta}{r} \left(\underbrace{\sin\theta \frac{\partial v_r}{\partial \theta} + \cos\theta v_r}_{B} + \cos\theta \frac{\partial v_\theta}{\partial \theta} - \underbrace{\sin\theta v_\theta}_{C} \right)$$

$$+ \frac{\partial v_{\hat{z}}}{\partial \hat{z}}. \tag{2.335}$$

When expanded, the terms labeled A, B, and C cancel in this expression. Then using the trigonometric identity $\sin^2\theta + \cos^2\theta = 1$, we arrive at the simple form

$$\nabla^T \cdot \mathbf{v} = \frac{\partial v_r}{\partial r} + \frac{v_r}{r} + \frac{1}{r}\frac{\partial v_\theta}{\partial \theta} + \frac{\partial v_{\hat{z}}}{\partial \hat{z}} = \frac{1}{r}\frac{\partial}{\partial r}(r v_r) + \frac{1}{r}\frac{\partial v_\theta}{\partial \theta} + \frac{\partial v_{\hat{z}}}{\partial \hat{z}}. \tag{2.336}$$

Using the same procedure, we can show that the Laplacian operator transforms to

$$\nabla^T \cdot \nabla = \nabla^2 = \frac{1}{r}\frac{\partial}{\partial r}\left(r\frac{\partial}{\partial r}\right) + \frac{1}{r^2}\frac{\partial^2}{\partial \theta^2} + \frac{\partial^2}{\partial \hat{z}^2}. \tag{2.337}$$

Alternate Derivations

Presented here are brief details of an alternate, more formal, derivation of the gradient and Laplacian operators in cylindrical coordinates. The general background was presented in Section 2.5. As before, one can transform from the Cartesian system with (x, y, z) as coordinates to the cylindrical system with $(r, \theta, \hat{z})$ as coordinates via $x = r\cos\theta$, $y = r\sin\theta$, $z = \hat{z}$. We will consider the domain $r \in [0, \infty)$, $\theta \in [0, 2\pi]$, $\hat{z} \in (-\infty, \infty)$. Then, with the exception of the origin $(x, y, z) = (0, 0, 0)$, every (x, y, z) will map to a unique $(r, \theta, \hat{z})$.

The Jacobian matrix of the transformation is

$$\mathbf{J} = \begin{pmatrix} \frac{\partial x}{\partial r} & \frac{\partial x}{\partial \theta} & \frac{\partial x}{\partial \hat{z}} \\ \frac{\partial y}{\partial r} & \frac{\partial y}{\partial \theta} & \frac{\partial y}{\partial \hat{z}} \\ \frac{\partial z}{\partial r} & \frac{\partial z}{\partial \theta} & \frac{\partial z}{\partial \hat{z}} \end{pmatrix} = \begin{pmatrix} \cos\theta & -r\sin\theta & 0 \\ \sin\theta & r\cos\theta & 0 \\ 0 & 0 & 1 \end{pmatrix}. \tag{2.338}$$

We have $J = \det \mathbf{J} = r$; this can be shown to tell us that the transformation is singular and thus nonunique when $r = 0$. It is orientation-preserving for $r > 0$, and it is volume-preserving only for $r = 1$; thus, in general, it does not preserve volume.

If we take $d\mathbf{x} = (dx, dy, dz)^T$, and $d\mathbf{r} = (dr, d\theta, d\hat{z})^T$, we have

$$d\mathbf{x} = \mathbf{J} \cdot d\mathbf{r}, \qquad \begin{pmatrix} dx \\ dy \\ dz \end{pmatrix} = \begin{pmatrix} \frac{\partial x}{\partial r} & \frac{\partial x}{\partial \theta} & \frac{\partial x}{\partial \hat{z}} \\ \frac{\partial y}{\partial r} & \frac{\partial y}{\partial \theta} & \frac{\partial y}{\partial \hat{z}} \\ \frac{\partial z}{\partial r} & \frac{\partial z}{\partial \theta} & \frac{\partial z}{\partial \hat{z}} \end{pmatrix} \begin{pmatrix} dr \\ d\theta \\ d\hat{z} \end{pmatrix}. \tag{2.339}$$

2.6 Cylindrical Coordinates

Now we want distance to be invariant in either coordinate system. We have the standard result for Cartesian systems that

$$ds^2 = d\mathbf{x}^T \cdot d\mathbf{x} = dx^2 + dy^2 + dz^2. \tag{2.340}$$

For invariance of distance in the transformed system, we thus require

$$ds^2 = d\mathbf{x}^T \cdot d\mathbf{x} = (\mathbf{J} \cdot d\mathbf{r})^T \cdot (\mathbf{J} \cdot d\mathbf{r}) = d\mathbf{r}^T \cdot \underbrace{\mathbf{J}^T \cdot \mathbf{J}}_{\mathbf{G}} \cdot d\mathbf{r} = d\mathbf{r}^T \cdot \mathbf{G} \cdot d\mathbf{r}. \tag{2.341}$$

Recall the metric tensor $\mathbf{G}$ is defined by Eq. (2.255) as

$$\mathbf{G} = \mathbf{J}^T \cdot \mathbf{J} = \begin{pmatrix} \cos\theta & \sin\theta & 0 \\ -r\sin\theta & r\cos\theta & 0 \\ 0 & 0 & 1 \end{pmatrix} \begin{pmatrix} \cos\theta & -r\sin\theta & 0 \\ \sin\theta & r\cos\theta & 0 \\ 0 & 0 & 1 \end{pmatrix} = \begin{pmatrix} 1 & 0 & 0 \\ 0 & r^2 & 0 \\ 0 & 0 & 1 \end{pmatrix}. \tag{2.342}$$

Because $\mathbf{G}$ is diagonal, the implication can be shown to be that new coordinates axes are also orthogonal. We have $g = \det \mathbf{G} = r^2$. So for our system

$$ds^2 = \begin{pmatrix} dr & d\theta & d\hat{z} \end{pmatrix} \begin{pmatrix} 1 & 0 & 0 \\ 0 & r^2 & 0 \\ 0 & 0 & 1 \end{pmatrix} \begin{pmatrix} dr \\ d\theta \\ d\hat{z} \end{pmatrix} = dr^2 + (r\,d\theta)^2 + d\hat{z}^2. \tag{2.343}$$

Now the gradient operator in the Cartesian system is related to that of the cylindrical system via the same analysis used to obtain Eq. (2.251):

$$\nabla = \begin{pmatrix} \frac{\partial}{\partial x} \\ \frac{\partial}{\partial y} \\ \frac{\partial}{\partial z} \end{pmatrix} = (\mathbf{J}^T)^{-1} \begin{pmatrix} \frac{\partial}{\partial r} \\ \frac{\partial}{\partial \theta} \\ \frac{\partial}{\partial \hat{z}} \end{pmatrix}, \tag{2.344}$$

$$= \begin{pmatrix} \cos\theta & -\frac{\sin\theta}{r} & 0 \\ \sin\theta & \frac{\cos\theta}{r} & 0 \\ 0 & 0 & 1 \end{pmatrix} \begin{pmatrix} \frac{\partial}{\partial r} \\ \frac{\partial}{\partial \theta} \\ \frac{\partial}{\partial \hat{z}} \end{pmatrix} = \begin{pmatrix} \cos\theta \frac{\partial}{\partial r} - \frac{\sin\theta}{r}\frac{\partial}{\partial \theta} \\ \sin\theta \frac{\partial}{\partial r} + \frac{\cos\theta}{r}\frac{\partial}{\partial \theta} \\ \frac{\partial}{\partial \hat{z}} \end{pmatrix}. \tag{2.345}$$

We expand this to express the gradient operator as

$$\nabla = \mathbf{i}\left(\cos\theta\frac{\partial}{\partial r} - \frac{\sin\theta}{r}\frac{\partial}{\partial \theta}\right) + \mathbf{j}\left(\sin\theta\frac{\partial}{\partial r} + \frac{\cos\theta}{r}\frac{\partial}{\partial \theta}\right) + \mathbf{k}\frac{\partial}{\partial \hat{z}}, \tag{2.346}$$

$$= (\cos\theta\mathbf{e}_r - \sin\theta\mathbf{e}_\theta)\left(\cos\theta\frac{\partial}{\partial r} - \frac{\sin\theta}{r}\frac{\partial}{\partial \theta}\right)$$
$$+ (\sin\theta\mathbf{e}_r + \cos\theta\mathbf{e}_\theta)\left(\sin\theta\frac{\partial}{\partial r} + \frac{\cos\theta}{r}\frac{\partial}{\partial \theta}\right) + \mathbf{e}_{\hat{z}}\frac{\partial}{\partial \hat{z}}, \tag{2.347}$$

$$= \mathbf{e}_r\left(\cos^2\theta + \sin^2\theta\right)\frac{\partial}{\partial r} + \mathbf{e}_\theta\left(\frac{\sin^2\theta + \cos^2\theta}{r}\right)\frac{\partial}{\partial \theta} + \mathbf{e}_{\hat{z}}\frac{\partial}{\partial \hat{z}}, \tag{2.348}$$

$$= \mathbf{e}_r\frac{\partial}{\partial r} + \mathbf{e}_\theta\frac{1}{r}\frac{\partial}{\partial \theta} + \mathbf{e}_{\hat{z}}\frac{\partial}{\partial \hat{z}}. \tag{2.349}$$

Then we find for the divergence operator

$$\nabla^T \cdot \mathbf{v} = \begin{pmatrix} \cos\theta \frac{\partial}{\partial r} - \frac{\sin\theta}{r}\frac{\partial}{\partial \theta} & \sin\theta \frac{\partial}{\partial r} + \frac{\cos\theta}{r}\frac{\partial}{\partial \theta} & \frac{\partial}{\partial \hat{z}} \end{pmatrix} \begin{pmatrix} v_r \cos\theta - v_\theta \sin\theta \\ v_r \sin\theta + v_\theta \cos\theta \\ v_{\hat{z}} \end{pmatrix}, \quad (2.350)$$

$$= \frac{1}{r}\frac{\partial}{\partial r}(rv_r) + \frac{1}{r}\frac{\partial v_\theta}{\partial \theta} + \frac{\partial v_{\hat{z}}}{\partial \hat{z}}. \quad (2.351)$$

Consider next the Laplacian operator, $\nabla^2 = \nabla^T \cdot \nabla$, which is

$$\nabla^2 = \nabla^T \cdot \nabla, \quad (2.352)$$

$$= \begin{pmatrix} \cos\theta \frac{\partial}{\partial r} - \frac{\sin\theta}{r}\frac{\partial}{\partial \theta} & \sin\theta \frac{\partial}{\partial r} + \frac{\cos\theta}{r}\frac{\partial}{\partial \theta} & \frac{\partial}{\partial \hat{z}} \end{pmatrix} \begin{pmatrix} \cos\theta \frac{\partial}{\partial r} - \frac{\sin\theta}{r}\frac{\partial}{\partial \theta} \\ \sin\theta \frac{\partial}{\partial r} + \frac{\cos\theta}{r}\frac{\partial}{\partial \theta} \\ \frac{\partial}{\partial \hat{z}} \end{pmatrix}. \quad (2.353)$$

Detailed expansion reveals that this reduces to

$$\nabla^T \cdot \nabla = \nabla^2 = \frac{1}{r}\frac{\partial}{\partial r}\left(r\frac{\partial}{\partial r}\right) + \frac{1}{r^2}\frac{\partial^2}{\partial \theta^2} + \frac{\partial^2}{\partial \hat{z}^2} = \frac{\partial^2}{\partial r^2} + \frac{1}{r}\frac{\partial}{\partial r} + \frac{1}{r^2}\frac{\partial^2}{\partial \theta^2} + \frac{\partial^2}{\partial \hat{z}^2}. \quad (2.354)$$

Example 2.25 Find $\nabla^T \cdot \mathbf{v}$ in cylindrical coordinates using the formalism of generalized coordinates, Eq. (2.304); adapted from Powers and Sen (2015, p. 50).

Solution
We take (ξ^1, ξ^2, ξ^3) to be a set of Cartesian coordinates and (x^1, x^2, x^3) to be the corresponding cylindrical coordinates. The orientation-preserving transformations are

$$x^1 = +\sqrt{(\xi^1)^2 + (\xi^2)^2}, \qquad x^2 = \tan^{-1}\left(\frac{\xi^2}{\xi^1}\right), \qquad x^3 = \xi^3. \quad (2.355)$$

Here we restrict attention to positive x^1. The inverse transformation is

$$\xi^1 = x^1 \cos x^2, \qquad \xi^2 = x^1 \sin x^2, \qquad \xi^3 = x^3. \quad (2.356)$$

Finding $\nabla^T \cdot \mathbf{v}$ corresponds to finding

$$\Delta_i v^i = \frac{\partial v^i}{\partial x^i} + \Gamma^i_{il} v^l = \frac{1}{\sqrt{g}}\frac{\partial}{\partial x^i}\left(\sqrt{g} v^i\right). \quad (2.357)$$

Now for $i = j$

$$\Gamma^i_{il} v^l = \frac{\partial^2 \xi^p}{\partial x^i \partial x^l}\frac{\partial x^i}{\partial \xi^p} v^l, \quad (2.358)$$

$$= \frac{\partial^2 \xi^1}{\partial x^i \partial x^l}\frac{\partial x^i}{\partial \xi^1} v^l + \frac{\partial^2 \xi^2}{\partial x^i \partial x^l}\frac{\partial x^i}{\partial \xi^2} v^l + \underbrace{\frac{\partial^2 \xi^3}{\partial x^i \partial x^l}}_{=0}\frac{\partial x^i}{\partial \xi^3} v^l. \quad (2.359)$$

Noting that all second partials of ξ^3 are zero,

$$\Gamma^i_{il} v^l = \frac{\partial^2 \xi^1}{\partial x^i \partial x^l}\frac{\partial x^i}{\partial \xi^1} v^l + \frac{\partial^2 \xi^2}{\partial x^i \partial x^l}\frac{\partial x^i}{\partial \xi^2} v^l. \quad (2.360)$$

Expanding the i summation,

$$\Gamma^i_{il}v^l = \frac{\partial^2\xi^1}{\partial x^1 \partial x^l}\frac{\partial x^1}{\partial \xi^1}v^l + \frac{\partial^2\xi^1}{\partial x^2 \partial x^l}\frac{\partial x^2}{\partial \xi^1}v^l + \frac{\partial^2\xi^1}{\partial x^3 \partial x^l}\underbrace{\frac{\partial x^3}{\partial \xi^1}}_{=0}v^l$$
$$+ \frac{\partial^2\xi^2}{\partial x^1 \partial x^l}\frac{\partial x^1}{\partial \xi^2}v^l + \frac{\partial^2\xi^2}{\partial x^2 \partial x^l}\frac{\partial x^2}{\partial \xi^2}v^l + \frac{\partial^2\xi^2}{\partial x^3 \partial x^l}\underbrace{\frac{\partial x^3}{\partial \xi^2}}_{=0}v^l. \tag{2.361}$$

Noting that partials of x^3 with respect to ξ^1 and ξ^2 are zero,

$$\Gamma^i_{il}v^l = \frac{\partial^2\xi^1}{\partial x^1 \partial x^l}\frac{\partial x^1}{\partial \xi^1}v^l + \frac{\partial^2\xi^1}{\partial x^2 \partial x^l}\frac{\partial x^2}{\partial \xi^1}v^l + \frac{\partial^2\xi^2}{\partial x^1 \partial x^l}\frac{\partial x^1}{\partial \xi^2}v^l + \frac{\partial^2\xi^2}{\partial x^2 \partial x^l}\frac{\partial x^2}{\partial \xi^2}v^l. \tag{2.362}$$

Expanding the l summation, we get

$$\Gamma^i_{il}v^l = \frac{\partial^2\xi^1}{\partial x^1 \partial x^1}\frac{\partial x^1}{\partial \xi^1}v^1 + \frac{\partial^2\xi^1}{\partial x^1 \partial x^2}\frac{\partial x^1}{\partial \xi^1}v^2 + \underbrace{\frac{\partial^2\xi^1}{\partial x^1 \partial x^3}}_{=0}\frac{\partial x^1}{\partial \xi^1}v^3$$
$$+ \frac{\partial^2\xi^1}{\partial x^2 \partial x^1}\frac{\partial x^2}{\partial \xi^1}v^1 + \frac{\partial^2\xi^1}{\partial x^2 \partial x^2}\frac{\partial x^2}{\partial \xi^1}v^2 + \underbrace{\frac{\partial^2\xi^1}{\partial x^2 \partial x^3}}_{=0}\frac{\partial x^2}{\partial \xi^1}v^3$$
$$+ \frac{\partial^2\xi^2}{\partial x^1 \partial x^1}\frac{\partial x^1}{\partial \xi^2}v^1 + \frac{\partial^2\xi^2}{\partial x^1 \partial x^2}\frac{\partial x^1}{\partial \xi^2}v^2 + \underbrace{\frac{\partial^2\xi^2}{\partial x^1 \partial x^3}}_{=0}\frac{\partial x^1}{\partial \xi^2}v^3$$
$$+ \frac{\partial^2\xi^2}{\partial x^2 \partial x^1}\frac{\partial x^2}{\partial \xi^2}v^1 + \frac{\partial^2\xi^2}{\partial x^2 \partial x^2}\frac{\partial x^2}{\partial \xi^2}v^2 + \underbrace{\frac{\partial^2\xi^2}{\partial x^2 \partial x^3}}_{=0}\frac{\partial x^2}{\partial \xi^2}v^3. \tag{2.363}$$

Again removing the x^3 variation, we get

$$\Gamma^i_{il}v^l = \frac{\partial^2\xi^1}{\partial x^1 \partial x^1}\frac{\partial x^1}{\partial \xi^1}v^1 + \frac{\partial^2\xi^1}{\partial x^1 \partial x^2}\frac{\partial x^1}{\partial \xi^1}v^2 + \frac{\partial^2\xi^1}{\partial x^2 \partial x^1}\frac{\partial x^2}{\partial \xi^1}v^1 + \frac{\partial^2\xi^1}{\partial x^2 \partial x^2}\frac{\partial x^2}{\partial \xi^1}v^2$$
$$+ \frac{\partial^2\xi^2}{\partial x^1 \partial x^1}\frac{\partial x^1}{\partial \xi^2}v^1 + \frac{\partial^2\xi^2}{\partial x^1 \partial x^2}\frac{\partial x^1}{\partial \xi^2}v^2 + \frac{\partial^2\xi^2}{\partial x^2 \partial x^1}\frac{\partial x^2}{\partial \xi^2}v^1 + \frac{\partial^2\xi^2}{\partial x^2 \partial x^2}\frac{\partial x^2}{\partial \xi^2}v^2. \tag{2.364}$$

Substituting for the partial derivatives, we find

$$\Gamma^i_{il}v^l = 0v^1 - \sin x^2 \cos x^2 v^2 - \sin x^2\left(\frac{-\sin x^2}{x^1}\right)v^1 - x^1\cos x^2\left(\frac{-\sin x^2}{x^1}\right)v^2$$
$$+ 0v^1 + \cos x^2 \sin x^2 v^2 + \cos x^2\left(\frac{\cos x^2}{x^1}\right)v^1 - x^1 \sin x^2\left(\frac{\cos x^2}{x^1}\right)v^2, \tag{2.365}$$
$$= \frac{v^1}{x^1}. \tag{2.366}$$

So, in cylindrical coordinates

$$\nabla^T \cdot \mathbf{v} = \frac{\partial v^1}{\partial x^1} + \frac{\partial v^2}{\partial x^2} + \frac{\partial v^3}{\partial x^3} + \frac{v^1}{x^1}. \tag{2.367}$$

With $g = (x^1)^2$, we have $\sqrt{g} = x^1$, and alternatively can say

$$\nabla^T \cdot \mathbf{v} = \frac{1}{\sqrt{g}}\frac{\partial}{\partial x_1}(\sqrt{g}v^i) = \frac{1}{x^1}\frac{\partial}{\partial x_1}(x^1 v^1) + \frac{\partial v^2}{\partial x^2} + \frac{\partial v^3}{\partial x^3}. \tag{2.368}$$

Note that in standard cylindrical notation, $x^1 = r, x^2 = \theta, x^3 = z$. Considering $\mathbf{v}$ to be a velocity vector, we get

$$\nabla^T \cdot \mathbf{v} = \frac{\partial}{\partial r}\left(\frac{dr}{dt}\right) + \frac{\partial}{\partial \theta}\left(\frac{d\theta}{dt}\right) + \frac{\partial}{\partial z}\left(\frac{dz}{dt}\right) + \frac{1}{r}\left(\frac{dr}{dt}\right), \tag{2.369}$$

$$= \frac{1}{r}\frac{\partial}{\partial r}\left(r\frac{dr}{dt}\right) + \frac{1}{r}\frac{\partial}{\partial \theta}\left(r\frac{d\theta}{dt}\right) + \frac{\partial}{\partial z}\left(\frac{dz}{dt}\right), \tag{2.370}$$

$$= \frac{1}{r}\frac{\partial}{\partial r}(rv_r) + \frac{1}{r}\frac{\partial v_\theta}{\partial \theta} + \frac{\partial v_z}{\partial z}. \tag{2.371}$$

This is fully equivalent to Eq. (2.336). Here we have also used the more traditional $v_\theta = r(d\theta/dt) = x^1 v^2$, along with $v_r = v^1, v_z = v^3$. For practical purposes, this ensures that v_r, v_θ, v_z all have the same dimensions.

2.7 Spherical Coordinates

To illuminate an important non-Cartesian coordinate system, consider the spherical coordinate system. The transformation and inverse transformation are

$$x = r\cos\theta\sin\phi, \quad y = r\sin\theta\sin\phi, \quad z = r\cos\phi, \tag{2.372}$$

$$r = +\sqrt{x^2 + y^2 + z^2}, \quad \theta = \tan^{-1}(y/x), \quad \phi = \cos^{-1}\frac{z}{\sqrt{x^2 + y^2 + z^2}}. \tag{2.373}$$

A diagram is shown in Fig. 2.11. We have used the convention for θ and ϕ commonly employed in the mathematics literature. With our convention, when $\phi = \pi/2$, we are in the $z = 0$ plane, and we recover the standard polar coordinate transformation $x = r\cos\theta$, $y = r\sin\theta$. It is common in the physics and fluid mechanics literatures to switch the roles of θ and ϕ. The

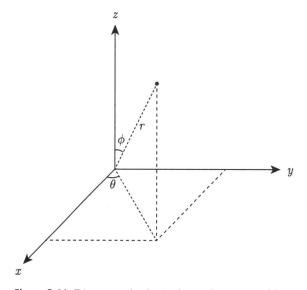

Figure 2.11 Diagram of spherical coordinate variables.

Jacobian matrix of the transformation is

$$\mathbf{J} = \begin{pmatrix} \frac{\partial x}{\partial r} & \frac{\partial x}{\partial \phi} & \frac{\partial x}{\partial \theta} \\ \frac{\partial y}{\partial r} & \frac{\partial y}{\partial \phi} & \frac{\partial y}{\partial \theta} \\ \frac{\partial z}{\partial r} & \frac{\partial z}{\partial \phi} & \frac{\partial z}{\partial \theta} \end{pmatrix} = \begin{pmatrix} \cos\theta \sin\phi & r\cos\theta\cos\phi & -r\sin\theta\sin\phi \\ \sin\theta\sin\phi & r\cos\phi\sin\theta & r\cos\theta\sin\phi \\ \cos\phi & -r\sin\phi & 0 \end{pmatrix}. \quad (2.374)$$

We have $J = \det \mathbf{J} = r^2 \sin\phi$, so the transformation is singular and thus nonunique when either $r = 0$, $\phi = 0$, or $\phi = \pi$. It is orientation-preserving for $r > 0$, $\phi \in [0, \pi]$, and it is volume-preserving only for $r^2 \sin\phi = 1$; thus, in general it does not preserve volume.

The metric tensor $\mathbf{G} = \mathbf{J}^T \cdot \mathbf{J}$, is

$$\mathbf{G} = \begin{pmatrix} \cos\theta\sin\phi & \sin\theta\sin\phi & \cos\phi \\ r\cos\theta\cos\phi & r\cos\phi\sin\theta & -r\sin\phi \\ -r\sin\theta\cos\phi & r\cos\theta\sin\phi & 0 \end{pmatrix} \begin{pmatrix} \cos\theta\sin\phi & r\cos\theta\cos\phi & -r\sin\theta\sin\phi \\ \sin\theta\sin\phi & r\cos\phi\sin\theta & r\cos\theta\sin\phi \\ \cos\phi & -r\sin\phi & 0 \end{pmatrix}, \quad (2.375)$$

$$= \begin{pmatrix} 1 & 0 & 0 \\ 0 & r^2 & 0 \\ 0 & 0 & (r\sin\phi)^2 \end{pmatrix}. \quad (2.376)$$

So we get

$$ds^2 = \begin{pmatrix} dr & d\phi & d\theta \end{pmatrix} \begin{pmatrix} 1 & 0 & 0 \\ 0 & r^2 & 0 \\ 0 & 0 & (r\sin\phi)^2 \end{pmatrix} \begin{pmatrix} dr \\ d\phi \\ d\theta \end{pmatrix} = dr^2 + (r\,d\phi)^2 + (r\sin\phi\,d\theta)^2. \quad (2.377)$$

Because $\mathbf{G}$ is diagonal, the new coordinates axes are also orthogonal. The gradient operator in the Cartesian system is related to that of the spherical system via

$$\nabla = \begin{pmatrix} \frac{\partial}{\partial x} \\ \frac{\partial}{\partial y} \\ \frac{\partial}{\partial z} \end{pmatrix} = (\mathbf{J}^T)^{-1} \begin{pmatrix} \frac{\partial}{\partial r} \\ \frac{\partial}{\partial \phi} \\ \frac{\partial}{\partial \theta} \end{pmatrix} = \begin{pmatrix} \cos\theta\sin\phi & \frac{\cos\theta\cos\phi}{r} & -\frac{\csc\phi\sin\theta}{r} \\ \sin\theta\sin\phi & \frac{\cos\phi\sin\theta}{r} & \frac{\cos\theta\csc\phi}{r} \\ \cos\phi & -\frac{\sin\phi}{r} & 0 \end{pmatrix} \begin{pmatrix} \frac{\partial}{\partial r} \\ \frac{\partial}{\partial \phi} \\ \frac{\partial}{\partial \theta} \end{pmatrix}, \quad (2.378)$$

$$= \begin{pmatrix} \cos\theta\sin\phi\frac{\partial}{\partial r} + \frac{\cos\theta\cos\phi}{r}\frac{\partial}{\partial \phi} - \frac{\csc\phi\sin\theta}{r}\frac{\partial}{\partial \theta} \\ \sin\theta\sin\phi\frac{\partial}{\partial r} + \frac{\cos\phi\sin\theta}{r}\frac{\partial}{\partial \phi} + \frac{\cos\theta\csc\phi}{r}\frac{\partial}{\partial \theta} \\ \cos\phi\frac{\partial}{\partial r} - \frac{\sin\phi}{r}\frac{\partial}{\partial \phi} \end{pmatrix}. \quad (2.379)$$

With a series of operations, we can find

$$\nabla = \mathbf{e}_r \frac{\partial}{\partial r} + \mathbf{e}_\phi \frac{1}{r}\frac{\partial}{\partial \phi} + \mathbf{e}_\theta \frac{1}{r\sin\phi}\frac{\partial}{\partial \theta}. \quad (2.380)$$

We perform an analogous series of operations for cylindrical coordinates in analysis that led to Eq. (2.330). Again, omitting details, one can find formulæ for the divergence of a vector field $\mathbf{v} = v_r \mathbf{e}_r + v_\phi \mathbf{e}_\phi + v_\theta \mathbf{e}_\theta$ as well as the Laplacian operator:

$$\nabla^T \cdot \mathbf{v} = \frac{1}{r^2}\frac{\partial}{\partial r}(r^2 v_r) + \frac{1}{r\sin\phi}\frac{\partial}{\partial \phi}(\sin\phi\, v_\phi) + \frac{1}{r\sin\phi}\frac{\partial v_\theta}{\partial \theta}, \quad (2.381)$$

$$\nabla^T \cdot \nabla = \nabla^2 = \frac{1}{r^2}\frac{\partial}{\partial r}\left(r^2 \frac{\partial}{\partial r}\right) + \frac{1}{r^2 \sin\phi}\frac{\partial}{\partial \phi}\left(\sin\phi \frac{\partial}{\partial \phi}\right) + \frac{1}{r^2 \sin^2\phi}\frac{\partial^2}{\partial \theta^2}. \quad (2.382)$$

An important limit in spherical coordinates is the limit in which changes with respect to θ are negligible. This yields a two-dimensional axisymmetric limit in which variation with r and ϕ is permitted. For this limit, $\nabla^2 \neq E^2$, and $\nabla^4 \neq E^4$. Panton (2013, p. 280) shows

$$E^2 = \frac{\partial^2}{\partial r^2} + \frac{\sin\phi}{r^2}\frac{\partial}{\partial \phi}\left(\frac{1}{\sin\phi}\frac{\partial}{\partial \phi}\right), \quad E^4 = \left(\frac{\partial^2}{\partial r^2} + \frac{\sin\phi}{r^2}\frac{\partial}{\partial \phi}\left(\frac{1}{\sin\phi}\frac{\partial}{\partial \phi}\right)\right)^2. \quad (2.383)$$

2.8 Quadratic Forms

The discussion of this section is adapted from Powers and Sen (2015, Section 7.14). At times, one may be given a polynomial for which one wants to determine conditions under which the expression is positive. For example, if we have

$$f(\xi_1, \xi_2, \xi_3) = 18\xi_1^2 - 16\xi_1\xi_2 + 5\xi_2^2 + 12\xi_1\xi_3 - 4\xi_2\xi_3 + 6\xi_3^2, \quad (2.384)$$

it is not obvious whether or not there exist (ξ_1, ξ_2, ξ_3) that will yield positive or negative values of f. However, it is easily verified that f can be rewritten as

$$f(\xi_1, \xi_2, \xi_3) = 2(\xi_1 - \xi_2 + \xi_3)^2 + 3(2\xi_1 - \xi_2)^2 + 4(\xi_1 + \xi_3)^2. \quad (2.385)$$

So in this case $f \geq 0$ for all (ξ_1, ξ_2, ξ_3). How to demonstrate positivity (or non-positivity or indeterminant positivity) of such expressions is the topic of this section.

A *quadratic form* is an expression such as

$$f(\xi_1, \ldots, \xi_N) = \sum_{j=1}^{N}\sum_{i=1}^{N} a_{ij}\xi_i\xi_j, \quad (2.386)$$

where a_{ij} is a component of a real, symmetric matrix that we call $\mathbf{A}$. The surface represented by the equation $\sum_{j=1}^{N}\sum_{i=1}^{N} a_{ij}\xi_i\xi_j = f_o$, where f_o is a constant, is a *quadric* surface. With the coefficient matrix defined, we can represent f as

$$f(\boldsymbol{\xi}) = \boldsymbol{\xi}^T \cdot \mathbf{A} \cdot \boldsymbol{\xi}. \quad (2.387)$$

Now, it is well known that the symmetric $\mathbf{A}$ can be decomposed as $\mathbf{Q} \cdot \boldsymbol{\Lambda} \cdot \mathbf{Q}^{-1}$, where $\mathbf{Q}$ is an orthogonal matrix populated by orthonormalized eigenvectors of $\mathbf{A}$, and $\boldsymbol{\Lambda}$ is the corresponding diagonal matrix of real eigenvalues. Thus, Eq. (2.387) becomes

$$f(\boldsymbol{\xi}) = \boldsymbol{\xi}^T \cdot \underbrace{\mathbf{Q} \cdot \boldsymbol{\Lambda} \cdot \mathbf{Q}^{-1}}_{\mathbf{A}} \cdot \boldsymbol{\xi}. \quad (2.388)$$

Because $\mathbf{Q}$ is orthogonal, $\mathbf{Q}^T = \mathbf{Q}^{-1}$, and we find

$$f(\boldsymbol{\xi}) = \boldsymbol{\xi}^T \cdot \mathbf{Q} \cdot \boldsymbol{\Lambda} \cdot \mathbf{Q}^T \cdot \boldsymbol{\xi}. \quad (2.389)$$

Define $\mathbf{x}$ so that $\mathbf{x} = \mathbf{Q}^T \cdot \boldsymbol{\xi} = \mathbf{Q}^{-1} \cdot \boldsymbol{\xi}$, giving $\boldsymbol{\xi} = \mathbf{Q} \cdot \mathbf{x}$. Thus, Eq. (2.389) becomes

$$f(\mathbf{x}) = (\mathbf{Q} \cdot \mathbf{x})^T \cdot \mathbf{Q} \cdot \boldsymbol{\Lambda} \cdot \mathbf{x} = \mathbf{x}^T \cdot \mathbf{Q}^T \cdot \mathbf{Q} \cdot \boldsymbol{\Lambda} \cdot \mathbf{x} = \mathbf{x}^T \cdot \boldsymbol{\Lambda} \cdot \mathbf{x}. \quad (2.390)$$

2.8 Quadratic Forms

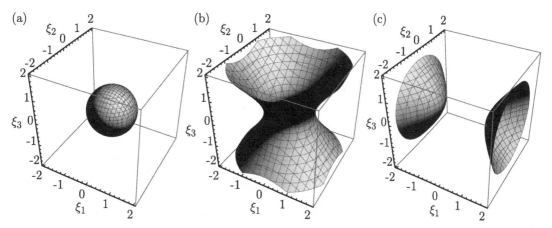

Figure 2.12 Plots of quadric surfaces for (a) $\xi_1^2+\xi_2^2+\xi_3^2=1$, a sphere; (b) $\xi_1^2+\xi_2^2-\xi_3^2=1$, a hyperboloid of revolution of one sheet; and (c) $\xi_1^2-\xi_2^2-\xi_3^2=1$, a hyperboloid of revolution of two sheets.

This canonical representation of a quadratic form is one in which the products of terms with different indices (i.e. $\xi_i \xi_j$, $i \neq j$) do not appear. If $\mathbf{Q}$ is the orthogonal matrix and $\lambda_1, \ldots, \lambda_N$ the eigenvalues of $\mathbf{A}$, a change in coordinates

$$\begin{pmatrix} \xi_1 \\ \vdots \\ \xi_N \end{pmatrix} = \mathbf{Q} \cdot \begin{pmatrix} x_1 \\ \vdots \\ x_N \end{pmatrix}, \tag{2.391}$$

will reduce the quadratic form, Eq. (2.386), to its canonical representation:

$$f(x_1, \ldots, x_N) = \lambda_1 x_1^2 + \lambda_2 x_2^2 + \cdots + \lambda_N x_N^2. \tag{2.392}$$

If $\lambda_i > 0$, $i = 1, \ldots, N$, $f > 0$, so f is positive definite. If $\lambda_i < 0$, $i = 1, \ldots, N$, $f < 0$, so f is negative definite. If some of the eigenvalues are positive and others negative, then f could be positive or negative. Eigenvalues of zero do not affect f. As an example we give surface plots for three classes of quadric surfaces in Fig. 2.12. They are for $f(\xi_1, \xi_2, \xi_3) = \xi_1^2 + \xi_2^2 + \xi_3^2 = 1$, a sphere; $f(\xi_1, \xi_2, \xi_3) = \xi_1^2 + \xi_2^2 - \xi_3^2 = 1$, a hyperboloid of revolution of one sheet; and $f(\xi_1, \xi_2, \xi_3) = \xi_1^2 - \xi_2^2 - \xi_3^2 = 1$, a hyperboloid of revolution of two sheets. Other classes exist.

Example 2.26 Change

$$f(\xi_1, \xi_2) = 2\xi_1^2 + 2\xi_1 \xi_2 + 2\xi_2^2, \tag{2.393}$$

to a canonical representation of its quadratic form.

Solution
For $N = 2$, Eq. (2.386) becomes

$$f(\xi_1, \xi_2) = a_{11} \xi_1^2 + (a_{12} + a_{21}) \xi_1 \xi_2 + a_{22} \xi_2^2. \tag{2.394}$$

We choose $\mathbf{A}$ to be symmetric. This gives $a_{11} = 2$, $a_{12} = 1$, $a_{21} = 1$, $a_{22} = 2$. So we get

$$\mathbf{A} = \begin{pmatrix} 2 & 1 \\ 1 & 2 \end{pmatrix}. \tag{2.395}$$

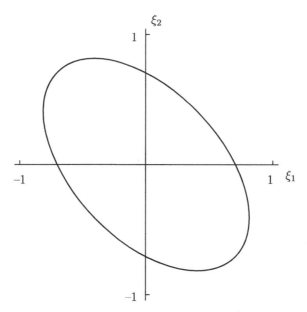

Figure 2.13 Quadric associated with $f(\xi_1,\xi_2) = (\xi_1 - \xi_2)^2/2 + 3(\xi_1 + \xi_2)^2/2 = 1$.

The eigenvalues of $\mathbf{A}$ are $\lambda = 1$, $\lambda = 3$. An orthogonal matrix, whose columns are populated by the normalized eigenvectors of $\mathbf{A}$, is

$$\mathbf{Q} = \begin{pmatrix} \frac{1}{\sqrt{2}} & \frac{1}{\sqrt{2}} \\ -\frac{1}{\sqrt{2}} & \frac{1}{\sqrt{2}} \end{pmatrix}, \qquad \mathbf{Q}^{-1} = \mathbf{Q}^T = \begin{pmatrix} \frac{1}{\sqrt{2}} & -\frac{1}{\sqrt{2}} \\ \frac{1}{\sqrt{2}} & \frac{1}{\sqrt{2}} \end{pmatrix}. \tag{2.396}$$

The transformation $\boldsymbol{\xi} = \mathbf{Q} \cdot \mathbf{x}$ gives

$$\xi_1 = \frac{1}{\sqrt{2}}(x_1 + x_2), \qquad \xi_2 = \frac{1}{\sqrt{2}}(-x_1 + x_2). \tag{2.397}$$

We have $\det \mathbf{Q} = 1$, so the transformation is orientation-preserving. The inverse transformation $\mathbf{x} = \mathbf{Q}^{-1} \cdot \boldsymbol{\xi} = \mathbf{Q}^T \cdot \boldsymbol{\xi}$ gives

$$x_1 = \frac{1}{\sqrt{2}}(\xi_1 - \xi_2), \qquad x_2 = \frac{1}{\sqrt{2}}(\xi_1 + \xi_2). \tag{2.398}$$

Using Eqs. (2.397) to eliminate ξ_1 and ξ_2 in Eq. (2.393), we get

$$f(x_1, x_2) = x_1^2 + 3x_2^2. \tag{2.399}$$

In terms of the original variables, we get

$$f(\xi_1, \xi_2) = \frac{1}{2}(\xi_1 - \xi_2)^2 + \frac{3}{2}(\xi_1 + \xi_2)^2. \tag{2.400}$$

For this two-variable problem, a quadric "surface" is actually an ellipse in the (ξ_1, ξ_2) plane. We give a plot of the quadric for $f(\xi_1, \xi_2) = 1$ in Fig. 2.13.

Example 2.27 Transform

$$f(\xi_1, \xi_2, \xi_3) = 18\xi_1^2 - 16\xi_1\xi_2 + 5\xi_2^2 + 12\xi_1\xi_3 - 4\xi_2\xi_3 + 6\xi_3^2, \tag{2.401}$$

to a canonical representation of its quadratic form.

Solution
For $N = 3$, Eq. (2.386) becomes

$$f(\xi_1, \xi_2, \xi_3) = \begin{pmatrix} \xi_1 & \xi_2 & \xi_3 \end{pmatrix} \begin{pmatrix} 18 & -8 & 6 \\ -8 & 5 & -2 \\ 6 & -2 & 6 \end{pmatrix} \begin{pmatrix} \xi_1 \\ \xi_2 \\ \xi_3 \end{pmatrix} = \boldsymbol{\xi}^T \cdot \mathbf{A} \cdot \boldsymbol{\xi}. \qquad (2.402)$$

The eigenvalues of $\mathbf{A}$ are $\lambda_1 = 1, \lambda_2 = 4, \lambda_3 = 24$. An orthogonal matrix induced by the eigenvectors of $\mathbf{A}$ is

$$\mathbf{Q} = \begin{pmatrix} -\frac{4}{\sqrt{69}} & -\frac{1}{\sqrt{30}} & \frac{13}{\sqrt{230}} \\ -\frac{7}{\sqrt{69}} & \sqrt{\frac{2}{15}} & -3\sqrt{\frac{2}{115}} \\ \frac{2}{\sqrt{69}} & \sqrt{\frac{5}{6}} & \sqrt{\frac{5}{46}} \end{pmatrix}, \qquad \mathbf{Q}^{-1} = \mathbf{Q}^T = \begin{pmatrix} -\frac{4}{\sqrt{69}} & -\frac{7}{\sqrt{69}} & \frac{2}{\sqrt{69}} \\ -\frac{1}{\sqrt{30}} & \sqrt{\frac{2}{15}} & \sqrt{\frac{5}{6}} \\ \frac{13}{\sqrt{230}} & -3\sqrt{\frac{2}{115}} & \sqrt{\frac{5}{46}} \end{pmatrix}. \qquad (2.403)$$

For this nonunique choice of $\mathbf{Q}$, we note that $\det \mathbf{Q} = -1$, so it is a reflection. It could easily be reformulated to be a rotation. The inverse transformation $\mathbf{x} = \mathbf{Q}^{-1} \cdot \boldsymbol{\xi} = \mathbf{Q}^T \cdot \boldsymbol{\xi}$ is

$$x_1 = \frac{-4}{\sqrt{69}} \xi_1 - \frac{7}{\sqrt{69}} \xi_2 + \frac{2}{\sqrt{69}} \xi_3, \qquad (2.404)$$

$$x_2 = -\frac{1}{\sqrt{30}} \xi_1 + \sqrt{\frac{2}{15}} \xi_2 + \sqrt{\frac{5}{6}} \xi_3, \qquad (2.405)$$

$$x_3 = \frac{13}{\sqrt{230}} \xi_1 - 3\sqrt{\frac{2}{115}} \xi_2 + \sqrt{\frac{5}{46}} \xi_3. \qquad (2.406)$$

Directly imposing, then, the canonical representation of the quadratic form of Eq. (2.392) onto Eq. (2.401), we get

$$f(x_1, x_2, x_3) = x_1^2 + 4x_2^2 + 24x_3^2. \qquad (2.407)$$

In terms of the original variables, we get

$$f(\xi_1, \xi_2, \xi_3) = \left(\frac{-4}{\sqrt{69}} \xi_1 - \frac{7}{\sqrt{69}} \xi_2 + \frac{2}{\sqrt{69}} \xi_3 \right)^2 + 4 \left(-\frac{1}{\sqrt{30}} \xi_1 + \sqrt{\frac{2}{15}} \xi_2 + \sqrt{\frac{5}{6}} \xi_3 \right)^2$$

$$+ 24 \left(\frac{13}{\sqrt{230}} \xi_1 - 3\sqrt{\frac{2}{115}} \xi_2 + \sqrt{\frac{5}{46}} \xi_3 \right)^2. \qquad (2.408)$$

It is clear that $f(\xi_1, \xi_2, \xi_3)$ is positive definite. Moreover, by performing the multiplications, it is seen that the original form is recovered. Further manipulation would also show that

$$f(\xi_1, \xi_2, \xi_3) = 2(\xi_1 - \xi_2 + \xi_3)^2 + 3(2\xi_1 - \xi_2)^2 + 4(\xi_1 + \xi_3)^2, \qquad (2.409)$$

so we see quadratic form is not unique. For this problem, a quadric surface is a triaxial ellipsoid in the (ξ_1, ξ_2, ξ_3) volume. We plot the quadric $f(\xi_1, \xi_2, \xi_3) = 1$ in Fig. 2.14.

SUMMARY

This chapter focused on geometry, the theory of place. Mathematical tools to describe geometry were reviewed, with special focus on scalars, vectors, and tensors. Tools necessary to describe transformations of coordinates, scalars, vectors, and tensors were discussed along with tools

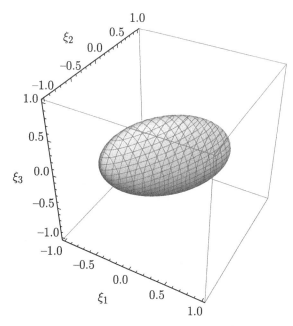

Figure 2.14 Quadric associated with $f(\xi_1, \xi_2, \xi_3) = 2(\xi_1 - \xi_2 + \xi_3)^2 + 3(2\xi_1 - \xi_2)^2 + 4(\xi_1 + \xi_3)^2 = 1$.

to describe derivatives in a multi-variable spatio-temporal environment required for the full description of the mechanics of continuum fluids.

PROBLEMS

2.1 Plot the quadric surfaces $\xi_1^2/2 + \xi_2^2 + \xi_3^2 = 1$, $\xi_1^2/2 + \xi_2^2 + 0\xi_3^2 = 1$, and $\xi_1^2/2 + \xi_2^2 - \xi_3^2 = 1$, for $\xi_1, \xi_2, \xi_3 \in [-2, 2]$.

2.2 Show using Cartesian index notation that
$$\mathbf{u}^T \cdot \mathbf{v} = \frac{1}{4}\left((\mathbf{u}+\mathbf{v})^T \cdot (\mathbf{u}+\mathbf{v}) - (\mathbf{u}-\mathbf{v})^T \cdot (\mathbf{u}-\mathbf{v})\right).$$

2.3 Show using Cartesian index notation that
$$\nabla \times (\mathbf{u} \times \mathbf{v}) = \left(\nabla^T \cdot \left(\mathbf{v}\mathbf{u}^T - \mathbf{u}\mathbf{v}^T\right)\right)^T.$$

2.4 Show using Cartesian index notation that
$$\nabla^T \cdot (\mathbf{u} \times \mathbf{v}) = (\nabla \times \mathbf{u})^T \cdot \mathbf{v} - (\nabla \times \mathbf{v})^T \cdot \mathbf{u}.$$

2.5 Show using Cartesian index notation that
$$\nabla \times (\mathbf{u} \times \mathbf{v}) = (\mathbf{v}^T \cdot \nabla)\mathbf{u} - (\mathbf{u}^T \cdot \nabla)\mathbf{v} + \mathbf{u}(\nabla^T \cdot \mathbf{v}) - \mathbf{v}(\nabla^T \cdot \mathbf{u}).$$

2.6 Show using Cartesian index notation that
$$\nabla \times (\nabla \times \mathbf{v}) = \nabla(\nabla^T \cdot \mathbf{v}) - \nabla^2 \mathbf{v}.$$

2.7 The following two-dimensional stretch mapping linear transformation exists between $\boldsymbol{\xi}$, a set of Cartesian coordinates, and $\mathbf{x}$, a set of deformed non-Cartesian coordinates: $\boldsymbol{\xi} = \mathbf{J} \cdot \mathbf{x}$. Here the Jacobian matrix is

$$\mathbf{J} = \begin{pmatrix} 1/2 & 0 \\ 0 & 1 \end{pmatrix}.$$

Determine if the transformation is area- and orientation-preserving and if it is either a rotation or reflection. Show how a unit square in $\boldsymbol{\xi}$ space is represented in the transformed $\mathbf{x}$ space by making an appropriate plot. Show how a unit circle in $\boldsymbol{\xi}$ space is represented in the transformed $\mathbf{x}$ space by developing an appropriate equation.

2.8 The following two-dimensional squeeze mapping linear transformation exists between $\boldsymbol{\xi}$, a set of Cartesian coordinates, and $\mathbf{x}$, a set of deformed non-Cartesian coordinates: $\boldsymbol{\xi} = \mathbf{J} \cdot \mathbf{x}$. Here the Jacobian matrix is

$$\mathbf{J} = \begin{pmatrix} 1/2 & 0 \\ 0 & 2 \end{pmatrix}.$$

Determine if the transformation is area- and orientation-preserving and if it is either a rotation or reflection. Show how a unit square in $\boldsymbol{\xi}$ space is represented in the transformed $\mathbf{x}$ space by making an appropriate plot. Show how a unit circle in $\boldsymbol{\xi}$ space is represented in the transformed $\mathbf{x}$ space by developing an appropriate equation.

2.9 The following two-dimensional linear transformation exists between $\boldsymbol{\xi}$, a set of Cartesian coordinates, and $\mathbf{x}$, a set of deformed non-Cartesian coordinates: $\boldsymbol{\xi} = \mathbf{J} \cdot \mathbf{x}$. Here the Jacobian matrix is

$$\mathbf{J} = \begin{pmatrix} -1 & 1 \\ -1 & 0 \end{pmatrix}.$$

Determine if the transformation is area-preserving and orientation-preserving and if it is either a rotation or reflection. Show how a unit circle in $\boldsymbol{\xi}$ space is represented as an ellipse in the transformed $\mathbf{x}$ space by making an appropriate plot. Show the lengths of the major and minor axes are the singular values of $\mathbf{J}^{-1}$, which also are the reciprocals of the singular values of $\mathbf{J}$.

2.10 Linear transformations exist between $\boldsymbol{\xi}$, a set of Cartesian coordinates, and $\mathbf{x}$, a set of deformed non-Cartesian coordinates: $\boldsymbol{\xi} = \mathbf{J} \cdot \mathbf{x}$. The Jacobian matrices are

$$\mathbf{J} = \begin{pmatrix} 2 & 1 & 1 \\ -1 & 2 & 0 \\ 1 & 0 & 2 \end{pmatrix}, \quad \mathbf{J} = \begin{pmatrix} 10 & 1 & 1 \\ -1 & 2 & 0 \\ 1 & 0 & 2 \end{pmatrix}, \quad \mathbf{J} = \begin{pmatrix} 10 & 1 & 1 \\ -1 & 10 & 0 \\ 1 & 0 & 2 \end{pmatrix}.$$

Determine if each transformation is volume- and orientation-preserving. Show how a unit sphere in $\boldsymbol{\xi}$ space is represented as an ellipsoid in the transformed $\mathbf{x}$ space by making appropriate plots. Find the lengths of the principal axes of the ellipsoids. Associate each ellipsoid with an approximation of the shape of either an egg, pancake, or spaghetti noodle.

2.11 Consider the matrix

$$\mathbf{A} = \begin{pmatrix} 1 & 0 & 0 \\ 1 & -1 & 1 \\ 0 & 1 & -1 \end{pmatrix}.$$

The operation $\mathbf{A} \cdot \mathbf{x} = \mathbf{b}$ may be considered one that maps a 3×1 vector $\mathbf{x}$ into the 3×1 vector $\mathbf{b}$. We allow $\mathbf{x}$ to span a space of dimension three. Using standard techniques from linear algebra, find the dimension of the space spanned by the vectors $\mathbf{b}$. If $\mathbf{b} = (1,0,0)^T$, determine if there is a vector $\mathbf{x}$ that maps to $\mathbf{b}$. If not, identify all vectors $\mathbf{x}$ that minimize the error $||\mathbf{A} \cdot \mathbf{x} - \mathbf{b}||$. Of those vectors that minimize the error, identify the one with the smallest magnitude. Repeat for $\mathbf{b} = (1,0,1)^T$. It may be advantageous to operate on the equation with the operator $\mathbf{A}^T$ and employ the robust method of Gaussian elimination.

2.12 Consider the two matrices

$$\mathbf{Q}_1 = \begin{pmatrix} \cos\alpha & -\sin\alpha & 0 \\ \sin\alpha & \cos\alpha & 0 \\ 0 & 0 & 1 \end{pmatrix}, \qquad \mathbf{Q}_2 = \begin{pmatrix} \cos\alpha & \sin\alpha & 0 \\ \sin\alpha & -\cos\alpha & 0 \\ 0 & 0 & 1 \end{pmatrix}.$$

Show that both $\mathbf{Q}_1$ and $\mathbf{Q}_2$ are volume-preserving, and that $\mathbf{Q}_1$ is also orientation-preserving and thus a rotation, while $\mathbf{Q}_2$ is orientation-reversing and thus a reflection. Show that $\mathbf{Q}_1^T \cdot \mathbf{Q}_1 = \mathbf{I}$, $\mathbf{Q}_2^T \cdot \mathbf{Q}_2 = \mathbf{I}$, so both matrices are orthogonal. Consider the transformed representation of a vector

$$\mathbf{x}' = \mathbf{Q}^T \cdot \mathbf{x}.$$

In contrast to continuum mechanics, which interprets the transformation as applied to the coordinate system (an alias transformation), consider the approach of an alibi transformation, where the action of an operator on a vector transforms the vector, but not the coordinate system. Find the action of $\mathbf{Q}_1$ and $\mathbf{Q}_2$ on $\mathbf{x} = (1,0,0)^T$ if $\alpha = \pi/6$. Interpret the result in terms of rotation and reflection. For the reflection, find the angle of inclination of the plane of reflection.

2.13 Consider $\mathbf{u} = (1,2,3)^T$, $\mathbf{v} = (4,5,6)^T$. Choose numerical values for a rotation matrix $\mathbf{Q}_1$, and show the representation of the dyadic product tensor $\mathbf{T} = \mathbf{u}\mathbf{v}^T$ transforms as expected under rotation:

$$(\mathbf{Q}_1^T \cdot \mathbf{T} \cdot \mathbf{Q}_1) = (\mathbf{Q}_1^T \cdot \mathbf{u})(\mathbf{Q}_1^T \cdot \mathbf{v})^T, \qquad \mathbf{T}' = \mathbf{u}'\mathbf{v}'^T.$$

Exchange two rows of $\mathbf{Q}_1$ so as to form a reflection matrix $\mathbf{Q}_2$. Show that

$$(\mathbf{Q}_2^T \cdot \mathbf{T} \cdot \mathbf{Q}_2) = (\mathbf{Q}_2^T \cdot \mathbf{u})(\mathbf{Q}_2^T \cdot \mathbf{v})^T, \qquad \mathbf{T}' = \mathbf{u}'\mathbf{v}'^T,$$

thus verifying tensorial nature of the dyadic product under rotation or reflection.

2.14 Consider $\mathbf{u} = (1,2,3)^T$, $\mathbf{v} = (4,5,6)^T$. Choose numerical values for a rotation matrix $\mathbf{Q}_1$, and show the cross product vector $\mathbf{w} = \mathbf{u} \times \mathbf{v}$ transforms as expected under rotation:

$$(\mathbf{Q}_1^T \cdot \mathbf{w}) = (\mathbf{Q}_1^T \cdot \mathbf{u}) \times (\mathbf{Q}_1^T \cdot \mathbf{v}), \qquad \mathbf{w}' = \mathbf{u}' \times \mathbf{v}'.$$

Exchange two rows of $\mathbf{Q}_1$ so as to form a reflection matrix $\mathbf{Q}_2$. Show that

$$(\mathbf{Q}_2^T \cdot \mathbf{w}) = -(\mathbf{Q}_2^T \cdot \mathbf{u}) \times (\mathbf{Q}_2^T \cdot \mathbf{v}), \qquad \mathbf{w}' = -\mathbf{u}' \times \mathbf{v}',$$

thus requiring us to interpret $\mathbf{w}$ as a pseudo-vector. Show the following is a valid rule for rotation or reflection transformations:

$$(\mathbf{Q}^T \cdot \mathbf{w}) = (\det \mathbf{Q})((\mathbf{Q}^T \cdot \mathbf{u}) \times (\mathbf{Q}^T \cdot \mathbf{v})), \qquad \mathbf{w}' = (\det \mathbf{Q})(\mathbf{u}' \times \mathbf{v}').$$

2.15 Consider $\mathbf{u} = (1, 2, 3)^T$, $\mathbf{v} = (4, 5, 6)^T$, $\mathbf{w} = (1, 2, 1)^T$. Choose numerical values for a rotation matrix $\mathbf{Q}_1$, and show the equation defining the *scalar triple product*, which is $V = \mathbf{w}^T \cdot (\mathbf{u} \times \mathbf{v})$, transforms as expected under rotation:

$$V = (\mathbf{Q}_1^T \cdot \mathbf{w})^T \cdot ((\mathbf{Q}_1^T \cdot \mathbf{u}) \times (\mathbf{Q}_1^T \cdot \mathbf{v})), \qquad V' = \mathbf{w}'^T \cdot (\mathbf{u}' \times \mathbf{v}'),$$

yielding $V' = V$, as desired for scalars. The volume of the parallelepiped defined by the three linearly independent vectors $\mathbf{u}$, $\mathbf{v}$, and $\mathbf{w}$ is given by the magnitude of V. Exchange two rows of $\mathbf{Q}_1$ so as to form a reflection matrix $\mathbf{Q}_2$. Show that

$$V = -(\mathbf{Q}_2^T \cdot \mathbf{w})^T \cdot ((\mathbf{Q}_2^T \cdot \mathbf{u}) \times (\mathbf{Q}_2^T \cdot \mathbf{v})), \qquad V' = -\mathbf{w}'^T \cdot (\mathbf{u}' \times \mathbf{v}'),$$

in order for $V' = V$. This requires us to interpret the scalar triple product as a *pseudoscalar*. Show the following is a valid rule for rotation or reflection transformations:

$$V = (\det \mathbf{Q})(\mathbf{Q}^T \cdot \mathbf{w})^T \cdot ((\mathbf{Q}^T \cdot \mathbf{u}) \times (\mathbf{Q}^T \cdot \mathbf{v})), \qquad V' = (\det \mathbf{Q})\mathbf{w}'^T \cdot (\mathbf{u}' \times \mathbf{v}').$$

Lastly, demonstrate for our choices of $\mathbf{u}$, $\mathbf{v}$, and $\mathbf{w}$ that

$$V = \mathbf{w}^T \cdot (\mathbf{u} \times \mathbf{v}) = \epsilon_{ijk} w_i u_j v_k = \begin{vmatrix} w_1 & w_2 & w_3 \\ u_1 & u_2 & u_3 \\ v_1 & v_2 & v_2 \end{vmatrix}.$$

2.16 Consider the matrix

$$\mathbf{A} = \begin{pmatrix} 1 & 2 & 3 \\ 2 & 3 & 1 \\ 3 & 2 & 1 \end{pmatrix}.$$

Decompose $\mathbf{A}$ into symmetric and anti-symmetric parts. Find a rotation matrix $\mathbf{Q}$ such that the symmetric part of $\mathbf{A}$ is $\mathbf{Q} \cdot \mathbf{\Lambda} \cdot \mathbf{Q}^T$, with $\mathbf{\Lambda}$ as a diagonal matrix with the eigenvalues of the symmetric part of $\mathbf{A}$ on its diagonal. Find the dual vector associated with the anti-symmetric part of $\mathbf{A}$.

2.17 Consider the tensor $\mathbf{T}$ and rotation matrix $\mathbf{Q}$:

$$\mathbf{T} = \begin{pmatrix} 1 & 2 \\ 3 & 4 \end{pmatrix}, \qquad \mathbf{Q} = \begin{pmatrix} \frac{3}{5} & \frac{4}{5} \\ -\frac{4}{5} & \frac{3}{5} \end{pmatrix}.$$

Find the representation $\mathbf{T}'$ of $\mathbf{T}$ in the rotated coordinate system defined by $\mathbf{Q}$. Show the matrix invariants of $\mathbf{T}$ and $\mathbf{T}'$ are identical.

2.18 By calculating both the relevant surface and volume integrals, show that Gauss's theorem is valid if

$$\mathbf{v} = x\mathbf{i} + y\mathbf{j} + 0\mathbf{k},$$

and A is the closed surface which consists of a circular base and the hemisphere of unit radius with center at the origin and $z \geq 0$, that is, $x^2 + y^2 + z^2 = 1$. A transformation to spherical coordinates will aid in the solution.

2.19 Consider the transformation from two-dimensional Cartesian coordinates to parabolic polar coordinates:
$$x = \frac{1}{2}(u^2 - v^2), \quad y = uv.$$
Plot curves of constant u and v in the (x, y) plane. Find the metric tensor $\mathbf{G}$ and Laplacian operator in parabolic polar coordinates. Consider the domain $u \in (-\infty, \infty)$, $v \in [0, \infty)$.

2.20 Determine if $f(\xi_1, \xi_2, \xi_3) = \xi_1^2 + 2\xi_2^2 + 2\xi_2\xi_3$ is positive definite by casting it in a canonical quadratic form.

2.21 Consider an orthogonal coordinate system in a three-dimensional space such that the differential distance ds obeys
$$ds^2 = g_{11}(x_1, x_2, x_3)\, dx_1^2 + g_{22}(x_1, x_2, x_3)\, dx_2^2 + g_{33}(x_1, x_2, x_3)\, dx_3^2,$$
where g_{ij} are the components of the metric tensor of the transformation. It can be shown that the gradient of a scalar function $\Phi(x_1, x_2, x_3)$ is
$$\nabla \Phi = \frac{1}{\sqrt{g_{11}}} \frac{\partial \Phi}{\partial x_1} \mathbf{e}_1 + \frac{1}{\sqrt{g_{22}}} \frac{\partial \Phi}{\partial x_2} \mathbf{e}_2 + \frac{1}{\sqrt{g_{33}}} \frac{\partial \Phi}{\partial x_3} \mathbf{e}_3.$$
Use this algorithm to find $\nabla \Phi$ in Cartesian, cylindrical, and spherical coordinates.

2.22 Consider an orthogonal coordinate system in a three-dimensional space such that the differential distance ds obeys
$$ds^2 = g_{11}(x_1, x_2, x_3)\, dx_1^2 + g_{22}(x_1, x_2, x_3)\, dx_2^2 + g_{33}(x_1, x_2, x_3)\, dx_3^2,$$
where g_{ij} are the components of the metric tensor of the transformation. It can be shown that the divergence of a vector function $\mathbf{v}(x_1, x_2, x_3)$ is
$$\nabla^T \cdot \mathbf{v} = \frac{1}{\sqrt{g_{11}g_{22}g_{33}}} \left(\frac{\partial}{\partial x_1}\left(\sqrt{g_{22}g_{33}}\, v_1\right) + \frac{\partial}{\partial x_2}\left(\sqrt{g_{11}g_{22}}\, v_2\right) + \frac{\partial}{\partial x_3}\left(\sqrt{g_{11}g_{22}}\, v_3\right) \right).$$
Use this algorithm to find $\nabla^T \cdot \mathbf{v}$ in Cartesian, cylindrical, and spherical coordinates.

2.23 Consider an orthogonal coordinate system in a three-dimensional space in which the metric tensor is at most of function of two variables, such that the differential distance ds obeys
$$ds^2 = g_{11}(x_1, x_2)\, dx_1^2 + g_{22}(x_1, x_2)\, dx_2^2 + g_{33}(x_1, x_2)\, dx_3^2,$$
where g_{ij} are the components of the metric tensor of the transformation. If the only nonzero components of a vector field in this space are taken to have the form $v_1(x_1, x_2)$ and $v_2(x_1, x_2)$, and the vector field is taken to be solenoidal, then it can be shown that a function $\psi(x_1, x_2)$ exists such that it induces vector field components
$$v_1 = \frac{1}{\sqrt{g_{22}g_{33}}} \frac{\partial \psi}{\partial x_2}, \quad v_2 = \frac{-1}{\sqrt{g_{11}g_{33}}} \frac{\partial \psi}{\partial x_1}.$$
For the Cartesian coordinate system, neglecting variation in z, and with $\psi = \psi(x, y)$, use this algorithm to find $v_x(x, y)$ and $v_y(x, y)$ in terms of $\psi(x, y)$ and show that $\nabla^T \cdot \mathbf{v} = 0$. For the cylindrical coordinate system, neglecting variation in $\hat{z}$, and with $\psi = \psi(r, \theta)$, use this algorithm to find $v_r(r, \theta)$ and $v_\theta(r, \theta)$ in terms of $\psi(r, \theta)$ and show that $\nabla^T \cdot$

$\mathbf{v} = 0$. For the spherical coordinate system, neglecting variation in θ, and with $\psi = \psi(r, \phi)$, use this algorithm to find $v_r(r, \phi)$ and $v_\phi(r, \phi)$ in terms of $\psi(r, \phi)$ and show that $\nabla^T \cdot \mathbf{v} = 0$.

2.24 Consider the transformation from Cartesian $(\xi^1, \xi^2)^T$ to $(x^1, x^2)^T$ via the transformation $\xi^1 = x^2$, $\xi^2 = x^1$. Determine if the transformation is area- and orientation-preserving. Use the method of quadratic forms to cast $(ds)^2 = (d\xi^1)^2 + (d\xi^2)^2$ in terms of dx^1 and dx^2 as a sum of squares. Show that $(ds)^2 \geq 0$ for all dx^1, dx^2.

2.25 Consider the transformation from Cartesian $(x, y)^T$ to non-Cartesian $(\xi, \eta)^T$ via the linear transformation $x = \xi + \eta$, $y = -\eta$. Determine if the transformation is area- and orientation-preserving. Use the method of quadratic forms to cast $ds^2 = dx^2 + dy^2$ in terms of $d\xi$ and $d\eta$ as a sum of squares. Show that $ds^2 \geq 0$ for all $d\xi, d\eta$. Plot curves of constant ξ and η in the (x, y) plane.

2.26 Three linearly independent, nonorthogonal basis vectors $\mathbf{e}_1, \mathbf{e}_2, \mathbf{e}_3$, have scalar triple product $\mathbf{e}_1^T \cdot (\mathbf{e}_2 \times \mathbf{e}_3) = V \neq 0$. Here $|V|$ is the volume of the parallelepiped defined by the basis vectors. A reciprocal basis is defined by the reciprocal basis vectors

$$\mathbf{e}^1 = \frac{\mathbf{e}_2 \times \mathbf{e}_3}{V}, \qquad \mathbf{e}^2 = \frac{\mathbf{e}_3 \times \mathbf{e}_1}{V}, \qquad \mathbf{e}^3 = \frac{\mathbf{e}_1 \times \mathbf{e}_2}{V}.$$

Take $\mathbf{e}_1 = (1, 0, 0)^T$, $\mathbf{e}_2 = (1, 2, 0)^T$, $\mathbf{e}_3 = (1, 2, 3)^T$. Find the reciprocal basis vectors, and show $\mathbf{e}^{iT} \cdot \mathbf{e}_j = \delta^i_j$. Find the representation of $\mathbf{v} = (0, 0, 1)^T$ in terms of the basis and reciprocal basis vectors.

2.27 The orientation of the Euler axis is associated with the argument of the complex eigenvalues of a rotation matrix $\mathbf{Q}$. Find the orientation of the Euler axis associated with the following rotation matrices:

$$\mathbf{Q} = \begin{pmatrix} 0 & -1 & 0 \\ 1 & 0 & 0 \\ 0 & 0 & 1 \end{pmatrix}, \qquad \mathbf{Q} = \begin{pmatrix} -\frac{1}{\sqrt{3}} & -\frac{1}{\sqrt{3}} & -\frac{1}{\sqrt{3}} \\ -\frac{1}{\sqrt{6}} & -\frac{1}{\sqrt{6}} & \sqrt{\frac{2}{3}} \\ -\frac{1}{\sqrt{2}} & \frac{1}{\sqrt{2}} & 0 \end{pmatrix}.$$

2.28 Using the alibi transformation approach, we may rotate a vector $\mathbf{x}$ to an arbitrary rotated orientation $\mathbf{x}' = \mathbf{Q} \cdot \mathbf{x}$ through the use of three so-called Euler angles, $\alpha_1, \alpha_2, \alpha_3$, which are associated with rotations about the 1, 2, and 3 axes, respectively. The associated rotation matrices are

$$\mathbf{Q}_1 = \begin{pmatrix} 1 & 0 & 0 \\ 0 & \cos\alpha_1 & -\sin\alpha_1 \\ 0 & \sin\alpha_1 & \cos\alpha_1 \end{pmatrix},$$

$$\mathbf{Q}_2 = \begin{pmatrix} \cos\alpha_2 & 0 & \sin\alpha_2 \\ 0 & 1 & 0 \\ -\sin\alpha_2 & 0 & \cos\alpha_2 \end{pmatrix}, \quad \mathbf{Q}_3 = \begin{pmatrix} \cos\alpha_3 & -\sin\alpha_3 & 0 \\ \sin\alpha_3 & \cos\alpha_3 & 0 \\ 0 & 0 & 1 \end{pmatrix}.$$

The order of application of the matrices is important, and we may consider a 3,2,1 rotation taking $\mathbf{x}$ into $\mathbf{x}'$ as

$$\mathbf{x}' = \mathbf{Q}_1 \cdot \mathbf{Q}_2 \cdot \mathbf{Q}_3 \cdot \mathbf{x}.$$

The first rotation is about the 3 axis, the second about the 2 axis, and the third about the 1 axis. If $\mathbf{x} = (1, 1, 1)^T$, and $\alpha_1 = \pi$, $\alpha_2 = \pi/2$, $\alpha_3 = \pi/3$, find the composite rotation matrix $\mathbf{Q} = \mathbf{Q}_1 \cdot \mathbf{Q}_2 \cdot \mathbf{Q}_3$, the rotated vector $\mathbf{x}'$, the orientation of the axis of composite rotation, and the angle of the composite rotation. Repeat for a 1,2,3 rotation, for which $\mathbf{Q} = \mathbf{Q}_3 \cdot \mathbf{Q}_2 \cdot \mathbf{Q}_1$.

2.29 Examine the literature, for example Feynman et al. (1963, Volume 1, Chapter 52-5) or Aris (1962, p. 36), and report on the distinction between so-called *polar vectors* and *axial vectors*, which are also known as pseudo-vectors. Examples of polar vectors include position $\mathbf{x}$, velocity $\mathbf{v}$, and acceleration $\mathbf{a}$. Examples of axial vectors are those whose value depends on the choice of a right-handed or left-handed coordinate system, such as might occur when a cross product or curl is involved. Examples include the vorticity $\boldsymbol{\omega}$, which is the curl of the velocity vector $\boldsymbol{\omega} = \nabla \times \mathbf{v}$. Discuss the invariance of both classes of vectors when they are subjected to transformations followed by inverse transformations. Consider rotation and reflection transformations and their inverses.

2.30 Consider the vector field, represented in Cartesian coordinates $\boldsymbol{\xi} = (\xi^1, \xi^2, \xi^3)^T$ as $\mathbf{v} = (2\xi^1 + \xi^2, 2\xi^1 - \xi^2, 0)^T$. At the point $(\xi^1, \xi^2, \xi^3)^T = (1, 1, 1)^T$, evaluate $\mathbf{v}$, $\vartheta = \nabla^T \cdot \mathbf{v}$, and $\boldsymbol{\omega} = \nabla \times \mathbf{v}$ in the Cartesian representation. Consider a coordinate transformation given by $\boldsymbol{\xi} = \mathbf{J} \cdot \mathbf{x}$. In the special case in which $\mathbf{J}$ is a rotation matrix with

$$\mathbf{J} = \begin{pmatrix} \frac{1}{\sqrt{2}} & -\frac{1}{\sqrt{2}} & 0 \\ \frac{1}{\sqrt{2}} & \frac{1}{\sqrt{2}} & 0 \\ 0 & 0 & 1 \end{pmatrix},$$

find the representation of $\boldsymbol{\xi}$, $\mathbf{v}$, and $\boldsymbol{\omega}$ in the rotated coordinate system at the point in question $\boldsymbol{\xi} = (1, 1, 1)^T$ and determine whether or not the vectors are invariant when transformed back to the unrotated system. Determine if the result is consistent with $\mathbf{v}$ as a polar vector and $\boldsymbol{\omega}$ as an axial vector. Determine $\nabla^T \cdot \mathbf{v}$ for both representations.

2.31 Consider the vector field, represented in Cartesian coordinates $\boldsymbol{\xi} = (\xi^1, \xi^2, \xi^3)^T$ as $\mathbf{v} = (2\xi^1 + \xi^2, 2\xi^1 - \xi^2, 0)^T$. At the point $(\xi^1, \xi^2, \xi^3)^T = (1, 1, 1)^T$, evaluate $\mathbf{v}$, $\vartheta = \nabla^T \cdot \mathbf{v}$, and $\boldsymbol{\omega} = \nabla \times \mathbf{v}$ in the Cartesian representation. Consider a linear coordinate transformation given by $\boldsymbol{\xi} = \mathbf{J} \cdot \mathbf{x}$. In the case in which $\mathbf{J}$ is a reflection matrix with

$$\mathbf{J} = \begin{pmatrix} \frac{1}{\sqrt{2}} & \frac{1}{\sqrt{2}} & 0 \\ \frac{1}{\sqrt{2}} & -\frac{1}{\sqrt{2}} & 0 \\ 0 & 0 & 1 \end{pmatrix},$$

find the representation of the vectors $\boldsymbol{\xi}$, $\mathbf{v}$, and $\boldsymbol{\omega}$ in the reflected coordinate system at $\boldsymbol{\xi} = (1, 1, 1)^T$, and determine whether or not the vectors are invariant when transformed back to the unreflected system. Determine if the result is consistent with $\mathbf{v}$ as a polar vector and $\boldsymbol{\omega}$ as an axial vector. Determine $\nabla^T \cdot \mathbf{v}$ for both representations.

FURTHER READING

Eringen, A. C. (1989). *Mechanics of Continua*, 2nd ed. Malabar, Florida: Krieger.

Hubbard, J., and Burke-Hubbard, B. (2015). *Vector Calculus, Linear Algebra, and Differential Forms: A Unified Approach*, 5th ed. Ithaca, New York: Matrix Editions.

Hughes, W. F., and Gaylord, E. W. (1964). *Basic Equations of Engineering Science*, Schaum's Outline Series. New York: McGraw-Hill.

Kay, D. C (2011). *Tensor Calculus*, Schaum's Outline Series. New York: McGraw-Hill.

Lovelock, D., and Rund, H. (1975). *Tensors, Differential Forms, and Variational Principles*. New York: Dover.

McConnell, A. J. (1957). *Applications of Tensor Analysis*. New York: Dover.

Paolucci, S. (2016). *Continuum Mechanics and Thermodynamics of Matter*. New York: Cambridge University Press.

Schey, H. M. (2004). *Div, Grad, Curl, and All That*, 4th ed. London: W. W. Norton.

Thorne, K. S., and Blandford, R. D. (2017). *Modern Classical Physics: Optics, Fluids, Plasmas, Elasticity, and Statistical Physics*. Princeton, New Jersey: Princeton University Press.

Truesdell, C. A. (1991). *A First Course in Rational Continuum Mechanics*, Vol. 1, 2nd ed. Boston, Massachusetts: Academic Press.

REFERENCES

Anderson, D. A., Tannehill, J. C., Pletcher, R. H., Munipalli, R., and Shankar, V. (2021). *Computational Fluid Mechanics and Heat Transfer*, 4th ed. Boca Raton, Florida: CRC Press.

Aris, R. (1962). *Vectors, Tensors, and the Basic Equations of Fluid Mechanics*. New York: Dover.

Feynman, R. P., Leighton, R. B., and Sands, M. (1963). *The Feynman Lectures on Physics*, 3 vols., Reading, Massachusetts: Addison-Wesley.

Kaplan, W. (2003). *Advanced Calculus*, 5th ed. Boston, Massachusetts: Addison-Wesley.

Karamcheti, K. (1980). *Principles of Ideal-Fluid Aerodynamics*, 2nd ed. Malabar, Florida: Krieger.

Kundu, P. K., Cohen, I. M., and Dowling, D. R. (2015). *Fluid Mechanics*, 6th ed. Amsterdam: Academic Press.

Liseikin, V. L. (2017). *Grid Generation Methods*, 3rd ed. Cham, Switzerland: Springer.

Panton, R. L. (2013). *Incompressible Flow*, 4th ed. New York: John Wiley.

Powers, J. M. (2016). *Combustion Thermodynamics and Dynamics*. New York: Cambridge University Press.

Powers, J. M., and Sen, M. (2015). *Mathematical Methods in Engineering*. New York: Cambridge University Press.

Riley, K. F., Hobson, M. P., and Bence, S. J. (2006). *Mathematical Methods for Physics and Engineering*, 3rd ed. Cambridge, UK: Cambridge University Press.

Strang, G. (2006). *Linear Algebra and its Applications*, 4th ed. Boston, Massachusetts: Cengage.

Whitaker, S. (1968). *Introduction to Fluid Mechanics*. Malabar, Florida: Krieger.

Yih, C.-S. (1977). *Fluid Mechanics*. East Hampton, Connecticut: West River Press.

3 Kinematics

Here we advance from geometry to consider *kinematics*, the study of motion in space. We will not yet consider forces that cause the motion. If we knew the position of every fluid particle as a function of time, we could also describe the velocity and acceleration of each particle. We could also make statements about how groups of particles translate, rotate, and deform. This is the essence of kinematics, the tool to describe the motion of an infinitesimally small fluid particle, as well as a continuum of such particles. Common notions developed from the kinematics of point masses apply to fluid particles.

Fluid motion is generally a nonlinear phenomenon. Here, we will develop tools, using a local linear analysis, to break down complex flows to a combination of fundamental motions. We then return to nonlinearity by showing how kinematics can be considered as a nonlinear dynamical system, a field for which a broad mathematical literature exists, for example Guckenheimer and Holmes (2002), Hirsch et al. (2013), or Perko (2006). Standard treatment of fluid kinematics is widespread; see Aris (1962), Batchelor (2000), Currie (2013), Kundu et al. (2015), Ottino (1989), or Panton (2013). Meneveau (2011) gives a good discussion linking kinematics, dynamical systems, and turbulence.

3.1 Motivating Problem

Before proceeding to a general analysis of fluid kinematics, let us study a simple motivating problem. Consider the two-dimensional configuration of Fig. 3.1. Attention is focused on a single fluid particle P residing within the space. The time-dependent position of the fluid particle is $\mathbf{x} = \mathbf{r}(t)$. At $t = 0$, the fluid particle is located at $\mathbf{r}(0) = \mathbf{x}^o$. The velocity vector $\mathbf{v}$

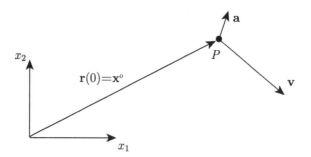

Figure 3.1 Configuration for the two-dimensional motion of a single fluid particle P.

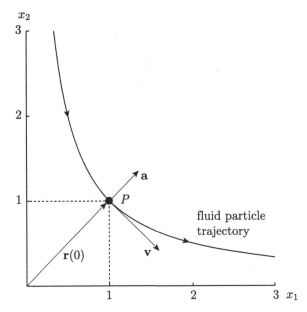

Figure 3.2 Position, velocity, acceleration, and trajectory of a fluid particle P for $x_1 = e^t$, $x_2 = e^{-t}$.

and acceleration vector **a** of the particle have their standard interpretations from Newtonian physics:

$$\mathbf{v} = \frac{d\mathbf{r}}{dt}, \qquad \mathbf{a} = \frac{d\mathbf{v}}{dt} = \frac{d^2\mathbf{r}}{dt^2}. \tag{3.1}$$

Let us consider a special case to illustrate the kinematics of a fluid particle. Consider

$$\mathbf{r}(t) = \begin{pmatrix} r_1(t) \\ r_2(t) \end{pmatrix} = \begin{pmatrix} x_1^o e^t \\ x_2^o e^{-t} \end{pmatrix}. \tag{3.2}$$

This gives us $r_1(t)$ and $r_2(t)$. For this simple flow, and assuming $x_1^o \neq 0$, $x_2^o \neq 0$, we can eliminate the explicit dependence on t to form

$$r_1(t) r_2(t) = x_1^o x_2^o. \tag{3.3}$$

And because $\mathbf{x} = \mathbf{r}(t)$, we can say

$$x_1 x_2 = x_1^o x_2^o. \tag{3.4}$$

This is a hyperbola in (x_1, x_2) space and represents the trajectory of P. By inspection, the velocity and acceleration vectors of this particle are

$$\mathbf{v}(t) = \frac{d\mathbf{r}}{dt} = \begin{pmatrix} x_1^o e^t \\ -x_2^o e^{-t} \end{pmatrix}, \qquad \mathbf{a}(t) = \frac{d^2\mathbf{r}}{dt^2} = \begin{pmatrix} x_1^o e^t \\ x_2^o e^{-t} \end{pmatrix}. \tag{3.5}$$

Taking $x_1^o = 1$, $x_2^o = 1$, we plot the trajectory of P, along with associated velocity and acceleration vectors in Fig. 3.2. For the special case for which $x_1^o = 0$, the trajectory in (x_1, x_2) space is the line $x_1 = 0$; similarly, for $x_2^o = 0$, the trajectory is $x_2 = 0$.

Now consider an ensemble of points, corresponding to selecting multiple values of x_1^o and x_2^o. We make similar plots for fluid particle trajectories, along with velocity and acceleration vector

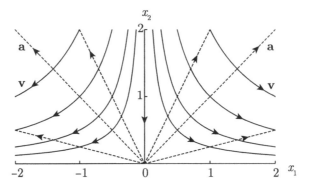

Figure 3.3 Trajectories, velocity and acceleration vector fields of an ensemble of fluid particles for $x_1 = x_1^o e^t$, $x_2 = x_2^o e^{-t}$.

fields in Fig. 3.3. This particular flow field will be considered in more detail in Section 8.3.5. Clearly, one can imagine it physically as a jet of fluid parallel to the x_2 axis directed down towards the origin. Any fluid particle that was initially on the x_2 axis ($x_1^o = 0$) will be driven to the origin, where it comes to rest. Fluid particles with $x_1^o > 0$ are swept to the right, and those with $x_1^o < 0$ are swept to the left. The acceleration vector field points radially outward from the origin. When dynamics are considered, the force field that is associated with such an acceleration can be examined.

3.2 Lagrangian Description

We move on now to a more general discussion. A Lagrangian[1] description is similar to a classical Newtonian description of motion in that each fluid particle is effectively labeled and tracked in terms of its initial position x_j^o and time $\hat{t}$. It should be recognized that we are treating x_j^o as a set of variables that fill the spatial continuum.

We take the position vector of a particle r_i to be

$$r_i = \tilde{r}_i(x_j^o, \hat{t}). \tag{3.6}$$

The velocity v_i of a particular particle is the time derivative of its position, holding x_j^o fixed:

$$v_i = \left.\frac{\partial \tilde{r}_i}{\partial \hat{t}}\right|_{x_j^o}. \tag{3.7}$$

Holding x_j^o fixed ensures we are able to identify the actual fluid particle under consideration. The acceleration a_i of a particular particle is the second time derivative of its position, holding x_j^o fixed:

$$a_i = \left.\frac{\partial^2 \tilde{r}_i}{\partial \hat{t}^2}\right|_{x_j^o}. \tag{3.8}$$

We can also write other variables as functions of time and initial position, for example, we could have for density $\rho(x_j^o, \hat{t})$.

---

[1] Joseph-Louis Lagrange (originally Giuseppe Luigi Lagrangia), 1736–1813, Italian-born, Italian-French mathematician. Worked on celestial mechanics and the three-body problem. Part of the committee that formulated the metric system.

The Lagrangian description has important value, especially in flows in which specific entities require tracking, such as flows with embedded small droplets. However, for the many applications in which knowledge of individual fluid particles is not required, it is typically not used. In solid mechanics, it is often critically important to know the location of each solid element, and it is the method of choice.

3.3 Eulerian Description

It is more common in fluid mechanics to use the Eulerian description of fluid motion. In this description, all variables are taken to be functions of time and *local* position, rather than initial position. We will first consider a one-dimensional limit, then extend to multi-dimensions.

3.3.1 One-Dimensional

Here, we will confine attention to a single spatial coordinate x and take the local position to be given by the position vector $x = r$. A general transformation from a spatially one-dimensional space-time coordinate system (t, x) to another $(\hat{t}, x^o)$ can take the form

$$t = t(\hat{t}, x^o), \qquad x = x(\hat{t}, x^o). \tag{3.9}$$

At this point, we can extend and adapt the analysis introduced in Section 2.5. While that discussion was focused on spatial coordinate transformations, there is no reason it cannot be extended to space-time systems. Forming the total differentials for dt and dx yields

$$\begin{pmatrix} dt \\ dx \end{pmatrix} = \underbrace{\begin{pmatrix} \frac{\partial t}{\partial \hat{t}}\big|_{x^o} & \frac{\partial t}{\partial x^o}\big|_{\hat{t}} \\ \frac{\partial x}{\partial \hat{t}}\big|_{x^o} & \frac{\partial x}{\partial x^o}\big|_{\hat{t}} \end{pmatrix}}_{\mathbf{J}} \begin{pmatrix} d\hat{t} \\ dx^o \end{pmatrix}. \tag{3.10}$$

This has the Jacobian matrix $\mathbf{J}$ of

$$\mathbf{J} = \begin{pmatrix} \frac{\partial t}{\partial \hat{t}}\big|_{x^o} & \frac{\partial t}{\partial x^o}\big|_{\hat{t}} \\ \frac{\partial x}{\partial \hat{t}}\big|_{x^o} & \frac{\partial x}{\partial x^o}\big|_{\hat{t}} \end{pmatrix}. \tag{3.11}$$

We will consider the transformation from Lagrangian coordinates to Eulerian coordinates given by the more specific form

$$t = \hat{t}, \qquad x = \tilde{r}(\hat{t}, x^o). \tag{3.12}$$

We also require, at the initial state $t = \hat{t} = 0$, that the Lagrangian and Eulerian coordinates coincide: $x = x^o = \tilde{r}(0, x^o)$. The more specific transformation has the more specific Jacobian matrix

$$\mathbf{J} = \begin{pmatrix} 1 & 0 \\ \frac{\partial \tilde{r}}{\partial \hat{t}}\big|_{x^o} & \frac{\partial \tilde{r}}{\partial x^o}\big|_{\hat{t}} \end{pmatrix}. \tag{3.13}$$

We have the Jacobian determinant J as

$$J = \det \mathbf{J} = \left.\frac{\partial \tilde{r}}{\partial x^o}\right|_{\hat{t}}. \tag{3.14}$$

As long as $J > 0$, the transformation is nonsingular and orientation-preserving. At $\hat{t} = 0$, we have $x = \tilde{r} = x^o$, Thus, at $\hat{t} = 0$, we get $J = 1$. So at the initial state, the transformation is length- and orientation-preserving.

3.3.2 Multi-Dimensional

Now, we will take the local position to be given by the position vector $x_i = r_i$. A general transformation from one coordinate system (t, x_i) to another $(\hat{t}, x_i^o)$ can take the general form

$$t = t(\hat{t}, x_j^o), \qquad x_i = x_i(\hat{t}, x_j^o). \tag{3.15}$$

We again extend and adapt the analysis introduced in Section 2.5. The definitions of the total differentials dt and dx_i tell us

$$\begin{pmatrix} dt \\ dx_i \end{pmatrix} = \underbrace{\begin{pmatrix} \left.\frac{\partial t}{\partial \hat{t}}\right|_{x_j^o} & \left.\frac{\partial t}{\partial x_j^o}\right|_{\hat{t}} \\ \left.\frac{\partial x_i}{\partial \hat{t}}\right|_{x_j^o} & \left.\frac{\partial x_i}{\partial x_j^o}\right|_{\hat{t}} \end{pmatrix}}_{\mathbf{J}} \begin{pmatrix} d\hat{t} \\ dx_j^o \end{pmatrix}. \tag{3.16}$$

This has the Jacobian matrix $\mathbf{J}$ of

$$\mathbf{J} = \begin{pmatrix} \left.\frac{\partial t}{\partial \hat{t}}\right|_{x_j^o} & \left.\frac{\partial t}{\partial x_j^o}\right|_{\hat{t}} \\ \left.\frac{\partial x_i}{\partial \hat{t}}\right|_{x_j^o} & \left.\frac{\partial x_i}{\partial x_j^o}\right|_{\hat{t}} \end{pmatrix}. \tag{3.17}$$

We will consider the transformation from Lagrangian coordinates to Eulerian coordinates given by the more specific form

$$t = \hat{t}, \qquad x_i = \tilde{r}_i(\hat{t}, x_j^o). \tag{3.18}$$

We also require, at the initial state $t = \hat{t} = 0$, that the Lagrangian and Eulerian coordinates coincide: $x_i = x_i^o = \tilde{r}_i(0, x_j^o)$. The more specific transformation has the more specific Jacobian matrix

$$\mathbf{J} = \begin{pmatrix} 1 & 0 \\ \left.\frac{\partial \tilde{r}_i}{\partial \hat{t}}\right|_{x_j^o} & \left.\frac{\partial \tilde{r}_i}{\partial x_j^o}\right|_{\hat{t}} \end{pmatrix}. \tag{3.19}$$

We have the Jacobian determinant J as

$$J = \det \mathbf{J} = \det \left.\frac{\partial \tilde{r}_i}{\partial x_j^o}\right|_{\hat{t}}. \tag{3.20}$$

As long as $J > 0$, the transformation is nonsingular and orientation-preserving. At $\hat{t} = 0$, we have $x_i = \tilde{r}_i = x_i^o$, Thus, at $\hat{t} = 0$, we get $J = \det \delta_{ij} = 1$. So at the initial state, the transformation is volume- and orientation-preserving.

3.4 Material Derivative

The *material derivative* is the derivative following a fluid particle. It is also known as the *substantial derivative*. It is trivial in Lagrangian coordinates, because by definition, a Lagrangian description keeps track of a fluid particle. It is not as straightforward in the Eulerian viewpoint. We will present several approaches to this, beginning with a simplified approach that suffices to capture most of the important features that will be needed for analysis of fluid motion. We will follow this with a more nuanced presentation based on coordinate transformations that will expose more of the features distinguishing the Lagrangian and Eulerian approaches.

3.4.1 Simple Approach

Consider a scalar fluid property such as density ρ that is a function of three Eulerian spatial coordinates and time: $\rho(t, x_1, x_2, x_3)$. Find its material derivative using a simplistic approach based on the definition of the derivative of a function of many variables. If $\rho = \rho(t, x_1, x_2, x_3)$, then by definition of the total differential, we have

$$d\rho = \frac{\partial \rho}{\partial t} dt + \frac{\partial \rho}{\partial x_1} dx_1 + \frac{\partial \rho}{\partial x_2} dx_2 + \frac{\partial \rho}{\partial x_3} dx_3. \tag{3.21}$$

Let us scale by dt and then say

$$\frac{d\rho}{dt} = \frac{\partial \rho}{\partial t} + \frac{\partial \rho}{\partial x_1}\frac{dx_1}{dt} + \frac{\partial \rho}{\partial x_2}\frac{dx_2}{dt} + \frac{\partial \rho}{\partial x_3}\frac{dx_3}{dt}. \tag{3.22}$$

If we select a path with known velocity $dx_1/dt = w_1$, $dx_2/dt = w_2$, $dx_3/dt = w_3$, we have

$$\frac{d\rho}{dt}_{tot} = \frac{\partial \rho}{\partial t} + \frac{\partial \rho}{\partial x_1}w_1 + \frac{\partial \rho}{\partial x_2}w_2 + \frac{\partial \rho}{\partial x_3}w_3 = \frac{\partial \rho}{\partial t} + w_1\frac{\partial \rho}{\partial x_1} + w_2\frac{\partial \rho}{\partial x_2} + w_3\frac{\partial \rho}{\partial x_3}, \tag{3.23}$$

$$= \partial_o \rho + w_i \partial_i \rho = \frac{\partial \rho}{\partial t} + \mathbf{w}^T \cdot \nabla \rho. \tag{3.24}$$

Here, we have defined the *total derivative operator* as

$$\frac{d}{dt}_{tot} = \partial_o + w_i \partial_i = \frac{\partial}{\partial t} + \mathbf{w}^T \cdot \nabla. \tag{3.25}$$

The total derivative can be applied on any path, not necessarily that of a fluid particle. It is only occasionally required. More importantly, on fluid particle path we select $\mathbf{w} = \mathbf{v}$ and have $dx_1/dt = v_1$, $dx_2/dt = v_2$, $dx_3/dt = v_3$, so we get the more important material derivative:

$$\frac{d\rho}{dt} = \frac{\partial \rho}{\partial t} + \frac{\partial \rho}{\partial x_1}v_1 + \frac{\partial \rho}{\partial x_2}v_2 + \frac{\partial \rho}{\partial x_3}v_3 = \frac{\partial \rho}{\partial t} + v_1\frac{\partial \rho}{\partial x_1} + v_2\frac{\partial \rho}{\partial x_2} + v_3\frac{\partial \rho}{\partial x_3}, \tag{3.26}$$

$$= \partial_o \rho + v_i \partial_i \rho = \frac{\partial \rho}{\partial t} + \mathbf{v}^T \cdot \nabla \rho. \tag{3.27}$$

The material derivative operator is

$$\frac{d}{dt} = \partial_o + v_i \partial_i = \frac{\partial}{\partial t} + \mathbf{v}^T \cdot \nabla. \tag{3.28}$$

From here on, d/dt is understood to be the material derivative.

3.4.2 Coordinate Transformation Approach

Let us achieve the same result now with a more formal approach based on coordinate transformations. We first look at a system with only one spatial dimension and then extend to three.

One-Dimensional

Consider a scalar fluid property, such as density ρ, that is a function of one spatial dimension and time. We seek to represent ρ as a function of either Eulerian space-time coordinates (x,t) or Lagrangian space-time coordinates $(x^o, \hat{t})$. We have from the chain rule

$$\begin{pmatrix} \left.\frac{\partial \rho}{\partial \hat{t}}\right|_{x^o} \\ \left.\frac{\partial \rho}{\partial x^o}\right|_{\hat{t}} \end{pmatrix} = \underbrace{\begin{pmatrix} \left.\frac{\partial t}{\partial \hat{t}}\right|_{x^o} & \left.\frac{\partial x}{\partial \hat{t}}\right|_{x^o} \\ \left.\frac{\partial t}{\partial x^o}\right|_{\hat{t}} & \left.\frac{\partial x}{\partial x^o}\right|_{\hat{t}} \end{pmatrix}}_{=\mathbf{J}^T} \begin{pmatrix} \left.\frac{\partial \rho}{\partial t}\right|_{x} \\ \left.\frac{\partial \rho}{\partial x}\right|_{t} \end{pmatrix}. \tag{3.29}$$

We recognize $\mathbf{J}^T$ as the transpose of the Jacobian matrix of the transformation. Invoking our transformation, Eq. (3.12), we get

$$\begin{pmatrix} \left.\frac{\partial \rho}{\partial \hat{t}}\right|_{x^o} \\ \left.\frac{\partial \rho}{\partial x^o}\right|_{\hat{t}} \end{pmatrix} = \begin{pmatrix} 1 & \left.\frac{\partial \tilde{r}}{\partial \hat{t}}\right|_{x^o} \\ 0 & \left.\frac{\partial \tilde{r}}{\partial x^o}\right|_{\hat{t}} \end{pmatrix} \begin{pmatrix} \left.\frac{\partial \rho}{\partial t}\right|_{x} \\ \left.\frac{\partial \rho}{\partial x}\right|_{t} \end{pmatrix} = \begin{pmatrix} 1 & v \\ 0 & \left.\frac{\partial \tilde{r}}{\partial x^o}\right|_{\hat{t}} \end{pmatrix} \begin{pmatrix} \left.\frac{\partial \rho}{\partial t}\right|_{x} \\ \left.\frac{\partial \rho}{\partial x}\right|_{t} \end{pmatrix}. \tag{3.30}$$

Thus, we get the one-dimensional equivalent of Eq. (3.27):

$$\left.\frac{\partial \rho}{\partial \hat{t}}\right|_{x^o} = \left.\frac{\partial \rho}{\partial t}\right|_{x} + v \left.\frac{\partial \rho}{\partial x}\right|_{t}. \tag{3.31}$$

Here it is unambiguous that this is the derivative following a specific material particle, because the partial derivative on the left-hand side demands that x^o be held fixed. So we are really saying the material derivative is such that

$$\frac{d}{dt} = \left.\frac{\partial}{\partial \hat{t}}\right|_{x^o}. \tag{3.32}$$

We also find, from inverting Eq. (3.29), that one gets the relationship

$$\begin{pmatrix} \left.\frac{\partial}{\partial t}\right|_{x} \\ \left.\frac{\partial}{\partial x}\right|_{t} \end{pmatrix} = (\mathbf{J}^T)^{-1} \begin{pmatrix} \left.\frac{\partial}{\partial \hat{t}}\right|_{x^o} \\ \left.\frac{\partial}{\partial x^o}\right|_{\hat{t}} \end{pmatrix} = \frac{1}{\left.\frac{\partial \tilde{r}}{\partial x^o}\right|_{\hat{t}}} \begin{pmatrix} 1 & -v \\ 0 & 1 \end{pmatrix} \begin{pmatrix} \left.\frac{\partial}{\partial \hat{t}}\right|_{x^o} \\ \left.\frac{\partial}{\partial x^o}\right|_{\hat{t}} \end{pmatrix} = \frac{1}{\left.\frac{\partial \tilde{r}}{\partial x^o}\right|_{\hat{t}}} \begin{pmatrix} \left.\frac{\partial}{\partial \hat{t}}\right|_{x^o} - v \left.\frac{\partial}{\partial x^o}\right|_{\hat{t}} \\ \left.\frac{\partial}{\partial x^o}\right|_{\hat{t}} \end{pmatrix}. \tag{3.33}$$

This extends the earlier-developed Eq. (2.251).

Example 3.1 The one-dimensional unsteady motion of a set of fluid particles is given by

$$\tilde{r}(x^o, \hat{t}) = x^o \left(1 + \left(\frac{\hat{t}}{\tau}\right)^2\right), \tag{3.34}$$

where τ is a constant. Particles are compressed in such a way that its density evolution is governed by the equation

$$\left.\frac{\partial \rho}{\partial \hat{t}}\right|_{x^o} = \frac{\rho_o x^o}{L\tau}, \qquad \rho(0, x^o) = \rho_o \left(\frac{x^o}{L}\right)\left(1 - \frac{x^o}{L}\right). \tag{3.35}$$

Here ρ_o and L are constant reference density and length, respectively. Analyze the fluid motion and density evolution in an Eulerian frame.

Solution

First note that when $\hat{t} = 0$ that $\tilde{r} = x^o$, as required. The transformation to Eulerian coordinates is given by

$$x = x^o \left(1 + \left(\frac{\hat{t}}{\tau}\right)^2\right), \qquad t = \hat{t}. \tag{3.36}$$

The velocity and acceleration of a fluid particle are

$$v = \left.\frac{\partial \tilde{r}}{\partial \hat{t}}\right|_{x^o} = 2x^o \frac{\hat{t}}{\tau^2}, \qquad a = \left.\frac{\partial^2 \tilde{r}}{\partial \hat{t}^2}\right|_{x^o} = \frac{2x^o}{\tau^2}. \tag{3.37}$$

We can integrate the density evolution equation to get

$$\rho(\hat{t}, x^o) = \frac{\rho_o x^o \hat{t}}{L\tau} + f(x^o). \tag{3.38}$$

Here $f(x^o)$ is an arbitrary function of x^o that can be evaluated with the initial condition so that

$$\rho(\hat{t}, x^o) = \rho_o \frac{x^o}{L} \frac{\hat{t}}{\tau} + \rho_o \left(\frac{x^o}{L}\right)\left(1 - \frac{x^o}{L}\right) = \rho_o \frac{x^o}{L}\left(\frac{\hat{t}}{\tau} + 1 - \frac{x^o}{L}\right). \tag{3.39}$$

Because of the simple nature of the flow field, the transformation to Eulerian coordinates is easy and seen by inspection to give

$$\rho(t, x) = \rho_o \frac{x/L}{1 + \left(\frac{t}{\tau}\right)^2}\left(\frac{t}{\tau} + 1 - \frac{x/L}{1 + \left(\frac{t}{\tau}\right)^2}\right). \tag{3.40}$$

For $\tau = 1$, $L = 1$, $\rho_o = 1$, plots of $\rho(x, t)$ and $\rho(x^o, \hat{t})$ are shown in Fig. 3.4.

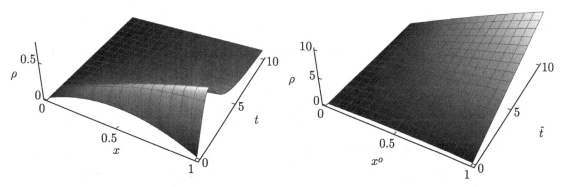

Figure 3.4 Eulerian plot $\rho(x, t)$ and Lagrangian plot of $\rho(x^o, \hat{t})$ for Example 3.1 originally described in Lagrangian coordinates, with $\tau = 1$, $L = 1$, $\rho_o = 1$.

Multi-Dimensional

Now let us perform the same exercise in multiple dimensions. The algebra is more complicated, but the general ideas are similar. From the chain rule, we have

$$\begin{pmatrix} \frac{\partial \rho}{\partial \hat{t}}\big|_{x_j^o} \\ \frac{\partial \rho}{\partial x_j^o}\big|_{\hat{t}} \end{pmatrix} = \underbrace{\begin{pmatrix} \frac{\partial t}{\partial \hat{t}}\big|_{x_j^o} & \frac{\partial x_i}{\partial \hat{t}}\big|_{x_j^o} \\ \frac{\partial t}{\partial x_j^o}\big|_{\hat{t}} & \frac{\partial x_i}{\partial x_j^o}\big|_{\hat{t}} \end{pmatrix}}_{=\mathbf{J}^T} \begin{pmatrix} \frac{\partial \rho}{\partial t}\big|_{x_i} \\ \frac{\partial \rho}{\partial x_i}\big|_{t} \end{pmatrix}. \tag{3.41}$$

We recognize $\mathbf{J}^T$ as the transpose of the Jacobian matrix of the transformation. Invoking our transformation, Eq. (3.18), we get

$$\begin{pmatrix} \frac{\partial \rho}{\partial \hat{t}}\big|_{x_j^o} \\ \frac{\partial \rho}{\partial x_j^o}\big|_{\hat{t}} \end{pmatrix} = \begin{pmatrix} 1 & \frac{\partial \tilde{r}_i}{\partial \hat{t}}\big|_{x_j^o} \\ 0 & \frac{\partial \tilde{r}_i}{\partial x_j^o}\big|_{\hat{t}} \end{pmatrix} \begin{pmatrix} \frac{\partial \rho}{\partial t}\big|_{x_i} \\ \frac{\partial \rho}{\partial x_i}\big|_{t} \end{pmatrix} = \begin{pmatrix} 1 & v_i \\ 0 & \frac{\partial \tilde{r}_i}{\partial x_j^o}\big|_{\hat{t}} \end{pmatrix} \begin{pmatrix} \frac{\partial \rho}{\partial t}\big|_{x_i} \\ \frac{\partial \rho}{\partial x_i}\big|_{t} \end{pmatrix}. \tag{3.42}$$

Thus, we get the equivalent, after again recognizing that holding x_j^o fixed is the same as holding x_i^o fixed:

$$\frac{\partial \rho}{\partial \hat{t}}\bigg|_{x_i^o} = \frac{\partial \rho}{\partial t}\bigg|_{x_i} + v_i \frac{\partial \rho}{\partial x_i}\bigg|_{t}. \tag{3.43}$$

We also find from inverting Eq. (3.41), that one gets the relationship

$$\begin{pmatrix} \frac{\partial}{\partial t}\big|_{x_i} \\ \frac{\partial}{\partial x_i}\big|_{t} \end{pmatrix} = \left(\mathbf{J}^T\right)^{-1} \begin{pmatrix} \frac{\partial}{\partial \hat{t}}\big|_{x_j^o} \\ \frac{\partial}{\partial x_j^o}\big|_{\hat{t}} \end{pmatrix}. \tag{3.44}$$

3.5 Streamlines

Streamlines are curves that are everywhere instantaneously parallel to velocity vectors. If a differential distance vector dx_k is parallel to a velocity vector v_j, then the cross product of the two vectors must be zero; hence, for a streamline we must have

$$\epsilon_{ijk} v_j \, dx_k = 0, \qquad \mathbf{v} \times d\mathbf{x} = \mathbf{0}. \tag{3.45}$$

Recalling from Eq. (2.108) that the cross product can be interpreted as a determinant, this condition reduces to

$$\begin{vmatrix} \mathbf{e}_1 & \mathbf{e}_2 & \mathbf{e}_3 \\ v_1 & v_2 & v_3 \\ dx_1 & dx_2 & dx_3 \end{vmatrix} = \mathbf{0}. \tag{3.46}$$

Expanding the determinant gives

$$\mathbf{e}_1(v_2 \, dx_3 - v_3 \, dx_2) + \mathbf{e}_2(v_3 \, dx_1 - v_1 \, dx_3) + \mathbf{e}_3(v_1 \, dx_2 - v_2 \, dx_1) = \mathbf{0}. \tag{3.47}$$

Because the basis vectors $\mathbf{e}_1$, $\mathbf{e}_2$, and $\mathbf{e}_3$ are linearly independent, the coefficient on each must be zero, giving rise to

$$v_2\, dx_3 = v_3\, dx_2, \quad \Rightarrow \quad \frac{dx_3}{v_3} = \frac{dx_2}{v_2}, \tag{3.48}$$

$$v_3\, dx_1 = v_1\, dx_3, \quad \Rightarrow \quad \frac{dx_1}{v_1} = \frac{dx_3}{v_3}, \tag{3.49}$$

$$v_1\, dx_2 = v_2\, dx_1, \quad \Rightarrow \quad \frac{dx_2}{v_2} = \frac{dx_1}{v_1}. \tag{3.50}$$

Combining, we get

$$\frac{dx_1}{v_1} = \frac{dx_2}{v_2} = \frac{dx_3}{v_3}. \tag{3.51}$$

At a fixed instant in time, $t = t_o$, we set the terms in Eq. (3.51) all equal to an arbitrary differential parameter $d\tau$ to obtain

$$\frac{dx_1}{v_1(x_1, x_2, x_3; t = t_o)} = \frac{dx_2}{v_2(x_1, x_2, x_3; t = t_o)} = \frac{dx_3}{v_3(x_1, x_2, x_3; t = t_o)} = d\tau. \tag{3.52}$$

Here τ should not be thought of as time, but just as a dummy parameter. Streamlines are only defined at a fixed time. While they will generally look different at different times, in the process of actually integrating to obtain them, time does not enter into the calculation. We then divide each equation by $d\tau$ and find they are equivalent to a system of ordinary differential equations of the *autonomous* form

$$\frac{dx_1}{d\tau} = v_1(x_1, x_2, x_3; t = t_o), \qquad x_1(\tau = 0) = x_{1o}, \tag{3.53}$$

$$\frac{dx_2}{d\tau} = v_2(x_1, x_2, x_3; t = t_o), \qquad x_2(\tau = 0) = x_{2o}, \tag{3.54}$$

$$\frac{dx_3}{d\tau} = v_3(x_1, x_2, x_3; t = t_o), \qquad x_3(\tau = 0) = x_{3o}. \tag{3.55}$$

The set is autonomous because the right-hand side has no explicit dependence on the independent variable, here τ. After integration, which in general must be done numerically, we find $x_1(\tau; t_o, x_{1o})$, $x_2(\tau; t_o, x_{2o})$, $x_3(\tau; t_o, x_{3o})$, letting the parameter τ vary over whatever domain we choose.

3.6 Pathlines

The pathlines are the locus of points traversed by a particular fluid particle. For an Eulerian description of motion in which the velocity field is known as a function of space and time $v_j(x_i, t)$, we can get the pathlines by integrating the following set of three *nonautonomous* ordinary differential equations, with the associated initial conditions:

$$\frac{dx_1}{dt} = v_1(x_1, x_2, x_3, t), \qquad x_1(t = t_o) = x_{1o}, \tag{3.56}$$

$$\frac{dx_2}{dt} = v_2(x_1, x_2, x_3, t), \qquad x_2(t = t_o) = x_{2o}, \tag{3.57}$$

$$\frac{dx_3}{dt} = v_3(x_1, x_2, x_3, t), \qquad x_3(t = t_o) = x_{3o}. \tag{3.58}$$

In general these are nonlinear equations, and often require numerical solution. Doing so gives us $x_1(t; x_{1o})$, $x_2(t; x_{2o})$, $x_3(t; x_{3o})$. They are nonautonomous because the right-hand side depends on the independent variable, here t.

3.7 Streaklines

A streakline is the locus of points that have passed through a particular point at some past time $t = \hat{t}$. Streaklines can be found by integrating a similar set of equations to those for pathlines.

$$\frac{dx_1}{dt} = v_1(x_1, x_2, x_3, t), \qquad x_1(t = \hat{t}) = x_{1o}, \qquad (3.59)$$

$$\frac{dx_2}{dt} = v_2(x_1, x_2, x_3, t), \qquad x_2(t = \hat{t}) = x_{2o}, \qquad (3.60)$$

$$\frac{dx_3}{dt} = v_3(x_1, x_2, x_3, t), \qquad x_3(t = \hat{t}) = x_{3o}. \qquad (3.61)$$

After integration, which is generally done numerically, we get $x_1(t; x_{1o}, \hat{t})$, $x_2(t; x_{2o}, \hat{t})$, $x_3(t; x_{3o}, \hat{t})$. Then, if we fix time t and the particular point in which we are interested $(x_{1o}, x_{2o}, x_{3o})^T$, we get a parametric representation of a streakline $x_1(\hat{t})$, $x_2(\hat{t})$, $x_3(\hat{t})$.

Example 3.2 If $v_1 = 2x_1 + t$, $v_2 = x_2 - 2t$, find (a) the streamline through the point $(1,1)^T$ at $t = 1$, (b) the pathline for the fluid particle that is at the point $(1,1)^T$ at $t = 1$, and (c) the streakline through the point $(1,1)^T$ at $t = 1$.

Solution

For the streamline, we have the following set of differential equations:

$$\frac{dx_1}{d\tau} = 2x_1 + t|_{t=1}, \qquad x_1(\tau = 0) = 1, \qquad (3.62)$$

$$\frac{dx_2}{d\tau} = x_2 - 2t|_{t=1}, \qquad x_2(\tau = 0) = 1. \qquad (3.63)$$

Here it is inconsequential where the parameter τ has its origin, as long as *some* value of τ corresponds to a streamline through $(1,1)^T$, so we have taken the origin for $\tau = 0$ to be the point $(1,1)^T$. These equations at $t = 1$ are

$$\frac{dx_1}{d\tau} = 2x_1 + 1, \qquad x_1(\tau = 0) = 1, \qquad (3.64)$$

$$\frac{dx_2}{d\tau} = x_2 - 2, \qquad x_2(\tau = 0) = 1. \qquad (3.65)$$

Solving, we get

$$x_1 = \frac{3}{2}e^{2\tau} - \frac{1}{2}, \qquad x_2 = -e^{\tau} + 2. \qquad (3.66)$$

Solving for τ, we find

$$\tau = \frac{1}{2}\ln\left(\frac{2}{3}\left(x_1 + \frac{1}{2}\right)\right). \qquad (3.67)$$

So, eliminating τ and writing $x_2(x_1)$, we get the streamline to be

$$x_2 = 2 - \sqrt{\frac{2}{3}\left(x_1 + \frac{1}{2}\right)}. \tag{3.68}$$

For the pathline we have the following equations:

$$\frac{dx_1}{dt} = 2x_1 + t, \qquad x_1(t=1) = 1, \tag{3.69}$$

$$\frac{dx_2}{dt} = x_2 - 2t, \qquad x_2(t=1) = 1. \tag{3.70}$$

These have solution

$$x_1 = \frac{7}{4}e^{2(t-1)} - \frac{t}{2} - \frac{1}{4}, \qquad x_2 = -3e^{t-1} + 2t + 2. \tag{3.71}$$

It is algebraically difficult to eliminate t so as to write $x_2(x_1)$ explicitly. However, the analysis certainly gives a parametric representation of the pathline, which can be plotted in x_1, x_2 space.

For the streakline we have the following equations:

$$\frac{dx_1}{dt} = 2x_1 + t, \qquad x_1(t=\hat{t}) = 1, \tag{3.72}$$

$$\frac{dx_2}{dt} = x_2 - 2t, \qquad x_2(t=\hat{t}) = 1. \tag{3.73}$$

These have solution

$$x_1 = \frac{5 + 2\hat{t}}{4}e^{2(t-\hat{t})} - \frac{t}{2} - \frac{1}{4}, \qquad x_2 = -(1 + 2\hat{t})e^{t-\hat{t}} + 2t + 2. \tag{3.74}$$

We evaluate the streakline at $t = 1$ and get

$$x_1 = \frac{5 + 2\hat{t}}{4}e^{2(1-\hat{t})} - \frac{3}{4}, \qquad x_2 = -(1 + 2\hat{t})e^{1-\hat{t}} + 4. \tag{3.75}$$

Once again, it is algebraically difficult to eliminate $\hat{t}$ so as to write $x_2(x_1)$ explicitly. However, the analysis gives a parametric representation of the streakline, which can be plotted in x_1, x_2 space. A plot of the streamline, pathline, and streakline for this problem is shown in Fig. 3.5. At the point $(1,1)^T$, all three intersect with the same slope. This can also be deduced from the equations governing streamlines, pathlines, and streaklines.

3.8 Kinematic Decomposition of Motion

In general the motion of a fluid is nonlinear in nearly all respects. Certainly, it is common for particle pathlines to have a nonlinear path; however, this is not actually a hallmark of nonlinearity in that linear theories of fluid motion routinely predict pathlines with finite curvature. More to the point, we cannot in general use the method of superposition to add one flow to another to generate a third. One fundamental source of nonlinearity is the nonlinear operator $v_i \partial_i$, which we will see appears in most of our governing equations.

However, the local behavior of fluids is nearly always dominated by linear effects. By analyzing only the linear effects induced by small changes in velocity that we will associate with the velocity gradient, we will learn much about fluid motion. In the linear analysis, we will see that a fluid particle's motion can be described as a summation of (1) linear translation, (2)

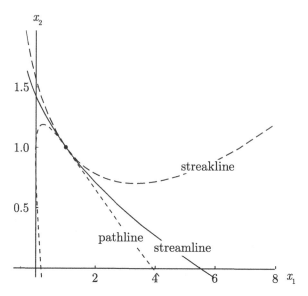

Figure 3.5 Streamlines, pathlines, and streaklines for unsteady flow of Example 3.2.

rotation as a rigid body, and (3) straining. The straining will be of two classes: (a) extensional, and (b) shear. Both types of straining can be thought of as deformation rates. We use the word "straining" in contrast to "strain" to distinguish fluid behavior from that of a flexible solid. Generally it is the rate of change of strain (that is the "straining") that has most relevance for a fluid, while it is the actual strain that has the most relevance for a flexible solid. This is because the stress in a flexible solid responds to strain, while the stress in a fluid responds to a strain rate. Nevertheless, while strain itself is associated with equilibrium configurations of a flexible solid, when its motion is decomposed, strain rate is relevant. In contrast, a rigid solid can be described by only a sum of linear translation and rotation. A point mass only translates; it cannot rotate or strain. This is summarized as follows:

$$\text{fluid motion} = \text{translation} + \text{rotation} + \underbrace{\text{extensional straining} + \text{shear straining}}_{\text{straining}},$$

$$\text{flexible solid motion} = \text{translation} + \text{rotation} + \underbrace{\text{extensional straining} + \text{shear straining}}_{\text{straining}},$$

$$\text{rigid solid motion} = \text{translation} + \text{rotation},$$

$$\text{point mass motion} = \text{translation}.$$

Let us consider in detail the configuration shown in Fig. 3.6. Here we have a fluid particle at point P with coordinates x_j and velocity v_j. A small distance $dr_j = dx_j$ away is the fluid particle at point P', with coordinates $x_j + dx_j$. This particle moves with velocity $v_j + dv_j$. We can describe the difference in location by the product of a unit vector α_j and a scalar differential distance magnitude $d\mathbf{s}$:

$$dr_j = dx_j = \alpha_j \, d\mathbf{s}, \qquad d\mathbf{r} = d\mathbf{x} = \boldsymbol{\alpha} \, d\mathbf{s}. \tag{3.76}$$

3.8 Kinematic Decomposition of Motion

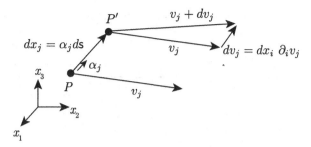

Figure 3.6 Diagram of fluid particle P in motion with velocity v_j and nearby neighbor particle P' with velocity $v_j + dv_j$.

Note that $\alpha_j = \boldsymbol{\alpha}$ is in general not aligned with the velocity vector, and the differential distance $d\mathsf{s}$ is not associated with the arc length along a particle path. Later in Section 3.14, we will select an alignment with the particle path, and thus choose $\boldsymbol{\alpha} = \boldsymbol{\alpha}_t$ and $d\mathsf{s} = ds$. Here $\boldsymbol{\alpha}_t$ is the unit tangent to the particle path, and ds is the arc length.

3.8.1 Translation

We have the motion at P' to be $v_j + dv_j$. Obviously, the first term v_j represents translation.

3.8.2 Rigid Body Rotation and Straining

What remains is dv_j, and we shall see that it is appropriate to characterize this term as a rigid body rotation combined with straining. We have from the definition of a total differential, ignoring time variations, that

$$dv_j = dx_i \, \partial_i v_j, \qquad d\mathbf{v}^T = d\mathbf{x}^T \cdot \nabla \mathbf{v}^T. \tag{3.77}$$

We could also interpret this as a scalar operator $(d\mathbf{x}^T \cdot \nabla)$ operating on a vector $\mathbf{v}^T$ and transpose both sides to get the fully equivalent

$$d\mathbf{v} = (d\mathbf{x}^T \cdot \nabla)\mathbf{v}. \tag{3.78}$$

It will be more useful, however, to focus on the combination $\nabla \mathbf{v}^T$ and say equivalently

$$d\mathbf{v} = \left(\nabla \mathbf{v}^T\right)^T \cdot d\mathbf{x} = \mathbf{L}^T \cdot d\mathbf{x}. \tag{3.79}$$

Here

$$\partial_i v_j = \nabla \mathbf{v}^T \equiv \mathbf{L}, \qquad \text{so} \qquad L_{ij} = \partial_i v_j = \frac{\partial v_j}{\partial x_i}, \tag{3.80}$$

is the *velocity gradient tensor*. Our convention, drawing on that introduced in Eq. (2.207), allows index ordering consistency with $L_{ij} = \partial_i v_j$, and this serves to maintain clarity for our nomenclature choices. As noted by Bird et al. (1987, Vol. 1, p. 306), the literature is not uniform on the notation for the velocity gradient in that many sources will use $\mathbf{L}^T$ to connote what we have defined as $\mathbf{L}$. To reconcile our results with those that have employed the alternate choice, one need only replace $\mathbf{L}$ by $\mathbf{L}^T$.

It is worth expanding the expression for $d\mathbf{v}^T$ to see the matrix structures:

$$\underbrace{\begin{pmatrix} dv_1 & dv_2 & dv_3 \end{pmatrix}}_{d\mathbf{v}^T} = \underbrace{\begin{pmatrix} dx_1 & dx_2 & dx_3 \end{pmatrix}}_{d\mathbf{x}^T} \underbrace{\begin{pmatrix} \partial_1 v_1 & \partial_1 v_2 & \partial_1 v_3 \\ \partial_2 v_1 & \partial_2 v_2 & \partial_2 v_3 \\ \partial_3 v_1 & \partial_3 v_2 & \partial_3 v_3 \end{pmatrix}}_{\mathbf{L}}. \qquad (3.81)$$

Transposing both sides, we get

$$\underbrace{\begin{pmatrix} dv_1 \\ dv_2 \\ dv_3 \end{pmatrix}}_{d\mathbf{v}} = \underbrace{\begin{pmatrix} \partial_1 v_1 & \partial_2 v_1 & \partial_3 v_1 \\ \partial_1 v_2 & \partial_2 v_2 & \partial_3 v_2 \\ \partial_1 v_3 & \partial_2 v_3 & \partial_3 v_3 \end{pmatrix}}_{\mathbf{L}^T} \underbrace{\begin{pmatrix} dx_1 \\ dx_2 \\ dx_3 \end{pmatrix}}_{d\mathbf{x}}. \qquad (3.82)$$

We can break $\partial_i v_j = \mathbf{L}$ into symmetric and anti-symmetric parts and say

$$\mathbf{L} = \mathbf{D} + \mathbf{R}, \qquad (3.83)$$

where

$$\mathbf{D} = D_{ij} \equiv \partial_{(i} v_{j)}, \qquad \mathbf{R} = R_{ij} \equiv \partial_{[i} v_{j]}, \qquad (3.84)$$

with the deformation tensor, $\mathbf{D}$, and the rotation tensor, $\mathbf{R}$. Then we get

$$dv_j = \underbrace{dx_i\, \partial_{(i} v_{j)}}_{\text{shear and extensional straining}} + \underbrace{dx_i\, \partial_{[i} v_{j]}}_{\text{rotation}}. \qquad (3.85)$$

Thus,

$$dv_j = dx_i\, D_{ij} + dx_i\, R_{ij} = (\alpha_i D_{ij} + \alpha_i R_{ij})\, d\mathbf{s}, \qquad (3.86)$$

$$d\mathbf{v}^T = d\mathbf{x}^T \cdot \mathbf{D} + d\mathbf{x}^T \cdot \mathbf{R} = \left(\boldsymbol{\alpha}^T \cdot \mathbf{D} + \boldsymbol{\alpha}^T \cdot \mathbf{R}\right) d\mathbf{s}, \qquad (3.87)$$

$$d\mathbf{v} = \mathbf{D} \cdot d\mathbf{x} + \mathbf{R}^T \cdot d\mathbf{x} = \left(\mathbf{D} \cdot \boldsymbol{\alpha} + \mathbf{R}^T \cdot \boldsymbol{\alpha}\right) d\mathbf{s}. \qquad (3.88)$$

Let

$$dv_j^{(s)} = dx_i\, \partial_{(i} v_{j)} = \alpha_i \partial_{(i} v_{j)}\, d\mathbf{s}, \qquad (3.89)$$

$$d\mathbf{v}^{(s)T} = d\mathbf{x}^T \cdot \mathbf{D} = \boldsymbol{\alpha}^T \cdot \mathbf{D}\, d\mathbf{s}, \qquad \text{so} \qquad d\mathbf{v}^{(s)} = \mathbf{D} \cdot d\mathbf{x} = \mathbf{D} \cdot \boldsymbol{\alpha}\, d\mathbf{s}. \qquad (3.90)$$

We will see this is associated with straining, both by shear and extension. Further, let

$$dv_j^{(r)} = dx_i\, \partial_{[i} v_{j]} = \alpha_i \partial_{[i} v_{j]}\, d\mathbf{s}, \qquad (3.91)$$

$$d\mathbf{v}^{(r)T} = d\mathbf{x}^T \cdot \mathbf{R} = \boldsymbol{\alpha}^T \cdot \mathbf{R}\, d\mathbf{s}, \qquad \text{so} \qquad d\mathbf{v}^{(r)} = \mathbf{R}^T \cdot d\mathbf{x} = \mathbf{R}^T \cdot \boldsymbol{\alpha}\, d\mathbf{s}. \qquad (3.92)$$

We will see this is associated with rotation as a rigid body. The total velocity difference is sum of the velocity difference due to straining and that due to rigid body rotation:

$$dv_j = dv_j^{(s)} + dv_j^{(r)}, \qquad d\mathbf{v} = d\mathbf{v}^{(s)} + d\mathbf{v}^{(r)}. \qquad (3.93)$$

Rigid Body Rotation

Let us examine $dv_j^{(r)}$. We define the *vorticity vector* ω_k as the curl of the velocity field:

$$\omega_k = \epsilon_{kij}\partial_i v_j, \qquad \boldsymbol{\omega} = \nabla \times \mathbf{v}. \tag{3.94}$$

For right-handed coordinate systems, calculation of vorticity from a velocity vector field in unrotated and rotated coordinate systems obeys ordinary vector transformation rules. However, calculation of vorticity from a velocity vector field in unreflected and reflected coordinate systems does not obey ordinary vector transformation rules. Because of this, it is better to consider the vorticity to be a pseudo-vector, that is, an axial vector.

Let us now split the velocity gradient $\partial_i v_j$ into its symmetric and anti-symmetric parts and recast the vorticity vector as

$$\omega_k = \underbrace{\epsilon_{kij}\partial_{(i}v_{j)}}_{=0} + \epsilon_{kij}\partial_{[i}v_{j]}. \tag{3.95}$$

The first term on the right-hand side is zero because it is the tensor inner product of an anti-symmetric and symmetric tensor. In what remains, we see that half of the vorticity ω_k is actually the dual vector, Ω_k, associated with the anti-symmetric $\partial_{[i}v_{j]}$ (see Section 2.1.5):

$$\omega_k = \epsilon_{kij}\partial_{[i}v_{j]} = \nabla \times \mathbf{v}, \tag{3.96}$$

$$\Omega_k = \frac{1}{2}\omega_k = \frac{1}{2}\epsilon_{kij}\partial_{[i}v_{j]} = \frac{1}{2}\nabla \times \mathbf{v}. \tag{3.97}$$

Using Eq. (2.90) to invert Eq. (3.97), we find

$$\partial_{[i}v_{j]} = \epsilon_{kij}\Omega_k = \frac{1}{2}\epsilon_{kij}\omega_k. \tag{3.98}$$

Thus, we have

$$dv_j^{(r)} = dx_i \frac{1}{2}\epsilon_{kij}\omega_k = \epsilon_{kij}\left(\frac{\omega_k}{2}\right)dx_i = \epsilon_{jki}\left(\frac{\omega_k}{2}\right)dx_i = \frac{1}{2}\boldsymbol{\omega} \times d\mathbf{r}, \tag{3.99}$$

and with $\boldsymbol{\Omega} = \boldsymbol{\omega}/2$, we get the rigid body rotation velocity of one point about another:

$$d\mathbf{v}^{(r)} = \boldsymbol{\Omega} \times d\mathbf{r}. \tag{3.100}$$

By introducing this definition for $\boldsymbol{\Omega}$, we see this term takes on the exact form for the differential velocity due to rigid body rotation of P' about P from classical rigid body kinematics. Hence, we give it the same interpretation. If $\boldsymbol{\omega} = \mathbf{0}$, the fluid is said to be *irrotational*. Irrotational velocity fields are sometimes described as *lamellar*.

Straining

Next we consider the remaining term, which we will associate with straining. First, let us further decompose this into what will be seen to be an extensional (*es*) straining and a shear straining (*ss*):

$$dv_k^{(s)} = \underbrace{dv_k^{(es)}}_{\text{extension}} + \underbrace{dv_k^{(ss)}}_{\text{shear}}, \qquad d\mathbf{v}^{(s)} = d\mathbf{v}^{(es)} + d\mathbf{v}^{(ss)}. \tag{3.101}$$

Extensional Straining Let us define the extensional straining to be the component of straining in the direction of dx_j. To do this, we need to project $dv_j^{(s)}$ onto the unit vector α_j, then point the result in the direction of that same unit vector:

$$dv_k^{(es)} = \underbrace{\left(\alpha_j dv_j^{(s)}\right)}_{\text{projection of straining}} \alpha_k. \tag{3.102}$$

Now using the definition of $dv_j^{(s)}$, Eq. (3.89), we get

$$dv_k^{(es)} = \left(\alpha_j \underbrace{\left(\alpha_i \partial_{(i} v_{j)}\, d\mathsf{s}\right)}_{=dv_j^{(s)}}\right) \alpha_k = \left(\alpha_i \partial_{(i} v_{j)} \alpha_j\right) \alpha_k\, d\mathsf{s}, \quad d\mathbf{v}^{(es)} = \left(\boldsymbol{\alpha}^T \cdot \mathbf{D} \cdot \boldsymbol{\alpha}\right) \boldsymbol{\alpha}\, d\mathsf{s}. \tag{3.103}$$

Now, because $\alpha_i \alpha_j$ is symmetric, we are led to a useful result. Consider the series of operations involving the velocity gradient, in general asymmetric, and a scalar quantity, $\mathcal{D}$:

$$\mathcal{D} = \boldsymbol{\alpha}^T \cdot \mathbf{L} \cdot \boldsymbol{\alpha} = \boldsymbol{\alpha}^T \cdot (\mathbf{D} + \mathbf{R}) \cdot \boldsymbol{\alpha} = \boldsymbol{\alpha}^T \cdot \mathbf{D} \cdot \boldsymbol{\alpha} + \underbrace{\boldsymbol{\alpha}^T \cdot \mathbf{R} \cdot \boldsymbol{\alpha}}_{=0} = \boldsymbol{\alpha}^T \cdot \mathbf{D} \cdot \boldsymbol{\alpha}. \tag{3.104}$$

Thus, we can recast Eq. (3.103) as

$$d\mathbf{v}^{(es)} = \left(\boldsymbol{\alpha}^T \cdot \mathbf{D} \cdot \boldsymbol{\alpha}\right) \boldsymbol{\alpha}\, d\mathsf{s} = \left(\boldsymbol{\alpha}^T \cdot \mathbf{L} \cdot \boldsymbol{\alpha}\right) \boldsymbol{\alpha}\, d\mathsf{s} = \mathcal{D} \boldsymbol{\alpha}\, d\mathsf{s}. \tag{3.105}$$

Shear Straining What straining that is not aligned with the axis connecting P and P' must then be normal to that axis, and can be visualized to represent a shearing between the two points. Hence, the shear straining is

$$dv_j^{(ss)} = dv_j^{(s)} - dv_j^{(es)} = \left(\partial_{(j} v_{i)} \alpha_i - \alpha_i \partial_{(i} v_{k)} \alpha_k \alpha_j\right)\, d\mathsf{s}, \tag{3.106}$$

$$= \left(\partial_{(j} v_{i)} \alpha_i - \alpha_p \partial_{(p} v_{k)} \alpha_k \underbrace{\delta_{ji} \alpha_i}_{\alpha_j}\right)\, d\mathsf{s}, \tag{3.107}$$

$$= \left(\partial_{(j} v_{i)} - \left(\alpha_p \partial_{(p} v_{k)} \alpha_k\right) \delta_{ji}\right) \alpha_i\, d\mathsf{s}, \tag{3.108}$$

$$d\mathbf{v}^{(ss)} = \left(\mathbf{D} - \left(\boldsymbol{\alpha}^T \cdot \mathbf{D} \cdot \boldsymbol{\alpha}\right) \mathbf{I}\right) \cdot \boldsymbol{\alpha}\, d\mathsf{s} = (\boldsymbol{\alpha} \times (\mathbf{D} \cdot \boldsymbol{\alpha})) \times \boldsymbol{\alpha}\, d\mathsf{s}. \tag{3.109}$$

Here Eq. (2.110) was employed to obtain the final form; if $\boldsymbol{\alpha}$ is an eigenvector of $\mathbf{D}$, $d\mathbf{v}^{(ss)} = \mathbf{0}$ because $\mathbf{D} \cdot \boldsymbol{\alpha} = \lambda \boldsymbol{\alpha}$ and $\boldsymbol{\alpha} \times \boldsymbol{\alpha} = \mathbf{0}$.

3.8.3 Principal Axes of Strain Rate

We recall from our earlier discussion of Section 2.3 that the principal axes of stress are those axes for which the force associated with a given axis points in the same direction as that axis. We can extend this idea to straining, but develop it in a slightly different fashion based on notions from linear algebra. We first recall that most arbitrary asymmetric square matrices $\mathbf{L}$ can be decomposed into a diagonal form as follows:

$$\mathbf{L} = \mathbf{S} \cdot \boldsymbol{\Lambda} \cdot \mathbf{S}^{-1}. \tag{3.110}$$

Here $\mathbf{S}$ is a matrix of the same dimension as $\mathbf{L}$ that has in its columns the right eigenvectors of $\mathbf{L}$. The matrix $\mathbf{\Lambda}$ also has the same dimension as $\mathbf{L}$; it is a diagonal matrix whose diagonal is populated by the eigenvalues, $\lambda_1, \lambda_2, \lambda_3$ of $\mathbf{L}$. It is important that the eigenvalues be ordered in such a fashion that they are associated with the correct eigenvectors in the columns of $\mathbf{S}$. Some matrices, such as those that do not have enough linearly independent eigenvectors, cannot be diagonalized; however, the argument can be extended through use of the singular value decomposition (SVD), to be considered soon in Section 3.8.5.

If $\mathbf{L}$ is symmetric, it can be shown that its eigenvalues are guaranteed to be real, and its eigenvectors are guaranteed to be orthogonal. Further, because the eigenvectors can always be scaled by a constant and remain eigenvectors, we can choose to scale them in such a way that they are all normalized. In such a case in which the matrix $\mathbf{S}$ has orthonormal columns, the matrix is orthogonal, and we call it $\mathbf{Q}$, as discussed in Section 2.1.2. So, when $\mathbf{L}$ is symmetric, such as when $\boldsymbol{\omega} = \mathbf{0}$, the velocity gradient is a symmetric tensor, $\mathbf{L} = \mathbf{D}$, and we have the following decomposition:

$$\mathbf{D} = \mathbf{Q} \cdot \mathbf{\Lambda} \cdot \mathbf{Q}^{-1}. \tag{3.111}$$

Orthogonal matrices have the property that their transpose is equal to their inverse, Eq. (2.14), and so we also have the even more useful

$$\mathbf{D} = \mathbf{Q} \cdot \mathbf{\Lambda} \cdot \mathbf{Q}^T. \tag{3.112}$$

Geometrically $\mathbf{Q}$ can be constructed to be equivalent to a matrix of direction cosines; as we have seen in Section 2.1.2. When so done, its transpose $\mathbf{Q}^T$ is a rotation matrix that rotates but does not stretch a vector when it operates on the vector.

Now let us consider the straining component of the velocity difference. Taking the symmetric $\partial_{(i} v_{j)} = \mathbf{D}$, which we further assume to be a constant for this analysis, we rewrite Eq. (3.90) using Gibbs notation as

$$d\mathbf{v}^{(s)} = \mathbf{D} \cdot d\mathbf{x} = \mathbf{Q} \cdot \mathbf{\Lambda} \cdot \mathbf{Q}^T \cdot d\mathbf{x}. \tag{3.113}$$

Now let us select what amounts to a special axes rotation. Perform matrix multiplication by the orthogonal matrix $\mathbf{Q}^T$:

$$\mathbf{Q}^T \cdot d\mathbf{v}^{(s)} = \mathbf{Q}^T \cdot \mathbf{Q} \cdot \mathbf{\Lambda} \cdot \mathbf{Q}^T \cdot d\mathbf{x} = \mathbf{Q}^{-1} \cdot \mathbf{Q} \cdot \mathbf{\Lambda} \cdot \mathbf{Q}^T \cdot d\mathbf{x} = \mathbf{\Lambda} \cdot \mathbf{Q}^T \cdot d\mathbf{x}, \tag{3.114}$$

$$d\underbrace{\left(\mathbf{Q}^T \cdot \mathbf{v}^{(s)}\right)}_{=\mathbf{v}'^{(s)}} = \mathbf{\Lambda} \cdot d\underbrace{\left(\mathbf{Q}^T \cdot \mathbf{x}\right)}_{=\mathbf{x}'}. \tag{3.115}$$

The last step is possible because $\mathbf{D}$ and thus $\mathbf{Q}^T$ are assumed constant. Recall from the definition of vectors, Eq. (2.41), that $\mathbf{Q}^T \cdot \mathbf{v}^{(s)} = \mathbf{v}'^{(s)}$ and $\mathbf{Q}^T \cdot \mathbf{x} = \mathbf{x}'$. That is, these are the representations of the vectors in a specially rotated coordinate system, so we have

$$d\mathbf{v}'^{(s)} = \mathbf{\Lambda} \cdot d\mathbf{x}'. \tag{3.116}$$

Because $\mathbf{\Lambda}$ is diagonal, we see that a perturbation in $\mathbf{x}'$ confined to any one of the rotated coordinate axes induces a change in velocity that lies in the same direction as that coordinate axis. For instance on the $1'$ axis, we have $dv_1'^{(s)} = \lambda_1 dx'_1$. That is to say that *in this specially rotated frame, all straining is extensional; there is no shear straining.*

3.8.4 Extensional Strain Rate Quadric

Let us study in some more detail the scalar that is the magnitude of the velocity difference due to extensional strain rate, given by Eq. (3.104):

$$\mathcal{D} = \boldsymbol{\alpha}^T \cdot \mathbf{D} \cdot \boldsymbol{\alpha}. \tag{3.117}$$

For a given $\mathbf{D}$, this is a quadratic equation for the components of $\boldsymbol{\alpha}$. However, because $\boldsymbol{\alpha}$ is a unit vector, it is subject to the constraint

$$\boldsymbol{\alpha}^T \cdot \boldsymbol{\alpha} = 1. \tag{3.118}$$

For a two-dimensional system, for example, Eq. (3.117) becomes

$$\mathcal{D} = D_{11}\alpha_1^2 + 2D_{12}\alpha_1\alpha_2 + D_{22}\alpha_2^2. \tag{3.119}$$

In (α_1, α_2) space and fixed $\mathcal{D}$, this may form an ellipse, hyperbola, or circle, depending on numerical values of D_{ij}. However, we have the constraint

$$\alpha_1^2 + \alpha_2^2 = 1. \tag{3.120}$$

One may imagine that as α_1 is varied for a given $\mathbf{D}$ that α_2 will also vary, as will $\mathcal{D}$, and that there could be extreme values of $\mathcal{D}$, depending on α_1.

Let us return to three-dimensional systems. Because $\mathbf{D}$ is symmetric, we can decompose it into a diagonal form and then say

$$\mathcal{D} = \boldsymbol{\alpha}^T \cdot \underbrace{\mathbf{Q} \cdot \boldsymbol{\Lambda} \cdot \mathbf{Q}^T}_{\mathbf{D}} \cdot \boldsymbol{\alpha} = (\mathbf{Q}^T \cdot \boldsymbol{\alpha})^T \cdot \boldsymbol{\Lambda} \cdot \mathbf{Q}^T \cdot \boldsymbol{\alpha}. \tag{3.121}$$

Then, defining a rotated coordinate system by $\boldsymbol{\alpha}' = \mathbf{Q}^T \cdot \boldsymbol{\alpha}$, we see

$$\mathcal{D} = \boldsymbol{\alpha}'^T \cdot \boldsymbol{\Lambda} \cdot \boldsymbol{\alpha}'. \tag{3.122}$$

Because of the diagonal form of $\boldsymbol{\Lambda}$, this can be written in full as

$$\mathcal{D} = \lambda^{(1)}\alpha_1'^2 + \lambda^{(2)}\alpha_2'^2 + \lambda^{(3)}\alpha_3'^2. \tag{3.123}$$

Equation (3.123) represents the *quadric of extensional strain rate*. It is subject to the constraint that $\boldsymbol{\alpha}'$ is a unit vector; thus,

$$1 = \alpha_1'^2 + \alpha_2'^2 + \alpha_3'^2. \tag{3.124}$$

One can formally show that $\mathcal{D}$ has a maximum given by the maximum eigenvalue and a minimum given by the minimum eigenvalue. Following details given by Fung (1965), we can adopt the ordering $\lambda^{(1)} \geq \lambda^{(2)} \geq \lambda^{(3)}$, and recognize $\mathcal{D}$ could be positive or negative. Then it can be shown that the quadric of extensional strain rate can take on many geometric forms:

- triaxial ellipsoid, if $\lambda^{(1)} > \lambda^{(2)} > \lambda^{(3)} > 0$, $\mathcal{D} > 0$, or $0 > \lambda^{(1)} > \lambda^{(2)} > \lambda^{(3)}$, $\mathcal{D} < 0$,
- ellipsoid of revolution (also called oblate or prolate spheroid), if $\lambda^{(i)}$ and $\mathcal{D}$ all share the same sign, and either $\lambda^{(1)} = \lambda^{(2)}$, $\lambda^{(2)} = \lambda^{(3)}$, or $\lambda^{(1)} = \lambda^{(3)}$,
- sphere, if $\lambda^{(1)} = \lambda^{(2)} = \lambda^{(3)}$ and share the same sign with $\mathcal{D}$,
- hyperboloid of one sheet, if $\lambda^{(1)} \geq \lambda^{(2)} > 0$, $\lambda^{(3)} < 0$, and $\mathcal{D} > 0$, or
- hyperboloid of two sheets, if $\lambda^{(1)} \geq \lambda^{(2)} > 0$, $\lambda^{(3)} < 0$, and $\mathcal{D} < 0$.

Special forms exist for either $\mathcal{D} = 0$ or any of the $\lambda^{(i)} = 0$. For strictly two-dimensional flows, we could expect the quadric of extensional strain rate to be either an ellipse, circle, hyperbola, or a line.

Example 3.3 Extremize $\mathcal{D}$ with the aid of the method of Lagrange multipliers for the case of unequal eigenvalues ordered such that $\lambda^{(1)} > \lambda^{(2)} > \lambda^{(3)}$.

Solution
Let us first define an auxiliary function $\hat{\mathcal{D}}$ as

$$\hat{\mathcal{D}} = \lambda^{(1)} \alpha_1'^2 + \lambda^{(2)} \alpha_2'^2 + \lambda^{(3)} \alpha_3'^2 - \sigma \left(\alpha_1'^2 + \alpha_2'^2 + \alpha_3'^2 - 1 \right). \tag{3.125}$$

Here σ is the Lagrange multiplier. We will also enforce the constraint $\alpha_1'^2 + \alpha_2'^2 + \alpha_3'^2 = 1$. Extremizing $\hat{\mathcal{D}}$ also extremizes $\mathcal{D}$. For extreme $\hat{\mathcal{D}}$, we require

$$\frac{\partial \hat{\mathcal{D}}}{\partial \alpha_1'} = 2\lambda^{(1)} \alpha_1' - 2\sigma \alpha_1' = 0, \tag{3.126}$$

$$\frac{\partial \hat{\mathcal{D}}}{\partial \alpha_2'} = 2\lambda^{(2)} \alpha_2' - 2\sigma \alpha_2' = 0, \tag{3.127}$$

$$\frac{\partial \hat{\mathcal{D}}}{\partial \alpha_3'} = 2\lambda^{(3)} \alpha_3' - 2\sigma \alpha_3' = 0. \tag{3.128}$$

If we allow $\alpha_1' \neq 0$, $\alpha_2' \neq 0$, $\alpha_3' \neq 0$, we are led to $\lambda^{(1)} = \lambda^{(2)} = \lambda^{(3)} = \sigma$. But this cannot be for our assumed ordering. In order to prevent this, at least two of the values of α_i' must be zero. There are then three cases to consider. If $\alpha_2' = \alpha_3' = 0$, then $\alpha_1' = 1$, and $\mathcal{D} = \lambda^{(1)}$. If $\alpha_1' = \alpha_3' = 0$, then $\alpha_2' = 1$, and $\mathcal{D} = \lambda^{(2)}$. If $\alpha_1' = \alpha_2' = 0$, then $\alpha_3' = 1$, and $\mathcal{D} = \lambda^{(3)}$. And from our assumed ordering, the maximum value is $\lambda^{(1)}$, and the minimum value is $\lambda^{(3)}$.

Example 3.4 Analyze $\mathcal{D}$, the magnitude of the velocity difference attributable to extensional straining in a selected direction $\boldsymbol{\alpha}$ for the deformation tensor

$$\mathbf{D} = \begin{pmatrix} 3 & 2 \\ 2 & 3 \end{pmatrix}. \tag{3.129}$$

Solution
We first note the eigenvalues of $\mathbf{D}$ are given by the roots of the characteristic polynomial

$$(3 - \lambda)(3 - \lambda) - 4 = 0. \tag{3.130}$$

This yields

$$\lambda^{(1)} = 5, \qquad \lambda^{(2)} = 1. \tag{3.131}$$

And as an aside, $I_{\dot{\epsilon}}^{(1)} = \text{tr } \mathbf{D} = 6$, $I_{\dot{\epsilon}}^{(3)} = \det \mathbf{D} = 5$. We have the quadric of strain rate

$$\mathcal{D} = D_{11} \alpha_1^2 + 2 D_{12} \alpha_1 \alpha_2 + D_{22} \alpha_2^2 = 3\alpha_1^2 + 4\alpha_1 \alpha_2 + 3\alpha_2^2. \tag{3.132}$$

We can plot contours for which $\mathcal{D}$ is constant in the (α_1, α_2) plane and get an infinite family of curves. However, we also have a constraint, namely $\alpha_1^2 + \alpha_2^2 = 1$.

Example 3.3 showed that the eigenvalues give extreme values of $\mathcal{D}$, and so we examine those two contours, $\mathcal{D} = 5$ and $\mathcal{D} = 1$:

$$5 = 3\alpha_1^2 + 4\alpha_1\alpha_2 + 3\alpha_2^2, \qquad 1 = 3\alpha_1^2 + 4\alpha_1\alpha_2 + 3\alpha_2^2. \qquad (3.133)$$

These two curves, along with the unit circle $\alpha_1^2 + \alpha_2^2 = 1$, are plotted in Fig. 3.7. For intersection of any contour of $\mathcal{D}$ with the unit circle, we require $\mathcal{D} \in [\lambda^{(min)}, \lambda^{(max)}]$, so $\mathcal{D} \in [1, 5]$.

The method of quadratic forms, see Section 2.8, can be applied. It can be shown by computing the eigenvectors of $\mathbf{D}$ and casting their normalized values into the columns of the orthogonal matrix $\mathbf{Q}$ that a diagonal decomposition of $\mathbf{D}$ is

$$\mathbf{D} = \mathbf{Q} \cdot \mathbf{\Lambda} \cdot \mathbf{Q}^T, \qquad (3.134)$$

$$\begin{pmatrix} 3 & 2 \\ 2 & 3 \end{pmatrix} = \begin{pmatrix} 1/\sqrt{2} & -1/\sqrt{2} \\ 1/\sqrt{2} & 1/\sqrt{2} \end{pmatrix} \begin{pmatrix} 5 & 0 \\ 0 & 1 \end{pmatrix} \begin{pmatrix} 1/\sqrt{2} & 1/\sqrt{2} \\ -1/\sqrt{2} & 1/\sqrt{2} \end{pmatrix}. \qquad (3.135)$$

We also have

$$\boldsymbol{\alpha}' = \mathbf{Q}^T \cdot \boldsymbol{\alpha} = \begin{pmatrix} 1/\sqrt{2} & 1/\sqrt{2} \\ -1/\sqrt{2} & 1/\sqrt{2} \end{pmatrix} \begin{pmatrix} \alpha_1 \\ \alpha_2 \end{pmatrix} = \begin{pmatrix} \frac{\alpha_1 + \alpha_2}{\sqrt{2}} \\ \frac{-\alpha_1 + \alpha_2}{\sqrt{2}} \end{pmatrix}. \qquad (3.136)$$

Our equation for $\mathcal{D}$ in rotated and then unrotated coordinates becomes

$$\mathcal{D} = \boldsymbol{\alpha}'^T \cdot \mathbf{\Lambda} \cdot \boldsymbol{\alpha}' = \lambda^{(1)} \alpha_1'^2 + \lambda^{(2)} \alpha_2'^2 = 5\alpha_1'^2 + \alpha_2'^2, \qquad (3.137)$$

$$= 5 \left(\frac{\alpha_1 + \alpha_2}{\sqrt{2}} \right)^2 + \left(\frac{-\alpha_1 + \alpha_2}{\sqrt{2}} \right)^2. \qquad (3.138)$$

By inspection, this is an ellipse with major and minor axes inclined relative to the (α_1, α_2) coordinate axes. Full expansion recovers our original

$$\mathcal{D} = 3\alpha_1^2 + 4\alpha_1\alpha_2 + 3\alpha_2^2. \qquad (3.139)$$

One should prefer Eq. (3.138) over the original, equivalent, and easy to obtain form of Eq. (3.139). That is because Eq. (3.138) is obviously an ellipse because of the positive coefficients on the quadratic terms. Moreover, the form gives the alignment of the major and minor axes via the rotation matrix $\mathbf{Q}^T$. Trigonometric analysis shows the principal axes are rotated clockwise by $\theta = 45°$. Lastly note that the matrix $\mathbf{D}$ could have one or more negative eigenvalues, in which case contours of $\mathcal{D}$ need not be ellipses.

Physically, one could imagine an initially spherical fluid particle centered at the origin straining in such a way that the strain rate for the direction inclined $45°$ to the α_1 axis was purely extensional with value 5, and that for the direction inclined at $-45°$ to the α_1 axis was purely extensional with value 1. Because the extensional straining on both principal axes is positive, one would not be surprised that the fluid particle's volume is expanding in this flow. Analysis tools in the upcoming Section 3.10 will show that the volume expansion rate in fact is proportional to the sum of the eigenvalues of $\mathbf{D}$, which here is $5 + 1 = 6$.

3.8.5 Singular Value Decomposition

Linear algebra provides a number of ways to decompose matrices, as discussed by Golub and Van Loan (2013) or Strang (2006). Each has a particular interpretation, and it is often geometric. Sometimes the implications of a particular decomposition for fluid kinematics are obvious and useful, other times less so. But they always exist and have some meaning, because they simply recast physically relevant matrices in different forms.

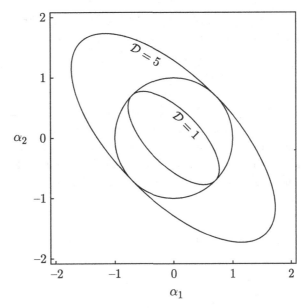

Figure 3.7 Two special contours of $\mathcal{D}$ along with the unit circle illustrating the extreme values of $\mathcal{D}$ and the orientation of the axes along with the extreme values of extensional straining are realized.

One of the most important decompositions that has had particular importance in recent decades in a wide range of science and technology is the so-called *singular value decomposition*. This builds on the singular values introduced in Sections 2.3 and 2.5. We present here a discussion that is an adaptation of that originally given by Powers and Sen (2015, Section 7.9.8). Any rectangular matrix has an SVD, and it is the most general form of diagonalization. For our discussion, however, it suffices to specialize to square matrices; these are relevant to velocity gradient tensors. Any square matrix $\mathbf{L}$ can be factored into the SVD

$$\mathbf{L} = \mathbf{Q}_2 \cdot \mathbf{\Sigma} \cdot \mathbf{Q}_1^T, \tag{3.140}$$

where $\mathbf{Q}_2$ and $\mathbf{Q}_1^T$ are orthogonal, square matrices, and $\mathbf{\Sigma}$ is a square matrix that has positive numbers σ_i, $(i = 1, 2, \ldots, r)$ in the first r positions on the main diagonal, and zero everywhere else. It turns out that r is the rank of $\mathbf{L}$. The columns of $\mathbf{Q}_2$ are the eigenvectors of $\mathbf{L} \cdot \mathbf{L}^T$. The columns of $\mathbf{Q}_1$ are the eigenvectors of $\mathbf{L}^T \cdot \mathbf{L}$. The values σ_i, $(i = 1, 2, \ldots, r) \in \mathbb{R}^1$ are called the singular values of $\mathbf{L}$. They are analogous to eigenvalues and are in fact the positive square roots of the eigenvalues of $\mathbf{L} \cdot \mathbf{L}^T$ or $\mathbf{L}^T \cdot \mathbf{L}$. Because the matrix from which the eigenvalues are drawn is symmetric, the eigenvalues, and thus the singular values, are guaranteed real. If $\mathbf{L}$ also symmetric, the absolute value of the eigenvalues of $\mathbf{L}$ will equal its singular values. If $\mathbf{L}$ is asymmetric, there is no simple relation between its eigenvalues and singular values. One notes that once more the product of a matrix and its transpose has importance.

The column vectors of $\mathbf{Q}_2$ and $\mathbf{Q}_1$ are even more than orthonormal: they also must be chosen in such a way that $\mathbf{L} \cdot \mathbf{Q}_1$ is a scalar multiple of $\mathbf{Q}_2$. This comes directly from post-multiplying the general form of the singular value decomposition, Eq. (3.140), by $\mathbf{Q}_2$. We find $\mathbf{L} \cdot \mathbf{Q}_1 = \mathbf{Q}_2 \cdot \mathbf{\Sigma}$. So in fact a more robust way of computing the singular value decomposition is to first compute one of the orthogonal matrices, and then compute the other orthogonal matrix with which the first one is consistent.

With regard to fluid kinematics, we recall Eq. (3.79). The transpose of the velocity gradient acting on a differential distance vector gives the differential velocity vector:

$$d\mathbf{v} = \mathbf{L}^T \cdot d\mathbf{x} = \left(\mathbf{Q}_2 \cdot \mathbf{\Sigma} \cdot \mathbf{Q}_1^T\right)^T \cdot d\mathbf{x} = \left(\mathbf{\Sigma} \cdot \mathbf{Q}_1^T\right)^T \cdot \mathbf{Q}_2^T \cdot d\mathbf{x} = \mathbf{Q}_1 \cdot \mathbf{\Sigma} \cdot \mathbf{Q}_2^T \cdot d\mathbf{x}.$$
(3.141)

This is a transformation of a distance to a velocity, so it is not a transformation such as a traditional coordinate transformation, which maps a distance into another distance. The cumulative effects of rotation, shear, and extension are seen to be combined into two rotations, $\mathbf{Q}_1$ and $\mathbf{Q}_2^T$, along with an eigen-stretching, embodied within $\mathbf{\Sigma}$. Physically, it is appropriate to think of each of the rotation matrices as dimensionless and the matrix $\mathbf{\Sigma}$ to have units of inverse time, so that the units of $d\mathbf{v}$ are those of velocity.

Example 3.5 Compute the singular value decomposition of the following velocity gradient tensor:

$$\mathbf{L} = \begin{pmatrix} \frac{157}{145} & \frac{713}{1740} \\ \frac{24}{145} & \frac{143}{145} \end{pmatrix},$$
(3.142)

and consider it in the context of fluid kinematics.

Solution
Here we have a square, invertible, asymmetric matrix with $\det \mathbf{L} = 1$. First let us compute

$$\mathbf{L} \cdot \mathbf{L}^T = \begin{pmatrix} \frac{157}{145} & \frac{713}{1740} \\ \frac{24}{145} & \frac{143}{145} \end{pmatrix} \begin{pmatrix} \frac{157}{145} & \frac{24}{145} \\ \frac{713}{1740} & \frac{143}{145} \end{pmatrix} = \begin{pmatrix} \frac{193}{144} & \frac{7}{12} \\ \frac{7}{12} & 1 \end{pmatrix}.$$
(3.143)

The eigenvalues of $\mathbf{L} \cdot \mathbf{L}^T$ are computed as $9/16, 16/9$. Their square roots are the singular values, $\sigma_1 = 3/4, \sigma_2 = 4/3$. These can be cast onto the diagonal of the diagonal matrix $\mathbf{\Sigma}$. The normalized eigenvectors of $\mathbf{L} \cdot \mathbf{L}^T$ may be computed and cast into the columns of $\mathbf{Q}_2$. We also choose the eigenvectors such that $\mathbf{Q}_2$ is a rotation matrix, with $\det \mathbf{Q}_2 = 1$. We find then that

$$\mathbf{Q}_2 = \begin{pmatrix} \frac{3}{5} & \frac{4}{5} \\ -\frac{4}{5} & \frac{3}{5} \end{pmatrix}.$$
(3.144)

The normalized eigenvectors of $\mathbf{L}^T \cdot \mathbf{L}$ may be computed and cast into the columns of $\mathbf{Q}_1$, taking care that they are ordered such that $\mathbf{L} \cdot \mathbf{Q}_1 = \mathbf{Q}_2 \cdot \mathbf{\Sigma}$. Alternatively, we could simply compute $\mathbf{Q}_1 = \mathbf{L}^{-1} \cdot \mathbf{Q}_2 \cdot \mathbf{\Sigma}$, provided $\mathbf{L}$ is invertible, which is the case here. We get

$$\mathbf{Q}_1 = \begin{pmatrix} \frac{20}{29} & \frac{21}{29} \\ -\frac{21}{29} & \frac{20}{29} \end{pmatrix}.$$
(3.145)

It can be confirmed that $\det \mathbf{Q}_1 = 1$, so $\mathbf{Q}_1$ is also a rotation. The singular value decomposition here is $\mathbf{L} = \mathbf{Q}_2 \cdot \mathbf{\Sigma} \cdot \mathbf{Q}_1^T$:

$$\mathbf{L} = \begin{pmatrix} \frac{157}{145} & \frac{713}{1740} \\ \frac{24}{145} & \frac{143}{145} \end{pmatrix} = \underbrace{\begin{pmatrix} \frac{3}{5} & \frac{4}{5} \\ -\frac{4}{5} & \frac{3}{5} \end{pmatrix}}_{\mathbf{Q}_2} \underbrace{\begin{pmatrix} \frac{3}{4} & 0 \\ 0 & \frac{4}{3} \end{pmatrix}}_{\mathbf{\Sigma}} \underbrace{\begin{pmatrix} \frac{20}{29} & -\frac{21}{29} \\ \frac{21}{29} & \frac{20}{29} \end{pmatrix}}_{\mathbf{Q}_1^T}.$$
(3.146)

Recalling Eq. (3.141), we can better envision the action of the velocity gradient on a chosen differential distance vector $d\mathbf{x}$ via

$$d\mathbf{v} = \mathbf{Q}_1 \cdot \mathbf{\Sigma} \cdot \mathbf{Q}_2^T \cdot d\mathbf{x}.$$
(3.147)

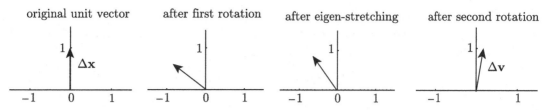

Figure 3.8 Unit position vector transforming to a velocity vector under the action of a velocity gradient tensor via rotation, stretching, and rotation of the singular value decomposition of the velocity gradient.

So a vector $d\mathbf{x}$ is first rotated by the action of $\mathbf{Q}_2^T$. Because $\mathbf{Q}_2^T$ is dimensionless, this rotated vector has units of distance. Then it is eigen-stretched by $\boldsymbol{\Sigma}$. Because the singular values have units of inverse time, this eigen-stretched vector now is a velocity. It is then rotated again by the dimensionless $\mathbf{Q}_1$ and remains a velocity vector. For finite vectors, we can approximate by replacing the d with a Δ and say

$$\Delta \mathbf{v} = \mathbf{Q}_1 \cdot \boldsymbol{\Sigma} \cdot \mathbf{Q}_2^T \cdot \Delta \mathbf{x}. \tag{3.148}$$

The action of this composition of matrix operations on a unit vector $\Delta \mathbf{x} = (0,1)^T$, is depicted in Fig. 3.8. The final vector, $\Delta \mathbf{v}$ is simply $\mathbf{L}^T \cdot \Delta \mathbf{x} = (24/145, 143/145)^T$.

Lastly, we note the singular value decomposition involves the eigenvalues of the matrix $\mathbf{L}^T \cdot \mathbf{L}$, which must be positive semi-definite real numbers. This can easily be related to the positive semi-definite nature of the eigenvalues of the metric tensor defined in Section 2.5, $\mathbf{G} = \mathbf{J}^T \cdot \mathbf{J}$, where $\mathbf{J}$ is the Jacobian matrix of a particular coordinate transformation and can be easily associated with ellipsoids of transformation described in Section 2.5 and the quadratic forms described in Section 2.8.

3.8.6 Polar Decompositions

Other decompositions exist. We next present an adaptation of a discussion given by Powers and Sen (2015, Section 7.9.9). A square invertible, potentially asymmetric matrix $\mathbf{L}$ with $\det \mathbf{L} > 0$, with $\mathbf{L}$ potentially representing the velocity gradient tensor, has the following two *polar decompositions*:

$$\mathbf{L} = \mathbf{Q} \cdot \mathbf{W}, \tag{3.149}$$

$$\mathbf{L} = \mathbf{V} \cdot \mathbf{Q}. \tag{3.150}$$

We take $\mathbf{Q}$ to be an orthogonal rotation matrix, and both $\mathbf{W}$ and $\mathbf{V}$ are symmetric positive definite square matrices. Equation (3.149) is known as the right polar decomposition, and Eq. (3.150) is known as the left polar decomposition. Both can be derived from the singular value decomposition.

Let us take the square, asymmetric $\mathbf{L}$ to have the singular value decomposition:

$$\mathbf{L} = \mathbf{Q}_2 \cdot \boldsymbol{\Sigma} \cdot \mathbf{Q}_1^T. \tag{3.151}$$

For convenience, we construct both $\mathbf{Q}_1$ and $\mathbf{Q}_2$ so they are rotation matrices. Because $\mathbf{L}$ is square, it has the same dimensions as $\mathbf{Q}_1$, $\mathbf{\Sigma}$, and $\mathbf{Q}_2$. If $\mathbf{L}$ is asymmetric, $\mathbf{Q}_1 \neq \mathbf{Q}_2$. Now, recalling that $\mathbf{Q}_1^T \cdot \mathbf{Q}_1 = \mathbf{I}$, we can recast Eq. (3.151) as

$$\mathbf{L} = \mathbf{Q}_2 \cdot \underbrace{\mathbf{Q}_1^T \cdot \mathbf{Q}_1}_{\mathbf{I}} \cdot \mathbf{\Sigma} \cdot \mathbf{Q}_1^T = \underbrace{\mathbf{Q}_2 \cdot \mathbf{Q}_1^T}_{=\mathbf{Q}} \cdot \underbrace{\mathbf{Q}_1 \cdot \mathbf{\Sigma} \cdot \mathbf{Q}_1^T}_{=\mathbf{W}}. \tag{3.152}$$

If we take

$$\mathbf{Q} = \mathbf{Q}_2 \cdot \mathbf{Q}_1^T, \qquad \mathbf{W} = \mathbf{Q}_1 \cdot \mathbf{\Sigma} \cdot \mathbf{Q}_1^T, \tag{3.153}$$

we recover Eq. (3.149). Because $\mathbf{Q}$ is formed from the product of two rotation matrices, it is itself a rotation matrix. And because $\mathbf{\Sigma}$ is diagonal with guaranteed positive entries, it and the similar matrix $\mathbf{W}$ are guaranteed positive definite. Thus, the action of $\mathbf{L}$ on a vector is the composition of the action of $\mathbf{W}$ followed by a rotation $\mathbf{Q}$. The action of $\mathbf{W}$ is first a rotation, followed by a stretch along the orthogonal principal axes, followed by a reverse of the first rotation.

Further note that

$$\mathbf{W} \cdot \mathbf{W} = \mathbf{Q}_1 \cdot \mathbf{\Sigma} \cdot \underbrace{\mathbf{Q}_1^T \cdot \mathbf{Q}_1}_{=\mathbf{I}} \cdot \mathbf{\Sigma} \cdot \mathbf{Q}_1^T = \mathbf{Q}_1 \cdot \mathbf{\Sigma} \cdot \mathbf{\Sigma} \cdot \mathbf{Q}_1^T. \tag{3.154}$$

Noting that $\mathbf{\Sigma} \cdot \mathbf{\Sigma}$ is diagonal, the right-hand side of Eq. (3.154) is nothing more than the diagonalization of $\mathbf{L}^T \cdot \mathbf{L}$; thus,

$$\mathbf{W} \cdot \mathbf{W} = \mathbf{L}^T \cdot \mathbf{L} \equiv \mathbf{C}. \tag{3.155}$$

We extend the notion of a square root to matrix operators which are positive semi-definite such as $\mathbf{L}^T \cdot \mathbf{L}$, which allows us to infer from Eq. (3.155) the equation

$$\mathbf{W} = \sqrt{\mathbf{L}^T \cdot \mathbf{L}}. \tag{3.156}$$

In other contexts, including coordinate transformations, the matrix $\mathbf{C}$ is known as the *right Cauchy[2]–Green[3] tensor*. In the context of coordinate transformations, it is analogous to the metric tensor defined in Eq. (2.255).

The analysis is similar for the left polar decomposition. Recalling that $\mathbf{Q}_2^T \cdot \mathbf{Q}_2 = \mathbf{I}$, we recast Eq. (3.151) as

$$\mathbf{L} = \mathbf{Q}_2 \cdot \mathbf{\Sigma} \cdot \underbrace{\mathbf{Q}_2^T \cdot \mathbf{Q}_2}_{\mathbf{I}} \cdot \mathbf{Q}_1^T = \underbrace{\mathbf{Q}_2 \cdot \mathbf{\Sigma} \cdot \mathbf{Q}_2^T}_{=\mathbf{V}} \cdot \underbrace{\mathbf{Q}_2 \cdot \mathbf{Q}_1^T}_{=\mathbf{Q}}. \tag{3.157}$$

If we take

$$\mathbf{V} = \mathbf{Q}_2 \cdot \mathbf{\Sigma} \cdot \mathbf{Q}_2^T, \tag{3.158}$$

---

[2] Augustin-Louis Cauchy, 1789–1857, French mathematician and military engineer, worked in complex analysis, optics, and theory of elasticity.
[3] George Green, 1793–1841, English baker and self-taught mathematical physicist.

and **Q** as in Eq. (3.153), we recover the left polar decomposition of Eq. (3.150). Similarly, it can be shown that

$$\mathbf{V} \cdot \mathbf{V} = \mathbf{L} \cdot \mathbf{L}^T \equiv \mathbf{B}. \tag{3.159}$$

When **L** is associated with coordinate transformations, the matrix **B** is known as the *left Cauchy–Green tensor*, and can be related to the metric tensor. We also see that

$$\mathbf{V} = \sqrt{\mathbf{L} \cdot \mathbf{L}^T}. \tag{3.160}$$

When **L** is symmetric, $\mathbf{Q}_1 = \mathbf{Q}_2$, $\mathbf{Q} = \mathbf{I}$, and $\mathbf{W} = \mathbf{V}$; that is, there is no distinction between the left and right polar decompositions. If $\det \mathbf{L} < 0$, a similar decomposition can be made; however, the matrix **Q** may additionally require reflection, $\det \mathbf{Q} = -1$, and is not a pure rotation.

Example 3.6 Find the right and left polar decompositions of the velocity gradient tensor

$$\mathbf{L} = \begin{pmatrix} \frac{157}{145} & \frac{713}{1740} \\ \frac{24}{145} & \frac{143}{145} \end{pmatrix}. \tag{3.161}$$

Solution

The relevant singular value decomposition has been identified in the earlier example. For the right polar decomposition, we can then form the rotation matrix

$$\mathbf{Q} = \mathbf{Q}_2 \cdot \mathbf{Q}_1^T = \begin{pmatrix} \frac{144}{145} & \frac{17}{145} \\ -\frac{17}{145} & \frac{144}{145} \end{pmatrix}. \tag{3.162}$$

We can also find

$$\mathbf{W} = \mathbf{Q}_1 \cdot \mathbf{\Sigma} \cdot \mathbf{Q}_1^T = \begin{pmatrix} \frac{888}{841} & \frac{245}{841} \\ \frac{245}{841} & \frac{10\,369}{10\,092} \end{pmatrix}. \tag{3.163}$$

The right Cauchy–Green tensor is

$$\mathbf{C} = \mathbf{W} \cdot \mathbf{W} = \mathbf{L}^T \cdot \mathbf{L} = \begin{pmatrix} \frac{1009}{841} & \frac{6125}{10\,092} \\ \frac{6125}{10\,092} & \frac{138\,121}{121\,104} \end{pmatrix}. \tag{3.164}$$

It can be verified that $\mathbf{Q} \cdot \mathbf{W} = \mathbf{L}$. With regard to fluid kinematics, we see that

$$d\mathbf{v} = \mathbf{L}^T \cdot d\mathbf{x} = (\mathbf{Q} \cdot \mathbf{W})^T \cdot d\mathbf{x} = \mathbf{W}^T \cdot \mathbf{Q}^T \cdot d\mathbf{x}. \tag{3.165}$$

So $d\mathbf{x}$ is first rotated by $\mathbf{Q}^T$ then further modulated by $\mathbf{W}^T$.

One can also find

$$\mathbf{V} = \mathbf{Q}_2 \cdot \mathbf{\Sigma} \cdot \mathbf{Q}_2^T = \begin{pmatrix} \frac{337}{300} & \frac{7}{25} \\ \frac{7}{25} & \frac{24}{25} \end{pmatrix}. \tag{3.166}$$

The left Cauchy–Green tensor is

$$\mathbf{B} = \mathbf{V} \cdot \mathbf{V} = \mathbf{L} \cdot \mathbf{L}^T = \begin{pmatrix} \frac{193}{144} & \frac{7}{12} \\ \frac{7}{12} & 1 \end{pmatrix}. \tag{3.167}$$

It can be confirmed that $\mathbf{V} \cdot \mathbf{Q} = \mathbf{L}$.

Other types of matrix decompositions of square matrices could be considered and include those named (a) Jordan, (b) Schur, (c) $\mathbf{L} \cdot \mathbf{U}$, and (d) $\mathbf{Q} \cdot \mathbf{R}$.

3.8.7 Decomposition of a Material Line Element

The decompositions of the previous sections were applied to the transformation of a differential distance vector $d\mathbf{x}$ to a differential velocity vector $d\mathbf{v}$. These vectors have different units, and so cannot be directly compared. To do a more direct comparison of the action of the velocity field on a fluid element, let us instead consider the related transformation of a differential distance vector to a translated, stretched, and rotated distance vector due to a spatially nonuniform velocity field. Let us then consider in detail the configuration shown in Fig. 3.9, an adaptation of Fig. 3.6. Here the fluid labeled P is initially located at position $\mathbf{x}^o$ and has velocity $\mathbf{v}$. Here, the superscript "o" connotes the initial configuration. The fluid labeled P' is nearby and separated by an infinitesimally small vector $d\mathbf{x}^o$. The spatially nonuniform velocity field induces the fluid at P and P' to move slightly differently, so that at an infinitesimally small time increment dt, they have a new displacement, $d\mathbf{x}$. Because we are considering infinitesimal increments in space and time, linear descriptions suffice. The fluid labeled P moves from $\mathbf{x}_P^o$ to $\mathbf{x}_P$ via

$$\mathbf{x}_P = \mathbf{x}_P^o + \mathbf{v}\, dt. \tag{3.168}$$

The fluid labeled P' moves from $\mathbf{x}_{P'}^o$ to $\mathbf{x}_{P'}$ via

$$\mathbf{x}_{P'} = \mathbf{x}_{P'}^o + (\mathbf{v} + d\mathbf{v})\, dt. \tag{3.169}$$

Using Eq. (3.79) and the fact that $\mathbf{x}_{P'}^o = \mathbf{x}_P^o + d\mathbf{x}^o$, we can recast Eq. (3.169) as

$$\mathbf{x}_{P'} = \mathbf{x}_P^o + d\mathbf{x}^o + (\mathbf{v} + \mathbf{L}^T \cdot d\mathbf{x}^o)\, dt. \tag{3.170}$$

Employing Eq. (3.168), we then substitute and rearrange to get

$$\mathbf{x}_{P'} = \mathbf{x}_P + d\mathbf{x}^o + (\mathbf{L}^T \cdot d\mathbf{x}^o)\, dt = \mathbf{x}_P + (\mathbf{I} + dt\, \mathbf{L}^T) \cdot d\mathbf{x}^o. \tag{3.171}$$

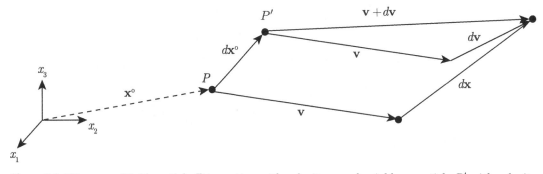

Figure 3.9 Diagram of fluid particle P in motion with velocity $\mathbf{v}$ and neighbor particle P' with velocity $\mathbf{v} + d\mathbf{v}$. The modulation of the connecting differential line segment $d\mathbf{x}^o$ into $d\mathbf{x}$ is depicted.

Realizing that $d\mathbf{x} = \mathbf{x}_{P'} - \mathbf{x}_P$, we then get

$$d\mathbf{x} = (\mathbf{I} + dt\ \mathbf{L}^T) \cdot d\mathbf{x}^o. \tag{3.172}$$

Thus, the initial differential distance vector $d\mathbf{x}^o$ is transformed by the matrix $\mathbf{I} + \Delta t\ \mathbf{L}^T$ to a new differential distance vector $d\mathbf{x}$. Recalling Eq. (3.83), $\mathbf{L} = \mathbf{D} + \mathbf{R}$, we can also say

$$d\mathbf{x} = (\mathbf{I} + dt\ \mathbf{D}) \cdot d\mathbf{x}^o + dt\ \mathbf{R}^T \cdot d\mathbf{x}^o. \tag{3.173}$$

The symmetric $\mathbf{I} + dt\ \mathbf{D}$ can be shown to have eigenvalues $1 + dt$ eig $\mathbf{D}$. Here "eig" is the operator that yields the eigenvalues. Because the eigenvalues of $\mathbf{D}$ are real, so too are those of $\mathbf{I} + dt\ \mathbf{D}$; this portion of the transformation can be considered as an eigen-stretching along the principal axes. The second term, $dt\ \mathbf{R}^T$, induces a rotation of $d\mathbf{x}^o$.

Example 3.7 Examine the time-evolution from $t = 0$ to $t = 0.2$ of the vector connecting fluid particles initially at $(0,0)^T$ and $(0,1)^T$ under the action of the velocity field, $v_1 = (157/145)x_1 + 24/145x_2$, $v_2 = (713/1740)x_1 + (143/145)x_2$. This is the same flow field described in Fig. 3.8, but here we will consider the transformation of the spatial vector into another spatial vector through the action of the velocity field.

Solution
Because we have finite values for increments in x and t, we will replace the infinitesimal d with the finite Δ. Here we have for Δx^o and velocity gradient $\mathbf{L}$

$$\Delta \mathbf{x}^o = \begin{pmatrix} 0 \\ 1 \end{pmatrix}, \quad \mathbf{L} = \begin{pmatrix} \partial_1 v_1 & \partial_1 v_2 \\ \partial_2 v_1 & \partial_2 v_2 \end{pmatrix} = \begin{pmatrix} \frac{157}{145} & \frac{713}{1740} \\ \frac{24}{145} & \frac{143}{145} \end{pmatrix}. \tag{3.174}$$

With this, we can specialize Eq. (3.172) as

$$\Delta \mathbf{x} \approx (\mathbf{I} + \Delta t\ \mathbf{L}^T) \cdot \Delta \mathbf{x}^o, \tag{3.175}$$

$$\begin{pmatrix} \Delta x_1 \\ \Delta x_2 \end{pmatrix} \approx \left(\begin{pmatrix} 1 & 0 \\ 0 & 1 \end{pmatrix} + 0.2 \begin{pmatrix} \frac{157}{145} & \frac{24}{145} \\ \frac{713}{1740} & \frac{143}{145} \end{pmatrix} \right) \begin{pmatrix} 0 \\ 1 \end{pmatrix} = \begin{pmatrix} \frac{24}{725} \\ \frac{868}{725} \end{pmatrix}. \tag{3.176}$$

Formally, because we are considering finite Δt, we can only require an approximation of equality. The action of the velocity vector field on the vector $\Delta \mathbf{x}^o$ at $t = \Delta t$ is shown in Fig. 3.10. The matrix that is taking action on $\Delta \mathbf{x}^o$, which we can call $\mathbf{A}$, is

$$\mathbf{A} = \mathbf{I} + \Delta t\ \mathbf{L}^T = \begin{pmatrix} \frac{882}{725} & \frac{24}{725} \\ \frac{713}{8700} & \frac{868}{725} \end{pmatrix} = \begin{pmatrix} 1.216\,55 & 0.033\,1034 \\ 0.081\,954 & 1.197\,24 \end{pmatrix}. \tag{3.177}$$

The matrix $\mathbf{A}$ is dimensionless and asymmetric. It can be decomposed in a variety of ways. Let us choose the SVD. Doing so, we find

$$\mathbf{A} = \begin{pmatrix} 0.756\,812 & -0.653\,633 \\ 0.653\,633 & 0.756\,812 \end{pmatrix} \begin{pmatrix} 1.265\,48 & 0 \\ 0 & 1.148\,81 \end{pmatrix} \begin{pmatrix} 0.769\,883 & 0.638\,186 \\ -0.638\,186 & 0.769\,883 \end{pmatrix}. \tag{3.178}$$

Considered with this decomposition, the vector is first rotated, then eigen-stretched, then rotated again. One can show that the vector $d\mathbf{x}^o$ has magnitude of unity, while $d\mathbf{x}$ has expanded to magnitude

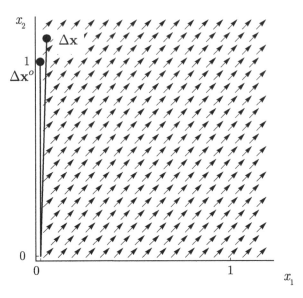

Figure 3.10 Plot of the modulation of the vector $\Delta \mathbf{x}^o$ into $\Delta \mathbf{x}$ under the action of a velocity field.

$4\sqrt{13/145} = 1.1977$. The SVD accounts for all aspects of the velocity field in a combined fashion. Other decompositions allow for other types of interpretations, for instance in terms of rigid body rotation, extensional strain, and shear strain. Such local linear analysis only makes sense for small Δt. For large time, the fluid particles move too far away from their initial configuration for the local linearization to be accurate.

3.9 Convected Derivative

Let us expand the notion introduced in the previous section of tracking the evolution of a material line element. Consider the expression for differential velocity, Eq. (3.78), $d\mathbf{v} = (d\mathbf{x}^T \cdot \nabla)\mathbf{v}$. As $\mathbf{v} = d\mathbf{x}/dt$, we can rewrite Eq. (3.78) and commute the differential operator d with the material derivative d/dt to get

$$d\left(\frac{d}{dt}\mathbf{x}\right) = (d\mathbf{x}^T \cdot \nabla)\mathbf{v}, \qquad \frac{d}{dt}d\mathbf{x} = (d\mathbf{x}^T \cdot \nabla)\mathbf{v}. \tag{3.179}$$

A similar formulation is also briefly introduced by Panton (2013, p. 67, p. 306). We can go on to expand the material derivative to say

$$\frac{\partial}{\partial t}d\mathbf{x} + (\mathbf{v}^T \cdot \nabla)d\mathbf{x} = (d\mathbf{x}^T \cdot \nabla)\mathbf{v}, \qquad \frac{\partial}{\partial t}d\mathbf{x} + (\mathbf{v}^T \cdot \nabla)d\mathbf{x} - (d\mathbf{x}^T \cdot \nabla)\mathbf{v} = \mathbf{0}. \tag{3.180}$$

Guided by Thorne and Blandford (2017, pp. 735–737), we define a new derivative operator, $\mathscr{D}/\mathscr{D}t$, such that

$$\frac{\mathscr{D}}{\mathscr{D}t}d\mathbf{x} \equiv \frac{d}{dt}d\mathbf{x} - (d\mathbf{x}^T \cdot \nabla)\mathbf{v} = \frac{\partial}{\partial t}d\mathbf{x} + (\mathbf{v}^T \cdot \nabla)d\mathbf{x} - (d\mathbf{x}^T \cdot \nabla)\mathbf{v}. \tag{3.181}$$

With our definition of the velocity gradient, Eq. (3.80) $\mathbf{L} = \nabla \mathbf{v}^T$, one can also say

$$\frac{\mathscr{D}}{\mathscr{D}t} d\mathbf{x} \equiv \frac{d}{dt} d\mathbf{x} - \mathbf{L}^T \cdot d\mathbf{x}. \tag{3.182}$$

Consistent with much of the continuum mechanics literature, we call the operator $\mathscr{D}/\mathscr{D}t$ the *convected derivative*; Thorne and Blandford call it the *fluid derivative*. The convected derivative $\mathscr{D}/\mathscr{D}t$ is different than the material derivative d/dt. It renders Eq. (3.78) to take the form

$$\frac{\mathscr{D}}{\mathscr{D}t}(d\mathbf{x}) = \mathbf{0}. \tag{3.183}$$

The combination

$$(\mathbf{v}^T \cdot \nabla) d\mathbf{x} - (d\mathbf{x}^T \cdot \nabla) \mathbf{v} \equiv \mathcal{L}_\mathbf{v} d\mathbf{x} \equiv [\mathbf{v}, d\mathbf{x}], \tag{3.184}$$

is described by Thorne and Blandford as a type of *Lie*[4] *derivative* $\mathcal{L}_\mathbf{v} d\mathbf{x}$, who also describe it as the *commutator* of $\mathbf{v}$ and $d\mathbf{x}$, which they call $[\mathbf{v}, d\mathbf{x}]$. It arises in differential geometry and group theory; see Cantwell (2002, p. 135). See also Marsden and Hughes (1983). A nonzero commutator indicates the described operation does not commute. What constitutes a Lie derivative and commutator may take a variety of forms depending on which branch of the literature one consults and the context of the physical problems considered.

The literature also discusses a variety of related derivatives that involve some form of commutator; the nomenclature is nonuniform. Brodkey (1967, p. 398) has a useful description. He uses, as do we, material derivative so as to refer to the time-variation detected by an observer who is translating, but not rotating or deforming, with a fluid particle. He defines the *Jaumann*[5] *derivative* so as to refer to the time-variation detected by an observer who is translating and rotating, but not deforming, with a fluid particle. Finally he defines, as do we, the *convected derivative* so as to refer to the time-variation detected by an observer who is translating, rotating, and deforming with a fluid particle. Paolucci (2016, p. 128) has a similar discussion. Bird et al. (1987, Vol. 1, pp. 306–312, Chapter 9) and Bird et al. (2007, p. 250) give further analysis of the Jaumann derivative, which they call the *corotational derivative*, and the convected derivative, which they call the *codeformational derivative*. We can finally recast Eq. (3.78) in any of the following forms:

$$\frac{\partial}{\partial t}(d\mathbf{x}) + \mathcal{L}_\mathbf{v}(d\mathbf{x}) = \mathbf{0}, \qquad \frac{\partial}{\partial t}(d\mathbf{x}) + [\mathbf{v}, d\mathbf{x}] = \mathbf{0}, \qquad \frac{\mathscr{D}}{\mathscr{D}t}(d\mathbf{x}) = \mathbf{0}. \tag{3.185}$$

We can use this to reinterpret Fig. 3.6 by considering Fig. 3.11. Two nearby fluid particles, P and Q, are initially separated by distance $d\mathbf{x}$. This vector is composed of a set of points in the line linking P and Q. The velocity field is nonuniform, so P, Q, and points between advect in nonparallel trajectories to P', Q', and points between after a time dt. Taylor[6] series tells us

---

[4] Marius Sophus Lie, 1842–1899, Norwegian mathematician.
[5] Gustav Andreas Johannes Jaumann, 1863–1924, Austrian physicist.
[6] Brook Taylor, 1685–1731, English mathematician and artist, published on capillary action, magnetism, and thermometers, adjudicated the dispute between Newton and Leibniz over priority in developing calculus, contributed to the method of finite differences, invented integration by parts, and has his name ascribed to the Taylor series of which variants were earlier discovered by Gregory, Newton, Leibniz, Johann Bernoulli, and de Moivre.

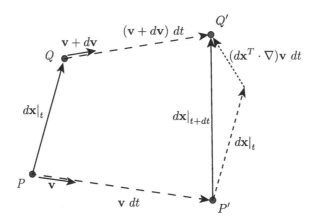

Figure 3.11 Evolution of a vector $d\mathbf{x}$ linking two material points P and Q in a nonuniform velocity field.

$$d\mathbf{x}|_{t+dt} = d\mathbf{x}|_t + \left(\frac{d}{dt}d\mathbf{x}\right)\bigg|_t dt + \cdots = d\mathbf{x}|_t + (d\mathbf{x}^T \cdot \nabla)\mathbf{v}\big|_t\, dt + \cdots . \quad (3.186)$$

This is the vector sum depicted in Fig. 3.11. Thus, the distance relative to the original between P' and Q' is $(d\mathbf{x}^T \cdot \nabla)\mathbf{v}\, dt$. And the material derivative $(d/dt)(d\mathbf{x}) = (d\mathbf{x}^T \cdot \nabla)\mathbf{v}$ gives the relative velocity between P and Q. One can consider $(\mathscr{D}/\mathscr{D}t)(d\mathbf{x}) = \mathbf{0}$ as a property of the convected derivative. Consider then an observer in a frame translating, rotating, and deforming precisely with the entire fluid. As time progresses, that observer will observe no change in the set of points $d\mathbf{x}$ linking P and Q; hence, $(\mathscr{D}/\mathscr{D}t)(d\mathbf{x}) = \mathbf{0}$. Scaling by dt and recognizing $\mathbf{v} = d\mathbf{x}/dt$, we recover what is depicted in Fig. 3.6:

$$\mathbf{v}|_{t+dt} = \mathbf{v}|_t + \left(\frac{d}{dt}d\mathbf{x}\right)\bigg|_t + \cdots = \mathbf{v}|_t + (d\mathbf{x}^T \cdot \nabla)\mathbf{v}\big|_t + \cdots . \quad (3.187)$$

We can also easily reconcile Eqs. (3.172) and (3.186). When evaluated at t, we have the o state, and at $t + dt$ the general state, so Eq. (3.186) becomes

$$d\mathbf{x} = d\mathbf{x}^o + (d\mathbf{x}^{oT} \cdot \nabla)\mathbf{v}\, dt. \quad (3.188)$$

Transpose both sides, recognizing the scalar operator $(d\mathbf{x}^{oT} \cdot \nabla)$ is invariant, to get

$$d\mathbf{x}^T = d\mathbf{x}^{oT} + (d\mathbf{x}^{oT} \cdot \nabla)\mathbf{v}^T\, dt. \quad (3.189)$$

Use the associative property and the definition of the velocity gradient, Eq. (3.80), to get

$$d\mathbf{x}^T = d\mathbf{x}^{oT} + d\mathbf{x}^{oT} \cdot (\nabla \mathbf{v}^T)\, dt = d\mathbf{x}^{oT} + d\mathbf{x}^{oT} \cdot \mathbf{L}\, dt. \quad (3.190)$$

Transpose both sides, and use the distributive property to recover Eq. (3.172):

$$d\mathbf{x} = d\mathbf{x}^o + \mathbf{L}^T \cdot d\mathbf{x}^o\, dt = \left(\mathbf{I} + dt\, \mathbf{L}^T\right) \cdot d\mathbf{x}^o. \quad (3.191)$$

Most of the applications for these special derivatives are for tensors used to model viscoelastic and other non-Newtonian flows. A good discussion in this context is given by Joseph (1990, pp. 7–14) as well as Morrison (2001). For tensors, such as the stress tensor $\mathbf{T}$, Paolucci (2016, p. 128) and many others describe how these various derivatives may be extended. For example, the convected derivative of the stress tensor, using our convention for $\mathbf{L}$, is given as

$$\frac{\mathscr{D}\mathbf{T}}{\mathscr{D}t} = \frac{d\mathbf{T}}{dt} - \mathbf{L}^T \cdot \mathbf{T} - \mathbf{T} \cdot \mathbf{L}. \tag{3.192}$$

Expanding the material derivative and velocity gradient tensor, this can be rewritten, independent of the ordering convention for $\mathbf{L}$, as

$$\frac{\mathscr{D}\mathbf{T}}{\mathscr{D}t} = \frac{\partial \mathbf{T}}{\partial t} + (\mathbf{v}^T \cdot \nabla)\mathbf{T} - (\nabla \mathbf{v}^T)^T \cdot \mathbf{T} - \mathbf{T} \cdot (\nabla \mathbf{v}^T). \tag{3.193}$$

In many sources, this is called the *upper convected derivative*. To provide a flavor of the many other types of derivatives, we provide one more. Using our convention for the rotation tensor $\mathbf{R}$, the Jaumann (i.e. corotational) derivative can be denoted by $\mathfrak{D}/\mathfrak{D}t$ and is given for a tensor, after recognizing $\mathbf{R}^T = -\mathbf{R}$, by

$$\frac{\mathfrak{D}\mathbf{T}}{\mathfrak{D}t} = \frac{d\mathbf{T}}{dt} + \mathbf{R} \cdot \mathbf{T} - \mathbf{T} \cdot \mathbf{R}. \tag{3.194}$$

One will also find a variety of notations for these derivatives. Common notations include

$$\dot{\mathbf{T}} \equiv \frac{d\mathbf{T}}{dt}, \quad \text{material}; \quad \overset{\triangledown}{\mathbf{T}} \equiv \frac{\mathscr{D}\mathbf{T}}{\mathscr{D}t}, \quad \text{convected}; \quad \overset{\circ}{\mathbf{T}} \equiv \frac{\mathfrak{D}\mathbf{T}}{\mathfrak{D}t}, \quad \text{Jaumann}. \tag{3.195}$$

3.10 Expansion Rate

The expansion rate can be determined for both a material volume as well as an arbitrary volume. We briefly consider both here.

3.10.1 Material Volume

Consider a small material volume of fluid, also called a particle of fluid. As introduced in Section 2.4.7, we define a material volume as a volume enclosed by a surface across which there is no flux of mass. We shall later see in Section 4.1, by invoking the mass conservation axiom for a nonrelativistic system, that the implication is that the mass of a material volume is constant, but we need not yet consider this. In general the volume containing this particle can increase or decrease. It is useful to quantify the rate of this increase or decrease. Additionally, this will give a flavor of the analysis to come for the evolution axioms in Chapter 4.

Taking $V_m(t)$ to denote the time-dependent finite material volume, we must have

$$V_m = \int_{V_m(t)} dV. \tag{3.196}$$

Using the Reynolds transport theorem, Eq. (2.241), we take the time derivative of both sides and obtain

$$\frac{dV_m}{dt} = \int_{V_m(t)} \underbrace{\partial_o(1)}_{=0} dV + \int_{A_m(t)} n_i v_i \, dA = \int_{A_m(t)} n_i v_i \, dA, \tag{3.197}$$

$$= \int_{V_m(t)} \partial_i v_i \, dV, \quad \text{by Gauss's theorem, Eq. (2.222),} \tag{3.198}$$

$$= \overline{\partial_i v_i} \, V_m, \quad \text{by the mean value theorem.} \tag{3.199}$$

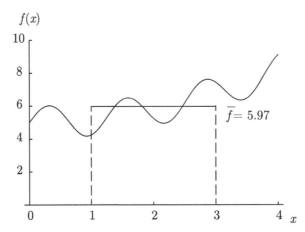

Figure 3.12 Plot illustrating the mean value theorem applied to $f(x) = 5 + x^2/5 + \sin 5x$ within the domain $x \in [1, 3]$. The mean value theorem holds that the mean value of the function is the area under the curve, that is, the integral, scaled by the domain length, here $3 - 1 = 2$. So the mean value theorem gives $\overline{f} = (\int_1^3 (5 + x^2/5 + \sin 5x) \, dx)/(3 - 1)$, reducing to $\overline{f} = (88 + 3(\sin 5)(\sin 10))/15$. So $\overline{f} = 5.97$.

We recall from calculus the mean value theorem, which states that for any integral, a mean value of the integrand can be defined, denoted by an over-bar, as for example $\int_a^b f(x) \, dx = \overline{f}(b-a)$. We give a simple example illustrating the mean value theorem in Fig. 3.12. As we shrink the size of the material volume to zero, the mean value approaches the local value, so we get

$$\frac{1}{V_m} \frac{dV_m}{dt} = \overline{\partial_i v_i}, \tag{3.200}$$

$$\lim_{V_m \to 0} \frac{1}{V_m} \frac{dV_m}{dt} = \partial_i v_i = \nabla^T \cdot \mathbf{v} = \text{div } \mathbf{v} = \text{tr } \mathbf{D} \equiv \vartheta. \tag{3.201}$$

Equation (3.201) describes the *relative expansion rate*, ϑ, also known as the *dilatation rate* of a material fluid particle. A fluid particle for which $\vartheta = \partial_i v_i = 0$ must have a relative expansion rate of zero, and satisfies conditions to be an *incompressible fluid*. The velocity field for an incompressible fluid is solenoidal.

3.10.2 Arbitrary Volume

For a time-dependent arbitrary volume, we have

$$V_a = \int_{V_a(t)} dV. \tag{3.202}$$

Using the general transport theorem, Eq. (2.238), we take the time derivative of both sides and obtain

$$\frac{dV_a}{dt} = \int_{V_a(t)} \underbrace{\partial_o(1)}_{=0} dV + \int_{A_a(t)} n_i w_i \, dA = \int_{A_a(t)} n_i w_i \, dA, \tag{3.203}$$

$$= \int_{V_a(t)} \partial_i w_i \, dV, \quad \text{by Gauss's theorem, Eq. (2.222)}, \tag{3.204}$$

$$= \overline{\partial_i w_i} \, V_a, \quad \text{by the mean value theorem.} \tag{3.205}$$

As we shrink the size of the arbitrary volume to zero, the mean value approaches the local value, so we get

$$\frac{1}{V_a}\frac{dV_a}{dt} = \overline{\partial_i w_i}, \tag{3.206}$$

$$\lim_{V_a \to 0} \frac{1}{V_a}\frac{dV_a}{dt} = \partial_i w_i = \nabla^T \cdot \mathbf{w} = \text{div } \mathbf{w}. \tag{3.207}$$

Equation (3.207) describes the *relative expansion rate* of an arbitrary volume.

Example 3.8 Consider a sphere whose initial radius is $r(0) = r_o$ and whose outer boundary is moving outwards, normal to the surface, with a uniform and constant velocity normal to the surface of magnitude w_o. Find $V(t)$.

Solution

The scenario is depicted in Fig. 3.13. Here the volume is formally no longer arbitrary, as we have specified it, so we now just call it $V(t)$. It is not, however, a material volume. And the bounding surface is $A(t)$. The velocity of the outer boundary is uniform and aligned with the outer surface normal $\mathbf{n}$,

$$\mathbf{w} = w_o \mathbf{n}, \quad \text{thus} \quad \mathbf{w}^T \cdot \mathbf{n} = w_o \underbrace{\mathbf{n}^T \cdot \mathbf{n}}_{=1} = +w_o. \tag{3.208}$$

Thus, Eq. (3.203) reduces to

$$\frac{dV}{dt} = \int_{A(t)} w_o \, dA = w_o \int_{A(t)} dA = w_o A(t). \tag{3.209}$$

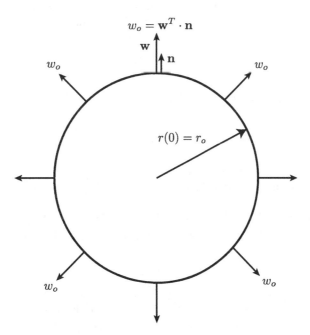

Figure 3.13 Diagram of uniformly expanding sphere.

For the sphere, we have $V = 4\pi r^3/3$ and $A = 4\pi r^2$, so we get

$$\frac{d}{dt}\left(\frac{4}{3}\pi r^3\right) = w_o 4\pi r^2, \quad \text{so} \quad \frac{1}{3}\frac{d}{dt}r^3 = w_o r^2, \quad \text{so} \quad r^2\frac{dr}{dt} = w_o r^2, \quad \frac{dr}{dt} = w_o. \tag{3.210}$$

This is not a surprise and likely could be inferred by intuition. Integrating and applying the initial condition $r(0) = r_o$, we get $r(t) = w_o t + r_o$. Thus,

$$V(t) = \frac{4}{3}\pi\left(w_o t + r_o\right)^3. \tag{3.211}$$

3.11 Invariants of the Strain Rate Tensor

The tensor associated with straining (also called the deformation rate tensor or strain rate tensor) $\partial_{(i}v_{j)}$ is symmetric. Consequently, it has three real eigenvalues, $\lambda_{\dot{\epsilon}}^{(i)}$, and an orientation for which the strain rate is aligned with the eigenvectors. As with stress, there are also three principal invariants of strain rate, the analog to Eqs. (2.138–2.140):

$$I_{\dot{\epsilon}}^{(1)} = \partial_{(i}v_{i)} = \partial_i v_i = \vartheta = \lambda_{\dot{\epsilon}}^{(1)} + \lambda_{\dot{\epsilon}}^{(2)} + \lambda_{\dot{\epsilon}}^{(3)} = \text{tr}\,\mathbf{D}, \tag{3.212}$$

$$I_{\dot{\epsilon}}^{(2)} = \frac{1}{2}(\partial_{(i}v_{i)}\partial_{(j}v_{j)} - \partial_{(i}v_{j)}\partial_{(j}v_{i)}) = \lambda_{\dot{\epsilon}}^{(1)}\lambda_{\dot{\epsilon}}^{(2)} + \lambda_{\dot{\epsilon}}^{(2)}\lambda_{\dot{\epsilon}}^{(3)} + \lambda_{\dot{\epsilon}}^{(3)}\lambda_{\dot{\epsilon}}^{(1)} = (\det \mathbf{D})(\text{tr}\,\mathbf{D}^{-1}), \tag{3.213}$$

$$I_{\dot{\epsilon}}^{(3)} = \epsilon_{ijk}\partial_{(1}v_{i)}\partial_{(2}v_{j)}\partial_{(3}v_{k)} = \lambda_{\dot{\epsilon}}^{(1)}\lambda_{\dot{\epsilon}}^{(2)}\lambda_{\dot{\epsilon}}^{(3)} = \det \mathbf{D}. \tag{3.214}$$

The physical interpretation for $I_{\dot{\epsilon}}^{(1)}$ is obvious in that it is equal to the relative rate of volume change for a material element, $(1/V)dV/dt$. Aris (1962, p. 92) discusses how $I_{\dot{\epsilon}}^{(2)}$ is related to $(1/V)d^2V/dt^2$ and $I_{\dot{\epsilon}}^{(3)}$ is related to $(1/V)d^3V/dt^3$.

3.12 Two-Dimensional Kinematics

Next, consider some important two-dimensional cases.

3.12.1 General Two-Dimensional Flows

For two-dimensional motion, we have the velocity vector as (v_1, v_2), and for the unit tangent to the vector separating two nearby particles (α_1, α_2).

Rotation

Recalling that $dx_i = \alpha_i\,ds$, we have from Eq. (3.91) for rotation

$$dv_j^{(r)} = \partial_{[i}v_{j]}\,dx_i = \alpha_i\partial_{[i}v_{j]}\,ds = \left(\alpha_1\partial_{[1}v_{j]} + \alpha_2\partial_{[2}v_{j]}\right)ds. \tag{3.215}$$

Thus,

$$dv_1^{(r)} = \left(\alpha_1 \underbrace{\partial_{[1}v_{1]}}_{=0} + \alpha_2 \partial_{[2}v_{1]}\right) d\mathsf{s} = \alpha_2 \partial_{[2}v_{1]}\, d\mathsf{s}, \tag{3.216}$$

$$dv_2^{(r)} = \left(\alpha_1 \partial_{[1}v_{2]} + \alpha_2 \underbrace{\partial_{[2}v_{2]}}_{=0}\right) d\mathsf{s} = \alpha_1 \partial_{[1}v_{2]}\, d\mathsf{s}. \tag{3.217}$$

Rewriting in terms of the actual derivatives, we get

$$dv_1^{(r)} = \frac{1}{2}\alpha_2\left(\partial_2 v_1 - \partial_1 v_2\right) d\mathsf{s}, \qquad dv_2^{(r)} = \frac{1}{2}\alpha_1\left(\partial_1 v_2 - \partial_2 v_1\right) d\mathsf{s}. \tag{3.218}$$

Also for the vorticity vector, we get

$$\omega_k = \epsilon_{kij}\partial_i v_j. \tag{3.219}$$

The only nonzero component is ω_3, which comes to

$$\omega_3 = \underbrace{\epsilon_{311}}_{=0}\partial_1 v_1 + \underbrace{\epsilon_{312}}_{=1}\partial_1 v_2 + \underbrace{\epsilon_{321}}_{=-1}\partial_2 v_1 + \underbrace{\epsilon_{322}}_{=0}\partial_2 v_2 = \partial_1 v_2 - \partial_2 v_1. \tag{3.220}$$

For a two-dimensional flow, vorticity is orthogonal to velocity. In general, a vector field that is orthogonal to its own curl so that $\mathbf{v}^T \cdot \nabla \times \mathbf{v} = \mathbf{v}^T \cdot \boldsymbol{\omega} = 0$ is known as a *complex lamellar vector field*, in contrast to a lamellar vector field, that has $\nabla \times \mathbf{v} = \boldsymbol{\omega} = \mathbf{0}$. A two-dimensional velocity field is a complex lamellar vector field, but need not be lamellar.

Extension

For extensional straining we have

$$dv_k^{(es)} = \alpha_k \alpha_i \alpha_j \partial_{(i} v_{j)}\, d\mathsf{s}, \tag{3.221}$$
$$= \alpha_k \left(\alpha_1\alpha_1\partial_{(1}v_{1)} + \alpha_1\alpha_2\partial_{(1}v_{2)} + \alpha_2\alpha_1\partial_{(2}v_{1)} + \alpha_2\alpha_2\partial_{(2}v_{2)}\right) d\mathsf{s}, \tag{3.222}$$
$$= \alpha_k \left(\alpha_1^2\partial_1 v_1 + \alpha_1\alpha_2\left(\partial_1 v_2 + \partial_2 v_1\right) + \alpha_2^2 \partial_2 v_2\right) d\mathsf{s}, \tag{3.223}$$
$$dv_1^{(es)} = \alpha_1 \left(\alpha_1^2\partial_1 v_1 + \alpha_1\alpha_2\left(\partial_1 v_2 + \partial_2 v_1\right) + \alpha_2^2 \partial_2 v_2\right) d\mathsf{s}, \tag{3.224}$$
$$dv_2^{(es)} = \alpha_2 \left(\alpha_1^2\partial_1 v_1 + \alpha_1\alpha_2\left(\partial_1 v_2 + \partial_2 v_1\right) + \alpha_2^2 \partial_2 v_2\right) d\mathsf{s}. \tag{3.225}$$

Shear

For shear straining, we have

$$dv_j^{(ss)} = dv_j^{(s)} - dv_j^{(es)} = \left(\alpha_i \partial_{(i}v_{j)} - \alpha_j \alpha_i \alpha_k \partial_{(i}v_{k)}\right) d\mathsf{s}, \tag{3.226}$$
$$dv_1^{(ss)} = \left(\alpha_1\partial_1 v_1 + \alpha_2\left(\frac{\partial_2 v_1 + \partial_1 v_2}{2}\right) - \alpha_1\left(\alpha_1^2\partial_1 v_1 + \alpha_1\alpha_2\left(\partial_1 v_2 + \partial_2 v_1\right) + \alpha_2^2 \partial_2 v_2\right)\right) d\mathsf{s}, \tag{3.227}$$

$$dv_2^{(ss)} = \left(\alpha_2\partial_2 v_2 + \alpha_1\left(\frac{\partial_1 v_2 + \partial_2 v_1}{2}\right) - \alpha_2\left(\alpha_1^2\partial_1 v_1 + \alpha_1\alpha_2(\partial_1 v_2 + \partial_2 v_1) + \alpha_2^2\partial_2 v_2\right)\right)d\mathbf{s}. \tag{3.228}$$

Expansion

For expansion, we have

$$\frac{1}{V}\frac{dV}{dt} = \partial_1 v_1 + \partial_2 v_2. \tag{3.229}$$

Invariants

Similar to two-dimensional invariants of $\mathbf{T}$ considered in Eqs. (2.143, 2.144), we have

$$I_{\dot{\epsilon}}^{(1)} = \partial_1 v_1 + \partial_2 v_2 = \lambda_{\dot{\epsilon}}^{(1)} + \lambda_{\dot{\epsilon}}^{(2)} = \operatorname{tr}\mathbf{D}, \tag{3.230}$$

$$I_{\dot{\epsilon}}^{(3)} = (\partial_1 v_1)(\partial_2 v_2) - (\partial_2 v_1 + \partial_1 v_2)^2/4 = \lambda_{\dot{\epsilon}}^{(1)}\lambda_{\dot{\epsilon}}^{(2)} = \det\mathbf{D}. \tag{3.231}$$

3.12.2 Relative Motion Along 1 and 2 Axes

Let us consider in detail the configuration shown in Fig. 3.14 in which the particle separation is along the 1 axis. Hence $\alpha_1 = 1$, $\alpha_2 = 0$, and $\alpha_3 = 0$.

Rotation

For rotation, we have

$$dv_1^{(r)} = 0, \qquad dv_2^{(r)} = \frac{1}{2}(\partial_1 v_2 - \partial_2 v_1)\ d\mathbf{s} = \frac{\omega_3}{2}\ d\mathbf{s}. \tag{3.232}$$

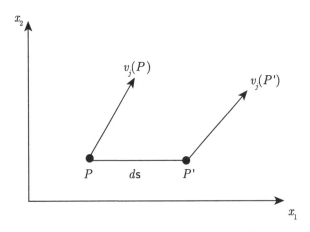

Figure 3.14 Diagram of fluid particle at P and P' in motion.

Extension

For extensional straining, we have

$$dv_1^{(es)} = \partial_1 v_1 \, ds, \qquad dv_2^{(es)} = 0. \tag{3.233}$$

Shear

For shear straining, we have

$$dv_1^{(ss)} = 0, \qquad dv_2^{(ss)} = \frac{1}{2} \left(\partial_1 v_2 + \partial_2 v_1 \right) ds = \partial_{(1} v_{2)} \, ds. \tag{3.234}$$

Expansion

For expansion, we have

$$\frac{1}{V} \frac{dV}{dt} = \partial_1 v_1 + \partial_2 v_2. \tag{3.235}$$

Next, consider in detail the configuration shown in Fig. 3.15 in which the particle separation is aligned with the 2 axis. Hence $\alpha_1 = 0$, $\alpha_2 = 1$, and $\alpha_3 = 0$.

Rotation

For rotation, we have

$$dv_1^{(r)} = \frac{1}{2} \left(\partial_2 v_1 - \partial_1 v_2 \right) ds = -\frac{\omega_3}{2} \, ds, \qquad dv_2^{(r)} = 0. \tag{3.236}$$

Extension

For extensional straining, we have

$$dv_1^{(es)} = 0, \qquad dv_2^{(es)} = \partial_2 v_2 \, ds. \tag{3.237}$$

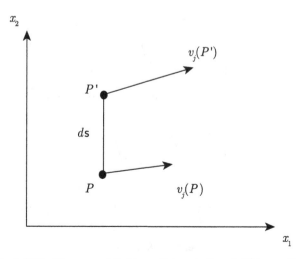

Figure 3.15 Diagram of fluid particles at P and P' in motion.

Shear

For shear straining, we have

$$dv_1^{(ss)} = \frac{1}{2}\left(\partial_2 v_1 + \partial_1 v_2\right) d\mathsf{s} = \partial_{(1} v_{2)} \, d\mathsf{s}, \qquad dv_2^{(ss)} = 0. \tag{3.238}$$

Expansion

For expansion, we have

$$\frac{1}{V}\frac{dV}{dt} = \partial_1 v_1 + \partial_2 v_2. \tag{3.239}$$

3.12.3 Uniform Flow

Consider the kinematics of a uniform two-dimensional flow in which

$$v_1 = k_1, \qquad v_2 = k_2, \tag{3.240}$$

as shown in Fig. 3.16.

- *Streamlines*: $\frac{dx_1}{v_1} = \frac{dx_2}{v_2}$, $\quad \frac{dx_1}{k_1} = \frac{dx_2}{k_2}$, $\quad x_1 = \left(\frac{k_1}{k_2}\right) x_2 + C$.
- *Rotation*: $\omega_3 = \partial_1 v_2 - \partial_2 v_1 = \partial_1(k_1) - \partial_2(k_2) = 0$.
- *Extension*:
  - on 1 axis, $\partial_1 v_1 = 0$,
  - on 2 axis: $\partial_2 v_2 = 0$.
- *Shear for unrotated element*: $\frac{1}{2}\left(\partial_1 v_2 + \partial_2 v_1\right) = 0$.
- *Expansion*: $\partial_1 v_1 + \partial_2 v_2 = 0$.
- *Acceleration*:

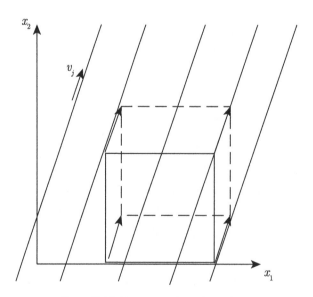

Figure 3.16 Diagram of uniform flow.

$$\frac{dv_1}{dt} = \partial_o v_1 + v_1 \partial_1 v_1 + v_2 \partial_2 v_1 = 0 + k_1 \partial_1(k_1) + k_2 \partial_2(k_1) = 0,$$
$$\frac{dv_2}{dt} = \partial_o v_2 + v_1 \partial_1 v_2 + v_2 \partial_2 v_2 = 0 + k_1 \partial_1(k_2) + k_2 \partial_2(k_2) = 0.$$

For this simple flow, the streamlines are straight lines, there is no rotation, no extension, no shear, no expansion, and no acceleration.

3.12.4 Pure Rigid Body Rotation

Consider the kinematics of a two-dimensional flow in which

$$v_1 = -kx_2, \qquad v_2 = kx_1, \tag{3.241}$$

as depicted in Fig. 3.17.

- *Streamlines*: $\frac{dx_1}{v_1} = \frac{dx_2}{v_2}$, $\quad \frac{dx_1}{-kx_2} = \frac{dx_2}{kx_1}$, $\quad x_1\,dx_1 = -x_2\,dx_2$, $\quad x_1^2 + x_2^2 = C$.
- *Rotation*: $\omega_3 = \partial_1 v_2 - \partial_2 v_1 = \partial_1(kx_1) - \partial_2(-kx_2) = 2k$.
- *Extension*:
  - on 1 axis: $\partial_1 v_1 = 0$,
  - on 2 axis: $\partial_2 v_2 = 0$.
- *Shear for unrotated element*: $\frac{1}{2}\left(\partial_1(kx_1) + \partial_2(-kx_2)\right) = k - k = 0$.
- *Expansion*: $\partial_1 v_1 + \partial_2 v_2 = 0 + 0 = 0$.
- *Acceleration*:
$$\frac{dv_1}{dt} = \partial_o v_1 + v_1 \partial_1 v_1 + v_2 \partial_2 v_1 = 0 - kx_2 \partial_1(-kx_2) + kx_1 \partial_2(-kx_2) = -k^2 x_1,$$
$$\frac{dv_2}{dt} = \partial_o v_2 + v_1 \partial_1 v_2 + v_2 \partial_2 v_2 = 0 - kx_2 \partial_1(kx_1) + kx_1 \partial_2(kx_1) = -k^2 x_2.$$

In this flow, the velocity magnitude grows linearly with distance from the origin. This is precisely how a rotating rigid body behaves. The streamlines are circles. The rotation is positive for positive k, hence counterclockwise. There is no deformation in extension or shear, and there is no expansion. The acceleration is pointed towards the origin.

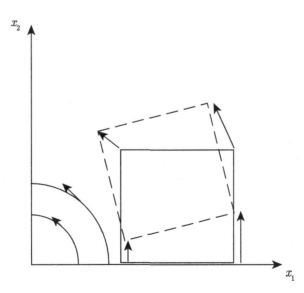

Figure 3.17 Diagram of pure rigid body rotation.

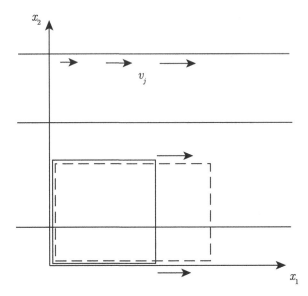

Figure 3.18 Diagram of extensional flow, a one-dimensional compressible flow.

3.12.5 Pure Extensional Motion (a Compressible Flow)

Consider the kinematics of a two-dimensional flow in which

$$v_1 = kx_1, \qquad v_2 = 0, \tag{3.242}$$

as presented in Fig. 3.18.

- *Streamlines*: $\frac{dx_1}{v_1} = \frac{dx_2}{v_2}$, $\quad v_2\, dx_1 = v_1\, dx_2$, $\quad 0 = kx_1\, dx_2$, $\quad x_2 = C$.
- *Rotation*: $\omega_3 = \partial_1 v_2 - \partial_2 v_1 = \partial_1(0) - \partial_2(kx_1) = 0$.
- *Extension*:
  - *on 1 axis*: $\partial_1 v_1 = k$,
  - *on 2 axis*: $\partial_2 v_2 = 0$.
- *Shear for unrotated element*: $\frac{1}{2}(\partial_1 v_2 + \partial_2 v_1) = \frac{1}{2}(\partial_1(0) + \partial_2(kx_1)) = 0$.
- *Expansion*: $\partial_1 v_1 + \partial_2 v_2 = k$.
- *Acceleration*:
  $\frac{dv_1}{dt} = \partial_o v_1 + v_1 \partial_1 v_1 + v_2 \partial_2 v_1 = 0 + kx_1 \partial_1(kx_1) + 0\partial_2(kx_1) = k^2 x_1$,
  $\frac{dv_2}{dt} = \partial_o v_2 + v_1 \partial_1 v_2 + v_2 \partial_2 v_2 = 0 + kx_1 \partial_1(0) + 0\partial_2(0) = 0$.

In this flow, the streamlines are straight lines; there is no fluid rotation; there is extension (stretching) deformation along the 1 axis, but no shear deformation along this axis. The relative expansion rate is positive for positive k, indicating a compressible flow. The acceleration is confined to the x_1 direction.

3.12.6 Pure Shear Straining

Consider the kinematics of a two-dimensional flow in which

$$v_1 = kx_2, \qquad v_2 = kx_1, \tag{3.243}$$

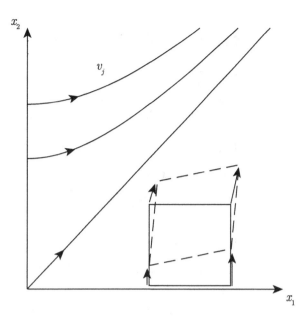

Figure 3.19 Diagram of pure shearing flow.

as given in Fig. 3.19.

- *Streamlines*: $\frac{dx_1}{v_1} = \frac{dx_2}{v_2}$, $\quad \frac{dx_1}{kx_2} = \frac{dx_2}{kx_1}$, $\quad x_1\, dx_1 = x_2\, dx_2$, $\quad x_1^2 = x_2^2 + C$.
- *Rotation*: $\omega_3 = \partial_1 v_2 - \partial_2 v_1 = \partial_1(kx_1) - \partial_2(kx_2) = k - k = 0$.
- *Extension*:
  - *on 1 axis*: $\partial_1 v_1 = \partial_1(kx_2) = 0$,
  - *on 2 axis*: $\partial_2 v_2 = \partial_2(kx_1) = 0$.
- *Shear for unrotated element*: $\frac{1}{2}(\partial_1 v_2 + \partial_2 v_1) = \frac{1}{2}(\partial_1(kx_1) + \partial_2(kx_2)) = k$.
- *Expansion*: $\partial_1 v_1 + \partial_2 v_2 = 0$.
- *Acceleration*:
  $\frac{dv_1}{dt} = \partial_o v_1 + v_1 \partial_1 v_1 + v_2 \partial_2 v_1 = 0 + kx_2 \partial_1(kx_2) + kx_1 \partial_2(kx_2) = k^2 x_1$,
  $\frac{dv_2}{dt} = \partial_o v_2 + v_1 \partial_1 v_2 + v_2 \partial_2 v_2 = 0 + kx_2 \partial_1(kx_1) + kx_1 \partial_2(kx_1) = k^2 x_2$.

In this flow, the streamlines are hyperbolas; there is no rotation or axial extension along the coordinate axes; there is positive shear deformation for an element aligned with the coordinate axes, and no expansion. So, the pure shear deformation preserves volume. The fluid is accelerating away from the origin.

3.12.7 Ideal Corner Flow

Consider the kinematics of a two-dimensional flow in which

$$v_1 = kx_1, \qquad v_2 = -kx_2, \qquad (3.244)$$

as depicted in Fig. 3.20.

- *Streamlines*: $\frac{dx_1}{v_1} = \frac{dx_2}{v_2}$, $\quad \frac{dx_1}{kx_1} = \frac{dx_2}{-kx_2}$, $\quad \ln x_1 = -\ln x_2 + C'$, $\quad x_1 x_2 = C$.
- *Rotation*: $\omega_3 = \partial_1 v_2 - \partial_2 v_1 = \partial_1(-kx_2) - \partial_2(kx_1) = 0$.

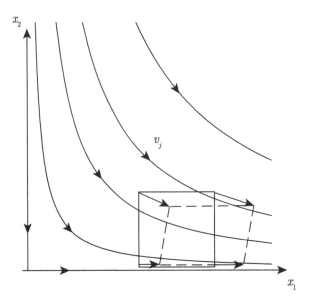

Figure 3.20 Diagram of an ideal corner flow.

- *Extension*:
  - on 1 axis: $\partial_1 v_1 = \partial_1(kx_1) = k$,
  - on 2 axis: $\partial_2 v_2 = \partial_2(-kx_2) = -k$.
- *Shear for unrotated element*: $\frac{1}{2}(\partial_1 v_2 + \partial_2 v_1) = \frac{1}{2}(\partial_1(-kx_2) + \partial_2(kx_1)) = 0$.
- *Expansion*: $\partial_1 v_1 + \partial_2 v_2 = k - k = 0$.
- *Acceleration*:
  $\frac{dv_1}{dt} = \partial_o v_1 + v_1 \partial_1 v_1 + v_2 \partial_2 v_1 = 0 + kx_1 \partial_1(kx_1) - kx_2 \partial_2(kx_1) = k^2 x_1$,
  $\frac{dv_2}{dt} = \partial_o v_2 + v_1 \partial_1 v_2 + v_2 \partial_2 v_2 = 0 + kx_1 \partial_1(-kx_2) - kx_2 \partial_2(-kx_2) = k^2 x_2$.

In this flow, the streamlines are hyperbolas; there is no rotation or shear along the coordinate axes; there is extensional strain for an element aligned with the coordinate axes, but no net expansion. So, the ideal corner flow preserves volume. The fluid is accelerating away from the origin.

3.12.8 Couette Flow: Shear + Rotation

Consider the kinematics of a two-dimensional flow in which

$$v_1 = kx_2, \qquad v_2 = 0, \tag{3.245}$$

as given in Fig. 3.21. This is known as a Couette[7] flow.

- *Streamlines*: $\frac{dx_1}{v_1} = \frac{dx_2}{v_2}$, $\quad \frac{dx_1}{kx_2} = \frac{dx_2}{0}$, $\quad 0 = kx_2\, dx_2$, $\quad x_2 = C$.
- *Rotation*: $\omega_3 = \partial_1 v_2 - \partial_2 v_1 = \partial_1(0) - \partial_2(kx_2) = -k$.
- *Extension*:
  - on 1 axis: $\partial_1 v_1 = \partial_1(kx_2) = 0$,
  - on 2 axis: $\partial_2 v_2 = \partial_2(0) = 0$.

---

[7] Maurice Marie Alfred Couette, 1858–1943, French fluid mechanician, rheologist, and teacher.

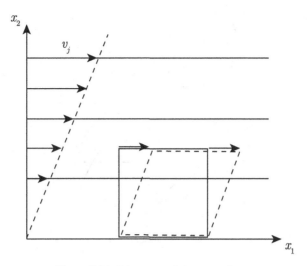

Figure 3.21 Diagram of Couette flow.

- *Shear for unrotated element*: $\frac{1}{2}(\partial_1 v_2 + \partial_2 v_1) = \frac{1}{2}(\partial_1(0) + \partial_2(kx_2)) = \frac{k}{2}$.
- *Expansion*: $\partial_1 v_1 + \partial_2 v_2 = 0$.
- *Acceleration*:
$\frac{dv_1}{dt} = \partial_o v_1 + v_1 \partial_1 v_1 + v_2 \partial_2 v_1 = 0 + kx_2 \partial_1(kx_2) + 0\partial_2(kx_2) = 0,$
$\frac{dv_2}{dt} = \partial_o v_2 + v_1 \partial_1 v_2 + v_2 \partial_2 v_2 = 0 + kx_2 \partial_1(0) + 0\partial_2(0) = 0.$

Here the streamlines are straight lines, and the flow is rotational! The rotation is clockwise because $\omega < 0$ for $k > 0$. The constant volume rotation is combined with a constant volume shear deformation for the element aligned with the coordinate axes. The fluid is not accelerating.

3.12.9 Ideal Irrotational Vortex: Extension + Shear

Consider the kinematics of a two-dimensional flow of Fig. 3.22.

$$v_1 = -k\frac{x_2}{x_1^2 + x_2^2}, \qquad v_2 = k\frac{x_1}{x_1^2 + x_2^2}. \tag{3.246}$$

- *Streamlines*: $\frac{dx_1}{v_1} = \frac{dx_2}{v_2}$, $\frac{dx_1}{-k\frac{x_2}{x_1^2+x_2^2}} = \frac{dx_2}{k\frac{x_1}{x_1^2+x_2^2}}$, $-\frac{dx_1}{x_2} = \frac{dx_2}{x_1}$, $x_1^2 + x_2^2 = C$.
- *Rotation*: $\omega_3 = \partial_1 v_2 - \partial_2 v_1 = \partial_1\left(k\frac{x_1}{x_1^2+x_2^2}\right) - \partial_2\left(-k\frac{x_2}{x_1^2+x_2^2}\right) = 0$.
- *Extension*:
  - on 1 axis: $\partial_1 v_1 = \partial_1\left(-k\frac{x_2}{x_1^2+x_2^2}\right) = 2k\frac{x_1 x_2}{(x_1^2+x_2^2)^2}$,
  - on 2 axis: $\partial_2 v_2 = \partial_2\left(k\frac{x_1}{x_1^2+x_2^2}\right) = -2k\frac{x_1 x_2}{(x_1^2+x_2^2)^2}$.
- *Shear for unrotated element*: $\frac{1}{2}(\partial_1 v_2 + \partial_2 v_1) = k\frac{x_2^2 - x_1^2}{(x_1^2+x_2^2)^2}$.
- *Expansion*: $\partial_1 v_1 + \partial_2 v_2 = 0$.
- *Acceleration*:
$\frac{dv_1}{dt} = \partial_o v_1 + v_1 \partial_1 v_1 + v_2 \partial_2 v_1 = -\frac{k^2 x_1}{(x_1^2+x_2^2)^2},$
$\frac{dv_2}{dt} = \partial_o v_2 + v_1 \partial_1 v_2 + v_2 \partial_2 v_2 = -\frac{k^2 x_2}{(x_1^2+x_2^2)^2}.$

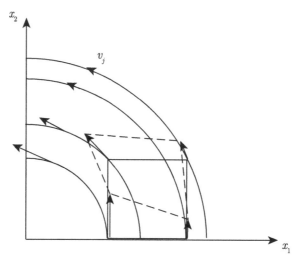

Figure 3.22 Diagram of ideal irrotational vortex.

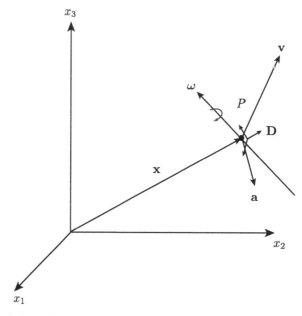

Figure 3.23 Diagram of three-dimensional kinematics features: location **x**, velocity **v**, acceleration **a**, vorticity $\boldsymbol{\omega}$, and deformation **D**.

The streamlines are circles, and the fluid element does not rotate about its own axis! It does rotate about the origin. It deforms by extension and shear in such a way that overall the volume is constant.

3.13 Three-Dimensional Kinematics: Summary

We graphically summarize three-dimensional kinematics in Fig. 3.23. Here we depict the motion of a fluid particle P, which has Eulerian position vector **x**, velocity vector **v**, and acceleration vector **a**. The particle P is also shown undergoing a rigid-body rotation about an axis aligned with the vorticity vector $\boldsymbol{\omega}$. It also undergoes an extensional deformation aligned with the principal axes of the deformation tensor **D**. As the position **x** is varied continuously, all field quantities, **v**, **a**, $\boldsymbol{\omega}$, and **D**, vary continuously as well.

3.14 Kinematics as a Dynamical System

Let us close this chapter by applying some standard notions from dynamical systems theory to fluid kinematics. In this section, we draw upon work presented previously by Powers (2016, Section 6.1), Powers and Sen (2015, Section 9.6), and Powers et al. (2015). See also Paolucci (2016, Section 3.2.6), Perry and Chong (1987), and Samelson and Wiggins (2006). Such notions have motivated a rich literature in so-called *chaotic advection* which through describing the tangled paths of motion of fluids allows one to better understand the fundamental phenomena of *mixing*; see Aref et al. (2017) for a detailed summary. While mixing is most often considered in limits where fluid inertial effects dominate viscous forces (see Chapter 14 on turbulence), mixing also occurs in the opposite limit in which fluid inertial effects are small relative to viscous forces; it is this limit that is most often associated with chaotic advection. We will defer detailed discussion of one important example of chaotic mixing to Section 13.4.4 and here simply focus on casting kinematics of fluid motion as a dynamical system.

Let us imagine that we are given a time-independent flow field, for which the fluid velocity is a function of position only. Then the motion of an individual fluid particle is governed by the following autonomous system of nonlinear ordinary differential equations:

$$\frac{d\mathbf{x}}{dt} = \mathbf{v}(\mathbf{x}(t)), \qquad \mathbf{x}(0) = \mathbf{x}^o. \tag{3.247}$$

Here, the initial position of the fluid particle is given by the constant vector $\mathbf{x}^o$. In the neighborhood of $\mathbf{x}^o$, this has a local linear behavior via Taylor series expansion in the neighborhood of $\mathbf{x}^o$, closely related to Eq. (3.172):

$$\frac{d\mathbf{x}}{dt} = \frac{d}{dt}(\mathbf{x} - \mathbf{x}^o) = \mathbf{v}(\mathbf{x}^o) + \mathbf{L}^T\big|_{\mathbf{x}^o} \cdot (\mathbf{x} - \mathbf{x}^o) + \cdots. \tag{3.248}$$

The solution of Eq. (3.247) can be expressed in general form

$$\mathbf{x} = \mathbf{x}(t; \mathbf{x}^o), \tag{3.249}$$

a function of time parameterized by the initial condition of the fluid particle. Such a solution is certainly a pathline, streamline, and streakline. It is also known as a trajectory in the dynamical systems literature.

Now, from the definition of the total differential for time-independent flow, see Eq. (3.79), we have

$$d\mathbf{v} = \underbrace{(\nabla\mathbf{v}^T)^T}_{\mathbf{L}^T} \cdot d\mathbf{x} = \mathbf{L}^T \cdot d\mathbf{x}. \tag{3.250}$$

Scaling by dt and employing Eq. (3.247) gives the acceleration vector $\mathbf{a}$ as

$$\mathbf{a} = \frac{d\mathbf{v}}{dt} = \mathbf{L}^T \cdot \frac{d\mathbf{x}}{dt} = \mathbf{L}^T \cdot \mathbf{v}. \qquad (3.251)$$

We will use this relation to aid in understanding particle acceleration in the following problem that considers the kinematics of a nontrivial velocity field. Consider the following nonlinear autonomous system, which could describe the steady three-dimensional kinematics of a fluid particle:

$$\frac{dx_1}{dt} = v_1(x_1, x_2, x_3) = 3 - x_1 - 2x_2, \qquad x_1^o = 0, \qquad (3.252)$$

$$\frac{dx_2}{dt} = v_2(x_1, x_2, x_3) = -3 + 3x_1 + x_2 - x_1 x_2, \qquad x_2^o = 0, \qquad (3.253)$$

$$\frac{dx_3}{dt} = v_3(x_1, x_2, x_3) = 1 - x_3, \qquad x_3^o = 0. \qquad (3.254)$$

Let us find its equilibria, the linear stability of the equilibria, the nonlinear fluid particle trajectory, and a representation of the fluid acceleration vector field. Because the forcing function for dx_2/dt includes $-x_1 x_2$, the system is nonlinear. When considering dynamical systems, one should always consider equilibrium points. Such points exist when $v_1 = v_2 = v_3 = 0$; in fluid mechanics, they are known as *stagnation points*. They are found by solving the nonlinear algebraic problem:

$$0 = v_1(x_1, x_2, x_3) = 3 - x_1 - 2x_2, \qquad (3.255)$$

$$0 = v_2(x_1, x_2, x_3) = -3 + 3x_1 + x_2 - x_1 x_2, \qquad (3.256)$$

$$0 = v_3(x_1, x_2, x_3) = 1 - x_3. \qquad (3.257)$$

For a nonlinear system of algebraic equations, there is no guarantee of a unique root. Algebraic analysis of this system reveals two roots:

$$\text{root 1: } x_{1o} = 1, \qquad x_{2o} = 1, \qquad x_{3o} = 1, \qquad (3.258)$$

$$\text{root 2: } x_{1o} = -3, \qquad x_{2o} = 3, \qquad x_{3o} = 1. \qquad (3.259)$$

The "o" as a subscript denotes a stagnation condition, in contrast to "o" as a superscript, which denotes an initial condition. Direct substitution into the equations for velocity confirm that both are stagnation points. Flow in the neighborhood of a stagnation point can be understood by considering the locally linear behavior. We take the initial condition to be very close to the stagnation condition $\mathbf{x}^o \sim \mathbf{x}_o$. Taylor series of the velocity in the neighborhood of an arbitrary stagnation point, Eq. (3.248) with $\mathbf{v}(\mathbf{x}^o) = \mathbf{0}$, reveals the local kinematics are given by the locally linear system

$$\underbrace{\begin{pmatrix} \frac{d}{dt}(x_1 - x_{1o}) \\ \frac{d}{dt}(x_2 - x_{2o}) \\ \frac{d}{dt}(x_3 - x_{3o}) \end{pmatrix}}_{d/dt(\mathbf{x}-\mathbf{x}_o)} = \underbrace{\begin{pmatrix} \frac{\partial v_1}{\partial x_1} & \frac{\partial v_1}{\partial x_2} & \frac{\partial v_1}{\partial x_3} \\ \frac{\partial v_2}{\partial x_1} & \frac{\partial v_2}{\partial x_2} & \frac{\partial v_2}{\partial x_3} \\ \frac{\partial v_3}{\partial x_1} & \frac{\partial v_3}{\partial x_2} & \frac{\partial v_3}{\partial x_3} \end{pmatrix}}_{=\mathbf{L}^T}\bigg|_{x_{io}} \underbrace{\begin{pmatrix} x_1 - x_{1o} \\ x_2 - x_{2o} \\ x_3 - x_{3o} \end{pmatrix}}_{\mathbf{x}-\mathbf{x}_o}, \qquad (3.260)$$

$$= \underbrace{\begin{pmatrix} -1 & -2 & 0 \\ 3 - x_2 & 1 - x_1 & 0 \\ 0 & 0 & -1 \end{pmatrix}}_{=\mathbf{L}^T}\bigg|_{x_{io}} \underbrace{\begin{pmatrix} x_1 - x_{1o} \\ x_2 - x_{2o} \\ x_3 - x_{3o} \end{pmatrix}}_{\mathbf{x}-\mathbf{x}_o}. \qquad (3.261)$$

Note that the local velocity gradient depends on the local position for this nonlinear system. Let us evaluate this in the neighborhood of root 1, $(x_{1o}, x_{2o}, x_{3o})^T = (1,1,1)^T$:

$$\begin{pmatrix} \frac{d}{dt}(x_1 - 1) \\ \frac{d}{dt}(x_2 - 1) \\ \frac{d}{dt}(x_3 - 1) \end{pmatrix} = \underbrace{\begin{pmatrix} -1 & -2 & 0 \\ 2 & 0 & 0 \\ 0 & 0 & -1 \end{pmatrix}}_{=\mathbf{L}^T} \begin{pmatrix} x_1 - 1 \\ x_2 - 1 \\ x_3 - 1 \end{pmatrix}. \qquad (3.262)$$

As discussed in standard texts on applied mathematics – for example Powers and Sen (2015, Section 9.6) – the local dynamics in the neighborhood of the stagnation point are dictated by the eigenvalues of the coefficient matrix, $\mathbf{L}^T$. Those are evaluated as $\lambda = (-1 + i\sqrt{15})/2$, $(-1 - i\sqrt{15})/2$, and -1. Because the real part of each eigenvalue is negative, root 1 is stable. Because two of the eigenvalues have imaginary components, there is a local oscillatory behavior. This stagnation point is a so-called *stable spiral node*. Had all eigenvalues been real and positive, it would have been an unstable *source node*. Had all eigenvalues been real and negative, it would have been a stable *sink node*. Had the eigenvalues been real, with some positive and others negative, it would have been an unstable *saddle node*. For root 2, one performs the same exercise and finds the eigenvalues are $\lambda = 4, -1, -1$. Because of the presence of a positive real eigenvalue, this is an unstable saddle point, and one would not expect it to be reached from a general initial condition.

Numerical solution of this nonlinear system of ordinary differential equations for initial condition $x_1(0) = x_2(0) = x_3(0) = 0$ yields $x_1(t), x_2(t), x_3(t)$, which for this time-independent velocity field induces the time-dependent particle pathlines plotted in Fig. 3.24. Had the initial condition been chosen to be near the unstable saddle at $(-3, 3, 1)$, we would have generated what is known as a *heteroclinic trajectory*. Instead, we have generated a generic trajectory. It is seen that the trajectory reaches the stable stagnation point at $(1,1,1)$, and that it approaches it in a spiral.

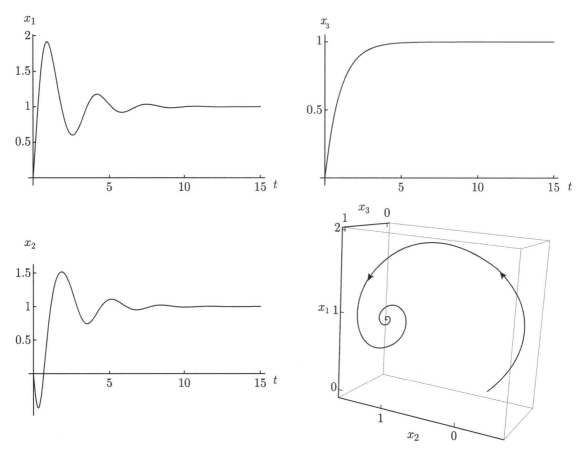

Figure 3.24 Plot of $x_1(t)$, $x_2(t)$, $x_3(t)$, along with the coincident pathline (which for the steady flow is also a streamline and streakline), for a steady three-dimensional fluid particle that commences at the origin.

We can use Eq. (3.251) to calculate the acceleration vector field:

$$\begin{pmatrix} \frac{dv_1}{dt} \\ \frac{dv_2}{dt} \\ \frac{dv_3}{dt} \end{pmatrix} = \begin{pmatrix} \frac{\partial v_1}{\partial x_1} & \frac{\partial v_1}{\partial x_2} & \frac{\partial v_1}{\partial x_3} \\ \frac{\partial v_2}{\partial x_1} & \frac{\partial v_2}{\partial x_2} & \frac{\partial v_2}{\partial x_3} \\ \frac{\partial v_3}{\partial x_1} & \frac{\partial v_3}{\partial x_2} & \frac{\partial v_3}{\partial x_3} \end{pmatrix} \begin{pmatrix} \frac{dx_1}{dt} \\ \frac{dx_2}{dt} \\ \frac{dx_3}{dt} \end{pmatrix}, \quad (3.263)$$

$$= \begin{pmatrix} -1 & -2 & 0 \\ 3-x_2 & 1-x_1 & 0 \\ 0 & 0 & -1 \end{pmatrix} \begin{pmatrix} 3 - x_1 - 2x_2 \\ -3 + 3x_1 + x_2 - x_1 x_2 \\ 1 - x_3 \end{pmatrix}, \quad (3.264)$$

$$= \begin{pmatrix} 3 - 5x_1 + 2x_1 x_2 \\ 6 + 3x_1 - 3x_1^2 - 8x_2 - x_1 x_2 + x_1^2 x_2 + 2x_2^2 \\ -1 + x_3 \end{pmatrix}. \quad (3.265)$$

If we know the kinematics of a fluid particle, we know everything about its motion, including its acceleration. We shall soon, in Section 4.2, discuss things like Newton's second law of motion that relates acceleration to forces. If we know the acceleration, it is possible to induce what the force was that generated it by simply multiplying the acceleration by the mass. Rarely is this the case, however. It is more common to know something about the forces and to use this to deduce what the motion is. We could also apply the complete mathematical theory of dynamical systems to understand the system better.

Near the equilibrium at $(1, 1, 1)^T$, we can decompose the velocity gradient $\mathbf{L}$ into symmetric ($\mathbf{D}$) and anti-symmetric ($\mathbf{R}$) parts:

$$\underbrace{\begin{pmatrix} -1 & 2 & 0 \\ -2 & 0 & 0 \\ 0 & 0 & -1 \end{pmatrix}}_{\mathbf{L}} = \underbrace{\begin{pmatrix} -1 & 0 & 0 \\ 0 & 0 & 0 \\ 0 & 0 & -1 \end{pmatrix}}_{\mathbf{D}} + \underbrace{\begin{pmatrix} 0 & 2 & 0 \\ -2 & 0 & 0 \\ 0 & 0 & 0 \end{pmatrix}}_{\mathbf{R}}. \quad (3.266)$$

The vorticity vector is found by $\omega_k = \epsilon_{kij} R_{ij}$. So $\omega_1 = \epsilon_{123} R_{23} + \epsilon_{132} R_{32} = R_{23} - R_{32} = 0$, $\omega_2 = \epsilon_{213} R_{13} + \epsilon_{231} R_{31} = -R_{13} + R_{31} = 0$, $\omega_3 = \epsilon_{312} R_{12} + \epsilon_{321} R_{21} = R_{12} - R_{21} = 4$. So the fluid has rotation as it approaches the equilibrium:

$$\omega = \begin{pmatrix} 0 \\ 0 \\ 4 \end{pmatrix}.$$

The eigenvalues of $\mathbf{D}$ near the equilibrium are

$$\lambda^{(1)} = -1, \quad \lambda^{(2)} = -1, \quad \lambda^{(3)} = 0.$$

Because there are no positive eigenvalues of $\mathbf{D}$ and one zero eigenvalue, there exists an eigen-direction for which the extensional strain rate is neutral and two other eigen-directions for which the extensional straining is towards the equilibrium. However, the complete dynamics are described by $\mathbf{L}$, not $\mathbf{D}$, and the rotation is sufficiently strong to counteract the neutrality in one of the directions. Because $\operatorname{tr} \mathbf{D} = \operatorname{tr} \mathbf{L} = -2$, there is an overall tendency for volumes near the origin to shrink. Now the quadric of extensional strain is written as $\mathcal{D} = \boldsymbol{\alpha}^T \cdot \mathbf{D} \cdot \boldsymbol{\alpha}$, which is

$$\mathcal{D} = -\alpha_1^2 - \alpha_3^2. \quad (3.267)$$

We also need to enforce

$$\alpha_1^2 + \alpha_2^2 + \alpha_3^2 = 1. \quad (3.268)$$

Optimization reveals a maximum extensional strain rate condition of

$$\mathcal{D} = 0, \quad \text{for} \quad \alpha_1 = 0, \quad \alpha_2 = \pm 1, \quad \alpha_3 = 0. \quad (3.269)$$

Similarly, a minimal extensional strain rate is found of

$$\mathcal{D} = -1, \quad \text{for} \quad \alpha_2 = 0, \quad (3.270)$$

for any set of α_1 and α_3 that are on the unit circle $\alpha_1^2 + \alpha_3^2 = 1$. For $\mathcal{D} = 0$, the quadric is a set of two parallel planes. For $\mathcal{D} = -1$, the quadric is a cylinder of unit radius. Both are plotted in Fig. 3.25. This case has many degeneracies, reflected in the fact the geometries of

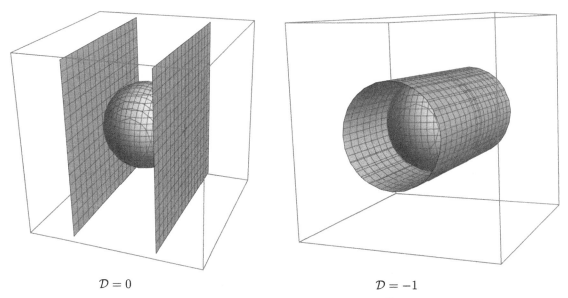

$\mathcal{D} = 0$ $\mathcal{D} = -1$

Figure 3.25 Plot of quadrics of extensional strain rate in $(\alpha_1, \alpha_2, \alpha_3)^T$ space for its extreme values for a three-dimensional velocity field near an equilibrium point. Also plotted is the unit sphere.

the quadrics are simple. For more general deformation tensors $\mathbf{D}$, more complicated geometries of the quadrics will arise.

Now, we seek to analyze a particular pathline. Note that the velocity vector is tangent to the fluid particle trajectory. Let us study a unit vector that happens to be tangent to the velocity field:

$$\boldsymbol{\alpha}_t = \frac{\mathbf{v}}{||\mathbf{v}||}. \tag{3.271}$$

Next, use the quotient rule to examine how the unit tangent vector evolves with time:

$$\frac{d\boldsymbol{\alpha}_t}{dt} = \frac{1}{||\mathbf{v}||}\frac{d\mathbf{v}}{dt} - \frac{\mathbf{v}}{||\mathbf{v}||^2}\frac{d||\mathbf{v}||}{dt}. \tag{3.272}$$

We can scale Eq. (3.251) by $||\mathbf{v}||$ to get $(1/||\mathbf{v}||)d\mathbf{v}/dt = \mathbf{L}^T \cdot \mathbf{v}/||\mathbf{v}|| = \mathbf{L}^T \cdot \boldsymbol{\alpha}_t$. Thus, Eq. (3.272) can be rewritten as

$$\frac{d\boldsymbol{\alpha}_t}{dt} = \mathbf{L}^T \cdot \boldsymbol{\alpha}_t - \frac{\mathbf{v}}{||\mathbf{v}||^2}\frac{d||\mathbf{v}||}{dt} = \mathbf{L}^T \cdot \boldsymbol{\alpha}_t - \boldsymbol{\alpha}_t \frac{1}{||\mathbf{v}||}\frac{d||\mathbf{v}||}{dt}. \tag{3.273}$$

Next consider the following series of operations starting with Eq. (3.251):

$$\mathbf{a} = \frac{d\mathbf{v}}{dt} = \mathbf{L}^T \cdot \mathbf{v}, \quad \text{so} \quad \mathbf{v}^T \cdot \frac{d\mathbf{v}}{dt} = \mathbf{v}^T \cdot \mathbf{L}^T \cdot \mathbf{v}, \tag{3.274}$$

$$\frac{d}{dt}\left(\frac{\mathbf{v}^T \cdot \mathbf{v}}{2}\right) = \mathbf{v}^T \cdot \mathbf{L}^T \cdot \mathbf{v}, \quad \text{so} \quad \frac{d}{dt}\left(\frac{||\mathbf{v}||^2}{2}\right) = \mathbf{v}^T \cdot \mathbf{L}^T \cdot \mathbf{v}, \tag{3.275}$$

3.14 Kinematics as a Dynamical System

$$||\mathbf{v}||\frac{d}{dt}(||\mathbf{v}||) = \mathbf{v}^T \cdot \mathbf{L}^T \cdot \mathbf{v}, \tag{3.276}$$

$$\frac{1}{||\mathbf{v}||}\frac{d}{dt}(||\mathbf{v}||) = \frac{\mathbf{v}^T}{||\mathbf{v}||} \cdot \mathbf{L}^T \cdot \frac{\mathbf{v}}{||\mathbf{v}||} = \boldsymbol{\alpha}_t^T \cdot \mathbf{L}^T \cdot \boldsymbol{\alpha}_t. \tag{3.277}$$

Now substitute Eq. (3.277) into Eq. (3.273) to get

$$\frac{d\boldsymbol{\alpha}_t}{dt} = \mathbf{L}^T \cdot \boldsymbol{\alpha}_t - \left(\boldsymbol{\alpha}_t^T \cdot \mathbf{L}^T \cdot \boldsymbol{\alpha}_t\right)\boldsymbol{\alpha}_t. \tag{3.278}$$

As an aside, take the dot product of Eq. (3.278) with $\boldsymbol{\alpha}_t$ to get

$$\boldsymbol{\alpha}_t^T \cdot \frac{d\boldsymbol{\alpha}_t}{dt} = \boldsymbol{\alpha}_t^T \cdot \mathbf{L}^T \cdot \boldsymbol{\alpha}_t - \left(\boldsymbol{\alpha}_t^T \cdot \mathbf{L}^T \cdot \boldsymbol{\alpha}_t\right)\underbrace{\boldsymbol{\alpha}_t^T \cdot \boldsymbol{\alpha}_t}_{=1}, \tag{3.279}$$

$$= \boldsymbol{\alpha}_t^T \cdot \mathbf{L}^T \cdot \boldsymbol{\alpha}_t - \boldsymbol{\alpha}_t^T \cdot \mathbf{L}^T \cdot \boldsymbol{\alpha}_t = 0. \tag{3.280}$$

This must be an identity, because $\boldsymbol{\alpha}_t^T \cdot \boldsymbol{\alpha}_t = 1$, and its time derivative gives $\boldsymbol{\alpha}_t^T \cdot d\boldsymbol{\alpha}_t/dt = 0$. Now recalling Eq. (3.83), and employing $\boldsymbol{\alpha}_t^T \cdot \mathbf{R}^T \cdot \boldsymbol{\alpha}_t = 0$, because of the anti-symmetry of $\mathbf{R}$, and $\mathbf{D}^T = \mathbf{D}$, because of the symmetry of $\mathbf{D}$, Eq. (3.278) can be rewritten as

$$\frac{d\boldsymbol{\alpha}_t}{dt} = \mathbf{L}^T \cdot \boldsymbol{\alpha}_t - \left(\boldsymbol{\alpha}_t^T \cdot \mathbf{D} \cdot \boldsymbol{\alpha}_t\right)\boldsymbol{\alpha}_t. \tag{3.281}$$

Let us consider how a volume stretches in a direction aligned with the velocity vector. We first specialize the general differential arc length to that found along the particle path: $d\mathbf{s} = ds$. Now, recall from geometry that the square of the differential arc length must be

$$ds^2 = d\mathbf{x}^T \cdot d\mathbf{x}, \tag{3.282}$$

where $d\mathbf{x}$ is also confined to the particle path. Consider now how this quantity changes with time when we move with the particle:

$$\frac{d}{dt}(ds)^2 = \frac{d}{dt}\left(d\mathbf{x}^T \cdot d\mathbf{x}\right) = d\mathbf{x}^T \cdot \frac{d}{dt}(d\mathbf{x}) + \left(\frac{d}{dt}(d\mathbf{x})\right)^T \cdot d\mathbf{x}, \tag{3.283}$$

$$= d\mathbf{x}^T \cdot d\left(\frac{d\mathbf{x}}{dt}\right) + \left(d\left(\frac{d\mathbf{x}}{dt}\right)\right)^T \cdot d\mathbf{x}, \tag{3.284}$$

$$= d\mathbf{x}^T \cdot d\mathbf{v} + d\mathbf{v}^T \cdot d\mathbf{x} = 2\, d\mathbf{x}^T \cdot d\mathbf{v} = 2\, d\mathbf{x}^T \cdot \mathbf{L}^T \cdot d\mathbf{x}, \tag{3.285}$$

$$2\, ds \frac{d}{dt}(ds) = 2\, d\mathbf{x}^T \cdot \mathbf{L}^T \cdot d\mathbf{x}, \tag{3.286}$$

$$\frac{1}{ds}\frac{d}{dt}(ds) = \frac{d\mathbf{x}^T}{ds} \cdot \mathbf{L}^T \cdot \frac{d\mathbf{x}}{ds}. \tag{3.287}$$

Recall now that

$$\boldsymbol{\alpha}_t = \frac{\mathbf{v}}{||\mathbf{v}||} = \frac{\frac{d\mathbf{x}}{dt}}{\frac{ds}{dt}} = \frac{d\mathbf{x}}{ds}. \tag{3.288}$$

So, Eq. (3.287) can be rewritten as

$$\frac{1}{ds}\frac{d}{dt}(ds) = \boldsymbol{\alpha}_t^T \cdot \mathbf{L}^T \cdot \boldsymbol{\alpha}_t, \tag{3.289}$$

$$\frac{d}{dt}(\ln ds) = \boldsymbol{\alpha}_t^T \cdot (\mathbf{D} + \mathbf{R})^T \cdot \boldsymbol{\alpha}_t = \boldsymbol{\alpha}_t^T \cdot \mathbf{D} \cdot \boldsymbol{\alpha}_t = \mathbf{D} : \boldsymbol{\alpha}_t \boldsymbol{\alpha}_t^T \equiv \mathcal{D}_t. \tag{3.290}$$

This relative tangential stretching rate is related to the result of Eq. (3.103) for extensional strain rate. Specializing Eq. (3.103) for a particle pathline, we can say

$$d\mathbf{v}^{(es)} = (\boldsymbol{\alpha}_t^T \cdot \mathbf{D} \cdot \boldsymbol{\alpha}_t)\boldsymbol{\alpha}_t \, ds, \tag{3.291}$$

$$\frac{d\mathbf{v}^{(es)}}{ds} = (\boldsymbol{\alpha}_t^T \cdot \mathbf{D} \cdot \boldsymbol{\alpha}_t)\boldsymbol{\alpha}_t, \tag{3.292}$$

$$\boldsymbol{\alpha}_t^T \cdot \frac{d\mathbf{v}^{(es)}}{ds} = (\boldsymbol{\alpha}_t^T \cdot \mathbf{D} \cdot \boldsymbol{\alpha}_t) \underbrace{\boldsymbol{\alpha}_t^T \cdot \boldsymbol{\alpha}_t}_{=1} = \boldsymbol{\alpha}_t^T \cdot \mathbf{D} \cdot \boldsymbol{\alpha}_t = \frac{1}{ds}\frac{d}{dt}(ds), \tag{3.293}$$

$$= \boldsymbol{\alpha}_t^T \cdot \mathbf{D} \cdot \boldsymbol{\alpha}_t = \frac{1}{ds}d\left(\frac{ds}{dt}\right) = \frac{d\|\mathbf{v}\|}{ds} = \mathbf{D}:\boldsymbol{\alpha}_t\boldsymbol{\alpha}_t^T = \mathcal{D}_t. \tag{3.294}$$

Here, we invoked Eq. (3.290) to obtain Eq. (3.294). The quantity $\boldsymbol{\alpha}_t^T \cdot \mathbf{D} \cdot \boldsymbol{\alpha}_t = \mathbf{D}:\boldsymbol{\alpha}_t\boldsymbol{\alpha}_t^T = \mathcal{D}_t$ is a measure of how the magnitude of the velocity changes with respect to arc length along the particle path.

We can gain further insight into how velocity magnitude changes by a diagonal decomposition of $\mathbf{D} = \mathbf{Q} \cdot \boldsymbol{\Lambda} \cdot \mathbf{Q}^T$. Here $\mathbf{Q}$ is an orthogonal rotation matrix with the normalized eigenvectors of $\mathbf{D}$ in its columns, and $\boldsymbol{\Lambda}$ is the diagonal matrix with the eigenvalues of $\mathbf{D}$ in its diagonal. Thus,

$$\frac{d\|\mathbf{v}\|}{ds} = \boldsymbol{\alpha}_t^T \cdot \underbrace{\mathbf{Q} \cdot \boldsymbol{\Lambda} \cdot \mathbf{Q}^T}_{\mathbf{D}} \cdot \boldsymbol{\alpha}_t = (\mathbf{Q}^T \cdot \boldsymbol{\alpha}_t)^T \cdot \boldsymbol{\Lambda} \cdot (\mathbf{Q}^T \cdot \boldsymbol{\alpha}_t). \tag{3.295}$$

The operation $\mathbf{Q}^T \cdot \boldsymbol{\alpha}_t \equiv \boldsymbol{\alpha}_s$ generates a new rotated unit vector $\boldsymbol{\alpha}_s = (\alpha_{s1}, \alpha_{s2}, \alpha_{s3})^T$. Thus, we can state

$$\frac{d\|\mathbf{v}\|}{ds} = \alpha_{s1}^2 \lambda^{(1)} + \alpha_{s2}^2 \lambda^{(2)} + \alpha_{s3}^2 \lambda^{(3)}, \qquad 1 = \alpha_{s1}^2 + \alpha_{s2}^2 + \alpha_{s3}^2. \tag{3.296}$$

The rate of change of the velocity magnitude along a particle pathline can be understood to be a weighted average of the eigenvalues of the deformation tensor $\mathbf{D}$. In the special case in which $\boldsymbol{\alpha}_t$ is the ith eigenvector of $\mathbf{D}$, we simply get $d\|\mathbf{v}\|/ds = \lambda^{(i)}$. Here $\lambda^{(i)}$ is the corresponding eigenvalue.

If we extend Eq. (3.201) to differential material volumes, we could say the relative expansion rate is

$$\frac{1}{dV}\frac{d}{dt}(dV) = \frac{d}{dt}(\ln dV) = \operatorname{tr}\mathbf{D} = \vartheta. \tag{3.297}$$

Now our differential volume can be formed by $dV = dA\, ds$, where dA is the cross-sectional area normal to the flow direction. Thus,

$$\ln dV = \ln dA + \ln ds, \quad \text{so} \quad \ln dA = \ln dV - \ln ds, \tag{3.298}$$

$$\frac{d}{dt}(\ln dA) = \frac{d}{dt}(\ln dV) - \frac{d}{dt}(\ln ds). \tag{3.299}$$

Substitute from Eqs. (3.290, 3.297) to get the relative rate of change of the differential area normal to the flow direction:

$$\frac{d}{dt}(\ln dA) = \operatorname{tr}\mathbf{D} - \boldsymbol{\alpha}_t^T \cdot \mathbf{D} \cdot \boldsymbol{\alpha}_t. \tag{3.300}$$

This relation, while not identical, is similar to the expression for shear strain rate, Eq. (3.109). We can also use Eq. (2.102) to rewrite Eq. (3.300) as

$$\frac{d}{dt}(\ln dA) = \mathbf{D} : \mathbf{I} - \mathbf{D} : \boldsymbol{\alpha}_t \boldsymbol{\alpha}_t^T = \mathbf{D} : \left(\mathbf{I} - \boldsymbol{\alpha}_t \boldsymbol{\alpha}_t^T\right). \tag{3.301}$$

Now the matrix $\mathbf{P} = \mathbf{I} - \boldsymbol{\alpha}_t \boldsymbol{\alpha}_t^T$ has some surprising properties. It is what is known as a *projection matrix*. One finds $\mathbf{P} \cdot \mathbf{P} = \mathbf{P}$, which renders it *idempotent*. It projects vectors $\mathbf{v}$ onto a plane yielding $\mathbf{v}_p = \mathbf{P} \cdot \mathbf{v}$. The vector $\boldsymbol{\alpha}_t$ is normal to the two vectors that span the plane of projection. One finds $\mathbf{P} \cdot \mathbf{v}_p = \mathbf{v}_p$. The matrix $\mathbf{P}$ is singular and has rank two. Because it is symmetric, it has a set of three orthogonal eigenvectors that can be normalized to form an orthonormal set. Its three eigenvalues are 1, 1, and 0. Thus, it has $\|\mathbf{P}\| = 1$. The eigenvector associated with the zero eigenvalue can be selected as $\boldsymbol{\alpha}_t$, the unit tangent to the curve. Thus, the other two eigenvectors can be thought of as unit normals to the curve, which we label $\boldsymbol{\alpha}_{n1}$ and $\boldsymbol{\alpha}_{n2}$. These eigenvectors are not unique; however, a set can always be found. We can summarize the decomposition:

$$\mathbf{P} = \mathbf{I} - \boldsymbol{\alpha}_t \boldsymbol{\alpha}_t^T = \mathbf{Q} \cdot \boldsymbol{\Lambda} \cdot \mathbf{Q}^T, \tag{3.302}$$

$$= \begin{pmatrix} \vdots & \vdots & \vdots \\ \boldsymbol{\alpha}_{n1} & \boldsymbol{\alpha}_{n2} & \boldsymbol{\alpha}_t \\ \vdots & \vdots & \vdots \end{pmatrix} \begin{pmatrix} 1 & 0 & 0 \\ 0 & 1 & 0 \\ 0 & 0 & 0 \end{pmatrix} \begin{pmatrix} \cdots & \boldsymbol{\alpha}_{n1}^T & \cdots \\ \cdots & \boldsymbol{\alpha}_{n2}^T & \cdots \\ \cdots & \boldsymbol{\alpha}_t^T & \cdots \end{pmatrix}, \tag{3.303}$$

$$= \boldsymbol{\alpha}_{n1} \boldsymbol{\alpha}_{n1}^T + \boldsymbol{\alpha}_{n2} \boldsymbol{\alpha}_{n2}^T. \tag{3.304}$$

As an aside, note that

$$\mathbf{I} = \boldsymbol{\alpha}_t \boldsymbol{\alpha}_t^T + \boldsymbol{\alpha}_{n1} \boldsymbol{\alpha}_{n1}^T + \boldsymbol{\alpha}_{n2} \boldsymbol{\alpha}_{n2}^T. \tag{3.305}$$

The two unit normals are orthogonal to each other, $\boldsymbol{\alpha}_{n1}^T \cdot \boldsymbol{\alpha}_{n2} = 0$. Thus, we have

$$\frac{d}{dt}(\ln dA) = \mathbf{D} : \left(\boldsymbol{\alpha}_{n1} \boldsymbol{\alpha}_{n1}^T + \boldsymbol{\alpha}_{n2} \boldsymbol{\alpha}_{n2}^T\right) = \boldsymbol{\alpha}_{n1}^T \cdot \mathbf{D} \cdot \boldsymbol{\alpha}_{n1} + \boldsymbol{\alpha}_{n2}^T \cdot \mathbf{D} \cdot \boldsymbol{\alpha}_{n2}. \tag{3.306}$$

Comparing to Eq. (3.290) that has one mode associated with $\boldsymbol{\alpha}_t$ available for stretching of the one-dimensional arc length in the streamwise direction, there are two modes associated with $\boldsymbol{\alpha}_{n1}, \boldsymbol{\alpha}_{n2}$ available for stretching the two-dimensional area.

The form $\boldsymbol{\alpha}_{n1}^T \cdot \mathbf{D} \cdot \boldsymbol{\alpha}_{n1}$ suggests it determines the relative normal stretching rate in the direction of $\boldsymbol{\alpha}_{n1}$; a similar rate exists for the other normal direction. One might imagine that there exists a normal direction that yields extreme values for relative normal stretching rates. It can be shown this achieved by the following. First, define a rectangular matrix, $\widehat{\mathbf{Q}}$, whose columns are populated by $\boldsymbol{\alpha}_{n1}$ and $\boldsymbol{\alpha}_{n2}$:

$$\widehat{\mathbf{Q}} = \begin{pmatrix} \vdots & \vdots \\ \boldsymbol{\alpha}_{n1} & \boldsymbol{\alpha}_{n2} \\ \vdots & \vdots \end{pmatrix}. \tag{3.307}$$

Then project the 3×3 matrix $\mathbf{D}$ onto this basis to form the 2×2 matrix $\widehat{\mathbf{D}}$ associated with stretching in the directions normal to the motion:

$$\widehat{\mathbf{D}} = \widehat{\mathbf{Q}}^T \cdot \mathbf{D} \cdot \widehat{\mathbf{Q}}. \tag{3.308}$$

The eigenvalues of $\widehat{\mathbf{D}}$ give the maximum and minimum values of the relative normal stretching rates, and the eigenvectors give the associated directions of extremal normal stretching.

Looked at another way and motivated by standard results from differential geometry, we can make special choices, $\boldsymbol{\alpha}_{n1} = \boldsymbol{\alpha}_{np}$, $\boldsymbol{\alpha}_{n2} = \boldsymbol{\alpha}_{nb}$. Here $\boldsymbol{\alpha}_{np}$ is the so-called "principal normal unit vector" and $\boldsymbol{\alpha}_{nb}$ is the so-called "binormal unit vector." The following results are described in more detail in many sources, for example Powers and Sen (2015, p. 89). One can obtain the binormal unit vector via

$$\boldsymbol{\alpha}_{nb} = \boldsymbol{\alpha}_t \times \boldsymbol{\alpha}_{np}. \tag{3.309}$$

After effort, we find the *Frenet–Serret*[8] *relations*:

$$\frac{d\boldsymbol{\alpha}_t}{ds} = \kappa \boldsymbol{\alpha}_{np}, \qquad \frac{d\boldsymbol{\alpha}_{np}}{ds} = -\kappa \boldsymbol{\alpha}_t + \tau \boldsymbol{\alpha}_{nb}, \qquad \frac{d\boldsymbol{\alpha}_{nb}}{ds} = -\tau \boldsymbol{\alpha}_{np}. \tag{3.310}$$

Here κ and τ are the *curvature* and *torsion*, respectively, of the curve. One can with more effort show that they are given by

$$\kappa = \frac{\sqrt{\left\|\frac{d^2\mathbf{x}}{dt^2}\right\|^2 \left\|\frac{d\mathbf{x}}{dt}\right\|^2 - \left(\frac{d\mathbf{x}}{dt}^T \cdot \frac{d^2\mathbf{x}}{dt^2}\right)^2}}{\left\|\frac{d\mathbf{x}}{dt}\right\|^3} = \frac{\left\|\frac{d\mathbf{x}}{dt} \times \frac{d^2\mathbf{x}}{dt^2}\right\|}{\left\|\frac{d\mathbf{x}}{dt}\right\|^3}, \tag{3.311}$$

$$\tau = \frac{\left(\frac{d\mathbf{x}}{dt} \times \frac{d^2\mathbf{x}}{dt^2}\right)^T \cdot \frac{d^3\mathbf{x}}{dt^3}}{\left\|\frac{d^2\mathbf{x}}{dt^2}\right\|^2 \left\|\frac{d\mathbf{x}}{dt}\right\|^2 - \left(\frac{d\mathbf{x}}{dt}^T \cdot \frac{d^2\mathbf{x}}{dt^2}\right)^2} = \frac{\det\left(\frac{d\mathbf{x}}{dt}, \frac{d^2\mathbf{x}}{dt^2}, \frac{d^3\mathbf{x}}{dt^3}\right)}{\left\|\frac{d\mathbf{x}}{dt} \times \frac{d^2\mathbf{x}}{dt^2}\right\|^2}. \tag{3.312}$$

We see $\kappa \geq 0$, while $\tau \in (-\infty, \infty)$. Powers and Sen use the opposite sign convention for τ; that used here is more common. Note κ and τ are expressed here as functions of time. This is the case for a particle moving along a path in time. But just as the intrinsic curvature of a mountain road is independent of the speed of the vehicle on the road, despite the traveling vehicle experiencing a time-dependency of curvature, the curvature and torsion can be considered more fundamentally to be functions of position only, given that the velocity field is known as a function of position. Analysis reveals that

$$\kappa = \frac{\sqrt{\left(\mathbf{v}^T \cdot \mathbf{L} \cdot \mathbf{L}^T \cdot \mathbf{v}\right)\left(\mathbf{v}^T \cdot \mathbf{v}\right) - \left(\mathbf{v}^T \cdot \mathbf{L}^T \cdot \mathbf{v}\right)^2}}{(\mathbf{v}^T \cdot \mathbf{v})^{3/2}}, \tag{3.313}$$

$$= \frac{1}{\|\mathbf{v}\|} \sqrt{\boldsymbol{\alpha}_t^T \cdot \mathbf{L} \cdot \mathbf{L}^T \cdot \boldsymbol{\alpha}_t - \left(\boldsymbol{\alpha}_t^T \cdot \mathbf{D} \cdot \boldsymbol{\alpha}_t\right)^2} = \frac{1}{\|\mathbf{v}\|} \sqrt{\left(\frac{\|\mathbf{a}\|}{\|\mathbf{v}\|}\right)^2 - \mathcal{D}_t^2}. \tag{3.314}$$

One could also develop an expression for torsion that is explicitly dependent on position. The expression is complicated and requires the use of third-order tensors to capture the higher-order spatial variations. We can also use this intrinsic orthonormal basis to get

---

[8] Jean Frédéric Frenet, 1816–1900, and Joseph Alfred Serret, 1819–1885, French mathematicians.

$$\frac{d}{dt}(\ln dA) = \mathbf{D} : \left(\boldsymbol{\alpha}_{np}\boldsymbol{\alpha}_{np}^T + \boldsymbol{\alpha}_{nb}\boldsymbol{\alpha}_{nb}^T\right) = \boldsymbol{\alpha}_{np}^T \cdot \mathbf{D} \cdot \boldsymbol{\alpha}_{np} + \boldsymbol{\alpha}_{nb}^T \cdot \mathbf{D} \cdot \boldsymbol{\alpha}_{nb}. \qquad (3.315)$$

To illustrate how kinematics illuminates the general field of nonlinear dynamical systems, consider the problem first presented by Mengers (2012), and later by Powers (2016, Section 6.2.2), Powers and Sen (2015, Section 9.6.3), and Powers et al. (2015). Take

$$\frac{dx_1}{dt} = v_1(x_1, x_2, x_3) = \frac{1}{20}(1 - x_1^2), \qquad (3.316)$$

$$\frac{dx_2}{dt} = v_2(x_1, x_2, x_3) = -2x_2 - \frac{35}{16}x_3 + 2(1 - x_1^2)x_3, \qquad (3.317)$$

$$\frac{dx_3}{dt} = v_3(x_1, x_2, x_3) = x_2 + x_3, \qquad (3.318)$$

and identify heteroclinic trajectories and their attractiveness. We can consider this system to be one of prescribed fluid motion that can be analyzed. There are only two finite equilibria for this system, a saddle with one unstable mode at $(-1, 0, 0)^T$ and a sink at $(1, 0, 0)$. Because the first equation is uncoupled from the second two and is sufficiently simple, it can be integrated exactly to form $x_1 = \tanh(t/20)$. This, coupled with $x_2 = 0$ and $x_3 = 0$, satisfies all differential equations and connects the equilibria, so the x_1 axis for $x_1 \in [-1, 1]$ is a heteroclinic trajectory. One then asks if nearby trajectories are attracted to it. This can be answered by a local geometry-based analysis. Our system is of the form $d\mathbf{x}/dt = \mathbf{v}(\mathbf{x})$. Let us consider its behavior in the neighborhood of a generic initial point $\mathbf{x}^o$ that is on the heteroclinic trajectory, but is far from equilibrium. We then locally linearize our system, similar to Eqs. (3.172, 3.248), as

$$\frac{d}{dt}(\mathbf{x} - \mathbf{x}^o) = \underbrace{\mathbf{v}(\mathbf{x}^o)}_{\text{translation}} + \underbrace{\mathbf{L}^T\big|_{\mathbf{x}^o} \cdot (\mathbf{x} - \mathbf{x}^o)}_{\text{deformation+rotation}} + \cdots, \qquad (3.319)$$

$$= \underbrace{\mathbf{v}(\mathbf{x}^o)}_{\text{translation}} + \underbrace{\mathbf{D}\big|_{\mathbf{x}^o} \cdot (\mathbf{x} - \mathbf{x}^o)}_{\text{deformation}} + \underbrace{\mathbf{R}^T\big|_{\mathbf{x}^o} \cdot (\mathbf{x} - \mathbf{x}^o)}_{\text{rotation}} + \cdots. \qquad (3.320)$$

Here, we have employed the local velocity gradient $\mathbf{L}$ as well as its symmetric ($\mathbf{D}$) and anti-symmetric ($\mathbf{R}$) parts:

$$\mathbf{L} = \partial_i v_j = \mathbf{D} + \mathbf{R}, \qquad \mathbf{D} = \frac{\mathbf{L} + \mathbf{L}^T}{2}, \qquad \mathbf{R} = \frac{\mathbf{L} - \mathbf{L}^T}{2}. \qquad (3.321)$$

The symmetry of $\mathbf{D}$ allows definition of a real orthonormal basis. For this three-dimensional system, the dual vector $\boldsymbol{\Omega}$ of the anti-symmetric $\mathbf{R}$ defines the axis of rotation, and its magnitude Ω describes the rotation rate. Now the relative volumetric stretching rate is given by $\operatorname{tr}\mathbf{L} = \operatorname{tr}\mathbf{D} = \operatorname{div}\mathbf{v} = \vartheta$. And one can recall Eq. (3.117), which shows that the relative linear stretching rate $\mathcal{D}$ associated with any direction with unit vector $\boldsymbol{\alpha}$ is $\mathcal{D} = \boldsymbol{\alpha}^T \cdot \mathbf{D} \cdot \boldsymbol{\alpha}$.

For our system, we have

$$\mathbf{L}^T = \begin{pmatrix} -\frac{x_1}{10} & 0 & 0 \\ -4x_1 x_3 & -2 & -\frac{35}{16} + 2(1 - x_1^2) \\ 0 & 1 & 1 \end{pmatrix}. \qquad (3.322)$$

We see that the relative volumetric expansion rate is

$$\vartheta = \operatorname{tr} \mathbf{L} = -1 - \frac{x_1}{10}. \tag{3.323}$$

Because on the heteroclinic trajectory $x_1 \in [-1, 1]$, we always have a locally shrinking volume on that trajectory. By inspection, the unit tangent vector to the heteroclinic trajectory is $\boldsymbol{\alpha}_t = (1, 0, 0)^T$. So the relative tangential stretching rate on the heteroclinic trajectory is

$$\mathcal{D}_t = \boldsymbol{\alpha}_t^T \cdot \mathbf{L} \cdot \boldsymbol{\alpha}_t = \boldsymbol{\alpha}_t^T \cdot \mathbf{D} \cdot \boldsymbol{\alpha}_t = -\frac{x_1}{10}. \tag{3.324}$$

So near the saddle we have $\mathcal{D}_t = 1/10$, and near the sink we have $\mathcal{D}_t = -1/10$. Now we are concerned with stretching in directions normal to the heteroclinic trajectory. Certainly two unit normal vectors are $\boldsymbol{\alpha}_{n1} = (0, 1, 0)^T$ and $\boldsymbol{\alpha}_{n2} = (0, 0, 1)^T$. But there are also infinitely many other unit normals. A detailed optimization calculation reveals, however, that if we (1) form the 3×2 matrix $\mathbf{Q}_n$ with $\boldsymbol{\alpha}_{n1}$ and $\boldsymbol{\alpha}_{n2}$ in its columns:

$$\mathbf{Q}_n = \begin{pmatrix} 0 & 0 \\ 1 & 0 \\ 0 & 1 \end{pmatrix}, \tag{3.325}$$

and (2) form the 2×2 matrices $\mathbf{D}_n$ and $\mathbf{R}_n$ associated with the plane normal to the heteroclinic trajectory

$$\mathbf{D}_n = \mathbf{Q}_n^T \cdot \mathbf{D} \cdot \mathbf{Q}_n, \qquad \mathbf{R}_n = \mathbf{Q}_n^T \cdot \mathbf{R} \cdot \mathbf{Q}_n, \tag{3.326}$$

that (a) the eigenvalues of $\mathbf{D}_n$ give the extreme values of the relative normal stretching rates $\mathcal{D}_{n1}$ and $\mathcal{D}_{n2}$, and the normalized eigenvectors give the associated directions for extreme normal stretching and (b) the magnitude of extremal rotation in the hyperplane normal to $\boldsymbol{\alpha}_t$ is given by $\Omega = \|\mathbf{R}_n\|$. On the heteroclinic trajectory, we find

$$\mathbf{D} = \begin{pmatrix} -\frac{x_1}{10} & 0 & 0 \\ 0 & -2 & -\frac{19}{32} + 1 - x_1^2 \\ 0 & -\frac{19}{32} + 1 - x_1^2 & 1 \end{pmatrix}. \tag{3.327}$$

The reduced deformation tensor associated with motion in the normal plane is

$$\mathbf{D}_n = \mathbf{Q}_n^T \cdot \mathbf{D} \cdot \mathbf{Q}_n = \begin{pmatrix} -2 & -\frac{19}{32} + 1 - x_1^2 \\ -\frac{19}{32} + 1 - x_1^2 & 1 \end{pmatrix}. \tag{3.328}$$

Its eigenvalues give the extremal relative normal stretching rates that are

$$\mathcal{D}_{n,1,2} = -\frac{1}{2} \pm \frac{\sqrt{2473 - 832 x_1^2 + 1024 x_1^4}}{32}. \tag{3.329}$$

For $x_1 \in [-1, 1]$, we have $\mathcal{D}_{n,1} \approx 1$ and $\mathcal{D}_{n,2} \approx -2$. *Because of the presence of a positive relative normal stretching rate, one cannot guarantee trajectories are attracted to the heteroclinic trajectory, even though volume of nearby points is shrinking.* Positive normal stretching does not guarantee divergence from the heteroclinic trajectory; it permits it. Rotation can orient a collection of nearby points into regions where there is either positive or negative normal stretching. There are two possibilities for the heteroclinic trajectory to be attracting: either (1) all normal stretching rates are negative, or (2) the rotation rate is sufficiently fast and the

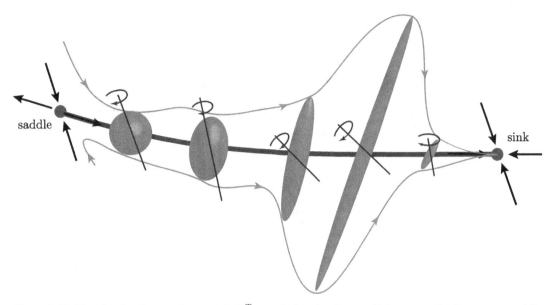

Figure 3.26 Sketch of a phase volume with $\nabla^T \cdot \mathbf{v} < 0$ showing heteroclinic connection between a saddle and sink equilibria along with the evolution of a set of points initially configured as a sphere as they move into regions with some positive normal stretching rates.

overall system is volume-decreasing, so that the integrated effect is relaxation to the heteroclinic trajectory. Such systems have $\nabla^T \cdot \mathbf{v} < 0$. In the dynamical systems literature, this is known as a dissipative system; however, in fluid mechanics we reserve the word "dissipative" for systems that have thermodynamic irreversibilities, as will be discussed in Section 4.5.

We illustrate these notions in the sketch of Fig. 3.26. Here we imagine a sphere of points as initial conditions near the saddle. We imagine that the system is such that the volume shrinks as the sphere moves. While the overall volume shrinks, one of the normal stretching rates is positive, admitting divergence of nearby trajectories from the heteroclinic trajectory. Rotation orients the volume into a region where negative normal stretching brings all points ultimately to the sink.

For our system, families of trajectories are shown in Fig. 3.27(a), and it is seen that there is divergence from the heteroclinic trajectory. This must be attributed to some points experiencing positive normal stretching away from the heteroclinic trajectory. For this case, the vorticity vector is $\boldsymbol{\omega} = \nabla \times \mathbf{v} = (51/16 - 2(1 - x_1^2), 0, -4x_1 x_3)^T$. On the heteroclinic trajectory, it reduces to $\boldsymbol{\omega} = (51/16 - 2(1 - x_1^2), 0, 0)^T$. The rotation rate on the heteroclinic trajectory, $\boldsymbol{\Omega} = \boldsymbol{\omega}/2 = (51/32 - (1 - x_1^2), 0, 0)^T$, has a magnitude Ω of near unity, and the time scales of rotation are close to the time scales of normal stretching.

We can modify the system to include more rotation. For instance, replacing Eq. (3.318) by $dx_3/dt = 10x_2 + x_3$ introduces a sufficient amount of rotation to render the heteroclinic trajectory to be attractive to nearby trajectories. Detailed analysis reveals that this small change (1) does not change the location of the two equilibria, (2) does not change the heteroclinic trajectory connecting the two equilibria, (3) modifies the dynamics near each

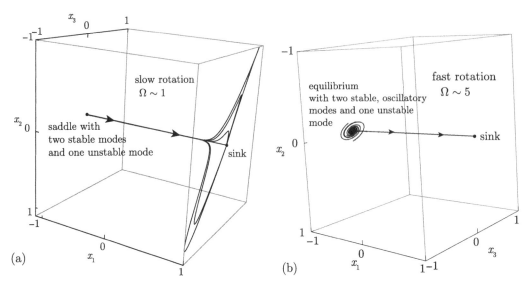

Figure 3.27 Plots of trajectories near the heteroclinic connection between equilibria with one unstable mode and a sink illustrating (a) divergence of nearby trajectories due to positive normal stretching with insufficiently rapid rotation and (b) convergence of nearby trajectories in the presence of positive normal stretching with sufficiently rapid rotation.

equilibrium such that both have two stable oscillatory modes, with the equilibrium at $(-1,0,0)^T$ also containing a third unstable mode and that at $(1,0,0)^T$ containing a third stable mode, (4) does not change that the system has a negative volumetric stretch rate on the heteroclinic trajectory, (5) does not change that a positive normal stretching mode exists on the heteroclinic trajectory, and (6) enhances the rotation such that the heteroclinic trajectory is locally attractive. This is illustrated in Fig. 3.27(b).

Had the local velocity gradient been purely symmetric, interpretation would be easier. It is the effect of a nonzero anti-symmetric part of $\mathbf{L}$ that induces the geometrical complexities of rotation. Such systems are often known as *non-normal dynamical systems* and have been considered in detail by Trefethen and Embree (2005).

SUMMARY

This chapter developed tools for describing the motion of a continuum fluid in space and time. First the Lagrangian description was developed because it is the natural extension of Newtonian mechanics of point masses. The more useful Eulerian description was then developed which does not focus on individual fluid particles. Standard descriptive tools were then introduced, including streamlines, pathlines, and streaklines. Importantly, local fluid motion was decomposed into a linear combination of translation, rigid body rotation, and deformation. The deformation was seen in general to decompose into extensional and shear deformation; however, special coordinate systems were seen to be available for which all of the deformation was aligned with special coordinate axes. In the special case in which the motion of a fluid was completely specified, it was seen how standard tools from dynamical system theory could be applied to better understand fluid motion.

PROBLEMS

3.1 A two-dimensional fluid particle trajectory is given by $x = t$, $y = t^4$. Determine its pathline, and plot it in x, y space. Include the velocity vectors and acceleration vectors on the pathline. Find the curvature of the pathline, and identify where it takes extreme values. Find the unit tangent and normal vectors, $\boldsymbol{\alpha}_t$, $\boldsymbol{\alpha}_{np}$, to the fluid pathline. Analyze how the particle's acceleration vector aligns with the particle trajectory and the direction normal to the trajectory.

3.2 A fluid particle trajectory is given by $x = t$, $y = t^2$, $z = t^3$. Find the velocity and acceleration vectors. At $t = 1$, evaluate the position, velocity, acceleration, trajectory curvature, trajectory torsion, and unit vectors $\boldsymbol{\alpha}_t$, $\boldsymbol{\alpha}_{np}$, and $\boldsymbol{\alpha}_{nb}$.

3.3 A fluid particle trajectory is given by

$$\mathbf{x}(t) = a \cos t \, \mathbf{i} + a \sin t \, \mathbf{j} + bt \, \mathbf{k}.$$

Find its curvature κ and torsion τ. Show that the acceleration vector is aligned with the principal unit normal to the trajectory.

3.4 Consider a steady flow field whose kinematics are described by $v_1 = (v_o/L)x_1$, $v_2 = 0$, $v_3 = 0$. Here v_o and L are constants. Find the acceleration vector, $d\mathbf{v}/dt$. Find the vorticity vector, $\boldsymbol{\omega}$. Find the deformation tensor, $\mathbf{D}$. Find the expansion rate, ϑ.

3.5 Consider the velocity field given by $v_1 = -x_2 - t$, $v_2 = x_1 + t$. Find the streamline through $(1, 0)^T$ at $t = 0$, the pathline for the fluid particle that is at $(1, 0)^T$ at $t = 0$, and the streakline through the point $(1, 0)^T$ at $t = 0$.

3.6 A steady two-dimensional velocity field is given by

$$v_1 = -x_1 + 4x_2, \qquad v_2 = -4x_1 - x_2.$$

Plot the velocity vector field. Determine the acceleration vector field and plot it. For a fluid particle that at $t = 0$ is at $(x_1, x_2)^T = (-2, 1)^T$, find a solution of $x_1(t)$, $x_2(t)$. Plot the pathline and the magnitudes of velocity and acceleration as functions of time.

3.7 Consider again the steady two-dimensional velocity field given by

$$v_1 = -x_1 + 4x_2, \qquad v_2 = -4x_1 - x_2.$$

Find the velocity gradient tensor $\mathbf{L}$, the deformation tensor $\mathbf{D}$, the rotation tensor $\mathbf{R}$, $||\mathbf{L}||$, $||\mathbf{D}||$, $||\mathbf{R}||$, the vorticity vector $\boldsymbol{\omega}$, the expansion rate ϑ, and the invariants of $\mathbf{L}$ and $\mathbf{D}$. Determine the quadric of extensional strain rate $\mathcal{D}$. Show $||\mathbf{D}|| = \max |\mathcal{D}|$. Show $||\mathbf{R}|| = ||\boldsymbol{\omega}||/2$.

3.8 A steady, two-dimensional velocity field is given by

$$v_1 = 1 + \frac{x_2^2 - x_1^2}{(x_1^2 + x_2^2)^2}, \qquad v_2 = -\frac{2x_1 x_2}{(x_1^2 + x_2^2)^2}.$$

Plot the velocity vector field. Determine and plot the acceleration vector field. For a fluid particle that at $t = 0$ is at $(x_1, x_2)^T = (-2, 1)^T$, find a numerical solution of $x_1(t), x_2(t)$, plot the pathline, and the magnitudes of velocity and acceleration as functions of time.

3.9 Consider again the steady, two-dimensional velocity field given by

$$v_1 = 1 + \frac{x_2^2 - x_1^2}{\left(x_1^2 + x_2^2\right)^2}, \qquad v_2 = -\frac{2x_1 x_2}{\left(x_1^2 + x_2^2\right)^2}.$$

Find the velocity gradient tensor $\mathbf{L}$, the deformation tensor $\mathbf{D}$, the vorticity vector $\boldsymbol{\omega}$, the expansion rate ϑ, and the invariants of $\mathbf{L}$ and $\mathbf{D}$. Examine and plot the quadric of extensional strain rate $\mathcal{D}$ at the point $(x_1, x_2)^T = (-2, 1)^T$.

3.10 Study the following nonlinear, autonomous system that describes the steady, three-dimensional kinematics of a fluid particle:

$$\frac{dx_1}{dt} = v_1(x_1, x_2, x_3) = -x_1 - 4x_2, \qquad x_1^o = 1,$$

$$\frac{dx_2}{dt} = v_2(x_1, x_2, x_3) = 2x_1 - x_1 x_2, \qquad x_2^o = 1,$$

$$\frac{dx_3}{dt} = v_3(x_1, x_2, x_3) = -x_3 - x_1, \qquad x_3^o = 1.$$

Find its equilibria (stagnation points), the linear stability of the equilibria, the fluid particle trajectory, and the acceleration vector field. Near the equilibrium point $(0, 0, 0)^T$, evaluate the deformation tensor $\mathbf{D}$, the rotation tensor $\mathbf{R}$, the vorticity vector $\boldsymbol{\omega}$, along with $\|\mathbf{D}\|$, $\|\mathbf{R}\|$, and $\|\boldsymbol{\omega}\|$. Explain why $(0, 0, 0)^T$ is a stable equilibrium despite the fact that $\mathbf{D}$ possesses an eigenvalue with a positive real part. Find the quadrics of extensional strain rate $\mathcal{D}$. Plot the quadrics for the extreme values of $\mathcal{D}$, and show that one is a hyperboloid of one sheet and the other a hyperboloid of two sheets. Near equilibrium, show $\|\mathbf{D}\| = \max |\mathcal{D}|$ and $\|\mathbf{R}\| = \|\boldsymbol{\omega}\|/2$.

3.11 Consider $\mathbf{A} \cdot \mathbf{x} = \mathbf{b}$, where

$$\mathbf{A} = \begin{pmatrix} 1 & 1 \\ 1 & 4 \\ 1 & 5 \end{pmatrix}, \qquad \mathbf{b} = \begin{pmatrix} 1 \\ 0 \\ 0 \end{pmatrix}.$$

An exact solution $\mathbf{x}$ does not exist, but by operating on both sides by $\mathbf{A}^T$, we can identify a best approximation. Find the best $\mathbf{x} = \mathbf{x}_p$ that minimizes $\|\mathbf{A} \cdot \mathbf{x} - \mathbf{b}\|$. Show that $\mathbf{x} = \mathbf{x}_p$ maps to $\mathbf{b}_p$ where $\mathbf{b}_p = \mathbf{P} \cdot \mathbf{b}$ with $\mathbf{P} = \mathbf{A} \cdot (\mathbf{A}^T \cdot \mathbf{A})^{-1} \cdot \mathbf{A}^T$. Show $\mathbf{P}$ is a projection matrix.

3.12 Consider what is known as a *Clebsch*[9] *decomposition* or a *complex lamellar decomposition*. A velocity field has a decomposition

$$\mathbf{v} = \mathbf{v}^{(\phi)} + \mathbf{v}^{(\omega)},$$

where $\mathbf{v}^{(\phi)}$ and $\mathbf{v}^{(\omega)}$ are the potential and vortical parts, respectively, defined by

$$\mathbf{v}^{(\phi)} = \nabla \phi, \qquad \mathbf{v}^{(\omega)} = \sigma \nabla \chi.$$

---

[9] Rudolf Friedrich Alfred Clebsch, (1833–1872), German mathematician.

Here ϕ, σ, and χ are nonunique scalar fields. Consider
$$\phi = x^2 + y^2 + z^2, \qquad \chi = x - y - z, \qquad \sigma = xyz.$$
Find $\mathbf{v}$, $\mathbf{v}^{(\phi)}$, $\mathbf{v}^{(\omega)}$, $\boldsymbol{\omega}$, and ϑ. Show $\mathbf{v}^{(\omega)}$ is a complex lamellar vector field as $\mathbf{v}^{(\omega)T} \cdot \boldsymbol{\omega} = 0$. Show $\boldsymbol{\omega} = \nabla \times \mathbf{v}^{(\omega)} = \nabla \sigma \times \nabla \chi$. Show surfaces of simultaneously constant χ and σ are vortex lines. Show $\mathbf{v}^{(\phi)}$ is irrotational but that $\mathbf{v}^{(\omega)}$ is not solenoidal.

3.13 Consider what is known as a *Helmholtz*[10] *decomposition*. A velocity field has a decomposition into potential and vortical parts:
$$\mathbf{v} = \mathbf{v}^{(\phi)} + \mathbf{v}^{(\omega)},$$
where $\mathbf{v}^{(\phi)}$ and $\mathbf{v}^{(\omega)}$ are the potential and vortical parts, respectively, defined by
$$\mathbf{v}^{(\phi)} = \nabla \phi, \qquad \mathbf{v}^{(\omega)} = \nabla \times \boldsymbol{\psi}.$$
Here ϕ is a nonunique scalar potential field, and $\boldsymbol{\psi}$ is a nonunique vector field. We require that $\boldsymbol{\psi}$ be solenoidal: $\nabla^T \cdot \boldsymbol{\psi} = 0$. Consider
$$\phi = x^2 + y^2 + z^2, \qquad \boldsymbol{\psi} = \left(x^2 z + xy + yz^2, xyz + y^2 + z, -\frac{3xz^2}{2} - 3yz \right)^T.$$
Find $\mathbf{v}$, $\mathbf{v}^{(\phi)}$, $\mathbf{v}^{(\omega)}$, $\boldsymbol{\omega}$, and ϑ. Show $\mathbf{v}^{(\omega)}$ is not a complex lamellar vector field. Show $\boldsymbol{\omega} = \nabla \times \mathbf{v}^{(\omega)} = -\nabla^2 \boldsymbol{\psi}$. Show $\mathbf{v}^{(\phi)}$ is irrotational and $\mathbf{v}^{(\omega)}$ is solenoidal.

3.14 Consider the Helmholtz decomposition of a velocity vector field as defined in the previous problem $\mathbf{v} = \mathbf{v}^{(\phi)} + \mathbf{v}^{(\omega)} = \nabla \phi + \nabla \times \boldsymbol{\psi}$. Show using standard identities from vector calculus and the definitions of the previous problem that the solenoidal condition $\nabla^T \cdot \boldsymbol{\psi} = 0$ and the expansion rate definition $\nabla^T \cdot \mathbf{v}^{(\phi)} = \vartheta$ allow us to admit Poisson equations for ϕ and $\boldsymbol{\psi}$:
$$\nabla^2 \phi = \vartheta, \qquad \nabla^2 \boldsymbol{\psi} = -\boldsymbol{\omega}.$$
See Wu et al. (2006) for discussion of decompositions related to fluid kinematics.

3.15 Given a velocity field $\mathbf{v}$, one can uniquely identify the vorticity field via $\boldsymbol{\omega} = \nabla \times \mathbf{v}$. Given the vorticity field, show that one cannot uniquely identify the velocity field. Identify the most general family of velocity fields that maps to the same vorticity field as $\mathbf{v}$. Examine the physics literature and describe this in the context of what is known as *gauge invariance* under a *gauge transformation*.

3.16 Two-dimensional Cartesian coordinates $\boldsymbol{\xi}$ are transformed to non-Cartesian coordinates $\mathbf{x}$ via the shear mapping linear transformation $\mathbf{x} = \mathbf{J}^{-1} \cdot \boldsymbol{\xi}$, with
$$\mathbf{J}^{-1} = \begin{pmatrix} 1 & -1 \\ 0 & 1 \end{pmatrix}.$$

---

[10] Hermann von Helmholtz, 1821–1894, Potsdam-born German physicist and philosopher, empiricist and refuter of the notion that scientific conclusions could be drawn from philosophical ideas, graduated from medical school, wrote convincingly on the science and physiology of music, and developed theories of vortex motion as well as thermodynamics and electrodynamics.

Show that the singular values from the SVD of $\mathbf{J}^{-1}$ determine the maximum and minimum unmodulated magnitudes of $\mathbf{x}$ induced by $\mathbf{J}^{-1}$ operating on unit vectors $\boldsymbol{\xi}$. Find unit vectors $\boldsymbol{\xi}$ that induce the maximum and minimum magnitudes of $\mathbf{x}$. Show this maximum magnitude of $\mathbf{x}$ induced by a unit vector $\boldsymbol{\xi}$ is $||\mathbf{J}^{-1}||$.

3.17 A fluid in motion has a locally constant rotation tensor

$$\mathbf{R} = \begin{pmatrix} 0 & \Omega_3 & -\Omega_2 \\ -\Omega_3 & 0 & \Omega_1 \\ \Omega_2 & -\Omega_1 & 0 \end{pmatrix}.$$

Find its dual vector $\boldsymbol{\Omega}$. Show $\boldsymbol{\Omega}$ points in the same direction as the eigenvector of $\mathbf{R}$ associated with its only real eigenvalue. Show $d\mathbf{v}^{(r)} = \mathbf{0}$ if $d\mathbf{x}$ is aligned with the axis of rotation. If $\boldsymbol{\Omega} = (1, 1, 1)^T$, use the SVD of $\mathbf{R}^T$ to interpret $d\mathbf{v}^{(r)}$.

3.18 Given the rotation matrix $\mathbf{Q} = (\boldsymbol{\alpha}_t, \boldsymbol{\alpha}_{np}, \boldsymbol{\alpha}_{nb})^T$, show that $d\mathbf{Q}/ds = \mathbf{A} \cdot \mathbf{Q}$ requires $\mathbf{A}$ to be anti-symmetric; compare it to the matrix induced by the Frenet–Serret relations.

FURTHER READING

Truesdell, C. (2018). *The Kinematics of Vorticity*. Mineola, New York: Dover.

REFERENCES

Aref, H., Blake, J. R., Budišić, M. et al. (2017). Frontiers of chaotic advection, *Reviews of Modern Physics*, **89**(2), 025007.

Aris, R. (1962). *Vectors, Tensors, and the Basic Equations of Fluid Mechanics*. New York: Dover.

Batchelor, G. K. (2000). *An Introduction to Fluid Dynamics*. Cambridge, UK: Cambridge University Press.

Bird, R. B., Armstrong, R. C., and Hassager, O. (1987). *Dynamics of Polymeric Liquids*, 2 vols. New York: John Wiley.

Bird, R. B., Stewart, W. E., and Lightfoot, E. N. (2007). *Transport Phenomena*, revised 2nd ed. New York: John Wiley.

Brodkey, R. S. (1967). *The Phenomena of Fluid Motions*. New York: Dover.

Cantwell, B. J. (2002). *Introduction to Symmetry Analysis*. New York: Cambridge University Press.

Currie, I. G. (2013). *Fundamental Mechanics of Fluids*, 4th ed. Boca Raton, Florida: CRC Press.

Golub, G. H., and Van Loan, C. F. (2013). *Matrix Computations*. Baltimore: Johns Hopkins University Press.

Guckenheimer, J., and Holmes, P. H. (2002). *Nonlinear Oscillations, Dynamical Systems, and Bifurcations of Vector Fields*. New York: Springer-Verlag.

Hirsch, M. W., Smale, S. and Devaney, R. L. (2013). *Differential Equations, Dynamical Systems, and an Introduction to Chaos*, 3rd ed. Waltham, Massachusetts: Academic Press.

Joseph, D. D. (1990). *Fluid Dynamics of Viscoelastic Liquids*. New York: Springer.

Kundu, P. K., Cohen, I. M., and Dowling, D. R. (2015). *Fluid Mechanics*, 6th ed. Amsterdam: Academic Press.

Marsden, J. E., and Hughes, T. J. R. (1983). *Mathematical Foundations of Elasticity*. Mineola, New York: Dover.

Meneveau, C. (2011). Lagrangian dynamics and models of the velocity gradient tensor in turbulent flows, *Annual Review of Fluid Mechanics*, **43**, 219–245.

Mengers, J. D. (2012). Slow invariant manifolds for reaction-diffusion systems, Ph.D. Dissertation, University of Notre Dame, Notre Dame, Indiana.

Morrison, F. A. (2001). *Understanding Rheology*. New York: Oxford University Press.

Ottino, J. M. (1989). *The Kinematics of Mixing: Stretching, Chaos, and Transport*. Cambridge, UK: Cambridge University Press.

Panton, R. L. (2013). *Incompressible Flow*, 4th ed. New York: John Wiley.

Paolucci, S. (2016). *Continuum Mechanics and Thermodynamics of Matter*. New York: Cambridge University Press

Perko, L. (2006). *Differential Equations and Dynamical Systems*, 3rd ed. Berlin: Springer.

Perry, A. E., and Chong, M. S. (1987). A description of eddying motions and flow patterns using critical-point concepts, *Annual Review of Fluid Mechanics*, **19**, 125–155.

Powers, J. M. (2016). *Combustion Thermodynamics and Dynamics*. New York: Cambridge University Press.

Powers, J. M., and Sen, M. (2015). *Mathematical Methods in Engineering*. New York: Cambridge University Press.

Powers, J. M., Paolucci, S., Mengers, J. D., and Al-Khateeb, A. N. (2015). Slow attractive canonical invariant manifolds for reactive systems, *Journal of Mathematical Chemistry*, **53**(2), 737–766.

Samelson, R. M., and Wiggins, S. (2006). *Lagrangian Transport in Geophysical Jets and Waves: the Dynamical Systems Approach*. New York: Springer.

Strang, G. (2006). *Linear Algebra and its Applications*, 4th ed. Boston, Massachusetts: Cengage.

Thorne, K. S., and Blandford, R. D. (2017). *Modern Classical Physics: Optics, Fluids, Plasmas, Elasticity, and Statistical Physics*. Princeton, New Jersey: Princeton University Press.

Trefethen, L. N., and Embree, M. (2005). *Spectra and Pseudospectra: The Behavior of Nonnormal Matrices and Operators*. Princeton, New Jersey: Princeton University Press.

Wu, J.-Z., Ma, H.-Y., and Zhou, M.-D. (2006). *Vorticity and Vortex Dynamics*. Berlin: Springer.

4 Evolution Axioms

The primary goal of this chapter is to convert verbal notions that embody the basic axioms of nonrelativistic continuum mechanics into usable mathematical expressions. They will have generality beyond fluid mechanics in that they apply to any continuum material, for example solids. First, we must list those axioms. These axioms will speak to the evolution in time of mass, linear momenta, angular momenta, energy, and entropy. The axioms themselves are simply principles that have been observed to have wide validity as long as the particle velocity is small relative to the speed of light and length scales are sufficiently large to contain many molecules. We give a simplistic mathematical formulation of the axioms we will use in Fig. 4.1. The meaning of the various symbols and equations will be examined in this chapter, and they will be extended to forms that better reflect the geometric richness of fluid mechanics. To many, they will be familiar from basic physics and thermodynamics courses. Many of these axioms can be applied to molecules or macroscale rigid bodies, but will not be here.

The axioms cannot be proven. They are simply statements that have been useful in describing the universe. In that sense they are analogous to axioms of geometry, such as Euclid's parallel postulate. Most texts discuss the axioms in one way or another, often implicitly, and often stirring in discussion of specific materials. Those that take the explicit material-independent approach as we do here include Aris (1962), Eringen (1989), Fung (1965), Kundu et al. (2015),

$$\frac{dm}{dt} = 0, \qquad \text{mass,}$$
$$\frac{d}{dt}(m\mathbf{v}) = \mathbf{F}, \qquad \text{linear momenta,}$$
$$\frac{d\hat{\mathbf{L}}}{dt} = \mathbf{r} \times \mathbf{F}, \qquad \text{angular momenta,}$$
$$\frac{dE}{dt} = \frac{dQ}{dt} - \frac{dW}{dt}, \qquad \text{energy,}$$
$$\frac{dS}{dt} \geq \frac{1}{T}\frac{dQ}{dt}, \qquad \text{entropy.}$$

Figure 4.1 Axioms for the evolution in time of mass, linear momenta, angular momenta, energy, and entropy for our continuum fluid in simplistic form.

Leal (2007), Paolucci (2016), Panton (2013), Truesdell (1991), Whitaker (1968), and Woods (1975).

A summary of the axioms in words is as follows:

- *Mass conservation principle*: The time rate of change of mass of a material volume is zero.
- *Linear momenta evolution principle*: The time rate of change of the linear momenta of a material volume is equal to the sum of forces acting on the volume. This is Euler's generalization of Newton's second law of motion.
- *Angular momenta evolution principle*: The time rate of change of the angular momenta of a material volume is equal to the sum of the torques acting on the volume.
- *Energy evolution principle*: The time rate of change of energy within a material volume is equal to the difference of rates at which heat is received and work is done. This is the first law of thermodynamics.
- *Entropy evolution inequality*: The time rate of change of entropy within a material volume is greater than or equal to the ratio of the rate of heat transferred to the volume and the absolute temperature of the volume. This is the second law of thermodynamics.

We call them axioms of *evolution* because in general the quantity of interest evolves in time. *Conservation* is a special case that implies no change. So for material volumes, mass is formally conserved. However, linear momenta, angular momenta, energy, and entropy of material volumes may evolve in time in response to, respectively, net forces, net torques, heat and work, and reversible heat transfer along with irreversibility-generating mechanisms. While it is more common to speak of momenta and energy conservation, it is more accurate to speak of their evolution, and so we do.

A less general case is realized when the same quantities are conserved. We give less useful words here to describe how our evolution axioms reduce in such special cases:

- *Mass conservation*: The time rate of change of mass of a material volume is zero.
- *Linear momenta conservation*: The time rate of change of linear momenta of a material volume is zero if there are no net forces acting on that volume.
- *Angular momenta conservation*: The time rate of change of angular momenta of a material volume is zero if there are no net torques acting on that volume.
- *Energy conservation*: The time rate of change of energy of a material volume is zero if there are no net combined heat and work transfers to that volume.
- *Entropy conservation*: The time rate of change of entropy of a material volume is zero if there is no heat transfer to that volume and no irreversible processes occurring within it.

Next we shall systematically convert these words into mathematical form.

4.1 Mass

The mass conservation axiom is simple to state mathematically. It is

$$\frac{d}{dt} m_{V_m(t)} = 0. \tag{4.1}$$

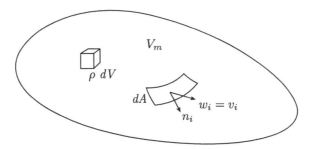

Figure 4.2 Diagram of finite material volume V_m, infinitesimal mass element $\rho\, dV$, and infinitesimal surface element dA with unit normal n_i, and general velocity w_i equal to fluid velocity v_i.

As introduced in Section 2.4.7, $V_m(t)$ stands for a material volume that can evolve in time, and $m_{V_m(t)}$ is the mass enclosed within. We interpret d/dt as the material derivative. A relevant material volume is depicted in Fig. 4.2. We define the mass of the material volume based upon the local value of density, $\rho(x_i, t)$, that may vary within $V_m(t)$:

$$m_{V_m(t)} = \int_{V_m(t)} \rho(x_i, t)\, dV. \tag{4.2}$$

From here on, we will typically forgo writing $\rho(x_i, t)$ in favor of ρ and understand that it is a function of position and time. So, the mass conservation axiom is

$$\frac{d}{dt} \int_{V_m(t)} \rho\, dV = 0. \tag{4.3}$$

Recalling Leibniz's rule, Eq. (2.238), $d/dt \int_{V_a(t)} [\ ]\, dV = \int_{V_a(t)} \partial_o [\ ]\, dV + \int_{A_a(t)} n_i w_i [\ ]\, dA$, we specialize the arbitrary velocity to the fluid velocity so that $w_i = v_i$. This is because we are considering a material volume, and thus the Reynolds transport theorem, Eq. (2.241). So we get

$$\frac{d}{dt} \int_{V_m(t)} \rho\, dV = \int_{V_m(t)} \partial_o \rho\, dV + \int_{A_m(t)} n_i v_i \rho\, dA = 0. \tag{4.4}$$

Now the integral form of Eq. (4.4) is in fact the most fundamental representation of the mass conservation principle. It applies for both continuous flows as well as for flows with embedded discontinuities such as the shock waves we will study in Section 9.6.

For this chapter, we will assume that there are no embedded discontinuities and proceed. We invoke Gauss's theorem, Eq. (2.222) $\int_{V_m(t)} \partial_i [\ ]\, dV = \int_{A_m(t)} n_i [\ ]\, dA$, to convert the surface integral to a volume integral to get the mass conservation axiom to read as

$$\int_{V_m(t)} \partial_o \rho\, dV + \int_{V_m(t)} \partial_i(\rho v_i)\, dV = \int_{V_m(t)} (\partial_o \rho + \partial_i(\rho v_i))\, dV = 0. \tag{4.5}$$

Now, in an important step, we realize that the only way for this integral – that is, for an arbitrary material volume – to always be zero is for the integrand itself to always be zero. Hence, we have

$$\partial_o \rho + \partial_i(\rho v_i) = 0. \tag{4.6}$$

We write this in expanded Cartesian and Gibbs notation as

$$\partial_o \rho + \partial_1(\rho v_1) + \partial_2(\rho v_2) + \partial_3(\rho v_3) = 0, \qquad \frac{\partial \rho}{\partial t} + \nabla^T \cdot (\rho \mathbf{v}) = 0. \qquad (4.7)$$

These equations, along with Eq. (4.6), are all in what is known as *conservative form*. The conservative form is also known as the *divergence form*. The conservative form shows mass (equivalently ρ) is conserved when mass fluxes, $\rho \mathbf{v}$, are in balance. There are several alternative forms for this axiom. Using the product rule, we can say also

$$\underbrace{\partial_o \rho + v_i \partial_i \rho}_{\text{material derivative of density}} + \rho \partial_i v_i = 0, \qquad (4.8)$$

or, writing in what is called the *nonconservative* form,

$$\frac{d\rho}{dt} + \rho \partial_i v_i = 0, \qquad \frac{d\rho}{dt} + \rho \nabla^T \cdot \mathbf{v} = 0, \qquad (4.9)$$

$$(\partial_o \rho + v_1 \partial_1 \rho + v_2 \partial_2 \rho + v_3 \partial_3 \rho) + \rho (\partial_1 v_1 + \partial_2 v_2 + \partial_3 v_3) = 0. \qquad (4.10)$$

For flows with no embedded discontinuities, the conservative and nonconservative forms give identical information. We can also say

$$\underbrace{\frac{1}{\rho}\frac{d\rho}{dt}}_{\text{relative rate of density increase}} = - \underbrace{\partial_i v_i}_{\text{relative rate of particle volume expansion}}. \qquad (4.11)$$

Recalling Eq. (3.201), we see the relative rate of density increase of a fluid particle is the negative of its relative rate of expansion, as expected. So, we also have

$$\frac{1}{\rho}\frac{d\rho}{dt} = -\frac{1}{V_m}\frac{dV_m}{dt}, \qquad \text{so} \qquad \rho \frac{dV_m}{dt} + V_m \frac{d\rho}{dt} = 0, \qquad (4.12)$$

$$\frac{d}{dt}(\rho V_m) = 0, \qquad \text{so} \qquad \frac{d}{dt}(m_{V_m}) = 0. \qquad (4.13)$$

This returns us to our original mass conservation statement, Eq. (4.1)!

Example 4.1 For a fluid in the velocity vector field

$$\mathbf{v} = \begin{pmatrix} v_1 \\ 0 \\ 0 \end{pmatrix} = \begin{pmatrix} v_o \frac{x_1}{L} \\ 0 \\ 0 \end{pmatrix}, \qquad (4.14)$$

find the time rate of change of its density as well as its acceleration.

Solution
For this one-dimensional flow that only varies in the x_1 direction, we have

$$\nabla^T \cdot \mathbf{v} = \frac{\partial v_1}{\partial x_1} + \frac{\partial v_2}{\partial x_2} + \frac{\partial v_3}{\partial x_3} = \frac{\partial}{\partial x_1}\left(v_o \frac{x_1}{L}\right) + 0 + 0 = \frac{v_o}{L}. \qquad (4.15)$$

We have from mass conservation that $d\rho/dt = -\rho \nabla^T \cdot \mathbf{v}$. This gives

$$\frac{d\rho}{dt} = -\frac{\rho v_o}{L}. \qquad (4.16)$$

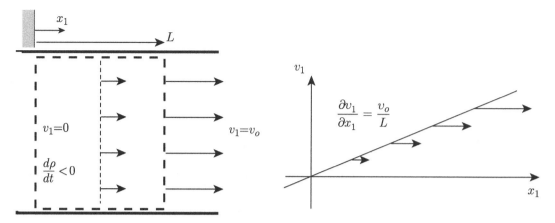

Figure 4.3 Diagram of a fluid element in a velocity field with nonzero divergence.

For $\rho, v_o, L > 0$, this flow has the density of a fluid particle decreasing with time. That is because the boundaries of the material volume are being stretched by the velocity field.

The acceleration of the fluid particle is $d\mathbf{v}/dt = \partial \mathbf{v}/\partial t + \mathbf{v}^T \cdot \nabla \mathbf{v}$. Because we have strictly one-dimensional motion, this reduces to

$$\frac{dv_1}{dt} = \frac{\partial v_1}{\partial t} + v_1 \frac{\partial v_1}{\partial x_1} = 0 + \left(v_o \frac{x_1}{L}\right)\left(v_o \frac{1}{L}\right) = \frac{v_o^2 x_1}{L^2}. \tag{4.17}$$

The fluid particle is accelerating to the right. Though the velocity field does not change with time, the position and velocity of an individual fluid particle change as it travels. The configuration is shown in Fig. 4.3. The fluid is stationary at $x_1 = 0$ and has velocity v_o at $x = L$. This stretches the material volume, causing the density of the material volume to drop. And the fluid particle accelerates as its x_1-coordinate increases. Also shown is the velocity as a function of x_1; the slope of this curve is the velocity gradient, which here is constant.

Let us consider a special case of the Reynolds transport theorem, Eq. (2.241), for a fluid that obeys mass conservation. The general tensor in Eq. (2.241) can be recast as

$$T_{jk...} = \rho \mathcal{T}_{jk...} \tag{4.18}$$

This is useful if $T_{jk...}$ is a property that has units of some quantity per unit mass. Then $\mathcal{T}_{jl...}$ is the same quantity per unit volume. Then the Reynolds transport theorem becomes

$$\frac{d}{dt}\int_{V_m(t)} \rho \mathcal{T}_{jk...}(x_i, t)\, dV = \int_{V_m(t)} \partial_o(\rho \mathcal{T}_{jk...})\, dV + \int_{A_m(t)} n_l \rho v_l \mathcal{T}_{jk...}\, dA, \tag{4.19}$$

$$= \int_{V_m(t)} \left(\partial_o(\rho \mathcal{T}_{jk...}) + \partial_l(\rho v_l \mathcal{T}_{jk...})\right)\, dV, \tag{4.20}$$

$$= \int_{V_m(t)} \mathcal{T}_{jk...} \underbrace{\left(\partial_o \rho + \partial_l(\rho v_l)\right)}_{=0}\, dV$$

$$+ \int_{V_m(t)} \rho \left(\underbrace{\partial_o \mathcal{T}_{jk\ldots} + v_l \partial_l \mathcal{T}_{jk\ldots}}_{= d\mathcal{T}_{jk\ldots}/dt} \right) dV, \qquad (4.21)$$

$$= \int_{V_m(t)} \rho \frac{d\mathcal{T}_{jk\ldots}}{dt} dV. \qquad (4.22)$$

Lastly, let us consider an approach to mass conservation that is probably more common in the solid mechanics literature, but is no less valid for a fluid. Here we follow the discussion given by Aris (1962). We consider a transformation from Eulerian coordinates x_i to Lagrangian coordinates x_j^o. The Jacobian matrix $\mathbf{J}$ and Jacobian determinant J of this transformation is

$$\mathbf{J} = \frac{\partial x_i}{\partial x_j^o}, \qquad J = \det \frac{\partial x_i}{\partial x_j^o}. \qquad (4.23)$$

In a detailed analysis, Aris shows that

$$\frac{1}{J} \frac{dJ}{dt} = \nabla^T \cdot \mathbf{v} = \partial_i v_i. \qquad (4.24)$$

Thus, the relative rate of change of J is the expansion rate from Eq. (3.201); so

$$\frac{1}{V_m} \frac{dV_m}{dt} = \frac{1}{J} \frac{dJ}{dt}. \qquad (4.25)$$

Aris also shows that

$$dV = J \, dV_o, \qquad (4.26)$$

where V_o is the original volume of the material element.

We can now operate on our mass conservation relation as follows. We first transform mass conservation, Eq. (4.3), to the Lagrangian frame to get

$$\frac{d}{dt} \int_{V_o} \rho(x_i(x_j^o,t),t) J \, dV_o = \int_{V_o} \frac{d}{dt} \left(\rho(x_i(x_j^o,t),t) J \right) dV_o = 0, \qquad (4.27)$$

$$\int_{V_o} \left(J \frac{d\rho}{dt} + \rho \frac{dJ}{dt} \right) dV_o = \int_{V_o} \left(\frac{d\rho}{dt} + \rho \partial_i v_i \right) J \, dV_o = 0. \qquad (4.28)$$

Transform back to the material frame to get

$$\int_{V_m(t)} \left(\frac{d\rho}{dt} + \rho \partial_i v_i \right) dV = 0, \quad \text{so} \quad \frac{d\rho}{dt} + \rho \partial_i v_i = 0, \qquad (4.29)$$

as expected.

4.2 Linear Momenta

We develop here the linear momenta axiom. The axiom speaks to the time rate of change of the linear momenta vector due to vectors representing body and surface forces. As such we first consider the vector form. However, we shall see that for a fluid the vector of surface forces requires further decomposition, and that necessitates the introduction of a stress tensor, first briefly discussed in Section 1.2.3. We then develop the stress tensor and finish by incorporating it into the final form of the linear momenta axiom.

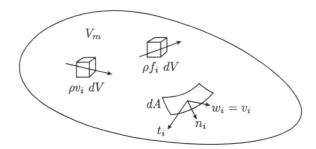

Figure 4.4 Diagram of finite material volume V_m, infinitesimal linear momenta element $\rho v_i\, dV$, infinitesimal body force element $\rho f_i\, dV$, and infinitesimal surface element dA with unit normal n_i, surface traction t_i, and general velocity w_i equal to fluid velocity v_i.

4.2.1 Preliminary Vector Form

The linear momenta evolution axiom is simple to state mathematically. It is

$$\underbrace{\frac{d}{dt}\int_{V_m(t)} \rho v_i\, dV}_{\text{rate of change of linear momenta}} = \underbrace{\int_{V_m(t)} \rho f_i\, dV}_{\text{body forces}} + \underbrace{\int_{A_m(t)} t_i\, dA}_{\text{surface forces}}. \qquad (4.30)$$

Again $V_m(t)$ stands for a material volume that can evolve in time. A relevant material volume is presented in Fig. 4.4. The term f_i represents a body force per unit mass. An example of such a force would be the gravitational force acting on a body, which when scaled by mass, yields gravitational acceleration g_i. The term t_i is a traction vector representing force per unit area; it is not to be confused with time t. We will need to express the traction vector in terms of the stress tensor.

Consider first the left-hand side, LHS, of the linear momenta principle

$$LHS = \int_{V_m(t)} \partial_o(\rho v_i)\, dV + \int_{A_m(t)} n_j v_j \rho v_i\, dA, \qquad \text{from Reynolds,} \qquad (4.31)$$

$$= \int_{V_m(t)} \left(\partial_o(\rho v_i) + \partial_j(\rho v_j v_i)\right)\, dV, \qquad \text{from Gauss.} \qquad (4.32)$$

So, the linear momenta principle is

$$\int_{V_m(t)} \left(\partial_o(\rho v_i) + \partial_j(\rho v_j v_i)\right)\, dV = \int_{V_m(t)} \rho f_i\, dV + \int_{A_m(t)} t_i\, dA. \qquad (4.33)$$

These are all expressed in terms of volume integrals except for the term involving surface forces. The following two sections describe how to cast all terms as volume integrals.

4.2.2 Surface Forces

The surface force per unit area is a vector we call the traction t_i. It has the units of stress, but it is not formally a stress, as stress is a tensor. The traction is a function of both position x_j and surface orientation n_k: $t_i = t_i(x_j, n_k)$.

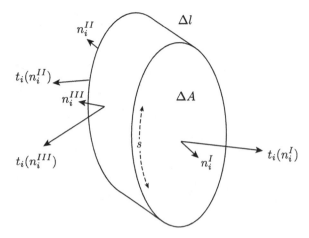

Figure 4.5 Diagram of pillbox element for stress analysis.

We intend to demonstrate that the traction can be stated in terms of a *stress tensor* T_{ji}, as written here:

$$t_i = n_j T_{ji}, \qquad \mathbf{t}^T = \mathbf{n}^T \cdot \mathbf{T}, \qquad \mathbf{t} = \mathbf{T}^T \cdot \mathbf{n}. \tag{4.34}$$

The following excursions are necessary to show this.

Show that the Force on One Side of a Surface is Equal and Opposite to that on the Opposite Side

Let us apply the principle of linear momenta to the material volume depicted in Fig. 4.5. Here we indicate the dependency of the traction on orientation by notation such as $t_i\left(n_i^{II}\right)$. This does not indicate multiplication, nor that i is a dummy index here. In Fig. 4.5, the thin pillbox has width Δl, circumference s, and a surface area for the circular region of ΔA. Surface I is a circular region; surface II is the opposite circular region, and surface III is the cylindrical side.

We apply the mean value theorem to the linear momenta principle and get

$$\overline{(\partial_o(\rho v_i) + \partial_j(\rho v_j v_i))}(\Delta A)(\Delta l) = \overline{\rho f_i}(\Delta A)(\Delta l) + \overline{t_i(n_i^I)}\Delta A + \overline{t_i(n_i^{II})}\Delta A + \overline{t_i(n_i^{III})}s(\Delta l). \tag{4.35}$$

Now we let $\Delta l \to 0$, holding for now s and ΔA fixed, to obtain

$$0 = \left(\overline{t_i(n_i^I)} + \overline{t_i(n_i^{II})}\right)\Delta A. \tag{4.36}$$

Now letting $\Delta A \to 0$, so that the mean value approaches the local value, and taking $n_i^I = -n_i^{II} \equiv n_i$, we get the useful result

$$t_i(n_i) = -t_i(-n_i). \tag{4.37}$$

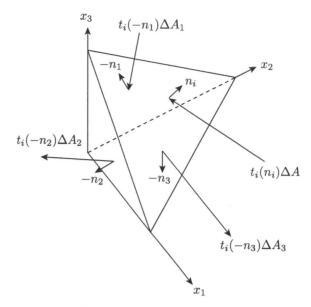

Figure 4.6 Diagram of tetrahedral element for stress analysis on an arbitrary plane.

At an infinitesimal length scale, the traction on one side of a surface is equal and opposite to that on the other. That is, there is a local force balance. This applies even if there is velocity and acceleration of the material on a macroscale. On the microscale, surface forces dominate inertial effects and body forces. This is a useful general principle that will be exploited in Section 11.1. It the fundamental reason why microorganisms have different propulsion systems than macroorganisms: they must overcome forces of a different nature.

Study the Stress on an Arbitrary Plane and Relate It to the Stress on Coordinate Planes

Now let us consider a rectangular parallelepiped aligned with the Cartesian axes that has been sliced at an oblique angle to form a tetrahedron. We will apply the linear momenta principle to this geometry and make a statement about the existence of a stress tensor. The described material volume is depicted in Fig. 4.6. Let ΔL be a characteristic length scale of the tetrahedron. Also let four unit normals n_j exist, one for each surface. They will be $-n_1, -n_2, -n_3$ for the surfaces associated with each coordinate direction. They are negative because the outer normal points opposite to the direction of the axes. Let n_i be the normal associated with the oblique face. Let ΔA denote the surface area of each face.

Now the volume of the tetrahedron must be of order L^3 and the surface area of order L^2. Thus, applying the mean value theorem to the linear momenta principle, we obtain the form

$$(\text{inertia}) \times (\Delta L)^3 = \sum (\text{body forces}) \times (\Delta L)^3 + \sum (\text{surface forces}) \times (\Delta L)^2. \qquad (4.38)$$

As before, for small volumes, $\Delta L \to 0$, and the linear momenta principle reduces to

$$\sum \text{surface forces} = 0. \qquad (4.39)$$

Applying this to the configuration of Fig. 4.6, we get

$$0 = \overline{t_i(n_i)}\Delta A + \overline{t_i(-n_1)}\Delta A_1 + \overline{t_i(-n_2)}\Delta A_2 + \overline{t_i(-n_3)}\Delta A_3. \qquad (4.40)$$

But we know that $t_i(n_i) = -t_i(-n_i)$, so

$$\overline{t_i(n_i)}\Delta A = \overline{t_i(n_1)}\Delta A_1 + \overline{t_i(n_2)}\Delta A_2 + \overline{t_i(n_3)}\Delta A_3. \qquad (4.41)$$

Now it is not a difficult geometry problem to show that $n_i \Delta A = \Delta A_i$, so we get

$$\overline{t_i(n_i)}\Delta A = n_1\overline{t_i(n_1)}\Delta A + n_2\overline{t_i(n_2)}\Delta A + n_3\overline{t_i(n_3)}\Delta A, \qquad (4.42)$$

$$\overline{t_i(n_i)} = n_1\overline{t_i(n_1)} + n_2\overline{t_i(n_2)} + n_3\overline{t_i(n_3)}. \qquad (4.43)$$

Letting the volume shrink to zero, mean values become local values, and we get

$$t_i(n_i) = n_1 t_i(n_1) + n_2 t_i(n_2) + n_3 t_i(n_3). \qquad (4.44)$$

Now we can consider terms like t_i to obviously be vectors, and the indicator, for example (n_1), tells us with which surface the vector is associated. This is precisely what a tensor does, and in fact we can say

$$t_i(n_i) = n_1 T_{1i} + n_2 T_{2i} + n_3 T_{3i}. \qquad (4.45)$$

In shorthand, we can say the same thing with

$$t_i = n_j T_{ji}, \qquad \text{or equivalently} \qquad t_j = n_i T_{ij}, \qquad \text{QED}. \qquad (4.46)$$

Here T_{ji} is the component of stress in the i direction associated with the surface whose normal is in the j direction.

Consider Pressure and the Viscous Stress Tensor

Pressure is a familiar concept from thermodynamics and fluid statics. It is often tempting and sometimes correct to think of the pressure as the force per unit area normal to a surface and the force tangential to a surface being somehow related to frictional forces. We shall see that in general this view is too simplistic.

First recall from thermodynamics that what we will call p, the *thermodynamic pressure*, is for a simple compressible substance a function of at most two intensive thermodynamic variables, say $p = f(\rho, e)$. Here e is the specific internal energy. Also recall that the thermodynamic pressure is defined as a normal stress for a static material.

To distinguish between thermodynamic stresses and other stresses, let us define the *viscous stress tensor* τ_{ji} as follows:

$$\tau_{ji} = T_{ji} + p\delta_{ji}. \qquad (4.47)$$

Recall that T_{ji} is the *total stress tensor*. We obviously also have

$$T_{ji} = -p\delta_{ji} + \tau_{ji}. \qquad (4.48)$$

With this definition, pressure is positive in compression, while T_{ji} and τ_{ji} are positive in tension. Let us also define the *mechanical pressure*, $p^{(m)}$, as the negative of the average normal surface stress

$$p^{(m)} \equiv -\frac{1}{3}T_{ii} = -\frac{1}{3}(T_{11} + T_{22} + T_{33}). \tag{4.49}$$

The often invoked *Stokes' assumption*, which remains a subject of widespread misunderstanding since it was first made by Stokes (1845), is often adopted for lack of a good alternative in answer to a question that will be addressed later in Section 5.4.3. It asserts that the thermodynamic pressure is equal to the mechanical pressure:

$$p = p^{(m)} = -\frac{1}{3}T_{ii}. \tag{4.50}$$

Presumably a pressure measuring device in a moving flow field would actually measure the mechanical pressure, and not necessarily the thermodynamic pressure, so it is important to have this issue clarified for proper reconciliation of theory and measurement. It will be seen that Stokes' assumption gives some minor æsthetic harmony in certain limits, but it is not well-established, and is more a convenience than a requirement for most materials. It is the case that various incarnations of more fundamental kinetic theory under the assumption of a dilute gas composed of inert hard spheres give rise to the conclusion that Stokes' assumption is valid. At moderate densities, these hard sphere kinetic theory models predict that Stokes' assumption is invalid. However, none of the common kinetic theory models are able to predict results from experiments, which nevertheless also give indication, albeit indirect, that Stokes' assumption is invalid. Kinetic theories and experiments that consider polyatomic molecules, which can suffer vibrational and rotational effects as well, show further deviation from Stokes' assumption. It is often plausibly argued that these so-called nonequilibrium effects, that is, molecular vibration and rotation, which are only important in high speed flow applications in which the flow velocity is on the order of the fluid sound speed, are the mechanisms that cause Stokes' assumption to be violated. Because they only are important in high speed applications, they are difficult to measure, though measurement of the decay of acoustic waves has provided some data. For liquids, there is little to no theory, and the limited data indicate that Stokes' assumption is invalid. Rosenhead (1954) summarizes a discussion of the basic physical questions. Gad-el-Hak (1995) provides additional insight to this still unsettled question.

Now contracting Eq. (4.48), we get

$$T_{ii} = -p\delta_{ii} + \tau_{ii}. \tag{4.51}$$

Using the fact that $\delta_{ii} = 3$ and inserting Eq. (4.50) in Eq. (4.51), we find *for a fluid that obeys Stokes' assumption* that

$$T_{ii} = \frac{1}{3}T_{ii}(3) + \tau_{ii}, \quad \text{so} \quad \tau_{ii} = 0. \tag{4.52}$$

That is to say, the trace of the viscous stress tensor is zero. Moreover, for a fluid that obeys Stokes' assumption, we can interpret the viscous stress as the deviation from the mean stress; that is, the viscous stress is a deviatoric stress:

$$\underbrace{T_{ji}}_{\text{total stress}} = \underbrace{\frac{1}{3}T_{kk}\delta_{ji}}_{\text{mean stress}} + \underbrace{\tau_{ji}}_{\text{deviatoric stress}}, \quad \text{valid only if Stokes' assumption holds.} \tag{4.53}$$

If Stokes' assumption does not hold, then a portion of τ_{ji} will also contribute to the mean stress; that is, the viscous stress is not then entirely deviatoric.

Finally, let us note what the traction vector is when the fluid is stationary. For a stationary fluid, there is no viscous stress, so $\tau_{ji} = 0$, and we have

$$T_{ji} = -p\delta_{ji}, \quad \text{stationary fluid.} \tag{4.54}$$

We get the traction vector for a stationary fluid on any surface with normal n_j by

$$t_i = n_j T_{ji} = -pn_j \delta_{ji} = -pn_i, \quad \text{stationary fluid.} \tag{4.55}$$

We see $t_i = -pn_i$; that is, the traction vector must be oriented in the same direction as the surface normal for a stationary fluid; all stresses are normal to any arbitrarily oriented surface.

4.2.3 Final Tensor Form

We are now prepared to write the linear momenta equation in final form. Substituting our expression for the traction vector, Eq. (4.46) into the linear momenta expression, Eq. (4.33), we get

$$\int_{V_m(t)} \left(\partial_o(\rho v_i) + \partial_j(\rho v_j v_i) \right) dV = \int_{V_m(t)} \rho f_i \, dV + \int_{A_m(t)} n_j T_{ji} \, dA. \tag{4.56}$$

Using Gauss's theorem, Eq. (2.222), to convert the surface integral into a volume integral, and combining all under one integral sign, we get

$$\int_{V_m(t)} \left(\partial_o(\rho v_i) + \partial_j(\rho v_j v_i) - \rho f_i - \partial_j T_{ji} \right) dV = 0. \tag{4.57}$$

Making the same argument as before regarding arbitrary material volumes, this must then require that the integrand be zero, so we obtain

$$\partial_o(\rho v_i) + \partial_j(\rho v_j v_i) - \rho f_i - \partial_j T_{ji} = 0. \tag{4.58}$$

We can do the same steps in Gibbs notation. We cast Eq. (4.33) as follows:

$$\int_{V_m(t)} \left(\frac{\partial}{\partial t}(\rho \mathbf{v}) + \left(\nabla^T \cdot \rho \mathbf{v} \mathbf{v}^T \right)^T \right) dV = \int_{V_m(t)} \rho \mathbf{f} \, dV + \int_{A_m(t)} \mathbf{t} \, dA, \tag{4.59}$$

$$= \int_{V_m(t)} \rho \mathbf{f} \, dV + \int_{A_m(t)} (\mathbf{t}^T)^T \, dA, \tag{4.60}$$

$$= \int_{V_m(t)} \rho \mathbf{f} \, dV + \int_{A_m(t)} (\mathbf{n}^T \cdot \mathbf{T})^T \, dA, \tag{4.61}$$

$$= \int_{V_m(t)} \rho \mathbf{f} \, dV + \int_{V_m(t)} (\nabla^T \cdot \mathbf{T})^T \, dV. \tag{4.62}$$

Combining under a single integral, we get

$$\int_{V_m(t)} \left(\frac{\partial}{\partial t}(\rho \mathbf{v}) + \left(\nabla^T \cdot \rho \mathbf{v} \mathbf{v}^T \right)^T - \rho \mathbf{f} - (\nabla^T \cdot \mathbf{T})^T \right) dV = \mathbf{0}, \tag{4.63}$$

$$\frac{\partial}{\partial t}(\rho \mathbf{v}) + \left(\nabla^T \cdot \rho \mathbf{v} \mathbf{v}^T \right)^T - \rho \mathbf{f} - (\nabla^T \cdot \mathbf{T})^T = \mathbf{0}. \tag{4.64}$$

Now let us replace $\mathbf{T}$ in favor of $\boldsymbol{\tau}$. Using then $T_{ji} = -p\delta_{ji} + \tau_{ji}$, we get in Cartesian index, Gibbs, and full notation

$$\partial_o(\rho v_i) + \partial_j(\rho v_j v_i) = \rho f_i - \partial_i p + \partial_j \tau_{ji}, \tag{4.65}$$

$$\frac{\partial}{\partial t}(\rho \mathbf{v}) + \left(\nabla^T \cdot (\rho \mathbf{v}\mathbf{v}^T)\right)^T = \rho \mathbf{f} - \nabla p + \left(\nabla^T \cdot \boldsymbol{\tau}\right)^T, \tag{4.66}$$

$$\partial_o(\rho v_1) + \partial_1(\rho v_1 v_1) + \partial_2(\rho v_2 v_1) + \partial_3(\rho v_3 v_1) = \rho f_1 - \partial_1 p + \partial_1 \tau_{11} + \partial_2 \tau_{21} + \partial_3 \tau_{31}, \tag{4.67}$$

$$\partial_o(\rho v_2) + \partial_1(\rho v_1 v_2) + \partial_2(\rho v_2 v_2) + \partial_3(\rho v_3 v_2) = \rho f_2 - \partial_2 p + \partial_1 \tau_{12} + \partial_2 \tau_{22} + \partial_3 \tau_{32}, \tag{4.68}$$

$$\partial_o(\rho v_3) + \partial_1(\rho v_1 v_3) + \partial_2(\rho v_2 v_3) + \partial_3(\rho v_3 v_3) = \rho f_3 - \partial_3 p + \partial_1 \tau_{13} + \partial_2 \tau_{23} + \partial_3 \tau_{33}. \tag{4.69}$$

The form is known as the linear momenta principle cast in conservative or divergence form. It is the first choice of forms for many numerical simulations, as discretizations of this form of the equation naturally preserve the correct values of global linear momenta, up to roundoff error. For the Gibbs notation, the transpose convention is particularly cumbersome and unfamiliar, though necessary for full consistency. One will more commonly see this equation written simply as $\partial/\partial t(\rho \mathbf{v}) + \nabla \cdot (\rho \mathbf{v}\mathbf{v}) = \rho \mathbf{f} - \nabla p + \nabla \cdot \boldsymbol{\tau}$.

There is a commonly used nonconservative form that makes some analysis and physical interpretation easier. Let us use the product rule to expand the linear momenta principle, then rearrange it, and use mass conservation, Eq. (4.6), and the definition of material derivative to rewrite the expression:

$$\rho \partial_o v_i + v_i \partial_o \rho + v_i \partial_j(\rho v_j) + \rho v_j \partial_j v_i = \rho f_i - \partial_i p + \partial_j \tau_{ji}, \tag{4.70}$$

$$\rho(\partial_o v_i + v_j \partial_j v_i) + v_i \underbrace{(\partial_o \rho + \partial_j(\rho v_j))}_{=0 \text{ by mass}} = \rho f_i - \partial_i p + \partial_j \tau_{ji}, \tag{4.71}$$

$$\rho \underbrace{(\partial_o v_i + v_j \partial_j v_i)}_{=\frac{dv_i}{dt}} = \rho f_i - \partial_i p + \partial_j \tau_{ji}, \tag{4.72}$$

$$\rho \frac{dv_i}{dt} = \rho f_i - \partial_i p + \partial_j \tau_{ji}, \qquad \rho \frac{d\mathbf{v}}{dt} = \rho \mathbf{f} - \nabla p + \left(\nabla^T \cdot \boldsymbol{\tau}\right)^T. \tag{4.73}$$

Despite the fact that ρ is now outside the derivative operator, this is valid for a fully compressible flow. This nonconservative form is sometimes known as *Cauchy's first law of motion*. Written in full, it becomes

$$\rho(\partial_o v_1 + v_1 \partial_1 v_1 + v_2 \partial_2 v_1 + v_3 \partial_3 v_1) = \rho f_1 - \partial_1 p + \partial_1 \tau_{11} + \partial_2 \tau_{21} + \partial_3 \tau_{31}, \tag{4.74}$$

$$\rho(\partial_o v_2 + v_1 \partial_1 v_2 + v_2 \partial_2 v_2 + v_3 \partial_3 v_2) = \rho f_2 - \partial_2 p + \partial_1 \tau_{12} + \partial_2 \tau_{22} + \partial_3 \tau_{32}, \tag{4.75}$$

$$\rho(\partial_o v_3 + v_1 \partial_1 v_3 + v_2 \partial_2 v_3 + v_3 \partial_3 v_3) = \rho f_3 - \partial_3 p + \partial_1 \tau_{13} + \partial_2 \tau_{23} + \partial_3 \tau_{33}. \tag{4.76}$$

We see that particles accelerate due to body forces and unbalanced surface forces. If the surface forces are nonzero but uniform, they will have no gradient or divergence, and hence will not contribute to accelerating a particle.

Example 4.2 For a fluid in the velocity vector field studied in Example 4.1,

$$\mathbf{v} = \begin{pmatrix} v_1 \\ 0 \\ 0 \end{pmatrix} = \begin{pmatrix} v_o \frac{x_1}{L} \\ 0 \\ 0 \end{pmatrix}, \tag{4.77}$$

and in a field of uniform stress, find the body force field that induces its motion.

Solution

We have seen in the previous example that this flow is accelerating and the density is decreasing with time. Here we seek to characterize the force that induces that acceleration. Because the stress field is uniform, it has no divergence. So while the stress may be nonzero, there is no surface force imbalance to induce acceleration. So the acceleration must be entirely due to a body force field.

We reduce the linear momenta axiom, Eq. (4.73) as follows:

$$\rho \frac{d\mathbf{v}}{dt} = \rho \mathbf{f} \underbrace{- \nabla p + \left(\nabla^T \cdot \boldsymbol{\tau}\right)^T}_{=\,0}. \tag{4.78}$$

So $d\mathbf{v}/dt = \mathbf{f}$. We recall that $\mathbf{f}$ is actually the body force per unit mass; we will use the common convention of simply calling this the body force. Even though we know from an earlier example that the flow is compressible, the density does not affect the body force calculation. The acceleration in the 2 and 3 directions is zero, so we must have the body forces in those directions as zero: $f_2 = f_3 = 0$. For our steady, spatially varying one-dimensional velocity field, the only nontrivial component of the linear momenta axiom is thus

$$v_1 \frac{\partial v_1}{\partial x_1} = f_1, \quad \text{so} \quad \left(\frac{v_o x_1}{L}\right) v_o \frac{1}{L} = f_1, \quad \text{so} \quad \frac{v_o^2 x_1}{L^2} = f_1. \tag{4.79}$$

So the body force is zero at the origin, and must increase linearly as x_1 increases. The body force vector field is thus given by

$$\mathbf{f} = \begin{pmatrix} \frac{v_o^2 x_1}{L^2} \\ 0 \\ 0 \end{pmatrix}. \tag{4.80}$$

Many body force fields, for example gravitational fields, can be characterized by a *body force potential field* $\varphi(x_1, x_2, x_3)$ for which

$$\mathbf{f} = -\nabla \varphi. \tag{4.81}$$

If a body force potential exists, it is said that we have a *conservative body force*. In contrast, a *nonconservative body force* has no associated potential. This will be taken up in more detail in Section 6.8.1. Here we have $\partial \varphi / \partial x_1 = -v_o^2 x_1/L^2$, $\partial \varphi / \partial x_2 = 0$, $\partial \varphi / \partial x_3 = 0$, so we see by inspection that the body force potential field is

$$\varphi = -\frac{1}{2} \left(\frac{v_o x_1}{L}\right)^2. \tag{4.82}$$

There is a constant of integration that we have set to zero for convenience. For this force field, equipotential curves are curves for which x_1 are constant, here vertical lines. The body force vector is orthogonal to equipotential lines and points in the direction of most rapid decrease of φ. Because the acceleration vector points in the same direction as the body force vector, the fluid accelerates from regions of high potential to low potential. This is shown in Fig. 4.7.

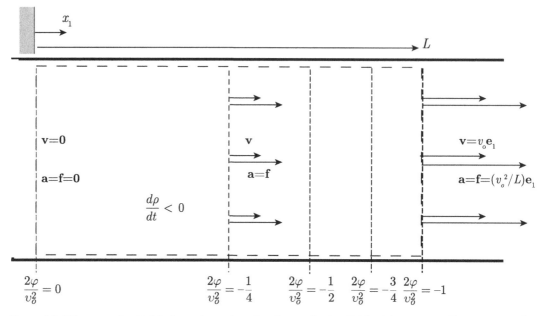

Figure 4.7 Diagram of a fluid element accelerating in a velocity field with nonzero divergence under the influence of a spatially nonuniform body force field and uniform stress field. Shown are curves of equipotential, and the acceleration, velocity, and body force vector fields.

Example 4.3 Show Newton's linear momenta principle in the limit of no viscous stress,

$$\frac{d\mathbf{v}}{dt} = \mathbf{f} - \frac{1}{\rho}\nabla p, \tag{4.83}$$

is invariant under the Galilean transformation of Section 1.2.1, written here as

$$x'_1 = x_1 - v_{1o}t, \quad x'_2 = x_2 - v_{2o}t, \quad x'_3 = x_3 - v_{3o}t, \quad t' = t. \tag{4.84}$$

Solution
We see $dt' = dt$. The spatial coordinates of fluid particles have time derivatives, with respect to the equivalent t or t', of

$$\frac{dx'_1}{dt'} = \frac{dx_1}{dt} - v_{1o}, \quad \frac{dx'_2}{dt'} = \frac{dx_2}{dt} - v_{2o}, \quad \frac{dx'_3}{dt'} = \frac{dx_3}{dt} - v_{3o}. \tag{4.85}$$

Defining the fluid particle velocities as usual, $dx_i/dt = v_i$, $dx'_i/dt' = v'_i$, we see

$$v'_1 = v_1 - v_{1o}, \quad v'_2 = v_2 - v_{2o}, \quad v'_3 = v_3 - v_{3o}. \tag{4.86}$$

Let us expand the equation set to be considered for transformation:

$$\frac{\partial v_1}{\partial t} + v_1\frac{\partial v_1}{\partial x_1} + v_2\frac{\partial v_1}{\partial x_2} + v_3\frac{\partial v_1}{\partial x_3} = f_1 - \frac{1}{\rho}\frac{\partial p}{\partial x_1}, \tag{4.87}$$

$$\frac{\partial v_2}{\partial t} + v_1\frac{\partial v_2}{\partial x_1} + v_2\frac{\partial v_2}{\partial x_2} + v_3\frac{\partial v_2}{\partial x_3} = f_2 - \frac{1}{\rho}\frac{\partial p}{\partial x_2}, \tag{4.88}$$

$$\frac{\partial v_3}{\partial t} + v_1\frac{\partial v_3}{\partial x_1} + v_2\frac{\partial v_3}{\partial x_2} + v_3\frac{\partial v_3}{\partial x_3} = f_3 - \frac{1}{\rho}\frac{\partial p}{\partial x_3}. \tag{4.89}$$

We need representations of the partial derivatives in the transformed coordinate system. Here it is advantageous to consider a so-called *space-time* formulation. Our original Cartesian system is obviously found by inverting the given transformation:

$$x_1 = x'_1 + v_{1o}t', \quad x_2 = x'_2 + v_{2o}t', \quad x_3 = x'_3 + v_{3o}t', \quad t = t'. \tag{4.90}$$

For a general space-time transformation, we have

$$\begin{pmatrix} dx_1 \\ dx_2 \\ dx_3 \\ dt \end{pmatrix} = \underbrace{\begin{pmatrix} \frac{\partial x_1}{\partial x'_1} & \frac{\partial x_1}{\partial x'_2} & \frac{\partial x_1}{\partial x'_3} & \frac{\partial x_1}{\partial t'} \\ \frac{\partial x_2}{\partial x'_1} & \frac{\partial x_2}{\partial x'_2} & \frac{\partial x_2}{\partial x'_3} & \frac{\partial x_2}{\partial t'} \\ \frac{\partial x_3}{\partial x'_1} & \frac{\partial x_3}{\partial x'_2} & \frac{\partial x_3}{\partial x'_3} & \frac{\partial x_3}{\partial t'} \\ \frac{\partial t}{\partial x'_1} & \frac{\partial t}{\partial x'_2} & \frac{\partial t}{\partial x'_3} & \frac{\partial t}{\partial t'} \end{pmatrix}}_{\mathbf{J}} \begin{pmatrix} dx'_1 \\ dx'_2 \\ dx'_3 \\ dt' \end{pmatrix}. \tag{4.91}$$

We have the Jacobian matrix $\mathbf{J}$, specialized for our Galilean transformation, as

$$\mathbf{J} = \begin{pmatrix} \frac{\partial x_1}{\partial x'_1} & \frac{\partial x_1}{\partial x'_2} & \frac{\partial x_1}{\partial x'_3} & \frac{\partial x_1}{\partial t'} \\ \frac{\partial x_2}{\partial x'_1} & \frac{\partial x_2}{\partial x'_2} & \frac{\partial x_2}{\partial x'_3} & \frac{\partial x_2}{\partial t'} \\ \frac{\partial x_3}{\partial x'_1} & \frac{\partial x_3}{\partial x'_2} & \frac{\partial x_3}{\partial x'_3} & \frac{\partial x_3}{\partial t'} \\ \frac{\partial t}{\partial x'_1} & \frac{\partial t}{\partial x'_2} & \frac{\partial t}{\partial x'_3} & \frac{\partial t}{\partial t'} \end{pmatrix} = \begin{pmatrix} 1 & 0 & 0 & v_{1o} \\ 0 & 1 & 0 & v_{2o} \\ 0 & 0 & 1 & v_{3o} \\ 0 & 0 & 0 & 1 \end{pmatrix}. \tag{4.92}$$

Here $J = \det \mathbf{J} = 1$, so the transformation is "volume"- and "orientation"-preserving, in the sense of space-time, and never singular. We also can show

$$\left(\mathbf{J}^T\right)^{-1} = \begin{pmatrix} 1 & 0 & 0 & 0 \\ 0 & 1 & 0 & 0 \\ 0 & 0 & 1 & 0 \\ -v_{1o} & -v_{2o} & -v_{3o} & 1 \end{pmatrix}, \tag{4.93}$$

and thus have from Eq. (2.251) that

$$\begin{pmatrix} \frac{\partial}{\partial x_1} \\ \frac{\partial}{\partial x_2} \\ \frac{\partial}{\partial x_3} \\ \frac{\partial}{\partial t} \end{pmatrix} = \begin{pmatrix} 1 & 0 & 0 & 0 \\ 0 & 1 & 0 & 0 \\ 0 & 0 & 1 & 0 \\ -v_{1o} & -v_{2o} & -v_{3o} & 1 \end{pmatrix} \begin{pmatrix} \frac{\partial}{\partial x'_1} \\ \frac{\partial}{\partial x'_2} \\ \frac{\partial}{\partial x'_3} \\ \frac{\partial}{\partial t'} \end{pmatrix}. \tag{4.94}$$

This gives, then, the simple

$$\frac{\partial}{\partial x_1} = \frac{\partial}{\partial x'_1}, \quad \frac{\partial}{\partial x_2} = \frac{\partial}{\partial x'_2}, \quad \frac{\partial}{\partial x_3} = \frac{\partial}{\partial x'_3}, \tag{4.95}$$

and the slightly more complicated

$$\frac{\partial}{\partial t} = \frac{\partial}{\partial t'} - v_{1o}\frac{\partial}{\partial x'_1} - v_{2o}\frac{\partial}{\partial x'_2} - v_{3o}\frac{\partial}{\partial x'_3}. \tag{4.96}$$

Let us apply the full Galilean transformation to one of the linear momenta equations, Eq. (4.87):

$$\frac{\partial v_1}{\partial t} + v_1\frac{\partial v_1}{\partial x_1} + v_2\frac{\partial v_1}{\partial x_2} + v_3\frac{\partial v_1}{\partial x_3} = f_1 - \frac{1}{\rho}\frac{\partial p}{\partial x_1}. \tag{4.97}$$

First, we replace all the velocities with their respective relative velocities:

$$\frac{\partial}{\partial t}(v'_1 + v_{1o}) + (v'_1 + v_{1o})\frac{\partial}{\partial x_1}(v'_1 + v_{1o})$$
$$+ (v'_2 + v_{2o})\frac{\partial}{\partial x_2}(v'_1 + v_{1o}) + (v'_3 + v_{3o})\frac{\partial}{\partial x_3}(v'_1 + v_{1o}) = f_1 - \frac{1}{\rho}\frac{\partial p}{\partial x_1}. \tag{4.98}$$

Because v_{io} are all constant, the equation simplifies to

$$\frac{\partial v_1'}{\partial t} + (v_1' + v_{1o})\frac{\partial v_1'}{\partial x_1} + (v_2' + v_{2o})\frac{\partial v_1'}{\partial x_2} + (v_3' + v_{3o})\frac{\partial v_1'}{\partial x_3} = f_1 - \frac{1}{\rho}\frac{\partial p}{\partial x_1}. \tag{4.99}$$

Because of how spatial derivatives transform, Eq. (4.95), we can say

$$\frac{\partial v_1'}{\partial t} + (v_1' + v_{1o})\frac{\partial v_1'}{\partial x_1'} + (v_2' + v_{2o})\frac{\partial v_1'}{\partial x_2'} + (v_3' + v_{3o})\frac{\partial v_1'}{\partial x_3'} = f_1 - \frac{1}{\rho}\frac{\partial p}{\partial x_1'}. \tag{4.100}$$

We next transform the time derivative via Eq. (4.96) to get

$$\frac{\partial v_1'}{\partial t'} - v_{1o}\frac{\partial v_1'}{\partial x_1'} - v_{2o}\frac{\partial v_1'}{\partial x_2'} - v_{3o}\frac{\partial v_1'}{\partial x_3'}$$
$$+ (v_1' + v_{1o})\frac{\partial v_1'}{\partial x_1'} + (v_2' + v_{2o})\frac{\partial v_1'}{\partial x_2'} + (v_3' + v_{3o})\frac{\partial v_1'}{\partial x_3'} = f_1 - \frac{1}{\rho}\frac{\partial p}{\partial x_1'}. \tag{4.101}$$

This simplifies significantly to yield an equation that is invariant in form from the untransformed Eq. (4.87):

$$\frac{\partial v_1'}{\partial t'} + v_1'\frac{\partial v_1'}{\partial x_1'} + v_2'\frac{\partial v_1'}{\partial x_2'} + v_3'\frac{\partial v_1'}{\partial x_3'} = f_1 - \frac{1}{\rho}\frac{\partial p}{\partial x_1'}. \tag{4.102}$$

This extends to the 2 and 3 linear momentum equations, yielding the general transformed linear momenta equation to be represented as

$$\frac{\partial v_i'}{\partial t'} + v_j'\frac{\partial v_i'}{\partial x_j'} = f_i - \frac{1}{\rho}\frac{\partial p}{\partial x_i'}. \tag{4.103}$$

In terms of the material derivative, we could say

$$\frac{dv_i'}{dt'} = f_i - \frac{1}{\rho}\frac{\partial p}{\partial x_i'}. \tag{4.104}$$

The invariance of the linear momenta principle under Galilean transformation is the linchpin of Newtonian mechanics. The invariance holds when viscous stresses are considered. To demonstrate this requires knowing the functional form of $\boldsymbol{\tau}$, and this awaits development in Section 5.4.

4.3 Angular Momenta

It is easy to overlook the angular momenta principle; for many problems, its consequence is so simple that it is often just asserted without proof. In classical rigid body mechanics, it is redundant with the linear momenta principle. It is, however, an independent axiom for continuous deformable media.

4.3.1 General Relation for Polar Fluids

Let us first briefly consider the angular momenta of a *polar fluid*. Such a fluid may rotate on a macroscale about an axis; additionally fluid particles may rotate about their own axes on a microscale. Many, including Aris (1962, p. 103), Panton (2013, p. 89), or Paolucci (2016), have additional discussion of polar and nonpolar fluids.

Let us recall some notions from classical rigid body mechanics, referring to Fig. 4.8. We have the angular momenta vector $\hat{\mathbf{L}}$ for the particle of Fig. 4.8:

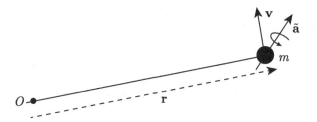

Figure 4.8 Diagram of particle of mass m with velocity $\mathbf{v}$ and intrinsic spin $\tilde{\mathbf{a}}$ rotating about an axis centered at point O, with radial distance vector $\mathbf{r}$.

$$\hat{\mathbf{L}} = \mathbf{r} \times (m\mathbf{v}) + m\tilde{\mathbf{a}}. \tag{4.105}$$

Here a portion of the particle's angular momenta is due to its rotation about O and another portion is due to it intrinsic rotation about its own axis, denoted by $\tilde{\mathbf{a}}$. Any force $\mathbf{F}$ that acts on m with lever arm $\mathbf{r}$ induces a torque $\hat{\mathbf{T}}$, that is,

$$\hat{\mathbf{T}} = \mathbf{r} \times \mathbf{F}. \tag{4.106}$$

Now let us apply these notions for an infinitesimal fluid particle with differential mass $\rho\, dV$:

$$\text{angular momenta} = (\mathbf{r} \times \mathbf{v} + \tilde{\mathbf{a}})(\rho\, dV) = \rho(\epsilon_{ijk} r_j v_k + \tilde{a}_i)\, dV, \tag{4.107}$$

$$\text{torque of body force} = \mathbf{r} \times (\rho\, dV)\mathbf{f} = \rho \epsilon_{ijk} r_j f_k\, dV, \tag{4.108}$$

$$\text{torque of surface force} = \mathbf{r} \times \mathbf{t}\, dA = \epsilon_{ijk} r_j t_k\, dA,$$

$$= \mathbf{r} \times (\mathbf{n}^T \cdot \mathbf{T})^T\, dA = \epsilon_{ijk} r_j n_p T_{pk}\, dA, \tag{4.109}$$

$$\text{angular momenta of surface couples} = \mathbf{n}^T \cdot \mathbf{H}\, dA = n_k H_{ki}\, dA, \tag{4.110}$$

$$\text{angular momenta of body couples} = \rho \mathbf{g}\, dV = \rho g_i\, dV. \tag{4.111}$$

Here we have introduced a so-called surface couple $\mathbf{H}$ and body couple $\mathbf{g}$ that may be a result of some magnetic effect or equivalent. Now the principle, which in words says the time rate of change of angular momenta of a material volume is equal to the sum of external couples (or torques) on the system, becomes mathematically,

$$\underbrace{\frac{d}{dt} \int_{V_m(t)} \rho(\epsilon_{ijk} r_j v_k + \tilde{a}_i)\, dV}_{\text{rate of change of angular momenta}} = \underbrace{\int_{V_m(t)} \rho\left(\epsilon_{ijk} r_j f_k + g_i\right) dV}_{\text{body force torques}}$$

$$+ \underbrace{\int_{A_m(t)} \left(\epsilon_{ijk} r_j n_p T_{pk} + n_k H_{ki}\right) dA}_{\text{surface force torques}}. \tag{4.112}$$

We apply Reynolds transport theorem and Gauss's theorem much as we have before and let the volume of the material volume shrink to zero now. First, with Reynolds, we get

$$\int_{V_m(t)} (\partial_o \rho \epsilon_{ijk} r_j v_k + \partial_o \rho \tilde{a}_i) \ dV + \int_{A_m(t)} (\epsilon_{ijk} \rho r_j v_k n_p v_p + \rho \tilde{a}_i n_p v_p) \ dA$$
$$= \int_{V_m(t)} \rho(\epsilon_{ijk} r_j f_k + g_i) \ dV + \int_{A_m(t)} (\epsilon_{ijk} r_j n_p T_{pk} + n_k H_{ki}) \ dA. \quad (4.113)$$

Next, using Gauss's theorem, we get

$$\int_{V_m(t)} (\partial_o \rho \epsilon_{ijk} r_j v_k + \partial_o \rho \tilde{a}_i) \ dV + \int_{V_m(t)} (\epsilon_{ijk} \partial_p (\rho r_j v_k v_p) + \partial_p (\rho \tilde{a}_i v_p)) \ dV$$
$$= \int_{V_m(t)} \rho(\epsilon_{ijk} r_j f_k + g_i) \ dV + \int_{V_m(t)} \epsilon_{ijk} \partial_p (r_j T_{pk}) \ dV + \int_{V_m(t)} \partial_k H_{ki} \ dV. \quad (4.114)$$

As the volume is arbitrary, the integrand formed by placing all terms under the same integral must be zero, which yields

$$\epsilon_{ijk} \left(\partial_o (\rho r_j v_k) + \partial_p (\rho r_j v_p v_k) - \rho r_j f_k - \partial_p (r_j T_{pk}) \right) + \partial_o (\rho \tilde{a}_i) + \partial_p (\rho v_p \tilde{a}_i) = \rho g_i + \partial_k H_{ki}. \quad (4.115)$$

Using the product rule to expand some of the derivatives, we get

$$\epsilon_{ijk} \left(r_j \partial_o (\rho v_k) + \rho v_k \underbrace{\partial_o r_j}_{= 0} + r_j \partial_p (\rho v_p v_k) + \rho v_p v_k \underbrace{\partial_p r_j}_{\delta_{pj}} - r_j \rho f_k - r_j \partial_p T_{pk} - T_{pk} \underbrace{\partial_p r_j}_{\delta_{pj}} \right)$$
$$+ \partial_o (\rho \tilde{a}_i) + \partial_p (\rho v_p \tilde{a}_i) = \rho g_i + \partial_k H_{ki}. \quad (4.116)$$

Applying the simplifications indicated and rearranging, we get

$$\epsilon_{ijk} r_j \underbrace{(\partial_o (\rho v_k) + \partial_p (\rho v_p v_k) - \rho f_k - \partial_p T_{pk})}_{=0 \text{ by linear momenta}} + \partial_o (\rho \tilde{a}_i) + \partial_p (\rho v_p \tilde{a}_i)$$
$$= \rho g_i + \partial_k H_{ki} - \underbrace{\rho \epsilon_{ijk} v_j v_k}_{=0} + \epsilon_{ijk} T_{jk}. \quad (4.117)$$

We then simplify using the mass equation to write

$$\rho \partial_o \tilde{a}_i + \rho v_p \partial_p \tilde{a}_i = \rho g_i + \partial_k H_{ki} + \epsilon_{ijk} T_{jk}. \quad (4.118)$$

Because ϵ_{ijk} is anti-symmetric in j, k, its tensor inner product with the symmetric part of T_{jk} is zero, leaving only the anti-symmetric part. Thus, we get

$$\rho \frac{d\tilde{a}_i}{dt} = \rho g_i + \partial_k H_{ki} + \epsilon_{ijk} T_{[jk]}. \quad (4.119)$$

This is sometimes known as *Cauchy's second law of motion*. The intrinsic spin of a fluid element is changed by body couples, surface couples, and anti-symmetric components of the stress tensor. In terms of the dual vector notation introduced in Eq. (2.88), we can say in Gibbs form

$$\rho \frac{d\tilde{\mathbf{a}}}{dt} = \rho \mathbf{g} + (\nabla^T \cdot \mathbf{H})^T + \mathbf{t}_\times. \quad (4.120)$$

4.3.2 Symmetry of the Stress Tensor for Nonpolar Fluids

From here on, our discussion will be mainly focused on so-called *nonpolar fluids* that lack intrinsic molecular angular momenta and under which no surface or body couples are applied.

So we take $d\tilde{a}_i/dt = 0$, $g_i = 0$, $H_{ki} = 0$, and the angular momenta principle reduces to the simple statement that

$$T_{[ij]} = 0. \tag{4.121}$$

That is, the anti-symmetric part of the stress tensor must be zero. Hence, the stress tensor, absent any surface couples, must be symmetric, and we get Cauchy's second law of motion to reduce to

$$T_{ij} = T_{ji}, \qquad \mathbf{T} = \mathbf{T}^T. \tag{4.122}$$

The symmetry of the stress tensor enables us to take a slight and useful diversion back to the stress tensor itself to consider some properties it also has because of the symmetry induced by the angular momenta principle in the limits we considered. We take these diversions in the next two sections.

4.3.3 Cauchy's Stress Quadric

We consider *Cauchy's stress quadric*. It is analogous to the extensional strain rate quadric considered in Section 3.8.4. Let us define σ as the scalar magnitude of stress in a chosen direction whose unit normal is $\mathbf{n}$. Analogous to Eq. (3.117), we then have

$$\sigma = \mathbf{n}^T \cdot \mathbf{T} \cdot \mathbf{n}, \tag{4.123}$$

subject to the constraint

$$\mathbf{n}^T \cdot \mathbf{n} = 1. \tag{4.124}$$

Repeating the analysis of Section 3.8.4, because $\mathbf{T}$ is symmetric, we can decompose it into a diagonal form and then say

$$\sigma = \mathbf{n}^T \cdot \underbrace{\mathbf{Q} \cdot \mathbf{\Lambda} \cdot \mathbf{Q}^T}_{\mathbf{T}} \cdot \mathbf{n} = (\mathbf{Q}^T \ \mathbf{n})^T \ \mathbf{\Lambda} \ (\mathbf{Q}^T \ \mathbf{n}). \tag{4.125}$$

Then, defining a rotated coordinate system by $\mathbf{n}' = \mathbf{Q}^T \cdot \mathbf{n}$, we see

$$\sigma = \mathbf{n}'^T \cdot \mathbf{\Lambda} \cdot \mathbf{n}'. \tag{4.126}$$

Because of the diagonal form of $\mathbf{\Lambda}$, this can be written in full as

$$\sigma = \lambda^{(1)} n_1'^2 + \lambda^{(2)} n_2'^2 + \lambda^{(3)} n_3'^2. \tag{4.127}$$

These are subject to the constraint that $\mathbf{n}'$ is a unit vector; thus,

$$1 = n_1'^2 + n_2'^2 + n_3'^2. \tag{4.128}$$

One can repeat the analysis of Section 3.8.4 to formally show that σ has a maximum given by the maximum eigenvalue and a minimum given by the minimum eigenvalue. Had $\mathbf{T}$ not been symmetric, it would have been possible for some of the eigenvalues to have been complex, and the physical interpretation would require considerable refinement. For the symmetric $\mathbf{T}$, this analysis is useful in that it provides the extreme values of stress along with the orientation of the planes on which those extreme values occur.

Example 4.4 A symmetric stress tensor is given by

$$\mathbf{T} = \begin{pmatrix} 1 & 0 & 0 \\ 0 & 3 & 1 \\ 0 & 1 & 3 \end{pmatrix}. \tag{4.129}$$

Analyze Cauchy's stress quadric.

Solution

Because $\mathbf{T}$ is symmetric, its eigenvalues must be real, and its eigenvectors orthogonal. It can be calculated that the three eigenvalues are

$$\lambda^{(1)} = 4, \qquad \lambda^{(2)} = 2, \qquad \lambda^{(3)} = 1. \tag{4.130}$$

It can be shown, by computing the eigenvectors of $\mathbf{T}$ and casting their normalized values into the columns of a rotation matrix $\mathbf{Q}$, that a diagonal decomposition of $\mathbf{T}$ is

$$\mathbf{T} = \mathbf{Q} \cdot \mathbf{\Lambda} \cdot \mathbf{Q}^T, \tag{4.131}$$

$$\begin{pmatrix} 1 & 0 & 0 \\ 0 & 3 & 1 \\ 0 & 1 & 3 \end{pmatrix} = \begin{pmatrix} 0 & 0 & 1 \\ 1/\sqrt{2} & -1/\sqrt{2} & 0 \\ 1/\sqrt{2} & 1/\sqrt{2} & 0 \end{pmatrix} \begin{pmatrix} 4 & 0 & 0 \\ 0 & 2 & 0 \\ 0 & 0 & 1 \end{pmatrix} \begin{pmatrix} 0 & 1/\sqrt{2} & 1/\sqrt{2} \\ 0 & -1/\sqrt{2} & 1/\sqrt{2} \\ 1 & 0 & 0 \end{pmatrix}. \tag{4.132}$$

It is easily verified that $\det \mathbf{Q} = 1$, so the transformation is a rotation. This allows us to define a rotated system, denoted with a prime; the vector $\mathbf{n}$ transforms via

$$\mathbf{n}' = \mathbf{Q}^T \cdot \mathbf{n}, \qquad \begin{pmatrix} n'_1 \\ n'_2 \\ n'_3 \end{pmatrix} = \begin{pmatrix} 0 & 1/\sqrt{2} & 1/\sqrt{2} \\ 0 & -1/\sqrt{2} & 1/\sqrt{2} \\ 1 & 0 & 0 \end{pmatrix} \begin{pmatrix} n_1 \\ n_2 \\ n_3 \end{pmatrix} = \begin{pmatrix} \frac{n_2+n_3}{\sqrt{2}} \\ \frac{n_3-n_2}{\sqrt{2}} \\ n_1 \end{pmatrix}. \tag{4.133}$$

We then find

$$\sigma = 4\left(\frac{n_2+n_3}{\sqrt{2}}\right)^2 + 2\left(\frac{n_3-n_2}{\sqrt{2}}\right)^2 + n_1^2. \tag{4.134}$$

This is subject to the constraint

$$n_1^2 + n_2^2 + n_3^2 = 1. \tag{4.135}$$

We certainly could analyze this with the method of Lagrange multipliers. This problem, however, is simple enough that intuition allows us to see the extreme values of σ. When $\mathbf{n} = (1,0,0)^T$, we obviously get

$$\sigma = \lambda^{(3)} = 1. \tag{4.136}$$

This is the minimum value of σ. When $n_1 = 0$ and $n_2 = \sqrt{2}/2$, $n_3 = -\sqrt{2}/2$, we get

$$\sigma = \lambda^{(2)} = 2. \tag{4.137}$$

This is an eigenvalue, but not an extreme value. When $n_1 = 0$ and $n_2 = n_3 = \sqrt{2}/2$, we get

$$\sigma = \lambda^{(1)} = 4. \tag{4.138}$$

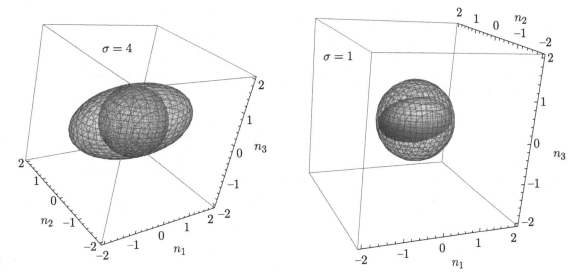

Figure 4.9 Plots of Cauchy's stress quadrics and unit spheres.

This is the maximum value of σ. Thus, as for the extensional strain rate quadric, it is the eigenvalues of the tensor that give the extreme values. It can be shown that the eigenvectors give the orientation of the element for which the extreme values are realized. In this case, they involve a rotation of $\pi/4$ about the x_1 axis. The stress quadrics and unit spheres are plotted in Fig. 4.9. Obviously, the quadrics here are triaxial ellipsoids.

4.3.4 Lamé Stress Ellipsoid

Related to Cauchy's stress quadric, for flows for which the stress tensor is symmetric, we can use the device of the Lamé[1] stress ellipsoid to better understand the nature of stress.

Because of its symmetry, the stress tensor $\mathbf{T}$ will possess three real eigenvalues, $\lambda^{(1)}$, $\lambda^{(2)}$, and $\lambda^{(3)}$, and three orthonormal eigenvectors, which we can call $\mathbf{n}^{(1)}$, $\mathbf{n}^{(2)}$, and $\mathbf{n}^{(3)}$. Now, in general, any traction vector is given by Eq. (4.34), $\mathbf{t} = \mathbf{T}^T \cdot \mathbf{n}$. Because here $\mathbf{T}$ is symmetric, any traction vector is also given by $\mathbf{t} = \mathbf{T} \cdot \mathbf{n}$. Because of the nature of the eigenvectors, we have three special traction vectors associated with each of the eigen-directions:

$$\mathbf{t}^{(1)} = \lambda^{(1)}\mathbf{n}^{(1)} = \mathbf{T} \cdot \mathbf{n}^{(1)}, \quad \mathbf{t}^{(2)} = \lambda^{(2)}\mathbf{n}^{(2)} = \mathbf{T} \cdot \mathbf{n}^{(2)}, \quad \mathbf{t}^{(3)} = \lambda^{(3)}\mathbf{n}^{(3)} = \mathbf{T} \cdot \mathbf{n}^{(3)}.$$
(4.139)

Let us now rotate our coordinate system so that it is aligned with the orthonormal eigenvectors. We will call this the primed system. In this coordinate system, the representation of the stress tensor, $\mathbf{T}'$ is

---

[1] Gabriel Lamé, 1795–1870, French mathematician.

$$\mathbf{T}' = \begin{pmatrix} \lambda^{(1)} & 0 & 0 \\ 0 & \lambda^{(2)} & 0 \\ 0 & 0 & \lambda^{(3)} \end{pmatrix}. \tag{4.140}$$

Now choose an arbitrary plane with unit normal $\mathbf{n} = (n_1, n_2, n_3)^T$. The traction vector associated with that plane is

$$\mathbf{t} = \mathbf{T}' \cdot \mathbf{n} = \begin{pmatrix} t_1 \\ t_2 \\ t_3 \end{pmatrix} = \begin{pmatrix} \lambda^{(1)} & 0 & 0 \\ 0 & \lambda^{(2)} & 0 \\ 0 & 0 & \lambda^{(3)} \end{pmatrix} \begin{pmatrix} n_1 \\ n_2 \\ n_3 \end{pmatrix} = \begin{pmatrix} \lambda^{(1)} n_1 \\ \lambda^{(2)} n_2 \\ \lambda^{(3)} n_3 \end{pmatrix}. \tag{4.141}$$

Now for our arbitrary unit normal vector $\mathbf{n}$, we must have

$$n_1^2 + n_2^2 + n_3^2 = 1. \tag{4.142}$$

Thus, we have

$$\left(\frac{t_1}{\lambda^{(1)}}\right)^2 + \left(\frac{t_2}{\lambda^{(2)}}\right)^2 + \left(\frac{t_3}{\lambda^{(3)}}\right)^2 = 1. \tag{4.143}$$

This is the equation of an ellipsoid in the space $(t_1, t_2, t_3)^T$. In contrast to Cauchy's stress quadric, which may be an ellipsoid, or a variety of hyperboloids, this construction must always yield an ellipsoid. It has an analog to the well-known Mohr's[2] circle, which is common in solid mechanics.

Example 4.5 Consider the symmetric stress tensor of the previous example:

$$\mathbf{T} = \begin{pmatrix} 1 & 0 & 0 \\ 0 & 3 & 1 \\ 0 & 1 & 3 \end{pmatrix}. \tag{4.144}$$

Determine and plot the Lamé stress ellipsoid.

Solution
The symmetric matrix possesses a set of orthonormal eigenvectors and real eigenvalues. The eigenvalues are found to be $\lambda^{(1)} = 4$, $\lambda^{(2)} = 2$, $\lambda^{(3)} = 1$. After rotation so that the coordinate axes are aligned with the eigenvectors, the stress tensor is represented as

$$\mathbf{T}' = \begin{pmatrix} 4 & 0 & 0 \\ 0 & 2 & 0 \\ 0 & 0 & 1 \end{pmatrix}. \tag{4.145}$$

By inspection, the Lamé stress ellipsoid for this system is triaxial and is given by

$$\left(\frac{t_1}{4}\right)^2 + \left(\frac{t_2}{2}\right)^2 + \left(\frac{t_3}{1}\right)^2 = 1. \tag{4.146}$$

The ellipsoid is plotted in Fig. 4.10. Note that the coordinate axes are not distances but tractions. The plot shows the available values of the components of $\mathbf{t}$ for all possible orientations $\mathbf{n}$. If we had chosen $\mathbf{n} = (1, 0, 0)^T$, we would for that orientation have found a maximum traction magnitude with $\mathbf{t} = (4, 0, 0)^T$. Had we chosen $\mathbf{n} = (0, 0, 1)^T$, we would have found a minimum traction magnitude with $\mathbf{t} = (0, 0, 1)^T$.

---

[2] Christian Otto Mohr, 1835–1918, German civil engineer.

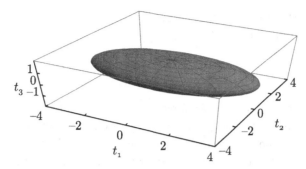

Figure 4.10 Plot of triaxial Lamé stress ellipsoid, $(t_1/4)^2 + (t_2/2)^2 + (t_3/1)^2 = 1$.

4.4 Energy

We recall the first law of thermodynamics, which states that the time rate of change of a material volume's internal and kinetic energy equals the rate of heat transferred to the material volume less the rate of work done by the material volume. Here we have adopted the engineering sign convention for heat and work, motivated by steam engine analysis, for which thermal energy came "in" and work came "out." Mathematically, this is stated as

$$\frac{d\mathcal{E}}{dt} = \frac{dQ}{dt} - \frac{dW}{dt}. \tag{4.147}$$

In this case (though this is not uniformly enforced in this book), the upper case letters denote extensive thermodynamic properties. For example, $\mathcal{E}$ is extensive *total energy*, inclusive of internal and kinetic,

$$\mathcal{E} = \rho V \left(e + \frac{1}{2} v_j v_j \right), \tag{4.148}$$

with SI units of J. The computational fluid dynamics literature often makes the unfortunate choice of defining the "total energy" as $E = e + v_j v_j/2$ with units of J/kg. Thus, it is really a specific energy and violates the thermodynamics convention that lower case variables are used for intensive properties. We will not use this nomenclature, and will generally reserve upper case variables for extensive properties. We will consider "total" to imply the sum of internal and kinetic, which could either be extensive or intensive. We take the extensive internal energy to be $E = \rho V e$. We could have included potential energy in $\mathcal{E}$, but will instead absorb it into the work term W. The corresponding intensive total energy ε with SI units of J/kg is $\varepsilon = e + v_j v_j/2$. Let us consider each term in the first law of thermodynamics in detail and then write the equation in final form.

4.4.1 Total Energy

For a fluid particle, the differential amount of extensive total energy is

$$d\mathcal{E} = \rho \left(e + \frac{1}{2} v_j v_j \right) dV = \underbrace{\rho\, dV}_{\text{mass}} \underbrace{\left(e + \frac{1}{2} v_j v_j \right)}_{\text{internal + kinetic}}. \tag{4.149}$$

The differential extensive total energy $d\mathcal{E}$ is the product of the differential mass and the sum of the specific internal and kinetic energies.

4.4.2 Work

Recall that work is done when a force acts through a distance, and a work rate arises when a force acts through a distance at a particular rate in time (hence, a velocity is involved). Recall also that work is the dot product of the force vector with the position. The dot product of the force vector and the velocity vector yields the the work rate. In shorthand, we have

$$dW = d\mathbf{x}^T \cdot \mathbf{F}, \qquad \frac{dW}{dt} = \frac{d\mathbf{x}^T}{dt} \cdot \mathbf{F} = \mathbf{v}^T \cdot \mathbf{F}. \tag{4.150}$$

Here W has the SI units of J, and the vector $\mathbf{F}$ has the SI units of N. We contrast this with our expression for body force per unit mass $\mathbf{f}$, which has SI units of N/kg = m/s^2. Now for the materials we consider, we must describe work done by two types of forces: (1) body, and (2) surface. First for the work rate done by a body force, we have

$$\text{work rate done } by \text{ force on fluid} = v_i(\rho \, dV)f_i, \tag{4.151}$$
$$\text{work rate done } by \text{ fluid} = -\rho v_i f_i \, dV. \tag{4.152}$$

Next for the work rate done by a surface force, we have

$$\text{work rate done } by \text{ force on fluid} = v_i(t_i \, dA) = v_i((n_j T_{ji}) \, dA), \tag{4.153}$$
$$\text{work rate done } by \text{ fluid} = -n_j T_{ji} v_i \, dA = -n_i T_{ij} v_j \, dA. \tag{4.154}$$

We focus attention on systems for which surface stresses are due to thermodynamic pressure and viscous effects and body forces due to gravity. Additional complexity arises when additional forces, such as those induced by electrical and magnetic effects, are considered.

4.4.3 Heat Transfer

The sign convention of the heat transfer rate must be considered with care. We recall that heat transfer *to* a body is associated with an increase in that body's energy. Now following the scenario depicted in the material volume of Fig. 4.11, we define the heat flux vector q_i as

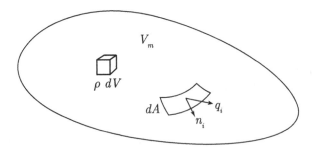

Figure 4.11 Diagram of finite material volume V_m, infinitesimal mass element $\rho \, dV$, and infinitesimal surface element dA with unit normal n_i, and heat flux vector q_i.

a vector that points in the direction of thermal energy flow that has units of energy per area per time; in SI this would be W/m². So, we have

$$\text{heat transfer rate } \textit{from} \text{ body through } dA = n_i q_i \, dA = \mathbf{n}^T \cdot \mathbf{q} \, dA, \quad (4.155)$$

$$\text{heat transfer rate } \textit{to} \text{ body through } dA = -n_i q_i \, dA = -\mathbf{n}^T \cdot \mathbf{q} \, dA. \quad (4.156)$$

4.4.4 Conservative Form

Putting the words of the first law into equation form, we get Eq. (4.147). We expand that into a more useful form to find

$$\underbrace{\frac{d}{dt} \int_{V_m(t)} \rho \left(e + \frac{1}{2} v_j v_j\right) dV}_{\frac{d\mathcal{E}}{dt},\text{ rate of change of energy}} = \underbrace{\int_{A_m(t)} (-n_i q_i) \, dA}_{\frac{dQ}{dt},\text{ heat transfer rate}}$$

$$\underbrace{- \left(\int_{A_m(t)} (-n_i T_{ij} v_j) \, dA + \int_{V_m(t)} (-\rho f_i v_i) \, dV \right)}_{\frac{dW}{dt},\text{ work rate}}, \quad (4.157)$$

$$\frac{d}{dt} \int_{V_m(t)} \rho \left(e + \frac{1}{2} v_j v_j\right) dV = -\int_{A_m(t)} n_i q_i \, dA + \int_{A_m(t)} n_i T_{ij} v_j \, dA + \int_{V_m(t)} \rho f_i v_i \, dV. \quad (4.158)$$

Omitting details of an identical application of the Reynolds transport theorem and Gauss's theorem, and shrinking the volume towards zero, we obtain the differential equation of energy evolution in conservative form (in Cartesian index then Gibbs notation):

$$\underbrace{\partial_o \left(\rho \left(e + \frac{1}{2} v_j v_j \right) \right)}_{\text{rate of change of total energy}} + \underbrace{\partial_i \left(\rho v_i \left(e + \frac{1}{2} v_j v_j \right) \right)}_{\text{advection of total energy}}$$

$$= - \underbrace{\partial_i q_i}_{\text{diffusive heat flux}} + \underbrace{\partial_i (T_{ij} v_j)}_{\text{surface force work rate}} + \underbrace{\rho v_i f_i}_{\text{body force work rate}}, \quad (4.159)$$

$$\frac{\partial}{\partial t} \left(\rho \left(e + \frac{1}{2} \mathbf{v}^T \cdot \mathbf{v} \right) \right) + \nabla^T \cdot \left(\rho \mathbf{v} \left(e + \frac{1}{2} \mathbf{v}^T \cdot \mathbf{v} \right) \right)$$
$$= -\nabla^T \cdot \mathbf{q} + \nabla^T \cdot (\mathbf{T} \cdot \mathbf{v}) + \rho \mathbf{v}^T \cdot \mathbf{f}. \quad (4.160)$$

This is a scalar equation as there are no free indices. We can segregate the work done by the surface forces into that done by pressure forces and that done by viscous forces by rewriting this in terms of p and τ_{ij} as follows:

$$\partial_o \left(\rho \left(e + \frac{1}{2} v_j v_j \right) \right) + \partial_i \left(\rho v_i \left(e + \frac{1}{2} v_j v_j \right) \right)$$
$$= -\partial_i q_i - \partial_i (p v_i) + \partial_i (\tau_{ij} v_j) + \rho v_i f_i, \quad (4.161)$$

$$\frac{\partial}{\partial t} \left(\rho \left(e + \frac{1}{2} \mathbf{v}^T \cdot \mathbf{v} \right) \right) + \nabla^T \cdot \left(\rho \mathbf{v} \left(e + \frac{1}{2} \mathbf{v}^T \cdot \mathbf{v} \right) \right)$$
$$= -\nabla^T \cdot \mathbf{q} - \nabla^T \cdot (p\mathbf{v}) + \nabla^T \cdot (\boldsymbol{\tau} \cdot \mathbf{v}) + \rho \mathbf{v}^T \cdot \mathbf{f}. \quad (4.162)$$

4.4.5 Secondary Forms

While the energy equation just derived is perfectly valid for all continuous materials, it is common to see other forms. Many will be described here.

Form Including Potential Energy

If the body force is conservative and time-independent, we can recover a familiar formulation in terms of the potential energy associated with the body force. So we adopt Eq. (4.81), $\mathbf{f} = -\nabla\varphi$ and further insist that φ is time-independent. The time-independence of φ is common for ordinary gravitational body forces. Let us then expand the body force term as follows, using the assumption that $\partial_o\varphi = 0$:

$$\rho v_i f_i = -\rho v_i \partial_i \varphi = -\rho \left(\underbrace{\partial_o \varphi}_{0} + v_i \partial_i \varphi \right) - \varphi \left(\underbrace{\frac{\partial \rho}{\partial t} + \partial_i(\rho v_i)}_{0} \right), \qquad (4.163)$$

$$= -(\partial_o(\rho\varphi) + \partial_i(\rho v_i \varphi)). \qquad (4.164)$$

We use this in Eq. (4.161) to get

$$\partial_o \left(\rho \left(e + \frac{1}{2} v_j v_j + \varphi \right) \right) + \partial_i \left(\rho v_i \left(e + \frac{1}{2} v_j v_j + \varphi \right) \right)$$
$$= -\partial_i q_i - \partial_i (p v_i) + \partial_i (\tau_{ij} v_j), \qquad (4.165)$$

$$\frac{\partial}{\partial t} \left(\rho \left(e + \frac{1}{2} \mathbf{v}^T \cdot \mathbf{v} + \varphi \right) \right) + \nabla^T \cdot \left(\rho \mathbf{v} \left(e + \frac{1}{2} \mathbf{v}^T \cdot \mathbf{v} + \varphi \right) \right)$$
$$= -\nabla^T \cdot \mathbf{q} - \nabla^T \cdot (p\mathbf{v}) + \nabla^T \cdot (\boldsymbol{\tau} \cdot \mathbf{v}). \qquad (4.166)$$

In this form, we explicitly see the presence of internal, kinetic, and potential energy. Adjustment is required for a time-dependent body force.

Enthalpy-based Conservative Form

It is common to bring the pressure-volume work term to the left-hand side to rewrite the conservative form of the energy equation, Eq. (4.161), as

$$\partial_o \left(\rho \left(e + \frac{1}{2} v_j v_j \right) \right) + \partial_i \left(\rho v_i \left(e + \frac{1}{2} v_j v_j + \frac{p}{\rho} \right) \right) = -\partial_i q_i + \partial_i (\tau_{ij} v_j)$$
$$+ \rho v_i f_i, \qquad (4.167)$$

$$\frac{\partial}{\partial t} \left(\rho \left(e + \frac{1}{2} \mathbf{v}^T \cdot \mathbf{v} \right) \right) + \nabla^T \cdot \left(\rho \mathbf{v} \left(e + \frac{1}{2} \mathbf{v}^T \cdot \mathbf{v} + \frac{p}{\rho} \right) \right) = -\nabla^T \cdot \mathbf{q} + \nabla^T \cdot (\boldsymbol{\tau} \cdot \mathbf{v})$$
$$+ \rho \mathbf{v}^T \cdot \mathbf{f}. \qquad (4.168)$$

Recall from elementary thermodynamics that the specific *enthalpy* h is defined as

$$h = e + \frac{p}{\rho}. \qquad (4.169)$$

Using this definition, the first law in conservative form can be rewritten as

$$\partial_o\left(\rho\left(e+\frac{1}{2}v_jv_j\right)\right)+\partial_i\left(\rho v_i\left(h+\frac{1}{2}v_jv_j\right)\right)=-\partial_i q_i+\partial_i(\tau_{ij}v_j)$$
$$+\rho v_i f_i, \quad (4.170)$$

$$\frac{\partial}{\partial t}\left(\rho\left(e+\frac{1}{2}\mathbf{v}^T\cdot\mathbf{v}\right)\right)+\nabla^T\cdot\left(\rho\mathbf{v}\left(h+\frac{1}{2}\mathbf{v}^T\cdot\mathbf{v}\right)\right)=-\nabla^T\cdot\mathbf{q}+\nabla^T\cdot(\boldsymbol{\tau}\cdot\mathbf{v})$$
$$+\rho\mathbf{v}^T\cdot\mathbf{f}. \quad (4.171)$$

Note that both e and h are present in this form.

Mechanical Energy Equation

The mechanical energy equation has no foundation in the first law of thermodynamics; instead, it is entirely a consequence of the linear momenta principle. It is the type of energy that is often considered in classical Newtonian particle mechanics, a world in which energy is either potential or kinetic but not thermal. We include it here because one needs to be able to distinguish mechanical from thermal energy and it will be useful in later analyses.

The mechanical energy equation, a consequence of the linear momenta principle, is obtained by taking the dot product of the velocity vector with the linear momenta principle:

$$\mathbf{v}^T\cdot\text{linear momenta}.$$

In detail, we get

$$v_j(\rho\partial_o v_j+\rho v_i\partial_i v_j)=\rho v_j f_j-v_j\partial_j p+(\partial_i\tau_{ij})v_j, \quad (4.172)$$

$$\rho\partial_o\left(\frac{v_jv_j}{2}\right)+\rho v_i\partial_i\left(\frac{v_jv_j}{2}\right)=\rho v_j f_j-v_j\partial_j p+(\partial_i\tau_{ij})v_j, \quad (4.173)$$

$$\frac{v_jv_j}{2}\text{ mass}:\frac{v_jv_j}{2}\partial_o\rho+\frac{v_jv_j}{2}\partial_i(\rho v_i)=0. \quad (4.174)$$

We add Eqs. (4.173) and (4.174) and use the product rule to get

$$\partial_o\left(\rho\frac{v_jv_j}{2}\right)+\partial_i\left(\rho v_i\frac{v_jv_j}{2}\right)=\rho v_j f_j-v_j\partial_j p+(\partial_i\tau_{ij})v_j, \quad (4.175)$$

$$\frac{\partial}{\partial t}\left(\rho\frac{\mathbf{v}^T\cdot\mathbf{v}}{2}\right)+\nabla^T\cdot\left(\rho\mathbf{v}\frac{\mathbf{v}^T\cdot\mathbf{v}}{2}\right)=\rho\mathbf{v}^T\cdot\mathbf{f}-\mathbf{v}^T\cdot\nabla p+\left(\nabla^T\cdot\boldsymbol{\tau}\right)\cdot\mathbf{v}. \quad (4.176)$$

The term $\rho v_j v_j/2$ represents the volume-averaged kinetic energy, with SI units J/m^3. The mechanical energy equation, Eq. (4.175), predicts the kinetic energy increases due to three effects: (1) fluid motion in the direction of a body force, (2) fluid motion in the direction of *decreasing* pressure, or (3) fluid motion in the direction of *increasing* viscous stress. Body forces themselves affect mechanical energy, while it is imbalances in surface forces that affect mechanical energy. We also present the nonconservative form of the mechanical energy equation, Eq. (4.173), as

$$\rho\frac{d}{dt}\left(\frac{v_jv_j}{2}\right)=\rho v_j f_j-v_j\partial_j p+(\partial_i\tau_{ij})v_j, \quad (4.177)$$

$$\rho\frac{d}{dt}\left(\frac{\mathbf{v}^T\cdot\mathbf{v}}{2}\right)=\rho\mathbf{v}^T\cdot\mathbf{f}-\mathbf{v}^T\cdot\nabla p+\left(\nabla^T\cdot\boldsymbol{\tau}\right)\cdot\mathbf{v}. \quad (4.178)$$

If the body force is conservative and time-independent, we can use Eq. (4.164) to recast the conservative form of the mechanical energy equation as

$$\partial_o \left(\rho \left(\frac{v_j v_j}{2} + \varphi\right)\right) + \partial_i \left(\rho v_i \left(\frac{v_j v_j}{2} + \varphi\right)\right) = -v_j \partial_j p + (\partial_i \tau_{ij}) v_j, \tag{4.179}$$

$$\frac{\partial}{\partial t}\left(\rho \left(\frac{\mathbf{v}^T \cdot \mathbf{v}}{2} + \varphi\right)\right) + \nabla^T \cdot \left(\rho \mathbf{v}\left(\frac{\mathbf{v}^T \cdot \mathbf{v}}{2} + \varphi\right)\right) = -\mathbf{v}^T \cdot \nabla p + \left(\nabla^T \cdot \boldsymbol{\tau}\right) \cdot \mathbf{v}. \tag{4.180}$$

In nonconservative form, this is

$$\rho \frac{d}{dt}\left(\frac{v_j v_j}{2} + \varphi\right) = -v_j \partial_j p + (\partial_i \tau_{ij}) v_j, \tag{4.181}$$

$$\rho \frac{d}{dt}\left(\frac{\mathbf{v}^T \cdot \mathbf{v}}{2} + \varphi\right) = -\mathbf{v}^T \cdot \nabla p + \left(\nabla^T \cdot \boldsymbol{\tau}\right) \cdot \mathbf{v}. \tag{4.182}$$

These forms allow one to see how the sum of the kinetic and potential energies evolves in response to surface forces.

Thermal Energy Equation

If we take the conservative form of the energy equation (4.161) and subtract from it the mechanical energy equation (4.175), we get an equation for the evolution of thermal energy:

$$\partial_o(\rho e) + \partial_i(\rho v_i e) = -\partial_i q_i - p \partial_i v_i + \tau_{ij} \partial_i v_j, \tag{4.183}$$

$$\frac{\partial}{\partial t}(\rho e) + \nabla^T \cdot (\rho \mathbf{v} e) = -\nabla^T \cdot \mathbf{q} - p \nabla^T \cdot \mathbf{v} + \boldsymbol{\tau} : \nabla \mathbf{v}^T. \tag{4.184}$$

Here ρe is the volume-averaged internal energy with SI units J/m^3. The thermal energy equation (4.183) predicts thermal energy (or internal energy) increases due to three effects: (1) negative divergence of the heat flux (more heat enters than leaves), (2) pressure force accompanied by a mean negative volumetric deformation (that is, a uniform compression; note that $\partial_i v_i$ is the relative expansion rate), or (3) viscous force associated with a deformation (we will worry about the sign later). For a general fluid, this includes a mean volumetric deformation as well as a deviatoric deformation. If the fluid satisfies Stokes' assumption, it is only the deviatoric deformation that induces a change in internal energy in the presence of viscous stress. Note that the term $\boldsymbol{\tau} : \nabla \mathbf{v}^T = \boldsymbol{\tau} : \mathbf{L} = \text{tr}(\boldsymbol{\tau} \cdot \mathbf{L}) = \text{tr}(\mathbf{L} \cdot \boldsymbol{\tau})$.

In contrast to mechanical energy, thermal energy changes do not require surface force imbalances; instead they require kinematic deformation. Moreover, body forces have no influence on thermal energy. The work done by a body force is partitioned entirely to the mechanical energy of a body.

Energy-based Nonconservative Form

We can obtain the commonly used nonconservative form of the energy equation, also known as the energy equation following a fluid particle, by the following operations. First, expand the thermal energy equation (4.183):

$$\rho \partial_o e + e \partial_o \rho + \rho v_i \partial_i e + e \partial_i(\rho v_i) = -\partial_i q_i - p \partial_i v_i + \tau_{ij} \partial_i v_j. \tag{4.185}$$

Then regroup and notice terms common from mass conservation, Eq. (4.6):

$$\rho \underbrace{(\partial_o e + v_i \partial_i e)}_{\frac{de}{dt}} + e \underbrace{(\partial_o \rho + \partial_i(\rho v_i))}_{=0 \text{ by mass}} = -\partial_i q_i - p \partial_i v_i + \tau_{ij} \partial_i v_j, \tag{4.186}$$

so we get

$$\rho \frac{de}{dt} = -\partial_i q_i - p \partial_i v_i + \tau_{ij} \partial_i v_j, \qquad \rho \frac{de}{dt} = -\nabla^T \cdot \mathbf{q} - p \nabla^T \cdot \mathbf{v} + \boldsymbol{\tau} : \nabla \mathbf{v}^T. \tag{4.187}$$

As an aside, we note for polar fluids that $\boldsymbol{\tau}$ may have an anti-symmetric component, which we call $\boldsymbol{\tau}_a$. If there is also an anti-symmetric component to the velocity gradient $\nabla \mathbf{v}^T$, which is $\mathbf{R}$, associated with rotation as a rigid body, then the combination $\boldsymbol{\tau}_a : \mathbf{R}$ contributes to change e. Now e, the internal energy, may be devised to include many effects, including translational kinetic energy of molecules as well as perhaps micro-spin of molecules. For such a fluid, a change in e could imply either a change in translational kinetic energy, reflected a change in the ordinary temperature T, or a change in the micro-spin of the molecules. To ascertain how such a change in e is partitioned requires appeal to either a more detailed theory than is presented here or to experiment. We will not consider it further. If $\boldsymbol{\tau}$ is symmetric, as it is for nonpolar fluids, the only contribution of $\boldsymbol{\tau} : \nabla \mathbf{v}^T$ is due to the symmetric part of the velocity gradient $\mathbf{D}$, so one could also say

$$\rho \frac{de}{dt} = -\partial_i q_i - p \partial_i v_i + \tau_{ij} \partial_{(i} v_{j)}, \qquad \rho \frac{de}{dt} = -\nabla^T \cdot \mathbf{q} - p \nabla^T \cdot \mathbf{v} + \boldsymbol{\tau} : \mathbf{D}. \tag{4.188}$$

Note that $\boldsymbol{\tau} : \mathbf{D} = \text{tr}(\boldsymbol{\tau} \cdot \mathbf{D}) = \text{tr}(\mathbf{D} \cdot \boldsymbol{\tau})$.

Enthalpy-based Nonconservative Form

Often the energy equation is cast in terms of enthalpy. This is generally valid, but especially useful in environments in which pressure changes are small, for example low Mach number flows. Now starting with the energy equation following a particle (4.187), we can use one form of the mass equation, Eq. (4.11), to eliminate the relative expansion rate $\partial_i v_i$ in favor of the material derivative of density to get

$$\rho \frac{de}{dt} = -\partial_i q_i + \frac{p}{\rho} \frac{d\rho}{dt} + \tau_{ij} \partial_i v_j. \tag{4.189}$$

Rearranging, we get

$$\rho \left(\frac{de}{dt} - \frac{p}{\rho^2} \frac{d\rho}{dt} \right) = -\partial_i q_i + \tau_{ij} \partial_i v_j. \tag{4.190}$$

Now differentiating Eq. (4.169), $h = e + p/\rho$, we find

$$dh = de - \frac{p}{\rho^2} d\rho + \frac{1}{\rho} dp, \tag{4.191}$$

$$\frac{dh}{dt} = \frac{de}{dt} - \frac{p}{\rho^2} \frac{d\rho}{dt} + \frac{1}{\rho} \frac{dp}{dt}, \tag{4.192}$$

$$\rho \frac{de}{dt} - \frac{p}{\rho} \frac{d\rho}{dt} = \rho \frac{dh}{dt} - \frac{dp}{dt}. \tag{4.193}$$

So, using Eq. (4.193) to eliminate de/dt in Eq. (4.190) in favor of dh/dt, the energy equation in terms of enthalpy becomes

$$\rho \frac{dh}{dt} = \frac{dp}{dt} - \partial_i q_i + \tau_{ij} \partial_i v_j, \qquad \rho \frac{dh}{dt} = \frac{dp}{dt} - \nabla^T \cdot \mathbf{q} + \boldsymbol{\tau} : \nabla \mathbf{v}^T. \qquad (4.194)$$

Entropy-based Forms

By using standard relations from thermodynamics, we can write the energy equation in terms of entropy. It is important to note that this is just an algebraic substitution. The physical principle that this equation will represent is still *energy* evolution from the first law of thermodynamics.

Recall the Gibbs equation from thermodynamics that serves to define entropy s:

$$T \, ds = de + p \, d\hat{v}. \qquad (4.195)$$

As an aside, we have employed the same symbol, s for both entropy and arc length; the context will allow the two to be distinguished. This Gibbs equation is valid for simple compressible substances, for which the only reversible work mode is pressure-volume work. Additional physics, such as electro-magnetic effects or chemical reactions, would require the Gibbs equation to be modified. Here T is the absolute temperature, and $\hat{v}$ is the specific volume, $\hat{v} = V/m = 1/\rho$. When we recall that $T \, ds$ is the reversible heat transfer and $p \, d\hat{v}$ is the reversible work, we see that the Gibbs equation, consistent with the first law of thermodynamics, tells us that the change in internal energy equals difference between the reversible heat transfer and the reversible work, $de = T \, ds - p \, d\hat{v}$. As an aside, there is also a second Gibbs equation. With $h = e + p\hat{v}$, we have $dh = de + p \, d\hat{v} + \hat{v} \, dp$. Eliminating de in the Gibbs equation in favor of dh gives the second Gibbs equation

$$dh = T \, ds + \hat{v} \, dp. \qquad (4.196)$$

In terms of ρ, the Gibbs equation is

$$T \, ds = de - \frac{p}{\rho^2} \, d\rho. \qquad (4.197)$$

Taking the material derivative of Eq. (4.197), which is operationally equivalent to dividing by dt, and solving for de/dt, we get

$$\frac{de}{dt} = T \frac{ds}{dt} + \frac{p}{\rho^2} \frac{d\rho}{dt}. \qquad (4.198)$$

This is still essentially a thermodynamic definition of s. Now use Eq. (4.198) in the nonconservative energy equation (4.187) to get an alternate expression for the first law:

$$\rho T \frac{ds}{dt} + \frac{p}{\rho} \frac{d\rho}{dt} = -\partial_i q_i - p \partial_i v_i + \tau_{ij} \partial_i v_j. \qquad (4.199)$$

Recalling Eq. (4.11), $-\partial_i v_i = (1/\rho)(d\rho/dt)$, we have

$$\rho T \frac{ds}{dt} = -\partial_i q_i + \tau_{ij} \partial_i v_j, \qquad \rho \frac{ds}{dt} = -\frac{1}{T} \partial_i q_i + \frac{1}{T} \tau_{ij} \partial_i v_j. \qquad (4.200)$$

Using the fact that from the quotient rule we have $\partial_i(q_i/T) = (1/T)\partial_i q_i - (q_i/T^2)\partial_i T$, we can then say

$$\rho \frac{ds}{dt} = -\partial_i \left(\frac{q_i}{T}\right) - \frac{1}{T^2}q_i \partial_i T + \frac{1}{T}\tau_{ij}\partial_i v_j, \tag{4.201}$$

$$= -\nabla^T \cdot \left(\frac{\mathbf{q}}{T}\right) - \frac{1}{T^2}\mathbf{q}^T \cdot \nabla T + \frac{1}{T}\boldsymbol{\tau} : \nabla \mathbf{v}^T. \tag{4.202}$$

From this statement, we can conclude from the *first law* of thermodynamics that the entropy of a fluid particle changes due to heat transfer and to deformation in the presence of viscous stress. We will make a more precise statement about entropy changes after we introduce the second law of thermodynamics.

The energy equation in terms of entropy can be written in conservative form by adding the product of s and the mass equation, $s\partial_o \rho + s\partial_i(\rho v_i) = 0$, to Eq. (4.202) to obtain

$$\partial_o(\rho s) + \partial_i(\rho v_i s) = -\partial_i \left(\frac{q_i}{T}\right) - \frac{1}{T^2}q_i \partial_i T + \frac{1}{T}\tau_{ij}\partial_i v_j, \tag{4.203}$$

$$\frac{\partial}{\partial t}(\rho s) + \nabla^T \cdot (\rho \mathbf{v} s) = -\nabla^T \cdot \left(\frac{\mathbf{q}}{T}\right) - \frac{1}{T^2}\mathbf{q}^T \cdot \nabla T + \frac{1}{T}\boldsymbol{\tau} : \nabla \mathbf{v}^T. \tag{4.204}$$

4.5 Entropy Inequality

Let us use a nonrigorous method to suggest a form of the entropy inequality that is consistent with classical thermodynamics. Recall the mathematical statement of the entropy inequality from classical thermodynamics:

$$dS \geq \frac{dQ}{T}. \tag{4.205}$$

Here S is the extensive entropy, with SI units J/K, and Q is the thermal energy transfer into a system with SI units of J. Notice that entropy can go up or down in a process, depending on the heat transferred. If the process is adiabatic, $dQ = 0$, and the entropy can either remain fixed or rise. Now for our continuous material we have

$$dS = \rho s \, dV, \qquad dQ = -q_i n_i \, dA \, dt. \tag{4.206}$$

Here we have used s for the specific entropy, which has SI units J/kg/K. Notice we must be careful with our sign convention. When the heat flux vector is aligned with the outward normal, heat leaves the system. Because we take positive dQ to represent heat into a system, we need the negative sign.

The second law becomes then

$$\rho s \, dV \geq -\frac{q_i}{T} n_i \, dA \, dt. \tag{4.207}$$

Now integrate over the finite geometry: on the left-hand side this is a volume integral and the right-hand side this is a surface integral.

$$\int_{V_m(t)} \rho s \, dV \geq \left(\int_{A_m(t)} -\frac{q_i}{T} n_i \, dA\right) dt. \tag{4.208}$$

Differentiating with respect to time and then applying our typical machinery to the second law gives rise to

$$\underbrace{\frac{d}{dt} \int_{V_m(t)} \rho s \, dV}_{\text{rate of change of entropy}} \geq \underbrace{\int_{A_m(t)} -\frac{q_i}{T} n_i \, dA,}_{\text{heat transfer to temperature ratio}} \qquad (4.209)$$

$$\int_{V_m(t)} \partial_o(\rho s) \, dV + \int_{A_m(t)} \rho s v_i n_i \, dA \geq \int_{A_m(t)} -\frac{q_i}{T} n_i \, dA, \qquad (4.210)$$

$$\int_{V_m(t)} (\partial_o(\rho s) + \partial_i(\rho s v_i)) \, dV \geq \int_{V_m(t)} -\partial_i \left(\frac{q_i}{T}\right) dV, \qquad (4.211)$$

$$= \int_{V_m(t)} -\partial_i \left(\frac{q_i}{T}\right) dV + \int_{V_m(t)} \dot{\mathcal{I}} \, dV. \qquad (4.212)$$

Here we have defined the *irreversibility production rate*, $\dot{\mathcal{I}} \geq 0$, as a positive semi-definite scalar with units W/K/m^3. It is simply a convenience to replace the inequality with an equality. Then we have the conservative form

$$\partial_o(\rho s) + \partial_i(\rho s v_i) = -\partial_i \left(\frac{q_i}{T}\right) + \dot{\mathcal{I}}. \qquad (4.213)$$

Invoking mass conservation, Eq. (4.6), we get the nonconservative form of the second law to be

$$\rho \frac{ds}{dt} = -\partial_i \left(\frac{q_i}{T}\right) + \dot{\mathcal{I}}. \qquad (4.214)$$

Now if we subtract from this the first law written in terms of entropy, Eq. (4.202), we get

$$\dot{\mathcal{I}} = -\frac{1}{T^2} q_i \partial_i T + \frac{1}{T} \underbrace{\tau_{ij} \partial_i v_j}_{\Phi}. \qquad (4.215)$$

With these definitions, we can state the second law as

$$\dot{\mathcal{I}} = -\frac{1}{T^2} q_i \partial_i T + \frac{1}{T} \underbrace{\tau_{ij} \partial_i v_j}_{\Phi} \geq 0. \qquad (4.216)$$

As an aside, we have defined the commonly used *viscous dissipation function* Φ as

$$\Phi \equiv \tau_{ij} \partial_i v_j. \qquad (4.217)$$

For symmetric stress tensors, we also have $\Phi = \tau_{ij} \partial_{(i} v_{j)}$. Now because $\dot{\mathcal{I}} \geq 0$, we can view the entirety of the second law as the following constraint, sometimes called the *weak form* of the *Clausius*[3]–*Duhem*[4] inequality:

$$-\frac{1}{T^2} q_i \partial_i T + \frac{1}{T} \tau_{ij} \partial_i v_j \geq 0, \qquad -\frac{1}{T^2} \mathbf{q}^T \cdot \nabla T + \frac{1}{T} \boldsymbol{\tau} : \nabla \mathbf{v}^T \geq 0. \qquad (4.218)$$

Recalling that τ_{ij} is symmetric by the angular momenta principle for no external couples, and, consequently, that its tensor inner product with the velocity gradient only has a contribution

---

[3] Rudolf Clausius, 1822–1888, Prussian-born German mathematical physicist, key figure in making thermodynamics a science, and author of the well-known statement of the second law of thermodynamics.

[4] Pierre Maurice Marie Duhem, 1861–1916, French physicist, mathematician, and philosopher.

from the symmetric part of the velocity gradient (that is, the deformation rate or strain rate tensor), the entropy inequality reduces to

$$-\frac{1}{T^2}q_i\partial_i T + \frac{1}{T}\tau_{ij}\partial_{(i}v_{j)} \geq 0, \qquad -\frac{1}{T^2}\mathbf{q}^T\cdot\nabla T + \frac{1}{T}\boldsymbol{\tau}:\mathbf{D} \geq 0. \qquad (4.219)$$

We shall see in upcoming sections that we will be able to specify q_i and τ_{ij} in such a fashion that is both consistent with experiment and satisfies the entropy inequality.

The more restrictive (and in some cases, *overly* restrictive) *strong form* of the Clausius–Duhem inequality requires each term to be greater than or equal to zero. For our system the strong form, realizing that the absolute temperature $T > 0$, is

$$-q_i\partial_i T \geq 0, \quad \Phi = \tau_{ij}\partial_{(i}v_{j)} \geq 0; \qquad -\mathbf{q}^T\cdot\nabla T \geq 0, \quad \Phi = \boldsymbol{\tau}:\mathbf{D} \geq 0. \qquad (4.220)$$

It is straightforward to show that terms that generate entropy due to viscous work also dissipate mechanical energy. This can be demonstrated by considering the mechanisms that cause mechanical energy to change within a finite fixed control volume V. First consider the nonconservative form of the mechanical energy equation, Eq. (4.177):

$$\rho\frac{d}{dt}\left(\frac{v_jv_j}{2}\right) = \rho v_j f_j - v_j\partial_j p + v_j\partial_i\tau_{ij}. \qquad (4.221)$$

Now use the product rule to restate the pressure and viscous work terms so as to achieve

$$\rho\frac{d}{dt}\left(\frac{v_jv_j}{2}\right) = \rho v_j f_j - \partial_j(v_j p) + p\partial_j v_j + \partial_i(\tau_{ij}v_j) - \underbrace{\tau_{ij}\partial_i v_j}_{=\Phi\geq 0}. \qquad (4.222)$$

So, here we see what induces *local* changes in mechanical energy. We see that body forces, pressure forces, and viscous forces in general can induce the mechanical energy to rise or fall. However, that part of the viscous stresses that is associated with the viscous dissipation, Φ, is guaranteed to induce a *local decrease* in mechanical energy. It is sometimes said that this is a transformation in which mechanical energy dissipates into thermal energy. There are additional nuances to this relation that we will consider in Section 14.1.1.

To study global changes in mechanical energy, we consider the conservative form of the mechanical energy equation, Eq. (4.175), here written in the same way that takes advantage of application of the product rule to the pressure and viscous terms:

$$\partial_o\left(\rho\frac{v_jv_j}{2}\right) + \partial_i\left(\rho v_i\frac{v_jv_j}{2}\right) = \rho v_j f_j - \partial_j(v_j p) + p\partial_j v_j + \partial_i(\tau_{ij}v_j) - \tau_{ij}\partial_i v_j. \qquad (4.223)$$

Now integrate over a fixed control volume with closed boundaries, so that

$$\int_V \partial_o\left(\rho\frac{v_jv_j}{2}\right)dV + \int_V \partial_i\left(\rho v_i\frac{v_jv_j}{2}\right)dV = \int_V \rho v_j f_j\,dV - \int_V \partial_j(v_j p)\,dV + \int_V p\partial_j v_j\,dV$$
$$+ \int_V \partial_i(\tau_{ij}v_j)\,dV - \int_V \tau_{ij}\partial_i v_j\,dV. \qquad (4.224)$$

Applying Leibniz's rule, Eq. (2.238), and Gauss's theorem, Eq. (2.222), we get

$$\frac{d}{dt}\int_V \rho\frac{v_jv_j}{2}\,dV + \int_A n_i\rho v_i\frac{v_jv_j}{2}\,dA = \int_V \rho v_j f_j\,dV - \int_A n_j v_j p\,dA + \int_V p\partial_j v_j\,dV$$
$$+ \int_A n_i(\tau_{ij}v_j)\,dA - \int_V \tau_{ij}\partial_i v_j\,dV. \qquad (4.225)$$

Now on the surface of the closed fixed volume, the velocity is zero, so we get

$$\frac{d}{dt}\int_V \rho \frac{v_j v_j}{2}\, dV = \int_V \rho v_j f_j\, dV + \int_V p\partial_j v_j\, dV - \int_V \underbrace{\tau_{ij}\partial_i v_j}_{\text{positive}}\, dV. \tag{4.226}$$

The strong form of the second law requires that $\tau_{ij}\partial_i v_j = \tau_{ij}\partial_{(i} v_{j)} \geq 0$. So, we see for a finite fixed closed volume of fluid that a body force and pressure force in conjunction with local volume changes can cause the global mechanical energy to either grow or decay. However, the viscous stress always induces a decay of global mechanical energy; in other words it is a *dissipative* effect, in that it dissipates mechanical energy.

The remaining term in the strong form of the Clausius–Duhem inequality, Eq. (4.220), namely $-\mathbf{q}^T \cdot \nabla T$, requires a more subtle interpretation. Some would think of it generically as a "dissipative" term; however, it does not dissipate mechanical energy. It certainly is an effect that is guaranteed to increase the entropy of a fluid particle as time increases, as it is associated with a portion of the entropy production that is irreversible.

4.6 Integral Forms

Our governing equations are formulated based upon laws that apply to a material element. We are not often interested in an actual material element but in some other fixed or moving volume in space. Rules for such systems can by formulated with Leibniz's rule in conjunction with the differential forms of our axioms.

Let us first recall Leibniz's rule (2.243) for an arbitrary scalar function f (that has no relation to our body force term) over a time-dependent arbitrary volume $V_a(t)$:

$$\frac{d}{dt}\int_{V_a(t)} f\, dV = \int_{V_a(t)} \partial_o f\, dV + \int_{A_a(t)} n_i w_i f\, dA. \tag{4.227}$$

Recall that w_i is the velocity of the arbitrary surface, not necessarily the particle velocity.

4.6.1 Mass

We rewrite the mass conservation, Eq. (4.6), as

$$\partial_o \rho = -\partial_i(\rho v_i). \tag{4.228}$$

Now let us use this, and let $f = \rho$ in Leibniz's rule, Eq. (4.227), to get

$$\frac{d}{dt}\int_{V_a(t)} \rho\, dV = \int_{V_a(t)} \partial_o \rho\, dV + \int_{A_a(t)} n_i w_i \rho\, dA, \tag{4.229}$$

$$= \int_{V_a(t)} (-\partial_i(\rho v_i))\, dV + \int_{A_a(t)} n_i w_i \rho\, dA. \tag{4.230}$$

Invoking Gauss's theorem, Eq. (2.222), we get

$$\frac{d}{dt}\int_{V_a(t)} \rho\, dV = \int_{A_a(t)} n_i \rho(w_i - v_i)\, dA. \tag{4.231}$$

Now consider three special cases.

Fixed Volume

We take $w_i = 0$. So the arbitrary volume that is a function of time, $V_a(t)$, becomes a fixed volume, V. It is bounded by a fixed surface A.

$$\frac{d}{dt}\int_V \rho \, dV = -\int_A n_i \rho v_i \, dA, \qquad \frac{d}{dt}\int_V \rho \, dV + \int_A n_i \rho v_i \, dA = 0. \tag{4.232}$$

Material Volume

Here we take $w_i = v_i$, and thus $V_a(t)$ becomes $V_m(t)$. We get the expected

$$\frac{d}{dt}\int_{V_m(t)} \rho \, dV = 0. \tag{4.233}$$

Moving Rigid Enclosure with Holes

Say the volume considered is a rigid enclosure with holes through with fluid can enter and exit. The our arbitrary surface $A_a(t)$ can be specified as

$$\begin{aligned}A_a(t) &= A_e(t) &&\text{area of entrances and exits}\\ &+ A_s(t) &&\text{rigid moving surface with } w_i = v_i\\ &+ A_s &&\text{fixed rigid surface with } w_i = v_i = 0.\end{aligned} \tag{4.234}$$

Then we get

$$\frac{d}{dt}\int_{V_a(t)} \rho \, dV + \int_{A_e(t)} \rho n_i(v_i - w_i) \, dA = 0. \tag{4.235}$$

Example 4.6 Consider the volume shown in Fig. 4.12. Water enters a circular hole of diameter $D_1 = 0.01\,\text{m}$ with velocity $v_1 = 3\,\text{m/s}$. Water enters another circular hole of diameter $D_2 = 0.03\,\text{m}$ with velocity $v_2 = 2\,\text{m/s}$. The cross-sectional area of the cylindrical tank is $A = 2\,\text{m}^2$. The tank has height H. Water at density ρ_w exists in the tank at height $h(t)$. Air at density ρ_a fills the remainder of the tank. Find the rate of rise of the water dh/dt.

Solution

Consider two control volumes: (1) V_1, the fixed volume enclosing the entire tank, and (2) $V_2(t)$, the material volume attached to the air.

First, let us write mass conservation for the material volume 2:

$$\frac{d}{dt}\int_{V_2} \rho_a \, dV = 0, \qquad \frac{d}{dt}\int_{h(t)}^{H} \rho_a A \, dz = 0. \tag{4.236}$$

Mass conservation for V_1 is

$$\frac{d}{dt}\int_{V_1} \rho \, dV + \int_{A_e} \rho v_i n_i \, dA = 0. \tag{4.237}$$

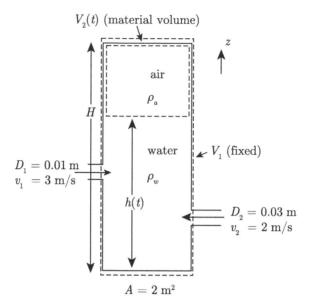

Figure 4.12 Sketch of volume with water and air being filled with water.

Now break up V_1 and write A_e explicitly

$$\frac{d}{dt}\int_0^{h(t)} \rho_w A\ dz + \underbrace{\frac{d}{dt}\int_{h(t)}^{H} \rho_a A\ dz}_{=0} = -\int_{A_1} \rho_w v_i n_i\ dA - \int_{A_2} \rho_w v_i n_i\ dA. \qquad (4.238)$$

Continuing, we get

$$\frac{d}{dt}\int_0^{h(t)} \rho_w A\ dz = \rho_w v_1 A_1 + \rho_w v_2 A_2 = \frac{\rho_w \pi}{4}(v_1 D_1^2 + v_2 D_2^2), \qquad (4.239)$$

$$\rho_w A \frac{d}{dt}\int_0^{h(t)} dz = \frac{\rho_w \pi}{4}(v_1 D_1^2 + v_2 D_2^2), \qquad (4.240)$$

$$\frac{dh}{dt} = \frac{\pi}{4A}(v_1 D_1^2 + v_2 D_2^2), \qquad (4.241)$$

$$= \frac{\pi}{4(2\,\text{m}^2)}\left(\left(3\,\frac{\text{m}}{\text{s}}\right)(0.01\,\text{m})^2 + \left(2\,\frac{\text{m}}{\text{s}}\right)(0.03\,\text{m})^2\right), \qquad (4.242)$$

$$= \frac{21\pi}{80\,000}\,\frac{\text{m}}{\text{s}} = 8.247 \times 10^{-4}\,\frac{\text{m}}{\text{s}}. \qquad (4.243)$$

4.6.2 Linear Momenta

Let us perform the same exercise for the linear momenta equation. First, in a strictly mathematical step, apply Leibniz's rule, Eq. (4.227), to linear momenta, $f = \rho v_i$:

$$\frac{d}{dt}\int_{V_a(t)} \rho v_i\ dV = \int_{V_a(t)} \partial_o(\rho v_i)\ dV + \int_{A_a(t)} n_j w_j \rho v_i\ dA. \qquad (4.244)$$

Now invoke the physical linear momenta axiom, Eq. (4.58). Here the axiom gives us an expression for $\partial_o(\rho v_i)$. We will also convert volume integrals to surface integrals via Gauss's theorem, Eq. (2.222), to get

$$\frac{d}{dt}\int_{V_a(t)} \rho v_i \, dV = -\int_{A_a(t)} (\rho n_j(v_j - w_j)v_i + n_i p - n_j \tau_{ij}) \, dA + \int_{V_a(t)} \rho f_i \, dV. \quad (4.245)$$

Now momenta flux terms only have values at entrances and exits (at solid surfaces, we get $v_i = w_i$), so we can say

$$\frac{d}{dt}\int_{V_a(t)} \rho v_i \, dV + \int_{A_e(t)} \rho n_j(v_j - w_j)v_i \, dA = -\int_{A_a(t)} n_i p \, dA + \int_{A_a(t)} n_j \tau_{ij} \, dA$$
$$+ \int_{V_a(t)} \rho f_i \, dV. \quad (4.246)$$

The surface forces are evaluated along all surfaces, not just entrances and exits.

4.6.3 Energy

Applying the same analysis to the energy equation, we obtain

$$\frac{d}{dt}\int_{V_a(t)} \rho\left(e + \frac{1}{2}v_j v_j\right) dV = -\int_{A_a(t)} \rho n_i(v_i - w_i)\left(e + \frac{1}{2}v_j v_j\right) dA$$
$$- \int_{A_a(t)} n_i q_i \, dA - \int_{A_a(t)} (n_i v_i p - n_i \tau_{ij} v_j) \, dA$$
$$+ \int_{V_a(t)} \rho v_i f_i \, dV. \quad (4.247)$$

4.6.4 General Expression

If we have a governing equation from a physical principle that is of the form

$$\partial_o f_j + \partial_i(v_i f_j) = \partial_i g_j + h_j, \quad (4.248)$$

then we can say for an arbitrary volume that

$$\underbrace{\frac{d}{dt}\int_{V_a(t)} f_j \, dV}_{\text{change of } f_j} = \underbrace{-\int_{A_a(t)} n_i f_j(v_i - w_i) \, dA}_{\text{flux of } f_j} + \underbrace{\int_{A_a(t)} n_i g_j \, dA}_{\text{effect of } g_j} + \underbrace{\int_{V_a(t)} h_j \, dV}_{\text{effect of } h_j}. \quad (4.249)$$

4.7 Summary of Axioms

Here we summarize the mathematical form of our axioms. We give the Cartesian index, Gibbs, and the full nonorthogonal index notation. Details of development of the nonorthogonal index notation are omitted. We will first present the conservative form and then the nonconservative form.

4.7.1 Conservative Form

Cartesian Index Form

$$\partial_o \rho + \partial_i(\rho v_i) = 0, \tag{4.250}$$

$$\partial_o(\rho v_i) + \partial_j(\rho v_j v_i) = \rho f_i - \partial_i p + \partial_j \tau_{ji}, \tag{4.251}$$

$$\tau_{ij} = \tau_{ji}, \tag{4.252}$$

$$\partial_o\left(\rho\left(e + \frac{1}{2}v_j v_j\right)\right) + \partial_i\left(\rho v_i\left(e + \frac{1}{2}v_j v_j\right)\right) = -\partial_i q_i - \partial_i(p v_i) + \partial_i(\tau_{ij} v_j) + \rho v_i f_i, \tag{4.253}$$

$$\partial_o(\rho s) + \partial_i(\rho s v_i) \geq -\partial_i\left(\frac{q_i}{T}\right). \tag{4.254}$$

Gibbs Form

$$\frac{\partial \rho}{\partial t} + \nabla^T \cdot (\rho \mathbf{v}) = 0, \tag{4.255}$$

$$\frac{\partial}{\partial t}(\rho \mathbf{v}) + \left(\nabla^T \cdot (\rho \mathbf{v} \mathbf{v}^T)\right)^T = \rho \mathbf{f} - \nabla p + \left(\nabla^T \cdot \boldsymbol{\tau}\right)^T, \tag{4.256}$$

$$\boldsymbol{\tau} = \boldsymbol{\tau}^T, \tag{4.257}$$

$$\frac{\partial}{\partial t}\left(\rho\left(e + \frac{1}{2}\mathbf{v}^T \cdot \mathbf{v}\right)\right) + \nabla^T \cdot \left(\rho \mathbf{v}\left(e + \frac{1}{2}\mathbf{v}^T \cdot \mathbf{v}\right)\right) = -\nabla^T \cdot \mathbf{q} - \nabla^T \cdot (p\mathbf{v})$$
$$+ \nabla^T \cdot (\boldsymbol{\tau} \cdot \mathbf{v}) + \rho \mathbf{v}^T \cdot \mathbf{f}, \tag{4.258}$$

$$\frac{\partial}{\partial t}(\rho s) + \nabla^T \cdot (\rho s \mathbf{v}) \geq -\nabla^T \cdot \left(\frac{\mathbf{q}}{T}\right). \tag{4.259}$$

Nonorthogonal Index Form

Here we present for completeness, but without detailed proof, the governing equations for general nonorthogonal coordinate systems. We use notation and concepts introduced in Section 2.5. Here we have extended the development of Vinokur (1974) to include the effects of momenta and energy diffusion. This extension has been guided by general notions found in standard works such as Aris (1962) as well as Liseikin (2017).

$$\frac{\partial}{\partial t}\left(\sqrt{g}\,\rho\right) + \frac{\partial}{\partial x^k}\left(\sqrt{g}\,\rho v^k\right) = 0, \tag{4.260}$$

$$\frac{\partial}{\partial t}\left(\sqrt{g}\,\rho v^j \frac{\partial \xi^i}{\partial x^j}\right) + \frac{\partial}{\partial x^k}\left(\sqrt{g}\,\rho v^j v^k \frac{\partial \xi^i}{\partial x^j}\right) = \sqrt{g}\,\rho f^j \frac{\partial \xi^i}{\partial x^j}$$
$$- \frac{\partial}{\partial x^k}\left(\sqrt{g}\,p g^{jk} \frac{\partial \xi^i}{\partial x^j}\right)$$
$$+ \frac{\partial}{\partial x^k}\left(\sqrt{g}\,\tau^{jk} \frac{\partial \xi^i}{\partial x^j}\right), \tag{4.261}$$

$$\tau^{jk} = \tau^{kj}, \tag{4.262}$$

$$\frac{\partial}{\partial t}\left(\sqrt{g}\,\rho\left(e+\frac{1}{2}g_{ij}v^iv^j\right)\right)+\frac{\partial}{\partial x^k}\left(\sqrt{g}\,\rho v^k\left(e+\frac{1}{2}g_{ij}v^iv^j\right)\right) = -\frac{\partial}{\partial x^k}\left(\sqrt{g}\,q^k\right)$$
$$-\frac{\partial}{\partial x^k}\left(\sqrt{g}\,pv^k\right)$$
$$+\frac{\partial}{\partial x^k}\left(\sqrt{g}\,g_{ij}v^j\tau^{ik}\right)$$
$$+\sqrt{g}\,\rho g_{ij}v^j f^i, \qquad (4.263)$$
$$\frac{\partial}{\partial t}\left(\sqrt{g}\,\rho s\right)+\frac{\partial}{\partial x^k}\left(\sqrt{g}\,\rho s v^k\right) \geq -\frac{\partial}{\partial x^k}\left(\sqrt{g}\,\frac{q^k}{T}\right). \qquad (4.264)$$

4.7.2 Nonconservative Form

Cartesian Index form

$$\frac{d\rho}{dt} = -\rho \partial_i v_i, \qquad (4.265)$$
$$\rho \frac{dv_i}{dt} = \rho f_i - \partial_i p + \partial_j \tau_{ji}, \qquad (4.266)$$
$$\tau_{ij} = \tau_{ji}, \qquad (4.267)$$
$$\rho \frac{de}{dt} = -\partial_i q_i - p \partial_i v_i + \tau_{ij}\partial_i v_j, \qquad (4.268)$$
$$\rho \frac{ds}{dt} \geq -\partial_i\left(\frac{q_i}{T}\right). \qquad (4.269)$$

Gibbs Form

$$\frac{d\rho}{dt} = -\rho \nabla^T \cdot \mathbf{v}, \qquad (4.270)$$
$$\rho \frac{d\mathbf{v}}{dt} = \rho \mathbf{f} - \nabla p + \left(\nabla^T \cdot \boldsymbol{\tau}\right)^T, \qquad (4.271)$$
$$\boldsymbol{\tau} = \boldsymbol{\tau}^T, \qquad (4.272)$$
$$\rho \frac{de}{dt} = -\nabla^T \cdot \mathbf{q} - p\nabla^T \cdot \mathbf{v} + \boldsymbol{\tau} : \nabla \mathbf{v}^T, \qquad (4.273)$$
$$\rho \frac{ds}{dt} \geq -\nabla^T \cdot \left(\frac{\mathbf{q}}{T}\right). \qquad (4.274)$$

Nonorthogonal Index Form

$$\frac{\partial \rho}{\partial t} + v^i \frac{\partial \rho}{\partial x^i} = -\frac{\rho}{\sqrt{g}}\frac{\partial}{\partial x^i}\left(\sqrt{g}\,v^i\right), \qquad (4.275)$$
$$\rho\left(\frac{\partial v^i}{\partial t} + v^j\left(\frac{\partial v^i}{\partial x^j}+\Gamma^i_{jl}v^l\right)\right) = \rho f^i - g^{ij}\frac{\partial p}{\partial x^j} + \frac{1}{\sqrt{g}}\frac{\partial}{\partial x^j}\left(\sqrt{g}\,\tau^{ij}\right) + \Gamma^i_{jk}\tau^{jk}, \qquad (4.276)$$
$$\tau^{jk} = \tau^{kj}, \qquad (4.277)$$

$$\rho\left(\frac{\partial e}{\partial t} + v^i \frac{\partial e}{\partial x^i}\right) = -\frac{1}{\sqrt{g}} \frac{\partial}{\partial x^i}\left(\sqrt{g}\, q^i\right) - \frac{p}{\sqrt{g}} \frac{\partial}{\partial x^i}\left(\sqrt{g}\, v^i\right)$$
$$+ g_{ik} \tau^{kj} \left(\frac{\partial v^i}{\partial x^j} + \Gamma^i_{jl} v^l\right), \qquad (4.278)$$

$$\rho\left(\frac{\partial s}{\partial t} + v^i \frac{\partial s}{\partial x^i}\right) \geq -\frac{1}{\sqrt{g}} \frac{\partial}{\partial x^i}\left(\sqrt{g}\, \frac{q^i}{T}\right). \qquad (4.279)$$

The term $\Gamma^i_{jl} v^l$ can be shown to represent the effects of non-Cartesian terms such as centripetal and Coriolis accelerations, as was introduced in Section 2.6.1.

4.7.3 Physical Interpretations

Each term in the governing axioms represents a physical mechanism. This approach is emphasized in the classical text by Bird et al. (2007) on transport processes. In general, the partial differential equations can be represented in the following form:

$$\text{local change} = \text{advection} + \text{diffusion} + \text{source}. \qquad (4.280)$$

Here we consider *advection* and *diffusion* to be types of transport phenomena. If we have a fixed volume of material, a property of that material, such as its thermal energy, can change because an outside flow sweeps energy in from outside due to bulk fluid motion. That is advection. Often the term *convection* is used in a similar fashion as advection. Both advection and convection are generally associated with bulk fluid motion that has $v_i \neq 0$. Such motion is macroscopic and identifiable. It is thus associated with information and thus order. Advection can apply to mass, momentum, or energy. Convection is sometimes restricted to advection of energy. Diffusion is associated with disordered, random motion of molecules. Diffusion is nonzero, even if there is no bulk fluid motion, $v_i = 0$. That is to say, a fluid at rest at the macroscale is generally in motion at the microscale, where the motion is random. The thermal energy can also change because random molecular motions allow slow leakage to the outside or leakage in from the outside. That is diffusion. Or the material can undergo intrinsic changes inside, such as viscous work, which converts kinetic energy into thermal energy.

Let us write the Gibbs form of the nonconservative equations of mass, linear momenta, and energy in a slightly different way to illustrate these mechanisms:

$$\begin{aligned}
\frac{\partial \rho}{\partial t} =& \quad \text{local change in mass} \\
& - \mathbf{v}^T \cdot \nabla \rho \quad \text{advection of mass} \\
& + 0 \quad \text{diffusion of mass} \\
& - \rho \nabla^T \cdot \mathbf{v}, \quad \text{volume expansion source,} \\
\rho \frac{\partial \mathbf{v}}{\partial t} =& \quad \text{local change in linear momenta} \\
& - \rho \left(\mathbf{v}^T \cdot \nabla\right) \mathbf{v} \quad \text{advection of linear momenta}
\end{aligned} \qquad (4.281)$$

$$
\begin{aligned}
&+ \left(\nabla^T \cdot \boldsymbol{\tau}\right)^T && \text{diffusion of linear momenta} && (4.282)\\
&+ \rho \mathbf{f} && \text{body force source of linear momenta}\\
&- \nabla p && \text{pressure gradient source of linear momenta,}
\end{aligned}
$$

$$
\begin{aligned}
\rho \frac{\partial e}{\partial t} =& \quad \text{local change in thermal energy}\\
&- \rho \mathbf{v}^T \cdot \nabla e && \text{advection of thermal energy}\\
&- \nabla^T \cdot \mathbf{q} && \text{diffusion of thermal energy}\\
&- p \nabla^T \cdot \mathbf{v} && \text{pressure work thermal energy source}\\
&+ \boldsymbol{\tau} : \nabla \mathbf{v}^T && \text{viscous work thermal energy source.} && (4.283)
\end{aligned}
$$

Briefly considering the second law of thermodynamics, we note that the irreversibility production rate $\hat{\mathcal{I}}$ is solely associated with diffusion of linear momenta and diffusion of energy. This makes sense in that diffusion is associated with random molecular motions and thus disorder. Advection is associated with an ordered motion of matter in that we retain knowledge of the position of the matter. Pressure–volume work is a reversible work and does not contribute to entropy changes. A portion of the heat transfer can be considered to be reversible. All of the work done by the viscous forces is irreversible work.

4.8 Incompleteness of the Axioms

These axioms are valid for any material that can be modeled as a continuum under the influence of the forces we have mentioned. Specifically, they are valid for both solid and fluid mechanics, which is remarkable.

While the axioms are complete, the equations are not! We have 23 unknowns here $\rho(1), v_i(3), f_i(3), p(1), \tau_{ij}(9), e(1), q_i(3), T(1), s(1)$, and only eight equations (one mass, three linear momenta, three independent angular momenta, one energy). We cannot count the second law as an equation, as it is an inequality. However, whatever result we get must be consistent with the second law of thermodynamics. In any case, we have a shortage of equations. We will see in Chapter 5 how we can use constitutive equations founded in empiricism for particular materials to complete our system.

SUMMARY

This chapter has presented the axioms of nonrelativistic continuum mechanics that may be applied to any continuum material, fluid or solid. Those axioms give principles that govern the evolution in time of mass, linear and angular momenta, energy, and entropy of a system. The axioms are designed to be consistent with principles of Newtonian mechanics and classical thermodynamics. They are incomplete in that to solve actual physical problems, one must specify properties of the material being studied.

PROBLEMS

4.1 A one-dimensional, steady velocity field is given in dimensionless form by $v_1 = x_1$, $v_2 = 0$, $v_3 = 0$. Find the most general steady dimensionless density field. Show there is a singularity at the plane defined by $x_1 = 0$, and discuss its problematic physical consequences.

4.2 A two-dimensional, steady velocity field is given in dimensionless form by $v_1 = x_1$, $v_2 = x_2$, $v_3 = 0$. Find the most general steady dimensionless density field. Transformation to cylindrical coordinates can aid the analysis. Show there is a singularity at the axis defined by $x_1 = x_2 = 0$, and discuss its problematic physical consequences.

4.3 A dimensionless velocity field is given by $v_1 = x_1 + x_2$, $v_2 = x_1 - x_2$. The stress tensor field describing surface forces is spatially uniform. Find the body force field that induces this motion. Show it to be conservative, and find the body force potential $\varphi(x_1, x_2)$. Show equipotential curves of the body force field are circles, and plot them along with the acceleration and velocity vector fields. Show the flow is irrotational, $\boldsymbol{\omega} = \nabla \times \mathbf{v} = \mathbf{0}$. Because of this, the velocity vector field is given by the gradient of a *velocity potential* ϕ, with $\mathbf{v} = \nabla \phi$. Find $\phi(x_1, x_2)$, and plot it with the velocity vector field. Find a compact algebraic form for $\phi(x_1, x_2)$ that demonstrates equipotential curves are hyperbolas; this analysis is aided by the method of quadratic forms.

4.4 A velocity field is given by $v_1 = x_1 + x_2$, $v_2 = x_1 - x_2$. The stress tensor field is

$$\mathbf{T} = \begin{pmatrix} x_1 & x_2 \\ x_2 & -x_1 \end{pmatrix}.$$

Take the fluid to obey Stokes' assumption. Assume the equations have been scaled to be in dimensionless form. Find the body force field that induces this motion. Because the velocity field has zero divergence, the flow is incompressible. Take $\rho = 1$. Show the body force field is conservative, and find the body force potential $\varphi(x_1, x_2)$. Find the pressure field. Plot equipotential body force field curves, the acceleration vector field, the velocity vector field, and the force vector field induced by surface stress.

4.5 A velocity field is given by $v_1 = x_1 + x_2$, $v_2 = x_1 - x_2$. The stress tensor field is

$$\mathbf{T} = \begin{pmatrix} x_1 & x_2 \\ x_1 & x_2 \end{pmatrix}.$$

Take the fluid to obey Stokes' assumption. Assume the equations have been scaled to dimensionless form. Find the body force field that induces this motion. Because the velocity field has zero divergence, the flow is incompressible, and one can take $\rho = 1$. Show the body force field is conservative, and find the body force potential $\varphi(x_1, x_2)$. Plot equipotential lines, the acceleration vector field, the velocity vector field, and isobars. Find a surface couple tensor $\mathbf{H}$ that could induce the asymmetry in the stress tensor, assuming zero body couple $\mathbf{g}$, and equilibrated micro-spin, $d\tilde{\mathbf{a}}/dt = \mathbf{0}$.

4.6 Consider the case of an incompressible fluid whose acceleration vector is a known constant $\mathbf{a}$, and the body force is a known constant $\mathbf{f}$. Also restrict attention to the case for which the stress tensor is diagonal: $\mathbf{T} = -p\mathbf{I}$, where p is the pressure. Find the most general expression for the pressure field.

4.7 Consider the first law of thermodynamics cast in terms of entropy:

$$\rho \frac{ds}{dt} = -\nabla^T \cdot \left(\frac{\mathbf{q}}{T}\right) - \frac{1}{T^2} \mathbf{q}^T \cdot \nabla T + \frac{1}{T} \boldsymbol{\tau} : \nabla \mathbf{v}^T.$$

Integrate this over a finite material volume $V_m(t)$ and determine necessary conditions for the fluid within the finite material volume to be isentropic.

4.8 A stress tensor is given by

$$\mathbf{T} = \begin{pmatrix} 1 & 1 & 1 \\ 1 & 2 & 1 \\ 1 & 1 & -3 \end{pmatrix}.$$

Find the traction vector $\mathbf{t}$ associated with the plane whose unit normal is $\mathbf{n} = (1/\sqrt{3}, 1/\sqrt{3}, 1/\sqrt{3})^T$. Determine and plot the Lamé stress ellipsoid. Determine and plot Cauchy's stress quadrics and associated unit spheres. Use the method of quadratic forms to write Cauchy's stress quadric in canonical form.

4.9 A compressible fluid exists within a finite fixed volume V. Heat flux is permitted at the boundary, but mass flux is not. A time-independent body force exists. Reduce the first law of thermodynamics to its simplest form for the integrated total energy, first for a nonconservative body force and second for a conservative body force. Give a physical interpretation in terms of fundamental thermodynamics.

4.10 Consider a steady flow in which all variables are dimensionless that is incompressible with $\rho = 1$ and with velocity field $v_1 = x_1$, $v_2 = -x_2$. The heat flux is zero, and the spatially uniform stress tensor is

$$\mathbf{T} = \begin{pmatrix} 1 & 2 \\ 2 & 3 \end{pmatrix}.$$

Identify the body force field that induces this motion. Find and interpret a solution for the internal energy field of the form $e = e(x_2/x_1)$.

4.11 Consider a steady flow in which all variables are dimensionless that is incompressible with $\rho = 1$ and with $v_1 = -x_2$, $v_2 = x_1$. The heat flux is zero, and the spatially uniform stress tensor is

$$\mathbf{T} = \begin{pmatrix} 0 & -1 \\ 1 & 0 \end{pmatrix}.$$

Identify the body force field that induces this motion. Find and interpret a solution for the internal energy field.

4.12 A two-dimensional Cartesian system has coordinates $(\xi^1, \xi^2)^T$. Consider the simple linear stretching transformation to the stretched system $(x^1, x^2)^T$ defined by $\xi^1 = 2x^1$, $\xi^2 = (1/2)x^2$. Here the "1" and "2" superscripts are not exponents, but simply identifiers. Determine if the transformation is orthogonal, area-preserving, and orientation-preserving. Formulate the mass conservation axiom in the transformed coordinate system. Transform the velocity vectors as well as the position coordinates.

4.13 Repeat the previous problem for the linear transformation to the non-Cartesian system $(x^1, x^2)^T$ defined by $\xi^1 = 2x^1 + 2x^2$, $\xi^2 = 2x^1 + 3x^2$.

4.14 Show that if the linear momenta equation is transformed to a coordinate system rotating with constant angular velocity $\mathbf{\Omega}$ about an axis passing through the origin, and $\mathbf{x}$ is the distance from the origin, that the linear momenta equation is

$$\frac{d\mathbf{v}}{dt} = \mathbf{f} - \frac{1}{\rho}\nabla p + \frac{1}{\rho}\left(\nabla^T \cdot \boldsymbol{\tau}\right)^T - \mathbf{\Omega} \times (\mathbf{\Omega} \times \mathbf{x}) - 2\mathbf{\Omega} \times \mathbf{v}.$$

Here $\mathbf{\Omega} \times (\mathbf{\Omega} \times \mathbf{x})$ is the centripetal acceleration, and $2\mathbf{\Omega} \times \mathbf{v}$ is the Coriolis acceleration. Show a further reduction is possible to achieve

$$\frac{d\mathbf{v}}{dt} = \mathbf{f} - \frac{1}{\rho}\nabla p + \frac{1}{\rho}\left(\nabla^T \cdot \boldsymbol{\tau}\right)^T + \frac{1}{2}\nabla\left((\mathbf{\Omega} \times \mathbf{x})^T \cdot (\mathbf{\Omega} \times \mathbf{x})\right) - 2\mathbf{\Omega} \times \mathbf{v}.$$

4.15 As described by Kundu et al. (2015, Chapter 13), *geostrophic flows* are those described by simplified models of atmospheric fluid dynamics near the surface of a planet in the limit in which the linear momenta principle reduces to a balance between Coriolis acceleration, gravity forces, and pressure gradient. Let us slightly simplify the spherical geometry and consider then a plane rotating about a central axis normal to the plane at constant angular velocity $\mathbf{\Omega}$. The balance between fluid inertial effects and forces is described by the following reduced linear momenta equation:

$$2\rho\mathbf{\Omega} \times \mathbf{v} = \rho\mathbf{f} - \nabla p,$$

where $\mathbf{\Omega} = (0, 0, \Omega_3)^T$ is the constant angular velocity, $\mathbf{f} = (0, 0, -g)^T$ is the constant gravitational acceleration, and density ρ is taken as constant. Take the curl of the linear momenta equation and show for this flow field that the *Taylor*[5]*–Proudman*[6] *theorem* holds: $\partial \mathbf{v}/\partial x_3 = \mathbf{0}$. Take the divergence of the linear momenta equation, and show that if the velocity field is also irrotational, that the pressure field is harmonic. Show for geostrophic flows with no component of vertical velocity, $v_3 = 0$, that the velocity vector is tangent to isobaric surfaces.

REFERENCES

Aris, R. (1962). *Vectors, Tensors, and the Basic Equations of Fluid Mechanics*. New York: Dover.

Bird, R. B., Stewart, W. E., and Lightfoot, E. N. (2007). *Transport Phenomena*, revised 2nd ed. New York: John Wiley.

Eringen, A. C. (1989). *Mechanics of Continua*, 2nd ed. Malabar, Florida: Krieger.

Fung, Y. C. (1965). *Foundations of Solid Mechanics*. Englewood Cliffs, New Jersey: Prentice-Hall.

Gad-el-Hak, M. (1995). Questions in fluid mechanics: Stokes' hypothesis for a Newtonian, isotropic fluid, *Journal of Fluids Engineering – Transactions of the ASME*, **117**(1), 3–5.

Kundu, P. K., Cohen, I. M., and Dowling, D. R. (2015). *Fluid Mechanics*, 6th ed. Amsterdam: Academic Press.

---

[5] Geoffrey Ingram Taylor, 1886–1975, English physicist.
[6] Joseph Proudman, 1888–1975, English mathematician and oceanographer.

Leal, L. G. (2007). *Advanced Transport Phenomena: Fluid Mechanics and Convective Transport Processes*. New York: Cambridge University Press.

Liseikin, V. L. (2017). *Grid Generation Methods*, 3rd ed. Cham, Switzerland: Springer.

Panton, R. L. (2013). *Incompressible Flow*, 4th ed. New York: John Wiley.

Paolucci, S. (2016). *Continuum Mechanics and Thermodynamics of Matter*. New York: Cambridge University Press.

Rosenhead, L. (1954). The second coefficient of viscosity: A brief review of fundamentals, *Proceedings of the Royal Society of London. Series A. Mathematical and Physical Sciences*, **226**(1164), 1–6.

Stokes, G. G. (1845). On the theories of internal friction of fluids in motion, *Transactions of the Cambridge Philosophical Society*, **8**, 287–305.

Truesdell, C. A. (1991). *A First Course in Rational Continuum Mechanics*, Vol. 1, 2nd ed. Boston, Massachusetts: Academic Press.

Vinokur, M. (1974). Conservation equations of gasdynamics in curvilinear coordinate systems, *Journal of Computational Physics*, **14**(2), 105–125.

Whitaker, S. (1968). *Introduction to Fluid Mechanics*. Malabar, Florida: Krieger.

Woods, L. C. (1975). *The Thermodynamics of Fluid Systems*. Oxford: Clarendon Press.

5 Constitutive Equations

In this chapter, we turn to the problem of completing the set of equations presented in Section 4.8 by introducing specific *constitutive equations*. They are material-specific and thus depend upon the *constitution* of the material. As an example constitutive equation, one can consider the well-known *Fourier's*[1] *law* for an isotropic material:

$$\mathbf{q} = -k\nabla T. \tag{5.1}$$

Here $\mathbf{q}$ is the heat flux vector, introduced in Section 4.4.3, and k is the thermal conductivity, introduced in Section 1.3. We will see in this chapter that for second law satisfaction $k \geq 0$. A depiction of a temperature field and the heat flux vector induced by Fourier's law is shown in Fig. 5.1. Thermal energy flows from regions of high temperature to regions of low temperature.

Constitutive equations are models based on experiment. They must satisfy principles of continuum mechanics. However, they are not as fundamental as the previously developed

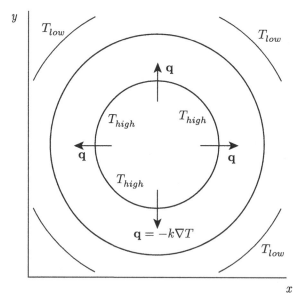

Figure 5.1 Schematic to illustrate Fourier's law for an isotropic material, $\mathbf{q} = -k\nabla T$.

---

[1] Jean-Baptiste-Joseph Fourier, 1768–1830, French mathematician and Egyptologist who studied the transfer of heat and the representation of mathematical functions by infinite series summations of other functions.

axioms. They will complete our continuum description. They can be useful ad hoc relations that in some sense model the sub-continuum nanostructure. In some cases, for example, the sub-continuum kinetic theory of gases, it can be shown that when the sub-continuum is formally averaged, that one obtains commonly used constitutive equations. In many cases, however, constitutive equations simply represent curve fits to basic experimental results, which can vary widely from material to material.

Importantly, constitutive equations are not completely arbitrary. Whatever is proposed must satisfy basic principles of mechanics. Aris (1962, pp. 190–192) summarizes an extensive list of principles for viability of a constitutive model, of which we will only focus on some in a simplified form. Among them are (i) they are consistent with the axioms, most importantly, the second law of thermodynamics; (ii) they are posed as tensorial quantities so that they may be correctly represented in transformed coordinate systems; (iii) when combined with the equations of motion, the number of equations should equal the number of unknowns; (iv) they have dimensional consistency; that is, the units must be correct; (v) the response of the system should be the same for all observers, this encompasses the principle of Galilean frame invariance and is known as *material indifference*; and (vi) the system have what is known as *equipresence*, which implies that certain dependencies on state variables or their gradients not be dismissed arbitrarily. It should be recognized that agreement is not universal for nuances of some of these principles; see for example Edelen and McLennan (1973) or Woods (1982). Let us briefly expand on the principle of equipresence. For example, we might hope to develop a constitutive equation for the heat flux vector $\mathbf{q}$. Experiment and intuition suggests it should be related to the gradient of temperature, $\mathbf{q} = -k\nabla T$. But continuum mechanics admits a broader range of constitutive models for $\mathbf{q}$. Naïvely, we might in general expect it to be a function of a large number of variables or their gradients:

$$\mathbf{q} = \mathbf{q}(\rho, T, \mathbf{v}, \nabla\rho, \nabla T, \dots). \tag{5.2}$$

One could envision a grand program to explore all such possibilities. Many dependencies could be dismissed, but some would remain. Practically, this is not feasible. Even linear couplings, such as

$$\mathbf{q} = \beta_1 \nabla\rho + \beta_2 \nabla T + \cdots \tag{5.3}$$

are difficult to evaluate. We will examine these briefly in this chapter for a few cases.

However, one should recognize that sometimes the physics demands that couplings be present. For example, when one considers simple exothermically reacting binary mixtures, with Y as the mass fraction of the products of reaction, it is well known that heat and mass flux vectors of the form

$$\mathbf{q} = -k\nabla T + \rho\mathcal{D}\hat{q}\nabla Y, \qquad \mathbf{j}_m = -\rho\mathcal{D}\nabla Y, \tag{5.4}$$

better capture observed physics. Here $\mathcal{D}$ is the mass diffusivity, $\hat{q}$ is the exothermic energy release per unit mass, and $\mathbf{j}_m$ is the diffusive mass flux, here specified by what is known as Fick's law; see Bird et al. (2007) or Powers (2016).

5.1 Frame Indifference

Our choice of a constitutive equation must be invariant under a Galilean transformation. Probably the most important consequence of this that arises in fluid mechanics is that *constitutive equations may depend on velocity differences or velocity derivatives, but not on velocity itself*. One must take care when models are posed involving fluid interactions with solid surfaces. Such solid surfaces in laboratory frames could be moving or stationary. If one transforms to a different steadily moving frame, an observer should record the same response. Special care is thus required when employing simple and common models of fluid friction such as the so-called Darcy[2] friction factor, which are often only valid in a laboratory frame and may lack frame invariance. We will not focus on the Darcy model, except briefly in Section 9.4.

Example 5.1 Examine the frame indifference under a Galilean transformation of Eq. (4.83),

$$\frac{d\mathbf{v}}{dt} = \mathbf{f} - \frac{1}{\rho}\nabla p, \tag{5.5}$$

for a body force of the form $\mathbf{f} = \beta\mathbf{v}$, where β is a constant.

Solution
Our model is thus

$$\frac{d\mathbf{v}}{dt} = \beta\mathbf{v} - \frac{1}{\rho}\nabla p. \tag{5.6}$$

Models of this type are occasionally used for drag for a fluid interacting with a solid wall or porous matrix of fixed solid particles. For example, we will discuss so-called Fanno flows in Section 9.4, and such models are sometimes formulated in ways that do not survive Galilean transformation. The models often presume the wall or porous matrix is stationary, but as we will see, to be frame invariant, the velocity of the solid needs to be built into the model, even if it is zero in the laboratory frame.

We have already seen in Eq. (4.104) that if $\mathbf{f}$ is constant, the system is frame invariant under Galilean transformation, and many details of the transformation are identical, so we will focus here only on those relevant to $\mathbf{f}$. The transformation is taken to be

$$\mathbf{x}' = \mathbf{x} - \mathbf{v}_o t, \qquad t' = t, \qquad \mathbf{v}' = \mathbf{v} - \mathbf{v}_o. \tag{5.7}$$

Here $\mathbf{v}_o$ is a constant velocity. We have already seen that the material derivative is frame invariant as is the gradient. So under this transformation, the linear momenta principle becomes

$$\frac{d\mathbf{v}'}{dt'} = \beta(\mathbf{v}' + \mathbf{v}_o) - \frac{1}{\rho}\nabla' p. \tag{5.8}$$

As this equation in the moving frame differs from that in the laboratory frame, it would predict different values of quantities such as pressure, which would not be observed, thus invalidating the model.

The original model can be fixed by having its constitutive equation for $\mathbf{f}$ depend on velocity differences. If we take instead

$$\frac{d\mathbf{v}}{dt} = \beta(\mathbf{v} - \mathbf{v}_w) - \frac{1}{\rho}\nabla p, \tag{5.9}$$

---

[2] Henry Philibert Gaspard Darcy, 1803–1858, French engineer.

where $\mathbf{v}_w$ is the velocity of the wall, under Galilean transformation, we get

$$\frac{d\mathbf{v}'}{dt'} = \beta(\mathbf{v}' - \mathbf{v}'_w) - \frac{1}{\rho}\nabla' p. \tag{5.10}$$

The corrected model is invariant. If the wall is stationary in the laboratory frame, there is no problem, and one can simply specify the additional relationship $\mathbf{v}_w = \mathbf{0}$.

5.2 Second Law Restrictions and Onsager Relations

Any proposed constitutive equation must not violate axioms of continuum mechanics. The axiom that requires the most attention is the second law of thermodynamics. We will see the entropy inequality provides additional restrictions on the form of constitutive equations. Recall the second law (equivalently, the weak form of the Clausius–Duhem inequality, Eq. (4.219)) tells us that

$$-\frac{1}{T^2}q_i\partial_i T + \frac{1}{T}\tau_{ij}\partial_{(i}v_{j)} \geq 0. \tag{5.11}$$

We would like to find forms of q_i and τ_{ij} that are consistent with Eq. (5.11).

5.2.1 Weak Form of the Clausius–Duhem Inequality

The weak form suggests that we may want to consider both q_i and τ_{ij} to be functions involving coupling of the temperature gradient $\partial_i T$ and the deformation tensor $\partial_{(i}v_{j)}$. By hypothesizing a coupling, we will be exploring the principle of equipresence. We will see that the coupling we hypothesize is one that is neither useful or used as it will violate other physical principles; however, its consideration is an excellent exercise that will expose methods that are useful and used when additional physical effects, such as anisotropy or coupled thermo-electric effects, are considered. The approach we will take is common in the field of *nonequilibrium thermodynamics*; summaries and relevant discussions of this extensive field are given by Callen (1985), de Groot and Mazur (1984), Gyarmati (1970), Hirschfelder et al. (1954), Kondepudi and Prigogine (1998), Lavenda (1978), Müller and Ruggeri (1998), Reynolds (1968), and Woods (1975).

Motivating Nonphysical Example

We can learn much about this method by studying a problem confined to fluid mechanics that the method will discern to be nonphysical. However, when the method is extended to include other physics, as shown by Reynolds (1968) for such devices as thermocouples, it is more than an exercise, but a tool to study and better describe nature. To illustrate the method, it suffices to consider a one-dimensional limit. In the one-dimensional limit, the weak form of the entropy inequality, Eq. (4.219), reduces to

$$-\frac{1}{T^2}q\frac{\partial T}{\partial x} + \frac{1}{T}\tau\frac{\partial u}{\partial x} \geq 0. \tag{5.12}$$

We can write the entropy inequality in a generalized matrix form as

$$\begin{pmatrix} -\frac{1}{T}\frac{\partial T}{\partial x} & \frac{1}{u}\frac{\partial u}{\partial x} \end{pmatrix} \begin{pmatrix} \frac{q}{T} \\ \frac{\tau u}{T} \end{pmatrix} \geq 0. \tag{5.13}$$

A factor of u/u was introduced to the viscous stress term. This allows for dimensional consistency in that q/T has the same units as $\tau u/T$.

Let us now hypothesize a coupled constitutive equation for q and τ. We speculate that a *linear relationship* exists between the generalized fluxes q/T and $\tau u/T$ and the generalized driving gradients $-(1/T)\partial T/\partial x$ and $(1/u)\partial u/\partial x$:

$$\frac{q}{T} = C_{11}\left(-\frac{1}{T}\frac{\partial T}{\partial x}\right) + C_{12}\frac{1}{u}\frac{\partial u}{\partial x}, \tag{5.14}$$

$$\frac{\tau u}{T} = C_{21}\left(-\frac{1}{T}\frac{\partial T}{\partial x}\right) + C_{22}\frac{1}{u}\frac{\partial u}{\partial x}. \tag{5.15}$$

This is perhaps the simplest nontrivial form that respects the coupling of equipresence. It is hypotheses of this type that are widely used and validated for more general systems.

For uncoupled systems with $C_{12} = C_{21} = 0$, we get $q = -C_{11}\partial T/\partial x$, which is physically defensible. However, we get an unusual relationship for τ linking it to $\partial u/\partial x$, but with extra terms that cannot be defended on physical grounds. Nevertheless, to illustrate the method, let us proceed and see where we are led by this hypothesis.

In matrix form, the hypothesized constitutive equation becomes

$$\begin{pmatrix} \frac{q}{T} \\ \frac{\tau u}{T} \end{pmatrix} = \begin{pmatrix} C_{11} & C_{12} \\ C_{21} & C_{22} \end{pmatrix} \begin{pmatrix} -\frac{1}{T}\frac{\partial T}{\partial x} \\ \frac{1}{u}\frac{\partial u}{\partial x} \end{pmatrix}. \tag{5.16}$$

We then substitute this hypothesized relationship into the entropy inequality to obtain

$$\begin{pmatrix} -\frac{1}{T}\frac{\partial T}{\partial x} & \frac{1}{u}\frac{\partial u}{\partial x} \end{pmatrix} \begin{pmatrix} C_{11} & C_{12} \\ C_{21} & C_{22} \end{pmatrix} \begin{pmatrix} -\frac{1}{T}\frac{\partial T}{\partial x} \\ \frac{1}{u}\frac{\partial u}{\partial x} \end{pmatrix} \geq 0. \tag{5.17}$$

We next segregate the matrix C_{ij} into a symmetric and anti-symmetric part to get

$$\begin{pmatrix} -\frac{1}{T}\frac{\partial T}{\partial x} & \frac{1}{u}\frac{\partial u}{\partial x} \end{pmatrix} \left(\begin{pmatrix} C_{11} & \frac{C_{12}+C_{21}}{2} \\ \frac{C_{21}+C_{12}}{2} & C_{22} \end{pmatrix} + \begin{pmatrix} 0 & \frac{C_{12}-C_{21}}{2} \\ \frac{C_{21}-C_{12}}{2} & 0 \end{pmatrix} \right) \begin{pmatrix} -\frac{1}{T}\frac{\partial T}{\partial x} \\ \frac{1}{u}\frac{\partial u}{\partial x} \end{pmatrix} \geq 0. \tag{5.18}$$

Distributing the multiplication, we find

$$\begin{pmatrix} -\frac{1}{T}\frac{\partial T}{\partial x} & \frac{1}{u}\frac{\partial u}{\partial x} \end{pmatrix} \begin{pmatrix} C_{11} & \frac{C_{12}+C_{21}}{2} \\ \frac{C_{21}+C_{12}}{2} & C_{22} \end{pmatrix} \begin{pmatrix} -\frac{1}{T}\frac{\partial T}{\partial x} \\ \frac{1}{u}\frac{\partial u}{\partial x} \end{pmatrix}$$

$$+ \underbrace{\begin{pmatrix} -\frac{1}{T}\frac{\partial T}{\partial x} & \frac{1}{u}\frac{\partial u}{\partial x} \end{pmatrix} \begin{pmatrix} 0 & \frac{C_{12}-C_{21}}{2} \\ \frac{C_{21}-C_{12}}{2} & 0 \end{pmatrix} \begin{pmatrix} -\frac{1}{T}\frac{\partial T}{\partial x} \\ \frac{1}{u}\frac{\partial u}{\partial x} \end{pmatrix}}_{=0} \geq 0. \tag{5.19}$$

The second term is identically zero for all values of temperature and velocity gradients. So what remains is the inequality involving only a symmetric matrix:

$$\begin{pmatrix} -\frac{1}{T}\frac{\partial T}{\partial x} & \frac{1}{u}\frac{\partial u}{\partial x} \end{pmatrix} \begin{pmatrix} C_{11} & \frac{C_{12}+C_{21}}{2} \\ \frac{C_{21}+C_{12}}{2} & C_{22} \end{pmatrix} \begin{pmatrix} -\frac{1}{T}\frac{\partial T}{\partial x} \\ \frac{1}{u}\frac{\partial u}{\partial x} \end{pmatrix} \geq 0. \tag{5.20}$$

Now in a well-known result from linear algebra that can easily be inferred from the method of quadratic forms, Section 2.8, a necessary and sufficient condition for satisfying this inequality is that the new coefficient matrix be positive semi-definite. Further, the matrix will be positive semi-definite if it has positive semi-definite eigenvalues. The eigenvalues of the new coefficient matrix can be shown to be

$$\lambda = \frac{1}{2}\left((C_{11} + C_{22}) \pm \sqrt{(C_{11} - C_{22})^2 + (C_{12} + C_{21})^2}\right). \tag{5.21}$$

Because the terms inside the radical are positive semi-definite, the eigenvalues must be real. This is a consequence of the parent matrix being symmetric. Now we require two positive semi-definite eigenvalues. First, if $C_{11} + C_{22} < 0$, we obviously have at least one negative eigenvalue, so we demand that $C_{11} + C_{22} \geq 0$. We then must have

$$C_{11} + C_{22} \geq \sqrt{(C_{11} - C_{22})^2 + (C_{12} + C_{21})^2}. \tag{5.22}$$

This gives rise to

$$(C_{11} + C_{22})^2 \geq (C_{11} - C_{22})^2 + (C_{12} + C_{21})^2. \tag{5.23}$$

Expanding and simplifying, one gets

$$C_{11} C_{22} \geq \left(\frac{C_{12} + C_{21}}{2}\right)^2. \tag{5.24}$$

Now the right-hand side is positive semi-definite, so the left-hand side must be also. Thus

$$C_{11} C_{22} \geq 0. \tag{5.25}$$

The only way for the sum and product of C_{11} and C_{22} to be positive semi-definite is to demand that $C_{11} \geq 0$ and $C_{22} \geq 0$. Thus, we arrive at the final set of conditions to satisfy the second law:

$$C_{11} \geq 0, \qquad C_{22} \geq 0, \qquad C_{11} C_{22} \geq \left(\frac{C_{12} + C_{21}}{2}\right)^2. \tag{5.26}$$

Now an important school of thought, founded by Onsager,[3] in twentieth-century thermodynamics takes an extra step and makes the *further* assertion that the original matrix C_{ij} itself must be symmetric. That is, $C_{12} = C_{21}$. This remarkable assertion is independent of the second law, and is, for other scenarios, consistent with experimental results. Consequently, the second law in combination with Onsager's independent demand requires that

$$C_{11} \geq 0, \qquad C_{22} \geq 0, \qquad C_{12} \leq \sqrt{C_{11} C_{22}}. \tag{5.27}$$

All this said, we must dismiss our hypothesis in this specific case on other physical grounds, namely that such a hypothesis results in an infinite shear stress for a fluid at rest! In the special case in which $\partial T/\partial x = 0$, our hypothesis predicts $\tau = C_{22}(T/u^2)(\partial u/\partial x)$. Obviously, this is inconsistent with any observation and so we reject this hypothesis. Additionally, this assumed form is not frame invariant because of the velocity dependency. So, why did we go to this trouble? First, we now have indication that we should not expect to find heat flux to depend on

---

[3] Lars Onsager, 1903–1976, Norwegian-born American physical chemist, earned the Ph.D. and taught at Yale, developed a systematic theory for irreversible chemical processes.

deformation. Second, it illustrates some general techniques in continuum mechanics. Moreover, the techniques we used have actually been applied to other more complicated phenomena which are physical, and of practical importance.

Real Physical Effects

That such a matrix such as we studied in the previous section was asserted to be symmetric is a manifestation of what is known as a general *Onsager relation*, developed by Onsager in 1931 for more general systems and for which he was awarded a Nobel Prize in chemistry in 1968. These actually describe a surprising variety of physical phenomena, and are described in detail many texts, including Fung (1965), Reynolds (1968), and Woods (1975). A well-known example is the Peltier[4] effect in which conduction of both heat and electrical charge is influenced by gradients of charge and temperature. This forms the basis of the operation of a thermocouple. Other relations that exist are the Soret[5] effect in which diffusive mass fluxes are induced by temperature gradients, the Dufour effect in which a diffusive energy flux is induced by a species concentration gradient, the Hall[6] effect for coupled electrical and magnetic effects (that explains the operation of an electric motor), the Seeback[7] effect in which electromotive forces are induced by different conducting elements at different temperatures, the Thomson[8] effect in which heat is transferred when electric current flows in a conductor in which there is a temperature gradient, and the principle of detailed balance for multi-species chemical reactions. One of the more important physical applications of Onsager's approach is that of conduction of thermal energy in an anisotropic material, and it is one we will explore using the same methods in Section 5.3.

5.2.2 Strong Form of the Clausius–Duhem Inequality

A less general way to satisfy the second law is to take the sufficient (but not necessary!) condition that each term in the entropy inequality be positive semi-definite:

$$-\frac{1}{T^2} q_i \partial_i T \geq 0, \qquad (5.28)$$

$$\frac{1}{T} \tau_{ij} \partial_{(i} v_{j)} \geq 0. \qquad (5.29)$$

Once again, this is called the strong form of the entropy inequality and is potentially overly restrictive.

---

[4] Jean-Charles-Athanase Peltier, 1785–1845, French clockmaker, retired at 30 to study science.
[5] Charles Soret, 1854–1904, Swiss physicist and chemist.
[6] Edwin Herbert Hall, 1855–1938, Maine-born American physicist.
[7] Thomas Johann Seebeck, 1770–1831, German medical doctor.
[8] William Thomson (Lord Kelvin), 1824–1907, Belfast-born British mathematician and physicist, key figure in nineteen-century engineering science including mathematics, thermodynamics, and electrodynamics.

5.3 Fourier's Law

Let us examine the restriction on q_i from the strong form of the entropy inequality to infer the common constitutive relation known as Fourier's law. The portion of the strong form of the entropy inequality with which we are concerned here is Eq. (5.28). Now *one* way to guarantee this inequality is satisfied is to specify the constitutive relation for the heat flux vector as

$$q_i = -k\partial_i T, \quad \text{with} \quad k \geq 0. \tag{5.30}$$

This is the well-known Fourier's law for an isotropic material, with k as the thermal conductivity. It has the proper behavior under Galilean transformations and rotations; more importantly, it is consistent with macroscale experiments for isotropic materials and can be justified from an underlying microscale theory. Substitution of Fourier's law for an isotropic material into the strong form of entropy inequality, Eq. (5.28), yields

$$\frac{1}{T^2}k(\partial_i T)(\partial_i T) \geq 0, \tag{5.31}$$

that for $k \geq 0$ is a true statement. The second law allows other forms as well. The expression $q_i = -k((\partial_j T)(\partial_j T))\partial_i T$ is consistent with the second law. It does not match experiments well for most materials, however.

Following Duhamel,[9] we can also generalize Fourier's law for an anisotropic material. This example was explored extensively by Onsager and many others. Additional background, including citation of original source material, is given by Powers (2004). Let us only consider anisotropic materials for which the thermal conductivity in any given direction is a constant. For such materials, the thermal conductivity is a tensor k_{ij}, and Fourier's law generalizes to

$$q_i = -k_{ij}\partial_j T. \tag{5.32}$$

This states that, for a fixed temperature gradient, the heat flux depends on the orientation. This is characteristic of anisotropic substances such as layered materials. Substitution of the generalized Fourier's law into the strong form of the entropy inequality, Eq. (5.28), gives now

$$\frac{1}{T^2}k_{ij}(\partial_j T)(\partial_i T) \geq 0, \quad \frac{1}{T^2}(\partial_i T)k_{ij}(\partial_j T) \geq 0, \quad \frac{1}{T^2}(\nabla T)^T \cdot \mathbf{K} \cdot \nabla T \geq 0. \tag{5.33}$$

Now $1/T^2 > 0$, so we must have $(\partial_i T)k_{ij}(\partial_j T) \geq 0$ for all possible values of ∇T. Now any possible anti-symmetric portion of k_{ij} cannot contribute to the inequality. We can see this by expanding k_{ij} in the entropy inequality to get

$$\partial_i T \left(\frac{1}{2}(k_{ij} + k_{ji}) + \frac{1}{2}(k_{ij} - k_{ji}) \right) \partial_j T \geq 0, \tag{5.34}$$

$$\partial_i T \left(k_{(ij)} + k_{[ij]} \right) \partial_j T \geq 0, \tag{5.35}$$

$$(\partial_i T)k_{(ij)}(\partial_j T) + \underbrace{(\partial_i T)k_{[ij]}(\partial_j T)}_{=0} \geq 0, \tag{5.36}$$

$$(\partial_i T)k_{(ij)}(\partial_j T) \geq 0. \tag{5.37}$$

---

[9] Jean-Marie Constant Duhamel, 1797–1872, scholar who applied mathematics to problems in heat transfer, mechanics, and acoustics.

The anti-symmetric part of k_{ij} makes no contribution to the entropy generation because it involves the tensor inner product of a symmetric tensor with an anti-symmetric tensor, which is identically zero.

Next, we again use the well-known result from linear algebra that the entropy inequality is satisfied if $k_{(ij)}$ is a positive semi-definite tensor. This will be the case if all the eigenvalues of $k_{(ij)}$ are nonnegative. That this is sufficient to satisfy the entropy inequality is made plausible if we consider $\partial_j T$ to be an eigenvector, so that $k_{(ij)}\partial_j T = \lambda \delta_{ij}\partial_j T$ giving rise to an entropy inequality of

$$(\partial_i T)\lambda \delta_{ij}(\partial_j T) \geq 0, \qquad \lambda(\partial_i T)(\partial_i T) \geq 0. \tag{5.38}$$

The inequality holds for all $\partial_i T$ as long as $\lambda \geq 0$.

Further now, when we consider the contribution of the heat flux vector to the energy equation, we see any possible anti-symmetric portion of the thermal conductivity tensor will be inconsequential. This is seen by the following analysis, which considers only relevant terms in the energy equation:

$$\rho \frac{de}{dt} = -\partial_i q_i + \cdots = \partial_i \left(k_{ij}\partial_j T\right) + \cdots = k_{ij}\partial_i \partial_j T + \cdots \tag{5.39}$$

$$= \left(k_{(ij)} + k_{[ij]}\right)\partial_i \partial_j T + \cdots = k_{(ij)}\partial_i \partial_j T + \underbrace{k_{[ij]}\partial_i \partial_j T}_{=0} + \cdots = k_{(ij)}\partial_i \partial_j T + \cdots . \tag{5.40}$$

So, it seems any possible anti-symmetric portion of k_{ij} will have no consequence as far as the first or second laws are concerned. However, an anti-symmetric portion of k_{ij} would induce a heat flux orthogonal to the direction of the temperature gradient. In a remarkable confirmation of Onsager's principle, experimental measurements on anisotropic crystalline materials demonstrate that there is no component of heat flux orthogonal to the temperature gradient, and thus, the thermal conductivity tensor k_{ij} in fact has zero anti-symmetric part, and thus is symmetric, $k_{ij} = k_{ji}$. For our particular case with a tensorial thermal conductivity, the competing effects are the heat fluxes in three directions, caused by temperature gradients in three directions:

$$\begin{pmatrix} q_1 \\ q_2 \\ q_3 \end{pmatrix} = - \begin{pmatrix} k_{11} & k_{12} & k_{13} \\ k_{21} & k_{22} & k_{23} \\ k_{31} & k_{32} & k_{33} \end{pmatrix} \begin{pmatrix} \partial_1 T \\ \partial_2 T \\ \partial_3 T \end{pmatrix}. \tag{5.41}$$

The symmetry condition, Onsager's principle, requires that $k_{12} = k_{21}$, $k_{13} = k_{31}$, and $k_{23} = k_{32}$. So, the experimentally verified Onsager's principle further holds that the heat flux for an anisotropic material is given by

$$\begin{pmatrix} q_1 \\ q_2 \\ q_3 \end{pmatrix} = - \begin{pmatrix} k_{11} & k_{12} & k_{13} \\ k_{12} & k_{22} & k_{23} \\ k_{13} & k_{23} & k_{33} \end{pmatrix} \begin{pmatrix} \partial_1 T \\ \partial_2 T \\ \partial_3 T \end{pmatrix}. \tag{5.42}$$

Now it is well known that the thermal conductivity tensor k_{ij} will be positive semi-definite if all its eigenvalues are nonnegative. The eigenvalues will be guaranteed real upon adopting Onsager symmetry. The characteristic polynomial for the eigenvalues is of the same form as Eq. (2.133) and is given by

$$\lambda^3 - I_k^{(1)}\lambda^2 + I_k^{(2)}\lambda - I_k^{(3)} = 0, \tag{5.43}$$

where the invariants of the thermal conductivity tensor k_{ij}, are given by the standard

$$I_k^{(1)} = k_{ii} = \operatorname{tr} \mathbf{K}, \tag{5.44}$$

$$I_k^{(2)} = \frac{1}{2}(k_{ii}k_{jj} - k_{ij}k_{ji}) = (\det \mathbf{K})\left(\operatorname{tr} \mathbf{K}^{-1}\right), \tag{5.45}$$

$$I_k^{(3)} = \epsilon_{ijk}k_{1j}k_{2j}k_{3j} = \det \mathbf{K}. \tag{5.46}$$

In a standard result from linear algebra, one can show that if all three invariants are positive semi-definite, then the eigenvalues are all positive semi-definite, and as a result, the matrix itself is positive semi-definite. Hence, in order for k_{ij} to be positive semi-definite, we demand that

$$I_k^{(1)} \geq 0, \qquad I_k^{(2)} \geq 0, \qquad I_k^{(3)} \geq 0, \tag{5.47}$$

which is equivalent to demanding that

$$k_{11} + k_{22} + k_{33} \geq 0, \tag{5.48}$$

$$k_{11}k_{22} + k_{11}k_{33} + k_{22}k_{33} - k_{12}^2 - k_{13}^2 - k_{23}^2 \geq 0, \tag{5.49}$$

$$k_{13}(k_{12}k_{23} - k_{22}k_{13}) + k_{23}(k_{12}k_{13} - k_{11}k_{23}) + k_{33}(k_{11}k_{22} - k_{12}k_{12}) \geq 0. \tag{5.50}$$

If $\det \mathbf{K} \neq 0$, the conditions reduce to

$$\operatorname{tr} \mathbf{K} \geq 0, \qquad \operatorname{tr} \mathbf{K}^{-1} \geq 0, \qquad \det \mathbf{K} > 0. \tag{5.51}$$

Now by considering $\partial_i T = (1, 0, 0)^T$, and demanding $(\partial_i T)k_{ij}(\partial_j T) \geq 0$, we conclude that $k_{11} \geq 0$. Similarly, by considering $\partial_i T = (0, 1, 0)^T$ and $\partial_i T = (0, 0, 1)^T$, we conclude that $k_{22} \geq 0$ and $k_{33} \geq 0$, respectively. Thus, $\operatorname{tr} \mathbf{K} \geq 0$ is automatically satisfied. In equation form, we then have

$$k_{11} \geq 0, \qquad k_{22} \geq 0, \qquad k_{33} \geq 0, \tag{5.52}$$

$$k_{11}k_{22} + k_{11}k_{33} + k_{22}k_{33} - k_{12}^2 - k_{13}^2 - k_{23}^2 \geq 0, \tag{5.53}$$

$$k_{13}(k_{12}k_{23} - k_{22}k_{13}) + k_{23}(k_{12}k_{13} - k_{11}k_{23}) + k_{33}(k_{11}k_{22} - k_{12}k_{12}) \geq 0. \tag{5.54}$$

While not a proof, numerical experimentation gives strong indication that the remaining conditions can be satisfied if, loosely stated, $k_{11}, k_{22}, k_{33} \gg |k_{12}|, |k_{23}|, |k_{13}|$. That is, for positive semi-definiteness, (1) *each* diagonal element must be positive semi-definite, (2) off-diagonal terms can be positive or negative, and (3) diagonal terms must have amplitudes that are, loosely speaking, larger than the amplitudes of off-diagonal terms.

Example 5.2 Let us consider heat conduction in the limit of two dimensions and a constant anisotropic thermal conductivity tensor, without imposing Onsager's conditions.

Solution
Let us take then

$$\begin{pmatrix} q_1 \\ q_2 \end{pmatrix} = - \begin{pmatrix} k_{11} & k_{12} \\ k_{21} & k_{22} \end{pmatrix} \begin{pmatrix} \partial_1 T \\ \partial_2 T \end{pmatrix}. \tag{5.55}$$

The second law demands that

$$\begin{pmatrix} \partial_1 T & \partial_2 T \end{pmatrix} \begin{pmatrix} k_{11} & k_{12} \\ k_{21} & k_{22} \end{pmatrix} \begin{pmatrix} \partial_1 T \\ \partial_2 T \end{pmatrix} \geq 0. \quad (5.56)$$

This is expanded as

$$\begin{pmatrix} \partial_1 T & \partial_2 T \end{pmatrix} \left(\begin{pmatrix} k_{11} & \frac{k_{12}+k_{21}}{2} \\ \frac{k_{21}+k_{12}}{2} & k_{22} \end{pmatrix} + \begin{pmatrix} 0 & \frac{k_{12}-k_{21}}{2} \\ \frac{k_{21}-k_{12}}{2} & 0 \end{pmatrix} \right) \begin{pmatrix} \partial_1 T \\ \partial_2 T \end{pmatrix} \geq 0. \quad (5.57)$$

As before, the anti-symmetric portion makes no contribution to the left-hand side, giving

$$\begin{pmatrix} \partial_1 T & \partial_2 T \end{pmatrix} \begin{pmatrix} k_{11} & \frac{k_{12}+k_{21}}{2} \\ \frac{k_{21}+k_{12}}{2} & k_{22} \end{pmatrix} \begin{pmatrix} \partial_1 T \\ \partial_2 T \end{pmatrix} \geq 0. \quad (5.58)$$

And, demanding that the eigenvalues of the symmetric part of the thermal conductivity tensor be positive gives rise to the conditions, identical to that of an earlier analysis, that

$$k_{11} \geq 0, \qquad k_{22} \geq 0, \qquad k_{11} k_{22} \geq \left(\frac{k_{12} + k_{21}}{2} \right)^2. \quad (5.59)$$

The energy equation becomes

$$\rho \frac{de}{dt} = -\partial_i q_i + \cdots, \quad (5.60)$$

$$= \begin{pmatrix} \partial_1 & \partial_2 \end{pmatrix} \begin{pmatrix} k_{11} & k_{12} \\ k_{21} & k_{22} \end{pmatrix} \begin{pmatrix} \partial_1 T \\ \partial_2 T \end{pmatrix} + \cdots, \quad (5.61)$$

$$= \begin{pmatrix} \partial_1 & \partial_2 \end{pmatrix} \begin{pmatrix} k_{11} \partial_1 T + k_{12} \partial_2 T \\ k_{21} \partial_1 T + k_{22} \partial_2 T \end{pmatrix} + \cdots, \quad (5.62)$$

$$= k_{11} \partial_1 \partial_1 T + (k_{12} + k_{21}) \partial_1 \partial_2 T + k_{22} \partial_2 \partial_2 T + \cdots, \quad (5.63)$$

$$= k_{11} \frac{\partial^2 T}{\partial x_1^2} + (k_{12} + k_{21}) \frac{\partial^2 T}{\partial x_1 \partial x_2} + k_{22} \frac{\partial^2 T}{\partial x_2^2} + \cdots. \quad (5.64)$$

One sees that the energy evolution depends only on the symmetric part of the thermal conductivity tensor.

Imposition of Onsager's relations gives simply $k_{12} = k_{21}$, giving rise to second law restrictions

$$k_{11} \geq 0, \qquad k_{22} \geq 0, \qquad k_{11} k_{22} \geq k_{12}^2, \quad (5.65)$$

and an energy equation of

$$\rho \frac{de}{dt} = k_{11} \frac{\partial^2 T}{\partial x_1^2} + 2 k_{12} \frac{\partial^2 T}{\partial x_1 \partial x_2} + k_{22} \frac{\partial^2 T}{\partial x_2^2} + \cdots. \quad (5.66)$$

Example 5.3 Consider the ramifications of a heat flux vector in violation of Onsager's principle: flux in which the anisotropic thermal conductivity is purely anti-symmetric. For simplicity consider an incompressible, isobaric, stationary liquid with constant specific heat c_p. For the heat flux, we take

$$\begin{pmatrix} q_1 \\ q_2 \end{pmatrix} = -\begin{pmatrix} 0 & -\beta \\ \beta & 0 \end{pmatrix} \begin{pmatrix} \partial_1 T \\ \partial_2 T \end{pmatrix}. \quad (5.67)$$

This holds that heat flux in the 1 direction is induced only by temperature gradients in the 2 direction and heat flux in the 2 direction is induced only by temperature gradients in the 1 direction.

Solution

The second law demands that

$$\begin{pmatrix} \partial_1 T & \partial_2 T \end{pmatrix} \begin{pmatrix} 0 & -\beta \\ \beta & 0 \end{pmatrix} \begin{pmatrix} \partial_1 T \\ \partial_2 T \end{pmatrix} \geq 0, \tag{5.68}$$

$$\begin{pmatrix} \partial_1 T & \partial_2 T \end{pmatrix} \begin{pmatrix} -\beta \partial_2 T \\ \beta \partial_1 T \end{pmatrix} \geq 0, \tag{5.69}$$

$$-\beta(\partial_1 T)(\partial_2 T) + \beta(\partial_1 T)(\partial_2 T) \geq 0, \tag{5.70}$$

$$0 \geq 0. \tag{5.71}$$

So, the second law holds.

For the incompressible stationary liquid with constant specific heat, the velocity field, and the energy equation, the upcoming Eq. (6.67), reduces to the simple

$$\rho c_p \frac{\partial T}{\partial t} = -\partial_i q_i. \tag{5.72}$$

Imposing our unusual expression for heat flux, we get

$$\rho c_p \frac{\partial T}{\partial t} = \begin{pmatrix} \partial_1 & \partial_2 \end{pmatrix} \begin{pmatrix} 0 & -\beta \\ \beta & 0 \end{pmatrix} \begin{pmatrix} \partial_1 T \\ \partial_2 T \end{pmatrix} = \begin{pmatrix} \partial_1 & \partial_2 \end{pmatrix} \begin{pmatrix} -\beta \partial_2 T \\ \beta \partial_1 T \end{pmatrix}, \tag{5.73}$$

$$= -\beta \partial_1 \partial_2 T + \beta \partial_1 \partial_2 T = 0. \tag{5.74}$$

So, this unusual heat flux vector is one that induces no change in temperature. In terms of the first law of thermodynamics, a net energy flux into a control volume in the 1 direction is exactly counterbalanced by a net energy flux out of the same control volume in the 2 direction. Thus, the first law holds as well.

Let us consider a temperature distribution for this unusual material. And let us consider it to apply to the domain $x \in [0,1]$, $y \in [0,1]$, $t \in [0,\infty)$. Take

$$T(x_1, x_2, t) = x_2. \tag{5.75}$$

Obviously, this satisfies the first law as $\partial T/\partial t = 0$. Let us check the heat flux.

$$q_1 = \beta \partial_2 T = \beta, \qquad q_2 = -\beta \partial_1 T = 0. \tag{5.76}$$

Now the lower boundary at $x_2 = 0$ has $T = 0$. The upper boundary has $x_2 = 1$ so $T = 1$. And this constant temperature gradient in the 2 direction is inducing a constant heat flux in the 1 direction, $q_1 = -\beta$. The energy flux that enters at $x_1 = 0$ departs at $x_1 = 1$, maintaining energy conservation. One can consider an equivalent problem in polar coordinates. Taking

$$x_1 = r \cos \theta, \qquad x_2 = r \sin \theta, \tag{5.77}$$

and applying the transformation rule,

$$\begin{pmatrix} \partial_1 \\ \partial_2 \end{pmatrix} = \begin{pmatrix} \frac{\partial r}{\partial x_1} & \frac{\partial \theta}{\partial x_1} \\ \frac{\partial r}{\partial x_2} & \frac{\partial \theta}{\partial x_2} \end{pmatrix} \begin{pmatrix} \frac{\partial}{\partial r} \\ \frac{\partial}{\partial \theta} \end{pmatrix}, \tag{5.78}$$

one finds

$$\begin{pmatrix} \partial_1 \\ \partial_2 \end{pmatrix} = \begin{pmatrix} \cos \theta & -\frac{\sin \theta}{r} \\ \sin \theta & \frac{\cos \theta}{r} \end{pmatrix} \begin{pmatrix} \frac{\partial}{\partial r} \\ \frac{\partial}{\partial \theta} \end{pmatrix}. \tag{5.79}$$

So, transforming $q_1 = \beta \partial_2 T$, and $q_2 = -\beta \partial_1 T$ gives

$$\begin{pmatrix} q_1 \\ q_2 \end{pmatrix} = \beta \begin{pmatrix} \sin\theta & \frac{\cos\theta}{r} \\ -\cos\theta & \frac{\sin\theta}{r} \end{pmatrix} \begin{pmatrix} \frac{\partial T}{\partial r} \\ \frac{\partial T}{\partial \theta} \end{pmatrix}. \tag{5.80}$$

Standard trigonometry gives

$$\begin{pmatrix} q_r \\ q_\theta \end{pmatrix} = \underbrace{\begin{pmatrix} \cos\theta & \sin\theta \\ -\sin\theta & \cos\theta \end{pmatrix}}_{\text{rotation matrix}} \begin{pmatrix} q_1 \\ q_2 \end{pmatrix}. \tag{5.81}$$

Applying the rotation matrix to both sides gives then

$$\begin{pmatrix} \cos\theta & \sin\theta \\ -\sin\theta & \cos\theta \end{pmatrix} \begin{pmatrix} q_1 \\ q_2 \end{pmatrix} = \beta \begin{pmatrix} \cos\theta & \sin\theta \\ -\sin\theta & \cos\theta \end{pmatrix} \begin{pmatrix} \sin\theta & \frac{\cos\theta}{r} \\ -\cos\theta & \frac{\sin\theta}{r} \end{pmatrix} \begin{pmatrix} \frac{\partial T}{\partial r} \\ \frac{\partial T}{\partial \theta} \end{pmatrix}, \tag{5.82}$$

$$\begin{pmatrix} q_r \\ q_\theta \end{pmatrix} = \beta \begin{pmatrix} 0 & \frac{1}{r} \\ -1 & 0 \end{pmatrix} \begin{pmatrix} \frac{\partial T}{\partial r} \\ \frac{\partial T}{\partial \theta} \end{pmatrix}, \tag{5.83}$$

or simply

$$q_r = \frac{\beta}{r} \frac{\partial T}{\partial \theta}, \qquad q_\theta = -\beta \frac{\partial T}{\partial r}. \tag{5.84}$$

Now the steady state temperature distribution in the annular region $1/2 < r < 1$, $T = r$, describes a domain with an inner boundary held at $T = 1/2$ and an outer boundary held at $T = 1$. Such a temperature distribution would induce a heat flux in the θ direction only, so that $q_r = 0$ and $q_\theta = -\beta$. That is, the heat goes round and round the domain, but never enters or exits at any boundary.

Now such a flux is counterintuitive precisely because it has never been observed or measured. It is for this reason that we can adopt Onsager's hypothesis and demand that, independent of the first and second laws of thermodynamics,

$$\beta = 0, \tag{5.85}$$

and the thermal conductivity tenser is purely symmetric.

5.4 Stress–Strain Rate Relation for a Newtonian Fluid

We now seek to satisfy the second part of the strong form of the entropy inequality, Eq. (5.29). Recalling that $T > 0$, this reduces to

$$\underbrace{\tau_{ij} \partial_{(i} v_{j)}}_{\Phi} \geq 0. \tag{5.86}$$

This form suggests that we seek a constitutive equation for the viscous stress tensor τ_{ij} that is a function of the deformation tensor $\partial_{(i} v_{j)}$. Unlike the ultimately nonphysical form hypothesized when we considered the weak form of the entropy inequality in Section 5.2.1, the strong form admits a physically viable model that agrees with macroscale experiments and microscale theories. Here we will focus on the simplest of such theories, for what is known as a *Newtonian fluid*, a fluid that is isotropic and whose viscous stress varies linearly with strain rate. In general, this is a discipline unto itself known as *rheology*. Additional background may be

5.4 Stress–Strain Rate Relation for a Newtonian Fluid

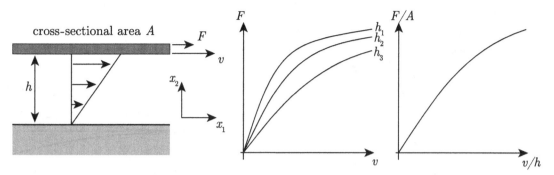

Figure 5.2 Diagram of simple Couette flow experiment with measurements of stress versus strain rate.

found in a variety of sources including Bird et al. (1987), Morrison (2001), Schowalter (1978), or Whitaker (1968).

5.4.1 Motivating Experiments

We can pull a flat plate over a fluid and measure the force necessary to maintain a specified velocity. This situation and some expected results are shown in Fig. 5.2. We observe the following. (1) At the upper and lower plate surfaces, the fluid has the same velocity of each plate. This is called the *no-slip* condition. (2) The faster the velocity v of the upper plate is, the higher the force necessary to pull the plate is. The increase can be linear or nonlinear. (3) When experiments are performed with different plate area and different gap width, a single universal curve results when F/A is plotted against v/h. (4) The velocity profile is linear with increasing x_2. In a way similar on a molecular scale to energy diffusion, this experiment is describing diffusion of momentum from the pulled plate into the fluid below it. The constitutive equation we develop for viscous stress, when combined with the governing axioms, will model momentum diffusion.

We can associate F/A with a shear stress: τ_{21}; this is stress on the 2 face in the 1 direction. We can associate v/h with a velocity gradient, here $\partial_2 v_1$. Considering the velocity gradient is equivalent to considering the deformation tensor, as far as the second law is concerned. As such, we will occasionally use the terms interchangeably in this discussion. We loosely define the coefficient of viscosity μ for this configuration as

$$\mu = \frac{\tau_{21}}{\partial_2 v_1} = \frac{\text{viscous stress}}{\text{strain rate}}. \tag{5.87}$$

We shall see in the next section that μ is better described as the *first coefficient of viscosity* and that a second coefficient is also required to fully characterize an isotropic fluid. The viscosity is the analog of solid mechanics' Young's[10] modulus: the ratio of stress to strain. In general, μ is a nonequilibrium thermodynamic property of a material. It is a nonequilibrium property

---

[10] Thomas Young, 1773–1829, English physician and physicist whose experiments in interferometry revived the wave theory of light, Egyptologist who helped decipher the Rosetta stone, worked on surface tension in fluids, gave the word "energy" scientific significance, and developed Young's modulus in elasticity.

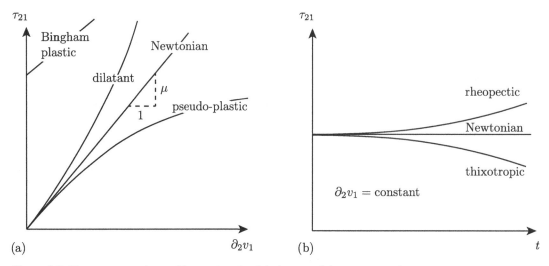

Figure 5.3 Newtonian and non-Newtonian fluid behavior: (a) variation of viscous stress with strain rate, (b) variation of viscous stress with time for constant strain rate.

because it cannot be measured in a fluid in a zero-velocity equilibrium state. It is often a strong function of temperature, but can vary with pressure as well. A Newtonian fluid has a viscosity that does not depend on strain rate, but could depend on temperature and pressure.

A non-Newtonian fluid could deviate from the Newtonian limit in several ways. The most common is that the viscosity coefficient could be a function of the strain rate. Some typical behavior is given in Fig. 5.3(a). Here the pseudo-plastic, also known as *shear thinning*, fluid has a viscosity that decreases with increasing strain rate. Fluids such as ketchup, molasses, or blood may be modeled as pseudo-plastic. The dilatant, also known as *shear thickening*, fluid has a viscosity that increases with strain rate. A suspension of corn starch in water is an example of a dilatant fluid. A Bingham plastic is a fluid that requires a finite shear stress before it moves. An example is toothpaste. For some fluids, it is observed that when subjected to a constant strain rate, the required shear stress varies with time. Fluids for with the stress decreases with time are known as *thixotropic*; those for which it increases with time are known as *rheopectic*. Both are shown in Fig. 5.3(b). Peanut butter may be thixotropic; printer ink can be rheopectic. Another deviation is that the stress could depend upon combinations of strain rate and strain, rendering the material to have both fluid- and solid-like features; such materials are known as *viscoelastic*. We will consider a few non-Newtonian problems in Section 10.3; however, our focus will be on Newtonian fluids. Much of our development will be valid for temperature- and pressure-dependent viscosity, while most examples will consider only constant viscosity.

5.4.2 Analysis for an Isotropic Newtonian Fluid

Here we present the method described by Whitaker (1968, pp. 139–145) to describe the viscous stress as a function of strain rate for an isotropic fluid with constant viscosity. An isotropic fluid has no directional dependencies. A fluid composed of aligned long chain polymers is an example of a fluid that is most likely not isotropic. Most gases and many common liquids are isotropic. Following Whitaker, we first *postulate* that stress is a function of deformation rate (strain rate) only. Thus, we are not allowing viscous stress to be a function of the rigid body rotation rate. While it seems intuitive that rigid body rotation should not induce viscous stress, Batchelor (2000, p. 144) mentions that there is no rigorous proof for this; hence, we describe our statement as a postulate:

$$\tau_{ij} = f_{ij}(\partial_{(k}v_{l)}). \tag{5.88}$$

Written in more detail, we have postulated a relationship of the form

$$\tau_{11} = f_{11}(\partial_{(1}v_{1)}, \partial_{(2}v_{2)}, \partial_{(3}v_{3)}, \partial_{(1}v_{2)}, \partial_{(2}v_{3)}, \partial_{(3}v_{1)}\partial_{(2}v_{1)}, \partial_{(3}v_{2)}, \partial_{(1}v_{3)}), \tag{5.89}$$

$$\tau_{12} = f_{12}(\partial_{(1}v_{1)}, \partial_{(2}v_{2)}, \partial_{(3}v_{3)}, \partial_{(1}v_{2)}, \partial_{(2}v_{3)}, \partial_{(3}v_{1)}\partial_{(2}v_{1)}, \partial_{(3}v_{2)}, \partial_{(1}v_{3)}), \tag{5.90}$$

$$\vdots$$

$$\tau_{33} = f_{33}(\partial_{(1}v_{1)}, \partial_{(2}v_{2)}, \partial_{(3}v_{3)}, \partial_{(1}v_{2)}, \partial_{(2}v_{3)}, \partial_{(3}v_{1)}\partial_{(2}v_{1)}, \partial_{(3}v_{2)}, \partial_{(1}v_{3)}). \tag{5.91}$$

We then require that $\tau_{ij} = 0$ if $\partial_{(i}v_{j)} = 0$; hence, no strain rate, no stress. We also require that viscous stress is *linearly* related to strain rate:

$$\tau_{ij} = \hat{C}_{ijkl}\partial_{(k}v_{l)}. \tag{5.92}$$

This is the imposition of the assumption of a Newtonian fluid. Here $\hat{C}_{ijkl}$ is a fourth-order tensor. We can say that the viscous stress is a *linear mapping* of the strain rate, in that zero strain rate implies zero viscous stress, and the relation between the two is linear. Thus, we have in matrix form

$$\begin{pmatrix} \tau_{11} \\ \tau_{22} \\ \tau_{33} \\ \tau_{12} \\ \tau_{23} \\ \tau_{31} \\ \tau_{21} \\ \tau_{32} \\ \tau_{13} \end{pmatrix} = \begin{pmatrix} \hat{C}_{1111} & \hat{C}_{1122} & \hat{C}_{1133} & \hat{C}_{1112} & \hat{C}_{1123} & \hat{C}_{1131} & \hat{C}_{1121} & \hat{C}_{1132} & \hat{C}_{1113} \\ \hat{C}_{2211} & \hat{C}_{2222} & \hat{C}_{2233} & \hat{C}_{2212} & \hat{C}_{2223} & \hat{C}_{2231} & \hat{C}_{2221} & \hat{C}_{2232} & \hat{C}_{2213} \\ \hat{C}_{3311} & \hat{C}_{3322} & \hat{C}_{3333} & \hat{C}_{3312} & \hat{C}_{3323} & \hat{C}_{3331} & \hat{C}_{3321} & \hat{C}_{3332} & \hat{C}_{3313} \\ \hat{C}_{1211} & \hat{C}_{1222} & \hat{C}_{1233} & \hat{C}_{1212} & \hat{C}_{1223} & \hat{C}_{1231} & \hat{C}_{1221} & \hat{C}_{1232} & \hat{C}_{1213} \\ \hat{C}_{2311} & \hat{C}_{2322} & \hat{C}_{2333} & \hat{C}_{2312} & \hat{C}_{2323} & \hat{C}_{2331} & \hat{C}_{2321} & \hat{C}_{2332} & \hat{C}_{2313} \\ \hat{C}_{3111} & \hat{C}_{3122} & \hat{C}_{3133} & \hat{C}_{3112} & \hat{C}_{3123} & \hat{C}_{3131} & \hat{C}_{3121} & \hat{C}_{3132} & \hat{C}_{3113} \\ \hat{C}_{2111} & \hat{C}_{2122} & \hat{C}_{2133} & \hat{C}_{2112} & \hat{C}_{2123} & \hat{C}_{2131} & \hat{C}_{2121} & \hat{C}_{2132} & \hat{C}_{2113} \\ \hat{C}_{3211} & \hat{C}_{3222} & \hat{C}_{3233} & \hat{C}_{3212} & \hat{C}_{3223} & \hat{C}_{3231} & \hat{C}_{3221} & \hat{C}_{3232} & \hat{C}_{3213} \\ \hat{C}_{1311} & \hat{C}_{1322} & \hat{C}_{1333} & \hat{C}_{1312} & \hat{C}_{1323} & \hat{C}_{1331} & \hat{C}_{1321} & \hat{C}_{1332} & \hat{C}_{1313} \end{pmatrix} \begin{pmatrix} \partial_{(1}v_{1)} \\ \partial_{(2}v_{2)} \\ \partial_{(3}v_{3)} \\ \partial_{(1}v_{2)} \\ \partial_{(2}v_{3)} \\ \partial_{(3}v_{1)} \\ \partial_{(2}v_{1)} \\ \partial_{(3}v_{2)} \\ \partial_{(1}v_{3)} \end{pmatrix}.$$

(5.93)

There are $3^4 = 81$ unknown coefficients $\hat{C}_{ijkl}$. We found one of them in our simple thought experiment in which we found

$$\tau_{21} = \tau_{12} = \mu\partial_2 v_1 = \mu(2\partial_{(1}v_{2)}). \tag{5.94}$$

Hence in this special case $\hat{C}_{1212} = 2\mu$.

Now we could do 81 separate experiments, or we could take advantage of the assumption that the fluid has no directional dependency. Imagine that observer A conducts an experiment to measure the stress tensor in reference frame $\mathcal{A}$. The observer begins with the "viscosity tensor" $\hat{C}_{ijkl}$. The experiment is conducted by varying strain rate and measuring stress. With complete knowledge A feels confident this knowledge could be used to predict the stress in rotated frame $\mathcal{A}'$.

Consider observer A' who is oriented to frame $\mathcal{A}'$. Oblivious to observer A, A' conducts the same experiment to measure what for her or him is τ'_{ij} The value that A' measures must be the same that A predicts in order for the system to be isotropic. This places restrictions on the viscosity tensor $\hat{C}_{ijkl}$. We intend to show that if the fluid is isotropic, only *two* of the 81 coefficients are distinct and nonzero.

We first use symmetry properties of the stress and strain rate tensor to reduce to 36 unknown coefficients. In actuality there are only six independent components of stress and six independent components of deformation because both are symmetric tensors. Consequently, we can write our linear stress–strain rate relation as

$$\begin{pmatrix}\tau_{11}\\\tau_{22}\\\tau_{33}\\\tau_{12}\\\tau_{23}\\\tau_{31}\end{pmatrix} = \begin{pmatrix}\hat{C}_{1111} & \hat{C}_{1122} & \hat{C}_{1133} & \hat{C}_{1112}+\hat{C}_{1121} & \hat{C}_{1123}+\hat{C}_{1132} & \hat{C}_{1131}+\hat{C}_{1113}\\\hat{C}_{2211} & \hat{C}_{2222} & \hat{C}_{2233} & \hat{C}_{2212}+\hat{C}_{2221} & \hat{C}_{2223}+\hat{C}_{2232} & \hat{C}_{2231}+\hat{C}_{2213}\\\hat{C}_{3311} & \hat{C}_{3322} & \hat{C}_{3333} & \hat{C}_{3312}+\hat{C}_{3321} & \hat{C}_{3323}+\hat{C}_{3332} & \hat{C}_{3331}+\hat{C}_{3313}\\\hat{C}_{1211} & \hat{C}_{1222} & \hat{C}_{1233} & \hat{C}_{1212}+\hat{C}_{1221} & \hat{C}_{1223}+\hat{C}_{1232} & \hat{C}_{1231}+\hat{C}_{1213}\\\hat{C}_{2311} & \hat{C}_{2322} & \hat{C}_{2333} & \hat{C}_{2312}+\hat{C}_{2321} & \hat{C}_{2323}+\hat{C}_{2332} & \hat{C}_{2331}+\hat{C}_{2313}\\\hat{C}_{3111} & \hat{C}_{3122} & \hat{C}_{3133} & \hat{C}_{3112}+\hat{C}_{3121} & \hat{C}_{3123}+\hat{C}_{3132} & \hat{C}_{3131}+\hat{C}_{3113}\end{pmatrix}\begin{pmatrix}\partial_{(1}v_{1)}\\\partial_{(2}v_{2)}\\\partial_{(3}v_{3)}\\\partial_{(1}v_{2)}\\\partial_{(2}v_{3)}\\\partial_{(3}v_{1)}\end{pmatrix}.$$
(5.95)

Now adopting Whitaker's notation for simplification, we define this matrix of $\hat{C}$'s as a new matrix of C's. Here, now C itself is not a tensor, while $\hat{C}$ is a tensor. We take equivalently

$$\begin{pmatrix}\tau_{11}\\\tau_{22}\\\tau_{33}\\\tau_{12}\\\tau_{23}\\\tau_{31}\end{pmatrix} = \begin{pmatrix}C_{11} & C_{12} & C_{13} & C_{14} & C_{15} & C_{16}\\C_{21} & C_{22} & C_{23} & C_{24} & C_{25} & C_{26}\\C_{31} & C_{32} & C_{33} & C_{34} & C_{35} & C_{36}\\C_{41} & C_{42} & C_{43} & C_{44} & C_{45} & C_{46}\\C_{51} & C_{52} & C_{53} & C_{54} & C_{55} & C_{56}\\C_{61} & C_{62} & C_{63} & C_{64} & C_{65} & C_{66}\end{pmatrix}\begin{pmatrix}\partial_{(1}v_{1)}\\\partial_{(2}v_{2)}\\\partial_{(3}v_{3)}\\\partial_{(1}v_{2)}\\\partial_{(2}v_{3)}\\\partial_{(3}v_{1)}\end{pmatrix}. \qquad (5.96)$$

Next, recalling that for tensorial quantities

$$\tau'_{ij} = \ell_{ki}\ell_{lj}\tau_{kl}, \qquad \partial'_{(i}v'_{j)} = \ell_{ki}\ell_{lj}\partial_{(k}v_{l)}, \qquad (5.97)$$

let us subject our fluid to a battery of rotations and see what can be concluded by enforcing material indifference.

180° Rotation About the x_3 Axis

For this rotation, depicted in Fig. 5.4, we have direction cosines

$$\mathbf{Q} = \ell_{ki} = \begin{pmatrix}\ell_{11}=-1 & \ell_{12}=0 & \ell_{13}=0\\\ell_{21}=0 & \ell_{22}=-1 & \ell_{23}=0\\\ell_{31}=0 & \ell_{32}=0 & \ell_{33}=1\end{pmatrix}. \qquad (5.98)$$

So, applying Eq. (2.9), we see

$$\begin{pmatrix}x'_1 & x'_2 & x'_3\end{pmatrix} = \begin{pmatrix}x_1 & x_2 & x_3\end{pmatrix}\begin{pmatrix}-1 & 0 & 0\\0 & -1 & 0\\0 & 0 & 1\end{pmatrix}, \qquad (5.99)$$

which yields

$$x'_1 = -x_1, \qquad x'_2 = -x_2, \qquad x'_3 = x_3, \qquad (5.100)$$

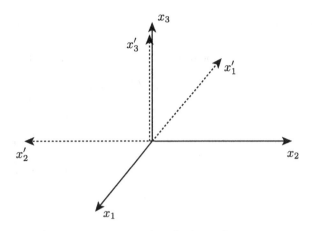

Figure 5.4 Rotation of 180° about the x_3 axis.

which is consistent with Fig. 5.4. Because $\det \ell_{ki} = 1$ and because its transpose equals its inverse, the transformation is a rotation. Applying the transformation rules to each term in the shear stress tensor, we get

$$\tau'_{11} = \ell_{k1}\ell_{l1}\tau_{kl} = (-1)^2\tau_{11} = \tau_{11}, \qquad \tau'_{22} = \ell_{k2}\ell_{l2}\tau_{kl} = (-1)^2\tau_{22} = \tau_{22}, \tag{5.101}$$

$$\tau'_{33} = \ell_{k3}\ell_{l3}\tau_{kl} = (1)^2\tau_{33} = \tau_{33}, \qquad \tau'_{12} = \ell_{k1}\ell_{l2}\tau_{kl} = (-1)^2\tau_{12} = \tau_{12}, \tag{5.102}$$

$$\tau'_{23} = \ell_{k2}\ell_{l3}\tau_{kl} = (-1)(1)\tau_{23} = -\tau_{23}, \quad \tau'_{31} = \ell_{k3}\ell_{l1}\tau_{kl} = (1)(-1)\tau_{31} = -\tau_{31}. \tag{5.103}$$

Likewise, we find that

$$\partial'_{(1}v'_{1)} = \partial_{(1}v_{1)}, \quad \partial'_{(2}v'_{2)} = \partial_{(2}v_{2)}, \qquad \partial'_{(3}v'_{3)} = \partial_{(3}v_{3)}, \tag{5.104}$$

$$\partial'_{(1}v'_{2)} = \partial_{(1}v_{2)}, \quad \partial'_{(2}v'_{3)} = -\partial_{(2}v_{3)}, \quad \partial'_{(3}v'_{1)} = -\partial_{(3}v_{1)}. \tag{5.105}$$

Now our observer A' who is in the rotated system would say that

$$\tau'_{11} = C_{11}\partial'_{(1}v'_{1)} + C_{12}\partial'_{(2}v'_{2)} + C_{13}\partial'_{(3}v'_{3)} + C_{14}\partial'_{(1}v'_{2)} + C_{15}\partial'_{(2}v'_{3)} + C_{16}\partial'_{(3}v'_{1)}, \tag{5.106}$$

while our observer A who used tensor algebra to predict τ'_{11} would say

$$\tau'_{11} = C_{11}\partial'_{(1}v'_{1)} + C_{12}\partial'_{(2}v'_{2)} + C_{13}\partial'_{(3}v'_{3)} + C_{14}\partial'_{(1}v'_{2)} - C_{15}\partial'_{(2}v'_{3)} - C_{16}\partial'_{(3}v'_{1)}. \tag{5.107}$$

Because we want both predictions to be the same, we must require that

$$C_{15} = C_{16} = 0. \tag{5.108}$$

In matrix form, our observer A would predict for the rotated frame that

$$\begin{pmatrix} \tau'_{11} \\ \tau'_{22} \\ \tau'_{33} \\ \tau'_{12} \\ -\tau'_{23} \\ -\tau'_{31} \end{pmatrix} = \begin{pmatrix} C_{11} & C_{12} & C_{13} & C_{14} & C_{15} & C_{16} \\ C_{21} & C_{22} & C_{23} & C_{24} & C_{25} & C_{26} \\ C_{31} & C_{32} & C_{33} & C_{34} & C_{35} & C_{36} \\ C_{41} & C_{42} & C_{43} & C_{44} & C_{45} & C_{46} \\ C_{51} & C_{52} & C_{53} & C_{54} & C_{55} & C_{56} \\ C_{61} & C_{62} & C_{63} & C_{64} & C_{65} & C_{66} \end{pmatrix} \begin{pmatrix} \partial'_{(1}v'_{1)} \\ \partial'_{(2}v'_{2)} \\ \partial'_{(3}v'_{3)} \\ \partial'_{(1}v'_{2)} \\ -\partial'_{(2}v'_{3)} \\ -\partial'_{(3}v'_{1)} \end{pmatrix}. \tag{5.109}$$

To retain material indifference between the predictions of our two observers, we require $C_{15} = C_{16} = C_{25} = C_{26} = C_{35} = C_{36} = C_{45} = C_{46} = C_{51} = C_{52} = C_{53} = C_{54} = C_{61} = C_{62} = C_{63} = C_{64} = 0$. This eliminates 16 coefficients; our viscosity matrix becomes

$$\begin{pmatrix} C_{11} & C_{12} & C_{13} & C_{14} & 0 & 0 \\ C_{21} & C_{22} & C_{23} & C_{24} & 0 & 0 \\ C_{31} & C_{32} & C_{33} & C_{34} & 0 & 0 \\ C_{41} & C_{42} & C_{43} & C_{44} & 0 & 0 \\ 0 & 0 & 0 & 0 & C_{55} & C_{56} \\ 0 & 0 & 0 & 0 & C_{65} & C_{66} \end{pmatrix}. \tag{5.110}$$

with only 20 independent coefficients.

180° Rotation About the x_1 Axis

This rotation is presented in Fig. 5.5. Applying Eq. (2.9), we see

$$\begin{pmatrix} x'_1 & x'_2 & x'_3 \end{pmatrix} = \begin{pmatrix} x_1 & x_2 & x_3 \end{pmatrix} \begin{pmatrix} 1 & 0 & 0 \\ 0 & -1 & 0 \\ 0 & 0 & -1 \end{pmatrix}. \tag{5.111}$$

This yields

$$x'_1 = x_1, \qquad x'_2 = -x_2, \qquad x'_3 = -x_3, \tag{5.112}$$

which is consistent with Fig. 5.5. Omitting the rest of the details of the previous calculation, this rotation with $\det \ell_{ki} = 1$ has a set of direction cosines of

$$\ell_{ki} = \begin{pmatrix} 1 & 0 & 0 \\ 0 & -1 & 0 \\ 0 & 0 & -1 \end{pmatrix}. \tag{5.113}$$

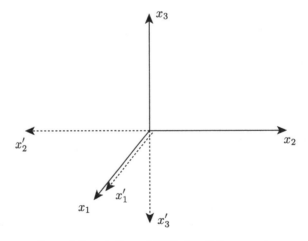

Figure 5.5 Rotation of 180° about the x_1 axis.

5.4 Stress–Strain Rate Relation for a Newtonian Fluid

Application of this rotation leads to the viscosity matrix

$$\begin{pmatrix} C_{11} & C_{12} & C_{13} & 0 & 0 & 0 \\ C_{21} & C_{22} & C_{23} & 0 & 0 & 0 \\ C_{31} & C_{32} & C_{33} & 0 & 0 & 0 \\ 0 & 0 & 0 & C_{44} & 0 & 0 \\ 0 & 0 & 0 & 0 & C_{55} & 0 \\ 0 & 0 & 0 & 0 & 0 & C_{66} \end{pmatrix}, \quad (5.114)$$

with only 12 independent coefficients.

180° Rotation About the x_2 Axis

One is tempted to perform this rotation as well, but nothing new is learned from it!

90° Rotation About the x_1 Axis

Having exhausted 180° rotations, let us turn to 90° rotations. We first rotate about the x_1 axis. This rotation is given in Fig. 5.6. This rotation has a set of direction cosines of

$$\ell_{ki} = \begin{pmatrix} 1 & 0 & 0 \\ 0 & 0 & -1 \\ 0 & 1 & 0 \end{pmatrix}. \quad (5.115)$$

Because $\det \ell_{ki} = 1$ and because its transpose equals its inverse, the transformation is a rotation. Application leads to the viscosity matrix

$$\begin{pmatrix} C_{11} & C_{12} & C_{12} & 0 & 0 & 0 \\ C_{21} & C_{22} & C_{23} & 0 & 0 & 0 \\ C_{21} & C_{23} & C_{22} & 0 & 0 & 0 \\ 0 & 0 & 0 & C_{44} & 0 & 0 \\ 0 & 0 & 0 & 0 & C_{55} & 0 \\ 0 & 0 & 0 & 0 & 0 & C_{66} \end{pmatrix}, \quad (5.116)$$

with only eight independent coefficients.

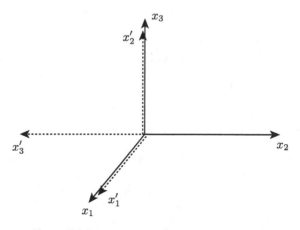

Figure 5.6 Rotation of 90° about the x_1 axis.

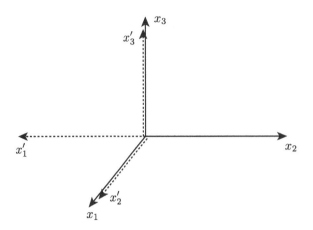

Figure 5.7 Rotation of $90°$ about the x_3 axis.

$90°$ Rotation About the x_3 Axis

This rotation is shown in Fig. 5.7. This rotation has a set of direction cosines of

$$\ell_{ki} = \begin{pmatrix} 0 & 1 & 0 \\ -1 & 0 & 0 \\ 0 & 0 & 1 \end{pmatrix}. \tag{5.117}$$

Because $\det \ell_{ki} = 1$ and because its transpose equals its inverse, the transformation is a rotation. Application leads to the conclusion that the viscosity matrix must be

$$\begin{pmatrix} C_{11} & C_{12} & C_{12} & 0 & 0 & 0 \\ C_{12} & C_{11} & C_{12} & 0 & 0 & 0 \\ C_{12} & C_{12} & C_{11} & 0 & 0 & 0 \\ 0 & 0 & 0 & C_{44} & 0 & 0 \\ 0 & 0 & 0 & 0 & C_{44} & 0 \\ 0 & 0 & 0 & 0 & 0 & C_{44} \end{pmatrix}, \tag{5.118}$$

with only three independent coefficients.

$90°$ Rotation About the x_2 Axis

We learn nothing from this rotation.

$45°$ Rotation About the x_3 Axis

This rotation is depicted in Fig. 5.8. This rotation has a set of direction cosines of

$$\ell_{ki} = \begin{pmatrix} \sqrt{2}/2 & -\sqrt{2}/2 & 0 \\ \sqrt{2}/2 & \sqrt{2}/2 & 0 \\ 0 & 0 & 1 \end{pmatrix}. \tag{5.119}$$

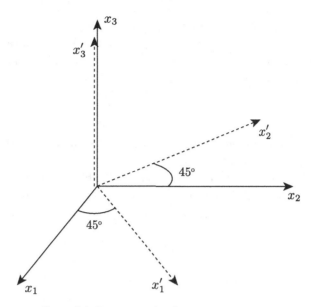

Figure 5.8 Rotation of $45°$ about the x_3 axis.

Because $\det \ell_{ki} = 1$ and because its transpose equals its inverse, the transformation is a rotation. After considerable algebra, application of this rotation leads to the conclusion that the viscosity matrix must be of the form

$$\begin{pmatrix} C_{44}+C_{12} & C_{12} & C_{12} & 0 & 0 & 0 \\ C_{12} & C_{44}+C_{12} & C_{12} & 0 & 0 & 0 \\ C_{12} & C_{12} & C_{44}+C_{12} & 0 & 0 & 0 \\ 0 & 0 & 0 & C_{44} & 0 & 0 \\ 0 & 0 & 0 & 0 & C_{44} & 0 \\ 0 & 0 & 0 & 0 & 0 & C_{44} \end{pmatrix}, \qquad (5.120)$$

with only two independent coefficients.

Try as we might, we cannot reduce this any further with more rotations. It can be proved more rigorously, as shown in most books on tensor analysis, that this is the furthest reduction that can be made. So, for an isotropic Newtonian fluid, we can expect two independent coefficients to parameterize the relation between strain rate and viscous stress. The relation between stress and strain rate can be expressed in detail as

$$\tau_{11} = C_{44}\partial_{(1}v_{1)} + C_{12}\left(\partial_{(1}v_{1)} + \partial_{(2}v_{2)} + \partial_{(3}v_{3)}\right), \qquad (5.121)$$

$$\tau_{22} = C_{44}\partial_{(2}v_{2)} + C_{12}\left(\partial_{(1}v_{1)} + \partial_{(2}v_{2)} + \partial_{(3}v_{3)}\right), \qquad (5.122)$$

$$\tau_{33} = C_{44}\partial_{(3}v_{3)} + C_{12}\left(\partial_{(1}v_{1)} + \partial_{(2}v_{2)} + \partial_{(3}v_{3)}\right), \qquad (5.123)$$

$$\tau_{12} = C_{44}\partial_{(1}v_{2)}, \qquad (5.124)$$

$$\tau_{23} = C_{44}\partial_{(2}v_{3)}, \qquad (5.125)$$

$$\tau_{31} = C_{44}\partial_{(3}v_{1)}. \qquad (5.126)$$

Using traditional notation, we take $C_{44} \equiv 2\mu$, with μ as the *first coefficient of viscosity*, and $C_{12} \equiv \lambda$, with λ as the *second coefficient of viscosity*. There are a variety of other nomenclatures for μ and λ. Following Paolucci (2016), we can also call μ the *shear viscosity* and λ the *dilatational viscosity*. We also can define the *bulk viscosity*, ζ, as

$$\zeta \equiv \lambda + \frac{2}{3}\mu, \tag{5.127}$$

which is a term in common usage. A similar analysis in solid mechanics leads one to conclude for an isotropic material in which the stress tensor is linearly related to the strain (rather than the strain rate) gives rise to two independent coefficients, the elastic modulus and the shear modulus. In solids, these both can be easily measured, and they are independent. In terms of our original fourth-order tensor, we can write the linear relationship $\tau_{ij} = \hat{C}_{ijkl}\partial_{(i}v_{j)}$ as

$$\begin{pmatrix}\tau_{11}\\\tau_{22}\\\tau_{33}\\\tau_{12}\\\tau_{23}\\\tau_{31}\\\tau_{21}\\\tau_{32}\\\tau_{13}\end{pmatrix} = \begin{pmatrix}2\mu+\lambda & \lambda & \lambda & 0 & 0 & 0 & 0 & 0 & 0\\\lambda & 2\mu+\lambda & \lambda & 0 & 0 & 0 & 0 & 0 & 0\\\lambda & \lambda & 2\mu+\lambda & 0 & 0 & 0 & 0 & 0 & 0\\0 & 0 & 0 & 2\mu & 0 & 0 & 0 & 0 & 0\\0 & 0 & 0 & 0 & 2\mu & 0 & 0 & 0 & 0\\0 & 0 & 0 & 0 & 0 & 2\mu & 0 & 0 & 0\\0 & 0 & 0 & 0 & 0 & 0 & 2\mu & 0 & 0\\0 & 0 & 0 & 0 & 0 & 0 & 0 & 2\mu & 0\\0 & 0 & 0 & 0 & 0 & 0 & 0 & 0 & 2\mu\end{pmatrix}\begin{pmatrix}\partial_{(1}v_{1)}\\\partial_{(2}v_{2)}\\\partial_{(3}v_{3)}\\\partial_{(1}v_{2)}\\\partial_{(2}v_{3)}\\\partial_{(3}v_{1)}\\\partial_{(2}v_{1)}\\\partial_{(3}v_{2)}\\\partial_{(1}v_{3)}\end{pmatrix}. \tag{5.128}$$

Because of the symmetry of $\partial_{(i}v_{j)}$, this representation is not unique. Note that the following, as well as other linear combinations, is an identically equivalent statement:

$$\begin{pmatrix}\tau_{11}\\\tau_{22}\\\tau_{33}\\\tau_{12}\\\tau_{23}\\\tau_{31}\\\tau_{21}\\\tau_{32}\\\tau_{13}\end{pmatrix} = \begin{pmatrix}2\mu+\lambda & \lambda & \lambda & 0 & 0 & 0 & 0 & 0 & 0\\\lambda & 2\mu+\lambda & \lambda & 0 & 0 & 0 & 0 & 0 & 0\\\lambda & \lambda & 2\mu+\lambda & 0 & 0 & 0 & 0 & 0 & 0\\0 & 0 & 0 & \mu & 0 & 0 & \mu & 0 & 0\\0 & 0 & 0 & 0 & \mu & 0 & 0 & \mu & 0\\0 & 0 & 0 & 0 & 0 & \mu & 0 & 0 & \mu\\0 & 0 & 0 & \mu & 0 & 0 & \mu & 0 & 0\\0 & 0 & 0 & 0 & \mu & 0 & 0 & \mu & 0\\0 & 0 & 0 & 0 & 0 & \mu & 0 & 0 & \mu\end{pmatrix}\begin{pmatrix}\partial_{(1}v_{1)}\\\partial_{(2}v_{2)}\\\partial_{(3}v_{3)}\\\partial_{(1}v_{2)}\\\partial_{(2}v_{3)}\\\partial_{(3}v_{1)}\\\partial_{(2}v_{1)}\\\partial_{(3}v_{2)}\\\partial_{(1}v_{3)}\end{pmatrix}. \tag{5.129}$$

The same result may be achieved by consideration of the isotropy properties $\hat{C}_{ijkl}$ and its decomposition into more fundamental forms involving combinations of the Kronecker delta as described by Aris (1962), Batchelor (2000), Paolucci (2016), or Yih (1977).

In shorthand Cartesian index and Gibbs notation, the viscous stress tensor is given by

$$\tau_{ij} = 2\mu\partial_{(i}v_{j)} + \lambda\partial_k v_k \delta_{ij}, \quad \boldsymbol{\tau} = 2\mu\left(\frac{\nabla\mathbf{v}^T + (\nabla\mathbf{v}^T)^T}{2}\right) + \lambda(\nabla^T\cdot\mathbf{v})\mathbf{I}. \tag{5.130}$$

By performing minor algebraic manipulations, the viscous stress tensor can be cast in a way that elucidates more of the physics of how strain rate influences stress. It can be verified by direct expansion that the viscous stress tensor can be written as

5.4 Stress–Strain Rate Relation for a Newtonian Fluid

$$\tau_{ij} = \underbrace{\left((2\mu + 3\lambda)\underbrace{\frac{\partial_k v_k}{3}}_{\text{mean strain rate}}\delta_{ij}\right)}_{\text{mean viscous stress}} + 2\mu \underbrace{\left(\partial_{(i} v_{j)} - \frac{1}{3}\partial_k v_k \delta_{ij}\right)}_{\text{deviatoric viscous stress}}, \quad (5.131)$$

$$\boldsymbol{\tau} = (2\mu + 3\lambda)\left(\frac{\nabla^T \cdot \mathbf{v}}{3}\right)\mathbf{I} + 2\mu\left(\frac{\nabla \mathbf{v}^T + (\nabla \mathbf{v}^T)^T}{2} - \frac{1}{3}\left(\nabla^T \cdot \mathbf{v}\right)\mathbf{I}\right). \quad (5.132)$$

Here it is seen that a mean strain rate, really a volumetric change, induces a mean viscous stress, as long as $\lambda \neq -(2/3)\mu$. If either $\lambda = -(2/3)\mu$ or $\partial_k v_k = 0$, all viscous stress is deviatoric. Further, for $\mu \neq 0$, a deviatoric strain rate induces a deviatoric viscous stress. Eliminating λ in favor of the bulk viscosity ζ, we can say

$$\tau_{ij} = \zeta\, \partial_k v_k \delta_{ij} + 2\mu\left(\partial_{(i} v_{j)} - \frac{1}{3}\partial_k v_k \delta_{ij}\right), \quad (5.133)$$

$$\boldsymbol{\tau} = \zeta\left(\nabla^T \cdot \mathbf{v}\right)\mathbf{I} + 2\mu\left(\frac{\nabla \mathbf{v}^T + (\nabla \mathbf{v}^T)^T}{2} - \frac{1}{3}\left(\nabla^T \cdot \mathbf{v}\right)\mathbf{I}\right). \quad (5.134)$$

Employing the expansion rate ϑ and deformation tensor $\mathbf{D}$, we could also say

$$\boldsymbol{\tau} = \zeta \vartheta \mathbf{I} + 2\mu\left(\mathbf{D} - \frac{1}{3}\vartheta \mathbf{I}\right). \quad (5.135)$$

We can form the mean viscous stress by contracting the viscous stress tensor:

$$\frac{1}{3}\tau_{ii} = \left(\frac{2}{3}\mu + \lambda\right)\partial_k v_k = \zeta\, \partial_k v_k. \quad (5.136)$$

The mean viscous stress is a scalar, and is thus independent of orientation; it is directly proportional to the first invariant of the viscous stress tensor. Obviously, the mean viscous stress is zero if $\lambda = -(2/3)\mu$, which occurs if the bulk viscosity is zero. Now the total stress tensor is given by

$$T_{ij} = -p\delta_{ij} + 2\mu\partial_{(i}v_{j)} + \lambda\partial_k v_k \delta_{ij}, \quad (5.137)$$

$$\mathbf{T} = -p\mathbf{I} + 2\mu\left(\frac{\nabla \mathbf{v}^T + (\nabla \mathbf{v}^T)^T}{2}\right) + \lambda(\nabla^T \cdot \mathbf{v})\mathbf{I}. \quad (5.138)$$

We notice the stress tensor has three components: (1) a uniform diagonal tensor with the hydrostatic pressure, (2) a tensor that is directly proportional to the strain rate tensor, and (3) a uniform diagonal tensor that is proportional to the first invariant of the strain rate tensor: $I_{\dot{\epsilon}}^{(1)} = \text{tr}(\partial_{(i}v_{k)}) = \partial_k v_k$. Consequently, the stress tensor can be written as

$$T_{ij} = \underbrace{\left(-p + \lambda I_{\dot{\epsilon}}^{(1)}\right)\delta_{ij}}_{\text{isotropic}} + \underbrace{2\mu\partial_{(i}v_{j)}}_{\text{linear in strain rate}}, \quad (5.139)$$

$$\mathbf{T} = \left(-p + \lambda I_{\dot{\epsilon}}^{(1)}\right)\mathbf{I} + 2\mu\left(\frac{\nabla \mathbf{v}^T + (\nabla \mathbf{v}^T)^T}{2}\right). \quad (5.140)$$

Recalling that $\delta_{ij} = \mathbf{I}$ as well as $I_{\dot{\epsilon}}^{(1)}$ are invariant under a rotation of coordinate axes, we deduce that the stress is related linearly to the strain rate. Moreover, when the axes are rotated to be aligned with the principal axes of strain rate, the stress is purely normal stress and has a magnitude given by the magnitude of the associated eigenvalue. Let us next consider two typical elements to aid in interpreting the relation between viscous stress and strain rate for a general Newtonian fluid.

Diagonal Component

Consider a typical diagonal component of the viscous stress tensor, say τ_{11}:

$$\tau_{11} = \underbrace{\left((2\mu + 3\lambda)\underbrace{\left(\frac{\partial_1 v_1 + \partial_2 v_2 + \partial_3 v_3}{3}\right)}_{\text{mean strain rate}}\right)}_{\text{mean viscous stress}} + \underbrace{2\mu\left(\underbrace{\partial_1 v_1 - \frac{1}{3}(\partial_1 v_1 + \partial_2 v_2 + \partial_3 v_3)}_{\text{deviatoric strain rate}}\right)}_{\text{deviatoric viscous stress}}. \tag{5.141}$$

If we choose our axes to be the principal axes of the strain rate tensor, then these terms will appear on the diagonal of the stress tensor, and there will be no off-diagonal elements. Thus, the fundamental physics of the stress–strain rate relationship are completely embodied in a natural way in this expression.

Off-Diagonal Component

If we are not aligned with the principal axes, then off-diagonal terms will be nonzero. A typical off-diagonal component, say τ_{12}, has the following form:

$$\tau_{12} = 2\mu \partial_{(1} v_{2)} + \lambda \partial_k v_k \underbrace{\delta_{12}}_{=0} = 2\mu \partial_{(1} v_{2)} = \mu(\partial_1 v_2 + \partial_2 v_1). \tag{5.142}$$

This is associated with shear deformation for elements aligned with the 1 and 2 axes; it is independent of the value of λ, which is only associated with the mean strain rate.

5.4.3 Stokes' Assumption

It is a straightforward matter to measure μ. It is not straightforward to measure λ. As discussed in Section 4.2.2, Stokes suggested to require that the mechanical pressure (that is, the average normal stress) be equal to the thermodynamic pressure; see Eq. (4.50) and the surrounding discussion. We have seen that the consequence of this is Eq. (4.52): $\tau_{ii} = 0$. If we enforce this on our expression for τ_{ij}, we get

$$\tau_{ii} = 0 = 2\mu \partial_{(i} v_{i)} + \lambda \partial_k v_k \delta_{ii} = 2\mu \partial_i v_i + 3\lambda \partial_k v_k = (2\mu + 3\lambda)\partial_i v_i. \tag{5.143}$$

Because in general $\partial_i v_i \neq 0$, Stokes' assumption implies that

$$\lambda = -\frac{2}{3}\mu, \qquad \zeta = 0, \qquad \text{iff Stokes' assumption satisfied.} \tag{5.144}$$

So, a Newtonian fluid satisfying Stokes' assumption has the following constitutive equation:

$$\tau_{ij} = 2\mu \underbrace{\left(\partial_{(i} v_{j)} - \frac{1}{3} \partial_k v_k \delta_{ij} \right)}_{\text{deviatoric strain rate}}, \quad \boldsymbol{\tau} = 2\mu \left(\frac{(\nabla \mathbf{v}^T + (\nabla \mathbf{v}^T)^T)}{2} - \frac{1}{3} (\nabla^T \cdot \mathbf{v}) \mathbf{I} \right). \quad (5.145)$$

$$\underbrace{\phantom{2\mu \left(\partial_{(i} v_{j)} - \frac{1}{3} \partial_k v_k \delta_{ij} \right)}}_{\text{deviatoric viscous stress}}$$

Incompressible flows have $\partial_i v_i = 0$; thus, λ plays no role in determining the viscous stress in such flows. For the fluid that obeys Stokes' assumption, the viscous stress is entirely deviatoric and is induced only by a deviatoric strain rate. Employing the expansion rate ϑ and deformation tensor $\mathbf{D}$, we have for a Newtonian fluid that satisfies Stokes' assumption

$$\boldsymbol{\tau} = 2\mu \left(\mathbf{D} - \frac{1}{3} \vartheta \mathbf{I} \right) = 2\mu \left(\mathbf{D} - \frac{1}{3} (\operatorname{tr} \mathbf{D}) \mathbf{I} \right). \quad (5.146)$$

Clearly, for such a fluid, the viscous stress is purely deviatoric with a mean value of zero and is proportional to the deviation of the deformation from its mean.

5.4.4 Second Law Restrictions

Recall that, in order for the constitutive equation for viscous stress to be consistent with the second law of thermodynamics, it is sufficient (but perhaps overly restrictive) to require that Eq. (5.29) hold:

$$\frac{1}{T} \underbrace{\tau_{ij} \partial_{(i} v_{j)}}_{=\Phi} \geq 0, \qquad \frac{1}{T} \Phi \geq 0. \quad (5.147)$$

Invoking our constitutive equation for viscous stress, and realizing that the absolute temperature $T > 0$, we have then that the viscous dissipation function Φ must satisfy

$$\Phi = (2\mu \partial_{(i} v_{j)} + \lambda \partial_k v_k \delta_{ij})(\partial_{(i} v_{j)}) \geq 0. \quad (5.148)$$

This reduces to the sum of two squares:

$$\Phi = 2\mu \partial_{(i} v_{j)} \partial_{(i} v_{j)} + \lambda \partial_k v_k \partial_i v_i \geq 0, \quad \Phi = 2\mu (\mathbf{D} : \mathbf{D}) + \lambda (\nabla^T \cdot \mathbf{v})^2 \geq 0. \quad (5.149)$$

We seek restrictions on μ and λ such that $\Phi \geq 0$. Requiring $\mu \geq 0$ and $\lambda \geq 0$ guarantees satisfaction of the second law. However, Stokes' assumption of $\lambda = -2\mu/3$ does not meet this criterion, and so we are motivated to check more carefully.

One-Dimensional Systems

Let us first check the criterion for a strictly one-dimensional system. For such a system, our second law restriction reduces to

$$2\mu \partial_{(1} v_{1)} \partial_{(1} v_{1)} + \lambda \partial_1 v_1 \partial_1 v_1 \geq 0, \quad (5.150)$$

$$(2\mu + \lambda) \partial_1 v_1 \partial_1 v_1 \geq 0, \quad \text{so} \quad 2\mu + \lambda \geq 0, \quad \text{so} \quad \lambda \geq -2\mu. \quad (5.151)$$

Obviously, if $\mu \geq 0$ and $\lambda = -2\mu/3$, the entropy inequality is satisfied. We also could satisfy the inequality for negative μ with sufficiently large positive λ, but there is no experimental evidence for this.

Two-Dimensional Systems

Extension to a two-dimensional system is more complicated. For such systems, expansion of our second law condition gives

$$2\mu\partial_{(1}v_{1)}\partial_{(1}v_{1)} + 2\mu\partial_{(1}v_{2)}\partial_{(1}v_{2)} + 2\mu\partial_{(2}v_{1)}\partial_{(2}v_{1)} + 2\mu\partial_{(2}v_{2)}\partial_{(2}v_{2)}$$
$$+ \lambda\left(\partial_{(1}v_{1)} + \partial_{(2}v_{2)}\right)\left(\partial_{(1}v_{1)} + \partial_{(2}v_{2)}\right) \geq 0. \quad (5.152)$$

Taking advantage of symmetry of the deformation tensor, we can say

$$2\mu\partial_{(1}v_{1)}\partial_{(1}v_{1)} + 4\mu\partial_{(1}v_{2)}\partial_{(1}v_{2)} + 2\mu\partial_{(2}v_{2)}\partial_{(2}v_{2)} + \lambda\left(\partial_{(1}v_{1)} + \partial_{(2}v_{2)}\right)\left(\partial_{(1}v_{1)} + \partial_{(2}v_{2)}\right) \geq 0. \quad (5.153)$$

Expanding the product and regrouping gives

$$(2\mu + \lambda)\partial_{(1}v_{1)}\partial_{(1}v_{1)} + 4\mu\partial_{(1}v_{2)}\partial_{(1}v_{2)} + (2\mu + \lambda)\partial_{(2}v_{2)}\partial_{(2}v_{2)} + 2\lambda\partial_{(1}v_{1)}\partial_{(2}v_{2)} \geq 0. \quad (5.154)$$

In matrix form, we can write this inequality using the method of quadratic forms; thermodynamics is satisfied see Section 2.8:

$$\Phi = \begin{pmatrix} \partial_{(1}v_{1)} & \partial_{(2}v_{2)} & \partial_{(1}v_{2)} \end{pmatrix} \begin{pmatrix} (2\mu + \lambda) & \lambda & 0 \\ \lambda & (2\mu + \lambda) & 0 \\ 0 & 0 & 4\mu \end{pmatrix} \begin{pmatrix} \partial_{(1}v_{1)} \\ \partial_{(2}v_{2)} \\ \partial_{(1}v_{2)} \end{pmatrix} \geq 0. \quad (5.155)$$

This may be verified by direct expansion. As we have discussed before, the condition that this hold for all values of the deformation is that the symmetric part of the coefficient matrix have eigenvalues that are greater than or equal to zero. In fact, here the coefficient matrix is symmetric. Let us find the eigenvalues κ of the coefficient matrix. The eigenvalues are found by evaluating the following equation:

$$\begin{vmatrix} (2\mu + \lambda) - \kappa & \lambda & 0 \\ \lambda & (2\mu + \lambda) - \kappa & 0 \\ 0 & 0 & 4\mu - \kappa \end{vmatrix} = 0. \quad (5.156)$$

We get the characteristic polynomial

$$(4\mu - \kappa)\left((2\mu + \lambda - \kappa)^2 - \lambda^2\right) = 0. \quad (5.157)$$

This has roots

$$\kappa = 4\mu, \quad \kappa = 2\mu, \quad \kappa = 2(\mu + \lambda). \quad (5.158)$$

For the two-dimensional system, we see now formally that we must satisfy both

$$\mu \geq 0, \quad \lambda \geq -\mu. \quad (5.159)$$

This is more restrictive than for the one-dimensional system, but we see that a fluid obeying Stokes' assumption $\lambda = -2\mu/3$ still satisfies this inequality.

Three-Dimensional Systems

For a full three-dimensional variation, the entropy inequality $(2\mu\partial_{(i}v_{j)} + \lambda\partial_k v_k \delta_{ij})(\partial_{(i}v_{j)}) \geq 0$, when expanded, is equivalent to the following quadratic form:

$$\Phi = \begin{pmatrix} \partial_{(1}v_{1)} & \partial_{(2}v_{2)} & \partial_{(3}v_{3)} & \partial_{(1}v_{2)} & \partial_{(2}v_{3)} & \partial_{(3}v_{1)} \end{pmatrix} \begin{pmatrix} \lambda+2\mu & \lambda & \lambda & 0 & 0 & 0 \\ \lambda & \lambda+2\mu & \lambda & 0 & 0 & 0 \\ \lambda & \lambda & \lambda+2\mu & 0 & 0 & 0 \\ 0 & 0 & 0 & 4\mu & 0 & 0 \\ 0 & 0 & 0 & 0 & 4\mu & 0 \\ 0 & 0 & 0 & 0 & 0 & 4\mu \end{pmatrix} \begin{pmatrix} \partial_{(1}v_{1)} \\ \partial_{(2}v_{2)} \\ \partial_{(3}v_{3)} \\ \partial_{(1}v_{2)} \\ \partial_{(2}v_{3)} \\ \partial_{(3}v_{1)} \end{pmatrix} \geq 0. \quad (5.160)$$

Again this must hold for arbitrary values of the deformation, so we must require that the eigenvalues κ of the interior matrix be greater than or equal to zero to satisfy the entropy inequality. One can show that the six eigenvalues for the interior matrix are

$$\kappa = 2\mu, \quad \kappa = 2\mu, \quad \kappa = 4\mu, \quad \kappa = 4\mu, \quad \kappa = 4\mu, \quad \kappa = 3\lambda + 2\mu. \quad (5.161)$$

Two of the eigenvalues are degenerate, but this is not a particular problem. We need now that $\kappa \geq 0$, so the entropy inequality requires that

$$\mu \geq 0, \quad \lambda \geq -\frac{2}{3}\mu. \quad (5.162)$$

Obviously, a fluid that satisfies Stokes' assumption does not violate the entropy inequality, but it does give rise to a minimum level of satisfaction. This does not mean the fluid is isentropic! It simply means one of the six eigenvalues is zero.

Now using standard techniques from linear algebra for quadratic forms, the entropy inequality can, after much effort, be manipulated into the form

$$\Phi = \frac{2}{3}\mu\left((\partial_{(1}v_{1)} - \partial_{(2}v_{2)})^2 + (\partial_{(2}v_{2)} - \partial_{(3}v_{3)})^2 + (\partial_{(3}v_{3)} - \partial_{(1}v_{1)})^2\right)$$
$$+ \left(\lambda + \frac{2}{3}\mu\right)(\partial_{(1}v_{1)} + \partial_{(2}v_{2)} + \partial_{(3}v_{3)})^2 + 4\mu((\partial_{(1}v_{2)})^2 + (\partial_{(2}v_{3)})^2 + (\partial_{(3}v_{1)})^2) \geq 0.$$
$$(5.163)$$

Obviously, this is a sum of perfect squares, and holds for all values of the strain rate tensor. It can be verified by direct expansion that this term is identical to the strong form of the entropy inequality for viscous stress. It can further be verified by direct expansion that the entropy inequality can also be written more compactly as

$$\Phi = 2\mu \underbrace{\left(\partial_{(i}v_{j)} - \frac{1}{3}\partial_k v_k \delta_{ij}\right)\left(\partial_{(i}v_{j)} - \frac{1}{3}\partial_k v_k \delta_{ij}\right)}_{\text{(deviatoric strain rate)}^2} + \left(\lambda + \frac{2}{3}\mu\right)\underbrace{(\partial_i v_i)(\partial_j v_j)}_{\text{(mean strain rate)}^2} \geq 0. \quad (5.164)$$

So, for a Newtonian fluid, we see that the increase in entropy due to viscous dissipation is attributable to two effects: deviatoric strain rate and mean strain rate. The terms involving both are perfect squares. As long as $\mu \geq 0$ and $\lambda \geq -2\mu/3$, the second law is satisfied. In the one-dimensional limit, we find

$$\Phi = (2\mu + \lambda)(\partial_1 v_1)^2, \tag{5.165}$$

and if Stokes' assumption is satisfied, we have

$$\Phi = \frac{4}{3}\mu(\partial_1 v_1)^2. \tag{5.166}$$

We can also write the strong form of the entropy inequality for a Newtonian fluid $(2\mu \partial_{(i} v_{j)} + \lambda \partial_k v_k \delta_{ij})(\partial_{(i} v_{j)}) \geq 0$, in terms of the principal invariants of strain rate. Leaving out details that can be verified by direct expansion, we find the following form:

$$\Phi = (2\mu + \lambda)\left(I_{\dot{\epsilon}}^{(1)}\right)^2 - 4\mu I_{\dot{\epsilon}}^{(2)} \geq 0. \tag{5.167}$$

Because this is in terms of the invariants, we are assured that it is independent of the coordinate system. It is, however, not obvious that this form is positive semi-definite. We can use the definitions of the invariants of strain rate to rewrite the inequality as

$$\Phi = 2\mu \left(\partial_{(i} v_{j)} \partial_{(j} v_{i)} - \frac{1}{3}(\partial_i v_i)(\partial_j v_j) \right) + \left(\lambda + \frac{2}{3}\mu\right)(\partial_i v_i)(\partial_j v_j) \geq 0. \tag{5.168}$$

In terms of the eigenvalues of the strain rate tensor, κ_1, κ_2, and κ_3, this becomes

$$\Phi = 2\mu \left(\kappa_1^2 + \kappa_2^2 + \kappa_3^2 - \frac{1}{3}(\kappa_1 + \kappa_2 + \kappa_3)^2 \right) + \left(\lambda + \frac{2}{3}\mu\right)(\kappa_1 + \kappa_2 + \kappa_3)^2 \geq 0. \tag{5.169}$$

This then reduces to a positive semi-definite form:

$$\Phi = \frac{2}{3}\mu \left((\kappa_1 - \kappa_2)^2 + (\kappa_1 - \kappa_3)^2 + (\kappa_2 - \kappa_3)^2 \right) + \left(\lambda + \frac{2}{3}\mu\right)(\kappa_1 + \kappa_2 + \kappa_3)^2 \geq 0. \tag{5.170}$$

Because the eigenvalues are invariant under rotation, this form is invariant.

We summarize by noting relations between mean and deviatoric stress and strain rates for Newtonian fluids. The influence of each on each has been seen or can be shown to be as follows. (1) A mean strain rate will induce a *time rate of change* in the mean thermodynamic stress via traditional thermodynamic relations (e.g. for an isothermal ideal gas $dp/dt = RT(d\rho/dt) = -\rho RT \partial_i v_i$) and will induce an additional mean viscous stress for fluids that do not obey Stokes' assumption. (2) A deviatoric strain rate will not directly induce a mean stress. (3) A deviatoric strain rate will directly induce a deviatoric stress. (4) A mean strain rate will induce entropy production only for a fluid that does not obey Stokes' assumption. (5) A deviatoric strain rate will always induce entropy production in a viscous fluid.

5.5 Irreversibility Production Rate

For our isotropic Newtonian fluid that satisfies Stokes' assumption and Fourier's law, we can now report the irreversibility production rate $\dot{\mathcal{I}}$ from Eq. (4.216):

$$\dot{\mathcal{I}} = \frac{k}{T^2} \partial_i T \partial_i T + \frac{2\mu}{T}\left(\partial_{(i} v_{j)} - \frac{1}{3}\partial_k v_k \delta_{ij}\right)\left(\partial_{(i} v_{j)} - \frac{1}{3}\partial_k v_k \delta_{ij}\right) \geq 0, \tag{5.171}$$

$$= \frac{k}{T^2}(\nabla T)^T \cdot \nabla T + \frac{2\mu}{T}\left(\mathbf{D} - \frac{1}{3}(\text{tr }\mathbf{D})\mathbf{I}\right) : \left(\mathbf{D} - \frac{1}{3}(\text{tr }\mathbf{D})\mathbf{I}\right) \geq 0. \tag{5.172}$$

As long as $k \geq 0$ and $\mu \geq 0$, we are guaranteed satisfaction of the second law.

5.6 Thermodynamic Equations of State

Thermodynamic equations of state provide algebraic relations between variables such as pressure, temperature, energy, and entropy. They do not involve velocity. They are formally valid for materials at rest. As long as the time scales of equilibration of the thermodynamic variables are much faster than the finest time scales of fluid dynamics, it is a valid assumption to use an ordinary equations of state. Such assumptions can be violated in high speed flows in which vibrational and rotational modes of oscillation become excited. They may also be invalid in highly rarefied flows such as might occur in the upper atmosphere.

Typically, we will require two types of equations, a *thermal equation of state* that for simple compressible substances with one reversible work mode gives the pressure as a function of two independent thermodynamic variables, for example

$$p = p(\rho, T), \tag{5.173}$$

and a *caloric equation of state* that gives the internal energy as a function of two independent thermodynamic variables, for example

$$e = e(\rho, T). \tag{5.174}$$

There are additional conditions regarding internal consistency of the equations of state; that is, just any stray functional forms will not do.

We outline here a method for generating equations of state with internal consistency based on satisfying the entropy inequality. First let us define a new thermodynamic property, $\hat{a}$, the Helmholtz free energy:

$$\hat{a} = e - Ts. \tag{5.175}$$

We can take the material time derivative of Eq. (5.175) to get

$$\frac{d\hat{a}}{dt} = \frac{de}{dt} - T\frac{ds}{dt} - s\frac{dT}{dt}. \tag{5.176}$$

It will be shown in Section 9.1.1 that there are a set of natural, "canonical," variables for describing $\hat{a}$, which are T and ρ, or equivalently T and $\hat{v}$. Here, we choose $\hat{a} = \hat{a}(T, \rho)$. Taking the total time derivative of this form of $\hat{a}$ yields another form for $d\hat{a}/dt$:

$$\frac{d\hat{a}}{dt} = \left.\frac{\partial \hat{a}}{\partial T}\right|_\rho \frac{dT}{dt} + \left.\frac{\partial \hat{a}}{\partial \rho}\right|_T \frac{d\rho}{dt}. \qquad (5.177)$$

Now we also have the energy equation, Eq. (4.187), and entropy inequality, Eq. (4.269):

$$\rho \frac{de}{dt} = -\partial_i q_i - p \partial_i v_i + \tau_{ij} \partial_i v_j, \qquad (5.178)$$

$$\rho \frac{ds}{dt} \geq -\partial_i \left(\frac{q_i}{T}\right). \qquad (5.179)$$

Using Eq. (5.176) to eliminate de/dt in favor of $d\hat{a}/dt$ in the energy equation, Eq. (5.178), gives a modified energy equation:

$$\rho \left(\frac{d\hat{a}}{dt} + T\frac{ds}{dt} + s\frac{dT}{dt}\right) = -\partial_i q_i - p \partial_i v_i + \tau_{ij} \partial_i v_j. \qquad (5.180)$$

Next, we use Eq. (5.177) to eliminate $d\hat{a}/dt$ in Eq. (5.180) to get

$$\rho \left(\left.\frac{\partial \hat{a}}{\partial T}\right|_\rho \frac{dT}{dt} + \left.\frac{\partial \hat{a}}{\partial \rho}\right|_T \frac{d\rho}{dt} + T\frac{ds}{dt} + s\frac{dT}{dt}\right) = -\partial_i q_i - p \partial_i v_i + \tau_{ij} \partial_i v_j. \qquad (5.181)$$

Now in this modified energy equation, we solve for $\rho\, ds/dt$ to get

$$\rho \frac{ds}{dt} = -\frac{1}{T}\partial_i q_i - \frac{p}{T}\partial_i v_i + \frac{1}{T}\tau_{ij}\partial_i v_j - \frac{\rho}{T}\left.\frac{\partial \hat{a}}{\partial T}\right|_\rho \frac{dT}{dt} - \frac{\rho}{T}\left.\frac{\partial \hat{a}}{\partial \rho}\right|_T \frac{d\rho}{dt} - \frac{\rho s}{T}\frac{dT}{dt}. \qquad (5.182)$$

Substituting this version of the energy equation into the second law, Eq. (5.179), gives

$$-\frac{1}{T}\partial_i q_i - \frac{p}{T}\partial_i v_i + \frac{1}{T}\tau_{ij}\partial_i v_j - \frac{\rho}{T}\left.\frac{\partial \hat{a}}{\partial T}\right|_\rho \frac{dT}{dt} - \frac{\rho}{T}\left.\frac{\partial \hat{a}}{\partial \rho}\right|_T \frac{d\rho}{dt} - \frac{\rho s}{T}\frac{dT}{dt} \geq -\partial_i\left(\frac{q_i}{T}\right). \qquad (5.183)$$

Rearranging and using the mass conservation relation to eliminate $\partial_i v_i$, we get

$$-\frac{q_i}{T^2}\partial_i T - \frac{p}{T}\left(-\frac{1}{\rho}\frac{d\rho}{dt}\right) + \frac{1}{T}\tau_{ij}\partial_i v_j - \frac{\rho}{T}\left.\frac{\partial \hat{a}}{\partial T}\right|_\rho \frac{dT}{dt} - \frac{\rho}{T}\left.\frac{\partial \hat{a}}{\partial \rho}\right|_T \frac{d\rho}{dt} - \frac{\rho s}{T}\frac{dT}{dt} \geq 0, \qquad (5.184)$$

$$-\frac{q_i}{T}\partial_i T + \frac{p}{\rho}\frac{d\rho}{dt} + \tau_{ij}\partial_i v_j - \rho\left.\frac{\partial \hat{a}}{\partial T}\right|_\rho \frac{dT}{dt} - \rho\left.\frac{\partial \hat{a}}{\partial \rho}\right|_T \frac{d\rho}{dt} - \rho s\frac{dT}{dt} \geq 0, \qquad (5.185)$$

$$-\frac{q_i}{T}\partial_i T + \tau_{ij}\partial_i v_j + \frac{1}{\rho}\frac{d\rho}{dt}\underbrace{\left(p - \rho^2 \left.\frac{\partial \hat{a}}{\partial \rho}\right|_T\right)}_{0} - \rho\frac{dT}{dt}\underbrace{\left(s + \left.\frac{\partial \hat{a}}{\partial T}\right|_\rho\right)}_{0} \geq 0. \qquad (5.186)$$

Now in our discussion of the strong form of the entropy inequality, we have already found forms for q_i and τ_{ij} for which the terms involving these phenomena are positive semi-definite. We can guarantee the remaining two terms are consistent with the second law, and are associated with reversible processes, by requiring that

$$p = \rho^2 \left.\frac{\partial \hat{a}}{\partial \rho}\right|_T, \qquad s = -\left.\frac{\partial \hat{a}}{\partial T}\right|_\rho. \qquad (5.187)$$

We insist on these conditions because $d\rho/dt$ and dT/dt can be positive, negative, or zero, and to guarantee second law satisfaction in weak form, we must then insist that derivatives of ρ or T not lead to second law violation.

If we take the nonobvious, but experimentally defensible, choice for $\hat{a}$ of

$$\hat{a}(T,\rho) = e_o + c_v(T - T_o) - T\left(s_o + c_v \ln\left(\frac{T}{T_o}\right) - R \ln\left(\frac{\rho}{\rho_o}\right)\right), \qquad (5.188)$$

then we get for pressure

$$p = \rho^2 \left.\frac{\partial \hat{a}}{\partial \rho}\right|_T = \rho^2 \left(\frac{RT}{\rho}\right) = \rho RT. \qquad (5.189)$$

This equation for pressure is a thermal equation of state for an *ideal gas*, and R is the engineering gas constant. As introduced in Eq. (1.18), it is the ratio of the universal gas constant and the molecular mass of the particular gas, $R = \Re/\mathcal{M}$.

Solving for entropy s, we get

$$s = -\left.\frac{\partial \hat{a}}{\partial T}\right|_\rho = s_o + c_v \ln\left(\frac{T}{T_o}\right) - R \ln\left(\frac{\rho}{\rho_o}\right). \qquad (5.190)$$

Then, we get for e

$$e = \hat{a} + Ts = e_o + c_v(T - T_o). \qquad (5.191)$$

We call this equation for energy a caloric equation of state for a *calorically perfect* gas. It is calorically perfect because the specific heat at constant volume c_v is assumed a true constant here. In general for ideal gases, it can be shown to be at most a function of temperature. For more general nonideal gases, one can also find general forms for $\hat{a}(\rho, T)$ and use that to find thermal and caloric state equations (see Section 9.1).

Reynolds (1968) and many other sources discuss the important topic of how to determine which equations of state are appropriate for a particular material under particular conditions. Typically, the ideal gas assumption is appropriate for gases for which intermolecular forces and the effect of finite molecular volume are negligible. One might expect significant intermolecular forces as a material cools to a near-liquid state at low temperatures. Or at high density, the finite volume of molecules affects the behavior of the gas. Monatomic inert gases such as helium and argon are well modeled as calorically perfect ideal gases over a wide range of temperatures. For multi-atomic molecules such as O_2 or CH_4, the effects of molecular vibration and rotation combine to render the specific heat to be a function of temperature at moderately high temperatures. And at higher temperatures, such molecules can dissociate and ionize. When this is realized, one must turn to models of chemically reacting mixing gases and ionized plasmas to capture system behavior.

Numerical values of physical properties that define relevant thermodynamic regimes vary from material to material. For example, at $p = 100\,\text{kPa}$, N_2 boils at $T = 77.24\,\text{K}$. At temperatures just above the boiling point, the ideal gas approximation is not good. But for $T = 180$ K, the ideal gas approximation has small error. Similarly for N_2 at $p = 100$ kPa, $T = 298$ K, the calorically perfect assumption is good, and the specific heat is approximated well as $c_p = 1.042$ kJ/kg/K. However, for $T = 500$ K, we find $c_p = 1.0575$ kJ/kg/K. For $T = 1000$ K, it is $c_p = 1.1700$ kJ/kg/K. The increase is due to the energy being diverted to rotational and vibrational modes at higher temperatures. At very high temperatures, for example, at $T = 6000$ K, significant dissociation at $p = 100$ kPa occurs, resulting in significant concentrations of both N and N_2, with respective mole fractions of 0.21 and 0.79.

SUMMARY

This chapter has developed mathematical models for the particular behavior of actual fluids. When coupled with the general axioms of mechanics, one has a complete system to describe fluid motion. General guiding principles were first presented such that the models are consistent with principles such as Galilean frame invariance, the second law of thermodynamics, and Onsager reciprocity. Special attention was given to development of a model to describe an isotropic fluid whose stress varies linearly with strain rate.

PROBLEMS

5.1 Subject the one-dimensional linear momentum equation for inviscid flow

$$\frac{\partial v_x}{\partial t} + v_x \frac{\partial v_x}{\partial x} = f_x - \frac{1}{\rho}\frac{\partial p}{\partial x},$$

to the coordinate transformation

$$x' = x - X(t), \qquad t' = t,$$

accompanied by the dependent variable transformation

$$\rho' = \rho, \qquad p' = p, \qquad v'_x = v_x - \frac{dX}{dt}.$$

Identify the class of transformations for which the linear momentum equation is invariant.

5.2 For a compressible, Newtonian fluid with constant viscosity that satisfies Stokes' assumption, show

$$(\nabla^T \cdot \boldsymbol{\tau})^T = \mu\left(\frac{4}{3}\nabla\vartheta - \nabla\times\boldsymbol{\omega}\right),$$

where $\vartheta = \nabla^T \cdot \mathbf{v}$ and $\boldsymbol{\omega} = \nabla \times \mathbf{v}$. Show that this is actually a Helmholtz decomposition (see Problem 3.13) of the form $(\nabla^T \cdot \boldsymbol{\tau})^T = \nabla\phi + \nabla\times\boldsymbol{\psi}$. Identify ϕ and $\boldsymbol{\psi}$.

5.3 The energy equation for a fluid with an internal source of heat $\mathcal{Q}(x,y,z,t)$ is

$$\rho\frac{de}{dt} = -\nabla^T \cdot \mathbf{q} - p\nabla^T \cdot \mathbf{v} + \boldsymbol{\tau}:\mathbf{D} + \mathcal{Q}.$$

Consider a time-independent, stationary fluid that only has temperature variation in the Cartesian coordinates x and y and that is subjected to an internal source of heating,

$\mathcal{Q}(x,y)$. The fluid has known temperature field $T(x,y) = xy(x-1)(y-1)$ and has anisotropic thermal conductivity with

$$\mathbf{q} = -\mathbf{K} \cdot \nabla T, \qquad \mathbf{K} = \begin{pmatrix} 2 & 1 \\ 1 & 2 \end{pmatrix}.$$

Show that the second law of thermodynamics is satisfied. Plot the temperature field, the heat flux vector field, and the heat source field $\mathcal{Q}(x,y)$. Plot separately the irreversibility production rate field $\dot{\mathcal{I}}(x,y)$ along with the heat flux vector field.

5.4 In contrast to a Newtonian fluid, whose stress tensor varies linearly with the deformation tensor, a Stokesian fluid may possess quadratic variation:

$$T_{ij} = -p\delta_{ij} + \beta \partial_{(i} v_{j)} + \gamma \partial_{(i} v_{k)} \partial_{(k} v_{j)}.$$

Find the mean and deviatoric stresses first for a Newtonian fluid satisfying Stokes' assumption and then for a Stokesian fluid. Determine if a Stokesian fluid satisfies Stokes' assumption.

5.5 A Newtonian fluid satisfying Stokes' assumption moves in the velocity field

$$v_1 = v_o \frac{x_1 + x_2 + x_3}{L}, \qquad v_2 = v_o \frac{x_1^2 - x_2^2}{L^2}, \qquad v_3 = v_o \frac{x_1 x_2 x_3}{L^3}.$$

Consider the special case for which $v_o = 1$, $L = 1$, $\mu = 1$. Find the viscous stress tensor, and find its eigenvalues at $x_1 = x_2 = x_3 = 1$. Show the eigenvalues sum to zero, and give an explanation for this. At the point $(1,1,1)^T$, find the viscous traction stress vector associated with the plane whose outer normal is $\mathbf{n} = (1/\sqrt{3}, 1/\sqrt{3}, 1/\sqrt{3})^T$.

5.6 Consider a Newtonian fluid satisfying Stokes' assumption and Fourier's law with constant μ and k moving in the velocity and temperature fields given by

$$v_1 = v_o \left(-2\frac{x_2}{L} + \left(\frac{x_1}{L}\right)^3 \right), \quad v_2 = v_o \left(\frac{x_1}{L} - \left(\frac{x_2}{L}\right)^3 \right), \quad v_3 = 0, \quad T = T_o \left(\left(\frac{x_1}{L}\right)^2 + 2\left(\frac{x_2}{L}\right)^2 \right).$$

Find an expression for the viscous dissipation function Φ and the irreversibility production rate $\dot{\mathcal{I}}$. For $k = 1$, $\mu = 1$, $T_o = 1$, $L = 1$, $v_o = 1$, plot the velocity vector field along with the temperature and irreversibility production rate fields.

5.7 A virial equation of state has Helmholtz free energy of

$$\hat{a}(T, \hat{v}) = c_v(T - T_o) - T \left(c_v \ln\left(\frac{T}{T_o}\right) + R \ln\left(\frac{\hat{v} - b}{\hat{v}_o - b}\right) \right),$$

with c_v, R, T_o, b, and $\hat{v}_o$ as constants. Recognizing $\hat{v} = 1/\rho$, find $p(\rho, T)$, $e(\rho, T)$. Find an expression for the enthalpy $h(\rho, T)$. Find an expression for the specific heat at constant pressure, defined as $c_p = \partial h / \partial T |_p$. With the Gibbs free energy defined as $g = h - Ts$, find an expression for the Gibbs free energy, $g(T, \hat{v})$.

REFERENCES

Aris, R. (1962). *Vectors, Tensors, and the Basic Equations of Fluid Mechanics.* New York: Dover.

Batchelor, G. K. (2000). *An Introduction to Fluid Dynamics.* Cambridge, UK: Cambridge University Press.

Bird, R. B., Armstrong, R. C., and Hassager, O. (1987). *Dynamics of Polymeric Liquids*, 2 vols. New York: John Wiley.

Bird, R. B., Stewart, W. E., and Lightfoot, E. N. (2007). *Transport Phenomena*, revised 2nd ed. New York: John Wiley.

Callen, H. B. (1985). *Thermodynamics and an Introduction to Thermostatistics*, 2nd ed. New York: John Wiley.

de Groot, S. R., and Mazur, P. (1984). *Non-Equilibrium Thermodynamics.* New York: Dover.

Edelen, D. G. B., and McLennan, J. A. (1973). Material indifference: A principle or a convenience, *International Journal of Engineering Science*, **11**(8), 813–817.

Fung, Y. C. (1965). *Foundations of Solid Mechanics.* Englewood Cliffs, New Jersey: Prentice-Hall.

Gyarmati, I. (1970). *Non-Equilibrium Thermodynamics. Field Theory and Variational Principles.* New York: Springer.

Hirschfelder, J. O., Curtiss, C. F., and Bird, R. B. (1954). *Molecular Theory of Gases and Liquids.* New York: John Wiley.

Kondepudi, D., and Prigogine, I. (1998). *Modern Thermodynamics: From Heat Engines to Dissipative Structures.* New York: John Wiley.

Lavenda, B. H. (1978). *Thermodynamics of Irreversible Processes.* New York: John Wiley.

Morrison, F. A. (2001). *Understanding Rheology.* New York: Oxford University Press.

Müller, I., and Ruggeri, T. (1998). *Rational Extended Thermodynamics*, 2nd ed. New York: Springer.

Paolucci, S. (2016). *Continuum Mechanics and Thermodynamics of Matter.* New York: Cambridge University Press.

Powers, J. M. (2004). On the necessity of positive semi-definite conductivity and Onsager reciprocity in modeling heat conduction in an anisotropic media, *Journal of Heat Transfer*, **126**(5), 670–675.

Powers, J. M. (2016). *Combustion Thermodynamics and Dynamics.* New York: Cambridge University Press.

Reynolds, W. C. (1968). *Thermodynamics*, 2nd ed. New York: McGraw-Hill.

Schowalter, W. R. (1978). *Mechanics of Non-Newtonian Fluids.* Oxford: Pergamon.

Whitaker, S. (1968). *Introduction to Fluid Mechanics.* Malabar, Florida: Krieger.

Woods, L. C. (1975). *The Thermodynamics of Fluid Systems.* Oxford: Clarendon Press.

Woods, L. C. (1982). Thermodynamic inequalities in continuum mechanics, *IMA Journal of Applied Mathematics*, **29**(3), 221–246.

Yih, C.-S. (1977). *Fluid Mechanics.* East Hampton, Connecticut: West River Press.

6 Governing Equations: Summary and Special Cases

In this chapter, we consider a variety of topics related to the governing equations as a system. We briefly discuss boundary and interface conditions, necessary for a complete system, summarize the partial differential equations in various forms, present some special cases of the governing equations, present the equations in a dimensionless form, and consider a few cases for which the linear momenta equation can be integrated once.

6.1 Boundary and Interface Conditions

We briefly mention a topic that is of critical importance in the modeling of fluid systems: boundary and interface conditions. For a fluid with moving surfaces, for example flow around deformable bubbles or for surfaces that contain cusps, this can present a significant challenge. We will mainly consider simple scenarios for which the topologies and analyses are straightforward.

We first consider conditions for viscous fluids. At fluid–solid interfaces, it is observed in the continuum regime that fluid sticks to a solid boundary, so that we can safely take the fluid and solid velocities to be identical at the interface. This is the no-slip condition. If the fluid is denoted by 1 and the solid by 2, we say at a no-slip boundary that

$$\mathbf{v}_1 = \mathbf{v}_2, \qquad \text{viscous, no-slip surface.} \tag{6.1}$$

As one approaches the molecular level, the no-slip assumption breaks down.

At the interface of two distinct, immiscible viscous fluids, labeled 1 and 2, with no surface tension, one requires continuity of temperature T, total stress $\mathbf{T}$, and energy flux $\mathbf{q}$ across the interface. This yields

$$T_1 = T_2, \qquad \mathbf{T}_1 = \mathbf{T}_2, \qquad \mathbf{q}_1 = \mathbf{q}_2. \tag{6.2}$$

These also hold if either 1 or 2 is a solid. In the important case of an adiabatic surface, one must have $\mathbf{q}_1 = \mathbf{q}_2 = \mathbf{0}$. Density need not be continuous in the absence of mass diffusion. Were mass diffusion present, the fluids would not be immiscible, and density would be a continuous variable. Additionally for such flows as those involving bubbles, the effect of surface tension needs to be considered. The effect of surface tension is to admit a discontinuity of $\mathbf{T}$ across an interface. Except for a short analysis in Section 12.2, we shall not consider surface tension; many texts give a more complete treatment. Surface geometry is considered extensively by Aris (1962), and surface physics by Leal (2007).

When one considers the inviscid limit, other possibilities arise. One can admit at an interface a fluid or solid with a tangential velocity component different than that of the fluid or solid on the opposite side of the interface. However, to satisfy mass conservation for an impermeable solid wall, one cannot allow the fluid to have a velocity normal to the solid surface. If the surface has a unit normal $\mathbf{n}$, we can characterize the *no-penetration boundary condition* as

$$\mathbf{v}_1^T \cdot \mathbf{n} = \mathbf{v}_2^T \cdot \mathbf{n}, \qquad \text{no-penetration for an inviscid, slip surface.} \tag{6.3}$$

If the solid is stationary, then the no-penetration condition at the slip surface is simply

$$\mathbf{v}^T \cdot \mathbf{n} = 0, \qquad \text{no-penetration for a stationary wall.} \tag{6.4}$$

One can allow a normal velocity for a porous surface that may admit injection of fluid into the domain.

For compressible inviscid fluids, as will be discussed in Section 9.6, these conditions need to be relaxed at interfaces that describe what are known as shocks and contact discontinuities. At such interfaces, one must enforce continuity of mass, momentum, and energy fluxes. However, at a shock discontinuity, this is achieved without requiring continuity of density, pressure, temperature, or fluid particle velocity. At a contact discontinuity, one finds continuity of pressure and fluid particle velocity, but not temperature or density.

6.2 Compressible Navier–Stokes Equations

Here write a complete set of equations, the compressible *Navier[1]–Stokes equations*, given here for a fluid that satisfies Stokes' assumption, but for which the viscosity μ (as well as thermal conductivity k) may be variable. They are an expanded version of those presented in Section 4.7. They are valid for a wide variety of flows, ranging from low speed to high speed, from inertia-dominated to viscous-dominated. A typical physical scenario in which such full equations might be required is in challenging flows such as a blunt body atmospheric re-entry problem at moderate supersonic velocity in which one can expect full compressibility, temperature-dependent diffusive transport properties, and fluid velocities that range from high freestream values to low values at the surface of the body. Viscous dissipation of mechanical energy creates a demanding thermal environment near the surface of the body, which must be accounted when designing for the survival of the re-entry vehicle. Our equations will be relevant for air at Mach numbers up to approximately five. Above that Mach number, previously neglected physics, such as molecular dissociation and ionization, become important.

6.2.1 Conservative Form

Cartesian Index Form

$$\partial_o \rho + \partial_i(\rho v_i) = 0, \tag{6.5}$$

---

[1] Claude Louis Marie Henri Navier, 1785–1836, Dijon-born French civil engineer and mathematician, specialist in road and bridge building.

$$\partial_o(\rho v_i) + \partial_j(\rho v_j v_i) = \rho f_i - \partial_i p + \partial_j \left(2\mu \left(\partial_{(j} v_{i)} - \frac{1}{3} \partial_k v_k \delta_{ji} \right) \right), \qquad (6.6)$$

$$\partial_o \left(\rho \left(e + \frac{1}{2} v_j v_j \right) \right) + \partial_i \left(\rho v_i \left(e + \frac{1}{2} v_j v_j \right) \right)$$
$$= \rho v_i f_i - \partial_i(p v_i) + \partial_i(k \partial_i T) + \partial_i \left(2\mu \left(\partial_{(i} v_{j)} - \frac{1}{3} \partial_k v_k \delta_{ij} \right) v_j \right), \qquad (6.7)$$

$$p = p(\rho, T), \quad e = e(\rho, T), \quad \mu = \mu(\rho, T), \quad k = k(\rho, T). \qquad (6.8)$$

Gibbs Form

$$\frac{\partial \rho}{\partial t} + \nabla^T \cdot (\rho \mathbf{v}) = 0, \qquad (6.9)$$

$$\frac{\partial}{\partial t}(\rho \mathbf{v}) + \left(\nabla^T \cdot (\rho \mathbf{v} \mathbf{v}^T) \right)^T$$
$$= \rho \mathbf{f} - \nabla p + \left(\nabla^T \cdot \left(2\mu \left(\frac{\nabla \mathbf{v}^T + (\nabla \mathbf{v}^T)^T}{2} - \frac{1}{3}(\nabla^T \cdot \mathbf{v}) \mathbf{I} \right) \right) \right)^T, \qquad (6.10)$$

$$\frac{\partial}{\partial t} \left(\rho \left(e + \frac{1}{2} \mathbf{v}^T \cdot \mathbf{v} \right) \right) + \nabla^T \cdot \left(\rho \mathbf{v} \left(e + \frac{1}{2} \mathbf{v}^T \cdot \mathbf{v} \right) \right)$$
$$= \rho \mathbf{v}^T \cdot \mathbf{f} - \nabla^T \cdot (p \mathbf{v}) + \nabla^T \cdot (k \nabla T) + \nabla^T \cdot \left(\left(2\mu \left(\frac{\nabla \mathbf{v}^T + (\nabla \mathbf{v}^T)^T}{2} - \frac{1}{3}(\nabla^T \cdot \mathbf{v}) \mathbf{I} \right) \right) \cdot \mathbf{v} \right), \qquad (6.11)$$

$$p = p(\rho, T), \quad e = e(\rho, T), \quad \mu = \mu(\rho, T), \quad k = k(\rho, T). \qquad (6.12)$$

6.2.2 Nonconservative Form

Cartesian Index Form

$$\frac{d\rho}{dt} = -\rho \partial_i v_i, \qquad (6.13)$$

$$\rho \frac{dv_i}{dt} = \rho f_i - \partial_i p + \partial_j \left(2\mu \left(\partial_{(j} v_{i)} - \frac{1}{3} \partial_k v_k \delta_{ji} \right) \right), \qquad (6.14)$$

$$\rho \frac{de}{dt} = -p \partial_i v_i + \partial_i(k \partial_i T) + \underbrace{2\mu \left(\partial_{(i} v_{j)} - \frac{1}{3} \partial_k v_k \delta_{ij} \right) \partial_i v_j}_{\Phi}, \qquad (6.15)$$

$$p = p(\rho, T), \quad e = e(\rho, T), \quad \mu = \mu(\rho, T), \quad k = k(\rho, T). \qquad (6.16)$$

Gibbs Form

$$\frac{d\rho}{dt} = -\rho \nabla^T \cdot \mathbf{v}, \qquad (6.17)$$

$$\rho \frac{d\mathbf{v}}{dt} = \rho \mathbf{f} - \nabla p + \left(\nabla^T \cdot \left(2\mu \left(\frac{\nabla \mathbf{v}^T + (\nabla \mathbf{v}^T)^T}{2} - \frac{1}{3}(\nabla^T \cdot \mathbf{v})\mathbf{I} \right) \right) \right)^T, \quad (6.18)$$

$$\rho \frac{de}{dt} = -p\nabla^T \cdot \mathbf{v} + \nabla^T \cdot (k\nabla T) + \underbrace{2\mu \left(\frac{\nabla \mathbf{v}^T + (\nabla \mathbf{v}^T)^T}{2} - \frac{1}{3}(\nabla^T \cdot \mathbf{v})\mathbf{I} \right) : \nabla \mathbf{v}^T}_{\Phi}, \quad (6.19)$$

$$p = p(\rho, T), \quad e = e(\rho, T), \quad \mu = \mu(\rho, T), \quad k = k(\rho, T). \quad (6.20)$$

We take μ, and k to be nonequilibrium thermodynamic properties of temperature and density. In practice, both dependencies are often weak, especially the dependency of μ and k on density. We also assume we know the form of the external body force per unit mass f_i. We also no longer formally present the angular momenta principle, as it has been absorbed into our constitutive equation for viscous stress. We also need not write the second law, as we can guarantee its satisfaction as long as $\mu \geq 0, k \geq 0$.

In summary, and returning to the formulation with e, we have nine unknowns: ρ, $v_i(3)$, p, e, T, μ, and k; and nine equations: mass, linear momenta (3), energy, thermal state, caloric state, and thermodynamic relations for viscosity and thermal conductivity. When coupled with initial, interface, and boundary conditions, all dependent variables can, in principle, be expressed as functions of position x_i and time t, and this knowledge utilized to design devices of practical importance.

A useful alternative form of the nonconservative form of the compressible energy equation is a version of Eq. (4.194):

$$\rho \frac{dh}{dt} = \frac{dp}{dt} + \nabla^T \cdot (k\nabla T) + \Phi, \quad (6.21)$$

where we have invoked Fourier's law, and Φ can be taken for a compressible, Newtonian fluid satisfying Stokes' assumption. Now consider the thermodynamics of enthalpy with the aim of writing the energy equation as a temperature evolution equation. So consider a general compressible material for which $h = h(T, p)$. Then we have

$$dh = \left.\frac{\partial h}{\partial T}\right|_p dT + \left.\frac{\partial h}{\partial p}\right|_T dp. \quad (6.22)$$

From classical thermodynamics, we have

$$\left.\frac{\partial h}{\partial T}\right|_p = c_p, \quad (6.23)$$

for a general material, where c_p is the specific heat at constant pressure. And from the second Gibbs equation, Eq. (4.196), $dh = T\,ds + \hat{v}\,dp$, we deduce

$$\left.\frac{\partial h}{\partial p}\right|_T = T \left.\frac{\partial s}{\partial p}\right|_T + \hat{v}. \quad (6.24)$$

To proceed further, we must first develop a necessary *Maxwell*[2] *relation*. The topic of Maxwell relations is nontrivial and draws upon a systematic knowledge of thermodynamic potentials.

---

[2] James Clerk Maxwell, 1831–1879, Scottish mathematical physicist.

Additional details are given by Callen (1985), Powers (2016), or Reynolds (1968). We will only develop here what is necessary for our purposes. With the Gibbs free energy g defined as

$$g = h - Ts, \tag{6.25}$$

we have by differentiating that $dg = dh - T\,ds - s\,dT$. Combining with the second Gibbs equation, $dh = T\,ds + \hat{v}\,dp$, we get

$$dg = \hat{v}\,dp - s\,dT. \tag{6.26}$$

This suggests the canonical variables for g are p and T, so we take $g = g(p,T)$. So

$$dg = \left.\frac{\partial g}{\partial p}\right|_T dp + \left.\frac{\partial g}{\partial T}\right|_p dT. \tag{6.27}$$

Comparing, we see that

$$\hat{v} = \left.\frac{\partial g}{\partial p}\right|_T, \qquad -s = \left.\frac{\partial g}{\partial T}\right|_p. \tag{6.28}$$

Now, take $\partial/\partial T$ of $\hat{v}$ and $\partial/\partial p$ of s and equate the mixed second partial derivatives to get the necessary Maxwell relation:

$$\left.\frac{\partial s}{\partial p}\right|_T = -\left.\frac{\partial \hat{v}}{\partial T}\right|_p. \tag{6.29}$$

Substitute this into Eq. (6.24) to get

$$\left.\frac{\partial h}{\partial p}\right|_T = \hat{v} - T\left.\frac{\partial \hat{v}}{\partial T}\right|_p. \tag{6.30}$$

Then Eq. (6.22) becomes

$$dh = c_p\,dT + \left(\hat{v} - T\left.\frac{\partial \hat{v}}{\partial T}\right|_p\right) dp. \tag{6.31}$$

With $\hat{v} = 1/\rho$, we see

$$dh = c_p\,dT + \left(\frac{1}{\rho} + \frac{1}{\rho^2}T\left.\frac{\partial \rho}{\partial T}\right|_p\right) dp. \tag{6.32}$$

Next define the *thermal expansion coefficient* β as

$$\beta = -\frac{1}{\rho}\left.\frac{\partial \rho}{\partial T}\right|_p, \tag{6.33}$$

to get

$$dh = c_p\,dT + \frac{1}{\rho}(1 - T\beta)\,dp. \tag{6.34}$$

We next scale by dt, multiply by ρ, and get the result for general thermodynamic equations of state that for fully compressible materials

$$\rho\frac{dh}{dt} = \rho c_p\frac{dT}{dt} + (1 - T\beta)\frac{dp}{dt}. \tag{6.35}$$

We use this to eliminate $\rho\, dh/dt$ in Eq. (6.21) to express the first law of thermodynamics as a temperature evolution equation for a fully compressible flow:

$$\rho c_p \frac{dT}{dt} = T\beta \frac{dp}{dt} + \nabla^T \cdot (k\nabla T) + \Phi. \tag{6.36}$$

6.3 Fluid Statics

In the limit of a stationary fluid, we have $\mathbf{v} = \mathbf{0}$, and the compressible Navier–Stokes equations in Cartesian and Gibbs form reduce to

$$\partial_o \rho = 0, \qquad \frac{\partial \rho}{\partial t} = 0, \tag{6.37}$$

$$0 = \rho f_i - \partial_i p, \qquad \mathbf{0} = \rho \mathbf{f} - \nabla p, \tag{6.38}$$

$$\rho \partial_o h = \partial_o p + \partial_i (k \partial_i T), \qquad \rho \frac{\partial h}{\partial t} = \frac{\partial p}{\partial t} + \nabla^T \cdot (k\nabla T), \tag{6.39}$$

$$p = p(\rho, T), \qquad h = h(\rho, T), \qquad k = k(\rho, T). \tag{6.40}$$

Here, in anticipation of low Mach number analysis that will be done in future chapters, we have formulated the energy equation in terms of enthalpy, drawing from Eq. (4.194). For such a flow, ρ is not time-varying, but may be a function of position. Other variables may be functions of both space and time.

Example 6.1 For a fluid that has constant density ρ_o, subjected to a constant body force $\mathbf{f} = -g\mathbf{k}$, find the pressure distribution. Assume the pressure at $z = 0$ is p_o.

Solution
In the limit of a constant density field with zero velocity, the mass equation is trivially satisfied, and the linear momenta equation may be directly solved for p. We have then

$$\nabla p = -\rho_o g \mathbf{k}. \tag{6.41}$$

Thus $\partial p/\partial x = 0$, $\partial p/\partial y = 0$, and

$$\frac{\partial p}{\partial z} = -\rho_o g. \tag{6.42}$$

We easily see that p may not be a function of x, y or t, so this becomes

$$\frac{dp}{dz} = -\rho_o g, \qquad p(0) = p_o, \tag{6.43}$$

$$p(z) = p_o - \rho_o g z. \tag{6.44}$$

The pressure increases as z decreases.

Example 6.2 Consider a steady, stationary, isentropic atmosphere described by a calorically perfect ideal gas with large variation in height. Consider variation in z only and find expressions for $T(z)$, $\rho(z)$, $p(z)$. Assume $k \sim 0$, which is required for an isentropic atmosphere. This is an example of a *stratified fluid*.

Solution

Assume a time-independent, fully compressible, stationary atmosphere with $\mathbf{v} = \mathbf{0}$. Assume also variation in z only and the presence of a constant downward gravitational body force of $\mathbf{f} = -g\mathbf{k}$. Then, the linear momenta principle, $\nabla p = \rho \mathbf{f}$, reduces to the simple

$$\frac{dp}{dz} = -\rho g. \tag{6.45}$$

As $\rho > 0$ and $g > 0$, we see that the pressure gradient $dp/dz < 0$. That is, the pressure goes down as we go up in height. If the density were constant, as it would be for liquid water, we could integrate this to get a linear relationship between p and z. However, for air, ρ is known to vary over large distances z. So a more complicated analysis is required.

Let us eliminate both p and ρ from Eq. (6.45) in favor of T. Let us first impose the calorically perfect ideal gas assumption. Thus,

$$p = \rho R T. \tag{6.46}$$

Now, invoke the isentropic relation, see Section 9.3.1, for a calorically perfect ideal gas:

$$p = p_o \left(\frac{T}{T_o}\right)^{\frac{\gamma}{\gamma-1}}, \tag{6.47}$$

in which γ is the constant ratio of specific heats, $\gamma = c_p/c_v$, which for air at moderate temperatures is $\gamma = 7/5$. This is a purely thermodynamic result and makes no appeal to fluid mechanics.

Now, take d/dz of Eq. (6.47), using the chain rule, to get

$$\frac{dp}{dz} = \frac{p_o}{T_o}\frac{\gamma}{\gamma-1}\left(\frac{T}{T_o}\right)^{\frac{\gamma}{\gamma-1}-1}\frac{dT}{dz}. \tag{6.48}$$

Now, use Eq. (6.46) to eliminate ρ in Eq. (6.45) and use Eq. (6.48) to eliminate dp/dz in Eq. (6.45), yielding

$$\underbrace{\frac{p_o}{T_o}\frac{\gamma}{\gamma-1}\left(\frac{T}{T_o}\right)^{\frac{1}{\gamma-1}}\frac{dT}{dz}}_{dp/dz} = -\underbrace{\frac{p}{RT}}_{\rho}g. \tag{6.49}$$

Now, use Eq. (6.47) to eliminate p to get

$$\frac{p_o}{T_o}\frac{\gamma}{\gamma-1}\left(\frac{T}{T_o}\right)^{\frac{1}{\gamma-1}}\frac{dT}{dz} = -\frac{p_o\left(\frac{T}{T_o}\right)^{\frac{\gamma}{\gamma-1}}}{RT}g, \tag{6.50}$$

$$\frac{\gamma}{\gamma-1}\left(\frac{T}{T_o}\right)^{\frac{1}{\gamma-1}}\frac{dT}{dz} = -\frac{\left(\frac{T}{T_o}\right)^{\frac{\gamma}{\gamma-1}}}{R\frac{T}{T_o}}g = -\frac{\left(\frac{T}{T_o}\right)^{\frac{1}{\gamma-1}}}{R}g, \tag{6.51}$$

$$\frac{dT}{dz} = -\frac{\gamma-1}{\gamma}\frac{g}{R} = -\frac{\frac{c_p}{c_v}-1}{\frac{c_p}{c_v}}\frac{g}{c_p-c_v} = -\frac{g}{c_p}. \tag{6.52}$$

So dT/dz is a constant. Let us use common numerical values to estimate the gradient:

$$\frac{dT}{dz} = -\frac{9.81 \frac{\text{m}}{\text{s}^2}}{1004.5 \frac{\text{J}}{\text{kg K}}} = -0.009\,766\,\frac{\text{K}}{\text{m}} = -9.766\,\frac{\text{K}}{\text{km}}. \tag{6.53}$$

Commercial airplanes fly at altitude near 10 km. So we might expect temperatures at this altitude to be about 100 K less than on the ground with this simple estimate.

Integrating, and requiring $T = T_o$ at $z = 0$, we get

$$T(z) = T_o - \frac{g}{c_p}z, \qquad (6.54)$$

valid for large and small z. Using this in the ideal gas law and isentropic relationship yields

$$\rho(z) = \rho_o\left(1 - \frac{g}{c_p T_o}z\right)^{\frac{1}{\gamma-1}}, \quad p(z) = p_o\left(1 - \frac{g}{c_p T_o}z\right)^{\frac{\gamma}{\gamma-1}}. \qquad (6.55)$$

Near $z = 0$, Taylor series expansion shows a locally linear decrease in ρ and p for small positive z:

$$\rho(z) \sim \rho_o\left(1 - \frac{g}{c_p T_o(\gamma-1)}z + \cdots\right), \quad p(z) \sim p_o\left(1 - \frac{\gamma g}{c_p T_o(\gamma-1)}z + \cdots\right). \qquad (6.56)$$

Pressure, density, and temperature all decrease as altitude z increases.

6.4 Incompressible Navier–Stokes Equations

While one can always employ the full compressible Navier–Stokes equations in simulation, there is often a significant advantage both in computational speed and in comprehension of results by removing model features that may be unimportant in particular limits. One such feature is compressibility, as it is observed in many flows that density variations are often small. If we make the assumption, which can be justified in the limit when fluid particle velocities are small relative to the velocity of sound waves in the fluid, that density changes following a particle are negligible (that is, $d\rho/dt \to 0$), the Navier–Stokes equations simplify considerably. *This does not require the density to be constant everywhere in the flow, although it could be.* Our incompressibility assumption allows for stratified flows, for which the density of individual particles still can remain constant, but may vary from pathline to pathline. Such can be the case in low speed flows in which a heated fluid acquires a lower density, inducing the fluid to rise in a gravitational field. We shall also assume that viscosity μ and thermal conductivity k are constants, though this is not necessary. Let us examine the mass, linear momenta, and energy equations in this limit.

6.4.1 Mass

Expanding the mass equation, $\partial_o \rho + \partial_i(\rho v_i) = 0$, we get

$$\underbrace{\partial_o \rho + v_i \partial_i \rho}_{\frac{d\rho}{dt} \to 0} + \rho \partial_i v_i = 0. \qquad (6.57)$$

We are assuming that the first two terms in this expression, which form $d\rho/dt$, go to zero; hence the mass equation becomes $\rho \partial_i v_i = 0$. Because $\rho > 0$, we can say

$$\partial_i v_i = 0, \qquad \nabla^T \cdot \mathbf{v} = 0. \qquad (6.58)$$

So, for an incompressible fluid, the relative expansion rate is $\vartheta = 0$, by Eq. (3.201).

6.4.2 Linear Momenta

We next examine the linear momenta axiom. We begin with presenting a simplified expression for the viscous stress tensor in the incompressible limit, $\nabla^T \cdot \mathbf{v} = 0$, for which Eq. (5.130) reduces to

$$\boldsymbol{\tau} = \mu \left(\nabla \mathbf{v}^T + (\nabla \mathbf{v}^T)^T \right) = \mu \begin{pmatrix} 2\partial_1 v_1 & \partial_1 v_2 + \partial_2 v_1 & \partial_1 v_3 + \partial_3 v_1 \\ \partial_2 v_1 + \partial_1 v_2 & 2\partial_2 v_2 & \partial_2 v_3 + \partial_3 v_2 \\ \partial_3 v_1 + \partial_1 v_3 & \partial_2 v_3 + \partial_3 v_2 & 2\partial_3 v_3 \end{pmatrix} = 2\mu \partial_{(i} v_{j)}. \tag{6.59}$$

The divergence of the viscous stress is

$$\partial_j \left(2\mu \left(\partial_{(j} v_{i)} - \frac{1}{3} \underbrace{\partial_k v_k}_{=0} \delta_{ij} \right) \right) = \partial_j \left(2\mu \left(\partial_{(j} v_{i)} \right) \right) = \partial_j \left(\mu \left(\partial_i v_j + \partial_j v_i \right) \right). \tag{6.60}$$

Because μ is constant here, we get

$$\mu \left(\partial_j \partial_i v_j + \partial_j \partial_j v_i \right) = \mu \left(\partial_i \underbrace{\partial_j v_j}_{=0} + \partial_j \partial_j v_i \right) = \mu \partial_j \partial_j v_i. \tag{6.61}$$

Everything else in the linear momenta equation is unchanged; hence, we get

$$\rho \partial_o v_i + \rho v_j \partial_j v_i = \rho f_i - \partial_i p + \mu \partial_j \partial_j v_i, \qquad \rho \frac{d\mathbf{v}}{dt} = \rho \mathbf{f} - \nabla p + \mu \nabla^2 \mathbf{v}. \tag{6.62}$$

In the incompressible, constant viscosity limit, the mass and linear momenta equations form a complete set of four equations in four unknowns: p, v_i. We will see in this limit the energy equation is coupled to mass and linear momenta, but it is only a one-way coupling.

6.4.3 Energy

We next take up the energy equation. For an incompressible fluid that is allowed to have density vary from pathline, there are some subtleties to this analysis involving the low Mach number limit that makes the results not obvious. Panton (2013, Chapter 10) gives an extended discussion.

Let us specialize the fully compressible Eq. (6.36) for the incompressible limit for a Newtonian fluid that satisfies Stokes' assumption. To that end, first consider the viscous dissipation function

$$\Phi = 2\mu \left(\partial_{(i} v_{j)} - \frac{1}{3} \underbrace{\partial_k v_k}_{=0} \delta_{ij} \right) \partial_i v_j = 2\mu \underbrace{\partial_{(i} v_{j)}}_{\text{sym.}} \left(\underbrace{\partial_{(i} v_{j)}}_{\text{sym.}} + \underbrace{\partial_{[i} v_{j]}}_{\text{anti-sym.}} \right) = 2\mu \partial_{(i} v_{j)} \partial_{(i} v_{j)}, \tag{6.63}$$

$$= 2\mu \left(\frac{\nabla \mathbf{v}^T + (\nabla \mathbf{v}^T)^T}{2} \right) : \left(\frac{\nabla \mathbf{v}^T + (\nabla \mathbf{v}^T)^T}{2} \right) = 2\mu \mathbf{D} : \mathbf{D}. \tag{6.64}$$

Here Φ is a scalar function and is obviously positive for $\mu > 0$ because it is a tensor inner product of a tensor with itself. So we can specialize Eq. (6.36) for the incompressible limit with constant k as

$$\rho c_p \frac{dT}{dt} = T\beta \frac{dp}{dt} + k\nabla^2 T + \underbrace{2\mu \left(\frac{\nabla \mathbf{v}^T + (\nabla \mathbf{v}^T)^T}{2}\right) : \left(\frac{\nabla \mathbf{v}^T + (\nabla \mathbf{v}^T)^T}{2}\right)}_{\Phi}. \quad (6.65)$$

For an ideal gas $p = \rho RT$, and we get $\beta = 1/T$. The incompressible first law of thermodynamics becomes

$$\rho c_p \frac{dT}{dt} = \frac{dp}{dt} + k\nabla^2 T + \Phi, \quad \text{ideal gas.} \quad (6.66)$$

For a liquid or solid, $\beta \sim 0$, and the incompressible first law of thermodynamics becomes

$$\rho c_p \frac{dT}{dt} = k\nabla^2 T + \Phi, \quad \text{liquid or solid.} \quad (6.67)$$

6.4.4 Poisson Equation for Pressure

In the numerical solution of incompressible Navier–Stokes equations in multiple dimensions, one is often required to solve a Poisson[3] equation for the pressure field. Let us see how such an equation arises. We will not solve this equation as there are complicated issues associated with the boundary conditions.

For simplicity, make two assumptions. First, consider the body force to be zero, $f_i = 0$. Second let us consider constant density flow, thus requiring that ρ be the same constant on all streamlines and not allowing stratified flow. Then take the divergence of the constant density, constant property linear momenta equation under these limits, Eq. (6.62):

$$\partial_i (\rho \partial_o v_i + \rho v_j \partial_j v_i) = \partial_i (-\partial_i p + \mu \partial_j \partial_j v_i), \quad (6.68)$$

$$\rho \partial_o \underbrace{\partial_i v_i}_{=0} + \rho \partial_i (v_j \partial_j v_i) = -\partial_i \partial_i p + \mu \partial_j \partial_j \underbrace{\partial_i v_i}_{=0}. \quad (6.69)$$

Because mass is conserved, we have $\partial_i v_i = 0$, by Eq. (6.58), so

$$\rho \partial_i (v_j \partial_j v_i) = -\partial_i \partial_i p, \quad (6.70)$$

$$\rho(v_j \partial_j \underbrace{\partial_i v_i}_{=0} + (\partial_i v_j)(\partial_j v_i)) = -\partial_i \partial_i p, \quad (6.71)$$

$$\partial_i \partial_i p = -\rho(\partial_i v_j)(\partial_j v_i), \qquad \nabla^2 p = -\rho \nabla \mathbf{v}^T : \nabla \mathbf{v}^T. \quad (6.72)$$

After splitting $\nabla \mathbf{v}^T$ into its symmetric and anti-symmetric parts, it can be verified by direct expansion that this is equivalent to the Poisson equation

$$\nabla^2 p = -\rho \left(\mathbf{D} : \mathbf{D} - \frac{\boldsymbol{\omega}^T \cdot \boldsymbol{\omega}}{2}\right). \quad (6.73)$$

---

[3] Siméon Denis Poisson, 1781–1840, French mathematician taught by Laplace, Lagrange, and Legendre, studied partial differential equations, potential theory, elasticity, and electrodynamics.

Expanding, we find

$$\frac{\partial^2 p}{\partial x^2} + \frac{\partial^2 p}{\partial y^2} + \frac{\partial^2 p}{\partial z^2} = -\rho\left(\left(\frac{\partial u}{\partial x}\right)^2 + \left(\frac{\partial v}{\partial y}\right)^2 + \left(\frac{\partial w}{\partial z}\right)^2 + 2\left(\frac{\partial u}{\partial y}\frac{\partial v}{\partial x} + \frac{\partial u}{\partial z}\frac{\partial w}{\partial x} + \frac{\partial v}{\partial z}\frac{\partial w}{\partial y}\right)\right). \tag{6.74}$$

For two-dimensional Cartesian flows, we have $w = 0$ and no variation with z. Using incompressibility, $\partial u/\partial x + \partial v/\partial y = 0$, the Poisson equation for pressure reduces in the two-dimensional limit to the much simpler

$$\frac{\partial^2 p}{\partial x^2} + \frac{\partial^2 p}{\partial y^2} = 2\rho\left(\frac{\partial u}{\partial x}\frac{\partial v}{\partial y} - \frac{\partial u}{\partial y}\frac{\partial v}{\partial x}\right). \tag{6.75}$$

It is interesting to note that the Laplacian of the pressure field is balanced by terms that arise from fluid inertia but not from viscous stresses. Of course the more primitive pressure and velocity fields do depend on viscosity. If we assume a nonzero conservative body force that can be written as the gradient of a potential, $\mathbf{f} = -\nabla\varphi$, then in the constant density limit, Eq. (6.73) becomes

$$\nabla^2(p + \rho\varphi) = -\rho\left(\mathbf{D} : \mathbf{D} - \frac{\boldsymbol{\omega}^T \cdot \boldsymbol{\omega}}{2}\right). \tag{6.76}$$

For a nonconservative body force, we simply have

$$\nabla^2 p = -\rho\left(\mathbf{D} : \mathbf{D} - \frac{\boldsymbol{\omega}^T \cdot \boldsymbol{\omega}}{2}\right) + \rho\nabla^T \cdot \mathbf{f}. \tag{6.77}$$

6.4.5 Summary

The incompressible constant property equations are summarized next in Gibbs notation:

$$\nabla^T \cdot \mathbf{v} = 0, \tag{6.78}$$

$$\rho\frac{d\mathbf{v}}{dt} = \rho\mathbf{f} - \nabla p + \mu\nabla^2\mathbf{v}, \tag{6.79}$$

$$\rho c_p \frac{dT}{dt} = \beta T \frac{dp}{dt} + k\nabla^2 T + \Phi. \tag{6.80}$$

For an ideal gas $\beta = 1/T$ and for a liquid or solid $\beta \sim 0$. In the limit of low Mach number gas, the incompressible energy equation is well modeled by neglecting viscous dissipation and pressure changes giving the widely used approximation of

$$\rho c_p \frac{dT}{dt} \approx k\nabla^2 T. \tag{6.81}$$

We will justify this more clearly in the analysis of Section 6.7.8.

In contrast, consider incompressible liquids, such as viscous lubricating oils operating in high strain rate environments such as rotating machinery. Such fluids acquire high temperatures due to conversion of frictional work to thermal energy; viscous dissipation cannot be neglected. For such flows, one commonly casts the energy equation as

$$\rho c_p \frac{dT}{dt} = k\nabla^2 T + \Phi. \tag{6.82}$$

Really, for the liquid we have taken $c_p \sim c_v \sim c$ and neglected any work done by fluid expansion, equivalent to taking $\beta = 0$. Such is the approach either discussed or applied by Arpaci and Larsen (1984, p. 89), Bejan (2013, p. 13), Bird et al. (2007, pp. 337–338, p. 342), Burmeister (1993, p. 37, p. 64), and White (2006, p. 97). We will effectively take this approach in some later analysis, for example Section 10.1.1.

6.4.6 Limits for One-Dimensional Diffusion

For a stationary, isobaric fluid ($v_i = 0$), we have $d/dt = \partial/\partial t$ and $\Phi = 0$; hence, for constant properties ρ, p, k, c_p, the energy equation, Eq. (6.80), can be written in a familiar form:

$$\frac{\partial T}{\partial t} = \alpha \nabla^2 T. \tag{6.83}$$

Here we have, from Eq. (1.22), $\alpha = k/\rho/c_p$ as the thermal diffusivity. In SI, thermal diffusivity has units of m^2/s. For one-dimensional cases for which all variation is in the x_2 direction, we get

$$\frac{\partial T}{\partial t} = \alpha \frac{\partial^2 T}{\partial x_2^2}. \tag{6.84}$$

Compare this to the momentum equation for a specific form of the velocity field, namely, $v_i(x_i) = v_1(x_2, t)$. When we also have no pressure gradient and no body force, the constant density linear momenta principle, Eq. (6.79), reduces to

$$\frac{\partial v_1}{\partial t} = \nu \frac{\partial^2 v_1}{\partial x_2^2}. \tag{6.85}$$

Here we have from Eq. (1.21) $\nu = \mu/\rho$ as the momentum diffusivity, more commonly known as the kinematic viscosity. In SI, momentum diffusivity has units of m^2/s; these are the same as for thermal diffusivity. This equation has an identical form to that for one-dimensional energy diffusion. In fact the physical mechanism governing both, random molecular collisions, is the same.

6.5 Boussinesq Approximation

We consider next the *Boussinesq*[4] *approximation*. It is sometimes known as the Oberbeck–Boussinesq approximation, as it is now recognized that Oberbeck[5] arrived at a similar set several years before Boussinesq. This approximation enables efficient modeling of such features of nature as warm air rising against a gravity force. Experience suggests that such flows fields have variable density accompanied by flow velocities that are much less than the speed of sound. This suggests the incompressibility assumption, valid in the limit of low Mach number, is viable.

---

[4] Joseph Valentin Boussinesq, 1842–1929, French mathematician and physicist.
[5] Anton Oberbeck, 1846–1900, German physicist.

6.5 Boussinesq Approximation

We will not rigorously derive these limits, but provide some limited supporting analysis. A more detailed discussion is given by many, including Currie (2013), Kundu et al. (2015), and Leal (2007). While these standard developments are useful, they do contain some ad hoc assumptions. More rigorous expositions are given by Mihaljan (1962) and Spiegel and Veronis (1960). A systematic treatment of the various limits involved in low Mach number flow in regimes where the Boussinesq approximation may or may not be appropriate requires a more sophisticated analysis, as described for example by Chenoweth and Paolucci (1986) or Paolucci (1982).

Confining our analysis to calorically perfect ideal gases with $p = \rho RT$, a key feature of the Boussinesq approximation is that pressure changes are sufficiently small that the ideal gas law is well approximated by a nearly isobaric state equation, $\rho T \approx \rho_o T_o$, where T_o is a reference state. So variable temperature induces variable density. However, the Mach number is sufficiently small such that $d\rho/dt \sim 0$; that is, the density change following a fluid particle is negligible. This enables us to require $\nabla^T \cdot \mathbf{v} = 0$ throughout the flow, while allowing ρ to vary from pathline to pathline. In these limits, viscous dissipation and dp/dt is also negligible; additional justification for neglecting these terms will be given in Section 6.7.8 using scaling analysis. We take kinematic viscosity $\nu = \mu/\rho_o$, and thermal diffusivity $\alpha = k/\rho_o/c_p$ to be constants. As constructed, the Boussinesq approximation recognizes temperature-dependent density variations as reflected in the body force term.

Drawing from the discussions of Currie (2013) and Kundu et al. (2015), let us begin with the full linear momenta equation for an incompressible, Newtonian fluid, Eq. (6.62):

$$\rho \frac{d\mathbf{v}}{dt} = \rho \mathbf{f} - \nabla p + \mu \nabla^2 \mathbf{v}. \tag{6.86}$$

Now if the fluid is stationary, we have $\mathbf{v} = \mathbf{0}$, and the force balance is between pressure and body forces:

$$\mathbf{0} = \rho \mathbf{f} - \nabla p. \tag{6.87}$$

Let us define solutions to this stationary problem as $\rho = \rho_o(x,y,z)$, $p = p_o(x,y,z)$, so that $\rho_o \mathbf{f} = \nabla p_o$. This result is identical to that obtained by the fully compressible Navier–Stokes equations in the limit of $\mathbf{v} = \mathbf{0}$. Note that both density and pressure fields may vary with space. One might expect this in the deep ocean, where the pressure certainly varies with depth, and the large pressure changes induce modest but measurable density changes. One might also expect this in the atmosphere if one is considering large changes in altitude, such as may be important in weather.

Subtract the stationary solution from the linear momenta equation to get

$$\rho \frac{d\mathbf{v}}{dt} = (\rho - \rho_o)\mathbf{f} - \nabla(p - p_o) + \mu \nabla^2 \mathbf{v}. \tag{6.88}$$

Define now $\rho' = \rho - \rho_o$ and $p' = p - p_o$. Using this and dividing by ρ_o renders the linear momenta equation to be

$$\left(1 + \frac{\rho'}{\rho_o}\right)\frac{d\mathbf{v}}{dt} = \frac{\rho'}{\rho_o}\mathbf{f} - \frac{1}{\rho_o}\nabla p' + \frac{\mu}{\rho_o}\nabla^2 \mathbf{v}. \tag{6.89}$$

So far, no approximation has been made beyond $\nabla^T \cdot \mathbf{v} = 0$. At this point, we invoke some plausible, but nevertheless ad hoc assumptions. As Spiegel and Veronis (1960) note, some of these "must be verified a posteriori from solutions of the problem." It is difficult to defend the internal consistency of some of these assumptions. We assume density changes from pathline to pathline are small, so that $\rho'/\rho_o \ll 1$; additionally we now approximate ρ_o as a constant, ignoring any spatial variation it might have. However, we choose to retain the body force effect, yielding, after taking $\nu = \mu/\rho_o$,

$$\frac{d\mathbf{v}}{dt} = \frac{\rho'}{\rho_o}\mathbf{f} - \frac{1}{\rho_o}\nabla p' + \nu \nabla^2 \mathbf{v}. \tag{6.90}$$

We treat ν as a constant, despite the fact that it has been scaled by $\rho_o(x, y, z)$. All these assumptions can break down for stratified flow where the density changes due to large changes in height may be large. This form, highlighting the deviations of density and pressure, ρ' and p', from their hydrostatic values, is often featured in the literature. Let us recover the form of Kundu et al. by returning to ρ and p:

$$\frac{d\mathbf{v}}{dt} = \frac{\rho - \rho_o}{\rho_o}\mathbf{f} - \frac{1}{\rho_o}\nabla (p - p_o) + \nu \nabla^2 \mathbf{v}. \tag{6.91}$$

Then, recognizing that $\nabla p_o = \rho_o \mathbf{f}$, we get

$$\frac{d\mathbf{v}}{dt} = \frac{\rho}{\rho_o}\mathbf{f} - \frac{1}{\rho_o}\nabla p + \nu \nabla^2 \mathbf{v}. \tag{6.92}$$

There is another common form in usage. Starting with the nearly isobaric state equation, we operate as follows:

$$\frac{\rho}{\rho_o} = \frac{T_o}{T} = \frac{T_o}{T - T_o + T_o} = \frac{1}{\frac{T - T_o}{T_o} + 1}. \tag{6.93}$$

Now we wish to consider temperature differences $T - T_o$ that are small relative to the absolute temperature T_o. In that limit, series expansion gives us

$$\frac{\rho}{\rho_o} = 1 - \frac{T - T_o}{T_o} + \cdots. \tag{6.94}$$

Thus, we could cast the momenta equation as

$$\frac{d\mathbf{v}}{dt} = \left(1 - \frac{T - T_o}{T_o}\right)\mathbf{f} - \frac{1}{\rho_o}\nabla p + \nu \nabla^2 \mathbf{v}. \tag{6.95}$$

One also sees that

$$\frac{\rho - \rho_o}{\rho_o} = -\frac{T - T_o}{T_o}. \tag{6.96}$$

So we could also write Eq. (6.91) as

$$\frac{d\mathbf{v}}{dt} = -\frac{T - T_o}{T_o}\mathbf{f} - \frac{1}{\rho_o}\nabla (p - p_o) + \nu \nabla^2 \mathbf{v}. \tag{6.97}$$

So, for example, if we are in a gravitational field for which $\mathbf{f} = -g\mathbf{k}$ and only consider the z momentum, we get

$$\frac{dw}{dt} = \frac{T - T_o}{T_o} g + \cdots. \tag{6.98}$$

A fluid that is locally warmer than T_o will be induced to accelerate upwards by the body force. We conclude by summarizing the complete Boussinesq approximation.

6.5.1 Cartesian Index Form

$$\partial_i v_i = 0, \tag{6.99}$$

$$\frac{dv_i}{dt} = \frac{\rho}{\rho_o} f_i - \frac{1}{\rho_o} \partial_i p + \nu \partial_j \partial_j v_i, \tag{6.100}$$

$$\frac{dT}{dt} = \alpha \partial_i \partial_i T, \tag{6.101}$$

$$\rho T = \rho_o T_o. \tag{6.102}$$

6.5.2 Gibbs Form

$$\nabla^T \cdot \mathbf{v} = 0, \tag{6.103}$$

$$\frac{d\mathbf{v}}{dt} = \frac{\rho}{\rho_o} \mathbf{f} - \frac{1}{\rho_o} \nabla p + \nu \nabla^2 \mathbf{v}, \tag{6.104}$$

$$\frac{dT}{dt} = \alpha \nabla^2 T, \tag{6.105}$$

$$\rho T = \rho_o T_o. \tag{6.106}$$

6.6 Euler Equations

The *Euler equations* are best described as the special case of the compressible Navier–Stokes equations in the limit in which diffusion is negligibly small relative to other effects; thus, we consider $\tau_{ij} \to 0$, $q_i \to 0$. For a Newtonian fluid that obeys Stokes' assumption and Fourier's law, we could also insist that $\mu \to 0$, $k \to 0$. If $\mu = 0$ and $k = 0$, the material is sometimes distinguished as a *perfect fluid*. Some sources, for example Landau and Lifshitz (1959), label a perfect fluid as an *ideal fluid*, not to be conflated with the thermodynamic notion of an ideal gas. Note that a perfect fluid cannot sustain any shear stress due to the absence of viscosity, and thus it does not fit neatly into our definition from Chapter 1 of a fluid as a material that continuously deforms when shear is applied. As Euler equations are typically used for compressible flows for which body forces are often negligible, we also take $f_i \to 0$, though this can be relaxed. A version of these equations was first presented by Euler (1757), although he only considered the mass and linear momenta principles. With the later nineteenth-century development of thermodynamics, the Euler equations have been taken to include the first law as well. For a complete set, they must be supplemented by appropriate thermodynamic equations of state; we leave these in a general form here.

Because the Euler equations neglect entropy-generating diffusive mechanisms, for continuous regions of flow, the second law tells us that the entropy is constant. However, we shall see later in Section 9.6 that shock discontinuities induce entropy changes. Also, because viscous stress is negligible, the angular momenta principle is irrelevant for the Euler equations. We summarize some of the various forms of the Euler equations next. They represent seven equations for the seven unknowns ρ, v_i, p, e, T.

6.6.1 Conservative Form

Cartesian Index Form

$$\partial_o \rho + \partial_i(\rho v_i) = 0, \tag{6.107}$$

$$\partial_o(\rho v_i) + \partial_j(\rho v_j v_i + p\delta_{ji}) = 0, \tag{6.108}$$

$$\partial_o \left(\rho \left(e + \frac{1}{2} v_j v_j \right) \right) + \partial_i \left(\rho v_i \left(e + \frac{1}{2} v_j v_j + \frac{p}{\rho} \right) \right) = 0, \tag{6.109}$$

$$p = p(\rho, T), \quad e = e(\rho, T). \tag{6.110}$$

Gibbs Form

$$\frac{\partial \rho}{\partial t} + \nabla^T \cdot (\rho \mathbf{v}) = 0, \tag{6.111}$$

$$\frac{\partial}{\partial t}(\rho \mathbf{v}) + \left(\nabla^T \cdot (\rho \mathbf{v} \mathbf{v}^T + p\mathbf{I}) \right)^T = \mathbf{0}, \tag{6.112}$$

$$\frac{\partial}{\partial t} \left(\rho \left(e + \frac{1}{2} \mathbf{v}^T \cdot \mathbf{v} \right) \right) + \nabla^T \cdot \left(\rho \mathbf{v} \left(e + \frac{1}{2} \mathbf{v}^T \cdot \mathbf{v} + \frac{p}{\rho} \right) \right) = 0, \tag{6.113}$$

$$p = p(\rho, T), \quad e = e(\rho, T). \tag{6.114}$$

6.6.2 Nonconservative Form

Cartesian Index Form

$$\frac{d\rho}{dt} = -\rho \partial_i v_i, \tag{6.115}$$

$$\rho \frac{dv_i}{dt} = -\partial_i p, \tag{6.116}$$

$$\rho \frac{de}{dt} = -p \partial_i v_i, \tag{6.117}$$

$$p = p(\rho, T), \quad e = e(\rho, T). \tag{6.118}$$

Gibbs Form

$$\frac{d\rho}{dt} = -\rho \nabla^T \cdot \mathbf{v}, \tag{6.119}$$

$$\rho \frac{d\mathbf{v}}{dt} = -\nabla p, \tag{6.120}$$

$$\rho \frac{de}{dt} = -p \nabla^T \cdot \mathbf{v}, \tag{6.121}$$

$$p = p(\rho, T), \quad e = e(\rho, T). \tag{6.122}$$

6.6.3 Alternate Forms of the Energy Equation

The neglect of entropy-generating mechanisms allows the energy equation in the Euler equations to be cast in some simple forms that clearly illuminate the fluid's behavior. One can specialize Eq. (4.189) to cast the energy equation as

$$\rho \frac{de}{dt} = \frac{p}{\rho} \frac{d\rho}{dt}, \quad \text{so} \quad \frac{de}{dt} - \frac{p}{\rho^2} \frac{d\rho}{dt} = 0. \tag{6.123}$$

In terms of differentials this is simply

$$de - \frac{p}{\rho^2} d\rho = 0, \tag{6.124}$$

which, when compared to the Gibbs equation, Eq. (4.197), tells us that this flow is isentropic, $ds = 0$, on a particle pathline. Moreover, using the definition of specific volume, $\hat{v} = 1/\rho$, the nonconservative form of the energy equation in the Euler equations is simply

$$de = -p \, d\hat{v}. \tag{6.125}$$

This result, also familiar from classical thermodynamics, says the change is energy within the context of the isentropic Euler equations is attributable solely to the reversible work done by a pressure force acting to change the volume. One can compare this isentropic equation to the Gibbs equation, Eq. (4.195), which allows for entropy changes, $de = T \, ds - p \, d\hat{v}$. In terms of the material derivative, one would say

$$\frac{de}{dt} = -p \frac{d\hat{v}}{dt}. \tag{6.126}$$

One could state the energy equation in terms of enthalpy by considering Eq. (4.194) in the limit of $q_i = 0$, $\tau_{ij} = 0$:

$$\rho \frac{dh}{dt} = \frac{dp}{dt}. \tag{6.127}$$

In terms of differentials, this gives the classical thermodynamics result

$$dh = \hat{v} \, dp. \tag{6.128}$$

One can compare this isentropic equation to Eq. (4.196) the second Gibbs equation, which allows for entropy changes, $dh = T \, ds + \hat{v} \, dp$. Equivalently, one could state the energy equation in terms of entropy by considering Eq. (4.202) in the limit of $q_i = 0$, $\tau_{ij} = 0$:

$$\frac{ds}{dt} = 0, \qquad \partial_o s + v_i \partial_i s = 0, \qquad \frac{\partial s}{\partial t} + \mathbf{v}^T \cdot \nabla s = 0. \qquad (6.129)$$

Integrating on a particle pathline, we get $s = C$; the constant C may vary from pathline to pathline. We adopt the following nomenclature: *Isentropic flow* connotes that the entropy s remains constant on a pathline, but may vary from pathline to pathline; *Homeoentropic flow* connotes that the entropy s is the same constant throughout all of the flow field.

For the special case in which the fluid is a calorically perfect ideal gas, we have $p = \rho R T$, $e = c_v T + \hat{e}$, and the first law, Eq. (6.124), reduces to

$$c_v \, dT - \frac{p}{\rho^2} \, d\rho = 0, \quad \text{so} \quad c_v \, d\left(\frac{p}{\rho R}\right) - \frac{p}{\rho^2} \, d\rho = 0, \qquad (6.130)$$

$$\frac{c_v}{R}\left(\frac{1}{\rho} dp - \frac{p}{\rho^2} d\rho\right) - \frac{p}{\rho^2} d\rho = 0, \quad \text{so} \quad \frac{1}{\gamma - 1}\left(dp - \frac{p}{\rho} d\rho\right) - \frac{p}{\rho} d\rho = 0, \qquad (6.131)$$

$$\frac{dp}{p} = \gamma \frac{d\rho}{\rho}, \quad \text{so} \quad \ln \frac{p}{p_o} = \gamma \ln \frac{\rho}{\rho_o}, \qquad (6.132)$$

$$\frac{p}{p_o} = \left(\frac{\rho}{\rho_o}\right)^\gamma, \quad \text{so} \quad \frac{p}{\rho^\gamma} = C. \qquad (6.133)$$

This is the well-known relation for the isentropic behavior of a calorically perfect ideal gas. Here C is a constant. We really confined ourselves to a particle pathline as we were considering the material time derivative. So the "constant" C actually can take on different values on different pathlines. Another way to cast this version of the energy equation (for an inviscid calorically perfect ideal gas) is

$$\frac{d}{dt}\left(\frac{p}{\rho^\gamma}\right) = 0, \quad \partial_o\left(\frac{p}{\rho^\gamma}\right) + v_i \partial_i\left(\frac{p}{\rho^\gamma}\right) = 0, \quad \frac{\partial}{\partial t}\left(\frac{p}{\rho^\gamma}\right) + \mathbf{v}^T \cdot \nabla\left(\frac{p}{\rho^\gamma}\right) = 0. \qquad (6.134)$$

6.7 Dimensionless Compressible Navier–Stokes Equations

Here we discuss how to scale the Navier–Stokes equations into a set of dimensionless equations. Panton (2013, Chapter 8) gives a general background for scaling. White (2006) gives a detailed discussion of the dimensionless form of the Navier–Stokes equations.

Consider the Navier–Stokes equations for a calorically perfect ideal gas that has Newtonian behavior, satisfies Stokes' assumption, and has constant viscosity, thermal conductivity, and specific heat at constant pressure:

$$\frac{d\rho}{dt} + \rho \partial_i v_i = 0, \qquad (6.135)$$

$$\rho \frac{dv_i}{dt} = \rho f_i - \partial_i p + \mu \partial_j \left(2\left(\partial_{(j} v_{i)} - \frac{1}{3}\partial_k v_k \delta_{ji}\right)\right), \qquad (6.136)$$

$$\rho \frac{dh}{dt} = \frac{dp}{dt} + k \partial_i \partial_i T + 2\mu \left(\partial_{(i} v_{j)} - \frac{1}{3}\partial_k v_k \delta_{ij}\right) \partial_i v_j, \qquad (6.137)$$

$$p = \rho R T, \quad h = c_p T + \hat{h}. \qquad (6.138)$$

6.7 Dimensionless Compressible Navier–Stokes Equations

Here R is the gas constant for the particular gas we are considering, which is the ratio of the universal gas constant $\Re$ and the gas's molecular mass $\mathcal{M}$: $R = \Re/\mathcal{M}$. Also $\hat{h}$ is a constant.

Now solutions to these equations, which may be of the form, for example, of $p(x_1, x_2, x_3, t)$, are necessarily parameterized by the constants from constitutive laws such as c_p, R, μ, k, f_i, in addition to parameters from initial and boundary conditions. That is, our solutions will really be of the form

$$p(x_1, x_2, x_3, t; c_p, R, \mu, k, f_i, \dots). \tag{6.139}$$

It is desirable for many reasons to reduce the number of parametric dependencies of these solutions. Some of these reasons include (1) identification of groups of terms that truly govern the features of the flow, (2) efficiency of presentation of results, and (3) efficiency of design of experiments. The Navier–Stokes equations (and nearly all sets of physically motivated equations) can be reduced in complexity by considering *scaled* versions of the same equations. For a given problem, the proper scales are *nonunique*, though some choices will be more helpful than others. One generally uses the following rules of thumb in choosing scales: (1) reduce variables so that their scaled value is near unity, (2) demonstrate that certain physical mechanisms may be negligible relative to other physical mechanisms, and (3) simplify initial and boundary conditions. In forming dimensionless equations, one must usually look for a characteristic length scale L, and a characteristic time scale t_c.

Often an ambient velocity or sound speed exists that can be used to form either a length or time scale; for example, given v_o and L we can take $t_c = L/v_o$; given v_o and t_c we can take $L = v_o t_c$. If, for example, our physical problem involves the flow over a body of length L (and whose other dimensions are of the same order as L), and freestream conditions are known to be $p = p_o$, $v_i = (v_o, 0, 0)^T$, $\rho = \rho_o$, as depicted in Fig. 6.1, knowledge of freestream pressure and density fixes all other freestream thermodynamic variables, for example e, T, via the thermodynamic relations. For this problem, let the $*$ subscript represent a dimensionless variable. Define the following scaled dependent variables:

$$\rho_* = \frac{\rho}{\rho_o}, \quad p_* = \frac{p}{p_o}, \quad v_{*i} = \frac{v_i}{v_o}, \quad T_* = \frac{\rho_o R}{p_o} T, \quad h_* = \frac{\rho_o}{p_o} h. \tag{6.140}$$

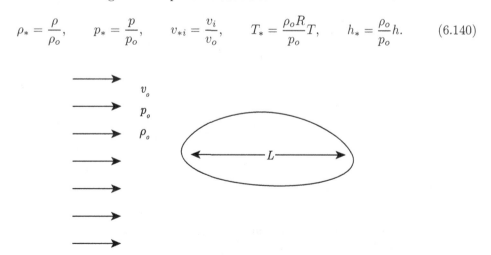

Figure 6.1 Diagram of known flow from infinity approaching body with characteristic length L.

Define the following scaled independent variables:

$$x_{*i} = \frac{x_i}{L}, \qquad t_* = \frac{v_o}{L}t. \tag{6.141}$$

Here we have scaled distance by the length scale of the body under study, and time has been scaled by the residence time a fluid particle has as it travels over the body. We can consider the time scale here to be *advection-based*. With these definitions, the operators must also be scaled, that is,

$$\partial_o = \frac{\partial}{\partial t} = \frac{dt_*}{dt}\frac{\partial}{\partial t_*} = \frac{v_o}{L}\frac{\partial}{\partial t_*} = \frac{v_o}{L}\partial_{*o}, \qquad \partial_{*o} = \frac{L}{v_o}\partial_o, \tag{6.142}$$

$$\partial_i = \frac{\partial}{\partial x_i} = \frac{dx_{*i}}{dx_i}\frac{\partial}{\partial x_{*i}} = \frac{1}{L}\frac{\partial}{\partial x_{*i}} = \frac{1}{L}\partial_{*i}, \qquad \partial_{*i} = L\partial_i. \tag{6.143}$$

6.7.1 Mass

Let us make these substitutions into the mass equation:

$$\frac{d\rho}{dt} + \rho\partial_i v_i = 0, \quad \text{so} \quad \frac{\rho_o v_o}{L}\left(\frac{d\rho_*}{dt_*} + \rho_*\partial_{*i}v_{*i}\right) = 0, \tag{6.144}$$

$$\frac{d\rho_*}{dt_*} + \rho_*\partial_{*i}v_{*i} = 0. \tag{6.145}$$

The mass equation is unchanged in form when we transform to a dimensionless version.

6.7.2 Linear Momenta

We have a similar analysis for the linear momenta equation:

$$\rho\frac{dv_i}{dt} = \rho f_i - \partial_i p + 2\mu\partial_j\left(\partial_{(j}v_{i)} - \frac{1}{3}\partial_k v_k \delta_{ji}\right), \tag{6.146}$$

$$\frac{\rho_o v_o^2}{L}\rho_*\frac{dv_{*i}}{dt_*} = \rho_o \rho_* f_i - \frac{p_o}{L}\partial_{*i}p_* + \frac{2\mu v_o}{L^2}\partial_{*j}\left(\partial_{(*j}v_{*i)} - \frac{1}{3}\partial_{*k}v_{*k}\delta_{ji}\right), \tag{6.147}$$

$$\rho_*\frac{dv_{*i}}{dt_*} = \frac{f_i L}{v_o^2}\rho_* - \frac{p_o}{\rho_o v_o^2}\partial_{*i}p_* + \frac{2\mu}{\rho_o v_o L}\partial_{*j}\left(\partial_{(*j}v_{*i)} - \frac{1}{3}\partial_{*k}v_{*k}\delta_{ji}\right). \tag{6.148}$$

With this scaling, we have generated three distinct dimensionless groups of terms that drive the linear momenta equation:

$$\frac{f_i L}{v_o^2}, \qquad \frac{p_o}{\rho_o v_o^2}, \quad \text{and} \quad \frac{\mu}{\rho_o v_o L}. \tag{6.149}$$

These groups are closely related to the following groups of terms, which have the associated interpretations indicated:

- *Froude number* Fr:[6] With the body force per unit mass $f_i = g\bar{g}_i$, with $g > 0$ is the gravitational acceleration magnitude and $\bar{g}_i$ is a unit vector pointing in the direction of gravitational acceleration,

---

[6] William Froude, 1810–1879, English engineer and naval architect.

$$Fr^2 \equiv \frac{v_o^2}{gL} = \frac{\text{flow kinetic energy}}{\text{gravitational potential energy}}. \tag{6.150}$$

- *Mach number* M_o: Using notions developed in Chapter 9, with the Mach number M_o defined as the ratio of the ambient velocity to the ambient sound speed, and recalling that for a calorically perfect ideal gas the square of the ambient sound speed, c_o^2, is $c_o^2 = \gamma p_o/\rho_o$, with γ as the ratio of specific heats $\gamma = c_p/c_v = (1 + R/c_v)$, we have

$$M_o^2 \equiv \frac{v_o^2}{c_o^2} = \frac{v_o^2}{\gamma \frac{p_o}{\rho_o}} = \frac{\rho_o v_o^2}{\gamma p_o} = \frac{v_o^2}{\gamma R T_o} = \frac{\text{flow kinetic energy}}{\text{thermal energy}}. \tag{6.151}$$

Here we have taken $T_o = p_o/\rho_o/R$.

- *Reynolds number* Re: We have

$$Re \equiv \frac{\rho_o v_o L}{\mu} = \frac{\rho_o v_o^2}{\mu \frac{v_o}{L}} = \frac{\text{dynamic pressure}}{\text{viscous stress}}. \tag{6.152}$$

With these definitions, we get

$$\rho_* \frac{dv_{*i}}{dt_*} = \frac{1}{Fr^2} \overline{g}_i \rho_* - \frac{1}{\gamma} \frac{1}{M_o^2} \partial_{*i} p_* + \frac{2}{Re} \partial_{*j} \left(\partial_{(*j} v_{*i)} - \frac{1}{3} \partial_{*k} v_{*k} \delta_{ji} \right). \tag{6.153}$$

The relative magnitudes of Fr, M_o, and Re play a crucial role in determining which physical mechanisms are most influential in changing the fluid's linear momenta.

6.7.3 Energy

The analysis is of the exact same form for the energy equation:

$$\rho \frac{dh}{dt} = \frac{dp}{dt} + k \partial_i \partial_i T + 2\mu \left(\partial_{(i} v_{j)} - \frac{1}{3} \partial_k v_k \delta_{ij} \right) \partial_i v_j, \tag{6.154}$$

$$\frac{v_o p_o}{L} \frac{dh_*}{dt_*} = \frac{k}{L^2} \frac{p_o}{\rho_o R} \partial_{*i} \partial_{*i} T_* + \frac{p_o v_o}{L} \frac{dp_*}{dt_*} + \frac{2\mu v_o^2}{L^2} \left(\partial_{(*i} v_{*j)} - \frac{1}{3} \partial_{*k} v_{*k} \delta_{ij} \right) \partial_{*i} v_{*j}, \tag{6.155}$$

$$\rho_* \frac{dh_*}{dt_*} = \frac{k}{L R \rho_o v_o} \partial_{*i} \partial_{*i} T_* + \frac{dp_*}{dt_*} + \frac{2\mu v_o^2}{L^2} \frac{L}{\rho_o v_o} \frac{1}{\frac{p_o}{\rho_o}} \left(\partial_{(*i} v_{*j)} - \frac{1}{3} \partial_{*k} v_{*k} \delta_{ij} \right) \partial_{*i} v_{*j}. \tag{6.156}$$

Now examining the dimensionless groups, we see that

$$\frac{k}{L R \rho_o v_o} = \frac{k}{c_p} \frac{c_p}{R} \frac{1}{L \rho_o v_o} = \frac{k}{\mu c_p} \frac{c_p}{c_p - c_v} \frac{\mu}{\rho_o v_o L} = \frac{1}{Pr} \frac{\gamma}{\gamma - 1} \frac{1}{Re}. \tag{6.157}$$

Here we have a dimensionless group, the Prandtl number, Pr, defined earlier in Eq. (1.20), with

$$Pr \equiv \frac{\mu c_p}{k} = \frac{\frac{\mu}{\rho_o}}{\frac{k}{\rho_o c_p}} = \frac{\text{momentum diffusivity}}{\text{energy diffusivity}} = \frac{\nu}{\alpha}. \tag{6.158}$$

This has employed definitions of diffusivities given earlier in Eqs. (1.22, 1.21). We also see that

$$\frac{2\mu v_o^2}{L^2} \frac{L}{\rho_o v_o} \frac{1}{\frac{p_o}{\rho_o}} = \frac{2\mu}{\rho_o v_o L} \frac{\gamma v_o^2}{\gamma \frac{p_o}{\rho_o}} = 2 \frac{1}{Re} \gamma M_o^2. \tag{6.159}$$

So, the dimensionless energy equation becomes

$$\rho_* \frac{dh_*}{dt_*} = \frac{\gamma}{\gamma-1} \frac{1}{Pr} \frac{1}{Re} \partial_{*i}\partial_{*i}T_* + \frac{dp_*}{dt_*} + 2\gamma \frac{M_o^2}{Re} \left(\partial_{(*i}v_{*j)} - \frac{1}{3}\partial_{*k}v_{*k}\delta_{ij} \right) \partial_{*i}v_{*j}. \tag{6.160}$$

6.7.4 Thermal State Equation

$$p_o p_* = \rho_o \rho_* R \left(\frac{p_o}{\rho_o R} \right) T_*, \quad \text{so} \quad p_* = \rho_* T_*. \tag{6.161}$$

6.7.5 Caloric State Equation

$$\frac{p_o}{\rho_o} h_* = c_p \left(\frac{p_o}{\rho_o R} \right) T_* + \hat{h}, \quad \text{so} \quad h_* = \frac{c_p}{R} T_* + \frac{\rho_o \hat{h}}{p_o} = \frac{\gamma}{\gamma-1} T_* + \underbrace{\frac{\rho_o \hat{h}}{p_o}}_{\text{unimportant}}. \tag{6.162}$$

For completeness, we retain the term $\rho_o \hat{h}/p_o$. It actually plays no role in this nonreactive flow because energy only enters via its derivatives. When flows with chemical reactions are modeled, this term may be important.

6.7.6 Upstream Conditions

Scaling the upstream conditions, we get

$$p_* = 1, \qquad \rho_* = 1, \qquad v_{*i} = (1,0,0)^T. \tag{6.163}$$

With this we then get secondary relationships

$$T_* = 1, \qquad h_* = \frac{\gamma}{\gamma-1} + \frac{\rho_o \hat{h}}{p_o}. \tag{6.164}$$

6.7.7 Reduction in Parameters

We lastly note that our original system had the following ten independent parameters:

$$\rho_o, \ p_o, \ c_p, \ R, \ L, \ v_o, \ \mu, \ k, \ f_i, \ \hat{h}. \tag{6.165}$$

Our scaled system, however, has only *six* independent parameters:

$$Re, \ Pr, \ M_o, \ Fr, \ \gamma, \ \frac{\rho_o \hat{h}}{p_o}. \tag{6.166}$$

We have lost no information, nor made any approximations, and we have a system with fewer dependencies.

6.7.8 Alternate Scaling

Other features may be highlighted by alternate scalings. For example, let us transform from our starred set of dimensionless variables to a hatted set of dimensionless variables via

$$x_* = \hat{x}, \quad t_* = \frac{v_o L}{\alpha}\hat{t}, \quad v_{*i} = \frac{\alpha}{v_o L}\hat{v}_i, \quad p_* = \frac{\rho_o \alpha^2}{p_o L^2}\hat{p}, \quad \rho_* = \hat{\rho}, \quad T_* = \hat{T}. \qquad (6.167)$$

In terms of the original dimensional variable, $\hat{t} = \alpha/L^2 t$. The time scale in this scaling is *diffusion-based*. After rescaling all the equations, the model transforms to

$$\frac{d\hat{\rho}}{d\hat{t}} + \hat{\rho}\hat{\partial}_i \hat{v}_i = 0, \qquad (6.168)$$

$$\hat{\rho}\frac{d\hat{v}_i}{d\hat{t}} = \frac{gL^3}{\alpha\nu}(Pr)\overline{g}_i\hat{\rho} - \hat{\partial}_i\hat{p} + 2(Pr)\hat{\partial}_j\left(\hat{\partial}_{(j}\hat{v}_{i)} - \frac{1}{3}\hat{\partial}_k\hat{v}_k\delta_{ji}\right), \qquad (6.169)$$

$$\hat{\rho}\frac{d\hat{T}}{d\hat{t}} = \hat{\partial}_i\hat{\partial}_i\hat{T} + (\gamma-1)\left(\frac{M_o}{RePr}\right)^2\left(\frac{d\hat{p}}{d\hat{t}} + 2Pr\left(\left(\hat{\partial}_{(i}\hat{v}_{j)} - \frac{1}{3}\hat{\partial}_k\hat{v}_k\delta_{ij}\right)\hat{\partial}_i\hat{v}_j\right)\right),$$

$$\gamma\left(\frac{M_o}{RePr}\right)^2\hat{p} = \hat{\rho}\hat{T}. \qquad (6.170)$$

These equations are fully compressible and have few embedded assumptions. In this scaling, the approximation of Eq. (6.81) for incompressible flow is seen to be justified in either the low Mach number or high Reynolds number limits. Later in Section 12.4 when we study *natural convection*, we will relate the term $gL^3/(\alpha\nu)$ to the *Rayleigh*[7] *number*, to be defined by Eq. (12.139), in the context of a specific problem.

6.8 First Integrals of Linear Momenta

Under special circumstances, we can integrate the linear momenta principle to obtain a simplified equation. We will consider two cases here: (1) the *Bernoulli*[8] *equation* and (2) *Crocco's*[9] *equation*. An expanded discussion of mathematical details of many of the topics of this section is given by Wu et al. (2006).

6.8.1 Bernoulli Equation

What is commonly called the Bernoulli equation is really a first integral of the linear momenta principle. Under different assumptions, we can get different flavors of the Bernoulli equation. A first integral of the linear momenta principle exists under the following conditions: (1) Viscous stresses are negligible relative to other terms, $\tau_{ij} \sim 0$, (2) The fluid is barotropic, $p = p(\rho)$ or $\rho = \rho(p)$. Three common barotropic conditions are (a) a constant density fluid, $\rho = C$, (b)

---

[7] John William Strutt (Lord Rayleigh), 1842–1919, English mathematician and physicist, who described correctly why the sky is blue, won the Nobel prize for the discovery of argon, and described traveling waves and solitons.

[8] Daniel Bernoulli, 1700–1782, Dutch-born Swiss mathematician of the prolific and mathematical Bernoulli family, son of Johann Bernoulli, who put forth his fluid mechanical principle in the 1738 *Hydrodynamica*, in competition with his father's 1738 *Hydraulica*.

[9] Luigi Crocco, 1909–1986, Sicilian-born, Italian applied mathematician and theoretical aerodynamicist and rocket engineer.

an isothermal ideal gas, for example $p = \rho(RT)$, with R and T constant, or (c) an isentropic calorically perfect ideal gas, for example $p/p_o = (\rho/\rho_o)^\gamma$, with γ as the ratio of specific heats. (3) Body forces are conservative (introduced earlier in Eq. (4.81)), so we can write $f_i = -\partial_i \varphi$, where φ is a known potential function, and (4) One of the following two conditions are satisfied: (a) the flow is irrotational, $\omega_k = \epsilon_{kij}\partial_i v_j = 0$, or (b) the flow is steady, $\partial_o = 0$.

First consider a version of the general linear momenta equation in nonconservative form, Eq. (4.72) scaled by ρ:

$$\partial_o v_i + v_j \partial_j v_i = -\frac{1}{\rho}\partial_i p + f_i + \frac{1}{\rho}\partial_j \tau_{ji}. \tag{6.171}$$

Now use our vector identity, Eq. (2.232), to rewrite the advective term, and impose our assumptions to arrive at

$$\partial_o v_i + \partial_i \left(\frac{1}{2}v_j v_j\right) - \epsilon_{ijk} v_j \omega_k = -\frac{1}{\rho}\partial_i p - \partial_i \varphi. \tag{6.172}$$

Now let us define, just for this particular analysis, a new function P. We will take P to be a function of pressure p, and thus implicitly, a function of x_i and t. For the barotropic fluid, we define P as

$$\mathsf{P}(p(x_i,t)) \equiv \int_{p_o}^{p(x_i,t)} \frac{d\hat{p}}{\rho(\hat{p})}. \tag{6.173}$$

In the special case of constant density flow with $1/\rho = 1/\rho_o$, we have

$$\mathsf{P} = \frac{p}{\rho_o} - \frac{p_o}{\rho_o}, \quad \text{constant density.} \tag{6.174}$$

In the special case of isothermal flow of an ideal gas with $1/\rho = RT/p$, we have

$$\mathsf{P} = \frac{p_o}{\rho_o} \ln \frac{p}{p_o}, \quad \text{isothermal, ideal gas.} \tag{6.175}$$

In the special case of isentropic flow of calorically perfect ideal gas with $1/\rho = (1/\rho_o)(p/p_o)^{-1/\gamma}$, see Eq. (6.133), we have

$$\mathsf{P} = \frac{\gamma}{\gamma - 1}\frac{p_o}{\rho_o}\left(\left(\frac{p}{p_o}\right)^{\frac{\gamma-1}{\gamma}} - 1\right), \quad \text{isentropic, calorically perfect ideal gas.} \tag{6.176}$$

Recalling Leibniz's rule for one spatial dimension, Eq. (2.244),

$$\frac{d}{dt}\int_{x=a(t)}^{x=b(t)} f(x,t)\,dx = \int_{x=a(t)}^{x=b(t)} \partial_o f\,dx + \frac{db}{dt}f(b(t),t) - \frac{da}{dt}f(a(t),t), \tag{6.177}$$

we let $\partial/\partial x_i$ play the role of d/dt to get

$$\frac{\partial}{\partial x_i}\mathsf{P} = \frac{\partial}{\partial x_i}\int_{p_o}^{p(x_i,t)} \frac{d\hat{p}}{\rho(\hat{p})} = \frac{1}{\rho(p(x_i,t))}\frac{\partial p}{\partial x_i} - \underbrace{\frac{1}{\rho(p_o)}\frac{\partial p_o}{\partial x_i}}_{=0} + \underbrace{\int_{p_o}^{p(x_i,t)} \frac{\partial}{\partial x_i}\left(\frac{1}{\rho(\hat{p})}\right) d\hat{p}}_{=0}. \tag{6.178}$$

As p_o is constant, and the integrand has no *explicit* dependency on x_i, we get

$$\frac{\partial}{\partial x_i}\mathsf{P} = \frac{1}{\rho(p(x_i,t))}\frac{\partial p}{\partial x_i}. \tag{6.179}$$

So, our linear momenta principle reduces to

$$\partial_o v_i + \partial_i \left(\frac{1}{2} v_j v_j\right) - \epsilon_{ijk} v_j \omega_k = -\partial_i \mathsf{P} - \partial_i \varphi, \tag{6.180}$$

$$\frac{\partial \mathbf{v}}{\partial t} + \nabla \left(\frac{\mathbf{v}^T \cdot \mathbf{v}}{2}\right) - \mathbf{v} \times \boldsymbol{\omega} = -\nabla \mathsf{P} - \nabla \varphi. \tag{6.181}$$

Consider now some special cases.

Irrotational Case

If the fluid is irrotational, which implies it is lamellar as discussed in Section 3.8.2, we have $\omega_k = \epsilon_{klm} \partial_l v_m = 0$. Consequently, we can write the velocity vector as the gradient of a potential function ϕ, defined earlier, p. 190, as the velocity potential:

$$\partial_m \phi = v_m. \tag{6.182}$$

If the velocity takes this form, then the vorticity is

$$\omega_k = \epsilon_{klm} \partial_l \partial_m \phi. \tag{6.183}$$

Because ϵ_{klm} is anti-symmetric and $\partial_l \partial_m$ is symmetric, their tensor inner product must be zero; hence, such a flow is irrotational: $\omega_k = \epsilon_{klm} \partial_l \partial_m \phi = 0$. So, the linear momenta principle, Eq. (6.180), reduces to

$$\partial_o \partial_i \phi + \partial_i \left(\frac{1}{2}(\partial_j \phi)(\partial_j \phi)\right) = -\partial_i \mathsf{P} - \partial_i \varphi, \tag{6.184}$$

$$\partial_i \left(\partial_o \phi + \frac{1}{2}(\partial_j \phi)(\partial_j \phi) + \mathsf{P} + \varphi\right) = 0. \tag{6.185}$$

We integrate to get

$$\partial_o \phi + \frac{1}{2}(\partial_j \phi)(\partial_j \phi) + \mathsf{P} + \varphi = f(t), \quad \frac{\partial \phi}{\partial t} + \frac{1}{2} \nabla^T \phi \cdot \nabla \phi + \mathsf{P} + \varphi = f(t). \tag{6.186}$$

Here $f(t)$ is an arbitrary function of time that can be chosen to match conditions in a given problem.

Steady Case

Streamline Integration In the steady case, we take $\partial_o = 0$, but $\omega_k \neq 0$. Rearranging the steady version of the linear momenta equation, Eq. (6.180), we get

$$\partial_i \left(\frac{1}{2} v_j v_j\right) + \partial_i \mathsf{P} + \partial_i \varphi = \epsilon_{ijk} v_j \omega_k, \quad \text{so} \quad \partial_i \left(\frac{1}{2} v_j v_j + \mathsf{P} + \varphi\right) = \epsilon_{ijk} v_j \omega_k. \tag{6.187}$$

Taking the dot product of both sides with v_i, we get

$$v_i \partial_i \left(\frac{1}{2} v_j v_j + \mathsf{P} + \varphi\right) = v_i \epsilon_{ijk} v_j \omega_k = \underbrace{\epsilon_{ijk} v_i v_j}_{=0} \omega_k = 0. \tag{6.188}$$

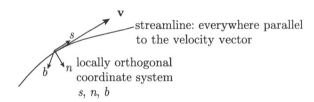

Figure 6.2 Local orthogonal intrinsic coordinate system oriented with local velocity field.

The term on the right-hand side is zero because it is the tensor inner product of a symmetric and anti-symmetric tensor.

For a local coordinate system that has component s aligned with the velocity vector v_i, and the other two directions n, and b, mutually orthogonal, we have $v_i = (v_s, 0, 0)^T$. Such a system is depicted in Fig. 6.2, we will get many simplifications. Our linear momenta principle then reduces to

$$(v_s, 0, 0) \begin{pmatrix} \partial_s \\ \partial_n \\ \partial_b \end{pmatrix} \left(\frac{1}{2} v_j v_j + \mathsf{P} + \varphi \right) = 0. \tag{6.189}$$

Forming this dot product yields

$$v_s \frac{\partial}{\partial s} \left(\frac{1}{2} v_j v_j + \mathsf{P} + \varphi \right) = 0. \tag{6.190}$$

For $v_s \neq 0$, we get that

$$\frac{1}{2} v_j v_j + \mathsf{P} + \varphi = C(n, b). \tag{6.191}$$

On a particular streamline, the function $C(n, b)$ will be a constant.

Lamb Surfaces We can extend the idea of integration along a streamline to describe what are known as *Lamb surfaces* by again considering the steady, inviscid linear momenta principle with conservative body forces, Eq. (6.187):

$$\partial_i \left(\frac{1}{2} v_j v_j + \mathsf{P} + \varphi \right) = \epsilon_{ijk} v_j \omega_k. \tag{6.192}$$

Now taking the quantity $\mathcal{B}$ to be

$$\mathcal{B} \equiv \frac{1}{2} v_j v_j + \mathsf{P} + \varphi, \tag{6.193}$$

the linear momenta principle, Eq. (6.187), becomes

$$\partial_i \mathcal{B} = \epsilon_{ijk} v_j \omega_k. \tag{6.194}$$

Now the vector $\epsilon_{ijk} v_j \omega_k$ is orthogonal to both velocity v_j and vorticity ω_k because of the nature of the cross product. Also the vector $\partial_i \mathcal{B}$ is orthogonal to a surface on which $\mathcal{B}$ is constant. Consequently, the surface on which $\mathcal{B}$ is constant must be tangent to both the velocity and vorticity vectors. Surfaces of constant $\mathcal{B}$ thus are composed of families of streamlines on which the Bernoulli constant has the same value. In addition they contain families of vortex lines. These are the Lamb surfaces of the flow.

Irrotational, Steady, Incompressible case

In the irrotational, steady, incompressible case, we recover the form most commonly used (and misused) of the Bernoulli equation, namely,

$$\frac{1}{2}v_j v_j + \mathsf{P} + \varphi = C. \tag{6.195}$$

The constant is truly constant throughout the flow field. With $\mathsf{P} = p/\rho_o - p_o/\rho_o$ and $\varphi = g_z z + \varphi_o$ (with $g_z > 0$, and rising z corresponding to rising distance from the earth's surface, we get $\mathbf{f} = -\nabla\varphi = -g_z \mathbf{k}$) for a constant gravitational field, we get

$$\frac{1}{2}v_j v_j + \frac{p}{\rho_o} + g_z z = C. \tag{6.196}$$

Here we have absorbed constants p_o/ρ_o and φ_o into C.

6.8.2 Crocco's Equation

It is common, especially in works on compressible flow, to present what is known as *Crocco's equation*, for example Crocco (1937). The many different versions presented in standard texts are nonuniform and often possess ambiguous regimes of validity. Its utility is confined mainly to providing an alternative way of expressing the linear momenta principle that provides some insight into the factors that influence fluid motion. In special cases, it can be integrated to form a more useful relationship, similar to the Bernoulli equation, between fundamental fluid variables. The heredity of this theorem is not always clear, though, as we shall see it is nothing more than a combination of the linear momenta principle coupled with some definitions from thermodynamics. Its derivation is often confined to inviscid flows. Here we will first present a result valid for general viscous flows for the evolution of *stagnation enthalpy* that is closely related to Crocco's equation. As introduced earlier in Section 3.14, a stagnation property, such as stagnation enthalpy, is the value the property acquires when the fluid is brought to rest at a stagnation point. Next we will show how one of the restrictions can be relaxed so as to obtain what we call the extended Crocco's equation. We then show how this reduces to a form that is similar to a form presented in many texts.

Stagnation Enthalpy Variation

First, again consider the general linear momenta equation, Eq. (6.171):

$$\partial_o v_i + v_j \partial_j v_i = -\frac{1}{\rho}\partial_i p + f_i + \frac{1}{\rho}\partial_j \tau_{ji}. \tag{6.197}$$

Now, as before in the development of the Bernoulli equation, use our vector identity, Eq. (2.232), to rewrite the advective term, but retain the viscous terms to get

$$\partial_o v_i + \partial_i\left(\frac{1}{2}v_j v_j\right) - \epsilon_{ijk} v_j \omega_k = -\frac{1}{\rho}\partial_i p + f_i + \frac{1}{\rho}\partial_j \tau_{ji}. \tag{6.198}$$

Taking the dot product with v_i and rearranging, we get

$$\partial_o \left(\frac{1}{2} v_i v_i\right) + v_i \partial_i \left(\frac{1}{2} v_j v_j\right) = \underbrace{\epsilon_{ijk} v_i v_j \omega_k}_{=0} - \frac{1}{\rho} v_i \partial_i p + v_i f_i + \frac{1}{\rho} v_i \partial_j \tau_{ji}. \tag{6.199}$$

Again, because ϵ_{ijk} is anti-symmetric and $v_i v_j$ is symmetric, their tensor inner product is zero, so we get

$$\partial_o \left(\frac{1}{2} v_i v_i\right) + v_i \partial_i \left(\frac{1}{2} v_j v_j\right) = -\frac{1}{\rho} v_i \partial_i p + v_i f_i + \frac{1}{\rho} v_i \partial_j \tau_{ji}. \tag{6.200}$$

Now recall the second Gibbs relation from thermodynamics, Eq. (4.196), which is rearranged to give

$$T \, ds = dh - \frac{1}{\rho} \, dp. \tag{6.201}$$

If we choose to apply this relation to the motion following a fluid particle, we can say then that

$$T \frac{ds}{dt} = \frac{dh}{dt} - \frac{1}{\rho} \frac{dp}{dt}. \tag{6.202}$$

Expanding, we get

$$T(\partial_o s + v_i \partial_i s) = \partial_o h + v_i \partial_i h - \frac{1}{\rho}(\partial_o p + v_i \partial_i p). \tag{6.203}$$

Rearranging, we get

$$T(\partial_o s + v_i \partial_i s) - (\partial_o h + v_i \partial_i h) + \frac{1}{\rho} \partial_o p = -\frac{1}{\rho} v_i \partial_i p. \tag{6.204}$$

We then use this identity to eliminate the pressure gradient term from the linear momenta equation in favor of enthalpy, entropy, and unsteady pressure terms:

$$\partial_o \left(\frac{1}{2} v_i v_i\right) + v_i \partial_i \left(\frac{1}{2} v_j v_j\right) = T(\partial_o s + v_i \partial_i s) - (\partial_o h + v_i \partial_i h) + \frac{1}{\rho} \partial_o p + v_i f_i + \frac{1}{\rho} v_i \partial_j \tau_{ji}. \tag{6.205}$$

Rearranging slightly, noting that $v_i v_i = v_j v_j$, and assuming the body force is conservative so that $f_i = -\partial_i \varphi$, we get

$$\partial_o \left(h + \frac{1}{2} v_j v_j + \varphi\right) + v_i \partial_i \left(h + \frac{1}{2} v_j v_j + \varphi\right) = T(\partial_o s + v_i \partial_i s) + \frac{1}{\rho} \partial_o p + \frac{1}{\rho} v_i \partial_j \tau_{ji}. \tag{6.206}$$

Here we have made the common assumption that the body force potential φ is independent of time, which allows us to absorb it within the time derivative. If we define, as is common, the *stagnation enthalpy* h_o as

$$h_o = h + \frac{1}{2} v_j v_j + \varphi, \tag{6.207}$$

we can then state

$$\partial_o h_o + v_i \partial_i h_o = T(\partial_o s + v_i \partial_i s) + \frac{1}{\rho} \partial_o p + \frac{1}{\rho} v_i \partial_j \tau_{ji}, \tag{6.208}$$

$$\frac{dh_o}{dt} = T \frac{ds}{dt} + \frac{1}{\rho} \frac{\partial p}{\partial t} + \frac{1}{\rho} \mathbf{v}^T \cdot \left(\nabla^T \cdot \boldsymbol{\tau}\right)^T. \tag{6.209}$$

The stagnation enthalpy is sometimes known as the *total enthalpy*. We can use the first law of thermodynamics written in terms of entropy, Eq. (4.200), $\rho(ds/dt) = -(1/T)\partial_i q_i + (1/T)\tau_{ij}\partial_i v_j$, to eliminate the entropy derivative in favor of those terms that generate entropy to arrive at

$$\rho \frac{dh_o}{dt} = \partial_i(\tau_{ij}v_j - q_i) + \partial_o p = \nabla^T \cdot (\boldsymbol{\tau} \cdot \mathbf{v} - \mathbf{q}) + \frac{\partial p}{\partial t}. \tag{6.210}$$

Thus, we see that the total enthalpy of a fluid particle is influenced by energy and momentum diffusion as well as an unsteady pressure field.

Extended Crocco's Equation

With a slight modification of the preceding analysis, we can arrive at the *extended Crocco's equation*. Begin once more with an earlier version of the linear momenta principle, Eq. (6.198):

$$\partial_o v_i + \partial_i \left(\frac{1}{2} v_j v_j\right) - \epsilon_{ijk} v_j \omega_k = -\frac{1}{\rho}\partial_i p + f_i + \frac{1}{\rho}\partial_j \tau_{ji}. \tag{6.211}$$

Now assume we have a functional representation of enthalpy in the form

$$h = h(s, p). \tag{6.212}$$

Then we get

$$dh = \left.\frac{\partial h}{\partial s}\right|_p ds + \left.\frac{\partial h}{\partial p}\right|_s dp. \tag{6.213}$$

We also thus deduce from the second Gibbs relation $dh = T\,ds + (1/\rho)\,dp$ that

$$\left.\frac{\partial h}{\partial s}\right|_p = T, \qquad \left.\frac{\partial h}{\partial p}\right|_s = \frac{1}{\rho}. \tag{6.214}$$

Now, because we have $h = h(s, p)$, we can take its derivative with respect to each and all of the coordinate directions to obtain

$$\frac{\partial h}{\partial x_i} = \left.\frac{\partial h}{\partial s}\right|_p \frac{\partial s}{\partial x_i} + \left.\frac{\partial h}{\partial p}\right|_s \frac{\partial p}{\partial x_i} \quad \text{or} \quad \partial_i h = \left.\frac{\partial h}{\partial s}\right|_p \partial_i s + \left.\frac{\partial h}{\partial p}\right|_s \partial_i p. \tag{6.215}$$

Substituting known values for the thermodynamic derivatives, we get from the second Gibbs equation that

$$\partial_i h = T\,\partial_i s + \frac{1}{\rho}\,\partial_i p. \tag{6.216}$$

We can use this to eliminate directly the pressure gradient term from the linear momenta equation to obtain then

$$\partial_o v_i + \partial_i \left(\frac{1}{2} v_j v_j\right) - \epsilon_{ijk} v_j \omega_k = T\partial_i s - \partial_i h + f_i + \frac{1}{\rho}\partial_j \tau_{ji}. \tag{6.217}$$

Rearranging slightly, and again assuming the body force is conservative so that $f_i = -\partial_i \varphi$, we get the extended Crocco's equation:

$$\partial_o v_i + \partial_i \left(h + \frac{1}{2} v_j v_j + \varphi\right) = T\partial_i s + \epsilon_{ijk} v_j \omega_k + \frac{1}{\rho}\partial_j \tau_{ji}. \tag{6.218}$$

Again, employing the total enthalpy, $h_o = h + \frac{1}{2}v_j v_j + \varphi$, we write the extended Crocco's equation as

$$\partial_o v_i + \partial_i h_o = T\partial_i s + \epsilon_{ijk} v_j \omega_k + \frac{1}{\rho}\partial_j \tau_{ji}, \tag{6.219}$$

$$\frac{\partial \mathbf{v}}{\partial t} + \nabla h_o = T\nabla s + \mathbf{v} \times \boldsymbol{\omega} + \frac{1}{\rho}(\nabla^T \cdot \boldsymbol{\tau})^T. \tag{6.220}$$

Traditional Crocco's Equation

For an unsteady, inviscid flow, the extended Crocco's equation reduces to what is usually called Crocco's equation, Liepmann and Roshko (2002), Anderson (2021):

$$\partial_o v_i + \partial_i h_o = T\partial_i s + \epsilon_{ijk} v_j \omega_k, \qquad \frac{\partial \mathbf{v}}{\partial t} + \nabla h_o = T\nabla s + \mathbf{v} \times \boldsymbol{\omega}. \tag{6.221}$$

In the steady state limit, this reduces to another version sometimes called Crocco's equation, Batchelor (2000):

$$\partial_i h_o = T\partial_i s + \epsilon_{ijk} v_j \omega_k, \qquad \nabla h_o = T\nabla s + \mathbf{v} \times \boldsymbol{\omega}. \tag{6.222}$$

If the flow is further required to be homeoentropic, we get

$$\partial_i h_o = \epsilon_{ijk} v_j \omega_k, \qquad \nabla h_o = \mathbf{v} \times \boldsymbol{\omega}. \tag{6.223}$$

Similar to Lamb surfaces, we find that surfaces on which h_o is constant are parallel to both the velocity and vorticity vector fields. Taking the dot product with v_i, we get

$$v_i \partial_i h_o = v_i \epsilon_{ijk} v_j \omega_k = \epsilon_{ijk} v_i v_j \omega_k = 0. \tag{6.224}$$

Integrating this along a streamline, as for the Bernoulli equation, we find

$$h_o = C(n, b), \qquad h + \frac{1}{2}v_j v_j + \varphi = C(n, b), \tag{6.225}$$

so we see that the stagnation enthalpy is constant along a streamline and varies from streamline to streamline. If the flow is steady, homeoentropic, and irrotational, the total enthalpy will be constant throughout the flow-field:

$$h + \frac{1}{2}v_j v_j + \varphi = C. \tag{6.226}$$

In terms of internal energy, we can rewrite this as

$$e + \frac{1}{2}v_j v_j + \frac{p}{\rho} + \varphi = C. \tag{6.227}$$

This is in a remarkably similar form to the Bernoulli equation for a steady, incompressible, irrotational fluid, Eq. (6.196). However, the assumptions for each are different. Bernoulli made no appeal to the first law of thermodynamics, while Crocco did. This version of the Bernoulli equation is restricted to incompressible flows, while this version of Crocco's equation is fully compressible.

SUMMARY

This chapter summarized the description of a fluid by various complete systems of equations, composed of a combination of axioms and constitutive laws. The most important of these are the compressible Navier–Stokes equations. Various special cases of these were considered including the incompressible Navier–Stokes equations, the Boussinesq approximation in which pressure changes are small, but moderate temperature and density changes are admitted, and the Euler equations, for which viscous effects are negligible, but pressure, density, and temperature changes may be large. Various scalings were applied to the equations, and dimensionless groups of parameters were identified that enable one to better understand what physics may dominate particular flow fields. Lastly, several special cases were identified in which some of the equations could be integrated; the most important of these yielding many versions of the well-known Bernoulli equation.

PROBLEMS

6.1 Write a complete set of one-dimensional, Cartesian, compressible Navier–Stokes equations for a Newtonian, calorically perfect ideal gas that satisfies Stokes' assumption, obeys Fourier's law, and has constant viscosity, thermal conductivity, and body force. Present them in conservative and then nonconservative forms. Show there are an equal number of equations and unknowns.

6.2 Write a complete set of steady, two-dimensional, Cartesian, incompressible Navier–Stokes equations for a Newtonian, calorically perfect ideal gas that obeys Fourier's law and has constant density, viscosity, thermal conductivity, and body force. Present them in nonconservative form. Show are there an equal number of equations and unknowns.

6.3 Consider the incompressible Navier–Stokes equations for a Newtonian fluid with constant viscosity and conservative body force. Show that the acceleration vector, composed of terms that represent the net force per unit mass due to pressure, body, and viscous effects, has a natural Helmholtz decomposition as the sum of the gradient of a scalar potential field and curl of a solenoidal vector field, analogous to that defined in Problem 3.13 for a velocity field.

6.4 Write the incompressible energy equation for temperature evolution for a compressible, viscous, nonideal gas that models finite intermolecular forces with the thermal state equation

$$p = \rho R T - a \rho^2.$$

Here R and a are constant. The gas is Newtonian and satisfies Stokes' assumption. This incompressible fluid has constant density on a pathline, but density may vary from pathline to pathline.

6.5 Consider the Euler equations for a calorically perfect ideal gas. Show the energy equation reduces to $ds/dt = 0$, then show this is satisfied by the spatially uniform $(p/p_o) = (\rho/\rho_o)^\gamma$. For such a flow, rewrite the Euler equations as a complete set in terms of v_i and ρ.

6.6 An inviscid flow has $u = y$, $v = -x$, $w = 0$, and $p = (x^2 + y^2)/2$. The flow has $\rho = 1$. There are no body forces. Confirm the flow is incompressible, rotational, and satisfies the linear momenta axiom. Find $\mathcal{B}$ and plot Lamb surfaces of $\mathcal{B}$ in the x, y, z volume. Confirm $\nabla \mathcal{B}$ is orthogonal to the velocity and vorticity vectors.

6.7 Consider a one-dimensional, compressible, isentropic, calorically perfect ideal gas undergoing steady motion from a stagnation state of zero velocity. Taking p_o and ρ_o to be the stagnation pressure and density, respectively, and ignoring body forces, show that the Bernoulli equation reduces to the equivalent equations

$$\frac{\gamma}{\gamma-1}\left(\frac{p}{p_o}\right)^{\frac{\gamma-1}{\gamma}}\frac{p_o}{\rho_o} + \frac{1}{2}u^2 = \frac{\gamma}{\gamma-1}\frac{p}{\rho} + \frac{1}{2}u^2 = \frac{\gamma}{\gamma-1}\frac{p_o}{\rho_o}.$$

Define the square of the Mach number to be $M^2 = u^2/(\gamma p/\rho)$, and find an expression for p/p_o in terms of M^2 and γ.

6.8 A steady, inviscid, compressible flow has dimensionless density, pressure, and velocity fields given by

$$\rho = \frac{3}{x^2+y^2+z^2+1}, \quad p = 1 - \frac{1}{(x^2+y^2+z^2+1)^3}, \quad \mathbf{v} = \begin{pmatrix} -\frac{2(y-xz)}{(x^2+y^2+z^2+1)^2} \\ \frac{2(x+yz)}{(x^2+y^2+z^2+1)^2} \\ \frac{-x^2-y^2+z^2+1}{(x^2+y^2+z^2+1)^2} \end{pmatrix}.$$

Show that this solution satisfies the mass and linear momenta equations from the Euler equations. This solution is a special case of a so-called *Hopf*[10] *fibration* that is studied in the field of differential topology. If the fluid is a calorically perfect ideal gas with $\gamma = 2$ and has no diffusive heat transfer, show an energy source $\mathcal{Q}$ is required for the steady, diffusion-free first law of thermodynamics, $\rho \mathbf{v}^T \cdot \nabla e + p \nabla^T \cdot \mathbf{v} = \mathcal{Q}$, to be satisfied. Find $\mathcal{Q}$. Numerically integrate the velocity field to find a family of 50 streamlines; consider streamlines commencing at $(x, y, z)^T = (x_o, 0, 0)^T$ with x_o uniformly distributed in the domain $x_o \in [1/2, 1]$. Plot the family of streamlines.

FURTHER READING

Hughes, W. F., and Gaylord, E. W. (1964). *Basic Equations of Engineering Science*, Schaum's Outline Series. New York: McGraw-Hill.

REFERENCES

Anderson, J. D. (2021). *Modern Compressible Flow with Historical Perspective*, 4th ed. New York: McGraw-Hill.

Aris, R. (1962). *Vectors, Tensors, and the Basic Equations of Fluid Mechanics*. New York: Dover.

Arpaci, V. S., and Larsen, P. S. (1984). *Convection Heat Transfer*. Englewood Cliffs, New Jersey: Prentice-Hall.

Batchelor, G. K. (2000). *An Introduction to Fluid Dynamics*. Cambridge, UK: Cambridge University Press.

Bejan, A. (2013). *Convection Heat Transfer*, 4th ed. Hoboken, New Jersey: John Wiley.

Bird, R. B., Stewart, W. E., and Lightfoot, E. N. (2007). *Transport Phenomena*, Revised 2nd ed. New York: John Wiley.

---

[10] Heinz Hopf, 1894–1971, German mathematician.

Burmeister, L. C. (1993). *Convective Heat Transfer*, 2nd ed. New York: John Wiley.

Callen, H. B. (1985). *Thermodynamics and an Introduction to Thermostatistics*, 2nd ed. New York: John Wiley.

Chenoweth, D. R., and Paolucci, S. (1986). Natural convection in an enclosed vertical air layer with large horizontal temperature differences, *Journal of Fluid Mechanics*, **169**, 173–210.

Crocco, L. (1937). Eine neue stromfunktion für die erforschung der bewegung der gase mit rotation, *Zeitschrift für Angewandte Mathematik und Mechanik*. **17**(1), 1–7.

Currie, I. G. (2013). *Fundamental Mechanics of Fluids*, 4th ed. Boca Raton, Florida: CRC Press.

Euler, L. (1757). Principes généraux du mouvement des fluides, *Mémoires de l'Académie des Sciences de Berlin*. **11**, 274–315.

Kundu, P. K., Cohen, I. M., and Dowling, D. R. (2015). *Fluid Mechanics*, 6th ed. Amsterdam: Academic Press.

Landau, L. D., and Lifshitz, E. M. (1959). *Fluid Mechanics*. Oxford: Pergamon Press.

Leal, L. G. (2007). *Advanced Transport Phenomena: Fluid Mechanics and Convective Transport Processes*. New York: Cambridge University Press.

Liepmann, H. W., and Roshko, A. (2002). *Elements of Gasdynamics*. New York: Dover.

Mihaljan, J. M. (1962). A rigorous exposition of the Boussinesq approximations applicable to a thin layer of fluid, *Astrophysical Journal*, **136**(3), 1126–1133.

Panton, R. L. (2013). *Incompressible Flow*, 4th ed. New York: John Wiley.

Paolucci, S. (1982). On the filtering of sound from the Navier–Stokes equations, SAND82-8257, Livermore, California: Sandia National Laboratories.

Powers, J. M. (2016). *Combustion Thermodynamics and Dynamics*. New York: Cambridge University Press.

Reynolds, W. C. (1968). *Thermodynamics*, 2nd ed. New York: McGraw-Hill.

Spiegel, E. A., and Veronis, G. (1960). On the Boussinesq approximations for a compressible fluid, *Astrophysical Journal*, **131**(2), 442–447.

White, F. M. (2006). *Viscous Fluid Flow*, 3rd ed. New York: McGraw-Hill.

Wu, J.-Z., Ma, H.-Y., and Zhou, M.-D. (2006). *Vorticity and Vortex Dynamics*. Berlin: Springer.

Part II

Solutions in Various Flow Regimes

7 Vortical Flow

In this chapter we will consider the kinematics and dynamics of fluid elements rotating about their centers of mass. Such an element is often described as a vortex, and is a commonly seen in fluids. However, a precise definition of a vortex is difficult to formulate. Rotating fluids may be observed, among other places, in weather patterns and airfoil wakes.

A rotating fluid element is depicted in Fig. 7.1. Here the center of mass is labeled G. While such a fluid element can also be translating and deforming, we focus here on rotation about an axis passing through G. The two most common quantities that are used to characterize rotating fluids are the vorticity vector, Eq. (3.94):

$$\boldsymbol{\omega} = \nabla \times \mathbf{v}, \tag{7.1}$$

and a new scalar quantity we define as the *circulation*, Γ:

$$\Gamma = \oint_C \mathbf{v}^T \cdot d\mathbf{r}. \tag{7.2}$$

Here $\oint_C$ is the integral about a closed contour C. Both concepts will be important in this chapter.

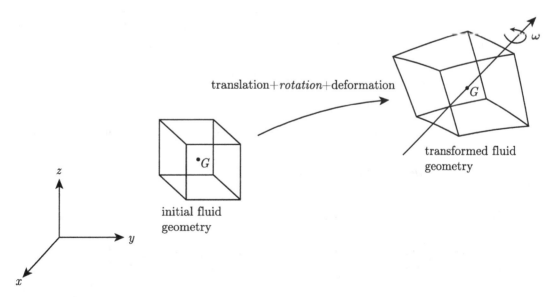

Figure 7.1 Fluid element whose motion includes rotation about its center of mass.

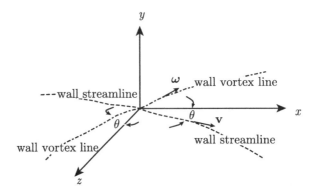

Figure 7.2 Wall streamlines and vortex lines at wall $y = 0$.

7.1 Streamlines and Vortex Lines

We defined a streamline in Section 3.5 as a curve that is everywhere tangent to the velocity vector **v**. Similarly, a *vortex line* is a curve that is everywhere tangent to the vorticity vector $\boldsymbol{\omega}$. For general three-dimensional flows away from walls, there is no restriction on the alignment of streamlines and vortex lines. We recall from Section 2.4.7 that for the unusual case of a Beltrami flow field, the velocity vector is everywhere a scalar multiple of its vorticity vector; $\boldsymbol{\omega} = \alpha \mathbf{v}$, with α as a scalar field variable; thus, $\boldsymbol{\omega} \times \mathbf{v} = \mathbf{0}$ for Beltrami flows. We contrast a Beltrami field with the previously defined (Section 3.12.1) complex lamellar vector field, for which $\mathbf{v}^T \cdot \boldsymbol{\omega} = 0$, rendering streamlines orthogonal to vortex lines. As a first step in understanding vortical flow, let us consider the behavior of fluid velocity and vorticity at and near a stationary no-slip wall. It seems odd that a streamline can be defined at a stationary wall where the velocity of a viscous fluid is zero, but in the neighborhood of the wall, the fluid velocity is small but nonzero. We can extrapolate the position of streamlines near the wall to the wall to define a wall streamline. Consider the geometry depicted in Fig. 7.2. Here the (x, z) plane is locally attached to a stationary wall at $y = 0$, and the y direction is normal to the wall. Wall streamlines and vortex lines are presented in the figure. Because the flow satisfies a no-slip condition, we have at the stationary wall

$$u(x, y=0, z) = 0, \qquad v(x, y=0, z) = 0, \qquad w(x, y=0, z) = 0. \tag{7.3}$$

Because of this, partial derivatives of all velocities with respect to either x or z will also be zero at $y = 0$:

$$\left.\frac{\partial u}{\partial x}\right|_{y=0} = \left.\frac{\partial u}{\partial z}\right|_{y=0} = \left.\frac{\partial v}{\partial x}\right|_{y=0} = \left.\frac{\partial v}{\partial z}\right|_{y=0} = \left.\frac{\partial w}{\partial x}\right|_{y=0} = \left.\frac{\partial w}{\partial z}\right|_{y=0} = 0. \tag{7.4}$$

Near the wall, the velocity is near zero, so the Mach number is small, and the flow is well modeled

as incompressible. So here, the mass conservation equation implies that $\nabla^T \cdot \mathbf{v} = 0$. Applying this at the wall, we get

7.1 Streamlines and Vortex Lines

$$\underbrace{\left.\frac{\partial u}{\partial x}\right|_{y=0}}_{=0} + \left.\frac{\partial v}{\partial y}\right|_{y=0} + \underbrace{\left.\frac{\partial w}{\partial z}\right|_{y=0}}_{=0} = 0, \quad \text{so} \quad \left.\frac{\partial v}{\partial y}\right|_{y=0} = 0. \tag{7.5}$$

Now let us examine the behavior of u, v, and w, as we leave the wall in the y direction. Consider a Taylor series of each:

$$u = \underbrace{u|_{y=0}}_{=0} + \left.\frac{\partial u}{\partial y}\right|_{y=0} y + \frac{1}{2}\left.\frac{\partial^2 u}{\partial y^2}\right|_{y=0} y^2 + \cdots, \tag{7.6}$$

$$v = \underbrace{v|_{y=0}}_{=0} + \underbrace{\left.\frac{\partial v}{\partial y}\right|_{y=0}}_{=0} y + \frac{1}{2}\left.\frac{\partial^2 v}{\partial y^2}\right|_{y=0} y^2 + \cdots, \tag{7.7}$$

$$w = \underbrace{w|_{y=0}}_{=0} + \left.\frac{\partial w}{\partial y}\right|_{y=0} y + \frac{1}{2}\left.\frac{\partial^2 w}{\partial y^2}\right|_{y=0} y^2 + \cdots. \tag{7.8}$$

So we get

$$u = \left.\frac{\partial u}{\partial y}\right|_{y=0} y + \cdots, \quad v = \frac{1}{2}\left.\frac{\partial^2 v}{\partial y^2}\right|_{y=0} y^2 + \cdots, \quad w = \left.\frac{\partial w}{\partial y}\right|_{y=0} y + \cdots. \tag{7.9}$$

Now for streamlines, we must have from Eq. (3.51) that

$$\frac{dx}{u} = \frac{dy}{v} = \frac{dz}{w}. \tag{7.10}$$

For the streamline near the wall, consider just $dx/u = dz/w$, and also tag the streamline as dz_s, so that the slope of the wall streamline, that is, the tangent of the angle θ between the wall streamline and the x axis, is

$$\tan\theta = \left.\frac{dz_s}{dx}\right|_{y=0} = \lim_{y \to 0} \frac{w}{u} = \frac{\left.\frac{\partial w}{\partial y}\right|_{y=0}}{\left.\frac{\partial u}{\partial y}\right|_{y=0}}. \tag{7.11}$$

Now consider the vorticity vector evaluated at the wall:

$$\omega_x|_{y=0} = \left.\frac{\partial w}{\partial y}\right|_{y=0} - \underbrace{\left.\frac{\partial v}{\partial z}\right|_{y=0}}_{=0} = \left.\frac{\partial w}{\partial y}\right|_{y=0}, \tag{7.12}$$

$$\omega_y|_{y=0} = \underbrace{\left.\frac{\partial u}{\partial z}\right|_{y=0}}_{=0} - \underbrace{\left.\frac{\partial w}{\partial x}\right|_{y=0}}_{=0} = 0, \tag{7.13}$$

$$\omega_z|_{y=0} = \underbrace{\left.\frac{\partial v}{\partial x}\right|_{y=0}}_{=0} - \left.\frac{\partial u}{\partial y}\right|_{y=0} = -\left.\frac{\partial u}{\partial y}\right|_{y=0}. \tag{7.14}$$

So we see that on the wall at $y = 0$, the vorticity vector has no component in the y direction. Hence, it must be parallel to the wall itself. Further, we can then define the slope of the vortex line, dz_v/dx, at the wall in the same fashion as we define a streamline:

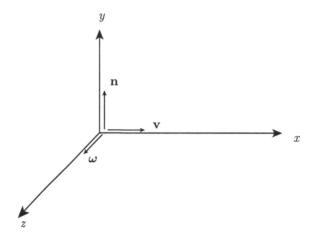

Figure 7.3 Coordinate system aligned with wall streamlines and vortex lines.

$$\left.\frac{dz_v}{dx}\right|_{y=0} = -\frac{\omega_z}{\omega_x} = -\frac{\left.\frac{\partial u}{\partial y}\right|_{y=0}}{\left.\frac{\partial w}{\partial y}\right|_{y=0}} = -\frac{1}{\left.\frac{dz_s}{dx}\right|_{y=0}}. \tag{7.15}$$

Because the slope of the vortex line is the negative reciprocal of the slope of the streamline, we have that at a no-slip wall, streamlines are orthogonal to vortex lines, and locally the field is a complex lamellar vector field. We also note that streamlines are orthogonal to vortex lines for flow with variation in the x and y directions only. This motivates a local coordinate system attached to the wall with the x axis is aligned with the wall streamline and the z axis aligned with the wall vortex line. As before the y axis is normal to the wall. The coordinate system aligned with the wall streamlines and vortex lines is given in Fig. 7.3. In the figure we take the direction **n** to be normal to the wall.

7.2 Incompressible Navier–Stokes Equations in Cylindrical Coordinates

Although it is possible to use Cartesian index notation to describe a rotating fluid, some of the ideas are better conveyed in a non-Cartesian system, such as the cylindrical coordinate system. For that reason, the Gibbs notation will often be used in this chapter. Additionally, when we wish to compare to Cartesian representations, we will mainly revert to the more traditional $(x, y, z)^T$ and $(u, v, w)^T$ for the position and velocity vectors, respectively. Using the transformations developed in Section 2.6, taking $\hat{z} = z$, it can be shown that the incompressible Navier–Stokes equations for a Newtonian fluid with constant viscosity and body force confined to the $-z$ direction are

$$0 = \frac{1}{r}\frac{\partial}{\partial r}(rv_r) + \frac{1}{r}\frac{\partial v_\theta}{\partial \theta} + \frac{\partial v_z}{\partial z}, \tag{7.16}$$

$$\left(\frac{\partial v_r}{\partial t} - \frac{v_\theta^2}{r}\right) + v_r\frac{\partial v_r}{\partial r} + \frac{v_\theta}{r}\frac{\partial v_r}{\partial \theta} + v_z\frac{\partial v_r}{\partial z} = -\frac{1}{\rho}\frac{\partial p}{\partial r} + \nu\left(\nabla^2 v_r - \frac{v_r}{r^2} - \frac{2}{r^2}\frac{\partial v_\theta}{\partial \theta}\right), \tag{7.17}$$

$$\left(\frac{\partial v_\theta}{\partial t} + \frac{v_r v_\theta}{r}\right) + v_r \frac{\partial v_\theta}{\partial r} + \frac{v_\theta}{r}\frac{\partial v_\theta}{\partial \theta} + v_z \frac{\partial v_\theta}{\partial z} = -\frac{1}{\rho}\frac{1}{r}\frac{\partial p}{\partial \theta} + \nu\left(\nabla^2 v_\theta + \frac{2}{r^2}\frac{\partial v_r}{\partial \theta} - \frac{v_\theta}{r^2}\right), \tag{7.18}$$

$$\frac{\partial v_z}{\partial t} + v_r \frac{\partial v_z}{\partial r} + \frac{v_\theta}{r}\frac{\partial v_z}{\partial \theta} + v_z \frac{\partial v_z}{\partial z} = -\frac{1}{\rho}\frac{\partial p}{\partial z} + \nu \nabla^2 v_z - g_z. \tag{7.19}$$

Here we recall Eq. (2.354) for the Laplacian operator:

$$\nabla^2 = \frac{\partial^2}{\partial r^2} + \frac{1}{r}\frac{\partial}{\partial r} + \frac{1}{r^2}\frac{\partial^2}{\partial \theta^2} + \frac{\partial^2}{\partial z^2}. \tag{7.20}$$

Within the acceleration terms, strictly unsteady terms, advective terms, as well as centripetal and Coriolis terms appear. For the body force, we have taken $\mathbf{f} = -g_z \mathbf{e}_z$. The viscous terms have additional complications that we have not presented in detail but arise because we must transform $\nabla^2 \mathbf{v}$; there are many non-intuitive terms that arise when expanded in full.

7.3 Ideal Rotational Vortex

Let us now consider the kinematics and dynamics of an ideal rotational vortex, which we define to be a fluid rotating as a rigid body. Let us assume incompressible flow, so $\nabla^T \cdot \mathbf{v} = 0$, assume a simple velocity field, and ask what forces could have given rise to that velocity field. Take

$$v_r = 0, \qquad v_\theta = \frac{\omega_o r}{2}, \qquad v_z = 0. \tag{7.21}$$

This velocity field was also considered in a Cartesian representation in Section 3.12.4. The kinematics of this flow are simple and depicted in Fig. 7.4, similar to Fig. 3.17. Here ω_o is a constant. The velocity is zero at the origin and grows linearly in amplitude with distance

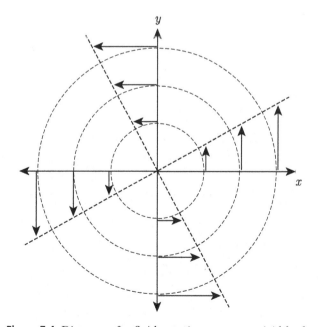

Figure 7.4 Diagram of a fluid rotating as a pure rigid body.

from the origin. The flow is steady, and the streamlines are circles centered about the origin. Obviously, as $r \to \infty$, compressibility effects would become important as the velocity approached the sound speed. That understood, one can still ask if this assumed velocity field satisfies incompressible mass conservation:

$$\frac{1}{r}\frac{\partial}{\partial r}(r(0)) + \frac{1}{r}\underbrace{\frac{\partial}{\partial \theta}\left(\frac{\omega_o r}{2}\right)}_{=0} + \frac{\partial}{\partial z}(0) \stackrel{?}{=} 0. \tag{7.22}$$

Obviously, it does.

Next let us consider the acceleration of an element of fluid and the forces that could give rise to that acceleration. First consider that portion of the acceleration that is neither centripetal nor Coriolis for this flow:

$$\underbrace{\frac{\partial}{\partial t}}_{=0} + \underbrace{v_r}_{=0}\frac{\partial}{\partial r} + \underbrace{\frac{v_\theta}{r}\frac{\partial}{\partial \theta}}_{=0} + \underbrace{v_z}_{=0}\frac{\partial}{\partial z} = 0. \tag{7.23}$$

As the only nonzero component of velocity, v_θ, has no dependency on θ, the unsteady and advective portions of the acceleration are zero for this flow. And because $v_r = 0$, there is no Coriolis acceleration. So the only acceleration is centripetal and is $-v_\theta^2/r = -\omega_o^2 r/4$. It points to the center, and thus is parallel to $\mathbf{e}_r$.

Consider now the viscous terms for this flow. We recall from Eq. (6.59) for an incompressible, Newtonian fluid that

$$\tau_{ij} = \tau_{ji} = 2\mu \partial_{(i}v_{j)} = \mu\left(\partial_i v_j + \partial_j v_i\right), \tag{7.24}$$

$$\partial_j \tau_{ji} = \mu\left(\partial_j \partial_i v_j + \partial_j \partial_j v_i\right) = \mu\left(\partial_i \underbrace{\partial_j v_j}_{=0} + \partial_j \partial_j v_i\right) = \mu \nabla^2 \mathbf{v}. \tag{7.25}$$

We also note that

$$\nabla \times \boldsymbol{\omega} = \epsilon_{ijk}\partial_j \omega_k = \epsilon_{ijk}\partial_j \epsilon_{kmn}\partial_m v_n = \epsilon_{kij}\epsilon_{kmn}\partial_j \partial_m v_n, \tag{7.26}$$

$$= (\delta_{im}\delta_{jn} - \delta_{in}\delta_{jm})\partial_j \partial_m v_n = \partial_j \partial_i v_j - \partial_j \partial_j v_i = \partial_i \underbrace{\partial_j v_j}_{=0} - \partial_j \partial_j v_i, \tag{7.27}$$

$$= -\partial_j \partial_j v_i = -\nabla^2 \mathbf{v}. \tag{7.28}$$

Comparing, we see that for this incompressible flow,

$$\left(\nabla^T \cdot \boldsymbol{\tau}\right)^T = -\mu(\nabla \times \boldsymbol{\omega}). \tag{7.29}$$

Now, using relations that can be developed for the curl in cylindrical coordinates, we have for this flow that

$$\omega_r = \frac{1}{r}\frac{\partial v_z}{\partial \theta} - \frac{\partial v_\theta}{\partial z} = 0, \tag{7.30}$$

$$\omega_\theta = \frac{\partial v_r}{\partial z} - \frac{\partial v_z}{\partial r} = 0, \tag{7.31}$$

$$\omega_z = \frac{1}{r}\frac{\partial}{\partial r}(rv_\theta) - \frac{1}{r}\frac{\partial v_r}{\partial \theta} = \frac{1}{r}\frac{\partial}{\partial r}\left(r\frac{\omega_o r}{2}\right) = \omega_o. \tag{7.32}$$

So the flow has a constant rotation rate, ω_o. Because it is constant, its curl is zero, and we have for this flow that $\left(\nabla^T \cdot \boldsymbol{\tau}\right)^T = \mathbf{0}$. We could just as well show for this flow that $\boldsymbol{\tau} = \mathbf{0}$. That is because the kinematics are those of pure rotation as a rigid body with no deformation. No deformation implies no viscous stress.

Hence, the three linear momenta equations in the cylindrical coordinate system reduce to the following:

$$-\frac{v_\theta^2}{r} = -\frac{1}{\rho}\frac{\partial p}{\partial r}, \qquad 0 = -\frac{1}{\rho}\frac{1}{r}\frac{\partial p}{\partial \theta}, \qquad 0 = -\frac{1}{\rho}\frac{\partial p}{\partial z} - g_z. \qquad (7.33)$$

The r momentum equation strikes a balance between centripetal inertia and radial pressure gradients. The θ momentum equation shows that as there is no acceleration in this direction, there can be no net pressure force to induce it. The z momentum equation enforces a balance between pressure forces and gravitational body forces.

If we take $p = p(r, \theta, z)$ and $p(r_o, \theta, z_o) = p_o$, then

$$dp = \frac{\partial p}{\partial r} dr + \frac{\partial p}{\partial \theta} d\theta + \frac{\partial p}{\partial z} dz = \frac{\rho v_\theta^2}{r} dr + 0\, d\theta - \rho g_z\, dz, \qquad (7.34)$$

$$= \frac{\rho \omega_o^2 r}{4} dr - \rho g_z\, dz, \qquad (7.35)$$

$$p(r, z) = p_o + \frac{\rho \omega_o^2}{8}(r^2 - r_o^2) - \rho g_z(z - z_o). \qquad (7.36)$$

Now on a surface of constant pressure we have $p(r, z) = \hat{p}$. So

$$\hat{p} = p_o + \frac{\rho \omega_o^2}{8}(r^2 - r_o^2) - \rho g_z(z - z_o), \qquad (7.37)$$

$$z = z_o + \frac{p_o - \hat{p}}{\rho g_z} + \frac{\omega_o^2}{8 g_z}(r^2 - r_o^2). \qquad (7.38)$$

So a surface of constant pressure is a parabola in r with a minimum at $r = 0$. This is consistent with what one observes upon spinning a bucket of water.

Now let us rearrange our general equation for the pressure field and eliminate ω_o using $v_\theta = \omega_o r/2$ and defining $v_{\theta o} = \omega_o r_o/2$:

$$p - \frac{1}{2}\rho v_\theta^2 + \rho g_z z = p_o - \frac{1}{2}\rho v_{\theta o}^2 + \rho g_z z_o = C. \qquad (7.39)$$

This looks similar to the steady, irrotational, incompressible Bernoulli equation, Eq. (6.196), in which $p + \frac{1}{2}\rho v^2 + \rho g_z z = K$. But there is a difference in the sign on one of the terms. Now add ρv_θ^2 to both sides of the equation to get

$$p + \frac{1}{2}\rho v_\theta^2 + \rho g_z z = C + \rho v_\theta^2. \qquad (7.40)$$

Because $v_\theta = \omega_o r/2$, $v_r = 0$, we have lines of constant r as streamlines, and v_θ is constant on those streamlines, so that we get

$$p + \frac{1}{2}\rho v_\theta^2 + \rho g_z z = C', \qquad \text{on a streamline}. \qquad (7.41)$$

Here C' varies from streamline to streamline, consistent with Eq. (6.191). We lastly note that the circulation for this system depends on position. If we choose our contour integral to be a circle of radius a about the origin, we find

$$\Gamma = \oint_C \mathbf{v}^T \cdot d\mathbf{r} = \oint_C v_\theta \mathbf{e}_\theta^T \cdot (a \, d\theta \, \mathbf{e}_\theta) = \int_0^{2\pi} \left(\frac{1}{2}\omega_o a\right)(a \, d\theta) = \pi a^2 \omega_o. \quad (7.42)$$

7.4 Ideal Irrotational Vortex

Now let us perform a similar analysis for the following velocity field:

$$v_r = 0, \qquad v_\theta = \frac{\Gamma_o}{2\pi r}, \qquad v_z = 0. \quad (7.43)$$

We have considered the same velocity field as represented in Cartesian coordinates, in Section 3.12.9. The kinematics of this flow are also simple and depicted in Fig. 7.5. We see once again that the streamlines are circles about the origin. But here, as opposed to the ideal rotational vortex, $v_\theta \to 0$ as $r \to \infty$ and $v_\theta \to \infty$ as $r \to 0$. The components of the vorticity vector for this flow are

$$\omega_r = \frac{1}{r}\frac{\partial v_z}{\partial \theta} - \frac{\partial v_\theta}{\partial z} = 0, \quad (7.44)$$

$$\omega_\theta = \frac{\partial v_r}{\partial z} - \frac{\partial v_z}{\partial r} = 0, \quad (7.45)$$

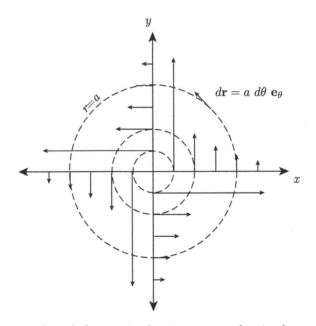

Figure 7.5 Diagram of an ideal irrotational point vortex and a circular contour of $r = a$.

$$\omega_z = \frac{1}{r}\frac{\partial}{\partial r}(rv_\theta) - \frac{1}{r}\frac{\partial v_r}{\partial \theta} = \frac{1}{r}\frac{\partial}{\partial r}\left(r\frac{\Gamma_o}{2\pi r}\right) = \frac{1}{r}\frac{\partial}{\partial r}\left(\frac{\Gamma_o}{2\pi}\right) = 0. \qquad (7.46)$$

This flow field, which seems to be the epitome of a rotating flow, is formally irrotational as it has zero vorticity! What is happening is that a fluid element not at the origin is actually undergoing severe deformation as it rotates about the origin; however, it does not rotate about its own center of mass (see Fig. 3.22). Therefore, the vorticity vector is zero, except at the origin, where it is undefined.

The circulation for this flow about a circle of radius a is

$$\Gamma = \oint_C \mathbf{v}^T \cdot d\mathbf{r} = \oint_C v_\theta \mathbf{e}_\theta^T \cdot (a\, d\theta\, \mathbf{e}_\theta) = \int_0^{2\pi} v_\theta(a\, d\theta) = \int_0^{2\pi} \frac{\Gamma_o}{2\pi a} a\, d\theta = \Gamma_o. \qquad (7.47)$$

So the circulation is nonzero and independent of the radius of the closed contour. In fact it can be shown that as long as the closed contour includes the origin in its interior that any closed contour will have this same circulation. We call Γ_o the ideal irrotational vortex strength, in that it is proportional to the magnitude of the velocity at any radius.

Let us once again consider the forces that could induce the motion of this vortex if the flow happens to be incompressible with constant properties and in a potential field in which the gravitational body force per unit mass is $-g_z\mathbf{k}$. Recall again that $\left(\nabla^T \cdot \boldsymbol{\tau}\right)^T = -\mu(\nabla \times \boldsymbol{\omega})$, and that because $\boldsymbol{\omega} = \mathbf{0}$ then $\left(\nabla^T \cdot \boldsymbol{\tau}\right)^T = \mathbf{0}$ for this flow. Because there is deformation here, $\boldsymbol{\tau}$ itself is not zero, but its divergence is zero. For example, if we consider one component of viscous stress $\tau_{r\theta}$ and use standard relations that can be derived for incompressible, Newtonian fluids, we find that

$$\tau_{r\theta} = \mu\left(r\frac{\partial}{\partial r}\left(\frac{v_\theta}{r}\right) + \frac{1}{r}\frac{\partial v_r}{\partial \theta}\right) = \mu r\frac{\partial}{\partial r}\left(\frac{\Gamma_o}{2\pi r^2}\right) = -\frac{\mu\Gamma_o}{\pi r^2}. \qquad (7.48)$$

The only nonzero component of acceleration is the centripetal acceleration; it is associated with the radial direction. The equations of motion reduce to the same ones as for the ideal rotational vortex:

$$-\frac{v_\theta^2}{r} = -\frac{1}{\rho}\frac{\partial p}{\partial r}, \qquad 0 = -\frac{1}{\rho}\frac{1}{r}\frac{\partial p}{\partial \theta}, \qquad 0 = -\frac{1}{\rho}\frac{\partial p}{\partial z} - g_z. \qquad (7.49)$$

Once more we can deduce a pressure field that is consistent with these and the same set of conditions at $r = r_o$, $z = z_o$, with $p = p_o$:

$$dp = \frac{\partial p}{\partial r}\, dr + \underbrace{\frac{\partial p}{\partial \theta}}_{=0}\, d\theta + \frac{\partial p}{\partial z}\, dz = \frac{\rho v_\theta^2}{r}\, dr - \rho g_z\, dz, \qquad (7.50)$$

$$= \frac{\rho \Gamma_o^2}{4\pi^2}\frac{dr}{r^3} - \rho g_z\, dz, \qquad (7.51)$$

$$p - p_o = -\frac{\rho \Gamma_o^2}{8\pi^2}\left(\frac{1}{r^2} - \frac{1}{r_o^2}\right) - \rho g_z(z - z_o), \qquad (7.52)$$

$$p + \frac{\rho \Gamma_o^2}{8\pi^2}\frac{1}{r^2} + \rho g_z z = p_o + \frac{\rho \Gamma_o^2}{8\pi^2}\frac{1}{r_o^2} + \rho g_z z_o, \qquad (7.53)$$

$$p + \frac{1}{2}\rho v_\theta^2 + \rho g_z z = p_o + \frac{1}{2}\rho v_{\theta o}^2 + \rho g_z z_o = C. \qquad (7.54)$$

This is fully equivalent to the Bernoulli equation for an irrotational, steady, incompressible flow, Eq. (6.196). Here, the Bernoulli constant C is truly constant for the entire flow field and not just along a streamline.

On isobars, we have $p = \hat{p}$, and that yields

$$\hat{p} - p_o = -\frac{\rho \Gamma_o^2}{8\pi^2}\left(\frac{1}{r^2} - \frac{1}{r_o^2}\right) - \rho g_z(z - z_o), \tag{7.55}$$

$$z = z_o + \frac{p_o - \hat{p}}{\rho g_z} + \frac{\Gamma_o^2}{8\pi^2 g_z}\left(\frac{1}{r^2} - \frac{1}{r_o^2}\right). \tag{7.56}$$

The pressure goes to negative infinity at the origin. One can show that actual forces, obtained by integrating pressure over area, are in fact bounded.

7.5 Helmholtz Vorticity Transport Equation

Here we will take the curl of the linear momenta principle to obtain a relationship, the Helmholtz vorticity transport equation, that shows what induces the vorticity field to evolve in a compressible, viscous fluid.

7.5.1 General Development

First, we recall some useful vector identities:

$$(\mathbf{v}^T \cdot \nabla)\mathbf{v} = \nabla\left(\frac{\mathbf{v}^T \cdot \mathbf{v}}{2}\right) + \boldsymbol{\omega} \times \mathbf{v}, \tag{7.57}$$

$$\nabla \times (\boldsymbol{\omega} \times \mathbf{v}) = (\mathbf{v}^T \cdot \nabla)\boldsymbol{\omega} - (\boldsymbol{\omega}^T \cdot \nabla)\mathbf{v} + \boldsymbol{\omega}(\nabla^T \cdot \mathbf{v}) - \mathbf{v}(\nabla^T \cdot \boldsymbol{\omega}), \tag{7.58}$$

$$\nabla \times (\nabla \phi) = \mathbf{0}, \tag{7.59}$$

$$\nabla^T \cdot (\nabla \times \mathbf{v}) = \nabla^T \cdot \boldsymbol{\omega} = 0. \tag{7.60}$$

The first is equivalent to Eq. (2.232); the others can be proved.

We start now with the linear momenta principle for a compressible, viscous fluid ; we recast Eq. (4.271) and write

$$\frac{\partial \mathbf{v}}{\partial t} + (\mathbf{v}^T \cdot \nabla)\mathbf{v} = \mathbf{f} - \frac{1}{\rho}\nabla p + \frac{1}{\rho}\left(\nabla^T \cdot \boldsymbol{\tau}\right)^T. \tag{7.61}$$

With Eq. (7.57), we expand the term $(\mathbf{v}^T \cdot \nabla)\mathbf{v}$ and then apply the curl operator to both sides to get

$$\nabla \times \left(\frac{\partial \mathbf{v}}{\partial t} + \nabla\left(\frac{\mathbf{v}^T \cdot \mathbf{v}}{2}\right) + \boldsymbol{\omega} \times \mathbf{v}\right) = \nabla \times \left(\mathbf{f} - \frac{1}{\rho}\nabla p + \frac{1}{\rho}\left(\nabla^T \cdot \boldsymbol{\tau}\right)^T\right). \tag{7.62}$$

This becomes, via the linearity of the various operators,

$$\frac{\partial}{\partial t}\underbrace{(\nabla \times \mathbf{v})}_{\boldsymbol{\omega}} + \underbrace{\nabla \times \left(\nabla\left(\frac{\mathbf{v}^T \cdot \mathbf{v}}{2}\right)\right)}_{=0} + \nabla \times \boldsymbol{\omega} \times \mathbf{v} = \nabla \times \mathbf{f} - \nabla \times \left(\frac{1}{\rho}\nabla p\right) + \nabla \times \left(\frac{1}{\rho}\left(\nabla^T \cdot \boldsymbol{\tau}\right)^T\right). \tag{7.63}$$

The second term is zero because of Eq. (7.59). Using our vector identity for the term with two cross products, Eq. (7.58), we get

$$\underbrace{\frac{\partial \boldsymbol{\omega}}{\partial t} + (\mathbf{v}^T \cdot \nabla)\boldsymbol{\omega}}_{=\frac{d\boldsymbol{\omega}}{dt}} - (\boldsymbol{\omega}^T \cdot \nabla)\mathbf{v} + \boldsymbol{\omega}(\underbrace{\nabla^T \cdot \mathbf{v}}_{=-\frac{1}{\rho}\frac{d\rho}{dt}}) - \mathbf{v}(\underbrace{\nabla^T \cdot \boldsymbol{\omega}}_{=0}) = \nabla \times \mathbf{f} - \nabla \times \left(\frac{1}{\rho}\nabla p\right)$$
$$+ \nabla \times \left(\frac{1}{\rho}\left(\nabla^T \cdot \boldsymbol{\tau}\right)^T\right). \quad (7.64)$$

The fifth term is zero because of Eq. (7.60). Rearranging, we have

$$\frac{d\boldsymbol{\omega}}{dt} - \frac{\boldsymbol{\omega}}{\rho}\frac{d\rho}{dt} = (\boldsymbol{\omega}^T \cdot \nabla)\mathbf{v} + \nabla \times \mathbf{f} - \nabla \times \left(\frac{1}{\rho}\nabla p\right) + \nabla \times \left(\frac{1}{\rho}\left(\nabla^T \cdot \boldsymbol{\tau}\right)^T\right), \quad (7.65)$$

$$\frac{1}{\rho}\frac{d\boldsymbol{\omega}}{dt} - \frac{\boldsymbol{\omega}}{\rho^2}\frac{d\rho}{dt} = \left(\frac{\boldsymbol{\omega}^T}{\rho} \cdot \nabla\right)\mathbf{v} + \frac{1}{\rho}\nabla \times \mathbf{f} - \frac{1}{\rho}\nabla \times \left(\frac{1}{\rho}\nabla p\right) + \frac{1}{\rho}\nabla \times \left(\frac{1}{\rho}\left(\nabla^T \cdot \boldsymbol{\tau}\right)^T\right), \quad (7.66)$$

$$\frac{d}{dt}\left(\frac{\boldsymbol{\omega}}{\rho}\right) = \left(\frac{\boldsymbol{\omega}^T}{\rho} \cdot \nabla\right)\mathbf{v} + \frac{1}{\rho}\nabla \times \mathbf{f} - \frac{1}{\rho}\nabla \times \left(\frac{1}{\rho}\nabla p\right) + \frac{1}{\rho}\nabla \times \left(\frac{1}{\rho}\left(\nabla^T \cdot \boldsymbol{\tau}\right)^T\right), \quad (7.67)$$

$$\rho\frac{d}{dt}\left(\frac{\boldsymbol{\omega}}{\rho}\right) = (\boldsymbol{\omega}^T \cdot \nabla)\mathbf{v} + \nabla \times \mathbf{f} - \nabla \times \left(\frac{1}{\rho}\nabla p\right) + \nabla \times \left(\frac{1}{\rho}\left(\nabla^T \cdot \boldsymbol{\tau}\right)^T\right). \quad (7.68)$$

Now consider the term $-\nabla \times ((1/\rho)\nabla p)$. In Cartesian index notation, we have

$$-\epsilon_{ijk}\partial_j\left(\frac{1}{\rho}\partial_k p\right) = -\epsilon_{ijk}\left(\frac{1}{\rho}\partial_j\partial_k p - \frac{1}{\rho^2}(\partial_j\rho)(\partial_k p)\right), \quad (7.69)$$

$$= -\frac{1}{\rho}\underbrace{\epsilon_{ijk}\partial_j\partial_k p}_{=0} + \frac{1}{\rho^2}\epsilon_{ijk}(\partial_j\rho)(\partial_k p) = \frac{1}{\rho^2}\nabla\rho \times \nabla p. \quad (7.70)$$

We write the final general form of the vorticity transport equation as

$$\rho\frac{d}{dt}\left(\frac{\boldsymbol{\omega}}{\rho}\right) = \underbrace{(\boldsymbol{\omega}^T \cdot \nabla)\mathbf{v}}_{A} + \underbrace{\nabla \times \mathbf{f}}_{B} + \underbrace{\frac{1}{\rho^2}\nabla\rho \times \nabla p}_{C} + \underbrace{\nabla \times \left(\frac{1}{\rho}\left(\nabla^T \cdot \boldsymbol{\tau}\right)^T\right)}_{D}. \quad (7.71)$$

Here we see the evolution of the vorticity scaled by the density is affected by four physical mechanisms, which we label A, B, C, and D:

- A: bending and stretching of vortex tubes,
- B: nonconservative body force effect (if $\mathbf{f} = -\nabla\varphi$, then $\mathbf{f}$ is conservative, and $\nabla \times \mathbf{f} = -\nabla \times \nabla\varphi = \mathbf{0}$),
- C: non-barotropic, also known as *baroclinic*, effects (if a fluid is barotropic, then $p = p(\rho)$ and $\nabla p = (dp/d\rho)\nabla\rho$ thus $\nabla\rho \times \nabla p = \nabla\rho \times (dp/d\rho)\nabla\rho = 0$), and
- D: viscous effects.

Mechanisms B and D have obvious physical interpretations. Mechanisms A and C are described next.

7.5.2 Bending and Stretching of Vortex Tubes

Let us consider generation of vorticity by three-dimensional effects. Such effects are commonly characterized as the bending and stretching of what are known as vortex tubes. Here we focus on the incompressible, barotropic, conservative body force limit of Eq. (7.71):

$$\frac{d\boldsymbol{\omega}}{dt} = (\boldsymbol{\omega}^T \cdot \nabla)\mathbf{v}. \tag{7.72}$$

If we consider a coordinate system that is oriented with the vorticity field as depicted in Fig. 7.6, we will get many simplifications. We take the following directions:

- w: the direction parallel to the vorticity vector,
- n: the principal normal direction, pointing towards the center of curvature,
- b: the binormal direction, orthogonal to w and n.

With this system, we can say that

$$(\boldsymbol{\omega}^T \cdot \nabla)\mathbf{v} = \left(\begin{pmatrix} \omega_w & 0 & 0 \end{pmatrix} \begin{pmatrix} \frac{\partial}{\partial w} \\ \frac{\partial}{\partial n} \\ \frac{\partial}{\partial b} \end{pmatrix} \right) \mathbf{v} = \omega_w \frac{\partial \mathbf{v}}{\partial w}. \tag{7.73}$$

So for the inviscid flow we have

$$\frac{d\boldsymbol{\omega}}{dt} = \omega_w \frac{\partial \mathbf{v}}{\partial w}. \tag{7.74}$$

We have in terms of components

$$\frac{d\omega_w}{dt} = \omega_w \frac{\partial v_w}{\partial w}, \quad \frac{d\omega_n}{dt} = \omega_w \frac{\partial v_n}{\partial w}, \quad \frac{d\omega_b}{dt} = \omega_w \frac{\partial v_b}{\partial w}. \tag{7.75}$$

The term $\partial v_w / \partial w$ we know from kinematics represents a local stretching or extensional straining. Just as a rotating figure skater increases his or her angular velocity by concentrating his or her mass about a vertical axis, so does a rotating fluid. The first of these expressions says that the component of rotation aligned with the present increases if there is stretching in that direction. This is depicted in Fig. 7.7.

The second and third terms enforce that if v_n or v_b are changing in the w direction, when accompanied by nonzero ω_w, that changes in the non-aligned components of $\boldsymbol{\omega}$ are induced. Hence the previously zero components ω_n, ω_b acquire nonzero values, and the lines parallel to the vorticity vector bend. Hence, we have the term *bending of vortex tubes*. It is generally

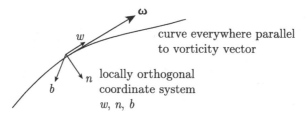

Figure 7.6 Local orthogonal intrinsic coordinate system oriented with local vorticity field.

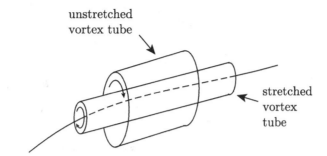

Figure 7.7 Increase in vorticity due to stretching of a vortex tube.

accepted that the bending and stretching of vortex tubes is an important mechanism in the transition from laminar to turbulent flow.

Lastly, we note that vortex stretching has been described as the critical physical process that induces fundamental mathematical challenges in solving the incompressible, three-dimensional Navier–Stokes equations. Whether or not these equations possess unique smooth solutions at high Reynolds number remains an open question; see Doering (2009) for a full discussion and review.

7.5.3 Baroclinic (Non-Barotropic) Effects

If a fluid is barotropic then we can write $p = p(\rho)$, or $\rho = \rho(p)$. As an example, an isentropic calorically perfect ideal gas has $p/p_o = (\rho/\rho_o)^\gamma$. Here γ is the ratio of specific heats, and the o subscript indicates a constant value. Such a gas is barotropic. For such a fluid, we must have by the chain rule that $\partial_i p = (dp/d\rho)\partial_i \rho$. Hence ∇p and $\nabla \rho$ are vectors that point in the same direction. Moreover, isobars (lines of constant pressure) must be parallel to isochores (lines of constant density). If, as depicted in Fig. 7.8, we calculate the resultant vector from the net pressure force, as well as the center of mass for a finite fluid volume, we would see that the resultant force had no lever arm with the center of gravity. Hence it would generate no torque, and no tendency for the fluid element to rotate about its center of mass; hence no vorticity would be generated by this force.

For a baroclinic fluid, we do not have $p = p(\rho)$; hence, we must expect that ∇p points in a different direction than $\nabla \rho$. If we examine this scenario, as depicted in Fig. 7.9, we discover that the resultant force from the pressure has a nonzero lever arm with the center of mass of the fluid element. Hence, it generates a torque, a tendency to rotate the fluid about G, and vorticity.

7.5.4 Incompressible, Conservative Body Force Limit

The Helmholtz vorticity transport equation (7.71) reduces significantly in special limiting cases involving incompressible flow in the limit of a conservative body force:

$$\frac{d\boldsymbol{\omega}}{dt} - (\boldsymbol{\omega}^T \cdot \nabla)\mathbf{v} = \frac{1}{\rho}\nabla \times (\nabla^T \cdot \boldsymbol{\tau})^T. \tag{7.76}$$

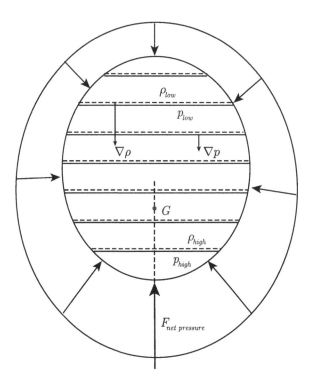

Figure 7.8 Isobars and isochores, center of mass G, and center of pressure for a barotropic fluid.

Isotropic, Newtonian, Constant Viscosity

Now if we further require that the fluid be isotropic and Newtonian with constant viscosity, the viscous term can be written as

$$\nabla \times (\nabla^T \cdot \boldsymbol{\tau})^T = \epsilon_{ijk}\partial_j\partial_m(2\mu(\partial_{(m}v_{k)} - (1/3)\underbrace{\partial_l v_l}_{=0}\delta_{mk})), \tag{7.77}$$

$$= \mu\epsilon_{ijk}\partial_j\partial_m(\partial_m v_k + \partial_k v_m) = \mu\epsilon_{ijk}\partial_j(\partial_m\partial_m v_k + \partial_m\partial_k v_m), \tag{7.78}$$

$$= \mu\epsilon_{ijk}\partial_j(\partial_m\partial_m v_k + \partial_k\underbrace{\partial_m v_m}_{=0}) = \mu\partial_m\partial_m\underbrace{\epsilon_{ijk}\partial_j v_k}_{\boldsymbol{\omega}} = \mu\nabla^2\boldsymbol{\omega}. \tag{7.79}$$

So we get, recalling that kinematic viscosity $\nu = \mu/\rho$,

$$\frac{d\boldsymbol{\omega}}{dt} - (\boldsymbol{\omega}^T \cdot \nabla)\mathbf{v} = \nu\nabla^2\boldsymbol{\omega}, \quad \frac{\partial\boldsymbol{\omega}}{\partial t} + (\mathbf{v}^T \cdot \nabla)\boldsymbol{\omega} - (\boldsymbol{\omega}^T \cdot \nabla)\mathbf{v} = \nu\nabla^2\boldsymbol{\omega}. \tag{7.80}$$

Here we have used the material derivative $d/dt = \partial/\partial t + \mathbf{v}^T \cdot \nabla$.

Following Thorne and Blandford (2017, pp. 735–737) and recalling the definitions of the Lie derivative, commutator, and convected derivative, Eqs. (3.181, 3.184), we recast Eq. (7.80), the Helmholtz vorticity transport equation for an incompressible, constant viscosity, Newtonian fluid, in any of the following forms:

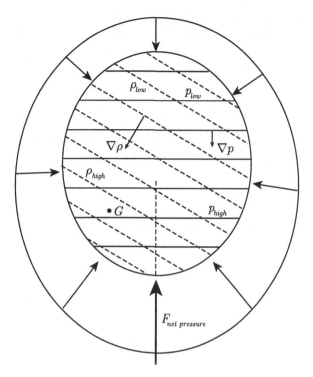

Figure 7.9 Isobars and isochores, center of mass G, and center of pressure for a baroclinic fluid.

$$\frac{\partial \boldsymbol{\omega}}{\partial t} + \mathcal{L}_{\mathbf{v}}\boldsymbol{\omega} = \nu\nabla^2\boldsymbol{\omega}, \qquad \frac{\partial \boldsymbol{\omega}}{\partial t} + [\mathbf{v}, \boldsymbol{\omega}] = \nu\nabla^2\boldsymbol{\omega}, \qquad \frac{\mathscr{D}\boldsymbol{\omega}}{\mathscr{D}t} = \nu\nabla^2\boldsymbol{\omega}. \qquad (7.81)$$

One can interpret the convected derivative of vorticity $\mathscr{D}\boldsymbol{\omega}/\mathscr{D}t$ as the rate of change of $\boldsymbol{\omega}$ relative to a vector such as $d\mathbf{x}$ that moves, rotates, and deforms with the fluid.

Two-Dimensional, Isotropic, Newtonian, Constant Viscosity

If we further require two-dimensionality, then we have $\boldsymbol{\omega} = (0, 0, \omega_3(x_1, x_2))^T$, and $\nabla = (\partial_1, \partial_2, 0)^T$, so $\boldsymbol{\omega}^T \cdot \nabla = 0$. Thus, we get the simple

$$\frac{d\omega_3}{dt} = \nu\nabla^2\omega_3 = \nu\left(\frac{\partial^2\omega_3}{\partial x_1^2} + \frac{\partial^2\omega_3}{\partial x_2^2}\right). \qquad (7.82)$$

If the flow is also inviscid $\nu = 0$, we get

$$\frac{d\omega_3}{dt} = 0, \qquad (7.83)$$

and we find that there is no tendency for vorticity to change along a streamline. If we further have an initially irrotational state, then we get $\boldsymbol{\omega} = \mathbf{0}$ for all space and time.

Example 7.1 Consider the flow of a viscous fluid with $\rho = \mu = \nu = 1$. Take the dimensionless velocity vector to be $\mathbf{v} = (y^3, 0, 0)^T$. Show the flow is incompressible, and find the vorticity vector $\boldsymbol{\omega}$. Then find any nonconservative body force vector $\mathbf{f}$ and a corresponding pressure field p that allow the Helmholtz vorticity transport equation and the linear momenta equation to be simultaneously satisfied.

Solution

The flow is incompressible because

$$\nabla^T \cdot \mathbf{v} = \frac{\partial}{\partial x} y^3 + \frac{\partial}{\partial y}(0) + \frac{\partial}{\partial z}(0) = 0.$$

The flow is rotational because

$$\boldsymbol{\omega} = \nabla \times \mathbf{v} = 0\mathbf{i} + 0\mathbf{j} + \left(\frac{\partial}{\partial x}(0) - \frac{\partial}{\partial y}(y^3)\right)\mathbf{k} = -3y^2 \mathbf{k}.$$

The steady Helmholtz vorticity transport equation, including body force effects, for this incompressible flow with $\nu = 1$ is

$$(\mathbf{v}^T \cdot \nabla)\boldsymbol{\omega} = \nabla^2 \boldsymbol{\omega} + \nabla \times \mathbf{f}. \tag{7.84}$$

The only nonzero portion of $\mathbf{v}^T \cdot \nabla$ is $y^3 \partial/\partial x$. And $\nabla^2 \boldsymbol{\omega} = (0, 0, -6)^T$. So we get

$$y^3 \frac{\partial}{\partial x} \begin{pmatrix} 0 \\ 0 \\ 3y^2 \end{pmatrix} = \begin{pmatrix} 0 \\ 0 \\ 6 \end{pmatrix} + \nabla \times \mathbf{f}, \tag{7.85}$$

$$\begin{pmatrix} 0 \\ 0 \\ 0 \end{pmatrix} = \begin{pmatrix} 0 \\ 0 \\ -6 \end{pmatrix} + \nabla \times \mathbf{f}. \tag{7.86}$$

So we need the $\mathbf{k}$ component of $\nabla \times \mathbf{f}$ to be equal to 6. There are an infinite number of solutions for $\mathbf{f}$ because if the solution for $\mathbf{f}$ includes any conservative potential $\nabla \varphi$, its curl will be zero. Let us take a simple, nonunique solution:

$$\mathbf{f} = \begin{pmatrix} -6y \\ 0 \\ 0 \end{pmatrix}. \tag{7.87}$$

This body force allows satisfaction of the Helmholtz vorticity transport equation. Now consider the linear momenta equation for $\rho = \mu = 1$:

$$\frac{d\mathbf{v}}{dt} = -\nabla p + \mathbf{f} + \nabla^2 \mathbf{v}. \tag{7.88}$$

For this flow $d\mathbf{v}/dt = \partial \mathbf{v}/\partial t + \mathbf{v}^T \cdot \nabla \mathbf{v} = \mathbf{0}$. The flow is not accelerating. There is an imbalance of viscous stress as $\nabla^2 \mathbf{v} = 6y\mathbf{i}$. We see that the force due to viscous terms exactly balances the body force $\mathbf{f}$, yielding $\nabla p = \mathbf{0}$. So we can take the pressure field to be constant:

$$p = p_o. \tag{7.89}$$

This also satisfies Eq. (6.77) for $\rho = 1$: $\nabla^2 p = -(\mathbf{D} : \mathbf{D} - \boldsymbol{\omega}^T \cdot \boldsymbol{\omega}/2) + \nabla^T \cdot \mathbf{f} = -(9y^4/2 - 9y^4/2) + 0 = 0$. Other choices can also satisfy. For instance, if we had chosen $\mathbf{f} = (0, 6x, 0)^T$ and $p = 6xy + p_o$, all equations would be satisfied. Nonlinear systems, such as the Navier–Stokes equations, need not have unique solutions.

7.6 Kelvin's Circulation Theorem

Kelvin's circulation theorem describes how the circulation of a material volume in a fluid changes with time. We first recall from Eq. (7.2) the definition of circulation: $\Gamma = \oint_C \mathbf{v}^T \cdot d\mathbf{x}$. Here C is a closed contour. We next take the material derivative of Γ to get

$$\frac{d\Gamma}{dt} = \frac{d}{dt} \oint_C \mathbf{v}^T \cdot d\mathbf{x} = \oint_C \frac{d\mathbf{v}^T}{dt} \cdot d\mathbf{x} + \oint_C \mathbf{v}^T \cdot \frac{d}{dt} d\mathbf{x}, \tag{7.90}$$

$$= \oint_C \left(\frac{d\mathbf{v}}{dt}\right)^T \cdot d\mathbf{x} + \oint_C \mathbf{v}^T \cdot d\left(\frac{d\mathbf{x}}{dt}\right) = \oint_C \left(\frac{d\mathbf{v}}{dt}\right)^T \cdot d\mathbf{x} + \oint_C \mathbf{v}^T \cdot d\mathbf{v}, \tag{7.91}$$

$$= \oint_C \left(\frac{d\mathbf{v}}{dt}\right)^T \cdot d\mathbf{x} + \underbrace{\oint_C d\left(\frac{1}{2}\mathbf{v}^T \cdot \mathbf{v}\right)}_{=0} = \oint_C \left(\frac{d\mathbf{v}}{dt}\right)^T \cdot d\mathbf{x}. \tag{7.92}$$

Because we have chosen a material volume for our closed contour, $d\mathbf{x}/dt$ must be the fluid particle velocity. This then allows us to write the second term as a perfect differential, that integrates fluid particle acceleration over the closed contour to be zero. We continue now by using the linear momenta principle to replace the particle acceleration with density-scaled forces to arrive at

$$\frac{d\Gamma}{dt} = \oint_C \left(\mathbf{f} - \frac{1}{\rho}\nabla p + \frac{1}{\rho}\left(\nabla^T \cdot \boldsymbol{\tau}\right)^T\right)^T \cdot d\mathbf{x}. \tag{7.93}$$

If now the fluid is inviscid ($\boldsymbol{\tau} = \mathbf{0}$), the body force is conservative ($\mathbf{f} = -\nabla \varphi$), and the fluid is barotropic ($(1/\rho)\nabla p = \nabla \mathsf{P}$), we then have

$$\frac{d\Gamma}{dt} = \oint_C (-\nabla\varphi - \nabla\mathsf{P})^T \cdot d\mathbf{x} = -\oint_C \nabla^T(\varphi + \mathsf{P}) \cdot d\mathbf{x} = -\underbrace{\oint_C d(\varphi + \mathsf{P})}_{=0}. \tag{7.94}$$

The integral on the right-hand side is zero because the contour is closed; hence, the integral is path-independent. Consequently, we arrive at the common version of Kelvin's circulation theorem that holds that for a fluid that is inviscid, barotropic, and subjected to conservative body forces, the circulation following a material volume does not change with time:

$$\frac{d\Gamma}{dt} = 0. \tag{7.95}$$

This is similar to the Helmholtz vorticity transport equation, which, when we make the additional stipulation of two-dimensionality and incompressibility, gives $d\boldsymbol{\omega}/dt = \mathbf{0}$. This is not surprising as the vorticity is closely linked to the circulation via Stokes' theorem, Eq. (2.231), which states

$$\Gamma = \oint_C \mathbf{v}^T \cdot d\mathbf{x} = \int_A (\nabla \times \mathbf{v})^T \cdot \mathbf{n} \, dA = \int_A \boldsymbol{\omega}^T \cdot \mathbf{n} \, dA. \tag{7.96}$$

7.7 Two-Dimensional Potential Flow of Ideal Point Vortices

This section will consider the fluid motion induced by the simultaneous interaction of a family of ideal *irrotational* point vortices in an incompressible flow field. We first briefly consider the three-dimensional case before turning attention to the two-dimensional limit. Because the flow is irrotational and incompressible, we have the following useful results: (1) Because $\nabla \times \mathbf{v} = \mathbf{0}$, we can adopt Eq. (6.182) and write the velocity vector as the gradient of a scalar potential ϕ:

$$\mathbf{v} = \nabla \phi, \qquad \text{if irrotational.} \tag{7.97}$$

We recall ϕ is the velocity potential, and the associated flow is known as a *potential flow*. (2) Because $\nabla^T \cdot \mathbf{v} = 0$, we then have

$$\nabla^T \cdot \nabla \phi = \nabla^2 \phi = 0, \tag{7.98}$$

or expanding, we have Laplace's equation, Eq. (2.217):

$$\frac{\partial^2 \phi}{\partial x^2} + \frac{\partial^2 \phi}{\partial y^2} + \frac{\partial^2 \phi}{\partial z^2} = 0, \tag{7.99}$$

and solution $\phi(x, y, z)$ is a harmonic function. (3) We notice that the equation for ϕ is linear; hence, the method of superposition is valid for the velocity potential. That is, we can add an arbitrary number of velocity potentials together and get an admissible flow field. (4) The irrotational, unsteady Bernoulli equation, Eq. (6.186), gives us the time- and space-dependent pressure field. This equation is not linear, so we do not expect pressures from elementary solutions to add to form total pressures.

Recalling that the incompressible, three-dimensional, constant viscosity Helmholtz vorticity transport equation can be written via Eq. (7.80) as

$$\frac{d\boldsymbol{\omega}}{dt} = (\boldsymbol{\omega}^T \cdot \nabla) \mathbf{v} + \nu \nabla^2 \boldsymbol{\omega}, \tag{7.100}$$

we see that a flow that is initially irrotational everywhere in an unbounded fluid will always be irrotational, as $d\boldsymbol{\omega}/dt = \mathbf{0}$. There is no mechanism to change the vorticity from its uniform initial value of zero. This even holds for a viscous flow. However, in a bounded medium, the no-slip boundary condition almost always tends to diffuse vorticity into the flow as we shall see. Further for inviscid, barotropic flow, from Kelvin's circulation theorem, Eq. (7.95), the circulation Γ has no tendency to change following a particle; that is, Γ advects unchanged along particle pathlines.

From here on in this chapter, we restrict attention to two-dimensional, inviscid, incompressible potential flows composed of ideal point vortices. The key equation will be the velocity field induced by a single ideal point vortex, Eq. (7.43) $v_\theta = \Gamma_o/(2\pi r)$. This can be considered a special two-dimensional limit of what is known as the *Biot*[1]–*Savart*[2] *law*, which can be developed for three-dimensional flows. A full discussion of the Biot–Savart law's derivation and reduction to the two-dimensional limit we use is given by Karamcheti (1980). Additional discussion is given by Cummings et al. (2015), Panton (2013), and many others.

---

[1] Jean-Baptiste Biot, 1774–1862, French physicist.
[2] Félix Savart, 1791–1841, French physicist.

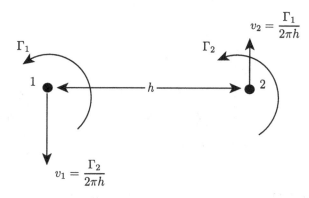

Figure 7.10 Diagram of the mutual influence of two ideal point vortices on each other.

Figure 7.11 Diagram showing the center of rotation G.

7.7.1 Two interacting Ideal Vortices

Let us apply these notions to two ideal counterrotating vortices 1 and 2, with respective strengths, Γ_1 and Γ_2, as shown in Fig. 7.10. Were it isolated, vortex 1 would have no tendency to move, but would induce a velocity at a distance h away from its center of $\Gamma_1/(2\pi h)$. This induced velocity in fact advects vortex 2 to satisfy Kelvin's circulation theorem. Similarly, vortex 2 induces a velocity of vortex 1 of $\Gamma_2/(2\pi h)$.

The center of rotation G is the point along the 1,2 axis for which the induced velocity is zero, as is illustrated in Fig. 7.11. To calculate it, we equate the induced velocities of each vortex

$$\frac{\Gamma_1}{2\pi h_G} = \frac{\Gamma_2}{2\pi (h - h_G)}, \qquad \text{yielding} \qquad h_G = h\frac{\Gamma_1}{\Gamma_1 + \Gamma_2}. \tag{7.101}$$

A pair of equal strength counterrotating vortices is illustrated in Fig. 7.12. Such vortices induce the same velocity in each other, so they will propagate as a pair at a fixed distance from one another.

7.7.2 Image Vortex

If we choose to model the fluid as inviscid, then we can no longer enforce the no-slip condition at a wall. However, at a slip wall, we must require that the velocity vector be parallel to the wall. This is the no-penetration condition, looser than a no-slip condition. No-penetration

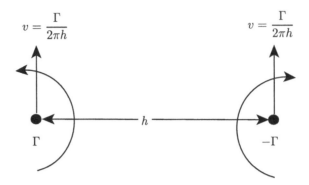

Figure 7.12 Diagram showing a pair of counterrotating vortices of equal strength.

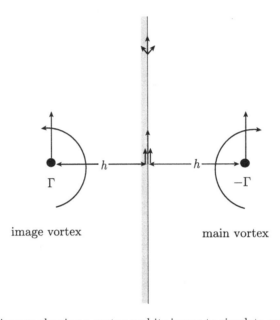

Figure 7.13 Diagram showing a vortex and its image to simulate an inviscid wall.

through a surface with outer normal $\mathbf{n}$ simply requires Eq. (6.4), $\mathbf{v}^T \cdot \mathbf{n} = 0$, and allows no mass to penetrate such a surface, while allowing slip at the surface because $\mathbf{v}^T \cdot \boldsymbol{\alpha}_t \neq 0$, with $\boldsymbol{\alpha}_t$ as the unit tangent vector to the surface. We can model the motion of an ideal vortex separated by a distance h from an inviscid slip wall by placing a so-called *image vortex* on the other side of the wall. The image vortex will induce a velocity that when superposed with the original vortex, renders the resultant velocity to be parallel to the wall. A vortex and its image vortex, which generates a straight streamline at a wall, is depicted in Fig. 7.13.

7.7.3 Vortex Sheets

We can model the slip line between two inviscid fluids moving at different velocities by what is known as a *vortex sheet*. A vortex sheet is depicted in Fig. 7.14. Here we have a distribution of infinitesimally small ideal vortices, each of strength $d\Gamma$, on the x axis. Each of these vortices induces an infinitesimally small velocity $d\mathbf{v}$ at an arbitrary point $(\tilde{x}, \tilde{y})$. The influence of the ideal point vortex at $(x, 0)$ is depicted in the figure. It generates an infinitesimally small velocity with magnitude

$$d\|\mathbf{v}\| = \frac{d\Gamma}{2\pi h} = \frac{d\Gamma}{2\pi\sqrt{(\tilde{x}-x)^2 + \tilde{y}^2}}. \quad (7.102)$$

Using basic trigonometry, we can deduce that the influence of the single vortex of differential strength on each velocity component is

$$du = \frac{-d\Gamma \tilde{y}}{2\pi((\tilde{x}-x)^2 + \tilde{y}^2)} = \frac{-\frac{d\Gamma}{dx}\tilde{y}}{2\pi((\tilde{x}-x)^2 + \tilde{y}^2)}\,dx, \quad (7.103)$$

$$dv = \frac{d\Gamma(\tilde{x}-x)}{2\pi((\tilde{x}-x)^2 + \tilde{y}^2)} = \frac{\frac{d\Gamma}{dx}(\tilde{x}-x)}{2\pi((\tilde{x}-x)^2 + \tilde{y}^2)}\,dx. \quad (7.104)$$

Here $d\Gamma/dx$ is a measure of the strength of the vortex sheet, and is taken here to be a constant. Let us account for the effects of *all* of the differential vortices by integrating from $x = -L$ to $x = L$ and then letting $L \to \infty$. We obtain then the total velocity components u and v at each point to be

$$u = \lim_{L\to\infty} -\frac{\frac{d\Gamma}{dx}}{2\pi}\left(\underbrace{\tan^{-1}\left(\frac{L-\tilde{x}}{\tilde{y}}\right)}_{\to \pm\frac{\pi}{2}} + \underbrace{\tan^{-1}\left(\frac{L+\tilde{x}}{\tilde{y}}\right)}_{\to \pm\frac{\pi}{2}}\right) = \begin{cases} -\frac{1}{2}\frac{d\Gamma}{dx}, & \text{if } \tilde{y} > 0, \\ \frac{1}{2}\frac{d\Gamma}{dx}, & \text{if } \tilde{y} < 0, \end{cases} \quad (7.105)$$

$$v = \lim_{L\to\infty} \frac{\frac{d\Gamma}{dx}}{4\pi}\ln\frac{(L-\tilde{x})^2 + \tilde{y}^2}{(L+\tilde{x})^2 + \tilde{y}^2} = 0. \quad (7.106)$$

So the vortex sheet generates no y component of velocity anywhere in the flow field and two uniform x components of velocity of opposite sign above and below the x axis. We can consider this flow to be an *inviscid shear layer*. We shall study viscous shear layers in Section 11.2.4, and the stability of inviscid shear layers in Section 12.2.

7.7.4 Potential of an Ideal Irrotational Vortex

Let us calculate the velocity potential function ϕ associated with a single ideal irrotational vortex. Consider an ideal irrotational vortex centered at the origin, and represent the velocity field here in cylindrical coordinates:

$$v_r = 0, \qquad v_\theta = \frac{\Gamma_o}{2\pi r}, \qquad v_z = 0. \quad (7.107)$$

We have considered the same velocity field in Section 7.4 and, as represented in Cartesian coordinates, in Section 3.12.9. Now in cylindrical coordinates the gradient operating on a scalar function gives

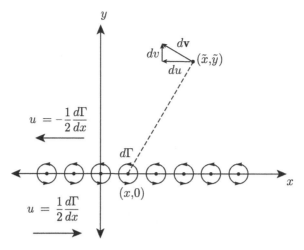

Figure 7.14 Diagram showing schematic of vortex sheet.

$$\nabla \phi = \mathbf{v}, \qquad (7.108)$$

$$\frac{\partial \phi}{\partial r}\mathbf{e}_r + \frac{1}{r}\frac{\partial \phi}{\partial \theta}\mathbf{e}_\theta + \frac{\partial \phi}{\partial z}\mathbf{e}_z = 0\mathbf{e}_r + \frac{\Gamma_o}{2\pi r}\mathbf{e}_\theta + 0\mathbf{e}_z, \qquad (7.109)$$

$$\frac{\partial \phi}{\partial r} = 0, \qquad \frac{1}{r}\frac{\partial \phi}{\partial \theta} = \frac{\Gamma_o}{2\pi r}, \qquad \frac{\partial \phi}{\partial z} = 0. \qquad (7.110)$$

Integrating and setting the arbitrary constant to zero, we get

$$\phi = \frac{\Gamma_o}{2\pi}\theta. \qquad (7.111)$$

In Cartesian coordinates, we have

$$\phi = \frac{\Gamma_o}{2\pi}\tan^{-1}\left(\frac{y}{x}\right). \qquad (7.112)$$

There are some additional nuances to the function $\tan^{-1}$ that will be considered in Section 8.2; for our purposes here, this standard usage will suffice. Lines of constant potential for the ideal vortex centered at the origin along with the velocity vector field are depicted in Fig. 7.15.

7.7.5 Interaction of Multiple Vortices

Here we will consider the interactions of a large number of vortices by using the method of superposition for the velocity potentials. If we have two vortices with strengths Γ_1 and Γ_2 centered at arbitrary locations (x_1, y_1) and (x_2, y_2), as depicted in Fig. 7.16, the potential for each is given by

$$\phi_1 = \frac{\Gamma_1}{2\pi}\tan^{-1}\left(\frac{y-y_1}{x-x_1}\right), \qquad \phi_2 = \frac{\Gamma_2}{2\pi}\tan^{-1}\left(\frac{y-y_2}{x-x_2}\right). \qquad (7.113)$$

Because the equation governing the velocity potential, $\nabla^2\phi = 0$, is linear, we can add the two potentials and still satisfy the overall equation so that

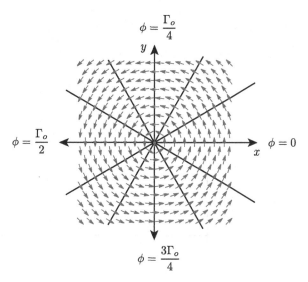

Figure 7.15 Lines of constant potential and velocity vector field for an ideal irrotational vortex.

$$\phi = \frac{\Gamma_1}{2\pi}\tan^{-1}\left(\frac{y-y_1}{x-x_1}\right) + \frac{\Gamma_2}{2\pi}\tan^{-1}\left(\frac{y-y_2}{x-x_2}\right), \quad (7.114)$$

is a solution. Taking the gradient of ϕ, we find

$$\nabla\phi = \left(-\left(\frac{\Gamma_1}{2\pi}\right)\frac{y-y_1}{(x-x_1)^2+(y-y_1)^2} - \left(\frac{\Gamma_2}{2\pi}\right)\frac{y-y_2}{(x-x_2)^2+(y-y_2)^2}\right)\mathbf{i}$$
$$+ \left(\left(\frac{\Gamma_1}{2\pi}\right)\frac{x-x_1}{(x-x_1)^2+(y-y_1)^2} + \left(\frac{\Gamma_2}{2\pi}\right)\frac{x-x_2}{(x-x_2)^2+(y-y_2)^2}\right)\mathbf{j}, \quad (7.115)$$

so that

$$u(x,y) = -\left(\frac{\Gamma_1}{2\pi}\right)\frac{y-y_1}{(x-x_1)^2+(y-y_1)^2} - \left(\frac{\Gamma_2}{2\pi}\right)\frac{y-y_2}{(x-x_2)^2+(y-y_2)^2}, \quad (7.116)$$

$$v(x,y) = \left(\frac{\Gamma_1}{2\pi}\right)\frac{x-x_1}{(x-x_1)^2+(y-y_1)^2} + \left(\frac{\Gamma_2}{2\pi}\right)\frac{x-x_2}{(x-x_2)^2+(y-y_2)^2}. \quad (7.117)$$

Extending this to a collection of N vortices located at (x_i, y_i) at a given time, we have the following for the velocity field:

$$u(x,y) = -\sum_{i=1}^{N}\left(\frac{\Gamma_i}{2\pi}\right)\frac{y-y_i}{(x-x_i)^2+(y-y_i)^2}, \quad (7.118)$$

$$v(x,y) = \sum_{i=1}^{N}\left(\frac{\Gamma_i}{2\pi}\right)\frac{x-x_i}{(x-x_i)^2+(y-y_i)^2}. \quad (7.119)$$

Now to advect (that is, to move) the kth vortex, we move it with the velocity induced by the other vortices, because vortices advect with the flow. Recalling that the velocity is the time derivative of the position $u_k = dx_k/dt, v_k = dy_k/dt$, we then get the following $2N$ nonlinear

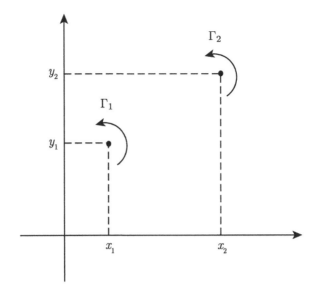

Figure 7.16 Two vortices at arbitrary locations.

ordinary differential equations for the $2N$ unknowns, the x and y positions of each of the N vortices:

$$\frac{dx_k}{dt} = \sum_{i=1, i \neq k}^{N} -\left(\frac{\Gamma_i}{2\pi}\right) \frac{y_k - y_i}{(x_k - x_i)^2 + (y_k - y_i)^2}, \quad x_k(0) = x_k^o, \quad k = 1, \ldots, N, \quad (7.120)$$

$$\frac{dy_k}{dt} = \sum_{i=1, i \neq k}^{N} \left(\frac{\Gamma_i}{2\pi}\right) \frac{x_k - x_i}{(x_k - x_i)^2 + (y_k - y_i)^2}, \quad y_k(0) = y_k^o, \quad k = 1, \ldots, N. \quad (7.121)$$

This set of equations, except for three or fewer point vortices, must be integrated numerically. As has already been demonstrated, the integration for two vortices is not difficult; the integration is not as straightforward for three vortices, as discussed by Aref (1979). These ordinary differential equations are highly nonlinear and typically give rise to chaotic motion of the point vortices. It is a similar calculation to the motion of point masses in a Newtonian gravitational field, except that the essential variation goes as $1/r$ for vortices and $1/r^2$ for Newtonian gravitational fields. Thus, the dynamics are different. Nevertheless just as calculations for large numbers of celestial bodies can give rise to solar systems, clusters of planets, and galaxies, similar "galaxies" of vortices can be predicted with the equations for vortex dynamics.

Example 7.2 Consider three ideal point vortices, each of strength $\Gamma = 1$, in a two-dimensional, incompressible, irrotational flow field. The vortices are initially at $(-1, -1)^T$, $(0, 0)^T$, and $(1, 1)^T$. Find the trajectory of the vortex initially at the origin for $t = [0, 1000]$.

Solution

The equations are

$$\frac{dx_1}{dt} = -\frac{y_1 - y_2}{2\pi\left((x_1-x_2)^2 + (y_1-y_2)^2\right)} - \frac{y_1 - y_3}{2\pi\left((x_1-x_3)^2 + (y_1-y_3)^2\right)}, \tag{7.122}$$

$$\frac{dx_2}{dt} = -\frac{y_2 - y_1}{2\pi\left((x_2-x_1)^2 + (y_2-y_1)^2\right)} - \frac{y_2 - y_3}{2\pi\left((x_2-x_3)^2 + (y_2-y_3)^2\right)}, \tag{7.123}$$

$$\frac{dx_3}{dt} = -\frac{y_3 - y_1}{2\pi\left((x_3-x_1)^2 + (y_3-y_1)^2\right)} - \frac{y_3 - y_2}{2\pi\left((x_3-x_2)^2 + (y_3-y_2)^2\right)}, \tag{7.124}$$

$$\frac{dy_1}{dt} = \frac{x_1 - x_2}{2\pi\left((x_1-x_2)^2 + (y_1-y_2)^2\right)} + \frac{x_1 - x_3}{2\pi\left((x_1-x_3)^2 + (y_1-y_3)^2\right)}, \tag{7.125}$$

$$\frac{dy_2}{dt} = \frac{x_2 - x_1}{2\pi\left((x_2-x_1)^2 + (y_2-y_1)^2\right)} + \frac{x_2 - x_3}{2\pi\left((x_2-x_3)^2 + (y_2-y_3)^2\right)}, \tag{7.126}$$

$$\frac{dy_3}{dt} = \frac{x_3 - x_1}{2\pi\left((x_3-x_1)^2 + (y_3-y_1)^2\right)} + \frac{x_3 - x_2}{2\pi\left((x_3-x_2)^2 + (y_3-y_2)^2\right)}, \tag{7.127}$$

$$x_1(0) = -1, \quad x_2(0) = 0, \quad x_3(0) = 1, \quad y_1(0) = -1, \quad y_2(0) = 0, \quad y_3(0) = 1. \tag{7.128}$$

Numerical solution is required. We do so, and plot the trajectory of the vortex that originated at the origin in Fig. 7.17.

7.7.6 Pressure Field

We have thus far focused on the kinematics of vortices. We have actually used dynamics in our incorporation of the Helmholtz vorticity transport equation and Kelvin's theorem, but their

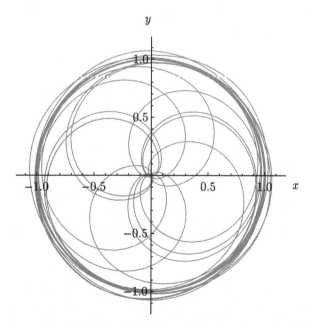

Figure 7.17 Trajectory for $t = [0, 1000]$ of an ideal point vortex initially at $(x,y)^T = (0,0)^T$ that evolves due to the presence of three equal strength ideal point vortices.

simple results really only justify the use of simple kinematics. Dynamics asks what are the forces that give rise to the motion. Here, we will assume there is no body force, that the fluid is inviscid, in which case it must be pressure forces that give rise to the motion, and that the fluid is at rest at infinity. We then have the proper conditions for which the Bernoulli equation can be used to give the pressure field. We consider two cases, a single stationary point vortex, and a group of N moving point vortices.

Single Stationary Vortex

If we take $p = p_\infty$ in the far field and $f_i = \mathbf{f} = \mathbf{0}$, this steady flow gives us

$$\frac{1}{2}\mathbf{v}^T \cdot \mathbf{v} + \frac{p}{\rho} = \frac{1}{2}\mathbf{v}_\infty^T \cdot \mathbf{v}_\infty + \frac{p_\infty}{\rho}, \tag{7.129}$$

$$\frac{1}{2}\left(\frac{\Gamma_o}{2\pi r}\right)^2 + \frac{p}{\rho} = 0 + \frac{p_\infty}{\rho}, \tag{7.130}$$

$$p(r) = p_\infty - \frac{\rho \Gamma_o^2}{8\pi^2}\frac{1}{r^2}. \tag{7.131}$$

The pressure goes to negative infinity at the origin. This is obviously unphysical. As will be shown in Section 10.2.6, it can be corrected by including viscous effects, which turn out not to substantially alter our main conclusions. It can be verified by direct substitution into the irrotational limit of Eq. (6.73) that

$$\nabla^2 p = -\rho \mathbf{D} : \mathbf{D} = -\frac{\rho \Gamma_o}{2\pi^2 r^4}. \tag{7.132}$$

Group of N Vortices

For a collection of N vortices, the flow is certainly not steady, and we must in general retain the time-dependent velocity potential in the Bernoulli equation, Eq. (6.186), yielding

$$\frac{\partial \phi}{\partial t} + \frac{1}{2}(\nabla \phi)^T \cdot \nabla \phi + \frac{p}{\rho} = f(t). \tag{7.133}$$

Now we require that as $r \to \infty$ that $p \to p_\infty$. We also know that as $r \to \infty$ that $\phi \to 0$, hence $\nabla \phi \to 0$ as well. Hence as $r \to \infty$, we have $p_\infty/\rho = f(t)$. So our final result is

$$p = p_\infty - \frac{1}{2}\rho(\nabla \phi)^T \cdot \nabla \phi - \frac{\partial \phi}{\partial t}. \tag{7.134}$$

So with a knowledge of the velocity field through ϕ, we can determine the pressure field that must have given rise to that velocity field.

Example 7.3 Consider the flow field generated by two ideal point vortices in an incompressible, irrotational flow field. We take $\rho = 1$, and each vortex has $\Gamma = 1$. One is located at $(x,y) = (-1,-1)^T$; the other is at $(1,1)$. Both are held stationary. Plot the potential field, the velocity vector field and the pressure field. Take $p_\infty = 0$.

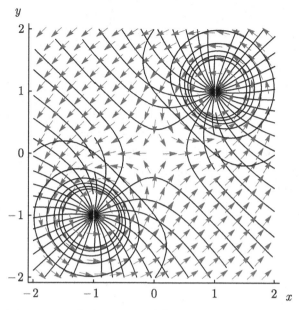

Figure 7.18 Equipotential curves, velocity vector field, and pressure field for two stationary point vortices in an incompressible, irrotational flow field.

Solution

The potential field is

$$\phi = \frac{1}{2\pi}\left(\tan^{-1}\frac{y+1}{x+1} + \tan^{-1}\frac{y-1}{x-1}\right). \tag{7.135}$$

The velocity vector field is

$$u(x,y) = \frac{\partial \phi}{\partial x} = -\frac{x^2 y - 2x + y^3}{\pi\left(x^4 + 2x^2 y^2 - 8xy + y^4 + 4\right)}, \tag{7.136}$$

$$v(x,y) = \frac{\partial \phi}{\partial y} = \frac{x^3 + xy^2 - 2y}{\pi\left(x^4 + 2x^2 y^2 - 8xy + y^4 + 4\right)}. \tag{7.137}$$

Because the vortices are held stationary, $\partial \phi/\partial t = 0$. The pressure field for $\rho = 1$ is

$$p(x,y) = -\frac{1}{2}(\nabla \phi)^T \cdot \nabla \phi = -\frac{x^2 + y^2}{2\pi^2\left(x^4 + 2x^2 y^2 - 8xy + y^4 + 4\right)}. \tag{7.138}$$

All is plotted in Fig. 7.18.

SUMMARY

This chapter considered aspects of rotating fluids, with a focus on those that are incompressible. The key property to characterize rotation is vorticity, and the linear momenta principle was employed to describe how and why vorticity evolves. The special case of an ideal point vortex was studied, and two-dimensional flows composed of collections of point vortices were considered in detail. For such flows, it is possible to associate the kinematics with a pressure field, thus identifying the dynamics that induce the kinematics.

PROBLEMS

7.1 A two-dimensional, incompressible, inviscid, irrotational flow field exists with an ideal point vortex of strength $\Gamma = -1$ held stationary at $(x,y)^T = (1,0)^T$. There is a vertical wall at $x = 0$. Introduce an image vortex; find the potential $\phi(x,y)$ and velocity vector fields $u(x,y)$, $v(x,y)$. Plot all on the domain $x \in [-2,2]$, $y \in [-2,2]$. Show that the wall is a slip surface with zero penetration.

7.2 Show for an incompressible, Newtonian fluid with constant viscosity that

$$(\nabla^T \cdot \boldsymbol{\tau})^T = -\mu \nabla \times \boldsymbol{\omega}.$$

7.3 Define the *enstrophy* as $\boldsymbol{\omega}^T \cdot \boldsymbol{\omega}/2$, a scalar that is half the square of the magnitude of the vorticity vector. For an incompressible fluid with constant viscosity and conservative body forces, take the dot product of $\boldsymbol{\omega}$ with the appropriate Helmholtz vorticity transport equation, $d\boldsymbol{\omega}/dt = (\boldsymbol{\omega}^T \cdot \nabla)\mathbf{v} + \nu \nabla^2 \boldsymbol{\omega}$, and show how to obtain the evolution equation for enstrophy:

$$\frac{d}{dt}\frac{\boldsymbol{\omega}^T \cdot \boldsymbol{\omega}}{2} = \boldsymbol{\omega}\boldsymbol{\omega}^T : \mathbf{D} + \nu \nabla^2 \frac{\boldsymbol{\omega}^T \cdot \boldsymbol{\omega}}{2} - \nu \nabla \boldsymbol{\omega} : (\nabla \boldsymbol{\omega})^T.$$

7.4 Define the *helicity* as $\mathbf{v}^T \cdot \boldsymbol{\omega}$, a scalar field related to the alignment of the velocity and vorticity vector fields. For an incompressible flow field with a conservative body force, combine the linear momenta and Helmholtz vorticity transport equations to show how to arrive at a helicity transport equation:

$$\frac{d}{dt}(\mathbf{v}^T \cdot \boldsymbol{\omega}) = \boldsymbol{\omega}^T \cdot \nabla \left(\frac{\mathbf{v}^T \cdot \mathbf{v}}{2} - \varphi - \frac{p}{\rho}\right) + \nu \nabla^2 (\mathbf{v}^T \cdot \boldsymbol{\omega}) - 2\nu \nabla \boldsymbol{\omega} : \mathbf{L}^T.$$

Helicity is better considered to be a pseudo-scalar because of its dependency on the vorticity pseudo-vector. It will suffer a sign reversal under a reflection transformation followed by an inverse of the same reflection transformation.

7.5 Consider a dimensionless two-dimensional, incompressible flow with conservative body forces with deviatoric stress tensor

$$\boldsymbol{\tau} = \begin{pmatrix} xy^2 & x^2 - 2y \\ -x^2 - y & -xy^2 \end{pmatrix}.$$

The anti-symmetric portion of the tensor is induced by a body couple, and the symmetric portion is driven by viscous effects. Show both effects induce fluid rotation.

7.6 Consider the inviscid flow field with zero body force given by

$$u = -\frac{x^2 y}{2} + x + y + z, \quad v = x^2 - y^2, \quad w = xyz + 2yz - z.$$

Show the flow is incompressible. Find $d\boldsymbol{\omega}/dt$.

7.7 A two-dimensional, incompressible, inviscid, irrotational flow field with zero body force exists with four point vortices, each held stationary. The first is at $(x,y)^T = (-1,-1)^T$ and has $\Gamma_1 = 1$. The second is at $(x,y)^T = (1,1)^T$ and has $\Gamma_2 = -1$. The third is at $(x,y)^T = (-1,1)^T$ and has $\Gamma_3 = 1$. The fourth is at $(x,y)^T = (1,-1)^T$ and has $\Gamma_4 = -1$. Find the potential $\phi(x,y)$, pressure $p(x,y)$, and velocity vector fields $u(x,y)$, $v(x,y)$, and plot all on the same plot. Take $\rho = 1$ and $p_\infty = 0$.

7.8 Consider a two-dimensional, incompressible, irrotational flow field seeded with a collection of N ideal point vortices, each with strength $\Gamma_i = 1$, $i = 1,\ldots,N$. At $t = 0$, the vortices are distributed along the x axis in the domain $x \in [-1,1]$ with equal distance between them. Numerically solve for the time-dependent trajectories. Plot the trajectory of the vortex initially at $(-1,0)^T$ for $N = 2, 3, 4, 10$, for $t = [0,200]$.

7.9 Consider a two-dimensional, incompressible, irrotational flow field seeded with a collection of $N = 51$ ideal point vortices, each with strength $\Gamma_i = 1$, $i = 1,\ldots,N$. At $t = 0$, the vortices are distributed along the x axis in the domain $x \in [-1,1]$ with equal distance between them. Numerically solve for the time-dependent trajectories. Plot the location of each vortex in a series of snapshots in time for $t = [0,10]$, those being $t = 0, 0.01, 0.1, 1, 10$. Make an animation of the vortex motion with software tools. Describe in words the behavior of the group of vortices.

7.10 Consider the flow field generated by two ideal point vortices in an incompressible, irrotational flow field with no body force. Take $\rho = 1$, and take each vortex to have $\Gamma = 1$. At $t = 0$, one is located at $(x,y)^T = (-1,0)^T$; the other is at $(1,0)^T$. Both are free to move under the influence of the other. At $t = 0$, plot the potential field, the velocity vector field, and the pressure field. Take $p_\infty = 0$. Compare the pressure field to that which is found if the vortices are held in place.

7.11 Show the Helmholtz vorticity transport equation for an incompressible, isotropic, Newtonian fluid with constant viscosity subject to a nonconservative body force is

$$\frac{\partial \boldsymbol{\omega}}{\partial t} + \nabla \times (\boldsymbol{\omega} \times \mathbf{v}) = \nabla \times \mathbf{f} + \nu \nabla^2 \boldsymbol{\omega}.$$

7.12 A Trkalian[3] flow field is one whose velocity vector is everywhere a constant scalar multiple of its vorticity vector; $\boldsymbol{\omega} = \alpha \mathbf{v}$, with α as a constant. For an incompressible flow of a Newtonian fluid with constant viscosity and conservative body force, show for a Trkalian flow that $\partial \boldsymbol{\omega}/\partial t = \nu \nabla^2 \boldsymbol{\omega}$. Show for a Trkalian flow that the pressure field is governed by $\nabla p = -\rho \mathbf{v}^T \cdot \nabla \mathbf{v}$.

7.13 Show that the velocity field

$$\mathbf{v} = e^{-\alpha^2 \nu t} \left(-\frac{\cos\left(\frac{\alpha x}{\sqrt{2}}\right) \sin\left(\frac{\alpha y}{\sqrt{2}}\right)}{\sqrt{2}}, \frac{\sin\left(\frac{\alpha x}{\sqrt{2}}\right) \cos\left(\frac{\alpha y}{\sqrt{2}}\right)}{\sqrt{2}}, \cos\left(\frac{\alpha x}{\sqrt{2}}\right) \cos\left(\frac{\alpha y}{\sqrt{2}}\right) \right)^T,$$

is Trkalian and that it satisfies the appropriate incompressibility condition and Helmholtz vorticity transport equation. Find an expression for the helicity field, $\mathbf{v}^T \cdot \boldsymbol{\omega}$. Find an

---

[3] Viktor Trkal 1888–1956, Czech physicist.

expression for the pressure field. For $\nu = 0$, $\alpha = 1$, determine numerically and plot the particle pathline of the particle that is at $(1, 1, 1)^T$ at $t = 0$.

7.14 As introduced in the discussion around Eq. (2.237), a Beltrami flow field is one whose velocity vector is everywhere a scalar multiple of its vorticity vector; $\boldsymbol{\omega} = \alpha \mathbf{v}$, with α as a scalar field variable; thus, $\boldsymbol{\omega} \times \mathbf{v} = \mathbf{0}$. A generalized Beltrami flow has the looser requirement that $\nabla \times (\boldsymbol{\omega} \times \mathbf{v}) = \mathbf{0}$. For an incompressible flow of a Newtonian fluid with constant viscosity and conservative body force, show for both Beltrami and generalized Beltrami flows that $\partial \boldsymbol{\omega} / \partial t = \nu \nabla^2 \boldsymbol{\omega}$. See Wang (1991) for background on Beltrami fields.

7.15 Show the velocity vector field $\mathbf{v} = (4y, -2x, 0)^T$ is a generalized Beltrami field. Show the flow field is incompressible. Plot the velocity vector field to show it can be considered to be an elliptic vortex.

7.16 Show the velocity vector field $\mathbf{v} = (1 + 2y + 2y/(x^2+y^2), -2x/(x^2+y^2), 0)^T$ is a generalized Beltrami field. Show the flow field is incompressible. Plot the velocity vector field to show it can be considered to be a vortex embedded within a shear flow.

7.17 Show for an incompressible, Newtonian fluid with constant viscosity and conservative body force whose velocity field is a generalized Beltrami field that

$$\nabla^2 \frac{\partial \mathbf{v}}{\partial t} = \nu \nabla^4 \mathbf{v}.$$

7.18 Consider an incompressible flow of a Newtonian fluid with constant viscosity and nonconservative body force with $\rho = \mu = \nu = 1$. For such a flow, the Helmholtz vorticity transport and linear momenta equations are

$$\frac{\partial \boldsymbol{\omega}}{\partial t} + (\mathbf{v}^T \cdot \nabla) \boldsymbol{\omega} = (\boldsymbol{\omega}^T \cdot \nabla) \mathbf{v} + \nabla^2 \boldsymbol{\omega} + \nabla \times \mathbf{f}, \quad \frac{\partial \mathbf{v}}{\partial t} + (\mathbf{v}^T \cdot \nabla) \mathbf{v} = -\nabla p + \mathbf{f} + \nabla^2 \mathbf{v}.$$

Consider next the steady velocity field

$$\mathbf{v} = \left(xyz, xy, -xz - \frac{yz^2}{2}\right)^T.$$

Find any $\mathbf{f}$ and p such that the steady linear momenta and Helmholtz vorticity transport equations are satisfied.

FURTHER READING

Batchelor, G. K. (2000). *An Introduction to Fluid Dynamics*. Cambridge, UK: Cambridge University Press.

Chorin, A. J. (1993). *Vorticity and Turbulence*. New York: Springer.

Chorin, A. J., and Marsden, J. E. (2000). *A Mathematical Introduction to Fluid Mechanics*, 3rd ed. New York: Springer.

Cottet, G.-H., and Koumoutsakos, P. D. (2000). *Vortex Methods: Theory and Practice*, 2nd ed. Cambridge, UK: Cambridge University Press.

Greenspan, H. P. (1968). *The Theory of Rotating Fluids*. Cambridge, UK: Cambridge University Press.

Kuethe, A. M., and Chow, C.-Y. (1998). *Foundations of Aerodynamics: Bases of Aerodynamic Design*, 5th ed. New York: John Wiley.

Majda, A. J., and Bertozzi, A. L. (2002). *Vorticity and Incompressible Flow*. Cambridge, UK: Cambridge University Press.

Newton, P. K. (2001). *The N-Vortex Problem: Analytical Techniques*. New York: Springer.
Saffman, P. G. (1992). *Vortex Dynamics*. Cambridge, UK: Cambridge University Press.
Ting, L., and Klein, R. (1991). *Viscous Vortical Flows*. Berlin: Springer.
Ting, L., Klein, R. and Knio, O. M. (2007). *Vortex Dominated Flows: Analysis and Computation for Multiple Scale Phenomena*. Berlin: Springer.
Tritton, D. J. (1988). *Physical Fluid Dynamics*, 2nd ed. Oxford: Oxford University Press.
Truesdell, C. (2018). *The Kinematics of Vorticity*. Mineola, New York: Dover.
Vanyo, J. P. (1993). *Rotating Fluids in Engineering and Science*. Boston, Massachusetts: Butterworth-Heinemann.
Voropayev, S. I., and Afanasyev, Y. D. (1994). *Vortex Structures in a Stratified Fluid: Order from Chaos*. London: Chapman and Hall.
Wu, J.-Z., Ma, H.-Y., and Zhou, M.-D. (2006). *Vorticity and Vortex Dynamics*. Berlin: Springer.
Wu, J.-Z., Ma, H.-Y., and Zhou, M.-D. (2015). *Vortical Flows*. Berlin: Springer.

REFERENCES

Aref, H. (1979). Motion of three vortices, *Physics of Fluids*, **22**(3), 393–400.
Cummings, R. M., Mason, W. H., Morton, S. A., and McDaniel, D. R. (2015). *Applied Computational Aerodynamics: A Modern Engineering Approach*. New York: Cambridge University Press.
Doering, C. R. (2009). The 3D Navier–Stokes problem, *Annual Review of Fluid Mechanics*, **41**, 109–128.
Karamcheti, K. (1980). *Principles of Ideal-Fluid Aerodynamics*, 2nd ed. Malabar, Florida: Krieger.
Kundu, P. K., Cohen, I. M., and Dowling, D. R. (2015). *Fluid Mechanics*, 6th ed. Amsterdam: Academic Press.
Panton, R. L. (2013). *Incompressible Flow*, 4th ed. New York: John Wiley.
Thorne, K. S., and Blandford, R. D. (2017). *Modern Classical Physics: Optics, Fluids, Plasmas, Elasticity, and Statistical Physics*. Princeton, New Jersey: Princeton University Press.
Wang, C. Y. (1991). Exact solutions of the steady-state Navier–Stokes equations, *Annual Review of Fluid Mechanics*, **23**, 159–177.

8 Potential Flow

This chapter will expand upon potential flow, introduced in Section 7.7, and will mainly be restricted to steady, two-dimensional planar, incompressible potential flow. Such flows can be characterized by a scalar potential field. An example of such a field along with associated streamlines is given in Fig. 8.1. This particular potential field can be shown to model the steady two-dimensional flow of an incompressible inviscid fluid over an elliptic body. Knowledge of this scalar field is sufficient to deduce all flow variables. As is typical for such fields, gradients of the potential induce flow. A beautiful mathematical theory was developed for potential flows in the nineteenth century and will be described in this chapter. Additionally, potential flow solutions can be applied in disparate fields, because the equations governing potential flow of a fluid are identical in form to those governing some forms of energy and mass diffusion, as well as electro-magnetics.

Despite its beauty, in some ways it is impractical for many engineering applications, though not all. As the theory necessarily ignores all vorticity generating mechanisms, it must ignore viscous effects. Consequently, the theory is incapable of predicting drag forces on solid bodies. As a result, those who needed to know the drag resorted in the nineteenth century to more empirically based methods. In the early twentieth century, Prandtl took steps to reconcile the practical viscous world of engineering with the more mathematical world of potential flow with his viscous boundary layer theory. He showed that indeed potential flow solutions could be of value away from no-slip walls, and provided a remedy to fix the solutions in the neighborhood of the wall. In so doing, he opened a new field of applied mathematics known as matched asymptotic analysis. We will consider such linkages to potential flow and viscous flow in the upcoming Chapter 11. To make such a connection, it is first necessary to understand potential

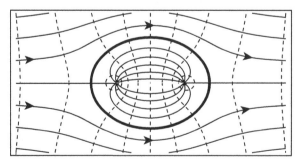

Figure 8.1 Equipotential lines (dotted) and streamlines (solid) in a potential field representing steady two-dimensional flow of an inviscid fluid over an ellipse.

flow. These and other reasons to study potential flow are summarized in the following points: (1) Portions of many real flow fields are captured well by potential theory, and those that are not can often be remedied by application of a viscous boundary layer theory. (2) Low speed aerodynamics are often well described by potential flow theory. (3) Study of potential flow solutions can give insight into fluid behavior and aid in the honing of a more precise intuition. (4) Potential flow solutions are useful as test cases for verification of numerical methods. (5) There is pedagogical and historical value in understanding potential flow.

We will focus on fundamental potential flows about simple geometries. It is possible to address potential flows about more complicated geometries using the method of *conformal mapping*, not covered here. In this method, one transforms a complicated geometry to a topologically equivalent simple geometry, solves in the simple domain, and then transforms back to the original coordinate system. This powerful method is explained well in such sources as Brown and Churchill (2014) or Panton (2013, p. 464).

8.1 Stream Functions and Velocity Potentials

We first consider *stream functions* and velocity potentials. We have seen velocity potentials before in Section 6.8.1 in the study of the Bernoulli equation and in Section 7.7 in study of ideal vortices. In this chapter, we will adopt the same assumption of irrotationality, and further require that the flow be two-dimensional planar. Recall if a flow velocity is confined to the (x, y) plane, then the vorticity vector is confined to the z direction and takes the form first shown in Eq. (3.220):

$$\boldsymbol{\omega} = \begin{pmatrix} 0 \\ 0 \\ \frac{\partial v}{\partial x} - \frac{\partial u}{\partial y} \end{pmatrix}. \tag{8.1}$$

So if the flow is two-dimensional planar and irrotational, we have

$$\frac{\partial v}{\partial x} - \frac{\partial u}{\partial y} = 0. \tag{8.2}$$

Moreover, because of irrotationality, we can express the velocity vector $\mathbf{v}$ as the gradient of a potential ϕ, the velocity potential, as earlier shown in Eq. (6.182):

$$\mathbf{v} = \nabla \phi. \tag{8.3}$$

With this definition, fluid flows from regions of low velocity potential to regions of high velocity potential. The scalar velocity components are

$$u = \frac{\partial \phi}{\partial x}, \qquad v = \frac{\partial \phi}{\partial y}. \tag{8.4}$$

Here we have adopted the common sign convention from aeronautics. In some sources, one finds the sign convention for ϕ inverted, for example Bird et al. (2007, p. 127). We see by substitution into the irrotationality condition, Eq. (8.2), that this is true identically:

$$\frac{\partial v}{\partial x} - \frac{\partial u}{\partial y} = \frac{\partial}{\partial x}\left(\frac{\partial \phi}{\partial y}\right) - \frac{\partial}{\partial y}\left(\frac{\partial \phi}{\partial x}\right) = 0. \tag{8.5}$$

Now for two-dimensional incompressible flows, we have by specializing Eq. (6.58) that

$$\frac{\partial u}{\partial x} + \frac{\partial v}{\partial y} = 0. \tag{8.6}$$

Substituting for u and v in favor of ϕ, we get Laplace's equation for ϕ, seen earlier in Eqs. (2.217, 7.98):

$$\frac{\partial}{\partial x}\left(\frac{\partial \phi}{\partial x}\right) + \frac{\partial}{\partial y}\left(\frac{\partial \phi}{\partial y}\right) = \nabla^2 \phi = 0. \tag{8.7}$$

Solutions $\phi(x,y)$ are harmonic functions. Now if the flow is also incompressible, we can define the stream function ψ as follows:

$$u = \frac{\partial \psi}{\partial y}, \qquad v = -\frac{\partial \psi}{\partial x}. \tag{8.8}$$

Here we have again adopted the common sign convention from aeronautics. In some disciplines – for example oceanography, meteorology, or chemical engineering – one finds the sign convention for ψ inverted, for example Bird et al. (2007, p. 127). Direct substitution of our stream function into the incompressible mass conservation equation, Eq. (8.6), yields an identity:

$$\frac{\partial u}{\partial x} + \frac{\partial v}{\partial y} = \frac{\partial}{\partial x}\left(\frac{\partial \psi}{\partial y}\right) + \frac{\partial}{\partial y}\left(-\frac{\partial \psi}{\partial x}\right) = 0. \tag{8.9}$$

Now, forming a pair of equations that will be critically important in the development of potential flow theory, we can set our definitions of u and v in terms of ϕ and ψ, Eqs. (8.4, 8.8), equal to each other, as they must be:

$$\underbrace{\frac{\partial \phi}{\partial x}}_{u} = \underbrace{\frac{\partial \psi}{\partial y}}_{u}, \qquad \underbrace{\frac{\partial \phi}{\partial y}}_{v} = \underbrace{-\frac{\partial \psi}{\partial x}}_{v}. \tag{8.10}$$

If we differentiate the first with respect to y and the second with respect to x, we see

$$\frac{\partial^2 \phi}{\partial y \partial x} = \frac{\partial^2 \psi}{\partial y^2}, \qquad \frac{\partial^2 \phi}{\partial x \partial y} = -\frac{\partial^2 \psi}{\partial x^2}. \tag{8.11}$$

Now subtract the second from the first to get Laplace's equation for ψ:

$$0 = \frac{\partial^2 \psi}{\partial y^2} + \frac{\partial^2 \psi}{\partial x^2} = \nabla^2 \psi. \tag{8.12}$$

Solutions $\psi(x,y)$ are harmonic functions.

Let us now examine lines of constant ϕ (equipotential lines) and lines of constant ψ (that we will see are streamlines). So take $\phi = C_1$, $\psi = C_2$. Because $\phi = \phi(x,y)$, we can take the total differential on a curve of constant ϕ and get

$$d\phi = \frac{\partial \phi}{\partial x}\,dx + \frac{\partial \phi}{\partial y}\,dy = u\,dx + v\,dy = 0, \quad \text{so} \quad \left.\frac{dy}{dx}\right|_{\phi=C_1} = -\frac{u}{v}. \tag{8.13}$$

Now for $\psi = \psi(x, y)$ we similarly get on curves of constant ψ that

$$d\psi = \frac{\partial \psi}{\partial x} dx + \frac{\partial \psi}{\partial y} dy = -v\, dx + u\, dy = 0, \quad \text{so} \quad \left.\frac{dy}{dx}\right|_{\psi=C_2} = \frac{v}{u}. \tag{8.14}$$

One notes

$$\left.\frac{dy}{dx}\right|_{\phi=C_1} = -1 \Big/ \left.\frac{dy}{dx}\right|_{\psi=C_2}; \tag{8.15}$$

hence, curves of constant ϕ are orthogonal to curves of constant ψ. We also note that on $\psi = C_2$, we have $dx/u = dy/v$; hence, by Eq. (3.51), lines of $\psi = C_2$ must be streamlines. As an aside, the definition of the stream function $u = \partial \psi / \partial y, v = -\partial \psi / \partial x$, can be rewritten as

$$\frac{dx}{dt} = \frac{\partial \psi}{\partial y}, \quad \frac{dy}{dt} = -\frac{\partial \psi}{\partial x}. \tag{8.16}$$

This is a common form from classical dynamics in which we can interpret ψ as the *Hamiltonian*[1] of the system. We shall not pursue this path, but note that a significant literature exists for Hamiltonian systems; see Powers and Sen (2015, Section 9.6.6), for an introduction, or Goldstein (1950, Chapter 7), for a detailed discussion.

Now the study of ϕ and ψ is essentially kinematics. The only incursion of dynamics is that we must have irrotational flow. Recalling the Helmholtz vorticity transport equation, Eq. (7.71), we realize that we can only have potential flow when the vorticity - generating mechanisms (three-dimensional effects, nonconservative body forces, baroclinic effects, and viscous effects) are suppressed. In that case, the dynamics, which is the driving force for the fluid motion, can be understood in the context of the unsteady Bernoulli equation, Eq. (6.186), taken for incompressible flow and negligible body force, in which limit, Eq. (6.173) reduces to $\mathsf{P} = p/\rho$:

$$\frac{\partial \phi}{\partial t} + \frac{1}{2} \nabla^T \phi \cdot \nabla \phi + \frac{p}{\rho} = f(t). \tag{8.17}$$

We do not have to require steady flow to have a potential flow field. It is also easy to correct for the presence of a conservative body force.

Solutions to the two key equations of two-dimensional planar potential flow $\nabla^2 \phi = 0, \nabla^2 \psi = 0$ are efficiently studied using methods involving complex variables. We will delay discussing solutions until we have presented the enabling mathematics.

8.2 Mathematics of Complex Variables

Here we briefly introduce relevant elements of complex variable theory. One can consult Brown and Churchill (2014) for a detailed presentation of the supporting mathematics. Recall that

---

[1] William Rowan Hamilton, 1805–1865, Dublin-based Anglo-Irish mathematician. Discovered the quaternion group, an extension of complex numbers, while walking along a canal and engraved it in the Broom Bridge. He invented the notion of the dot and cross products, coined the terms "scalar" and "tensor," and was the first to use "vector" in the modern sense. Introduced the ∇ operator in 1837, albeit in a rotated form, ◁, based on the harp.

the imaginary number i is defined such that

$$i^2 = -1, \qquad i = \sqrt{-1}. \tag{8.18}$$

8.2.1 Euler's Formula

We can get the useful *Euler's formula* by considering the following Taylor series expansions of common functions about $s = 0$:

$$e^s = 1 + s + \frac{1}{2!}s^2 + \frac{1}{3!}s^3 + \frac{1}{4!}s^4 + \frac{1}{5!}s^5 \cdots, \tag{8.19}$$

$$\sin s = 0 + s + 0\frac{1}{2!}s^2 - \frac{1}{3!}s^3 + 0\frac{1}{4!}s^4 + \frac{1}{5!}s^5 \cdots, \tag{8.20}$$

$$\cos s = 1 + 0s - \frac{1}{2!}s^2 + 0\frac{1}{3!}s^3 + \frac{1}{4!}s^4 + 0\frac{1}{5!}s^5 \cdots. \tag{8.21}$$

With these expansions, now consider the following combinations: $(\cos s + i \sin s)_{s=\theta}$ and $e^s|_{s=i\theta}$:

$$\cos\theta + i\sin\theta = 1 + i\theta - \frac{1}{2!}\theta^2 - i\frac{1}{3!}\theta^3 + \frac{1}{4!}\theta^4 + i\frac{1}{5!}\theta^5 + \cdots, \tag{8.22}$$

$$e^{i\theta} = 1 + i\theta + \frac{1}{2!}(i\theta)^2 + \frac{1}{3!}(i\theta)^3 + \frac{1}{4!}(i\theta)^4 + \frac{1}{5!}(i\theta)^5 + \cdots, \tag{8.23}$$

$$= 1 + i\theta - \frac{1}{2!}\theta^2 - i\frac{1}{3!}\theta^3 + \frac{1}{4!}\theta^4 + i\frac{1}{5!}\theta^5 + \cdots. \tag{8.24}$$

As the two series are identical, we have Euler's formula:

$$e^{i\theta} = \cos\theta + i\sin\theta. \tag{8.25}$$

8.2.2 Polar and Cartesian Representations

We take $x \in \mathbb{R}^1$, $y \in \mathbb{R}^1$ and define the complex number z to be

$$z = x + iy. \tag{8.26}$$

We say that $z \in \mathbb{C}^1$. We define the operator $\Re$ as selecting the real part of a complex number and $\Im$ as selecting the imaginary part of a complex number. The operator $\Re$ should not be conflated with the universal gas constant, for which we use the same symbol, Eq. (1.17). For Eq. (8.26), we see $\Re(z) = x$, $\Im(z) = y$. Both operators $\Re$ and $\Im$ take $\mathbb{C}^1 \to \mathbb{R}^1$. We can multiply and divide Eq. (8.26) by $\sqrt{x^2 + y^2}$ to obtain

$$z = \underbrace{\sqrt{x^2 + y^2}}_{r} \left(\underbrace{\frac{x}{\sqrt{x^2 + y^2}}}_{\cos\theta} + i \underbrace{\frac{y}{\sqrt{x^2 + y^2}}}_{\sin\theta} \right). \tag{8.27}$$

Noting the similarities between this and the transformation between Cartesian and polar coordinates suggests we adopt

$$r = \sqrt{x^2 + y^2}, \qquad \cos\theta = \frac{x}{\sqrt{x^2 + y^2}}, \qquad \sin\theta = \frac{y}{\sqrt{x^2 + y^2}}. \tag{8.28}$$

Table 8.1 Comparison of the action of arg, Tan^{-1}, and arctan

x	y	$\arg(x+iy)$	$\text{Tan}^{-1}(x,y)$	$\tan^{-1}(y/x)$
1	1	$\pi/4$	$\pi/4$	$\pi/4$
-1	1	$3\pi/4$	$3\pi/4$	$-\pi/4$
-1	-1	$-3\pi/4$	$-3\pi/4$	$\pi/4$
1	-1	$-\pi/4$	$-\pi/4$	$-\pi/4$

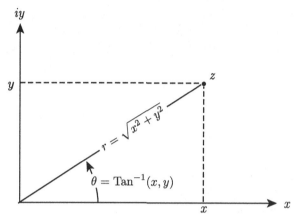

Figure 8.2 Polar and Cartesian representation of a complex number z.

Thus, we have

$$z = r(\cos\theta + i\sin\theta) = re^{i\theta}. \tag{8.29}$$

We often say that a complex number can be characterized by its magnitude $|z|$ and its argument, θ, we say then

$$r = |z|, \qquad \theta = \arg z. \tag{8.30}$$

Here, $r \in \mathbb{R}^1$ and $\theta \in \mathbb{R}^1$. Note that $|e^{i\theta}| = 1$. If $x > 0$, the function $\arg z$ is identical to $\tan^{-1}(y/x)$ from Eq. (2.306) and is suggested by the polar and Cartesian representation of z as shown in Fig. 8.2. However, we recognize that the ordinary $\tan^{-1}$ (also known as arctan) function maps onto the range $[-\pi/2, \pi/2]$, while we would like arg to map onto $[-\pi, \pi]$. For example, to capture the entire unit circle if $r = 1$, we need $\theta \in [-\pi, \pi]$. This can be achieved if we define arg, also known as Tan^{-1}, as follows:

$$\arg z = \arg(x+iy) = \text{Tan}^{-1}(x,y) = 2\tan^{-1}\left(\frac{y}{x+\sqrt{x^2+y^2}}\right). \tag{8.31}$$

Iff $x > 0$, this reduces to the more typical

$$\arg z = \arg(x+iy) = \text{Tan}^{-1}(x,y) = \arctan\left(\frac{y}{x}\right) = \tan^{-1}\left(\frac{y}{x}\right), \qquad x > 0. \tag{8.32}$$

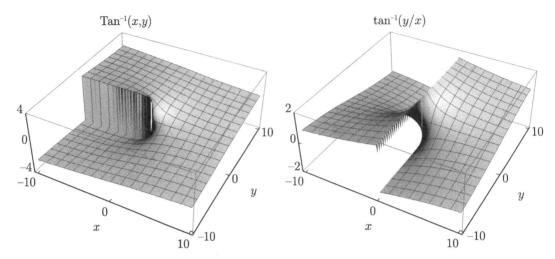

Figure 8.3 Comparison of $\text{Tan}^{-1}(x,y)$ and $\tan^{-1}(y/x)$.

The preferred and more general form is Eq. (8.31). We give simple function evaluations involving $\tan^{-1}$ and Tan^{-1} for selected values of x and y in Table 8.1. Use of Tan^{-1} effectively captures the correct quadrant of the complex plane corresponding to different positive and negative values of x and y. The function is sometimes known as Arctan or atan2. A comparison of $\text{Tan}^{-1}(x,y)$ and $\tan^{-1}(y/x)$ is given in Fig. 8.3.

Now we can define the *complex conjugate* $\overline{z}$ as

$$\overline{z} = x - iy = \sqrt{x^2 + y^2}\left(\frac{x}{\sqrt{x^2+y^2}} - i\frac{y}{\sqrt{x^2+y^2}}\right), \tag{8.33}$$

$$= r\left(\cos\theta - i\sin\theta\right) = r\left(\cos(-\theta) + i\sin(-\theta)\right) = re^{-i\theta}. \tag{8.34}$$

Note now that

$$z\overline{z} = (x+iy)(x-iy) = x^2 + y^2 = |z|^2 = re^{i\theta}re^{-i\theta} = r^2 = |z|^2. \tag{8.35}$$

We also have

$$\sin\theta = \frac{e^{i\theta} - e^{-i\theta}}{2i}, \qquad \cos\theta = \frac{e^{i\theta} + e^{-i\theta}}{2}. \tag{8.36}$$

Example 8.1 Use the polar representation of z to find all roots to the algebraic equation

$$z^4 = 1. \tag{8.37}$$

Solution
We know that $z = re^{i\theta}$. We also note that the constant 1 can be represented as

$$1 = e^{2n\pi i}, \qquad n = 0, 1, 2, \ldots. \tag{8.38}$$

This will be useful in finding all roots to our equation. With this representation, Eq. (8.37) becomes

$$r^4 e^{4i\theta} = e^{2n\pi i}, \qquad n = 0, 1, 2, \ldots. \tag{8.39}$$

We have a solution when
$$r = 1, \qquad \theta = \frac{n\pi}{2}, \qquad n = 0, 1, 2, \ldots. \tag{8.40}$$
There are distinct solutions for $n = 0, 1, 2, 3$. For larger n, the solutions repeat. So we have four distinct solutions
$$z = e^{0i}, \quad z = e^{i\pi/2}, \quad z = e^{i\pi}, \quad z = e^{3i\pi/2}. \tag{8.41}$$
In Cartesian form, the four distinct solutions are
$$z = \pm 1, \quad z = \pm i. \tag{8.42}$$

Example 8.2 Find all roots to
$$z^3 = i. \tag{8.43}$$

Solution
We proceed in a similar fashion as for the previous example. We know that
$$i = e^{i(\pi/2 + 2n\pi)}, \qquad n = 0, 1, 2, \ldots. \tag{8.44}$$
Substituting this into Eq. (8.43), we get
$$r^3 e^{3i\theta} = e^{i(\pi/2 + 2n\pi)}, \qquad n = 0, 1, 2, \ldots. \tag{8.45}$$
Solving, we get
$$r = 1, \qquad \theta = \frac{\pi}{6} + \frac{2n\pi}{3}. \tag{8.46}$$
There are only three distinct values of θ, those being $\theta = \pi/6$, $\theta = 5\pi/6$, $\theta = 3\pi/2$. So the three distinct roots are
$$z = e^{i\pi/6}, \quad z = e^{5i\pi/6}, \quad z = e^{3i\pi/2}. \tag{8.47}$$
In Cartesian form these roots are
$$z = \frac{\sqrt{3} + i}{2}, \quad z = \frac{-\sqrt{3} + i}{2}, \quad z = -i. \tag{8.48}$$
Diagrams of the solutions to this and the previous example are shown in Fig. 8.4. For both examples, the roots are uniformly distributed about the unit circle, with four roots for the quartic equation and three for the cubic.

8.2.3 Cauchy–Riemann Equations

It is possible to define complex functions of complex variables $W(z)$. For example, take a complex function to be defined as

$$W(z) = z^2 + z = (x + iy)^2 + (x + iy), \tag{8.49}$$
$$= x^2 + 2xyi - y^2 + x + iy = \underbrace{(x^2 + x - y^2)}_{\phi(x,y)} + i \underbrace{(2xy + y)}_{\psi(x,y)}. \tag{8.50}$$

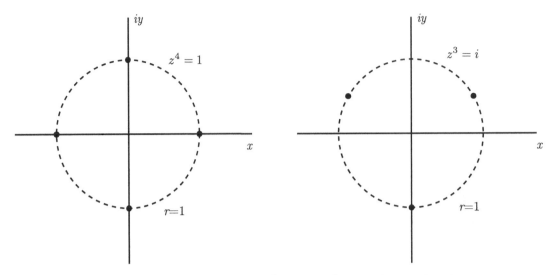

Figure 8.4 Diagram of solutions to $z^4 = 1$ and $z^3 = i$ in the complex plane.

In general, we can say
$$W(z) = \phi(x, y) + i\psi(x, y). \tag{8.51}$$

Here ϕ and ψ are *real* functions of *real* variables, which for our example function are $\phi = x^2 + x - y^2$ and $\psi = 2xy + y$.

Now $W(z)$ is defined as *analytic* at z_o if dW/dz exists at z_o and is independent of the direction in which it was calculated. We recall the definition of the derivative:
$$\left.\frac{dW}{dz}\right|_{z_o} = \lim_{\Delta z \to 0} \frac{W(z_o + \Delta z) - W(z_o)}{\Delta z}. \tag{8.52}$$

Now there are many paths that we can choose to evaluate the derivative. Let us consider two distinct paths, $y = C_1$ and $x = C_2$. We will get a result that can be shown to be valid for arbitrary paths. For $y = C_1$, we have $\Delta z = \Delta x$, so
$$\left.\frac{dW}{dz}\right|_{z_o} = \lim_{\Delta x \to 0} \frac{W(x_o + iy_o + \Delta x) - W(x_o + iy_o)}{\Delta x} = \left.\frac{\partial W}{\partial x}\right|_y. \tag{8.53}$$

For $x = C_2$, we have $\Delta z = i\Delta y$, so
$$\left.\frac{dW}{dz}\right|_{z_o} = \lim_{\Delta y \to 0} \frac{W(x_o + iy_o + i\Delta y) - W(x_o + iy_o)}{i\Delta y} = \left.\frac{1}{i}\frac{\partial W}{\partial y}\right|_x = \left.-i\frac{\partial W}{\partial y}\right|_x. \tag{8.54}$$

Now for an analytic function, we need
$$\left.\frac{\partial W}{\partial x}\right|_y = \left.-i\frac{\partial W}{\partial y}\right|_x, \tag{8.55}$$

or, expanding, we need
$$\frac{\partial \phi}{\partial x} + i\frac{\partial \psi}{\partial x} = -i\left(\frac{\partial \phi}{\partial y} + i\frac{\partial \psi}{\partial y}\right) = \frac{\partial \psi}{\partial y} - i\frac{\partial \phi}{\partial y}. \tag{8.56}$$

For equality, and thus path-independence of the derivative, we require

$$\frac{\partial \phi}{\partial x} = \frac{\partial \psi}{\partial y}, \qquad \frac{\partial \phi}{\partial y} = -\frac{\partial \psi}{\partial x}. \qquad (8.57)$$

These are the well-known *Cauchy–Riemann* equations for analytic functions of complex variables. *They are identical to our kinematic equations, Eq. (8.10), for incompressible irrotational fluid mechanics. Consequently, any analytic complex function is guaranteed to be a physical solution.* There are an infinite number of functions from which to choose!

We define the *complex velocity potential* as

$$W(z) = \phi(x, y) + i\psi(x, y), \qquad (8.58)$$

and taking a derivative of this potential, we have

$$\frac{dW}{dz} = \frac{\partial \phi}{\partial x} + i\frac{\partial \psi}{\partial x} = u - iv. \qquad (8.59)$$

Because the direction of the derivative does not matter, we can equivalently say

$$\frac{dW}{dz} = -i\left(\frac{\partial \phi}{\partial y} + i\frac{\partial \psi}{\partial y}\right) = \left(\frac{\partial \psi}{\partial y} - i\frac{\partial \phi}{\partial y}\right) = u - iv. \qquad (8.60)$$

We can associate the velocity magnitude with the magnitude of dW/dz:

$$\left|\frac{dW}{dz}\right|^2 = \overline{\frac{dW}{dz}} \frac{dW}{dz} = (u+iv)(u-iv) = u^2 + v^2, \quad \text{so} \quad \left|\frac{dW}{dz}\right| = \sqrt{u^2 + v^2}. \qquad (8.61)$$

Most common functions can be shown to be analytic. For example, for the function of Eq. (8.49)

$$W(z) = z^2 + z, \qquad (8.62)$$

which we have seen can be expressed as

$$W(z) = (x^2 + x - y^2) + i(2xy + y), \qquad (8.63)$$

we have

$$\phi(x, y) = x^2 + x - y^2, \qquad \psi(x, y) = 2xy + y, \qquad (8.64)$$

$$\frac{\partial \phi}{\partial x} = 2x + 1, \qquad \frac{\partial \psi}{\partial x} = 2y, \qquad (8.65)$$

$$\frac{\partial \phi}{\partial y} = -2y, \qquad \frac{\partial \psi}{\partial y} = 2x + 1. \qquad (8.66)$$

The fields of ψ and ϕ are plotted in Fig. 8.5. The Cauchy–Riemann equations are satisfied because $\partial \phi/\partial x = \partial \psi/\partial y$ and $\partial \phi/\partial y = -\partial \psi/\partial x$. Moreover,

$$\nabla^2 \phi = \frac{\partial}{\partial x}(2x+1) + \frac{\partial}{\partial y}(-2y) = 2 - 2 = 0, \qquad (8.67)$$

$$\nabla^2 \psi = \frac{\partial}{\partial x}(2y) + \frac{\partial}{\partial y}(2x+1) = 0 + 0 = 0, \qquad (8.68)$$

$$\mathbf{v} = \begin{pmatrix} u \\ v \end{pmatrix} = \begin{pmatrix} 2x+1 \\ -2y \end{pmatrix}. \qquad (8.69)$$

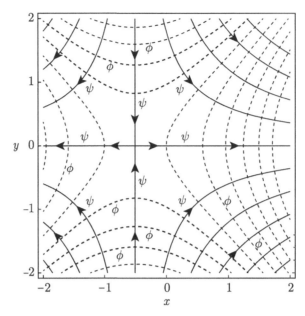

Figure 8.5 Fields of $\psi(x,y)$, $\phi(x,y)$ corresponding to the complex potential $W(z) = z^2 + 1$.

So the derivative is independent of direction, and we can say

$$\frac{dW}{dz} = \left.\frac{\partial W}{\partial x}\right|_y = (2x+1) + i(2y) = 2(x+iy) + 1 = 2z + 1. \tag{8.70}$$

We could get this result by ordinary rules of derivatives for real functions.

For an example of a non-analytic function consider $W(z) = \bar{z}$. Thus,

$$W(z) = x - iy. \tag{8.71}$$

So $\phi = x$ and $\psi = -y$, $\partial\phi/\partial x = 1$, $\partial\phi/\partial y = 0$, and $\partial\psi/\partial x = 0$, $\partial\psi/\partial y = -1$. Because $\partial\phi/\partial x \neq \partial\psi/\partial y$, the Cauchy–Riemann equations are not satisfied, and the derivative depends on direction. In fact, any complex function, $W(z)$, that has an explicit dependency on $\bar{z}$ will not be analytic.

8.3 Elementary Complex Potentials

Let us examine some simple analytic functions and see the fluid mechanics to which they correspond.

8.3.1 Uniform Flow

Take

$$W(z) = Uz, \quad \text{with} \quad U \in \mathbb{C}^1. \tag{8.72}$$

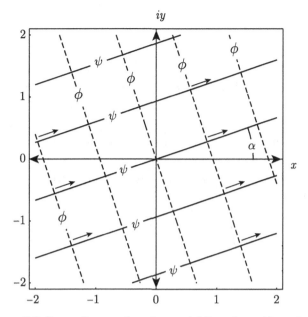

Figure 8.6 Streamlines and equipotential lines for uniform flow.

Here U is a complex constant with units of velocity. Then

$$\frac{dW}{dz} = U = u - iv. \tag{8.73}$$

Because U is complex, we can say

$$U = U_o e^{-i\alpha} = U_o \cos\alpha - iU_o \sin\alpha. \tag{8.74}$$

Thus, we get

$$u = U_o \cos\alpha, \qquad v = U_o \sin\alpha. \tag{8.75}$$

This represents a spatially uniform flow with streamlines inclined at angle α to the x axis. The flow is shown in Fig. 8.6.

8.3.2 Sources and Sinks

Take

$$W(z) = B \ln z, \qquad \text{with} \qquad B \in \mathbb{R}^1. \tag{8.76}$$

Here B is a real constant with units of area per time. With $z = re^{i\theta}$, we have $\ln z = \ln r + i\theta$. So

$$W(z) = B \ln r + iB\theta. \tag{8.77}$$

Consequently, we have for the velocity potential and stream function

$$\phi = B \ln r, \qquad \psi = B\theta. \tag{8.78}$$

8 Potential Flow

Now $\mathbf{v} = \nabla \phi$, so

$$v_r = \frac{\partial \phi}{\partial r} = \frac{B}{r}, \qquad v_\theta = \frac{1}{r}\frac{\partial \phi}{\partial \theta} = 0. \tag{8.79}$$

So the velocity is all radial, and becomes infinite at $r = 0$. We can show that the volumetric flow rate per unit depth is bounded, and is in fact a constant. For this two-dimensional flow, we really want to consider the volumetric flow rate per unit depth. We shall call this Q and recognize it has units of m²/s. (It is also common to interpret Q as a volumetric flow rate with units of m³/s. Our notation would have to be adjusted if we took this interpretation.) Then with dA as a differential area per unit depth, we have $dA = r\, d\theta$, and the volumetric flow rate per unit depth Q through a surface is

$$Q = \int_A \mathbf{v}^T \cdot \boldsymbol{\alpha}_n \, dA = \int_0^{2\pi} v_r r \, d\theta = \int_0^{2\pi} \frac{B}{r} r \, d\theta = 2\pi B. \tag{8.80}$$

The volumetric flow rate per unit depth is a constant. If $B > 0$, we have a *source*. If $B < 0$, we have a *sink*. The potential for a source/sink is often written as

$$W(z) = \frac{Q}{2\pi} \ln z. \tag{8.81}$$

For a source located at a point z_o that is not at the origin, we can say

$$W(z) = \frac{Q}{2\pi} \ln(z - z_o). \tag{8.82}$$

The flow is depicted in Fig. 8.7.

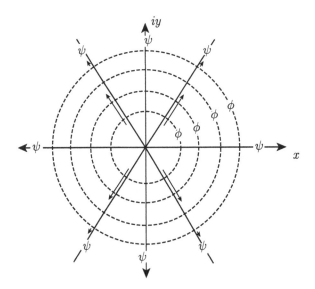

Figure 8.7 Streamlines and equipotential lines for source flow.

8.3.3 Point Vortices

For an ideal point vortex, identical to what was considered in Section 7.7.4, we have

$$W(z) = iB \ln z, \quad \text{with} \quad B \in \mathbb{R}^1. \tag{8.83}$$

Here B is a real constant with units of area per time. So

$$W(z) = iB(\ln r + i\theta) = -B\theta + iB \ln r. \tag{8.84}$$

Consequently,

$$\phi = -B\theta, \quad \psi = B \ln r. \tag{8.85}$$

We get the velocity field from

$$v_r = \frac{\partial \phi}{\partial r} = 0, \quad v_\theta = \frac{1}{r}\frac{\partial \phi}{\partial \theta} = -\frac{B}{r}. \tag{8.86}$$

So we see that the streamlines are circles about the origin, and there is no radial component of velocity. Consider the circulation of this flow, Eq. (7.2):

$$\Gamma = \oint_C \mathbf{v}^T \cdot d\mathbf{r} = \int_0^{2\pi} -\frac{B}{r} r \, d\theta = -2\pi B. \tag{8.87}$$

So we often write the complex potential in terms of the ideal vortex strength Γ_o:

$$W(z) = -\frac{i\Gamma_o}{2\pi} \ln z. \tag{8.88}$$

For an ideal vortex not at $z = z_o$, we say

$$W(z) = -\frac{i\Gamma_o}{2\pi} \ln(z - z_o). \tag{8.89}$$

The point vortex flow is shown in Fig. 8.8.

8.3.4 Superposition of Sources

Because the equation for velocity potential is linear, we can use the method of superposition to create new solutions as summations of elementary solutions. Say we want to model the effect of a wall on a source as shown in Fig. 8.9. At the wall we want $u(0, y) = 0$. That is,

$$\Re\left(\frac{dW}{dz}\right) = \Re(u - iv) = 0, \quad \text{on} \quad z = iy. \tag{8.90}$$

Now let us place a source at $z = a$ and superpose an image source at $z = -a$. Here a is a real scalar; $a \in \mathbb{R}^1$. So we have for the complex potential

$$W(z) = \underbrace{\frac{Q}{2\pi}\ln(z-a)}_{\text{original}} + \underbrace{\frac{Q}{2\pi}\ln(z+a)}_{\text{image}} = \frac{Q}{2\pi}\ln((z-a)(z+a)) = \frac{Q}{2\pi}\ln(z^2 - a^2), \tag{8.91}$$

$$\frac{dW}{dz} = \frac{Q}{2\pi}\frac{2z}{z^2 - a^2}. \tag{8.92}$$

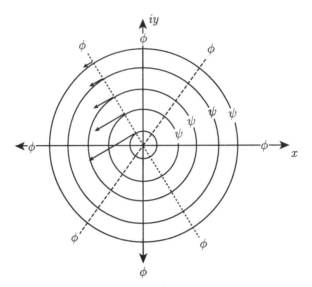

Figure 8.8 Streamlines, equipotential, and velocity vectors for a point vortex.

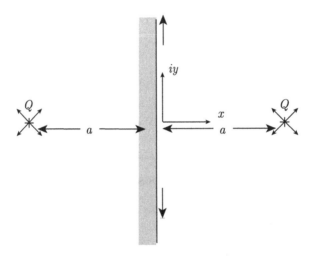

Figure 8.9 Diagram for source–wall interaction.

Now on $z = iy$, which is the location of the wall, we have

$$\frac{dW}{dz} = \frac{Q}{2\pi}\left(\frac{2iy}{-y^2 - a^2}\right) = u - iv. \tag{8.93}$$

The term is purely imaginary; hence, the real part is zero, and we have $u = 0$ on the wall, as desired. On the wall we do have a nonzero y component of velocity. Hence the wall is not a no-slip wall. On the wall we have then

$$v = \frac{Q}{\pi}\frac{y}{y^2 + a^2}. \tag{8.94}$$

We find the location on the wall of the maximum v velocity by setting the derivative with respect to y to be zero,

$$\frac{\partial v}{\partial y} = \frac{Q}{\pi} \frac{(y^2 + a^2) - y(2y)}{(y^2 + a^2)^2} = 0. \tag{8.95}$$

Solving, we find critical points at $y = \pm a$. It can be shown that v is a local maximum at $y = a$ and a local minimum at $y = -a$. So on the wall we have

$$\frac{1}{2}(u^2 + v^2) = \frac{1}{2}\frac{Q^2}{\pi^2} \frac{y^2}{(y^2 + a^2)^2}. \tag{8.96}$$

We can use the Bernoulli equation to find the pressure field, assuming steady flow and that $p \to p_o$ as $r \to \infty$. So the Bernoulli equation in this limit

$$\frac{1}{2}\nabla^T \phi \cdot \nabla \phi + \frac{p}{\rho} = \frac{p_o}{\rho}, \tag{8.97}$$

reduces to

$$p = p_o - \frac{1}{2}\rho \frac{Q^2}{\pi^2} \frac{y^2}{(y^2 + a^2)^2}. \tag{8.98}$$

The pressure is p_o at $y = 0$ and is p_o as $y \to \infty$. By integrating the pressure over the wall surface, one would find the net force on the wall induced by the source.

8.3.5 Flow in Corners

Flow in or around a corner can be modeled by the complex potential, with $B \in \mathbb{R}^1$:

$$W(z) = Bz^n = B\left(re^{i\theta}\right)^n = Br^n e^{in\theta} = Br^n(\cos(n\theta) + i\sin(n\theta)). \tag{8.99}$$

Here B is a real constant with units of L^{2-n}/t_c. Here L is a length, and t_c is a time. So we have

$$\phi = Br^n \cos n\theta, \qquad \psi = Br^n \sin n\theta. \tag{8.100}$$

Now recall that lines on which ψ is constant are streamlines. Examining the stream function, we obviously have streamlines when $\psi = 0$, which occurs whenever $\theta = 0$ or $\theta = \pi/n$.

For example, if $n = 2$, we model a stream striking a flat wall; kinematics of this have been previously described in Section 3.12.7. For this flow, we have

$$W(z) = Bz^2 = B(x + iy)^2 = B((x^2 - y^2) + i(2xy)), \tag{8.101}$$
$$\phi = B(x^2 - y^2), \qquad \psi = B(2xy). \tag{8.102}$$

For $n = 2$, the constant B has units of inverse time. The streamlines here are hyperbolas. For the velocity field, we take

$$\frac{dW}{dz} = 2Bz = 2B(x + iy) = u - iv, \tag{8.103}$$
$$u = 2Bx, \qquad v = -2By. \tag{8.104}$$

This flow actually represents flow in a corner formed by a right angle or flow striking a flat plate, or the impingement of two streams. For $n = 2$, streamlines are depicted in Fig. 8.10.

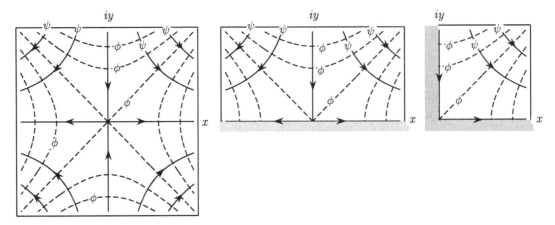

Figure 8.10 Diagram for impingement flow, stagnation flow, and flow in a corner, $W(z) = z^2$.

Let us explore kinematics and dynamics of the flow field induced by the complex potential $W(z) = Bz^2$. We have

$$W(z) = Bz^2 = B(x+iy)^2 = B\left((x^2 - y^2) + 2xyi\right). \tag{8.105}$$

Thus, we have

$$\phi(x,y) = B\left(x^2 - y^2\right), \qquad \psi(x,y) = 2Bxy. \tag{8.106}$$

Because B has units of inverse time, and x and y of length, both u and v have units of length per time. We see that the velocity field is given by

$$u = \frac{\partial \phi}{\partial x} = 2Bx, \qquad v = \frac{\partial \phi}{\partial y} = -2By. \tag{8.107}$$

We can also get velocity from the stream function

$$u = \frac{\partial \psi}{\partial y} = 2Bx, \qquad v = -\frac{\partial \psi}{\partial x} = -2By. \tag{8.108}$$

We also can get the velocity field directly via

$$\frac{dW}{dz} = 2Bz = u - iv. \tag{8.109}$$

Thus

$$2B(x+iy) = u - iv. \tag{8.110}$$

Comparing, we see that

$$u = 2Bx, \qquad v = -2By. \tag{8.111}$$

So the velocity vector is

$$\mathbf{v} = B \begin{pmatrix} 2x \\ -2y \end{pmatrix}. \tag{8.112}$$

The velocity vector is zero at the origin, which is a stagnation point. We also see Laplace's equations are satisfied because

$$\frac{\partial^2 \phi}{\partial x^2} + \frac{\partial^2 \phi}{\partial y^2} = 2B - 2B = 0, \qquad \frac{\partial^2 \psi}{\partial x^2} + \frac{\partial^2 \psi}{\partial y^2} = 0 + 0 = 0. \qquad (8.113)$$

Incompressibility is satisfied as

$$\nabla^T \cdot \mathbf{v} = \frac{\partial u}{\partial x} + \frac{\partial v}{\partial y} = 2B - 2B = 0. \qquad (8.114)$$

Irrotationality is satisfied because

$$\omega_z = \frac{\partial v}{\partial x} - \frac{\partial u}{\partial y} = 0 - 0 = 0. \qquad (8.115)$$

Let us examine the deformation tensor that is the symmetric part of the velocity gradient tensor:

$$\mathbf{D} = \begin{pmatrix} \frac{\partial u}{\partial x} & \frac{1}{2}\left(\frac{\partial u}{\partial y} + \frac{\partial v}{\partial x}\right) \\ \frac{1}{2}\left(\frac{\partial u}{\partial y} + \frac{\partial v}{\partial x}\right) & \frac{\partial v}{\partial y} \end{pmatrix} = \begin{pmatrix} 2B & 0 \\ 0 & -2B \end{pmatrix}. \qquad (8.116)$$

The deformation tensor has units of inverse time, which is appropriate for a strain rate. Employing Eqs. (2.143, 2.144), the principal invariants of $\mathbf{D}$ are $I_{\dot{\epsilon}}^{(1)} = \mathrm{tr}\,\mathbf{D} = 0$, $I_{\dot{\epsilon}}^{(3)} = \det \mathbf{D} = -4B^2$. The deformation tensor is already in diagonal form, so there is no necessity to rotate the axes to identify the principal axes. The principal axes are the eigenvectors associated with the tensor. We can normalize them and take them to be the unit vectors $\mathbf{i}$ and $\mathbf{j}$. There is positive extensional straining aligned with $\mathbf{i}$. This extensional straining is exactly counterbalanced by negative extensional straining aligned with $\mathbf{j}$. So the volume of the deforming fluid particle is preserved, as required by incompressibility. For the fluid element aligned with the coordinate axes, there is no shear deformation.

The local acceleration vector of a fluid particle is given by

$$\mathbf{a} = \frac{d\mathbf{v}}{dt} = \begin{pmatrix} \frac{\partial u}{\partial t} + u\frac{\partial u}{\partial x} + v\frac{\partial u}{\partial y} \\ \frac{\partial v}{\partial t} + u\frac{\partial v}{\partial x} + v\frac{\partial v}{\partial y} \end{pmatrix} = B^2 \begin{pmatrix} 4x \\ 4y \end{pmatrix}. \qquad (8.117)$$

The acceleration vector has units of length per time squared, as appropriate. The acceleration vector is zero at the origin; it points outward from the origin when away from the origin. A fluid particle on a stagnation streamline has no curvature, and its acceleration vector is parallel to its velocity vector. Streamlines that are not stagnation streamlines have nonzero curvature. The acceleration of such a fluid particle has a centripetal component that points towards the local instantaneous center of curvature.

Dynamics tells us that acceleration vectors must be induced by a net force. In our problem, in which we neglect viscous and body forces, the only net force in play is that induced by the gradient of pressure. Let us find the pressure field associated with this flow field. The Bernoulli equation gives us

$$p_o = p + \frac{1}{2}\rho(u^2 + v^2) = p + \frac{1}{2}\rho B^2(4x^2 + 4y^2), \qquad (8.118)$$

$$p - p_o = -2\rho B^2(x^2 + y^2). \qquad (8.119)$$

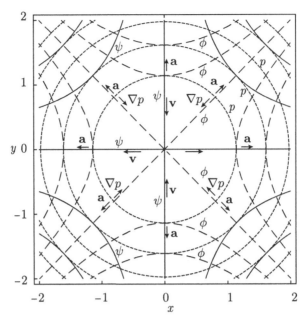

Figure 8.11 Diagram for $W(z) = z^2$ of streamlines, equipotential lines, isobars, velocity vectors, acceleration vectors, pressure gradient vectors.

By inspection, isobars are circles. The peak pressure is at the origin; the pressure decreases with the square of the distance from the origin. Fluid accelerates from regions of high pressure to regions of low pressure, and the acceleration vector points in the opposite direction of the pressure gradient vector:

$$\nabla p = -4\rho B^2 \left(x\mathbf{i} + y\mathbf{j} \right) = -\rho \mathbf{a}. \tag{8.120}$$

Thus, we recover our linear momenta equation

$$\rho \frac{d\mathbf{v}}{dt} = -\nabla p. \tag{8.121}$$

A diagram of streamlines, equipotential lines, isobars, velocity vectors, and acceleration vectors is given in Fig. 8.11. Lastly, for this irrotational flow, Eq. (6.73) reduces to

$$\nabla^2 p = -\rho \mathbf{D} : \mathbf{D}. \tag{8.122}$$

Direct calculation of both terms reveals this is satisfied as

$$\nabla^2 p = -\rho \mathbf{D} : \mathbf{D} = -8\rho B^2. \tag{8.123}$$

Thus the pressure field is anharmonic and satisfies a Poisson equation.

Let us take the case for which $B = 1$ with arbitrary units of inverse time and focus on one particular streamline, that for which $\psi = 2$. The equation of this streamline is

$$y = \frac{1}{x}. \tag{8.124}$$

We can use standard notions from calculus to define the curvature, κ. Specializing the more general three-dimensional form presented in Eq. (3.311) by taking $x(t) = t$, $y(t) = y(x)$, $z(t) = 0$, the general formula for curvature for two-dimensional curves is

$$\kappa = \frac{\frac{d^2 y}{dx^2}}{\left(1 + \left(\frac{dy}{dx}\right)^2\right)^{3/2}}. \tag{8.125}$$

For this particular $\psi = 2$ streamline, we find

$$\kappa = \frac{2}{x^3 \left(1 + \frac{1}{x^4}\right)^{3/2}}. \tag{8.126}$$

Taylor series reveals that

$$\lim_{x \to 0} \kappa = 0, \qquad \lim_{x \to \infty} \kappa = 0. \tag{8.127}$$

This makes sense as the streamline $\psi = 1$ in the first quadrant approaches the x axis for large x and the y axis for small positive x. These axes have no curvature. For $x \in [0, +\infty)$, one can show with calculus that κ has a maximum value given by

$$\kappa_{max} = \kappa(x = 1) = \frac{\sqrt{2}}{2}. \tag{8.128}$$

For this streamline, when $x = 1$, we have $y = 1/x = 1/1 = 1$. So the point of maximum streamline curvature is at an angle of $\pi/4$ from the x axis.

The velocity magnitude is given by

$$||\mathbf{v}|| = \sqrt{4x^2 + 4y^2}. \tag{8.129}$$

For the $y = 1/x$ streamline, this gives

$$||\mathbf{v}|| = \sqrt{4x^2 + \frac{4}{x^2}}. \tag{8.130}$$

Calculus reveals this has a local minimum of $2\sqrt{2}$ at the point $(x, y)^T = (1, 1)^T$.

From this analysis, one can see the following. On the x axis, with $y = 0$ and for $x \to \infty$, $\mathbf{v}$ and $\mathbf{a}$ are parallel and point in positive x direction. The fluid acceleration is parallel to the pressure gradient vector, and both are parallel to the streamwise direction. On the y axis, with $x = 0$ and for $y \to \infty$, $\mathbf{v}$ and $\mathbf{a}$ are parallel and point in opposite directions. The fluid acceleration is parallel to the pressure gradient vector, and both are parallel to the streamwise direction. For points on the curve $\theta = \pi/4$, $\mathbf{v}$ is orthogonal to $\mathbf{a}$. The fluid acceleration is parallel to the pressure gradient vector, and both are in the stream-normal direction. Here the acceleration is all centripetal and due to streamline curvature. And at such points the velocity magnitude has a local minimum. For intermediate points, $\mathbf{v}$ is neither parallel nor orthogonal to $\mathbf{a}$. The acceleration is parallel to the pressure gradient vector, and there are nonzero components of acceleration in the streamwise and stream-normal directions.

We can get the streamlines by direct integration. Because we know the velocity field, $u = 2x$, $v = -2y$, we then have the system of differential equations for the streamlines, pathlines, and

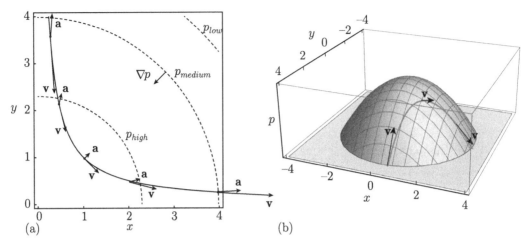

Figure 8.12 Kinematics and dynamics on the streamline $y = 1/x$ for the flow field defined by the complex potential $W(z) = Bz^2$, $B = 1$: (a) planar rendering, (b) three-dimensional rendering.

streaklines for this steady flow. Let us find the streamline that passes through $(x, y)^T = (1, 1)^T$. We have the kinematics cast as a dynamical system as discussed in Section 3.14:

$$\frac{dx}{dt} = 2x, \qquad x(0) = 1, \qquad \frac{dy}{dt} = -2y, \qquad y(0) = 1. \tag{8.131}$$

Integrating, we get

$$x(t) = e^{2t}, \qquad y(t) = e^{-2t}. \tag{8.132}$$

This is a parametric solution for the streamline, streakline, and pathline that passes through $(1, 1)^T$ at $t = 0$. Obviously, $1/x = e^{-2t} = y$, so this streamline is given, as expected, by

$$y = \frac{1}{x}. \tag{8.133}$$

The streamline $y = 1/x$ along with the local velocity and acceleration vectors and a few pressure contours are plotted in Fig. 8.12(a). The same information is plotted in Fig. 8.12(b), but in a three-dimensional rendering. The scale for pressure is arbitrary and not given. Clearly one can envision this as an analog of a particle climbing a hill in a gravitational field, with the pressure serving the role of elevation. At the bottom of the hill, the velocity is high. Forces induce the fluid to decelerate until it reaches its local maximum, which is not the apex of the hill, and then begins to accelerate as it moves down the hill.

The unit tangent to the streamline is

$$\boldsymbol{\alpha}_t = \frac{\mathbf{v}}{||\mathbf{v}||} = \frac{2x\mathbf{i} - 2y\mathbf{j}}{\sqrt{4x^2 + 4y^2}} = \frac{2x\mathbf{i} - \frac{2}{x}\mathbf{j}}{\sqrt{4x^2 + \frac{4}{x^2}}} = \frac{x^2}{\sqrt{1 + x^4}}\mathbf{i} - \frac{1}{\sqrt{1 + x^4}}\mathbf{j} = \begin{pmatrix} \frac{x^2}{\sqrt{1+x^4}} \\ -\frac{1}{\sqrt{1+x^4}} \end{pmatrix}. \tag{8.134}$$

We can specialize Eq. (3.117) to find the stretching rate $\mathcal{D}_t$ in the streamwise direction along the streamline $y = 1/x$. This is

$$\mathcal{D}_t = \boldsymbol{\alpha}_t^T \cdot \mathbf{D} \cdot \boldsymbol{\alpha}_t = \begin{pmatrix} \frac{x^2}{\sqrt{1+x^4}} & -\frac{1}{\sqrt{1+x^4}} \end{pmatrix} \begin{pmatrix} 2 & 0 \\ 0 & -2 \end{pmatrix} \begin{pmatrix} \frac{x^2}{\sqrt{1+x^4}} \\ -\frac{1}{\sqrt{1+x^4}} \end{pmatrix} = 2\frac{x^4 - 1}{x^4 + 1}. \qquad (8.135)$$

There is no streamwise stretching at $x = 1$. As $x \to \infty$, $\mathcal{D}_t \to 2$, and as $x \to 0$, $\mathcal{D}_t \to -2$.

We see by inspection that the unit vector normal to the streamline $\boldsymbol{\alpha}_n$ must be

$$\boldsymbol{\alpha}_n = \begin{pmatrix} \frac{1}{\sqrt{1+x^4}} \\ \frac{x^2}{\sqrt{1+x^4}} \end{pmatrix}. \qquad (8.136)$$

This vector obviously has $\boldsymbol{\alpha}_n^T \cdot \boldsymbol{\alpha}_t = 0$, and $\boldsymbol{\alpha}_n^T \cdot \boldsymbol{\alpha}_n = 1$. Moreover, $\boldsymbol{\alpha}_n$ points toward the center of curvature of the streamline. We can also find the stretching rate $\mathcal{D}_n$ in the stream-normal direction along the streamline $y = 1/x$. This is

$$\mathcal{D}_n = \boldsymbol{\alpha}_n^T \cdot \mathbf{D} \cdot \boldsymbol{\alpha}_n = \begin{pmatrix} \frac{1}{\sqrt{1+x^4}} & \frac{x^2}{\sqrt{1+x^4}} \end{pmatrix} \begin{pmatrix} 2 & 0 \\ 0 & -2 \end{pmatrix} \begin{pmatrix} \frac{1}{\sqrt{1+x^4}} \\ \frac{x^2}{\sqrt{1+x^4}} \end{pmatrix} = 2\frac{1 - x^4}{x^4 + 1}. \qquad (8.137)$$

There is no stream-normal stretching at $x = 1$. As $x \to \infty$, $\mathcal{D}_n \to -2$, and as $x \to 0$, $\mathcal{D}_n \to 2$. Moreover, because the two-dimensional flow is incompressible and thus area-preserving, streamwise stretching is balanced by stream-normal stretching so that overall one has

$$\mathcal{D}_t + \mathcal{D}_n = 2\frac{x^4 - 1}{x^4 + 1} + 2\frac{1 - x^4}{x^4 + 1} = 0. \qquad (8.138)$$

Now let us perform an analysis for extreme values for $\mathcal{D}$ similar to that performed in generating Fig. 3.7. For a general direction $\boldsymbol{\alpha}$, we have

$$\mathcal{D} = \boldsymbol{\alpha}^T \cdot \mathbf{D} \cdot \boldsymbol{\alpha} = \begin{pmatrix} \alpha_1 & \alpha_2 \end{pmatrix} \begin{pmatrix} 2 & 0 \\ 0 & -2 \end{pmatrix} \begin{pmatrix} \alpha_1 \\ \alpha_2 \end{pmatrix} = 2\alpha_1^2 - 2\alpha_2^2. \qquad (8.139)$$

As before, we could plot contours for which $\mathcal{D}$ is constant in the (α_1, α_2) plane and get an infinite family of curves. In contrast to the ellipses of Fig. 3.7, here we have hyperbolas as contours. Once more, we also have the constraint of $\alpha_1^2 + \alpha_2^2 = 1$. And once more, the eigenvalues of $\mathbf{D}$ are the special contours of $\mathcal{D}$, suggesting we examine the two contours

$$2 = 2\alpha_1^2 - 2\alpha_2^2, \qquad -2 = 2\alpha_1^2 - 2\alpha_2^2. \qquad (8.140)$$

These two curves, along with the unit circle $\alpha_1^2 + \alpha_2^2 = 1$ are plotted in Fig. 8.13 An infinite family of contours of $\mathcal{D}$ exist. Many of them will also intersect the unit circle, and so are candidate solutions. However, the special contours we selected are extreme values. For intersection with the unit circle, we require

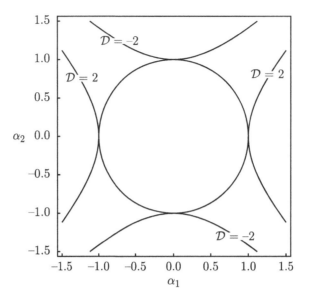

Figure 8.13 Two special contours of $\mathcal{D}$ along with the unit circle illustrating the extreme values of $\mathcal{D}$ for a simple stagnation potential flow defined by $W(z) = z^2$.

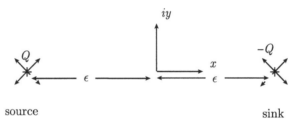

Figure 8.14 Source–sink pair.

$$\mathcal{D} \in [\lambda_{min}, \lambda_{max}], \quad \text{thus} \quad \mathcal{D} \in [-2, 2]. \tag{8.141}$$

Because $\mathbf{D}$ is already diagonal, the eigenvectors are aligned with the unrotated coordinate axes. It is straightforward to show for $\mathcal{D} = -2$, that $\boldsymbol{\alpha} = (0, 1)^T$ and for $\mathcal{D} = 2$, that $\boldsymbol{\alpha} = (1, 0)^T$.

8.3.6 Doublets

We can form what is known as a *doublet* flow by considering the superposition of a source and sink and let the two approach each other. Consider a source and sink of equal and opposite strength straddling the y axis, each separated from the origin by a distance ϵ as depicted in Fig. 8.14. The complex velocity potential is

$$W(z) = \frac{Q}{2\pi} \ln(z + \epsilon) - \frac{Q}{2\pi} \ln(z - \epsilon) = \frac{Q}{2\pi} \ln\left(\frac{z + \epsilon}{z - \epsilon}\right). \tag{8.142}$$

It can be shown by synthetic division that as $\epsilon \to 0$, that

$$\frac{z + \epsilon}{z - \epsilon} = 1 + \epsilon \frac{2}{z} + \epsilon^2 \frac{2}{z^2} + \cdots. \tag{8.143}$$

So the potential approaches

$$W(z) \sim \frac{Q}{2\pi} \ln\left(1 + \epsilon \frac{2}{z} + \epsilon^2 \frac{2}{z^2} + \cdots\right). \tag{8.144}$$

Now because $\ln(1+x) \to x$ as $x \to 0$, we get for small ϵ that

$$W(z) \sim \frac{Q}{2\pi} \epsilon \frac{2}{z} \sim \frac{Q\epsilon}{\pi z}. \tag{8.145}$$

Now let us require that

$$\lim_{\epsilon \to 0} \frac{Q\epsilon}{\pi} \to \mu. \tag{8.146}$$

Here, we use a standard notation of μ to denote doublet strength; in this context, μ has no relation to viscosity. Then we have

$$W(z) = \frac{\mu}{z} = \frac{\mu}{x+iy} \frac{x-iy}{x-iy} = \frac{\mu(x-iy)}{x^2+y^2}. \tag{8.147}$$

So

$$\phi(x,y) = \mu \frac{x}{x^2+y^2}, \qquad \psi(x,y) = -\mu \frac{y}{x^2+y^2}. \tag{8.148}$$

In polar coordinates, we then say

$$\phi = \mu \frac{\cos\theta}{r}, \qquad \psi = -\mu \frac{\sin\theta}{r}. \tag{8.149}$$

Streamlines and equipotential lines for a doublet are plotted in Fig. 8.15. Notice because the sink is infinitesimally to the right of the source, there exists a directionality. This can be considered a type of *dipole moment*; in this case, the direction of the dipole is $-\mathbf{i}$.

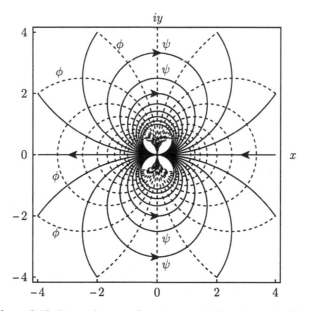

Figure 8.15 Streamlines and equipotential lines for a doublet.

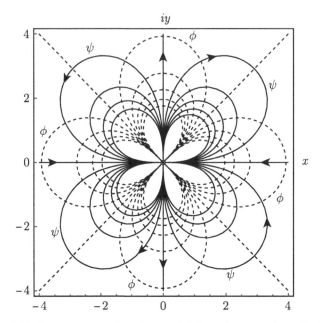

Figure 8.16 Streamlines and equipotential lines for a quadrupole, $k = 1$.

8.3.7 Quadrupoles

It is natural to examine a higher-order potentials. Let us look at one known as a *quadrupole*:

$$W(z) = \frac{k}{z^2} = k\frac{1}{(x+iy)(x+iy)} = k\frac{(x-iy)^2}{(x^2+y^2)^2} = k\frac{x^2 - y^2 - 2ixy}{(x^2+y^2)^2}. \quad (8.150)$$

This gives

$$\phi(x,y) = k\frac{x^2 - y^2}{(x^2+y^2)^2}, \qquad \psi(x,y) = k\frac{-2xy}{(x^2+y^2)^2}. \quad (8.151)$$

Streamlines and equipotential lines for a quadrupole with $k = 1$ are plotted in Fig. 8.16.

8.3.8 Rankine Half Body

Now consider the superposition of a uniform stream and a source, which we define to be a *Rankine*[2] *half body*, with $U, Q \in \mathbb{R}^1$:

$$W(z) = Uz + \frac{Q}{2\pi}\ln z = Ure^{i\theta} + \frac{Q}{2\pi}(\ln r + i\theta), \quad (8.152)$$

$$= Ur(\cos\theta + i\sin\theta) + \frac{Q}{2\pi}(\ln r + i\theta), \quad (8.153)$$

$$= \left(Ur\cos\theta + \frac{Q}{2\pi}\ln r\right) + i\left(Ur\sin\theta + \frac{Q}{2\pi}\theta\right). \quad (8.154)$$

---

[2] William John Macquorn Rankine, 1820–1872, Scottish engineer and mechanician, pioneer of thermodynamics and steam engine theory, studied fatigue in railway engine axles.

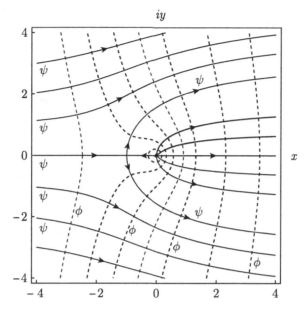

Figure 8.17 Streamlines and equipotential lines for a Rankine half body.

So

$$\phi = Ur\cos\theta + \frac{Q}{2\pi}\ln r, \qquad \psi = Ur\sin\theta + \frac{Q}{2\pi}\theta. \tag{8.155}$$

Streamlines for a Rankine half body are plotted in Fig. 8.17. Now for the Rankine half body, it is clear that there is a stagnation point somewhere on the x axis, along $\theta = \pi$. With the velocity given by

$$\frac{dW}{dz} = U + \frac{Q}{2\pi z} = u - iv, \tag{8.156}$$

we get

$$U + \frac{Q}{2\pi}\frac{1}{r}e^{-i\theta} = u - iv, \tag{8.157}$$

$$U + \frac{Q}{2\pi}\frac{1}{r}(\cos\theta - i\sin\theta) = u - iv, \tag{8.158}$$

$$u = U + \frac{Q}{2\pi r}\cos\theta, \qquad v = \frac{Q}{2\pi r}\sin\theta. \tag{8.159}$$

For $\theta = \pi$, we get $u = 0$ if

$$0 = U + \frac{Q}{2\pi r}(-1), \quad \text{so} \quad r = \frac{Q}{2\pi U}. \tag{8.160}$$

8.3.9 Flow over a Cylinder

We can model flow past a cylinder without circulation by superposing a uniform flow with a doublet. Defining $a^2 = \mu/U$, we write

$$W(z) = Uz + \frac{\mu}{z} = U\left(z + \frac{a^2}{z}\right) = U\left(re^{i\theta} + \frac{a^2}{re^{i\theta}}\right), \qquad (8.161)$$

$$= U\left(r(\cos\theta + i\sin\theta) + \frac{a^2}{r}(\cos\theta - i\sin\theta)\right), \qquad (8.162)$$

$$= U\left(\left(r\cos\theta + \frac{a^2}{r}\cos\theta\right) + i\left(r\sin\theta - \frac{a^2}{r}\sin\theta\right)\right), \qquad (8.163)$$

$$= Ur\left(\cos\theta\left(1 + \frac{a^2}{r^2}\right) + i\sin\theta\left(1 - \frac{a^2}{r^2}\right)\right). \qquad (8.164)$$

So

$$\phi = Ur\cos\theta\left(1 + \frac{a^2}{r^2}\right), \qquad \psi = Ur\sin\theta\left(1 - \frac{a^2}{r^2}\right). \qquad (8.165)$$

Now on $r = a$, we have $\psi = 0$. Because the stream function is constant here, the curve $r = a$, a circle, must be a streamline through which no mass can pass. A diagram of the streamlines and equipotential lines is plotted in Fig. 8.18.

For the velocities, we have

$$v_r = \frac{\partial \phi}{\partial r} = U\cos\theta\left(1 + \frac{a^2}{r^2}\right) + Ur\cos\theta\left(-2\frac{a^2}{r^3}\right) = U\cos\theta\left(1 - \frac{a^2}{r^2}\right), \qquad (8.166)$$

$$v_\theta = \frac{1}{r}\frac{\partial \phi}{\partial \theta} = -U\sin\theta\left(1 + \frac{a^2}{r^2}\right). \qquad (8.167)$$

So on $r = a$, we have $v_r = 0$, and $v_\theta = -2U\sin\theta$. Thus, on the surface, we have

$$\nabla^T\phi \cdot \nabla\phi = 4U^2\sin^2\theta. \qquad (8.168)$$

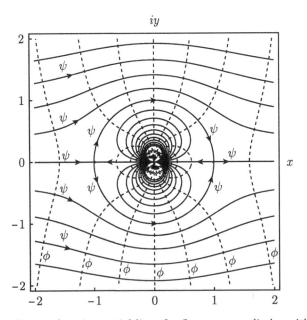

Figure 8.18 Streamlines and equipotential lines for flow over a cylinder without circulation.

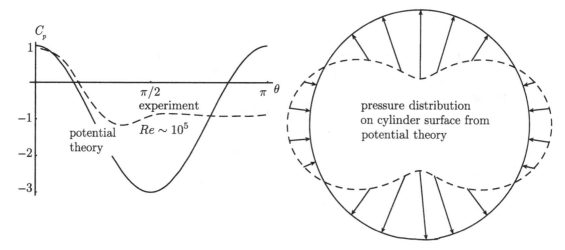

Figure 8.19 Pressure distribution for ideal flow over a cylinder without circulation.

The Bernoulli equation, Eq. (6.196), for a steady flow with $p \to p_\infty$ as $r \to \infty$, then gives

$$\frac{p}{\rho} + \frac{1}{2}\nabla^T \phi \cdot \nabla \phi = \frac{p_\infty}{\rho} + \frac{U^2}{2}, \quad \text{so} \quad p = p_\infty + \frac{1}{2}\rho U^2 (1 - 4\sin^2 \theta). \tag{8.169}$$

The pressure coefficient C_p, defined next, then is

$$C_p \equiv \frac{p - p_\infty}{\frac{1}{2}\rho U^2} = 1 - 4\sin^2 \theta. \tag{8.170}$$

A diagram of the pressure distribution, both predicted and experimentally observed, is plotted in Fig. 8.19. One can see that $C_p = 0$ when $\theta = \pm \pi/6, \pm 5\pi/6$. Potential theory predicts the pressure well on the front surface of the cylinder, but not so well on the back surface. This is because in real fluids at sufficiently high Reynolds number, a phenomenon known as flow separation manifests itself in regions of negative pressure gradients. Correct modeling of separation events requires a re-introduction of viscous stresses; thus, a potential theory cannot predict separation. This will be considered in Section 11.2.2 when we consider viscous *Falkner*[3]–*Skan*[4] solutions. The sketched experimental curve in Fig. 8.19 is for a viscous flow with $Re \sim 10^5$. Different Reynolds numbers would induce different curves, and this is a reflection of some of the complicated flow physics involved in viscous, separated flows.

Lastly, for this irrotational flow, we find that Eq. (6.73) reduces to

$$\nabla^2 p = -\rho \mathbf{D} : \mathbf{D}. \tag{8.171}$$

Direct calculation of both $\nabla^2 p$ and $-\rho \mathbf{D} : \mathbf{D}$ reveals both are given by

$$\nabla^2 p = -\rho \mathbf{D} : \mathbf{D} = -\frac{8\rho U^2}{r^6}. \tag{8.172}$$

---

[3] Victor Montague Falkner, 1897–1965, English mathematician and aerodynamicist.
[4] Sylvia Winifred Skan, 1897–1972, English applied mathematician and aerodynamicist.

The pressure field is anharmonic.

Example 8.3 For a cylinder of radius c at rest in an accelerating potential flow field with a far field velocity of $U = a + bt$, find the pressure on the stagnation point of the cylinder.

Solution
The velocity potential and velocities for this flow are

$$\phi(r, \theta, t) = (a + bt) r \cos\theta \left(1 + \frac{c^2}{r^2}\right), \tag{8.173}$$

$$v_r = \frac{\partial \phi}{\partial r} = (a + bt) \cos\theta \left(1 - \frac{c^2}{r^2}\right), \tag{8.174}$$

$$v_\theta = \frac{1}{r}\frac{\partial \phi}{\partial \theta} = -(a + bt) \sin\theta \left(1 + \frac{c^2}{r^2}\right), \tag{8.175}$$

$$\frac{1}{2}\nabla^T\phi \cdot \nabla\phi = \frac{1}{2}(a + bt)^2 \left(\cos^2\theta \left(1 - \frac{c^2}{r^2}\right)^2 + \sin^2\theta \left(1 + \frac{c^2}{r^2}\right)^2\right), \tag{8.176}$$

$$= \frac{1}{2}(a + bt)^2 \left(1 + \frac{c^4}{r^4} + \frac{2c^2}{r^2}\left(\sin^2\theta - \cos^2\theta\right)\right). \tag{8.177}$$

Also, because the flow is unsteady, we will need $\partial \phi / \partial t$:

$$\frac{\partial \phi}{\partial t} = br \cos\theta \left(1 + \frac{c^2}{r^2}\right). \tag{8.178}$$

In the limit as $r \to \infty$,

$$\frac{\partial \phi}{\partial t} \to br \cos\theta, \quad \frac{1}{2}\nabla^T\phi \cdot \nabla\phi \to \frac{1}{2}(a + bt)^2. \tag{8.179}$$

We also have that on the surface of the cylinder

$$v_r(r = c, \theta, t) = 0. \tag{8.180}$$

The unsteady Bernoulli equation, the incompressible, zero-body force version of Eq. (6.186), gives us

$$\frac{\partial \phi}{\partial t} + \frac{1}{2}\nabla^T\phi \cdot \nabla\phi + \frac{p}{\rho} = f(t). \tag{8.181}$$

We use the far field behavior to evaluate $f(t)$:

$$br \cos\theta + \frac{1}{2}(a + bt)^2 + \frac{p}{\rho} = f(t). \tag{8.182}$$

Now if we make the non-intuitive choice of $f(t) = \frac{1}{2}(a + bt)^2 + p_o/\rho$, we get

$$br \cos\theta + \frac{1}{2}(a + bt)^2 + \frac{p}{\rho} = \frac{1}{2}(a + bt)^2 + \frac{p_o}{\rho}. \tag{8.183}$$

So

$$p = p_o - \rho br \cos\theta = p_o - \rho bx. \tag{8.184}$$

Because the flow at infinity is accelerating, there must be a far field pressure gradient to induce this acceleration. Consider the x momentum equation in the far field

$$\rho \frac{du}{dt} = -\frac{\partial p}{\partial x}, \quad \text{giving} \quad \rho(b) = -(-\rho b). \tag{8.185}$$

So for the pressure field, we have

$$\underbrace{br\cos\theta\left(1+\frac{c^2}{r^2}\right)}_{\partial\phi/\partial t}+\underbrace{\frac{1}{2}(a+bt)^2\left(1+\frac{c^4}{r^4}+\frac{2c^2}{r^2}(\sin^2\theta-\cos^2\theta)\right)}_{\nabla^T\phi\cdot\nabla\phi/2}+\frac{p}{\rho}=\underbrace{\frac{1}{2}(a+bt)^2+\frac{p_o}{\rho}}_{f(t)}, \quad (8.186)$$

which can be solved for p to yield

$$p(r,\theta,t)=p_o-\rho br\cos\theta\left(1+\frac{c^2}{r^2}\right)-\frac{1}{2}\rho(a+bt)^2\left(\frac{c^4}{r^4}+\frac{2c^2}{r^2}(\sin^2\theta-\cos^2\theta)\right). \quad (8.187)$$

For the stagnation point, we evaluate as

$$p(c,\pi,t)=p_o-\rho bc(-1)(1+1)-\frac{1}{2}\rho(a+bt)^2(1+2(1)(0-1)), \quad (8.188)$$

$$=p_o+\frac{1}{2}\rho(a+bt)^2+2\rho bc=p_o+\frac{1}{2}\rho U^2+2\rho bc. \quad (8.189)$$

The first two terms would be predicted by a naïve extension of the steady Bernoulli equation, Eq. (6.196). The final term, however, is not intuitive and is a purely unsteady effect.

8.4 Forces Induced by Potential Flow

There are more basic ways to describe the force on bodies using complex variables directly in contrast to the method employing the Bernoulli equation. We shall present those methods in this section, following a necessary discussion of the motivating complex variable theory. Mathematical background is given by Brown and Churchill (2014).

8.4.1 Contour Integrals

Consider the closed contour integral of a complex function in the complex plane. For such integrals, there is a useful theory that we will not prove, but will demonstrate here. Consider contour integrals enclosing the origin with a circle in the complex plane for four functions. The contour in each is

$$C: z=\hat{R}e^{i\theta}, \quad \theta\in[0,2\pi). \quad (8.190)$$

For such a contour, $dz=i\hat{R}e^{i\theta}\,d\theta$.

Simple Pole

We describe a simple pole with the complex potential

$$W(z)=\frac{a}{z}. \quad (8.191)$$

The contour integral is

$$\oint_C W(z)\,dz=\oint_C \frac{a}{z}\,dz=\int_{\theta=0}^{\theta=2\pi}\frac{a}{\hat{R}e^{i\theta}}i\hat{R}e^{i\theta}\,d\theta=ai\int_0^{2\pi}d\theta=2\pi ia. \quad (8.192)$$

Constant Potential

We describe a constant with the complex potential

$$W(z) = b. \tag{8.193}$$

The contour integral is

$$\oint_C W(z)\, dz = \oint_C b\, dz = \int_{\theta=0}^{\theta=2\pi} bi\hat{R}e^{i\theta}\, d\theta = \left.\frac{bi\hat{R}}{i}e^{i\theta}\right|_0^{2\pi} = 0, \tag{8.194}$$

because $e^{0i} = e^{2\pi i} = 1$.

Uniform Flow

We describe a constant with the complex potential

$$W(z) = cz. \tag{8.195}$$

The contour integral is

$$\oint_C W(z)\, dz = \oint_C cz\, dz = \int_{\theta=0}^{\theta=2\pi} c\hat{R}e^{i\theta} i\hat{R}e^{i\theta}\, d\theta, \tag{8.196}$$

$$= ic\hat{R}^2 \int_0^{2\pi} e^{2i\theta}\, d\theta = \left.\frac{ic\hat{R}^2}{2i}e^{2i\theta}\right|_0^{2\pi} = 0. \tag{8.197}$$

because $e^{0i} = e^{4\pi i} = 1$.

Quadrupole

A quadrupole potential is described by

$$W(z) = \frac{k}{z^2}. \tag{8.198}$$

Taking the contour integral, we find

$$\oint_C \frac{k}{z^2}\, dz = k \int_0^{2\pi} \frac{i\hat{R}e^{i\theta}}{\hat{R}^2 e^{2i\theta}}\, d\theta = \frac{ki}{\hat{R}} \int_0^{2\pi} e^{-i\theta}\, d\theta = \left.\frac{ki}{\hat{R}}\frac{1}{-i}e^{-i\theta}\right|_0^{2\pi} = 0. \tag{8.199}$$

So the only nonzero contour integral is for functions of the form $W(z) = a/z$. We find that all polynomial powers of z have a zero contour integral about the origin for arbitrary contours except this special one.

8.4.2 Laurent Series

Now it can be shown that any function can be expanded, much as for a Taylor series, as a *Laurent series*:[5]

$$W(z) = \cdots + C_{-2}(z-z_o)^{-2} + C_{-1}(z-z_o)^{-1} + C_0(z-z_o)^0 + C_1(z-z_o)^1 + C_2(z-z_o)^2 + \cdots. \tag{8.200}$$

In compact summation notation, we can say

$$W(z) = \sum_{n=-\infty}^{n=\infty} C_n(z-z_o)^n. \tag{8.201}$$

Taking the contour integral of both sides we get

$$\oint_C W(z)\, dz = \oint_C \sum_{n=-\infty}^{n=\infty} C_n(z-z_o)^n\, dz = \sum_{n=-\infty}^{n=\infty} C_n \oint_C (z-z_o)^n\, dz. \tag{8.202}$$

From our just completed analysis, this has value $2\pi i$ only when $n = -1$, so

$$\oint_C W(z)\, dz = C_{-1} 2\pi i. \tag{8.203}$$

Here C_{-1} is known as the *residue* of the Laurent series. In general, we have the *Cauchy integral theorem*, which holds that if $W(z)$ is analytic within and on a closed curve C except for a finite number of singular points, then

$$\oint_C W(z)\, dz = 2\pi i \sum \text{residues}. \tag{8.204}$$

Let us get a simple formula for C_n. We first exchange m for n in Eq. (8.201) and say

$$W(z) = \sum_{m=-\infty}^{m=\infty} C_m(z-z_o)^m. \tag{8.205}$$

Then we operate as follows:

$$\frac{W(z)}{(z-z_o)^{n+1}} = \sum_{m=-\infty}^{m=\infty} C_m(z-z_o)^{m-n-1}, \tag{8.206}$$

$$\oint_C \frac{W(z)}{(z-z_o)^{n+1}}\, dz = \oint_C \sum_{m=-\infty}^{m=\infty} C_m(z-z_o)^{m-n-1}\, dz, \tag{8.207}$$

$$= \sum_{m=-\infty}^{m=\infty} C_m \oint_C (z-z_o)^{m-n-1}\, dz. \tag{8.208}$$

Here C is any closed contour that has z_o in its interior. The contour integral on the right-hand side only has a nonzero value when $n = m$. Let us then insist that $n = m$, giving

$$\oint_C \frac{W(z)}{(z-z_o)^{n+1}}\, dz = C_n \underbrace{\oint_C (z-z_o)^{-1}\, dz}_{=2\pi i}. \tag{8.209}$$

---

[5] Pierre Alphonse Laurent, 1813–1854, Parisian engineer.

We know from earlier analysis that the contour integral enclosing a simple pole such as found on the right-hand side has a value of $2\pi i$. Solving, we find then that

$$C_n = \frac{1}{2\pi i} \oint_C \frac{W(z)}{(z-z_o)^{n+1}} \, dz. \tag{8.210}$$

If the closed contour C encloses no poles, then

$$\oint_C W(z) \, dz = 0. \tag{8.211}$$

8.4.3 Pressure Distribution for Steady Flow

For steady, irrotational, incompressible flow with no body force present, we have the Bernoulli equation. We recast Eq. (6.196) as

$$\frac{p}{\rho} + \frac{1}{2}\nabla^T \phi \cdot \nabla \phi = \frac{p_\infty}{\rho} + \frac{1}{2}U_\infty^2. \tag{8.212}$$

We can write this in terms of the complex potential in a simple fashion. First, recall that

$$\nabla^T \phi \cdot \nabla \phi = u^2 + v^2. \tag{8.213}$$

We also have $dW/dz = u - iv$, so $\overline{dW/dz} = u + iv$. Consequently,

$$\frac{dW}{dz}\overline{\frac{dW}{dz}} = u^2 + v^2 = \nabla^T \phi \cdot \nabla \phi. \tag{8.214}$$

So we get the pressure field from the Bernoulli equation to be

$$p = p_\infty + \frac{1}{2}\rho\left(U_\infty^2 - \frac{dW}{dz}\overline{\frac{dW}{dz}}\right). \tag{8.215}$$

The pressure coefficient C_p is

$$C_p = \frac{p - p_\infty}{\frac{1}{2}\rho U_\infty^2} = 1 - \frac{1}{U_\infty^2}\frac{dW}{dz}\overline{\frac{dW}{dz}}. \tag{8.216}$$

8.4.4 Blasius Force Theorem

For steady flows, we can find the net contribution of a pressure force on an arbitrarily shaped solid body with the *Blasius*[6] *force theorem*. Consider the geometry shown in Fig. 8.20. The surface of the arbitrarily shaped body is described by A_b, and C is a closed contour containing A_b. First consider the linear momenta equation for steady flow, no body forces, and no viscous forces:

$$\rho\left(\mathbf{v}^T \cdot \nabla\right)\mathbf{v} = -\nabla p, \qquad \text{add mass to get conservative form,} \tag{8.217}$$

$$\left(\nabla^T \cdot (\rho \mathbf{v}\mathbf{v}^T)\right)^T = -\nabla p, \qquad \text{integrate over } V, \tag{8.218}$$

---

[6] Paul Richard Heinrich Blasius, 1883–1970, who gave mathematical description of a similarity solution to the boundary layer problem.

8.4 Forces Induced by Potential Flow

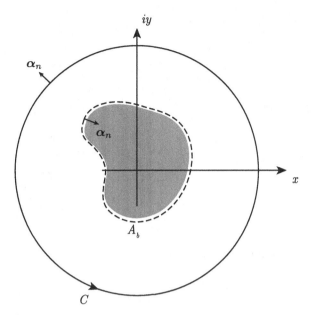

Figure 8.20 Potential flow about an arbitrarily shaped two-dimensional body with fluid control volume indicated.

$$\int_V \left(\nabla^T \cdot (\rho \mathbf{v}\mathbf{v}^T)\right)^T \, dV = -\int_V \nabla p \, dV, \qquad \text{use Gauss}, \tag{8.219}$$

$$\int_A \rho \mathbf{v}(\mathbf{v}^T \cdot \boldsymbol{\alpha}_n) \, dA = -\int_A p \boldsymbol{\alpha}_n \, dA. \tag{8.220}$$

Now the surface integral here is really a line integral with unit depth b, $dA = b \, ds$. Moreover, the surface enclosing the *fluid* has an inner contour A_b and an outer contour C. Now on C, which we prescribe, we will know $x(s)$ and $y(s)$. Here s is arc length. So on C we also get the unit tangent $\boldsymbol{\alpha}_t$ and unit outward normal $\boldsymbol{\alpha}_n$:

$$\boldsymbol{\alpha}_t = \begin{pmatrix} \frac{dx}{ds} \\ \frac{dy}{ds} \end{pmatrix}, \qquad \boldsymbol{\alpha}_n = \begin{pmatrix} \frac{dy}{ds} \\ -\frac{dx}{ds} \end{pmatrix}, \qquad \text{on } C. \tag{8.221}$$

A loose analysis that can be verified more rigorously shows $\boldsymbol{\alpha}_t$ and $\boldsymbol{\alpha}_n$ are unit vectors. Standard geometry tells us $ds^2 = dx^2 + dy^2$. So $\boldsymbol{\alpha}_t = (dx/\sqrt{dx^2+dy^2}, dy/\sqrt{dx^2+dy^2})^T$. By inspection $||\boldsymbol{\alpha}_t|| = 1$. A similar result holds for $\boldsymbol{\alpha}_n$.

On A_b we have, because it is a solid surface,

$$\mathbf{v}^T \cdot \boldsymbol{\alpha}_n = 0, \qquad \text{on } A_b. \tag{8.222}$$

Now let the force on the body due to fluid pressure be $\mathbf{F}$:

$$\int_{A_b} p \boldsymbol{\alpha}_n \, dA = \mathbf{F}. \tag{8.223}$$

Now return to our linear momenta equation,

$$\int_A \rho \mathbf{v}\mathbf{v}^T \cdot \boldsymbol{\alpha}_n \, dA = -\int_A p\boldsymbol{\alpha}_n \, dA. \tag{8.224}$$

Break this up to get

$$\oint_{A_b} \rho \underbrace{\mathbf{v} \, \mathbf{v}^T \cdot \boldsymbol{\alpha}_n}_{=0} \, dA + \oint_C \rho \mathbf{v}\mathbf{v}^T \cdot \boldsymbol{\alpha}_n \, dA = -\underbrace{\oint_{A_b} p\boldsymbol{\alpha}_n \, dA}_{=\mathbf{F}} - \oint_C p\boldsymbol{\alpha}_n \, dA, \tag{8.225}$$

$$\oint_C \rho \mathbf{v}\mathbf{v}^T \cdot \boldsymbol{\alpha}_n \, dA = -\mathbf{F} - \oint_C p\boldsymbol{\alpha}_n \, dA. \tag{8.226}$$

We can break this into x and y components:

$$\oint_C \rho u \underbrace{\left(u\frac{dy}{ds} - v\frac{dx}{ds}\right)}_{\mathbf{v}^T \cdot \boldsymbol{\alpha}_n} b \, ds = -F_x - \oint_C p \underbrace{\frac{dy}{ds}}_{\alpha_{nx}} b \, ds, \tag{8.227}$$

$$\oint_C \rho v \underbrace{\left(u\frac{dy}{ds} - v\frac{dx}{ds}\right)}_{\mathbf{v}^T \cdot \boldsymbol{\alpha}_n} b \, ds = -F_y - \oint_C p \underbrace{\left(-\frac{dx}{ds}\right)}_{\alpha_{ny}} b \, ds. \tag{8.228}$$

Solving for F_x and F_y per unit depth, we get

$$\frac{F_x}{b} = \oint_C -p \, dy - \rho u^2 \, dy + \rho u v \, dx, \quad \frac{F_y}{b} = \oint_C p \, dx + \rho v^2 \, dx - \rho u v \, dy. \tag{8.229}$$

Now Bernoulli gives us $p = p_o - (1/2)\rho(u^2 + v^2)$, where p_o is some constant. So the x force per unit depth becomes

$$\frac{F_x}{b} = \oint_C -p_o \, dy + \frac{1}{2}\rho(u^2 + v^2) \, dy - \rho u^2 \, dy + \rho u v \, dx. \tag{8.230}$$

Because the integral over a closed contour of a constant p_o is zero, we get

$$\frac{F_x}{b} = \oint_C \frac{1}{2}\rho(-u^2 + v^2) \, dy + \rho u v \, dx = \frac{1}{2}\rho \oint_C (-u^2 + v^2) \, dy + 2uv \, dx. \tag{8.231}$$

Similarly for the y direction, we get

$$\frac{F_y}{b} = \oint_C p_o \, dx - \frac{1}{2}\rho(u^2 + v^2) \, dx + \rho v^2 \, dx - \rho u v \, dy, \tag{8.232}$$

$$= \frac{1}{2}\rho \oint_C (-u^2 + v^2) \, dx - 2uv \, dy. \tag{8.233}$$

Now consider the linear combination $(F_x - iF_y)/b$:

$$\frac{F_x - iF_y}{b} = \frac{1}{2}\rho \oint_C (-u^2 + v^2)\, dy + 2uv\, dx - (-u^2 + v^2)i\, dx + 2uvi\, dy, \quad (8.234)$$

$$= \frac{1}{2}\rho \oint_C (i(u^2 - v^2) + 2uv)\, dx + ((-u^2 + v^2) + 2uvi)\, dy, \quad (8.235)$$

$$= \frac{1}{2}\rho \oint_C (i(u^2 - v^2) + 2uv)\, dx + (i(u^2 - v^2) + 2uv)i\, dy, \quad (8.236)$$

$$= \frac{1}{2}\rho \oint_C (i(u^2 - v^2) + 2uv)(dx + i\, dy), \quad (8.237)$$

$$= \frac{1}{2}\rho \oint_C i(u - iv)^2(dx + i\, dy) = \frac{1}{2}\rho i \oint_C \left(\frac{dW}{dz}\right)^2 dz. \quad (8.238)$$

So if we have the complex potential, we can get the force on a body.

8.4.5 Kutta–Zhukovsky Lift Theorem

We consider here the *Kutta*[7]–*Zhukovsky*[8] *lift theorem*. Consider the geometry presented in Fig. 8.21. Here we consider a flow with a freestream constant velocity of U. We take an arbitrary

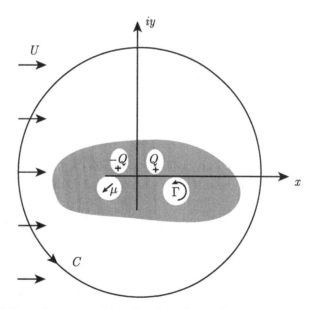

Figure 8.21 Potential flow about an arbitrarily shaped two-dimensional body with distribution of sources, sinks, vortices, and dipoles.

---

[7] Martin Wilhelm Kutta, 1867–1944, Silesian-born German mechanician, co-developer of the Runge–Kutta method for integrating ordinary differential equations.
[8] Nikolai Egorovich Zhukovsky, 1847–1921, Russian applied mathematician and mechanician, father of Russian aviation, who purchased a glider from Lilienthal, and developed the lift theorem independently of Kutta.

body shape to enclose a distribution of canceling source–sink pairs, doublets, point vortices, quadruples, and any other non-mass adding potential flow term. This combination gives rise to some surface that is a streamline.

Now far from the body surface, a contour sees all of these features as effectively concentrated at the origin. Then, the potential can be written as

$$W(z) \sim \underbrace{Uz}_{\text{uniform flow}} + \underbrace{\frac{Q}{2\pi}\ln z - \frac{Q}{2\pi}\ln z}_{\text{canceling source–sink pair}} + \underbrace{\frac{i\Gamma}{2\pi}\ln z}_{\text{clockwise! vortex}} + \underbrace{\frac{\mu}{z}}_{\text{doublet}} + \cdots. \tag{8.239}$$

The sign convention for Γ has been violated here, by tradition. Now let us take D to be the so-called drag force per unit depth and L to be the so-called lift force per unit depth, so in terms of F_x and F_y, we have

$$\frac{F_x}{b} = D, \qquad \frac{F_y}{b} = L. \tag{8.240}$$

Now by the Blasius force theorem, we have

$$D - iL = \frac{1}{2}\rho i \oint_C \left(\frac{dW}{dz}\right)^2 dz = \frac{1}{2}\rho i \oint_C \left(U + \frac{i\Gamma}{2\pi z} - \frac{\mu}{z^2} + \cdots\right)^2 dz, \tag{8.241}$$

$$= \frac{1}{2}\rho i \oint_C \left(U^2 + \frac{i\Gamma U}{\pi z} - \frac{1}{z^2}\left(\frac{\Gamma^2}{4\pi^2} + 2U\mu\right) + \cdots\right) dz. \tag{8.242}$$

Now the Cauchy integral theorem, Eq. (8.204), gives us the contour integral to be $2\pi i \sum \text{residues}$. Here the residue is $i\Gamma U/\pi$. So we get

$$D - iL = \frac{1}{2}\rho i \left(2\pi i \left(\frac{i\Gamma U}{\pi}\right)\right) = -i\rho\Gamma U. \tag{8.243}$$

We see that

$$D = 0, \qquad L = \rho U \Gamma. \tag{8.244}$$

This is a remarkably simple and elegant result! Note the following. (1) The term Γ is associated with *clockwise* circulation here. This is something of a tradition in aerodynamics. (2) Because for airfoils $\Gamma \sim U$, we get the lift force $L \sim \rho U^2$. (3) For steady inviscid flow, there is no drag force. Consideration of either unsteady or viscous effects would lead to a nonzero x component of force. This force is often described as inducing the *Magnus*[9] *effect* on spinning bodies moving through fluids.

Example 8.4 Consider the flow over a cylinder of radius a with clockwise circulation Γ.

Solution
To do so, we can superpose a point vortex onto the potential for flow over a cylinder in the following fashion:

$$W(z) = U\left(z + \frac{a^2}{z}\right) + \frac{i\Gamma}{2\pi}\ln\left(\frac{z}{a}\right). \tag{8.245}$$

[9] Heinrich Gustav Magnus, 1802–1870, German experimental physicist.

Breaking this up as before into real and imaginary parts, we get

$$W(z) = \left(Ur\cos\theta\left(1+\frac{a^2}{r^2}\right)\right) + i\left(Ur\sin\theta\left(1-\frac{a^2}{r^2}\right)\right) + \frac{i\Gamma}{2\pi}\left(\ln\left(\frac{r}{a}\right)+i\theta\right). \quad (8.246)$$

So, we find

$$\psi = \Im(W(z)) = Ur\sin\theta\left(1-\frac{a^2}{r^2}\right) + \frac{\Gamma}{2\pi}\ln\left(\frac{r}{a}\right). \quad (8.247)$$

On $r = a$, we find that $\psi = 0$, so the addition of the circulation in the way we have proposed maintains the cylinder surface as a streamline. It is important to note that this is valid for *arbitrary* Γ. That is, the potential flow solution for flow over a cylinder is *nonunique*. In aerodynamics, this is used to advantage to add just enough circulation to enforce the so-called *Kutta condition*. The Kutta condition is an experimental observation that for a steady flow, the trailing edge of an airfoil is a stagnation point.

The Kutta–Zhukovsky lift theorem tells us that whenever we add circulation, a lift force $L = \rho U \Gamma$ is induced. This is consistent with the phenomena observed in baseball that the "fastball" rises. The fastball leaves the pitcher's hand traveling towards the batter and rotating towards the pitcher. The induced aerodynamic force is opposite to the force of gravity.

Let us get the lift force the hard way and confirm the Kutta–Zhukovsky theorem. We can get the velocity field from the velocity potential:

$$\phi = \Re(W(z)) = Ur\cos\theta\left(1+\frac{a^2}{r^2}\right) - \frac{\Gamma\theta}{2\pi}. \quad (8.248)$$

Thus, we differentiate ϕ appropriately to find v_r and v_θ:

$$v_r = \frac{\partial \phi}{\partial r} = Ur\cos\theta\left(-\frac{2a^2}{r^3}\right) + U\cos\theta\left(1+\frac{a^2}{r^2}\right), \quad (8.249)$$

$$v_r|_{r=a} = U\cos\theta\left(-\frac{2a^3}{a^3}+1+\frac{a^2}{a^2}\right) = 0, \quad (8.250)$$

$$v_\theta = \frac{1}{r}\frac{\partial \phi}{\partial \theta} = \frac{1}{r}\left(-Ur\sin\theta\left(1+\frac{a^2}{r^2}\right) - \frac{\Gamma}{2\pi}\right), \quad (8.251)$$

$$v_\theta|_{r=a} = -U\sin\theta\left(1+\frac{a^2}{a^2}\right) - \frac{\Gamma}{2\pi a} = -2U\sin\theta - \frac{\Gamma}{2\pi a}. \quad (8.252)$$

We get the pressure on the cylinder surface from the Bernoulli equation:

$$p = p_\infty + \frac{1}{2}\rho U^2 - \frac{1}{2}\rho \nabla^T \phi \cdot \nabla \phi = p_\infty + \frac{1}{2}\rho U^2 - \frac{1}{2}\rho\left(-2U\sin\theta - \frac{\Gamma}{2\pi a}\right)^2. \quad (8.253)$$

Now for a small element of the cylinder at $r = a$, the surface area is $dA = br\, d\theta = ba\, d\theta$. This is depicted in Fig. 8.22. We also note that the x and y forces depend on the orientation of the element, given by θ. Elementary trigonometry shows that the elemental x and y forces per depth are

$$\frac{dF_x}{b} = -p(\cos\theta)a\, d\theta, \qquad \frac{dF_y}{b} = -p(\sin\theta)a\, d\theta. \quad (8.254)$$

So integrating over the entire cylinder, we obtain,

$$\frac{F_x}{b} = \int_0^{2\pi} -\underbrace{\left(p_\infty + \frac{1}{2}\rho U^2 - \frac{1}{2}\rho\left(-2U\sin\theta - \frac{\Gamma}{2\pi a}\right)^2\right)}_{p}(\cos\theta)a\, d\theta, \quad (8.255)$$

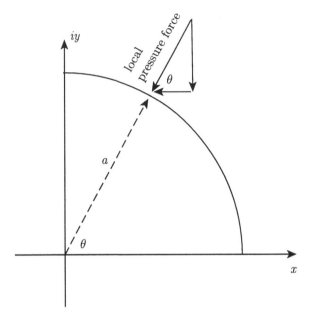

Figure 8.22 Pressure force on a differential area element of cylindrical surface.

$$\frac{F_y}{b} = \int_0^{2\pi} -\underbrace{\left(p_\infty + \frac{1}{2}\rho U^2 - \frac{1}{2}\rho\left(-2U\sin\theta - \frac{\Gamma}{2\pi a}\right)^2\right)}_{p}(\sin\theta)a\,d\theta. \qquad (8.256)$$

Integration via computer algebra gives

$$\frac{F_x}{b} = 0, \qquad \frac{F_y}{b} = \rho U \Gamma. \qquad (8.257)$$

This is identical to the result we expect from the Kutta–Zhukovsky lift theorem, Eq. (8.244).

SUMMARY

This chapter has focused on flow fields that are steady, two-dimensional planar, irrotational, and incompressible. These assumptions allow the use of two scalar functions, a velocity potential and stream function, to efficiently describe complicated fluid kinematics. The dynamics can be understood with consideration of a Bernoulli principle. Analysis of such flows is aided by the remarkable theory of analytic complex functions, which can be directly applied to fluid mechanics.

PROBLEMS

8.1 A complex potential is given by $W(z) = z^4$. Find and plot the associated velocity potential $\phi(x,y)$, stream function $\psi(x,y)$, and velocity vector field $\mathbf{v} = u(x,y)\mathbf{i} + v(x,y)\mathbf{j}$. Confirm the flow is incompressible and irrotational.

8.2 A complex potential is given by $W(z) = \sin z$. Show $\sin z = \sin x \cosh y + i \cos x \sinh y$ by use of standard trigonometric identities and Taylor series expansions of sin, cos, sinh, and cosh. Find and plot the associated velocity potential $\phi(x, y)$, stream function $\psi(x, y)$, and velocity vector field $\mathbf{v} = u(x, y)\mathbf{i} + v(x, y)\mathbf{j}$. Confirm the flow is incompressible and irrotational.

8.3 For potential flow over a cylinder of radius $a = 1$ and freestream velocity $U = 1$ with density $\rho = 1$ and freestream pressure $p_\infty = 0$, find expressions for $u(x, y)$, $v(x, y)$, and $p(x, y)$. Plot isobars on the same plot with the stream function, velocity potential, and velocity vector fields. Take the body force to be zero.

8.4 Add the potential of an ideal point vortex, $W(z) = -i\Gamma_o/(2\pi) \ln z$, to the potential of flow over a cylinder of radius $a = 1$ and freestream velocity $U = 1$ with density $\rho = 1$ and freestream pressure $p_\infty = 0$. Show $r = 1$ remains a streamline even in the presence of the ideal vortex. Find expressions for $u(x, y)$, $v(x, y)$, and $p(x, y)$. Plot isobars on the same plot with the stream function and velocity potential for $\Gamma_o = -2\pi, -4\pi, -9\pi/2$. Describe how the location of the stagnation points evolves as Γ_o evolves. Find the lift force. Neglect body forces.

8.5 Consider potential flow over a cylinder with circulation given by

$$W(z) = z + \frac{1}{z} + i \ln z.$$

Find an expression for the streamline curvature $\kappa(x, y)$, and plot contours of κ superposed on the velocity vector field and the stream function.

8.6 Consider the quadrupole given by $W(z) = 1/z^2$. Find $u(x, y)$, $v(x, y)$, and with $\rho = 1$, $p_\infty = 0$, find $p(x, y)$. Find the acceleration field $a_x(x, y)$, $a_y(x, y)$. Plot curves of constant ϕ superposed onto the velocity vector field. Then plot curves of constant p superposed onto the acceleration vector field. Neglect body forces.

8.7 Consider again the quadrupole given by $W(z) = 1/z^2$. Find an expression for the deformation tensor $\mathbf{D}$. Find an expression for the only nonzero invariant of the deformation tensor and give a contour plot of it in the (x, y) plane superposed onto the velocity vector field.

8.8 Consider an incompressible flow that when cast in spherical coordinates has no velocity or variation in the θ direction. In this limit, the continuity equation, $\nabla^T \cdot \mathbf{v} = 0$, is

$$\frac{1}{r^2}\frac{\partial}{\partial r}(r^2 v_r) + \frac{1}{r \sin \phi}\frac{\partial}{\partial \phi}(\sin \phi \, v_\phi) = 0.$$

With

$$v_r = \frac{1}{r^2 \sin \phi}\frac{\partial \psi}{\partial \phi}, \qquad v_\phi = -\frac{1}{r \sin \phi}\frac{\partial \psi}{\partial r},$$

show that this definition guarantees that mass conservation is satisfied for the incompressible flow and that

$$\psi(r, \phi) = \frac{1}{2}Ur^2 \sin^2 \phi \left(1 - \frac{a^3}{r^3}\right),$$

represents the flow of a uniform freestream with velocity U over a sphere of radius a. Show $\nabla^2 \psi \neq 0$, but that when the E^2 operator, Eq. (2.383), is applied, that

$$E^2 \psi = \frac{\partial^2 \psi}{\partial r^2} + \frac{\sin \phi}{r^2} \frac{\partial}{\partial \phi} \left(\frac{1}{\sin \phi} \frac{\partial \psi}{\partial \phi} \right) = 0.$$

In this problem ϕ is a variable for the spherical coordinate system and has no relation to a velocity potential.

8.9 Consider again an incompressible flow that when cast in spherical coordinates has no velocity or variation in the θ direction. With the velocity potential, here denoted as Φ, take

$$v_r = \frac{\partial \Phi}{\partial r}, \qquad v_\phi = \frac{1}{r} \frac{\partial \Phi}{\partial \phi}.$$

This yields $\mathbf{v} = \nabla \Phi$. Show that this guarantees that mass conservation is satisfied for the incompressible flow. The potential field

$$\Phi(r, \phi) = U r \cos \phi \left(1 + \frac{a^3}{2 r^3} \right),$$

represents the flow of a uniform freestream with velocity U over a sphere of radius a. Show $\nabla^2 \Phi = 0$. Show the velocity field generated by this velocity potential is identical to that generated by the stream function from the previous problem. Show this potential field is irrotational by direct calculation of the only possible nonzero component of vorticity:

$$\omega_\theta = \frac{1}{r} \frac{\partial}{\partial r} (r v_\phi) - \frac{1}{r} \frac{\partial v_r}{\partial \phi}.$$

8.10 Recall the Helmholtz decomposition (see Problem 3.13). A velocity field has a nonunique decomposition into potential and vortical parts:

$$\mathbf{v} = \mathbf{v}^{(\phi)} + \mathbf{v}^{(\omega)},$$

where $\mathbf{v}^{(\phi)}$ and $\mathbf{v}^{(\omega)}$ are the potential and vortical parts, respectively, defined by

$$\mathbf{v}^{(\phi)} = \nabla \phi, \qquad \mathbf{v}^{(\omega)} = \nabla \times \boldsymbol{\psi}.$$

Here ϕ is a nonunique scalar potential field, and $\boldsymbol{\psi}$ is a nonunique vector field. We require that $\boldsymbol{\psi}$ be solenoidal: $\nabla^T \cdot \boldsymbol{\psi} = 0$. Apply this to a compressible, rotational two-dimensional planar flow field with

$$\phi = x^2 + y^2, \qquad \boldsymbol{\psi} = \left(0, 0, x^2 + y^2 \right)^T.$$

Find $\mathbf{v}, \mathbf{v}^{(\phi)}, \mathbf{v}^{(\omega)}, \boldsymbol{\omega}$, and ϑ. Show $\nabla^2 \boldsymbol{\psi} = -\boldsymbol{\omega}$ and $\nabla^2 \phi = \vartheta$. Then, apply the Helmholtz decomposition to an incompressible, irrotational, two-dimensional planar source flow field:

$$\phi = 0, \qquad \boldsymbol{\psi} = \left(0, 0, \ln \frac{1}{\sqrt{x^2 + y^2}} \right)^T.$$

Find $\mathbf{v}, \mathbf{v}^{(\phi)}, \mathbf{v}^{(\omega)}, \boldsymbol{\omega}$, and ϑ. Show $\nabla^2 \boldsymbol{\psi} = \mathbf{0}$ and $\nabla^2 \phi = 0$. Then show

$$\phi = \mathrm{Tan}^{-1}(x, y), \qquad \boldsymbol{\psi} = \mathbf{0},$$

yields an identical velocity field.

FURTHER READING

Basset, A. B. (1888). *A Treatise on Hydrodynamics with Numerous Examples*, 2 vols. Cambridge, UK: Deighton, Bell, and Co.

Batchelor, G. K. (2000). *An Introduction to Fluid Dynamics*. Cambridge, UK: Cambridge University Press.

Brodkey, R. S. (1967). *The Phenomena of Fluid Motions*. New York: Dover.

Currie, I. G. (2013). *Fundamental Mechanics of Fluids*, 4th ed. Boca Raton, Florida: CRC Press.

Karamcheti, K. (1980). *Principles of Ideal-Fluid Aerodynamics*, 2nd ed. Malabar, Florida: Krieger.

Katz, J., and Plotkin, A. (2001). *Low-Speed Aerodynamics*, 2nd ed. New York: Cambridge University Press.

Kuethe, A. M., and Chow, C.-Y. (1998). *Foundations of Aerodynamics: Bases of Aerodynamic Design*, 5th ed. New York: John Wiley.

Kundu, P. K., Cohen, I. M., and Dowling, D. R. (2015). *Fluid Mechanics*, 6th ed. Amsterdam: Academic Press.

Lamb, H. (1993). *Hydrodynamics*, 6th ed. New York: Dover.

Leal, L. G. (2007). *Advanced Transport Phenomena: Fluid Mechanics and Convective Transport Processes*. New York: Cambridge University Press.

Yih, C.-S. (1977). *Fluid Mechanics*. East Hampton, Connecticut: West River Press.

REFERENCES

Bird, R. B., Stewart, W. E., and Lightfoot, E. N. (2007). *Transport Phenomena*, revised 2nd ed. New York: John Wiley.

Brown, J. W., and Churchill, R. V. (2014). *Complex Variables and Applications*, 9th ed. New York: McGraw-Hill.

Goldstein, H. (1950). *Classical Mechanics*. Reading, Massachusetts: Addison-Wesley.

Panton, R. L. (2013). *Incompressible Flow*, 4th ed. New York: John Wiley.

Powers, J. M., and Sen, M. (2015). *Mathematical Methods in Engineering*. New York: Cambridge University Press.

9 One-Dimensional Compressible Flow

This chapter will focus on one-dimensional flow of a compressible fluid. The emphasis will be on inviscid problems, with one brief excursion into viscous compressible flow that will serve as a transition to a study of viscous flow in following chapters. The compressibility we will study here is that which is induced when the fluid particle velocity is of similar magnitude to the fluid sound speed. For such flows, it will be seen that kinetic energy changes are often as important as thermal energy changes, and we can expect to find large variation in nearly all flow variables, for example density, temperature, pressure and velocity. Such flows have widespread physical relevance in jet and rocket propulsion, aerodynamics, hypersonics, acoustics, explosions, and astrophysics. We will only consider one-dimensional problems, which are sufficient to expose most of the relevant flow physics. A paradigm problem that will well illustrate remarkable features of compressible flow is that of a piston-driven shock wave, depicted in Fig. 9.1. Here a piston with velocity $\hat{u}_p$ is driving a shock wave with velocity U into a fluid at rest at ambient pressure p_1. In a thin zone, usually modeled as infinitely thin in the inviscid limit, the pressure rapidly increases to p_2. When modeled in the inviscid limit, mathematical discontinuities are admitted, and one must return to the original integral form of the governing equations (e.g. Eqs. (4.4, 4.30, 4.158)) in order to correctly analyze the shock. Understanding shock waves is essential for describing atmospheric re-entry of space vehicles, supersonic flight, detonation, and many other physically important problems.

There are interesting problems in compressible flow that will be deferred to later chapters. Later in Section 11.2.6 we will consider flow with variable density induced by local heating in a nearly constant pressure environment known as natural convection. For such flows, body forces are relevant, and it is common to induce fluid velocities that are well below the sound speed. Such flows are of practical importance in weather prediction, heating, ventilation, and air conditioning. While formally a compressible flow, methods of analysis for natural convection flows are dramatically different than for high speed compressible flows, so they are deferred. Also in Section 11.2.7, we will study compressible viscous boundary layers. Such flows are best studied after incompressible viscous boundary layers have been considered, which will be done in Section 11.2.1.

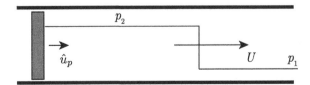

Figure 9.1 Schematic of a piston-driven shock wave.

The flows we will study in this chapter provide excellent examples of the importance of the coupling of the complete system of mass, momentum, and energy evolution axioms along with constitutive equations, especially thermodynamic state relations. This stands in contrast to the many incompressible flow solutions considered in Chapters 7 and 8. For most incompressible flows, it is possible to solve the mass and momentum equations without considering either the energy equation or thermodynamic equations of state. Additionally, recalling from Section 4.7.3 how the governing equations can be taken as a balance of local change with advection, diffusion, and source terms, Eq. (4.280), we will see examples here of such balances. Most of the balances will be such that local change is balanced by advection. However, we will consider in Section 9.12 an important example in which advection balances diffusion in a compressible flow environment.

We will assume for this chapter a one-dimensional flow model so that $v \equiv 0$, $w \equiv 0$, $\partial/\partial y \equiv 0$, and $\partial/\partial z \equiv 0$. Except for Section 9.12 on viscous shock waves, friction and heat transfer will not be modeled rigorously. Instead, they will be modeled in a fashion that loosely captures the relevant physics and retains analytic tractability. Mathematically, we will not model friction and heat transfer as a classical diffusion processes; consequently, we will consider $\mu \equiv 0$ and $k \equiv 0$. However, we will introduce simpler, less rigorous, new terms to model friction and heat transfer. They will will have a different mathematical character. As a consequence, some of our solutions will not represent rational limiting cases of the more fundamental Navier–Stokes equations. Direct comparison of results using our modeling approximations will never completely agree with equivalent (and expensive) predictions of compressible Navier–Stokes equations. Further, we will ignore the influences of an external body force, $f_i = 0$. Except for Section 9.12, our model will best be seen as an adaptation of the Euler equations of Section 6.6. It will have the advantage of yielding rapid and non-intuitive insight into how actual fluids behave under the extreme conditions of flow at velocities near or above the speed of sound.

9.1 Thermodynamics of Compressible Fluids

To understand a compressible flowing fluid, it is essential to first understand the thermodynamics of a compressible stationary fluid. Here let us briefly consider compressible fluids with general nonideal equations of state. Consideration of nonideal equations of state is the proper framework for studying compressible flow and shock propagation in dense fluids with significant intermolecular forces. This is relevant for a gas near the vapor dome, for example a van der Waals[1] gas, or for solids that have been shocked far beyond their yield stress. Most aeronautical applications are well captured by an ideal gas equation of state, and many common results are restricted to that limit. Our results will be more general, but will require a more detailed consideration of thermodynamics that is typical.

We presume here an understanding of fundamental classical thermodynamics and focus on specialized aspects we will need to aid in the study of compressible mechanics of fluids. Here, we will focus attention on the speed of sound as well as how a typical thermodynamic analysis is affected by nonideal equations of state.

---

[1] Johannes Diderik van der Waals, 1837–1923, Dutch thermodynamicist and 1910 Nobel laureate in physics for his work in developing his celebrated equation of state.

9.1.1 Maxwell Relation

To best understand the thermodynamics of sound speed for general materials, we must develop another Maxwell relation, similar to one found in Section 6.4.3. Consider first the Helmholtz free energy, $\hat{a}$. It is a thermodynamic property and was introduced in Eq. (5.175): $\hat{a} = e - Ts$. Differentiating, we get

$$d\hat{a} = de - T\,ds - s\,dT. \tag{9.1}$$

Now use the Gibbs equation, Eq. (4.195), to eliminate de to get

$$d\hat{a} = \underbrace{(-p\,d\hat{v} + T\,ds)}_{de} - T\,ds - s\,dT = -p\,d\hat{v} - s\,dT. \tag{9.2}$$

We have $\hat{a} = \hat{a}(\hat{v}, T)$, because intensive thermodynamic properties are functions of two independent state variables for simple compressible substances. Then calculus tells us

$$d\hat{a} = \left.\frac{\partial \hat{a}}{\partial \hat{v}}\right|_T d\hat{v} + \left.\frac{\partial \hat{a}}{\partial T}\right|_{\hat{v}} dT. \tag{9.3}$$

Comparing to Eq. (9.2), we see that we must have

$$-p = \left.\frac{\partial \hat{a}}{\partial \hat{v}}\right|_T, \quad -s = \left.\frac{\partial \hat{a}}{\partial T}\right|_{\hat{v}}. \tag{9.4}$$

These are fully equivalent to the earlier Eq. (5.187) after using the chain rule with $\hat{v} = 1/\rho$. Differentiating the first with respect to T and the second with respect to $\hat{v}$ gives

$$-\left.\frac{\partial p}{\partial T}\right|_{\hat{v}} = \frac{\partial^2 \hat{a}}{\partial T \partial \hat{v}}, \quad -\left.\frac{\partial s}{\partial \hat{v}}\right|_T = \frac{\partial^2 \hat{a}}{\partial \hat{v} \partial T}. \tag{9.5}$$

Assuming $\hat{a}$ is continuous and sufficiently differentiable, the order of differentiation of the mixed second partials does not matter, thus giving the Maxwell relation

$$\left.\frac{\partial p}{\partial T}\right|_{\hat{v}} = \left.\frac{\partial s}{\partial \hat{v}}\right|_T. \tag{9.6}$$

This is useful because $\partial p/\partial T|_{\hat{v}}$ is available from the thermal equation of state, and it will be required in analysis of the next section in which we find caloric equations of state that are consistent with a given thermal equation of state.

9.1.2 Internal Energy from Thermal Equation of State

Let us first find the internal energy $e(T, \hat{v})$ for a general material whose thermal equation of state, $p = p(T, \hat{v})$, is known. Starting then with e, we have

$$e = e(T, \hat{v}), \quad \text{thus} \quad de = \left.\frac{\partial e}{\partial T}\right|_{\hat{v}} dT + \left.\frac{\partial e}{\partial \hat{v}}\right|_T d\hat{v}. \tag{9.7}$$

Specific heat at constant volume, c_v, for a general material is defined as

$$c_v = \left.\frac{\partial e}{\partial T}\right|_{\hat{v}}. \tag{9.8}$$

Using this, Eq. (9.7) becomes

$$de = c_v\, dT + \left.\frac{\partial e}{\partial \hat{v}}\right|_T d\hat{v}. \tag{9.9}$$

Now from the Gibbs equation, Eq. (4.195), we find

$$de = T\, ds - p\, d\hat{v}, \quad \text{so} \quad \frac{de}{d\hat{v}} = T\frac{ds}{d\hat{v}} - p, \quad \text{so} \quad \left.\frac{\partial e}{\partial \hat{v}}\right|_T = T\left.\frac{\partial s}{\partial \hat{v}}\right|_T - p. \tag{9.10}$$

Substitute from the Maxwell relation, Eq. (9.6), to get

$$\left.\frac{\partial e}{\partial \hat{v}}\right|_T = T\left.\frac{\partial p}{\partial T}\right|_{\hat{v}} - p. \tag{9.11}$$

Thus, Eq. (9.9) can be rewritten as

$$de = c_v\, dT + \underbrace{\left(T\left.\frac{\partial p}{\partial T}\right|_{\hat{v}} - p\right)}_{\left.\frac{\partial e}{\partial \hat{v}}\right|_T} d\hat{v}. \tag{9.12}$$

Integrating, we then get the caloric equation of state:

$$\int_{e_o}^{e} d\hat{e} = \int_{T_o}^{T} c_v(\hat{T})\, d\hat{T} + \int_{\hat{v}_o}^{\hat{v}} \left(T\left.\frac{\partial p}{\partial T}\right|_{\tilde{v}} - p\right) d\tilde{v}, \tag{9.13}$$

$$e(T, \hat{v}) = e_o + \int_{T_o}^{T} c_v(\hat{T})\, d\hat{T} + \int_{\hat{v}_o}^{\hat{v}} \left(T\left.\frac{\partial p}{\partial T}\right|_{\tilde{v}} - p\right) d\tilde{v}. \tag{9.14}$$

This is the caloric equation of state that is thermodynamically consistent with the given thermal equation of state. Here $\hat{T}$ and $\tilde{v}$ are dummy variables of integration.

Example 9.1 Find a general expression for $e(T, \hat{v})$ if we have an ideal gas:

$$p(T, \hat{v}) = \frac{RT}{\hat{v}}. \tag{9.15}$$

Solution
Proceed as follows:

$$\left.\frac{\partial p}{\partial T}\right|_v = \frac{R}{\hat{v}}, \tag{9.16}$$

$$T\left.\frac{\partial p}{\partial T}\right|_{\hat{v}} - p = \frac{RT}{\hat{v}} - p = \frac{RT}{\hat{v}} - \frac{RT}{\hat{v}} = 0. \tag{9.17}$$

Thus, e is

$$e(T) = e_o + \int_{T_o}^{T} c_v(\hat{T})\, d\hat{T}. \tag{9.18}$$

Iff c_v is a constant, the caloric equation of state is

$$e(T) = e_o + c_v(T - T_o). \tag{9.19}$$

Example 9.2 Find a general expression for $e(T, \hat{v})$ for a van der Waals gas:

$$p(T, \hat{v}) = \frac{RT}{\hat{v} - b} - \frac{a}{\hat{v}^2}. \qquad (9.20)$$

Solution

Proceed as before:

$$\left.\frac{\partial p}{\partial T}\right|_{\hat{v}} = \frac{R}{\hat{v} - b}, \qquad (9.21)$$

$$T \left.\frac{\partial p}{\partial T}\right|_{\hat{v}} - p = \frac{RT}{\hat{v} - b} - p = \frac{RT}{\hat{v} - b} - \left(\frac{RT}{\hat{v} - b} - \frac{a}{\hat{v}^2}\right) = \frac{a}{\hat{v}^2}. \qquad (9.22)$$

Thus, the caloric equation of state for e is

$$e(T, \hat{v}) = e_o + \int_{T_o}^{T} c_v(\hat{T}) \, d\hat{T} + \int_{\hat{v}_o}^{\hat{v}} \frac{a}{\tilde{v}^2} \, d\tilde{v} = e_o + \int_{T_o}^{T} c_v(\hat{T}) \, d\hat{T} + a \left(\frac{1}{\hat{v}_o} - \frac{1}{\hat{v}}\right). \qquad (9.23)$$

If c_v is constant, the caloric equation of state for the van der Waals gas reduces to

$$e(T, \hat{v}) = e_o + c_v(T - T_o) - a \left(\frac{1}{\hat{v}} - \frac{1}{\hat{v}_o}\right). \qquad (9.24)$$

Example 9.3 Analyze the canonical form of the van der Waals gas in which we have the Helmholtz free energy $\hat{a}(T, \hat{v})$ as

$$\hat{a}(T, \hat{v}) = e_o + c_v(T - T_o) - a\left(\frac{1}{\hat{v}} - \frac{1}{\hat{v}_o}\right) - T\left(s_o + c_v \ln \frac{T}{T_o} - R \ln \frac{\hat{v}_o - b}{\hat{v} - b}\right). \qquad (9.25)$$

Here e_o, $\hat{v}_o$, T_o, and s_o are reference state constants, and we take the specific heat c_v to be a constant. Also note the van der Waals constant a is distinct from the Helmholtz free energy $\hat{a}$.

Solution

The advantage of the canonical form is that it contains all thermodynamic information about the material. Let us see that it is consistent with standard forms of the van der Waals gas. First let us use Eq. (9.4) to get the pressure and entropy:

$$p = -\left.\frac{\partial \hat{a}}{\partial \hat{v}}\right|_T = \frac{RT}{\hat{v} - b} - \frac{a}{\hat{v}^2}, \quad s = -\left.\frac{\partial \hat{a}}{\partial T}\right|_{\hat{v}} = s_o + c_v \ln \frac{T}{T_o} - R \ln \frac{\hat{v}_o - b}{\hat{v} - b}. \qquad (9.26)$$

Then with $e = \hat{a} + Ts$ and $h = e + p\hat{v}$, we get e and h as functions of T and $\hat{v}$ as

$$e = e_o + c_v(T - T_o) - a\left(\frac{1}{\hat{v}} - \frac{1}{\hat{v}_o}\right), \qquad (9.27)$$

$$h = e_o + c_v(T - T_o) + \left(\frac{RT}{\hat{v} - b} - \frac{a}{\hat{v}^2}\right)\hat{v} - a\left(\frac{1}{\hat{v}} - \frac{1}{\hat{v}_o}\right). \qquad (9.28)$$

Example 9.4 As shown in Fig. 9.2, a van der Waals gas with $R = 200\,\text{J/kg/K}$, $a = 150\,\text{Pa}\,\text{m}^6/\text{kg}^2$, $b = 0.001\,\text{m}^3/\text{kg}$, $c_v = (350\,\text{J/kg/K}) + (0.2\,\text{J/kg/K}^2)(T - 300\,\text{K})$ begins at $T_1 = 300\,\text{K}$, $p_1 = 10^5\,\text{Pa}$. It is isothermally compressed to state 2 for which $p_2 = 10^6\,\text{Pa}$. It is then isochorically heated to state 3 for which $T_3 = 1000\,\text{K}$. Find w_{13}, q_{13}, and $s_3 - s_1$. Assume the surroundings are at $1000\,\text{K}$.

Solution

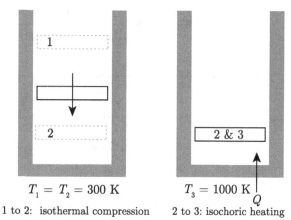

Figure 9.2 Two step thermodynamic process (isothermal compression then isochoric heating) applied to a van der Waals gas.

Recall

$$p(T, \hat{v}) = \frac{RT}{\hat{v} - b} - \frac{a}{\hat{v}^2}. \tag{9.29}$$

So at state 1

$$10^5 \text{ Pa} = \frac{\left(200 \, \frac{\text{J}}{\text{kg K}}\right)(300 \, \text{K})}{\hat{v}_1 - 0.001 \, \frac{\text{m}^3}{\text{kg}}} - \frac{150 \, \text{Pa} \, \text{m}^6/\text{kg}^2}{\hat{v}_1^2}. \tag{9.30}$$

Expanding and ignoring the units, we get

$$-0.15 + 150\hat{v} - 60\,100\hat{v}^2 + 100\,000\hat{v}^3 = 0. \tag{9.31}$$

This is a cubic equation, and it thus has three solutions:

$$\hat{v}_1 = 0.598 \, \frac{\text{m}^3}{\text{kg}}, \qquad \hat{v}_1 = 0.001\,25 \pm 0.0097i \, \frac{\text{m}^3}{\text{kg}}. \tag{9.32}$$

Only the first root is physical. At state 2 we know p_2 and T_2, so we can determine $\hat{v}_2$:

$$10^6 \text{ Pa} = \frac{\left(200 \, \frac{\text{J}}{\text{kg K}}\right)(300 \, \text{K})}{\hat{v}_2 - 0.001 \, \frac{\text{m}^3}{\text{kg}}} - \frac{150 \, \text{Pa} \, \text{m}^6/\text{kg}^2}{\hat{v}_2^2}. \tag{9.33}$$

The physical solution is $\hat{v}_2 = 0.0585 \, \text{m}^3/\text{kg}$. Now at state 3 we know $\hat{v}_3 = \hat{v}_2$ and T_3. Determine p_3:

$$p_3 = \frac{\left(200 \, \frac{\text{J}}{\text{kg K}}\right)(1000 \, \text{K})}{0.0585 \, \frac{\text{m}^3}{\text{kg}} - 0.001 \, \frac{\text{m}^3}{\text{kg}}} - \frac{150 \, \text{Pa} \, \text{m}^6/\text{kg}^2}{\left(0.0585 \, \frac{\text{m}^3}{\text{kg}}\right)^2} = 3\,434\,430 \, \text{Pa}. \tag{9.34}$$

Now $w_{13} = w_{12} + w_{23} = \int_1^2 p \, d\hat{v} + \int_2^3 p \, d\hat{v} = \int_1^2 p \, d\hat{v}$ because $2 \to 3$ is at constant volume. So

$$w_{13} = \int_{\hat{v}_1}^{\hat{v}_2} \left(\frac{RT}{\hat{v} - b} - \frac{a}{\hat{v}^2}\right) d\hat{v}, \tag{9.35}$$

$$= RT_1 \int_{\hat{v}_1}^{\hat{v}_2} \frac{d\hat{v}}{\hat{v} - b} - a \int_{\hat{v}_1}^{\hat{v}_2} \frac{d\hat{v}}{\hat{v}^2} = RT_1 \ln\left(\frac{\hat{v}_2 - b}{\hat{v}_1 - b}\right) + a\left(\frac{1}{\hat{v}_2} - \frac{1}{\hat{v}_1}\right), \tag{9.36}$$

$$= \left(200 \, \frac{\text{J}}{\text{kg K}}\right)(300 \, \text{K}) \ln \left(\frac{\left(0.0585 \, \frac{\text{m}^3}{\text{kg}}\right) - \left(0.001 \, \frac{\text{m}^3}{\text{kg}}\right)}{\left(0.598 \, \frac{\text{m}^3}{\text{kg}}\right) - \left(0.001 \, \frac{\text{m}^3}{\text{kg}}\right)} \right)$$
$$+ \left(150 \, \text{Pa} \, \text{m}^6/\text{kg}\right) \left(\frac{1}{\left(0.0585 \, \frac{\text{m}^3}{\text{kg}}\right)} - \frac{1}{\left(0.598 \, \frac{\text{m}^3}{\text{kg}}\right)} \right) = -138\,095 \, \frac{\text{J}}{\text{kg}}. \quad (9.37)$$

The gas is compressed, so the work is negative. Because e is a state property, we have

$$e_3 - e_1 = \int_{T_1}^{T_3} c_v(T) \, dT + a \left(\frac{1}{\hat{v}_1} - \frac{1}{\hat{v}_3} \right). \quad (9.38)$$

Now

$$c_v = \left(350 \, \frac{\text{J}}{\text{kg K}}\right) + \left(0.2 \, \frac{\text{J}}{\text{kg K}^2}\right)(T - (300 \, \text{K})), \quad (9.39)$$

$$= \left(290 \, \frac{\text{J}}{\text{kg K}}\right) + \left(0.2 \, \frac{\text{J}}{\text{kg K}^2}\right) T, \quad (9.40)$$

so

$$e_3 - e_1 = \int_{T_1}^{T_3} \left(\left(290 \, \frac{\text{J}}{\text{kg K}}\right) + \left(0.2 \, \frac{\text{J}}{\text{kg K}^2}\right) T \right) dT + a \left(\frac{1}{\hat{v}_1} - \frac{1}{\hat{v}_3} \right), \quad (9.41)$$

$$= \left(290 \, \frac{\text{J}}{\text{kg K}}\right)(T_3 - T_1) + \left(0.1 \, \frac{\text{J}}{\text{kg K}^2}\right)(T_3^2 - T_1^2) + a \left(\frac{1}{\hat{v}_1} - \frac{1}{\hat{v}_3} \right), \quad (9.42)$$

$$= \left(290 \, \frac{\text{J}}{\text{kg K}}\right)((1000 \, \text{K}) - (300 \, \text{K})) + \left(0.1 \, \frac{\text{J}}{\text{kg K}^2}\right)((1000 \, \text{K})^2 - (300 \, \text{K})^2)$$

$$+ (150 \, \text{Pa} \, \text{m}^6/\text{kg}) \left(\frac{1}{\left(0.598 \, \frac{\text{m}^3}{\text{kg}}\right)} - \frac{1}{\left(0.0585 \, \frac{\text{m}^3}{\text{kg}}\right)} \right) = 291\,687 \, \frac{\text{J}}{\text{kg}}. \quad (9.43)$$

Now from the first law, we have

$$e_3 - e_1 = q_{13} - w_{13}, \quad (9.44)$$

$$q_{13} = e_3 - e_1 + w_{13} = \left(291\,687 \, \frac{\text{J}}{\text{kg}}\right) - \left(138\,095 \, \frac{\text{J}}{\text{kg}}\right) = 153\,592 \, \frac{\text{J}}{\text{kg}}. \quad (9.45)$$

The heat transfer is positive as heat was added to the system.

Now find the entropy change. Manipulate the Gibbs equation, Eq. (4.197):

$$T \, ds = de + p \, d\hat{v}, \quad (9.46)$$

$$ds = \frac{1}{T} de + \frac{p}{T} d\hat{v} = \frac{1}{T} \left(c_v(T) \, dT + \frac{a}{\hat{v}^2} d\hat{v} \right) + \frac{p}{T} d\hat{v}, \quad (9.47)$$

$$= \frac{1}{T} \left(c_v(T) \, dT + \frac{a}{\hat{v}^2} d\hat{v} \right) + \frac{1}{T} \left(\frac{RT}{\hat{v} - b} - \frac{a}{\hat{v}^2} \right) d\hat{v}, \quad (9.48)$$

$$= \frac{c_v(T)}{T} dT + \frac{R}{\hat{v} - b} d\hat{v}, \quad (9.49)$$

$$s_3 - s_1 = \int_{T_1}^{T_3} \frac{c_v(T)}{T} dT + R \ln \frac{\hat{v}_3 - b}{\hat{v}_1 - b}, \quad (9.50)$$

$$= \int_{300 \, \text{K}}^{1000 \, \text{K}} \left(\frac{\left(290 \, \frac{\text{J}}{\text{kg K}}\right)}{T} + \left(0.2 \, \frac{\text{J}}{\text{kg K}^2}\right) \right) dT + R \ln \frac{\hat{v}_3 - b}{\hat{v}_1 - b}, \quad (9.51)$$

$$= \left(290 \, \frac{\text{J}}{\text{kg K}}\right) \ln \frac{1000 \, \text{K}}{300 \, \text{K}} + \left(0.2 \, \frac{\text{J}}{\text{kg K}^2}\right) ((1000 \, \text{K}) - (300 \, \text{K}))$$

$$+ \left(200 \, \frac{\text{J}}{\text{kg K}}\right) \ln \frac{\left(0.0585 \, \frac{\text{m}^3}{\text{kg}}\right) - \left(0.001 \, \frac{\text{m}^3}{\text{kg}}\right)}{\left(0.598 \, \frac{\text{m}^3}{\text{kg}}\right) - \left(0.001 \, \frac{\text{m}^3}{\text{kg}}\right)} = 21 \, \frac{\text{J}}{\text{kg K}}. \tag{9.52}$$

Is the second law satisfied for each portion of the process? First look at $1 \to 2$:

$$e_2 - e_1 = q_{12} - w_{12}, \tag{9.53}$$

$$q_{12} = e_2 - e_1 + w_{12}, \tag{9.54}$$

$$= \left(\int_{T_1}^{T_2} c_v(T) \, dT + a \left(\frac{1}{\hat{v}_1} - \frac{1}{\hat{v}_2}\right)\right) + \left(RT_1 \ln \left(\frac{\hat{v}_2 - b}{\hat{v}_1 - b}\right) + a \left(\frac{1}{\hat{v}_2} - \frac{1}{\hat{v}_1}\right)\right). \tag{9.55}$$

Because $T_1 = T_2$ and the fact that we can cancel the terms in a, we get

$$q_{12} = RT_1 \ln \left(\frac{\hat{v}_2 - b}{\hat{v}_1 - b}\right), \tag{9.56}$$

$$= \left(200 \, \frac{\text{J}}{\text{kg K}}\right) (300 \, \text{K}) \ln \frac{\left(0.0585 \, \frac{\text{m}^3}{\text{kg}}\right) - \left(0.001 \, \frac{\text{m}^3}{\text{kg}}\right)}{\left(0.598 \, \frac{\text{m}^3}{\text{kg}}\right) - \left(0.001 \, \frac{\text{m}^3}{\text{kg}}\right)} = -140408 \, \frac{\text{J}}{\text{kg}}. \tag{9.57}$$

Because the process is isothermal, we find

$$s_2 - s_1 = R \ln \frac{\hat{v}_2 - b}{\hat{v}_1 - b}, \tag{9.58}$$

$$= \left(200 \, \frac{\text{J}}{\text{kg K}}\right) \ln \frac{\left(0.0585 \, \frac{\text{m}^3}{\text{kg}}\right) - \left(0.001 \, \frac{\text{m}^3}{\text{kg}}\right)}{\left(0.598 \, \frac{\text{m}^3}{\text{kg}}\right) - \left(0.001 \, \frac{\text{m}^3}{\text{kg}}\right)} = -468.0 \, \frac{\text{J}}{\text{kg K}}. \tag{9.59}$$

Entropy *drops* because heat was transferred *out* of the system.

Check the second law. Note that in this portion of the process in which the heat is transferred out of the system, that the surroundings must have $T_{surr} \leq 300 \, \text{K}$. For this portion of the process let us take $T_{surr} = 300 \, \text{K}$.

$$s_2 - s_1 \geq \frac{q_{12}}{T}? \tag{9.60}$$

$$-468.0 \, \frac{\text{J}}{\text{kg K}} \geq \frac{-140\,408 \, \frac{\text{J}}{\text{kg}}}{300 \, \text{K}}, \tag{9.61}$$

$$\geq -468.0 \, \frac{\text{J}}{\text{kg K}}, \quad \text{ok}. \tag{9.62}$$

Next look at $2 \to 3$

$$q_{23} = e_3 - e_2 + w_{23} = \left(\int_{T_2}^{T_3} c_v(T) \, dT + a \left(\frac{1}{\hat{v}_2} - \frac{1}{\hat{v}_3}\right)\right) + \left(\int_{\hat{v}_2}^{\hat{v}_3} p \, d\hat{v}\right). \tag{9.63}$$

Because the process is isochoric, we have

$$q_{23} = \int_{T_2}^{T_3} c_v(T) \, dT, \tag{9.64}$$

$$= \int_{300 \, \text{K}}^{1000 \, \text{K}} \left(\left(290 \, \frac{\text{J}}{\text{kg K}}\right) + \left(0.2 \, \frac{\text{J}}{\text{kg K}^2}\right) T\right) dT = 294\,000 \, \frac{\text{J}}{\text{kg}}. \tag{9.65}$$

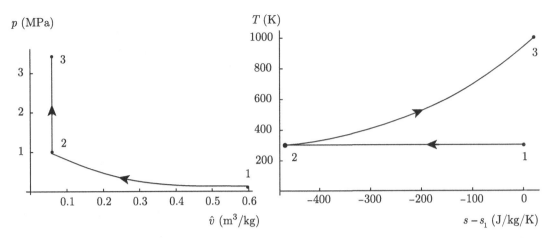

Figure 9.3 $(P, \hat{v})$ and (T, s) plots for two step thermodynamic process (isothermal compression then isochoric heating) applied to a van der Waals gas.

Now look at the entropy change for the isochoric process:

$$s_3 - s_2 = \int_{T_2}^{T_3} \frac{c_v(T)}{T} dT = \int_{T_2}^{T_3} \left(\frac{\left(290 \frac{\text{J}}{\text{kg K}}\right)}{T} + \left(0.2 \frac{\text{J}}{\text{kg K}^2}\right) \right) dT, \quad (9.66)$$

$$= \left(290 \frac{\text{J}}{\text{kg K}}\right) \ln \frac{1000\,\text{K}}{300\,\text{K}} + \left(0.2 \frac{\text{J}}{\text{kg K}^2}\right) ((1000\,\text{K}) - (300\,\text{K})), \quad (9.67)$$

$$= 489 \frac{\text{J}}{\text{kg K}}. \quad (9.68)$$

Entropy *rises* because heat was transferred *into* the system.

In order to transfer heat into the system, we must have a different thermal reservoir. This one must have $T_{surr} \geq 1000\,\text{K}$. Assume here that the heat transfer was from a reservoir held at $1000\,\text{K}$ to assess the influence of the second law.

$$s_3 - s_2 \geq \frac{q_{23}}{T}? \quad (9.69)$$

$$489 \frac{\text{J}}{\text{kg K}} \geq \frac{294\,000 \frac{\text{J}}{\text{kg}}}{1000\,\text{K}}, \quad (9.70)$$

$$\geq 294 \frac{\text{J}}{\text{kg K}}, \quad \text{ok.} \quad (9.71)$$

Diagrams of the two step process in both the $(P, \hat{v})$ and (T, s) planes are shown in Fig. 9.3. We recall the work for the process is the area under the curve in the $(P, \hat{v})$ plane, and the heat transfer is the area under the curve in the (T, s) plane.

9.1.3 Sound Speed

Let us find the sound speed $c(T, \rho)$ for a general material with known thermal equation of state $p(\rho, T)$. Later, in Section 9.2.3, we will see how one can obtain the sound speed from a general caloric equation of state $e(p, \rho)$. At this point, c is best thought of as a thermodynamic property. In Section 9.6.7, we will see how it represents the speed of propagation of small acoustic disturbances. Let us define c as

$$c = \sqrt{\left.\frac{\partial p}{\partial \rho}\right|_s}, \qquad c^2 = \left.\frac{\partial p}{\partial \rho}\right|_s. \tag{9.72}$$

Because $\hat{v} = 1/\rho$, one can show

$$c^2 = \frac{d\hat{v}}{d\rho} \left.\frac{\partial p}{\partial \hat{v}}\right|_s = -\hat{v}^2 \left.\frac{\partial p}{\partial \hat{v}}\right|_s. \tag{9.73}$$

Take the Gibbs relation, Eq. (4.195), $T\,ds = de + p\,d\hat{v}$, and use Eq. (9.12) to eliminate de:

$$T\,ds = \underbrace{\left(c_v\,dT + \left(T \left.\frac{\partial p}{\partial T}\right|_{\hat{v}} - p\right) d\hat{v}\right)}_{de} + p\,d\hat{v} = c_v\,dT + T \left.\frac{\partial p}{\partial T}\right|_{\hat{v}} d\hat{v}, \tag{9.74}$$

$$= c_v\,dT - \frac{T}{\rho^2} \left.\frac{\partial p}{\partial T}\right|_\rho d\rho. \tag{9.75}$$

Because $p = p(T, \hat{v})$, we can also say $p = p(T, \rho)$, and then we get the differential

$$dp = \left.\frac{\partial p}{\partial T}\right|_\rho dT + \left.\frac{\partial p}{\partial \rho}\right|_T d\rho, \quad \text{so} \quad dT = \frac{dp - \left.\frac{\partial p}{\partial \rho}\right|_T d\rho}{\left.\frac{\partial p}{\partial T}\right|_\rho}. \tag{9.76}$$

Thus, substituting for dT in Eq. (9.75), we find

$$T\,ds = c_v \left(\frac{dp - \left.\frac{\partial p}{\partial \rho}\right|_T d\rho}{\left.\frac{\partial p}{\partial T}\right|_\rho}\right) - \frac{T}{\rho^2} \left.\frac{\partial p}{\partial T}\right|_\rho d\rho. \tag{9.77}$$

Grouping terms in dp and $d\rho$, we get

$$T\,ds = \left(\frac{c_v}{\left.\frac{\partial p}{\partial T}\right|_\rho}\right) dp - \left(c_v \frac{\left.\frac{\partial p}{\partial \rho}\right|_T}{\left.\frac{\partial p}{\partial T}\right|_\rho} + \frac{T}{\rho^2} \left.\frac{\partial p}{\partial T}\right|_\rho\right) d\rho. \tag{9.78}$$

For the isentropic sound speed, we must have $ds \equiv 0$; we thus obtain

$$c^2 = \left.\frac{\partial p}{\partial \rho}\right|_s = \frac{1}{c_v} \left.\frac{\partial p}{\partial T}\right|_\rho \left(c_v \frac{\left.\frac{\partial p}{\partial \rho}\right|_T}{\left.\frac{\partial p}{\partial T}\right|_\rho} + \frac{T}{\rho^2} \left.\frac{\partial p}{\partial T}\right|_\rho\right) = \left.\frac{\partial p}{\partial \rho}\right|_T + \frac{T}{c_v \rho^2} \left(\left.\frac{\partial p}{\partial T}\right|_\rho\right)^2. \tag{9.79}$$

So the isentropic sound speed for a general thermal equation of state is

$$c(T, \rho) = \sqrt{\left.\frac{\partial p}{\partial \rho}\right|_s} = \sqrt{\left.\frac{\partial p}{\partial \rho}\right|_T + \frac{T}{c_v \rho^2} \left(\left.\frac{\partial p}{\partial T}\right|_\rho\right)^2}. \tag{9.80}$$

Without the benefit of modern understanding of thermodynamics, Newton (1999, pp. 776–778, originally presented 1687) concluded the speed of sound was $\sqrt{\partial p/\partial \rho|_T}$, though he used an approach that was more opaque. So in an era in which Boyle's[2] law relating pressure and density for an isothermal gas was known, Newton could effectively predict an isothermal sound speed of $\sqrt{p/\rho}$. He then went to considerable effort to measure the sound speed, but could not reconcile the discrepancy of his measurements with his theory. Newton's approach was corrected by Laplace in 1816, who generated what amounts to our adiabatic prediction, long before notions of thermodynamics were settled. Laplace's notions rested on an uncertain theoretical foundation; he in fact adjusted his theory often, and it was not until thermodynamics was well established several decades later that our understanding of sound speed was clarified. The interested reader can consult Finn (1964).

Example 9.5 Find the sound speed for a calorically perfect ideal gas:

$$p(T, \rho) = \rho RT. \tag{9.81}$$

Solution

The necessary partial derivatives are

$$\left.\frac{\partial p}{\partial \rho}\right|_T = RT, \qquad \left.\frac{\partial p}{\partial T}\right|_\rho = \rho R. \tag{9.82}$$

So

$$c(T, \rho) = \sqrt{RT + \frac{T}{c_v \rho^2}(\rho R)^2} = \sqrt{RT + \frac{R^2 T}{c_v}}, \tag{9.83}$$

$$= \sqrt{RT\left(1 + \frac{R}{c_v}\right)} = \sqrt{RT\left(1 + \frac{c_p - c_v}{c_v}\right)} = \sqrt{\gamma RT}. \tag{9.84}$$

Sound speed depends on temperature alone for the calorically perfect ideal gas. For calorically perfect air with $\gamma = 7/5$, $R = 287\,\text{J/kg/K}$, $T = 300\,\text{K}$, we find $c = 347.189\,\text{m/s}$. Contrast this with an isothermal sound speed of $c_T = \sqrt{RT} = 293.428\,\text{m/s}$; this can be compared with Newton's (1999, p. 776) prediction of $979\,\text{ft/s} = 298.399\,\text{m/s}$.

Example 9.6 Find the sound speed of a so-called *virial gas*:

$$p(T, \rho) = \rho RT(1 + b\rho). \tag{9.85}$$

Solution

The necessary partial derivatives are

$$\left.\frac{\partial p}{\partial \rho}\right|_T = RT + 2b\rho RT, \qquad \left.\frac{\partial p}{\partial T}\right|_\rho = \rho R(1 + b\rho). \tag{9.86}$$

Thus,

$$c(T, \rho) = \sqrt{RT + 2b\rho RT + \frac{T}{c_v \rho^2}(\rho R(1+b\rho))^2} = \sqrt{RT\left(1 + 2b\rho + \frac{R}{c_v}(1+b\rho)^2\right)}. \tag{9.87}$$

The sound speed of a virial gas depends on both temperature and density.

[2] Robert Boyle, 1627–1691, Anglo-Irish natural philosopher.

In Section 9.11.1, we shall need to consider $p = p(\rho, s)$, taking advantage of the fact that in thermodynamics, one can cast any intensive thermodynamic variable in terms of two other independent intensive thermodynamic variables.

Example 9.7 For a calorically perfect ideal gas, find $p = p(\rho, s)$.

Solution
Start with the Gibbs equation, Eq. (4.195), $T\,ds = de + p\,d\hat{v}$. For a calorically perfect ideal gas, we have $de = c_v\,dT$, $p = RT/\hat{v}$, so the Gibbs equation reduces to

$$ds = c_v \frac{dT}{T} + R \frac{d\hat{v}}{\hat{v}} = c_v \frac{dT}{T} + R \frac{-\frac{1}{\rho^2} d\rho}{\frac{1}{\rho}} = c_v \frac{dT}{T} - R \frac{d\rho}{\rho}. \tag{9.88}$$

Now because $p = \rho RT$, we also have

$$dp = \rho R\,dT + RT\,d\rho, \quad \text{so} \quad \frac{dT}{T} = \frac{dp}{p} - \frac{d\rho}{\rho}. \tag{9.89}$$

Substitute this into Eq. (9.88) to get

$$ds = c_v \left(\frac{dp}{p} - \frac{d\rho}{\rho} \right) - R \frac{d\rho}{\rho} = c_v \frac{dp}{p} - (c_v + R) \frac{d\rho}{\rho} = c_v \frac{dp}{p} - c_p \frac{d\rho}{\rho}, \tag{9.90}$$

$$\frac{ds}{c_v} = \frac{dp}{p} - \gamma \frac{d\rho}{\rho}, \quad \text{yielding} \quad \frac{s - s_o}{c_v} = \ln \frac{p}{p_o} - \gamma \ln \frac{\rho}{\rho_o}, \tag{9.91}$$

$$\ln \frac{p}{p_o} = \ln \left(\frac{\rho}{\rho_o} \right)^\gamma + \frac{s - s_o}{c_v}, \quad \text{so} \quad p(\rho, s) = p_o \left(\frac{\rho}{\rho_o} \right)^\gamma \exp \left(\frac{s - s_o}{c_v} \right). \tag{9.92}$$

Note that for an isentropic case, we have $s = s_o$, yielding the barotropic relation $p/p_o = (\rho/\rho_o)^\gamma$.

9.2 Generalized One-Dimensional Flow

Now let us introduce the evolution axioms of mass, momentum, and energy to augment our thermodynamics. This allows us to address the dynamics of compressible fluid flow. Here we will derive in a conventional way the one-dimensional equations of compressible flow with area change. The development parallels that given by Shapiro (1953, Chapter 8). Although for the geometry we use, it will appear that we should be using at least two-dimensional equations, our results will be approximately correct when we interpret them as yielding average values at a given x location. Our results will be valid as long as the area changes slowly relative to how fast the flow can adjust to area changes. We can be guided by our equations from Section 6.2. However, the common equations we will employ are not easily cast as limiting cases of previously derived equations. Instead, the ad hoc nature of common area change, friction, and heat transfer models we will employ makes a fresh derivation essential.

We consider flow with area change, heat transfer, and wall friction, depicted in Fig. 9.4. We adopt the following conventions. (1) Surfaces 1 and 2 are open and allow advective fluxes of mass, momentum, and energy. (2) Surface w is a closed wall; no mass flux through the wall is allowed. (3) External heat flux q_w (energy/area/time, W/m^2) *through* the wall is allowed: q_w is a known function, often taken as a constant. (4) Diffusive, longitudinal heat transfer is ignored, $q_x = 0$; thus, thermal conductivity $k = 0$. (5) Wall shear τ_w (force/area, N/m^2) is allowed: τ_w

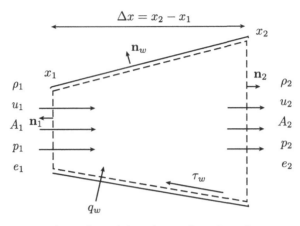

Figure 9.4 Control volume diagram for one-dimensional compressible flow with area change, heat transfer, and wall friction.

is a known function of fluid properties and system geometry. (6) Diffusive viscous stress is not allowed, $\tau_{xx} = 0$; thus, viscosity $\mu = 0$. (7) The cross-sectional area is a known function, $A(x)$.

9.2.1 Mass

Take an over-bar notation to indicate a volume-averaged quantity. The amount of mass in a control volume after a time increment Δt is equal to the original amount of mass plus that which came in minus that which left:

$$\underbrace{\bar{\rho}\bar{A}\Delta x\big|_{t+\Delta t}}_{\text{new mass}} = \underbrace{\bar{\rho}\bar{A}\Delta x\big|_{t}}_{\text{old mass}} + \underbrace{\rho_1 A_1 (u_1 \Delta t)}_{\text{mass in}} - \underbrace{\rho_2 A_2 (u_2 \Delta t)}_{\text{mass out}}. \tag{9.93}$$

Rearrange and divide by $\Delta x \Delta t$:

$$\frac{\bar{\rho}\bar{A}\big|_{t+\Delta t} - \bar{\rho}\bar{A}\big|_{t}}{\Delta t} + \frac{\rho_2 A_2 u_2 - \rho_1 A_1 u_1}{\Delta x} = 0. \tag{9.94}$$

Taking the limit as $\Delta t \to 0, \Delta x \to 0$, and thus allowing volume-averaged values to take on local values, we get

$$\frac{\partial}{\partial t}(\rho A) + \frac{\partial}{\partial x}(\rho A u) = 0. \tag{9.95}$$

If the flow is steady, then

$$\frac{d}{dx}(\rho A u) = 0, \tag{9.96}$$

$$Au \frac{d\rho}{dx} + \rho u \frac{dA}{dx} + \rho A \frac{du}{dx} = 0, \tag{9.97}$$

$$\frac{1}{\rho}\frac{d\rho}{dx} + \frac{1}{A}\frac{dA}{dx} + \frac{1}{u}\frac{du}{dx} = 0. \tag{9.98}$$

Now integrate Eq. (9.96) from x_1 to x_2 to get

$$\int_{x_1}^{x_2} \frac{d}{dx}(\rho A u)\, dx = \int_{x_1}^{x_2} 0\, dx, \quad \text{so} \quad \int_1^2 d(\rho A u) = 0, \tag{9.99}$$

$$\rho_2 u_2 A_2 - \rho_1 u_1 A_1 = 0, \quad \text{so} \quad \rho_2 u_2 A_2 = \rho_1 u_1 A_1 \equiv \dot{m}. \tag{9.100}$$

Here $\dot{m}$ is the *mass flow rate* with units kg/s. For steady flow, it is a constant.

9.2.2 Linear Momentum

Newton's second law of motion says the time rate of change of linear momentum of a body equals the sum of the forces acting on the body. In the x direction this yields for a system:

$$\frac{d}{dt}(mu) = \sum F_x. \tag{9.101}$$

In discrete form this becomes

$$\frac{mu|_{t+\Delta t} - mu|_t}{\Delta t} = \sum F_x, \tag{9.102}$$

$$mu|_{t+\Delta t} = mu|_t + \left(\sum F_x\right)\Delta t. \tag{9.103}$$

For a control volume containing fluid, we must also account for the momentum that enters and leaves the control volume. The amount of momentum in a control volume after a time increment Δt is equal to the original amount of momentum plus that which came in minus that which left plus that introduced by the impulse of the forces acting on the control volume. Note the following. (1) The pressure force at surface 1 *pushes* the fluid. (2) The pressure force at surface 2 *restrains* the fluid. (3) The force due to the reaction of the wall to the pressure force *pushes* the fluid if the area change is positive. (4) The force due to the reaction of the wall to the shear force *restrains* the fluid. We write the linear momentum principle as

$$\underbrace{(\bar{\rho}\bar{A}\Delta x)\,\bar{u}|_{t+\Delta t}}_{\text{new momentum}} = \underbrace{(\bar{\rho}\bar{A}\Delta x)\,\bar{u}|_t}_{\text{old momentum}} + \underbrace{(\rho_1 A_1 (u_1 \Delta t))\, u_1}_{\text{momentum in}} - \underbrace{(\rho_2 A_2 (u_2 \Delta t))\, u_2}_{\text{momentum out}}$$

$$+ \underbrace{(p_1 A_1)\,\Delta t - (p_2 A_2)\,\Delta t + (\bar{p}(A_2 - A_1))\,\Delta t}_{\text{net impulse from pressure forces}}$$

$$- \underbrace{(\tau_w \bar{L}_p \Delta x)\,\Delta t}_{\text{impulse from friction}}. \tag{9.104}$$

Rearrange and divide by $\Delta x \Delta t$ to get

$$\frac{\bar{\rho}\bar{A}\bar{u}|_{t+\Delta t} - \bar{\rho}\bar{A}\bar{u}|_t}{\Delta t} + \frac{\rho_2 A_2 u_2^2 - \rho_1 A_1 u_1^2}{\Delta x} = -\frac{p_2 A_2 - p_1 A_1}{\Delta x} + \bar{p}\frac{A_2 - A_1}{\Delta x} - \tau_w \bar{L}_p. \tag{9.105}$$

In the limit $\Delta x \to 0, \Delta t \to 0$ we get

$$\frac{\partial}{\partial t}(\rho A u) + \frac{\partial}{\partial x}(\rho A u^2) = -\frac{\partial}{\partial x}(pA) + p\frac{\partial A}{\partial x} - \tau_w L_p. \tag{9.106}$$

In steady state, we find

$$\frac{d}{dx}(\rho A u^2) = -\frac{d}{dx}(pA) + p\frac{dA}{dx} - \tau_w L_p, \quad (9.107)$$

$$\rho A u \frac{du}{dx} + u\underbrace{\frac{d}{dx}(\rho A u)}_{=0} = -p\frac{dA}{dx} - A\frac{dp}{dx} + p\frac{dA}{dx} - \tau_w L_p, \quad (9.108)$$

$$\rho u \frac{du}{dx} = -\frac{dp}{dx} - \tau_w \frac{L_p}{A}, \quad (9.109)$$

$$\rho u\, du + dp = -\tau_w \frac{L_p}{A}\, dx, \quad (9.110)$$

$$\rho\, d\left(\frac{u^2}{2}\right) + dp = -\tau_w \frac{L_p}{A}\, dx. \quad (9.111)$$

If there is no wall shear, then Eq. (9.111) reduces to

$$dp = -\rho\, d\left(\frac{u^2}{2}\right), \quad \text{no wall shear.} \quad (9.112)$$

An increase in velocity magnitude decreases the pressure. We can equivalently rewrite Eq. (9.110) for $\tau_w = 0$ as

$$\rho u \frac{du}{dx} + \frac{dp}{dx} = 0, \quad \text{no wall shear.} \quad (9.113)$$

For flow with no area change, $dA/dx = 0$. For that limit, along with the limit of $\tau_w = 0$, Eq. (9.107) reduces to

$$\frac{d}{dx}(\rho A u^2) = -\frac{d}{dx}(pA), \quad \text{so} \quad \frac{d}{dx}(\rho u^2 + p) = 0, \quad (9.114)$$

$$\rho u^2 + p = \rho_1 u_1^2 + p_1, \quad \text{no wall shear, no area change.} \quad (9.115)$$

Note that Eq. (9.115) resembles many of the Bernoulli equations studied in Section 6.8.1, but it is not a traditional Bernoulli equation: there is no factor of $1/2$ multiplying ρu^2. Let us see why this is. First, recall in deriving Eq. (9.115) we made no appeal to the first law of thermodynamics, any equation of state, or any assumption such as isentropic flow or barotropic flow. It is simply a first integral of the conservative form of the linear momentum principle. Typical Bernoulli equations arise from first integrals of the nonconservative version of the linear momentum principle with additional side conditions. For example, the one-dimensional steady, zero body force limit of Eq. (6.172) is $\rho\, d/dx(u^2/2) + dp/dx = 0$. This nonconservative form cannot be integrated exactly without further knowledge of how p varies with ρ. If one had additional information, for example, that the flow was isentropic, one could form a relationship between p and ρ and then integrate the nonconservative form of the linear momentum principle. But given that the result of Eq. (9.115) requires no such additional assumptions, it is to be preferred. A rigorous derivation of how the compressible flow equations for a calorically perfect ideal gas in the isentropic low Mach number limit reduce to the incompressible Bernoulli equation is manifested in the upcoming Eq. (9.214).

9.2.3 Energy

The first law of thermodynamics states that the change of total energy of a body equals the heat transferred to the body minus the work done by the body:

$$E_2 - E_1 = Q - W, \quad \text{thus} \quad E_2 = E_1 + Q - W. \tag{9.116}$$

So for our control volume this becomes the following when we also account for the energy flux in and out of the control volume in addition to the work and heat transfer:

$$\underbrace{(\bar{\rho}\bar{A}\Delta x)\left(\bar{e}+\frac{\bar{u}^2}{2}\right)\Big|_{t+\Delta t}}_{\text{new energy}} = \underbrace{(\bar{\rho}\bar{A}\Delta x)\left(\bar{e}+\frac{\bar{u}^2}{2}\right)\Big|_{t}}_{\text{old energy}}$$

$$+\underbrace{\rho_1 A_1 (u_1 \Delta t)\left(e_1+\frac{u_1^2}{2}\right)}_{\text{energy in}} - \underbrace{\rho_2 A_2 (u_2 \Delta t)\left(e_2+\frac{u_2^2}{2}\right)}_{\text{energy out}}$$

$$+\underbrace{q_w\left(\bar{L}_p \Delta x\right)\Delta t}_{\text{heat in}} - \underbrace{\left((-p_1 A_1)(u_1 \Delta t)-(-p_2 A_2)(u_2 \Delta t)\right)}_{\text{net work by pressure forces at inlets/exits}}. \tag{9.117}$$

Note that the work done by mean pressure times the area difference is not included because it is acting on a stationary boundary. Also note that the work done by the wall shear force is not included. In neglecting work done by the wall shear force, we have taken an approach that is nearly universal, but fundamentally difficult to defend. It is likely unproductive to advance a more rigorous approach for a common model that already contains many compromises and probably confuse the reader; consequently, results for flow with friction will be consistent with those of other sources. The argument typically used to justify this is that the real fluid satisfies no-slip at the boundary; thus, the wall shear actually does no work. However, one can argue that within the context of the one-dimensional model that has been posed, the shear force behaves as an external force that reduces the fluid's mechanical energy. Moreover, it is possible to show that neglect of this term results in the loss of frame invariance under Galilean transformation, a serious defect indeed!

Returning to the energy equation, we rearrange and divide by $\Delta t \Delta x$:

$$\frac{\bar{\rho}\bar{A}\left(\bar{e}+\frac{\bar{u}^2}{2}\right)\Big|_{t+\Delta t} - \bar{\rho}\bar{A}\left(\bar{e}+\frac{\bar{u}^2}{2}\right)\Big|_{t}}{\Delta t} + \frac{\rho_2 A_2 u_2\left(e_2+\frac{u_2^2}{2}+\frac{p_2}{\rho_2}\right) - \rho_1 A_1 u_1\left(e_1+\frac{u_1^2}{2}+\frac{p_1}{\rho_1}\right)}{\Delta x}$$
$$= q_w \bar{L}_p. \tag{9.118}$$

In differential form as $\Delta x \to 0, \Delta t \to 0$, we get

$$\frac{\partial}{\partial t}\left(\rho A\left(e+\frac{u^2}{2}\right)\right) + \frac{\partial}{\partial x}\left(\rho A u\left(e+\frac{u^2}{2}+\frac{p}{\rho}\right)\right) = q_w L_p. \tag{9.119}$$

In steady state, we find

$$\frac{d}{dx}\left(\rho A u \left(e + \frac{u^2}{2} + \frac{p}{\rho}\right)\right) = q_w L_p, \tag{9.120}$$

$$\rho A u \frac{d}{dx}\left(e + \frac{u^2}{2} + \frac{p}{\rho}\right) + \left(e + \frac{u^2}{2} + \frac{p}{\rho}\right)\underbrace{\frac{d}{dx}(\rho A u)}_{=0} = q_w L_p, \tag{9.121}$$

$$\rho u \frac{d}{dx}\left(e + \frac{u^2}{2} + \frac{p}{\rho}\right) = \frac{q_w L_p}{A}, \tag{9.122}$$

$$\rho u \left(\frac{de}{dx} + u\frac{du}{dx} + \frac{1}{\rho}\frac{dp}{dx} - \frac{p}{\rho^2}\frac{d\rho}{dx}\right) = \frac{q_w L_p}{A}. \tag{9.123}$$

Now consider the product of velocity and momentum from Eq. (9.109) to get an equation for the mechanical energy:

$$\rho u^2 \frac{du}{dx} + u\frac{dp}{dx} = -\frac{\tau_w L_p u}{A}. \tag{9.124}$$

Subtract this, the mechanical energy, from Eq. (9.123) to get an equation for the thermal energy:

$$\rho u \frac{de}{dx} - \frac{p u}{\rho}\frac{d\rho}{dx} = \frac{q_w L_p}{A} + \frac{\tau_w L_p u}{A}, \quad \text{so} \quad \frac{de}{dx} - \frac{p}{\rho^2}\frac{d\rho}{dx} = \frac{(q_w + \tau_w u) L_p}{\dot{m}}. \tag{9.125}$$

Because $e = e(p, \rho)$, we have

$$de = \left.\frac{\partial e}{\partial \rho}\right|_p d\rho + \left.\frac{\partial e}{\partial p}\right|_\rho dp, \quad \text{so} \quad \frac{de}{dx} = \left.\frac{\partial e}{\partial \rho}\right|_p \frac{d\rho}{dx} + \left.\frac{\partial e}{\partial p}\right|_\rho \frac{dp}{dx}. \tag{9.126}$$

Thus, Eq. (9.125) becomes

$$\underbrace{\left.\frac{\partial e}{\partial \rho}\right|_p \frac{d\rho}{dx} + \left.\frac{\partial e}{\partial p}\right|_\rho \frac{dp}{dx}}_{\frac{de}{dx}} - \frac{p}{\rho^2}\frac{d\rho}{dx} = \frac{(q_w + \tau_w u) L_p}{\dot{m}}, \tag{9.127}$$

$$\frac{dp}{dx} - \underbrace{\left(\frac{\frac{p}{\rho^2} - \left.\frac{\partial e}{\partial \rho}\right|_p}{\left.\frac{\partial e}{\partial p}\right|_\rho}\right)}_{\equiv c^2} \frac{d\rho}{dx} = \frac{(q_w + \tau_w u) L_p}{\dot{m} \left.\frac{\partial e}{\partial p}\right|_\rho}. \tag{9.128}$$

Now let us consider the term in parentheses, which we label c^2. It will be seen to be the square of the sound speed. We can put that term in a more common form by considering the Gibbs equation, Eq. (4.197), $T\,ds = de - (p/\rho^2)\,d\rho$, along with a general caloric equation of state $e = e(p, \rho)$, from which we can infer

$$de = \left.\frac{\partial e}{\partial p}\right|_\rho dp + \left.\frac{\partial e}{\partial \rho}\right|_p d\rho. \tag{9.129}$$

Substituting into the Gibbs equation, we get

$$T\,ds = \underbrace{\left.\frac{\partial e}{\partial p}\right|_\rho dp + \left.\frac{\partial e}{\partial \rho}\right|_p d\rho}_{de} - \frac{p}{\rho^2}\,d\rho. \tag{9.130}$$

Taking s to be constant and dividing by $d\rho$, we get

$$0 = \left.\frac{\partial e}{\partial p}\right|_\rho \left.\frac{\partial p}{\partial \rho}\right|_s + \left.\frac{\partial e}{\partial \rho}\right|_p - \frac{p}{\rho^2}. \tag{9.131}$$

Rearranging, we get

$$\left.\frac{\partial p}{\partial \rho}\right|_s = c^2 = \frac{\frac{p}{\rho^2} - \left.\frac{\partial e}{\partial \rho}\right|_p}{\left.\frac{\partial e}{\partial p}\right|_\rho}. \tag{9.132}$$

This general expression for sound speed is useful in situations in which the temperature is difficult to measure. So we get

$$\frac{dp}{dx} - c^2\frac{d\rho}{dx} = \frac{(q_w + \tau_w u)L_p}{\dot{m}\left.\frac{\partial e}{\partial p}\right|_\rho} = \frac{(q_w + \tau_w u)L_p}{\rho u A \left.\frac{\partial e}{\partial p}\right|_\rho}. \tag{9.133}$$

Here c is the isentropic sound speed, a thermodynamic property of the material. We shall see later in Section 9.6.7 why it is appropriate to interpret this property as the propagation speed of small disturbances. At this point, it should simply be thought of as a thermodynamic state property.

Example 9.8 Find the speed of sound for a calorically perfect ideal gas using the approach of this section.

Solution
For a calorically perfect ideal gas, we have

$$p = \rho R T, \qquad e = c_v T + e_o, \qquad c_p - c_v = R. \tag{9.134}$$

So

$$e = c_v \frac{p}{\rho R} + e_o = c_v \frac{p}{(c_p - c_v)\rho} + e_o = \frac{1}{\frac{c_p}{c_v} - 1}\frac{p}{\rho} + e_o = \frac{1}{\gamma - 1}\frac{p}{\rho} + e_o. \tag{9.135}$$

The relevant partial derivatives are

$$\left.\frac{\partial e}{\partial \rho}\right|_p = -\frac{1}{\gamma - 1}\frac{p}{\rho^2}, \qquad \left.\frac{\partial e}{\partial p}\right|_\rho = \frac{1}{\gamma - 1}\frac{1}{\rho}. \tag{9.136}$$

So by Eq. (9.132), we have

$$c^2 = \frac{\frac{p}{\rho^2} - \left(-\frac{1}{\gamma-1}\frac{p}{\rho^2}\right)}{\frac{1}{\gamma-1}\frac{1}{\rho}} = \frac{(\gamma-1)\frac{p}{\rho^2} + \frac{p}{\rho^2}}{\frac{1}{\rho}} = \gamma\frac{p}{\rho} = \gamma R T, \tag{9.137}$$

$$c = \sqrt{\gamma R T}. \tag{9.138}$$

This is identical to that obtained by a different method in Eq. (9.84).

Example 9.9 For $q_w = 0$, $\tau_w = 0$, find a relation between p and ρ for the steady flow of a calorically perfect ideal gas.

Solution
Start with the energy equation, Eq. (9.133), in the limit as $q_w = 0$, $\tau_w = 0$:

$$\frac{dp}{dx} = c^2 \frac{d\rho}{dx}. \tag{9.139}$$

Dividing, we could also say that $dp/d\rho = c^2$, which tells us that suppressing the q_w and τ_w entropy-generating mechanisms because in general $\partial p/\partial \rho|_s = c^2$. Now from the previous example, we know for a calorically perfect ideal gas that $c^2 = \gamma p/\rho$, so

$$\frac{dp}{dx} = \gamma \frac{p}{\rho} \frac{d\rho}{dx}, \quad \frac{1}{p}\frac{dp}{dx} = \gamma \frac{1}{\rho}\frac{d\rho}{dx}, \quad \frac{dp}{p} = \gamma \frac{d\rho}{\rho}, \tag{9.140}$$

$$\ln \frac{p}{p_o} = \gamma \ln \frac{\rho}{\rho_o} = \ln \left(\frac{\rho}{\rho_o}\right)^\gamma, \quad \frac{p}{p_o} = \left(\frac{\rho}{\rho_o}\right)^\gamma, \quad \frac{p}{\rho^\gamma} = \frac{p_o}{\rho_o^\gamma}. \tag{9.141}$$

This is equivalent to what we have earlier derived in Eq. (6.133). In terms of specific volume, we could say

$$p\hat{v}^\gamma = p_o \hat{v}_o^\gamma. \tag{9.142}$$

This is the equation for a so-called *polytropic process* in which the polytropic exponent is γ. This particular polytropic process is isentropic.

Consider now the special case of flow with no heat transfer $q_w \equiv 0$. We still allow area change and wall friction is allowed (see earlier discussion, p. 353):

$$\rho u \frac{d}{dx}\left(e + \frac{u^2}{2} + \frac{p}{\rho}\right) = 0, \tag{9.143}$$

$$e + \frac{u^2}{2} + \frac{p}{\rho} = e_1 + \frac{u_1^2}{2} + \frac{p_1}{\rho_1}, \quad h + \frac{u^2}{2} = h_1 + \frac{u_1^2}{2}. \tag{9.144}$$

In contrast to Eq. (9.115), Eq. (9.144) can take on the form of a typical Bernoulli equation, once one invokes a thermodynamic equation of state. As an example, for a calorically perfect ideal gas, one has $h = (\gamma/(\gamma - 1))(p/\rho)$. And for this case, Eq. (9.144) reduces to

$$\frac{\gamma}{\gamma - 1}\frac{p}{\rho} + \frac{u^2}{2} = \frac{\gamma}{\gamma - 1}\frac{p_1}{\rho_1} + \frac{u_1^2}{2}. \tag{9.145}$$

This is of the form given by Eq. (6.191).

9.2.4 Summary of Equations

We can summarize the one-dimensional compressible flow equations in various forms here. In the equations given next, we assume $A(x)$, τ_w, q_w, and L_p are all known.

Unsteady Conservative Form

$$\frac{\partial}{\partial t}(\rho A) + \frac{\partial}{\partial x}(\rho A u) = 0, \tag{9.146}$$

$$\frac{\partial}{\partial t}(\rho A u) + \frac{\partial}{\partial x}(\rho A u^2 + pA) = p\frac{dA}{dx} - \tau_w L_p, \tag{9.147}$$

$$\frac{\partial}{\partial t}\left(\rho A \left(e + \frac{u^2}{2}\right)\right) + \frac{\partial}{\partial x}\left(\rho A u \left(e + \frac{u^2}{2} + \frac{p}{\rho}\right)\right) = q_w L_p, \tag{9.148}$$

$$e = e(\rho, p), \qquad p = p(\rho, T). \tag{9.149}$$

Unsteady Nonconservative Form

$$\frac{d\rho}{dt} = -\frac{\rho}{A}\frac{\partial}{\partial x}(Au), \tag{9.150}$$

$$\rho \frac{du}{dt} = -\frac{\partial p}{\partial x} - \frac{\tau_w L_p}{A}, \tag{9.151}$$

$$\rho \frac{de}{dt} - \frac{p}{\rho}\frac{d\rho}{dt} = \frac{(q_w + \tau_w u)L_p}{A}, \tag{9.152}$$

$$e = e(\rho, p), \qquad p = p(\rho, T). \tag{9.153}$$

Steady Conservative Form

$$\frac{d}{dx}(\rho A u) = 0, \tag{9.154}$$

$$\frac{d}{dx}(\rho A u^2 + pA) = p\frac{dA}{dx} - \tau_w L_p, \tag{9.155}$$

$$\frac{d}{dx}\left(\rho A u \left(e + \frac{u^2}{2} + \frac{p}{\rho}\right)\right) = q_w L_p, \tag{9.156}$$

$$e = e(\rho, p), \qquad p = p(\rho, T). \tag{9.157}$$

Steady Nonconservative Form

$$u\frac{d\rho}{dx} = -\frac{\rho}{A}\frac{d}{dx}(Au), \tag{9.158}$$

$$\rho u \frac{du}{dx} = -\frac{dp}{dx} - \frac{\tau_w L_p}{A}, \tag{9.159}$$

$$\rho u \frac{de}{dx} - \frac{pu}{\rho}\frac{d\rho}{dx} = \frac{(q_w + \tau_w u)L_p}{A}, \tag{9.160}$$

$$e = e(\rho, p), \qquad p = p(\rho, T). \tag{9.161}$$

In whatever form we consider, we have five equations in five unknown dependent variables: ρ, u, p, e, and T. We can always use the thermal and caloric state equations to eliminate e and T to give rise to three equations in three unknowns.

Example 9.10 Let us consider the flow of air with heat addition depicted in Fig. 9.5.

We are given air initially at $p_1 = 100\,\text{kPa}$, $T_1 = 300\,\text{K}$, $u_1 = 10\,\text{m/s}$ flowing in a duct of length 100 m. The duct has a constant circular cross-sectional area of $A = 0.02\,\text{m}^2$ and is isobarically heated with a constant heat flux q_w along the entire surface of the duct. At the end of the duct, the flow has $p_2 =$

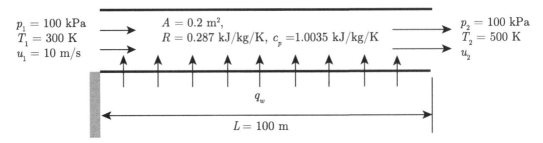

Figure 9.5 Schematic for a flow of calorically perfect ideal air with heat addition.

$100\,\text{kPa}$, $T_2 = 500\,\text{K}$. Find the mass flow rate $\dot{m}$, the wall heat flux q_w, and the entropy change $s_2 - s_1$; check for satisfaction of the second law. Assume a calorically perfect ideal gas, $R = 0.287\,\text{kJ}/(\text{kg}\,\text{K})$, $c_p = 1.0035\,\text{kJ}/(\text{kg}\,\text{K})$.

Solution

We begin by considering the geometry.

$$A = \pi r^2, \quad r = \sqrt{\frac{A}{\pi}}, \quad L_p = 2\pi r = 2\sqrt{\pi A} = 2\sqrt{\pi (0.02\,\text{m}^2)} = 0.501\,\text{m}. \tag{9.162}$$

Now get the mass flow rate. Because $p_1 = \rho_1 R T_1$, we have

$$\rho_1 = \frac{p_1}{RT_1} = \frac{100\,\text{kPa}}{\left(0.287\,\frac{\text{kJ}}{\text{kg}\,\text{K}}\right)(300\,\text{K})} = 1.161\,\frac{\text{kg}}{\text{m}^3}. \tag{9.163}$$

So

$$\dot{m} = \rho_1 u_1 A_1 = \left(1.161\,\frac{\text{kg}}{\text{m}^3}\right)\left(10\,\frac{\text{m}}{\text{s}}\right)(0.02\,\text{m}^2) = 0.2322\,\frac{\text{kg}}{\text{s}}. \tag{9.164}$$

Now get the flow variables at state 2:

$$\rho_2 = \frac{p_2}{RT_2}, = \frac{100\,\text{kPa}}{\left(0.287\,\frac{\text{kJ}}{\text{kg}\,\text{K}}\right)(500\,\text{K})} = 0.6969\,\frac{\text{kg}}{\text{m}^3}, \tag{9.165}$$

$$\rho_2 u_2 A_2 = \rho_1 u_1 A_1, \tag{9.166}$$

$$u_2 = \frac{\rho_1 u_1 A_1}{\rho_2 A_2} = \frac{\rho_1 u_1}{\rho_2} = \frac{\left(1.161\,\frac{\text{kg}}{\text{m}^3}\right)\left(10\,\frac{\text{m}}{\text{s}}\right)}{0.6969\,\frac{\text{kg}}{\text{m}^3}} = 16.67\,\frac{\text{m}}{\text{s}}. \tag{9.167}$$

Now consider the energy equation:

$$\rho u \frac{d}{dx}\left(e + \frac{u^2}{2} + \frac{p}{\rho}\right) = \frac{q_w L_p}{A}, \quad \text{so} \quad \frac{d}{dx}\left(h + \frac{u^2}{2}\right) = \frac{q_w L_p}{\dot{m}}, \tag{9.168}$$

$$\int_0^L \frac{d}{dx}\left(h + \frac{u^2}{2}\right) dx = \int_0^L \frac{q_w L_p}{\dot{m}}\,dx, \quad \text{so} \quad h_2 + \frac{u_2^2}{2} - h_1 - \frac{u_1^2}{2} = \frac{q_w L L_p}{\dot{m}}, \tag{9.169}$$

$$c_p(T_2 - T_1) + \frac{u_2^2}{2} - \frac{u_1^2}{2} = \frac{q_w L L_p}{\dot{m}}. \tag{9.170}$$

Solving for q_w, we get

$$q_w = \left(\frac{\dot{m}}{LL_p}\right)\left(c_p(T_2 - T_1) + \frac{u_2^2}{2} - \frac{u_1^2}{2}\right), \quad (9.171)$$

$$= \left(\frac{0.2322\,\frac{\text{kg}}{\text{s}}}{(100\,\text{m})(0.501\,\text{m})}\right)$$

$$\times \left(\left(1003.5\,\frac{\text{J}}{\text{kg K}}\right)(500\,\text{K} - 300\,\text{K}) + \frac{(16.67\,\frac{\text{m}}{\text{s}})^2}{2} - \frac{(10\,\frac{\text{m}}{\text{s}})^2}{2}\right), \quad (9.172)$$

$$= 0.004\,635\,\frac{\text{kg}}{\text{m}^2\,\text{s}}\left(200\,700\,\frac{\text{J}}{\text{kg}} - 88.9\,\frac{\text{m}^2}{\text{s}^2}\right) = 930\,\frac{\text{W}}{\text{m}^2}. \quad (9.173)$$

Note the unit J/kg is equal to the unit m^2/s^2. The heat flux is positive, which indicates a transfer of thermal energy *into* the air.

Now find the entropy change.

$$s_2 - s_1 = c_p \ln\left(\frac{T_2}{T_1}\right) - R \ln\left(\frac{p_2}{p_1}\right), \quad (9.174)$$

$$= \left(1003.5\,\frac{\text{J}}{\text{kg K}}\right)\ln\left(\frac{500\,\text{K}}{300\,\text{K}}\right) - \left(287\,\frac{\text{J}}{\text{kg K}}\right)\ln\left(\frac{100\,\text{kPa}}{100\,\text{kPa}}\right), \quad (9.175)$$

$$= 512.6\,\frac{\text{J}}{\text{kg K}}. \quad (9.176)$$

Is the second law satisfied? Assume the heat transfer takes place from a reservoir held at 500 K. The reservoir would have to be *at least* at 500 K in order to bring the fluid to its final state of 500 K. It could be greater than 500 K and still satisfy the second law.

$$S_2 - S_1 \geq \frac{Q_{12}}{T}, \quad \text{so} \quad \dot{S}_2 - \dot{S}_1 \geq \frac{\dot{Q}_{12}}{T}, \quad (9.177)$$

$$\dot{m}(s_2 - s_1) \geq \frac{\dot{Q}_{12}}{T}, \quad (9.178)$$

$$\geq \frac{q_w A_{tot}}{T}, \quad (9.179)$$

$$\geq \frac{q_w L L_p}{T}, \quad (9.180)$$

$$s_2 - s_1 \geq \frac{q_w L L_p}{\dot{m} T}, \quad (9.181)$$

$$512.6\,\frac{\text{J}}{\text{kg K}} \geq \frac{\left(930\,\frac{\text{J}}{\text{s m}^2}\right)(100\,\text{m})(0.501\,\text{m})}{\left(0.2322\,\frac{\text{kg}}{\text{s}}\right)(500\,\text{K})}, \quad (9.182)$$

$$512.6\,\frac{\text{J}}{\text{kg K}} \geq 401.3\,\frac{\text{J}}{\text{kg K}}. \quad (9.183)$$

9.2.5 Dynamical System Form

Now, let us uncouple the steady one-dimensional equations. First let us summarize again, in a slightly different manner than before, by rearranging the mass, Eq. (9.97), momentum, Eq. (9.109), and energy, Eq. (9.133), equations:

$$u\frac{d\rho}{dx} + \rho\frac{du}{dx} = -\frac{\rho u}{A}\frac{dA}{dx}, \quad (9.184)$$

$$\rho u \frac{du}{dx} + \frac{dp}{dx} = -\frac{\tau_w L_p}{A}, \tag{9.185}$$

$$\frac{dp}{dx} - c^2 \frac{d\rho}{dx} = \frac{(q_w + \tau_w u) L_p}{\rho u A \left.\frac{\partial e}{\partial p}\right|_\rho}. \tag{9.186}$$

In matrix form, these are recast as

$$\begin{pmatrix} u & \rho & 0 \\ 0 & \rho u & 1 \\ -c^2 & 0 & 1 \end{pmatrix} \begin{pmatrix} \frac{d\rho}{dx} \\ \frac{du}{dx} \\ \frac{dp}{dx} \end{pmatrix} = \begin{pmatrix} -\frac{\rho u}{A} \frac{dA}{dx} \\ -\frac{\tau_w L_p}{A} \\ \frac{(q_w + \tau_w u) L_p}{\rho u A \left.\frac{\partial e}{\partial p}\right|_\rho} \end{pmatrix}. \tag{9.187}$$

Use Cramer's rule, Example 2.14, to solve for the derivatives. First calculate the determinant of the coefficient matrix:

$$u\left((\rho u)(1) - (1)(0)\right) - \rho\left((0)(1) - (-c^2)(1)\right) = \rho\left(u^2 - c^2\right). \tag{9.188}$$

Implementing Cramer's rule, we find after detailed calculation:

$$\frac{d\rho}{dx} = \frac{\rho u \left(-\frac{\rho u}{A} \frac{dA}{dx}\right) - \rho \left(-\frac{\tau_w L_p}{A}\right) + \rho \left(\frac{(q_w + \tau_w u) L_p}{\rho u A \left.\frac{\partial e}{\partial p}\right|_\rho}\right)}{\rho\left(u^2 - c^2\right)}, \tag{9.189}$$

$$\frac{du}{dx} = \frac{-c^2 \left(-\frac{\rho u}{A} \frac{dA}{dx}\right) + u\left(-\frac{\tau_w L_p}{A}\right) - u\left(\frac{(q_w + \tau_w u) L_p}{\rho u A \left.\frac{\partial e}{\partial p}\right|_\rho}\right)}{\rho\left(u^2 - c^2\right)}, \tag{9.190}$$

$$\frac{dp}{dx} = \frac{\rho u c^2 \left(-\frac{\rho u}{A} \frac{dA}{dx}\right) - \rho c^2 \left(-\frac{\tau_w L_p}{A}\right) + \rho u^2 \left(\frac{(q_w + \tau_w u) L_p}{\rho u A \left.\frac{\partial e}{\partial p}\right|_\rho}\right)}{\rho\left(u^2 - c^2\right)}. \tag{9.191}$$

Simplify to find

$$\frac{d\rho}{dx} = \frac{1}{A} \frac{-\rho u^2 \frac{dA}{dx} + \tau_w L_p + \frac{(q_w + \tau_w u) L_p}{\rho u \left.\frac{\partial e}{\partial p}\right|_\rho}}{u^2 - c^2}, \tag{9.192}$$

$$\frac{du}{dx} = \frac{1}{A} \frac{c^2 \rho u \frac{dA}{dx} - u \tau_w L_p - \frac{(q_w + \tau_w u) L_p}{\rho \left.\frac{\partial e}{\partial p}\right|_\rho}}{\rho\left(u^2 - c^2\right)}, \tag{9.193}$$

$$\frac{dp}{dx} = \frac{1}{A} \frac{-c^2 \rho u^2 \frac{dA}{dx} + c^2 \tau_w L_p + \frac{(q_w + \tau_w u) L_p u}{\rho \left.\frac{\partial e}{\partial p}\right|_\rho}}{u^2 - c^2}. \tag{9.194}$$

We have here a system of coupled nonautonomous, nonlinear ordinary differential equations in standard form for dynamical system analysis: $d\mathbf{u}/dx = \mathbf{f}(\mathbf{u}(x), x)$. They are valid for *general* equations of state, and are *singular* when the fluid particle velocity is sonic $u = c$. This sonic singularity is the key to understanding important and non-intuitive physically observed behavior. When the velocity is sonic, the denominator is zero, and physical solutions can only be realized when the numerator simultaneously goes to zero. A useful way to analyze behavior near such singularities is to define a new independent variable, and solve the equivalent nonsingular autonomous system of the form $d\mathbf{u}/d\xi = \mathbf{f}(\mathbf{u}(\xi))$:

$$\frac{d\rho}{d\xi} = \frac{1}{A}\left(-\rho u^2 \frac{dA}{dx} + \tau_w L_p + \frac{(q_w + \tau_w u) L_p}{\rho u \left.\frac{\partial e}{\partial p}\right|_\rho}\right), \tag{9.195}$$

$$\frac{du}{d\xi} = \frac{1}{\rho A}\left(c^2 \rho u \frac{dA}{dx} - u\tau_w L_p - \frac{(q_w + \tau_w u) L_p}{\rho \left.\frac{\partial e}{\partial p}\right|_\rho}\right), \tag{9.196}$$

$$\frac{dp}{d\xi} = \frac{1}{A}\left(-c^2 \rho u^2 \frac{dA}{dx} + c^2 \tau_w L_p + \frac{(q_w + \tau_w u) L_p u}{\rho \left.\frac{\partial e}{\partial p}\right|_\rho}\right), \tag{9.197}$$

$$\frac{dx}{d\xi} = u^2 - c^2. \tag{9.198}$$

This formulation enables straightforward analysis of a sonic point in the flow. At such a point, one can easily ascertain the equilibrium by requiring all the terms on the right-hand side to be zero. One can then linearize about this equilibrium and discern the local behavior near the sonic point.

9.3 Isentropic Flow with Area Change

This section will consider isentropic flow with area change.

9.3.1 Isentropic Mach Number Relations

Take the special case of a calorically perfect ideal gas with $\tau_w = 0$, $q_w = 0$. Then we can recast Eqs. (9.154–9.156) as

$$\frac{d}{dx}(\rho u A) = 0, \quad \rho u \frac{du}{dx} + \frac{dp}{dx} = 0, \quad \frac{d}{dx}\left(e + \frac{u^2}{2} + \frac{p}{\rho}\right) = 0. \tag{9.199}$$

Integrate the energy equation using Eq. (4.169), $h = e + p/\rho$, to get

$$h + \frac{u^2}{2} = h_o + \frac{u_o^2}{2}. \tag{9.200}$$

If we define the "o" condition to be a condition of rest, then $u_o \equiv 0$. This is a stagnation condition, as introduced in Sections 3.14 and 6.82. In an unfortunate choice of nomenclature, properties evaluated for the fluid in motion are commonly named *static* properties, and are distinct from stagnation properties. We have

$$h + \frac{u^2}{2} = h_o, \quad (h - h_o) + \frac{u^2}{2} = 0. \tag{9.201}$$

Because we have a calorically perfect ideal gas,

$$c_p(T - T_o) + \frac{u^2}{2} = 0, \quad \text{so} \quad T - T_o + \frac{u^2}{2c_p} = 0, \tag{9.202}$$

$$1 - \frac{T_o}{T} + \frac{u^2}{2c_p T} = 0. \tag{9.203}$$

Now note that

$$c_p = c_p \frac{c_p - c_v}{c_p - c_v} = \frac{c_p}{c_v} \frac{c_p - c_v}{\frac{c_p}{c_v} - 1} = \frac{\gamma R}{\gamma - 1}, \tag{9.204}$$

so

$$1 - \frac{T_o}{T} + \frac{\gamma - 1}{2} \frac{u^2}{\gamma RT} = 0, \quad \text{giving} \quad \frac{T_o}{T} = 1 + \frac{\gamma - 1}{2} \frac{u^2}{\gamma RT}. \tag{9.205}$$

Recall the sound speed and Mach number for a calorically perfect ideal gas, $c^2 = \gamma RT$, $M^2 = (u/c)^2$. Then we get

$$\frac{T_o}{T} = 1 + \frac{\gamma - 1}{2} M^2, \quad \frac{T}{T_o} = \left(1 + \frac{\gamma - 1}{2} M^2\right)^{-1}. \tag{9.206}$$

Now if the flow is isentropic and for a calorically perfect ideal gas, it can be inferred from Eq. (9.141) that ratios of static to stagnation values are

$$\frac{T}{T_o} = \left(\frac{\rho}{\rho_o}\right)^{\gamma - 1} = \left(\frac{p}{p_o}\right)^{\frac{\gamma - 1}{\gamma}}. \tag{9.207}$$

Thus,

$$\frac{\rho}{\rho_o} = \left(1 + \frac{\gamma - 1}{2} M^2\right)^{-\frac{1}{\gamma - 1}}, \quad \frac{p}{p_o} = \left(1 + \frac{\gamma - 1}{2} M^2\right)^{-\frac{\gamma}{\gamma - 1}}. \tag{9.208}$$

For air $\gamma = 7/5$, so

$$\frac{T}{T_o} = \left(1 + \frac{1}{5} M^2\right)^{-1}, \quad \frac{\rho}{\rho_o} = \left(1 + \frac{1}{5} M^2\right)^{-\frac{5}{2}}, \quad \frac{p}{p_o} = \left(1 + \frac{1}{5} M^2\right)^{-\frac{7}{2}}. \tag{9.209}$$

Figures 9.6–9.8 show the variation of T, ρ, and p with M^2 for isentropic flow. Other thermodynamic properties can be determined from these, for example the sound speed:

$$\frac{c}{c_o} = \sqrt{\frac{\gamma RT}{\gamma RT_o}} = \sqrt{\frac{T}{T_o}} = \left(1 + \frac{\gamma - 1}{2} M^2\right)^{-1/2}. \tag{9.210}$$

Example 9.11 Show that how the isentropic relation for the ratio of stagnation to static pressures reduces to the incompressible Bernoulli equation in the limit of low Mach number, and show how it deviates from this as the Mach number rises.

Solution
Begin with the reciprocal of Eq. (9.208):

$$\frac{p_o}{p} = \left(1 + \frac{\gamma - 1}{2} M^2\right)^{\frac{\gamma}{\gamma - 1}}. \tag{9.211}$$

9.3 Isentropic Flow with Area Change

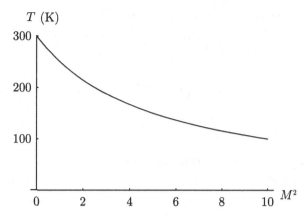

Figure 9.6 Static temperature versus Mach number squared for a calorically perfect ideal gas with $R = 0.287\,\text{kJ/kg/K}$, $\gamma = 7/5$, and $T_o = 300\,\text{K}$.

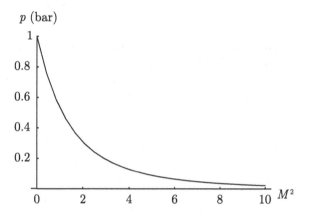

Figure 9.7 Static pressure versus Mach number squared for a calorically perfect ideal gas with $R = 0.287\,\text{kJ/kg/K}$, $\gamma = 7/5$, and $p_o = 1\,\text{bar}$.

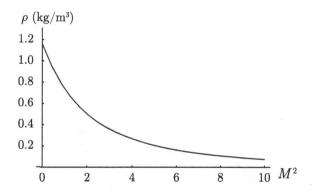

Figure 9.8 Static density versus Mach number squared for a calorically perfect ideal gas with $R = 0.287\,\text{kJ/kg/K}$, $\gamma = 7/5$, and $\rho_o = 1.16\,\text{kg/m}^3$.

Taylor series of this equation about $M = 0$ yields

$$\frac{p_o}{p} = 1 + \frac{\gamma}{2}M^2 + \frac{\gamma}{8}M^4 + \cdots = 1 + \frac{\gamma}{2}M^2\left(1 + \frac{1}{4}M^2 + \cdots\right), \quad (9.212)$$

$$= 1 + \frac{\gamma}{2}\frac{\rho u^2}{\gamma p}\left(1 + \frac{1}{4}M^2 + \cdots\right) = 1 + \frac{\rho u^2}{2p}\left(1 + \frac{1}{4}M^2 + \cdots\right), \quad (9.213)$$

$$\frac{p_o}{\rho} = \frac{p}{\rho} + \frac{u^2}{2}\left(1 + \frac{1}{4}M^2 + \cdots\right). \quad (9.214)$$

For $M = 0$, we recover the incompressible Bernoulli equation, see Eq. (6.196), and the correction at small finite Mach number to the incompressible Bernoulli equation estimate for stagnation pressure is evident.

Example 9.12 We are given an airplane flying into still air at $\hat{u}_p = 200$ m/s as sketched in Fig. 9.9(a). The ambient air is at 288 K and 101.3 kPa. Find the temperature, pressure, and density at the nose of the airplane. Assume steady one-dimensional isentropic flow of a calorically perfect ideal gas.

Solution
Because of Galilean invariance, we can formulate the problem in any convenient reference frame. Leaving out formalities of the transformation, we can see that the laboratory frame of Fig. 9.9(a) will induce an unsteady response, but if we formulate in the frame in which the plane is at rest, Fig. 9.9(b), a time-independent steady result will be found. In the steady wave frame, the ambient conditions are *static* while the nose conditions are *stagnation*:

$$M = \frac{u}{c} = \frac{u}{\sqrt{\gamma RT}} = \frac{200\,\frac{\text{m}}{\text{s}}}{\sqrt{\frac{7}{5}\left(287\,\frac{\text{J}}{\text{kg K}}\right)288\,\text{K}}} = 0.588, \quad (9.215)$$

so

$$T_o = T\left(1 + \frac{1}{5}M^2\right) = (288\,\text{K})\left(1 + \frac{1}{5}0.588^2\right) = 307.9\,\text{K}, \quad (9.216)$$

$$\rho_o = \rho\left(1 + \frac{1}{5}M^2\right)^{\frac{5}{2}} = \frac{101.3\,\text{kPa}}{\left(0.287\,\frac{\text{kJ}}{\text{kg K}}\right)(288\,\text{K})\left(1 + \frac{1}{5}0.588^2\right)^{\frac{5}{2}}} = 1.45\,\frac{\text{kg}}{\text{m}^3}, \quad (9.217)$$

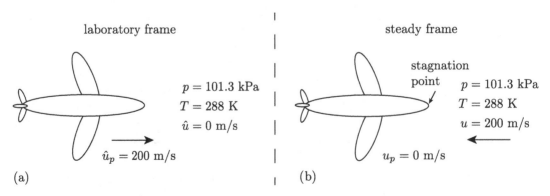

Figure 9.9 Configurations for estimating stagnation conditions at an airplane's nose.

$$p_o = p\left(1 + \frac{1}{5}M^2\right)^{\frac{7}{2}} = (101.3\,\text{kPa})\left(1 + \frac{1}{5}0.588^2\right)^{\frac{7}{2}} = 128\,\text{kPa}. \tag{9.218}$$

The temperature, pressure, and density all rise in the isentropic process. In this wave frame, the kinetic energy of the flow is being converted isentropically to thermal energy.

9.3.2 Sonic Properties

Let "*" denote a property at the sonic state $M^2 \equiv 1$. Then we get

$$\frac{T_*}{T_o} = \frac{2}{\gamma+1}, \quad \frac{\rho_*}{\rho_o} = \left(\frac{2}{\gamma+1}\right)^{\frac{1}{\gamma-1}}, \quad \frac{p_*}{p_o} = \left(\frac{2}{\gamma+1}\right)^{\frac{\gamma}{\gamma-1}}, \tag{9.219}$$

$$u_* = c_* = \sqrt{\gamma R T_*} = \sqrt{\frac{2\gamma}{\gamma+1}RT_o}, \quad \frac{c_*}{c_o} = \sqrt{\frac{2}{\gamma+1}}. \tag{9.220}$$

If the fluid is air modeled as a calorically perfect ideal gas with $\gamma = 7/5$, we get

$$\frac{T_*}{T_o} = 0.8333, \quad \frac{\rho_*}{\rho_o} = 0.6339, \quad \frac{p_*}{p_o} = 0.5283, \quad \frac{c_*}{c_o} = 0.9129. \tag{9.221}$$

9.3.3 Effect of Area Change

To understand the effect of area change, the influence of the mass equation must be considered. So far we have really only looked at energy. In the isentropic limit the mass, momentum, and energy equations for a calorically perfect ideal gas, Eqs. (9.98, 9.113, 9.140), reduce to

$$\frac{d\rho}{\rho} + \frac{du}{u} + \frac{dA}{A} = 0, \quad \rho u\,du + dp = 0, \quad \frac{dp}{p} = \gamma\frac{d\rho}{\rho}. \tag{9.222}$$

Substitute energy, then mass, into momentum:

$$\rho u\,du + \underbrace{\gamma\frac{p}{\rho}\,d\rho}_{dp} = 0, \quad \text{so} \quad \rho u\,du + \gamma\frac{p}{\rho}\underbrace{\left(-\frac{\rho}{u}\,du - \frac{\rho}{A}\,dA\right)}_{d\rho} = 0, \tag{9.223}$$

$$du + \gamma\frac{p}{\rho}\left(-\frac{1}{u^2}\,du - \frac{1}{uA}\,dA\right) = 0, \quad \text{so} \quad du\left(1 - \frac{\gamma p/\rho}{u^2}\right) = \gamma\frac{p}{\rho}\frac{dA}{uA}, \tag{9.224}$$

$$\frac{du}{u}\left(1 - \frac{\gamma p/\rho}{u^2}\right) = \frac{\gamma p/\rho}{u^2}\frac{dA}{A}, \quad \text{so} \quad \frac{du}{u}\left(1 - \frac{1}{M^2}\right) = \frac{1}{M^2}\frac{dA}{A}, \tag{9.225}$$

$$\frac{du}{u}(M^2 - 1) = \frac{dA}{A}, \quad \text{so} \quad \frac{du}{u} = \frac{1}{M^2-1}\frac{dA}{A}. \tag{9.226}$$

This is a useful result that yields some surprising and important physical insights. We note the following: (1) For subsonic flow, if fluid moves into a region of diverging area, its velocity goes down, consistent with our intuition. (2) For subsonic flow, if fluid moves into a region of converging area, its velocity goes up, consistent with our intuition. (3) For supersonic flow, if fluid moves into a region of diverging area, its velocity goes up, inconsistent with our intuition. (4) For supersonic flow, if fluid moves into a region of converging area, its velocity goes down, inconsistent with our intuition. (5) There is a singularity when $M^2 = 1$. (6) If $M^2 = 1$, we need $dA = 0$ to retain non-singular du. (7) An area minimum is necessary to induce a transition from subsonic to supersonic flow. Figure 9.10 depicts the performance of a fluid in a variable area duct.

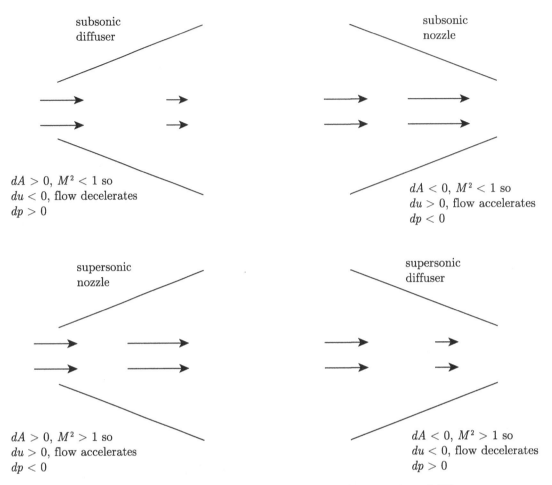

Figure 9.10 Behavior of a fluid with $u > 0$ in sub- and supersonic and diffusers.

Consider the area A at a sonic state. From the mass equation, Eq. (9.100):

$$\rho u A = \rho_* u_* A_* = \rho_* c_* A_*, \tag{9.227}$$

$$\frac{A}{A_*} = \frac{\rho_*}{\rho} c_* \frac{1}{u} = \frac{\rho_*}{\rho} \sqrt{\gamma R T_*} \frac{1}{u} = \frac{\rho_*}{\rho} \frac{\sqrt{\gamma R T_*}}{\sqrt{\gamma R T}} \frac{\sqrt{\gamma R T}}{u}, \tag{9.228}$$

$$= \frac{\rho_*}{\rho} \sqrt{\frac{T_*}{T}} \frac{1}{M} = \frac{\rho_*}{\rho_o} \frac{\rho_o}{\rho} \sqrt{\frac{T_* T_o}{T_o T}} \frac{1}{M}. \tag{9.229}$$

Substitute from earlier-developed relations and get

$$\frac{A}{A_*} = \frac{1}{M} \left(\frac{2}{\gamma+1} \left(1 + \frac{\gamma-1}{2} M^2 \right) \right)^{\frac{1}{2}\frac{\gamma+1}{\gamma-1}}. \tag{9.230}$$

Figure 9.11 shows the performance of A and M in a variable area duct. Note that A/A_* has a minimum value of 1 at $M = 1$, for each $A/A_* > 1$, there exist *two* values of M, and $A/A_* \to \infty$ as $M \to 0$ or $M \to \infty$.

9.3 Isentropic Flow with Area Change

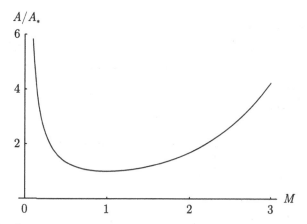

Figure 9.11 Ratio of local area to sonic throat area versus Mach number for a calorically perfect ideal gas with $R = 0.287\,\text{kJ/kg/K}$, $\gamma = 7/5$.

9.3.4 Choking

Consider mass flow rate variation with pressure difference. We know that a small pressure difference gives small velocity and small mass flow. As pressure difference grows, velocity and mass flow rate grow. The velocity is limited to sonic at a particular duct location. Thus, this provides fundamental restriction on mass flow rate. It can be shown that the sonic state corresponds to a maximum mass flow rate, $\dot{m}_{max}$, which is

$$\dot{m}_{max} = \rho_* u_* A_*. \tag{9.231}$$

If the material is a calorically perfect ideal gas then,

$$\dot{m}_{max} = \rho_o \left(\frac{2}{\gamma+1}\right)^{\frac{1}{\gamma-1}} \left(\sqrt{\frac{2\gamma}{\gamma+1} RT_o}\right) A_*, \tag{9.232}$$

$$= \rho_o \left(\frac{2}{\gamma+1}\right)^{\frac{1}{2}\frac{\gamma+1}{\gamma-1}} \sqrt{\gamma RT_o} A_*. \tag{9.233}$$

A flow that has a maximum mass flow rate is known as *choked* flow. Flows will choke at area minima in a duct.

Example 9.13 Consider an isentropic flow with area change and choking, as shown in Fig. 9.12. We are given air effectively at rest in a large tank with stagnation conditions $p_o = 200\,\text{kPa}$, $T_o = 600\,\text{K}$ flowing through a throat to an exit Mach number of 2.5. The desired mass flow rate is $2\,\text{kg/s}$. Find the (a) throat area, (b) exit pressure, (c) exit temperature, (d) exit velocity, and (e) exit area. Assume we have isentropic flow of a calorically perfect ideal gas with $\gamma = 7/5$.

Solution
While the air in the tank cannot have a velocity that is exactly zero and still have flow, the tank is so large that the velocity in the tank interior is small. First find the stagnation density via the ideal

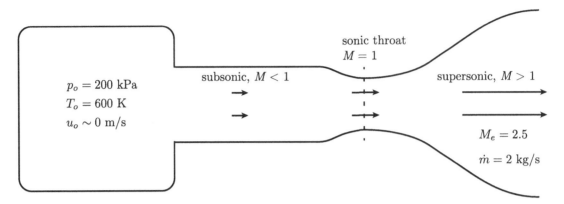

Figure 9.12 Isentropic expansion of air from a stagnation state to a supersonic exit.

gas law:

$$\rho_o = \frac{p_o}{RT_o} = \frac{200\,\text{kPa}}{\left(0.287\,\frac{\text{kJ}}{\text{kg K}}\right)(600\,\text{K})} = 1.161\,44\,\frac{\text{kg}}{\text{m}^3}. \tag{9.234}$$

Because the air necessarily flows through a sonic throat:

$$\dot{m}_{max} = \rho_o \left(\frac{2}{\gamma+1}\right)^{\frac{1}{2}\frac{\gamma+1}{\gamma-1}} \sqrt{\gamma RT_o}\, A_*, \quad \text{so} \quad A_* = \frac{\dot{m}_{max}}{\rho_o \left(\frac{2}{\gamma+1}\right)^{\frac{1}{2}\frac{\gamma+1}{\gamma-1}} \sqrt{\gamma RT_o}}, \tag{9.235}$$

$$A_* = \frac{2\,\frac{\text{kg}}{\text{s}}}{(1.161\,44\,\frac{\text{kg}}{\text{m}^3})(0.5787)\sqrt{1.4\left(287\,\frac{\text{J}}{\text{kg K}}\right)(600\,\text{K})}} = 0.006\,060\,33\,\text{m}^2. \tag{9.236}$$

Because we know M_e, we can use isentropic relations to find other exit conditions:

$$p_e = p_o \left(1 + \frac{\gamma-1}{2}M_e^2\right)^{-\frac{\gamma}{\gamma-1}} = (200\,\text{kPa})\left(1 + \frac{1}{5}2.5^2\right)^{-3.5} = 11.7055\,\text{kPa}, \tag{9.237}$$

$$T_e = T_o \left(1 + \frac{\gamma-1}{2}M_e^2\right)^{-1} = (600\,\text{K})\left(1 + \frac{1}{5}2.5^2\right)^{-1} = 266.667\,\text{K}. \tag{9.238}$$

Note that

$$\rho_e = \frac{p_e}{RT_e} = \frac{11.7055\,\text{kPa}}{\left(0.287\,\frac{\text{kJ}}{\text{kg K}}\right)(266.667\,\text{K})} = 0.152\,947\,\frac{\text{kg}}{\text{m}^3}. \tag{9.239}$$

The pressure, temperature, and density at the exit are all considerably lower than their values at the stagnation state. Now the exit velocity is simply

$$u_e = M_e c_e = M_e \sqrt{\gamma RT_e} = 2.5\sqrt{1.4\left(287\,\frac{\text{J}}{\text{kg K}}\right)(266.667\,\text{K})} = 818.332\,\frac{\text{m}}{\text{s}}. \tag{9.240}$$

Now determine the exit area:

$$A = \frac{A_*}{M_e}\left(\frac{2}{\gamma+1}\left(1 + \frac{\gamma-1}{2}M_e^2\right)\right)^{\frac{1}{2}\frac{\gamma+1}{\gamma-1}}, \tag{9.241}$$

$$= \frac{0.006\,060\,33\,\text{m}^2}{2.5}\left(\frac{5}{6}\left(1 + \frac{1}{5}2.5^2\right)\right)^3 = 0.015\,9794\,\text{m}^2. \tag{9.242}$$

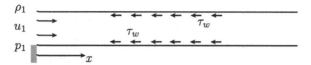

Figure 9.13 Diagram of Fanno flow.

9.4 Fanno Flow

Adiabatic flow with friction through a constant area duct is known as *Fanno*[3] *flow*. We give a depiction in Fig. 9.13. Wall friction is typically considered by taking the wall shear stress τ_w to be a constant. Wall friction is usually correlated with what is known as the Darcy friction factor, f, with

$$f \equiv \frac{8\tau_w}{\rho u^2}. \tag{9.243}$$

Now in common engineering analysis f is related to the local flow Reynolds number based on pipe diameter d, $Re_d = \rho u d/\mu$, and relative roughness of the duct ϵ/d, in which ϵ is the average surface roughness:

$$f = f\left(Re_d, \frac{\epsilon}{d}\right). \tag{9.244}$$

For steady laminar duct flow, the friction factor is independent of ϵ. It turns out the Poiseuille[4] flow solution, to be studied in Section 10.1.1, gives the friction factor for laminar pipe flow, which can be shown to be $f = 64/Re_d$.

If the flow is steady and turbulent, the friction factor is described by the following empirical formula known as the Colebrook equation:

$$\frac{1}{f^{1/2}} = -2.0 \log_{10}\left(\frac{\epsilon/d}{3.7} + \frac{2.51}{Re_d f^{1/2}}\right). \tag{9.245}$$

Often one needs to iterate to find f for turbulent flows. Alternatively, one can use the Moody[5] chart to estimate f. This is simply a graphical representation of the Colebrook equation and is found in most undergraduate fluid mechanics texts. While in principle f varies with a host of variables, in practice for Fanno flow, it is often estimated as a constant. Lastly, embedded within most presentations of Fanno flow is the implicit assumption that the bounding wall is stationary. Were the equations of Fanno flow subjected to a Galilean transformation, one would need to guarantee the wall friction model depended on velocity differences, not simply velocity.

To get a grasp on the effects of wall friction, consider a special case of generalized one-dimensional flow for which the flow is steady, one-dimensional, adiabatic, constant area, modeled by a Darcy friction model, and a calorically perfect ideal gas. Considering our general equations, Eqs. (9.192–9.194) in the limit $dA/dx = 0$, $q_w = 0$, we get

---

[3] Gino Girolamo Fanno, 1882–1962, Italian mechanical engineer.
[4] Jean Louis Poiseuille, 1799–1869, French physician.
[5] Lewis Ferry Moody, 1880–1953, American engineer and educator.

$$\frac{d\rho}{dx} = \frac{\tau_w L_p}{A} \frac{1 + \frac{1}{\rho \left.\frac{\partial e}{\partial p}\right|_\rho}}{u^2 - c^2}, \quad (9.246)$$

$$\frac{du}{dx} = -\frac{u\tau_w L_p}{A} \frac{1 + \frac{1}{\rho \left.\frac{\partial e}{\partial p}\right|_\rho}}{\rho(u^2 - c^2)}, \quad (9.247)$$

$$\frac{dp}{dx} = \frac{\tau_w L_p}{A} \frac{c^2 + \frac{u^2}{\rho \left.\frac{\partial e}{\partial p}\right|_\rho}}{u^2 - c^2}. \quad (9.248)$$

Now for a circular duct

$$L_p = 2\pi r, \quad A = \pi r^2, \quad \frac{L_p}{A} = \frac{2\pi r}{\pi r^2} = \frac{2}{r} = \frac{4}{d}. \quad (9.249)$$

For a calorically perfect ideal gas

$$e = \frac{1}{\gamma - 1}\frac{p}{\rho}, \quad \left.\frac{\partial e}{\partial p}\right|_\rho = \frac{1}{\gamma - 1}\frac{1}{\rho}, \quad 1 + \frac{1}{\rho \left.\frac{\partial e}{\partial p}\right|_\rho} = \gamma. \quad (9.250)$$

Making these substitutions yields

$$\frac{d\rho}{dx} = \frac{4\tau_w}{d}\frac{\gamma}{u^2 - c^2}, \quad (9.251)$$

$$\frac{du}{dx} = -\frac{4u\tau_w}{d}\frac{\gamma}{\rho(u^2 - c^2)}, \quad (9.252)$$

$$\frac{dp}{dx} = \frac{4\tau_w}{d}\frac{c^2 + u^2(\gamma - 1)}{u^2 - c^2}. \quad (9.253)$$

Substituting for τ_w gives

$$\frac{d\rho}{dx} = \frac{f\rho u^2}{2d}\frac{\gamma}{u^2 - c^2}, \quad (9.254)$$

$$\frac{du}{dx} = -\frac{f\rho u^2 u}{2d}\frac{\gamma}{\rho(u^2 - c^2)}, \quad (9.255)$$

$$\frac{dp}{dx} = \frac{f\rho u^2}{2d}\frac{c^2 + u^2(\gamma - 1)}{u^2 - c^2}. \quad (9.256)$$

Multiply each equation by c^2/c^2 to cast the model in terms of M^2:

$$\frac{d\rho}{dx} = \frac{f\rho M^2}{2d}\frac{\gamma}{M^2 - 1}, \quad (9.257)$$

$$\frac{du}{dx} = -\frac{fM^2 u}{2d}\frac{\gamma}{M^2 - 1}, \quad (9.258)$$

$$\frac{dp}{dx} = \frac{f\rho u^2}{2d}\frac{1 + M^2(\gamma - 1)}{M^2 - 1}. \quad (9.259)$$

Now with the definition of M^2 for the calorically perfect ideal gas, $M^2 = \rho u^2/(\gamma p)$, one differentiates and gets

$$\frac{dM^2}{dx} = \frac{u^2}{\gamma p}\frac{d\rho}{dx} + \frac{2\rho u}{\gamma p}\frac{du}{dx} - \frac{\rho u^2}{\gamma p^2}\frac{dp}{dx}, \qquad (9.260)$$

$$= \frac{u^2}{\gamma p}\frac{f\rho M^2}{2d}\frac{\gamma}{M^2-1} + \frac{2\rho u}{\gamma p}\left(-\frac{fM^2 u}{2d}\frac{\gamma}{M^2-1}\right) - \frac{\rho u^2}{\gamma p^2}\frac{f\rho u^2}{2d}\frac{1+M^2(\gamma-1)}{M^2-1}, \qquad (9.261)$$

$$= \frac{fM^4}{2d}\frac{\gamma}{M^2-1} - \frac{2fM^4}{2d}\frac{\gamma}{M^2-1} - \frac{\gamma fM^4}{2d}\frac{1+M^2(\gamma-1)}{M^2-1}, \qquad (9.262)$$

$$= \frac{\gamma fM^4}{d(1-M^2)}\left(1 + \frac{\gamma-1}{2}M^2\right). \qquad (9.263)$$

Rearranging gives

$$\frac{1-M^2}{\gamma (M^2)^2 \left(1+\frac{\gamma-1}{2}M^2\right)}\,dM^2 = \frac{f}{d}\,dx. \qquad (9.264)$$

This can be integrated. Of particular interest is the length of pipe necessary to induce a locally sonic, choked flow, condition. Integrate this expression from $x = 0$ to $x = L_*$. Here L_* is defined as the length at which the flow becomes sonic, so $M^2 = 1$ at $x = L_*$:

$$\int_{M_1^2}^1 \frac{1-M^2}{\gamma (M^2)^2 \left(1+\frac{\gamma-1}{2}M^2\right)}\,dM^2 = \int_0^{L_*} \frac{f}{d}\,dx. \qquad (9.265)$$

An analytic solution for this integral is

$$\frac{1-M_1^2}{\gamma M_1^2} + \frac{1+\gamma}{2\gamma}\ln\frac{(1+\gamma)M_1^2}{2+(\gamma-1)M_1^2} = \frac{f}{d}L_*. \qquad (9.266)$$

If one has $M_1 < 1$ at $x = 0$, one can also integrate to an arbitrary $M \leq 1$ to get

$$x = \frac{d}{f}g(M;\gamma,M_1). \qquad (9.267)$$

While we might prefer to invert to have $M(x)$, this is difficult. We also can examine the steady conservative form of the general equations, Eqs. (9.154–9.156) and determine for the conditions of Fanno flow, constant A, $q_w = 0$, that

$$\rho u = \rho_1 u_1, \qquad \frac{d}{dx}\left(\rho u^2 + p\right) = -\tau_w L_p, \qquad h + \frac{u^2}{2} = h_1 + \frac{u_1^2}{2}. \qquad (9.268)$$

Here one can see that for Fanno flow, mass flux is constant, total enthalpy is constant, but that $\rho u^2 + p$ decays due to frictional wall stress τ_w.

Example 9.14 Consider Fanno flow of a calorically perfect ideal gas with $\gamma = 7/5$, $M_1 = 0.1$, $f = 0.04$, $d = 0.01$ m.

Solution
Integrating Eq. (9.264) from 0 to arbitrary x, we find

$$\frac{1}{\gamma M_1^2} - \frac{1}{\gamma M^2} + \frac{1+\gamma}{2\gamma}\ln\left(\frac{M_1^2\left((\gamma-1)M^2+2\right)}{M^2\left((\gamma-1)M_1^2+2\right)}\right) = \frac{fx}{d}. \qquad (9.269)$$

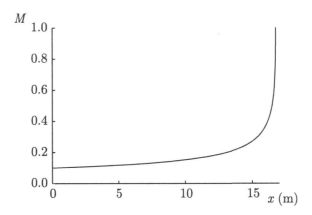

Figure 9.14 Mach number versus x for Fanno flow with $\gamma = 7/5$, $M_1 = 0.1$, $f = 0.04$, $d = 0.01$ m.

Figure 9.14 shows the variation of fluid Mach number with distance for this case. The Mach number rises as the flow moves down the duct. When it reaches $x = 16.73$ m, $M = 1$, and the flow is *frictionally choked*. It is somewhat counterintuitive that friction tends to increase a subsonic Mach number towards a sonic condition. Despite the fact that the Mach number is increasing, the mass flux ρu and total enthalpy $h + u^2/2$ remain constant; however, it can be seen from Eq. (9.147) that the quantity $\rho u^2 + p$ decreases. Were we to study a supersonic M_1, we would find that friction would tend to decrease it towards a $M = 1$ choked condition.

While with considerable effort, it is possible to get exact solutions for all state variables such as $\rho(x)$, $u(x)$, $p(x)$, it is straightforward to get them by numerical integration of the system of ordinary differential equations, Eqs. (9.257–9.259). We do so and show results in Fig. 9.15; we also plot the temperature. As the flow chokes, the density, pressure, and temperature steeply fall, while the velocity steeply rises. Again, as in much of compressible flow, non-intuitive results are obtained. The temperature goes down in the presence of friction! The velocity goes up in the presence of friction! Such is the nature of a coupled system. All of the conservation principles have been satisfied along with the state equations. Friction has done what it has been asked to do: decay the value of $\rho u^2 + p$.

9.5 Rayleigh Flow

Frictionless flow with heat addition through a constant area duct is known as *Rayleigh flow*. We give a diagram in Fig. 9.16. We could analyze in a similar fashion that we did for Fanno flow. Let us take a different approach. For frictionless flow through a constant area duct of a calorically perfect ideal gas, Eqs. (9.154–9.157), after recognizing for a circular duct that $L_p/A = 4/d$, integrating, and applying the freestream conditions, reduce to

$$\rho u = \rho_1 u_1, \tag{9.270}$$

$$\rho u^2 + p = \rho_1 u_1^2 + p_1, \tag{9.271}$$

9.5 Rayleigh Flow

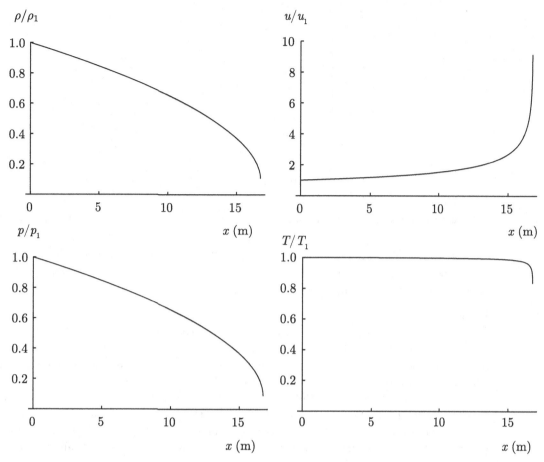

Figure 9.15 ρ/ρ_1, u/u_1, p/p_1, T/T_1 versus x for Fanno flow with $\gamma = 7/5$, $M_1 = 0.1$, $f = 0.04$, $d = 0.01$ m.

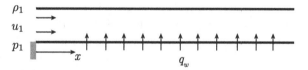

Figure 9.16 Diagram of Rayleigh flow.

$$h + \frac{u^2}{2} = h_1 + \frac{u_1^2}{2} + \frac{4q_w}{\rho_1 u_1 d}x, \tag{9.272}$$

$$h = \frac{\gamma}{\gamma - 1}\frac{p}{\rho}. \tag{9.273}$$

While these are four algebraic equations in the four unknowns $\rho(x)$, $u(x)$, $p(x)$, $h(x)$, their analysis is lengthy, even more so than a similar lengthy analysis that will soon be given in Section 9.6, which will be for the case in which $q_w = 0$. For $q_w > 0$, we have heat addition, and all flow variables will be functions of x. They can, however, be rapidly solved using methods of computer algebra.

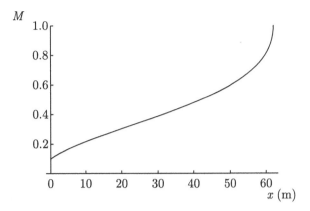

Figure 9.17 M versus x for Rayleigh flow with $\gamma = 7/5$, $M_1 = 0.1$, $q_w = 10\,000\,\text{W/m}^2$, $d = 0.01\,\text{m}$.

Example 9.15 Consider Rayleigh flow of a calorically perfect ideal gas with $\gamma = 7/5$, $M_1 = 0.1$, $q_w = 10\,000\,\text{W/m}^2$, $d = 0.01\,\text{m}$. Take $p_1 = 10^5\,\text{Pa}$, $T_1 = 300\,\text{K}$, $R = 287\,\text{J/kg/K}$. Plot Mach number, density, velocity, pressure, and temperature as functions of distance.

Solution
We solve the algebraic equations for Rayleigh flow with computer algebra. Two solutions are found. One is obviously nonphysical as it induces negative pressure. Had the initial condition been supersonic, two physical roots would have been found, and we would need to consider results more carefully. The other is physical, and we and plot these solutions. Figure 9.17 shows the variation of a fluid Mach number in a constant area duct with heat transfer. As for Fanno flow, for the Rayleigh flow, the Mach number rises as the flow moves down the duct. When it reaches $x = 62.03\,\text{m}$, $M = 1$, and the flow is *thermally choked*. Heat transfer tends to increase a subsonic Mach number towards a sonic condition. Despite the fact that the Mach number is increasing, the mass flux ρu and the total momentum $\rho u^2 + p$ remain constant; however, the heat transfer causes the total enthalpy $h + u^2/2$ to increase. Were we to study a supersonic M_1, we would find heat transfer would tend to decrease it towards a $M = 1$ choked condition. It is possible to get exact solutions for all state variables such as $\rho(x)$, $u(x)$, $p(x)$, and we do so again by computer algebra. Results are given in Fig. 9.18. We also plot the temperature. As the flow chokes, density and pressure steeply fall, while the velocity steeply rises. The temperature, after initially rising, suffers a small but steep fall near the thermal choking point.

Example 9.16 Consider a problem similar to the previous example. Take again flow of a calorically perfect ideal gas with $\gamma = 7/5$, $M_1 = 0.1$. But now, restrict the heat addition to a finite region of the duct, and include effects of area change:

$$q_w = \begin{cases} 10\,000\,\text{W/m}^2, & x \in [0, 60\,\text{m}], \\ 0\,\text{W/m}^2 & x > 60\,\text{m} \end{cases}, \quad d = (0.01\,\text{m})\left(1 - 0.12\exp\left(\frac{x - (40\,\text{m})}{4\,\text{m}}\right)^2\right). \quad (9.274)$$

Take $p_1 = 10^5\,\text{Pa}$, $T_1 = 300\,\text{K}$, $R = 287\,\text{J/kg/K}$. Prepare a plot of Mach number as a function of distance and compare to the previous example.

Solution
Here, the heat addition is suppressed for $x > 60\,\text{m}$, and a converging–diverging nozzle has been added with a throat at $x = 40\,\text{m}$. For this problem, the simultaneous presence of heat transfer and area

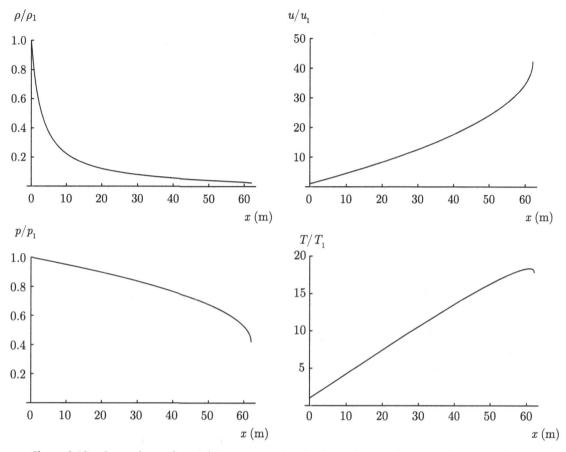

Figure 9.18 ρ/ρ_1, u/u_1, p/p_1, T/T_1 versus x for Rayleigh flow with $\gamma = 7/5$, $M_1 = 0.1$, $q_w = 10\,000\,\text{W/m}^2$, $d = 0.01\,\text{m}$.

change renders a strictly algebraic solution to be difficult. It is thus better to return to the ordinary differential equations of generalized one-dimensional flow, Eqs. (9.195–9.198). We simplify them here for the calorically perfect ideal gas with $\tau_w = 0$:

$$\frac{d\rho}{d\xi} = \frac{1}{A}\left(-\rho u^2 \frac{dA}{dx} + \frac{(\gamma-1)q_w(x)L_p}{u}\right), \tag{9.275}$$

$$\frac{du}{d\xi} = \frac{1}{\rho A}\left(\gamma p u \frac{dA}{dx} - (\gamma-1)q_w(x)L_p\right), \tag{9.276}$$

$$\frac{dp}{d\xi} = \frac{1}{A}\left(-\gamma p u^2 \frac{dA}{dx} + (\gamma-1)q_w(x)L_p u\right), \tag{9.277}$$

$$\frac{dx}{d\xi} = u^2 - \gamma\frac{p}{\rho}. \tag{9.278}$$

This is an autonomous system of four nonlinear ordinary differential equations in the four unknowns $\rho(\xi), u(\xi), p(\xi), x(\xi)$. One could use mass conservation, $\rho u A = \rho_1 u_1 A_1$, to reduce this to three nonlinear ordinary differential equations. Alternatively, one can integrate the full four equations and check the solution by verifying after calculation that mass is conserved.

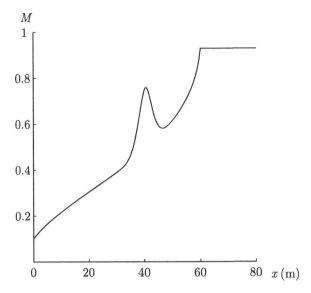

Figure 9.19 M versus x for flow with heat addition and area change $\gamma = 7/5$, $M_1 = 0.1$.

Substitution of functional forms for $q_w(x)$ and $A(x)$ leads to a system of ordinary differential equations that is difficult to write in a compact form. They may be solved numerically and results obtained. Initial conditions are

$$\rho(0) = 1.16144 \, \frac{\text{kg}}{\text{m}^3}, \quad u(0) = 34.7189 \, \frac{\text{m}}{\text{s}}, \quad p(0) = 10^5 \, \text{Pa}, \quad x(0) = 0 \, \text{m}. \qquad (9.279)$$

Results are shown in Fig. 9.19.

Near $x = 0$ m, the heat addition dominates as the Mach number increases. Near $x = 40$ m, the effect of area change is apparent, as the Mach number increases more with x. At the throat, where $x = 40$ m, the Mach number has a local maximum. Heat addition continues and soon dominates the effects of area change. At $x = 60$ m, the heat addition ceases, and there is no driving mechanism to further influence the flow, which equilibrates. Had the flow reached $M = 1$, the flow would have choked, and its analysis would be significantly more complicated. One could also include the effects of friction, and the results would be further modified. For this flow, it can be verified that the mass flow rate $\rho u A$ is a constant with value $\dot{m} = 0.00316703$ kg/s.

9.6 Normal Shock Waves

This section will develop relations for normal shock waves in fluids with general equations of state. It will be specialized to a calorically perfect ideal gas to illustrate common features of the waves. For this section, it is assumed we have one-dimensional steady flow, no area change, negligible viscous and wall friction effects, and negligible thermal conduction and wall heat transfer. We will first give an analysis for general one-dimensional unsteady flux-conservative systems. We will then specialize to the one-dimensional steady Euler equations for fluid mechanics. In this limit, the equations can be integrated, and one is led to a complicated system of nonlinear algebraic equations which are solved and interpreted physically.

9.6.1 Analysis for a General Flux-Conservative System

As described by LeVeque (1992), the proper way to arrive at the shock jump equations is to use a more primitive form of the conservation laws, expressed in terms of integrals of conserved quantities balanced by fluxes of those quantities, for example Eq. (4.4). Consider the scenario of Fig. 9.20. In both Fig. 9.20(a) and (b), we have a fixed volume bounded in the x direction by x_1 and x_2. If $\mathbf{q}$ is a set of conserved variables, and $\mathbf{f}(\mathbf{q})$ is the flux of $\mathbf{q}$ (e.g. for mass conservation, ρ is a conserved variable and ρu is the flux), then the primitive form of the conservation law can be written as

$$\frac{d}{dt} \int_{x_1}^{x_2} \mathbf{q}(x,t)\, dx = \mathbf{f}(\mathbf{q}(x_1,t)) - \mathbf{f}(\mathbf{q}(x_2,t)). \tag{9.280}$$

Here we have considered flow into and out of a one-dimensional box for $x \in [x_1, x_2]$. For the Euler equations, we have

$$\mathbf{q} = \begin{pmatrix} \rho \\ \rho u \\ \rho\left(e + \tfrac{1}{2}u^2\right) \end{pmatrix}, \qquad \mathbf{f}(\mathbf{q}) = \begin{pmatrix} \rho u \\ \rho u^2 + p \\ \rho u\left(e + \tfrac{1}{2}u^2 + \tfrac{p}{\rho}\right) \end{pmatrix}. \tag{9.281}$$

For a calorically perfect ideal gas with $e = p/\rho/(\gamma - 1)$, we have $q_1 = \rho$, $q_2 = \rho u$, $q_3 = p/(\gamma - 1) + \rho u^2/2$, it can be shown that $f_1 = q_2$, $f_2 = q_3(\gamma - 1) - q_2^2(\gamma - 3)/(2q_1)$, $f_3 = ((1 - \gamma)q_2^3 + 2\gamma q_1 q_2 q_3)/(2q_1^2)$. For nonideal gases, the form is more complicated.

If the solution is free of discontinuities, we study the limit $\Delta x = x_2 - x_1 \to 0$ and use Taylor series to rewrite Eq. (9.280) as

$$\int_{x_1}^{x_1+\Delta x} \frac{\partial \mathbf{q}}{\partial t}\, dx = \mathbf{f}(\mathbf{q}(x_1,t)) - \left(\mathbf{f}(\mathbf{q}(x_1,t)) + \frac{\partial \mathbf{f}}{\partial x}\Delta x + \cdots\right). \tag{9.282}$$

As we let $\Delta x \to 0$, the local values approach the mean values, and we can say for smooth flows that

$$\frac{\partial \mathbf{q}}{\partial t}\Delta x = -\frac{\partial \mathbf{f}}{\partial x}\Delta x + \cdots, \qquad \text{so} \qquad \frac{\partial \mathbf{q}}{\partial t} + \frac{\partial \mathbf{f}}{\partial x} = \mathbf{0}. \tag{9.283}$$

However, the partial differential equation form is not suitable for analysis of solutions that have embedded discontinuities, and can lead to error.

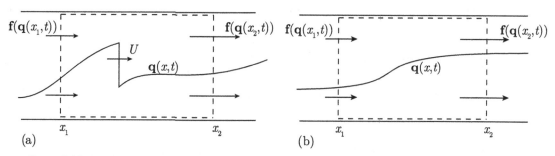

Figure 9.20 Schematic of general flux $\mathbf{f}$ into and out of a finite volume in which a general variable set $\mathbf{q}$ evolves: (a) internal discontinuity present, (b) smooth flow.

Returning to flows with embedded discontinuities, we consider the more primitive integral form. If we assume there is a discontinuity in the region $x \in [x_1, x_2]$ propagating at speed U, such as in Fig. 9.20(a), we can break up the integral in Eq. (9.280) into the form

$$\frac{d}{dt} \int_{x_1}^{x_1+Ut^-} \mathbf{q}(x,t) \, dx + \frac{d}{dt} \int_{x_1+Ut^+}^{x_2} \mathbf{q}(x,t) \, dx = \mathbf{f}(\mathbf{q}(x_1,t)) - \mathbf{f}(\mathbf{q}(x_2,t)). \tag{9.284}$$

Here $x_1 + Ut^-$ lies just before the discontinuity and $x_1 + Ut^+$ lies just past the discontinuity. Using Leibniz's rule, Eq. (2.244), we get

$$\mathbf{q}(x_1+Ut^-,t)D + 0 + \int_{x_1}^{x_1+Ut^-} \frac{\partial \mathbf{q}}{\partial t} \, dx + 0 - \mathbf{q}(x_1+Ut^+,t)D + \int_{x_1+Ut^+}^{x_2} \frac{\partial \mathbf{q}}{\partial t} \, dx$$
$$= \mathbf{f}(\mathbf{q}(x_1,t)) - \mathbf{f}(\mathbf{q}(x_2,t)). \tag{9.285}$$

Now if we assume that on either side of the discontinuity the volume of integration is sufficiently small so that the time and space variation of $\mathbf{q}$ is negligibly small, we get

$$\mathbf{q}(x_1)U - \mathbf{q}(x_2)U = \mathbf{f}(\mathbf{q}(x_1)) - \mathbf{f}(\mathbf{q}(x_2)), \tag{9.286}$$

$$U\left(\mathbf{q}(x_1) - \mathbf{q}(x_2)\right) = \mathbf{f}(\mathbf{q}(x_1)) - \mathbf{f}(\mathbf{q}(x_2)). \tag{9.287}$$

Defining next the notation for a jump as

$$[\![\mathbf{q}(x)]\!] \equiv \mathbf{q}(x_2) - \mathbf{q}(x_1), \tag{9.288}$$

the jump conditions are rewritten as

$$U [\![\mathbf{q}(x)]\!] = [\![\mathbf{f}(\mathbf{q}(x))]\!]. \tag{9.289}$$

If $U = 0$, as is the case when we transform to the frame where the wave is at rest, we simply recover

$$\mathbf{0} = \mathbf{f}(\mathbf{q}(x_1)) - \mathbf{f}(\mathbf{q}(x_2)), \quad \mathbf{f}(\mathbf{q}(x_1)) = \mathbf{f}(\mathbf{q}(x_2)), \quad [\![\mathbf{f}(\mathbf{q}(x))]\!] = \mathbf{0}. \tag{9.290}$$

That is, the fluxes on either side of the discontinuity are equal. We also get a more general result for $U \neq 0$, which is the well known

$$U = \frac{\mathbf{f}(\mathbf{q}(x_2)) - \mathbf{f}(\mathbf{q}(x_1))}{\mathbf{q}(x_2) - \mathbf{q}(x_1)} = \frac{[\![\mathbf{f}(\mathbf{q}(x))]\!]}{[\![\mathbf{q}(x)]\!]}. \tag{9.291}$$

The general *Rankine–Hugoniot*[6] *jump equations* then for the one-dimensional Euler equations across a non-stationary jump are given by

$$U \begin{pmatrix} \rho_2 - \rho_1 \\ \rho_2 u_2 - \rho_1 u_1 \\ \rho_2 \left(e_2 + \frac{1}{2}u_2^2\right) - \rho_1 \left(e_1 + \frac{1}{2}u_1^2\right) \end{pmatrix} = \begin{pmatrix} \rho_2 u_2 - \rho_1 u_1 \\ \rho_2 u_2^2 + p_2 - \rho_1 u_1^2 - p_1 \\ \rho_2 u_2 \left(e_2 + \frac{1}{2}u_2^2 + \frac{p_2}{\rho_2}\right) - \rho_1 u_1 \left(e_1 + \frac{1}{2}u_1^2 + \frac{p_1}{\rho_1}\right) \end{pmatrix}. \tag{9.292}$$

---

[6] Pierre Henri Hugoniot, 1851–1887, French engineer.

If we happen to choose a reference frame for which $U = 0$, then the jump equations reduce to

$$\begin{pmatrix} 0 \\ 0 \\ 0 \end{pmatrix} = \begin{pmatrix} \rho_2 u_2 - \rho_1 u_1 \\ \rho_2 u_2^2 + p_2 - \rho_1 u_1^2 - p_1 \\ \rho_2 u_2 \left(e_2 + \tfrac{1}{2} u_2^2 + \tfrac{p_2}{\rho_2} \right) - \rho_1 u_1 \left(e_1 + \tfrac{1}{2} u_1^2 + \tfrac{p_1}{\rho_1} \right) \end{pmatrix}. \quad (9.293)$$

These are the well-known Rankine–Hugoniot jump equations for a general material. In the following section, we will use a slightly less rigorous method to arrive at the same equations, and then analyze them.

9.6.2 Rankine–Hugoniot Equations

We will consider the problem in the context of the piston problem as shown in Fig. 9.21. The physical problem is as follows. Drive a piston with known velocity $\hat{u}_p$ into a fluid at rest ($\hat{u}_1 = 0$) with known properties, p_1, ρ_1 in the $\hat{x}$ laboratory frame. Determine the disturbance speed U. Determine the disturbance properties $\hat{u}_2, p_2, \rho_2$. In this frame of reference we have an *unsteady* problem, and it is difficult to solve. Let us employ a Galilean transformation so that we can study the problem as a *steady* problem. The transformation takes the laboratory frame $(\hat{x}, \hat{t})$ with velocity $\hat{u}$ to the transformed frame (x, t) with velocity u via

$$x = \hat{x} - U\hat{t}, \qquad t = \hat{t}, \qquad u = \hat{u} - U. \quad (9.294)$$

For the transformed problem, we will assume no variation with t so that all partial differential equations become ordinary differential equations in x. We will then integrate the ordinary differential equations to arrive at algebraic Rankine–Hugoniot jump equations. Next we solve as though U were known to get downstream "2" conditions: $u_2(U), p_2(U)$, etc. Then we invert to solve for U as function of u_2, the transformed piston velocity, $U(u_2)$. We then back transform to get all variables as function of $\hat{u}_2$, the laboratory piston velocity: $U(\hat{u}_2), p_2(\hat{u}_2), \rho_2(\hat{u}_2)$, etc.

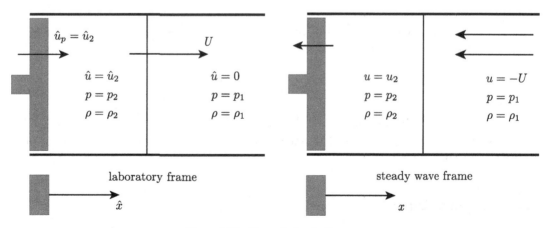

Figure 9.21 Normal shock diagram.

Under these assumptions, we can recast the evolution axioms and equation of state in the steady frame, Eqs. (9.154–9.157), as follows:

$$\frac{d}{dx}(\rho u) = 0, \qquad (9.295)$$

$$\frac{d}{dx}(\rho u^2 + p) = 0, \qquad (9.296)$$

$$\frac{d}{dx}\left(\rho u \left(h + \frac{u^2}{2}\right)\right) = 0, \qquad (9.297)$$

$$h = h(p, \rho). \qquad (9.298)$$

Upstream conditions are $\rho = \rho_1$, $p = p_1$, $u = -U$. With knowledge of the equation of state, we get $h = h_1$. In what is a natural, but in fact naïve, step we can integrate the equations from upstream to state "2" to give the correct Rankine–Hugoniot jump equations, which are effectively identical to the equations for Rayleigh flow in the limit as $q_w = 0$, Eqs.(9.270–9.273):

$$\rho_2 u_2 = -\rho_1 U, \qquad (9.299)$$

$$\rho_2 u_2^2 + p_2 = \rho_1 U^2 + p_1, \qquad (9.300)$$

$$h_2 + \frac{u_2^2}{2} = h_1 + \frac{U^2}{2}, \qquad (9.301)$$

$$h_2 = h(p_2, \rho_2). \qquad (9.302)$$

It is seen that these are fully equivalent to those obtained by the more nuanced analysis of the previous section, Eq. (9.293).

9.6.3 Rayleigh Line

We operate on the momentum equation, Eq. (9.300) as follows:

$$p_2 = p_1 + \rho_1 U^2 - \rho_2 u_2^2, \quad \text{so} \quad p_2 = p_1 + \frac{\rho_1^2 U^2}{\rho_1} - \frac{\rho_2^2 u_2^2}{\rho_2}. \qquad (9.303)$$

Because mass conservation, Eq. (9.299), gives us $\rho_2^2 u_2^2 = \rho_1^2 U^2$ we get an equation for the *Rayleigh Line*, a line in $(p, 1/\rho)$ space:

$$p_2 = p_1 + \rho_1^2 U^2 \left(\frac{1}{\rho_1} - \frac{1}{\rho_2}\right). \qquad (9.304)$$

Note that the Rayleigh line passes through the ambient state, has a *negative* slope, has a slope with magnitude proportional to square of the wave speed, and is independent of state and energy equations.

9.6.4 Hugoniot Curve

Let us now operate on the energy equation, using both mass and momentum to eliminate velocity. First eliminate u_2 via the mass equation, Eq. (9.299):

$$h_2 + \frac{u_2^2}{2} = h_1 + \frac{U^2}{2}, \tag{9.305}$$

$$h_2 + \frac{1}{2}\left(\frac{\rho_1 U}{\rho_2}\right)^2 = h_1 + \frac{U^2}{2}, \tag{9.306}$$

$$h_2 - h_1 + \frac{U^2}{2}\left(\left(\frac{\rho_1}{\rho_2}\right)^2 - 1\right) = 0, \tag{9.307}$$

$$h_2 - h_1 + \frac{U^2}{2}\left(\frac{(\rho_1 - \rho_2)(\rho_1 + \rho_2)}{\rho_2^2}\right) = 0. \tag{9.308}$$

Now use the Rayleigh line, Eq. (9.304), to eliminate U^2:

$$U^2 = (p_2 - p_1)\left(\frac{1}{\rho_1^2}\right)\left(\frac{1}{\rho_1} - \frac{1}{\rho_2}\right)^{-1} = (p_2 - p_1)\left(\frac{1}{\rho_1^2}\right)\left(\frac{\rho_1 \rho_2}{\rho_2 - \rho_1}\right). \tag{9.309}$$

So the energy equation, Eq. (9.308), becomes

$$h_2 - h_1 + \frac{1}{2}(p_2 - p_1)\left(\frac{1}{\rho_1^2}\right)\left(\frac{\rho_1 \rho_2}{\rho_2 - \rho_1}\right)\left(\frac{(\rho_1 - \rho_2)(\rho_1 + \rho_2)}{\rho_2^2}\right) = 0. \tag{9.310}$$

Straightforward algebra shows this reduces to

$$h_2 - h_1 = (p_2 - p_1)\left(\frac{1}{2}\right)\left(\frac{1}{\rho_2} + \frac{1}{\rho_1}\right) = \underbrace{\left(\frac{\hat{v}_2 + \hat{v}_1}{2}\right)}_{\hat{v}_{mean}} \underbrace{(p_2 - p_1)}_{\Delta p}. \tag{9.311}$$

So $\Delta h = \hat{v}_{mean} \Delta p$. This equation is the *Hugoniot* equation. It holds that enthalpy change equals the product of the mean volume, and the pressure difference, is independent of wave speed U and velocity u_2, and is independent of the equation of state. Note the similarity here between a common result for reversible thermodynamics. Using the definition of enthalpy, $h = e + p\hat{v}$ in the Gibbs equation gives $T\,ds = dh - \hat{v}\,dp$. For an isentropic change, we get $dh = \hat{v}\,dp$.

9.6.5 Solution Procedure for General Equations of State

The shocked state can be determined by the following procedure. (1) Specify the equation of state $h(p, \rho)$. (2) Substitute the equation of state into the Hugoniot, Eq. (9.311), to get a second relation between p_2 and ρ_2. (3) Use the Rayleigh line, Eq. (9.304), to eliminate p_2 in the Hugoniot so that the Hugoniot is a single equation in ρ_2. (4) Solve for ρ_2 as functions of "1" and U. (5) Back substitute to solve for p_2, u_2, h_2, T_2 as functions of "1" and U. (6) Invert to find U as function of "1" state and u_2. (7) Back transform to laboratory frame to get U as function of "1" state and piston velocity $\hat{u}_2 = \hat{u}_p$.

9.6.6 Calorically Perfect Ideal Gas Solutions

Let us follow this procedure for the special case of a calorically perfect ideal gas, for which $h = c_p(T - T_o) + \hat{h}$, and $p = \rho RT$. Thus,

$$h = c_p \left(\frac{p}{R\rho} - \frac{p_o}{R\rho_o}\right) + \hat{h} = \frac{c_p}{R}\left(\frac{p}{\rho} - \frac{p_o}{\rho_o}\right) + \hat{h} = \frac{\gamma}{\gamma - 1}\left(\frac{p}{\rho} - \frac{p_o}{\rho_o}\right) + \hat{h}. \quad (9.312)$$

Evaluate at states 1 and 2 and substitute into the Hugoniot equation, Eq. (9.311):

$$\left(\frac{\gamma}{\gamma-1}\left(\frac{p_2}{\rho_2} - \frac{p_o}{\rho_o}\right) + \hat{h}\right) - \left(\frac{\gamma}{\gamma-1}\left(\frac{p_1}{\rho_1} - \frac{p_o}{\rho_o}\right) + \hat{h}\right) = (p_2 - p_1)\left(\frac{1}{2}\right)\left(\frac{1}{\rho_2} + \frac{1}{\rho_1}\right). \quad (9.313)$$

Rearranging, we find

$$\frac{\gamma}{\gamma-1}\left(\frac{p_2}{\rho_2} - \frac{p_1}{\rho_1}\right) - (p_2 - p_1)\left(\frac{1}{2}\right)\left(\frac{1}{\rho_2} + \frac{1}{\rho_1}\right) = 0. \quad (9.314)$$

This reduces to a final form of

$$p_2 = p_1 \frac{\frac{\gamma+1}{\gamma-1}\frac{1}{\rho_1} - \frac{1}{\rho_2}}{\frac{\gamma+1}{\gamma-1}\frac{1}{\rho_2} - \frac{1}{\rho_1}}. \quad (9.315)$$

We see the Hugoniot for a calorically perfect ideal gas is a *hyperbola* in $(p, 1/\rho)$ space. As $1/\rho_2 \to (\gamma-1)/(\gamma+1)(1/\rho_1)$, we find $p_2 \to \infty$. For $\gamma = 7/5$, we get $\rho_2 \to 6\rho_1$ for infinite pressure. The Hugoniot has as $1/\rho_2 \to \infty, p_2 \to -p_1(\gamma-1)/(\gamma+1)$; note the nonphysical negative pressure.

The Rayleigh line and Hugoniot curve are depicted in Fig. 9.22. Note that the intersections of the two curves are solutions to the equations; the ambient state "1" is one solution; the other solution "2" is known as the shock solution; the shock solution has higher pressure and higher density; a higher wave speed implies higher pressure and higher density. A minimum wave speed exists. It occurs when the Rayleigh line is tangent to the Hugoniot curve, occurs for infinitesimally small pressure changes, corresponds to a sonic wave speed, and has disturbances that are *acoustic*. If pressure increases, it can be shown that entropy increases. If pressure decreases (for wave speeds that are less than sonic), entropy decreases; this is nonphysical.

Substitute the Rayleigh line into the Hugoniot equation to get a single equation for ρ_2:

$$p_1 + \rho_1^2 U^2 \left(\frac{1}{\rho_1} - \frac{1}{\rho_2}\right) = p_1 \frac{\frac{\gamma+1}{\gamma-1}\frac{1}{\rho_1} - \frac{1}{\rho_2}}{\frac{\gamma+1}{\gamma-1}\frac{1}{\rho_2} - \frac{1}{\rho_1}}. \quad (9.316)$$

This equation is quadratic in $1/\rho_2$ and factorizable. Use computer algebra to solve and get two solutions: one ambient solution, $1/\rho_2 = 1/\rho_1$, and one shocked solution. The shocked solution is

$$\frac{1}{\rho_2} = \frac{1}{\rho_1}\frac{\gamma-1}{\gamma+1}\left(1 + \frac{2\gamma}{(\gamma-1)U^2}\frac{p_1}{\rho_1}\right). \quad (9.317)$$

We perform a set of calculations for calorically perfect ideal air with $\gamma = 7/5$, $R = 0.287\,\text{kJ/kg/K}$, $p_1 = 10^5\,\text{Pa}$, $\rho_1 = 1.161\,44\,\text{kg/m}^3$. For this, the ideal gas law gives $T_1 = 300\,\text{K}$. The shocked density ρ_2 is plotted against wave speed U for calorically perfect ideal air in Fig. 9.23(a). Note the density solution allows allows all wave speeds $0 < U < \infty$. The range, however, is $U \in [c_1, \infty)$. The Rayleigh line and Hugoniot show $U \geq c_1$. The solution for $U = U(\hat{u}_p)$, (to be shown), requires $U \geq c_1$. For the *strong shock limit*: $U^2 \to \infty$, $\rho_2 \to (\gamma+1)/(\gamma-1)$. For the *acoustic limit*: $U^2 \to \gamma p_1/\rho_1$, $\rho_2 \to \rho_1$. For the *nonphysical limit*: $U^2 \to 0$, $\rho_2 \to 0$.

9.6 Normal Shock Waves

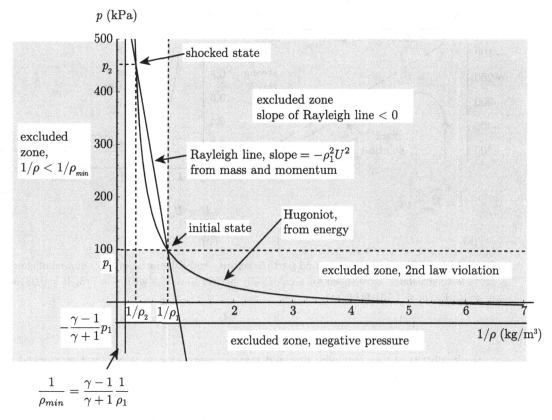

Figure 9.22 Rayleigh line and Hugoniot curve for a typical calorically perfect ideal gas.

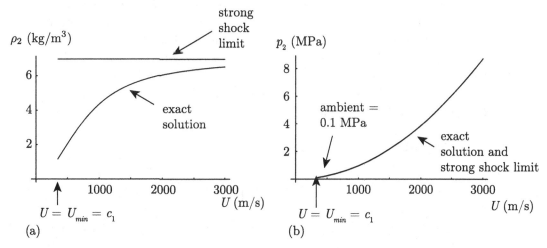

Figure 9.23 Shock (a) density and (b) pressure versus shock wave speed for a calorically perfect ideal gas with $\gamma = 7/5$, $R = 0.287\,\text{kJ/kg/K}$, $p_1 = 10^5$ Pa, $\rho_1 = 1.161\,44\,\text{kg/m}^3$.

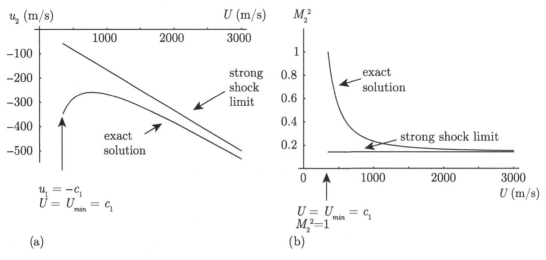

Figure 9.24 (a) Shock wave frame fluid particle velocity and (b) Mach number squared of shocked fluid particle versus shock wave speed for a calorically perfect ideal gas with $\gamma = 7/5$, $R = 0.287\,\text{kJ/kg/K}$, $p_1 = 10^5\,\text{Pa}$, $\rho_1 = 1.161\,44\,\text{kg/m}^3$.

Back substitute into Rayleigh line and mass conservation to solve for the shocked pressure and the fluid velocity in the shocked wave frame:

$$p_2 = \frac{2}{\gamma+1}\rho_1 U^2 - \frac{\gamma-1}{\gamma+1}p_1, \quad u_2 = -U\frac{\gamma-1}{\gamma+1}\left(1 + \frac{2\gamma}{(\gamma-1)U^2}\frac{p_1}{\rho_1}\right). \tag{9.318}$$

The shocked pressure p_2 is plotted against wave speed U for calorically perfect ideal air in Fig. 9.23(b) including both the exact solution and the solution in the strong shock limit. For these parameters, the results are indistinguishable. The shocked wave frame fluid particle velocity u_2 is plotted against wave speed U for calorically perfect ideal air in Fig. 9.24(a). The shocked wave frame fluid particle Mach number, $M_2^2 = \rho_2 u_2^2/(\gamma p_2)$, is plotted against wave speed U for calorically perfect ideal air in Fig. 9.24(b). In the steady frame, the Mach number of the undisturbed flow is (and must be) supersonic, and the shocked flow is (and must be) subsonic.

Transform back to the laboratory frame $u = \hat{u} - U$:

$$\hat{u}_2 - U = -U\frac{\gamma-1}{\gamma+1}\left(1 + \frac{2\gamma}{(\gamma-1)U^2}\frac{p_1}{\rho_1}\right). \tag{9.319}$$

Manipulate this equation and solve the resulting quadratic equation for U and get

$$U = \frac{\gamma+1}{4}\hat{u}_2 \pm \sqrt{\frac{\gamma p_1}{\rho_1} + \hat{u}_2^2\left(\frac{\gamma+1}{4}\right)^2}. \tag{9.320}$$

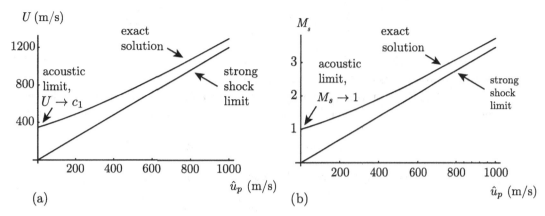

Figure 9.25 (a) Shock speed and (b) shock Mach number versus piston velocity for a calorically perfect ideal gas with $\gamma = 7/5$, $R = 0.287\,\text{kJ/kg/K}$, $p_1 = 10^5\,\text{Pa}$, $\rho_1 = 1.161\,44\,\text{kg/m}^3$.

Now if $\hat{u}_2 > 0$, we expect $U > 0$ so take positive root; also set the velocity equal to the piston velocity $\hat{u}_2 = \hat{u}_p$ and get

$$U = \frac{\gamma+1}{4}\hat{u}_p + \sqrt{\frac{\gamma p_1}{\rho_1} + \hat{u}_p^2\left(\frac{\gamma+1}{4}\right)^2}. \tag{9.321}$$

Note for the *acoustic limit*, as $\hat{u}_p \to 0$, $U \to c_1$, the shock speed approaches the sound speed. For the *strong shock limit*, as $\hat{u}_p \to \infty$, $U \to \hat{u}_p(\gamma+1)/2$.

The shock speed U is plotted against piston velocity $\hat{u}_p$ for calorically perfect ideal air in Fig. 9.25(a). Both the exact solution and strong shock limit are shown. If we define the Mach number of the shock as $M_s \equiv U/c_1$, we get

$$M_s = \frac{\gamma+1}{4}\frac{\hat{u}_p}{\sqrt{\gamma R T_1}} + \sqrt{1 + \frac{\hat{u}_p^2}{\gamma R T_1}\left(\frac{\gamma+1}{4}\right)^2}. \tag{9.322}$$

The shock Mach number M_s is plotted against piston velocity $\hat{u}_p$ for calorically perfect ideal air in Fig. 9.25(b). Both the exact solution and strong shock limit are shown. It is straightforward to recast Eq. (9.317), using for the calorically perfect ideal gas that $M_s^2 = \rho_1 U^2/\gamma/p_1$, as

$$\frac{\rho_1}{\rho_2} = \frac{(\gamma-1)M_s^2 + 2}{(\gamma+1)M_s^2} = 1 - \frac{2}{\gamma+1}\left(1 - \frac{1}{M_s^2}\right). \tag{9.323}$$

For the weak shock limit as $M_s \to 1$, the density ratio is $\rho_1/\rho_2 \sim 1 - 4(M_s - 1)/(\gamma + 1)$. And for the strong shock limit as $M_s \to \infty$, the density ratio approaches a constant, $(\gamma-1)/(\gamma+1)$.

Similarly, Eq. (9.318) can be recast as

$$\frac{p_2}{p_1} = 1 + \frac{2\gamma(M_s^2 - 1)}{\gamma+1}. \tag{9.324}$$

For the weak shock limit as $M_s \to 1$, the pressure ratio is $p_2/p_1 \sim 1 + 4\gamma(M_s - 1)/(\gamma+1)$, and for the strong shock limit the pressure ratio goes as $2\gamma M_s^2/(\gamma+1)$. Using the ideal gas law, it is easy to show that

$$\frac{T_2}{T_1} = 1 + \frac{2(\gamma-1)}{(\gamma+1)^2}\left(1 - \frac{1}{M_s^2}\right)(1 + \gamma M_s^2). \tag{9.325}$$

In the weak shock limit, $T_2/T_1 \sim 1 + 4(\gamma-1)(M_s-1)/(\gamma+1)$. For the strong shock limit, the temperature ratio goes as $2\gamma(\gamma-1)M_s^2/(\gamma+1)^2$.

Let us find the entropy change induced by a shock for a calorically perfect ideal gas. We first need an expression for the entropy change. Begin with the Gibbs equation, Eq. (4.195):

$$T\,ds = de + p\,d\hat{v}, \quad \text{so} \quad ds = \frac{de}{T} + \frac{p}{T}\,d\hat{v}. \tag{9.326}$$

Now invoke the calorically perfect ideal gas assumption to get

$$ds = c_v \frac{dT}{T} + R\frac{d\hat{v}}{\hat{v}}. \tag{9.327}$$

Now for the ideal gas with $p\hat{v} = RT$, we get

$$p\,d\hat{v} + \hat{v}\,dp = R\,dT. \tag{9.328}$$

Divide the left-hand side by $p\hat{v}$ and the right-hand side by the equivalent RT to get

$$\frac{d\hat{v}}{\hat{v}} + \frac{dp}{p} = \frac{dT}{T}. \tag{9.329}$$

Use this to eliminate temperature in Eq. (9.327) to get

$$ds = c_v\left(\frac{d\hat{v}}{\hat{v}} + \frac{dp}{p}\right) + R\frac{d\hat{v}}{\hat{v}} = (c_v + R)\frac{d\hat{v}}{\hat{v}} + c_v\frac{dp}{p}, \tag{9.330}$$

$$= c_p \frac{d\hat{v}}{\hat{v}} + c_v \frac{dp}{p} = c_v\left(\gamma\frac{d\hat{v}}{\hat{v}} + \frac{dp}{p}\right), \tag{9.331}$$

$$s_2 - s_1 = c_v\left(\gamma \ln\frac{\hat{v}_2}{\hat{v}_1} + \ln\frac{p_2}{p_1}\right) = c_v\left(\ln\left(\frac{\hat{v}_2}{\hat{v}_1}\right)^\gamma + \ln\frac{p_2}{p_1}\right), \tag{9.332}$$

$$= c_v\left(\ln\left(\frac{\rho_1}{\rho_2}\right)^\gamma + \ln\frac{p_2}{p_1}\right) = c_v \ln\left(\left(\frac{\rho_1}{\rho_2}\right)^\gamma \frac{p_2}{p_1}\right). \tag{9.333}$$

Then we use Eqs. (9.323) and (9.324) to eliminate the density and pressure ratios in favor of M_s, and follow this with algebraic reduction to arrive at

$$\frac{s_2 - s_1}{c_v} = \ln\left(\left(\frac{(\gamma-1)M_s^2 + 2}{(\gamma+1)M_s^2}\right)^\gamma \left(1 + \frac{2\gamma\left(M_s^2 - 1\right)}{\gamma+1}\right)\right). \tag{9.334}$$

For $\gamma = 7/5$, we plot $(s_2-s_1)/c_v$ as a function of M_s in Fig. 9.26. Clearly for $M_s = 1$, we have $(s_2 - s_1)/c_v = 0$, so the sonic wave is isentropic. And clearly for $M_s > 1$, the entropy rises, thus satisfying the second law for what is an adiabatic irreversible compression. For $M_s < 1$, the entropy is predicted to fall for an adiabatic expansion. This is not observed in nature for calorically perfect ideal gases and violates the second law of thermodynamics. So we must have $M_s \geq 1$ for a propagating discontinuity. And because $M_s \geq 1$, we must have $p_2 \geq p_1$, $T_2 \geq T_1$, $\rho_2 \geq \rho_1$; that is, we are restricted to compression shocks for calorically perfect ideal gases. This does not generalize to nonideal gases. Remarkably, we will find in Section 9.13 a special case for a nonideal gas that admits a supersonic rarefaction discontinuity.

9.6 Normal Shock Waves

Table 9.1 Solutions to the jump equations for a calorically perfect ideal gas, $\gamma = 7/5$, $R = 0.287$ kJ/kg/K

Root	ρ_2 (kg/m³)	u_2 (m/s)	p_2 (Pa)	h_2 (J/kg)	T_2 (K)	$\hat{v}_2$ (m³/kg)	Δs (J/kg/K)
1	1.161 44	−1000	$1.000\,00 \times 10^5$	86 100	300.000	0.861	0
2	4.348 06	−267.117	$9.512\,00 \times 10^5$	550 424	762.244	0.229 987	290.198

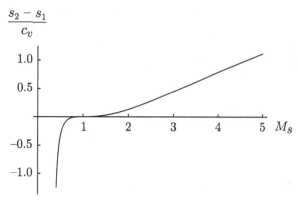

Figure 9.26 Scaled entropy jump through a discontinuity as a function of M_s for a calorically perfect ideal gas with $\gamma = 7/5$.

The equation for entropy jump is complicated. We can better understand it by performing a Taylor series expansion in the neighborhood of $M_s = 1$. Doing so yields

$$\frac{s_2 - s_1}{c_v} = \frac{16\gamma(\gamma-1)}{3(\gamma+1)^2}(M_s - 1)^3 + \mathcal{O}\left((M_s - 1)^4\right). \qquad (9.335)$$

Clearly for general $\gamma > 1$, the entropy rises for $M_s > 1$ and falls for $M_s < 1$, and the local behavior is cubic in the deviation of M_s from unity.

Lastly, we give shock values for a particular value of $U = 1000$ m/s. For this shock velocity, we get $\rho_2 = 4.348\,06$ kg/m³, $u_2 = -267.117$ m/s, $p_2 = 951\,200$ Pa, $T_2 = 762.244$ K, $M_2^2 = 0.232\,969$, $\Delta s/c_v = 0.404\,457$. As the entropy change is positive, the solution satisfies the second law of thermodynamics. The inert state is also a solution, presented as Root 1. The shock state is presented as Root 2. Both solutions are summarized in Table 9.1.

9.6.7 Weak Shock Limit

Let us now study shock waves in calorically perfect ideal gases that are sufficiently weak that linear approximations work well. This is the acoustic limit. Consider that state 2 is a small perturbation of state 1 so that

$$\rho_2 = \rho_1 + \Delta\rho, \qquad u_2 = u_1 + \Delta u, \qquad p_2 = p_1 + \Delta p. \qquad (9.336)$$

Assume $\Delta\rho \ll \rho_1$, $|\Delta u| \ll |u_1|$, $\Delta p \ll p_1$. Substituting into the Rankine–Hugoniot jump equations, Eqs. (9.299–9.301) restricted to a calorically perfect ideal gas, we get

$$(\rho_1 + \Delta\rho)(u_1 + \Delta u) = \rho_1 u_1, \tag{9.337}$$

$$(\rho_1 + \Delta\rho)(u_1 + \Delta u)^2 + (p_1 + \Delta p) = \rho_1 u_1^2 + p_1, \tag{9.338}$$

$$\frac{\gamma}{\gamma-1}\frac{p_1 + \Delta p}{\rho_1 + \Delta\rho} + \frac{1}{2}(u_1 + \Delta u)^2 = \frac{\gamma}{\gamma-1}\frac{p_1}{\rho_1} + \frac{1}{2}u_1^2. \tag{9.339}$$

Expanding, we get

$$\rho_1 u_1 + u_1(\Delta\rho) + \rho_1(\Delta u) + (\Delta\rho)(\Delta u) = \rho_1 u_1, \tag{9.340}$$

$$\left(\rho_1 u_1^2 + 2\rho_1 u_1(\Delta u) + u_1^2(\Delta\rho) + \rho_1(\Delta u)^2 + 2u_1(\Delta u)(\Delta\rho) + (\Delta\rho)(\Delta u)^2\right) + (p_1 + \Delta p)$$
$$= \rho_1 u_1^2 + p_1, \tag{9.341}$$

$$\frac{\gamma}{\gamma-1}\left(\frac{p_1}{\rho_1} + \frac{1}{\rho_1}\Delta p - \frac{p_1}{\rho_1^2}\Delta\rho + \cdots\right) + \frac{1}{2}\left(u_1^2 + 2u_1(\Delta u) + (\Delta u)^2\right)$$
$$= \frac{\gamma}{\gamma-1}\frac{p_1}{\rho_1} + \frac{1}{2}u_1^2. \tag{9.342}$$

Subtracting the base state and eliminating products of small quantities yields

$$u_1(\Delta\rho) + \rho_1(\Delta u) = 0, \tag{9.343}$$

$$2\rho_1 u_1(\Delta u) + u_1^2(\Delta\rho) + \Delta p = 0, \tag{9.344}$$

$$\frac{\gamma}{\gamma-1}\left(\frac{1}{\rho_1}\Delta p - \frac{p_1}{\rho_1^2}\Delta\rho\right) + u_1(\Delta u) = 0. \tag{9.345}$$

In matrix form this is

$$\begin{pmatrix} u_1 & \rho_1 & 0 \\ u_1^2 & 2\rho_1 u_1 & 1 \\ -\frac{\gamma}{\gamma-1}\frac{p_1}{\rho_1^2} & u_1 & \frac{\gamma}{\gamma-1}\frac{1}{\rho_1} \end{pmatrix} \begin{pmatrix} \Delta\rho \\ \Delta u \\ \Delta p \end{pmatrix} = \begin{pmatrix} 0 \\ 0 \\ 0 \end{pmatrix}. \tag{9.346}$$

As the right-hand side is zero, the determinant must be zero, and there must be a linear dependency of the solution. First check the determinant:

$$u_1\left(\frac{2\gamma}{\gamma-1}u_1 - u_1\right) - \rho_1\left(\frac{\gamma}{\gamma-1}\frac{u_1^2}{\rho_1} + \frac{\gamma}{\gamma-1}\frac{p_1}{\rho_1^2}\right) = 0, \tag{9.347}$$

$$\frac{u_1^2}{\gamma-1}(2\gamma - (\gamma-1)) - \frac{1}{\gamma-1}\left(\gamma u_1^2 + \gamma\frac{p_1}{\rho_1}\right) = 0, \tag{9.348}$$

$$u_1^2(\gamma+1) - \left(\gamma u_1^2 + \gamma\frac{p_1}{\rho_1}\right) = 0, \tag{9.349}$$

$$u_1^2 = \gamma\frac{p_1}{\rho_1} = c_1^2. \tag{9.350}$$

So the velocity is necessarily sonic for a small disturbance, with $u_1 = \pm c_1$. This validates the assertion of Section 9.1.3 that the thermodynamic property we studied represents the speed of small disturbances.

Take Δu to be known and solve a resulting 2×2 system:

$$\begin{pmatrix} u_1 & 0 \\ -\frac{\gamma}{\gamma-1}\frac{p_1}{\rho_1^2} & \frac{\gamma}{\gamma-1}\frac{1}{\rho_1} \end{pmatrix} \begin{pmatrix} \Delta\rho \\ \Delta p \end{pmatrix} = \begin{pmatrix} -\rho_1 \Delta u \\ -u_1 \Delta u \end{pmatrix}. \tag{9.351}$$

Solving, and choosing $u_1 = -c_1$ yields

$$\Delta \rho = -\frac{\rho_1 \Delta u}{u_1} = \rho_1 \frac{\Delta u}{c_1} = \rho_1 M_p, \quad \Delta p = -\rho_1 u_1 \Delta u = \rho_1 c_1^2 \frac{\Delta u}{c_1} = \rho_1 c_1^2 M_p. \quad (9.352)$$

Here we have taken the driving piston Mach number to be $M_p = \Delta u/c_1$. Note that this is consistent with the configuration of Fig. 9.21 when we choose the root $u_1 = -c_1$, with $c_1 > 0$. In this configuration $\Delta u > 0$, and the positive fluid particle velocity change induced by the shock induces a density and pressure rise.

9.7 Contact Discontinuities

Analysis of contact discontinuities is straightforward. One must enforce $u_1 = u_2$ and $p_1 = p_2$, but allow variation in T and ρ. Similar to the shock wave, let us consider a contact discontinuity propagating to the right with speed U into a fluid at rest at ρ_1, p_1. Then in the reference frame in which the contact discontinuity is at rest, we have $u_1 = -U$. Let us consider a jump in which the density jump is specified, so ρ_2 is considered known. Then the jump equations for a calorically perfect ideal gas are simply

$$u_2 = -U, \qquad p_2 = p_1, \qquad T_2 = \frac{\rho_1}{\rho_2} T_1. \quad (9.353)$$

The density and temperature jump over a contact discontinuity admitted by the Euler equations induces an entropy jump. The second law, however, places no restriction on whether such an entropy jump is positive or negative. This is because fluid particles do not cross the jump, as the jump and fluid on both sides of it move with the same velocity. The jump in entropy for the calorically perfect ideal gas is given by

$$s_2 - s_1 = c_p \ln \frac{T_2}{T_1} - R \underbrace{\ln \frac{p_2}{p_1}}_{=0} = c_p \ln \frac{T_2}{T_1} = c_p \ln \frac{\rho_1}{\rho_2}. \quad (9.354)$$

It can be positive or negative, depending on whether ρ_1/ρ_2 is greater or less than unity. One can also imagine performing a Galilean transformation to bring the velocity of the contact discontinuity to zero. This would then bring the velocity on either side of the jump to zero. As the pressures are equal on either side, there is no net force present to induce a change in any velocity. The temperatures on either side may be different, but because thermal conductivity is zero, there is no mechanism for diffusion of thermal energy, and the temperature jump may persist in equilibrium. This then admits a density and entropy jump to persist.

Example 9.17 Analyze a contact discontinuity moving with $U = 10$ m/s into an ambient calorically perfect ideal gas at rest with $\gamma = 7/5$, $R = 0.287$ kJ/kg/K, with $T_1 = 300$ K, $\rho_1 = 1.16144$ kg/m^3, $\rho_2 = \rho_1/2$.

Solution
The jump conditions are straightforward. We have $u_2 = u_1 = -U = -10$ m/s. We have $p_2 = p_1 = \rho_1 R T_1 = (1.16144 \text{ kg/m}^3)(0.287 \text{ kJ/kg/K})(300 \text{ K}) = 100$ kPa. We are given $\rho_2 = \rho_1/2 = 0.58072$ kg/m^3. The isobaric ideal gas law gives $T_2 = T_1(\rho_1/\rho_2) = 2T_1 = 600$ K. The entropy change for the calorically perfect ideal gas is easily shown to be

$$s_2 - s_1 = c_p \ln \frac{T_2}{T_1} - R \underbrace{\ln \frac{p_2}{p_1}}_{=0} = c_p \ln \frac{T_2}{T_1} = \frac{\gamma}{\gamma-1} R \ln \frac{T_2}{T_1}, \tag{9.355}$$

$$= \frac{7/5}{7/5 - 1} \left(0.287 \, \frac{\text{kJ}}{\text{kg K}}\right) \ln 2 = 696.266 \, \frac{\text{kJ}}{\text{kg K}}. \tag{9.356}$$

This right traveling wave induces a positive entropy jump. However, had the wave been left traveling, it would have had the same jump.

9.8 Throttling Device

A flow is throttled when, for example, it flows through a partially open valve. When it does so, we notice that there can be a significant pressure loss from one side of the partially open valve to the other. A sketch of a throttling device is given in Fig. 9.27. We model a throttling device as steady with one entrance and exit, with no work or heat transfer. We neglect changes in area as well as potential energy. Mass conservation tells us that

$$\rho_1 v_1 = \rho_2 v_2. \tag{9.357}$$

Energy conservation tells us that

$$h_1 + \frac{v_1^2}{2} = h_2 + \frac{v_2^2}{2}. \tag{9.358}$$

This is supplemented by an equation of state, and using this, one can determine state 1 in terms of state 2, given that one knows the thermodynamic state and velocity at 1 and the pressure drop to 2. The analysis is similar to that of a normal shock.

Example 9.18 Consider calorically perfect ideal air flowing in a duct at $p_1 = 100\,000$ Pa, $T_1 = 300$ K, $v_1 = 10$ m/s. Take $c_p = 1000$ J/kg/K. The air is throttled by a valve to $p_2 = 90\,000$ Pa. Consider fully the effects of compressibility and kinetic energy, find the downstream state after throttling, and show the assumption that $h_2 \sim h_1$ (giving for the calorically perfect ideal gas $T_2 \sim T_1$) is a good approximation for this flow.

Solution
From the ideal gas law, we get

$$\rho_1 = \frac{p_1}{RT_1} = \frac{100\,000 \, \text{Pa}}{\left(287 \, \frac{\text{J}}{\text{kg K}}\right)(300 \, \text{K})} = 1.161\,44 \, \frac{\text{kg}}{\text{m}^3}. \tag{9.359}$$

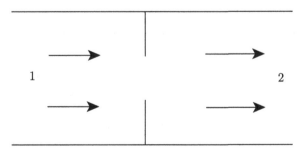

Figure 9.27 Diagram of throttling device.

We write the mass, energy, and thermal state equation as

$$\rho_2 v_2 = \rho_1 v_1, \quad c_p T_2 + \frac{v_2^2}{2} = c_p T_1 + \frac{v_1^2}{2}, \quad p_2 = \rho_2 R T_2. \tag{9.360}$$

Substituting numbers, we get the system of nonlinear algebraic equations

$$\rho_2 v_2 = \left(1.161\,44\,\frac{\text{kg}}{\text{m}^3}\right)\left(10\,\frac{\text{m}}{\text{s}}\right) = 11.6144\,\frac{\text{kg}}{\text{m}^2\,\text{s}}, \tag{9.361}$$

$$\left(1000\,\frac{\text{J}}{\text{kg K}}\right) T_2 + \frac{v_2^2}{2} = \left(1000\,\frac{\text{J}}{\text{kg K}}\right)(300\,\text{K}) + \frac{\left(10\,\frac{\text{m}}{\text{s}}\right)^2}{2} = 300\,050\,\frac{\text{J}}{\text{kg}}, \tag{9.362}$$

$$90\,000\,\text{Pa} = \rho_2 \left(287\,\frac{\text{J}}{\text{kg K}}\right) T_2. \tag{9.363}$$

This forms three equations in the three unknowns ρ_2, T_2, v_2. Detailed manipulation can reduce this to a quadratic equation, with two roots. The first root yields

$$\rho_2 = 1.045\,33\,\frac{\text{kg}}{\text{m}^3}, \quad T_2 = 299.988\,\text{K}, \quad v_2 = 11.1107\,\frac{\text{m}}{\text{s}}. \tag{9.364}$$

This is the physical root. We see that

$$T_2 \sim T_1. \tag{9.365}$$

Thus, from the ideal gas law

$$\rho_2 = \rho_1 \frac{T_1 p_2}{T_2 p_1} \sim \rho_1 \frac{p_2}{p_1} \sim \rho_1 \frac{90\,000\,\text{Pa}}{100\,000\,\text{Pa}} \sim 0.9 \left(1.161\,44\,\frac{\text{kg}}{\text{m}^3}\right) \sim 1.0453\,\frac{\text{kg}}{\text{m}^3}. \tag{9.366}$$

Then from mass conservation, we get

$$v_2 = v_1 \frac{\rho_1}{\rho_2} \sim v_1 \frac{1}{0.9} \sim \left(10\,\frac{\text{m}}{\text{s}}\right)\frac{1}{0.9} \sim 11.11\,\frac{\text{m}}{\text{s}}. \tag{9.367}$$

There is also an obviously nonphysical root: $\rho_2 = -0.000\,215\,037\,\text{kg/m}^3$, $T_2 = -1\,458\,300\,\text{K}$, $v_2 = -54011.1\,\text{m/s}$. Negative density and absolute temperature are physically unacceptable.

As an aside, detailed calculation reveals the exact solution for v_2 is

$$v_2 = \frac{c_p p_2 T_1 \left(\sqrt{\frac{p_1^2 v_1^2 (2 c_p T_1 + v_1^2)}{c_p^2 p_2^2 T_1^2} + 1} - 1\right)}{p_1 v_1}. \tag{9.368}$$

This is a complicated expression that is difficult to interpret. But Taylor series expansion for $v_1^2/(c_p T_1) \ll 1$ yields

$$v_2 = v_1 \left(\frac{p_1}{p_2}\right)\left(1 + \left(1 - \left(\frac{p_1}{p_2}\right)^2\right)\frac{v_1^2}{2 c_p T_1} + \cdots\right). \tag{9.369}$$

This is an excellent approximation that yields results fully equivalent to what was obtained earlier. Similarly, Taylor series of the solution for T_2 and ρ_2 in the same limit yields the approximations

$$T_2 = T_1 \left(1 + \left(1 - \left(\frac{p_1}{p_2}\right)^2\right)\frac{v_1^2}{2 c_p T_1} + \cdots\right), \tag{9.370}$$

$$\rho_2 = \rho_1 \left(\frac{p_2}{p_1}\right)\left(1 - \left(1 - \left(\frac{p_1}{p_2}\right)^2\right)\frac{v_1^2}{2 c_p T_1} + \cdots\right). \tag{9.371}$$

Direct substitution of the parameter values reveals that the approximations, relative to the exact solution, agree up to the fifth decimal place for v_2, T_2, and ρ_2.

9.9 Flow with Area Change and Normal Shocks

This section will consider flow from a reservoir with the fluid at stagnation conditions to a constant pressure environment. The pressure of the environment is commonly known as the *back pressure*: p_b. The key problem is given $A(x)$, stagnation conditions and p_b, find the pressure, temperature, density at all points in the duct and the mass flow rate.

9.9.1 Converging Nozzle

A converging nozzle operating at several different values of p_b is presented in Fig. 9.28. Also shown are possible pressure profiles and mass flow rates. The flow through the duct can be solved using the following steps. Check if $p_b \geq p_*$. If so, set $p_e = p_b$. Determine M_e from the isentropic flow relations. Then determine A_* from the A/A_* relation. Lastly at any point in the flow where A is known, compute A/A_* and then invert the A/A_* relation to find the local M. These flows are subsonic throughout and correspond to points a and b in Fig. 9.28. If $p_b = p_*$ then the flow is sonic at the exit and just choked. This corresponds to point c in Fig. 9.28.

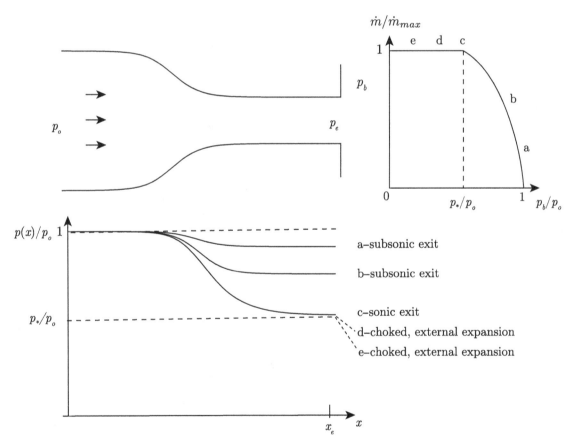

Figure 9.28 Diagrams of converging nozzle, pressure profiles for various back pressures, and mass flow rates for various back pressures.

9.9.2 Converging–Diverging Nozzle

A converging–diverging nozzle operating at several different values of p_b is depicted in Fig. 9.29. Also shown are several pressure profiles and possible mass flow rates. The flow through the duct can be solved using the a similar following procedure as for a converging nozzle. First, set $A_t = A_*$. With this assumption, calculate A_e/A_*. Then, determine M_{esub}, M_{esup}, both supersonic and subsonic, from the A/A_* relation. Next, determine p_{esub}, p_{esup}, from M_{esub}, M_{esup}; these are the supersonic and subsonic design pressures. If $p_b > p_{esub}$, the flow is subsonic throughout and the throat is not sonic. Use same procedure as for converging duct: Determine M_e by setting $p_e = p_b$ and using isentropic relations. If $p_{esub} > p_b > p_{esup}$, the procedure is complicated. First, estimate the pressure with a normal shock at the end of the duct, p_{esh}. If $p_b \geq p_{esh}$, there is a normal shock inside the duct. If $p_b < p_{esh}$, the duct flow is shockless, and there may be compression outside the duct. Now, if $p_{esup} = p_b$, the flow is at supersonic design conditions and the flow is shockless. If $p_b < p_{esup}$, the flow in the duct is isentropic and there is expansion outside the duct.

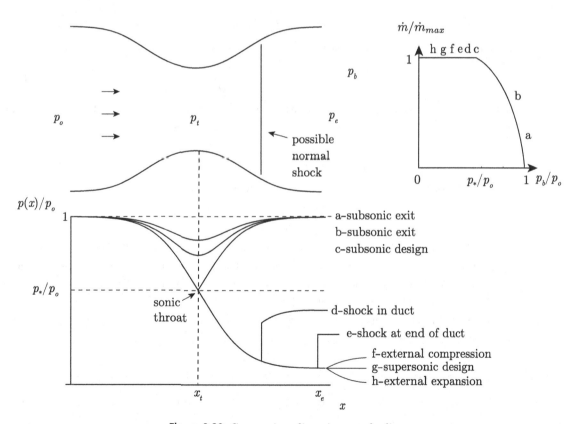

Figure 9.29 Converging–diverging nozzle diagram.

9.10 Acoustics

Let us consider in more detail the acoustic limit that was introduced in Section 9.6.7. We can examine the one-dimensional homeoentropic flow of a calorically perfect ideal gas in the limit in which the fluid velocity is small relative to the sound speed. These assumptions yield $p/\rho^\gamma = p_o/\rho_o^\gamma$. And the sound speed is $c = \sqrt{\gamma p/\rho}$. We thus need not explicitly consider the energy equation, because it is equivalent to requiring $p/\rho^\gamma = p_o/\rho_o^\gamma$. Because of homeoentropy, there can be no viscous forces or energy diffusion. So it suffices to consider the one-dimensional limit of the mass and linear momentum equations from the Euler equations presented in Section 6.6:

$$\frac{\partial \rho}{\partial t} + u\frac{\partial \rho}{\partial x} + \rho\frac{\partial u}{\partial x} = 0, \qquad \rho\frac{\partial u}{\partial t} + \rho u\frac{\partial u}{\partial x} + \frac{\partial p}{\partial x} = 0. \qquad (9.372)$$

Now from the definition of sound speed, we have

$$\left.\frac{\partial p}{\partial \rho}\right|_s = c^2, \qquad \frac{\partial p}{\partial x} = c^2\frac{\partial \rho}{\partial x}. \qquad (9.373)$$

Thus, our equations become

$$\frac{\partial \rho}{\partial t} + u\frac{\partial \rho}{\partial x} + \rho\frac{\partial u}{\partial x} = 0, \qquad \rho\frac{\partial u}{\partial t} + \rho u\frac{\partial u}{\partial x} + c^2\frac{\partial \rho}{\partial x} = 0. \qquad (9.374)$$

Now similar to Section 9.6.7, we consider small deviations from stagnation values with

$$\rho = \rho_o + \epsilon\rho_1, \qquad u = \epsilon u_1, \qquad p = p_o + \epsilon p_1. \qquad (9.375)$$

We consider $0 < \epsilon \ll 1$. We substitute for these, neglect higher-order terms, and obtain

$$\frac{\partial \rho_1}{\partial t} + \rho_o\frac{\partial u_1}{\partial x} = 0, \qquad \rho_o\frac{\partial u_1}{\partial t} + c_o^2\frac{\partial \rho_1}{\partial x} = 0. \qquad (9.376)$$

Here we have $c_o^2 = \gamma p_o/\rho_o$. Next differentiate the mass equation with respect to t and the linear momentum equation with respect to x and multiply the mass equation by c_o^2 to get

$$c_o^2\frac{\partial^2 \rho_1}{\partial x\partial t} + \rho_o c_o^2\frac{\partial^2 u_1}{\partial x^2} = 0, \qquad \rho_o\frac{\partial^2 u_1}{\partial t^2} + c_o^2\frac{\partial^2 \rho_1}{\partial t\partial x} = 0. \qquad (9.377)$$

Finally, subtract the mass equation from the linear momentum equation and scale by ρ_o to get

$$\frac{\partial^2 u_1}{\partial t^2} = c_o^2\frac{\partial^2 u_1}{\partial x^2}. \qquad (9.378)$$

This is the wave equation, which has been well studied. It is a linear partial differential equations of the type that is known as *hyperbolic*. Such problems, in contrast to say Laplace's equation, which requires boundary conditions, require initial data, and more limited boundary data. The wave equation has a number of well-known solutions, see Garabedian (1998) or Goodwine (2010). Once u_1 is known, it is straightforward to find ρ_1 and p_1. We present a few examples for solutions of the wave equation.

Example 9.19 Show the wave equation is satisfied by the well-known d'Alembert[a] solution:

$$u_1(x,t) = f(x - c_o t) + g(x + c_o t), \qquad (9.379)$$

where f and g are arbitrary functions.

Solution

We prove this by direct substitution of the solution into the wave equation. The relevant partial derivatives are

$$\frac{\partial u_1}{\partial t} = -c_o f' + c_o g', \qquad \frac{\partial^2 u_1}{\partial t^2} = c_o^2 f'' + c_o^2 g'', \tag{9.380}$$

$$\frac{\partial u_1}{\partial x} = f' + g', \qquad \frac{\partial^2 u_1}{\partial x^2} = f'' + g''. \tag{9.381}$$

By inspection, the wave equation is satisfied because

$$\frac{\partial^2 u_1}{\partial t^2} = c_o^2 \frac{\partial^2 u_1}{\partial x^2}. \tag{9.382}$$

We can think of f and g as representing left- and right-running signals. The signals can be shown to propagate with speed c_o. Moreover, they do not decay; thus they retain all their information. This is consistent with the entropy being preserved with no loss of information due to viscous diffusion or heat transfer.

By direct substitution into the governing partial differential equations and state equations, one can show by differentiating and integrating that

$$\rho_1(x,t) = \rho_o \frac{f(x - c_o t) - g(x + c_o t)}{c_o}, \tag{9.383}$$

$$p_1(x,t) = \rho_o c_o^2 \frac{f(x - c_o t) - g(x + c_o t)}{c_o}, \tag{9.384}$$

$$T_1(x,t) = \frac{\gamma - 1}{\gamma} \frac{c_o^2}{R} \frac{f(x - c_o t) - g(x + c_o t)}{c_o}. \tag{9.385}$$

Acoustic disturbances induce space and time variation for all state variables.

[a] Jean le Rond d'Alembert, 1717–1783, French mathematician.

Other solution approaches may be used. Let us solve the wave equation, Eq. (9.378),

$$\frac{\partial^2 u_1}{\partial t^2} = c_o^2 \frac{\partial^2 u_1}{\partial x^2}, \tag{9.386}$$

subject to boundary and initial conditions

$$u_1(0,t) = u_1(L,t) = 0, \qquad u_1(x,0) = f(x), \qquad \frac{\partial u_1}{\partial t}(x,0) = 0. \tag{9.387}$$

We will generate solutions for four sets of initial conditions:

$$f(x) = u_{1o} \sin \frac{\pi x}{L}, \qquad \text{mono-modal}, \tag{9.388}$$

$$= u_{1o} \left(\sin \frac{\pi x}{L} + \frac{1}{10} \sin \frac{10 \pi x}{L} \right), \qquad \text{bi-modal}, \tag{9.389}$$

$$= u_{1o} \left(\frac{x}{L} \right) \left(1 - \frac{x}{L} \right), \qquad \text{poly-modal}, \tag{9.390}$$

$$= u_{1o} \left(H\left(\frac{x}{L} - \frac{2}{5} \right) - H\left(\frac{x}{L} - \frac{3}{5} \right) \right), \qquad \text{poly-modal}. \tag{9.391}$$

Physically, the boundary conditions imply the fluid is confined to a tube closed at both ends and it is given an initial spatial distribution of velocity, which at $t = 0$ is stationary in time.

This might be difficult to actually realize in an experiment, but it is useful for illustrating the solution and its properties.

We assume solutions of the type

$$u_1(x,t) = A(x)B(t). \tag{9.392}$$

With this assumption, known as *separation of variables*, Eq. (9.386) becomes

$$A(x)\frac{d^2 B}{dt^2} = c_o^2 B(t)\frac{d^2 A}{dx^2}, \quad \text{so} \quad \frac{1}{c_o^2 B(t)}\frac{d^2 B}{dt^2} = \frac{1}{A(x)}\frac{d^2 A}{dx^2} = -\lambda^2. \tag{9.393}$$

For an arbitrary function of t to be equal to an arbitrary function of x, both functions must be the same constant, which we have selected to be $-\lambda^2$. This induces two second-order linear ordinary differential equations:

$$\frac{d^2 A}{dx^2} + \lambda^2 A = 0, \quad \frac{d^2 B}{dt^2} + c_o^2 \lambda^2 B = 0. \tag{9.394}$$

Consider the equation for A first. In order to satisfy the boundary conditions $u_1(0,t) = u_1(L,t) = 0$, we must have $A(0) = A(L) = 0$. As is done here, if u_1 is specified on a boundary it is known as a *Dirichlet*[7] boundary condition. Had the derivative $\partial u_1/\partial x$ been specified, the boundary condition would have been called a *Neumann*[8] boundary condition. When a linear combination of u_1 and $\partial u_1/\partial x$ is specified on a boundary, it is known as a *Robin*[9] boundary condition. Along with these boundary conditions, the first of Eq. (9.394) can be recast as an eigenvalue problem:

$$-\frac{d^2}{dx^2}A = \lambda^2 A, \quad A(0) = A(L) = 0. \tag{9.395}$$

With $\mathcal{L} = -d^2/dx^2$, a *self-adjoint positive definite linear operator* (see Powers and Sen, 2015, Chapter 6), this takes the form

$$\mathcal{L}A = \lambda^2 A. \tag{9.396}$$

Importantly, the linear operator $\mathcal{L}$ operating on A yields a scalar multiple of A. We recall that self-adjoint linear operators have *orthogonal* eigenfunctions and real eigenvalues. Because it can be shown that our $\mathcal{L}$ is positive definite, the eigenvalues are also positive, which is why we describe the eigenvalue as λ^2.

Solving Eq. (9.394), we see that

$$A(x) = C_1 \sin \lambda x + C_2 \cos \lambda x. \tag{9.397}$$

For $A(0) = 0$, we get

$$A(0) = 0 = C_2 \cos 0 = C_2. \tag{9.398}$$

Thus,

$$A(x) = C_1 \sin \lambda x. \tag{9.399}$$

---

[7] Peter Gustav Lejeune Dirichlet, 1805–1859, German mathematician.
[8] Carl Gottfried Neumann, 1832–1925, German mathematician.
[9] Victor Gustave Robin, 1855–1897, French mathematician.

Now at $x = L$, we have

$$A(L) = 0 = C_1 \sin \lambda L. \tag{9.400}$$

To guarantee this condition is satisfied, we must require that

$$\lambda L = n\pi, \quad \text{so} \quad \lambda = \frac{n\pi}{L}, \quad n = 1, 2, \ldots. \tag{9.401}$$

With this, we have

$$A(x) = C_n \sin \frac{n\pi x}{L}, \quad n = 1, 2, \ldots. \tag{9.402}$$

We note that the eigenvalues λ^2 are real and positive, and the eigenfunctions $C_n \sin n\pi x/L$ have an arbitrary amplitude. It will also soon be useful to employ the orthogonality property, $\int_0^L (\sin m\pi x/L)(\sin n\pi x/L)\, dx = 0$ if $m \neq n$ when m and n are integers, and that the integral is nonzero if $n = m$.

We now turn to solution of the second of Eq. (9.394), which is now restated as

$$\frac{d^2 B}{dt^2} + \left(\frac{n\pi c_o}{L}\right)^2 B = 0. \tag{9.403}$$

This has solution

$$B(t) = C_3 \sin \frac{n\pi c_o t}{L} + C_4 \cos \frac{n\pi c_o t}{L}. \tag{9.404}$$

Now for $\partial u_1/\partial t$ to be everywhere zero at $t = 0$, we must insist that $dB/dt(0) = 0$. Enforcing this gives

$$\frac{dB}{dt} = \frac{C_3 n\pi c_o}{L} \cos \frac{n\pi c_o t}{L} - \frac{C_4 n\pi c_o}{L} \sin \frac{n\pi c_o t}{L}, \tag{9.405}$$

$$\frac{dB}{dt}(t=0) = \frac{C_3 n\pi c_o}{L} \cos 0 - \frac{C_4 n\pi c_o}{L} \sin 0 = 0, \tag{9.406}$$

$$= \frac{C_3 n\pi c_o}{L} = 0. \tag{9.407}$$

We thus insist that $C_3 = 0$. Taking $\hat{C}_4 = C_1 C_4$, our solution combines to form

$$u_1(x,t) = \hat{C}_4 \cos \frac{n\pi c_o t}{L} \sin \frac{n\pi x}{L}. \tag{9.408}$$

We next recognize that this solution is valid for arbitrary positive integer n. Because the original equation is linear, the principle of superposition applies, and arbitrary linear combinations also are valid solutions. We can express this by generalizing to

$$u_1(x,t) = \sum_{n=1}^{\infty} C_n \cos \frac{n\pi c_o t}{L} \sin \frac{n\pi x}{L}. \tag{9.409}$$

We can use standard trigonometric reductions to recast Eq. (9.409) as

$$u_1(x,t) = \sum_{n=1}^{\infty} \frac{C_n}{2}\left(\sin\left(\frac{n\pi}{L}(x+c_o t)\right) - \sin\left(\frac{n\pi}{L}(x-c_o t)\right)\right). \tag{9.410}$$

Importantly, we note that the solution is consistent with the d'Alembert solution, Eq. (9.379). The solution can be thought of an infinite sum of left- and right-propagating waves. All modes travel at the same velocity magnitude c_o; formally, such waves are *nondispersive*. The amplitude of each mode does not decay with time; formally, such waves are *nondiffusive*. Were we to model momentum diffusion with nonzero viscosity, we would find it would induce amplitude decay.

We can fix the various values of C_n by applying the initial condition for $u_1(x,0) = f(x)$:

$$u_1(x,0) = f(x) = \sum_{n=1}^{\infty} C_n \sin \frac{n\pi x}{L}. \tag{9.411}$$

This amounts to finding the Fourier sine series expansion of $f(x)$. We get this by taking advantage of the orthogonality properties of $\sin n\pi x/L$ on the domain $x \in [0, L]$ by the following series of operations:

$$f(x) = \sum_{n=1}^{\infty} C_n \sin \frac{n\pi x}{L}, \tag{9.412}$$

$$\sin \frac{m\pi x}{L} f(x) = \sum_{n=1}^{\infty} C_n \sin \frac{m\pi x}{L} \sin \frac{n\pi x}{L}, \tag{9.413}$$

$$\int_0^L \sin \frac{m\pi x}{L} f(x)\, dx = \int_0^L \sum_{n=1}^{\infty} C_n \sin \frac{m\pi x}{L} \sin \frac{n\pi x}{L}\, dx, \tag{9.414}$$

$$= \sum_{n=1}^{\infty} C_n \underbrace{\int_0^L \sin \frac{m\pi x}{L} \sin \frac{n\pi x}{L}\, dx}_{=L\delta_{mn}/2}. \tag{9.415}$$

Because of orthogonality, the integral has value of 0 for $n \neq m$ and $L/2$ for $n = m$. Employing the Kronecker delta, Eq. (2.19), we get

$$\int_0^L \sin \frac{m\pi x}{L} f(x)\, dx = \frac{L}{2} \sum_{n=1}^{\infty} C_n \delta_{nm} = \frac{L}{2} C_m, \tag{9.416}$$

$$C_n = \frac{2}{L} \int_0^L \sin \frac{n\pi x}{L} f(x)\, dx. \tag{9.417}$$

This combined with Eq. (9.409) forms the solution for arbitrary $f(x)$.

If we have the mono-modal $f(x) = u_{1o} \sin(\pi x/L)$, the full solution is particularly simple. In this case the initial condition has exactly the functional form of the eigenfunction, and there is thus only a one-term Fourier series. The solution is, by inspection,

$$u_1(x,t) = u_{1o} \cos \frac{\pi c_o t}{L} \sin \frac{\pi x}{L}. \tag{9.418}$$

The solution is a single fundamental mode, given by half of a sine wave pinned at $x = 0$ and $x = L$. At any given point x, the velocity u_1 oscillates. For example, at $x = L/2$, we have

$$u_1(L/2, t) = u_{1o} \cos \frac{\pi c_o t}{L}. \tag{9.419}$$

We call this a *standing wave*. Because there is only one Fourier mode, it is also known as *mono-modal*.

Trigonometric expansion shows that Eq. (9.418) can be expanded as

$$u_1(x,t) = \frac{u_{1o}}{2}\left(\sin\left(\frac{\pi}{L}(x - c_o t)\right) + \sin\left(\frac{\pi}{L}(x + c_o t)\right)\right). \tag{9.420}$$

This form illustrates that the standing wave can be considered as a sum of two propagating signals, one moving to the left with speed c_o, the other moving to the right at speed c_o. This is consistent with the d'Alembert solution, Eq. (9.379). A plot of the single mode standing wave is shown in Fig. 9.30(a) for parameter values shown in the caption.

The solution is almost as simple for the bi-modal second initial condition. We must have

$$u_1(x,0) = u_{1o}\sin\frac{\pi x}{L} + \frac{u_{1o}}{10}\sin\frac{10\pi x}{L}. \tag{9.421}$$

By inspection again, the solution is

$$u_1(x,t) = u_{1o}\cos\frac{\pi c_o t}{L}\sin\frac{\pi x}{L} + \frac{u_{1o}}{10}\cos\frac{10\pi c_o t}{L}\sin\frac{10\pi x}{L}. \tag{9.422}$$

A plot of the bi-modal standing wave is shown in Fig. 9.30(c) for parameter values shown in the caption.

Next let us consider the poly-modal third initial distribution:

$$u_1(x,0) = f(x) = u_{1o}\frac{x}{L}\left(1 - \frac{x}{L}\right). \tag{9.423}$$

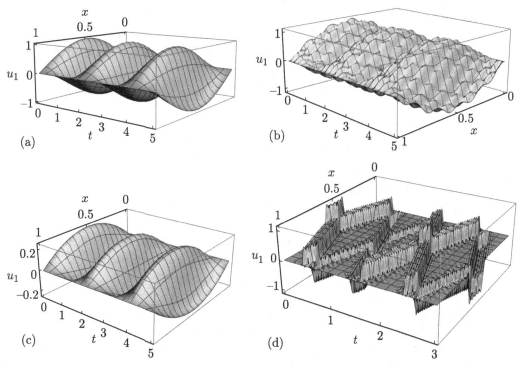

Figure 9.30 Response $u_1(x,t)$ for a solution to the wave equation with (a) a single Fourier mode (mono-modal), (b) two Fourier modes (bi-modal), (c) multiple Fourier modes (poly-modal), $f(x) = u_{1o}(x/L)(1 - x/L)$, and (d) a poly-modal "top-hat" initial condition, $f(x) = u_{1o}(H((x/L - 2/5) - H(x/L - 3/5))$, all with $u_{1o} = 1$, $c_o = 1$, $L = 1$.

For this $f(x)$, evaluation of C_n via Eq. (9.417) gives the set of C_n as

$$C_n = \frac{8u_{1o}}{\pi^3}\left\{1, 0, \frac{1}{27}, 0, \frac{1}{125}, 0\ldots\right\}, \qquad n = 1,\ldots,\infty. \tag{9.424}$$

Obviously, every odd term in the series is zero. This is a result of $f(x)$ having symmetry about $x = L/2$. It is possible to get a simple expression for C_n:

$$C_n = \begin{cases} \frac{8u_{1o}}{n^3\pi^3}, & n \text{ odd}, \\ 0 & n \text{ even}. \end{cases} \tag{9.425}$$

It is then possible to show that the solution can be expressed as the infinite series

$$u_1(x,t) = \frac{8u_{1o}}{\pi^3}\sum_{m=1}^{\infty}\frac{1}{(2m-1)^3}\cos\frac{(2m-1)\pi c_o t}{L}\sin\frac{(2m-1)\pi x}{L}. \tag{9.426}$$

A plot of the poly-modal standing wave is shown in Fig. 9.30(c) for parameter values shown in the caption. The plot looks similar to that for the mono-modal initial condition; this is because the quadratic polynomial initial condition is well modeled by a single Fourier mode. Recognize, however, that an infinite number smaller amplitude modes are present across infinite spectrum of frequencies. Note also that the frequencies of the modes are discretely separated. This is known as a *discrete spectrum* of frequencies. This feature is a consequence of the fact that for each sine wave to fit within the finite domain and still match the boundary conditions, only discretely separated frequencies are admitted. If we were to remove the boundary conditions at $x = 0$ and $x = L$, we would find instead a *continuous spectrum*.

We lastly consider the poly-modal initial condition which is a so-called "top-hat" function:

$$f(x) = u_{1o}\left(H\left(\frac{x}{L} - \frac{2}{5}\right) - H\left(\frac{x}{L} - \frac{3}{5}\right)\right). \tag{9.427}$$

Evaluation of C_n via Eq. (9.417) gives the set of C_n as

$$C_n = u_{1o}\left\{\frac{\sqrt{5}-1}{\pi}, 0, -\frac{1+\sqrt{5}}{3\pi}, 0, \frac{4}{5\pi}, 0, -\frac{1+\sqrt{5}}{7\pi}, 0, \frac{\sqrt{5}-1}{9\pi}, 0,\ldots\right\}. \tag{9.428}$$

A plot of the solution is shown in Fig. 9.30(c) for parameter values shown in the caption. Here 50 nonzero terms have been retained in the series. We note several important features of Fig. 9.30(c). The initial "top hat" signal immediately breaks into two distinct waveforms. One propagates to the right and the other to the left. This is consistent with the d'Alembert nature of the solution to the wave equation. When either waveform strikes the boundary at either $x = 0$ or $x = L$, there is a reflection, with the sign of u_1 changing. After a second reflection, both waves recombine to recover the initial waveform at a particular time. The pattern repeats, and there is no loss of information in the signal. Lastly, due to the finite number of terms in the series, there is a choppiness in the plotted approximate solution.

A so-called (x,t) diagram can be useful in understanding wave phenomena. In such a diagram, either contouring or shading is used to show how the dependent variable varies in the (x,t) plane. Figure 9.31 gives such a diagram for solution to the wave equation with the "top-hat" function as an initial condition. Here dark and light regions correspond to small and large u_1, respectively. Clearly signals are propagating with angle $\pm\pi/4$ in this plane, which

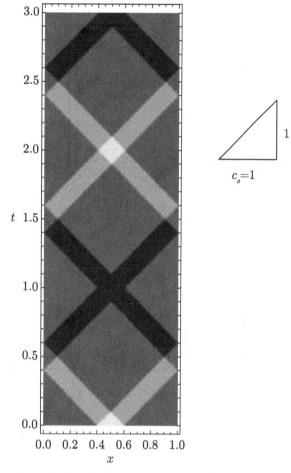

Figure 9.31 (x, t) diagram for solution to the wave equation with a "top-hat" initial condition, $f(x) = u_{1o}(H((x/L - 2/5) - H(x/L - 3/5))$, all with $u_{1o} = 1$, $c_o = 1$, $L = 1$.

corresponds to a wave speed of $c_o = 1$. We also clearly see the reflection process at $x = 0$ and $x = 1$.

Lastly we examine the variation of the amplitudes $|C_n|$ with n for each of the four cases. A plot is shown in Fig. 9.32. Figures such as this are related to the so-called power spectral density of a signal; in other contexts it is known as the "energy" spectral density. This is not obviously our energy with units of J; however, it is a common parlance. One can see how "energy" is partitioned into various modes of oscillation. One can show the frequency of oscillation is proportional to n.

9.11 Method of Characteristics

The previous section considered the propagation of waves for compressible flow in the acoustic limit. We consider the considerably more complicated problem of wave propagation away from

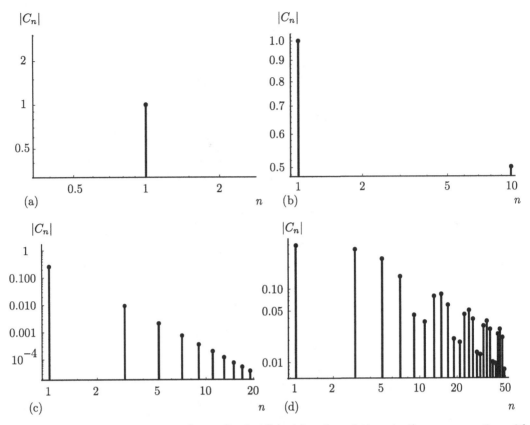

Figure 9.32 Variation of Fourier mode amplitude $|C_n|$ with n for solutions to the wave equation with (a) mono-modal signal (a single Fourier mode), (b) a bi-modal signal (two Fourier modes), (c) a poly-modal signal (multiple Fourier modes), $f(x) = u_{1o}(x/L)(1 - x/L)$, and (d) a poly-modal "top-hat" initial condition, $f(x) = u_{1o}(H((x/L - 2/5) - H(x/L - 3/5))$, all with $u_{1o} = 1$, $c_o = 1$, $L = 1$.

the acoustic limit in this section. Here we develop the so-called *method of characteristics* to model a one-dimensional unsteady, inviscid, non-heat conducting fluid. The emphasis will be on rarefaction waves. This analysis requires more rigor than much of traditional one-dimensional gas dynamics, and draws upon some challenging mathematical methods.

In neglecting diffusive transport, we have eliminated all mechanisms for entropy generation; consequently, we will be able to model the process as isentropic. Even without diffusion, shocks can generate entropy. However, expansion waves for ideal gases are inherently continuous, and do remain isentropic. We will consider a general equation of state, and later specialize to a calorically perfect ideal gas. The problem is inherently nonlinear and is also hyperbolic. A discussion of the mathematics as well as application to fluid mechanics is given by Garabedian (1998). We will later consider an unusual discontinuous expansion, which is admitted for nonideal gases, in Section 9.13.

9.11.1 Inviscid One-Dimensional Equations

The equations to be considered are shown here in nonconservative form. These are one-dimensional limits of the system developed in Section 6.6:

$$\frac{\partial \rho}{\partial t} + u\frac{\partial \rho}{\partial x} + \rho\frac{\partial u}{\partial x} = 0, \qquad (9.429)$$

$$\rho\frac{\partial u}{\partial t} + \rho u\frac{\partial u}{\partial x} + \frac{\partial p}{\partial x} = 0, \qquad (9.430)$$

$$\frac{\partial s}{\partial t} + u\frac{\partial s}{\partial x} = 0, \qquad (9.431)$$

$$p = p(\rho, s). \qquad (9.432)$$

Here we have written the energy equation in terms of entropy. The development of this was shown in Section 4.4.5. We have also utilized the general result from thermodynamics that any intensive property can be written as a function of two other independent thermodynamic properties. Here we have chosen to write pressure as a function of density and entropy, as we did in Eq. (9.92) for a special case. Thus, we have four equations for the four unknowns, ρ, u, p, s.

Because $p = p(\rho, s)$, we have

$$dp = \left.\frac{\partial p}{\partial \rho}\right|_s d\rho + \left.\frac{\partial p}{\partial s}\right|_\rho ds, \quad \text{so} \quad \left.\frac{\partial p}{\partial x}\right|_t = \left.\frac{\partial p}{\partial \rho}\right|_s \left.\frac{\partial \rho}{\partial x}\right|_t + \left.\frac{\partial p}{\partial s}\right|_\rho \left.\frac{\partial s}{\partial x}\right|_t. \qquad (9.433)$$

Now, let us define thermodynamic properties c^2 and ζ as follows:

$$c^2 \equiv \left.\frac{\partial p}{\partial \rho}\right|_s, \qquad \zeta \equiv \left.\frac{\partial p}{\partial s}\right|_\rho. \qquad (9.434)$$

Of course c^2 is the same as was earlier defined by Eq. (9.72). We will see that ζ will be unimportant. With analysis of this section, we will be able to ascribe to c the physical significance of the speed of propagation of small disturbances, the so-called sound speed, that we have already encountered in acoustics. If we know the equation of state, then we can think of c^2 and ζ as known thermodynamic functions of ρ and s. Our definitions give us

$$\frac{\partial p}{\partial x} = c^2\frac{\partial \rho}{\partial x} + \zeta\frac{\partial s}{\partial x}. \qquad (9.435)$$

Substituting into our governing equations, we see that pressure can be eliminated to give three equations in three unknowns:

$$\frac{\partial \rho}{\partial t} + u\frac{\partial \rho}{\partial x} + \rho\frac{\partial u}{\partial x} = 0, \qquad (9.436)$$

$$\rho\frac{\partial u}{\partial t} + \rho u\frac{\partial u}{\partial x} + \underbrace{c^2\frac{\partial \rho}{\partial x} + \zeta\frac{\partial s}{\partial x}}_{\frac{\partial p}{\partial x}} = 0, \qquad (9.437)$$

$$\frac{\partial s}{\partial t} + u\frac{\partial s}{\partial x} = 0. \qquad (9.438)$$

Now we can say that if $s = s(x,t)$,

$$ds = \frac{\partial s}{\partial t} dt + \frac{\partial s}{\partial x} dx, \quad \text{so} \quad \frac{ds}{dt} = \frac{\partial s}{\partial t} + \frac{dx}{dt}\frac{\partial s}{\partial x} = \frac{\partial s}{\partial t} + u\frac{\partial s}{\partial x}. \qquad (9.439)$$

Thus, on curves where $dx/dt = u$ (that by definition are particle pathlines), we have from substituting Eq. (9.439) into the energy equation (9.438) that

$$\frac{ds}{dt} = 0. \qquad (9.440)$$

Thus, we have converted the partial differential equation for the first law of thermodynamics into an ordinary differential equation. This can be integrated to give

$$s = C, \quad \text{on a particle pathline,} \quad \frac{dx}{dt} = u. \qquad (9.441)$$

This scenario is depicted in the so-called (x,t) diagram of Fig. 9.33.

This result is satisfying, but incomplete, as we do not in general know where the pathlines are. Let us try to apply this technique to the system in general. Consider our equations in matrix form:

$$\begin{pmatrix} 1 & 0 & 0 \\ 0 & \rho & 0 \\ 0 & 0 & 1 \end{pmatrix} \begin{pmatrix} \frac{\partial \rho}{\partial t} \\ \frac{\partial u}{\partial t} \\ \frac{\partial s}{\partial t} \end{pmatrix} + \begin{pmatrix} u & \rho & 0 \\ c^2 & \rho u & \zeta \\ 0 & 0 & u \end{pmatrix} \begin{pmatrix} \frac{\partial \rho}{\partial x} \\ \frac{\partial u}{\partial x} \\ \frac{\partial s}{\partial x} \end{pmatrix} = \begin{pmatrix} 0 \\ 0 \\ 0 \end{pmatrix}. \qquad (9.442)$$

These equations are of the form

$$A_{ij}\frac{\partial u_j}{\partial t} + B_{ij}\frac{\partial u_j}{\partial x} = f_i. \qquad (9.443)$$

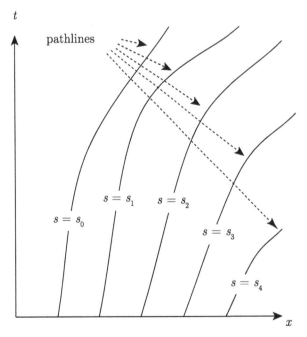

Figure 9.33 (x,t) diagram showing maintenance of entropy s along particle pathlines $dx/dt = u$ for isentropic flow.

As described by Whitham[10] (1974), there is a general technique to analyze such equations. First pre-multiply both sides of the equation by a yet to be determined vector of variables ℓ_i:

$$\ell_i A_{ij} \frac{\partial u_j}{\partial t} + \ell_i B_{ij} \frac{\partial u_j}{\partial x} = \ell_i f_i. \tag{9.444}$$

Now, this method will work if we can choose ℓ_i to render this product to be of the form similar to $\partial/\partial t + u(\partial/\partial x)$. Let us take

$$\ell_i A_{ij} \frac{\partial u_j}{\partial t} + \ell_i B_{ij} \frac{\partial u_j}{\partial x} = m_j \left(\frac{\partial u_j}{\partial t} + \lambda \frac{\partial u_j}{\partial x} \right), \tag{9.445}$$

$$= m_j \frac{du_j}{dt} \quad \text{on} \quad \frac{dx}{dt} = \lambda. \tag{9.446}$$

So comparing terms, we see that

$$\ell_i A_{ij} = m_j, \qquad \ell_i B_{ij} = \lambda m_j, \tag{9.447}$$

$$\lambda \ell_i A_{ij} = \lambda m_j, \tag{9.448}$$

so, we get by eliminating m_j that

$$\ell_i \left(\lambda A_{ij} - B_{ij} \right) = 0. \tag{9.449}$$

This is a left eigenvalue problem. We set the determinant of $\lambda A_{ij} - B_{ij}$ to zero for a nontrivial solution and find

$$\begin{vmatrix} \lambda - u & -\rho & 0 \\ -c^2 & \rho(\lambda - u) & -\zeta \\ 0 & 0 & \lambda - u \end{vmatrix} = 0. \tag{9.450}$$

Evaluating, we get

$$(\lambda - u)\left(\rho(\lambda - u)^2\right) + \rho(\lambda - u)(-c^2) = 0, \quad \text{so} \quad \rho(\lambda - u)\left((\lambda - u)^2 - c^2\right) = 0. \tag{9.451}$$

Solving we get

$$\lambda = u, \qquad \lambda = u \pm c. \tag{9.452}$$

Now the left eigenvectors ℓ_i give us the actual equations. First for $\lambda = u$, we get

$$\begin{pmatrix} \ell_1 & \ell_2 & \ell_3 \end{pmatrix} \begin{pmatrix} u - u & -\rho & 0 \\ -c^2 & \rho(u - u) & -\zeta \\ 0 & 0 & u - u \end{pmatrix} = \begin{pmatrix} 0 & 0 & 0 \end{pmatrix}, \tag{9.453}$$

$$\begin{pmatrix} \ell_1 & \ell_2 & \ell_3 \end{pmatrix} \begin{pmatrix} 0 & -\rho & 0 \\ -c^2 & 0 & -\zeta \\ 0 & 0 & 0 \end{pmatrix} = \begin{pmatrix} 0 & 0 & 0 \end{pmatrix}. \tag{9.454}$$

Two of the equations require that $\ell_1 = 0$ and $\ell_2 = 0$. There is no restriction on ℓ_3. We will select a normalized solution so that

$$\ell_i = (0, 0, 1). \tag{9.455}$$

---

[10] Gerald Beresford Whitham, 1927–2014, applied mathematician and developer of theory for nonlinear wave propagation.

Thus, $\ell_i A_{ij}(\partial u_j/\partial t) + \ell_i B_{ij}(\partial u_j/\partial x) = \ell_i f_i$ gives

$$\begin{pmatrix} 0 & 0 & 1 \end{pmatrix} \begin{pmatrix} 1 & 0 & 0 \\ 0 & \rho & 0 \\ 0 & 0 & 1 \end{pmatrix} \begin{pmatrix} \frac{\partial \rho}{\partial t} \\ \frac{\partial u}{\partial t} \\ \frac{\partial s}{\partial t} \end{pmatrix} + \begin{pmatrix} 0 & 0 & 1 \end{pmatrix} \begin{pmatrix} u & \rho & 0 \\ c^2 & \rho u & \zeta \\ 0 & 0 & u \end{pmatrix} \begin{pmatrix} \frac{\partial \rho}{\partial x} \\ \frac{\partial u}{\partial x} \\ \frac{\partial s}{\partial x} \end{pmatrix} = \begin{pmatrix} 0 & 0 & 1 \end{pmatrix} \begin{pmatrix} 0 \\ 0 \\ 0 \end{pmatrix},$$
(9.456)

$$\begin{pmatrix} 0 & 0 & 1 \end{pmatrix} \begin{pmatrix} \frac{\partial \rho}{\partial t} \\ \frac{\partial u}{\partial t} \\ \frac{\partial s}{\partial t} \end{pmatrix} + \begin{pmatrix} 0 & 0 & u \end{pmatrix} \begin{pmatrix} \frac{\partial \rho}{\partial x} \\ \frac{\partial u}{\partial x} \\ \frac{\partial s}{\partial x} \end{pmatrix} = 0,$$
(9.457)

$$\frac{\partial s}{\partial t} + u \frac{\partial s}{\partial x} = 0.$$
(9.458)

So as before with $s = s(x,t)$, we have $ds = (\partial s/\partial t)dt + (\partial s/\partial x)dx$, and $ds/dt = \partial s/\partial t + (dx/dt)(\partial s/\partial x)$. Now *if* we require dx/dt to be a particle pathline, $dx/dt = u$, then our energy equation, Eq. (9.458), gives us

$$\frac{ds}{dt} = 0, \quad \text{on} \quad \frac{dx}{dt} = u.$$
(9.459)

The special case in which the pathlines are straight in (x,t) space, corresponding to a uniform velocity field of $u(x,t) = u_o$, is shown in the (x,t) diagram of Fig. 9.34.

Now let us look at the remaining eigenvalues, $\lambda = u \pm c$:

$$\begin{pmatrix} \ell_1 & \ell_2 & \ell_3 \end{pmatrix} \begin{pmatrix} u \pm c - u & -\rho & 0 \\ -c^2 & \rho(u \pm c - u) & -\zeta \\ 0 & 0 & u \pm c - u \end{pmatrix} = \begin{pmatrix} 0 & 0 & 0 \end{pmatrix},$$
(9.460)

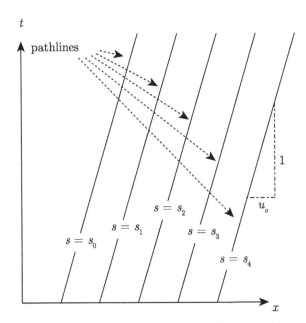

Figure 9.34 (x,t) diagram showing maintenance of entropy s along particle pathlines $dx/dt = u_o$ for isentropic flow.

$$\begin{pmatrix} \ell_1 & \ell_2 & \ell_3 \end{pmatrix} \begin{pmatrix} \pm c & -\rho & 0 \\ -c^2 & \pm \rho c & -\zeta \\ 0 & 0 & \pm c \end{pmatrix} = \begin{pmatrix} 0 & 0 & 0 \end{pmatrix}. \qquad (9.461)$$

As one of the components of the left eigenvector should be arbitrary, we will take $\ell_1 = 1$; we then arrive at the following equations:

$$\pm c - c^2 \ell_2 = 0, \Longrightarrow \ell_2 = \pm \frac{1}{c}, \qquad (9.462)$$

$$-\rho \pm \rho c \ell_2 = 0, \Longrightarrow \ell_2 = \pm \frac{1}{c}, \qquad (9.463)$$

$$-\zeta \ell_2 \pm c \ell_3 = 0, \Longrightarrow \ell_3 = \frac{\zeta}{c^2}. \qquad (9.464)$$

Thus, $\ell_i A_{ij}(\partial u_j/\partial t) + \ell_i B_{ij}(\partial u_j/\partial x) = \ell_i f_i$ gives

$$\begin{pmatrix} 1 & \pm \frac{1}{c} & \frac{\zeta}{c^2} \end{pmatrix} \begin{pmatrix} 1 & 0 & 0 \\ 0 & \rho & 0 \\ 0 & 0 & 1 \end{pmatrix} \begin{pmatrix} \frac{\partial \rho}{\partial t} \\ \frac{\partial u}{\partial t} \\ \frac{\partial s}{\partial t} \end{pmatrix} + \begin{pmatrix} 1 & \pm \frac{1}{c} & \frac{\zeta}{c^2} \end{pmatrix} \begin{pmatrix} u & \rho & 0 \\ c^2 & \rho u & \zeta \\ 0 & 0 & u \end{pmatrix} \begin{pmatrix} \frac{\partial \rho}{\partial x} \\ \frac{\partial u}{\partial x} \\ \frac{\partial s}{\partial x} \end{pmatrix}$$

$$= \begin{pmatrix} 1 \pm \frac{1}{c} \frac{\zeta}{c^2} \end{pmatrix} \begin{pmatrix} 0 \\ 0 \\ 0 \end{pmatrix}, \qquad (9.465)$$

$$\begin{pmatrix} 1 & \pm \frac{\rho}{c} & \frac{\zeta}{c^2} \end{pmatrix} \begin{pmatrix} \frac{\partial \rho}{\partial t} \\ \frac{\partial u}{\partial t} \\ \frac{\partial s}{\partial t} \end{pmatrix} + \begin{pmatrix} u \pm c & \rho \pm \rho \frac{u}{c} & \pm \frac{\zeta}{c} + \frac{\zeta u}{c^2} \end{pmatrix} \begin{pmatrix} \frac{\partial \rho}{\partial x} \\ \frac{\partial u}{\partial x} \\ \frac{\partial s}{\partial x} \end{pmatrix} = 0, \qquad (9.466)$$

$$\frac{\partial \rho}{\partial t} + (u \pm c)\frac{\partial \rho}{\partial x} \pm \frac{\rho}{c}\frac{\partial u}{\partial t} + \rho\left(1 \pm \frac{u}{c}\right)\frac{\partial u}{\partial x} + \frac{\zeta}{c^2}\frac{\partial s}{\partial t} + \left(\frac{\zeta u}{c^2} \pm \frac{\zeta}{c}\right)\frac{\partial s}{\partial x} = 0, \qquad (9.467)$$

$$\left(\frac{\partial \rho}{\partial t} + (u \pm c)\frac{\partial \rho}{\partial x}\right) \pm \frac{\rho}{c}\left(\frac{\partial u}{\partial t} + (u \pm c)\frac{\partial u}{\partial x}\right) + \frac{\zeta}{c^2}\left(\frac{\partial s}{\partial t} + (u \pm c)\frac{\partial s}{\partial x}\right) = 0, \qquad (9.468)$$

$$c^2 \underbrace{\left(\frac{\partial \rho}{\partial t} + (u \pm c)\frac{\partial \rho}{\partial x}\right)}_{d\rho/dt} \pm \rho c \underbrace{\left(\frac{\partial u}{\partial t} + (u \pm c)\frac{\partial u}{\partial x}\right)}_{du/dt} + \zeta \underbrace{\left(\frac{\partial s}{\partial t} + (u \pm c)\frac{\partial s}{\partial x}\right)}_{ds/dt} = 0. \qquad (9.469)$$

Now on lines where $dx/dt = u \pm c$, we get a transformation of the partial differential equations to ordinary differential equations:

$$c^2 \frac{d\rho}{dt} \pm \rho c \frac{du}{dt} + \zeta \frac{ds}{dt} = 0, \qquad \text{on} \qquad \frac{dx}{dt} = u \pm c. \qquad (9.470)$$

A plot of the *characteristics*, the lines on which the differential equations are obtained, is given in the (x, t) diagram of Fig. 9.35.

9.11.2 Homeoentropic Flow of a Calorically Perfect Ideal Gas

The equations developed so far are valid for a general equation of state. Here let us now consider the flow of a calorically perfect ideal gas, so $p = \rho R T$ and $e = c_v T + \hat{e}$. Consequently, we have the standard relation for the square of the sound speed of a calorically perfect ideal

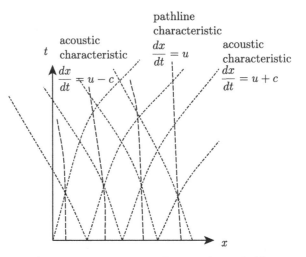

Figure 9.35 (x,t) diagram showing characteristics for pathlines $dx/dt = u$ and acoustic waves $dx/dt = u \pm c$.

gas, Eq. (9.137), $c^2 = \gamma p/\rho$. Further, let us take the flow to be homeoentropic (as introduced in Section 6.6.3), that is to say, not only does the entropy remain constant on pathlines, which is isentropic, but it has the same value on each pathline. That is, the entropy field is a constant. Consequently, we have the standard relation for a calorically perfect ideal gas:

$$\frac{p}{\rho^\gamma} = \frac{p_o}{\rho_o^\gamma}. \tag{9.471}$$

Because of homeoentropy, we no longer need consider the energy equation, and the linear combination of mass and linear momentum equations, Eq. (9.470), reduces to

$$c^2 \frac{d\rho}{dt} \pm \rho c \frac{du}{dt} = 0, \quad \text{on} \quad \frac{dx}{dt} = u \pm c. \tag{9.472}$$

Rearranging, we get

$$\frac{du}{dt} = \mp \frac{d\rho}{dt} \frac{c}{\rho}, \quad \text{on} \quad \frac{dx}{dt} = u \pm c. \tag{9.473}$$

Now $c^2 = \gamma p/\rho = c_o^2 (\rho/\rho_o)^{\gamma-1}$; thus, $c = c_o(\rho/\rho_o)^{(\gamma-1)/2}$, so

$$\frac{du}{dt} = \mp c_o(\rho/\rho_o)^{(\gamma-1)/2} \rho^{-1} \frac{d\rho}{dt} = \mp \frac{2}{\gamma-1} \frac{d}{dt}\left(c_o\left(\frac{\rho}{\rho_o}\right)^{\frac{\gamma-1}{2}}\right) = \mp \frac{2}{\gamma-1} \frac{dc}{dt}. \tag{9.474}$$

Regrouping, we find

$$\frac{d}{dt}\left(u \pm \frac{2}{\gamma-1} c\right) = 0. \tag{9.475}$$

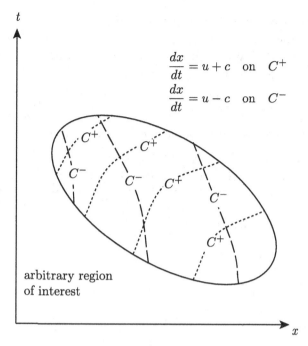

Figure 9.36 (x,t) diagram showing C^+ and C^- characteristics $dx/dt = u \pm c$.

Adapting notation used by Courant[11] and Friedrichs[12] (1976), we then integrate each of these equations, both of which are homogeneous, along characteristics to obtain algebraic relations

$$u + \frac{2}{\gamma - 1}c = 2\tilde{r}, \quad \text{on} \quad \frac{dx}{dt} = u + c, \quad C^+ \text{ characteristic}, \tag{9.476}$$

$$u - \frac{2}{\gamma - 1}c = -2\tilde{s}, \quad \text{on} \quad \frac{dx}{dt} = u - c, \quad C^- \text{ characteristic}. \tag{9.477}$$

Courant and Friedrichs' $\tilde{s}$ has no relation to entropy; it is just a new variable introduced for convenience. Also the presence of ± 2 is simply an arbitrary choice made by Courant and Friedrichs. A plot of the characteristics is given in the (x,t) diagram of Fig. 9.36. Now $\tilde{r}$ and $\tilde{s}$ can take on different values, depending on which characteristic we are on. On a given characteristic, they remain constant. Let us define additional parameters α and β to identify which characteristic we are on. So we have

$$u + \frac{2}{\gamma - 1}c = 2\tilde{r}(\beta), \quad \text{on} \quad \frac{dx}{dt} = u + c, \quad C^+ \text{ characteristic}, \tag{9.478}$$

$$u - \frac{2}{\gamma - 1}c = -2\tilde{s}(\alpha), \quad \text{on} \quad \frac{dx}{dt} = u - c, \quad C^- \text{ characteristic}. \tag{9.479}$$

---

[11] Richard Courant, 1888–1972, Prussian-born German mathematician, founder of the Courant Institute of Mathematical Sciences at New York University, author of a classic mathematical text on supersonic fluid mechanics.

[12] Kurt Otto Friedrichs, 1901–1982, German-born mathematician who emigrated to the United States in 1937, worked on partial differential equations of mathematical physics and fluid mechanics.

These quantities are known as *Riemann invariants*.[13]

Let us use the method of characteristics in the acoustic limit to recover the d'Alembert solution, Eq. (9.379). We recall our two first-order equations for acoustics in a calorically perfect ideal gas, Eqs. (9.376):

$$\frac{\partial \rho_1}{\partial t} + \rho_o \frac{\partial u_1}{\partial x} = 0, \qquad \rho_o \frac{\partial u_1}{\partial t} + c_o^2 \frac{\partial \rho_1}{\partial x} = 0. \qquad (9.480)$$

We next cast these in the general form of Eq. (9.443) to get

$$\underbrace{\begin{pmatrix} 1 & 0 \\ 0 & \rho_o \end{pmatrix}}_{\mathbf{A}} \underbrace{\begin{pmatrix} \frac{\partial \rho_1}{\partial t} \\ \frac{\partial u_1}{\partial t} \end{pmatrix}}_{\frac{\partial \mathbf{u}}{\partial t}} + \underbrace{\begin{pmatrix} 0 & \rho_o \\ c_o^2 & 0 \end{pmatrix}}_{\mathbf{B}} \underbrace{\begin{pmatrix} \frac{\partial \rho_1}{\partial x} \\ \frac{\partial u_1}{\partial x} \end{pmatrix}}_{\frac{\partial \mathbf{u}}{\partial x}} = \underbrace{\begin{pmatrix} 0 \\ 0 \end{pmatrix}}_{\mathbf{f}}. \qquad (9.481)$$

Here our vector of dependent variables is

$$\mathbf{u} = \begin{pmatrix} \rho_1 \\ u_1 \end{pmatrix}. \qquad (9.482)$$

The associated eigenvalue problem is

$$\det(\lambda \mathbf{A} - \mathbf{B}) = \begin{vmatrix} \lambda & -\rho_o \\ -c_o^2 & \rho_o \lambda \end{vmatrix} = 0. \qquad (9.483)$$

Solving gives

$$\rho_o \lambda^2 - \rho_o c_o^2 = 0, \qquad \text{so} \qquad \lambda = \pm c_o. \qquad (9.484)$$

We have two real and distinct eigenvalues. Let us find the eigenvectors.

$$\boldsymbol{\ell}^T \cdot (\lambda \mathbf{A} - \mathbf{B}) = \mathbf{0}^T, \qquad (9.485)$$

$$\begin{pmatrix} \ell_1 & \ell_2 \end{pmatrix} \begin{pmatrix} \lambda & -\rho_o \\ -c_o^2 & \rho_o \lambda \end{pmatrix} = \begin{pmatrix} 0 & 0 \end{pmatrix}, \qquad (9.486)$$

$$\begin{pmatrix} \ell_1 & \ell_2 \end{pmatrix} \begin{pmatrix} \pm c_o & -\rho_o \\ -c_o^2 & \pm \rho_o c_o \end{pmatrix} = \begin{pmatrix} 0 & 0 \end{pmatrix}. \qquad (9.487)$$

This yields two linearly dependent equations:

$$\pm c_o \ell_1 - c_o^2 \ell_2 = 0, \qquad -\rho_o \ell_1 \pm \rho_o c_o \ell_2 = 0. \qquad (9.488)$$

If we multiply the second by $\mp c_o / \rho_o$, we get the first. It is obvious the solution is not unique. If we take $\ell_2 = \tau$, with τ as any constant, then $\ell_1 = \pm c_o \tau$. Let us take $\tau = 1$, and thus take the eigenvectors to be

$$\boldsymbol{\ell} = \begin{pmatrix} \pm c_o \\ 1 \end{pmatrix}. \qquad (9.489)$$

Importantly, not only do we have two distinct and real eigenvalues, but we also have two linearly independent eigenvectors. This is the mathematical condition for our wave equation to be hyperbolic.

---

[13] Georg Friedrich Bernhard Riemann, 1826–1866, German mathematician and geometer whose work in non-Euclidean geometry was critical to Einstein's theory of general relativity, produced the first major study of shock waves.

We lastly use the eigenvalues and eigenvectors to recast our original system. Multiplying both sides of Eq. (9.481) by $\boldsymbol{\ell}^T$, we get

$$\underbrace{(\pm c_o \quad 1)}_{\boldsymbol{\ell}^T} \underbrace{\begin{pmatrix} 1 & 0 \\ 0 & \rho_o \end{pmatrix}}_{\mathbf{A}} \underbrace{\begin{pmatrix} \frac{\partial \rho_1}{\partial t} \\ \frac{\partial u_1}{\partial t} \end{pmatrix}}_{\frac{\partial \mathbf{u}}{\partial t}} + \underbrace{(\pm c_o \quad 1)}_{\boldsymbol{\ell}^T} \underbrace{\begin{pmatrix} 0 & \rho_o \\ c_o^2 & 0 \end{pmatrix}}_{\mathbf{B}} \underbrace{\begin{pmatrix} \frac{\partial \rho_1}{\partial x} \\ \frac{\partial u_1}{\partial x} \end{pmatrix}}_{\frac{\partial \mathbf{u}}{\partial x}} = \underbrace{(\pm c_o \quad 1)}_{\boldsymbol{\ell}^T} \underbrace{\begin{pmatrix} 0 \\ 0 \end{pmatrix}}_{\mathbf{f}}, \quad (9.490)$$

$$(\pm c_o \quad \rho_o) \begin{pmatrix} \frac{\partial \rho_1}{\partial t} \\ \frac{\partial u_1}{\partial t} \end{pmatrix} + (c_o^2 \quad \pm \rho_o c_o) \begin{pmatrix} \frac{\partial \rho_1}{\partial x} \\ \frac{\partial u_1}{\partial x} \end{pmatrix} = 0, \quad (9.491)$$

$$\pm c_o \frac{\partial \rho_1}{\partial t} + \rho_o \frac{\partial u_1}{\partial t} + c_o^2 \frac{\partial \rho_1}{\partial x} \pm \rho_o c_o \frac{\partial u_1}{\partial x} = 0, \quad (9.492)$$

$$\pm c_o \left(\frac{\partial \rho_1}{\partial t} \pm c_o \frac{\partial \rho_1}{\partial x} \right) + \rho_o \left(\frac{\partial u_1}{\partial t} \pm c_o \frac{\partial u_1}{\partial x} \right) = 0, \quad (9.493)$$

$$c_o^2 \left(\frac{\partial \rho_1}{\partial t} \pm c_o \frac{\partial \rho_1}{\partial x} \right) \pm \rho_o c_o \left(\frac{\partial u_1}{\partial t} \pm c_o \frac{\partial u_1}{\partial x} \right) = 0. \quad (9.494)$$

This reduces to two sets of differential equations valid on two different sets of characteristic lines:

$$c_o^2 \frac{d\rho_1}{dt} \pm \rho_o c_o \frac{du_1}{dt} = 0 \quad \text{on} \quad \frac{dx}{dt} = \pm c_o. \quad (9.495)$$

This can be compared to the full nonlinear version of Eq. (9.470). These can be rearranged to form equations analogous to Eq. (9.475):

$$\frac{d}{dt}\left(u_1 \pm c_o \frac{\rho_1}{\rho_o} \right) = 0 \quad \text{on} \quad \frac{dx}{dt} = \pm c_o. \quad (9.496)$$

Integrating, we find

$$u_1 + c_o \frac{\rho_1}{\rho_o} = C_1 \quad \text{on} \quad x = x_o + c_o t, \quad C^+ \text{ characteristic}, \quad (9.497)$$

$$u_1 - c_o \frac{\rho_1}{\rho_o} = C_2 \quad \text{on} \quad x = x_o - c_o t, \quad C^- \text{ characteristic}. \quad (9.498)$$

That is to say, the combinations of $u_1 \pm c_o \rho_1/\rho_o$ are preserved on lines for which $x = x_o \pm c_o t$. In this solution signals are propagated in two distinct directions, and those signals are preserved as they propagate. The constants C_1 and C_2 are the Riemann invariants for the system. The Riemann invariants are only invariant on a given characteristic and may vary from one characteristic to another. We depict a typical characteristic net in the (x, t) diagram of Fig. 9.37. In contrast to characteristics for the nonlinear Euler equations, which may themselves be nonlinear, those for the equations of linear acoustics are straight lines.

The just-completed analysis is common, and is often described as converting the partial differential equation to ordinary differential equations valid along so-called characteristic lines in x, t space. This is unsatisfying as the variation of C_1 and C_2 reflects the fact that we really are considering partial differential equations. Motivated by the existence of characteristic lines on which linear combinations of ρ_1 and u_1 must retain a constant value, let us seek a coordinate transformation to clarify this. More importantly, the formal coordinate transform will show us how we really are considering a partial differential equation in a more easily analyzed space.

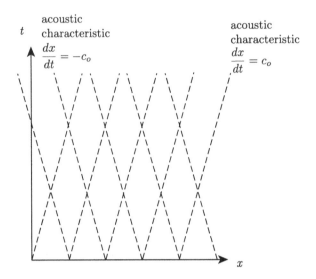

Figure 9.37 (x,t) diagram showing characteristics for linear acoustic waves; $dx/dt = \pm c_o$.

Motivated by the form of the characteristic lines, we take

$$\begin{pmatrix} \xi \\ \eta \end{pmatrix} = \begin{pmatrix} 1 & -c_o \\ 1 & c_o \end{pmatrix} \begin{pmatrix} x \\ t \end{pmatrix}. \qquad (9.499)$$

Thus,

$$\xi(x,t) = x - c_o t, \qquad \eta(x,t) = x + c_o t. \qquad (9.500)$$

Inverting, we get

$$\begin{pmatrix} x \\ t \end{pmatrix} = \underbrace{\begin{pmatrix} \frac{1}{2} & \frac{1}{2} \\ -\frac{1}{2c_o} & \frac{1}{2c_o} \end{pmatrix}}_{\mathbf{J}} \begin{pmatrix} \xi \\ \eta \end{pmatrix}. \qquad (9.501)$$

Here the Jacobian matrix $\mathbf{J}$ of the transformation is

$$\mathbf{J} = \begin{pmatrix} \frac{1}{2} & \frac{1}{2} \\ -\frac{1}{2c_o} & \frac{1}{2c_o} \end{pmatrix}. \qquad (9.502)$$

We find

$$J = |\mathbf{J}| = \det \mathbf{J} = \frac{1}{2c_o}. \qquad (9.503)$$

As an aside, we see the transformation is only area-preserving when $c_o = \pm 1/2$, and is orientation-preserving when $c_o > 0$.

We need rules for how the partial derivatives transform. The transformation rule tells us

$$\begin{pmatrix} \frac{\partial \rho_1}{\partial \xi} \\ \frac{\partial \rho_1}{\partial \eta} \end{pmatrix} = \underbrace{\begin{pmatrix} \frac{\partial x}{\partial \xi} & \frac{\partial t}{\partial \xi} \\ \frac{\partial x}{\partial \eta} & \frac{\partial t}{\partial \eta} \end{pmatrix}}_{\mathbf{J}^T} \begin{pmatrix} \frac{\partial \rho_1}{\partial x} \\ \frac{\partial \rho_1}{\partial t} \end{pmatrix} = \begin{pmatrix} \frac{1}{2} & -\frac{1}{2c_o} \\ \frac{1}{2} & \frac{1}{2c_o} \end{pmatrix} \begin{pmatrix} \frac{\partial \rho_1}{\partial x} \\ \frac{\partial \rho_1}{\partial t} \end{pmatrix}. \qquad (9.504)$$

Inverting, we find

$$\begin{pmatrix} \frac{\partial \rho_1}{\partial x} \\ \frac{\partial \rho_1}{\partial t} \end{pmatrix} = \underbrace{\begin{pmatrix} 1 & 1 \\ -c_o & c_o \end{pmatrix}}_{(\mathbf{J}^T)^{-1}} \begin{pmatrix} \frac{\partial \rho_1}{\partial \xi} \\ \frac{\partial \rho_1}{\partial \eta} \end{pmatrix}. \quad (9.505)$$

This is, in short

$$\frac{\partial}{\partial x} = \frac{\partial}{\partial \xi} + \frac{\partial}{\partial \eta}, \qquad \frac{\partial}{\partial t} = -c_o \frac{\partial}{\partial \xi} + c_o \frac{\partial}{\partial \eta}. \quad (9.506)$$

Employing these transformed operators on our original equation, Eq. (9.378), we get

$$\underbrace{\left(-c_o \frac{\partial}{\partial \xi} + c_o \frac{\partial}{\partial \eta}\right)}_{\partial/\partial t} \underbrace{\left(-c_o \frac{\partial}{\partial \xi} + c_o \frac{\partial}{\partial \eta}\right)}_{\partial/\partial t} u_1 = c_o^2 \underbrace{\left(\frac{\partial}{\partial \xi} + \frac{\partial}{\partial \eta}\right)}_{\partial/\partial x} \underbrace{\left(\frac{\partial}{\partial \xi} + \frac{\partial}{\partial \eta}\right)}_{\partial/\partial x} u_1, \quad (9.507)$$

$$c_o^2 \left(\frac{\partial^2 u_1}{\partial \xi^2} - 2 \frac{\partial^2 u_1}{\partial \xi \partial \eta} + \frac{\partial^2 u_1}{\partial \eta^2}\right) = c_o^2 \left(\frac{\partial^2 u_1}{\partial \xi^2} + 2 \frac{\partial^2 u_1}{\partial \xi \partial \eta} + \frac{\partial^2 u_1}{\partial \eta^2}\right), \quad (9.508)$$

$$-2 \frac{\partial^2 u_1}{\partial \xi \partial \eta} = 2 \frac{\partial^2 u_1}{\partial \xi \partial \eta}, \quad \text{so} \quad \frac{\partial^2 u_1}{\partial \xi \partial \eta} = 0. \quad (9.509)$$

We integrate this equation first with respect to ξ to get

$$\frac{\partial u_1}{\partial \eta} = h(\eta). \quad (9.510)$$

Note that when we integrate homogeneous partial differential equations, we must include an arbitrary function rather than the arbitrary constant we get for ordinary differential equations. We next integrate with respect to η to get

$$u_1 = \underbrace{\int_0^\eta h(\hat{\eta}) \, d\hat{\eta}}_{=f(\eta)} + g(\xi). \quad (9.511)$$

The integral of $h(\eta)$ simply yields another function of η, which we call $f(\eta)$. Thus, the general solution to $\partial^2 u_1 / \partial \xi \partial \eta = 0$ is

$$u_1(\xi, \eta) = f(\eta) + g(\xi). \quad (9.512)$$

We might say that we have separated the solution into two functions of two independent variables. Here the separated functions were combined as a sum. In other problems, the separated functions will combine as a product. In terms of our original coordinates, we can say

$$u_1(x, t) = f(x + c_o t) + g(x - c_o t). \quad (9.513)$$

Here f and g are completely arbitrary functions. This is the d'Alembert solution, Eq. (9.379).

9.11.3 Simple Waves

Return now to the full nonlinear Euler equations. Simple waves are defined to exist when either $\tilde{r}(\beta)$ or $\tilde{s}(\alpha)$ are constant everywhere in x, t space and not just on characteristics. For example, say $\tilde{s}(\alpha) = \tilde{s}_o$. Then the Riemann invariant

$$u - \frac{2}{\gamma - 1}c = -2\tilde{s}_o, \qquad \text{everywhere}, \tag{9.514}$$

is actually invariant over all of (x, t) space. Now the other Riemann invariant,

$$u + \frac{2}{\gamma - 1}c = 2\tilde{r}(\beta), \qquad \text{on } C^+, \tag{9.515}$$

takes on many values depending on β. However, it can be shown that for the simple wave, the characteristics have a constant slope in the (x, t) plane, as depicted in the (x, t) diagram of Fig. 9.38.

Now consider a rarefaction with a *prescribed* piston motion $u = u_p(t)$. A plot is given in the (x, t) diagram of Fig. 9.39. For this configuration, the Riemann invariant $u - 2c/(\gamma - 1) = -2\tilde{s}_o$ is valid everywhere. Let us evaluate $\tilde{s}_o$ in terms of more fundamental variables. For the piston problem we are considering, when $t = 0$, we have $u = 0$, $c = c_o$, so

$$u - \frac{2}{\gamma - 1}c = -\frac{2}{\gamma - 1}c_o, \qquad \text{everywhere.} \tag{9.516}$$

Thus, $\tilde{s}_o = c_o/(\gamma - 1)$.

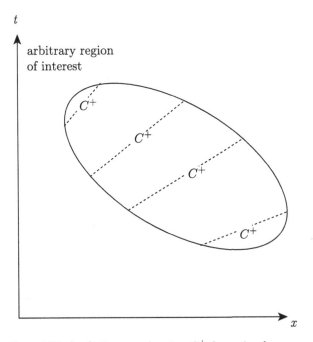

Figure 9.38 (x, t) diagram showing C^+ for a simple wave.

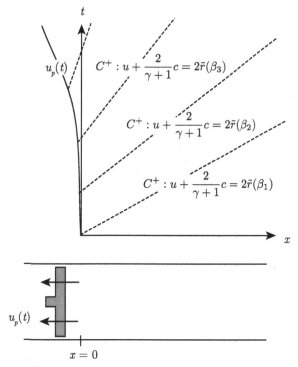

Figure 9.39 (x, t) diagram showing C^+ characteristics for isentropic rarefaction problem, along with piston cylinder arrangement.

Consider now a special characteristic $\hat{C}^+$ at $t = \hat{t}$. (see Fig. 9.40). At this time the piston moves with velocity $\hat{u}_p$, and the fluid velocity at the piston face is

$$u_{face}(\hat{t}) = \hat{u}_p. \qquad (9.517)$$

We get $c_{face}(\hat{t})$ from Eq. (9.516), which must be valid everywhere, including the face of the piston:

$$\underbrace{u_{face}}_{\hat{u}_p} - \frac{2}{\gamma - 1} c_{face} = -\frac{2}{\gamma - 1} c_o, \quad \text{so} \quad c_{face}(t = \hat{t}) = c_o + \frac{\gamma - 1}{2} \hat{u}_p. \qquad (9.518)$$

Also from Eq. (9.516), we have

$$c = c_o + \frac{\gamma - 1}{2} u, \quad \text{everywhere}, \qquad (9.519)$$

which is valid everywhere.

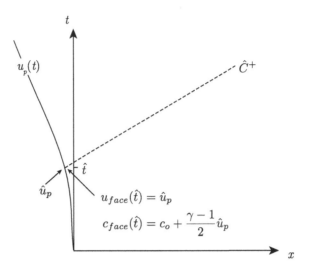

Figure 9.40 (x,t) diagram showing $\hat{C}^+$ for our rarefaction problem.

Now on $\hat{C}^+$, we have from Eq. (9.515) that

$$u + \frac{2}{\gamma - 1}c = \left(u_{face} + \frac{2}{\gamma - 1}c_{face}\right)_{t=\hat{t}}, \tag{9.520}$$

$$u + \frac{2}{\gamma - 1}\underbrace{\left(c_o + \frac{\gamma - 1}{2}u\right)}_{c} = \hat{u}_p + \frac{2}{\gamma - 1}\underbrace{\left(c_o + \frac{\gamma - 1}{2}\hat{u}_p\right)}_{c_{face}}, \tag{9.521}$$

$$2u + \frac{2}{\gamma - 1}c_o = 2\hat{u}_p + \frac{2}{\gamma - 1}c_o, \tag{9.522}$$

$$u = \hat{u}_p \quad \text{on} \quad \hat{C}^+. \tag{9.523}$$

So on $\hat{C}^+$, we have from Eq. (9.519) that

$$c = c_o + \frac{\gamma - 1}{2}\hat{u}_p. \tag{9.524}$$

So for $\hat{C}^+$, we get

$$\frac{dx}{dt} = u + c = \hat{u}_p + \underbrace{c_o + \frac{\gamma - 1}{2}\hat{u}_p}_{c} = \frac{\gamma + 1}{2}\hat{u}_p + c_o, \tag{9.525}$$

for a particular characteristic, this slope is a constant, as was earlier suggested. For our *prescribed* motion, $\hat{u}_p$ decreases with time and becomes more negative; hence the slope of our $\hat{C}^+$ characteristic decreases, and the characteristics *diverge* in x,t space. The slope of the leading characteristic is c_o, the ambient sound speed. The characteristic we consider, $\hat{C}^+$, is depicted in the (x,t) diagram of Fig. 9.40.

We can use our Riemann invariant along with isentropic relations to obtain other flow variables. From Eq. (9.516), we get

$$\frac{c}{c_o} = 1 + \frac{\gamma - 1}{2}\frac{u}{c_o}. \qquad (9.526)$$

Because the flow is homeoentropic, we have $c/c_o = (\rho/\rho_o)^{\frac{\gamma-1}{2}}$ and $p/p_o = (\rho/\rho_o)^\gamma$, so

$$\frac{p}{p_o} = \left(1 + \frac{\gamma - 1}{2}\frac{u}{c_o}\right)^{\frac{2\gamma}{\gamma-1}}, \qquad \frac{\rho}{\rho_o} = \left(1 + \frac{\gamma - 1}{2}\frac{u}{c_o}\right)^{\frac{2}{\gamma-1}}. \qquad (9.527)$$

9.11.4 Centered Rarefaction

If the piston is *suddenly* accelerated to a constant velocity, then a family of characteristics clusters at the origin on the (x,t) diagram and fans out in a *centered rarefaction fan*. This can also be studied using the similarity transformation $\eta = x/t$ that reduces the partial differential equations to ordinary differential equations. Centered and non-centered rarefactions are compared in the (x,t) diagram of Fig. 9.41.

Example 9.20 Analyze a centered rarefaction fan propagating into calorically perfect ideal air for a piston suddenly accelerated from rest to $u_p = -100$ m/s. Take the ambient air to be at $p_o = 10^5$ Pa, $T_o = 300$ K.

Solution
The ideal gas law gives $\rho_o = p_o/RT_o = (10^5 \text{ Pa})/((287 \text{ J/kg/K})(300 \text{ K})) = 1.16 \text{ kg/m}^3$. Now

$$c_o = \sqrt{\gamma R T_o} = \sqrt{\frac{7}{5}\left(287 \frac{\text{J}}{\text{kg K}}\right)(300 \text{ K})} = 347 \frac{\text{m}}{\text{s}}. \qquad (9.528)$$

On the final characteristic of the fan, C_f^+: $u = u_p = -100$ m/s. So

$$c = c_o + \frac{\gamma - 1}{2} u_p = \left(347 \frac{\text{m}}{\text{s}}\right) + \frac{7/5 - 1}{2}\left(-100 \frac{\text{m}}{\text{s}}\right) = 327 \frac{\text{m}}{\text{s}}. \qquad (9.529)$$

Now the final pressure is

$$\frac{p_f}{p_o} = \left(1 + \frac{\gamma - 1}{2}\frac{u_f}{c_o}\right)^{\frac{2\gamma}{\gamma-1}} = \left(1 + \frac{7/5 - 1}{2}\frac{(-100 \frac{\text{m}}{\text{s}})}{347 \frac{\text{m}}{\text{s}}}\right)^{\frac{2(7/5)}{7/5-1}} = 0.660. \qquad (9.530)$$

Hence $p_f = 6.6 \times 10^4$ Pa. Because the flow is homeoentropic, we get

$$\rho_f = \rho_o \left(\frac{p_f}{p_o}\right)^{\frac{1}{\gamma}} = \left(1.16 \frac{\text{kg}}{\text{m}^3}\right)(0.660)^{5/7} = 0.863 \frac{\text{kg}}{\text{m}^3}. \qquad (9.531)$$

And the final temperature is

$$T_f = \frac{p_f}{\rho_f R} = \frac{66.0 \times 10^3 \text{ Pa}}{(0.863 \frac{\text{kg}}{\text{m}^3})(287 \frac{\text{J}}{\text{kg K}})} = 266.4 \text{ K}. \qquad (9.532)$$

From linear acoustic theory, with $\Delta u = -100$ m/s, Section 9.6.7, we estimate that

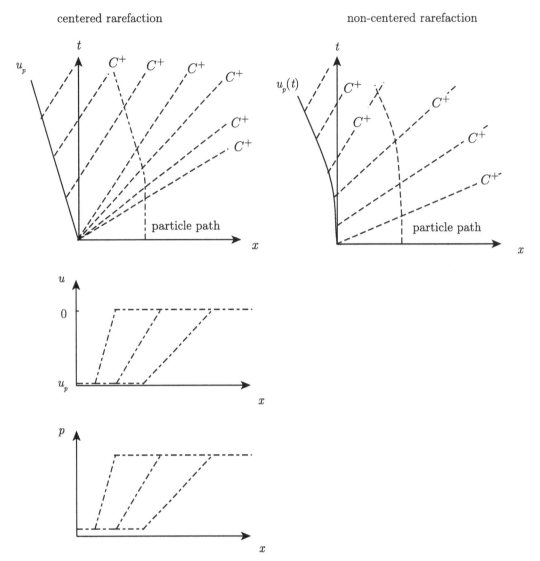

Figure 9.41 (x,t) diagrams for centered and non-centered rarefactions, along with pressure and velocity profiles for centered rarefaction fans.

$$\Delta \rho \sim \rho_o \frac{\Delta u}{c_o}, \qquad \Delta p \sim \rho_o c_o^2 \frac{\Delta u}{c_o}, \qquad \Delta T \sim (\gamma - 1) T_o \frac{\Delta u}{c_o}. \tag{9.533}$$

We compare the results of this problem with the estimates of linear acoustic theory, and see

$$\Delta \rho_{exact} = -0.298 \,\frac{\text{kg}}{\text{m}^3}, \qquad \Delta \rho_{linear} = -0.335 \,\frac{\text{kg}}{\text{m}^3}, \tag{9.534}$$

$$\Delta p_{exact} = -34.0 \times 10^3 \,\text{Pa}, \qquad \Delta p_{linear} = -40.3 \times 10^3 \,\text{Pa}, \tag{9.535}$$

$$\Delta T_{exact} = -33.6 \,\text{K}, \qquad \Delta T_{linear} = -34.6 \,\text{K}. \tag{9.536}$$

Here the linear equation for ΔT may be derived based on the equation of state.

9.11.5 Unsteady Inviscid Bateman–Burgers Shock Formation

We show here how the method of characteristics can be used to predict the onset of shock formation. To this point, we have described a common and traditional approach to the method of characteristics. Using common notation, we have written what began as partial differential equations in the form of ordinary differential equations, and it is often said that the method of characteristics is a way to *transform* partial into ordinary differential equations. However, the equations that result are certainly not in a standard form for ordinary differential equations; they are burdened with unusual side conditions.

It is in fact more sound to state that the method of characteristics transforms the partial differential equations in (x, t) space to another set of partial differential equations in a new space (ξ, τ) in which the integration is easier. Consider, for example, a model equation that is hyperbolic, the inviscid Bateman[14]–Burgers[15] equation:

$$\frac{\partial u}{\partial t} + u \frac{\partial u}{\partial x} = 0. \tag{9.537}$$

The viscous version of the model equation, $\partial u/\partial t + u \partial u/\partial x = \nu \partial^2 u/\partial x^2$, is widely known as Burgers' equation and is often cited as originating from Burgers (1948). However, the viscous version was given earlier by Bateman (1915). The inviscid Bateman–Burgers equation does not apply directly to a fluid; however, it contains one of the key mathematical features of fluid models: advective nonlinearity that has profound effect on the dynamics.

Now consider a general transformation $(x, t) \to (\xi, \tau)$:

$$x = x(\xi, \tau), \qquad t = t(\xi, \tau). \tag{9.538}$$

We assume the transformation to be unique and invertible. The definition of the total differentials dx and dt gives

$$\begin{pmatrix} dx \\ dt \end{pmatrix} = \underbrace{\begin{pmatrix} \frac{\partial x}{\partial \xi} & \frac{\partial x}{\partial \tau} \\ \frac{\partial t}{\partial \xi} & \frac{\partial t}{\partial \tau} \end{pmatrix}}_{\mathbf{J}} \begin{pmatrix} d\xi \\ d\tau \end{pmatrix}. \tag{9.539}$$

The Jacobian matrix of the transformation is

$$\mathbf{J} = \begin{pmatrix} \frac{\partial x}{\partial \xi} & \frac{\partial x}{\partial \tau} \\ \frac{\partial t}{\partial \xi} & \frac{\partial t}{\partial \tau} \end{pmatrix}. \tag{9.540}$$

And we have for the Jacobian determinant J:

$$J = \det \mathbf{J} = \frac{\partial x}{\partial \xi} \frac{\partial t}{\partial \tau} - \frac{\partial x}{\partial \tau} \frac{\partial t}{\partial \xi}. \tag{9.541}$$

Now from Eq. (2.251), we can deduce

$$\begin{pmatrix} \frac{\partial}{\partial x} \\ \frac{\partial}{\partial t} \end{pmatrix} = (\mathbf{J}^T)^{-1} \begin{pmatrix} \frac{\partial}{\partial \xi} \\ \frac{\partial}{\partial \tau} \end{pmatrix} = \frac{1}{J} \begin{pmatrix} \frac{\partial t}{\partial \tau} & -\frac{\partial t}{\partial \xi} \\ -\frac{\partial x}{\partial \tau} & \frac{\partial x}{\partial \xi} \end{pmatrix} \begin{pmatrix} \frac{\partial}{\partial \xi} \\ \frac{\partial}{\partial \tau} \end{pmatrix} = \frac{1}{J} \begin{pmatrix} \frac{\partial t}{\partial \tau} \frac{\partial}{\partial \xi} - \frac{\partial t}{\partial \xi} \frac{\partial}{\partial \tau} \\ -\frac{\partial x}{\partial \tau} \frac{\partial}{\partial \xi} + \frac{\partial x}{\partial \xi} \frac{\partial}{\partial \tau} \end{pmatrix}. \tag{9.542}$$

---

[14] Harry Bateman, 1882–1946, English mathematician.
[15] Johannes Martinus Burgers, 1895–1981, Dutch physicist.

With these transformation rules, Eq. (9.537) is rewritten as

$$\underbrace{\frac{1}{J}\left(-\frac{\partial x}{\partial \tau}\frac{\partial u}{\partial \xi}+\frac{\partial x}{\partial \xi}\frac{\partial u}{\partial \tau}\right)}_{\partial u/\partial t}+u\underbrace{\frac{1}{J}\left(\frac{\partial t}{\partial \tau}\frac{\partial u}{\partial \xi}-\frac{\partial t}{\partial \xi}\frac{\partial u}{\partial \tau}\right)}_{\partial u/\partial x}=0. \qquad (9.543)$$

Now by assumption, $J \neq 0$, so we can multiply by J to get

$$-\frac{\partial x}{\partial \tau}\frac{\partial u}{\partial \xi}+\frac{\partial x}{\partial \xi}\frac{\partial u}{\partial \tau}+u\frac{\partial t}{\partial \tau}\frac{\partial u}{\partial \xi}-u\frac{\partial t}{\partial \xi}\frac{\partial u}{\partial \tau}=0. \qquad (9.544)$$

Let us now restrict our transformation so as to satisfy the following requirements:

$$\frac{\partial x}{\partial \tau}=u\frac{\partial t}{\partial \tau}, \qquad t(\xi,\tau)=\tau. \qquad (9.545)$$

The first says that if we insist that ξ is held fixed, that the ratio of the change in x to the change in t will be u; this is equivalent to the more standard statement that on a characteristic line we have $dx/dt = u$. The second is a convenience simply equating τ to t. Applying the second restriction to the first, we can also say

$$\frac{\partial x}{\partial \tau}=u. \qquad (9.546)$$

With these restrictions, our inviscid Bateman–Burgers equation becomes

$$\underbrace{-\frac{\partial x}{\partial \tau}\frac{\partial u}{\partial \xi}}_{u}+\frac{\partial x}{\partial \xi}\frac{\partial u}{\partial \tau}+u\underbrace{\frac{\partial t}{\partial \tau}\frac{\partial u}{\partial \xi}}_{1}-u\underbrace{\frac{\partial t}{\partial \xi}\frac{\partial u}{\partial \tau}}_{0}=0, \qquad (9.547)$$

$$-u\frac{\partial u}{\partial \xi}+\frac{\partial x}{\partial \xi}\frac{\partial u}{\partial \tau}+u\frac{\partial u}{\partial \xi}=0, \qquad (9.548)$$

$$\frac{\partial x}{\partial \xi}\frac{\partial u}{\partial \tau}=0. \qquad (9.549)$$

Let us further require that $\partial x/\partial \xi \neq 0$. Then we have

$$\frac{\partial u}{\partial \tau}=0, \qquad \text{giving} \qquad u=f(\xi). \qquad (9.550)$$

Here f is an arbitrary function. Substitute this into Eq. (9.545) to get

$$\frac{\partial x}{\partial \tau}=f(\xi)\frac{\partial t}{\partial \tau}. \qquad (9.551)$$

We can integrate Eq. (9.551) to get

$$x=f(\xi)t+g(\xi). \qquad (9.552)$$

Here $g(\xi)$ is an arbitrary function. The coordinate transformation can be chosen for our convenience. To this end, remove t in favor of τ and set $g(\xi)=\xi$ so that x maps to ξ when $t=\tau=0$, giving

$$x(\xi,\tau)=f(\xi)\tau+\xi. \qquad (9.553)$$

We can then state the solution to the Bateman–Burgers equation, Eq. (9.537), parametrically as

$$u(\xi,\tau)=f(\xi), \qquad x(\xi,\tau)=f(\xi)\tau+\xi, \qquad t(\xi,\tau)=\tau. \qquad (9.554)$$

For this transformation, we have from Eq. (9.540) that

$$\mathbf{J} = \begin{pmatrix} 1 + \frac{df}{d\xi}\tau & f(\xi) \\ 0 & 1 \end{pmatrix}. \tag{9.555}$$

Thus,

$$J = \det \mathbf{J} = 1 + \frac{df}{d\xi}\tau. \tag{9.556}$$

We have a singularity in the coordinate transformation whenever $J = 0$, implying a difficulty when

$$\tau = -\frac{1}{\frac{df}{d\xi}}. \tag{9.557}$$

Let us solve the Bateman–Burgers equation, $\partial u/\partial t + u\partial u/\partial x = 0$, Eq. (9.537), if

$$u(x,0) = 1 + \sin \pi x, \qquad x \in [0,1]. \tag{9.558}$$

Let us not be concerned with that portion of u which at $t = 0$ has $x < 0$ or $x > 1$. The analysis can be modified to address this. We know the solution is given in general by Eqs. (9.554). At $t = 0$, we have $\tau = 0$, and thus $x = \xi$. And we have

$$f(\xi) = 1 + \sin \pi \xi. \tag{9.559}$$

Thus, we can say by inspection that the solution is

$$u(\xi,\tau) = 1 + \sin \pi \xi, \qquad x(\xi,\tau) = (1 + \sin \pi \xi)\tau + \xi, \qquad t(\xi,\tau) = \tau. \tag{9.560}$$

Results for $u(x,t)$ are plotted in Fig. 9.42. Another way to view the results is in the (x,t) diagram that gives contours of u in Fig. 9.43 These contours are generated before the transformation has become singular. The curves of constant u are the characteristics. Clearly those on the right of the maximum of u are coalescing, while those to the left are diverging. The coalescence corresponds to shock formation, and the divergence corresponds to a rarefaction.

One notes the following. The signal propagates to the right; this is a consequence of $u > 0$ in the domain we consider. Portions of the signal with higher u propagate faster. The signal distorts as t increases. The wave appears to "break" at $t = t_s$, with $1/4 \lesssim t_s \lesssim 1/2$. For $t > t_s$, it is possible to find multiple values of u at a given x and t. If u were a physical variable, we would not expect to see such multivaluedness in nature.

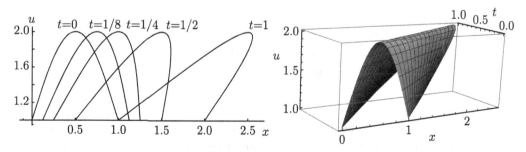

Figure 9.42 Solution to $\partial u/\partial t + u\partial u/\partial x = 0$ with $u(x,0) = 1 + \sin \pi x$.

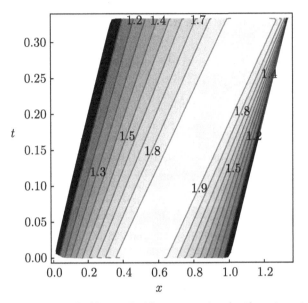

Figure 9.43 Early time solution to $\partial u/\partial t + u\partial u/\partial x = 0$ with $u(x,0) = 1 + \sin \pi x$ in the form of a contour plot in x, t space giving contours of constant u.

It appears to be challenging to write an explicit formula for $u(x,t)$. However, for small ξ, one can write a useful approximation. Taylor series expansion of Eq. (9.560) for small ξ yields

$$x(\xi, \tau) \sim (1 + \pi\tau)\xi + \tau + \cdots. \tag{9.561}$$

We invert this and use $\tau = t$ to find

$$\xi(x,t) \sim \frac{x-t}{1+\pi t} + \cdots. \tag{9.562}$$

Then, because $u = f(\xi)$, we get

$$u(x,t) \sim 1 + \sin\frac{\pi(x-t)}{1+\pi t} + \cdots. \tag{9.563}$$

This itself has a series expansion for small x and t of

$$u(x,t) \sim 1 + \pi(x-t) + \cdots. \tag{9.564}$$

Figure 9.44 shows how one can envision the portion of the initial sine wave with $x > 1/2$ steepening, while that portion with $x < 1/2$ flattens. We place arrows whose magnitude is proportional to the local value of u on the plot itself.

For our value of $f(\xi)$, we have from from Eq. (9.556) that

$$J = 1 + \pi\tau \cos \pi\xi. \tag{9.565}$$

Clearly, there exist values of (ξ, τ) for which $J = 0$. At such points, we can expect difficulties in our solution. In Fig. 9.45, we plot a portion of the locus of points for which $J = 0$ in the (ξ, τ) plane. We also see portions of this plane where the transformation is orientation-preserving, for

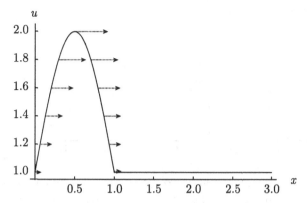

Figure 9.44 Diagram of response of u which satisfies the Bateman–Burgers equation $\partial u/\partial t + u\partial u/\partial x$ with $u(x,0) = 1 + \sin \pi x$.

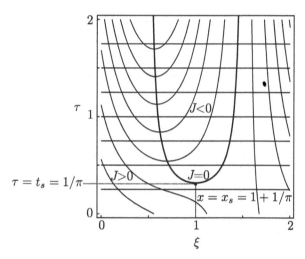

Figure 9.45 Curves where $J = 0$ and of constant x and t in the (ξ, τ) plane for our coordinate transformation.

which $J > 0$, and orientation-reversing, for which $J < 0$. Also shown in Fig. 9.45 are contours of constant x and t. Clearly when $J = 0$, the contours of constant x are parallel to those of constant t, and there are not enough linearly independent vectors to form a basis.

From Eq. (9.557), we can expect a singular coordinate transformation when

$$\tau = -\frac{1}{\frac{df}{d\xi}} = -\frac{1}{\pi \cos \pi \xi}. \tag{9.566}$$

We then substitute this into Eqs. (9.560) to get a parametric curve for when the transformation is singular, $x_s(\xi)$, $t_s(\xi)$:

$$x_s(\xi) = -\frac{1 + \sin \pi \xi}{\pi \cos \pi \xi} + \xi, \qquad t_s(\xi) = -\frac{1}{\pi \cos \pi \xi}. \tag{9.567}$$

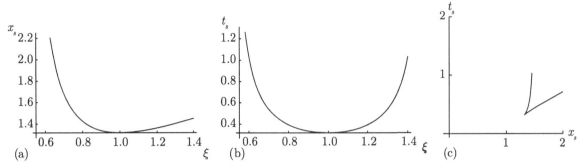

Figure 9.46 Plots indicating where the coordinate transformation of Eqs. (9.560) is singular: (a) $x_s(\xi)$ from Eq. (9.567), (b) $t_s(\xi)$ from Eq. (9.567), and (c) representation of the curve of singularity in (x,t) space.

A portion of this curve for where the transformation is singular is shown in Fig. 9.46. Figure 9.46(a) plots $x_s(\xi)$ from Eq. (9.567). Figure 9.46(b) plots $t_s(\xi)$ from Eq. (9.567). We see a parametric plot of the same quantities in Fig. 9.46(c). At early time the system is free of singularities. It can be shown that both $x_s(\xi)$ and $t_s(\xi)$ have a local minimum at $\xi = 1$, at which point, we have

$$x_s(1) = 1 + \frac{1}{\pi}, \qquad t_s(1) = \frac{1}{\pi}. \tag{9.568}$$

Examining Fig. 9.42, this appears to be the point at which the solution becomes multivalued. Examining Fig. 9.45, this is the point on the curve $J = 0$ that is a local minimum. So while x_s and t_s are well-behaved as functions of ξ for the domain considered, when the curves are projected into the (x,t) plane, there is a cusp at $(x,t) = (x_s(1), t_s(1)) = (1 + 1/\pi, 1/\pi)$.

Let us examine with Taylor series the behavior of $\partial u/\partial x$ in the neighborhood of the singularity. Our expectation is that the slope approaches infinity as the singularity is approached. From Eq. (9.543), we see that

$$\frac{\partial u}{\partial x} = \frac{1}{J}\left(\frac{\partial t}{\partial \tau}\frac{\partial u}{\partial \xi} - \frac{\partial t}{\partial \xi}\frac{\partial u}{\partial \tau}\right). \tag{9.569}$$

For our transformation, we have $\partial t/\partial \tau = 1$, $\partial t/\partial \xi = 0$, so

$$\frac{\partial u}{\partial x} = \frac{1}{J}\frac{\partial u}{\partial \xi}. \tag{9.570}$$

Now use our solution for u, Eq. (9.554), and for J, Eq. (9.556), to say

$$\frac{\partial u}{\partial x} = \frac{1}{1 + \frac{df}{d\xi}\tau}\frac{df}{d\xi}. \tag{9.571}$$

With $f(\xi)$ for our example from Eq. (9.559), we get

$$\frac{\partial u}{\partial x} = \frac{\pi \cos(\pi\xi)}{1 + \pi\tau\cos(\pi\xi)}. \tag{9.572}$$

We use computer algebra to perform a Taylor series expansion of $\partial u/\partial \xi$ about $\xi = 1$, $\tau = 1/\pi$ to find the behavior near the singularity to be

$$\frac{\partial u}{\partial x} = \frac{1}{\tau - \frac{1}{\pi}} + \left(\frac{\pi}{2\left(\tau - \frac{1}{\pi}\right)^2}\right)(\xi - 1)^2 + \cdots. \tag{9.573}$$

By inspection, we thus see that

$$\lim_{\tau \to \frac{1}{\pi}} \frac{\partial u}{\partial x} = -\infty. \tag{9.574}$$

It is necessary for τ to be increasing towards $1/\pi$ for the slope to be negative. This is the physically relevant approach as τ begins at zero and is increasing.

This procedure can be extended to the Euler equations, though it is more complicated. For the Euler equations, Courant and Friedrichs (1976) give some special solutions for rarefactions.

9.12 Viscous Shock Waves

Let us here consider further the effect of viscosity on one-dimensional shock structure. This challenging problem will also serve as a bridge to Chapter 10, which will focus on more commonly considered and simpler one-dimensional viscous flows.

We will first extend the Bateman–Burgers equation to include the effects of viscosity and compare to the results of the previous section. This study of shocks admitted by the inviscid and viscous Bateman–Burgers equations suggests we examine the analogs for the Euler and Navier–Stokes equations. We will extend the inviscid analysis of normal shock waves given in Section 9.6 to account for the effects of diffusion. The inviscid analysis admits infinitely thin shock waves. This is actually an approximation to what is observed in nature where shock waves have small but finite thickness. Within the thin shock wave, significant velocity and temperature gradients exist, rendering the neglected mechanisms of momentum and energy diffusion again relevant. This analysis will represent a case for which advection balances diffusion. Because of the presence of advective acceleration, this flow will include the effects of inertia. It will be seen that, in the limit as the diffusion coefficients approach zero, we recover identical results to the inviscid analysis.

9.12.1 Unsteady Viscous Bateman–Burgers Shock Formation

Our predictions of $u(x, t)$ change when diffusion is introduced. Consider the Bateman–Burgers equation:

$$\frac{\partial u}{\partial t} + u\frac{\partial u}{\partial x} = \nu \frac{\partial^2 u}{\partial x^2}. \tag{9.575}$$

When we discretize the partial differential equations and simulate via standard numerical methods the same problem whose diffusion-free solution is plotted in Fig. 9.42 for which $u(x, 0) = 1 + \sin \pi x$, we obtain the results plotted in Fig. 9.47 for four different values of $\nu = 1/1000$, $1/100$, $1/10$, and 1. While an exact solution to the Bateman–Burgers equation

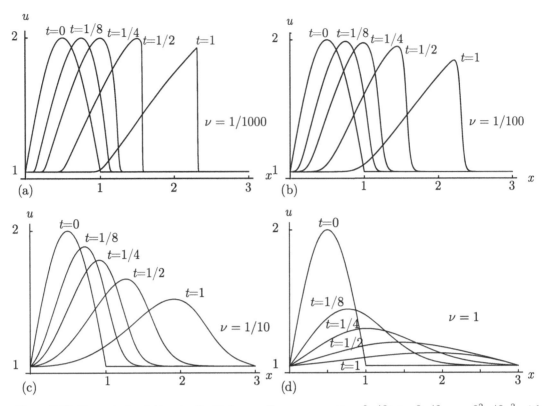

Figure 9.47 Numerical solution to the Bateman–Burgers equation $\partial u/\partial t + u\partial u/\partial x = \nu \partial^2 u/\partial x^2$ with $u(x,0) = 1 + \sin \pi x$ and various values of ν: (a) $1/1000$, (b) $1/100$, (c) $1/10$, (d) 1.

is available, in practice, it is complicated. It is often easier to obtain results by numerical discretization, and that is what we did here. The scheme used was sufficiently resolved to capture the thin zones present when ν was small. For the case in which $\nu = 1/100$, we plot the (x,t) diagram; the shading is proportional to the local value of u, in Fig. 9.48.

We note the following. We restricted our study to positive values of ν, which can be shown to be necessary for a stable solution as $t \to \infty$. If $\nu = 0$, our viscous Bateman–Burgers equation reduces to the inviscid Bateman–Burgers equation. As $\nu \to 0$, solutions to the viscous Bateman–Burgers equation seem to relax to a solution with an infinitely thin discontinuity; they do not relax to those solutions displayed in Fig. 9.42. A common tool to ascertain whether a solution is viable is to require that a viscous calculation of the equivalent case relax to the inviscid solution in the limit as viscosity approaches zero. In that case, it is said that the correct *vanishing viscosity solution* has been identified. LeVeque (1992, 2002) describes these in general terms and how they are a type of *weak solution* to the integral form of the conservation laws. For all values of ν, the solution $u(x,t)$ at a given time has a single value of u for a single value of x, in contrast to multivalued solutions exhibited by the diffusion-free analog. Thus, we conclude the discontinuous solution is the correct vanishing viscosity solution. As $\nu \to 0$, the peaks retain a larger magnitude. Thus, one can conclude that enhancing ν smears peaks.

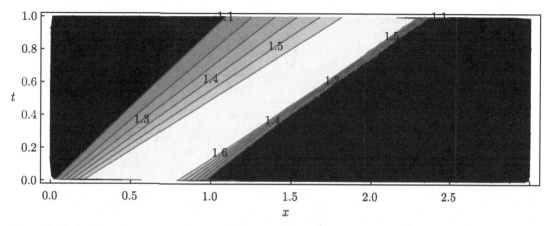

Figure 9.48 (x, t) diagram with contours of u for solution to the Bateman–Burgers equation $\partial u/\partial t + u\partial u/\partial x = \nu \partial^2 u/\partial x^2$ with $u(x,0) = 1 + \sin \pi x$, $\nu = 1/100$.

At early time the solutions to the viscous Bateman–Burgers equation resemble those of the Bateman–Burgers equation.

Let us try to understand this behavior. Fundamentally, it will be seen that in many cases, nonlinearity manifested in $u\partial u/\partial x$ can serve to steepen a waveform. If that steepening is unchecked by diffusion, either a formal discontinuity is admitted, or multivalued solutions. Now diffusion acts most strongly when gradients are steep, that is, when $\partial u/\partial x$ has large magnitude. As a wave steepens due to nonlinear effects, diffusion, which many have been initially unimportant, can reassert its importance and serve to suppress the growth due to the nonlinearity.

9.12.2 Steady Viscous Bateman–Burgers Shocks

Let us now consider a simpler problem for viscous shocks, the viscous Bateman–Burgers equation for a steady propagating wave. Let us examine solutions to Eq. (9.575) that can link a constant state where $u(-\infty, t) = u_1$ to a second constant state where $u(\infty, t) = u_2$. We shall see that this can be achieved by what is known as a steadily propagating wave solution. For such solutions, a waveform is maintained as the wave translates with a given velocity.

We can employ both a coordinate transformation and a change of variables. Let us take as new coordinates

$$\xi = x - Ut, \qquad \tau = t. \tag{9.576}$$

Here U is a constant which we will specify later. From this, we get in matrix form

$$\begin{pmatrix} \xi \\ \tau \end{pmatrix} = \underbrace{\begin{pmatrix} 1 & -U \\ 0 & 1 \end{pmatrix}}_{\mathbf{J}^{-1}} \begin{pmatrix} x \\ t \end{pmatrix}. \tag{9.577}$$

The inverse transformation is

$$\begin{pmatrix} x \\ t \end{pmatrix} = \underbrace{\begin{pmatrix} 1 & U \\ 0 & 1 \end{pmatrix}}_{\mathbf{J}} \begin{pmatrix} \xi \\ \tau \end{pmatrix}. \tag{9.578}$$

Here we have $J = \det \mathbf{J} = 1$. We get

$$\begin{pmatrix} \frac{\partial u}{\partial x} \\ \frac{\partial u}{\partial t} \end{pmatrix} = (\mathbf{J}^T)^{-1} \begin{pmatrix} \frac{\partial u}{\partial \xi} \\ \frac{\partial u}{\partial \tau} \end{pmatrix} = \begin{pmatrix} 1 & 0 \\ -U & 1 \end{pmatrix} \begin{pmatrix} \frac{\partial u}{\partial \xi} \\ \frac{\partial u}{\partial \tau} \end{pmatrix}. \tag{9.579}$$

Thus, we see

$$\frac{\partial}{\partial x} = \frac{\partial}{\partial \xi}, \qquad \frac{\partial}{\partial t} = -U \frac{\partial}{\partial \xi} + \frac{\partial}{\partial \tau}. \tag{9.580}$$

We apply this coordinate transformation to Eq. (9.575) to get

$$\underbrace{-U \frac{\partial u}{\partial \xi} + \frac{\partial u}{\partial \tau}}_{\frac{\partial u}{\partial t}} + \underbrace{u \frac{\partial u}{\partial \xi}}_{u \frac{\partial u}{\partial x}} = \underbrace{\nu \frac{\partial^2 u}{\partial \xi^2}}_{\nu \frac{\partial^2 u}{\partial x^2}}, \quad \text{so} \quad \frac{\partial u}{\partial \tau} + (u - U) \frac{\partial u}{\partial \xi} = \nu \frac{\partial^2 u}{\partial \xi^2}. \tag{9.581}$$

This form suggests we will realize further simplification by defining

$$w = u - U. \tag{9.582}$$

This is a Galilean transformation, with w being the relative velocity. Doing so, we get

$$\frac{\partial}{\partial \tau}(w + U) + w \frac{\partial}{\partial \xi}(w + U) = \nu \frac{\partial^2}{\partial \xi^2}(w + U), \quad \text{so} \quad \frac{\partial w}{\partial \tau} + w \frac{\partial w}{\partial \xi} = \nu \frac{\partial^2 w}{\partial \xi^2}. \tag{9.583}$$

Remarkably, Eq. (9.583) has precisely the same form as Eq. (9.575), with w standing in for u. Leaving aside for now any concerns about initial and boundary conditions, we say that our Galilean transformation, which transforms both the dependent and independent variables, has mapped Eq. (9.575) into itself.

Our boundary conditions transform to

$$w(-\infty, \tau) = u_1 - U \equiv w_1, \quad w(\infty, \tau) = u_2 - U \equiv w_2. \tag{9.584}$$

We shall see there is an additional requirement for U, to be determined.

We trivially note that if we seek solutions that are independent of ξ, Eq. (9.583) reduces to $dw/d\tau = 0$, which gives us $w = C$. The boundary conditions are only satisfied in the special case when $w_1 = w_2$, giving $w = w_1$. This is not particularly useful. We find nontrivial results when we seek solutions that are independent of τ; that is, we seek $w = w(\xi)$. Then Eq. (9.583) reduces to

$$w \frac{dw}{d\xi} = \nu \frac{d^2 w}{d\xi^2}, \quad \text{so} \quad \frac{d}{d\xi}\left(\frac{w^2}{2}\right) = \nu \frac{d^2 w}{d\xi^2}. \tag{9.585}$$

This gives

$$\frac{w^2}{2} + C_1 = \nu \frac{dw}{d\xi}. \tag{9.586}$$

Now as $\xi \to -\infty$, we expect $w \to w_1$ and $dw/d\xi \to 0$. This gives $C_1 = -w_1^2/2$. Thus,

$$\nu \frac{dw}{d\xi} = \frac{1}{2}(w^2 - w_1^2). \tag{9.587}$$

Separating variables and solving, we get

$$w(\xi) = w_1 \tanh\left(-\frac{w_1}{2\nu}\xi + C_2 w_1\right). \tag{9.588}$$

Examination of this solution reveals that $\lim_{\xi \to -\infty} w(\xi) = w_1$ and $\lim_{\xi \to \infty} w(\xi) = -w_1$ and $w(\xi) = 0$ when $\xi = 2\nu C_2$. Let us make the convenient assumption that $C_2 = 0$ to place the arbitrary zero-crossing at $\xi = 0$. Other choices would simply translate the zero-crossing and not otherwise affect the solution. Now we see we have satisfied the boundary condition at $\xi \to -\infty$ but not at $\xi = +\infty$. We can satisfy both boundary conditions at $\pm\infty$ by making the correct choice of the as of yet unspecified wave speed U. We thus would like to choose U such that

$$-w_1 = w_2. \tag{9.589}$$

Using our definitions, Eqs. (9.584), we get

$$-(u_1 - U) = u_2 - U, \quad \text{so} \quad U = \frac{u_1 + u_2}{2}. \tag{9.590}$$

Then

$$w_1 = u_1 - U = \frac{u_1 - u_2}{2}, \quad w_2 = u_2 - U = -\frac{u_1 - u_2}{2} = -w_1. \tag{9.591}$$

Our solution is then

$$w(\xi) = \frac{u_1 - u_2}{2} \tanh\left(-\frac{u_1 - u_2}{4\nu}\xi\right). \tag{9.592}$$

In terms of our untransformed variables, we have

$$u(x,t) = \frac{u_1 + u_2}{2} + \frac{u_1 - u_2}{2} \tanh\left(-\frac{u_1 - u_2}{4\nu}\left(x - \frac{u_1 + u_2}{2}t\right)\right). \tag{9.593}$$

One can confirm by direct calculation that Eq. (9.593) satisfies the viscous Bateman–Burgers equation, Eq. (9.575). Additionally, it satisfies both boundary conditions at $x = \pm\infty$. By inspection of the solution, the thickness ℓ of the zone were u adjusts from u_1 to u_2 is given by

$$\ell = \left|\frac{4\nu}{u_1 - u_2}\right|. \tag{9.594}$$

In the limit as $\nu \to 0$, we see $\ell \to 0$, and $u(x,t)$ suffers a jump from $u = u_1$ to $u = u_2$ at $x = (u_1 + u_2)t/2$.

This is fully consistent with an inviscid jump analysis. We propose the proper conservative form of the Bateman–Burgers equation to be

$$\frac{\partial u}{\partial t} + \frac{\partial}{\partial x}\left(\frac{u^2}{2}\right) = 0. \tag{9.595}$$

Recalling our general jump analysis from Eq. (9.291), we have $q = u$ and $f = u^2/2$ and taking $U = a$, giving

$$a = \frac{[\![f]\!]}{[\![q]\!]} = \frac{\left[\!\left[\frac{u^2}{2}\right]\!\right]}{[\![u]\!]} = \frac{1}{2}\frac{u_2^2 - u_1^2}{u_2 - u_1} = \frac{u_1 + u_2}{2}. \tag{9.596}$$

Because our independent analysis of the viscous Bateman–Burgers equation revealed that the propagation speed is $(u_1 + u_2)/2$, we conclude that the appropriate form of the Bateman–Burgers equation is that of Eq. (9.595), and not one of the many others, for example $\partial/\partial t(u^2/2) + \partial/\partial x(u^3/3) = 0$, which would yield a different propagation speed. Results are plotted in Fig. 9.49 for three different values of $\nu = 1/1000$, $1/100$, and $1/10$ for $u_1 = 3/2$, $u_2 = 1/2$ and $t = 2$. Clearly, all solutions relax at $\pm\infty$ to the correct values of u_1 and u_2. The only effect of ν is the thickness of the zone where u relaxes from u_1 to u_2. Also the propagation speed $a = (u_1 + u_2)/2 = 1$. Because the wave was centered at $x = 0$ at $t = 0$, we see at $t = 2$ its "center" has propagated to $x = 2$.

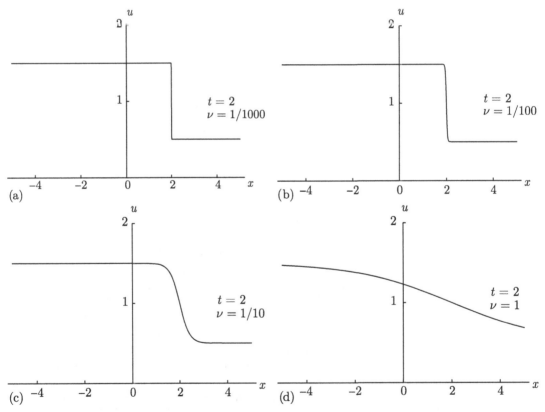

Figure 9.49 Propagating steady wave solution at $t = 2$ to the viscous Bateman–Burgers equation $\partial u/\partial t + u\partial u/\partial x = \nu \partial^2 u/\partial x^2$ with $u(-\infty, t) = u_1 = 3/2$, $u(\infty, t) = u_2 = 1/2$ and various values of ν: (a) $1/1000$, (b) $1/100$, (c) $1/10$, (d) 1.

9.12.3 Steady Navier–Stokes Shocks

We now consider a specific problem for which we can find an exact algebraic solution for a one-dimensional steady viscous shock propagating in a calorically perfect ideal gas with constant dynamic viscosity and Prandtl number of 3/4. It has been known for over a century, see Becker (1922) or Taylor (1910), but has often been overlooked in recent decades. One of the more lucid discussions is given by Morduchow and Libby (1949). Other relevant studies from this era include those of Gilbarg and Paolucci (1953), Ludford (1951), and von Mises (1950). A convincing argument for employing the continuum limit for this problem along with a presentation of the sub-continuum theory is given by Liepmann and Roshko (2002, Chapter 14) and Vincenti and Kruger (1965, pp. 412–415). In short, the viscous shock has a thickness that is approaching a few mean free paths. This is at the frontier near where continuum approximation fails. And it gives a remarkably successful prediction of the experimentally observed shock thickness as described in detail by Liepmann and Roshko (2002, Chapter 13) or Müller and Ruggeri (1998, pp. 277–308).

Here we derive a solution for the structure of a steady viscous shock propagating in a calorically perfect ideal gas with constant viscosity and constant Prandtl number $Pr = 3/4$. There are some more modern discussions of this problem, for example Bird et al. (2007, pp. 350–353), Iannelli (2013), and the one from which we draw extensively here, Ghia et al. (2021). The governing equations are best presented in a mixed conservative and nonconservative form:

$$\frac{\partial \rho}{\partial t} + \frac{\partial}{\partial x}(\rho u) = 0, \tag{9.597}$$

$$\rho \frac{\partial u}{\partial t} + \rho u \frac{\partial u}{\partial x} + \frac{\partial}{\partial x}(p - \tau) = 0, \tag{9.598}$$

$$\frac{\partial}{\partial t}\left(\rho\left(e + \frac{u^2}{2}\right)\right) + \frac{\partial}{\partial x}\left(\rho u \left(e + \frac{u^2}{2} + \frac{p}{\rho}\right) - (\tau u - q)\right) = 0, \tag{9.599}$$

$$p = \rho R T, \qquad e = c_v T, \tag{9.600}$$

$$q = -k \frac{\partial T}{\partial x}, \qquad \tau = \frac{4}{3} \mu \frac{\partial u}{\partial x}. \tag{9.601}$$

Due to the presence of diffusive transport terms, it is not necessary for the system to be in either conservative or nonconservative form. So for convenience, we have used the conservative forms of the mass and energy equations, Eqs. (9.597, 9.599), and the nonconservative form of linear momentum evolution, Eq. (9.598). The parameters μ, k, R, and c_v are taken to be constant. It is possible to relax these assumptions to include temperature-dependent properties. Secondary constant parameters are the specific heat at constant pressure $c_p = c_v + R$, the ratio of specific heats $\gamma = c_p/c_v$, and the Prandtl number $Pr = \mu c_p/k$. We could, as done in Section 9.6, use a Galilean transformation to better understand the nature of a propagating wave. Instead, we can learn nearly as much by simply taking the steady state limit of Eqs. (9.597–9.599), after taking the enthalpy, $h = e + p/\rho = c_p T$, to give

$$\frac{d}{dx}(\rho u) = 0, \tag{9.602}$$

$$\rho u \frac{du}{dx} + \frac{dp}{dx} - \frac{d\tau}{dx} = 0, \tag{9.603}$$

$$\frac{d}{dx}\left(\rho u\left(h+\frac{u^2}{2}\right)\right) - \frac{d}{dx}(\tau u - q) = 0. \tag{9.604}$$

We specify the upstream unshocked condition as $x \to -\infty$ to be

$$p \to p_1, \quad T \to T_1, \quad u \to u_1, \quad q \to 0, \quad \tau \to 0. \tag{9.605}$$

As a consequence, we get as $x \to -\infty$ that $\rho_1 = p_1/R/T_1$, $h_1 = c_p T_1$. We have that $\rho_1, T_1, h_1, u_1, p_1$ are all positive constants. All variables are required to be bounded for $x \in (-\infty, \infty)$.

Our energy equation, Eq. (9.604), is rearranged using Eqs. (9.601), and the definition of Pr in the following steps:

$$\frac{d}{dx}\left(\rho u\left(h+\frac{u^2}{2}\right)\right) - \frac{d}{dx}\bigg(\underbrace{\frac{4}{3}\mu\frac{du}{dx}}_{\tau}u + \underbrace{k\frac{dT}{dx}}_{-q}\bigg) = 0, \tag{9.606}$$

$$\frac{d}{dx}\left(\rho u\left(h+\frac{u^2}{2}\right)\right) - \frac{d}{dx}\left(\frac{4}{3}\mu\frac{du}{dx}u + \frac{\mu c_p}{Pr}\frac{dT}{dx}\right) = 0. \tag{9.607}$$

Now take $Pr = 3/4$, utilize $h = c_p T$, and bring u within the derivative operator, so to obtain

$$\frac{d}{dx}\left(\rho u\left(h+\frac{u^2}{2}\right)\right) - \frac{4}{3}\mu\frac{d}{dx}\left(\frac{d}{dx}\left(\frac{u^2}{2}\right) + \frac{dh}{dx}\right) = 0. \tag{9.608}$$

The assumption of $Pr = 3/4$ is not unreasonable for many gases; importantly, it allows us to obtain a closed form solution. One can consider other values of Pr, at the expense of requiring some additional numerical analysis. Now if we integrate the mass equation, Eq. (9.602), and apply the boundary conditions, we get

$$\rho u = \rho_1 u_1. \tag{9.609}$$

Next use Eq. (9.609) and the assumption of constant μ to simplify Eq. (9.608) to get

$$\rho_1 u_1 \frac{d}{dx}\left(h+\frac{u^2}{2}\right) - \frac{4}{3}\mu\frac{d^2}{dx^2}\left(h+\frac{u^2}{2}\right) = 0. \tag{9.610}$$

Integrate to get

$$\rho_1 u_1 \left(h+\frac{u^2}{2}\right) - \frac{4}{3}\mu\frac{d}{dx}\left(h+\frac{u^2}{2}\right) = C_1. \tag{9.611}$$

This can be rearranged to find

$$\frac{d}{dx}\left(h+\frac{u^2}{2}\right) = -\frac{3C_1}{4\mu} + \frac{3\rho_1 u_1}{4\mu}\left(h+\frac{u^2}{2}\right). \tag{9.612}$$

This is a linear first-order ordinary differential equation whose solution is

$$h+\frac{u^2}{2} = \frac{C_1}{\rho_1 u_1} + C_2 \exp\left(\frac{3\rho_1 u_1 x}{4\mu}\right). \tag{9.613}$$

To suppress unbounded growth as $x \to \infty$, one can take $C_2 = 0$. To satisfy the condition at $x \to -\infty$, one can take $C_1 = \rho_1 u_1(h_1 + u_1^2/2)$, yielding

$$h+\frac{u^2}{2} = h_1 + \frac{u_1^2}{2}. \tag{9.614}$$

Equation (9.614) is a result for viscous, heat conducting fluids, restricted to $Pr = 3/4$. Remarkably it is equivalent to two results obtained in the inviscid limit: (1) the steady, homeoentropic, irrotational, zero body-force version of the traditional Crocco's equation, Eq. (6.226), as well as (2) the Rankine–Hugoniot energy jump condition, Eq. (9.301). Were the Prandtl number to have a different value, energy and momentum would diffuse at different rates, and a more general result would have been obtained.

We can then apply the caloric state equation, Eq. (9.600), to Eq. (9.614) to give

$$c_p T + \frac{u^2}{2} = c_p T_1 + \frac{u_1^2}{2}, \quad \text{so} \quad T = T_1 + \frac{u_1^2 - u^2}{2c_p}. \tag{9.615}$$

The integrated mass equation, Eq. (9.609), can be solved for ρ to give

$$\rho = \frac{\rho_1 u_1}{u}. \tag{9.616}$$

The linear momentum equation, Eq. (9.603), can be recast as

$$\underbrace{\rho_1 u_1}_{\rho u} \frac{du}{dx} + \frac{dp}{dx} - \underbrace{\frac{d}{dx}\left(\frac{4}{3}\mu \frac{du}{dx}\right)}_{\tau} = 0, \quad \text{so} \quad \frac{d}{dx}\left(\rho_1 u_1 u + p - \frac{4}{3}\mu \frac{du}{dx}\right) = 0. \tag{9.617}$$

We integrate this and apply boundary conditions at $x \to -\infty$ to get

$$\rho_1 u_1 u + p - \frac{4}{3}\mu \frac{du}{dx} = \rho_1 u_1^2 + p_1. \tag{9.618}$$

We may then use the ideal gas law, Eq. (9.600), along with Eqs. (9.615, 9.616) to arrive at

$$\rho_1 u_1 u + \underbrace{\underbrace{\left(\frac{\rho_1 u_1}{u}\right)}_{\rho} R \underbrace{\left(T_1 + \frac{u_1^2 - u^2}{2c_p}\right)}_{T}}_{p} - \frac{4}{3}\mu \frac{du}{dx} = \rho_1 u_1^2 + p_1. \tag{9.619}$$

With some algebra, we can solve for du/dx to rewrite the linear momentum equation as

$$\frac{du}{dx} = -\frac{3(\gamma+1)\rho_1 u_1}{8\gamma\mu} \frac{(u_1 - u)(u - u_2)}{u}. \tag{9.620}$$

In this expression u_2 is the fully shocked value of u. It is

$$u_2 = u_1 \frac{\gamma-1}{\gamma+1}\left(1 + \frac{2}{(\gamma-1)M_1^2}\right). \tag{9.621}$$

Here, the freestream Mach number M_1 is given by $M_1^2 = u_1^2/(\gamma R T_1)$. More generally, the Mach number M is given by $M = u/\sqrt{\gamma R T}$. The solution for u_2 is identical to that obtained by solving the inviscid Rankine–Hugoniot jump equations. Because $\gamma > 1$, we have that $u_1 > u_2$. Within the shock we will expect that $u_1 > u(x) > u_2 > 0$. We also have $\rho_1 > 0$, $\mu > 0$. Because of these inequalities, we can expect from Eq. (9.620) that $du/dx < 0$. Note that $du/dx = 0$ if either $u = u_1$ or $u = u_2$, so the end states from analysis of Rankine–Hugoniot jump equations are equilibrium states.

We can rearrange Eq. (9.620) to get

$$\frac{\mu u}{(u_1 - u)(u - u_2)} \frac{du}{dx} = -\frac{3(\gamma + 1)\rho_1 u_1}{8\gamma}. \tag{9.622}$$

Integrating, we find

$$\int \frac{\mu u \, du}{(u_1 - u)(u - u_2)} = -\frac{3(\gamma + 1)\rho_1 u_1}{8\gamma}(x + C), \tag{9.623}$$

where C is a constant. This solution is invariant under translation in distance. Although we could apply boundary conditions at infinity, this poses numerical issues when one tries to represent $u(x)$. To alleviate these, we can select an initial condition as follows:

$$u(x_{in}) = u_{in}. \tag{9.624}$$

The value x_{in} can be selected arbitrarily, and it will be useful to choose u_{in} to be near to the unshocked value u_1. With this, we find that

$$\frac{\mu u_2}{u_1 - u_2} \ln\left(\frac{u - u_2}{u_{in} - u_2}\right) - \frac{\mu u_1}{u_1 - u_2} \ln\left(\frac{u_1 - u}{u_1 - u_{in}}\right) = -\frac{3(\gamma + 1)\rho_1 u_1}{8\gamma}(x - x_{in}) \tag{9.625}$$

satisfies the differential equation and boundary condition. We rearrange Eq. (9.625) to get $x(u)$ as

$$x(u) = x_{in} - \frac{8\gamma\mu}{3(\gamma + 1)\rho_1 u_1}\left(\frac{u_2}{u_1 - u_2} \ln\left(\frac{u - u_2}{u_{in} - u_2}\right) - \frac{u_1}{u_1 - u_2} \ln\left(\frac{u_1 - u}{u_1 - u_{in}}\right)\right), \tag{9.626}$$

$$= x_{in} - \frac{8\gamma\mu}{3(\gamma + 1)\rho_1(u_1 - u_2)}\left(\frac{u_2}{u_1} \ln\left(\frac{u - u_2}{u_{in} - u_2}\right) - \ln\left(\frac{u_1 - u}{u_1 - u_{in}}\right)\right), \tag{9.627}$$

$$= x_{in} - \frac{8\gamma\mu}{3(\gamma + 1)\rho_1(u_1 - u_2)} \ln\left(\frac{\left(\frac{u - u_2}{u_{in} - u_2}\right)^{\frac{u_2}{u_1}}}{\left(\frac{u_1 - u}{u_1 - u_{in}}\right)}\right). \tag{9.628}$$

We get an estimate of the shock thickness by the formula

$$\ell_{shock} \sim \frac{8\gamma\mu}{3(\gamma + 1)\rho_1(u_1 - u_2)}. \tag{9.629}$$

We can thus relate the shock thickness to the mean free path λ using Eq. (1.17) and taking $\rho \sim \rho_1$ to get

$$\ell_{shock} \sim \frac{8\gamma}{3(\gamma + 1)(u_1 - u_2)} \sqrt{\frac{\Re T}{3\mathcal{M}}} \lambda. \tag{9.630}$$

This is of course a crude estimate. If we interpret T as a mean temperature, $\sqrt{\Re T/3/\mathcal{M}}$ is of the same order of magnitude as the sound speed. And $u_1 - u_2$ is of the order of magnitude of the sound speed. And we know $\gamma = \mathcal{O}(1)$. So the shock thickness ℓ_{shock} is the same order of magnitude as the mean free path λ.

Equation (9.628) takes the form $x(u) = f(u)$. It is possible to invert f via numerical root-finding methods to form its inverse, f^{-1}, so $u(x) = f^{-1}(x)$. It is possible to use Eqs. (9.616, 9.615, 9.600) to get the density, temperature, pressure, and Mach number:

$$\rho(x) = \frac{\rho_1 u_1}{u(x)}, \tag{9.631}$$

$$T(x) = T_1 + \frac{u_1^2 - u(x)^2}{2c_p}, \tag{9.632}$$

$$p(x) = \underbrace{\frac{\rho_1 u_1}{u(x)}}_{\rho(x)} R \underbrace{\left(T_1 + \frac{u_1^2 - u(x)^2}{2c_p} \right)}_{T(x)}, \tag{9.633}$$

$$M(x) = \frac{u(x)}{\sqrt{\gamma R T(x)}}. \tag{9.634}$$

We can use Eq. (9.333) to get an equation for the entropy:

$$\frac{s(x) - s_1}{c_v} = \ln\left(\left(\frac{\rho_1}{\rho(x)}\right)^\gamma \frac{p(x)}{p_1} \right). \tag{9.635}$$

We give a shock structure for parameters that well approximate air: $u_1 = 905 \, \text{m/s}$, $p_1 = 10^5 \, \text{Pa}$, $T_1 = 300 \, \text{K}$, $R = 287 \, \text{J/kg/K}$, $c_p = 1004.5 \, \text{J/kg/K}$, $\mu = 19 \times 10^{-6} \, \text{Pa s}$, $x_{in} = 0 \, \text{m}$, $u_{in} = 904 \, \text{m/s}$. With these parameters, we get $\rho_1 = 1.16144 \, \text{kg/m}^3$, $\gamma = 7/5$, $M_1 = 2.60665$, and find from Eq. (9.621) that $u_2 = 261.828 \, \text{m/s}$. With these numbers, we get from Eq. (9.630) an estimate of the ratio of shock thickness to mean free path of

$$\frac{\ell_{shock}}{\lambda} \sim \frac{8\gamma}{3(\gamma+1)(u_1 - u_2)} \sqrt{\frac{\Re T}{3\mathcal{M}}}, \tag{9.636}$$

$$= \frac{8\left(\frac{7}{5}\right)}{3\left(\frac{7}{5}+1\right)\left((905\,\frac{\text{m}}{\text{s}}) - (281.828\,\frac{\text{m}}{\text{s}})\right)} \sqrt{\frac{(287\,\frac{\text{J}}{\text{kg K}})(300\,\text{K})}{3}} = 0.42. \tag{9.637}$$

Substituting the same numbers into Eq. (9.628), we get for $x(u)$

$$x(u) = -\left(3.95654 \times 10^{-8}\,\text{m}\right) \ln\left(\frac{\left(\frac{u - (261.828\,\text{m/s})}{(904\,\text{m/s}) - (261.828\,\text{m/s})}\right)^{0.289312}}{\left(\frac{(905\,\text{m/s}) - u}{(905\,\text{m/s}) - (904\,\text{m/s})}\right)} \right). \tag{9.638}$$

The shock thickness of $0.0395654\,\mu\text{m}$ is about half a mean free path. The mean free path length, as estimated by Eq. (1.14) with $\rho = \rho_1 = 1.16144\,\text{kg/m}^3$ is

$$\lambda = \frac{7.8895 \times 10^{-8}\,\frac{\text{kg}}{\text{m}^2}}{1.16144\,\frac{\text{kg}}{\text{m}^3}} = 6.792 \times 10^{-8}\,\text{m} = 0.06792\,\mu\text{m}. \tag{9.639}$$

Analytic inversion of Eq. (9.638) is impossible, but numerical root finding allows one to obtain $u(x)$. With $u(x)$, one can use Eqs. (9.631–9.634) to get density $\rho(x)$, temperature $T(x)$, pressure $p(x)$, and Mach number $M(x)$. Results for $u(x)$ and $\rho(x)$ are obtained from Eqs. (9.638, 9.631) and are given in Fig. 9.50. As $x \to -\infty$, the flow is unshocked. In the laboratory frame, we can think of the shock as propagating leftward. The predictions of the algebraic

Table 9.2 Steady viscous shock structure for calorically perfect ideal air

x (m)	u (m/s)	ρ (kg/m^3)	T (K)	p (Pa)	M
$-\infty$	905.000	1.161 44	300.000	100 000	2.606 65
0.0×10^{-7}	904.000	1.162 72	300.900	100 411	2.599 87
0.5×10^{-7}	901.465	1.165 99	303.178	101 456	2.582 83
1.0×10^{-7}	892.543	1.177 65	311.146	105 163	2.524 31
1.5×10^{-7}	861.558	1.220 00	338.199	118 417	2.337 19
2.0×10^{-7}	759.364	1.384 19	420.653	167 110	1.847 07
2.5×10^{-7}	492.440	2.134 48	586.973	359 577	1.014 00
3.0×10^{-7}	274.491	3.829 28	670.174	736 523	0.528 97
3.5×10^{-7}	262.000	4.011 85	673.510	775 480	0.503 64
4.0×10^{-7}	261.830	4.014 45	673.554	776 033	0.503 30
$+\infty$	261.828	4.014 48	673.555	776 040	0.503 30

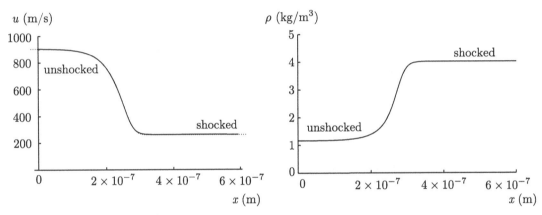

Figure 9.50 Viscous shock structure in air for $u(x)$, $\rho(x)$. The results plotted were predicted using three methods: (1) an exact algebraic solution, (2) a numerical solution of the corresponding steady ordinary differential equations, and (3) numerical solution of the corresponding unsteady partial differential equations relaxed to a time-independent limit. The same results were obtained for each method.

exact solution, Eq. (9.638), are indistinguishable from those obtained by two independent discrete numerical methods: numerical integration of the ordinary differential equation of Eq. (9.620) with $u(0) = 904$ m/s, and numerical integration of the unsteady partial differential equations of Eqs. (9.597–9.601) to sufficiently long time for transients to have relaxed with a standard numerical finite difference strategy. Because this is a useful problem for verification of computational fluid dynamics algorithms, in Table 9.2 we give numerical values for the exact viscous shock structure as reported in Fig. 9.50 for the calorically perfect ideal gas for the parameters chosen. Visual examination of the shock structure of Fig. 9.50 suggests a nearly complete relaxation over a 0.1 μm distance.

We plot the scaled entropy in Fig. 9.51. The unshocked gas is in a low entropy state. As fluid particle encounters the shock, its entropy rises, until it reaches a maximum, then falls. Clearly the shocked entropy is higher than the unshocked entropy. And the far field shocked entropy is

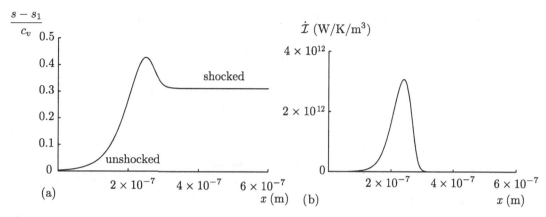

Figure 9.51 Viscous shock structure for (a) scaled entropy $(s(x) - s_1)/c_v$, and (b) irreversibility production rate $\dot{\mathcal{I}}(x)$.

precisely the value one would obtain in an inviscid analysis using the Rankine–Hugoniot jump equations. *Even though the entropy attains a local maximum within the viscous shock structure, the second law is not violated.* It may be recalled that the second law only requires $ds \geq dq/T$, and dq may be positive or negative. Because there are temperature gradients in the flow, a fluid particle may have $ds < 0$ if $dq < 0$. The region of the flow for which the entropy is falling may be attributed to heat transfer. We can confirm the second law is satisfied by examining the irreversibility production rate, $\dot{\mathcal{I}}$. In the one-dimensional limit, Eq. (5.171) reduces to

$$\dot{\mathcal{I}} = \frac{k}{T^2}\left(\frac{\partial T}{\partial x}\right)^2 + \frac{4\mu}{3T}\left(\frac{\partial u}{\partial x}\right)^2 = \frac{\mu}{T^2}\left(\underbrace{\frac{k}{\mu c_p}}_{1/Pr = 4/3} c_p \left(\frac{\partial T}{\partial x}\right)^2 + \frac{4T}{3}\left(\frac{\partial u}{\partial x}\right)^2\right), \quad (9.640)$$

$$= \frac{4\mu}{3T^2}\left(c_p \left(\frac{\partial T}{\partial x}\right)^2 + T\left(\frac{\partial u}{\partial x}\right)^2\right). \quad (9.641)$$

It is seen in Fig. 9.51(b) that $\dot{\mathcal{I}}$ is positive semi-definite and that it has a large magnitude within the viscous structure of the shock where gradients of u and T are steep.

9.13 Discontinuous Rarefactions for Nonideal Gases

The behavior of discontinuities in inviscid nonideal gases is often similar to those in ideal gases, with small corrections. However, for some gases in the vicinity of the thermodynamic vapor dome, one can find atypical behavior. In ideal gases, compression shock discontinuities may form by the following process. A driving piston begins to move and induces a small acoustic disturbance, traveling at the ambient speed of sound. This small acoustic wave slightly heats the material due to adiabatic compression, and the next compression wave travels slightly faster due to the temperature-sensitivity of the sound speed. It catches and strengthens the lead wave.

Subsequent waves do the same. The compression waves accumulate and can form a shock, which is discontinuous in the inviscid limit. For the same reason, for ideal gases, rarefaction waves cool the gas, and subsequent rarefactions travel slower than the lead rarefaction. Thus, for ideal gases, rarefactions disperse and are continuous. We have seen in Section 9.6.6 that compression shocks in ideal gases increase the entropy, and we have seen that rarefactions in ideal gases are isentropic so there is no violation of the second law of thermodynamics.

However, it is possible for a nonideal gas to exist in a region of thermodynamic space for which the sound speed decreases as the gas is adiabatically compressed. In such a region, compression waves will disperse and form a continuous isentropic wave, and rarefaction waves will accumulate and form a discontinuity in an inviscid gas. Neither of these violate the entropy inequality. Thus, it must be understood that the common notion that discontinuous rarefactions violate the second law of thermodynamics is typically based on the assumption of an ideal gas. We will demonstrate for a material with a common nonideal equation of state, a van der Waals gas, that rarefaction discontinuities may be admitted and satisfy the second law of thermodynamics. Further background is given by Emanuel (2016, pp. 49–50), Colonna and Guardone (2006) or Thompson (1971).

Example 9.21 Consider the fluorocarbon gas $C_{13}F_{22}$, known as PP10. It is well modeled as a van der Waals gas near the vapor dome with $R = 14.4843 \text{ J/kg/K}$, $c_v = 1131.588 \text{ J/kg/K}$, $a = 22.3889 \text{ m}^5/\text{kg/s}^2$, $b = 0.0007244 \text{ m}^3/\text{kg}$. Using the procedure of Section 9.6.5, evaluate the state that arises through a discontinuity for $\hat{v}_1 = 0.002700 \text{ m}^3/\text{kg}$, $T_1 = 656.225 \text{ K}$, $U = 37.89 \text{ m/s}$. The reference values, which will not affect anything of physical importance, can be taken as $e_o = 0 \text{ J/kg}$, $s_o = 0 \text{ J/kg/K}$, $\hat{v}_o = 1 \text{ m}^3/\text{kg}$, $T_o = 300 \text{ K}$. If the flow included chemical reaction, the reference values would play a role in determining physical quantities.

Solution

The numbers for this problem have been chosen carefully so as to demonstrate the unusual behavior that is possible for nonideal gases. It turns out the wave speed U is just above the ambient sound speed, which can be shown to be 37.3659 m/s, so the wave is weakly supersonic. We use the various state equations, Eqs. (9.26, 9.28), to evaluate thermodynamic properties at the undisturbed state, giving $\rho_1 = 370.370 \text{ kg/m}^3$, $p_1 = 1.7400 \times 10^6 \text{ Pa}$, and $h_1 = 3.99529 \times 10^5 \text{ J/kg/K}$. Notice that this is a dense gas. We have taken c_v be constant; however, it would be straightforward to include a dependency on temperature.

The Rankine–Hugoniot jump equations, Eqs. (9.299–9.302), after expanding the state equations, become

$$\rho_2 u_2 = -\rho_1 U, \tag{9.642}$$

$$\rho_2 u_2^2 + p_2 = \rho_1 U^2 + p_1, \tag{9.643}$$

$$h_2 + \frac{u_2^2}{2} = h_1 + \frac{U^2}{2}, \tag{9.644}$$

$$h_2 = e_o + c_v(T_2 - T_o) - a\left(\frac{1}{\hat{v}_2} - \frac{1}{\hat{v}_o}\right) + p_2 \hat{v}_2, \tag{9.645}$$

$$p_2 = \frac{RT_2}{\hat{v}_2 - b} - \frac{a}{\hat{v}_2^2}, \tag{9.646}$$

$$\hat{v}_2 = \frac{1}{\rho_2}. \tag{9.647}$$

9.13 Discontinuous Rarefactions for Nonideal Gases

Table 9.3 Solutions to the jump equations for a van der Waals gas

Root	ρ_2 (kg/m^3)	u_2 (m/s)	p_2 (Pa)	h_2 (J/kg)	T_2 (K)	$\hat{v}_2$ (m^3/kg)	Δs (J/kg/K)
1	370.370	−37.8900	1.7400×10^6	399 529	656.225	0.002 700 0	0
2	299.464	−46.8615	1.6141×10^6	399 149	653.875	0.003 339 3	0.000 493 251
3	445.797	−31.4792	1.82997×10^6	399 751	658.438	0.002 243 17	0.000 642 36
4	247.151	−56.778 04	1.4749×10^6	398 634	651.875	0.004 046 11	−0.001 375 9

These represent six equations for the six unknowns, $\rho_2, u_2, p_2, h_2, T_2, \hat{v}_2$. The undisturbed 1 state and U are fully specified.

The equations are solved, and four distinct roots are found; all are listed in Table 9.3. Root 1 is obviously the ambient state, which must be a solution of the Rankine–Hugoniot jump equations. Root 2 is a rarefaction in that the density, pressure, enthalpy, and temperature decrease. It is a candidate for a physical solution because *this rarefaction discontinuity does not violate the second law of thermodynamics as the entropy increases for this adiabatic jump*. Root 3 has a density, pressure, enthalpy, and temperature increase. It too is a candidate for a physical solution, as the entropy increases. Root 4 is a nonphysical rarefaction. With an entropy decrease, it violates the second law of thermodynamics.

The jump analysis has revealed two candidate nonambient solutions. Both satisfy all axioms of the mechanics of fluids. However, only one is the correct vanishing viscosity solution, which requires a viscous calculation to ascertain. For this case, such a calculation was performed, and it was found that Root 2, the rarefaction discontinuity, was the appropriate vanishing viscosity solution to the Navier–Stokes equations. The simulation was for a shock tube with two chambers, both with fluid initially at rest, and the left chamber at $p = 1.74 \times 10^6$ Pa, $\hat{v} = 0.0027$ m^3/kg, and the right chamber at $p = 1.5 \times 10^6$ Pa, $\hat{v} = 0.0035$ m^3/kg. Values of $\mu = 11.0 \times 10^{-6}$ Pa s and $k = 12.61 \times 10^{-3}$ W/m/K were used. The Navier–Stokes equations were discretized with a second-order central differencing scheme and a first-order Euler time-advance method was used. To resolve the thin viscous structures, fine resolution was necessary. A spatially uniform grid with $\Delta x = 1 \times 10^{-9}$ m was employed, and was verified to be sufficiently small that the results were insensitive to the grid. Indeed, this is likely smaller than the mean free path, but once the continuum assumption has been invoked, there is no lower bound for discretization size. However, one should not expect predictions at length scales below the mean free path to have physical relevance.

Results are shown in Fig. 9.52. In Fig. 9.52(a), one sees a disperse continuous compression fan propagating to the right and a rarefaction "shock," slightly thickened by viscosity, propagating to the left. The end state pressures for the rarefaction are identical to those predicted by the inviscid Rankine–Hugoniot analysis. In Fig. 9.52(b), one sees the second law is satisfied, as the irreversibility production rate $\dot{\mathcal{I}}$ is positive semi-definite. The bulk of irreversibility production is near the contact region, which does not appear in the pressure plot. In contrast to an ideal equation of state, there is more irreversibility production in the rarefaction than the compression. Had the calculation been done in the inviscid limit, there would have been no irreversibility production associated with the continuous compression fan.

The unusual behavior may be understood by considering the behavior of the sound speed for the nonideal gas. First, recall from Eq. (9.72) that the sound speed increases when the

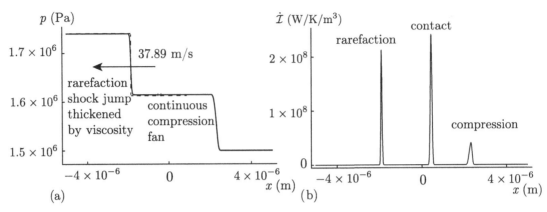

Figure 9.52 Results for an unsteady Navier–Stokes numerical simulation of a shock tube filled with a viscous van der Waals gas at $t = 50$ ns: (a) pressure profile, (b) irreversibility production rate profile. Calculations by A. M. Davies, University of Notre Dame, with analysis aided by K. R. Pielemeier, University of Notre Dame.

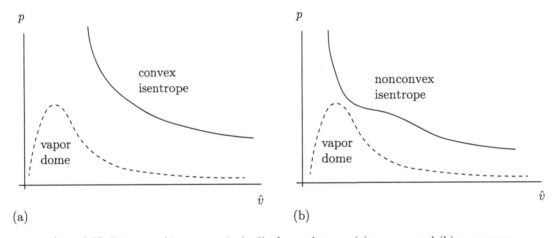

Figure 9.53 Diagram of isentropes in $(p, \hat{v})$ planes that are (a) convex, and (b) nonconvex.

magnitude of the slope of the isentrope $\partial p/\partial \hat{v}|_s$ increases. As most known materials have $\partial p/\partial \hat{v}|_s < 0$ and $\hat{v} > 0$, c^2 is guaranteed positive. Consider now Fig. 9.53, which displays two isentropes in the $(p, \hat{v})$ plane along with a vapor dome. For Fig. 9.53(a), as $\hat{v}$ increases on an isentrope far from the vapor dome, p decreases. Additionally, the magnitude of the slope $\partial p/\partial \hat{v}|_s$ decreases, inducing the sound speed to decrease with expansion. For this curve, we have $\partial^2 p/\partial \hat{v}^2|_s > 0$. That is to say, as $\hat{v}$ increases, the slope is becoming less negative, so the slope of the slope is positive. Curves with $\partial^2 p/\partial \hat{v}^2|_s \geq 0$ are known as *convex*.

Closer to the vapor dome, intermolecular forces can warp the isentrope as shown in Fig. 9.53(b). Thus, there are regions where expansion induces an increase in magnitude of the slope $\partial p/\partial \hat{v}|_s$ and thus an increase in the sound speed. For such regions, one has $\partial^2 p/\partial \hat{v}^2|_s < 0$, and isentropes there are described as *nonconvex*. Thompson (1971) encapsulated this with a dimensionless parameter called $\mathcal{G}$:

$$\mathcal{G} = \frac{\hat{v}^3}{2c^2} \left.\frac{\partial^2 p}{\partial \hat{v}^2}\right|_s. \tag{9.648}$$

For convexity, we require $\mathcal{G} \geq 0$. If $\mathcal{G} < 0$, the isentrope is nonconvex. Detailed calculation reveals that for our van der Waals gas

$$\mathcal{G}(\hat{v}, s) = \frac{\hat{v}^3}{2c^2}\left(-\frac{6a}{\hat{v}^4} + \frac{R T_o}{c_v}\left(\frac{R}{c_v} + 2\right)\frac{R + c_v}{(\hat{v} - b)^3}\exp\left(\frac{s - s_o}{c_v}\right)\left(\frac{\hat{v} - b}{\hat{v}_o - b}\right)^{-\frac{R}{c_v}}\right). \tag{9.649}$$

For a calorically perfect ideal gas, we have $a = b = 0$, and we find

$$\mathcal{G} = 1 + \frac{R}{2c_v} = \frac{\gamma + 1}{2} > 0, \tag{9.650}$$

where $\gamma = 1 + R/c_v = c_p/c_v$. So for an ideal gas, convexity is guaranteed, and one cannot find physically viable discontinuous rarefactions. In contrast, for the parameters studied in the van der Waals example that yielded a physically viable discontinuous rarefaction, one finds at the ambient state that $\mathcal{G} = -0.104\,354$, thus rendering it nonconvex, and thus allowing expansion waves to catch and reinforce each other so as to form a nearly discontinuous jump.

9.14 Taylor–Sedov Blast Waves

We close this chapter by considering an important unsteady inviscid problem: a blast wave that results from a packet of energy that is initially infinitely concentrated. We will study the Taylor–Sedov[16] blast wave solution. We will follow most closely two papers of Taylor (1950a, 1950b). Sedov's (1946) complementary study is also of interest. We shall follow Taylor's analysis and obtain what is known as *self-similar solutions*. There are more general approaches which may in fact expose more details of how self-similar solutions are obtained; we will consider these later in Section 10.2.1. Here, we will confine ourselves to Taylor's approach and use his notation.

The self-similar solution will be enabled by studying the equations for the motion of a diffusion-free ideal compressible fluid in what is known as the *strong shock limit* for a spherical shock wave. Now, a shock wave will raise both the internal and kinetic energy of the ambient fluid into which it is propagating. We would like to consider a scenario in which the total energy, kinetic and internal, enclosed by the strong spherical shock wave is a constant. The ambient fluid, a calorically perfect ideal gas with gas constant R (with special notation employed here to distinguish the gas constant R from the shock radius R) and ratio of specific heats γ, is initially at rest, and a point source of energy, E, exists at $r = 0$. For $t > 0$, this point source of energy is distributed to the mechanical and thermal energy of the surrounding fluid.

Let us follow now Taylor's (1950a) analysis. We shall write the governing inert, inviscid one-dimensional unsteady equations in spherical coordinates. Then we reduce the partial differential equations in r and t to ordinary differential equations in an appropriate similarity variable. Next, we solve the ordinary differential equations numerically and show our transformation

---

[16] Leonid Ivanovitch Sedov, 1907–1999, Soviet physicist.

guarantees constant total energy in the region $r \in (0, R(t)]$, where $R(t)$ is the locus of the moving shock wave. We shall also refer to specific equations given by Taylor (1950a).

9.14.1 Governing Equations

We rewrite the nonconservative form of the Euler equations, Eqs. (6.119–6.122):

$$\frac{d\rho}{dt} + \rho \nabla^T \cdot \mathbf{v} = 0, \tag{9.651}$$

$$\rho \frac{d\mathbf{v}}{dt} + \nabla p = \mathbf{0}, \tag{9.652}$$

$$\rho \frac{de}{dt} + p \nabla^T \cdot \mathbf{v} = 0, \tag{9.653}$$

$$p = p(\rho, T), \qquad e = e(\rho, T). \tag{9.654}$$

Now $\mathbf{v} = v_r \mathbf{e}_r + v_\theta \mathbf{e}_\theta + v_\phi \mathbf{e}_\phi$. Let us confine attention to the one-dimensional limit in which $v_\theta = v_\phi = 0$ and $v_r = v_r(r, t)$. The gradient operator, Eq. (2.380), in this limit gives

$$\nabla = \frac{\partial}{\partial r} \mathbf{e}_r. \tag{9.655}$$

The material derivative reduces as follows:

$$\frac{d}{dt} = \frac{\partial}{\partial t} + v_r \frac{\partial}{\partial r} + \underbrace{\frac{v_\phi}{r} \frac{\partial}{\partial \phi} + \frac{v_\theta}{r \sin \phi} \frac{\partial}{\partial \theta}}_{0} = \frac{\partial}{\partial t} + v_r \frac{\partial}{\partial r}. \tag{9.656}$$

The divergence is given by Eq. (2.381). Following traditional notation, we shall take $v_r(r, t) = u(r, t)$, where u is understood to be the velocity in the radial direction. Focus on the mass equation. In our one-dimensional limit, it becomes

$$\frac{\partial \rho}{\partial t} + u \frac{\partial \rho}{\partial r} + \rho \left(\frac{1}{r^2} \frac{\partial}{\partial r} \left(r^2 u \right) \right) = 0, \tag{9.657}$$

$$\frac{\partial \rho}{\partial t} + u \frac{\partial \rho}{\partial r} + \rho \left(\frac{\partial u}{\partial r} + \frac{2u}{r} \right) = 0. \tag{9.658}$$

The other equations are straightforwardly reduced, giving our governing equations, after adopting the calorically perfect ideal gas limit, to be

$$\frac{\partial \rho}{\partial t} + u \frac{\partial \rho}{\partial r} + \rho \frac{\partial u}{\partial r} = -\frac{2\rho u}{r}, \tag{9.659}$$

$$\frac{\partial u}{\partial t} + u \frac{\partial u}{\partial r} + \frac{1}{\rho} \frac{\partial p}{\partial r} = 0, \tag{9.660}$$

$$\left(\frac{\partial e}{\partial t} + u \frac{\partial e}{\partial r} \right) - \frac{p}{\rho^2} \left(\frac{\partial \rho}{\partial t} + u \frac{\partial \rho}{\partial r} \right) = 0, \tag{9.661}$$

$$e = \frac{1}{\gamma - 1} \frac{p}{\rho}, \tag{9.662}$$

$$p = \rho \mathrm{R} T. \tag{9.663}$$

Here we take the gas constant to be R, as following Taylor, we will reserve R for the shock radius. Note that for sufficiently large r, the equations reduce to the Cartesian limit. The conservative version, not shown here, can also be written in the form of Eq. (9.280). The conservative form induces a set of shock jump equations in the form of Eq. (9.290). Taking the subscript s to denote the shocked state and the subscript o to denote the unshocked state, the shock velocity to be dR/dt, and the shock Mach number $M_s = (dR/dt)/\sqrt{\gamma p_o/\rho_o}$, their solution gives the jump over a shock discontinuity via the Rankine–Hugoniot jump equations:

$$\frac{\rho_s}{\rho_o} = \frac{\gamma+1}{\gamma-1}\left(1 + \frac{2}{(\gamma-1)}\frac{1}{M_s^2}\right)^{-1}, \tag{9.664}$$

$$\frac{p_s}{p_o} = \frac{2\gamma}{\gamma+1}M_s^2 - \frac{\gamma-1}{\gamma+1}, \tag{9.665}$$

$$\frac{dR}{dt} = \frac{\gamma+1}{4}u_s + \sqrt{\frac{\gamma p_o}{\rho_o} + u_s^2\left(\frac{\gamma+1}{4}\right)^2}. \tag{9.666}$$

These yield identical results as those for Cartesian coordinate system studied earlier.

Let us look at the energy equation, Eq. (9.661) in more detail. With the material derivative as $d/dt = \partial/\partial t + u\,\partial/\partial r$, Eq. (9.661) can be rewritten as

$$\frac{de}{dt} - \frac{p}{\rho^2}\frac{d\rho}{dt} = 0. \tag{9.667}$$

As an aside, we recall that the specific volume $\hat{v}$ is defined as $\hat{v} = 1/\rho$. Thus, we have $d\hat{v}/dt = -(1/\rho^2)d\rho/dt$. Thus the energy equation can be rewritten as $de/dt + p\,d\hat{v}/dt = 0$, or $de/dt = -p\,d\hat{v}/dt$. In differential form, this is $de = -p\,d\hat{v}$. This says the change in energy is solely due to reversible work done by a pressure force. We might recall the Gibbs equation from thermodynamics, $de = T\,ds - p\,d\hat{v}$, where s is the entropy. For our system, we have $ds = 0$; thus, the flow is isentropic, at least behind the shock. It is isentropic because away from the shock, we have neglected all entropy-producing mechanisms like diffusion.

Let us now substitute the caloric state equation, Eq. (9.662), into the energy equation, Eq. (9.667):

$$\frac{1}{\gamma-1}\frac{d}{dt}\left(\frac{p}{\rho}\right) - \frac{p}{\rho^2}\frac{d\rho}{dt} = 0. \tag{9.668}$$

Straightforward analysis shows this reduces to

$$\frac{d}{dt}\left(\frac{p}{\rho^\gamma}\right) = 0. \tag{9.669}$$

This says that following a fluid particle, p/ρ^γ is a constant. In terms of specific volume, this says $P\hat{v}^\gamma = C$, which is a well-known isentropic relation for a calorically perfect ideal gas.

9.14.2 Similarity Transformation

We shall next make some nonintuitive and nonobvious choices for a transformed coordinate system and transformed dependent variables. These choices can be systematically studied with the techniques of group theory, not discussed here.

Independent Variables

Let us transform the independent variables $(r, t) \to (\eta, \tau)$ with

$$\eta = \frac{r}{R(t)}, \qquad \tau = t. \tag{9.670}$$

We will seek solutions such that the dependent variables are functions of η, the distance relative to the time-dependent shock, only. We will have little need for the transformed time τ because it is equivalent to the original time t.

Dependent Variables

Let us also define new dependent variables as

$$\frac{p}{p_o} = y = R^{-3} f_1(\eta), \qquad \frac{\rho}{\rho_o} = \psi(\eta), \qquad u = R^{-3/2} \phi_1(\eta). \tag{9.671}$$

These amount to definitions of a scaled pressure f_1, a scaled density ψ, and a scaled velocity ϕ_1, with the assumption that each is a function of η only. Here, p_o and ρ_o are constant ambient values of pressure and density, respectively. Note that ψ and ϕ are Taylor's notations, and have nothing to do with stream functions or velocity potentials.

We also assume the shock velocity to be of the form

$$U(t) = \frac{dR}{dt} = A R^{-3/2}. \tag{9.672}$$

The constant A is to be determined.

Derivative Transformations

By the chain rule we have

$$\frac{\partial}{\partial t} = \frac{\partial \eta}{\partial t} \frac{\partial}{\partial \eta} + \frac{\partial \tau}{\partial t} \frac{\partial}{\partial \tau}. \tag{9.673}$$

Now, by Eq. (9.670) we get

$$\frac{\partial \eta}{\partial t} = -\frac{r}{R^2} \frac{dR}{dt} = -\frac{\eta}{R(t)} \frac{dR}{dt} = -\frac{\eta}{R} A R^{-3/2} = -\frac{A\eta}{R^{5/2}}. \tag{9.674}$$

From Eq. (9.670) we simply get $\partial \tau / \partial t = 1$. Thus, the chain rule, Eq. (9.673), can be written as

$$\frac{\partial}{\partial t} = -\frac{A\eta}{R^{5/2}} \frac{\partial}{\partial \eta} + \frac{\partial}{\partial \tau}. \tag{9.675}$$

As we are insisting the $\partial/\partial \tau = 0$, we get

$$\frac{\partial}{\partial t} = -\frac{A\eta}{R^{5/2}} \frac{d}{d\eta}. \tag{9.676}$$

In the same way, we get

$$\frac{\partial}{\partial r} = \frac{\partial \eta}{\partial r} \frac{\partial}{\partial \eta} + \underbrace{\frac{\partial \tau}{\partial r} \frac{\partial}{\partial \tau}}_{=0} = \frac{1}{R} \frac{d}{d\eta}. \tag{9.677}$$

9.14.3 Transformed Equations

Let us now apply our rules for derivative transformation, Eqs. (9.674, 9.677), and our transformed dependent variables, Eqs. (9.671), to the governing equations.

Mass

First, we shall consider the mass equation, Eq. (9.659). We get

$$-\underbrace{\frac{A\eta}{R^{5/2}}\frac{d}{d\eta}}_{=\partial/\partial t}\underbrace{(\rho_o\psi)}_{=\rho} + \underbrace{R^{-3/2}\phi_1}_{=u}\underbrace{\frac{1}{R}\frac{d}{d\eta}}_{=\partial/\partial r}\underbrace{(\rho_o\psi)}_{=\rho} + \underbrace{\rho_o\psi}_{=\rho}\underbrace{\frac{1}{R}\frac{d}{d\eta}\left(R^{-3/2}\phi_1\right)}_{=\partial/\partial r}_{=u} = -\underbrace{\frac{2}{r}}_{=2/(\eta R)}\underbrace{\rho_o\psi}_{=\rho}\underbrace{R^{-3/2}\phi_1}_{=u}. \tag{9.678}$$

Realizing that $R(t) = R(\tau)$ is not a function of η, canceling the common factor of ρ_o, and eliminating r with Eq. (9.670), we can write

$$-\frac{A\eta}{R^{5/2}}\frac{d\psi}{d\eta} + \frac{\phi_1}{R^{5/2}}\frac{d\psi}{d\eta} + \frac{\psi}{R^{5/2}}\frac{d\phi_1}{d\eta} = -\frac{2}{\eta}\frac{\psi\phi_1}{R^{5/2}}, \tag{9.679}$$

$$-A\eta\frac{d\psi}{d\eta} + \phi_1\frac{d\psi}{d\eta} + \psi\frac{d\phi_1}{d\eta} = -\frac{2}{\eta}\psi\phi_1, \tag{9.680}$$

$$-A\eta\frac{d\psi}{d\eta} + \phi_1\frac{d\psi}{d\eta} + \psi\left(\frac{d\phi_1}{d\eta} + \frac{2}{\eta}\phi_1\right) = 0, \qquad \text{mass}. \tag{9.681}$$

Equation (9.681) is number 9 in Taylor's paper, which we will call here Eq. T(9).

Linear Momentum

Now, consider the linear momentum equation, Eq. (9.660), and apply the same transformations:

$$\frac{\partial}{\partial t}\underbrace{\left(R^{-3/2}\phi_1\right)}_{=u} + \underbrace{R^{-3/2}\phi_1}_{=u}\frac{\partial}{\partial r}\underbrace{\left(R^{-3/2}\phi_1\right)}_{=u} + \underbrace{\frac{1}{\rho_o\psi}}_{=1/\rho}\frac{\partial}{\partial r}\underbrace{\left(p_o R^{-3}f_1\right)}_{=P} = 0, \tag{9.682}$$

$$\underbrace{R^{-3/2}\frac{\partial\phi_1}{\partial t} - \frac{3}{2}R^{-5/2}\frac{dR}{dt}\phi_1}_{=\partial u/\partial t} + R^{-3/2}\phi_1\frac{\partial}{\partial r}\left(R^{-3/2}\phi_1\right) + \frac{1}{\rho_o\psi}\frac{\partial}{\partial r}\left(p_o R^{-3}f_1\right) = 0, \tag{9.683}$$

$$R^{-3/2}\left(-\frac{A\eta}{R^{5/2}}\right)\frac{d\phi_1}{d\eta} - \frac{3}{2}R^{-5/2}\left(AR^{-3/2}\right)\phi_1$$
$$+ R^{-3/2}\phi_1\frac{\partial}{\partial r}\left(R^{-3/2}\phi_1\right) + \frac{1}{\rho_o\psi}\frac{\partial}{\partial r}\left(p_o R^{-3}f_1\right) = 0, \tag{9.684}$$

$$-\frac{A\eta}{R^4}\frac{d\phi_1}{d\eta} - \frac{3}{2}\frac{A}{R^4}\phi_1 + R^{-3/2}\phi_1\frac{\partial}{\partial r}\left(R^{-3/2}\phi_1\right) + \frac{1}{\rho_o\psi}\frac{\partial}{\partial r}\left(p_o R^{-3}f_1\right) = 0, \tag{9.685}$$

$$-\frac{A\eta}{R^4}\frac{d\phi_1}{d\eta} - \frac{3}{2}\frac{A}{R^4}\phi_1 + R^{-3/2}\phi_1\frac{1}{R}\frac{d}{d\eta}\left(R^{-3/2}\phi_1\right) + \frac{1}{\rho_o\psi}\frac{1}{R}\frac{d}{d\eta}\left(p_o R^{-3}f_1\right) = 0, \tag{9.686}$$

$$-\frac{A\eta}{R^4}\frac{d\phi_1}{d\eta} - \frac{3}{2}\frac{A}{R^4}\phi_1 + \frac{\phi_1}{R^4}\frac{d\phi_1}{d\eta} + \frac{p_o}{\rho_o\psi}\frac{1}{R^4}\frac{df_1}{d\eta} = 0, \tag{9.687}$$

$$-A\eta\frac{d\phi_1}{d\eta} - \frac{3}{2}A\phi_1 + \phi_1\frac{d\phi_1}{d\eta} + \frac{p_o}{\rho_o\psi}\frac{df_1}{d\eta} = 0. \tag{9.688}$$

Our final form is

$$-A\left(\frac{3}{2}\phi_1 + \eta\frac{d\phi_1}{d\eta}\right) + \phi_1\frac{d\phi_1}{d\eta} + \frac{p_o}{\rho_o}\frac{1}{\psi}\frac{df_1}{d\eta} = 0, \qquad \text{linear momentum.} \tag{9.689}$$

Equation (9.689) is T(7).

Energy

Let us now consider the energy equation. It is best to begin with a form in which the equation of state has already been imposed. So, we will start by expanding Eq. (9.669) in terms of partial derivatives:

$$\underbrace{\frac{\partial p}{\partial t} + u\frac{\partial p}{\partial r}}_{=dp/dt} - \gamma\frac{p}{\rho}\underbrace{\left(\frac{\partial \rho}{\partial t} + u\frac{\partial \rho}{\partial r}\right)}_{=d\rho/dt} = 0, \tag{9.690}$$

$$\frac{\partial}{\partial t}\left(p_o R^{-3} f_1\right) + R^{-3/2}\psi_1\frac{\partial}{\partial r}\left(p_o R^{-3} f_1\right) - \gamma\frac{p_o R^{-3} f_1}{\rho_o\psi}\left(\frac{\partial}{\partial t}\left(\rho_o\psi\right) + R^{-3/2}\psi_1\frac{\partial}{\partial r}\left(\rho_o\psi\right)\right) = 0, \tag{9.691}$$

$$\frac{\partial}{\partial t}\left(R^{-3} f_1\right) + R^{-3/2}\phi_1\frac{\partial}{\partial r}\left(R^{-3} f_1\right) - \gamma\frac{R^{-3} f_1}{\psi}\left(\frac{\partial\psi}{\partial t} + R^{-3/2}\phi_1\frac{\partial\psi}{\partial r}\right) = 0, \tag{9.692}$$

$$R^{-3}\frac{\partial f_1}{\partial t} - 3R^{-4}\frac{dR}{dt}f_1 + R^{-3/2}\phi_1\frac{\partial}{\partial r}\left(R^{-3} f_1\right) - \gamma\frac{R^{-3} f_1}{\psi}\left(\frac{\partial\psi}{\partial t} + R^{-3/2}\phi_1\frac{\partial\psi}{\partial r}\right) = 0, \tag{9.693}$$

$$R^{-3}\left(-\frac{A\eta}{R^{5/2}}\right)\frac{df_1}{d\eta} - 3R^{-4}(AR^{-3/2})f_1 + R^{-3/2}\phi_1\frac{\partial}{\partial r}\left(R^{-3} f_1\right)$$
$$- \gamma\frac{R^{-3} f_1}{\psi}\left(\frac{\partial\psi}{\partial t} + R^{-3/2}\phi_1\frac{\partial\psi}{\partial r}\right) = 0. \tag{9.694}$$

Carrying on, we have

$$-\frac{A\eta}{R^{11/2}}\frac{df_1}{d\eta} - 3\frac{A}{R^{11/2}}f_1 + R^{-3/2}\phi_1 R^{-3}\frac{1}{R}\frac{df_1}{d\eta}$$
$$- \gamma\frac{R^{-3} f_1}{\psi}\left(\left(-\frac{A\eta}{R^{5/2}}\right)\frac{d\psi}{d\eta} + R^{-3/2}\phi_1\frac{1}{R}\frac{d\psi}{d\eta}\right) = 0, \tag{9.695}$$

$$-\frac{A\eta}{R^{11/2}}\frac{df_1}{d\eta} - 3\frac{A}{R^{11/2}}f_1 + \frac{\phi_1}{R^{11/2}}\frac{df_1}{d\eta} - \gamma\frac{f_1}{\psi R^{11/2}}\left(-A\eta\frac{d\psi}{d\eta} + \phi_1\frac{d\psi}{d\eta}\right) = 0, \tag{9.696}$$

$$-A\eta\frac{df_1}{d\eta} - 3Af_1 + \phi_1\frac{df_1}{d\eta} - \gamma\frac{f_1}{\psi}\left(-A\eta\frac{d\psi}{d\eta} + \phi_1\frac{d\psi}{d\eta}\right) = 0. \tag{9.697}$$

Our final form is

$$A\left(3f_1 + \eta\frac{df_1}{d\eta}\right) + \gamma\frac{f_1}{\psi}(-A\eta + \phi_1)\frac{d\psi}{d\eta} - \phi_1\frac{df_1}{d\eta} = 0, \qquad \text{energy.} \qquad (9.698)$$

Equation (9.698) is T(11), correcting for a typographical error replacing a r with γ.

9.14.4 Dimensionless Equations

Let us now write our conservation principles in dimensionless form. We take the constant ambient sound speed c_o to be defined for our gas as

$$c_o^2 \equiv \gamma\frac{p_o}{\rho_o}. \qquad (9.699)$$

We have used our notation for sound speed here; Taylor uses a instead. Let us also define

$$f \equiv \left(\frac{c_o}{A}\right)^2 f_1, \qquad \phi \equiv \frac{\phi_1}{A}. \qquad (9.700)$$

Mass

With these definitions, the mass equation, Eq. (9.681), becomes

$$-A\eta\frac{d\psi}{d\eta} + A\phi\frac{d\psi}{d\eta} + \psi\left(A\frac{d\phi}{d\eta} + \frac{2}{\eta}A\phi\right) = 0, \qquad (9.701)$$

$$-\eta\frac{d\psi}{d\eta} + \phi\frac{d\psi}{d\eta} + \psi\left(\frac{d\phi}{d\eta} + \frac{2}{\eta}\phi\right) = 0, \qquad (9.702)$$

$$\frac{d\psi}{d\eta}(\phi - \eta) = -\psi\left(\frac{d\phi}{d\eta} + \frac{2}{\eta}\phi\right), \qquad (9.703)$$

$$\frac{1}{\psi}\frac{d\psi}{d\eta} = \frac{\frac{d\phi}{d\eta} + \frac{2\phi}{\eta}}{\eta \quad \phi}, \qquad \text{mass.} \qquad (9.704)$$

Equation (9.704) is T(9a).

Linear Momentum

With the same definitions, the momentum equation, Eq. (9.689) becomes

$$-A\left(\frac{3}{2}A\phi + A\eta\frac{d\phi}{d\eta}\right) + A^2\phi\frac{d\phi}{d\eta} + \frac{p_o}{\rho_o}\frac{1}{\psi}\frac{A^2}{c_o^2}\frac{df}{d\eta} = 0, \qquad (9.705)$$

$$-\left(\frac{3}{2}\phi + \eta\frac{d\phi}{d\eta}\right) + \phi\frac{d\phi}{d\eta} + \frac{1}{\gamma}\frac{1}{\psi}\frac{df}{d\eta} = 0, \qquad (9.706)$$

$$\frac{d\phi}{d\eta}(\phi - \eta) - \frac{3}{2}\phi + \frac{1}{\gamma\psi}\frac{df}{d\eta} = 0, \qquad (9.707)$$

$$\frac{d\phi}{d\eta}(\eta - \phi) = \frac{1}{\gamma\psi}\frac{df}{d\eta} - \frac{3}{2}\phi, \qquad \text{momentum.} \qquad (9.708)$$

Equation (9.708) is T(7a).

Energy

The energy equation, Eq. (9.698) becomes

$$A\left(3\frac{A^2}{c_o^2}f + \eta\frac{A^2}{c_o^2}\frac{df}{d\eta}\right) + \gamma\frac{f}{\psi}\frac{A^2}{c_o^2}(-A\eta + A\phi)\frac{d\psi}{d\eta} - A\frac{A^2}{c_o^2}\phi\frac{df}{d\eta} = 0, \qquad (9.709)$$

$$3f + \eta\frac{df}{d\eta} + \gamma\frac{f}{\psi}(-\eta + \phi)\frac{d\psi}{d\eta} - \phi\frac{df}{d\eta} = 0, \qquad (9.710)$$

$$3f + \eta\frac{df}{d\eta} + \gamma\frac{1}{\psi}\frac{d\psi}{d\eta}f(-\eta + \phi) - \phi\frac{df}{d\eta} = 0, \quad \text{energy}. \qquad (9.711)$$

Equation (9.711) is T(11a).

9.14.5 Reduction to Nonautonomous Form

Let us eliminate $d\psi/d\eta$ and $d\phi/d\eta$ from Eq. (9.711) with use of Eqs. (9.704, 9.708):

$$3f + \eta\frac{df}{d\eta} + \gamma f\left(\frac{\frac{d\phi}{d\eta} + \frac{2\phi}{\eta}}{\eta - \phi}\right)(-\eta + \phi) - \phi\frac{df}{d\eta} = 0, \qquad (9.712)$$

$$3f + \eta\frac{df}{d\eta} + \gamma f\left(\frac{\frac{1}{\gamma\psi}\frac{df}{d\eta} - \frac{3}{2}\psi}{\eta - \phi} + \frac{2\phi}{\eta}\right)(-\eta + \phi) - \phi\frac{df}{d\eta} = 0, \qquad (9.713)$$

$$3f + (\eta - \phi)\frac{df}{d\eta} - \gamma f\left(\frac{\frac{1}{\gamma\psi}\frac{df}{d\eta} - \frac{3}{2}\phi}{\eta - \phi} + \frac{2\phi}{\eta}\right) = 0, \qquad (9.714)$$

$$3f(\eta - \phi) + (\eta - \phi)^2\frac{df}{d\eta} - \gamma f\left(\frac{1}{\gamma\psi}\frac{df}{d\eta} - \frac{3}{2}\phi + \frac{2\phi}{\eta}(\eta - \phi)\right) = 0, \qquad (9.715)$$

$$\left((\eta - \phi)^2 - \frac{f}{\psi}\right)\frac{df}{d\eta} - f\left(-3(\eta - \phi) - \frac{3}{2}\gamma\phi + \frac{2\gamma\phi}{\eta}(\eta - \phi)\right) = 0, \qquad (9.716)$$

$$\left((\eta - \phi)^2 - \frac{f}{\psi}\right)\frac{df}{d\eta} + f\left(3\eta - 3\phi + \frac{3}{2}\gamma\phi - 2\gamma\phi + \frac{2\gamma\phi^2}{\eta}\right) = 0, \qquad (9.717)$$

$$\left((\eta - \phi)^2 - \frac{f}{\psi}\right)\frac{df}{d\eta} + f\left(3\eta - \phi\left(3 + \frac{1}{2}\gamma\right) + \frac{2\gamma\phi^2}{\eta}\right) = 0. \qquad (9.718)$$

Rearranging, we get

$$\left((\eta - \phi)^2 - \frac{f}{\psi}\right)\frac{df}{d\eta} = f\left(-3\eta + \phi\left(3 + \frac{1}{2}\gamma\right) - \frac{2\gamma\phi^2}{\eta}\right). \qquad (9.719)$$

Equation (9.719) is T(14).

We can thus write an explicit nonautonomous ordinary differential equation for the evolution of f in terms of the state variables f, ψ, and ϕ, as well as the independent variable η:

$$\frac{df}{d\eta} = \frac{f\left(-3\eta + \phi\left(3 + \frac{1}{2}\gamma\right) - \frac{2\gamma\phi^2}{\eta}\right)}{(\eta - \phi)^2 - \frac{f}{\psi}}. \qquad (9.720)$$

Equation (9.720) can be directly substituted into the momentum equation, Eq. (9.708) to get

$$\frac{d\phi}{d\eta} = \frac{\frac{1}{\gamma\psi}\frac{df}{d\eta} - \frac{3}{2}\phi}{\eta - \phi}. \tag{9.721}$$

Then, Eq. (9.721) can be substituted into Eq. (9.704) to get

$$\frac{d\psi}{d\eta} = \psi\frac{\frac{d\phi}{d\eta} + \frac{2\phi}{\eta}}{\eta - \phi}. \tag{9.722}$$

Equations (9.720–9.722) form a nonautonomous system of first-order differential equations of the form

$$\frac{df}{d\eta} = g_1(f,\phi,\psi,\eta), \qquad \frac{d\phi}{d\eta} = g_2(f,\phi,\psi,\eta), \qquad \frac{d\psi}{d\eta} = g_3(f,\phi,\psi,\eta). \tag{9.723}$$

They can be integrated with standard numerical methods. One must of course provide conditions of all state variables at a particular point. We apply conditions not at $\eta = 0$, but at $\eta = 1$, the locus of the shock front. Following Taylor, the conditions are taken from the Rankine–Hugoniot jump equations, Eqs. (9.664–9.666), applied in the limit of a strong shock ($M_s \to \infty$). We omit the details of this analysis. We take the subscript s to denote the shock state at $\eta = 1$. For the density, one finds

$$\frac{\rho_s}{\rho_o} = \frac{\gamma+1}{\gamma-1}, \qquad \frac{\rho_o\psi_s}{\rho_o} = \frac{\gamma+1}{\gamma-1}, \qquad \psi_s = \psi(\eta=1) = \frac{\gamma+1}{\gamma-1}. \tag{9.724}$$

For the pressure, leaving out some details, one finds that

$$\frac{\frac{dR^2}{dt}}{c_o^2} = \frac{\gamma+1}{2\gamma}\frac{p_s}{p_o}, \tag{9.725}$$

$$\frac{A^2 R^{-3}}{c_o^2} = \frac{\gamma+1}{2\gamma} R^{-3} f_{1s} = \frac{\gamma+1}{2\gamma} R^{-3} \frac{A^2}{c_o^2} f_s, \tag{9.726}$$

$$1 = \frac{\gamma+1}{2\gamma} f_s, \quad \text{so} \quad f_s = f(\eta=1) = \frac{2\gamma}{\gamma+1}. \tag{9.727}$$

For the velocity, leaving out details, one finds

$$\frac{u_s}{\frac{dR}{dt}} = \frac{2}{\gamma+1}, \tag{9.728}$$

$$\frac{R^{-3/2}\phi_{1s}}{AR^{-3/2}} = \frac{2}{\gamma+1}, \tag{9.729}$$

$$\frac{R^{-3/2}A\phi_s}{AR^{-3/2}} = \frac{2}{\gamma+1}, \quad \text{so} \quad \phi_s = \phi(\eta=1) = \frac{2}{\gamma+1}. \tag{9.730}$$

Equations (9.724, 9.727, 9.730) form the appropriate set of initial conditions for the integration of Eqs. (9.720–9.722).

9.14.6 Numerical Solution

Solutions for $f(\eta)$, $\phi(\eta)$ and $\psi(\eta)$ are shown for $\gamma = 7/5$ in Figs. 9.54–9.56, respectively. So, we now have a similarity solution for the scaled variables. We need to relate this to physical

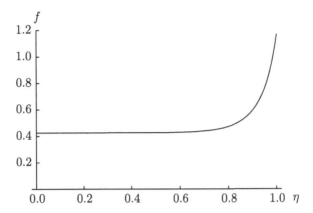

Figure 9.54 Scaled pressure f versus similarity variable η for $\gamma = 7/5$ in a Taylor–Sedov blast wave.

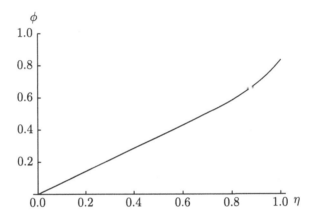

Figure 9.55 Scaled velocity ϕ versus similarity variable η for $\gamma = 7/5$ in a Taylor–Sedov blast wave.

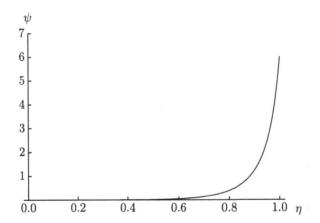

Figure 9.56 Scaled density ψ versus similarity variable η for $\gamma = 7/5$ in a Taylor–Sedov blast wave.

dimensional quantities. Let us assign some initial conditions for $t = 0$, $r > 0$; that is, away from the point source. Take

$$u(r,0) = 0, \quad \rho(r,0) = \rho_o, \quad p(r,0) = p_o. \tag{9.731}$$

We also have from Eq. (9.663) that

$$T(r,0) = \frac{p_o}{\rho_o R} = T_o. \tag{9.732}$$

Using Eq. (9.662), we further have

$$e(r,0) = \frac{1}{\gamma - 1} \frac{p_o}{\rho_o} = e_o. \tag{9.733}$$

Calculation of Total Energy

Now, as the point source expands, it will generate a strong shock wave. Material which has not been shocked is oblivious to the presence of the shock. Material which the shock wave has reached has been influenced by it. It stands to reason from the first law of thermodynamics that we want the total energy, internal plus kinetic, to be *constant* in the shocked domain, $r \in (0, R(t)]$, where $R(t)$ is the shock front location.

Let us recall some spherical geometry so the first law of thermodynamics can be properly formulated. Consider a thin differential spherical shell of thickness dr located somewhere in the shocked region: $r \in (0, R(t)]$. The volume of the thin shell is

$$dV = \underbrace{4\pi r^2}_{\text{(surface area)}} \underbrace{dr}_{\text{(thickness)}}. \tag{9.734}$$

The differential mass dm of this shell is

$$dm = \rho \, dV = 4\pi r^2 \rho \, dr. \tag{9.735}$$

Now, recall the mass-specific internal energy is e and the mass-specific kinetic energy is $u^2/2$. So, the total differential energy, internal plus kinetic, in the differential shell is

$$dE = \left(e + \frac{1}{2}u^2\right) dm = 4\pi \rho \left(e + \frac{1}{2}u^2\right) r^2 \, dr. \tag{9.736}$$

Now, the total energy E within the shock is the integral through the entire sphere,

$$E = \int_0^{R(t)} dE = \int_0^{R(t)} 4\pi \rho \left(e + \frac{1}{2}u^2\right) r^2 \, dr, \tag{9.737}$$

$$= \int_0^{R(t)} 4\pi \rho \left(\frac{1}{\gamma - 1}\frac{p}{\rho} + \frac{1}{2}u^2\right) r^2 \, dr, \tag{9.738}$$

$$= \underbrace{\frac{4\pi}{\gamma - 1} \int_0^{R(t)} P r^2 \, dr}_{\text{thermal energy}} + \underbrace{2\pi \int_0^{R(t)} \rho u^2 r^2 \, dr}_{\text{kinetic energy}}. \tag{9.739}$$

We introduce variables from our similarity transformations next:

$$E = \frac{4\pi}{\gamma - 1}\int_0^1 \underbrace{p_o R^{-3} f_1}_{P} \underbrace{R^2 \eta^2}_{r^2} \underbrace{R \, d\eta}_{dr} + 2\pi \int_0^1 \underbrace{\rho_o \psi}_{\rho} \underbrace{R^{-3}\phi_1^2}_{u^2} \underbrace{R^2\eta^2}_{r^2} \underbrace{R \, d\eta}_{dr}, \tag{9.740}$$

$$= \frac{4\pi}{\gamma-1}\int_0^1 p_o f_1 \eta^2\, d\eta + 2\pi \int_0^1 \rho_o \psi \phi_1^2 \eta^2\, d\eta, \tag{9.741}$$

$$= \frac{4\pi}{\gamma-1}\int_0^1 \frac{p_o A^2}{c_o^2} f \eta^2\, d\eta + 2\pi \int_0^1 \rho_o \psi A^2 \phi^2 \eta^2\, d\eta, \tag{9.742}$$

$$= 4\pi A^2 \left(\frac{p_o}{c_o^2(\gamma-1)}\int_0^1 f\eta^2\, d\eta + \frac{\rho_o}{2}\int_0^1 \psi\phi^2\eta^2\, d\eta \right), \tag{9.743}$$

$$= 4\pi A^2 \rho_o \underbrace{\left(\frac{1}{\gamma(\gamma-1)}\int_0^1 f\eta^2\, d\eta + \frac{1}{2}\int_0^1 \psi\phi^2\eta^2\, d\eta \right)}_{\text{dependent on } \gamma \text{ only}}. \tag{9.744}$$

The term inside the parentheses is dependent on γ only. So, if we consider air with $\gamma = 7/5$, we can – using our knowledge of $f(\eta)$, $\psi(\eta)$, and $\phi(\eta)$, which only depend on γ – calculate once and for all the value of the integrals. For $\gamma = 7/5$, we obtain via numerical quadrature

$$E = 4\pi A^2 \rho_o \left(\frac{1}{(7/5)(2/5)}(0.185\,194) + \frac{1}{2}(0.185\,168) \right) = 5.3192 \rho_o A^2. \tag{9.745}$$

Now, from Eqs. (9.671, 9.699, 9.700, 9.745) with $\gamma = 7/5$, we get

$$p = p_o R^{-3} f \frac{A^2}{c_o^2} = p_o R^{-3} f \frac{\rho_o}{\gamma p_o} A^2 = R^{-3} f \frac{1}{\gamma}\rho_o A^2 = R^{-3} f \frac{1}{\frac{7}{5}}\frac{E}{5.3192}, \tag{9.746}$$

$$p(r,t) = 0.1343 \frac{E}{R^3(t)} f\left(\frac{r}{R(t)}\right). \tag{9.747}$$

The peak pressure occurs at $\eta = 1$, where $r = R(t)$, and where

$$f(\eta = 1) = \frac{2\gamma}{\gamma + 1} = \frac{2(1.4)}{1.4+1} = 1.167. \tag{9.748}$$

So, at $\eta = 1$, where $r = R(t)$, we have

$$p(t) = (0.1343)(1.167)R^{-3}E = 0.1567\frac{E}{R^3}. \tag{9.749}$$

The peak pressure decays at a rate proportional to $1/R^3$ in the strong shock limit.

Now, from Eqs. (9.671, 9.700, 9.745) we get for u:

$$u = R^{-3/2} A\phi = R^{-3/2}\sqrt{\frac{E}{5.319\rho_o}}\,\phi, \tag{9.750}$$

$$u(r,t) = \sqrt{\frac{E}{5.319\rho_o}}\frac{1}{R^{3/2}(t)}\phi\left(\frac{r}{R(t)}\right). \tag{9.751}$$

Let us now explicitly solve for the shock position $R(t)$ and the shock velocity dR/dt. We have from Eqs. (9.672, 9.745) that

$$\frac{dR}{dt} = AR^{-3/2} = \sqrt{\frac{E}{5.319\rho_o}}\frac{1}{R^{3/2}(t)}, \tag{9.752}$$

$$R^{3/2}dR = \sqrt{\frac{E}{5.319\rho_o}}\, dt, \tag{9.753}$$

$$\frac{2}{5}R^{5/2} = \sqrt{\frac{E}{5.319\rho_o}}\, t + C. \tag{9.754}$$

Now, because $R(0) = 0$, we get $C = 0$, so

$$\frac{2}{5}R^{5/2} = \sqrt{\frac{E}{5.319\rho_o}}\, t, \tag{9.755}$$

$$t = \frac{2}{5}R^{5/2}\sqrt{5.319\rho_o}\, E^{-1/2} = 0.9225 R^{5/2} \rho_o^{1/2} E^{-1/2}. \tag{9.756}$$

Equation (9.756) is T(38). Solving for R, we get

$$R^{5/2} = \frac{1}{0.9225} t \rho_o^{-1/2} E^{1/2}, \quad \text{so} \quad R(t) = 1.03279 \rho_o^{-1/5} E^{1/5} t^{2/5}. \tag{9.757}$$

Thus, we have a prediction for the shock location as a function of time t, as well as point source energy E. If we know the position as a function of time, we can get the shock velocity by direct differentiation:

$$\frac{dR}{dt} = 0.4131 \rho_o^{-1/5} E^{1/5} t^{-3/5}. \tag{9.758}$$

If we can make a measurement of the blast wave location R at a given known time t, and we know the ambient density ρ_o, we can estimate the point source energy E. Let us invert Eq. (9.757) to solve for E and get

$$E = \frac{\rho_o R^5}{(1.03279)^5 t^2} = 0.85102 \frac{\rho_o R^5}{t^2}. \tag{9.759}$$

Comparison with Experimental Data

Taylor (1950b) gives data for the July 19, 1945, atomic bomb explosion at the Trinity site in New Mexico. We choose one point from the photographic record which finds the shock from the blast to be located at $R = 185\,\text{m}$ when $t = 62\,\text{ms}$. Let us assume the ambient air has a density of $\rho_o = 1.161\,\text{kg/m}^3$. Then, we can estimate the energy of the device by Eq. (9.759) as

$$E = 0.85102 \frac{\left(1.161\,\frac{\text{kg}}{\text{m}^3}\right)(185\,\text{m})^5}{(0.062\,\text{s})^2} = 55.7 \times 10^{12}\,\text{J}. \tag{9.760}$$

Now, 1 ton of the high explosive TNT (more specifically, 2,4,6-trinitrotoluene, $C_6H_2(NO_2)_3CH_3$) is known to contain 4.25×10^9 J of chemical energy. So, the estimated energy of the Trinity site device in terms of a TNT equivalent is

$$\text{TNT}_{equivalent} = \frac{55.7 \times 10^{12}\,\text{J}}{4.25 \times 10^9\,\frac{\text{J}}{\text{ton}}} = 13.1 \times 10^3\,\text{ton}. \tag{9.761}$$

In common parlance, the Trinity site device was a 13 kiloton bomb by this simple estimate. Taylor provides some nuanced corrections to this estimate. Modern estimates are now around 20 kiloton.

9.14.7 Contrast with Acoustic Limit

We saw in Eq. (9.749) that in the expansion associated with a strong shock, the pressure decays as $1/R^3$. Let us see how that compares with the decay of pressure in the limit of a weak shock.

Let us first rewrite the governing equations. Here, we (1) rewrite Eq. (9.659) in a conservative form, using the chain rule to absorb the source term inside the derivative, (2) repeat the linear momentum equation, Eq. (9.660), and (3) re-cast the energy equation for a calorically perfect ideal gas, Eq. (9.669) in terms of the full partial derivatives:

$$\frac{\partial \rho}{\partial t} + \frac{1}{r^2}\frac{\partial}{\partial r}\left(r^2 \rho u\right) = 0, \tag{9.762}$$

$$\frac{\partial u}{\partial t} + u\frac{\partial u}{\partial r} + \frac{1}{\rho}\frac{\partial p}{\partial r} = 0, \tag{9.763}$$

$$\frac{\partial p}{\partial t} + u\frac{\partial p}{\partial r} - \gamma\frac{p}{\rho}\left(\frac{\partial \rho}{\partial t} + u\frac{\partial \rho}{\partial r}\right) = 0. \tag{9.764}$$

Now, let us consider the acoustic limit, which corresponds to perturbations of a fluid at rest. Taking $0 < \epsilon \ll 1$, we recast the dependent variables ρ, p, and u as

$$\rho = \rho_o + \epsilon\rho_1 + \cdots, \quad p = p_o + \epsilon p_1 + \cdots, \quad u = \underbrace{u_o}_{=0} + \epsilon u_1 + \cdots. \tag{9.765}$$

Here, ρ_o and p_o are taken to be constants. The ambient velocity $u_o = 0$. Strictly speaking, we should nondimensionalize the equations before we introduce an asymptotic expansion. However, so doing would not change the essence of the argument to be made.

We next introduce our expansions into the governing equations:

$$\frac{\partial}{\partial t}(\rho_o + \epsilon\rho_1) + \frac{1}{r^2}\frac{\partial}{\partial r}\left(r^2(\rho_o + \epsilon\rho_1)(\epsilon u_1)\right) = 0, \tag{9.766}$$

$$\frac{\partial}{\partial t}(\epsilon u_1) + (\epsilon u_1)\frac{\partial}{\partial r}(\epsilon u_1) + \frac{1}{\rho_o + \epsilon\rho_1}\frac{\partial}{\partial r}(p_o + \epsilon p_1) = 0, \tag{9.767}$$

$$\frac{\partial}{\partial t}(p_o + \epsilon p_1) + (\epsilon u_1)\frac{\partial}{\partial r}(p_o + \epsilon p_1)$$
$$-\gamma\frac{p_o + \epsilon p_1}{\rho_o + \epsilon\rho_1}\left(\frac{\partial}{\partial t}(\rho_o + \epsilon\rho_1) + (\epsilon u_1)\frac{\partial}{\partial r}(\rho_o + \epsilon\rho_1)\right) = 0. \tag{9.768}$$

Now, derivatives of constants are all zero, and so at leading order the constant state satisfies the governing equations. At $\mathcal{O}(\epsilon)$, the equations reduce to

$$\frac{\partial \rho_1}{\partial t} + \frac{1}{r^2}\frac{\partial}{\partial r}(r^2 \rho_o u_1) = 0, \tag{9.769}$$

$$\frac{\partial u_1}{\partial t} + \frac{1}{\rho_o}\frac{\partial p_1}{\partial r} = 0, \tag{9.770}$$

$$\frac{\partial p_1}{\partial t} - \gamma\frac{p_o}{\rho_o}\frac{\partial \rho_1}{\partial t} = 0. \tag{9.771}$$

Now, adopt as before $c_o^2 = \gamma p_o/\rho_o$, so the energy equation, Eq. (9.771), becomes

$$\frac{\partial p_1}{\partial t} = c_o^2 \frac{\partial \rho_1}{\partial t}. \tag{9.772}$$

Now, substitute Eq. (9.772) into the mass equation, Eq. (9.769), to get

$$\frac{1}{c_o^2}\frac{\partial p_1}{\partial t} + \frac{1}{r^2}\frac{\partial}{\partial r}(r^2 \rho_o u_1) = 0. \tag{9.773}$$

We take the time derivative of Eq. (9.773) to get

$$\frac{1}{c_o^2}\frac{\partial^2 p_1}{\partial t^2} + \frac{\partial}{\partial t}\left(\frac{1}{r^2}\frac{\partial}{\partial r}\left(r^2 \rho_o u_1\right)\right) = 0, \qquad (9.774)$$

$$\frac{1}{c_o^2}\frac{\partial^2 p_1}{\partial t^2} + \frac{1}{r^2}\frac{\partial}{\partial r}\left(r^2 \rho_o \frac{\partial u_1}{\partial t}\right) = 0. \qquad (9.775)$$

We next use the momentum equation, Eq. (9.770), to eliminate $\partial u_1/\partial t$ in Eq. (9.775):

$$\frac{1}{c_o^2}\frac{\partial^2 p_1}{\partial t^2} + \frac{1}{r^2}\frac{\partial}{\partial r}\left(r^2 \rho_o \left(-\frac{1}{\rho_o}\frac{\partial p_1}{\partial r}\right)\right) = 0, \qquad (9.776)$$

$$\frac{1}{c_o^2}\frac{\partial^2 p_1}{\partial t^2} - \frac{1}{r^2}\frac{\partial}{\partial r}\left(r^2 \frac{\partial p_1}{\partial r}\right) = 0, \qquad (9.777)$$

$$\frac{1}{c_o^2}\frac{\partial^2 p_1}{\partial t^2} = \frac{1}{r^2}\frac{\partial}{\partial r}\left(r^2 \frac{\partial p_1}{\partial r}\right). \qquad (9.778)$$

This second-order linear partial differential equation has a well-known solution of the d'Alembert form:

$$p_1 = \frac{1}{r}g\left(t - \frac{r}{c_o}\right) + \frac{1}{r}h\left(t + \frac{r}{c_o}\right). \qquad (9.779)$$

Here, g and h are arbitrary functions which are chosen to match the initial conditions. Let us check this solution for g; the procedure can be repeated for h.

If $p_1 = (1/r)g(t - r/c_o)$, then

$$\frac{\partial p_1}{\partial t} = \frac{1}{r}g'\left(t - \frac{r}{c_o}\right), \qquad \frac{\partial^2 p_1}{\partial t^2} = \frac{1}{r}g''\left(t - \frac{r}{c_o}\right), \qquad (9.780)$$

and

$$\frac{\partial p_1}{\partial r} = -\frac{1}{c_o}\frac{1}{r}g'\left(t - \frac{r}{c_o}\right) - \frac{1}{r^2}g\left(t - \frac{r}{c_o}\right). \qquad (9.781)$$

With these results, let us substitute into Eq. (9.778) to see if it is satisfied:

$$\frac{1}{c_o^2}\frac{1}{r}g''\left(t - \frac{r}{c_o}\right) = \frac{1}{r^2}\frac{\partial}{\partial r}\left(r^2\left(-\frac{1}{c_o}\frac{1}{r}g'\left(t - \frac{r}{c_o}\right) - \frac{1}{r^2}g\left(t - \frac{r}{c_o}\right)\right)\right), \qquad (9.782)$$

$$= -\frac{1}{r^2}\frac{\partial}{\partial r}\left(\frac{r}{c_o}g'\left(t - \frac{r}{c_o}\right) + g\left(t - \frac{r}{c_o}\right)\right), \qquad (9.783)$$

$$= -\frac{1}{r^2}\left(-\frac{r}{c_o^2}g''\left(t - \frac{r}{c_o}\right) + \frac{1}{c_o}g'\left(t - \frac{r}{c_o}\right) - \frac{1}{c_o}g'\left(t - \frac{r}{c_o}\right)\right), \qquad (9.784)$$

$$= \frac{1}{r^2}\left(\frac{r}{c_o^2}g''\left(t - \frac{r}{c_o}\right)\right), \qquad (9.785)$$

$$= \frac{1}{c_o^2}\frac{1}{r}g''\left(t - \frac{r}{c_o}\right). \qquad (9.786)$$

Indeed, our form of $p_1(r,t)$ satisfies the governing partial differential equation. Moreover, we can see by inspection of Eq. (9.779) that the pressure decays as $1/r$ in the limit of acoustic disturbances. This is a much slower rate of decay than for the blast wave, which goes as $1/r^3$.

SUMMARY

This chapter focused on one-dimensional, inviscid, compressible flows. To understand such flows, one must consider fully coupled models of the evolution of mass, linear momentum, and energy along with thermodynamic equations of state. Isentropic and entropy-generating flows were considered. The absence of viscosity, along with nonlinearity gives rise to the possibility of flows with embedded discontinuities. These are observed in nature as shock waves. Real shock waves have a thin viscous structure, encroaching on the limit in which the continuum assumption is invalid; such waves were studied in the context of the Navier–Stokes equations. Other types of flow that were considered include flow with friction, heat transfer, and area change. Pressure disturbances ranging from the weak acoustic limit to the strong blast wave limit were considered. It was also shown for systems in which strong intermolecular forces were present that unusual behavior could be predicted.

PROBLEMS

9.1 A gas obeys the thermal state equation $p = \rho R T - b\rho^4$. Taking the specific heat at constant volume c_v to be a constant, find an expression for the sound speed, c, and the Helmholtz free energy, $\hat{a}$.

9.2 We are given calorically perfect ideal air, $\gamma = 7/5$, $R = 287\,\text{J/kg/K}$, with static conditions $p = 1\,\text{MPa}$, $T = 300\,\text{K}$, $u = 1000\,\text{m/s}$. The air comes to rest isentropically. Find the stagnation temperature and pressure.

9.3 We are given calorically perfect ideal air, $\gamma = 7/5$, $R = 287\,\text{J/kg/K}$, with stagnation conditions $p_o = 1\,\text{MPa}$, $T_o = 300\,\text{K}$, flowing isentropically through a throat into a diverging region with an exit Mach number of 3. The desired mass flow rate is $10\,\text{kg/s}$. Find the throat area, the exit pressure, the exit temperature, the exit velocity, and the exit area.

9.4 We are given calorically perfect ideal air in a thermally insulated tank with $p_o = 700\,\text{kPa}$, $T_o = 20°\text{C}$, $V = 1.5\,\text{m}^3$. The throat area in a converging nozzle is $0.65\,\text{cm}^2$. The air exhausts to a $101.3\,\text{kPa}$ environment. Find the time for the stagnation pressure in the tank to decrease to $500\,\text{kPa}$.

9.5 Consider Fanno flow of a calorically perfect ideal gas with $\gamma = 7/5$, $M_1 = 3$, $f = 0.04$, $d = 0.01\,\text{m}$. Plot $M(x)$, and identify the distance for which the flow chokes.

9.6 Consider Rayleigh flow of a calorically perfect ideal gas with $\gamma = 7/5$, $R = 287\,\text{J/kg/K}$, $M_1 = 2$, $q_w = 10\,000\,\text{W/m}^2$, $d = 0.01\,\text{m}$, $p_1 = 10^5\,\text{Pa}$, $T_1 = 300\,\text{K}$. Prepare a plot of Mach number as a function of distance including the two different branches of the solution. Identify the distance where the flow is thermally choked. Examine the combustion literature, and discuss the solution in the context of Chapman–Jouguet detonation theory.

9.7 For air, modeled as a calorically perfect ideal gas with $\gamma = 7/5$, $R = 287\,\text{J/kg/K}$, with ambient temperature and pressure of $T_1 = 300\,\text{K}$, $p_1 = 10^5\,\text{Pa}$, and shock speed $U = 500$ m/s, identify the shock state's density, velocity, pressure, and temperature. Find the jump in entropy. Find the jump in the square of the Mach number. Assess if the shocked state is physical.

9.8 For air, modeled as a calorically perfect ideal gas with $\gamma = 7/5$, $R = 287\,\text{J/kg/K}$, with ambient temperature and pressure of $T_1 = 300\,\text{K}$, $p_1 = 10^5\,\text{Pa}$, and shock speed $U = 300\,\text{m/s}$, identify the shock state's density, velocity, pressure, and temperature. Find the jump in entropy. Find the jump in the square of the Mach number. Assess if the shocked state is physical.

9.9 Introduce a small change in initial temperature to an example problem in the text, and consider the fluorocarbon gas $C_{13}F_{22}$, PP10. It is well modeled as a van der Waals gas near the vapor dome with constants $R = 14.4843\,\text{J/kg/K}$, $c_v = 1131.588\,\text{J/kg/K}$, $a = 22.3889\,\text{m}^5/\text{kg}/\text{s}^2$, $b = 0.000\,7244\,\text{m}^3/\text{kg}$. Evaluate the state that arises through a discontinuity for $\hat{v}_1 = 0.002\,700\,\text{m}^3/\text{kg}$, $T_1 = 656\,\text{K}$, $U = 37.89\,\text{m/s}$. The reference values can be taken as $e_o = 0\,\text{J/kg}$, $s_o = 0\,\text{J/kg/K}$, $\hat{v}_o = 1\,\text{m}^3/\text{kg}$, $T_o = 300\,\text{K}$. Identify all possible states admitted by analysis of Rankine–Hugoniot jump equations and describe the physical viability of each.

9.10 Introduce two small changes to the previous problem. Take instead $T_1 = 700\,\text{K}$ and $U = 100\,\text{m/s}$. Identify all possible states admitted by analysis of Rankine–Hugoniot jump equations and describe the physical viability of each.

9.11 Consider the one-dimensional, unsteady mass and linear momentum equations for compressible, inviscid flow with no body force of an isothermal, calorically perfect ideal gas. Write these equations in characteristic form.

9.12 Consider the inviscid Bateman–Burgers equation. Via coordinate stretching, show an acceptable similarity variable is $\eta = x/t$. Cast the Bateman–Burgers equation as an ordinary differential equation in this single similarity variable, and solve to find two solutions. For the solution that varies with x and t, plot contours of u in the (x, t) plane. Show this solution is equivalent to that found with the method of characteristics.

9.13 Consider the inviscid Bateman–Burgers equation. For an initial condition of $u(x,0) = H(-x)$, where H is the Heaviside unit step function, determine and plot $u(x, t = 1)$. Repeat for $u(x,0) = H(x)$.

9.14 Consider the Euler equations for a one-dimensional, unsteady, homeoentropic flow of a calorically perfect ideal gas. Show that the mass and linear momentum equations are self-similar and can be reduced to a pair of ordinary differential equations after employing the similarity variable $\eta = x/t$. Identify possible solutions $\rho(\eta)$, $u(\eta)$ and the corresponding $\rho(x,t)$, $u(x,t)$. Confirm, for $\rho_o = 1$, $\gamma = 2$, $c_o = 10$, that $\rho(x,t) = (x/t)^2/900$, $u(x,t) = (2/3)(x/t)$ is a solution.

9.15 Consider an exact solution for viscous shock structure in helium. Take $u_1 = 3000$ m/s, $p_1 = 10^5\,\text{Pa}$, $T_1 = 300\,\text{K}$, $x_{in} = 0\,\text{m}$, $u_{in} = 2999\,\text{m/s}$. Examine the literature to find appropriate values for R, γ, c_p, and μ. To enable the use of an exact solution, take $Pr = 3/4$; this is not far from the accepted value for helium of $Pr = 0.680$ at $300\,\text{K}$. Find and plot $u(x)$, $(s(x) - s_1)/c_v$, $\dot{\mathcal{I}}(x)$.

9.16 Formulate the problem of one-dimensional acoustics in cylindrical coordinates. Separate variables, and find the most general solution for the pressure perturbation $p_1(r,t)$.

9.17 Find an appropriate scaling and show for homeoentropic flow of a calorically perfect ideal gas with ratio of specific heats γ that the mass and inviscid linear momenta equations form a dimensionless system given by

$$\frac{\partial \rho}{\partial t} + (\mathbf{v}^T \cdot \nabla)\rho + \rho \nabla^T \cdot \mathbf{v} = 0,$$

$$\frac{\partial \mathbf{v}}{\partial t} + (\mathbf{v}^T \cdot \nabla)\mathbf{v} + \rho^{\gamma-2} \nabla \rho = \mathbf{0}.$$

9.18 For the dimensionless one-dimensional Cartesian version of the homeoentropic Euler equations for a calorically perfect ideal gas, taking $\gamma = 3$, show a solution is given by

$$\rho = \frac{\sqrt{1 - \frac{x^2}{t^2+1}}}{\sqrt{t^2+1}}, \qquad u = \frac{tx}{t^2+1}.$$

Plot $\rho(x)$ at various times, and show how this corresponds to a gas expanding into a vacuum. Provide an (x, t) diagram giving a contour plot of $\rho(x, t)$. See Cantwell (2002) for more general discussion of symmetry methods for compressible flows that motivate this and other solutions.

9.19 For the dimensionless two-dimensional Cartesian version of the homeoentropic Euler equations for a calorically perfect ideal gas, taking $\gamma = 2$, show a solution is given by

$$\rho = \frac{1 - \frac{x^2+y^2}{t^2+1}}{2(t^2+1)}, \qquad u = \frac{tx}{t^2+1}, \qquad v = \frac{ty}{t^2+1}.$$

Plot $\rho(x, y = 0)$ at various times and show how this corresponds to a gas expanding into a vacuum.

9.20 For the dimensionless three-dimensional Cartesian version of the homeoentropic Euler equations for a calorically perfect ideal gas, taking $\gamma = 5/3$, show a solution is given by

$$\rho = \frac{\left(1 - \frac{x^2+y^2+z^2}{t^2+1}\right)^{3/2}}{3\sqrt{3}(t^2+1)^{3/2}}, \qquad u = \frac{tx}{t^2+1}, \qquad v = \frac{ty}{t^2+1}, \qquad w = \frac{tz}{t^2+1}.$$

Plot $\rho(x, y = 0, z = 0)$ at various times and show how this corresponds to a gas expanding into a vacuum.

9.21 Repeat the numerical simulations for a Taylor–Sedov blast wave with $\gamma = 7/5$ and reproduce the plots for scaled pressure, velocity, and density, $f(\eta)$, $\phi(\eta)$, and $\psi(\eta)$.

FURTHER READING

Anderson, J. D. (2021). *Modern Compressible Flow with Historical Perspective*, 4th ed. New York: McGraw-Hill.

Brodkey, R. S. (1967). *The Phenomena of Fluid Motions*. New York: Dover.

Emanuel, G. (1986). *Gasdynamics: Theory and Applications*. New York: AIAA.

Hayes, W. D., and Probstein, R. F. (1959). *Hypersonic Flow Theory*. New York: Academic Press.

Lighthill, J. (1978). *Waves in Fluids*. Cambridge, UK: Cambridge University Press.

Novotný, A., and Straškraba, I. (2004). *Introduction to the Mathematical Theory of Compressible Flow*. Oxford: Oxford University Press.

Ramm, H. J. (1990). *Fluid Dynamics for the Study of Transonic Flow*. Oxford: Oxford University Press.

Sasoh, A. (2020). *Compressible Fluid Dynamics and Shock Waves*. Singapore: Springer.

Taylor, G. I. (1963). *The Scientific Papers of Sir Geoffrey Ingram Taylor*. 4 vols. Cambridge, UK: Cambridge University Press.

Thorne, K. S., and Blandford, R. D. (2017). *Modern Classical Physics: Optics, Fluids, Plasmas, Elasticity, and Statistical Physics*. Princeton, New Jersey: Princeton University Press.

von Mises, R. (2004). *Mathematical Theory of Compressible Flow*. Mineola, New York: Dover.

von Neumann, J. (1961). *Collected Works*. New York: Pergamon.

Zel'dovich, Y. B., and Raizer, Y. P. (2002). *Physics of Shock Waves and High-Temperature Hydrodynamic Phenomena*. Mineola, New York: Dover.

Zucrow, M. J., and Hoffman, J. D. (1976). *Gas Dynamics*, 2 vols. New York: John Wiley.

REFERENCES

Bateman, H. (1915). Some recent researches in the motion of fluids, *Monthly Weather Review*, **43**(4), 163–170.

Becker, R. (1922). Stoßwelle und Detonation, *Zeitschrift für Physik*, **8**, 321–362, English translation in NACA-TM-505, 1929.

Bird, R. B., Stewart, W. E., and Lightfoot, E. N. (2007). *Transport Phenomena*, revised 2nd ed. New York: John Wiley.

Burgers, J. M. (1948). A mathematical model illustrating the theory of turbulence, *Advances in Applied Mathematics*, **1**, 171–199.

Cantwell, B. J. (2002). *Introduction to Symmetry Analysis*. New York: Cambridge University Press.

Colonna, P., and Guardone, A. (2006). Molecular interpretation of nonclassical gas dynamics of dense vapors under the van der Waals model, *Physics of Fluids*, **19**(8), 086102.

Courant, R., and Friedrichs, K. O. (1976). *Supersonic Flow and Shock Waves*. New York: Springer.

Emanuel, G. (2016). *Analytical Fluid Dynamics*, 3rd ed. Boca Raton, Florida: CRC Press.

Finn, B. S. (1964). Laplace and the speed of sound, *Isis*, **55**(1), 7–19.

Garabedian, P. R. (1998). *Partial Differential Equations*. Providence, Rhode Island: American Mathematical Society.

Ghia, U., Bayyuk, S., Benek, J. et al. (2021). Recommended Practice for Code Verification in Computational Fluid Dynamics, AIAA R-141-2021; portions reprinted by permission of the American Institute of Aeronautics and Astronautics, Inc.

Gilbarg, D., and Paolucci, D. (1953). The structure of shock waves in the continuum theory of fluids, *Journal of Rational Mechanics and Analysis*, **2**(5), 617–642.

Goodwine, B. (2010). *Engineering Differential Equations: Theory and Applications*. New York: Springer.

Iannelli, J. (2013). An exact non-linear Navier–Stokes compressible-flow solution for CFD code verification, *International Journal for Numerical Methods in Fluids*, **72**(2), 157–176.

LeVeque, R. J. (1992). *Numerical Methods for Conservation Laws*. Basel: Birkhäuser.

LeVeque, R. J. (2002). *Finite Volume Methods for Hyperbolic Problems*. Cambridge, UK: Cambridge University Press.

Liepmann, H. W., and Roshko, A. (2002). *Elements of Gasdynamics*. New York: Dover.

Ludford, G. S. S. (1951). The classification of one-dimensional flows and the general problem of a compressible, viscous, heat-conducting fluid, *Journal of the Aeronautical Sciences*, **18**(12), 830–834.

Morduchow, M., and Libby, P. A. (1949). On a complete solution of the one-dimensional flow equations of a viscous, heat conducting, compressible gas, *Journal of the Aeronautical Sciences*, **16**(11), 674–684.

Müller, I., and Ruggeri, T. (1998). *Rational Extended Thermodynamics*, 2nd ed. New York: Springer.

Newton, I. (1999). *The Principia: Mathematical Principles of Natural Philosophy*. Cohen, I. B. and Whitman, A., trans. Berkeley, California: University of California Press.

Powers, J. M., and Sen, M. (2015). *Mathematical Methods in Engineering*. New York: Cambridge University Press.

Sedov, L. I. (1946). "Rasprostraneniya sil'nykh vzryvnykh voln," *Prikladnaya Matematika i Mekhanika*, **10**: 241–250.

Shapiro, A. H. (1953, 1954). *The Dynamics and Thermodynamics of Compressible Fluid Flow*, Vol. 1. New York: John Wiley; Vol. 2. New York: Ronald.

Taylor, G. I. (1910). The conditions necessary for discontinuous motion in gases, *Proceedings of the Royal Society of London Series A*, **84**(571), 271–377.

Taylor, G. I. (1950a). The formation of a blast wave by a very intense explosion. I. Theoretical discussion, *Proceedings of the Royal Society of London. Series A, Mathematical and Physical Sciences*, **201**(1065): 159–174.

Taylor, G. I. (1950b). The formation of a blast wave by a very intense explosion. II. The atomic explosion of 1945, *Proceedings of the Royal Society of London. Series A, Mathematical and Physical Sciences*, **201**(1065): 175–186.

Thompson, P. A. (1971). A fundamental derivative in gasdynamics, *Physics of Fluids*, **14**(9), 1843–1849.

Vincenti, W. G., and Kruger, C. H. (1965). *Introduction to Physical Gas Dynamics*. New York: John Wiley.

von Mises, R. (1950). On the thickness of a steady shock wave, *Journal of the Aeronautical Sciences*, **17**(9), 551–554.

Whitham, G. B. (1974). *Linear and Nonlinear Waves*. New York: John Wiley.

10 One-Dimensional Viscous Flow

Here we consider some basic problems in one-dimensional viscous flow. Application areas range from ordinary pipe flow to microscale fluid mechanics, such as found in micro-electronic or biological systems. A typical scenario is shown in Fig. 10.1. We will select this and various problems that illustrate the effects of advection, diffusion, and unsteady effects. Problems with constant fluid properties as well as variable properties will be considered. We will not consider body forces, so $\mathbf{f} = \mathbf{0}$. And we will typically insist that the velocity and temperature gradients in the x and z directions are zero, $\partial \mathbf{v}/\partial x = \mathbf{0}$, $\partial \mathbf{v}/\partial z = \mathbf{0}$, $\partial T/\partial x = 0$, $\partial T/\partial z = 0$. This will result typically in a flow that has at most one nonzero velocity component, $\mathbf{v} = u(y,t)\mathbf{i}$ in the x direction, and that varies in the y direction and perhaps in t. In Section 10.2.6, we will also consider a problem in cylindrical coordinates that has a single nonzero velocity component with variation in time and one spatial dimension: $\mathbf{v} = v_\theta(r,t)\mathbf{e}_\theta$.

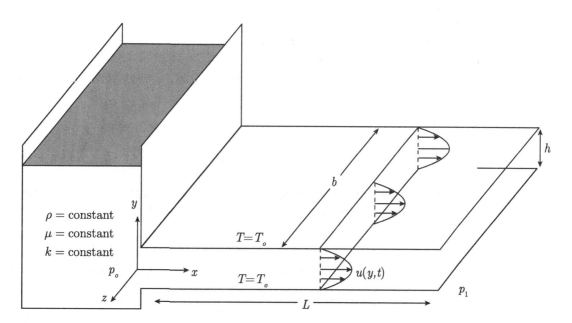

Figure 10.1 Pressure gradient-driven (Poiseuille) flow of liquid fluid in a slot with isothermal walls.

10.1 Flow with No Effects of Inertia

The first type of one-dimensional flow we will consider is flow with with negligible effects of inertia. For a fluid with finite density to have no inertial effects, it must be steady ($\partial/\partial t = 0$) and be what is known as *fully developed*. Fully developed flows will be seen to have no advective acceleration. We will see that these assumptions give rise to flows with a nonzero x velocity u that varies in the y direction, and that other velocities v, and w, will be zero.

10.1.1 Incompressible Poiseuille Flow in a Slot

Consider the flow depicted in Fig. 10.1. Here we have a large reservoir of liquid fluid with constant properties ρ, μ, k with a long narrow slot located around $y = 0$. We take the length of the slot in the z direction, b, to be long relative to the slot width in the y direction h. Attached to the slot are two parallel plates, separated by distance in the y direction h. The length of the plates in the x direction is L. We take $L \gg h$. Because of gravity forces, which we neglect in the slot, the pressure at the entrance of the slot p_o is higher than atmospheric. At the end of the slot, the fluid expels to the atmosphere at p_1. Hence, there is a pressure gradient in the x direction, which drives the flow in the slot. We will see that the flow is resisted by viscous stresses. An analogous flow in a circular duct is defined as a *Hagen*[1]*–Poiseuille*[2] *flow*. For convenience, we will call the simpler flow in a slot Poiseuille flow.

Near $x = 0$, the flow accelerates in what is known as the *entrance length*. If L is sufficiently long, we observe that sufficiently downstream of $x = 0$, the fluid particles no longer accelerate. It is at this point where the viscous shear forces exactly balance the pressure forces to give rise to the fully developed velocity field.

For this flow, let us make the additional assumptions that (1) there is no imposed pressure gradient in the z direction, and (2) the walls are held at a constant temperature, T_o. Incorporating some of these assumptions, we recast the incompressible constant property Navier–Stokes equations of Section 6.4.5 as

$$\partial_i v_i = 0, \tag{10.1}$$

$$\rho \partial_o v_i + \rho v_j \partial_j v_i = -\partial_i p + \mu \partial_j \partial_j v_i, \tag{10.2}$$

$$\rho c_p \partial_o T + \rho c_p v_j \partial_j T = k \partial_i \partial_i T + 2\mu \partial_{(i} v_{j)} \partial_{(i} v_{j)}. \tag{10.3}$$

Here we have five equations in five unknowns, v_i, p, and T. Because we model a liquid here, we use the form of the energy equation for $\beta = 0$, Eq. (6.82); thus, there is no dp/dt term.

---

[1] Gotthilf Ludwig Hagen, 1797–1884, German engineer who measured velocity of water in small diameter tubes.
[2] Jean Léonard Marie Poiseuille, 1797–1869, French physicist and physiologist.

10.1 Flow with No Effects of Inertia

As for all incompressible flows with constant properties, we can get the velocity field by only considering the mass and momenta equations. To get the temperature field, we will need the velocity field, however. The mass equation, recalling that gradients in x and z are zero, gives us

$$\underbrace{\frac{\partial}{\partial x} u}_{=0} + \frac{\partial}{\partial y} v + \underbrace{\frac{\partial}{\partial z} w}_{=0} = 0. \tag{10.4}$$

So the mass equation gives us

$$\frac{\partial v}{\partial y} = 0. \tag{10.5}$$

Now, from our assumptions of steady and fully developed flow, we know that v cannot be a function of x, z, or t. So the partial becomes a ordinary derivative, and mass conservation holds that $dv/dy = 0$. Integrating, we find that $v(y) = C$. The constant C must be zero, because we must satisfy a no-penetration boundary condition at either wall that $v(y = h/2) = v(y = -h/2) = 0$. Hence, mass conservation, coupled with the no-penetration boundary condition, gives us

$$v = 0. \tag{10.6}$$

Now consider the x momentum equation:

$$\rho \underbrace{\frac{\partial}{\partial t} u}_{=0} + \rho u \underbrace{\frac{\partial}{\partial x} u}_{=0} + \rho \underbrace{v}_{=0} \frac{\partial}{\partial y} u + \rho w \underbrace{\frac{\partial}{\partial z} u}_{=0} = -\frac{\partial p}{\partial x} + \mu \left(\underbrace{\frac{\partial^2}{\partial x^2} u}_{=0} + \frac{\partial^2}{\partial y^2} u + \underbrace{\frac{\partial^2}{\partial z^2} u}_{=0} \right), \tag{10.7}$$

$$0 = -\frac{\partial p}{\partial x} + \mu \frac{\partial^2 u}{\partial y^2}. \tag{10.8}$$

For this fully developed flow the acceleration, that is, the material derivative of velocity, is formally zero, and the equation gives rise to a balance of pressure and viscous surface forces.

For the y momentum equation, we get

$$\rho \underbrace{\frac{\partial}{\partial t} \underbrace{v}_{=0}}_{=0} + \rho u \underbrace{\frac{\partial}{\partial x} \underbrace{v}_{=0}}_{=0} + \rho \underbrace{v}_{=0} \underbrace{\frac{\partial}{\partial y} \underbrace{v}_{=0}}_{=0} + \rho w \underbrace{\frac{\partial}{\partial z} \underbrace{v}_{=0}}_{=0} = -\frac{\partial p}{\partial y} \tag{10.9}$$

$$+ \mu \left(\frac{\partial^2}{\partial x^2} v + \frac{\partial^2}{\partial y^2} \underbrace{v}_{=0} + \frac{\partial^2}{\partial z^2} \underbrace{v}_{=0} \right),$$

$$0 = \frac{\partial p}{\partial y}. \tag{10.10}$$

Hence, $p = p(x, z)$, but because we have assumed there is no pressure gradient in the z direction, we have at most that $p = p(x)$.

For the z momentum equation we get

$$\rho \underbrace{\frac{\partial}{\partial t} w}_{=0} + \rho u \underbrace{\frac{\partial}{\partial x} w}_{=0} + \rho \underbrace{v}_{=0} \frac{\partial}{\partial y} w + \rho w \underbrace{\frac{\partial}{\partial z} w}_{=0} = -\underbrace{\frac{\partial p}{\partial z}}_{=0} \quad (10.11)$$

$$+ \mu \left(\underbrace{\frac{\partial^2}{\partial x^2} w}_{=0} + \frac{\partial^2}{\partial y^2} w + \underbrace{\frac{\partial^2}{\partial z^2} w}_{=0} \right),$$

$$0 = \frac{\partial^2 w}{\partial y^2}. \quad (10.12)$$

Solution of this partial differential equation gives us

$$w = f(x,z)y + g(x,z). \quad (10.13)$$

Now to satisfy the no-slip condition, we must have $w = 0$ at $y = \pm h/2$. This leads us to two linear equations for f and g:

$$\begin{pmatrix} \frac{h}{2} & 1 \\ -\frac{h}{2} & 1 \end{pmatrix} \begin{pmatrix} f(x,z) \\ g(x,z) \end{pmatrix} = \begin{pmatrix} 0 \\ 0 \end{pmatrix}. \quad (10.14)$$

Because the determinant of the coefficient matrix, $h/2 + h/2 = h$, is nonzero, the only solution is the trivial solution $f(x,z) = g(x,z) = 0$. Hence,

$$w = 0. \quad (10.15)$$

Next consider how the energy equation reduces:

$$\rho c_p \underbrace{\frac{\partial}{\partial t} T}_{=0} + \rho c_p \left(u \underbrace{\frac{\partial}{\partial x} T}_{=0} + \underbrace{v}_{=0} \frac{\partial}{\partial y} T + \underbrace{w}_{=0} \frac{\partial}{\partial z} T \right) = k \left(\underbrace{\frac{\partial^2}{\partial x^2} T}_{=0} + \frac{\partial^2}{\partial y^2} T + \underbrace{\frac{\partial^2}{\partial z^2} T}_{=0} \right)$$
$$+ 2\mu \partial_{(i} v_{j)} \partial_{(i} v_{j)}. \quad (10.16)$$

The energy equation with our assumptions yields

$$0 = k \frac{\partial^2 T}{\partial y^2} + 2\mu \partial_{(i} v_{j)} \partial_{(i} v_{j)}. \quad (10.17)$$

There is no tendency for a particle's temperature to increase. There is a balance between thermal energy generated by viscous dissipation and that conducted away by energy diffusion. Thus, the energy path is (1) viscous work is done to generate thermal energy, (2) thermal energy diffuses throughout the channel and out the boundary. Now consider the viscous dissipation term for this flow:

$$\partial_i v_j = \begin{pmatrix} \underbrace{\partial_1 v_1}_{=0} & \underbrace{\partial_1 v_2}_{=0} & \underbrace{\partial_1 v_3}_{=0} \\ \partial_2 v_1 & \underbrace{\partial_2 v_2}_{=0} & \underbrace{\partial_2 v_3}_{=0} \\ \underbrace{\partial_3 v_1}_{=0} & \underbrace{\partial_3 v_2}_{=0} & \underbrace{\partial_3 v_3}_{=0} \end{pmatrix} = \begin{pmatrix} 0 & 0 & 0 \\ \partial_2 v_1 & 0 & 0 \\ 0 & 0 & 0 \end{pmatrix}, \tag{10.18}$$

$$\partial_{(i} v_{j)} = \begin{pmatrix} 0 & \frac{1}{2}\left(\partial_2 v_1 + \underbrace{\partial_1 v_2}_{=0}\right) & 0 \\ \frac{1}{2}\left(\partial_2 v_1 + \underbrace{\partial_1 v_2}_{=0}\right) & 0 & 0 \\ 0 & 0 & 0 \end{pmatrix} = \begin{pmatrix} 0 & \frac{1}{2}\frac{\partial u}{\partial y} & 0 \\ \frac{1}{2}\frac{\partial u}{\partial y} & 0 & 0 \\ 0 & 0 & 0 \end{pmatrix}. \tag{10.19}$$

Further,

$$\partial_{(i} v_{j)} \partial_{(i} v_{j)} = \left(\frac{1}{2}\frac{\partial u}{\partial y}\right)^2 + \left(\frac{1}{2}\frac{\partial u}{\partial y}\right)^2 = \frac{1}{2}\left(\frac{\partial u}{\partial y}\right)^2. \tag{10.20}$$

So the energy equation becomes finally

$$0 = k\frac{\partial^2 T}{\partial y^2} + \mu\left(\frac{\partial u}{\partial y}\right)^2. \tag{10.21}$$

At this point we have the x momentum and energy equations as the only two that seem to have any substance:

$$0 = -\frac{\partial p}{\partial x} + \mu\frac{\partial^2 u}{\partial y^2}, \quad 0 = k\frac{\partial^2 T}{\partial y^2} + \mu\left(\frac{\partial u}{\partial y}\right)^2. \tag{10.22}$$

This looks like two equations in three unknowns. One peculiarity of incompressible equations is that there is always some side condition, that ultimately hinges on the mass equation, that really gives a third equation. Without going into details, it involves for general flows solving a Poisson equation for pressure that is of the form $\nabla^2 p = f(u,v)$. Section 6.4.4 has described this in general terms. The Poisson equation involves second derivatives of pressure. Here we can obtain a simple form of this general equation by taking the partial derivative with respect to x of the x momentum equation:

$$0 = -\frac{\partial^2 p}{\partial x^2} + \mu\frac{\partial}{\partial x}\frac{\partial^2 u}{\partial y^2} = -\frac{\partial^2 p}{\partial x^2} + \frac{\partial^2}{\partial y^2}\underbrace{\frac{\partial u}{\partial x}}_{=0}. \tag{10.23}$$

The viscous term here is zero because of our assumption of fully developed flow. Moreover, because $p = p(x)$ only, we then get

$$\frac{d^2 p}{dx^2} = 0, \quad p(0) = p_o, \; p(L) = p_1, \tag{10.24}$$

which has a solution showing the pressure field must be linear in x:

$$p(x) = p_o - \frac{p_o - p_1}{L}x, \quad \text{so} \quad \frac{dp}{dx} = -\frac{p_o - p_1}{L}. \tag{10.25}$$

Now, because u is at most a function of y, we can convert partial derivatives to ordinary derivatives, and write the x momentum equation and energy equation as two ordinary differential equations in two unknowns with appropriate boundary conditions at the wall $y = \pm h/2$:

$$\frac{d^2 u}{dy^2} = -\frac{p_o - p_1}{\mu L}, \qquad u\left(\frac{h}{2}\right) = 0, \quad u\left(-\frac{h}{2}\right) = 0, \tag{10.26}$$

$$\frac{d^2 T}{dy^2} = -\frac{\mu}{k}\left(\frac{du}{dy}\right)^2, \qquad T\left(\frac{h}{2}\right) = T_o, \quad T\left(-\frac{h}{2}\right) = T_o. \tag{10.27}$$

We could solve these equations directly, but instead let us first cast them in dimensionless form. This will give our results some universality and efficiency. Moreover, it will reveal more fundamental groups of terms that govern the fluid behavior. Let us select scales such that dimensionless variables, denoted by a * subscript, are as follows:

$$y_* = \frac{y}{h}, \qquad T_* = \frac{T - T_o}{T_o}, \qquad u_* = \frac{u}{u_c}. \tag{10.28}$$

We have yet to determine the characteristic velocity u_c. The dimensionless temperature has been chosen to render it zero at the boundaries. With these choices, the x momentum equation becomes

$$\frac{u_c}{h^2}\frac{d^2 u_*}{dy_*^2} = -\frac{p_o - p_1}{\mu L}, \quad \text{so} \quad \frac{d^2 u_*}{dy_*^2} = -\frac{(p_o - p_1)h^2}{\mu L u_c}, \tag{10.29}$$

$$u_c u_*(x_* h = h/2) = u_c u_*(x_* h = -h/2) = 0, \quad \text{so} \quad u_*(x_* = 1/2) = u_*(x_* = -1/2) = 0. \tag{10.30}$$

Let us now choose the characteristic velocity to render the x momentum equation to have a simple form:

$$u_c \equiv \frac{(p_o - p_1)h^2}{\mu L}. \tag{10.31}$$

Now scale the energy equation:

$$\frac{T_o}{h^2}\frac{d^2 T_*}{dy_*^2} = -\frac{\mu u_c^2}{k h^2}\left(\frac{du_*}{dy_*}\right)^2, \tag{10.32}$$

$$\frac{d^2 T_*}{dy_*^2} = -\frac{\mu u_c^2}{k T_o}\left(\frac{du_*}{dy_*}\right)^2 = -\frac{\mu c_p}{k}\frac{u_c^2}{c_p T_o}\left(\frac{du_*}{dy_*}\right)^2 = -Pr Ec\left(\frac{du_*}{dy_*}\right)^2, \tag{10.33}$$

$$T_*\left(-\frac{1}{2}\right) = T_*\left(\frac{1}{2}\right) = 0. \tag{10.34}$$

Here we have grouped terms so that the Prandtl number, Eq. (6.158), $Pr = \mu c_p/k$, explicitly appears. Further, we have defined the Eckert[3] number Ec as

$$Ec = \frac{u_c^2}{c_p T_o} = \frac{\left(\frac{(p_o - p_1)h^2}{\mu L}\right)^2}{c_p T_o}. \tag{10.35}$$

---

[3] Ernst R. G. Eckert, 1904–2004, scholar of convective heat transfer.

10.1 Flow with No Effects of Inertia

In summary our dimensionless differential equations and boundary conditions are

$$\frac{d^2 u_*}{dy_*^2} = -1, \quad u\left(\pm\frac{1}{2}\right) = 0, \tag{10.36}$$

$$\frac{d^2 T_*}{dy_*^2} = -PrEc\left(\frac{du_*}{dy_*}\right)^2, \quad T_*\left(\pm\frac{1}{2}\right) = 0. \tag{10.37}$$

These boundary conditions are homogeneous; hence, they do not contribute to a nontrivial solution. The pressure gradient is an inhomogeneous forcing term in the momentum equation, and the viscous dissipation is a forcing term in the energy equation.

The solution for the velocity field that satisfies the differential equation and boundary conditions is quadratic in y_* and is

$$u_* = \frac{1}{2}\left(\left(\frac{1}{2}\right)^2 - y_*^2\right). \tag{10.38}$$

The maximum velocity occurs at $y_* = 0$ and has value

$$u_{*max} = \frac{1}{8}. \tag{10.39}$$

The mean velocity is found through integrating the velocity field to arrive at

$$u_{*mean} = \int_{-1/2}^{1/2} u_*(y_*)\, dy_* = \int_{-1/2}^{1/2} \frac{1}{2}\left(\left(\frac{1}{2}\right)^2 - y_*^2\right) dy_*, \tag{10.40}$$

$$= \frac{1}{2}\left(\frac{1}{4}y_* - \frac{1}{3}y_*^3\right)\bigg|_{-1/2}^{1/2} = \frac{1}{12}. \tag{10.41}$$

We could have scaled the velocity field in such a fashion that either the maximum or the mean velocity was unity. The scaling we chose gave rise to a non-unity value of both. In dimensional terms we could say

$$\frac{u}{\frac{(p_o - p_1)h^2}{\mu L}} = \frac{1}{2}\left(\left(\frac{1}{2}\right)^2 - \left(\frac{y}{h}\right)^2\right). \tag{10.42}$$

The velocity profile is shown in Fig. 10.2. This flow is rotational. For the two-dimensional flow, the only component of vorticity is in the z_* direction, and we have

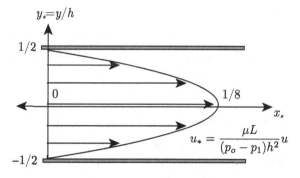

Figure 10.2 Velocity profile for Poiseuille flow in a slot.

$$\omega_{z_*} = \underbrace{\frac{\partial v_*}{\partial x_*}}_{=0} - \frac{\partial u_*}{\partial y_*} = y_*. \tag{10.43}$$

The vorticity magnitude is maximum at the solid walls at $y_* = \pm 1/2$, and zero at the centerline, $y_* = 0$. The dimensionless deformation tensor is

$$\mathbf{D}_* = \begin{pmatrix} \frac{\partial u_*}{\partial x_*} & \frac{1}{2}\left(\frac{\partial u_*}{\partial y_*} + \frac{\partial v_*}{\partial x_*}\right) \\ \frac{1}{2}\left(\frac{\partial u_*}{\partial y_*} + \frac{\partial v_*}{\partial x_*}\right) & \frac{\partial v_*}{\partial y_*} \end{pmatrix} = \begin{pmatrix} 0 & -\frac{y_*}{2} \\ -\frac{y_*}{2} & 0 \end{pmatrix}. \tag{10.44}$$

One can show the eigenvalues of $\mathbf{D}_*$ are given by $\lambda = \pm y_*/2$ and the eigenvectors are at angles of $\pi/4$ and $3\pi/4$ to the horizontal. So on these axes exists the rate of extreme extensional straining.

Now let us get the temperature field:

$$\frac{d^2 T_*}{dy_*^2} = -PrEc \left(\frac{d}{dy_*}\left(\frac{1}{2}\left(\left(\frac{1}{2}\right)^2 - y_*^2\right)\right)\right)^2 = -PrEc\, y_*^2, \tag{10.45}$$

$$\frac{dT_*}{dy_*} = -\frac{1}{3} PrEc\, y_*^3 + C_1, \tag{10.46}$$

$$T_* = -\frac{1}{12} PrEc\, y_*^4 + C_1 y_* + C_2, \tag{10.47}$$

$$0 = -\frac{1}{12} PrEc \frac{1}{16} + C_1 \frac{1}{2} + C_2, \quad y_* = \frac{1}{2}, \tag{10.48}$$

$$0 = -\frac{1}{12} PrEc \frac{1}{16} - C_1 \frac{1}{2} + C_2, \quad y_* = -\frac{1}{2}, \tag{10.49}$$

$$C_1 = 0, \quad C_2 = \frac{PrEc}{192}. \tag{10.50}$$

Regrouping, we find that

$$T_* = \frac{PrEc}{12}\left(\left(\frac{1}{2}\right)^4 - y_*^4\right). \tag{10.51}$$

In terms of dimensional quantities, we can say

$$\frac{T - T_o}{T_o} = \frac{(p_o - p_1)^2 h^4}{12\mu L^2 k T_o}\left(\left(\frac{1}{2}\right)^4 - \left(\frac{y}{h}\right)^4\right). \tag{10.52}$$

A diagram of the dimensionless temperature profile is given in Fig. 10.3. The maximum temperature occurs at the centerline, $y_* = 0$, and has dimensional value

$$T_{max} = T_o + \frac{(p_o - p_1)^2 h^4}{192 \mu L^2 k}. \tag{10.53}$$

From knowledge of the velocity and temperature field, we can calculate other quantities of interest. Let us calculate the field of shear stress and heat flux, and then evaluate both at the wall. First for the shear stress, recall that in dimensional form we have

$$\tau_{ij} = 2\mu \partial_{(i} v_{j)} + \lambda \underbrace{\partial_k v_k}_{=0} \delta_{ij} = 2\mu \partial_{(i} v_{j)}. \tag{10.54}$$

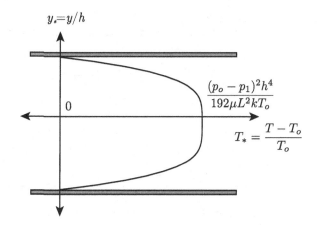

Figure 10.3 Temperature profile for Poiseuille flow in a slot.

We have already seen that the only nonzero components of the symmetric part of the velocity gradient tensor are the 12 and 21 components. Thus, the 21 stress component is

$$\tau_{21} = 2\mu \partial_{(2}v_{1)} = 2\mu \left(\frac{\partial_2 v_1 + \overbrace{\partial_1 v_2}^{=0}}{2} \right) = \mu \partial_2 v_1. \tag{10.55}$$

In (x, y) space, we then say here that

$$\tau_{yx} = \mu \frac{du}{dy}. \tag{10.56}$$

This is a stress on the y (tangential) face that points in the x direction; hence, it is certainly a shearing stress. In dimensionless terms, we can define a characteristic shear stress τ_c, so that the scale shear is $\tau_* = \tau_{yx}/\tau_c$. Thus, our equation for shear becomes

$$\tau_c \tau_* = \frac{\mu u_c}{h} \frac{du_*}{dy_*}. \tag{10.57}$$

Now take

$$\tau_c \equiv \frac{\mu u_c}{h} = \frac{\mu(p_o - p_1)h^2}{h\mu L} = (p_o - p_1)\left(\frac{h}{L}\right). \tag{10.58}$$

With this definition, we get

$$\tau_* = \frac{du_*}{dy_*}. \tag{10.59}$$

Evaluating for the velocity profile of the pressure gradient-driven flow, we find

$$\tau_* = -y_*. \tag{10.60}$$

The stress is zero at the centerline $y_* = 0$ and has maximum magnitude of $1/2$ at either wall, $y_* = \pm 1/2$. In dimensional terms, the wall shear stress τ_w is

$$\tau_w = -\frac{1}{2}(p_o - p_1)\left(\frac{h}{L}\right). \tag{10.61}$$

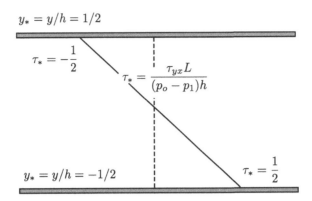

Figure 10.4 Shear stress profile for Poiseuille flow in a slot.

The wall shear stress is governed by the pressure difference and not the viscosity. However, the viscosity plays a determining role in selecting the maximum fluid velocity. The shear profile is presented in Fig. 10.4.

Next, let us calculate the heat flux vector. Recall that, for this flow, with no x or z variation of T, we have the only nonzero component of the heat flux vector as

$$q_y = -k\frac{dT}{dy}. \tag{10.62}$$

Now define scale the heat flux by a characteristic heat flux q_c, to be determined, to obtain a dimensionless heat flux:

$$q_* = \frac{q_y}{q_c}, \quad \text{so} \quad q_c q_* = -\frac{kT_o}{h}\frac{dT_*}{dy_*}, \quad q_* = -\frac{kT_o}{hq_c}\frac{dT_*}{dt_*}. \tag{10.63}$$

Let $q_c \equiv kT_o/h$, so

$$q_* = -\frac{dT_*}{dy_*} = \frac{1}{3}PrEc\, y_*^3. \tag{10.64}$$

For our flow, we have a cubic variation of the heat flux vector magnitude. There is no heat flux at the centerline, which corresponds to this being a region of no shear. The magnitude of the heat flux is maximum at the wall, the region of maximum shear. At the upper wall, we have

$$q_*|_{y_*=1/2} = \frac{1}{24}PrEc. \tag{10.65}$$

The heat flux profile is presented in Fig. 10.5. In dimensional terms we have

$$\frac{q_w}{\frac{kT_o}{h}} = \frac{1}{24}\frac{(p_o-p_1)^2 h^4}{\mu L^2 kT_o}, \quad \text{so} \quad q_w = \frac{1}{24}\frac{(p_o-p_1)^2 h^3}{\mu L^2}. \tag{10.66}$$

10.1.2 Incompressible Couette Flow

We next consider Couette flow for a flow with constant properties ρ, μ, and k. Couette flow implies that there is a moving plate at one boundary and a fixed plate at the other. It is a common experimental configuration, and used often to actually measure a fluid's viscosity.

10.1 Flow with No Effects of Inertia

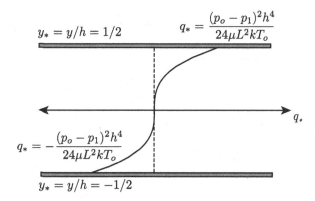

Figure 10.5 Heat flux profile for Poiseuille flow in a slot.

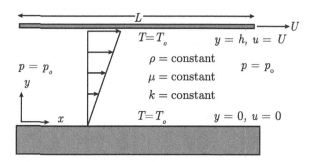

Figure 10.6 Configuration for incompressible Couette flow with isothermal walls.

Here we will take the same assumptions as for pressure gradient-driven flow in a slot, expect for (1) the boundary condition at the upper surface, which we will require to have a constant velocity U, and (2) there will be no pressure gradient. We will also shift the coordinates so that $y = 0$ matches the lower plate surface and $y = h$ matches the upper plate surface. The configuration for this flow is shown in Fig. 10.6.

Our equations governing this flow are similar to Eqs. (10.26, 10.27), except for the absence of the pressure gradient and a slight adjustment of the boundary conditions:

$$\frac{d^2 u}{dy^2} = 0, \qquad u(0) = 0, \quad u(h) = U, \tag{10.67}$$

$$\frac{d^2 T}{dy^2} = -\frac{\mu}{k}\left(\frac{du}{dy}\right)^2, \qquad T(0) = T_o, \quad T(h) = T_o. \tag{10.68}$$

In the momentum equation, there is no acceleration, and the gradient of shear stress must be zero. In the energy equation, there is no energy increase, and generation of thermal energy due to viscous work is balanced by diffusion of the thermal energy, ultimately out of the system through the boundaries. Here there are inhomogeneities in both the forcing terms and the boundary conditions. In terms of work, the pulling of the plate induces work.

Once again, let us scale the equations. This time, we have a natural velocity scale, U, the upper plate velocity. So take

$$y_* = \frac{y}{h}, \qquad T_* = \frac{T - T_o}{T_o}, \qquad u_* = \frac{u}{U}. \qquad (10.69)$$

The momentum equation becomes

$$\frac{d^2 u_*}{dy_*^2} = 0. \qquad (10.70)$$

with $u_*(0) = 0$, $u_*(1) = 1$. This has solution $u_* = C_1 y_* + C_2$. Applying the boundary conditions, we get

$$u_* = y_*. \qquad (10.71)$$

The velocity profile is shown in Fig. 10.6.

Let us now calculate the shear stress profile. With $\tau = \mu(du/dy)$, and taking $\tau_* = \tau/\tau_c$, we get

$$\tau_c \tau_* = \frac{\mu U}{h} \frac{du_*}{dy_*}, \qquad \text{so} \qquad \tau_* = \frac{\mu U}{h \tau_c} \frac{du_*}{dy_*}. \qquad (10.72)$$

Now taking $\tau_c \equiv \mu U/h$, we get

$$\tau_* = \frac{du_*}{dy_*} = 1. \qquad (10.73)$$

The shear is constant.

We now consider the heat transfer problem. Scaling, we get

$$\frac{T_o}{h^2} \frac{d^2 T_*}{dy_*^2} = -\frac{\mu U^2}{k h^2} \left(\frac{du_*}{dy_*}\right)^2, \qquad T_*(0) = T_*(1) = 0, \qquad (10.74)$$

$$\frac{d^2 T_*}{dy_*^2} = -\frac{\mu U^2}{k T_o} \left(\frac{du_*}{dy_*}\right)^2 = -\frac{\mu c}{k} \frac{U^2}{c T_o} \left(\frac{du_*}{dy_*}\right)^2 = -PrEc. \qquad (10.75)$$

Integrating, we get

$$\frac{dT_*}{dy_*} = -PrEc\, y_* + C_1, \qquad \text{so} \qquad T_* = -PrEc \frac{y_*^2}{2} + C_1 y_* + C_2. \qquad (10.76)$$

For $T_*(0) = 0$, we must have $C_2 = 0$, giving $T_* = -PrEc\, y_*^2/2 + C_1 y_*$. For $T_*(1) = 0$, we then get $0 = -PrEc/2 + C_1$, giving $C_1 = PrEc/2$, giving the temperature field as

$$T_*(y_*) = \frac{PrEc}{2} y_*(1 - y_*). \qquad (10.77)$$

For the wall heat transfer, recall $q_y = -k(dT/dy)$. Scaling, we get

$$q_c q_* = -\frac{k T_o}{h} \frac{dT_*}{dy_*}, \qquad \text{so} \qquad q_* = -\frac{k T_o}{h q_c} \frac{dT_*}{dy_*}. \qquad (10.78)$$

Now choose $q_c = kT_o/h$ so that

$$q_* = -\frac{dT_*}{dy_*} = PrEc\left(y_* - \frac{1}{2}\right). \qquad (10.79)$$

There is no heat flux at the midpoint of the channel. At the bottom wall $y_* = 0$, we get for the wall heat flux

$$q_*|_{y_*=0} = -\frac{PrEc}{2}. \qquad (10.80)$$

10.1.3 Incompressible Couette Flow with Pressure Gradient

We next consider Couette flow with a pressure gradient. Here we will take the same assumptions as made in the previous section, except that we will now allow for a pressure gradient. The configuration for this flow is shown in Fig. 10.7.

Our equations governing this flow are

$$\frac{d^2 u}{dy^2} = -\frac{p_o - p_1}{\mu L}, \qquad u(0) = 0, \quad u(h) = U, \qquad (10.81)$$

$$\frac{d^2 T}{dy^2} = -\frac{\mu}{k}\left(\frac{du}{dy}\right)^2, \qquad T(0) = T_o, \quad T(h) = T_o. \qquad (10.82)$$

Once again in the momentum equation, there is no acceleration, and viscous stress gradients balance pressure gradients. In the energy equation, there is no energy increase, and generation of thermal energy due to viscous work is balanced by diffusion of the thermal energy, ultimately out of the system through the boundaries. There are inhomogeneities in both the forcing terms and the boundary conditions. In terms of work, both the pressure gradient and the pulling of the plate induce work.

Once again let us scale the equations. This time, we have a natural velocity scale, U, the upper plate velocity. So take

$$y_* = \frac{y}{h}, \qquad T_* = \frac{T - T_o}{T_o}, \qquad u_* = \frac{u}{U}. \qquad (10.83)$$

The momentum equation becomes

$$\frac{U}{h^2}\frac{d^2 u_*}{dy_*^2} = -\frac{p_o - p_1}{\mu L}, \quad \text{so} \quad \frac{d^2 u_*}{dy_*^2} = -\frac{(p_o - p_1)h^2}{\mu U L}. \qquad (10.84)$$

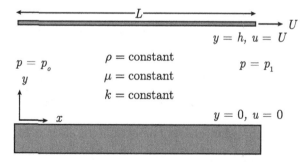

Figure 10.7 Configuration for incompressible Couette flow with pressure gradient.

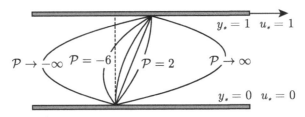

Figure 10.8 Velocity profiles for various values of $\mathcal{P}$ for Couette flow with pressure gradient.

With dimensionless pressure gradient

$$\mathcal{P} \equiv \frac{(p_o - p_1)h^2}{\mu U L}, \tag{10.85}$$

we get

$$\frac{d^2 u_*}{dy_*^2} = -\mathcal{P}, \quad u_*(0) = 0, \quad u_*(1) = 1. \tag{10.86}$$

This has solution

$$u_* = \underbrace{\frac{1}{2}\mathcal{P} y_*(1 - y_*)}_{\text{pressure effect}} + \underbrace{y_*}_{\text{Couette effect}}. \tag{10.87}$$

We see that the pressure gradient generates a velocity profile that is quadratic in y_*. This is distinguished from the Couette effect, which is the effect of the upper plate's motion, which gives a linear profile. Because our governing equation here is linear, it is appropriate to think of these as superposed solutions. Velocity profiles for various values of $\mathcal{P}$ are shown in Fig. 10.8.

Now calculate the shear stress. With $\tau = \mu(du/dy)$, and taking $\tau_* = \tau/\tau_c$, we get

$$\tau_c \tau_* = \frac{\mu U}{h} \frac{du_*}{dy_*}, \quad \text{so} \quad \tau_* = \frac{\mu U}{h \tau_c} \frac{du_*}{dy_*}. \tag{10.88}$$

Taking $\tau_c \equiv \mu U / h$, we get

$$\tau_* = \frac{du_*}{dy_*}, \quad \text{so} \quad \tau_* = -\mathcal{P} y_* + \frac{1}{2}\mathcal{P} + 1. \tag{10.89}$$

At the boundaries, we get

$$\tau_*|_{y_*=0} = \frac{1}{2}\mathcal{P} + 1, \quad \tau_*|_{y_*=1} = -\frac{1}{2}\mathcal{P} + 1. \tag{10.90}$$

The wall shear has pressure gradient and Couette effects. In fact we can select a pressure gradient to balance the Couette effect at one or the other wall, but not both.

We can also calculate the dimensionless volumetric flow rate Q_*, which for incompressible flow is directly proportional to the mass flux. Ignoring how the scaling would be done, we arrive at

$$Q_* = \int_0^1 u_* \, dy_* = \int_0^1 \left(-\frac{1}{2}\mathcal{P}y_*^2 + \left(1+\frac{1}{2}\mathcal{P}\right)y_*\right) dy_*, \tag{10.91}$$

$$= \left(-\frac{1}{6}\mathcal{P}y_*^3\right)_0^1 + \left(1+\frac{1}{2}\mathcal{P}\right)\frac{y_*^2}{2}\bigg|_0^1 = \frac{\mathcal{P}}{12} + \frac{1}{2}. \tag{10.92}$$

Again there is a pressure gradient and a Couette contribution. We could select $\mathcal{P}$ to give no net volumetric flow rate. We summarize some of the special cases as follows:

- $\mathcal{P} \to -\infty$: $u_* = (1/2)\mathcal{P}y_*(1-y_*)$; $\tau_* = \mathcal{P}(1/2 - y_*)$, $Q_* = \mathcal{P}/12$. Here the fluid flows in the opposite direction as driven by the plate because of the large pressure gradient.
- $\mathcal{P} = -6$. Here we get no net mass flow and $u_* = 3y_*^2 - 2y_*$, $\tau_* = 2y_*$, $Q_* = 0$.
- $\mathcal{P} = -2$. Here we get no shear at the bottom wall and $u_* = y_*^2$, $\tau_* = 2y_*$, $Q_* = 1/3$.
- $\mathcal{P} = 0$. Here we have no pressure gradient and $u_* = y_*$, $\tau_* = 1$, $Q_* = 1/2$.
- $\mathcal{P} = 2$. Here we get no shear at the top wall and $u_* = -y_*^2 + 2y_*$, $\tau* = -2y_* + 2$, $Q_* = 2/3$.
- $\mathcal{P} \to \infty$: $u_* = (1/2)\mathcal{P}y_*(1-y_*)$; $\tau_* = \mathcal{P}(1/2 - y_*)$, $Q_* = \mathcal{P}/12$. Here the fluid flows in the same direction as driven by the plate.

We now consider the heat transfer problem. Scaling, we get

$$\frac{T_o}{h^2}\frac{d^2 T_*}{dy_*^2} = -\frac{\mu U^2}{kh^2}\left(\frac{du_*}{dy_*}\right)^2, \qquad T_*(0) = T_*(1) = 0, \tag{10.93}$$

$$\frac{d^2 T_*}{dy_*^2} = -\frac{\mu U^2}{kT_o}\left(\frac{du_*}{dy_*}\right)^2 = -\frac{\mu c_p}{k}\frac{U^2}{c_p T_o}\left(\frac{du_*}{dy_*}\right)^2 = -Pr\,Ec\left(\frac{du_*}{dy_*}\right)^2, \tag{10.94}$$

$$= -Pr\,Ec\,\tau_*^2 = -Pr\,Ec\left(-\mathcal{P}y_* + \frac{1}{2}\mathcal{P} + 1\right)^2. \tag{10.95}$$

Detailed calculation, which can be verified by direct substitution, yields

$$T_* = \frac{Pr\,Ec}{24}y_*(1-y_*)(12 + 4\mathcal{P} + \mathcal{P}^2 - 8\mathcal{P}y_* - 2\mathcal{P}^2 y_* + 2\mathcal{P}^2 y_*^2). \tag{10.96}$$

For the wall heat transfer, recall $q_y = -k(dT/dy)$. Scaling, we get

$$q_c q_* = -\frac{kT_o}{h}\frac{dT_*}{dy_*}, \quad \text{so} \quad q_* = -\frac{kT_o}{hq_c}\frac{dT_*}{dy_*}. \tag{10.97}$$

As before choose $q_c = kT_o/h$, so $q_* = -dT_*/dy_*$. So

$$q_* = Pr\,Ec\left(\frac{\mathcal{P}^2}{3}y_*^3 - \mathcal{P}\left(\frac{1}{2}\mathcal{P} + 1\right)y_*^2 + \left(1+\frac{1}{2}\mathcal{P}\right)^2 y_* - \frac{1}{2} - \frac{\mathcal{P}}{6} - \frac{\mathcal{P}^2}{24}\right). \tag{10.98}$$

At the bottom wall $y_* = 0$, we get for the wall heat flux

$$q_*|_{y_*=0} = -Pr\,Ec\left(\frac{1}{2} + \frac{\mathcal{P}}{6} + \frac{\mathcal{P}^2}{24}\right). \tag{10.99}$$

10.1.4 Compressible Couette Flow

Next consider Couette flow for a flow with variable properties ρ, μ, and k. This introduces a significant complication, and it will be necessary to consider a fully coupled system involving mass, momentum, and energy evolution equations along with thermodynamic state equations and constitutive models for momentum and energy diffusion.

The configuration for this flow is shown in Fig. 10.9. We will make several of the same assumptions used in our incompressible Couette flow model, namely, the flow is steady, $\partial/\partial t = 0$; the flow is fully developed, $\partial/\partial x = \partial/\partial z = 0$; there are no imposed pressure gradients, $\partial p/\partial x = \partial p/\partial z = 0$, body forces are negligible; the upper wall has a constant velocity of U; the lower wall is stationary. However, we adopt the following modifications: variable density is allowed; the viscosity and thermal conductivity are modeled as having a temperature-dependency; the lower wall is taken to be adiabatic (while the upper wall is again taken to be isothermal); the fluid is modeled as a calorically perfect ideal gas. Delaying adoption of some of our assumptions, we take our governing compressible Navier–Stokes equations to be

$$\partial_o \rho + \partial_i(\rho v_i) = 0, \tag{10.100}$$

$$\rho \partial_o v_i + \rho v_j \partial_j v_i = -\partial_i p + \partial_j \tau_{ji}, \tag{10.101}$$

$$\rho \partial_o e + \rho v_j \partial_j e = -\partial_i q_i - p \partial_i v_i + \tau_{ij} \partial_i v_j, \tag{10.102}$$

$$\tau_{ij} = 2\mu \partial_{(i} v_{j)} - \frac{2}{3}\mu \partial_k v_k \delta_{ij}, \qquad q_i = -k \partial_i T, \tag{10.103}$$

$$e = c_v T + \hat{e}, \qquad p = \rho R T, \tag{10.104}$$

$$\mu = \mu_o \left(\frac{T}{T_o}\right)^n, \qquad k = k_o \left(\frac{T}{T_o}\right)^n. \tag{10.105}$$

We now find how they simplify by imposing our assumptions. With $\partial/\partial t = \partial/\partial x = \partial/\partial z = 0$, the mass equation simplifies considerably to yield

$$\frac{\partial}{\partial y}(\rho v) = 0. \tag{10.106}$$

Integrating, this yields

$$\rho v = f(x, z). \tag{10.107}$$

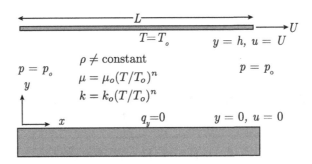

Figure 10.9 Configuration for compressible Couette flow with variable properties, one isothermal wall, and one adiabatic wall.

However, $v = 0$ at both $y = 0$ and $y = h$ because there can be no flux through these solid walls. So because of this, we must insist that to satisfy mass conservation, we must have

$$v(x, y, z, t) = 0. \tag{10.108}$$

This result induces considerable simplification in the remaining equations. First consider the viscous stress tensor. With $v = 0$ and $\partial/\partial x = \partial/\partial z = 0$, it simplifies to the following:

$$\tau_{ij} = \mu \begin{pmatrix} 0 & \frac{\partial u}{\partial y} & 0 \\ \frac{\partial u}{\partial y} & 0 & \frac{\partial w}{\partial y} \\ 0 & \frac{\partial w}{\partial y} & 0 \end{pmatrix}. \tag{10.109}$$

Thus the only nonzero terms are $\tau_{xy} = \tau_{yx} = \mu \partial u/\partial y$ and $\tau_{yz} = \tau_{zy} = \mu \partial w/\partial y$.

We now impose $\partial/\partial t = \partial/\partial x = \partial/\partial z = 0$, $v = 0$, $\tau_{xx} = \tau_{zx} = 0$ to the x momentum equation. This yields

$$0 = \frac{\partial \tau_{yx}}{\partial y}. \tag{10.110}$$

So with no inertial effects and no pressure gradient, we simply insist that there is no gradient in shear stress to maintain a force balance in the x direction. Integrating, we get

$$\tau_{yx} = f(x, z). \tag{10.111}$$

Here $f(x, z)$ is some function. However, because we allow no gradients with respect to x or z, the function at most can be a constant, which we call τ_o. Thus, our relation between stress and strain rate tells us

$$\tau_o = \mu \frac{\partial u}{\partial y}. \tag{10.112}$$

This cannot yet be integrated because μ is not constant. Also, while τ_o is a constant, we do not yet know its value.

We now impose $\partial/\partial t = \partial/\partial x = \partial/\partial z = 0$, $v = 0$, $\tau_{yy} = 0$ on the y momentum equation to yield the simple result

$$\frac{\partial p}{\partial y} = 0. \tag{10.113}$$

Now we already have $\partial p/\partial x = \partial p/\partial z = 0$ by assumption. Combining this with the result from the y momentum equation, as well as $\partial/\partial t = 0$, we see that

$$p(x, y, z, t) = p_o. \tag{10.114}$$

Here p_o is a constant.

We now impose $\partial/\partial t = \partial/\partial x = \partial/\partial z = 0$, $v = 0$, $p = p_o$ on the z momentum equation to yield the simple result

$$0 = \frac{\partial \tau_{yz}}{\partial y}. \tag{10.115}$$

We integrate this, and recognizing that there can be no variation with x, z, or t, we get

$$\tau_{yz} = \tau_1. \tag{10.116}$$

Here τ_1 is a constant. From our constitutive relation for τ_{yz}, recognizing that y is the only variable, we infer

$$\tau_1 = \mu \frac{dw}{dy}. \tag{10.117}$$

Now, we have $\mu = \mu(T)$, and we expect to find $T = T(y)$, so we can think of $\mu = \mu(y)$, which must be positive. Solving for w, and imposing the no-slip condition $w(0) = 0$, we get

$$w(y) = \tau_1 \int_0^y \frac{d\hat{y}}{\mu(\hat{y})}. \tag{10.118}$$

Here $\hat{y}$ is a dummy variable. The fluid velocity must also match the moving plate velocity, so we insist that $w(h) = 0$, giving

$$w(h) = \tau_1 \underbrace{\int_0^h \frac{d\hat{y}}{\mu(\hat{y})}}_{>0} = 0. \tag{10.119}$$

Because the integrand is strictly positive, the only way to satisfy is to insist that $\tau_1 = 0$. So the only nonzero components of the stress tensor are $\tau_{xy} = \tau_{yx} = \tau_1$.

We now impose $\partial/\partial t = \partial/\partial x = \partial/\partial z = 0$, $v = 0$, $p = p_o$, $\tau_{xy} = \tau_{yx} = \tau_o$ on the energy equation to yield

$$0 = -\frac{\partial q_y}{\partial y} + \tau_o \frac{\partial u}{\partial y}. \tag{10.120}$$

Mechanical energy is converted to thermal energy via viscous dissipation, and this energy is diffused in the y direction via thermal conduction. We can integrate this, and recognizing once more that there can be no variation with x, z, or t, recover

$$C = -q_y + \tau_o u. \tag{10.121}$$

Here C is a constant. Now at the bottom surface $y = 0$, we have $q_y = 0$ and $u = 0$ for our adiabatic, no-slip solid wall. Thus $C = 0$ and

$$\tau_o u = q_y. \tag{10.122}$$

Imposing our result, Eq. (10.112), this expands to

$$\mu u \frac{du}{dy} = q_y, \quad \text{so} \quad \frac{d}{dy}\left(\frac{u^2}{2}\right) = \frac{q_y}{\mu}. \tag{10.123}$$

Using the only surviving component of Eq. (10.103) to eliminate q_y, we get

$$\frac{d}{dy}\left(\frac{u^2}{2}\right) = -\frac{k(T)}{\mu(T)} \frac{dT}{dy}. \tag{10.124}$$

Let us integrate this with respect to y. While it may seem natural to apply the operator $\int_0^y$ to the equation, the Dirichlet conditions at the moving plate will be easier to impose if we apply the operator $\int_y^h$. Doing this yields

$$\int_y^h \frac{d}{d\hat{y}}\left(\frac{u^2}{2}\right) d\hat{y} = -\int_y^h \frac{k(T)}{\mu(T)} \frac{dT}{d\hat{y}} d\hat{y}. \tag{10.125}$$

Using basic calculus to change the independent variable on the left to u^2 and on the right to T and transforming the boundary conditions, we get

$$\int_{u^2/2}^{U^2/2} d\left(\frac{\hat{u}^2}{2}\right) = -\int_T^{T_o} \frac{k(\hat{T})}{\mu(\hat{T})} d\hat{T}, \quad \text{so} \quad \frac{1}{2}u^2 - \frac{1}{2}U^2 = -\int_{T_o}^{T} \frac{k(\hat{T})}{\mu(\hat{T})} d\hat{T}. \quad (10.126)$$

This gives us a useful relation between u and T. We still need to see how both vary with y. To do so, we can first recast Eq. (10.112) as

$$\mu(T)\, du = \tau_o\, dy. \quad (10.127)$$

Recognizing that we can use Eq. (10.126) to eliminate T in favor of u, we can say

$$dy = \frac{\mu(T(u))}{\tau_o}\, du. \quad (10.128)$$

Integrating and recognizing that $u = 0$ at $y = 0$, we get

$$y = \frac{1}{\tau_o} \int_0^u \mu(T(\hat{u}))\, d\hat{u}, \quad (10.129)$$

yielding a relationship between u and y. At the upper wall, $y = h$, we must have $u = U$, so

$$h = \frac{1}{\tau_o} \int_0^U \mu(T(\hat{u}))\, d\hat{u}. \quad (10.130)$$

Now we have sufficient information to determine the constant τ_o:

$$\tau_o = \frac{1}{h} \int_0^U \mu(T(\hat{u}))\, d\hat{u}. \quad (10.131)$$

So we rewrite Eq. (10.129) as

$$y = \frac{h \int_0^u \mu(T(\hat{u}))\, d\hat{u}}{\int_0^U \mu(T(\hat{u}))\, d\hat{u}}. \quad (10.132)$$

Our results thus far are valid for general $\mu(T)$ and $k(T)$. Let us impose our model choices of Eqs. (10.105) onto Eq. (10.126) to get

$$\frac{1}{2}u^2 - \frac{1}{2}U^2 = -\int_{T_o}^T \frac{k_o(\hat{T}/T_o)^n}{\mu_o(\hat{T}/T_o)^n} d\hat{T} = -\frac{k_o}{\mu_o}\int_{T_o}^T d\hat{T} = -\frac{k_o}{\mu_o}(T - T_o), \quad (10.133)$$

$$T(u) = T_o + \frac{2\mu_o}{k_o}(U^2 - u^2). \quad (10.134)$$

We now have enough information to identify the constant shear stress. Substituting into Eq. (10.131), we get

$$\tau_o = \frac{\mu_o}{h} \int_0^U \left(1 + \frac{\mu_o}{2k_o T_o}(U^2 - \hat{u}^2)\right)^n d\hat{u}. \quad (10.135)$$

For the special case of temperature-independent viscosity, we have $n = 0$, and get

$$\tau_o = \frac{\mu_o U}{h}, \quad \text{for } n = 0. \quad (10.136)$$

This is the same result we get for incompressible flow with constant properties. For $n = 1$, we allow temperature-dependent properties, and the integration can be performed exactly to yield

$$\tau_o = \frac{\mu_o U}{h}\left(1 + \frac{\mu_o U^2}{3 k_o T_o}\right). \tag{10.137}$$

Note that relative to the $n = 0$ case, the shear stress here is higher. That is because the higher temperature induced a higher viscosity, which is the case for common gases. We can also get the velocity profile. With our choice of viscosity, Eq. (10.132) becomes

$$y = \frac{h \int_0^u \left(1 + \frac{\mu_o}{2 k_o T_o}(U^2 - \hat{u}^2)\right)^n d\hat{u}}{\int_0^U \left(1 + \frac{\mu_o}{2 k_o T_o}(U^2 - \hat{u}^2)\right)^n d\hat{u}}. \tag{10.138}$$

For general n, the u varies nonlinearly with y. For the special case of temperature-independent properties, $n = 0$, we get the linear profile

$$\frac{u}{U} = \frac{y}{h}, \qquad n = 0. \tag{10.139}$$

For simple temperature dependence, with $n = 1$, we get

$$\frac{y}{h} = \frac{u}{U}\frac{\left(1 + \frac{U^2 \mu_o}{2 k_o T_o}\left(1 - \frac{1}{3}\left(\frac{u}{U}\right)^2\right)\right)}{\left(1 + \frac{U^2 \mu_o}{3 k_o T_o}\right)}. \tag{10.140}$$

Clearly, the dimensionless ratio $U^2 \mu_o/(2 k_o T_o)$, which can be related to the Prandtl, Mach, and Eckert numbers, dictates how far the velocity profile deviates from linearity. Let us examine this ratio in more detail. It is

$$\frac{U^2 \mu_o}{2 k_o T_o} = \frac{1}{2}\frac{U^2}{c_p T_o}\frac{\mu_o c_p}{k_o} = \frac{\gamma - 1}{2}\frac{U^2}{\gamma R T_o}\frac{\mu_o c_p}{k_o} = \frac{\gamma - 1}{2} M_o^2 Pr = \frac{1}{2} Ec Pr. \tag{10.141}$$

We give velocity profiles for various values of this ratio in Fig. 10.9. It is not difficult to generate a temperature profile, as we know the relation between T and u. And because the pressure field is constant, we can use knowledge of T to determine the density field $\rho(y)$. One could get a variety of other fields as well; for example, the vorticity field is given here by finding $\omega_z = -du/dy$.

10.2 Flow with Effects of Inertia

Let us consider incompressible, isobaric, one-dimensional flow for which unsteady effects render inertia to be relevant. However, we shall consider flows for which acceleration due to advection is negligible, thus allowing us to avoid nonlinear effects.

10.2.1 Stokes' First Problem

The first problem we will consider is known as Stokes' first problem, as Stokes (1851) addressed it in his work that also developed the Navier–Stokes equations. The problem is shown in Fig.

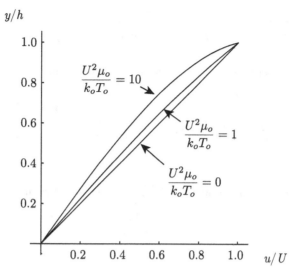

Figure 10.10 Velocity profiles for compressible Couette flow with variable properties, one isothermal wall, and one adiabatic wall for $n = 1$, and various values of $U^2\mu_o/k_o/T_o$.

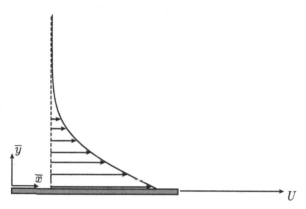

Figure 10.11 Schematic for Stokes' first problem of a suddenly accelerated plate diffusing linear momentum into a fluid at rest.

10.11. Because we will extensively employ dimensionless variables, and the $*$ subscript can become cumbersome, take the over-bar notation to denote a dimensional variable; for example, we will take dimensional distance to be $\bar{x}$, while dimensionless distance will be x. Consider a flat plate of infinite extent lying at rest for $\bar{t} < 0$ on the $\bar{y} = 0$ plane in $(\bar{x}, \bar{y}, \bar{z})$ space. In the volume described by $\bar{y} > 0$ exists a fluid of semi-infinite extent that is at rest at time $\bar{t} < 0$. At $\bar{t} = 0$, the plate is suddenly accelerated to a constant velocity of U, entirely in the $\bar{x}$ direction. Because the no-slip condition is satisfied for the viscous flow, this induces the fluid at the plate surface to acquire an instantaneous velocity of $\bar{u}(0) = U$. Because of diffusion of linear $\bar{x}$ momentum via tangential viscous shear forces, the fluid in the region above the plate begins to acquire a positive velocity in the $\bar{x}$ direction. We will use the Navier–Stokes equations to quantify this behavior. Let us make identical assumptions as we did in the previous section, except that (1) we will not neglect time derivatives, and (2) we will assume all pressure gradients are zero.

Under these assumptions, the $\overline{x}$ momentum equation,

$$\rho \frac{\partial}{\partial \overline{t}}\overline{u} + \underbrace{\rho \overline{u} \frac{\partial}{\partial \overline{x}}\overline{u}}_{=0} + \rho \underbrace{\overline{v}}_{=0} \frac{\partial}{\partial \overline{y}}\overline{u} + \rho \overline{w} \underbrace{\frac{\partial}{\partial \overline{z}}\overline{u}}_{=0} = -\underbrace{\frac{\partial \overline{p}}{\partial \overline{x}}}_{=0} + \mu \left(\underbrace{\frac{\partial^2}{\partial \overline{x}^2}\overline{u}}_{=0} + \frac{\partial^2}{\partial \overline{y}^2}\overline{u} + \underbrace{\frac{\partial^2}{\partial \overline{z}^2}\overline{u}}_{=0} \right), \tag{10.142}$$

is the only relevant component of linear momenta, and reduces to

$$\underbrace{\rho \frac{\partial \overline{u}}{\partial \overline{t}}}_{\text{(mass)(acceleration)}} = \underbrace{\mu \frac{\partial^2 \overline{u}}{\partial \overline{y}^2}}_{\text{shear force}}. \tag{10.143}$$

The energy equation reduces as follows:

$$\rho c_p \frac{\partial}{\partial \overline{t}}\overline{T} + \rho c_p \left(\underbrace{\overline{u} \frac{\partial}{\partial \overline{x}}\overline{T}}_{=0} + \underbrace{\overline{v}}_{=0} \frac{\partial}{\partial \overline{y}}\overline{T} + \underbrace{\overline{w}}_{=0} \frac{\partial}{\partial \overline{z}}\overline{T} \right) = \underbrace{\frac{d\overline{p}}{dt}}_{=0}$$

$$+ k \left(\underbrace{\frac{\partial^2}{\partial \overline{x}^2}\overline{T}}_{=0} + \frac{\partial^2}{\partial \overline{y}^2}\overline{T} + \underbrace{\frac{\partial^2}{\partial \overline{z}^2}\overline{T}}_{=0} \right)$$

$$+ 2\mu \partial_{(i}\overline{v}_{j)} \partial_{(i}\overline{v}_{j)}, \tag{10.144}$$

$$\underbrace{\rho c_p \frac{\partial \overline{T}}{\partial \overline{t}}}_{\text{energy increase}} = \underbrace{k \frac{\partial^2 \overline{T}}{\partial \overline{y}^2}}_{\text{energy diffusion}} + \underbrace{\mu \left(\frac{\partial \overline{u}}{\partial \overline{y}} \right)^2}_{\text{viscous work source}}. \tag{10.145}$$

Let us first consider the $\overline{x}$ momentum equation. Recalling the momentum diffusivity definition, Eq. (1.21), $\nu = \mu/\rho$, we get the following partial differential equation, initial and boundary conditions:

$$\frac{\partial \overline{u}}{\partial \overline{t}} = \nu \frac{\partial^2 \overline{u}}{\partial \overline{y}^2}, \quad \overline{u}(\overline{y}, 0) = 0, \quad \overline{u}(0, \overline{t}) = U, \quad \overline{u}(\infty, \overline{t}) = 0. \tag{10.146}$$

Now let us scale the equations. Choose

$$u = \frac{\overline{u}}{U}, \quad t = \frac{\overline{t}}{t_c}, \quad y = \frac{\overline{y}}{y_c}. \tag{10.147}$$

Here u, t, and y are dimensionless. We have yet to specify the characteristic length, y_c, and time, t_c. The equations become

$$\frac{U}{t_c} \frac{\partial u}{\partial t} = \frac{\nu U}{y_c^2} \frac{\partial^2 u}{\partial y^2}, \quad \text{so} \quad \frac{\partial u}{\partial t} = \frac{\nu t_c}{y_c^2} \frac{\partial^2 u}{\partial y^2}. \tag{10.148}$$

We choose
$$y_c \equiv \frac{\nu}{U} = \frac{\mu}{\rho U}. \tag{10.149}$$

Examining the SI units, we see $\mu/(\rho U)$ has units of length: $(\text{N s/m}^2)(\text{m}^3/\text{kg})(\text{s/m}) = (\text{kg m/s}^2)(\text{s/m}^2)(\text{m}^3/\text{kg})(\text{s/m}) = \text{m}$. With this choice, we get

$$\frac{\nu t_c}{y_c^2} = \frac{\nu t_c U^2}{\nu^2} = \frac{t_c U^2}{\nu}. \tag{10.150}$$

This suggests we choose $t_c = \nu/U^2$. With these choices, the complete system can be written as

$$\frac{\partial u}{\partial t} = \frac{\partial^2 u}{\partial y^2}, \quad u(y,0) = 0, \quad u(0,t) = 1, \quad u(\infty,t) = 0. \tag{10.151}$$

Now for *self-similarity*, we seek a transformation that reduces this partial differential equation, as well as its initial and boundary conditions, into an ordinary differential equation with suitable boundary conditions. If this transformation does not exist, no similarity solution exists. In this, but not all cases, the transformation does exist.

Let us first consider a general transformation from a (y,t) coordinate system to a new $(\eta, \hat{t})$ coordinate system. We assume then a general transformation

$$\eta = \eta(y,t), \quad \hat{t} = \hat{t}(y,t). \tag{10.152}$$

Following the procedure of Section 2.5, this induces

$$\begin{pmatrix} d\eta \\ d\hat{t} \end{pmatrix} = \underbrace{\begin{pmatrix} \frac{\partial \eta}{\partial y} & \frac{\partial \eta}{\partial t} \\ \frac{\partial \hat{t}}{\partial y} & \frac{\partial \hat{t}}{\partial t} \end{pmatrix}}_{\mathbf{J}} \begin{pmatrix} dy \\ dt \end{pmatrix}, \tag{10.153}$$

with Jacobian matrix

$$\mathbf{J} = \begin{pmatrix} \frac{\partial \eta}{\partial y} & \frac{\partial \eta}{\partial t} \\ \frac{\partial \hat{t}}{\partial y} & \frac{\partial \hat{t}}{\partial t} \end{pmatrix}. \tag{10.154}$$

The transformation rule for partial differentiation gives

$$\begin{pmatrix} \frac{\partial}{\partial y} \\ \frac{\partial}{\partial t} \end{pmatrix} = \underbrace{\begin{pmatrix} \frac{\partial \eta}{\partial y} & \frac{\partial \hat{t}}{\partial y} \\ \frac{\partial \eta}{\partial t} & \frac{\partial \hat{t}}{\partial t} \end{pmatrix}}_{\mathbf{J}^T} \begin{pmatrix} \frac{\partial}{\partial \eta} \\ \frac{\partial}{\partial \hat{t}} \end{pmatrix}. \tag{10.155}$$

This expands as:

$$\left.\frac{\partial}{\partial y}\right|_t = \left.\frac{\partial \eta}{\partial y}\right|_t \left.\frac{\partial}{\partial \eta}\right|_{\hat{t}} + \left.\frac{\partial \hat{t}}{\partial y}\right|_t \left.\frac{\partial}{\partial \hat{t}}\right|_\eta, \quad \left.\frac{\partial}{\partial t}\right|_y = \left.\frac{\partial \eta}{\partial t}\right|_y \left.\frac{\partial}{\partial \eta}\right|_{\hat{t}} + \left.\frac{\partial \hat{t}}{\partial t}\right|_y \left.\frac{\partial}{\partial \hat{t}}\right|_\eta. \tag{10.156}$$

Now we will restrict ourselves to the transformation

$$\hat{t} = t, \tag{10.157}$$

so we have $\partial \hat{t}/\partial t|_y = 1$ and $\partial \hat{t}/\partial y|_t = 0$. Thus, our rules for differentiation reduce to

$$\left.\frac{\partial}{\partial y}\right|_t = \left.\frac{\partial \eta}{\partial y}\right|_t \left.\frac{\partial}{\partial \eta}\right|_{\hat{t}}, \quad \left.\frac{\partial}{\partial t}\right|_y = \left.\frac{\partial \eta}{\partial t}\right|_y \left.\frac{\partial}{\partial \eta}\right|_{\hat{t}} + \left.\frac{\partial}{\partial \hat{t}}\right|_\eta. \tag{10.158}$$

The next assumption is key for a similarity solution to exist. We restrict ourselves to transformations for which all state variables are at most a function of η. That is, we allow no dependence on $\hat{t}$. Hence, we must require that $\partial/\partial\hat{t}|_\eta = 0$. Moreover, partial derivatives with respect to η become ordinary derivatives, giving us a final form of transformations for the derivatives

$$\left.\frac{\partial}{\partial y}\right|_t = \left.\frac{\partial \eta}{\partial y}\right|_t \frac{d}{d\eta}, \qquad \left.\frac{\partial}{\partial t}\right|_y = \left.\frac{\partial \eta}{\partial t}\right|_y \frac{d}{d\eta}. \tag{10.159}$$

Now returning to Stokes' first problem, let us assume that a similarity solution exists of the form $u(y,t) = u(\eta)$. It is not always possible to find a similarity variable η. One of the more robust ways to find a similarity variable, if it exists, comes from group theory and is explained in detail by Cantwell (2002). Group theory, which is too detailed to explicate in full here, relies on a generalized *symmetry* of equations to find simpler forms. In the same sense that a snowflake, subjected to rotations of $\pi/3$, $2\pi/3$, π, $4\pi/3$, $5\pi/3$, or 2π, is transformed into a form that is indistinguishable from its original form, we seek transformations of the variables in our partial differential equation that map the equation into a form that is indistinguishable from the original. When systems are subject to such transformations, known as group operators, they are said to exhibit symmetry.

Let us subject our governing partial differential equation along with initial and boundary conditions to an additional type of transformation, a simple stretching of space, time, and velocity:

$$\tilde{t} = e^a t, \qquad \tilde{y} = e^b y, \qquad \tilde{u} = e^c u. \tag{10.160}$$

Here the "∼" variables are stretched variables, and a, b, and c are constant parameters. The exponential will be seen to be a convenience that is not absolutely necessary. Note for $a \in (-\infty, \infty)$, $b \in (-\infty, \infty)$, $c \in (-\infty, \infty)$, that $e^a \in (0, \infty)$, $e^b \in (0, \infty)$, $e^c \in (0, \infty)$. So the stretching does not change the direction of the variable; that is, it is not a reflection transformation. With this stretching, the domain of the problem remains unchanged; that is, $t \in [0, \infty)$ maps into $\tilde{t} \in [0, \infty)$; $y \in [0, \infty)$ maps into $\tilde{y} \in [0, \infty)$. The range is also unchanged if we allow $u \in [0, \infty)$, which maps into $\tilde{u} \in [0, \infty)$. Direct substitution of the transformation shows that in the stretched space, the system becomes

$$e^{a-c}\frac{\partial \tilde{u}}{\partial \tilde{t}} = e^{2b-c}\frac{\partial^2 \tilde{u}}{\partial \tilde{y}^2}, \qquad e^{-c}\tilde{u}(\tilde{y},0) = 0, \quad e^{-c}\tilde{u}(0,\tilde{t}) = 1, \quad e^{-c}\tilde{u}(\infty,\tilde{t}) = 0. \tag{10.161}$$

In order that the stretching transformation map the system into a form indistinguishable from the original so that the transformation exhibits symmetry, we must take $c = 0$, $a = 2b$. So our symmetry transformation is

$$\tilde{t} = e^{2b}t, \qquad \tilde{y} = e^b y, \qquad \tilde{u} = u, \tag{10.162}$$

giving in transformed space

$$\frac{\partial \tilde{u}}{\partial \tilde{t}} = \frac{\partial^2 \tilde{u}}{\partial \tilde{y}^2}, \qquad \tilde{u}(\tilde{y},0) = 0, \quad \tilde{u}(0,\tilde{t}) = 1, \quad \tilde{u}(\infty,\tilde{t}) = 0. \tag{10.163}$$

Now both the original and transformed systems are the same, and the remaining stretching parameter b does not enter directly into either formulation, so we cannot expect it in the solution of either form. That is, we expect a solution to be independent of the stretching parameter b. This can be achieved if we take both u and $\tilde{u}$ to be functions of special combinations of the independent variables, combinations that are formed such that b does not appear. Eliminating b via $e^b = \tilde{y}/y$, we get

$$\frac{\tilde{t}}{t} = \left(\frac{\tilde{y}}{y}\right)^2, \quad \text{or after rearrangement} \quad \frac{y}{\sqrt{t}} = \frac{\tilde{y}}{\sqrt{\tilde{t}}}. \tag{10.164}$$

We thus expect $u = u\left(y/\sqrt{t}\right)$ or equivalently $\tilde{u} = \tilde{u}\left(\tilde{y}/\sqrt{\tilde{t}}\right)$. This form also allows $u = u\left(\alpha y/\sqrt{t}\right)$. Here α is any constant. Let us then define our similarity variable η as

$$\eta = \frac{y}{2\sqrt{t}}. \tag{10.165}$$

Here the factor of $1/2$ is simply a convenience adopted so that the solution takes on a traditional form. We would find that any constant in the similarity transformation would induce a self-similar result.

Let us rewrite the differential equation, boundary, and initial conditions in terms of the similarity variable η. We first must use the chain rule to get expressions for the derivatives. Applying the general results just developed, we get

$$\frac{\partial u}{\partial t} = \frac{\partial \eta}{\partial t}\frac{du}{d\eta} = -\frac{1}{2}\frac{y}{2}t^{-3/2}\frac{du}{d\eta} = -\frac{\eta}{2t}\frac{du}{d\eta}, \tag{10.166}$$

$$\frac{\partial u}{\partial y} = \frac{\partial \eta}{\partial y}\frac{du}{d\eta} = \frac{1}{2\sqrt{t}}\frac{du}{d\eta}, \tag{10.167}$$

$$\frac{\partial^2 u}{\partial y^2} = \frac{\partial}{\partial y}\left(\frac{\partial u}{\partial y}\right) = \frac{\partial}{\partial y}\left(\frac{1}{2\sqrt{t}}\frac{du}{d\eta}\right), \tag{10.168}$$

$$= \frac{1}{2\sqrt{t}}\frac{\partial}{\partial y}\left(\frac{du}{d\eta}\right) = \frac{1}{2\sqrt{t}}\left(\frac{1}{2\sqrt{t}}\frac{d^2u}{d\eta^2}\right) = \frac{1}{4t}\frac{d^2u}{d\eta^2}. \tag{10.169}$$

Thus, applying these rules to our governing linear momenta equation, we recover

$$-\frac{\eta}{2t}\frac{du}{d\eta} = \frac{1}{4t}\frac{d^2u}{d\eta^2}, \quad \text{so} \quad \underbrace{-2\eta\frac{du}{d\eta}}_{\text{acceleration}} = \underbrace{\frac{d^2u}{d\eta^2}}_{\text{viscous force imbalance}}, \tag{10.170}$$

$$\frac{d^2u}{d\eta^2} + 2\eta\frac{du}{d\eta} = 0. \tag{10.171}$$

Our governing equation has a singularity at $t = 0$. As it appears on both sides of the equation, we cancel it on both sides, but we shall see that this point is associated with special behavior of the similarity solution. The important result is that the reduced equation has dependency on η only. If this did not occur, we could not have a similarity solution.

Now consider the initial and boundary conditions. They transform as follows:

$$y = 0, \Longrightarrow \eta = 0, \quad y \to \infty, \Longrightarrow \eta \to \infty, \quad t \to 0, \Longrightarrow \eta \to \infty. \tag{10.172}$$

The three important points for t and y collapse into two corresponding points in η. This is also necessary for the similarity solution to exist. Consequently, our conditions in η space reduce to

$$u(0) = 1, \quad \text{no-slip;} \quad u(\infty) = 0, \quad \text{initial and far field.} \tag{10.173}$$

We solve the second-order differential equation by the method of reduction of order, noticing that it is really two first-order equations in disguise:

$$\frac{d}{d\eta}\left(\frac{du}{d\eta}\right) + 2\eta\left(\frac{du}{d\eta}\right) = 0. \tag{10.174}$$

We multiply by the integrating factor e^{η^2}:

$$e^{\eta^2}\frac{d}{d\eta}\left(\frac{du}{d\eta}\right) + 2\eta e^{\eta^2}\left(\frac{du}{d\eta}\right) = 0, \quad \text{so} \quad \frac{d}{d\eta}\left(e^{\eta^2}\frac{du}{d\eta}\right) = 0, \tag{10.175}$$

$$e^{\eta^2}\frac{du}{d\eta} = A, \quad \text{so} \quad \frac{du}{d\eta} = Ae^{-\eta^2}, \tag{10.176}$$

$$u = B + A\int_0^\eta e^{-s^2}\, ds. \tag{10.177}$$

Now applying the condition $u = 1$ at $\eta = 0$ gives

$$1 = B + A\underbrace{\int_0^0 e^{-s^2}\, ds}_{=0}, \quad \text{so} \quad B = 1. \tag{10.178}$$

So we have

$$u = 1 + A\int_0^\eta e^{-s^2}\, ds. \tag{10.179}$$

Now applying the condition $u = 0$ at $\eta \to \infty$, we get

$$0 = 1 + A\underbrace{\int_0^\infty e^{-s^2}\, ds}_{=\sqrt{\pi}/2}, \quad \text{thus,} \quad A = -\frac{2}{\sqrt{\pi}}. \tag{10.180}$$

Though not immediately obvious, it can be shown by a simple variable transformation to a polar coordinate system that the integral from 0 to ∞ has the value $\sqrt{\pi}/2$. It is not surprising that this integral has finite value over the semi-infinite domain as the integrand is bounded between zero and one, and decays rapidly to zero as $s \to \infty$.

Let us divert to evaluate this integral. To do so, consider the related integral I_2 defined over the first quadrant in $s, \hat{s}$ space, for which

$$I_2 \equiv \int_0^\infty \int_0^\infty e^{-s^2 - \hat{s}^2}\, ds\, d\hat{s}, \tag{10.181}$$

$$= \int_0^\infty e^{-\hat{s}^2}\int_0^\infty e^{-s^2}\, ds\, d\hat{s} = \left(\int_0^\infty e^{-s^2}\, ds\right)\left(\int_0^\infty e^{-\hat{s}^2}\, d\hat{s}\right), \tag{10.182}$$

$$= \left(\int_0^\infty e^{-s^2}\, ds\right)^2, \quad \text{so} \quad \sqrt{I_2} = \int_0^\infty e^{-s^2}\, ds. \tag{10.183}$$

Now transform to polar coordinates with $s = r\cos\theta$, $\hat{s} = r\sin\theta$. With this, we can show $ds\, d\hat{s} = r\, dr\, d\theta$ and $s^2 + \hat{s}^2 = r^2$. Substituting this into Eq. (10.181) and changing the limits of integration appropriately, we get

$$I_2 = \int_0^{\pi/2} \int_0^\infty e^{-r^2} r \, dr \, d\theta = \int_0^{\pi/2} \left(-\frac{1}{2}e^{-r^2}\right)_0^\infty d\theta = \int_0^{\pi/2} \left(\frac{1}{2}\right) d\theta = \frac{\pi}{4}. \quad (10.184)$$

Comparing with Eq. (10.183), we deduce

$$\sqrt{I_2} = \int_0^\infty e^{-s^2} \, ds = \frac{\sqrt{\pi}}{2}. \quad (10.185)$$

With this verified, we can return to our original analysis and say that the velocity profile can be written as

$$u(\eta) = 1 - \frac{2}{\sqrt{\pi}} \int_0^\eta e^{-s^2} \, ds, \quad (10.186)$$

$$u(y,t) = 1 - \frac{2}{\sqrt{\pi}} \int_0^{\frac{y}{2\sqrt{t}}} e^{-s^2} \, ds = \text{erfc}\left(\frac{y}{2\sqrt{t}}\right). \quad (10.187)$$

In the last form here, we have introduced the so-called error function complement, "erfc." Plots for the velocity profile in terms of both η and y, t are given in Fig. 10.12. We see that in similarity space, the curve is a single curve in which u has a value of unity at $\eta = 0$ and has nearly relaxed to zero when $\eta = 1$. In dimensionless physical space, we see that at early time, there is a thin momentum layer near the surface. At later time, more momentum is present in the fluid. We can say in fact that momentum is diffusing into the fluid.

We define the momentum diffusion length as the length for which significant momentum has diffused into the fluid. This is well estimated by taking $\eta = 1$. In terms of physical variables, we have

$$\frac{y}{2\sqrt{t}} = 1, \quad \text{so} \quad y = 2\sqrt{t}, \quad (10.188)$$

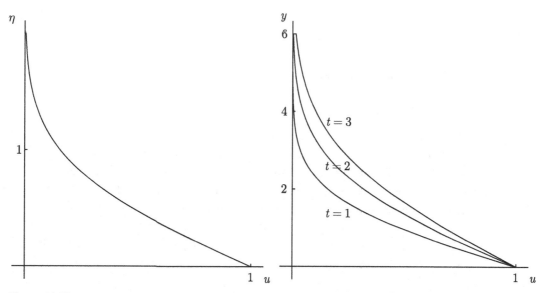

Figure 10.12 Diagram of velocity field solution for Stokes' first problem in both similarity coordinate η and primitive coordinates (y, t).

$$\frac{\overline{y}}{\frac{\nu}{U}} = 2\sqrt{\frac{\overline{t}}{\frac{\nu}{U^2}}}, \quad \text{so} \quad \overline{y} = \frac{2\nu}{U}\sqrt{\frac{U^2\overline{t}}{\nu}} = 2\sqrt{\nu\overline{t}}. \qquad (10.189)$$

We can in fact define this as a boundary layer thickness. That is to say, the momentum boundary layer thickness in Stokes' first problem grows at a rate proportional to the square root of momentum diffusivity and time. This class of result is a hallmark of all diffusion processes, be it diffusion of mass, momentum, or energy. Taking standard properties of air at 300 K, one atmosphere pressure, we find after one minute that its boundary layer thickness is 0.06 m. For engine oil under the same conditions, we get a thickness of 0.36 m. Lastly, using a related analysis, one can write the solution, Eq. (10.187) in terms of dimensional variables as

$$\frac{\overline{u}(\overline{y},\overline{t})}{U} = \text{erfc}\left(\frac{\overline{y}}{2\sqrt{\nu\overline{t}}}\right). \qquad (10.190)$$

We next consider the shear stress field. For this problem, the dimensional shear stress reduces to simply

$$\overline{\tau} = \mu \frac{\partial \overline{u}}{\partial \overline{y}}. \qquad (10.191)$$

Scaling as before by a characteristic stress τ_c, we get

$$\tau\tau_c = \frac{\mu U}{\frac{\nu}{U}} \frac{\partial u}{\partial y}, \quad \text{so} \quad \tau = \frac{\mu U^2}{\nu} \frac{1}{\tau_c} \frac{\partial u}{\partial y}. \qquad (10.192)$$

Taking $\tau_c = \mu U^2/\nu = \mu U^2/(\mu/\rho) = \rho U^2$, we get

$$\tau = \frac{\partial u}{\partial y} = \frac{1}{2\sqrt{t}} \frac{du}{d\eta} = \frac{1}{2\sqrt{t}}\left(-\frac{2}{\sqrt{\pi}}e^{-\eta^2}\right) = -\frac{1}{\sqrt{\pi t}}e^{-\eta^2} = -\frac{1}{\sqrt{\pi t}}\exp\left(-\frac{y}{2\sqrt{t}}\right)^2. \qquad (10.193)$$

Now at the wall, $y = 0$, and we get

$$\tau|_{y=0} = -\frac{1}{\sqrt{\pi t}}. \qquad (10.194)$$

So the shear stress does not have a similarity solution, but is directly related to time variation. The equation holds that the stress is infinite at $t = 0$, and decreases as time increases. This is because the velocity gradient flattens as time progresses. It can also be shown that while the stress is unbounded at a single point in time, the impulse over a finite time span is finite, even when the time span includes $t = 0$. It can also be shown that the flow corresponds to a pulse of vorticity being introduced at the wall, which subsequently diffuses into the fluid.

In dimensional terms, we can say

$$\frac{\overline{\tau}}{\rho U^2} = -\frac{1}{\sqrt{\pi \frac{U^2\overline{t}}{\nu}}}, \quad \text{so} \quad \overline{\tau} = -\frac{\rho U^2}{\sqrt{\pi \frac{U}{\nu}\sqrt{\overline{t}}}} = -\frac{\rho U \sqrt{\nu}}{\sqrt{\pi \overline{t}}} = -\frac{\mu U}{\sqrt{\pi \nu \overline{t}}}. \qquad (10.195)$$

For this two-dimensional flow, the dimensionless vorticity vector is confined to the z direction and has magnitude

$$\omega_z = \underbrace{\frac{\partial v}{\partial x}}_{=0} - \frac{\partial u}{\partial y} = \frac{1}{\sqrt{\pi t}} \exp\left(-\frac{y}{2\sqrt{t}}\right)^2. \tag{10.196}$$

The dimensionless acceleration vector is confined to the x direction and has magnitude

$$a_x = \frac{\partial u}{\partial t} = \frac{1}{2\sqrt{\pi} t^{3/2}} \exp\left(-\frac{y}{2\sqrt{t}}\right)^2. \tag{10.197}$$

The dimensionless deformation tensor reduces to

$$\mathbf{D} = \begin{pmatrix} 0 & \frac{1}{2}\frac{\partial u}{\partial y} \\ \frac{1}{2}\frac{\partial u}{\partial y} & 0 \end{pmatrix} = \begin{pmatrix} 0 & -\frac{1}{2\sqrt{\pi t}}\exp\left(-\frac{y}{2\sqrt{t}}\right)^2 \\ -\frac{1}{2\sqrt{\pi t}}\exp\left(-\frac{y}{2\sqrt{t}}\right)^2 & 0 \end{pmatrix}. \tag{10.198}$$

The deformation tensor has eigenvalues that represent the extreme values of extensional strain, and they are

$$\mathcal{D} = \pm\frac{1}{2\sqrt{\pi t}} \exp\left(-\frac{y}{2\sqrt{t}}\right)^2. \tag{10.199}$$

It is straightforward to show that the eigenvectors of $\mathcal{D}$ are aligned with coordinates axes that have been rotated through an angle of $\pi/4$.

Now let us consider the heat transfer problem. Recall the dimensional governing equation, initial and boundary conditions are

$$\rho c_p \frac{\partial \overline{T}}{\partial \overline{t}} = k \frac{\partial^2 \overline{T}}{\partial \overline{y}^2} + \mu \left(\frac{\partial \overline{u}}{\partial \overline{y}}\right)^2, \quad \overline{T}(\overline{y},0) = T_o, \quad \overline{T}(0,\overline{t}) = T_o, \quad \overline{T}(\infty,\overline{t}) = T_o. \tag{10.200}$$

We will adopt the same time t_c and length y_c scales as before. Take the dimensionless temperature to be $T = (\overline{T} - T_o)/T_o$. So we get

$$\frac{\rho c_p T_o}{t_c} \frac{\partial T}{\partial \overline{t}} = \frac{k T_o}{y_c^2} \frac{\partial^2 T}{\partial \overline{y}^2} + \frac{\mu U^2}{y_c^2} \left(\frac{\partial \overline{u}}{\partial \overline{y}}\right)^2, \tag{10.201}$$

$$\frac{\partial T}{\partial \overline{t}} = \frac{k T_o}{y_c^2} \frac{t_c}{\rho c_p T_o} \frac{\partial^2 T}{\partial \overline{y}^2} + \frac{\mu U^2}{y_c^2} \frac{t_c}{\rho c_p T_o}\left(\frac{\partial \overline{u}}{\partial \overline{y}}\right)^2. \tag{10.202}$$

Now

$$\frac{k}{y_c^2} \frac{t_c}{\rho c_p} = \frac{kU^2}{\nu^2} \frac{\nu}{U^2} \frac{1}{\rho c_p} = \frac{k}{\rho c_p \nu} = \frac{k}{\mu c_p} = \frac{1}{Pr}, \tag{10.203}$$

$$\frac{\mu U^2}{y_c^2} \frac{t_c}{\rho c_p T_o} = \frac{\mu U^2 U^2}{\nu^2} \frac{\nu}{U^2} \frac{1}{\rho c_p T_o} = \frac{\mu U^2}{\frac{\mu}{\rho}\rho c_p T_o} = \frac{U^2}{c_p T_o} = Ec. \tag{10.204}$$

So we have in dimensionless form

$$\frac{\partial T}{\partial t} = \frac{1}{Pr}\frac{\partial^2 T}{\partial y^2} + Ec\left(\frac{\partial u}{\partial y}\right)^2, \quad T(y,0) = 0, \quad T(0,t) = 0, \quad T(\infty,t) = 0. \tag{10.205}$$

Notice that the only driving inhomogeneity is the viscous work. Now we know from our solution of the linear momentum equation that

$$\frac{\partial u}{\partial y} = -\frac{1}{\sqrt{\pi t}} \exp\left(-\frac{y^2}{4t}\right). \quad (10.206)$$

So we can rewrite the equation for temperature variation as

$$\frac{\partial T}{\partial t} = \frac{1}{Pr}\frac{\partial^2 T}{\partial y^2} + \frac{Ec}{\pi t} \exp\left(-\frac{y^2}{2t}\right), \quad T(y,0) = 0, \quad T(0,t) = 0, \quad T(\infty,t) = 0. \quad (10.207)$$

Before considering the general solution, let us consider some limiting cases.

$Ec \to 0$

In the limit as $Ec \to 0$, we get a trivial solution, $T(y,t) = 0$.

$Pr \to \infty$

We recall that the Prandtl number is the ratio of momentum diffusivity to thermal diffusivity, Eq. (6.158); this limit corresponds to materials for which momentum diffusivity is much greater than thermal diffusivity. For example, for SAE 30 oil, the Prandtl number is around 3500. Naïvely assuming that we can simply neglect conduction, we write the energy equation in this limit as

$$\frac{\partial T}{\partial t} = \frac{Ec}{\pi t} \exp\left(-\frac{y^2}{2t}\right). \quad (10.208)$$

and with $T = T(\eta)$ and $\eta = y/(2\sqrt{t})$, we get the transformed partial time derivative to be

$$\frac{\partial T}{\partial t} = -\frac{\eta}{2t}\frac{dT}{d\eta}. \quad (10.209)$$

So the governing equation reduces to

$$-\frac{\eta}{2t}\frac{dT}{d\eta} = \frac{Ec}{\pi t}e^{-2\eta^2}, \quad \text{so} \quad \frac{dT}{d\eta} = -\frac{2Ec}{\pi}\frac{1}{\eta}e^{-2\eta^2}, \quad (10.210)$$

$$T = \frac{2Ec}{\pi}\int_\eta^\infty \frac{1}{s}e^{-2s^2}\, ds. \quad (10.211)$$

The equation has been solved so as to satisfy the boundary condition in the far field of $T(\infty) = 0$. Unfortunately, we notice that we cannot satisfy the boundary condition at $\eta = 0$. We simply do not have enough degrees of freedom. In actuality, what we have found is an outer solution, and to match the boundary condition at 0 we would have to reintroduce conduction, which has a higher derivative.

First let us see how the outer solution behaves near $\eta = 0$. Expanding the differential equation in a Taylor series about $\eta = 0$ and solving gives

$$\frac{dT}{d\eta} = -\frac{2Ec}{\pi}\left(\frac{1}{\eta} - 2\eta + 2\eta^3 + \cdots\right), \tag{10.212}$$

$$T = -\frac{2Ec}{\pi}\left(\ln\eta - \eta^2 + \frac{1}{2}\eta^4 + \cdots\right). \tag{10.213}$$

It turns out that solving the inner layer problem and the matching is of about the same difficulty as solving the full general problem, so we will defer this until later in this section.

$Pr \to 0$

In this limit, we get

$$\frac{\partial^2 T}{\partial y^2} = 0. \tag{10.214}$$

The solution that satisfies the boundary conditions is $T = 0$. In this limit, momentum diffuses slowly relative to energy. So we can interpret the results as follows. In the boundary layer, momentum is generated in a thin layer. Viscous dissipation in this layer gives rise to a local change in temperature in the layer that rapidly diffuses throughout the entire flow. The effect of smearing a localized finite thermal energy input over a semi-infinite domain has a negligible influence on the temperature of the global domain.

So let us bring back diffusion and study solutions for finite Prandtl number. Our governing equation in similarity variables then becomes

$$-\frac{\eta}{2t}\frac{dT}{d\eta} = \frac{1}{Pr}\frac{1}{4t}\frac{d^2T}{d\eta^2} + \frac{Ec}{\pi t}e^{-2\eta^2}, \tag{10.215}$$

$$\underbrace{-2\eta\frac{dT}{d\eta}}_{\text{temperature evolution}} = \underbrace{\frac{1}{Pr}\frac{d^2T}{d\eta^2}}_{\text{energy diffusion}} + \underbrace{\frac{4Ec}{\pi}e^{-2\eta^2}}_{\text{dissipation}}, \tag{10.216}$$

$$\frac{d^2T}{d\eta^2} + 2Pr\,\eta\frac{dT}{d\eta} = -\frac{4}{\pi}EcPr\,e^{-2\eta^2}, \quad T(0) = 0, \quad T(\infty) = 0. \tag{10.217}$$

The second-order differential equation is really two first-order differential equations in disguise. There is an integrating factor of $e^{Pr\,\eta^2}$. Multiplying by the integrating factor and operating on the system, we find

$$e^{Pr\,\eta^2}\frac{d^2T}{d\eta^2} + 2Pr\,\eta e^{Pr\,\eta^2}\frac{dT}{d\eta} = -\frac{4}{\pi}EcPr\,e^{(Pr-2)\eta^2}, \tag{10.218}$$

$$\frac{d}{d\eta}\left(e^{Pr\,\eta^2}\frac{dT}{d\eta}\right) = -\frac{4}{\pi}EcPr\,e^{(Pr-2)\eta^2}, \tag{10.219}$$

$$e^{Pr\,\eta^2}\frac{dT}{d\eta} = -\frac{4}{\pi}EcPr\int_0^\eta e^{(Pr-2)s^2}\,ds + C_1, \tag{10.220}$$

$$\frac{dT}{d\eta} = -\frac{4}{\pi} EcPr\, e^{-Pr\,\eta^2} \int_0^\eta e^{(Pr-2)s^2}\, ds + C_1 e^{-Pr\,\eta^2}, \tag{10.221}$$

$$T = -\frac{4}{\pi} EcPr \int_0^\eta e^{-Pr\,\alpha^2} \int_0^\alpha e^{(Pr-2)s^2}\, ds\, d\alpha$$
$$+ C_1 \int_0^\eta e^{-Pr\,s^2}\, ds + C_2. \tag{10.222}$$

The boundary condition $T(0) = 0$ gives us $C_2 = 0$. The boundary condition at ∞ gives us then

$$0 = -\frac{4}{\pi} EcPr \int_0^\infty e^{-Pr\,\alpha^2} \int_0^\alpha e^{(Pr-2)s^2}\, ds\, d\alpha + C_1 \underbrace{\int_0^\infty e^{-Pr\,s^2}\, ds}_{\frac{1}{2}\sqrt{\frac{\pi}{Pr}}}. \tag{10.223}$$

Therefore, we get

$$\frac{4}{\pi} EcPr \int_0^\infty e^{-Pr\,\alpha^2} \int_0^\alpha e^{(Pr-2)s^2}\, ds\, d\alpha = \frac{C_1}{2}\sqrt{\frac{\pi}{Pr}}, \tag{10.224}$$

$$C_1 = \frac{8}{\pi^{3/2}} EcPr^{3/2} \int_0^\infty e^{-Pr\,\alpha^2} \int_0^\alpha e^{(Pr-2)s^2}\, ds\, d\alpha. \tag{10.225}$$

So finally, we have for the temperature profile

$$T(\eta) = -\frac{4}{\pi} EcPr \int_0^\eta e^{-Pr\,\alpha^2} \int_0^\alpha e^{(Pr-2)s^2}\, ds\, d\alpha$$
$$+ \left(\frac{8}{\pi^{3/2}} EcPr^{3/2} \int_0^\infty e^{-Pr\,\alpha^2} \int_0^\alpha e^{(Pr-2)s^2}\, ds\, d\alpha\right) \int_0^\eta e^{-Pr\,s^2}\, ds. \tag{10.226}$$

This simplifies to

$$T(\eta) = -\frac{2EcPr}{\sqrt{\pi(2-Pr)}}$$
$$\times \left(\int_0^\eta e^{-Pr\,\alpha^2} \mathrm{erf}\left(\sqrt{2-Pr}\,\alpha\right) d\alpha - \mathrm{erf}\left(\sqrt{Pr}\,\eta\right) \int_0^\infty e^{-Pr\,\alpha^2} \mathrm{erf}\left(\sqrt{2-Pr}\,\alpha\right) d\alpha\right). \tag{10.227}$$

This analysis simplifies considerably in the limit of $Pr = 1$, which is when momentum and energy diffuse at the same rate. This is a close to reality for many gases. In this case, the temperature profile becomes

$$T(\eta) = -\frac{4}{\pi} Ec \int_0^\eta e^{-\alpha^2} \int_0^\alpha e^{-s^2}\, ds\, d\alpha + C_1 \int_0^\eta e^{-s^2}\, ds. \tag{10.228}$$

Now if $h(\alpha) = \int_0^\alpha e^{-s^2} ds$, we get $dh/d\alpha = e^{-\alpha^2}$. Using this, we can rewrite the temperature profile as

$$T(\eta) = -\frac{4}{\pi} Ec \int_0^\eta h(\alpha) \frac{dh}{d\alpha} d\alpha + C_1 \int_0^\eta e^{-s^2} ds, \qquad (10.229)$$

$$= -\frac{4Ec}{\pi} \int_0^\eta d\left(\frac{h^2}{2}\right) + C_1 \int_0^\eta e^{-s^2} ds, \qquad (10.230)$$

$$= -\frac{4Ec}{\pi} \left(\frac{1}{2}\right) \left(\int_0^\eta e^{-s^2} ds\right)^2 + C_1 \int_0^\eta e^{-s^2} ds, \qquad (10.231)$$

$$= \left(-\frac{2Ec}{\pi} \int_0^\eta e^{-s^2} ds + C_1\right) \int_0^\eta e^{-s^2} ds. \qquad (10.232)$$

Now for $T(\infty) = 0$, we get

$$0 = \left(-\frac{2Ec}{\pi} \int_0^\infty e^{-s^2} ds + C_1\right) \int_0^\infty e^{-s^2} ds, \qquad (10.233)$$

$$0 = \left(-\frac{2Ec}{\pi} \frac{\sqrt{\pi}}{2} + C_1\right) \frac{\sqrt{\pi}}{2}, \quad \text{so} \quad C_1 = \frac{Ec}{\sqrt{\pi}}. \qquad (10.234)$$

So the temperature profile can be expressed as

$$T(\eta) = \frac{Ec}{\sqrt{\pi}} \left(\int_0^\eta e^{-s^2} ds\right) \left(1 - \frac{2}{\sqrt{\pi}} \int_0^\eta e^{-s^2} ds\right). \qquad (10.235)$$

We notice that we can write this directly in terms of the velocity as

$$T(\eta) = \frac{Ec}{2} u(\eta) \left(1 - u(\eta)\right). \qquad (10.236)$$

This is a consequence of what is known as *Reynolds analogy* that holds for $Pr = 1$ that the temperature field can be directly related to the velocity field. The temperature field for Stokes' first problem for $Pr = 1$, $Ec = 1$ is plotted in Fig. 10.13.

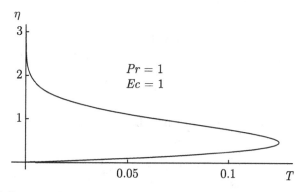

Figure 10.13 Plot of dimensionless temperature field for Stokes' first problem for $Pr = 1$, $Ec = 1$.

10.2.2 General Solution

Let us generalize the results of Section 10.2.1 for Stokes' first problem for a general set of initial conditions. Consider the following initial-boundary value problem:

$$\frac{\partial u}{\partial t} = \nu \frac{\partial^2 u}{\partial y^2}, \quad u(y \to \pm\infty, t) \to 0, \quad u(y, 0) = f(y). \tag{10.237}$$

Here we have returned to considering u, t, and y to be dimensional variables. For the solution we will present, it is important that the momentum vanish at infinity at all times. Thus constrained, we can allow for an otherwise arbitrary initial distribution of momentum given by the function $f(y)$. This partial differential equation has been widely studied and is known as the "heat equation," because its thermal analog was studied first. A well-known solution exists. Mei (1997) shows how it may be obtained via the technique of the Fourier transformation. Further background is given by the classic monograph of Carslaw and Jaeger (1986). The solution is

$$u(y, t) = \frac{1}{\sqrt{4\pi \nu t}} \int_{-\infty}^{\infty} f(\xi) \exp\left(-\frac{(y-\xi)^2}{4\nu t}\right) d\xi. \tag{10.238}$$

10.2.3 Momentum Pulse

Consider the special case in which a pulse of finite strength of momentum is initially concentrated at $y = 0$. This can be achieved by taking

$$u(y, 0) = f(y) = \mathcal{M}\delta(y). \tag{10.239}$$

Here $\delta(y)$ is the so-called *Dirac*[4] *delta function*, and $\mathcal{M}$ is a constant with units m²/s representing the momentum pulse strength. The Dirac delta function is defined such that

$$\int_{\alpha}^{\beta} g(y)\delta(y-a)\, dy = \begin{cases} 0 & a \notin [\alpha, \beta], \\ g(a) & a \in [\alpha, \beta]. \end{cases} \tag{10.240}$$

Importantly, if $g(y) = 1$, $\alpha \to -\infty$, $\beta \to \infty$, and $a = 0$, we get

$$\delta(y) = 0, \quad y \neq 0, \quad \text{and} \quad \int_{-\infty}^{\infty} \delta(y)\, dy = 1. \tag{10.241}$$

We can think of the Dirac delta function, $\delta(y)$, as a function of zero everywhere except at $y = 0$, where the function has infinite "height", with a "width" that is infinitely thin, such that the product of the height and width is unity.

Now for the solution substitute Eq. (10.239) into Eq. (10.238) to get

$$u(y, t) = \frac{\mathcal{M}}{\sqrt{4\pi\nu t}} \int_{-\infty}^{\infty} \delta(\xi) \exp\left(-\frac{(y-\xi)^2}{4\nu t}\right) d\xi = \frac{\mathcal{M}}{\sqrt{4\pi\nu t}} \exp\left(-\frac{y^2}{4\nu t}\right). \tag{10.242}$$

It can be verified that $u(y, t)$ satisfies the governing partial differential equation. Computer mathematical software verifies that

$$\int_{-\infty}^{\infty} u(y, t)\, dy = \int_{-\infty}^{\infty} \frac{\mathcal{M}}{\sqrt{4\pi\nu t}} \exp\left(-\frac{y^2}{4\nu t}\right) dy = \mathcal{M}. \tag{10.243}$$

---

[4] Paul Adrien Maurice Dirac, 1902–1984, British mathematical physicist and 1933 Nobel Laureate.

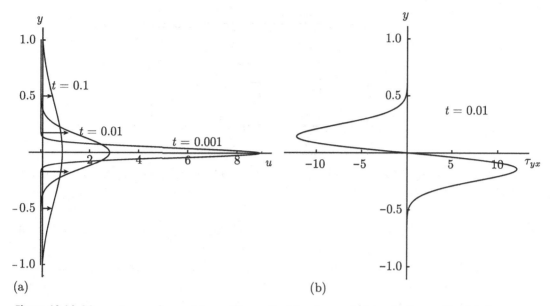

Figure 10.14 Momentum pulse problem with $\nu = 1$, $\mathcal{M} = 1$, $\rho = 1$: (a) velocity profile, (b) shear stress profile.

This is independent of time, so the amount of momentum in the domain is conserved. Plots of velocity profiles at $t = 0.001, 0.01, 0.1$ for $\nu = 1$, $\mathcal{M} = 1$ are given in Fig. 10.14(a). Clearly the pulse is concentrated at early time near $y = 0$ and diffuses into the quiescent flow as time advances.

The shear stress is given by $\tau_{yx} = \tau_{xy} = \mu \partial u/\partial y$. Recognizing that $\nu = \mu/\rho$, the time-dependent shear stress field is given by

$$\tau_{yx} = -\frac{\mathcal{M}\rho y}{2\sqrt{4\pi\nu t^3}} \exp\left(-\frac{y^2}{4\nu t}\right). \tag{10.244}$$

The stress is anti-symmetric about $y = 0$, decays to zero in the far field, and is singular near $t = 0$. A shear stress profile at $t = 0.01$ for $\nu = 1$, $\mathcal{M} = 1$, $\rho = 1$ is given in Fig. 10.14(b).

10.2.4 Unsteady Couette Flow

Consider now the unsteady extension of the steady Couette flow considered in Section 10.1.2. The configuration for this flow is shown in Fig. 10.15. We assume constant ρ, μ, $\partial/\partial x = \partial/\partial z = 0$, and that the fluid is initially at rest. We also have then that ν is constant. The fluid at $y = 0$ adheres to a stationary wall; the fluid at $y = h$ adheres to a flat plate suddenly accelerated to a constant velocity U. One can see how this is similar to Stokes' first problem, except the spatial domain is finite instead of semi-infinite. Our equations governing this flow are

$$\frac{\partial u}{\partial t} = \nu \frac{\partial^2 u}{\partial y^2}, \qquad u(y,0) = 0, \quad u(0,t) = 0, \quad u(h,t) = U. \tag{10.245}$$

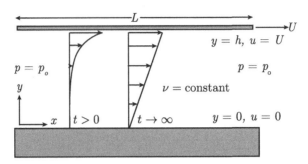

Figure 10.15 Configuration for incompressible Couette flow with isothermal walls.

All variables are dimensional here. This is a standard problem in linear partial differential equations that may be addressed with the method of separation of variables. It will prove useful in this to transform the dependent variable. We know from our earlier analysis that $u = Uy/h$ is a solution to the time-independent version of this problem. Let us pose the problem as one in which we study the deviation of u from its time-independent value. So let us define

$$\hat{u}(y,t) = u(y,t) - U\frac{y}{h}. \tag{10.246}$$

Our governing equations then transform to

$$\frac{\partial}{\partial t}\left(\hat{u} + U\frac{y}{h}\right) = \nu\frac{\partial^2}{\partial y^2}\left(\hat{u} + U\frac{y}{h}\right), \tag{10.247}$$

$$\hat{u}(y,0) + U\frac{y}{h} = 0, \quad \hat{u}(0,t) + U\frac{0}{h} = 0, \quad \hat{u}(h,t) + U\frac{h}{h} = U. \tag{10.248}$$

Simplifying, we get

$$\frac{\partial \hat{u}}{\partial t} = \nu\frac{\partial^2 \hat{u}}{\partial y^2}, \quad \hat{u}(y,0) = -U\frac{y}{h}, \quad \hat{u}(0,t) = 0, \quad \hat{u}(h,t) = 0. \tag{10.249}$$

The effect of the transformation has been to move the inhomogeneity from the $y = h$ boundary to the initial condition. The boundary conditions in space are now homogeneous, which will be a key for the success of the method we employ.

We next assume that solutions are of the form

$$\hat{u}(y,t) = A(y)B(t). \tag{10.250}$$

This is the same separation of variables approach we used for problems in acoustics; see Eq. (9.392). It often enables the solution of linear partial differential equations, but there is no guarantee that it will work. Let us see if it works for this problem. Substituting into our governing equation, we get

$$A(y)\frac{dB}{dt} = \nu B(t)\frac{d^2 A}{dy^2}, \quad \text{so} \quad \frac{1}{\nu B(t)}\frac{dB}{dt} = \frac{1}{A(y)}\frac{d^2 A}{dy^2} = -\lambda^2. \tag{10.251}$$

The leftmost term is a function of t only, and the term to its right is a function of y only. Because y and t are independent variables, the only way to maintain equality is if both terms are the same constant. We choose for convenience to call that constant $-\lambda^2$, though any choice would suffice. We then recover two ordinary differential equations:

$$\frac{dB}{dt} + \lambda^2 \nu B = 0, \qquad \frac{d^2 A}{dy^2} + \lambda^2 A = 0. \tag{10.252}$$

The second is of the form

$$-\frac{d^2}{dy^2} A = \lambda^2 A, \tag{10.253}$$

in which the linear operator $-d^2/dy^2$ operates on a function A to yield a scalar multiple of A. As seen earlier, this is an eigenvalue problem defined by the positive definite self-adjoint operator $-d^2/dy^2$. Such problems are guaranteed to have a set of real and positive eigenvalues λ_n^2 and a set of eigenfunctions $A_n(y)$. Here we are anticipating the existence of multiple eigenvalues and eigenfunctions. The eigenfunctions are guaranteed to have the remarkable property of orthogonality, which holds that

$$\int_0^h A_n(y) A_m(y)\, dy = \begin{cases} 0 & n \neq m, \\ K^2 & n = m. \end{cases} \tag{10.254}$$

Here K^2 is some positive definite scalar. This property of orthogonality will enable us to find the exact solution.

These both can be solved to get

$$B(t) = C_1 \exp\left(-\lambda^2 \nu t\right), \qquad A(y) = C_2 \sin \lambda y + C_3 \cos \lambda y. \tag{10.255}$$

We do not yet know λ. In order to satisfy the homogeneous spatial boundary conditions on $\hat{u}$ at $y = 0$ and $y = h$, we must insist that $A(0) = A(h) = 0$. Taking $A(0) = 0$ gives $0 = C_3$. Thus,

$$A(y) = C_2 \sin \lambda y. \tag{10.256}$$

Applying $A(h) = 0$ gives

$$0 = C_2 \sin \lambda h. \tag{10.257}$$

A trivial way to satisfy this is by insisting that $C_2 = 0$. But this yields no interesting solution. Another way to satisfy is by insisting that

$$\lambda h = n\pi, \quad n = 1, 2, 3, \ldots. \tag{10.258}$$

So $\lambda_n = n\pi/h$, $n = 1, 2, 3, \ldots$. We make this choice and say recast our solution as

$$\hat{u}(y, t) = C_1 C_2 \exp\left(-\left(\frac{n\pi}{h}\right)^2 \nu t\right) \sin\left(\frac{n\pi y}{h}\right). \tag{10.259}$$

Now this applies for any n. Because our equation is linear, linear combinations of solutions are also solutions. We can take then $C_n = C_1 C_2$ and sum solutions to say

$$\hat{u}(y, t) = \sum_{n=1}^{\infty} C_n \exp\left(-\left(\frac{n\pi}{h}\right)^2 \nu t\right) \sin\left(\frac{n\pi y}{h}\right). \tag{10.260}$$

Direct substitution into the original partial differential equation verifies this is a solution. It also can be seen to satisfy the homogeneous boundary conditions at $y = 0$ and $y = h$. This form confirms our decision to restrict n to positive integers. Obviously, the $n = 0$ term cannot contribute. And for negative integer n, we get a linearly dependent result because $\sin(n\pi y) = -\sin(-n\pi y)$. So this is not a linearly independent eigenfunction.

We must select values of C_n such that the initial condition is satisfied. At $t = 0$, we thus must have

$$\hat{u}(y, 0) = -U\frac{y}{h} = \sum_{n=1}^{\infty} C_n \sin\left(\frac{n\pi y}{h}\right). \tag{10.261}$$

This amounts to finding the Fourier-sine coefficients for the function $-Uy/h$.

Let us first confirm that our $A_n(y)$ are orthogonal. Doing so, we find

$$\int_0^h \sin\left(\frac{n\pi y}{h}\right) \sin\left(\frac{m\pi y}{h}\right) dy = \begin{cases} 0 & n \neq m, \\ \frac{h}{2} & n = m. \end{cases} \tag{10.262}$$

Now multiply both sides of Eq. (10.261) by $\sin(m\pi y/h)$ and integrate from 0 to h:

$$-\frac{U}{h} \int_0^h y \sin\left(\frac{m\pi y}{h}\right) dy = \sum_{n=1}^{\infty} C_n \int_0^h \sin\left(\frac{m\pi y}{h}\right) \sin\left(\frac{n\pi y}{h}\right) dy. \tag{10.263}$$

For $n \neq m$ the integral on the right-hand side is zero, and for $n = m$, it is $h/2$ by our earlier developed relations for the orthogonal eigenfunctions. Employing that, integrating the left-hand side, and exchanging n for m, we obtain

$$-\frac{U}{h}\left(\frac{(-1)^{m+1}h^2}{m\pi}\right) = C_m \frac{h}{2}, \quad \text{so} \quad C_n = \frac{2U(-1)^n}{n\pi}. \tag{10.264}$$

Thus, our solution for the deviation of velocity from its steady state distribution is

$$\hat{u}(y, t) = \frac{2U}{\pi} \sum_{n=1}^{\infty} \frac{(-1)^n}{n} \exp\left(-(n\pi)^2 \frac{\nu t}{h^2}\right) \sin\left(\frac{n\pi y}{h}\right). \tag{10.265}$$

Transforming from $\hat{u}$ to u, we see that

$$\frac{u(y, t)}{U} = \frac{y}{h} + \frac{2}{\pi} \sum_{n=1}^{\infty} \frac{(-1)^n}{n} \exp\left(-(n\pi)^2 \frac{\nu t}{h^2}\right) \sin\left(\frac{n\pi y}{h}\right). \tag{10.266}$$

The solution for this flow for various times is shown in Fig. 10.16. At sufficiently early times, $t \ll h^2/\nu$, the effect of the boundary at $y = 0$ is not strong, and the flow is well modeled as Stokes' first problem with the suddenly accelerated plate at $y = h$. Then, Eq. (10.190) can be modified to give the early time approximate solution:

$$\frac{u(y, t)}{U} = \text{erfc}\left(\frac{h - y}{2\sqrt{\nu t}}\right), \quad t \ll h^2/\nu. \tag{10.267}$$

The approximation is remarkably accurate at early time. For example, for $U = 1$, $h = 1$, $\nu = 1$, $t = 0.01$, the maximum error is about 10^{-12}. Later at $t = 0.1$, the maximum error is 0.025. When $t = 1$, the error is large and is near 0.5.

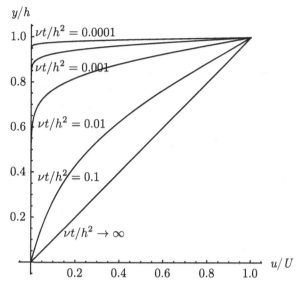

Figure 10.16 Solution at various times for unsteady Couette flow problem.

10.2.5 Stokes' Second Problem

Let us consider Stokes' (1851) second problem for a viscous fluid moving under the influence of an oscillating flat plate. We adopt the same linear momentum equation, Eq. (10.146), used for Stokes' first problem, but consider a different set of boundary conditions:

$$\frac{\partial u}{\partial t} = \nu \frac{\partial^2 u}{\partial y^2}, \quad u(y,0) = 0, \quad u(0,t) = U \sin \Omega t, \quad u(\infty,t) < \infty. \tag{10.268}$$

We can imagine this problem physically as one in which a flat plate at $y = 0$ is sinusoidally oscillating with time. We may recall Euler's formula, Eq. (8.25) and adapt it to give:

$$e^{i\Omega t} = \cos \Omega t + i \sin \Omega t. \tag{10.269}$$

We also recall that the real and imaginary parts are defined as

$$\Re\left(e^{i\Omega t}\right) = \cos \Omega t, \qquad \Im\left(e^{i\Omega t}\right) = \sin \Omega t. \tag{10.270}$$

Let us define a related auxiliary problem, with u defined as a complex variable whose imaginary part is u: $\Im(\mathsf{u}) = u$. We then take our extended problem to be

$$\frac{\partial \mathsf{u}}{\partial t} = \nu \frac{\partial^2 \mathsf{u}}{\partial y^2}, \quad \mathsf{u}(y,0) = 0, \quad \mathsf{u}(0,t) = U e^{i\Omega t}, \quad |\mathsf{u}(\infty,t)| < \infty. \tag{10.271}$$

Next let us seek a solution which is valid at long time. That is to say, we will not require our solution to satisfy any initial condition but will require it to satisfy the partial differential equation and boundary conditions at $y = 0$ and $y \to \infty$. We will gain many useful insights even though we will not capture the initial condition, which, with extra effort, we could.

Let us separate variables in the following fashion. Assume that

$$u(y,t) = f(y)e^{i\Omega t}. \tag{10.272}$$

Here $f(y)$ is a function to be determined. Ultimately we will only be concerned with the imaginary portion of this solution, which is the portion we will need to match the boundary condition. With this assumption, we find formulæ for the various partial derivatives to be

$$\frac{\partial u}{\partial t} = i\Omega f(y)e^{i\Omega t}, \quad \frac{\partial u}{\partial y} = \frac{df}{dy}e^{i\Omega t}, \quad \frac{\partial^2 u}{\partial y^2} = \frac{d^2 f}{dy^2}e^{i\Omega t}. \tag{10.273}$$

Then Eq. (10.271) becomes

$$i\Omega f e^{i\Omega t} = \nu \frac{d^2 f}{dy^2}e^{i\Omega t}, \quad \text{so} \quad i\Omega f = \nu \frac{d^2 f}{dy^2}. \tag{10.274}$$

Now assume that $f(y) = Ae^{ay}$, giving

$$i\Omega A e^{ay} = Aa^2 \nu e^{ay}, \quad \text{so} \quad i = a^2 \frac{\nu}{\Omega}. \tag{10.275}$$

Now in a polar representation, one has $i = e^{i\pi/2}$. More generally, we could say

$$i = e^{i(\pi/2 + 2n\pi)}, \quad n = 0, 1, 2, \dots. \tag{10.276}$$

Thus, Eq. (10.275) can be re-expressed as

$$e^{i(\pi/2 + 2n\pi)} = a^2 \frac{\nu}{\Omega}, \quad \text{so} \quad \sqrt{\frac{\Omega}{\nu}} e^{i(\pi/4 + n\pi)} = a. \tag{10.277}$$

Using Euler's formula, Eq. (8.25), we could then say

$$a = \sqrt{\frac{\Omega}{\nu}} \left(\cos\left(\frac{\pi}{4} + n\pi\right) + i \sin\left(\frac{\pi}{4} + n\pi\right) \right), \tag{10.278}$$

$$= \pm\sqrt{\frac{\Omega}{\nu}} \left(\frac{1}{\sqrt{2}} + \frac{i}{\sqrt{2}} \right) = \pm\sqrt{\frac{\Omega}{2\nu}} (1+i). \tag{10.279}$$

When n is even, we have the "plus" root; when odd, we have the "minus" root. For each root, we can have a solution; thus, we form linear combinations to get

$$f(y) = C_1 \exp\left(\sqrt{\frac{\Omega}{2\nu}}(1+i)y \right) + C_2 \exp\left(-\sqrt{\frac{\Omega}{2\nu}}(1+i)y \right). \tag{10.280}$$

Now because we take $\Omega > 0$, $\nu > 0$ and $y > 0$, we will need $C_1 = 0$ in order to keep $|u|$ bounded as $y \to \infty$. So we have

$$f(y) = C_2 \exp\left(-\sqrt{\frac{\Omega}{2\nu}}(1+i)y \right). \tag{10.281}$$

Then recombining, we find that

$$u(y,t) = C_2 \exp\left(-\sqrt{\frac{\Omega}{2\nu}}(1+i)y \right) \exp(i\Omega t). \tag{10.282}$$

Now at $y = 0$, we must have

$$U \exp(i\Omega t) = C_2 \exp(i\Omega t). \tag{10.283}$$

We thus need to take $C_2 = U$, giving

$$\mathsf{u}(y,t) = U \exp\left(-\sqrt{\frac{\Omega}{2\nu}}(1+i)y\right) \exp(i\Omega t). \tag{10.284}$$

We then find u by considering only the imaginary portion of u, giving

$$u(y,t) = \Im\left(U \exp\left(-\sqrt{\frac{\Omega}{2\nu}}(1+i)y\right) \exp(i\Omega t)\right), \tag{10.285}$$

$$= \Im\left(U \exp\left(-\sqrt{\frac{\Omega}{2\nu}}(1+i)y + i\Omega t\right)\right), \tag{10.286}$$

$$= \Im\left(U \exp\left(-\sqrt{\frac{\Omega}{2\nu}}y + i\left(\Omega t - \sqrt{\frac{\Omega}{2\nu}}y\right)\right)\right), \tag{10.287}$$

$$= U \exp\left(-\sqrt{\frac{\Omega}{2\nu}}y\right) \sin\left(\Omega t - \sqrt{\frac{\Omega}{2\nu}}y\right). \tag{10.288}$$

By inspection, the boundary condition is satisfied. Direct substitution reveals the solution also satisfies the heat equation. And clearly, as $y \to \infty$, $u \to 0$.

Now the amplitude of this wave-like solution has decayed to roughly $U/100$ at a point where

$$\sqrt{\frac{\Omega}{2\nu}}y = 4.5, \quad \text{so} \quad y = 4.5\sqrt{\frac{2\nu}{\Omega}}. \tag{10.289}$$

Thus the penetration depth of the wave into the domain is enhanced by high ν and low Ω. And below this depth, the material is ambivalent to the disturbance at the boundary. With regards to the oscillatory portion of the solution, we see the angular frequency is Ω and the wave number is $k = \sqrt{\Omega/(2\nu)}$.

The phase of the wave is given by

$$\phi = \Omega t - \sqrt{\frac{\Omega}{2\nu}}y. \tag{10.290}$$

Let us get the phase speed. If the phase itself is constant, we differentiate to get

$$\frac{d\phi}{dt} = 0 = \Omega - \sqrt{\frac{\Omega}{2\nu}}\frac{dy}{dt}, \quad \text{so} \quad \frac{dy}{dt} = \sqrt{2\nu\Omega}. \tag{10.291}$$

For $\nu = 1$, $U = 1$, $\Omega = 1$, we plot $u(y,t)$ in Fig. 10.17. Clearly, for $y \approx 4.5\sqrt{(2)(1)/1} = 6.4$, the effect of the sinusoidal velocity variation at $y = 0$ has small effect.

10.2.6 Decay of an Ideal Vortex

Here we consider how viscous effects modulate a basic inviscid flow solution. We presented solutions for ideal irrotational vortices in Section 7.4. Such a vortex is shown in Fig. 10.18. The steady velocity field was seen to be for the inviscid flow

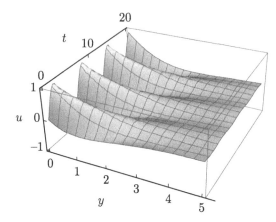

Figure 10.17 Solution to Stokes' second problem with $\nu = 1$, $U = 1$, $\Omega = 1$.

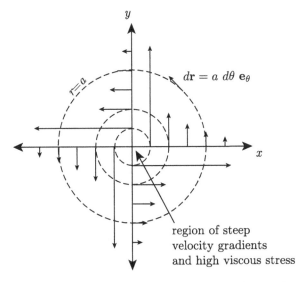

Figure 10.18 Diagram of an ideal irrotational point vortex and a circular contour of $r = a$.

$$v_r = 0, \qquad v_\theta = \frac{\Gamma_o}{2\pi r}, \qquad v_z = 0. \tag{10.292}$$

The pressure was seen to have singular behavior at $r = 0$. And the circulation $\Gamma = \oint_C \mathbf{v}^T \cdot d\mathbf{r}$ was seen to be a constant Γ_o. Examining Fig. 10.18, it seems obvious that near $r = 0$ there exist strong velocity gradients. Such gradients will induce viscous stresses that had been neglected. Let us here consider a unsteady viscous flow field whose initial condition is given by an ideal irrotational vortex. The presence of viscous stress will induce the field to decay with time, and we will study those dynamics.

To analyze this decay, let us assume that $v_r = v_z = 0$, and that $v_\theta = v_\theta(r,t)$. The only momentum equation we need consider is the θ momentum equation, which for an incompressible fluid with constant viscosity is

$$\rho \left(\frac{\partial v_\theta}{\partial t} + \underbrace{v_r \frac{\partial v_\theta}{\partial r} + \frac{v_\theta}{r} \frac{\partial v_\theta}{\partial \theta} + v_z \frac{\partial v_\theta}{\partial z} + \frac{v_r v_\theta}{r}}_{=0} \right)$$

$$= -\underbrace{\frac{1}{r} \frac{\partial p}{\partial \theta}}_{=0} + \mu \left(\frac{\partial}{\partial r} \left(\frac{1}{r} \frac{\partial}{\partial r} (r v_\theta) \right) + \underbrace{\frac{1}{r^2} \frac{\partial^2 v_\theta}{\partial \theta^2}}_{} + \frac{\partial^2 v_\theta}{\partial z^2} + \underbrace{\frac{2}{r^2} \frac{\partial v_r}{\partial \theta}}_{=0} \right). \quad (10.293)$$

This yields

$$\frac{\partial v_\theta}{\partial t} = \nu \frac{\partial}{\partial r} \left(\frac{1}{r} \frac{\partial}{\partial r} (r v_\theta) \right). \quad (10.294)$$

Expanding and adopting initial and boundary conditions, we pose the problem as

$$\frac{\partial v_\theta}{\partial t} = -\frac{\nu}{r^2} \frac{\partial}{\partial r} (r v_\theta) + \frac{\nu}{r} \frac{\partial^2}{\partial r^2} (r v_\theta), \quad (10.295)$$

$$v_\theta(0,t) = 0, \quad v_\theta(\infty, t) = \frac{\Gamma_o}{2\pi r}, \quad v_\theta(r, 0) = \frac{\Gamma_o}{2\pi r}. \quad (10.296)$$

Here we force the velocity to zero at the origin, in contrast to its infinite value for the inviscid case. We take the ideal irrotational vortex solution as an initial condition, and insist that in the far field, the solution relax to the inviscid solution.

Let us choose to study a scaled velocity; however, we shall choose a scaling that has spatial dependence. Better stated, we are choosing to study a transformed dependent variable. Let us take

$$v_* = \frac{v_\theta}{\Gamma_o/(2\pi r)}. \quad (10.297)$$

This transformation simplifies the initial and far field boundary condition, but complicates the governing partial differential equation. We substitute into the governing θ momentum equation to get

$$\frac{\partial}{\partial t} \left(\frac{\Gamma_o}{2\pi} \frac{1}{r} v_* \right) = -\frac{\nu}{r^2} \frac{\partial}{\partial r} \left(\frac{\Gamma_o}{2\pi} v_* \right) + \frac{\nu}{r} \frac{\partial^2}{\partial r^2} \left(\frac{\Gamma_o}{2\pi} v_* \right), \quad (10.298)$$

$$\frac{\partial v_*}{\partial t} = -\frac{\nu}{r} \frac{\partial v_*}{\partial r} + \nu \frac{\partial^2 v_*}{\partial r^2}, \quad v_*(0,t) = 0, \quad v_*(\infty, t) = 1, \quad v_*(r,0) = 1. \quad (10.299)$$

We seek self-similar solutions. We could use notions used previously involving coordinate stretching to identify a similarity variable. We could also use intuition and experience. This problem has a similar mathematical structure to Stokes' first problem of Section 10.2.1. Let us guess what amounts to the same transformation:

$$\eta = \frac{r}{\sqrt{\nu t}}. \quad (10.300)$$

Let us seek to find $v_* = v_*(\eta)$. The chain rule gives us

$$\frac{\partial v_*}{\partial t} = \frac{\partial \eta}{\partial t}\frac{dv_*}{d\eta} = -\frac{1}{2}\frac{r}{\sqrt{\nu t}}\frac{1}{t}\frac{dv_*}{d\eta} = -\frac{1}{2t}\eta\frac{dv_*}{d\eta}, \tag{10.301}$$

$$\frac{\partial v_*}{\partial r} = \frac{\partial \eta}{\partial r}\frac{dv_*}{d\eta} = \frac{1}{\sqrt{\nu t}}\frac{dv_*}{d\eta}, \tag{10.302}$$

$$\frac{\partial^2 v_*}{\partial r^2} = \frac{1}{\nu t}\frac{d^2 v_*}{d\eta^2}. \tag{10.303}$$

Thus, our governing equation transforms to

$$-\frac{1}{2t}\eta\frac{dv_*}{d\eta} = -\frac{\nu}{r}\frac{1}{\sqrt{\nu t}}\frac{dv_*}{d\eta} + \frac{\nu}{\nu t}\frac{d^2 v_*}{d\eta^2}, \tag{10.304}$$

$$\eta\frac{dv_*}{d\eta} = \frac{2}{r}\sqrt{\nu t}\frac{dv_*}{d\eta} - 2\frac{d^2 v_*}{d\eta^2}, \quad \text{so} \quad \eta\frac{dv_*}{d\eta} = \frac{2}{\eta}\frac{dv_*}{d\eta} - 2\frac{d^2 v_*}{d\eta^2}. \tag{10.305}$$

The differential equation is self-similar, with only dependence on η. The three conditions (initial and two boundary) collapse to two:

$$v_*(0) = 0, \qquad v_*(\infty) = 1. \tag{10.306}$$

We recast the differential equation as

$$\frac{d^2 v_*}{d\eta^2} + \left(\frac{\eta}{2} - \frac{1}{\eta}\right)\frac{dv_*}{d\eta} = 0. \tag{10.307}$$

This is a first-order ordinary differential equation for $dv_*/d\eta$. Multiplying by the integrating factor of $e^{\eta^2/4}/\eta$, we get

$$\frac{d}{d\eta}\left(\frac{1}{\eta}e^{\eta^2/4}\frac{dv_*}{d\eta}\right) = 0, \quad \text{so} \quad \frac{1}{\eta}e^{\eta^2/4}\frac{dv_*}{d\eta} = A, \tag{10.308}$$

$$\frac{dv_*}{d\eta} = A\eta e^{-\eta^2/4}, \quad \text{so} \quad v_* = -2Ae^{-\eta^2/4} + B. \tag{10.309}$$

To satisfy the far field condition, we take $B = 1$, giving

$$v_* = -2Ae^{-\eta^2/4} + 1. \tag{10.310}$$

For the condition at $\eta = 0$, we get $0 = -2A + 1$, giving $A = 1/2$, so the scaled and unscaled velocity solutions are

$$v_*(\eta) = 1 - e^{-\eta^2/4}, \qquad v_\theta(r,t) = \frac{\Gamma_o}{2\pi r}\left(1 - e^{-r^2/(4\nu t)}\right). \tag{10.311}$$

Clearly, when either $r \to \infty$ or $t \to 0$, we get the ideal irrotational vortex, $v_\theta = \Gamma_o/(2\pi r)$. For $t > 0$ and small r, Taylor series expansion of v_θ gives

$$v_\theta(r,t) \sim \frac{\Gamma_o r}{8\pi \nu t}\left(1 - \frac{r^2}{8\nu t} + \cdots\right). \tag{10.312}$$

The vorticity for this flow is

$$\omega_z = \frac{1}{r}\frac{\partial}{\partial r}(r v_\theta) = \frac{1}{r}\frac{\partial}{\partial r}\left(\frac{\Gamma_o}{2\pi}\left(1 - e^{-r^2/(4\nu t)}\right)\right) = \frac{\Gamma_o}{4\pi\nu t}e^{-r^2/(4\nu t)}. \tag{10.313}$$

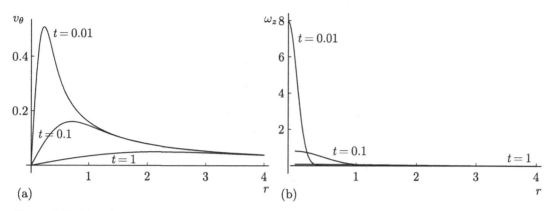

Figure 10.19 Plots for decay of an ideal vortex under the influence of viscous stress for an incompressible fluid: (a) velocity, (b) vorticity, for the case of $\Gamma_o = 1$, $\nu = 1$.

Plots of $v_\theta(r,t)$ and $\omega_z(r,t)$ for the case of $\Gamma_o = 1$, $\nu = 1$ are given in Fig. 10.19. The vorticity is initially concentrated in a pulse of finite strength at the origin; it diffuses with time into the flow. By $t = 1$ the vorticity is nearly zero throughout the domain. However, when we integrate ω_z throughout the domain, we see that is a constant, independent of time:

$$\int_A \omega_z \, dA = \int_0^\infty \omega_z \, (2\pi r) \, dr = \int_0^\infty \frac{\Gamma_o}{4\pi \nu t} e^{-r^2/(4\nu t)} (2\pi r) \, dr = \Gamma_o. \tag{10.314}$$

So the total vorticity simply redistributes by diffusion as time progresses.

10.3 Non-Newtonian Flow

Here we consider a few basic flows of some non-Newtonian fluids. Such fluids typically are composed of complicated molecules with more complicated intermolecular forces than found in simple gases. We will not consider the molecular origin of the non-Newtonian behavior, but will examine its consequences on some simple flow configurations. Additional background may be found in a variety of sources including Bird et al. (1987), Fung (1965), Joseph (1990), Morrison (2001), Paolucci (2016), and Schowalter (1978).

10.3.1 Strain Rate Dependent Viscosity

We start by considering a fluid whose viscosity depends on strain rate, as depicted in Fig. 10.20. This gives a plot of viscous stress versus strain rate. We can consider the slope of these curves to be the viscosity. Obviously, the slope of both the pseudo-plastic and dilatant fluids varies with strain rate, rendering them non-Newtonian. Let us imagine a general model for the viscosity:

$$\mu = \mu\left(p, T, I_{\dot{\epsilon}}^{(1)}, I_{\dot{\epsilon}}^{(2)}, I_{\dot{\epsilon}}^{(3)}\right) \geq 0. \tag{10.315}$$

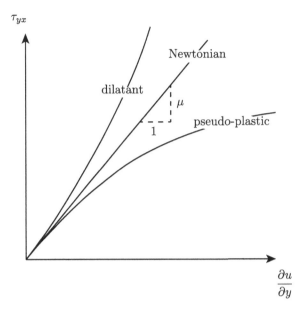

Figure 10.20 Variation of viscous stress with strain rate for Newtonian and non-Newtonian fluids.

In order to guarantee satisfaction of the second law – see Eq. (5.162) and the surrounding discussion – we insist that $\mu \geq 0$. Here, we think of μ as a thermodynamic transport property that has variation with p and T. Additionally, we allow dependence on strain rate through dependence on the principal invariants of the strain rate tensor, $I_\epsilon^{(1)}$, $I_\epsilon^{(2)}$, $I_\epsilon^{(3)}$, as introduced in Section 3.11. Because the strain rate dependency is embedded within tensor invariants, it is independent of the coordinate system chosen.

Let us now make some choices to enable a simple analysis. Let us ignore any dependency on p or T. Let us also consider an incompressible fluid. Thus, $\partial_i v_i = 0$, yielding $I_\epsilon^{(1)} = 0$. Let us only consider flows with $w = 0$ and $\partial/\partial z = 0$, so we have only two dimensions active. Following analysis of Eqs. (3.230, 3.231) we only consider dependency on $I_\epsilon^{(3)}$, giving us

$$\mu = \mu\left(I_\epsilon^{(3)}\right) \geq 0. \tag{10.316}$$

With this assumption, Eq. (3.213) reduces to

$$I_\epsilon^{(3)} = \det \mathbf{D} = \frac{\partial u}{\partial x}\frac{\partial v}{\partial y} - \frac{1}{4}\left(\frac{\partial u}{\partial y} + \frac{\partial v}{\partial x}\right)^2. \tag{10.317}$$

Note that $I_\epsilon^{(3)}$ could be positive or negative. Now for incompressible flow, the relation between viscous stress and strain rate, Eq. (5.145), reduces for our fluid to

$$\tau_{ij} = \mu\left(I_\epsilon^{(3)}\right)(\partial_i v_j + \partial_j v_i). \tag{10.318}$$

In the two-dimensional limit, we could say

$$\begin{pmatrix} \tau_{xx} & \tau_{xy} \\ \tau_{yx} & \tau_{yy} \end{pmatrix} = \mu\left(I_{\dot{\epsilon}}^{(3)}\right) \begin{pmatrix} 2\frac{\partial u}{\partial x} & \frac{\partial v}{\partial x}+\frac{\partial u}{\partial y} \\ \frac{\partial u}{\partial y}+\frac{\partial v}{\partial x} & 2\frac{\partial v}{\partial y} \end{pmatrix}. \tag{10.319}$$

Let us now select a specific functional form for μ:

$$\mu\left(I_{\dot{\epsilon}}^{(3)}\right) = \mu_o \left(\frac{\left|I_{\dot{\epsilon}}^{(3)}\right|}{\frac{U_c^2}{4h^2}}\right)^{\frac{n-1}{2}}. \tag{10.320}$$

Here empirically determined positive semi-definite constants are μ_o, n, U_c, and h. We require μ_o to have the dimensions of viscosity, n to be dimensionless, U_c to have dimensions of velocity, and h to have units of distance. We might think of U_c as some characteristic velocity and h to be a gap width.

Substituting for $I_{\dot{\epsilon}}^{(3)}$, we get

$$\mu = \mu_o \left(\frac{\left|\frac{\partial u}{\partial x}\frac{\partial v}{\partial y} - \frac{1}{4}\left(\frac{\partial u}{\partial y}+\frac{\partial v}{\partial x}\right)^2\right|}{\frac{U_c^2}{4h^2}}\right)^{\frac{n-1}{2}}. \tag{10.321}$$

Let us now make our standard assumptions for fully developed, incompressible flow, which leads to $\partial/\partial x = 0$, $v = 0$. This gives a much simpler relation for μ as

$$\mu = \mu_o \left(\left(\frac{h}{U_c}\frac{\partial u}{\partial y}\right)^2\right)^{\frac{n-1}{2}}. \tag{10.322}$$

Recognizing that $\mu \geq 0$, and accounting for the fact that $\partial u/\partial y \in (-\infty, \infty)$, we can rewrite this as

$$\mu = \mu_o \left(\frac{h}{U_c}\right)^{n-1} \left|\frac{\partial u}{\partial y}\right|^{n-1}. \tag{10.323}$$

Under these conditions, the elements of the viscous stress tensor are $\tau_{xx} = \tau_{yy} = 0$, and

$$\tau_{yx} = \tau_{xy} = \mu_o \left(\frac{h}{U_c}\right)^{n-1} \left|\frac{\partial u}{\partial y}\right|^{n-1} \frac{\partial u}{\partial y}. \tag{10.324}$$

For convenience, let us restrict study to cases for which $\partial u/\partial y \geq 0$. Then we have

$$\tau_{yx} = \tau_{xy} = \mu_o \left(\frac{h}{U_c}\right)^{n-1} \left(\frac{\partial u}{\partial y}\right)^n, \qquad \frac{\partial u}{\partial y} \geq 0. \tag{10.325}$$

For $n < 1$, the fluid is pseudo-plastic; for $n = 1$, it is Newtonian, and for $n > 1$, it is dilatant.

Couette Flow

Let us first study the simplest of non-Newtonian flows, a Couette flow. For steady, isobaric, zero body force flow of such a non-Newtonian fluid whose only nonzero velocity is $u(y)$, the x

momentum equation reduces to

$$\frac{d\tau_{yx}}{dy} = 0, \tag{10.326}$$

$$\frac{d}{dy}\left(\mu_o \left(\frac{h}{U_c}\right)^{n-1} \left(\frac{du}{dy}\right)^n\right) = 0, \tag{10.327}$$

$$\left(\frac{du}{dy}\right)^n = C_1. \tag{10.328}$$

With Couette flow boundary conditions as $u(0) = 0$, $u(h) = U$, the solution is obviously

$$u(y) = U\frac{y}{h}. \tag{10.329}$$

This is identical to the solution for Newtonian flow! The shear stress is

$$\tau_{yx} = \tau_{xy} = \mu_o \left(\frac{h}{U_c}\right)^{n-1} \left(\frac{U}{h}\right)^n = \mu_o \frac{U_c}{h}\left(\frac{U}{U_c}\right)^n. \tag{10.330}$$

Poiseuille Flow

Let us consider a non-Newtonian extension of the incompressible Poiseuille flow studied in Section 10.1.1 for a pseudo-plastic fluid with $n = 1/2$. While general n can be considered, the algebra is cumbersome, with considerable care required to account for raising negative numbers to noninteger powers. In contrast to Couette flow, here the non-Newtonian effects will alter the velocity field. The configuration is shown in Fig. 10.21. We make the typical assumptions of a steady fully developed incompressible flow with $u = u(y)$. A constant negative pressure gradient is assumed to exist in the x direction that induces a flow in the positive x direction. The fluid satisfies a no-slip condition at $y = \pm h/2$. Under these conditions, there are no effects of inertia, and the x momentum equation reduces to a force balance between pressure gradient and shear gradient:

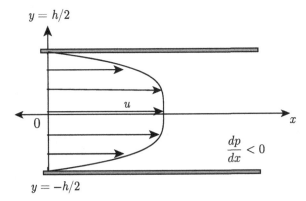

Figure 10.21 Diagram of velocity profile for pseudo-plastic non-Newtonian pressure gradient-driven flow in a slot.

10.3 Non-Newtonian Flow

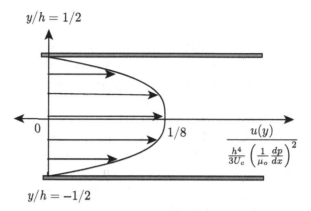

Figure 10.22 Plot of scaled velocity profile for pseudo-plastic, $n = 1/2$, non-Newtonian pressure gradient-driven flow in a slot.

$$0 = -\frac{dp}{dx} + \frac{d\tau_{yx}}{dy}, \quad (10.331)$$

$$0 = -\frac{dp}{dx} + \frac{d}{dy}\left(\mu_o \left(\frac{h}{U_c}\right)^{-1/2} \left(\frac{du}{dy}\right)^{1/2}\right), \quad y \in [-h/2, 0]. \quad (10.332)$$

To ensure $du/dy > 0$, we restrict attention to $y \in [-h/2, 0]$. A similar analysis can be done for $y > 0$, and the results have symmetry about $y = 0$. Recognizing that dp/dx is a constant, we can integrate to obtain

$$\mu_o \left(\frac{h}{U_c}\right)^{-1/2} \left(\frac{du}{dy}\right)^{1/2} = \frac{dp}{dx} y + C_1. \quad (10.333)$$

Because we expect symmetry at $y = 0$, we also expect $du/dy|_{y=0} = 0$ and use this condition to replace the boundary condition at $y = h/2$. This yields $C_1 = 0$, so

$$\mu_o \left(\frac{h}{U_c}\right)^{-1/2} \left(\frac{du}{dy}\right)^{1/2} = \frac{dp}{dx} y, \quad (10.334)$$

$$\frac{du}{dy} = \left(\frac{1}{\mu_o}\left(\frac{U_c}{h}\right)^{-1/2} \frac{dp}{dx}\right)^2 y^2, \quad (10.335)$$

$$\frac{u(y)}{\frac{h^4}{3U_c}\left(\frac{1}{\mu_o}\frac{dp}{dx}\right)^2} = \left(\frac{y}{h}\right)^3 + \left(\frac{1}{2}\right)^3, \quad y \in [-h/2, 0]. \quad (10.336)$$

Here we have applied the no-slip boundary condition requiring $u(-h/2) = 0$. Scaled results for $y \in [-h/2, h/2]$ are plotted in Fig. 10.22. It is straightforward to show that the shear stress profile is linear with y, whatever the value for n, with peak values at either wall and zero at $y = 0$.

10.3.2 Viscoelastic Flow

Let us consider briefly the complicated subject of *viscoelastic flow*. Viscoelastic materials often behave in an unusual manner relative to Newtonian fluids and may exhibit significant time-dependency. Their mathematical analysis requires some approaches not commonly employed for Newtonian fluids. We will consider just a small sample of this deep subject so as to give some insights into how they may be modeled.

Recall we defined a fluid in Chapter 1 as a material that flows when a shear stress is applied, contrasting it with a solid which relaxes to a new equilibrium in response to a shear stress. If the solid's response is linear, it is known as an *elastic solid*. Here let us consider a material that has both fluid and solid properties in that its response to an applied shear stress is that of both a fluid and a solid. There are numerous mathematical models for this type of behavior, and most are empirical curve fits.

We recall that for a Newtonian fluid, stress is proportional to strain rate and for an elastic solid, stress is proportional to strain; it is seen that we must introduce strain. To do so, we need to introduce *displacement*. A material element at any point $\mathbf{x}$ may be displaced a distance $\boldsymbol{\delta}$ from its initial point. The material velocity vector then is simply the time derivative of the displacement vector:

$$\mathbf{v} = \frac{\partial \boldsymbol{\delta}}{\partial t}. \tag{10.337}$$

As described in many standard texts in solid mechanics, for example Fung (1965), we can define the *strain tensor* $\boldsymbol{\gamma}$ as a symmetric tensor:

$$\boldsymbol{\gamma} = \frac{\nabla \boldsymbol{\delta}^T + (\nabla \boldsymbol{\delta}^T)^T}{2}. \tag{10.338}$$

Let us define the *strain rate* tensor $\dot{\boldsymbol{\gamma}}$ as

$$\dot{\boldsymbol{\gamma}} = \frac{\partial \boldsymbol{\gamma}}{\partial t} = \frac{1}{2}\left(\nabla \frac{\partial \boldsymbol{\delta}^T}{\partial t} + \left(\nabla \frac{\partial \boldsymbol{\delta}^T}{\partial t}\right)^T\right) = \frac{1}{2}\left(\nabla \mathbf{v}^T + (\nabla \mathbf{v}^T)^T\right). \tag{10.339}$$

Note that $\dot{\boldsymbol{\gamma}} = \mathbf{D}$. Here $\mathbf{D}$ is the deformation tensor defined by Eq. (3.84).

Next, let us consider a simple configuration known as a *Maxwellian element*, as shown in Fig. 10.23. Here the material element is subject to a normal stress σ and that stress is simultaneously resisted by an elastic spring with constant modulus G and a viscous dashpot, characterized by constant modulus μ. Indeed, the stress depicted is normal, which is the only possibility for a one-dimensional model. It can be extended to multiple dimensions, in which case tangential shear is admitted. We will first focus on a one-dimensional problem with scalar strain: $\gamma = \partial \delta_x / \partial x$. The element suffers two strains: γ_1 associated with the spring and γ_2 associated with the dashpot.

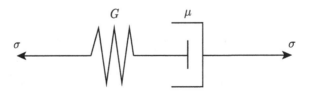

Figure 10.23 Maxwellian element for a viscoelastic material.

The stress is constant at any section of the configuration. From linear elasticity, we must have $\sigma = G\gamma_1$. And from fluid mechanics, we must have $\sigma = \mu \partial \gamma_2/\partial t$. For this element in series, the strains sum, giving

$$\gamma_1 + \gamma_2 = \gamma. \tag{10.340}$$

Taking time derivatives, we get

$$\frac{\partial \gamma_1}{\partial t} + \frac{\partial \gamma_2}{\partial t} = \frac{\partial \gamma}{\partial t}. \tag{10.341}$$

Eliminating γ_1 and γ_2, we get

$$\frac{\partial}{\partial t}\left(\frac{\sigma}{G}\right) + \frac{\sigma}{\mu} = \frac{\partial \gamma}{\partial t}, \quad \text{so} \quad \frac{\mu}{G}\frac{\partial \sigma}{\partial t} + \sigma = \mu\frac{\partial \gamma}{\partial t}. \tag{10.342}$$

Here we define the *relaxation time* as $\mu/G \equiv \lambda$. Thus,

$$\lambda \frac{\partial \sigma}{\partial t} + \sigma = \mu \frac{\partial \gamma}{\partial t}. \tag{10.343}$$

This is a first-order differential equation in time for σ. If, for example, we had a constant strain $\gamma \neq 0$, $\partial \gamma/\partial t = 0$, with $\sigma(t=0) = \sigma_o$, we would get

$$\lambda \frac{\partial \sigma}{\partial t} + \sigma = 0, \quad \text{so} \quad \sigma(t) = \sigma_o \exp\left(\frac{-t}{\lambda}\right). \tag{10.344}$$

That is to say, a suddenly strained element sees an initial value of stress, and then that relaxes exponentially to zero with time with a time constant $\lambda = \mu/G$. The material is behaving first as a solid, then as a fluid. Alternatively, if we have a constant strain rate $\dot{\gamma}$, with $\sigma(t=0) = \sigma_o$, we find

$$\sigma(t) = \mu\dot{\gamma} + e^{-t/\lambda}(\sigma_o - \mu\dot{\gamma}). \tag{10.345}$$

Here the stress relaxes in time to that of a viscous fluid, $\sigma(t \to \infty) = \mu\dot{\gamma}$.

For a more general solution, let us rearrange and multiply by the integrating factor $\exp(t/\lambda)$ to get

$$\exp\left(\frac{t}{\lambda}\right)\frac{\partial \sigma}{\partial t} + \frac{1}{\lambda}\exp\left(\frac{t}{\lambda}\right)\sigma = \frac{\mu}{\lambda}\exp\left(\frac{t}{\lambda}\right)\frac{\partial \gamma}{\partial t}, \tag{10.346}$$

$$\frac{\partial}{\partial t}\left(\exp\left(\frac{t}{\lambda}\right)\sigma\right) = \frac{\mu}{\lambda}\exp\left(\frac{t}{\lambda}\right)\frac{\partial \gamma}{\partial t}, \tag{10.347}$$

$$\exp\left(\frac{t}{\lambda}\right)\sigma = \frac{\mu}{\lambda}\int_{-\infty}^{t}\exp\left(\frac{\hat{t}}{\lambda}\right)\frac{\partial \gamma}{\partial \hat{t}}\,d\hat{t}, \tag{10.348}$$

$$\sigma(t) = \frac{\mu}{\lambda}\int_{-\infty}^{t}\exp\left(\frac{\hat{t}-t}{\lambda}\right)\frac{\partial \gamma}{\partial \hat{t}}\,d\hat{t}. \tag{10.349}$$

Here we have assumed that $\sigma \to 0$ as $t \to -\infty$.

Now for one-dimensional motion, we have

$$\gamma = \frac{\partial \delta_x}{\partial x}. \tag{10.350}$$

Then taking a time derivative, we get

$$\frac{\partial \gamma}{\partial t} = \frac{\partial}{\partial x}\frac{\partial \delta_x}{\partial t}. \tag{10.351}$$

Specializing Eq. (10.337) for the one-dimensional limit, we have $u = \partial \delta_x/\partial t$, so we get

$$\frac{\partial \gamma}{\partial t} = \frac{\partial u}{\partial x}. \tag{10.352}$$

The strain rate is the velocity gradient. Thus we can rewrite our constitutive equation for the viscoelastic material, Eq. (10.343), as

$$\lambda \frac{\partial \sigma}{\partial t} + \sigma = \mu \frac{\partial u}{\partial x}. \tag{10.353}$$

The linear momentum principle – neglecting advection, body forces – and pressure forces becomes

$$\rho \frac{\partial u}{\partial t} = \frac{\partial \sigma}{\partial x}. \tag{10.354}$$

Now take the space derivative of Eq. (10.353) and the time derivative of Eq. (10.354) to get

$$\lambda \frac{\partial^2 \sigma}{\partial x \partial t} + \frac{\partial \sigma}{\partial x} = \mu \frac{\partial^2 u}{\partial x^2}, \qquad \rho \frac{\partial^2 u}{\partial t^2} = \frac{\partial^2 \sigma}{\partial t \partial x}. \tag{10.355}$$

Eliminate the derivatives of σ and take $\nu = \mu/\rho$ to get a single equation:

$$\frac{\partial^2 u}{\partial t^2} + \frac{1}{\lambda}\frac{\partial u}{\partial t} = \frac{\nu}{\lambda}\frac{\partial^2 u}{\partial x^2}. \tag{10.356}$$

This is often known as the *telegraph equation*, and has application beyond fluid mechanics. Now if $\mu \to \infty$, we recover the traditional hyperbolic wave equation for elastic waves in solids

$$\frac{\partial^2 u}{\partial t^2} = \frac{G}{\rho}\frac{\partial^2 u}{\partial x^2}, \tag{10.357}$$

which has a d'Alembert solution of the form

$$u(x,t) = f\left(x + \sqrt{\frac{G}{\rho}}t\right) + g\left(x - \sqrt{\frac{G}{\rho}}t\right). \tag{10.358}$$

Here we see the so-called elastic wave speed is the well-known $\sqrt{G/\rho}$. Note that the wave speed makes no appeal to thermodynamics, in contrast to models of compressible fluid flow. For short relaxation time $\lambda \to 0$, we recover the traditional equation for fluid-like behavior:

$$\frac{\partial u}{\partial t} = \nu \frac{\partial^2 u}{\partial x^2}. \tag{10.359}$$

Example 10.1 Consider the telegraph equation and find solutions of the form $u = \Re(U(t)\exp(ikx))$. Evaluate for the special case of $\lambda = \mu = \rho = k = 1$.

Solution
Let us begin by considering the full function $u = U(t)\exp(ikx)$. We can take the real part later. We get

$$\frac{\partial u}{\partial t} = \exp(ikx)\frac{dU}{dt}, \tag{10.360}$$

$$\frac{\partial^2 u}{\partial t^2} = \exp(ikx)\frac{d^2 U}{dt^2}, \tag{10.361}$$

$$\frac{\partial^2 u}{\partial x^2} = -k^2 \exp(ikx) U(t). \tag{10.362}$$

Substitute into Eq. (10.356) to get

$$\exp(ikx)\frac{d^2 U}{dt^2} + \exp(ikx)\frac{1}{\lambda}\frac{dU}{dt} = \frac{-k^2 \nu}{\lambda}\exp(ikx)U(t), \tag{10.363}$$

$$\frac{d^2 U}{dt^2} + \frac{1}{\lambda}\frac{dU}{dt} + \frac{k^2 \nu}{\lambda}U(t) = 0. \tag{10.364}$$

This yields

$$U(t) = C_1 e^{\frac{1}{2}t\left(-\frac{\sqrt{1-4k^2\lambda\nu}}{\lambda}-\frac{1}{\lambda}\right)} + C_2 e^{\frac{1}{2}t\left(\frac{\sqrt{1-4k^2\lambda\nu}}{\lambda}-\frac{1}{\lambda}\right)}, \tag{10.365}$$

$$u(x,t) = C_1 e^{\frac{1}{2}t\left(-\frac{\sqrt{1-4k^2\lambda\nu}}{\lambda}-\frac{1}{\lambda}\right)+ikx} + C_2 e^{\frac{1}{2}t\left(\frac{\sqrt{1-4k^2\lambda\nu}}{\lambda}-\frac{1}{\lambda}\right)+ikx}. \tag{10.366}$$

Clearly for large k, $k > \sqrt{1/(4\lambda\nu)}$, there will be an oscillatory component in time, and for $k < \sqrt{1/(4\lambda\nu)}$, there will be no oscillation in time. For all $k > 0$, $\nu > 0$, $\lambda > 0$, the amplitude decays as $t \to \infty$. Extracting the real part of $u(x,t)$ is cumbersome. Let us study the special case for which $C_1 = 1$, $C_2 = 0$, $k = \lambda = \nu = 1$:

$$u(x,t) = \Re\left(\exp\left(-(1+\sqrt{3}i)t/2 + ix\right)\right), \tag{10.367}$$

$$= e^{-t/2}\cos\left(x - \frac{\sqrt{3}}{2}t\right). \tag{10.368}$$

Direct substitution into the telegraph equation $\partial^2 u/\partial t^2 + \partial u/\partial t = \partial^2 u/\partial x^2$ confirms this is a solution. This particular mode represents a signal propagating with speed $\sqrt{3}/2$ with amplitude decaying to zero as $t \to \infty$. Because the telegraph equation is linear, this solution could with effort be superposed with solutions for other modes to satisfy a set of initial and boundary conditions. Changing to a smaller wave number, $k = 1/2$, yields a solution with a standing wave whose amplitude decays with time:

$$u(x,t) = e^{-t/2}\cos\left(\frac{x}{2}\right). \tag{10.369}$$

A spatially homogeneous solution with $k = 0$ yields

$$u(x,t) = e^{-t}. \tag{10.370}$$

The discussion of viscoelastic materials based on simple springs and dashpots is useful for understanding, but does not engage with the full range of multi-dimensional, time-dependent continuum mechanics. Full appreciation requires significant effort, such as given by Bird et al. (1987). We will not dive deeply into this, but instead present an example which is a version of one given by Bird et al., which is itself an adaptation of Stokes' second problem.

Example 10.2 Consider an incompressible, isobaric, Maxwellian, viscoelastic material in the domain $y \in [0,\infty)$. The fluid velocity vector is confined to the x direction with variation in y and so is $u(y,t)$. At the surface $y = 0$, we have an oscillating plate inducing the no-slip condition $u(0,t) = U\sin\Omega t$. Find the behavior of the fluid at long time and compare it to a Newtonian fluid for the special case in which $\omega = \lambda = \nu = U = 1$.

Solution

Our x momentum equation is the same as for Stokes' first and second problems:

$$\rho \frac{\partial u}{\partial t} = \frac{\partial \tau_{yx}}{\partial y}. \qquad (10.371)$$

We adapt Eq. (10.353), taking as before the relaxation time $\lambda = \mu/G$, and say

$$\lambda \frac{\partial \tau_{yx}}{\partial t} + \tau_{yx} = \mu \frac{\partial u}{\partial y}. \qquad (10.372)$$

We reduce as before and obtain the telegraph equation, similar to Eq. (10.356), in the form

$$\frac{\partial^2 u}{\partial t^2} + \frac{1}{\lambda}\frac{\partial u}{\partial t} = \frac{\nu}{\lambda}\frac{\partial^2 u}{\partial y^2}. \qquad (10.373)$$

Here, we have taken $\nu = \mu/\rho$, the constant kinematic viscosity.

Let not be concerned with any initial condition, but require the following boundary conditions:

$$u(0,t) = U \sin \Omega t, \qquad u(\infty, t) < \infty. \qquad (10.374)$$

Recalling Euler's formula, $e^{i\Omega t} = \cos \Omega t + i \sin \Omega t$, we have $\Re(e^{i\Omega t}) = \cos \Omega t$, $\Im(e^{i\Omega t}) = \sin \Omega t$. Let us define a new complex variable $\mathsf{u}(y,t)$. Here $\Im(\mathsf{u}) = u(y,t)$. Then we study the related problem

$$\frac{\partial^2 \mathsf{u}}{\partial t^2} + \frac{1}{\lambda}\frac{\partial \mathsf{u}}{\partial t} = \frac{\nu}{\lambda}\frac{\partial^2 \mathsf{u}}{\partial y^2}. \qquad (10.375)$$

Now take

$$\mathsf{u}(y,t) = f(y) e^{i\Omega t}. \qquad (10.376)$$

This gives $\partial \mathsf{u}/\partial t = i\Omega f(y) e^{i\Omega t}$, $\partial^2 \mathsf{u}/\partial t = -\Omega^2 f(y) e^{i\Omega t}$, $\partial^2 \mathsf{u}/\partial t = (d^2 f/dy^2) e^{i\Omega t}$. Substitute, and cancel $e^{i\Omega t}$ to get

$$\left(-\Omega^2 + \frac{i\Omega}{\lambda}\right) f(y) = \frac{\nu}{\lambda} \frac{d^2 f}{dy^2}. \qquad (10.377)$$

Solving, we get

$$f(y) = C_1 e^{\frac{y\sqrt{-\lambda\Omega^2 + i\Omega}}{\sqrt{\nu}}} + C_2 e^{-\frac{y\sqrt{-\lambda\Omega^2 + i\Omega}}{\sqrt{\nu}}}, \qquad (10.378)$$

and recomposing to get u, we have

$$\mathsf{u}(y,t) = C_1 e^{i\Omega t + \frac{y\sqrt{-\lambda\Omega^2 + i\Omega}}{\sqrt{\nu}}} + C_2 e^{i\Omega t - \frac{y\sqrt{-\lambda\Omega^2 + i\Omega}}{\sqrt{\nu}}}. \qquad (10.379)$$

Finding the imaginary part of this is possible, but algebraically complicated. At this point let us invoke the special case of $\Omega = \lambda = \nu = U = 1$ to get

$$\mathsf{u}(y,t) = C_1 e^{it + y\sqrt{-1+i}} + C_2 e^{it - y\sqrt{-1+i}}. \qquad (10.380)$$

We use Euler's formula to recast $\sqrt{-1+i} = 2^{1/4}(\cos(3i\pi/8) + i\sin(3i\pi/8))$. Then, we get

$$\mathsf{u}(y,t) = e^{it}\left(C_1 e^{\sqrt[4]{2}y\left(\sin\left(\frac{\pi}{8}\right) + i\cos\left(\frac{\pi}{8}\right)\right)} + C_2 e^{-\sqrt[4]{2}y\left(\sin\left(\frac{\pi}{8}\right) + i\cos\left(\frac{\pi}{8}\right)\right)}\right). \qquad (10.381)$$

For a bounded flow as $y \to \infty$, we insist that $C_1 = 0$, giving

$$\mathsf{u}(y,t) = C_2 e^{it - \sqrt[4]{2}y\sin\left(\frac{\pi}{8}\right) - i\sqrt[4]{2}y\cos\left(\frac{\pi}{8}\right)}. \qquad (10.382)$$

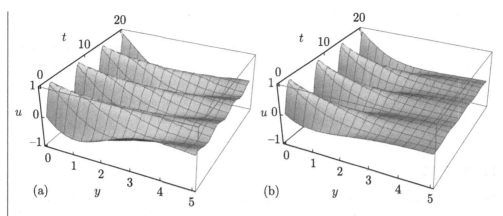

Figure 10.24 Velocity field time-evolution for Stokes' second problem for (a) Maxwellian and (b) Newtonian fluids.

Further analysis, including taking $C_2 = 1$ to satisfy the boundary condition, reveals the imaginary part of this to be

$$u(y,t) = e^{-\sqrt[4]{2}y \sin\left(\frac{\pi}{8}\right)} \sin\left(t - \sqrt[4]{2}y \cos\left(\frac{\pi}{8}\right)\right), \quad \text{Maxwellian.} \tag{10.383}$$

We compare this to the solution to the equivalent problem for a Newtonian fluid, which satisfies $\partial u/\partial t = \partial^2 u/\partial y^2$, $u(0,t) = \sin t$:

$$u(y,t) = e^{-y/\sqrt{2}} \sin\left(t - \frac{y}{\sqrt{2}}\right), \quad \text{Newtonian.} \tag{10.384}$$

Numerical evaluation allows a more direct comparison:

$$u(y,t) = e^{-0.455y} \sin(t - 1.098y), \quad \text{Maxwellian,} \tag{10.385}$$

$$u(y,t) = e^{-0.707y} \sin(t - 0.707y), \quad \text{Newtonian.} \tag{10.386}$$

There is a more rapid decay of velocity magnitude with y for a Newtonian fluid. And the phase speed for waves in the Newtonian fluid is greater. Plots for Maxwellian and Newtonian fluids are given in Fig. 10.24. One sees that the region of large amplitude velocity is confined to a smaller zone for the Newtonian fluid relative to the Maxwellian fluid.

It is possible to show with the aid of mathematical software that the full solution is

$$u(y,t) = U \exp\left(-y\sqrt[4]{\lambda^2 \Omega^2 + 1} \sqrt{\frac{\Omega}{\nu}} \cos\left(\frac{1}{2}\operatorname{Tan}^{-1}(-\lambda\Omega, 1)\right)\right)$$
$$\times \sin\left(\Omega t - y\sqrt[4]{\lambda^2 \Omega^2 + 1} \sqrt{\frac{\Omega}{\nu}} \sin\left(\frac{1}{2}\operatorname{Tan}^{-1}(-\lambda\Omega, 1)\right)\right). \tag{10.387}$$

Taylor series expansion of this for small relaxation time λ shows how the solution is that for Stokes' second problem, Eq. (10.288) with a correction term for small finite λ:

$$u(y,t) \sim U e^{-\sqrt{\frac{\Omega}{2\nu}}y} \sin\left(\Omega t - \sqrt{\frac{\Omega}{2\nu}}y\right)$$
$$\times \left(1 + \frac{\lambda\Omega}{2}\sqrt{\frac{\Omega}{2\nu}}y\left(1 - \cot\left(\Omega t - \sqrt{\frac{\Omega}{2\nu}}y\right)\right) + O\left(\lambda^2\right)\right). \tag{10.388}$$

SUMMARY

This chapter has focused on the one-dimensional flow of incompressible viscous fluids. First, important paradigm problems in which inertia played no role were considered, including Poiseuille and Couette flows. Attention was then given to simple unsteady flows for which fluid inertia played a critical role. Lastly some consideration was given to non-Newtonian flows with either strain rate-dependent viscosity or viscoelastic effects.

PROBLEMS

10.1 An incompressible fluid with density ρ and constant viscosity μ exists in the annular region between two infinitely long cylinders. Body forces are negligible. The inner cylinder has radius R_1 and is stationary. The outer cylinder has radius R_2 and rotates such that its velocity magnitude is v_2. Assume the field has no time-dependency and has a velocity field of the form $\mathbf{v} = v_\theta(r)\mathbf{e}_\theta$. Find the velocity field $v_\theta(r)$. Find the pressure field $p(r)$, assuming the constant of integration to be zero.

10.2 A Newtonian fluid with constant density ρ flows in a slot of width b. The slot is modeled as having infinite depth. The slot is at an angle of θ to the horizontal, and gravity acts vertically downward with the gravitational acceleration of g. There is no driving pressure gradient in the flow direction. Construct a coordinate system rotated at an angle θ to the horizontal, with x in the flow direction and y normal to the flow direction. If the viscosity μ is a constant, find the steady state velocity profile. Find the difference in pressure between the top and bottom of the channel, measured normal to the channel walls. If the viscosity μ varies with distance as $\mu(y) = \mu_o + \mu_1(y/b)$, find the steady state velocity profile.

10.3 For the constant viscosity limit of the previous problem, if the fluid is stationary at $t = 0$, find the unsteady response, and give an estimate for the time to relax to a steady state. Make plots of $u(y,t)$ for the parameters $g = 1$, $\theta = \pi/4$, $b = 1$, $\nu = \mu/\rho = 1$.

10.4 Solve a generalization of Stokes' first problem:

$$\frac{\partial u}{\partial t} = \nu \frac{\partial^2 u}{\partial y^2}, \qquad u(y \to \pm\infty) \to 0, \quad u(y,0) = U(H(y+L) - H(y-L)),$$

in terms of the error function. Here $H(\xi)$ is the Heaviside unit step function with $H(\xi) = 0$, for $\xi < 0$ and $H(\xi) = 1$, for $\xi \geq 0$. The initial distribution is sometimes called a "top hat" function for its shape. Plot the solution for $U = 1$, $\nu = 1$, $L = 1$ for $y \in [-5,5]$, $t \in (0,2]$.

10.5 Show the one-dimensional, incompressible Navier–Stokes equations for a Newtonian fluid with constant viscosity and constant body force, $\partial u/\partial x = 0$, $du/dt = -(1/\rho)\partial p/\partial x + f_x + \nu \partial^2 u/\partial x^2$, are invariant under the transformations

$$\tilde{x} = x + a(t), \quad \tilde{t} = t, \quad \tilde{u} = u + \frac{da}{dt}, \quad \tilde{p} = p - \rho x \frac{d^2 a}{dt^2} + b(t).$$

Here $a(t)$ and $b(t)$ are arbitrary functions of time. The result can be extended to multiple dimensions.

10.6 Show the incompressible Navier–Stokes equations for a Newtonian fluid with constant viscosity and no body force, $\nabla^T \cdot \mathbf{v} = 0$, $d\mathbf{v}/dt = -(1/\rho)\nabla p + \nu \nabla^2 \mathbf{v}$, are invariant under the stretching transformations

$$\tilde{\mathbf{x}} = e^s \mathbf{x}, \quad \tilde{t} = e^{2s} t, \quad \tilde{\mathbf{v}} = e^{-s} \mathbf{v}, \quad \tilde{p} = e^{-2s} p.$$

10.7 Show the pulse of x momentum described by the velocity field

$$u(y,z,t) = \frac{M}{4\pi \nu t} \exp\left(-\frac{y^2}{4\nu t} - \frac{z^2}{4\nu t}\right),$$

satisfies the incompressible, isobaric x momentum equation

$$\frac{\partial u}{\partial t} = \nu \left(\frac{\partial^2 u}{\partial y^2} + \frac{\partial^2 u}{\partial z^2}\right),$$

and describes the diffusion of a constant pulse of x momentum through the (y,z) plane. Find the total value of x momentum in the domain $y \in (-\infty, \infty)$, $z \in (-\infty, \infty)$. For $M = 1$, $\nu = 1$, plot $u(y, z, t = 1/10)$, $u(y, z, t = 1/2)$. Confirm the flow field is incompressible. Find an expression for the stress tensor, $\tau(y, z, t)$ if the density is ρ. For $\rho = \nu = M = y = z = t = 1$, find the eigenvalues of stress and the unit vectors of orientation α. Plot Cauchy's viscous stress quadrics.

10.8 For a non-Newtonian fluid with

$$\tau_{yx} = \mu_o \left(\frac{h}{U_c}\right)^{n-1} \left(\frac{\partial u}{\partial y}\right)^n, \quad \frac{\partial u}{\partial y} > 0,$$

formulate a modified Stokes' first problem for an incompressible flow with density ρ. To maintain $\partial u/\partial y > 0$, take $u(0,t) = -U$, with $U > 0$. Selecting $n = 1/2$ for a pseudoplastic fluid, taking $\nu_o = \mu_o/\rho$, choosing an appropriate scaling, and taking dimensionless parameter $\beta \equiv (4(U/U_c)(hU/\nu_o))^{-1/2}$, show the problem formulation can be reduced to

$$\frac{\partial u_*}{\partial t_*} = \beta \left(\frac{\partial u_*}{\partial y_*}\right)^{-1/2} \frac{\partial^2 u_*}{\partial y_*^2}, \quad u_*(y_*, 0) = 0, \quad u_*(0, t_*) = -1, \quad u_*(\infty, t_*) = 0.$$

Stretch the variables to find a self-similar formulation in terms of a similarity variable η. Find a solution, either numerical or analytic, that satisfies the no-slip and far field conditions. Plot results.

10.9 Repeat the previous problem for a dilatant fluid. For a non-Newtonian fluid with

$$\tau_{yx} = \mu_o \left(\frac{h}{U_c}\right)^{n-1} \left(\frac{\partial u}{\partial y}\right)^n, \quad \frac{\partial u}{\partial y} > 0,$$

formulate a modified Stokes' first problem for an incompressible flow with density ρ. To maintain $\partial u/\partial y > 0$, take $u(0,t) = -U$, with $U > 0$. Selecting $n = 3/2$ for a dilatant fluid, taking $\nu_o = \mu_o/\rho$, choosing an appropriate scaling, and taking dimensionless parameter $\beta \equiv \sqrt{(9/4)(U/U_c)(hU/\nu_o)}$, show the problem can be reduced to

$$\frac{\partial u_*}{\partial t_*} = \beta \left(\frac{\partial u_*}{\partial y_*}\right)^{1/2} \frac{\partial^2 u_*}{\partial y_*^2}, \quad u_*(y_*, 0) = 0, \quad u_*(0, t_*) = -1, \quad u_*(\infty, t_*) = 0.$$

Stretch the variables to find a self-similar formulation in terms of a similarity variable η. Find a solution that satisfies the no-slip condition. Identify issues with satisfying a far field condition.

10.10 An incompressible, Newtonian fluid with density ρ and viscosity μ is initially at rest in a channel between two stationary parallel plates of infinite extent. The plates are at $y = \pm h$. At $t = 0$ a pressure gradient of strength $\nabla p = (\Delta p/L)\mathbf{i}$ is suddenly applied and maintained. There is no body force. Assuming the velocity is restricted to $\mathbf{v} = u(y,t)\mathbf{i}$, find $u(y,t)$. Find numerical values for a three-term series expansion with $h = 1$, $\Delta p = -1$, $\nu = 1$, $L = 1$, and plot $u(y, t = 0)$, $u(y, t = 1/10)$, $u(y, t = 1)$.

10.11 An incompressible, Newtonian fluid with constant viscosity is contained between two infinite parallel plates separated by distance $2h$. There is no body force. At $t = 0$ the fluid is at rest, and the plate walls acquire a velocity U. Find the unsteady response and the time constant of relaxation of the slowest mode.

10.12 An incompressible, Newtonian fluid with constant viscosity is contained in a cylindrical tube of radius R. There is no body force. At $t = 0$ the fluid is at rest, and the tube walls acquire a velocity U. Making all appropriate assumptions, find the unsteady response. Find the time constant of relaxation of the slowest mode. Compare to the planar equivalent of the previous problem.

10.13 Consider Stokes' first problem for a type of incompressible, Oldroyd viscoelastic fluid:

$$\rho \frac{\partial u}{\partial t} = \frac{\partial \tau_{yx}}{\partial y}, \qquad \tau_{yx} + \lambda \frac{\partial \tau_{yx}}{\partial t} = \mu \left(\frac{\partial u}{\partial y} + \lambda \frac{\partial}{\partial t} \frac{\partial u}{\partial y} \right),$$
$$u(0,t) = U, \qquad u(\infty, t) = 0, \qquad u(y,0) = 0.$$

Here λ is a known time constant; if $\lambda = 0$, the fluid is Newtonian. Combine to form a single partial differential equation for u, and find the solution in terms of the error function complement.

FURTHER READING

Bird, R. B., Stewart, W. E., and Lightfoot, E. N. (2007). *Transport Phenomena*, revised 2nd ed. New York: John Wiley.

Brodkey, R. S. (1967). *The Phenomena of Fluid Motions*. New York: Dover.

Currie, I. G. (2013). *Fundamental Mechanics of Fluids*, 4th ed. Boca Raton, Florida: CRC Press.

Huilgol, R. R. (2015). *Fluid Mechanics of Viscoplasticity*. Berlin: Springer.

Joseph, D., Funada, T., and Wang, J. (2008). *Potential Flows of Viscous and Viscoelastic Fluids*. New York: Cambridge University Press.

Kundu, P. K., Cohen, I. M., and Dowling, D. R. (2015). *Fluid Mechanics*, 6th ed. Amsterdam: Academic Press.

Langlois, W. E., and Deville, M. O. (2014). *Slow Viscous Flow*, 2nd ed. Heidelberg: Springer.

Leal, L. G. (2007). *Advanced Transport Phenomena: Fluid Mechanics and Convective Transport Processes*. New York: Cambridge University Press.

Ockendon, H., and Ockendon, J. R. (1995). *Viscous Flow*. Cambridge, UK: Cambridge University Press.

Panton, R. L. (2013). *Incompressible Flow*, 4th ed. New York: John Wiley.

Schlichting, H., and Gersten, K. (2017). *Boundary Layer Theory*, 9th ed. New York: McGraw-Hill.

Sherman, F. S. (1990). *Viscous Flow*. New York: McGraw-Hill.
Whitaker, S. (1968). *Introduction to Fluid Mechanics*. Malabar, Florida: Krieger.
White, F. M. (2006). *Viscous Fluid Flow*, 3rd ed. New York: McGraw-Hill.
Yih, C.-S. (1977). *Fluid Mechanics*. East Hampton, Connecticut: West River Press.

REFERENCES

Bird, R. B., Armstrong, R. C., and Hassager, O. (1987). *Dynamics of Polymeric Liquids*, 2 vols. New York: John Wiley.
Cantwell, B. J. (2002). *Introduction to Symmetry Analysis*. New York: Cambridge University Press.
Carslaw, H. S., and Jaeger, J. C. (1986). *Conduction of Heat in Solids*, 2nd ed. Oxford: Oxford University Press.
Fung, Y. C. (1965). *Foundations of Solid Mechanics*. Englewood Cliffs, New Jersey: Prentice-Hall.
Joseph, D. D. (1990). *Fluid Dynamics of Viscoelastic Liquids*. New York: Springer.
Mei, C. C. (1997). *Mathematical Analysis in Engineering*. Cambridge, UK: Cambridge University Press.
Morrison, F. A. (2001). *Understanding Rheology*. New York: Oxford University Press.
Paolucci, S. (2016). *Continuum Mechanics and Thermodynamics of Matter*. New York: Cambridge University Press.
Schowalter, W. R. (1978). *Mechanics of Non-Newtonian Fluids*. Oxford: Pergamon.
Stokes, G. G. (1851). On the effect of the internal friction of fluids on the motion of pendulums, *Transactions of the Cambridge Philosophical Society*, **9**(2), 8–106.

11 Multi-Dimensional Viscous Flow

Here we consider some standard problems in multi-dimensional viscous flow. As for one-dimensional viscous flow, application areas are widespread and can include ordinary pipe flows as well as microscale fluid mechanics. We will restrict attention to problems that are steady and laminar. Most of the problems will be incompressible, except for one dealing with a problem in natural convection, Section 11.2.6, and another in compressible boundary layers, Section 11.2.7. The first section will consider problems in which inertia plays no role and solutions are a consequence of a force balance. The second section will reintroduce inertia by allowing steady advective acceleration, which is associated with a force imbalance.

A paradigm flow for this chapter, to be considered in Section 11.2.1, is the so-called *Blasius boundary layer*, depicted in Fig. 11.1. Here we have presented a fluid with uniform velocity in the x direction approaching a horizontal flat plate. The diagram shows the upper half plane, and one should realize that the results are symmetric about $y = 0$ for a thin flat plate. When the fluid strikes the flat plate, it sticks to its surface, thus inducing a velocity gradient in the y direction and an associated shear stress. The fluid's x velocity component dominates over the y component, and the x component varies strongly with both x and y. In the far field as $y \to \pm\infty$, the fluid velocity relaxes to its freestream value. This flow field has significant multi-dimensional variation, and the fluid particle decelerates significantly as the freestream fluid encounters the flat plate near $y = 0$. The curved dashed line indicates the outer edge of the boundary layer. Within the boundary layer, the flow is rotational and net viscous forces balance fluid acceleration. Outside of the boundary layer, velocity gradients are small, viscous forces are small, and the flow is well modeled by inviscid equations. The flow for the Blasius boundary layer is such that the pressure field is spatially uniform.

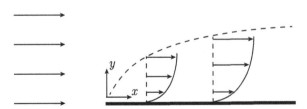

Figure 11.1 Diagram of a paradigm two-dimensional steady viscous flow, the Blasius boundary layer, for which a uniform freestream encounters a flat plate.

11.1 Flow with No Effects of Inertia

We begin with a treatment of flows for which effects of inertia are negligible, thus leading to force balances. We will not consider effects of body forces. So the force balances we study will be balances of surface forces: pressure and viscous. This flow regime is relevant to many microscale flows such as flow around small particles or in narrow channels. The flows will be taken to be incompressible. Such flows will induce spatially variable velocity fields, which formally have advective acceleration. However, the magnitude of this acceleration will be small relative to the surface forces.

11.1.1 Stokes Equations

Because we will mainly deal with dimensionless variables, let us return to an alternate notation for which dimensional variables have an over-bar. Consider the dimensional incompressible mass and linear momenta equations for a fluid with constant viscosity and no body force:

$$\overline{\nabla}^T \cdot \overline{\mathbf{v}} = 0, \qquad \frac{d\overline{\mathbf{v}}}{d\overline{t}} = -\frac{1}{\rho}\overline{\nabla}\overline{p} + \nu(\overline{\nabla}^T \cdot \overline{\nabla})\overline{\mathbf{v}}. \qquad (11.1)$$

In Section 6.7, we scaled the compressible version of these equations in a common fashion. That scaling was done for the scenario in which inertia-based terms were of comparable magnitude to pressure forces. We also considered an alternate scaling in Section 6.7.8 that highlighted conditions under which pressure variation and viscous dissipation could be neglected. Let us make yet another an alternate choice here and consider the scenario in which inertia terms are small relative to both pressure and shear forces, both of which are of comparable magnitude. Choosing a scaling under which to examine the equations is comparable to choosing a magnification setting on a microscope; different choices allow different features to be revealed.

Let us then take the following scaling:

$$\mathbf{x} = \frac{\overline{\mathbf{x}}}{L}, \quad \mathbf{v} = \frac{\overline{\mathbf{v}}}{v_o}, \quad t = \frac{v_o \overline{t}}{L}, \quad \frac{\partial}{\partial t} = \frac{L}{v_o}\frac{\partial}{\partial \overline{t}}, \quad \nabla = L\overline{\nabla}, \quad p = \frac{\overline{p} - p_o}{\mu v_o / L}. \qquad (11.2)$$

The key distinction from what was done in Section 6.7 is that pressure has been scaled differently. Defining the Reynolds number as $Re = v_o L/\nu$, this scaling leads us to

$$\nabla^T \cdot \mathbf{v} = 0, \qquad \frac{d\mathbf{v}}{dt} = \frac{1}{Re}\left(-\nabla p + (\nabla^T \cdot \nabla)\mathbf{v}\right). \qquad (11.3)$$

Now as $Re \to 0$, the momenta equation reduces to a balance between the surface forces of pressure and shear:

$$\nabla p = (\nabla^T \cdot \nabla)\mathbf{v}. \qquad (11.4)$$

Often these are written in dimensional form and are taken to be

$$\overline{\nabla}^T \cdot \overline{\mathbf{v}} = 0, \qquad \overline{\nabla}\overline{p} = \mu(\overline{\nabla}^T \cdot \overline{\nabla})\overline{\mathbf{v}}. \qquad (11.5)$$

These are known as *Stokes equations*, and flow governed by them is known as *Stokes flow*.

We return to the dimensionless form now. We get useful relations by taking the divergence and curl of the linear momenta equation. First we take the divergence to find

$$\nabla^T \cdot \nabla p = \nabla^T \cdot (\nabla^T \cdot \nabla)\mathbf{v} = (\nabla^T \cdot \nabla)\underbrace{\nabla^T \cdot \mathbf{v}}_{=0} = 0. \tag{11.6}$$

So the pressure satisfies Laplace's equation ($\nabla^2 p = 0$) and must be a harmonic function. Recall from Section 6.4.4 that the Laplacian of pressure is relevant for incompressible flow analysis. Next take the curl of the dimensionless linear momenta equation:

$$\underbrace{\nabla \times \nabla p}_{=0} = \nabla \times (\nabla^T \cdot \nabla)\mathbf{v}. \tag{11.7}$$

The curl of the gradient of any scalar must be zero, and we can commute the operators on the right-hand side to get

$$\mathbf{0} = (\nabla^T \cdot \nabla)\underbrace{\nabla \times \mathbf{v}}_{=\boldsymbol{\omega}} = (\nabla^T \cdot \nabla)\boldsymbol{\omega}. \tag{11.8}$$

Here the dimensionless vorticity vector is $\boldsymbol{\omega} = \nabla \times \mathbf{v}$, and we see that each component of the vorticity field satisfies Laplace's equation ($\nabla^2 \boldsymbol{\omega} = \mathbf{0}$) and must also be a harmonic function. Comparing to the incompressible, constant viscosity Helmholtz vorticity transport equation, Eq. (7.80), we see for Stokes flow, the transport of vorticity is restricted to diffusive transport, as all unsteady and advective terms are negligible. Lastly, in dimensionless form, we can relate the viscous stress tensor to the deformation tensor via

$$\boldsymbol{\tau} = 2\mathbf{D}. \tag{11.9}$$

Let us consider three important two-dimensional limits: (a) Cartesian, (b) polar, and (c) spherical axisymmetric. In each, we will be able to define a stream function and arrive at a compact formulation of the equations of motion. Introduction of a stream function for two-dimensional flow has some subtle and nonobvious consequences. One must be careful in extending this to general coordinate systems. Panton (2013, Chapters 12, 21, Appendix D) gives a detailed discussion that we will draw upon. In short, for some coordinate systems, we will need to replace the Laplacian operator ∇^2, with the E^2 operator; see Sections 2.4.4, 2.7. For systems for which neglect of a third dimension confines the geometry to a plane, such as (x, y) Cartesian systems or (r, θ) polar systems, we will find $\nabla^2 = E^2$. But for systems such as spherical coordinate systems in which neglect of variation in any coordinate does not confine the system to a plane, we will need the E^2 operator, which is slightly different.

Cartesian

First, we examine a Cartesian formulation. For two-dimensional incompressible flow, we can define the dimensionless stream function ψ, analogous to Eq. (8.8), as

$$u = \frac{\partial \psi}{\partial y}, \qquad v = -\frac{\partial \psi}{\partial x}. \tag{11.10}$$

And for this two-dimensional flow, we adapt Eq. (8.1) to get

$$\omega_z = \frac{\partial v}{\partial x} - \frac{\partial u}{\partial y}. \tag{11.11}$$

Eliminating velocities, we get

$$\omega_z = \frac{\partial}{\partial x}\left(-\frac{\partial \psi}{\partial x}\right) - \frac{\partial}{\partial y}\left(\frac{\partial \psi}{\partial y}\right), \qquad \frac{\partial^2 \psi}{\partial x^2} + \frac{\partial^2 \psi}{\partial y^2} = -\omega_z. \tag{11.12}$$

We now apply the two-dimensional Laplacian operator to both sides, giving

$$\left(\frac{\partial^2}{\partial x^2} + \frac{\partial^2}{\partial y^2}\right)\left(\frac{\partial^2 \psi}{\partial x^2} + \frac{\partial^2 \psi}{\partial y^2}\right) = -\left(\frac{\partial^2}{\partial x^2} + \frac{\partial^2}{\partial y^2}\right)\omega_z. \tag{11.13}$$

The right-hand side must be zero because of Eq. (11.8), and we can expand the left-hand side to get

$$\left(\frac{\partial^2}{\partial x^2} + \frac{\partial^2}{\partial y^2}\right)\left(\frac{\partial^2}{\partial x^2} + \frac{\partial^2}{\partial y^2}\right)\psi = \frac{\partial^4 \psi}{\partial x^4} + 2\frac{\partial^4 \psi}{\partial x^2 \partial y^2} + \frac{\partial^4 \psi}{\partial y^4} = 0. \tag{11.14}$$

This is often called the *biharmonic equation*. Sometimes it is abbreviated as $\nabla^4 \psi = 0$. Sometimes it is written as $\nabla^2 \nabla^2 \psi = 0$. Though uncommon, one could write it as $(\nabla^T \cdot \nabla)(\nabla^T \cdot \nabla)\psi = 0$. In index notation, one would simply say $\partial_i \partial_i \partial_j \partial_j \psi = 0$. And for this Cartesian system $E^2 = \nabla^2$, so we could also cast the biharmonic equation as

$$E^4 \psi = \nabla^4 \psi = 0, \qquad \text{Cartesian coordinates.} \tag{11.15}$$

Example 11.1 Analyze for the small Reynolds number limit the flow field given by the dimensionless stream function

$$\psi(x,y) = 2xy - \frac{\Omega}{6}(x^3 + y^3). \tag{11.16}$$

Solution

Here Ω is a dimensionless number. When $\Omega = 0$, we recover the potential flow field studied in Section 8.3.5 for $n = 2$, $B = 1$. The $\Omega = 0$ stream function describes the stagnation flow of two impinging streams, as well as the flow in a corner of angle $\pi/2$. By inspection, for $\Omega \neq 0$, we see that the flow satisfies the biharmonic equation, Eq. (11.13). So it is a solution to the Stokes equations. The velocity field is given by

$$u = \frac{\partial \psi}{\partial y} = 2x - \frac{\Omega}{2}y^2, \qquad v = -\frac{\partial \psi}{\partial x} = -2y + \frac{\Omega}{2}x^2. \tag{11.17}$$

Mass conservation for the incompressible flow is satisfied because

$$\frac{\partial u}{\partial x} + \frac{\partial v}{\partial y} = 2 - 2 = 0. \tag{11.18}$$

The flow is rotational because

$$\omega_z = \frac{\partial v}{\partial x} - \frac{\partial u}{\partial y} = \Omega x - (-\Omega y) = \Omega(x + y). \tag{11.19}$$

So for this simple flow, the vorticity varies over the plane and is zero if $y = -x$. By inspection, $\nabla^2 \omega_z = 0$, a condition for Stokes flow. It can be verified that $\nabla^2 \psi = -\omega_z$. We also see that $\nabla^2 u = -\Omega$ and $\nabla^2 v = \Omega$. Thus $\partial p/\partial x = -\Omega$ and $\partial p/\partial y = \Omega$. Taking $p(0,0) = 0$, we then have the pressure field as

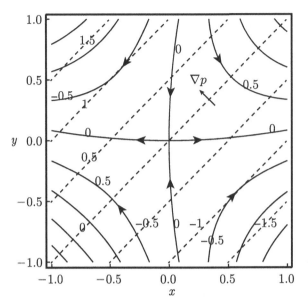

Figure 11.2 Streamlines and isobars for a simple Stokes flow for two impinging streams. Solid lines are streamlines; dashed lines are isobars.

$$p(x,y) = \Omega(y - x). \tag{11.20}$$

Obviously $\nabla^2 p = 0$, as required, and the pressure field is harmonic. The shear stress tensor field is given by

$$\boldsymbol{\tau} = \begin{pmatrix} 2\frac{\partial u}{\partial x} & \frac{\partial v}{\partial x} + \frac{\partial u}{\partial y} \\ \frac{\partial u}{\partial y} + \frac{\partial v}{\partial x} & 2\frac{\partial v}{\partial y} \end{pmatrix} = \begin{pmatrix} 4 & \Omega(x - y) \\ \Omega(x - y) & -4 \end{pmatrix}. \tag{11.21}$$

The relevant divergence of the stress tensor is

$$\nabla^T \cdot \boldsymbol{\tau} = \begin{pmatrix} \frac{\partial}{\partial x} & \frac{\partial}{\partial y} \end{pmatrix} \begin{pmatrix} 4 & \Omega(x - y) \\ \Omega(x - y) & -4 \end{pmatrix} = \begin{pmatrix} -\Omega & \Omega \end{pmatrix}. \tag{11.22}$$

The transpose of this vector is exactly the gradient of the pressure field, as required by the force balance in the zero inertial effect limit. Streamlines and isobars are plotted for $\Omega = 1$ in Fig. 11.2. The streamlines are similar to those for the stagnation potential flow depicted in Fig. 8.11; here, however, the stagnation streamlines with $\psi = 0$ have weak curvature. This Stokes flow has isobars that are straight lines inclined at $\pi/4$ to the horizontal; in a strong distinction, the potential flow has circular isobars with a pressure maximum at the origin.

The deformation tensor field, $\mathbf{D} = \boldsymbol{\tau}/2$, is given by

$$\mathbf{D} = \begin{pmatrix} \frac{\partial u}{\partial x} & (1/2)(\frac{\partial v}{\partial x} + \frac{\partial u}{\partial y}) \\ (1/2)(\frac{\partial u}{\partial y} + \frac{\partial v}{\partial x}) & \frac{\partial v}{\partial y} \end{pmatrix} = \begin{pmatrix} 2 & \Omega(x - y)/2 \\ \Omega(x - y)/2 & -2 \end{pmatrix}. \tag{11.23}$$

The strain rate can be characterized by its two principal invariants. They are

$$I_{\dot{\epsilon}}^{(1)} = \operatorname{tr} \mathbf{D} = 0, \qquad I_{\dot{\epsilon}}^{(3)} = \det \mathbf{D} = -4 - \frac{\Omega^2}{4}(x - y)^2. \tag{11.24}$$

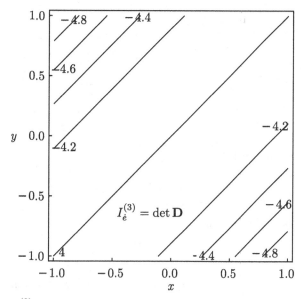

Figure 11.3 Contours of $I_{\dot{\epsilon}}^{(3)} = \det \mathbf{D}$, one of the principal invariants of strain rate, for a simple Stokes flow for two impinging streams.

For $\Omega = 0$, the field of $I_{\dot{\epsilon}}^{(3)}$ is uniform, and for $\Omega \neq 0$, its amplitude increases with distance from the line $y = x$. Contours of $I_{\dot{\epsilon}}^{(3)}$ are plotted for $\Omega = 1$ in Fig. 11.3.

As an aside, there is a nonzero advective acceleration:

$$(\mathbf{v}^T \cdot \nabla)\mathbf{v} = \frac{1}{2}\begin{pmatrix} 8x + \Omega y(2y - x^2\Omega) \\ 8y + \Omega x(2x - y^2\Omega) \end{pmatrix}. \qquad (11.25)$$

In the linear momenta equation, this term is multiplied by Re, which we are assuming is negligibly small. Nevertheless for small but finite Re for sufficiently large x or y, the advective acceleration for our flow will be sufficiently large that our assumptions for Stokes flow are no longer valid, and a more complete theory is required to predict the flow dynamics. Lastly, while the analysis of this flow field is straightforward, it may be difficult to actually construct an experiment with these features.

Polar

We now consider the solutions to the Stokes equations in the polar coordinate system, which is a two-dimensional limit of the cylindrical coordinate system introduced in Section 2.6. Here we neglect any component in the z direction and specialize the transformation of Section 2.6 to

$$x = r\cos\theta, \qquad r = \sqrt{x^2 + y^2}, \qquad (11.26)$$
$$y = r\sin\theta, \qquad \theta = \mathrm{Tan}^{-1}(x, y). \qquad (11.27)$$

We consider x and y and thus r and θ to all be dimensionless. Here we have employed the slightly more robust Tan^{-1} function, as described in Section 8.2.2. The polar limit of Eqs.

(2.330, 2.336, 2.337) are

$$\nabla = \frac{\partial}{\partial r}\mathbf{e}_r + \frac{1}{r}\frac{\partial}{\partial \theta}\mathbf{e}_\theta, \tag{11.28}$$

$$\nabla^T \cdot \mathbf{v} = \frac{1}{r}\frac{\partial}{\partial r}(rv_r) + \frac{1}{r}\frac{\partial v_\theta}{\partial \theta}, \tag{11.29}$$

$$\nabla^T \cdot \nabla = \nabla^2 = \frac{1}{r}\frac{\partial}{\partial r}\left(r\frac{\partial}{\partial r}\right) + \frac{1}{r^2}\frac{\partial^2}{\partial \theta^2}. \tag{11.30}$$

The biharmonic operator is

$$\nabla^4 = E^4 = \left(\frac{1}{r}\frac{\partial}{\partial r}\left(r\frac{\partial}{\partial r}\right) + \frac{1}{r^2}\frac{\partial^2}{\partial \theta^2}\right)^2. \tag{11.31}$$

The vorticity in polar coordinates is

$$\omega_z = \frac{1}{r}\frac{\partial}{\partial r}(rv_\theta) - \frac{1}{r}\frac{\partial v_r}{\partial \theta}. \tag{11.32}$$

The representation of the mass equation in polar coordinates is

$$\frac{1}{r}\frac{\partial}{\partial r}(rv_r) + \frac{1}{r}\frac{\partial v_\theta}{\partial \theta} = 0. \tag{11.33}$$

This suggests the definition of a stream function ψ as follows:

$$v_r = \frac{1}{r}\frac{\partial \psi}{\partial \theta}, \qquad v_\theta = -\frac{\partial \psi}{\partial r}. \tag{11.34}$$

Direct substitution into the mass equation shows this guarantees satisfaction.

Example 11.2 Analyze for the small Reynolds number limit the flow field given by the dimensionless stream function discussed in more detail by Panton (2013, p. 617), known as the Taylor (1962) scraper:

$$\psi(r,\theta) = -\frac{r}{\sin^2 \alpha - \alpha^2}\left(\alpha^2 \sin \theta - (\sin^2 \alpha)\theta \cos \theta + (\sin \alpha \cos \alpha - \alpha)\theta \sin \theta\right). \tag{11.35}$$

Here α is a constant.

Solution

This solution has remarkable features. Direct substitution into $E^4\psi = \nabla^4\psi = 0$ and detailed calculation aided by computer software verifies that $\psi(r,\theta)$ satisfies the biharmonic equation. Direct substitution also verifies that

$$\psi(r,0) = 0, \qquad \psi(r,\alpha) = 0, \tag{11.36}$$

$$v_r(r,0) = \frac{1}{r}\left.\frac{\partial \psi}{\partial \theta}\right|_{\theta=0} = 0, \qquad v_r(r,\alpha) = \frac{1}{r}\left.\frac{\partial \psi}{\partial \theta}\right|_{\theta=\alpha} = 1, \tag{11.37}$$

$$v_\theta(r,0) = -\left.\frac{\partial \psi}{\partial r}\right|_{\theta=0} = 0, \qquad v_\theta(r,\alpha) = -\left.\frac{\partial \psi}{\partial r}\right|_{\theta=\alpha} = 0. \tag{11.38}$$

Consider the wedge defined by $\theta = 0$ and $\theta = \alpha$. The wedge surface is a streamline through which no mass can penetrate; thus, it may represent a wall. The wall at $\theta = 0$ has a constant velocity of $v_r = 1$, $v_\theta = 0$, so it is a moving surface to which the fluid sticks. The wall at $\theta = \alpha$ is stationary and because of the no-slip condition, the fluid is also stationary there. Contours of the stream function for various values of α are plotted in Fig. 11.4.

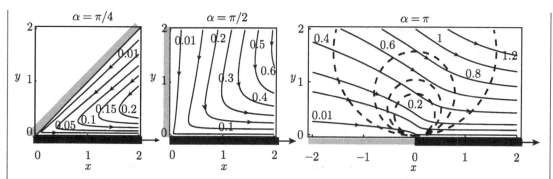

Figure 11.4 Streamlines for a Taylor scraper Stokes flow for various wedge angles $\alpha = \pi/4$, $\alpha = \pi/2$, $\alpha = \pi$. The wall at $y = 0$, $x \in [0, \infty)$ is pulled to the right with constant velocity of unity. Isobars are plotted in dashed lines for the $\alpha = \pi$ case.

For the case of $\alpha = \pi$, the solution has a simple representation in Cartesian coordinates of

$$\psi(x, y) = y\left(1 - \frac{1}{\pi}\text{Tan}^{-1}(x, y)\right). \tag{11.39}$$

For this flow, we find

$$u(x, y) = \frac{\partial \psi}{\partial y} = 1 - \frac{xy}{\pi(x^2 + y^2)} - \frac{1}{\pi}\text{Tan}^{-1}(x, y), \tag{11.40}$$

$$v(x, y) = -\frac{\partial \psi}{\partial x} = -\frac{y^2}{\pi(x^2 + y^2)}, \tag{11.41}$$

$$\omega_z = \frac{\partial v}{\partial x} - \frac{\partial u}{\partial y} = \frac{2x}{\pi(x^2 + y^2)}. \tag{11.42}$$

The velocities are not singular, but the vorticity is singular near the origin. Transforming back to polar coordinates, the vorticity is $\omega_z = 2\cos\theta/(\pi r)$, which is singular at $r = 0$. This is not surprising as there is a jump in velocity at the origin. It can be verified that $E^2\psi = \nabla^2\psi = -\omega_z$.

We can get the pressure field by considering $\nabla p = \nabla^2 \mathbf{v} = (\nabla^T \cdot \boldsymbol{\tau})^T$:

$$\begin{pmatrix} \frac{\partial p}{\partial x} \\ \frac{\partial p}{\partial y} \end{pmatrix} = \begin{pmatrix} \frac{\partial^2 u}{\partial x^2} + \frac{\partial^2 u}{\partial y^2} \\ \frac{\partial^2 v}{\partial x^2} + \frac{\partial^2 v}{\partial y^2} \end{pmatrix} = \begin{pmatrix} \frac{4xy}{\pi(x^2+y^2)^2} \\ \frac{2(-x^2+y^2)}{\pi(x^2+y^2)^2} \end{pmatrix}. \tag{11.43}$$

Integrating, and neglecting the constant of integration, we find

$$p(x, y) = \frac{-2y}{\pi(x^2 + y^2)}. \tag{11.44}$$

It can be verified that p is harmonic: $\nabla^2 p = 0$. In polar coordinates this is

$$p(r, \theta) = \frac{-2\sin\theta}{r}. \tag{11.45}$$

Obviously, the pressure is singular at the origin, unless we also have $\theta = 0, \pi$. Even though the pressure could be singular at the origin, its integrated effect is finite. One can show that

$$\lim_{y \to 0^+} \int_{-\infty}^{\infty} p(x, y) \, dx = \lim_{y \to 0^+} \int_{-\infty}^{\infty} \frac{-2y}{\pi(x^2 + y^2)} \, dx = -2. \tag{11.46}$$

This indicates the pressure acts as lifting force at $y \to 0^+$. Isobars for the $\alpha = 0$ case are plotted in Fig. 11.4.

We should recall, however, that both viscous and pressure forces act on the surface. Recalling Eq. (4.34) and the fact that the stress tensor is symmetric, the traction vector associated with a surface whose outer normal vector is **n** is

$$\mathbf{t} = (\boldsymbol{\tau} - p\mathbf{I}) \cdot \mathbf{n}. \tag{11.47}$$

At $y = 0$, the outer normal for the fluid is $\mathbf{n} = -\mathbf{j}$, and the traction is

$$\begin{pmatrix} t_x \\ t_y \end{pmatrix} = \begin{pmatrix} \tau_{xx} - p & \tau_{yx} \\ \tau_{xy} & \tau_{yy} - p \end{pmatrix} \begin{pmatrix} 0 \\ -1 \end{pmatrix} = \begin{pmatrix} -\tau_{yx} \\ -\tau_{yy} + p \end{pmatrix}. \tag{11.48}$$

For this wall at $y = 0$, the tangential (shear) traction is $-\tau_{yx}$. And the normal traction is $-\tau_{yy} + p$. Now we have from direct substitution for u and v

$$\boldsymbol{\tau} = \begin{pmatrix} 2\frac{\partial u}{\partial x} & \frac{\partial v}{\partial x} + \frac{\partial u}{\partial y} \\ \frac{\partial v}{\partial x} + \frac{\partial u}{\partial y} & 2\frac{\partial u}{\partial y} \end{pmatrix} = \begin{pmatrix} \frac{4x^2 y}{\pi(x^2+y^2)^2} & -\frac{2x(x-y)(x+y)}{\pi(x^2+y^2)^2} \\ -\frac{2x(x-y)(x+y)}{\pi(x^2+y^2)^2} & -\frac{4x^2 y}{\pi(x^2+y^2)^2} \end{pmatrix}. \tag{11.49}$$

Then we see, after including the pressure term, that we get

$$\begin{pmatrix} t_x \\ t_y \end{pmatrix} = \begin{pmatrix} \frac{2(x-y)(x+y)x}{\pi(x^2+y^2)^2} \\ \frac{2(x-y)(x+y)y}{\pi(x^2+y^2)^2} \end{pmatrix}. \tag{11.50}$$

At $y = 0$, we get

$$\begin{pmatrix} t_x \\ t_y \end{pmatrix} = \begin{pmatrix} \frac{2}{\pi x} \\ 0 \end{pmatrix}. \tag{11.51}$$

The surface traction at $y = 0$ is confined to the x direction. While it is singular at the origin, the function is odd, so that if we integrate from $x = -L$ to $x = L$, there will be a cancelation and thus a force balance.

The deformation tensor $\mathbf{D} = \boldsymbol{\tau}/2$:

$$\mathbf{D} = \begin{pmatrix} \frac{\partial u}{\partial x} & \frac{\frac{\partial v}{\partial x} + \frac{\partial u}{\partial y}}{2} \\ \frac{\frac{\partial v}{\partial x} + \frac{\partial u}{\partial y}}{2} & \frac{\partial u}{\partial y} \end{pmatrix} = \begin{pmatrix} \frac{2x^2 y}{\pi(x^2+y^2)^2} & -\frac{x(x-y)(x+y)}{\pi(x^2+y^2)^2} \\ -\frac{x(x-y)(x+y)}{\pi(x^2+y^2)^2} & -\frac{2x^2 y}{\pi(x^2+y^2)^2} \end{pmatrix}. \tag{11.52}$$

Again, the strain rate can be characterized by its two principal invariants. They are

$$I_{\dot{\epsilon}}^{(1)} = \operatorname{tr} \mathbf{D} = 0, \qquad I_{\dot{\epsilon}}^{(3)} = \det \mathbf{D} = -\left(\frac{x}{\pi(x^2+y^2)}\right)^2. \tag{11.53}$$

Contours of $I_{\dot{\epsilon}}^{(3)}$ are plotted in Fig. 11.5. The amplitude of $I_{\dot{\epsilon}}^{(3)}$ is singular at the origin and decays at $\mathcal{O}(r^{-2})$ with distance from the origin.

Spherical

We now consider solutions to the Stokes equations in an axisymmetric spherical coordinate system. We consider the limit in which variation in θ is negligible, so that all variables at most are a function of r and ϕ. Considering Eqs. (2.381, 2.382) in this limit, we obtain

$$\nabla^T \cdot \mathbf{v} = \frac{1}{r^2}\frac{\partial}{\partial r}(r^2 v_r) + \frac{1}{r\sin\phi}\frac{\partial}{\partial \phi}(\sin\phi\, v_\phi), \tag{11.54}$$

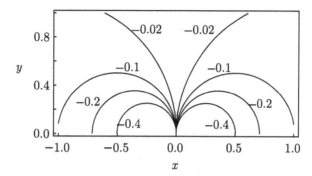

Figure 11.5 Contours of $I_{\dot{\epsilon}}^{(3)} = \det \mathbf{D}$, one of the principal invariants of strain rate, for a simple Stokes flow that is a Taylor scraper with $\alpha = \pi$.

$$\nabla^T \cdot \nabla = \nabla^2 = \frac{1}{r^2}\frac{\partial}{\partial r}\left(r^2 \frac{\partial}{\partial r}\right) + \frac{1}{r^2 \sin\phi}\frac{\partial}{\partial \phi}\left(\sin\phi \frac{\partial}{\partial \phi}\right). \quad (11.55)$$

In contrast to two-dimensional Cartesian and polar coordinates, $\nabla^2 \neq E^2$. We recall Eq. (2.383):

$$E^2 = \frac{\partial^2}{\partial r^2} + \frac{\sin\phi}{r^2}\frac{\partial}{\partial \phi}\left(\frac{1}{\sin\phi}\frac{\partial}{\partial \phi}\right), \quad E^4 = \left(\frac{\partial^2}{\partial r^2} + \frac{\sin\phi}{r^2}\frac{\partial}{\partial \phi}\left(\frac{1}{\sin\phi}\frac{\partial}{\partial \phi}\right)\right)^2. \quad (11.56)$$

The vorticity in axisymmetric spherical coordinates is

$$\omega_\theta = \frac{1}{r}\frac{\partial}{\partial r}(rv_\phi) - \frac{1}{r}\frac{\partial v_r}{\partial \phi}. \quad (11.57)$$

The representation of the mass equation in axisymmetric spherical coordinates is

$$\frac{1}{r^2}\frac{\partial}{\partial r}(r^2 v_r) + \frac{1}{r\sin\phi}\frac{\partial}{\partial \phi}(\sin\phi\, v_\phi) = 0. \quad (11.58)$$

This suggests the definition of a stream function ψ as follows:

$$v_r = \frac{1}{r^2 \sin\phi}\frac{\partial \psi}{\partial \phi}, \quad v_\phi = -\frac{1}{r\sin\phi}\frac{\partial \psi}{\partial r}. \quad (11.59)$$

Direct substitution into the mass equation shows this guarantees satisfaction.

And with these definitions, we recast the vorticity as

$$\omega_\theta = -\frac{1}{r}\frac{\partial}{\partial r}\left(r\frac{1}{r\sin\phi}\frac{\partial \psi}{\partial r}\right) - \frac{1}{r}\frac{\partial}{\partial \phi}\left(\frac{1}{r^2 \sin\phi}\frac{\partial \psi}{\partial \phi}\right), \quad (11.60)$$

$$= -\frac{1}{r\sin\phi}\frac{\partial^2 \psi}{\partial r^2} - \frac{1}{r^3}\frac{\partial}{\partial \phi}\left(\frac{1}{\sin\phi}\frac{\partial \psi}{\partial \phi}\right), \quad (11.61)$$

$$= -\frac{1}{r\sin\phi}\left(\frac{\partial^2 \psi}{\partial r^2} + \frac{\sin\phi}{r^2}\frac{\partial}{\partial \phi}\left(\frac{1}{\sin\phi}\frac{\partial \psi}{\partial \phi}\right)\right) = -\frac{1}{r\sin\phi}E^2\psi. \quad (11.62)$$

Example 11.3 Analyze for the small Reynolds number limit the well-known flow studied by Stokes (1851) in which a fluid flowing in the far field at constant velocity $v_z = -1$ encounters a sphere of unit radius centered at the origin.

Solution

It can be verified that a uniform flow with unit velocity directed in the negative z direction is described by the stream function

$$\hat{\psi} = -\frac{1}{2}r^2 \sin^2 \phi. \tag{11.63}$$

For this flow the velocity components are

$$v_r = \frac{1}{r^2 \sin \phi} \frac{\partial \hat{\psi}}{\partial \phi} = -\cos \phi, \qquad v_\phi = -\frac{1}{r \sin \phi} \frac{\partial \hat{\psi}}{\partial r} = \sin \phi. \tag{11.64}$$

In Cartesian coordinates, one can see that this flow is uniform with $v_z = -1$. Intuitively, for flow over a sphere, this suggests we seek solutions to the biharmonic equation of the form

$$\psi(r,\phi) = -\frac{1}{2}r^2 \sin^2 \phi \, F(r). \tag{11.65}$$

This is a type of separation of variables. We seek $F(r)$ so that (a) it goes to unity as $r \to \infty$, thus allowing ψ to satisfy a far field condition of $v_z = -1$, (b) the surface of the sphere at $r = 1$ is a streamline, thus satisfying a no-penetration boundary condition, and (c) there is no slip at the surface of the sphere.

Proceeding, we get

$$E^4 \psi = \left(\frac{\partial^2}{\partial r^2} + \frac{\sin \phi}{r^2} \frac{\partial}{\partial \phi} \left(\frac{1}{\sin \phi} \frac{\partial}{\partial \phi} \right) \right)^2 \left(-\frac{1}{2} r^2 \sin^2 \phi \, F(r) \right) = 0. \tag{11.66}$$

Evaluating yields

$$-8 \frac{dF}{dr} + 8r \left(\frac{d^2 F}{dr^2} + r \frac{d^3 F}{dr^3} \right) + r^3 \frac{d^4 F}{dr^4} = 0. \tag{11.67}$$

Assuming a polynomial solution of the form $F(r) = Cr^n$ leads to the equation $n(n-2)(n+1)(n+3) = 0$, giving $n = 0, 2, -1, -3$. Because of linearity, we can superpose solutions to get the form

$$F(r) = \frac{C_1}{r^3} + \frac{C_2}{r} + C_3 r^2 + C_4. \tag{11.68}$$

To prevent unbounded behavior for $r \to \infty$, we take $C_3 = 0$. And to satisfy the far field condition that $v_z = -1$, we take $C_4 = 1$. This gives

$$F(r) = \frac{C_1}{r^3} + \frac{C_2}{r} + 1. \tag{11.69}$$

For the surface of the sphere to be a streamline, we can insist that $F(1) = 0$, giving $C_2 = -1 - C_1$, so

$$F(r) = \frac{C_1}{r^3} + \frac{-1 - C_1}{r} + 1. \tag{11.70}$$

We also must enforce the surface of the sphere to be a no-slip surface giving $v_\phi(1, \phi) = 0$. Enforcing this yields $C_1 = 1/2$, giving then

$$F(r) = \frac{1}{2r^3} - \frac{3}{2r} + 1. \tag{11.71}$$

$$\psi(r,\phi) = -\frac{1}{2}r^2 \sin^2 \phi \left(\frac{1}{2r^3} - \frac{3}{2r} + 1 \right) = -\frac{1}{2}r^2 \sin^2 \phi \left(1 - \frac{1}{r}\right)^2 \left(1 + \frac{1}{2r}\right). \tag{11.72}$$

Direct substitution into $E^4 \psi = 0$ and detailed calculation verifies that $\psi(r,\phi)$ satisfies the equation. As an aside, $\nabla^4 \psi \neq 0$ for this flow. Direct substitution also verifies that

$$\psi(r,0) = 0, \qquad \psi(r,\pi) = 0, \qquad \psi(1,\phi) = 0. \tag{11.73}$$

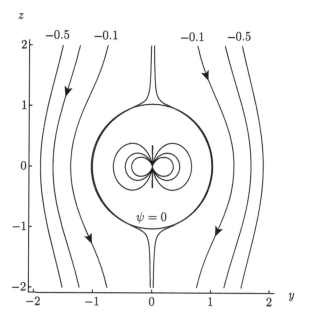

Figure 11.6 Plot of stream function contours $\psi(y,z)$ in the $x=0$ plane for Stokes flow over a stationary sphere with unit radius.

So the surface of the sphere is a streamline. The velocity components are

$$v_r = \frac{1}{r^2 \sin\phi}\frac{\partial \psi}{\partial \phi} = -\left(\frac{1}{2r^3} - \frac{3}{2r} + 1\right)\cos\phi, \tag{11.74}$$

$$v_\phi = -\frac{1}{r \sin\phi}\frac{\partial \psi}{\partial r} = \frac{(r-1)(4r^2 + r + 1)\sin\phi}{4r^3}. \tag{11.75}$$

Then we see that no-penetration and no-slip conditions are satisfied at the surface of the sphere, $r=1$:

$$v_r(1,\phi) = 0, \qquad v_\phi(1,\phi) = 0. \tag{11.76}$$

A plot of contours of ψ is shown in Fig. 11.6.

With knowledge of the velocities, we can calculate

$$\omega_\theta = \frac{1}{r}\frac{\partial}{\partial r}(rv_\phi) - \frac{1}{r}\frac{\partial v_r}{\partial \phi} = \frac{3\sin\phi}{2r^2}. \tag{11.77}$$

Then it can be verified that

$$-\frac{1}{r\sin\phi}E^2\psi = -\frac{1}{r\sin\phi}\left(\frac{\partial^2\psi}{\partial r^2} + \frac{\sin\phi}{r^2}\frac{\partial}{\partial \phi}\left(\frac{1}{\sin\phi}\frac{\partial \psi}{\partial \phi}\right)\right) = \frac{3\sin\phi}{2r^2} = \omega_\theta. \tag{11.78}$$

In the far field, $r \to \infty$, the velocity field reduces to

$$\lim_{r\to\infty} v_r = -\cos\phi, \qquad \lim_{r\to\infty} v_\phi = \sin\phi. \tag{11.79}$$

Referring to Fig. 11.7, this corresponds to a far field flow with $v_z = -1$, a downward directed uniform flow that approaches a stationary sphere. Because the operator E^4 is linear, we can use superposition. If we subtract the far field portion of the flow, we are effectively performing a Galilean transformation to bring the fluid to rest at infinity, while allowing the sphere to move. It can be seen that this is achieved by subtraction of a simple term from our stream function to give

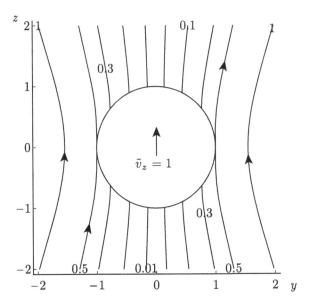

Figure 11.7 Plot of stream function contours $\tilde{\psi}(y,z)$ in the $x=0$ plane for Stokes flow over a sphere moving with positive unit velocity in the z direction.

$$\tilde{\psi}(r,\phi) = -\frac{1}{2}r^2\sin^2\phi\left(\frac{1}{2r^3} - \frac{3}{2r}\right). \tag{11.80}$$

For this system

$$\tilde{v}_r = \frac{1}{r^2\sin\phi}\frac{\partial\tilde{\psi}}{\partial\phi} = -\left(\frac{1}{2r^3} - \frac{3}{2r}\right)\cos\phi, \tag{11.81}$$

$$\tilde{v}_\phi = -\frac{1}{r\sin\phi}\frac{\partial\tilde{\psi}}{\partial r} = -\frac{(3r^2+1)\sin\phi}{4r^3}. \tag{11.82}$$

In the far field, $r \to \infty$, $\tilde{v}_r, \tilde{v}_\phi \to 0$. And at the surface of the sphere $r=1$, we get

$$\tilde{v}_r|_{r=1} = \cos\phi, \qquad \tilde{v}_\phi|_{r=1} = -\sin\phi, \qquad \tilde{\psi}|_{r=1} = \frac{\sin^2\phi}{2}. \tag{11.83}$$

Note that the surface of the sphere is not a streamline in this frame. This corresponds to the surface of the sphere uniformly rising with unit velocity in the z direction. A plot of contours of $\tilde{\psi}$ is shown in Fig. 11.7.

With knowledge of the stream function and thus the velocity field, one can get $\nabla p = \nabla^2 \mathbf{v}$, and the integrate to find the pressure. With knowledge of the velocity field, one can also obtain the viscous stress field. One can then evaluate the combined effect of these forces at the surface of the sphere, and integrate them over the entire surface. Returning then to a dimensional value, one obtains the well-known relation for drag on a sphere in the limit of low Re:

$$F = 6\pi\mu r_o v_o. \tag{11.84}$$

Here r_o is the radius of the sphere and v_o is the freestream velocity. We can get a dimensionless drag coefficient C_D via scaling by the dynamic pressure, $\rho v_o^2/2$ and the area πr_o^2:

$$C_D = \frac{F}{\frac{1}{2}\rho v_o^2\pi r_o^2} = \frac{6\pi\mu r_o v_o}{\frac{1}{2}\rho v_o^2\pi r_o^2} = \frac{12\mu}{\rho v_o r_o} = \frac{24\mu}{\rho v_o D_o} = \frac{24}{Re_D}. \tag{11.85}$$

Here Re_D is the Reynolds number based on the diameter. This formula matches experiments well for $Re_D < 0.5$. For ordinary fluids like air, the particle size must be small to be in the Stokes flow regime. A particle of diameter $1\,\mu\text{m}$ moving at $1\,\text{m/s}$ in $300\,\text{K}$ air with $\nu = 15.89 \times 10^{-6}\,\text{m}^2/\text{s}$ will have $Re_D = 0.0629$ and thus be in the regime in which inertial effects are negligible. Many important applications exist especially in fields such as microbiology in which particle sizes are small, and the particles may be embedded in a viscous fluid.

Lastly, we mention the well-known Oseen correction. In the far field, it is possible to show that advective inertia terms again become as important as viscous terms. Oseen re-introduced a simplified advective inertia term of the form $v_o \partial \mathbf{v}/\partial z$ and obtained an improved approximate solution. The end result is a modified formula for the drag force:

$$F = 6\pi \mu r_o v_o \left(1 + \frac{3}{8} Re_D\right). \tag{11.86}$$

As discussed by Yih (1977), there exist problems with Oseen's approximation near the sphere; an improvement was given by Proudman and Pearson (1957), who obtained the formula

$$F = 6\pi \mu r_o v_o \left(1 + \frac{3}{8} Re_D + \frac{9}{40} Re_D^2 \ln Re_D + \mathcal{O}(Re_D^2)\right). \tag{11.87}$$

11.1.2 Generalized Poiseuille Flow

A straightforward example of an internal flow in which Stokes equations may be applied is given by steady, multi-dimensional Poiseuille flow, an extension of the flow considered earlier in Section 10.1.1. We again utilize an over-bar notation to indicate a dimensional quantity. Consider the schematic of steady, fully developed ($\partial/\partial \overline{z} = 0$), channel flow with $\overline{u} = \overline{v} = 0$, $\overline{w} = \overline{w}(\overline{x}, \overline{y})$, driven by a pressure gradient $d\overline{p}/d\overline{z}$ in the $\overline{z}$ direction as shown in Fig. 11.8. We neglect body forces and consider the flow to be incompressible with constant properties ρ and μ. Because $\overline{u} = \overline{v} = 0$, the $\overline{x}$ and $\overline{y}$ momentum equations reduce to $\partial \overline{p}/\partial \overline{x} = 0$, and $\partial \overline{p}/\partial \overline{y} = 0$, respectively. And because we are considering steady flow, we can have at most $\overline{p} = \overline{p}(\overline{z})$. Because the flow is also fully developed, inertial effects are not present, and the $\overline{z}$ momentum equation reduces to a balance between pressure and shear gradients, giving

$$\rho \underbrace{\left(\frac{\partial \overline{w}}{\partial \overline{t}} + \overline{u}\frac{\partial \overline{w}}{\partial \overline{x}} + \overline{v}\frac{\partial \overline{w}}{\partial \overline{y}} + \overline{w}\frac{\partial \overline{w}}{\partial \overline{z}}\right)}_{=0} = -\frac{d\overline{p}}{d\overline{z}} + \mu \left(\frac{\partial^2 \overline{w}}{\partial \overline{x}^2} + \frac{\partial^2 \overline{w}}{\partial \overline{y}^2} + \underbrace{\frac{\partial^2 \overline{w}}{\partial \overline{z}^2}}_{=0}\right), \tag{11.88}$$

$$0 = -\frac{d\overline{p}}{d\overline{z}} + \mu \left(\frac{\partial^2 \overline{w}}{\partial \overline{x}^2} + \frac{\partial^2 \overline{w}}{\partial \overline{y}^2}\right). \tag{11.89}$$

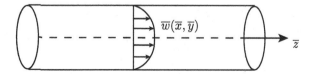

Figure 11.8 Schematic for steady, multi-dimensional Poiseuille flow.

This the equivalent of Stokes equation, Eq. (11.5), applied to this channel flow. If we take $\partial/\partial \bar{z}$ and recall that we have assumed $\bar{w} = \bar{w}(\bar{x}, \bar{y})$, we obtain $0 = -d^2\bar{p}/d\bar{z}^2$. This is consistent with our requirement for Stokes flow that $\nabla^2 \bar{p} = 0$. This then requires that

$$\frac{d\bar{p}}{d\bar{z}} = \text{constant}. \tag{11.90}$$

Now at a stationary wall, we will have $\bar{w} = 0$.

Let us scale the equations. We take L to be a characteristic length associated with the cross-sectional area and w_c to be a characteristic velocity. Then we scale by

$$x = \frac{\bar{x}}{L}, \qquad y = \frac{\bar{y}}{L}, \qquad w = \frac{\bar{w}}{w_c}. \tag{11.91}$$

Then Eq. (11.89) becomes

$$0 = -\frac{d\bar{p}}{d\bar{z}} \frac{L^2}{\mu w_c} + \frac{\partial^2 w}{\partial x^2} + \frac{\partial^2 w}{\partial y^2}. \tag{11.92}$$

Let us then choose $w_c = -L^2 d\bar{p}/d\bar{z}/\mu$. This choice guarantees that a negative pressure gradient will induce a positive characteristic velocity. This gives the Poisson equation

$$\nabla^2 w = \frac{\partial^2 w}{\partial x^2} + \frac{\partial^2 w}{\partial y^2} = -1. \tag{11.93}$$

Elliptic Cross Section

We can consider this for a variety of cross-sectional geometries. Let us first try an ellipse with the cross-sectional geometry defined by

$$\left(\frac{\bar{x}}{a}\right)^2 + \left(\frac{\bar{y}}{b}\right)^2 = 1. \tag{11.94}$$

This is depicted in Fig. 11.9. Eliminating $\bar{x}$ and $\bar{y}$ for x and y, we get

$$\left(\frac{xL}{a}\right)^2 + \left(\frac{yL}{b}\right)^2 = 1. \tag{11.95}$$

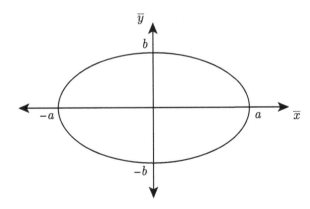

Figure 11.9 Elliptic cross section for a generalized Poiseuille flow.

Let us select $L = a$, giving us

$$x^2 + \left(\frac{a}{b}\right)^2 y^2 = 1. \tag{11.96}$$

Let us now change variables so as to transform the Poisson equation to Laplace's equation. We need the second derivative of a function to go to a constant. Some combination of quadratic polynomials may work. Let us guess that the appropriate transformed dependent variable is

$$W = w + C_1 x^2 + C_2 y^2, \tag{11.97}$$

and see if C_1 and C_2 can be chosen to remove the inhomogeneity in the Poisson equation. Take the Laplacian of both sides to get

$$\nabla^2 W = \nabla^2 w + 2C_1 + 2C_2. \tag{11.98}$$

Let us demand that $2C_1 + 2C_2 = 1$ so that $\nabla^2 W = \nabla^2 w + 1$. Then our governing equation is

$$\nabla^2 W = 0. \tag{11.99}$$

Now on the wall $w = 0$, so on the wall, denoted by a subscript w, we must have

$$W|_w = C_1 x^2 |_w + C_2 y^2 |_w = C_1 \left(x^2 |_w + \frac{C_2}{C_1} y^2 |_w \right). \tag{11.100}$$

Now let us select $C_2/C_1 = (a/b)^2$, so that

$$W|_w = C_1 \underbrace{\left(x^2 |_w + \left(\frac{a}{b}\right)^2 y^2 |_w \right)}_{=1} = C_1. \tag{11.101}$$

So on the elliptic wall boundary, W is a constant. We can now solve for C_1 and C_2 to get

$$C_1 = \frac{1}{2\left(1 + \left(\frac{a}{b}\right)^2\right)}, \quad C_2 = \frac{\left(\frac{a}{b}\right)^2}{2\left(1 + \left(\frac{a}{b}\right)^2\right)}. \tag{11.102}$$

Now our problem is reduced to

$$\nabla^2 W = 0, \quad W = C_1 \quad \text{on the boundary.} \tag{11.103}$$

By inspection, the solution over the entire domain is

$$W(x, y) = C_1 = \frac{1}{2\left(1 + \left(\frac{a}{b}\right)^2\right)}. \tag{11.104}$$

Because the equation is linear, we are guaranteed a unique solution, and this one clearly satisfies Laplace's equation and the boundary condition. Transforming back to the original dependent variable, we get

$$w(x,y) + C_1 x^2 + C_2 y^2 = C_1, \qquad (11.105)$$

$$w(x,y) = C_1 \left(1 - x^2 - \frac{C_2}{C_1} y^2\right), \qquad (11.106)$$

$$= \frac{1}{2\left(1 + \left(\frac{a}{b}\right)^2\right)} \left(1 - x^2 - \left(\frac{a}{b}\right)^2 y^2\right). \qquad (11.107)$$

In dimensional terms, we have

$$\frac{\overline{w}}{\frac{L^2\left(-\frac{d\overline{p}}{d\overline{z}}\right)}{\mu}} = \frac{1}{2\left(1 + \left(\frac{a}{b}\right)^2\right)} \left(1 - \left(\frac{\overline{x}}{a}\right)^2 - \left(\frac{\overline{y}}{b}\right)^2\right). \qquad (11.108)$$

Note that on any ellipse $(\overline{x}/a)^2 + (\overline{y}/b)^2 = \alpha^2$, $\overline{w}$ is a constant. The velocity takes on its maximum value for $\overline{x} = \overline{y} = 0$. Choosing $a/b = 2$, we plot contours of $w = (1/10)(1 - x^2 - 4y^2)$ in Fig. 11.10. The maximum is $w(0,0) = 1/10$.

Rectangular Cross Section

Let us next try a rectangular cross section for a rectangle centered at the origin with length $2L$ and height $2\beta L$. Here β is a dimensionless aspect ratio. This is depicted in Fig. 11.11. We adopt the same scaling $x = \overline{x}/L$, $y = \overline{y}/L$, $w = \overline{w}/(L^2(-d\overline{p}/d\overline{z})/\mu)$ and arrive at the following

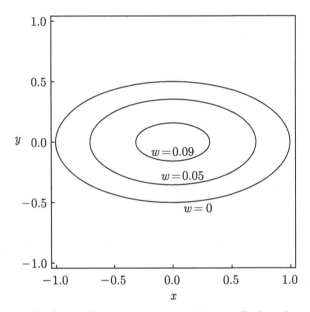

Figure 11.10 Contours of $w(x,y)$ for $a/b = 2$ in a generalized Poiseuille flow through a pipe with elliptic cross section.

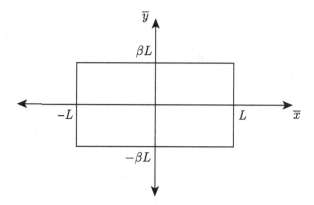

Figure 11.11 Rectangular cross section for a generalized Poiseuille flow.

z momentum equation and no-slip boundary conditions:

$$\frac{\partial^2 w}{\partial x^2} + \frac{\partial^2 w}{\partial y^2} = -1, \quad w(1,y) = w(-1,y) = w(x,\beta) = w(x,-\beta) = 0. \quad (11.109)$$

Because of the symmetry of the domain and boundary conditions, we can expect to have some zero-gradient conditions at $x = 0$ and $y = 0$. We will employ these soon.

Let us transform w to remove the inhomogeneity in the Poisson equation. Let us motivate this by solving the equivalent problem, assuming only y variation, giving

$$\frac{d^2 w}{dy^2} = -1, \quad w(y=\beta) = w(y=-\beta) = 0. \quad (11.110)$$

This yields $w = -y^2/2 + Ay + B$. Applying both boundary conditions then gives $w = (\beta^2 - y^2)/2$. We use this to motivate a new transformed dependent variable:

$$W = w - \frac{1}{2}\left(\beta^2 - y^2\right). \quad (11.111)$$

Obviously, $\nabla^2 W = \nabla^2 w + 1$, so in the transformed variable, we have

$$\frac{\partial^2 W}{\partial x^2} + \frac{\partial^2 W}{\partial y^2} = 0. \quad (11.112)$$

Let us now transform the boundary conditions; while doing so, we shall replace two of them by symmetry:

$$W(1,y) = \underbrace{w(1,y)}_{=0} - \frac{1}{2}(\beta^2 - y^2) = -\frac{1}{2}(\beta^2 - y^2), \quad (11.113)$$

$$\frac{\partial w}{\partial x}(0,y) = 0, \quad \Rightarrow \quad \frac{\partial W}{\partial x}(0,y) = 0, \quad (11.114)$$

$$W(x,\beta) = \underbrace{w(x,\beta)}_{=0} - \frac{1}{2}(\beta^2 - \beta^2) = 0, \quad (11.115)$$

$$\frac{\partial w}{\partial y}(x,0) = 0, \quad \Rightarrow \quad \frac{\partial W}{\partial y}(x,0) - y\big|_{y=0} = 0. \quad (11.116)$$

We had an inhomogeneous partial differential equation with homogeneous boundary conditions. With our transformation, we have a homogeneous partial differential equation with one inhomogeneous boundary condition.

Let us now separate variables. Assume $W = A(x)B(y)$. This gives us

$$A(x)\frac{d^2 B}{dy^2} + B(y)\frac{d^2 A}{dx^2} = 0, \tag{11.117}$$

$$\frac{1}{B(y)}\frac{d^2 B}{dy^2} = -\frac{1}{A(x)}\frac{d^2 A}{dx^2} = -\alpha^2. \tag{11.118}$$

This gives two ordinary differential equations. For B, we have

$$\frac{d^2 B}{dy^2} + \alpha^2 B = 0. \tag{11.119}$$

This has solution

$$B(y) = C_1 \sin \alpha y + C_2 \cos \alpha y. \tag{11.120}$$

In order for $\partial W/\partial y = 0$ at $y = 0$, we must demand $dB/dy = 0$ at $y = 0$. We get

$$\left.\frac{dB}{dy}\right|_{y=0} = \alpha C_1 \cos \alpha y|_{y=0} - \alpha C_2 \sin \alpha y|_{y=0}. \tag{11.121}$$

This yields $C_1 = 0$, and $B = C_2 \cos \alpha y$. For the no-slip condition at $y = \beta$, we must have $0 = C_2 \cos \alpha \beta$. This requires that

$$\alpha = \frac{(2n-1)\pi}{2\beta}, \quad n = 1, 2, 3, \ldots. \tag{11.122}$$

Our differential equation and solution for $A(x)$ is

$$\frac{d^2 A}{dx^2} - \alpha^2 A = 0, \quad A(x) = C_3 \sinh \alpha x + C_4 \cosh \alpha x. \tag{11.123}$$

For our symmetry boundary condition, we insist that $dA/dx = 0$ at $x = 0$. We find then that

$$\frac{dA}{dx} = \alpha C_3 \cosh \alpha x + \alpha C_4 \sinh \alpha x. \tag{11.124}$$

Enforcing the condition yields $C_3 = 0$, so

$$A(x) = C_4 \cosh \alpha x. \tag{11.125}$$

Now recompose and use superposition to get

$$W(x,y) = \sum_{n=1}^{\infty} C_n \cos\left(\frac{(2n-1)\pi}{2\beta}y\right) \cosh\left(\frac{(2n-1)\pi}{2\beta}x\right). \tag{11.126}$$

This solution can be verified to satisfy Laplace's equation and three of the four boundary conditions. We have one more boundary condition, Eq. (11.113), to satisfy, and we can use standard methods of Fourier series to do so. We thus need to select C_n such that

$$-\frac{1}{2}(\beta^2 - y^2) = \sum_{n=1}^{\infty} C_n \cos\left(\frac{(2n-1)\pi}{2\beta}y\right) \cosh\left(\frac{(2n-1)\pi}{2\beta}\right). \tag{11.127}$$

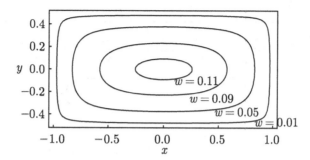

Figure 11.12 Contours of $w(x,y)$ for generalized Poiseuille flow through a pipe with rectangular cross section with $\beta = 1/2$.

Let us multiply both sides by $\cos((2m-1)\pi y/(2\beta))$, integrate from 0 to β, and take advantage of orthogonality:

$$-\frac{1}{2}\int_0^\beta \cos\left(\frac{(2m-1)\pi}{2\beta}y\right)(\beta^2 - y^2)\,dy$$
$$= \sum_{n=1}^\infty C_n \cosh\left(\frac{(2n-1)\pi}{2\beta}\right)\underbrace{\int_0^\beta \cos\left(\frac{(2m-1)\pi}{2\beta}y\right)\cos\left(\frac{(2n-1)\pi}{2\beta}y\right)dy}_{=\delta_{mn}\beta/2}. \quad (11.128)$$

Simplifying both sides and trading m for n, we get

$$(-1)^n \left(\frac{2\beta}{\pi(2n-1)}\right)^3 = C_n \frac{\beta}{2}\cosh\left(\frac{(2n-1)\pi}{2\beta}\right), \quad (11.129)$$

$$C_n = (-1)^n \left(\frac{2\beta}{\pi(2n-1)}\right)^3 \frac{2}{\beta \cosh\left(\frac{(2n-1)\pi}{2\beta}\right)}. \quad (11.130)$$

Returning to our original variable w, we get

$$w(x,y) = \frac{1}{2}(\beta^2 - y^2) + \sum_{n=1}^\infty (-1)^n \left(\frac{2\beta}{\pi(2n-1)}\right)^3 \frac{2}{\beta \cosh\left(\frac{(2n-1)\pi}{2\beta}\right)}$$
$$\times \cos\left(\frac{(2n-1)\pi}{2\beta}y\right)\cosh\left(\frac{(2n-1)\pi}{2\beta}x\right). \quad (11.131)$$

The maximum velocity is at the origin and is a function of the aspect ratio β:

$$w(0,0) = \frac{1}{2}\beta^2 + \sum_{n=1}^\infty (-1)^n \left(\frac{2\beta}{\pi(2n-1)}\right)^3 \frac{2}{\beta \cosh\left(\frac{(2n-1)\pi}{2\beta}\right)}. \quad (11.132)$$

For $\beta = 1/2$, a contour plot of $w(x,y)$ is given in Fig. 11.12.

11.1.3 Lubrication Theory

Here we will briefly introduce equations used in *lubrication theory*, often used to approximate the flow of a viscous fluid in a thin gap such as might exist in a bearing. Some assumptions of

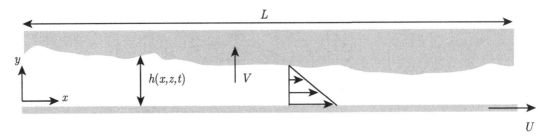

Figure 11.13 Schematic for lubrication theory.

this theory are not rigorous, but are plausible. We can consider the scenario depicted in Fig. 11.13. For this section, all variables are dimensional. Here we have two plates: a smooth plate at $y = 0$ moving horizontally with constant velocity U, and a rough plate moving vertically with variable velocity V. The gap width is $h(x, z, t)$. The key assumption of the theory is that the gap width is small relative to the length scale L over which we are concerned: $h(x, z, t)/L \ll 1$. We also assume incompressibility, constant pressure gradients in the x and z directions, constant material properties, negligible inertial effects, negligible body force, and typical boundary layer approximations (see Section 11.2.1). These lead us to the following mass and momenta equations:

$$\frac{\partial u}{\partial x} + \frac{\partial v}{\partial y} + \frac{\partial w}{\partial z} = 0, \tag{11.133}$$

$$0 = -\frac{\partial p}{\partial x} + \mu \frac{\partial^2 u}{\partial y^2}, \quad 0 = -\frac{\partial p}{\partial y}, \quad 0 = -\frac{\partial p}{\partial z} + \mu \frac{\partial^2 w}{\partial y^2}. \tag{11.134}$$

Let us take as boundary conditions

$$u(y = 0) = U, \quad u(y = h) = 0, \tag{11.135}$$

$$v(y = 0) = 0, \quad v(y = h) = V = \frac{\partial h}{\partial t}, \tag{11.136}$$

$$w(y = 0) = 0, \quad w(y = h) = 0. \tag{11.137}$$

The conditions at $y = h$ are approximations that do not account for the roughness of the upper wall. We can integrate the x and z momenta equations and apply the boundary conditions to get

$$u = \frac{h^2}{2\mu} \frac{\partial p}{\partial x} \left(\frac{y}{h}\right)\left(\left(\frac{y}{h}\right) - 1\right) + \left(1 - \frac{y}{h}\right) U, \quad w = \frac{h^2}{2\mu} \frac{\partial p}{\partial z} \left(\frac{y}{h}\right)\left(\left(\frac{y}{h}\right) - 1\right). \tag{11.138}$$

Now let us integrate the mass equation, Eq. (11.133), over the gap width:

$$\int_0^{h(x,z,t)} \frac{\partial u}{\partial x} \, dy + \int_0^{h(x,z,t)} \frac{\partial v}{\partial y} \, dy + \int_0^{h(x,z,t)} \frac{\partial w}{\partial z} \, dy = \int_0^{h(x,z,t)} 0 \, dy, \tag{11.139}$$

$$\int_0^{h(x,z,t)} \frac{\partial u}{\partial x} \, dy + \underbrace{v(y = h)}_{V} - \underbrace{v(y = 0)}_{0} + \int_0^{h(x,z,t)} \frac{\partial w}{\partial z} \, dy = 0, \tag{11.140}$$

$$\int_0^{h(x,z,t)} \frac{\partial u}{\partial x} \, dy + \int_0^{h(x,z,t)} \frac{\partial w}{\partial z} \, dy = -V. \tag{11.141}$$

From Leibniz's rule, Eq. (2.238), we can say

$$\frac{\partial}{\partial x}\int_0^{h(x,z,t)} u\,dy = \int_0^{h(x,z,t)} \frac{\partial u}{\partial x}\,dy + \underbrace{u(y=h)}_{0}\frac{\partial h}{\partial x} - u(y=0)\frac{\partial}{\partial x}0, \qquad (11.142)$$

$$\frac{\partial}{\partial x}\int_0^{h(x,z,t)} u\,dy = \int_0^{h(x,z,t)} \frac{\partial u}{\partial x}\,dy. \qquad (11.143)$$

Applying a version this also to the z term, and taking $V = \partial h/\partial t$, we can rewrite Eq. (11.141) as

$$\frac{\partial}{\partial x}\int_0^{h(x,z,t)} u\,dy + \frac{\partial}{\partial z}\int_0^{h(x,z,t)} w\,dy = -\frac{\partial h}{\partial t}. \qquad (11.144)$$

This result is simply a statement of mass conservation. We now substitute our results from the x and z linear momenta equations to get

$$\frac{\partial}{\partial x}\int_0^{h(x,z,t)} \left(\frac{h^2}{2\mu}\frac{\partial p}{\partial x}\left(\frac{y}{h}\right)\left(\left(\frac{y}{h}\right)-1\right) + \left(1-\frac{y}{h}\right)U\right)dy$$

$$+\frac{\partial}{\partial z}\int_0^{h(x,z,t)} \left(\frac{h^2}{2\mu}\frac{\partial p}{\partial z}\left(\frac{y}{h}\right)\left(\left(\frac{y}{h}\right)-1\right)\right)dy = -\frac{\partial h}{\partial t}. \qquad (11.145)$$

Integrating, we get

$$\frac{\partial}{\partial x}\left(\frac{hU}{2} - \frac{1}{12\mu}\frac{\partial p}{\partial x}h^3\right) + \frac{\partial}{\partial z}\left(-\frac{1}{12\mu}\frac{\partial p}{\partial z}h^3\right) = -\frac{\partial h}{\partial t}, \qquad (11.146)$$

$$\frac{1}{\mu}\left(\frac{\partial}{\partial x}\left(\frac{\partial p}{\partial x}h^3\right) + \frac{\partial}{\partial z}\left(\frac{\partial p}{\partial z}h^3\right)\right) = 12\frac{\partial h}{\partial t} + 6U\frac{\partial h}{\partial x}. \qquad (11.147)$$

This is the well-known *Reynolds equation* for flow in a channel whose lower surface moves at speed U and has gap height $h(x,z,t)$. In the limit in which h is constant, it reduces to $\nabla^2 p = 0$, and the model is relevant for what is known as *Hele-Shaw flow*. Recall again from Section 6.4.4 that the Laplacian of pressure often is relevant for incompressible flow analysis.

Let us consider a special case. Take $h = h_o(1 - x/(2L))$ and require no variation in z. And let us take $p(0) = p(L) = p_o$. While there is no overall pressure difference between $x = 0$ and $x = L$, there will be local pressure gradients in the region $x \in [0, L]$. This is because there is no inertia, and the nonzero viscous stress induced by the moving plate must be balanced by a pressure gradient. The Reynolds equation reduces to

$$\frac{d}{dx}\left(h^3\frac{dp}{dx}\right) = -\frac{3Uh_o\mu}{L}, \qquad p(0) = p(L) = p_o. \qquad (11.148)$$

Integration yields

$$p(x) = p_o + \frac{8\mu U}{h_o^2}\frac{1-x/L}{(2-x/L)^2}x. \qquad (11.149)$$

Clearly the boundary conditions at $x = 0$ and $x = L$ are satisfied. Straightforward methods from calculus reveal that the pressure takes on a maximum value of

$$p_{max} = p_o + \frac{LU\mu}{h_o^2}, \qquad (11.150)$$

at the location $x = 2L/3$. Thin gaps induce high pressure.

11.2 Flow with Effects of Inertia

In this section, we reintroduce effects of inertia. We will consider problems that can be addressed by what is known as a similarity transformation. The problems themselves will be fundamental ones that have variation in either time and one spatial coordinate, or with two spatial coordinates. This is in contrast with solutions of the previous section that varied only with one spatial coordinate.

Because two coordinates are involved, we must resort to solving partial differential equations. The similarity transformation actually reveals a hidden symmetry of the partial differential equations by defining a new independent variable that is a grouping of the original independent variables, under which the partial differential equations transform into ordinary differential equations. We then solve the resulting ordinary differential equations by standard techniques.

11.2.1 Blasius Boundary Layer

We consider the well-known problem of the flow of a viscous fluid over a flat plate. This problem forms the foundation for a variety of viscous flows over more complicated geometries. It also illustrates some important features of viscous flow physics, as well as giving the original motivating problem for the mathematical technique of matched asymptotic expansions. Here we will consider, as given in Fig. 11.14, the incompressible flow of viscous fluid of constant viscosity and thermal conductivity over a flat plate. In the far field, the fluid will be a uniform stream with constant velocity. At the plate surface, the no-slip condition must be enforced, which will give rise to a zone of adjustment where the fluid's velocity changes from zero at the plate surface to its freestream value. This zone is called the boundary layer. The diagram has a small inaccuracy as it only gives the u velocity component variation. Actually there is also a v component of the velocity vector, and that induces streamline curvature, not apparent in Fig. 11.14. We will focus on the domain $y > 0$, as the domain $y < 0$ yields symmetric results.

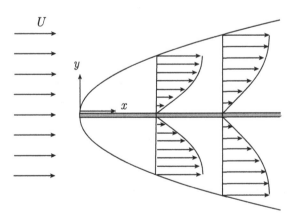

Figure 11.14 Schematic for flat plate boundary layer problem.

Considering first the velocity field, we find that the dimensionless steady two-dimensional Navier–Stokes equations are as follows:

$$\frac{\partial u}{\partial x} + \frac{\partial v}{\partial y} = 0, \qquad (11.151)$$

$$u\frac{\partial u}{\partial x} + v\frac{\partial u}{\partial y} = -\frac{\partial p}{\partial x} + \frac{1}{Re}\left(\frac{\partial^2 u}{\partial x^2} + \frac{\partial^2 u}{\partial y^2}\right), \qquad (11.152)$$

$$u\frac{\partial v}{\partial x} + v\frac{\partial v}{\partial y} = -\frac{\partial p}{\partial y} + \frac{1}{Re}\left(\frac{\partial^2 v}{\partial x^2} + \frac{\partial^2 v}{\partial y^2}\right). \qquad (11.153)$$

The dimensionless boundary conditions are

$$u(x, y \to \infty) = 1, \quad p(x, y \to \infty) = 0, \quad u(x, 0) = 0, \quad v(x, 0) = 0. \qquad (11.154)$$

We have assumed a scaling of the following form, where an over-bar denotes a dimensional variable:

$$u = \frac{\overline{u}}{U}, \quad v = \frac{\overline{v}}{U}, \quad x = \frac{\overline{x}}{L}, \quad y = \frac{\overline{y}}{L}, \quad p = \frac{\overline{p} - p_o}{\rho U^2}. \qquad (11.155)$$

For our flat plate of semi-infinite extent, we do not have a natural length scale. This suggests that we may find a similarity solution that removes the effect of L. It also may be appropriate to think of making the requirement that

$$L \gg \frac{\nu}{U}. \qquad (11.156)$$

This will ensure that $Re = UL/\nu \gg 1$. This is not entirely satisfying as we really want our domain to be semi-infinite with $L \to \infty$.

Now let us consider that for $Re \to \infty$, we have an outer solution of $u = 1$ to be valid for most of the flow field sufficiently far away from the plate surface. In fact, the so-called *outer solution*, $u = 1$, $v = 0$, $p = 0$, satisfies all of the governing equations and boundary conditions except for the no-slip condition at $y = 0$. Because in the limit as $Re \to \infty$, we effectively ignore the high-order derivatives found in the viscous terms, we cannot expect to satisfy all boundary conditions for the full problem. The outer solution is also an inviscid solution, allowing for a slip condition at the boundary.

Let us *rescale* our equations near the plate surface $y = 0$ to (a) bring back the effect of the viscous terms, (b) bring back the no-slip condition, and (c) match our inviscid outer solution to a viscous inner solution. This is *the* first example of the use of the method of matched asymptotic expansions as introduced by Prandtl and his student Blasius in the early twentieth century.

With some difficulty, we could show how to choose the scaling. Instead, let us simply adopt a scaling and show that it indeed achieves our desired end. So let us take a scaled y distance and velocity, denoted by a tilde, to be

$$\tilde{v} = \sqrt{Re}\, v, \qquad \tilde{y} = \sqrt{Re}\, y. \qquad (11.157)$$

With this scaling, assuming the Reynolds number is large, when we examine small y or v, both of $\mathcal{O}(1/\sqrt{Re})$, we are examining an order unity $\tilde{y}$ or $\tilde{v}$. Our equations rescale as

$$\frac{\partial u}{\partial x} + \frac{1/\sqrt{Re}}{1/\sqrt{Re}}\frac{\partial \tilde{v}}{\partial \tilde{y}} = 0, \tag{11.158}$$

$$u\frac{\partial u}{\partial x} + \frac{1/\sqrt{Re}}{1/\sqrt{Re}}\tilde{v}\frac{\partial u}{\partial \tilde{y}} = -\frac{\partial p}{\partial x} + \frac{1}{Re}\left(\frac{\partial^2 u}{\partial x^2} + Re\frac{\partial^2 u}{\partial \tilde{y}^2}\right), \tag{11.159}$$

$$\frac{1}{\sqrt{Re}}u\frac{\partial \tilde{v}}{\partial x} + \frac{(1/\sqrt{Re})(1/\sqrt{Re})}{1/\sqrt{Re}}\tilde{v}\frac{\partial \tilde{v}}{\partial \tilde{y}} = -\frac{1}{1/\sqrt{Re}}\frac{\partial p}{\partial \tilde{y}}$$
$$+ \frac{1}{Re}\left(\frac{1}{\sqrt{Re}}\frac{\partial^2 \tilde{v}}{\partial x^2} + \frac{1/\sqrt{Re}}{1/Re}\frac{\partial^2 \tilde{v}}{\partial \tilde{y}^2}\right). \tag{11.160}$$

Simplifying, this reduces to

$$\frac{\partial u}{\partial x} + \frac{\partial \tilde{v}}{\partial \tilde{y}} = 0, \tag{11.161}$$

$$u\frac{\partial u}{\partial x} + \tilde{v}\frac{\partial u}{\partial \tilde{y}} = -\frac{\partial p}{\partial x} + \frac{1}{Re}\frac{\partial^2 u}{\partial x^2} + \frac{\partial^2 u}{\partial \tilde{y}^2}, \tag{11.162}$$

$$u\frac{\partial \tilde{v}}{\partial x} + \tilde{v}\frac{\partial \tilde{v}}{\partial \tilde{y}} = -Re\frac{\partial p}{\partial \tilde{y}} + \frac{1}{Re}\frac{\partial^2 \tilde{v}}{\partial x^2} + \frac{\partial^2 \tilde{v}}{\partial \tilde{y}^2}. \tag{11.163}$$

Now in the limit as $Re \to \infty$, the rescaled equations reduce to the well-known *boundary layer equations*:

$$\frac{\partial u}{\partial x} + \frac{\partial \tilde{v}}{\partial \tilde{y}} = 0, \tag{11.164}$$

$$u\frac{\partial u}{\partial x} + \tilde{v}\frac{\partial u}{\partial \tilde{y}} = -\frac{\partial p}{\partial x} + \frac{\partial^2 u}{\partial \tilde{y}^2}, \tag{11.165}$$

$$0 = \frac{\partial p}{\partial \tilde{y}}. \tag{11.166}$$

To match the outer solution, we take boundary conditions

$$u(x, \tilde{y} \to \infty) = 1, \quad p(x, \tilde{y} \to \infty) = 0, \quad u(x, 0) = 0, \quad \tilde{v}(x, 0) = 0. \tag{11.167}$$

The $\tilde{y}$ momentum equation gives us

$$p = p(x). \tag{11.168}$$

In general, we can consider this to be an imposed pressure gradient that is supplied by the outer inviscid solution. For general flows, $dp/dx \neq 0$. For the Blasius problem, we will choose to study problems for which there is *no pressure gradient*. That is, we take

$$p(x) = 0, \quad \text{for a Blasius flat plate boundary layer}. \tag{11.169}$$

So called Falkner–Skan solutions, to be studied in Section 11.2.2, consider flows over nonhorizontal plates, for which the outer inviscid solution does not have a constant pressure. This ultimately affects the behavior of the fluid in the boundary layer, giving results that differ in important features from the Blasius problem.

With our assumptions, the Blasius problem reduces to

$$\frac{\partial u}{\partial x} + \frac{\partial \tilde{v}}{\partial \tilde{y}} = 0, \tag{11.170}$$

$$u\frac{\partial u}{\partial x} + \tilde{v}\frac{\partial u}{\partial \tilde{y}} = \frac{\partial^2 u}{\partial \tilde{y}^2}. \tag{11.171}$$

The boundary conditions are now only on velocity and are

$$u(x, \tilde{y} \to \infty) = 1, \quad u(x, 0) = 0, \quad \tilde{v}(x, 0) = 0. \tag{11.172}$$

To simplify, we invoke the stream function ψ, which allows us to satisfy continuity automatically and eliminate u and $\tilde{v}$ at the expense of raising the order of the differential equation. So taking

$$u = \frac{\partial \psi}{\partial \tilde{y}}, \quad \tilde{v} = -\frac{\partial \psi}{\partial x}, \tag{11.173}$$

we find that mass conservation reduces to $\partial^2 \psi / \partial x \partial \tilde{y} - \partial^2 \psi / \partial \tilde{y} \partial x = 0$. The x momentum equation and associated boundary conditions become

$$\frac{\partial \psi}{\partial \tilde{y}}\frac{\partial^2 \psi}{\partial x \partial \tilde{y}} - \frac{\partial \psi}{\partial x}\frac{\partial^2 \psi}{\partial \tilde{y}^2} = \frac{\partial^3 \psi}{\partial \tilde{y}^3}, \tag{11.174}$$

$$\frac{\partial \psi}{\partial \tilde{y}}(x, \tilde{y} \to \infty) = 1, \quad \frac{\partial \psi}{\partial \tilde{y}}(x, 0) = 0, \quad \frac{\partial \psi}{\partial x}(x, 0) = 0. \tag{11.175}$$

Let us try stretching all the variables of this system to see if there are stretching transformations under which the system exhibits symmetry; that is, we seek a stretching transformation under which the system is invariant. Take

$$\hat{x} = e^a x, \quad \hat{y} = e^b \tilde{y}, \quad \hat{\psi} = e^c \psi. \tag{11.176}$$

The x momentum equation and boundary conditions transform to

$$e^{a+2b-2c}\frac{\partial \hat{\psi}}{\partial \hat{y}}\frac{\partial^2 \hat{\psi}}{\partial \hat{x} \partial \hat{y}} - e^{a+2b-2c}\frac{\partial \hat{\psi}}{\partial \hat{x}}\frac{\partial^2 \hat{\psi}}{\partial \hat{y}^2} = e^{3b-c}\frac{\partial^3 \hat{\psi}}{\partial \hat{y}^3}, \tag{11.177}$$

$$e^{b-c}\frac{\partial \hat{\psi}}{\partial \hat{y}}(\hat{x}, \hat{y} \to \infty) = 1, \quad e^{b-c}\frac{\partial \hat{\psi}}{\partial \hat{y}}(\hat{x}, 0) = 0, \quad e^{a-c}\frac{\partial \hat{\psi}}{\partial \hat{x}}(\hat{x}, 0) = 0. \tag{11.178}$$

If we demand $b = c$ and $a = 2c$, then the transformation is invariant, yielding

$$\frac{\partial \hat{\psi}}{\partial \hat{y}}\frac{\partial^2 \hat{\psi}}{\partial \hat{x} \partial \hat{y}} - \frac{\partial \hat{\psi}}{\partial \hat{x}}\frac{\partial^2 \hat{\psi}}{\partial \hat{y}^2} = \frac{\partial^3 \hat{\psi}}{\partial \hat{y}^3}, \tag{11.179}$$

$$\frac{\partial \hat{\psi}}{\partial \hat{y}}(\hat{x}, \hat{y} \to \infty) = 1, \quad \frac{\partial \hat{\psi}}{\partial \hat{y}}(\hat{x}, 0) = 0, \quad \frac{\partial \hat{\psi}}{\partial \hat{x}}(\hat{x}, 0) = 0. \tag{11.180}$$

Now our transformation is reduced to

$$\hat{x} = e^{2c} x, \quad \hat{y} = e^c \tilde{y}, \quad \hat{\psi} = e^c \psi. \tag{11.181}$$

Because c does not appear explicitly in either the original equation set nor the transformed equation set, the solution must not depend on this stretching. Eliminating c from the transformation by $e^c = \sqrt{\hat{x}/x}$ we find that

$$\frac{\hat{y}}{\tilde{y}} = \sqrt{\frac{\hat{x}}{x}}, \qquad \frac{\hat{\psi}}{\psi} = \sqrt{\frac{\hat{x}}{x}}, \qquad (11.182)$$

or

$$\frac{\hat{y}}{\sqrt{\hat{x}}} = \frac{\tilde{y}}{\sqrt{x}}, \qquad \frac{\hat{\psi}}{\sqrt{\hat{x}}} = \frac{\psi}{\sqrt{x}}. \qquad (11.183)$$

Thus motivated, let us seek solutions of the form

$$\frac{\psi}{\sqrt{x}} = f\left(\frac{\tilde{y}}{\sqrt{x}}\right). \qquad (11.184)$$

That is, taking $\eta = \tilde{y}/\sqrt{x}$, we seek

$$\psi = \sqrt{x} f(\eta). \qquad (11.185)$$

Let us check that our similarity variable is independent of L, the unknown length scale:

$$\eta = \frac{\tilde{y}}{\sqrt{x}} = \frac{\sqrt{Re}\, y}{\sqrt{x}} = \frac{\sqrt{Re}\, \overline{y}/L}{\sqrt{\overline{x}/L}} = \sqrt{\frac{UL}{\nu}} \frac{\overline{y}}{L} \frac{\sqrt{L}}{\sqrt{\overline{x}}} = \sqrt{\frac{U}{\nu}} \frac{\overline{y}}{\sqrt{\overline{x}}}. \qquad (11.186)$$

So indeed, our similarity variable is independent of any arbitrary length scale we happen to have chosen.

With our similarity transformation, we have

$$\frac{\partial \eta}{\partial x} = -\frac{1}{2}\tilde{y} x^{-3/2} = -\frac{1}{2}\frac{\eta}{x}, \qquad \frac{\partial \eta}{\partial \tilde{y}} = \frac{1}{\sqrt{x}}. \qquad (11.187)$$

Now we need expressions for $\partial \psi/\partial x$, $\partial \psi/\partial \tilde{y}$, $\partial^2 \psi/\partial x \partial \tilde{y}$, $\partial^2 \psi/\partial \tilde{y}^2$, and $\partial^3 \psi/\partial \tilde{y}^3$. First, consider the partial derivatives of the stream function ψ. Operating on each partial derivative, we find

$$\frac{\partial \psi}{\partial x} = \frac{\partial}{\partial x}\left(\sqrt{x} f(\eta)\right) = \sqrt{x} \frac{df}{d\eta} \frac{\partial \eta}{\partial x} + \frac{1}{2}\frac{1}{\sqrt{x}} f, \qquad (11.188)$$

$$= \sqrt{x}\left(-\frac{1}{2}\right) \frac{\eta}{x} \frac{df}{d\eta} + \frac{1}{2}\frac{1}{\sqrt{x}} f = \frac{1}{2\sqrt{x}}\left(f - \eta \frac{df}{d\eta}\right). \qquad (11.189)$$

So we get for $\tilde{v}$ that

$$\tilde{v} = -\frac{\partial \psi}{\partial x} = \frac{1}{2\sqrt{x}}\left(\eta \frac{df}{d\eta} - f\right). \qquad (11.190)$$

And then we find

$$\frac{\partial \psi}{\partial \tilde{y}} = \frac{\partial}{\partial \tilde{y}}\left(\sqrt{x} f(\eta)\right) = \sqrt{x}\frac{\partial}{\partial \tilde{y}}(f(\eta)) = \sqrt{x}\frac{df}{d\eta}\frac{\partial \eta}{\partial \tilde{y}} = \sqrt{x}\frac{df}{d\eta}\frac{1}{\sqrt{x}} = \frac{df}{d\eta}. \qquad (11.191)$$

Thus, for u, we get

$$u = \frac{\partial \psi}{\partial \tilde{y}} = \frac{df}{d\eta}. \qquad (11.192)$$

So

$$\frac{\partial^2 \psi}{\partial x \partial \tilde{y}} = \frac{\partial}{\partial x}\left(\frac{\partial \psi}{\partial \tilde{y}}\right) = \frac{\partial}{\partial x}\left(\frac{df}{d\eta}\right) = \frac{d^2 f}{d\eta^2}\frac{\partial \eta}{\partial x} = -\frac{1}{2x}\eta\frac{d^2 f}{d\eta^2}, \quad (11.193)$$

$$\frac{\partial^2 \psi}{\partial \tilde{y}^2} = \frac{\partial}{\partial \tilde{y}}\left(\frac{\partial \psi}{\partial \tilde{y}}\right) = \frac{\partial}{\partial \tilde{y}}\left(\frac{df}{d\eta}\right) = \frac{d^2 f}{d\eta^2}\frac{\partial \eta}{\partial \tilde{y}} = \frac{1}{\sqrt{x}}\frac{d^2 f}{d\eta^2}, \quad (11.194)$$

$$\frac{\partial^3 \psi}{\partial \tilde{y}^3} = \frac{\partial}{\partial \tilde{y}}\left(\frac{\partial^2 \psi}{\partial \tilde{y}^2}\right) = \frac{\partial}{\partial \tilde{y}}\left(\frac{1}{\sqrt{x}}\frac{d^2 f}{d\eta^2}\right) = \frac{1}{\sqrt{x}}\frac{\partial}{\partial \tilde{y}}\left(\frac{d^2 f}{d\eta^2}\right) = \frac{1}{\sqrt{x}}\frac{d^3 f}{d\eta^3}\frac{\partial \eta}{\partial \tilde{y}}, \quad (11.195)$$

$$= \frac{1}{x}\frac{d^3 f}{d\eta^3}. \quad (11.196)$$

Now we substitute each of these expressions into the x momentum equation and get

$$\underbrace{\frac{df}{d\eta}}_{u}\underbrace{\left(-\frac{1}{2x}\eta\frac{d^2 f}{d\eta^2}\right)}_{\frac{\partial u}{\partial x}} + \underbrace{\frac{1}{2\sqrt{x}}\left(\eta\frac{df}{d\eta} - f\right)}_{\tilde{v}}\underbrace{\frac{1}{\sqrt{x}}\frac{d^2 f}{d\eta^2}}_{\frac{\partial u}{\partial \tilde{y}}} = \underbrace{\frac{1}{x}\frac{d^3 f}{d\eta^3}}_{\frac{\partial^2 u}{\partial \tilde{y}^2}}, \quad (11.197)$$

$$-\eta\frac{df}{d\eta}\frac{d^2 f}{d\eta^2} + \left(\eta\frac{df}{d\eta} - f\right)\frac{d^2 f}{d\eta^2} = 2\frac{d^3 f}{d\eta^3}, \quad (11.198)$$

$$\underbrace{-f\frac{d^2 f}{d\eta^2}}_{\text{advection}} = \underbrace{2\frac{d^3 f}{d\eta^3}}_{\text{momentum diffusion}}, \quad (11.199)$$

$$\frac{d^3 f}{d\eta^3} + \frac{1}{2}f\frac{d^2 f}{d\eta^2} = 0. \quad (11.200)$$

This is a third-order nonlinear ordinary differential equation for $f(\eta)$. We need three boundary conditions. Now at the surface $\tilde{y} = 0$, we have $\eta = 0$. And as $\tilde{y} \to \infty$, we have $\eta \to \infty$. To satisfy the no-slip condition on u at the plate surface, we require

$$\left.\frac{df}{d\eta}\right|_{\eta=0} = 0. \quad (11.201)$$

For no-slip on $\tilde{v}$, we require

$$\tilde{v}(0) = 0 = \frac{1}{2\sqrt{x}}\left(\eta\frac{df}{d\eta} - f\right), \quad \text{so} \quad 0 = \underbrace{0\left.\frac{df}{d\eta}\right|_{\eta=0}}_{=0} - f(0), \quad \text{so} \quad f(0) = 0. \quad (11.202)$$

And to satisfy the freestream condition on u as $\eta \to \infty$, we need

$$\left.\frac{df}{d\eta}\right|_{\eta\to\infty} = 1. \quad (11.203)$$

The most standard way to solve nonlinear ordinary differential equations of this type is to reduce them to systems of first-order ordinary differential equations and use some numerical technique, such as a Runge[1]–Kutta integration. We recall that Runge–Kutta methods, as well

---

[1] Carl David Tolmè Runge, 1856–1927, German mathematician and physicist who studied the spectral line elements of non-Hydrogen molecules.

as most other common techniques, require a well-defined set of initial conditions to predict the final state. To achieve the desired form, we define

$$g \equiv \frac{df}{d\eta}, \qquad h \equiv \frac{d^2 f}{d\eta^2}. \tag{11.204}$$

Thus, the x momentum equation becomes

$$\frac{dh}{d\eta} + \frac{1}{2}fh = 0. \tag{11.205}$$

But this is one equation in three unknowns. We need to write our equations as a system of three first-order equations, along with associated initial conditions. They are

$$\frac{df}{d\eta} = g, \quad \frac{dg}{d\eta} = h, \quad \frac{dh}{d\eta} = -\frac{1}{2}fh, \qquad f(0) = 0, \; g(0) = 0, \; h(0) = ? \tag{11.206}$$

Everything is well-defined except that we do not have an initial condition on h. We do, however, have a far field condition on g that is $g(\infty) = 1$. One viable option we have for getting a final solution is to use a numerical trial and error procedure, guessing $h(0)$ until we find that $g(\infty) \to 1$. We will use a slightly more efficient method here that only requires one guess.

To do this, let us first demonstrate the following lemma: If $F(\eta)$ is a solution to the Blasius equation $d^3 f/d\eta^3 + \frac{1}{2}f(d^2 f/d\eta^2) = 0$, then $f = aF(a\eta)$ is also a solution. The proof is as follows. Consider the transformation $f = aF$ and $\xi = a\eta$. Then we have $d/d\eta = (d\xi/d\eta)d/d\xi = a\,d/d\xi$. Then the Blasius equation transform to

$$a^4 \frac{d^3 F}{d\xi^3} + \frac{1}{2}a^4 F \frac{d^2 F}{d\xi^2} = 0, \qquad \frac{d^3 F}{d\xi^3} + \frac{1}{2}F \frac{d^2 F}{d\xi^2} = 0. \tag{11.207}$$

This is again the Blasius equation, and $F(\xi)$ is its solution. Hence $f = aF(a\eta)$ is a solution. This is true for any boundary conditions.

So to solve our nonlinear system, let us first solve the following related system:

$$\frac{dF}{d\eta} = G, \quad \frac{dG}{d\eta} = H, \quad \frac{dH}{d\eta} = -\frac{1}{2}FH, \quad F(0) = 0, \; G(0) = 0, \; H(0) = 1. \tag{11.208}$$

After one numerical integration, we find that with this guess for $H(0)$ that

$$G(\infty) = 2.08540918\ldots. \tag{11.209}$$

Now our numerical solution also gives us F, and so we know that $f = aF(a\eta)$ is also a solution. Moreover,

$$\frac{df}{d\eta} = a^2 \frac{dF(a\eta)}{d\eta}, \quad \text{which is} \quad g(\eta) = a^2 G(a\eta). \tag{11.210}$$

Now we want $g(\infty) = 1$, so take $1 = a^2 G(\infty)$, so $a^2 = 1/G(\infty)$. So

$$a = \frac{1}{\sqrt{G(\infty)}}. \tag{11.211}$$

Now

$$\frac{d^2 f}{d\eta^2} = a^3 \frac{d^2 F(a\eta)}{d\eta^2}, \quad \text{so} \quad \left.\frac{d^2 f}{d\eta^2}\right|_{\eta=0} = a^3 \left.\frac{d^2 F(a\eta)}{d\eta^2}\right|_{\eta=0}, \tag{11.212}$$

$$h(0) = a^3 H(0) = a^3(1) = G^{-3/2}(\infty) = (2.085\,409\,18)^{-3/2} = 0.332\,0573. \tag{11.213}$$

This is the proper choice for the initial condition on h. Numerically integrating once more, we get the behavior of f, g, and h as functions of η that indeed satisfies the condition at ∞. A plot of $u = df/d\eta$ as a function of η is shown in Fig. 11.15. From Fig. 11.15, we see that when $\eta = 5$, the velocity has nearly acquired the freestream value of $u = 1$. We can plot streamlines and the velocity vector field as well using the transformations to acquire ψ, u and $\tilde{v}$ as functions of x and $\tilde{y}$. They are plotted in Fig. 11.16. Notice that the streamlines have curvature, consistent with the velocity vector having nonzero components in the x and $\tilde{y}$ directions.

The associated kinematic topics of acceleration and vorticity vector fields, deformation tensors, are not straightforward and need to be carefully interpreted in light of the fact that we have scaled our equations in a particular fashion. Examination of the streamlines and velocity vector field of Fig. 11.16 suggests that the flow is decelerating in the streamwise direction, and that streamline curvature induces a stream-normal component of centripetal acceleration. The only forces available to induce such an acceleration are imbalanced viscous shear forces. Certainly one can visualize that a fluid element is both rotating as well as deforming in a volume-preserving fashion.

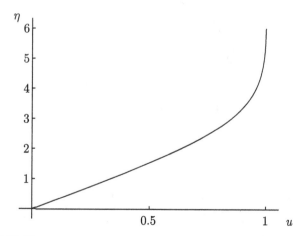

Figure 11.15 Velocity component profile for u for a Blasius boundary layer.

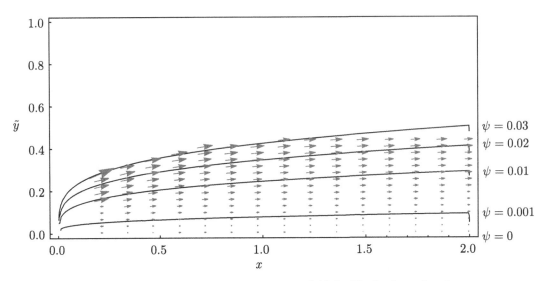

Figure 11.16 Streamlines and velocity vector field for Blasius boundary layer.

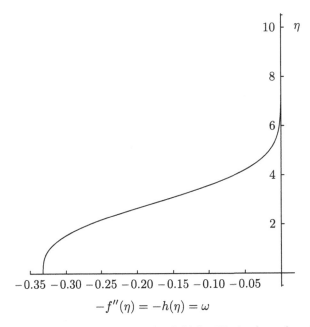

Figure 11.17 Dimensionless vorticity field for Blasius boundary layer.

Panton (2013, Chapter 20) shows that, to leading order, the dimensionless vorticity field is given by

$$\omega = -\frac{d^2 f}{d\eta^2} = -h(\eta). \tag{11.214}$$

A plot is given in Fig. 11.17. The vorticity is maximum at the no-slip boundary at $\eta = 0$. As $\eta \to \infty$, the flow becomes irrotational, consistent with the uniform freestream in the far field.

Examination of the numerical results shows that when $\eta = 4.9$, the u component of velocity has 0.99 of its freestream value. As the velocity only reaches its freestream value at ∞, we define the *boundary layer thickness*, $\delta_{0.99}$, as that value of $\bar{y}$ for which the velocity has 0.99 of its freestream value. Recalling that

$$\eta = \sqrt{\frac{U}{\nu}} \frac{\bar{y}}{\sqrt{\bar{x}}}, \tag{11.215}$$

we say that

$$4.9 = \sqrt{\frac{U}{\nu}} \frac{\delta_{0.99}}{\sqrt{\bar{x}}}. \tag{11.216}$$

Rearranging, we get

$$\frac{\delta_{0.99}}{\bar{x}} = 4.9\sqrt{\frac{\nu}{U\bar{x}}} = 4.9 Re_{\bar{x}}^{-1/2}. \tag{11.217}$$

Here we have taken a Reynolds number based on local distance to be

$$Re_{\bar{x}} = \frac{U\bar{x}}{\nu}. \tag{11.218}$$

This formula is valid for laminar flows, and has been seen to be valid for $Re_{\bar{x}} < 3 \times 10^6$. For greater lengths, there can be a transition to turbulent flow. For saturated liquid water at $T = 300\,\text{K}$ ($\nu = 0.858 \times 10^{-6}\,\text{m}^2/\text{s}$) flowing at $1\,\text{m/s}$ and a downstream distance of $1\,\text{m}$, we find $\delta_{0.99} = 0.45\,\text{cm}$. For one atmosphere air at $300\,\text{K}$ (yielding $\nu = 15.89 \times 10^{-6}\,\text{m}^2/\text{s}$), we find $\delta_{0.99} = 1.95\,\text{cm}$. We also note that the boundary layer grows with the square root of distance along the plate. We further note that higher kinematic viscosity leads to thicker boundary layers, while lower kinematic viscosity lead to thinner boundary layers.

The velocity $\tilde{v}$ has some nonintuitive behavior. It is plotted in Fig. 11.18. As seen from its definition in Eq. (11.190), scaling by $2\sqrt{x}$ is required to capture the exclusive dependency on η. And the calculation reveals that

$$\lim_{\eta \to \infty} 2\sqrt{x}\,\tilde{v} = 1.72. \tag{11.219}$$

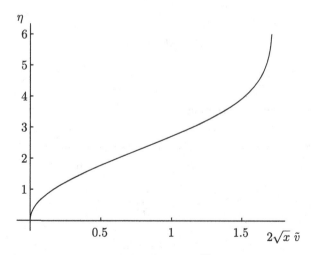

Figure 11.18 Velocity component profile for $2\sqrt{x}\,\tilde{v}$ in Blasius boundary layer.

We can unravel the various scalings then to get the following as $\eta \to \infty$:

$$0.86 = \sqrt{\tilde{x}}\,\tilde{v} = \sqrt{\frac{\overline{x}}{L}}\sqrt{Re}\,v = \sqrt{\frac{\overline{x}}{L}}\sqrt{\frac{UL}{\nu}}\frac{\overline{v}}{U} = \sqrt{\frac{U\overline{x}}{\nu}}\frac{\overline{v}}{U}, \qquad (11.220)$$

$$\frac{\overline{v}}{U} = 0.86 Re_{\overline{x}}^{-1/2}. \qquad (11.221)$$

We might expect our theory to force $\overline{v} \to 0$ as $\overline{y} \to \infty$. This is obviously not the case for our approximation, though it does so at large $\overline{x}$. The remedy is a complicated problem in asymptotic analysis as outlined by Van Dyke (1982, Chapter 7).

Now let us determine the shear stress at the wall, and the viscous force acting on the wall. So let us find

$$\tau_w = \mu \left.\frac{\partial \overline{u}}{\partial \overline{y}}\right|_{\overline{y}=0}. \qquad (11.222)$$

Consider

$$\frac{\partial u}{\partial \tilde{y}} = \frac{\partial^2 \psi}{\partial \tilde{y}^2} = \frac{1}{\sqrt{\tilde{x}}}\frac{d^2 f}{d\eta^2}, \quad \text{so} \quad \frac{\partial\left(\frac{\overline{u}}{U}\right)}{\partial\left(\sqrt{\frac{UL}{\nu}}\frac{\overline{y}}{L}\right)} = \frac{1}{\sqrt{\frac{\overline{x}}{L}}}\frac{d^2 f}{d\eta^2}, \qquad (11.223)$$

$$\frac{\partial \overline{u}}{\partial \overline{y}} = U\sqrt{\frac{\rho U}{\mu}}\frac{1}{\sqrt{\overline{x}}}\frac{d^2 f}{d\eta^2}, \quad \text{so} \quad \tau = \mu\frac{\partial \overline{u}}{\partial \overline{y}} = U\sqrt{\frac{\rho U \mu}{\overline{x}}}\frac{d^2 f}{d\eta^2}, \qquad (11.224)$$

$$\frac{\tau(0)}{\frac{1}{2}\rho U^2} = C_f = 2\sqrt{\frac{\mu}{\rho U \overline{x}}}\frac{d^2 f}{d\eta^2}(0), \quad \text{so} \quad C_f = 2Re_{\overline{x}}^{-1/2}\frac{d^2 f}{d\eta^2}(0) = \frac{0.664\ldots}{\sqrt{Re_{\overline{x}}}}. \qquad (11.225)$$

We notice at $\overline{x} = 0$, the stress is infinite. This seeming problem is seen not to be one when we consider the actual viscous force on a finite length of plate. Consider a plate of length L and width b. Then the viscous force acting on the plate is

$$F = \int_0^L \tau\, dA = \int_0^L \tau(\overline{x},0)b\, d\overline{x} = b\int_0^L f''(0)U\sqrt{\rho U \mu}\frac{1}{\sqrt{\overline{x}}}\, d\overline{x}, \qquad (11.226)$$

$$= bf''(0)U\sqrt{\rho U \mu}\int_0^L \frac{d\overline{x}}{\sqrt{\overline{x}}} = bf''(0)U\sqrt{\rho U \mu}\left(2\sqrt{\overline{x}}\right)\Big|_0^L, \qquad (11.227)$$

$$= 2bf''(0)U\sqrt{\rho U \mu}\sqrt{L}, \qquad (11.228)$$

$$\frac{F}{\frac{1}{2}\rho U^2 Lb} = C_D = 4f''(0)\sqrt{\frac{\mu}{\rho U L}} = 4f''(0)Re_L^{-1/2} = 1.328 Re_L^{-1/2}. \qquad (11.229)$$

Now let us consider the thermal boundary layer. Here we will take the boundary conditions so that the wall and far field are held at a constant fixed temperature $\overline{T} = T_o$. We need to scale the energy equation, so let us start with the steady incompressible two-dimensional dimensional energy equation, retaining the effects of viscous dissipation:

$$\rho c_p \left(\overline{u}\frac{\partial \overline{T}}{\partial \overline{x}} + \overline{v}\frac{\partial \overline{T}}{\partial \overline{y}}\right) = k\left(\frac{\partial^2 \overline{T}}{\partial \overline{x}^2} + \frac{\partial^2 \overline{T}}{\partial \overline{y}^2}\right) \qquad (11.230)$$

$$+ \mu\left(2\left(\frac{\partial \overline{u}}{\partial \overline{x}}\right)^2 + 2\left(\frac{\partial \overline{v}}{\partial \overline{y}}\right)^2 + \left(\frac{\partial \overline{u}}{\partial \overline{y}} + \frac{\partial \overline{v}}{\partial \overline{x}}\right)^2\right).$$

Taking as before,

$$x = \frac{\bar{x}}{L}, \quad y = \frac{\bar{y}}{L}, \quad T = \frac{\bar{T} - T_o}{T_o}, \quad u = \frac{\bar{u}}{U}, \quad v = \frac{\bar{v}}{U}. \quad (11.231)$$

Making these substitutions, we get

$$\frac{\rho c_p U T_o}{L}\left(u\frac{\partial T}{\partial x} + v\frac{\partial T}{\partial y}\right) = \frac{kT_o}{L^2}\left(\frac{\partial^2 T}{\partial x^2} + \frac{\partial^2 T}{\partial y^2}\right) \quad (11.232)$$
$$+ \frac{\mu U^2}{L^2}\left(2\left(\frac{\partial u}{\partial x}\right)^2 + 2\left(\frac{\partial v}{\partial y}\right)^2 + \left(\frac{\partial u}{\partial y} + \frac{\partial v}{\partial x}\right)^2\right),$$

$$u\frac{\partial T}{\partial x} + v\frac{\partial T}{\partial y} = \frac{k}{\rho c_p U L}\left(\frac{\partial^2 T}{\partial x^2} + \frac{\partial^2 T}{\partial y^2}\right) \quad (11.233)$$
$$+ \frac{\mu U}{\rho c_p L T_o}\left(2\left(\frac{\partial u}{\partial x}\right)^2 + 2\left(\frac{\partial v}{\partial y}\right)^2 + \left(\frac{\partial u}{\partial y} + \frac{\partial v}{\partial x}\right)^2\right).$$

Now we have

$$\frac{k}{\rho c_p U L} = \frac{k}{c_p \mu}\frac{\mu}{\rho U L} = \frac{1}{Pr}\frac{1}{Re}, \quad \frac{\mu U}{\rho c_p L T_o} = \frac{\mu}{\rho U L}\frac{U^2}{c_p T_o} = \frac{Ec}{Re}. \quad (11.234)$$

So the dimensionless energy equation with boundary conditions can be written as

$$u\frac{\partial T}{\partial x} + v\frac{\partial T}{\partial y} = \frac{1}{Pr Re}\left(\frac{\partial^2 T}{\partial x^2} + \frac{\partial^2 T}{\partial y^2}\right) \quad (11.235)$$
$$+ \frac{Ec}{Re}\left(2\left(\frac{\partial u}{\partial x}\right)^2 + 2\left(\frac{\partial v}{\partial y}\right)^2 + \left(\frac{\partial u}{\partial y} + \frac{\partial v}{\partial x}\right)^2\right),$$
$$T(x,0) = 0, \quad T(x,\infty) = 0. \quad (11.236)$$

Now as $Re \to \infty$, we see that $T = 0$ is a solution that satisfies the energy equation and all boundary conditions. For finite Reynolds number, nonzero velocity gradients generate a temperature field. Once again, we rescale in the boundary layer using $\tilde{v} = \sqrt{Re}\, v$, and $\tilde{y} = \sqrt{Re}\, y$. This gives

$$u\frac{\partial T}{\partial x} + \frac{1}{\sqrt{Re}}\frac{1}{1/\sqrt{Re}}\tilde{v}\frac{\partial T}{\partial \tilde{y}} = \frac{1}{Pr Re}\left(\frac{\partial^2 T}{\partial x^2} + Re\frac{\partial^2 T}{\partial \tilde{y}^2}\right) \quad (11.237)$$
$$+ \frac{Ec}{Re}\left(2\left(\frac{\partial u}{\partial x}\right)^2 + 2\left(\frac{\partial \tilde{v}}{\partial \tilde{y}}\right)^2 + \left(\sqrt{Re}\frac{\partial u}{\partial \tilde{y}} + \frac{1}{\sqrt{Re}}\frac{\partial \tilde{v}}{\partial x}\right)^2\right)$$

$$u\frac{\partial T}{\partial x} + \tilde{v}\frac{\partial T}{\partial \tilde{y}} = \frac{1}{Pr}\left(\frac{1}{Re}\frac{\partial^2 T}{\partial x^2} + \frac{\partial^2 T}{\partial \tilde{y}^2}\right)$$
$$+ Ec\left(\frac{2}{Re}\left(\frac{\partial u}{\partial x}\right)^2 + \frac{2}{Re}\left(\frac{\partial \tilde{v}}{\partial \tilde{y}}\right)^2 + \left(\frac{\partial u}{\partial \tilde{y}} + \frac{1}{Re}\frac{\partial \tilde{v}}{\partial x}\right)^2\right).$$
$$(11.238)$$

Now as $Re \to \infty$,

$$u\frac{\partial T}{\partial x} + \tilde{v}\frac{\partial T}{\partial \tilde{y}} = \frac{1}{Pr}\frac{\partial^2 T}{\partial \tilde{y}^2} + Ec\left(\frac{\partial u}{\partial \tilde{y}}\right)^2. \tag{11.239}$$

Now take $T = T(\eta)$ with $\eta = \tilde{y}/\sqrt{x}$ as well as $u = df/d\eta$, $\tilde{v} = (1/(2\sqrt{x}))(\eta(df/d\eta) - f)$ and $\partial u/\partial \tilde{y} = (1/\sqrt{x})(d^2 f/d\eta^2)$. We also have for derivatives, that

$$\frac{\partial T}{\partial x} = \frac{dT}{d\eta}\frac{\partial \eta}{\partial x} = \frac{dT}{d\eta}\left(-\frac{1}{2}\frac{\eta}{x}\right), \tag{11.240}$$

$$\frac{\partial T}{\partial \tilde{y}} = \frac{dT}{d\eta}\frac{\partial \eta}{\partial \tilde{y}} = \frac{dT}{d\eta}\frac{1}{\sqrt{x}}, \tag{11.241}$$

$$\frac{\partial^2 T}{\partial \tilde{y}^2} = \frac{\partial}{\partial \tilde{y}}\left(\frac{\partial T}{\partial \tilde{y}}\right) = \frac{\partial}{\partial \tilde{y}}\left(\frac{1}{\sqrt{x}}\frac{dT}{d\eta}\right) = \frac{1}{\sqrt{x}}\frac{\partial}{\partial \tilde{y}}\frac{dT}{d\eta} = \frac{1}{x}\frac{d^2 T}{d\eta^2}. \tag{11.242}$$

The energy equation is then rendered as

$$\underbrace{\frac{df}{d\eta}}_{u}\underbrace{\left(-\frac{1}{2}\frac{\eta}{x}\frac{dT}{d\eta}\right)}_{\frac{\partial T}{\partial x}} + \underbrace{\frac{1}{2\sqrt{x}}\left(\eta\frac{df}{d\eta} - f\right)}_{\tilde{v}}\underbrace{\frac{1}{\sqrt{x}}\frac{dT}{d\eta}}_{\frac{\partial T}{\partial \tilde{y}}} = \underbrace{\frac{1}{Pr}\frac{1}{x}\frac{d^2 T}{d\eta^2}}_{\frac{1}{Pr}\frac{\partial^2 T}{\partial \tilde{y}^2}} + \underbrace{\frac{Ec}{x}\left(\frac{d^2 f}{d\eta^2}\right)^2}_{Ec\left(\frac{\partial u}{\partial \tilde{y}}\right)^2}, \tag{11.243}$$

$$-\frac{1}{2}\eta\frac{df}{d\eta}\frac{dT}{d\eta} + \frac{1}{2}\left(\eta\frac{df}{d\eta} - f\right)\frac{dT}{d\eta} = \frac{1}{Pr}\frac{d^2 T}{d\eta^2} + Ec\left(\frac{d^2 f}{d\eta^2}\right)^2, \tag{11.244}$$

$$\underbrace{-\frac{1}{2}f\frac{dT}{d\eta}}_{\text{advection}} = \underbrace{\frac{1}{Pr}\frac{d^2 T}{d\eta^2}}_{\text{energy diffusion}} + \underbrace{Ec\left(\frac{d^2 f}{d\eta^2}\right)^2}_{\text{dissipation}}, \tag{11.245}$$

$$\frac{d^2 T}{d\eta^2} + \frac{Pr}{2}\frac{f}{d\eta}\frac{dT}{d\eta} = -Pr Ec\left(\frac{d^2 f}{d\eta^2}\right)^2, \quad T(0) = 0, \quad T(\infty) = 0. \tag{11.246}$$

Now for $Ec \to 0$, we get $T = 0$ as a solution that satisfies the governing differential equation and boundary conditions. Let us consider a solution for nontrivial Ec, but for $Pr = 1$. We could extend this for general values of Pr as well. Here, following the Reynolds analogy, when thermal diffusivity equals momentum diffusivity, we expect the temperature field to be directly related to the velocity field. For $Pr = 1$, the energy equation reduces to

$$\underbrace{\frac{d^2 T}{d\eta^2}}_{\text{energy diffusion}} + \underbrace{\frac{1}{2}f\frac{dT}{d\eta}}_{\text{advection}} = \underbrace{-Ec\left(\frac{d^2 f}{d\eta^2}\right)^2}_{\text{dissipation}}, \tag{11.247}$$

$$T(0) = 0, \quad T(\infty) = 0. \tag{11.248}$$

Multiplying the energy equation by the integrating factor, $e^{\int_0^\eta \frac{1}{2}f(t)\,dt}$, gives

$$e^{\int_0^\eta \frac{1}{2}f(t)\,dt}\frac{d^2 T}{d\eta^2} + \frac{1}{2}fe^{\int_0^\eta \frac{1}{2}f(t)\,dt}\frac{dT}{d\eta} = -Ec\, e^{\int_0^\eta \frac{1}{2}f(t)\,dt}\left(\frac{d^2 f}{d\eta^2}\right)^2, \tag{11.249}$$

$$\frac{d}{d\eta}\left(e^{\int_0^\eta \frac{1}{2}f(t)\,dt}\frac{dT}{d\eta}\right) = -Ec\, e^{\int_0^\eta \frac{1}{2}f(t)\,dt}\left(\frac{d^2 f}{d\eta^2}\right)^2. \tag{11.250}$$

Now from the x momentum equation, $f''' + \frac{1}{2}ff'' = 0$, we have

$$f = -2\frac{f'''}{f''}. \tag{11.251}$$

So we can rewrite the integrating factor as

$$e^{\int_0^\eta \frac{1}{2}f(t)\,dt} = e^{\int_0^\eta \frac{1}{2}\frac{(-2)f'''}{f''}\,dt} = e^{-\ln\left(\frac{f''(\eta)}{f''(0)}\right)} = \frac{f''(0)}{f''(\eta)}. \tag{11.252}$$

So the energy equation can be written as

$$\frac{d}{d\eta}\left(\frac{f''(0)}{f''(\eta)}\frac{dT}{d\eta}\right) = -Ec\left(\frac{f''(0)}{f''(\eta)}\right)\left(\frac{d^2f}{d\eta^2}\right)^2 = -Ecf''(0)\frac{d^2f}{d\eta^2}, \tag{11.253}$$

$$\frac{f''(0)}{f''(\eta)}\frac{dT}{d\eta} = -Ecf''(0)\int_0^\eta \frac{d^2f}{ds^2}\,ds + C_1, \tag{11.254}$$

$$\frac{dT}{d\eta} = -Ec\frac{d^2f}{d\eta^2}\int_0^\eta \frac{d^2f}{ds^2}\,ds + C_1\frac{d^2f}{d\eta^2}, \tag{11.255}$$

$$= -Ec\frac{d^2f}{d\eta^2}\left(\frac{df}{d\eta} - \underbrace{f'(0)}_{=0}\right) + C_1\frac{d^2f}{d\eta^2}, \tag{11.256}$$

$$= -Ec\frac{d^2f}{d\eta^2}\frac{df}{d\eta} + C_1\frac{d^2f}{d\eta^2}, \tag{11.257}$$

$$= -Ec\frac{d}{d\eta}\left(\frac{1}{2}\left(\frac{df}{d\eta}\right)^2\right) + C_1\frac{d^2f}{d\eta^2}, \tag{11.258}$$

$$T = -\frac{Ec}{2}\left(\frac{df}{d\eta}\right)^2 + C_1\frac{df}{d\eta} + C_2, \tag{11.259}$$

$$T(0) = 0 = -\frac{Ec}{2}\underbrace{(f'(0))^2}_{=0} + C_1\underbrace{f'(0)}_{=0} + C_2, \quad \text{so} \quad C_2 = 0, \tag{11.260}$$

$$T(\infty) = 0 = -\frac{Ec}{2}\underbrace{(f'(\infty))^2}_{=1} + C_1\underbrace{f'(\infty)}_{=1}, \quad \text{so} \quad C_1 = \frac{Ec}{2}, \tag{11.261}$$

$$T(\eta) = \frac{Ec}{2}\frac{df}{d\eta}\left(1 - \frac{df}{d\eta}\right) = \frac{Ec}{2}u(\eta)(1 - u(\eta)). \tag{11.262}$$

A plot of the temperature profile for $Pr = 1$ and $Ec = 1$ is given in Fig. 11.19.

Now it is common and more physically defensible in the low Mach number limit to neglect viscous dissipation as small. This corresponds to $Ec \to 0$. Let us slightly alter the boundary conditions so that we can find a nontrivial solution for $Ec = 0$. For simplicity, we also consider $Pr = 1$, but this is not difficult to generalize. We take now the far field to be at $\overline{T} = T_o$, as before, but the plate surface to be at $\overline{T} = T_1$. Then we choose the scaling $T = (\overline{T} - T_o)/(T_1 - T_o)$ and recast Eqs. (11.246) for $Pr = 1$, $Ec = 0$ as

$$\frac{d^2T}{d\eta^2} + \frac{1}{2}f\frac{dT}{d\eta} = 0, \qquad T(0) = 1, \quad T(\infty) = 0. \tag{11.263}$$

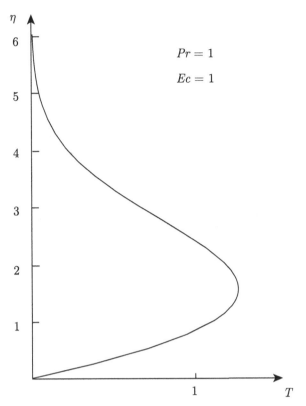

Figure 11.19 Temperature profile for Blasius boundary layer, $Ec = 1$, $Pr = 1$; wall temperature held equal to the freestream temperature.

Now the inhomogeneity is in the boundary conditions rather than the forcing term. The previously completed analysis is identical up to Eq. (11.259), which we write for $Ec = 0$ as

$$T = C_1 \frac{df}{d\eta} + C_2. \tag{11.264}$$

Now at the plate surface $\eta = 0$ and the far field $\eta \to \infty$, we get, respectively

$$1 = C_1 \underbrace{f'(0)}_{0} + C_2, \qquad 0 = C_1 \underbrace{f'(\infty)}_{1} + C_2. \tag{11.265}$$

Thus $C_1 = -1$, $C_2 = 1$, so

$$T(\eta) = 1 - \frac{df}{d\eta} = 1 - u(\eta). \tag{11.266}$$

Again for the $Pr = 1$ problem, the Reynolds analogy holds. We plot this temperature field in Fig. 11.20.

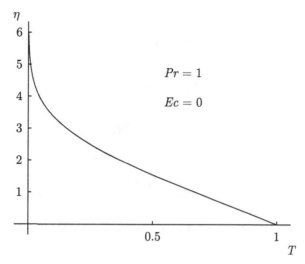

Figure 11.20 Temperature profile for Blasius boundary layer, $Ec = 0$, $Pr = 1$; wall temperature held distinct from the freestream temperature.

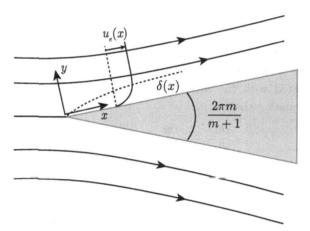

Figure 11.21 Falkner–Skan flow of an incompressible viscous fluid over a wedge.

11.2.2 Falkner–Skan Flow

Let us extend the analysis of the Blasius boundary layer to a flow with a nonzero pressure gradient. We will consider the flow of an incompressible viscous fluid over a wedge as depicted in Fig. 11.21. The complete wedge angle is taken to be $\pi(2m)/(m+1)$. As indicated in the diagram, we expect something similar to a Blasius boundary layer at the wedge surface. Away from the boundary layer, the flow is well modeled as inviscid. We will consider that to be well modeled by a potential flow solution that will be dictated by a balance of pressure gradient and advective fluid acceleration.

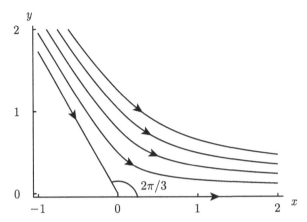

Figure 11.22 Streamlines for potential flow solution of corner flow with $n = 3/2$, giving a turning angle of $2\pi/3$.

Let us then adopt the boundary layer equations, Eqs. (11.164–11.166). After integrating Eq. (11.166), we recognize once more that $p = p(x)$, and thus our equation set can be cast as

$$\frac{\partial u}{\partial x} + \frac{\partial \tilde{v}}{\partial \tilde{y}} = 0, \qquad u\frac{\partial u}{\partial x} + \tilde{v}\frac{\partial u}{\partial \tilde{y}} = -\frac{dp}{dx} + \frac{\partial^2 u}{\partial \tilde{y}^2}. \tag{11.267}$$

In the freestream away from the boundary layer, we expect the flow to be entirely in the x direction with variation in only the x direction with negligible viscous forces and an exact solution, denoted with the subscript e, to be available from potential flow theory. The x momentum equation there is

$$u_e \frac{\partial u_e}{\partial x} + \underbrace{v_e \frac{\partial u_e}{\partial y}}_{\sim 0} = -\frac{dp}{dx} + \underbrace{\frac{1}{Re}\left(\frac{\partial^2 u_e}{\partial x^2} + \frac{\partial^2 u_e}{\partial y^2}\right)}_{\sim 0}, \tag{11.268}$$

$$u_e \frac{du_e}{dx} = -\frac{dp}{dx}. \tag{11.269}$$

We will replace $-dp/dx$ in our boundary layer equations by $u_e du_e/dx$, where u_e will come from a potential flow solution. Doing that, and re-introducing the stream function via Eq. (11.173), the x momentum equation for the boundary layer, Eq. (11.267), can be recast as

$$\frac{\partial \psi}{\partial \tilde{y}}\frac{\partial^2 \psi}{\partial x \partial \tilde{y}} - \frac{\partial \psi}{\partial x}\frac{\partial^2 \psi}{\partial \tilde{y}^2} = u_e \frac{du_e}{dx} + \frac{\partial^3 \psi}{\partial \tilde{y}^3}. \tag{11.270}$$

At this juncture, we step back to see how to relate the outer solution u_e to inviscid flow solutions studied earlier. Recall that potential flow in a corner can be described by Eqs. (8.100):

$$\phi = Br^n \cos n\theta, \qquad \psi = Br^n \sin n\theta. \tag{11.271}$$

Clearly the $\psi = 0$ streamline is found at $\theta = 0$, $\theta = \pi/n$. Note that n need not be an integer. For $n = 3/2$, contours of ψ are given in Fig. 11.22. We apply then Eq. (11.34) to give us

$$v_r = \frac{1}{r}\frac{\partial \psi}{\partial \theta} = Bnr^{n-1}\cos n\theta, \qquad v_\theta = -\frac{\partial \psi}{\partial r} = -Bnr^{n-1}\sin n\theta. \tag{11.272}$$

We can consider the wedge surface to be at $\theta = 0$, aligned with the x axis. So on the wedge surface, we have $u = v_r$, $v = v_\theta$ and

$$u(x) = Bnx^{n-1}, \qquad v = 0. \tag{11.273}$$

We need to correlate the parameter n with the wedge angle of Fig. 11.21. The diagram of Fig. 11.22 aids in this. Here the angle between the freestream flow and the wedge is π/n. Thus, the wedge half angle is $\pi(1 - 1/n)$, as indicated. The full wedge angle is $2\pi(1 - 1/n)$. Now, let us define m such that $m = n - 1$. Thus, the angle between the flow and the wedge is $\pi/(m+1)$. The full wedge angle transforms to $2\pi m/(m+1)$, and the wedge half-angle is $\pi m/(m+1)$. And the velocity at the surface of the wedge becomes, after choosing $B = 1/(m+1)$,

$$u(x) = x^m. \tag{11.274}$$

We take this to be $u_e(x)$. So the advective acceleration, which is equated with the pressure gradient and serves as a source term in the boundary layer equations, is

$$u_e \frac{du_e}{dx} = x^m(mx^{m-1}) = mx^{2m-1}. \tag{11.275}$$

This lets us consider some important flows. The diagram of Fig. 11.24 aids in this. For $m < 0$, we have flow into a diverging section. For $m = 0$, we have flow over a flat plate. For $m \in (0, 1)$, we have flow into a converging section, that is, a wedge. For $m = 1$, we have stagnation point flow. And for $m \in (1, 2)$, we have flow into a corner.

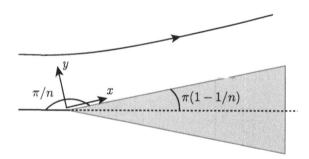

Figure 11.23 Diagram of Falkner–Skan flow.

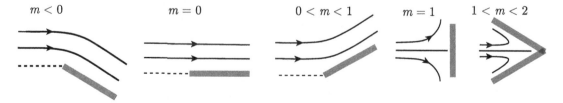

Figure 11.24 Geometries modeled by Falkner–Skan equations.

We now recast Eq. (11.270) as

$$\frac{\partial \psi}{\partial \tilde{y}} \frac{\partial^2 \psi}{\partial x \partial \tilde{y}} - \frac{\partial \psi}{\partial x} \frac{\partial^2 \psi}{\partial \tilde{y}^2} = mx^{2m-1} + \frac{\partial^3 \psi}{\partial \tilde{y}^3}. \tag{11.276}$$

The similarity variables for flow over a flat plate do not work. We guess, however, that

$$\eta = \tilde{y} x^a, \qquad \psi = x^b f(\eta). \tag{11.277}$$

A detailed calculation reveals self-similarity is achieved if $a = (m-1)/2$ and $b = (m+1)/2$. The result is a nonlinear third-order differential equation with conditions at the wedge surface and far field given by

$$\frac{d^3 f}{d\eta^3} + \frac{m+1}{2} f \frac{d^2 f}{d\eta^2} + m\left(1 - \left(\frac{df}{d\eta}\right)^2\right) = 0, \tag{11.278}$$

$$f(0) = 0, \qquad f'(0) = 0, \qquad f'(\infty) = 1. \tag{11.279}$$

When $m = 0$, we have a flat plate, and the equation reduces to that for a Blasius boundary layer, Eq. (11.200).

Let us examine the dimensionless similarity variable in terms of dimensional variables. With $a = (m-1)/2$ and $u_e = x^m$, we have

$$\eta = \frac{\tilde{y}}{\sqrt{x^{1-m}}} = \frac{\sqrt{Re}\, y}{\sqrt{\frac{x}{u_e}}} = \frac{y}{\sqrt{\frac{\nu}{\overline{u}_o L} \frac{x}{u_e}}} = \frac{\overline{y}/L}{\sqrt{\frac{\nu}{\overline{u}_o L} \frac{\overline{x}}{L} \frac{\overline{u}_o}{\overline{u}_e}}} = \frac{\overline{y}}{\sqrt{\nu \overline{x}/\overline{u}_e}} = \frac{\overline{y}}{\overline{x}} Re_{\overline{x}}^{-1/2}. \tag{11.280}$$

The undefined term L does not enter into the final form, as should be expected. It is noted that $\overline{u}_e$ is a function of $\overline{x}$ as prescribed by the outer potential flow solution. Plots of $f(\eta)$ for various values of m are given in Fig. 11.25. Now m can be thought of as representing the strength of the

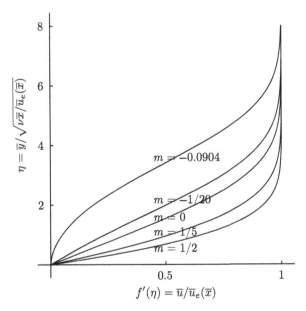

Figure 11.25 Solutions of Falkner–Skan boundary layer equations for various values of m.

pressure gradient. Increasing m corresponds to increasing the wedge angle associated with the converging cross section. It renders the pressure gradient to become more negative and causes the boundary layer to become thinner. For positive pressure gradients corresponding to flow in a diverging cross section, we can obtain physically defensible solutions for $m \in (0, -0.0904)$. When $m = -0.0904$, we see f' approaches $\eta = 0$ with $f'' = 0$. Physically, this implies zero stress at the wall. This is often correlated with the onset of flow reversal. In experiments, one does observe flow reversal in such a diverging portion of a duct. As the Falkner–Skan equations are nonlinear, there is no guarantee of a unique solution; Panton (2013, pp. 543–546) has a more complete discussion of how this affects interpretation of solutions.

11.2.3 Jets

Let us consider here the laminar flow of a jet of incompressible viscous fluid into a quiescent region. The fluid will be modeled as a jet with a finite fixed x momentum issuing from an infinitely thin slot at $(x, y) = (0, 0)$. We will restrict attention to steady two-dimensional flow; see Fig. 11.26. As depicted, we might expect the flow to be primarily in the x direction and for momentum diffusion to smear the initially concentrated pulse as it propagates in the positive x direction. We will model this with our boundary layer equations and neglect all pressure gradients. Following Schlichting and Gersten (2017), here we will not be as formal with scaling, but simply commence with the dimensional equations neglecting terms that we know from our more rigorous analysis should be negligible. So we take the dimensional analog of Eqs. (11.170, 11.171):

$$\frac{\partial u}{\partial x} + \frac{\partial v}{\partial \tilde{y}} = 0, \qquad u\frac{\partial u}{\partial x} + v\frac{\partial u}{\partial y} = \nu \frac{\partial^2 u}{\partial y^2}. \tag{11.281}$$

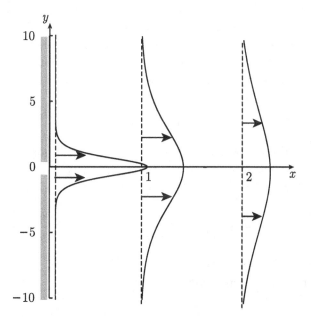

Figure 11.26 Plot of velocity profiles of $u(y)$ of a laminar incompressible jet, $\nu = 1$, $A = 1$, $\rho = 1$, at $x = 0.1$, $x = 1$, $x = 2$.

As boundary conditions, we take

$$v(x,0) = 0, \qquad \frac{\partial u}{\partial y}(x,0) = 0, \qquad u(x,\infty) = 0. \tag{11.282}$$

Recall the mass flux in the x direction is understood to be ρu: the mass per unit volume of the fluid, ρ, is carried by its velocity u. Similarly, the linear momentum flux in the x direction has the linear momentum per unit volume, ρu, carried by its velocity, u, giving the momentum flux to be ρu^2. At a given x, let us integrate the momentum flux through the entire y domain and call it J:

$$J = \rho \int_{-\infty}^{\infty} u^2 \, dy. \tag{11.283}$$

We will insist that J is a constant independent of x.

Now we take the dimensional version of the standard stream function, Eq. (11.173):

$$u = \frac{\partial \psi}{\partial y}, \qquad v = -\frac{\partial \psi}{\partial x}. \tag{11.284}$$

With this, mass conservation is guaranteed, and the x momentum equation and boundary conditions become

$$\frac{\partial \psi}{\partial y}\frac{\partial^2 \psi}{\partial x \partial y} - \frac{\partial \psi}{\partial x}\frac{\partial^2 \psi}{\partial y^2} = \nu \frac{\partial^3 \psi}{\partial y^3}, \tag{11.285}$$

$$\frac{\partial \psi}{\partial x}(x,0) = 0, \qquad \frac{\partial^2 \psi}{\partial y^2}(x,0) = 0, \qquad \frac{\partial \psi}{\partial y}(x,\infty) = 0. \tag{11.286}$$

Notice that $\psi(x,y) = 0$ is a solution that satisfies the governing equations and all boundary conditions; however, it has zero momentum flux J, so we seek another nontrivial solution. Because the equations are nonlinear, we need not be surprised that other solutions exist.

We now seek self-similar solutions. Through previously employed methods involving coordinate stretching, one can determine that the following are appropriate set of new variables:

$$\eta = \frac{1}{3\nu^{1/2}}\frac{y}{x^{2/3}}, \qquad \psi = \nu^{1/2}x^{1/3}f(\eta). \tag{11.287}$$

Let us confirm that these transformations are able to transform our partial differential equation to an ordinary differential equation. We use the chain rule to obtain the necessary partial derivatives on η:

$$\frac{\partial \eta}{\partial x} = -\frac{2}{3}\frac{1}{3\nu^{1/2}}\frac{y}{x^{2/3}}\frac{1}{x} = -\frac{2}{3}\frac{\eta}{x}, \qquad \frac{\partial \eta}{\partial y} = \frac{1}{3\nu^{1/2}}\frac{1}{x^{2/3}}. \tag{11.288}$$

We now turn to the necessary derivatives of ψ; we employ a "prime" notation for a derivative with respect to η, for example $df/d\eta = f'$:

$$\frac{\partial \psi}{\partial x} = \frac{\partial}{\partial x}\left(\nu^{1/2}x^{1/3}f(\eta)\right) = \nu^{1/2}\left(\frac{1}{3}x^{-2/3}f + x^{1/3}\left(-\frac{2}{3}\frac{\eta}{x}f'\right)\right), \tag{11.289}$$

11.2 Flow with Effects of Inertia

$$= \frac{\nu^{1/2}}{3x^{2/3}}(f - 2\eta f'), \qquad (11.290)$$

$$\frac{\partial \psi}{\partial y} = \nu^{1/2} x^{1/3} \frac{1}{3\nu^{1/2} x^{2/3}} f' = \frac{1}{3x^{1/3}} f', \qquad (11.291)$$

$$\frac{\partial^2 \psi}{\partial y^2} = \frac{1}{3x^{1/3}} \frac{1}{3\nu^{1/2} x^{2/3}} f'' = \frac{1}{9\nu^{1/2} x} f'', \qquad (11.292)$$

$$\frac{\partial^3 \psi}{\partial y^3} = \frac{1}{9\nu^{1/2} x} \frac{1}{3\nu^{1/2} x^{2/3}} f''' = \frac{1}{27\nu x^{5/3}} f''', \qquad (11.293)$$

$$\frac{\partial^2 \psi}{\partial x \partial y} = \frac{1}{3}\left(-\frac{1}{3x^{4/3}} f' + \frac{1}{x^{1/3}}\left(-\frac{2}{3}\frac{\eta}{x}\right)f''\right) = -\frac{1}{9x^{4/3}}(f' + 2\eta f''). \qquad (11.294)$$

We substitute all of these into our x momentum equation, Eq. (11.285), to get

$$\underbrace{\frac{1}{3x^{1/3}} f'}_{\frac{\partial \psi}{\partial y}} \underbrace{\left(\frac{-1}{9x^{4/3}}(f' + 2\eta f'')\right)}_{\frac{\partial^2 \psi}{\partial x \partial y}} - \underbrace{\left(\frac{\nu^{1/2}}{3x^{2/3}}(f - 2\eta f')\right)}_{\frac{\partial \psi}{\partial x}} \underbrace{\left(\frac{1}{9\nu^{1/2} x} f''\right)}_{\frac{\partial^2 \psi}{\partial y^2}} = \nu \underbrace{\left(\frac{1}{27\nu x^{5/3}} f'''\right)}_{\frac{\partial^3 \psi}{\partial y^3}},$$

$$(11.295)$$

This simplifies considerably:

$$-f'(f' + 2\eta f'') - (f - 2\eta f')f'' = f''', \qquad (11.296)$$

$$f''' + f f'' + f'^2 = 0. \qquad (11.297)$$

The partial differential equation indeed reduced to an ordinary differential equation under the transformation. We must also check the boundary conditions. For $v(x, 0) = 0$, we have at $\eta = 0$

$$f - 2\eta f' = 0. \qquad (11.298)$$

This yields $f(0) = 0$. The condition $\partial^2 \psi / \partial y^2 (x, 0)$ gives $f''(0) = 0$. And the condition $\partial \psi / \partial y (x, \infty) = 0$ gives $f'(\infty) = 0$. Now let us recast Eq. (11.297) as

$$\frac{d^3 f}{d\eta^3} + \frac{d}{d\eta}\left(f \frac{df}{d\eta}\right) = 0. \qquad (11.299)$$

We can integrate this once to form

$$\frac{d^2 f}{d\eta^2} + f \frac{df}{d\eta} = C. \qquad (11.300)$$

Now $f''(0) = f(0) = 0$, so $C = 0$, giving

$$\frac{d^2 f}{d\eta^2} + f \frac{df}{d\eta} = 0. \qquad (11.301)$$

We can recast this as

$$\frac{d^2 f}{d\eta^2} + \frac{1}{2} \frac{d}{d\eta} f^2 = 0. \qquad (11.302)$$

This may be integrated to give

$$\frac{df}{d\eta} + \frac{1}{2} f^2 = A. \qquad (11.303)$$

Solving this and enforcing $f(0) = 0$ gives

$$f(\eta) = \sqrt{2A} \tanh\left(\sqrt{\frac{A}{2}}\eta\right). \tag{11.304}$$

Now the velocity u can be determined. It must be

$$u = \frac{1}{3x^{1/3}}\frac{df}{d\eta} = \frac{A}{3x^{1/3}}\operatorname{sech}^2\left(\sqrt{\frac{A}{2}}\eta\right) = \frac{A}{3x^{1/3}}\operatorname{sech}^2\left(\sqrt{\frac{A}{2\nu}}\frac{y}{3x^{2/3}}\right). \tag{11.305}$$

Now let us determine J:

$$J = \rho \int_{-\infty}^{\infty} u^2\, dy = \rho \int_{-\infty}^{\infty} \left(\frac{A}{3x^{1/3}}\operatorname{sech}^2\left(\sqrt{\frac{A}{2\nu}}\frac{y}{3x^{2/3}}\right)\right)^2 dy = \frac{4\rho}{9}\sqrt{2A^3\nu}. \tag{11.306}$$

The key result here is that J is a constant; importantly, it has no dependence on x. If one knows the value of J, one can then pick A. Plots of the velocity profile for a fluid with $\nu = 1$, $\rho = 1$, and $A = 1$ are given at $x = 0.1$, $x = 1$, and $x = 2$ in Fig. 11.26.

We have plotted u in Fig. 11.26; one should remember that there is a small, nonzero component $v(x,y)$ available as well. Thus the actual streamlines have curvature. The stream function is

$$\psi(x,y) = \sqrt{2A}x^{1/3}\tanh\left(\sqrt{\frac{A}{2\nu}}\frac{y}{3x^{2/3}}\right). \tag{11.307}$$

For $\nu = 1$, $A = 1$, streamlines are plotted in Fig. 11.27. Clearly near $x = 0$ there is significant entrainment from $y = \pm\infty$. And for these parameter choices, there is obvious curvature of the streamlines.

11.2.4 Shear Layers

One of the most important flow configurations is the shear layer; see Fig. 11.29. Here two streams – one at U_1, the other at U_2 – come into contact. Because of momentum diffusion, a layer of adjustment develops during which the velocity adjusts from U_1 to U_2. This is the paradigm mixing problem, of fundamental importance in numerous physical applications. Here the mixing is of linear momentum; equivalents exist for mass and energy mixing. While on large scales mixing is enhanced by advection, true molecular mixing always occurs at scales where diffusion dominates.

Using methods that have been demonstrated in previous problems, one can study the incompressible, two-dimensional mixing of two streams possessing a fluid with the same viscosity. One adopts the similarity variables

$$\eta = y\sqrt{\frac{U_1}{\nu x}}, \qquad \psi = \sqrt{\nu U_1 x}f(\eta), \tag{11.308}$$

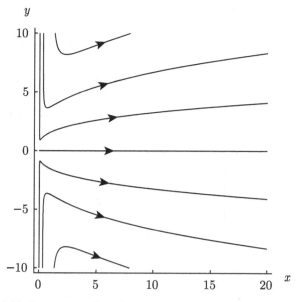

Figure 11.27 Streamlines for a laminar incompressible jet, $\nu = 1$, $A = 1$.

and is led to

$$f\frac{d^2 f}{d\eta^2} + 2\frac{d^3 f}{d\eta^3} = 0, \qquad (11.309)$$

$$f'(\infty) = 1, \qquad f'(-\infty) = \frac{U_2}{U_1}, \qquad f(0) = 0. \qquad (11.310)$$

This must be integrated numerically. We do so after choosing $\nu = 1$, $U_1 = 2$, $U_2 = 1$, and plot results in Fig. 11.28. Here, we have $u = U_1 f'(\eta)$.

11.2.5 von Kármán's Viscous Pump

Let us consider a problem first addressed by von Kármán (1921) in which a disk rotating at constant angular velocity entrains the motion of a viscous fluid, as depicted in Fig. 11.29. We will follow the development of Panton (2013). We consider the disk to be infinite and residing in the plane where $z = 0$. It is rotating about the z axis with constant angular velocity Ω. We will study a flow that is steady and incompressible, and assume the pressure gradient in the radial direction is zero: $\partial p/\partial r = 0$. In contrast, were the rotating fluid within a finite cylinder, we might expect a nonzero radial pressure gradient. This lack of a confining radial pressure gradient will induce $v_r \neq 0$ as will be seen upon examination of the radial momentum equation. At the plate surface, the no-slip condition will induce a known nonzero value of v_θ. To achieve mass and momentum balances, we shall see a nonzero v_z is required. The rotating plate will be seen to pull the fluid from infinite z and fling it towards infinite r. Importantly, all three velocity components can be expected to be nontrivial in this flow, so it is truly a three-dimensional flow field. We seek some simplification and suppose that none of the three

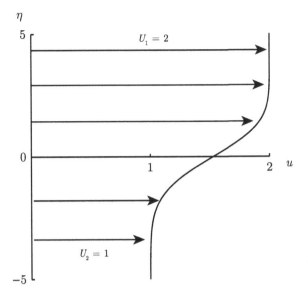

Figure 11.28 Solution for a laminar shear layer with $\nu = 1$, $U_1 = 2$, $U_2 = 1$.

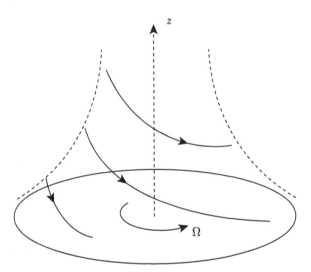

Figure 11.29 Schematic for von Kármán's viscous pump.

velocity components has any dependence on θ. So we seek $v_r(r,z)$, $v_\theta(r,z)$, $v_z(r,z)$. We assume the following form and will test if it leads to a solution:

$$v_r(r,z) = r\Omega F(z), \qquad v_\theta(r,z) = r\Omega G(z), \qquad v_z(r,z) = \sqrt{\nu\Omega}H(z). \qquad (11.311)$$

Here F, G, and H are all dimensionless functions. Let us also define a dimensionless z coordinate:

$$z_* = \frac{z}{\sqrt{\nu/\Omega}}. \qquad (11.312)$$

We now check if and how these assumptions are able to reduce the mass and momenta equations to more amenable forms. First we reduce the incompressible mass conservation equation:

$$\frac{1}{r}\frac{\partial}{\partial r}(rv_r) + \frac{1}{r}\frac{\partial v_\theta}{\partial \theta} + \frac{\partial v_z}{\partial z} = 0, \tag{11.313}$$

$$\frac{1}{r}\frac{\partial}{\partial r}\left(r^2\Omega F(z)\right) + \frac{1}{r}\frac{\partial}{\partial \theta}(r\Omega G(z)) + \frac{\partial}{\partial z}\left(\sqrt{\nu\Omega}H(z)\right) = 0, \tag{11.314}$$

$$\frac{1}{r}\Omega F(z)(2r) + 0 + \sqrt{\nu\Omega}\frac{dH}{dz} = 0, \tag{11.315}$$

$$2F(z) + 0 + \sqrt{\nu/\Omega}\frac{dH}{dz} = 0. \tag{11.316}$$

Eliminating z in favor of z_*, we get

$$2F(z_*) + \frac{dH}{dz_*} = 0. \tag{11.317}$$

Next we turn to the r momentum equation, which we have assumed has no pressure gradient:

$$v_r\frac{\partial v_r}{\partial r} + \frac{v_\theta}{r}\frac{\partial v_r}{\partial \theta} - \frac{v_\theta^2}{r} + v_z\frac{\partial v_r}{\partial z} = \nu\left(\frac{\partial}{\partial r}\left(\frac{1}{r}\frac{\partial}{\partial r}(rv_r)\right) + \frac{1}{r^2}\frac{\partial^2 v_r}{\partial \theta^2} + \frac{\partial^2 v_r}{\partial z^2} - \frac{2}{r^2}\frac{\partial v_\theta}{\partial \theta}\right). \tag{11.318}$$

Removing all $\partial/\partial\theta$ terms, we get

$$v_r\frac{\partial v_r}{\partial r} - \frac{v_\theta^2}{r} + v_z\frac{\partial v_r}{\partial z} = \nu\left(\frac{\partial}{\partial r}\left(\frac{1}{r}\frac{\partial}{\partial r}(rv_r)\right) + \frac{\partial^2 v_r}{\partial z^2}\right). \tag{11.319}$$

Substituting our assumed forms, we find

$$r\Omega F(z)\frac{\partial}{\partial r}(r\Omega F(z)) - \frac{(r\Omega G(z))^2}{r} + \sqrt{\nu\Omega}H(z)\frac{\partial}{\partial z}(r\Omega F(z))$$
$$= \nu\left(\frac{\partial}{\partial r}\left(\frac{1}{r}\frac{\partial}{\partial r}(r^2\Omega F(z))\right) + \frac{\partial^2}{\partial z^2}(r\Omega F(z))\right). \tag{11.320}$$

This simplifies to

$$r\Omega^2 F^2(z) - r\Omega^2 G^2(z) + \sqrt{\nu\Omega}\, r\Omega H(z)\frac{dF}{dz} = \nu r\Omega \frac{d^2 F}{dz^2}, \tag{11.321}$$

$$F^2(z) - G^2(z) + \sqrt{\frac{\nu}{\Omega}}H(z)\frac{dF}{dz} = \frac{\nu}{\Omega}\frac{d^2 F}{dz^2}, \tag{11.322}$$

$$\underbrace{F^2(z_*) + H(z_*)\frac{dF}{dz_*}}_{\text{advective acceleration, radial and axial}} - \underbrace{G^2(z_*)}_{\text{centripetal acceleration}} = \underbrace{\frac{d^2 F}{dz_*^2}}_{\text{diffusion}}. \tag{11.323}$$

When a similar exercise is done for the θ and z momenta equations, we get, after defining $\mathcal{P} = p/(\rho\Omega\nu)$, that

$$2F(z_*)G(z_*) + H(z_*)\frac{dG}{dz_*} = \frac{d^2 G}{dz_*^2}, \tag{11.324}$$

$$2H(z_*)F(z_*) - 2\frac{dF}{dz_*} = \frac{d\mathcal{P}}{dz_*}. \tag{11.325}$$

Our boundary conditions are seen to be, first at the plate surface $z_* = 0$

$$v_r(z_* = 0) = 0, \quad F(0) = 0, \tag{11.326}$$
$$v_\theta(z_* = 0) = r\Omega, \quad G(0) = 1, \tag{11.327}$$
$$v_z(z_* = 0) = 0, \quad H(0) = 0, \tag{11.328}$$
$$p(z_* = 0) = 0, \quad \mathcal{P}(0) = 0. \tag{11.329}$$

In the far field, we need

$$v_r(z_* \to \infty) = 0, \quad F(\infty) = 0, \tag{11.330}$$
$$v_\theta(z_* \to \infty) = 0, \quad G(\infty) = 0. \tag{11.331}$$

Note that we do not require v_z to vanish at infinity, as it will need to be nonzero so that fluid can be entrained.

We now seek a numerical solution. This is best enabled by casting the equations as a system of first-order differential equations. We take $I = dF/dz_*$ and $J = dG/dz_*$. Our system becomes

$$\frac{dF}{dz_*} = I, \quad F(0) = 0, \tag{11.332}$$
$$\frac{dG}{dz_*} = J, \quad G(0) = 0, \tag{11.333}$$
$$\frac{dH}{dz_*} = -2F, \quad H(0) = 0, \tag{11.334}$$
$$\frac{dI}{dz_*} = F^2 + HI - G^2, \quad I(0) = ?, \tag{11.335}$$
$$\frac{dJ}{dz_*} = 2FG + HG, \quad J(0) = ?, \tag{11.336}$$
$$\frac{d\mathcal{P}}{dz_*} = 2FH - 2I, \quad \mathcal{P}(0) = 0. \tag{11.337}$$

We do not know the initial conditions for I and J. However, we can guess until the far field conditions on F and G are satisfied. An accurate estimate is given by Rogers and Lance (1960), who provide many additional details. With the use of their values of $I(0) = 0.510\,233$, $J(0) = -0.615\,922$, the solution for F, G, and $-H$ is shown in Fig. 11.30. Clearly the far field conditions on F and G appear to be satisfied here.

We plot the individual terms from the z_* momentum equation in Fig. 11.31. The advective acceleration, $H\,dH/dz_*$, is always positive. It is near zero in the far field, reaches a maximum near $z_* = 2$, then decreases to zero at the $z_* = 0$ surface. The z_* velocity in the far field is $H(\infty) \to -0.885$. Its increase to zero at $z_* = 0$ is a consequence of the always positive acceleration.

Consistent with Newton's theory, the acceleration is driven by forces: here those due to imbalanced pressure and shear forces. In the far field, the net viscous force dominates the net pressure force. Here the viscous effect is positive to match the positive advective acceleration. In the far field, the pressure effect is small and negative. Near $z_* = 1$, the situation reverses: the viscous effect becomes negative, while the pressure effect become positive. As $z_* = 0$ is approached, the pressure and viscous effects balance, leading to zero advective acceleration.

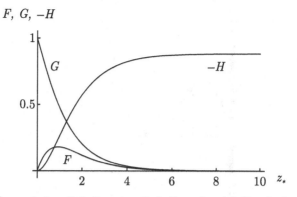

Figure 11.30 Solution for scaled radial F, azimuthal G, and axial H velocity components for von Kármán's viscous pump.

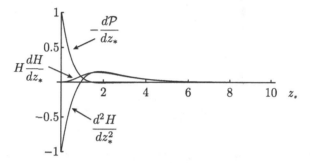

Figure 11.31 Advective z_* acceleration $H\,dH/dz_*$, viscous effect, d^2H/dz_*^2, and pressure gradient effect, $-d\mathcal{P}/dz_*$ as a function of z_* for von Kármán's viscous pump.

Lastly, we note the dimensional axial velocity achieved by the pump, a device whose purpose is to move fluid from one place to another. With our scaling:

$$v_z(z \to \infty) = \sqrt{\nu\Omega}\,H(\infty) = -0.885\sqrt{\nu\Omega}. \tag{11.338}$$

For example, if the fluid is water with $\nu \sim 10^{-6}\,\text{m}^2/\text{s}$, and the disk rotates at $\Omega = 100\,\text{rad/s}$ (roughly the speed of a lightly loaded automobile engine), then the velocity of the water in the far field is $\sqrt{(10^{-6}\,\text{m}^2/\text{s})(10^2\,\text{s}^{-1})}(-0.885) = -0.008\,85\,\text{m/s} = -8.885\,\text{mm/s}$.

11.2.6 Natural Convection

We next examine a fully coupled thermal-fluid problem, natural convection, in which warm air rises in a gravitational field due to density gradients. Let us consider a vertical flat plate, and follow much of the analysis and notation of Yih (1977), which is similar to that of Schlichting and Gersten (2017) and other sources. As shown in Fig. 11.32, we take the x axis to be vertical and the y axis to be horizontal. The stationary flat plate is held at temperature $T = T_1$. In the region $y > 0$, there exists a calorically perfect ideal gas with constant viscosity μ and constant thermal conductivity k. At $y = 0$, the fluid is stationary, so $u(x,0) = v(x,0) = 0$ and has $T(x,0) = T_1$. In the far field as $y \to \infty$, the fluid is at $T = T_o$, with $T_1 > T_o$, $\rho = \rho_o$ with zero

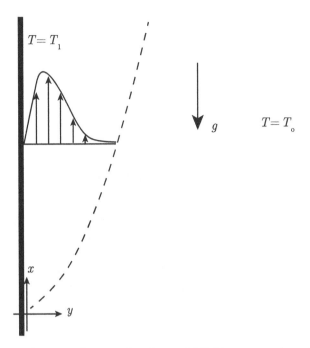

Figure 11.32 Schematic for natural convection for a cold fluid encountering a hot vertical plate.

velocity. The constant gravitational acceleration vector is $\mathbf{f} = -g\mathbf{i}$; that is, it points vertically downwards. We will consider flows for which the velocities are well below the speed of sound, rendering the pressure to be nearly a constant. Thus, the high temperature wall will induce the fluid to have low density near the wall. This low density fluid will rise as it is displaced by falling higher density fluid from the far field.

Consider the Boussinesq approximation for the steady Navier–Stokes equations, Eqs. (6.103–6.106) in the two-dimensional limit for a gas in a gravity field with $\mathbf{f} = -g\mathbf{i}$. Taking the typical assumptions for a boundary layer, we get the following dimensional equations:

$$\frac{\partial u}{\partial x} + \frac{\partial v}{\partial y} = 0, \tag{11.339}$$

$$u\frac{\partial u}{\partial x} + v\frac{\partial u}{\partial y} = -\frac{\rho}{\rho_o}g - \frac{1}{\rho_o}\frac{\partial p}{\partial x} + \nu\frac{\partial^2 u}{\partial y^2}, \tag{11.340}$$

$$0 = -\frac{1}{\rho_o}\frac{\partial p}{\partial y}, \tag{11.341}$$

$$u\frac{\partial T}{\partial x} + v\frac{\partial T}{\partial y} = \alpha\frac{\partial^2 T}{\partial y^2}, \tag{11.342}$$

$$\rho T = \rho_o T_o. \tag{11.343}$$

Now we wish to consider temperature differences $T - T_o$ that are small relative to the absolute far field temperature T_o. Recall that in that limit, series expansion gave us Eq. (6.94):

$$\frac{\rho}{\rho_o} = 1 - \frac{T - T_o}{T_o} + \cdots . \tag{11.344}$$

We use this to rewrite the x momentum equation, Eq. (11.340), as

$$u\frac{\partial u}{\partial x} + v\frac{\partial u}{\partial y} = -\left(1 - \frac{T - T_o}{T_o}\right)g - \frac{1}{\rho_o}\frac{dp}{dx} + \nu\frac{\partial^2 u}{\partial y^2}. \quad (11.345)$$

Here, because $\partial p/\partial y = 0$, we have employed dp/dx. Now, similar to the Falkner–Skan problem of Section 11.2.2, we can think of the pressure gradient as imposed from the far field. In the far field we have $u = v = 0$, $T = T_o$, so the x momentum equation reduces to

$$0 = -g - \frac{1}{\rho_o}\frac{dp}{dx}. \quad (11.346)$$

We impose this condition into the x momentum equation and get

$$u\frac{\partial u}{\partial x} + v\frac{\partial u}{\partial y} = \frac{T - T_o}{T_o}g + \nu\frac{\partial^2 u}{\partial y^2}. \quad (11.347)$$

Now let us define a convenient dimensionless temperature, θ:

$$\theta = \frac{T - T_o}{T_1 - T_o}. \quad (11.348)$$

With this definition, we have at the wall $\theta(y = 0) = 1$, and in the far field $\theta(y \to \infty) = 0$. Our equation system then can be rewritten as

$$\frac{\partial u}{\partial x} + \frac{\partial v}{\partial y} = 0, \quad (11.349)$$

$$u\frac{\partial u}{\partial x} + v\frac{\partial u}{\partial y} = \frac{T_1 - T_o}{T_o}g\theta + \nu\frac{\partial^2 u}{\partial y^2}, \quad (11.350)$$

$$u\frac{\partial \theta}{\partial x} + v\frac{\partial \theta}{\partial y} = \alpha\frac{\partial^2 \theta}{\partial y^2}. \quad (11.351)$$

These equations can be transformed to ordinary differential equations as follows. We define dimensional constant C as

$$C = \left(\frac{g(T_1 - T_o)}{4\nu^2 T_o}\right)^{1/4}. \quad (11.352)$$

Here C has SI units m$^{-3/4}$. We then take

$$\eta = C\frac{y}{x^{1/4}}, \quad \psi(x,y) = 4\nu C x^{3/4} f(\eta), \quad \theta(x,y) = \theta(\eta). \quad (11.353)$$

The employment of the stream function ψ guarantees satisfaction of the mass equation. The x momentum and energy equations can be shown to transform to

$$\frac{d^3 f}{d\eta^3} + 3f\frac{d^2 f}{d\eta^2} - 2\left(\frac{df}{d\eta}\right)^2 + \theta = 0, \quad (11.354)$$

$$\frac{d^2\theta}{d\eta^2} + 3(Pr)f\frac{d\theta}{d\eta} = 0, \quad (11.355)$$

$$f(0) = 0, \quad f'(0) = 0, \quad \theta(0) = 1, \quad f'(\infty) = 0, \quad \theta(\infty) = 0. \quad (11.356)$$

Here $Pr = \nu/\alpha = \mu c_p/k$ is the Prandtl number. These can be solved numerically. Scaled temperature, $\theta(\eta)$, and x velocity, $f'(\eta)$, are plotted Fig. 11.33 for $Pr = 1, 10$. For $Pr = 1$, $C = 1$, $\nu = 1/4$, Fig. 11.34 gives a plot of the stream function $\psi(x, y)$.

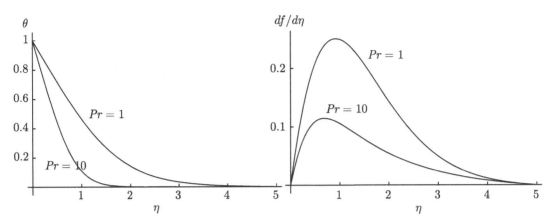

Figure 11.33 Scaled temperature $\theta(\eta)$ and x velocity $f'(\eta)$ for natural convection of a cold fluid encountering a hot vertical plate with $Pr = 1$, $Pr = 10$.

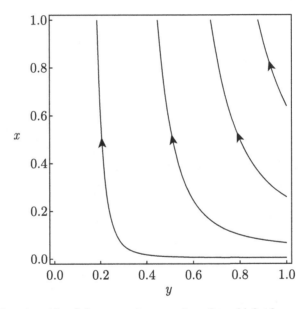

Figure 11.34 Stream function $\psi(x,y)$ for natural convection of a cold fluid encountering a hot vertical plate with $Pr = 1$, $C = 1$, $\nu = 1/4$.

The literature often defines C in terms of a so-called *Grashof[2] number*, a dimensionless number. The Grashof number based on the variable length x is defined as

$$Gr_x = \frac{g\beta(T_1 - T_o)x^3}{\nu^2}. \tag{11.357}$$

We recall β is the thermal expansion coefficient, Eq. (6.33). For an ideal gas $\beta = 1/T$. So for an ideal gas

$$Gr_x = \frac{g(T_1 - T_o)x^3}{T\nu^2}. \tag{11.358}$$

---

[2] Franz Grashof, 1826–1893, German engineer.

For small temperature changes, we can approximate the Grashof number using $\beta \sim 1/T_o$, giving

$$Gr_x \sim \frac{g(T_1 - T_o)x^3}{T_o \nu^2}. \tag{11.359}$$

With this, our similarity variable η can be expressed as

$$\eta = C \frac{y}{x^{1/4}} = \left(\frac{Gr_x}{4}\right)^{1/4} \frac{y}{x}. \tag{11.360}$$

11.2.7 Compressible Boundary Layer

Let us study a compressible boundary layer flow of a fluid over a flat plate. In contrast to the previous section, the fluid velocity here is allowed to be of the same order of magnitude as the speed of sound. Thus we will not be able to neglect density changes in any part of the flow, and we will have $\nabla^T \cdot \mathbf{v} \neq 0$. We will make several simplifying assumptions to render the problem more amenable to compact analysis. All of these could be relaxed, at the expense of considerably more complicated algebra. First, we will take the pressure to be constant. Let us also consider a calorically perfect ideal gas with μ and k having a simple linear temperature dependency. We will study the special case for which $Pr = \mu c_p/k = 1$. This is close to values seen in common gases. We consider steady, two-dimensional flow. We will be guided by the analysis of Yih (1977). Taking the typical boundary layer assumptions, our compressible mass, momentum, and energy equations reduce to

$$\frac{\partial}{\partial x}(\rho u) + \frac{\partial}{\partial y}(\rho v) = 0, \tag{11.361}$$

$$\rho\left(u\frac{\partial u}{\partial x} + v\frac{\partial u}{\partial y}\right) = \frac{\partial}{\partial y}\left(\mu \frac{\partial u}{\partial y}\right), \tag{11.362}$$

$$\mu c_p \left(u\frac{\partial T}{\partial x} + v\frac{\partial T}{\partial y}\right) = \frac{\partial}{\partial y}\left(k\frac{\partial T}{\partial y}\right) + \mu\left(\frac{\partial u}{\partial y}\right)^2, \tag{11.363}$$

$$\rho T = \rho_o T_o, \qquad \mu = \mu_o \left(\frac{T}{T_o}\right), \qquad k = k_o \left(\frac{T}{T_o}\right). \tag{11.364}$$

Adding the product of u and the momentum equation, Eq. (11.362), to the energy equation, Eq. (11.363), gives the form

$$\rho u \frac{\partial}{\partial x}\left(c_p T + \frac{u^2}{2}\right) + \rho v \frac{\partial}{\partial y}\left(c_p T + \frac{u^2}{2}\right) = \frac{\partial}{\partial y}\left(k\frac{\partial T}{\partial y} + u\mu\frac{\partial u}{\partial y}\right), \tag{11.365}$$

$$= \frac{\partial}{\partial y}\left(\mu\left(\frac{k}{\mu c_p}\frac{\partial}{\partial y}(c_p T) + \frac{\partial}{\partial y}\left(\frac{u^2}{2}\right)\right)\right). \tag{11.366}$$

Note that we have utilized the fact that c_p is constant for the calorically perfect ideal gas. Now because of the form of our simple temperature-dependent transport properties, we have $\mu c_p/k = \mu_o c_p/k_o = Pr$. And we are restricting attention to $Pr = 1$. This gives, after recognizing $d/dt = u\partial/\partial x + v\partial/\partial y$, the result

$$\rho \frac{d}{dt}\left(c_p T + \frac{u^2}{2}\right) = \frac{\partial}{\partial y}\left(\mu \frac{\partial}{\partial y}\left(c_p T + \frac{u^2}{2}\right)\right). \tag{11.367}$$

Now if the total enthalpy $c_p T + u^2/2$ is initially spatially uniform, and there is no flux of total enthalpy at the wall surface $y = 0$ or in the far field, then we are guaranteed that the total enthalpy will remain constant in the flow field:

$$c_p T + \frac{u^2}{2} = c_p T_o + \frac{U^2}{2}. \tag{11.368}$$

The isobaric assumption gives $\rho T = \rho_o T_o$, which then allows us to form a relation between ρ and u:

$$\frac{\rho_o}{\rho} + \frac{u^2}{2c_p T_o} = 1 + \frac{U^2}{2c_p T_o}. \tag{11.369}$$

Next, let us extend the notion of a stream function to compressible flows. Define the compressible stream function ψ such that

$$\rho u = \rho_o \frac{\partial \psi}{\partial y}, \qquad \rho v = -\rho_o \frac{\partial \psi}{\partial x}. \tag{11.370}$$

With this definition, the mass equation is guaranteed to be satisfied.

Next let us perform what is known as a *von Mises*[3] *transformation* so that instead of (x, y), we are in $(\hat{x}, \psi)$ space. This transformation could also be adapted for incompressible flows. The chain rule gives us

$$\left.\frac{\partial}{\partial x}\right|_y = \frac{\partial \hat{x}}{\partial x}\left.\frac{\partial}{\partial \hat{x}}\right|_\psi + \frac{\partial \psi}{\partial x}\left.\frac{\partial}{\partial \psi}\right|_{\hat{x}}, \qquad \left.\frac{\partial}{\partial y}\right|_x = \frac{\partial \hat{x}}{\partial y}\left.\frac{\partial}{\partial \hat{x}}\right|_\psi + \frac{\partial \psi}{\partial y}\left.\frac{\partial}{\partial \psi}\right|_{\hat{x}}. \tag{11.371}$$

Now we restrict our transformation to $\hat{x} = x$, giving

$$\left.\frac{\partial}{\partial x}\right|_y = \left.\frac{\partial}{\partial \hat{x}}\right|_\psi + \frac{\partial \psi}{\partial x}\left.\frac{\partial}{\partial \psi}\right|_{\hat{x}} = \left.\frac{\partial}{\partial \hat{x}}\right|_\psi - \frac{\rho v}{\rho_o}\left.\frac{\partial}{\partial \psi}\right|_{\hat{x}}, \tag{11.372}$$

$$\left.\frac{\partial}{\partial y}\right|_x = \frac{\partial \psi}{\partial y}\left.\frac{\partial}{\partial \psi}\right|_{\hat{x}} = \frac{\rho u}{\rho_o}\left.\frac{\partial}{\partial \psi}\right|_{\hat{x}}, \tag{11.373}$$

Then the x momentum equation, Eq. (11.362), becomes

$$\rho u\left(\frac{\partial u}{\partial \hat{x}} - \frac{\rho v}{\rho_o}\frac{\partial u}{\partial \psi}\right) + \rho v \frac{\rho u}{\rho_o}\frac{\partial u}{\partial \psi} = \frac{\rho u}{\rho_o}\frac{\partial}{\partial \psi}\left(\mu \frac{\rho u}{\rho_o}\frac{\partial u}{\partial \psi}\right), \tag{11.374}$$

$$\rho_o^2 \frac{\partial u}{\partial \hat{x}} = \frac{\partial}{\partial \psi}\left(\mu \rho u \frac{\partial u}{\partial \psi}\right). \tag{11.375}$$

Because of the special nature of our viscosity model, $\mu = \mu_o T/T_o$, and our isobaric ideal gas with $\rho = \rho_o T_o/T$, we must have $\rho \mu = \rho_o \mu_o$, yielding, after taking $\nu_o = \mu_o/\rho_o$,

$$\frac{\partial u}{\partial \hat{x}} = \nu_o \frac{\partial}{\partial \psi}\left(u \frac{\partial u}{\partial \psi}\right). \tag{11.376}$$

---

[3] Richard Elder von Mises, 1883–1953, Austrian mathematician and mechanician.

11.2 Flow with Effects of Inertia

Let us now transform independent variables from $(\hat{x}, \psi)$ to $(\tilde{x}, \eta)$:

$$\eta = \frac{\psi}{\sqrt{\nu_o U \hat{x}}}, \qquad \tilde{x} = \hat{x}. \tag{11.377}$$

This gives from the chain rule

$$\frac{\partial}{\partial \hat{x}} = \frac{\partial \eta}{\partial \hat{x}} \frac{\partial}{\partial \eta} + \frac{\partial \tilde{x}}{\partial \hat{x}} \frac{\partial}{\partial \tilde{x}} = -\frac{\eta}{2\tilde{x}} \frac{\partial}{\partial \eta} + \frac{\partial}{\partial \tilde{x}}, \tag{11.378}$$

$$\frac{\partial}{\partial \psi} = \frac{\partial \eta}{\partial \psi} \frac{\partial}{\partial \eta} + \frac{\partial \tilde{x}}{\partial \psi} \frac{\partial}{\partial \tilde{x}} = \frac{1}{\sqrt{\nu_o U \tilde{x}}} \frac{\partial}{\partial \eta}. \tag{11.379}$$

The $\hat{x}$ momentum equation becomes

$$-\frac{\eta}{2\tilde{x}} \frac{\partial u}{\partial \eta} + \frac{\partial u}{\partial \tilde{x}} = \frac{\nu_o}{\sqrt{\nu_o U \tilde{x}}} \frac{\partial}{\partial \eta} \left(u \frac{1}{\sqrt{\nu_o U \tilde{x}}} \frac{\partial u}{\partial \eta} \right) = \frac{1}{U \tilde{x}} \frac{\partial}{\partial \eta} \left(u \frac{\partial u}{\partial \eta} \right). \tag{11.380}$$

Now, let us assume that $u(\eta, \tilde{x}) = U f(\eta)$, yielding the nonlinear ordinary differential equation

$$-\frac{\eta}{2} \frac{df}{d\eta} = \frac{d}{d\eta} \left(f \frac{df}{d\eta} \right). \tag{11.381}$$

We have as boundary conditions $f(0) = 0$ and $f(\infty) = 1$. We could address this directly, but let us continue to transform. We define a new independent variable ξ and new dependent variable $F(\xi)$ such that

$$\frac{d\eta}{d\xi} = f(\eta), \qquad \eta = F(\xi). \tag{11.382}$$

Thus, we can also say

$$\frac{d\eta}{d\xi} = \frac{dF}{d\xi} = f. \tag{11.383}$$

And the chain rule tells us

$$\frac{d}{d\eta} = \frac{d\xi}{d\eta} \frac{d}{d\xi} = \frac{1}{\frac{d\eta}{d\xi}} \frac{d}{d\xi} = \frac{1}{\frac{dF}{d\xi}} \frac{d}{d\xi}. \tag{11.384}$$

Make these substitutions into the x momentum equation, Eq. (11.381), to recast it as

$$-\frac{F}{2} \frac{1}{\frac{dF}{d\xi}} \frac{d}{d\xi} \left(\frac{dF}{d\xi} \right) = \frac{1}{\frac{dF}{d\xi}} \frac{d}{d\xi} \left(\frac{dF}{d\xi} \frac{1}{\frac{dF}{d\xi}} \frac{d}{d\xi} \left(\frac{dF}{d\xi} \right) \right), \tag{11.385}$$

$$-\frac{F}{2} \frac{d^2 F}{d\xi^2} = \frac{d^3 F}{d\xi^3}, \tag{11.386}$$

$$\frac{d^3 F}{d\xi^3} + \frac{1}{2} F \frac{d^2 F}{d\xi^2} = 0. \tag{11.387}$$

Remarkably, this has exactly the same form as the Blasius equation for incompressible flow, Eq. (11.200)! One can further show that the boundary conditions are the same as those for the incompressible Blasius equation. Thus we have the solution for $F(\xi)$ via the same numerical integration as done previously. A long sequence of inverse transformations, not done here, is required to return to physical variables, for example $u(x, y)$.

SUMMARY

This chapter focused on multi-dimensional, viscous flow of a Newtonian fluid. First, consideration was given to flows with negligible inertial effects. Problems were studied in planar, cylindrical, and spherical geometries, including the important problem of viscous flow over a sphere. Consideration of inertial effects led to the important boundary layer theory. The boundary layer equations allow consideration of flows over flat plates, jets, shear layers, and other important physical problems, including one that involved a full coupling of mass, momenta, energy, and thermodynamics: natural convection induced by a hot vertical plate in contact with a cold fluid. One predicts for such flows that warm air rises.

PROBLEMS

11.1 As shown by Kundu et al. (2015, Chapter 13), *Ekman*[4] *transport* is described by simplified models of geophysical fluid dynamics in the limit in which the linear momenta principle reduces to a balance between Coriolis acceleration and viscous stress gradients. Consider then a plane rotating about a central axis normal to the plane at constant angular velocity Ω. The balance between fluid inertia effects and viscous forces is described by the following reduced linear momenta equation for an incompressible Newtonian fluid with constant kinematic viscosity:

$$2\Omega \times \mathbf{v} = \nu \nabla^2 \mathbf{v},$$

where $\Omega = (0, 0, \Omega_z)^T$ is the constant angular velocity. Consider the x and y momentum equations, and assume there is only variation in the z direction. Assume at $z = 0$ that known stress in the x direction $\tau = \rho\nu du/dz|_{z=0}$ is applied with no stress in the y direction so $0 = dv/dz|_{z=0}$. Let $u, v \to 0$ as $z \to -\infty$. Find $u(z)$ and $v(z)$ and interpret.

11.2 Consider Stokes flow of a Newtonian fluid with constant viscosity with a dimensionless stream function given by

$$\psi(x, y) = (x + y)y.$$

Neglect body forces. Show the flow is incompressible. Find expressions for the velocity, vorticity, pressure, and viscous stress fields, taking $\mu = 1$. Find all stagnation points. Plot the streamlines and velocity vector fields in the domain $x \in [-2, 2]$, $y \in [-2, 2]$. While inertial effects are neglected in the analysis of forces, the flow is accelerating. Find all points in the flow field where the acceleration is zero. Plot the acceleration vector field superposed onto the streamlines in the domain $x \in [-2, 2]$, $y \in [-2, 2]$. Determine if the condition for a generalized Beltrami field, $\nabla \times (\boldsymbol{\omega} \times \mathbf{v}) = \mathbf{0}$, has been met. Give a brief physical description of the flow.

11.3 Repeat the previous problem for

$$\psi(x, y) = y + y^2 + \tan^{-1}\frac{y}{x}.$$

---

[4] Vagn Walfrid Ekman, 1874–1954, Swedish oceanographer.

11.4 Repeat the previous problem for

$$\psi(x, y) = y^2 + e^{-2y} \cos 2x.$$

11.5 Repeat the previous problem for

$$\psi(x, y) = 2xy + x^3 - y^3.$$

11.6 Consider flow of a Newtonian fluid with constant viscosity with a dimensionless stream function given by

$$\psi(x, y) = (x + y)y.$$

Do not neglect inertial effects. Neglect all body forces. Show the flow is incompressible. Find expressions for the velocity, vorticity, pressure, and viscous stress fields, taking $\mu = 1$, $\rho = 1$. Compare to the case for which inertia is negligible.

11.7 Consider flow of a Newtonian fluid with constant viscosity with a dimensionless stream function given by

$$\psi(x, y) = 2xy + x^3 - y^3.$$

Do not neglect inertial effects. Do not neglect body forces, which may be nonconservative. Show the flow is incompressible. Find expressions for the velocity, vorticity, pressure, body force, and viscous stress fields, taking $\mu = 1$, $\rho = 1$. Compare to the case for which inertial effects are negligible.

11.8 Consider the Taylor scraper problem for $\alpha = \pi/6$ and a Stokes flow model with constant viscosity. Find $\psi(x, y)$, $u(x, y)$, $v(x, y)$, $\omega_z(x, y)$, $p(x, y)$, $I_\epsilon^{(3)}(x, y)$, and form relevant plots. Assume the equations have been scaled, or equivalently that $\mu = 1$.

11.9 For the problem of Stokes flow over a stationary sphere with the parameters employed in the text, consider the $x = 0$ plane and present a formula for $\omega_\theta(y, z)$. Then plot contours of ω_θ in the $x = 0$ plane.

11.10 Consider steady fully developed flow of an incompressible fluid with constant viscosity in a square channel of cross-sectional area L^2. There are no imposed pressure gradients or body forces. Three of the walls are stationary, and the fourth has a constant velocity of W. This is a multi-dimensional analog to Couette flow. Scale the problem appropriately. Find and plot the scaled velocity profile $w(x, y)$.

11.11 Consider the Reynolds equation with no variation in z or t in the special case in which $h(x) = h_o(1 - (1/2)(x/L)^2)$ with $p(0) = p(L) = p_o$ and $L = 1$, $p_o = 1$, $U = 1$, $\mu = 1$, $h_o = 1$. Find an analytic expression and plot for $p(x)$. Find the maximum value of $p(x)$ and the location x where it is realized.

11.12 Consider the incompressible, unsteady Navier–Stokes equations for a Newtonian fluid with constant viscosity and no body force. Confine attention to two-dimensional systems in Cartesian coordinates. Introduce the stream function such that $u = \partial \psi / \partial y$, $v =$

$-\partial\psi/\partial x$, and show the Navier–Stokes equations reduce to a single equation

$$\frac{\partial}{\partial t}\left(\frac{\partial^2\psi}{\partial x^2}+\frac{\partial^2\psi}{\partial y^2}\right)+\frac{\partial\psi}{\partial y}\frac{\partial}{\partial x}\left(\frac{\partial^2\psi}{\partial x^2}+\frac{\partial^2\psi}{\partial y^2}\right)-\frac{\partial\psi}{\partial x}\frac{\partial}{\partial y}\left(\frac{\partial^2\psi}{\partial x^2}+\frac{\partial^2\psi}{\partial y^2}\right)$$
$$=\nu\left(\frac{\partial^4\psi}{\partial x^4}+2\frac{\partial^4\psi}{\partial x^2\partial y^2}+\frac{\partial^4\psi}{\partial y^4}\right),$$

that can be interpreted as

$$\frac{d}{dt}\nabla^2\psi=\nu\nabla^4\psi.$$

Show further that this can be rewritten as

$$\frac{\partial}{\partial t}\nabla^2\psi+\frac{\partial(\nabla^2\psi,\psi)}{\partial(x,y)}=\nu\nabla^4\psi,$$

where one has a type of Jacobian determinant for advection-based terms:

$$\frac{\partial(\nabla^2\psi,\psi)}{\partial(x,y)}=\begin{vmatrix}\frac{\partial}{\partial x}\nabla^2\psi & \frac{\partial}{\partial y}\nabla^2\psi \\ \frac{\partial}{\partial x}\psi & \frac{\partial}{\partial y}\psi\end{vmatrix}.$$

See Bird et al. (2007, p. 123) for the nontrivial extension to two-dimensional non-Cartesian coordinate systems, while noting the alternate sign convention for ψ employed therein.

11.13 Show the linear momenta equation of the previous problem, $d/dt\nabla^2\psi=\nu\nabla^4\psi$, is invariant under the transformation to a coordinate system translating in a known time-dependent fashion

$$\tilde{x}=x+a(t),\quad \tilde{y}=y+b(t),\quad \tilde{t}=t,\quad \tilde{\psi}=\psi-x\frac{db}{dt}+y\frac{da}{dt}.$$

Here $a(t)$ and $b(t)$ are known functions. If $\psi(x,y,t)=(x+y)y+xt$ and $\rho=\mu=\nu=1$, find a pressure field $p(x,y,t)$ that guarantees satisfaction of the governing equations. See Cantwell (2002) for more general discussion of symmetry methods for viscous flows.

11.14 Consider a thermal boundary layer for a flow whose velocity field is described by a Blasius boundary layer. The flat plate and far field are held at a constant temperature of $T_o=300$ K. The fluid is liquid water moving at $U=100$ m/s. Thermal properties of liquid water are known to be $Pr=5.83$, $c_p=4179$ J/kg/K. Find the temperature field via numerical solution of the thermal boundary layer equation. Plot the dimensional temperature profile as a function of the similarity variable: $\overline{T}(\eta)$.

11.15 Consider a Falkner–Skan problem for $m=1/3$ and $m=-9/100$. Assume all variables to be dimensionless. Solve numerically. Plot $u(x=1,\tilde{y})$ and streamlines superposed onto the velocity vector field $(u,\tilde{v})$ for both values of m. Align the axes of the streamline and vector field plots with the proper angle of inclination to the horizontal.

11.16 Consider a natural convection problem as described in the text for the Boussinesq approximation. Take parameter values appropriate for air, $T_o=300$ K, $T_1=400$ K, $g=9.81$ m/s^2, $\nu=15.89\times10^{-6}$ m^2/s, $Pr=0.707$. Give a contour plot of $T(x,y)$ with labeled isotherms for a carefully chosen (x,y) domain that illustrates the spatial relaxation of temperature from its value at the wall to that in the far field.

FURTHER READING

Brodkey, R. S. (1967). *The Phenomena of Fluid Motions*. New York: Dover.

Currie, I. G. (2013). *Fundamental Mechanics of Fluids*, 4th ed. Boca Raton, Florida: CRC Press.

Joseph, D., Funada, T., and Wang, J. (2008). *Potential Flows of Viscous and Viscoelastic Fluids*. New York: Cambridge University Press.

Langlois, W. E., and Deville, M. O. (2014). *Slow Viscous Flow*, 2nd ed. Heidelberg: Springer.

Leal, L. G. (2007). *Advanced Transport Phenomena: Fluid Mechanics and Convective Transport Processes*. New York: Cambridge University Press.

Ockendon, H., and Ockendon, J. R. (1995). *Viscous Flow*. Cambridge, UK: Cambridge University Press.

Rogers, M. H., and Lance, G. N. (1960). The rotationally symmetric flow of a viscous fluid in the presence of an infinite rotating disk, *Journal of Fluid Mechanics*, **7**(4), 617–631.

Sherman, F. S. (1990). *Viscous Flow*. New York: McGraw-Hill.

Truesdell, C. A. (1991). *A First Course in Rational Continuum Mechanics*, Vol. 1, 2nd ed. Boston, Massachusetts: Academic Press.

Wang, C. Y. (1991). Exact solutions of the steady-state Navier–Stokes equations, *Annual Review of Fluid Mechanics*, **23**, 159–177.

White, F. M. (2006). *Viscous Fluid Flow*, 3rd ed. New York: McGraw-Hill.

REFERENCES

Bird, R. B., Stewart, W. E., and Lightfoot, E. N. (2007). *Transport Phenomena*, revised 2nd ed. New York: John Wiley.

Cantwell, B. J. (2002). *Introduction to Symmetry Analysis*. New York: Cambridge University Press.

Kundu, P. K., Cohen, I. M., and Dowling, D. R. (2015). *Fluid Mechanics*, 6th ed. Amsterdam: Academic Press.

Panton, R. L. (2013). *Incompressible Flow*, 4th ed. New York: John Wiley.

Proudman, I., and Pearson, J. R. A. (1957). Expansions at small Reynolds numbers for the flow past a sphere and a circular cylinder, *Journal of Fluid Mechanics*, **2**(3), 237–262.

Schlichting, H., and Gersten, K. (2017). *Boundary Layer Theory*, 9th ed. New York: McGraw-Hill.

Stokes, G. G. (2009). *Mathematical and Physical Papers*, 5 vols. Cambridge, UK: Cambridge University Press.

Taylor, G. I. (1962). On scraping viscous fluid from a plane surface. In: G. K. Batchelor (ed.), *The Scientific Papers of Sir Geoffrey Ingram Taylor*. Cambridge, UK: Cambridge University Press (1971), **4**, 410–413.

Van Dyke, M. (1982). *An Album of Fluid Motion*. Stanford, California: Parabolic Press.

von Kármán, T. (1921). Über laminare und turbulente Reibung, *Zeitschrift für Angewandte Mathematik und Mechanik*, **1**(4), 233–252.

Yih, C.-S. (1977). *Fluid Mechanics*. East Hampton, Connecticut: West River Press.

12 Linearly Unstable Flow

Here we study flows that possess steady solutions that may not persist in time if subjected to small perturbations. Often the behavior of a fluid with no time-dependency is dramatically different than one with time-dependency. Understanding what type of perturbation induces persistent time-dependency is essential for scientific and practical understanding of fluid behavior. An example that we will consider here as well as in later chapters is that of warm air rising or not rising; see Fig. 12.1. A fluid confined between a hot surface at the bottom and a cold surface at the top may, for small temperature differences, remain stationary and transfer thermal energy via ordinary conduction from bottom to top. In such a case, the tendency for the warm air to rise is suppressed by viscous forces. As the temperature difference increases, the fluid at the bottom acquires a lower density, and buoyancy forces may overcome viscous forces. This induces the fluid to accelerate from bottom to the top. This is the onset of a convective instability, which enhances heat transfer from bottom to top, as the moving fluid carries thermal energy.

In this chapter, we will restrict attention to problems in which the perturbations are sufficiently small that linear equations describe the dynamics. This is a broad and important subject, and we will illustrate it with only a few problems. We will study the dynamics of systems that have been linearized about their equilibrium states. Systems that are *stable* see an initial perturbation decay or not grow with time. Systems that are *asymptotically stable* see a perturbation decay with time. Systems that are *unstable* see a perturbation grow with time. If the linear system exhibits instability, it will likely grow into a state for which the assumptions of the linearization are violated. In such cases, one must reintroduce the neglected nonlinearities to capture the full system dynamics. We will not consider such nonlinear dynamics in any detail in this chapter, but will do so in Chapters 13 and 14.

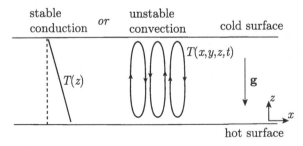

Figure 12.1 Stable conduction or unstable convection for a viscous fluid confined between a hot and cold surfaces in the presence of a gravity field.

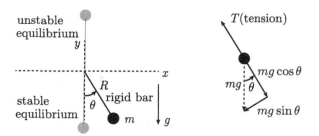

Figure 12.2 Configuration for simple pendulum problem.

12.1 Motivating Problem from Dynamics

Let us illustrate our approach to stability analysis by considering a simple example from dynamics: the simple pendulum problem depicted in Fig. 12.2. Here a mass m is at the end of a rigid bar that is hinged at the origin. The mass moves under the influence of two forces: a tension force T from the rigid bar and the gravity force mg. The gravitational constant is g, and the gravity force acts in the negative y direction.

Our approach for this and all problems in this chapter will be: (1) formulate the nonlinear governing equations; (2) identify all time-independent solutions to the governing equations; these are the equilibrium states of the system; (3) linearize the governing time-dependent equations about each of the equilibria; (4) subject the linearized system to an infinitesimally small perturbation; and (5) solve the linear system, and ascertain the conditions under which the perturbation grows or decays.

Let us first prepare the governing equations for this nonlinear pendulum problem. We require Newton's second law of motion to hold

$$m\frac{d^2 \mathbf{x}}{dt^2} = \sum \mathbf{F}, \tag{12.1}$$

where $\mathbf{x}$ is the position, and $\mathbf{F}$ is any force acting on m. There is an advantage to working in polar coordinates, in which case, we must take care to include the proper centripetal and Coriolis acceleration terms. The free body diagram in Fig. 12.2 aids in this. For the r momentum we have

$$m\left(\underbrace{\frac{d^2 R}{dt^2}}_{0} - \underbrace{R\left(\frac{d\theta}{dt}\right)^2}_{\text{centripetal}}\right) = -T + mg\cos\theta. \tag{12.2}$$

Note this completely consistent with Eq. (2.317) when we recognize $v_r = dR/dt$, $v_\theta = R\,d\theta/dt$. Because the bar is rigid, $d^2R/dt^2 = 0$, so

$$-mR\left(\frac{d\theta}{dt}\right)^2 = -T + mg\cos\theta. \tag{12.3}$$

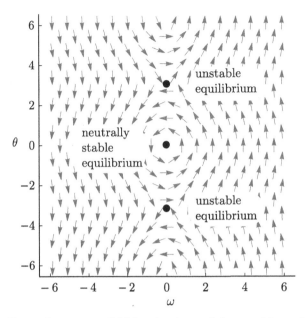

Figure 12.3 Phase plane vector field for simple pendulum problem with $g/R = 1$.

The θ momentum equation gives

$$m\left(R\frac{d^2\theta}{dt^2} + 2\underbrace{\overbrace{\frac{dR}{dt}}^{0}\frac{d\theta}{dt}}_{\text{Coriolis}}\right) = -mg\sin\theta. \qquad (12.4)$$

Again this is consistent with Eq. (2.317). Because the bar is rigid, there is no Coriolis acceleration, and θ momentum reduces to

$$\frac{d^2\theta}{dt^2} + \frac{g}{R}\sin\theta = 0. \qquad (12.5)$$

This is a *nonlinear* equation that completely determines the dynamics $\theta(t)$. With knowledge of $\theta(t)$, one can use the r momentum equation, Eq. (12.3), to determine the time-dependent tension force T. Let us focus on the θ momentum equation. We can write it as a system of two nonlinear first-order differential equations by defining the angular velocity $\omega = d\theta/dt$, yielding the nonlinear system

$$\frac{d\omega}{dt} = -\frac{g}{R}\sin\theta, \qquad \frac{d\theta}{dt} = \omega. \qquad (12.6)$$

The vector field defined in the phase plane for $g/R = 1$ is plotted in Fig. 12.3. The equilibrium states may now be ascertained. For equilibrium, we require the system not to change in time; therefore, we require

$$0 = -\frac{g}{R}\sin\theta, \qquad 0 = \omega. \qquad (12.7)$$

This is an algebra problem for the equilibrium state. For partial differential equations of fluid mechanics, we will be led to steady differential equations to solve for the equilibrium state. Our system here has multiple equilibria:

$$\theta = n\pi, \quad (n = 0, 1, 2, \ldots), \qquad \omega = 0. \tag{12.8}$$

Only two of these equilibria are distinct, $(\theta, \omega) = (0, 0)$ and $(\theta, \omega) = (\pi, 0)$. Physically, the pendulum is at equilibrium at either "top dead center" or "bottom dead center."

Let us now linearize the full nonlinear equation about each of the equilibria and examine system dynamics. First consider behavior near $\theta = 0$. Taylor series shows

$$\sin \theta = \theta - \frac{\theta^3}{3!} + \frac{\theta^5}{5!} - \cdots . \tag{12.9}$$

For small θ, the dominant term is $\sin \theta \sim \theta$, which is linear. We thus can approximate Eq. (12.5) by

$$\frac{d^2\theta}{dt^2} + \frac{g}{R}\theta = 0. \tag{12.10}$$

This has solution

$$\theta(t) = A \sin \sqrt{\frac{g}{R}} t + B \cos \sqrt{\frac{g}{R}} t, \tag{12.11}$$

where A and B are from the initial conditions. Because both sin and cos have amplitude of unity, the solution to the linear problem is bounded for all time. Thus, it is stable. It is not asymptotically stable, as it does not decay to zero. If we provided initial conditions, say $\theta(0) = \theta_o$, $d\theta/dt(0) = \omega(0) = 0$, we get

$$\theta(t) = \theta_o \cos \sqrt{\frac{g}{R}} t, \qquad \omega(t) = \frac{d\theta}{dt} = -\sqrt{\frac{g}{R}} \theta_o \sin \sqrt{\frac{g}{R}} t. \tag{12.12}$$

We insist that $\theta_o \ll 1$ so as to prevent the neglected nonlinear terms from having influence.

We can also cast this in terms of eigenvalue analysis, which will be the method of choice for fluid stability problems. Our first-order nonlinear system, Eqs. (12.6), after linearization, takes the form

$$\frac{d}{dt}\begin{pmatrix} \omega \\ \theta \end{pmatrix} = \begin{pmatrix} 0 & -\frac{g}{R} \\ 1 & 0 \end{pmatrix} \begin{pmatrix} \omega \\ \theta \end{pmatrix}. \tag{12.13}$$

Assuming $\omega = C_1 e^{\lambda t}$, $\theta = C_2 e^{\lambda t}$ gives

$$\begin{pmatrix} C_1 \lambda e^{\lambda t} \\ C_2 \lambda e^{\lambda t} \end{pmatrix} = \begin{pmatrix} 0 & -\frac{g}{R} \\ 1 & 0 \end{pmatrix} \begin{pmatrix} C_1 e^{\lambda t} \\ C_2 e^{\lambda t} \end{pmatrix}, \quad \text{so} \quad \lambda \begin{pmatrix} C_1 \\ C_2 \end{pmatrix} = \begin{pmatrix} 0 & -\frac{g}{R} \\ 1 & 0 \end{pmatrix} \begin{pmatrix} C_1 \\ C_2 \end{pmatrix}. \tag{12.14}$$

This is clearly an eigenvalue problem. We rearrange and get

$$\begin{pmatrix} -\lambda & -\frac{g}{R} \\ 1 & -\lambda \end{pmatrix} \begin{pmatrix} C_1 \\ C_2 \end{pmatrix} = \begin{pmatrix} 0 \\ 0 \end{pmatrix}. \tag{12.15}$$

Nontrivial solutions for $(C_1, C_2)^T$ require that the determinant of the coefficient matrix be zero:

$$\begin{vmatrix} -\lambda & -\frac{g}{R} \\ 1 & -\lambda \end{vmatrix} = 0, \tag{12.16}$$

giving

$$\lambda^2 + \frac{g}{R} = 0, \quad \text{so} \quad \lambda = \pm i\sqrt{\frac{g}{R}}. \tag{12.17}$$

Because $|e^{i\alpha t}| = 1$ for all real α, our solutions – which are of the form $e^{\lambda t}$ – are guaranteed not to grow in time. Had λ had a nonzero real part, growth (associated with a positive real part) or decay (associated with a negative real part) would have been seen. Each eigenvalue induces a nonunique eigenvector $(C_1, C_2)^T$. For $\lambda = \pm i\sqrt{g/R}$, we get $(C_1, C_2)^T = s(\pm i\sqrt{g/R}, 1)^T$, with s an arbitrary constant. However, the eigenvectors play no role in determining the stability.

Next, let us consider the behavior of the system near the equilibrium at $\theta = \pi$. Let us define a new variable $\hat{\theta} = \theta - \pi$. Thus, when θ is near π, $\hat{\theta}$ will be small. Our nonlinear θ momentum equation, Eq. (12.5), transforms, after recognizing $d^2\theta/dt^2 = d^2\hat{\theta}/dt^2$ and $\sin(\hat{\theta} + \pi) = -\sin\hat{\theta}$, to

$$\frac{d^2\hat{\theta}}{dt^2} - \frac{g}{R}\sin\hat{\theta} = 0. \tag{12.18}$$

Now we wish to consider small $\hat{\theta}$. As before, Taylor series gives $\sin\hat{\theta} \sim \hat{\theta} - \hat{\theta}^3/3! + \cdots$. We retain only the linear term and get

$$\frac{d^2\hat{\theta}}{dt^2} - \frac{g}{R}\hat{\theta} = 0. \tag{12.19}$$

This has solution

$$\hat{\theta} = A\exp\left(\sqrt{\frac{g}{R}}t\right) + B\exp\left(-\sqrt{\frac{g}{R}}t\right). \tag{12.20}$$

For $A \neq 0$, but otherwise arbitrary initial condition, $\hat{\theta}$ is guaranteed to grow exponentially in time, no matter how small the nonzero perturbation is. The "top dead center" equilibrium is unstable. We could have also performed an eigenvalue analysis on the linear system and would have obtained $\lambda = \pm\sqrt{g/R}$. The eigenvalues are real, with one stable mode and one unstable mode.

12.2 Planar Two-Dimensional Inviscid Instabilities

Now let us take up a stability problem in fluid mechanics. Consider the scenario shown in Fig. 12.4. Here we have two distinct fluids, labeled "1" and "2," in two distinct regions. Region 1 is mainly $y > 0$ and region 2 is mainly $y < 0$. We neglect all variation in z; so this is a two-dimensional flow. The two regions are segregated by an interface described by $y = \eta(x,t)$. We will consider η to be initially small, and ask whether it grows or decays.

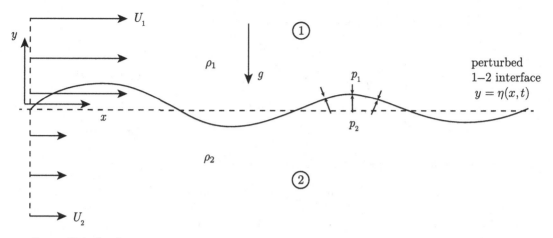

Figure 12.4 Configuration for simple inviscid shear flow associated with varieties of instability.

Importantly, both fluids are taken to be inviscid, so there is no momentum diffusion. Had we allowed momentum diffusion, we would have been obliged to also include mass diffusion, as they are both consequences of the same molecular collision processes. And one could certainly expect the mass of fluid 1 to diffuse into the region where fluid 2 exists and vice versa. As we neglect all molecular diffusion processes, we can expect our theory to maintain a well-defined interface segregating fluids 1 and 2. Indeed the interface could become contorted with steep spatial gradients, but as long as we neglect diffusion, such a condition is admitted. In reality, when the gradients become too steep, diffusion becomes important, and molecular mixing dominates at the small scales.

Initially, the fluids in regions 1 and 2 have uniform velocities $\mathbf{v}_1 = U_1\mathbf{i}$, $\mathbf{v}_2 = U_2\mathbf{i}$. Both fluids are incompressible, with respective densities ρ_1 and ρ_2. We allow for a nonzero body force dictated by the gravitational acceleration g, which points in the negative y direction. We allow for nonzero surface tension at the 1,2 interface. A consequence of this is that there will be a pressure jump at the interface, and a force balance will reveal that surface tension supplies the force to counter the pressure imbalance.

Now the Helmholtz vorticity transport equation for a two-dimensional incompressible, inviscid fluid, Eq. (7.83), tells us that a flow that is initially irrotational is always irrotational. In both regions 1 and 2, the flows are irrotational at $x = 0$; thus, as a fluid particle propagates in x, it will do so without rotation. Thus, we can associate the velocity vector with a velocity potential $\mathbf{v} = \nabla \phi$, and because of incompressibility, $\nabla^T \cdot \mathbf{v} = 0$, we get $\nabla^T \cdot \nabla \phi = 0$. This amounts to capturing the kinematics of our system. We can account for the dynamics of the system with the linear momenta equation in the form of the appropriate version of the unsteady Bernoulli relation, Eq. (6.186):

$$\frac{\partial \phi}{\partial t} + \frac{1}{2}\nabla^T \phi \cdot \nabla \phi + \frac{p}{\rho} + gy = f(t). \qquad (12.21)$$

Here we have used the appropriate $\mathsf{P} = p/\rho$ for an incompressible fluid and gravitational potential $\varphi = gy$ for the configuration of Fig. 12.4. Now because we have two distinct fluids and regions, we need two sets of equations. We thus write

$$\nabla^2 \phi_1 = 0, \qquad \text{kinematics, region 1,} \qquad (12.22)$$

$$\frac{\partial \phi_1}{\partial t} + \frac{1}{2}\nabla^T \phi_1 \cdot \nabla \phi_1 + \frac{p_1}{\rho_1} + gy = f_1(t), \qquad \text{dynamics, region 1,} \qquad (12.23)$$

$$\nabla^2 \phi_2 = 0, \qquad \text{kinematics, region 2,} \qquad (12.24)$$

$$\frac{\partial \phi_2}{\partial t} + \frac{1}{2}\nabla^T \phi_2 \cdot \nabla \phi_2 + \frac{p_2}{\rho_2} + gy = f_2(t), \qquad \text{dynamics, region 2.} \qquad (12.25)$$

In the far field we expect uniform flow, thus

$$\lim_{y \to \infty} \phi_1 = U_1 x, \qquad \lim_{y \to -\infty} \phi_2 = U_2 x. \qquad (12.26)$$

Next, let us consider conditions at the interface. We have, for convenience, described the interface by the curve $y = \eta(x,t)$. We do not yet know η. Let us, again for convenience, define a type of *level set function* $F(x, y, t)$ such that

$$F(x, y, t) = y - \eta(x, t). \qquad (12.27)$$

While definitions of this type are common in classical fluid mechanics, the notion of a level set function has broader utility in computational mechanics; see Sethian (1996).

At the interface, where we have $y = \eta(x,t)$, we must have $F = 0$. Focusing only on the interface, because $F = 0$, so must be all its derivatives, including its material derivative:

$$\frac{dF}{dt} = \frac{\partial F}{\partial t} + u\frac{\partial F}{\partial x} + v\frac{\partial F}{\partial y} = 0, \qquad (12.28)$$

$$\frac{\partial}{\partial t}(y - \eta(x,t)) + \frac{\partial \phi}{\partial x}\frac{\partial}{\partial x}(y - \eta(x,t)) + \frac{\partial \phi}{\partial y}\frac{\partial}{\partial y}(y - \eta(x,t)) = 0, \qquad (12.29)$$

$$-\frac{\partial \eta}{\partial t} + \frac{\partial \phi}{\partial x}\left(-\frac{\partial \eta}{\partial x}\right) + \frac{\partial \phi}{\partial y} = 0, \qquad (12.30)$$

$$\frac{\partial \phi}{\partial y} = \frac{\partial \eta}{\partial t} + \frac{\partial \phi}{\partial x}\frac{\partial \eta}{\partial x}. \qquad (12.31)$$

We call this the *kinematic interface condition*.

We also must consider dynamics at the interface. To do so requires a discussion of the topic of *surface tension*. To aid in understanding this topic, one may consider a balloon filled with high pressure gas, residing at rest in a low pressure atmosphere. The balloon membrane may be an elastic surface under tension that segregates the high pressure interior from the low pressure exterior. A diagram of this configuration is shown in Fig. 12.5. While a balloon is a three-dimensional object, the analysis is similar for two dimensions. We may consider then a cylindrical balloon, similar to an air mattress, in which the end effects are negligible. Experiments show that the pressure difference is inversely proportional to the radius of the balloon:

$$p_1 - p_2 = \frac{\sigma}{R}. \qquad (12.32)$$

Here $\sigma > 0$ is a material constant known at the *surface tension coefficient*, and R is the local radius of curvature of the surface. Cylinders of smaller radius support larger pressure

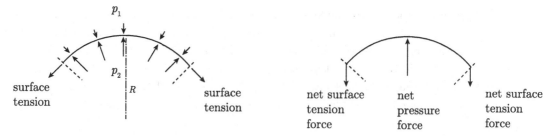

Figure 12.5 Configuration analyzing surface tension at an interface segregating two inviscid fluids.

differences. For three-dimensional flows, it would be necessary to introduce two local radii of curvature to properly characterize the surface.

We show a cutaway section in Fig. 12.5 that focuses on the net forces. For this section, because $p_2 > p_1$, and because of cancelation of horizontal forces, the net pressure force is vertical and points upward. Now the surface tension forces are tangent to the surface. The horizontal components of surface tension cancel, leaving the net effect as vertically downward force. Similar to what we saw in Section 4.2.2, in which inertia played no role in the analysis of surfaces without surface tension, it plays no role when surface tension is present. Therefore, we must expect a force balance.

Recall the curvature $\kappa = 1/R$; it is the reciprocal of the radius of curvature. Thus, our dynamical interface condition, Eq. (12.32), becomes

$$p_1 - p_2 = \sigma \kappa. \tag{12.33}$$

Now let us cast the curvature κ in terms of the surface $y = \eta(x,t)$. Adapting the formula for curvature of a two-dimensional curve given in Eq. (8.125), we have

$$\kappa = \frac{\frac{\partial^2 \eta}{\partial x^2}}{\left(1 + \left(\frac{\partial \eta}{\partial x}\right)^2\right)^{3/2}}. \tag{12.34}$$

Thus, we expect

$$p_1 - p_2 = \sigma \frac{\frac{\partial^2 \eta}{\partial x^2}}{\left(1 + \left(\frac{\partial \eta}{\partial x}\right)^2\right)^{3/2}}. \tag{12.35}$$

This is our *dynamic interface condition*.

Let us pause to check if our formula makes sense for a simple curve. For the configuration of Fig. 12.5, we expect $p_2 > p_1$. Let us say the curve is given by $y = \eta(x,t) = -x^2$. The shape of this parabola resembles that of Fig. 12.5. This parabola has curvature that is always negative $\kappa = -2/(1 + 4x^2)^{3/2}$. With $p_2 = p_1 - \sigma\kappa$ and $\sigma > 0$, $\kappa < 0$, a curve of this type guarantees $p_2 > p_1$.

Now pose a time-independent equilibrium solution that can be verified by substitution:

$$\phi_1 = U_1 x, \quad \phi_2 = U_2 x, \quad \eta(x,t) = 0, \quad p_1 = p_o - \rho_1 g y, \quad p_2 = p_o - \rho_2 g y. \tag{12.36}$$

Clearly the solution is time-independent and is thus at equilibrium. We see that in region 1, $u = \partial\phi_1/\partial x = U_1$, and region 2, $u = \partial\phi_2/\partial x = U_2$. So the velocity field is time-independent and spatially uniform within regions 1 and 2. The interface is time-independent, and located at $y = 0$; thus, it has no curvature, $\kappa = 0$. Because it has no curvature, surface tension is zero, and at the interface, we must have $p_1 = p_2$. That is in fact the case, as our pressure fields give $p_1 = p_2 = p_o$ at $y = 0$. Now for our Bernoulli equations, we have assigned

$$f_1(t) = \frac{p_o}{\rho_1} + \frac{U_1^2}{2}, \qquad f_2(t) = \frac{p_o}{\rho_2} + \frac{U_2^2}{2}. \tag{12.37}$$

With these choices for pressure and velocity fields, both Bernoulli equations are satisfied. The pressure field has a hydrostatic character; it increases linearly as depth decreases. So we have established what must be an equilibrium solution. It is steady with two uniform velocity fields. Indeed, the velocity gradient $\partial u/\partial y$ is infinite at $y = 0$; however, because we have neglected viscous stress, this velocity slip line is permitted.

Next, let us subject this steady flow field to a small perturbation. For convenience, we take $0 < \epsilon \ll 1$ to be a small dimensionless parameter, and define variables with "prime" notation as $\mathcal{O}(1)$. We thus take

$$\eta(x,t) = \epsilon \eta'(x,t) + \cdots, \quad \phi_1(x,y,t) = U_1 x + \epsilon \phi_1' + \cdots, \quad \phi_2(x,y,t) = U_2 x + \epsilon \phi_2' + \cdots, \tag{12.38}$$

$$p_1 = p_o - \rho_1 g y + \epsilon p_1' + \cdots, \quad p_2 = p_o - \rho_2 g y + \epsilon p_2' + \cdots. \tag{12.39}$$

Returning to the full equations and applying Eq. (12.37), our full dynamics equations, Eqs. (12.23, 12.25), can be cast as

$$\frac{\partial \phi_1}{\partial t} + \frac{1}{2}\nabla^T \phi_1 \cdot \nabla \phi_1 + \frac{p_1}{\rho_1} + gy = \frac{p_o}{\rho_1} + \frac{U_1^2}{2}, \tag{12.40}$$

$$\frac{\partial \phi_2}{\partial t} + \frac{1}{2}\nabla^T \phi_2 \cdot \nabla \phi_2 + \frac{p_2}{\rho_2} + gy = \frac{p_o}{\rho_2} + \frac{U_2^2}{2}. \tag{12.41}$$

Rearranging yields

$$p_1 = p_o + \frac{\rho_1 U_1^2}{2} - \rho_1 \left(\frac{\partial \phi_1}{\partial t} + \frac{1}{2}\nabla^T \phi_1 \cdot \nabla \phi_1 + gy \right), \tag{12.42}$$

$$p_2 = p_o + \frac{\rho_2 U_2^2}{2} - \rho_2 \left(\frac{\partial \phi_2}{\partial t} + \frac{1}{2}\nabla^T \phi_2 \cdot \nabla \phi_2 + gy \right). \tag{12.43}$$

Now at the perturbed interface, Eq. (12.35) becomes

$$p_1 - p_2 = \sigma \frac{\epsilon \frac{\partial^2 \eta'}{\partial x^2}}{\left(1 + \left(\epsilon \frac{\partial \eta'}{\partial x}\right)^2\right)^{3/2}} \sim \epsilon \sigma \frac{\partial^2 \eta'}{\partial x^2}. \tag{12.44}$$

Equating this with pressure differences calculated from Eqs. (12.42, 12.43), we get

$$\epsilon\sigma\frac{\partial^2 \eta'}{\partial x^2} = \frac{\rho_1 U_1^2}{2} - \frac{\rho_2 U_2^2}{2} - \rho_1\left(\frac{\partial \phi_1}{\partial t} + \frac{1}{2}\nabla^T\phi_1 \cdot \nabla\phi_1 + gy\right)$$
$$+ \rho_2\left(\frac{\partial \phi_2}{\partial t} + \frac{1}{2}\nabla^T\phi_2 \cdot \nabla\phi_2 + gy\right). \quad (12.45)$$

Now for $j = 1, 2$, we can say

$$\nabla^T\phi_j \cdot \nabla\phi_j = |\nabla\phi_j|^2 = |\nabla(U_j x + \epsilon\phi_j')|^2 = \left|\left(U_j + \epsilon\frac{\partial \phi_j'}{\partial x}\right)\mathbf{i} + \epsilon\frac{\partial \phi_j'}{\partial y}\mathbf{j}\right|^2, \quad (12.46)$$

$$= U_j^2 + 2\epsilon\frac{\partial \phi_j'}{\partial x}U_j + \mathcal{O}(\epsilon^2). \quad (12.47)$$

Similarly for $\partial\phi_j/\partial t$, we can say

$$\frac{\partial \phi_j}{\partial t} = \frac{\partial}{\partial t}\left(U_j x + \epsilon\phi_j'\right) = \epsilon\frac{\partial \phi_j'}{\partial t}. \quad (12.48)$$

These can be used to simplify our dynamic interface condition, Eq. (12.45), after neglecting $\mathcal{O}(\epsilon^2)$ and lesser terms, to

$$\sigma\frac{\partial^2 \eta'}{\partial x^2} = -\rho_1\left(\frac{\partial \phi_1'}{\partial t} + U_1\frac{\partial \phi_1'}{\partial x} + g\eta'\right) + \rho_2\left(\frac{\partial \phi_2'}{\partial t} + U_2\frac{\partial \phi_2'}{\partial x} + g\eta'\right). \quad (12.49)$$

Rearrange to give the dynamic interface condition incorporating the Bernoulli principle:

$$\rho_1\left(\frac{\partial \phi_1'}{\partial t} + U_1\frac{\partial \phi_1'}{\partial x}\right) - \rho_2\left(\frac{\partial \phi_2'}{\partial t} + U_2\frac{\partial \phi_2'}{\partial x}\right) + (\rho_1 - \rho_2)g\eta' + \sigma\frac{\partial^2 \eta'}{\partial x^2} = 0. \quad (12.50)$$

Now apply our approximations to the kinematic interface condition, Eq. (12.31). We get

$$\frac{\partial}{\partial y}\left(U_j x + \epsilon\phi_j'\right) = \epsilon\frac{\partial \eta'}{\partial t} + \epsilon\left(\frac{\partial}{\partial x}\left(U_j x + \epsilon\phi_j'\right)\right)\frac{\partial \eta'}{\partial x}, \quad (12.51)$$

$$\frac{\partial \phi_j'}{\partial y} = \frac{\partial \eta'}{\partial t} + U_j\frac{\partial \eta'}{\partial x}, \quad j = 1, 2. \quad (12.52)$$

The kinematic field equations become

$$\nabla^2\left(U_j x + \epsilon\phi_j'\right) = 0, \quad \text{so} \quad \nabla^2\phi_j' = 0. \quad (12.53)$$

In the far field, we require $\phi_j' = 0$.

We now seek solutions to our linear system of partial differential equations in space and time. Let us assume solutions of a particular exponential form:

$$\eta'(x,t) = Ae^{ik(x-ct)}, \quad \phi_1'(x,y,t) = Be^{ik(x-ct)}e^{-ky}, \quad \phi_2'(x,y,t) = Ce^{ik(x-ct)}e^{ky}. \quad (12.54)$$

Here k and c are assumed constants, as are the amplitudes A, B, and C. We could have started with a more general form, and would be led to the same conclusions. The form assumed here allows for some easy interpretations. First, we note that the presumed dependence on $x - ct$ suggest we are seeking traveling wave solutions with wave speed c. For example, the interface may be thought of as having a sinusoidal shape that propagates as a wave. Second, for $k > 0$, we see that $\phi_1' \to 0$ as $y \to +\infty$ and that $\phi_2' \to 0$ as $y \to -\infty$, as we desire.

Let us check to see if the field equations are satisfied for ϕ_j. Consider $\nabla^2 \phi_1 = 0$:

$$\frac{\partial^2}{\partial x^2}\left(Be^{ik(x-ct)}e^{-ky}\right) + \frac{\partial^2}{\partial y^2}\left(Be^{ik(x-ct)}e^{-ky}\right) = 0, \tag{12.55}$$

$$-k^2 Be^{-ky}e^{-ikct}e^{ikx} + k^2 Be^{-ky}e^{-ikct}e^{ikx} = 0. \tag{12.56}$$

Indeed, the field equation for ϕ_1' is satisfied. The same analysis shows the same for ϕ_2'.

Now let us apply our assumed form to the dynamic interface condition, Eq. (12.50):

$$\begin{aligned}0 = {}& \rho_1 \left(Be^{ik(x-ct)}e^{-ky}(-ick) + U_1 Be^{ik(x-ct)}e^{-ky}(ik)\right) \\ & - \rho_2 \left(Ce^{ik(x-ct)}e^{ky}(-ick) + U_2 Ce^{ik(x-ct)}e^{ky}(ik)\right) \\ & + (\rho_1 - \rho_2)gAe^{ik(x-ct)} - k^2 \sigma Ae^{ik(x-ct)}.\end{aligned} \tag{12.57}$$

At leading order, the interface is at $y = 0$; thus, at the interface the terms $e^{\pm ky} \sim 1$. Using that and dividing common factors, the dynamic interface condition reduces to

$$0 = \rho_1\left(B(-ick) + U_1 B(ik)\right) - \rho_2\left(C(-ick) + U_2 C(ik)\right) + (\rho_1 - \rho_2)gA - k^2\sigma A. \tag{12.58}$$

Rearranging, we get

$$A\left((\rho_1 - \rho_2)g - \sigma k^2\right) + Bik\rho_1(U_1 - c) - Cik\rho_2(U_2 - c) = 0. \tag{12.59}$$

Let us do a similar exercise for the linearized kinematic interface condition, Eq. (12.52). For $j = 1$, we get

$$-kBe^{ik(x-ct)}e^{-ky} = (-ick)Ae^{ik(x-ct)} + U_1 A(ik)e^{ik(x-ct)}. \tag{12.60}$$

At $y = 0$, this simplifies to

$$-kB = Aik(U_1 - c). \tag{12.61}$$

In the same way, for $j = 2$, we get

$$kC = Aik(U_2 - c). \tag{12.62}$$

Now, Eqs. (12.59, 12.61, 12.62) form a linear system for the constants A, B, and C:

$$\begin{pmatrix} ik(U_1 - c) & k & 0 \\ ik(U_2 - c) & 0 & -k \\ (\rho_1 - \rho_2)g - \sigma k^2 & ik\rho_1(U_1 - c) & -ik\rho_2(U_2 - c) \end{pmatrix}\begin{pmatrix} A \\ B \\ C \end{pmatrix} = \begin{pmatrix} 0 \\ 0 \\ 0 \end{pmatrix}. \tag{12.63}$$

Just as in the simple pendulum problem in which we were led to an eigenvalue problem in Eq. (12.14), so too are we here. The vector $(A, B, C)^T$ is an eigenvector with arbitrary magnitude. For a nontrivial solution, we enforce the condition that the coefficient matrix has determinant of zero. This will provide a relationship between c and k. We can imagine for a disturbance with a given wave number k that we will find the speed c at which it propagates by setting the determinant to zero:

$$\begin{vmatrix} ik(U_1 - c) & k & 0 \\ ik(U_2 - c) & 0 & -k \\ (\rho_1 - \rho_2)g - \sigma k^2 & ik\rho_1(U_1 - c) & -ik\rho_2(U_2 - c) \end{vmatrix} = 0. \qquad (12.64)$$

A detailed calculation with the aid of computer-based mathematical software verifies the values of c that force the determinant to zero are

$$c = \frac{\rho_1 U_1 + \rho_2 U_2}{\rho_1 + \rho_2} \pm \frac{1}{\rho_1 + \rho_2}\sqrt{(\rho_1 + \rho_2)\left(\sigma k - (\rho_1 - \rho_2)\frac{g}{k}\right) - \rho_1\rho_2(U_2 - U_1)^2}. \qquad (12.65)$$

Importantly, if the quantity within the square root is negative, then c will have an imaginary component. Let us examine the consequences of this. We could possibly have $c = c_R \pm ic_I$, where c_R and c_I are real numbers that are the real and imaginary parts of c. Then our interface position would be

$$\eta' = Ae^{ik(x - c_R t \mp c_I it)} = Ae^{ik(x - c_R t)}e^{\pm kc_I t}. \qquad (12.66)$$

We see the presence of a nonzero c_I induces the amplitude to grow exponentially with time. Such a system would be *unstable*. We can identify the *stability boundary* by setting the quantity inside the square root to be zero:

$$(\rho_1 + \rho_2)\left(\sigma k - (\rho_1 - \rho_2)\frac{g}{k}\right) - \rho_1\rho_2(U_2 - U_1)^2 = 0. \qquad (12.67)$$

Detailed calculation for $\sigma \neq 0$ reveals this stability boundary is reached if

$$k = \frac{1}{2\sigma}\left(\frac{\rho_1\rho_2}{\rho_1 + \rho_2}(U_1 - U_2)^2 \pm \sqrt{\left(\frac{\rho_1\rho_2}{\rho_1 + \rho_2}\right)^2 (U_1 - U_2)^4 - 4\sigma g(\rho_2 - \rho_1)}\right). \qquad (12.68)$$

If $\sigma = 0$, the stability threshold exists at

$$k = \frac{\rho_2^2 - \rho_1^2}{\rho_1\rho_2}\frac{g}{(U_1 - U_2)^2}. \qquad (12.69)$$

These formulæ are complicated. Let us simplify by studying some special cases.

12.2.1 Stationary Fluids with No Surface Tension: Rayleigh–Taylor Instability

Consider the Rayleigh–Taylor instability. If we take $U_1 = U_2 = 0$ and $\sigma = 0$, Eq. (12.65) simplifies considerably to

$$c = \pm\sqrt{\frac{g}{k}\frac{\rho_2 - \rho_1}{\rho_1 + \rho_2}}. \qquad (12.70)$$

Here we recover an obvious result. The stability threshold, where an imaginary component of c appears, is clearly at $\rho_1 = \rho_2$. And clearly, for $\rho_2 > \rho_1$, the system is stable; for $\rho_2 < \rho_1$, the system is unstable. A dense fluid resting over a light fluid will not remain at rest if subjected to perturbation. This has implications in a variety of applications, including the weather. The interface is located at

$$y = \eta'(x,t) = Ae^{ik\left(x-\left(\pm\sqrt{\frac{g}{k}\frac{\rho_2-\rho_1}{\rho_1+\rho_2}}\right)t\right)}. \qquad (12.71)$$

This describes the interface for the stable case $\rho_2 > \rho_1$.

For the unstable case, $\rho_1 > \rho_2$, a better description is

$$y = \eta'(x,t) = Ae^{ik\left(x-\left(\pm i\sqrt{\frac{g}{k}\frac{\rho_1-\rho_2}{\rho_1+\rho_2}}\right)t\right)} = Ae^{ikx}e^{\pm\sqrt{gk\frac{\rho_1-\rho_2}{\rho_1+\rho_2}}t}. \qquad (12.72)$$

The time constant of growth is

$$\tau = \sqrt{\frac{\rho_1+\rho_2}{gk(\rho_1-\rho_2)}}. \qquad (12.73)$$

Modes with high k (short wavelength) have a fast time constant; those with small k (long wavelength) have a slow time constant. Large density differences promote fast time constants.

12.2.2 Stationary Fluids with Surface Tension

If we take $U_1 = U_2 = 0$, Eq. (12.65) simplifies to

$$c = \pm\sqrt{\frac{1}{\rho_1+\rho_2}\left(k\sigma - \frac{(\rho_1-\rho_2)g}{k}\right)}. \qquad (12.74)$$

Clearly if $\rho_1 < \rho_2$, the argument of the square root is guaranteed positive, and we have a stable interface. But if $\rho_1 > \rho_2$, some modes will be stable and others unstable. If k is sufficiently large (corresponding to a short wavelength disturbance), surface tension will counterbalance the tendency of the heavy fluid to fall into the light fluid. For small k (long wavelength disturbances), the interface will be unstable. This could guide the design of a device. If the device width is smaller than the critical wavelength associated with the critical k, one might expect surface tension to be sufficient to maintain a stationary heavy fluid above a light fluid. The interface is located at

$$y = \eta'(x,t) = Ae^{ik\left(x-\left(\pm\sqrt{\frac{1}{\rho_1+\rho_2}\left(k\sigma - \frac{(\rho_1-\rho_2)g}{k}\right)}\right)t\right)}. \qquad (12.75)$$

Let us write this in a more convenient form for the unstable case with ρ_1 sufficiently greater than ρ_2 to overcome surface tension:

$$y = \eta'(x,t) = Ae^{ikx}e^{\pm\sqrt{\frac{k^2}{\rho_1+\rho_2}\left(-k\sigma + \frac{(\rho_1-\rho_2)g}{k}\right)}\,t}. \qquad (12.76)$$

Let us focus on the unstable mode, which corresponds to the positive root. The constant multiplying t is a good measure of the growth rate, which we will call α:

$$\alpha = \sqrt{\frac{k^2}{\rho_1+\rho_2}\left(-k\sigma + \frac{(\rho_1-\rho_2)g}{k}\right)}. \qquad (12.77)$$

The growth rate depends on k. We plot $\alpha(k)$ for parameters $\sigma = 1$, $\rho_1 = 2$, $\rho_2 = 1$, $g = 10$ in Fig. 12.6. It is seen there exists a k for which the growth rate is maximized. This has important

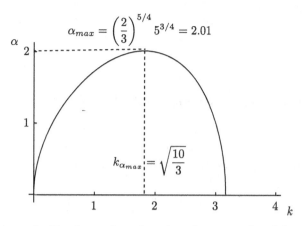

Figure 12.6 Growth rate α of a Fourier mode as function of wave number k for a stationary fluid with surface tension and body force effects, $\sigma = 1$, $\rho_1 = 2$, $\rho_2 = 1$, $g = 10$.

physical significance. An arbitrary initial condition may be expected to have contributions from Fourier modes at many values of k. While many of those modes could be unstable, the ones that would be expected to be seen first in an experiment are those with the largest growth rate. Using the tools of calculus, we find for this problem that the wave number at which the maximum growth rate occurs is

$$k_{\alpha_{max}} = \sqrt{\frac{g(\rho_1 - \rho_2)}{3\sigma}}. \tag{12.78}$$

At that value of k, the maximum growth rate is

$$\alpha_{max} = \sqrt{\frac{2\left(\frac{(\rho_1-\rho_2)g}{3\sigma^{1/3}}\right)^{3/2}}{\rho_1 + \rho_2}}. \tag{12.79}$$

For the parameters used for Fig. 12.6, we have $k_{\alpha_{max}} = \sqrt{10/3}$ and $\alpha_{max} = (2/3)^{5/4} 5^{3/4}$. For $k > \sqrt{g(\rho_1 - \rho_2)/\sigma}$, surface tension dominates such that there is no instability.

12.2.3 Surface Waves

Let us consider the limit in which $\rho_1 \to 0$ that can describe a surface wave. This might be appropriate for an air–water interface such as in an ocean. In that limit, Eq. (12.65) simplifies considerably to

$$c = U_2 \pm \sqrt{\frac{g}{k}\left(1 + \frac{k^2\sigma}{\rho_2 g}\right)}. \tag{12.80}$$

The interface is located at

$$y = \eta'(x,t) = Ae^{ik\left(x - \left(U_2 \pm \sqrt{\frac{g}{k}\left(1 + \frac{k^2\sigma}{\rho_2 g}\right)}\right)t\right)}, \tag{12.81}$$

and it is stable because there is no growth. Nor is there decay. The wave speed c is real for all k. For $k \ll \sqrt{\rho_2 g/\sigma}$, gravity effects dominate. For $k \gg \sqrt{\rho_2 g/\sigma}$, surface tension effects dominate.

12.2.4 Inviscid Shear Layer: Kelvin–Helmholtz Instability

Let us focus on the effect of distinct velocities for the special case in which both fluids are identical $\rho_1 = \rho_2$ and there is no surface tension, $\sigma = 0$. This is the inviscid limit of the viscous shear layer considered in Section 11.2.4. In these limits, Eq. (12.65) simplifies to

$$c = \frac{U_1 + U_2}{2} \pm i \frac{U_2 - U_1}{2}. \tag{12.82}$$

Because $c_I = \pm(U_2 - U_1)/2 \neq 0$, the flow is unstable for all k. The interface is located at

$$y = \eta'(x, t) = A e^{ik\left(x - \left(\frac{U_1+U_2}{2} \pm i \frac{U_2-U_1}{2}\right)t\right)}, \tag{12.83}$$

which simplifies to

$$y = \eta'(x, t) = A e^{ikx} e^{-ik\left(\frac{U_1+U_2}{2}\right)t} e^{\pm k\left(\frac{U_2-U_1}{2}\right)t}. \tag{12.84}$$

As long as $U_1 \neq U_2$, there will instability for all k. The time constant of the growth is

$$\tau = \frac{2}{k|U_2 - U_1|}. \tag{12.85}$$

Fast growth is promoted by large k (short wavelength) and large velocity difference.

12.2.5 Kelvin–Helmholtz Instability with Surface Tension

Let us see how nonzero surface tension stabilizes some modes of the Kelvin–Helmholtz instability. We study $\rho_1 = \rho_2 = \rho$ for $\sigma > 0$. Here, Eq. (12.65) simplifies to

$$c = \frac{U_1 + U_2}{2} \pm \sqrt{\frac{k\sigma}{2\rho} - \left(\frac{U_1 - U_2}{2}\right)^2}. \tag{12.86}$$

This gives stability if $k\sigma/2/\rho > ((U_1 - U_2)/2)^2$; thus, high k tends toward stability. If $k\sigma/2/\rho < ((U_1 - U_2)/2)^2$, we can say

$$c = \frac{U_1 + U_2}{2} \pm i \sqrt{\left(\frac{U_1 - U_2}{2}\right)^2 - \frac{k\sigma}{2\rho}}. \tag{12.87}$$

This yields for the interface

$$y = \eta'(x, t) = A e^{ikx} e^{-ik\left(\frac{U_1+U_2}{2}\right)t} e^{\pm k \sqrt{\left(\frac{U_1-U_2}{2}\right)^2 - \frac{k\sigma}{2\rho}}\, t}. \tag{12.88}$$

We see then that the growth rate of an unstable mode is

$$\alpha = k \sqrt{\left(\frac{U_1 - U_2}{2}\right)^2 - \frac{k\sigma}{2\rho}}. \tag{12.89}$$

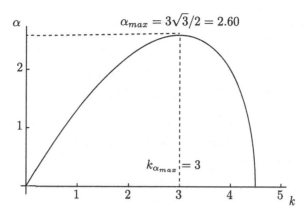

Figure 12.7 Growth rate α of a Fourier mode as function of wave number k for moving fluids with surface tension effects, $U_1 = 1$, $U_2 = 4$, $\sigma = 1$, $\rho_1 = \rho_2 = 1$.

Calculus reveals the value of k for which α is maximized is

$$k_{\alpha_{max}} = \frac{\rho}{2\sigma}(U_1 - U_2)^2, \tag{12.90}$$

and that the maximum value of growth rate α is

$$\alpha_{max} = \frac{\rho|U_1 - U_2|^3}{6\sqrt{3}\sigma}. \tag{12.91}$$

Velocity differences promote growth, and surface tension impedes growth. We plot $\alpha(k)$ for parameters $U_1 = 1$, $U_2 = 4$, $\sigma = 1$, $\rho_1 = \rho_2 = 1$ in Fig. 12.7. For these values $k_{\alpha_{max}} = 3$, $\alpha_{max} = 3\sqrt{3}/2 = 2.60$.

12.3 Parallel Viscous Flow Instability

The stability of what is known as parallel viscous flows is an important and complicated problem. It is treated in detail in sources such as Drazin and Reid (1981). We consider a parallel viscous flow to be one in which there is a dominant component of velocity that may vary in the normal direction, for example $\mathbf{v} \approx u(y)\mathbf{i}$. Such flows include Couette and Poiseuille flows, Blasius and Falkner–Skan boundary layers, shear layers, and others.

To get a flavor of the analysis, let us address a focused aspect of this with a standard approximation technique. Consider the linear stability of a two-dimensional boundary layer for incompressible flow over a flat plate with imposed pressure gradient. For example, we have an available similarity solution from Section 11.2.2 in the form of a Falkner–Skan boundary layer, which gives us $u(x, y)$. The Falkner–Skan solution also provides $v(x, y)$, but it is small compared to u. For a given $x = \hat{x}$ then, let us assume we have the Falkner–Skan solution, which gives us $u(x = \hat{x}, y) = \hat{u}(y)$ for an imposed known $p = \hat{p}(x)$. We will make the useful approximation that $v(x = \hat{x}, y) = 0$ for our boundary layer. If we were studying the stability of a Couette flow, we would take $\hat{u}(y) = y$ and $\hat{p}(x) = p_o$.

Now let us introduce perturbations in space and time to our solution. Take

$$u(x, y, t) = \hat{u}(y) + u'(x, y, t), \tag{12.92}$$
$$v(x, y, t) = v'(x, y, t), \tag{12.93}$$
$$p(x, y, t) = \hat{p}(x) + p'(x, y, t), \tag{12.94}$$

for which the prime notation indicates a small perturbation from the base state. These are introduced into the dimensionless two-dimensional unsteady Navier–Stokes equations, a straightforward extension of Eqs. (11.151–11.153), to obtain

$$\frac{\partial u'}{\partial x} + \frac{\partial v'}{\partial y} = 0, \tag{12.95}$$

$$\frac{\partial u'}{\partial t} + (\hat{u} + u')\frac{\partial u'}{\partial x} + v'\left(\frac{d\hat{u}}{dy} + \frac{\partial u'}{\partial y}\right) = -\frac{d\hat{p}}{dx} - \frac{\partial p'}{\partial x} + \frac{1}{Re}\left(\frac{\partial^2 u'}{\partial x^2} + \frac{d^2\hat{u}}{dy^2} + \frac{\partial^2 u'}{\partial y^2}\right), \tag{12.96}$$

$$\frac{\partial v'}{\partial t} + (\hat{u} + u')\frac{\partial v'}{\partial x} + v'\frac{\partial v'}{\partial y} = -\frac{\partial p'}{\partial y} + \frac{1}{Re}\left(\frac{\partial^2 v'}{\partial x^2} + \frac{\partial^2 v'}{\partial y^2}\right). \tag{12.97}$$

Now for the unperturbed flow under these assumptions, we must have $0 = -d\hat{p}/dx + (1/Re)d^2\hat{u}/dy^2$. We thus remove this as well as products of small terms to simplify to

$$\frac{\partial u'}{\partial x} + \frac{\partial v'}{\partial y} = 0, \tag{12.98}$$

$$\frac{\partial u'}{\partial t} + \hat{u}\frac{\partial u'}{\partial x} + v'\frac{d\hat{u}}{dy} = -\frac{\partial p'}{\partial x} + \frac{1}{Re}\left(\frac{\partial^2 u'}{\partial x^2} + \frac{\partial^2 u'}{\partial y^2}\right), \tag{12.99}$$

$$\frac{\partial v'}{\partial t} + \hat{u}\frac{\partial v'}{\partial x} = -\frac{\partial p'}{\partial y} + \frac{1}{Re}\left(\frac{\partial^2 v'}{\partial x^2} + \frac{\partial^2 v'}{\partial y^2}\right). \tag{12.100}$$

Next introduce the stream function so as to guarantee satisfaction of mass conservation:

$$u' = \frac{\partial \psi}{\partial y}, \qquad v' = -\frac{\partial \psi}{\partial x}. \tag{12.101}$$

The momenta equations become

$$\frac{\partial^2 \psi}{\partial y \partial t} + \hat{u}\frac{\partial^2 \psi}{\partial y \partial x} - \frac{\partial \psi}{\partial x}\frac{d\hat{u}}{dy} = -\frac{\partial p'}{\partial x} + \frac{1}{Re}\left(\frac{\partial^3 \psi}{\partial y \partial x^2} + \frac{\partial^3 \psi}{\partial y^3}\right), \tag{12.102}$$

$$-\frac{\partial^2 \psi}{\partial x \partial t} - \hat{u}\frac{\partial^2 \psi}{\partial x^2} = -\frac{\partial p'}{\partial y} - \frac{1}{Re}\left(\frac{\partial^3 \psi}{\partial x^3} + \frac{\partial^3 \psi}{\partial y^2 \partial x}\right). \tag{12.103}$$

Now we take $\partial/\partial y$ of the x momentum equation and subtract from it $\partial/\partial x$ of the y momentum equation to get

$$\left(\frac{\partial}{\partial t} + \hat{u}\frac{\partial}{\partial x}\right)\nabla^2 \psi - \frac{\partial \psi}{\partial x}\frac{d^2\hat{u}}{dy^2} = \frac{1}{Re}\nabla^4 \psi. \tag{12.104}$$

Let us now assume that

$$\psi(x, y, t) = \Psi(y)e^{i\alpha(x-ct)}. \tag{12.105}$$

With this assumption, we see that

$$\nabla^2 \psi = \left(\frac{d^2\Psi}{dy^2} - \alpha^2 \Psi\right) e^{i\alpha(x-ct)}, \qquad (12.106)$$

$$\nabla^4 \psi = \left(\frac{d^4\Psi}{dy^4} - \alpha^2 \frac{d^2\Psi}{dy^2} + \alpha^4 \Psi\right) e^{i\alpha(x-ct)}, \qquad (12.107)$$

$$\left(\frac{\partial}{\partial t} + \hat{u}\frac{\partial}{\partial x}\right) \nabla^2 \psi = \left(-i\alpha c \left(\frac{d^2\Psi}{dy^2} - \alpha^2 \Psi\right) + \hat{u} i\alpha \left(\frac{d^2\Psi}{dy^2} - \alpha^2 \Psi\right)\right) e^{i\alpha(x-ct)}. \qquad (12.108)$$

Applying the derivative operators and rearranging, the y momentum equation reduces to the well-known *Orr*[1]*–Sommerfeld*[2] *equation*:

$$(\hat{u}(y) - c)\left(\frac{d^2}{dy^2} - \alpha^2\right)\Psi - \frac{d^2\hat{u}}{dy^2}\Psi = \frac{1}{i\alpha}\frac{1}{Re}\left(\frac{d^2}{dy^2} - \alpha^2\right)^2 \Psi. \qquad (12.109)$$

For a Falkner–Skan boundary layer, the boundary conditions are such that the stream function and its derivatives must vanish at $y = 0$ and $y \to \infty$, yielding

$$\Psi(0) = \Psi'(0) = 0, \qquad \lim_{y \to \infty} \Psi(y) = \Psi'(y) = 0. \qquad (12.110)$$

For other problems with finite domains, such as Couette or Poiseuille flows, the boundary conditions are applied at finite values of y. This is a highly challenging eigenvalue problem; it will not be solved here. The problem is a fourth-order linear ordinary differential equation with variable coefficients. The boundary conditions are homogeneous, but two are at infinity. One can proceed as follows. Select a Falkner–Skan solution for a chosen imposed pressure gradient, thus giving $\hat{u}(y)$. Then select a numerical value of Re and disturbance wave number α to consider. This leaves only one free parameter in the problem, c. One could then approximate the semi-infinite domain as a finite domain, $y \in [0, y_{max}]$, with the selection of y_{max} guided by the Falkner–Skan results and chosen so the boundary layer has nearly relaxed to freestream conditions. At that point it is possible to discretize the domain and consider only a finite number of points N. At each point, the various continuous differential operators may be replaced by any consistent discrete analog, such as a finite difference operator. The local value of $\hat{u}(y)$ will be available at each discrete point. This will yield a large system of linear algebraic equations of the form

$$\mathbf{A}(c) \cdot \mathbf{\Psi} = \mathbf{0}. \qquad (12.111)$$

Here $\mathbf{A}$ is an $N \times N$ matrix that has in it numerical constants based on Re, α, $\hat{u}$, and the unknown parameter c. Accuracy will be increased as N increases. If a low-order finite difference is used to approximate the derivative operators, the matrix will be diagonally dominant. High-order finite difference approximations will reduce the sparsity of $\mathbf{A}$. The vector $\mathbf{\Psi}$ contains N discrete values, one for each spatial point, $(\Psi_1, \ldots, \Psi_N)^T$. While the trivial $\mathbf{\Psi} = \mathbf{0}$ obviously satisfies, one seeks an eigenvalue c for which nontrivial $\mathbf{\Psi}$ may exist. This can be achieved

---

[1] William McFadden Orr, 1866–1934, Irish mathematician.
[2] Arnold Johannes Wilhelm Sommerfeld, 1868–1951, German physicist.

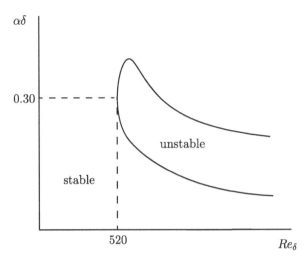

Figure 12.8 Sketch of marginal stability curve for a Blasius boundary layer.

by an iterative procedure. Background and solutions are given by sources such as Jordinson (1970) or Osborne (1967). The eigenvalue c could be real or complex. If the imaginary part of c is positive, the solution will be unstable. One can repeat this for a variety of α and Re and, with effort, identify $c = c(\alpha, Re)$. One can summarize the results with a sketch based on results for a zero pressure gradient Blasius boundary layer presented by Panton (2013, p. 750), shown in Fig. 12.8. Here $Re_\delta = U\delta/\nu$, where U is the free velocity and δ is the thickness of the boundary layer. For $Re_\delta < 520$, the zero pressure gradient Blasius boundary layer is stable. At $Re_\delta = 520$, a single mode of instability is reached for $\alpha\delta = 0.30$. For larger Reynolds numbers there exists a continuum of wave numbers α that induce instability. When one includes the effect of nonzero pressure gradient, the marginal stability curve shifts. For an adverse pressure gradient, $d\hat{p}/dx < 0$, the critical Reynolds number decreases.

Let us examine an interesting limit in which $Re \to \infty$. Then the Orr–Sommerfeld equation reduces to

$$(\hat{u}(y) - c)\left(\frac{d^2}{dy^2} - \alpha^2\right)\Psi - \frac{d^2\hat{u}}{dy^2}\Psi = 0. \tag{12.112}$$

With small rearrangement, one obtains what is known as the *Rayleigh equation*:

$$\frac{d^2\Psi}{dy^2} - \left(\alpha^2 + \frac{\frac{d^2\hat{u}}{dy^2}}{\hat{u}(y) - c}\right)\Psi = 0. \tag{12.113}$$

What this equation addresses is inviscid perturbations to a viscous flow. We seek a c at the threshold of stability. So if $c = c_R + ic_I$, we seek conditions for which $c_I = 0$. Recalling that for a linear second-order equation with constant coefficients, such as $\Psi'' - \beta\Psi = 0$, we could expect exponential growth for $\beta > 0$ and bounded oscillations for $\beta < 0$. Thus, we might expect some condition on $d^2\hat{u}/dy^2$ to be involved.

Let us see how such an analysis proceeds for the Rayleigh equation. Following Yih (1977), consider a flow between two boundaries at $y = y_1$ and $y = y_2$. We have ignored viscous effects, so we must sacrifice the no-slip condition. We require no mass penetration, which we enforce by taking $\Psi(y_1) = \Psi(y_2) = 0$. Now multiply the equation by $\overline{\Psi}$, where the bar connotes the complex conjugate, and integrate from y_1 to y_2:

$$\int_{y_1}^{y_2} \overline{\Psi} \frac{d^2\Psi}{dy^2} \, dy - \int_{y_1}^{y_2} \left(\alpha^2 + \frac{\frac{d^2\hat{u}}{dy^2}}{\hat{u}(y) - c} \right) \overline{\Psi}\Psi \, dy = 0. \qquad (12.114)$$

Integrating by parts, and recognizing $\overline{\Psi}\Psi = |\Psi|^2$, we get

$$\underbrace{\overline{\Psi} \frac{d\Psi}{dy} \bigg|_{y_1}^{y_2}}_{0} - \int_{y_1}^{y_2} \frac{d\Psi}{dy} \frac{d\overline{\Psi}}{dy} \, dy - \int_{y_1}^{y_2} \left(\alpha^2 + \frac{\frac{d^2\hat{u}}{dy^2}}{\hat{u}(y) - c} \right) |\Psi|^2 \, dy = 0, \qquad (12.115)$$

$$\underbrace{\int_{y_1}^{y_2} \left(\left|\frac{d\Psi}{dy}\right|^2 + \alpha^2 |\Psi|^2 \right) dy}_{\text{real}} + \int_{y_1}^{y_2} \frac{\frac{d^2\hat{u}}{dy^2}}{\hat{u}(y) - c} |\Psi|^2 \, dy = 0. \qquad (12.116)$$

Because the first term is real and cannot contribute to c_I, it suffices to study the second term. Let us multiply the integrand of the second part of the equation by $\overline{(\hat{u}-c)}/\overline{(\hat{u}-c)} = (\hat{u} - \overline{c})/(\hat{u} - \overline{c})$ to get

$$\underbrace{\int_{y_1}^{y_2} \left(\left|\frac{d\Psi}{dy}\right|^2 + \alpha^2 |\Psi|^2 \right) dy}_{\text{real}} + \int_{y_1}^{y_2} (\hat{u}(y) - \overline{c}) \frac{\frac{d^2\hat{u}}{dy^2}}{|\hat{u}(y) - c|^2} |\Psi|^2 \, dy = 0. \qquad (12.117)$$

The imaginary part of this is

$$c_I \int_{y_1}^{y_2} \frac{\frac{d^2\hat{u}}{dy^2}}{|\hat{u}(y) - c|^2} |\Psi|^2 \, dy = 0. \qquad (12.118)$$

If $d^2\hat{u}/dy^2$ is either always positive or always negative for $y \in [y_1, y_2]$, then we must have $c_I = 0$, and the flow is stable. If on the other hand there is a point of inflection where $d^2\hat{u}/dy^2 = 0$, within $y = [y_1, y_2]$, then it is possible to satisfy the equation for $c_I \neq 0$. So for a sufficient condition of stability, we have what is sometimes called Rayleigh's theorem:

$$\frac{d^2\hat{u}}{dy^2} \neq 0, \qquad y \in [y_1, y_2] \qquad \text{implies stability.} \qquad (12.119)$$

The converse is a necessary but insufficient condition for instability. A depiction of three common velocity profiles and an indication of their stability as predicted by the Rayleigh equation is given in Fig. 12.9.

Interestingly, our two-dimensional stability results have relevance for three-dimensional flows. As discussed in detail by Panton (2013, p. 745) and others, *Squire's*[3] *theorem* proves for the incompressible Navier–Stokes equations that for any unstable three-dimensional disturbance,

---

[3] Herbert Brian Squire, 1909–1961, British aerospace engineer.

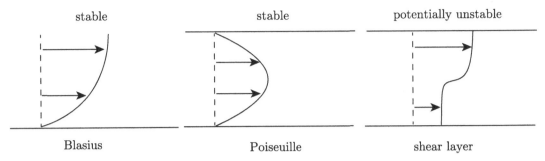

Figure 12.9 Diagram of stable and potentially unstable configurations of $\hat{u}(y)$ for high Reynolds number viscous flow governed by the Rayleigh equation.

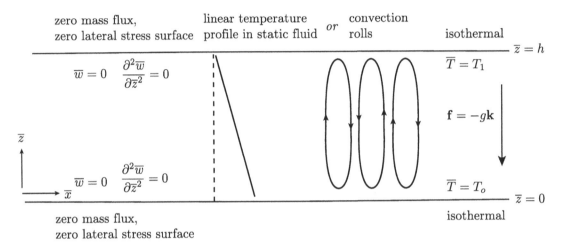

Figure 12.10 Configuration for Rayleigh–Bénard flow.

one can identify a corresponding two-dimensional disturbance whose instability is stronger. That is to say, the least stable disturbances are two-dimensional. Thus, it suffices to consider two-dimensional stability problems in order to ascertain the stability of qualifying three-dimensional flows.

12.4 Thermal Convection: Rayleigh–Bénard Instability

We next consider an important problem in thermal-fluid science: instability of buoyancy-driven flows. Consider the configuration shown in Fig. 12.10. We are guided here by the discussion of Drazin (2002); many other sources may be consulted, for example Koschmieder (1993). We have two surfaces, with the bottom plate held at a high temperature and the top plate held at a cold temperature. We expect heat to be transferred between the two plates. We might expect that when the flow is highly viscous, the fluid will be stationary, and the heat transfer will be by ordinary conduction. For flow that is not so viscous, we might expect the warm light fluid at the bottom to rise due to gravitational forces and be displaced by the dense fluid at the top. We might then expect the heat transfer rate to be enhanced by the advection of thermal energy. Often the advection of thermal energy is called convection. We study this here and seek to understand the stability of a stationary configuration in the presence of viscous and body forces.

Let us adopt the Boussinesq approximation for an ideal gas, Eqs. (6.103, 6.95, 6.105, 6.106), taking $\bar{\mathbf{f}} = -g\mathbf{k}$:

$$\overline{\nabla}^T \cdot \overline{\mathbf{v}} = 0, \tag{12.120}$$

$$\frac{d\overline{\mathbf{v}}}{d\bar{t}} = -\left(1 - \frac{\overline{T} - T_o}{T_o}\right) g\mathbf{k} - \frac{1}{\rho_o}\overline{\nabla}\bar{p} + \nu \overline{\nabla}^2 \overline{\mathbf{v}}, \tag{12.121}$$

$$\frac{d\overline{T}}{d\bar{t}} = \alpha \overline{\nabla}^2 \overline{T}, \tag{12.122}$$

$$\bar{\rho}\overline{T} = \rho_o T_o. \tag{12.123}$$

Here we use the over-bar notation to indicate the variable or operator is dimensional. Let us consider the region between two planes. The bottom plane is at $\bar{z} = 0$, and the top plane is at $\bar{z} = h$. We allow no mass flux through either of these planes so

$$\overline{w}(\bar{x}, \bar{y}, 0, \bar{t}) = \overline{w}(\bar{x}, \bar{y}, h, \bar{t}) = 0. \tag{12.124}$$

Following Rayleigh (1916), let us choose to model these planes as free surfaces with zero lateral stress. Other boundary conditions could be considered, but the ones chosen here admit a more straightforward solution. We thus insist that $\bar{\tau}_{zx} = \bar{\tau}_{zy} = 0$ at $\bar{z} = 0, h$, which from Eq. (6.59) gives us

$$\frac{\partial \overline{u}}{\partial \bar{z}} + \frac{\partial \overline{w}}{\partial \bar{x}} = 0, \qquad \frac{\partial \overline{w}}{\partial \bar{y}} + \frac{\partial \overline{v}}{\partial \bar{z}} = 0, \qquad \text{at } \bar{z} = 0, h. \tag{12.125}$$

Because $\overline{w}$ is always zero at both planar boundaries, independent of $\bar{x}$ or $\bar{y}$, we have at such boundaries $\partial \overline{w}/\partial \bar{x} = 0$, $\partial \overline{w}/\partial \bar{y} = 0$. Thus, the zero lateral stress condition gives

$$\frac{\partial \overline{u}}{\partial \bar{z}} = 0, \qquad \frac{\partial \overline{v}}{\partial \bar{z}} = 0, \qquad \text{at } \bar{z} = 0, h. \tag{12.126}$$

We take the temperature at the planar boundaries to be isothermal at two different temperatures:

$$\overline{T}(\bar{x}, \bar{y}, 0, \bar{t}) = T_o, \qquad \overline{T}(\bar{x}, \bar{y}, h, \bar{t}) = T_1. \tag{12.127}$$

When $T_o > T_1$, we expect the less dense fluid at $\bar{z} = 0$ to have a tendency to rise in the presence of the gravitational field with $\mathbf{f} = -g\mathbf{k}$. For $T_o < T_1$, there would be no tendency for the denser fluid at $\bar{z} = 0$ to rise. Note that this tendency to rise is counteracted by viscous forces that tend to suppress motion. If the viscous forces are sufficiently strong, there will be no motion. Because both boundaries are isothermal, there is no restriction on the boundary heat flux, and one can expect a thermal energy flux at both walls; thus, the problem is not adiabatic. For situations in which the gravitational forces overcome the viscous forces, thus inducing fluid acceleration, the energy necessary to maintain the kinetic energy of the fluid must come from fluxes of thermal energy at the boundary and conversion of this thermal energy into mechanical energy.

Consider the hydrostatic solution with $\overline{\mathbf{v}} = \mathbf{0}$. The energy equation reduces to $\overline{\nabla}^2 \overline{T} = 0$, and the obvious solution that satisfies both boundary conditions on $\overline{T}$ is the linear profile

$$\tilde{T}(\overline{z}) = T_o + (T_1 - T_o)\frac{\overline{z}}{h}. \tag{12.128}$$

Interestingly, this result shares features of a result of our our earlier analysis, Eq. (6.54): both predict linear variation of temperature with distance, albeit with different slopes. However, the assumptions that led to Eq. (6.54) included requiring the fluid to be isentropic. Here, the presence of energy diffusion renders the fluid to have variable entropy.

The momenta equation, Eq. (12.121), in the hydrostatic limit becomes

$$\mathbf{0} = -\left(1 - \frac{T_1 - T_o}{T_o}\frac{\overline{z}}{h}\right)g\mathbf{k} - \frac{1}{\rho_o}\overline{\nabla}\overline{p}. \tag{12.129}$$

Solving, and taking $\overline{p}(\overline{z} = 0) = p_o$, and calling the solution $\tilde{p}(\overline{z})$, we get

$$\tilde{p}(\overline{z}) = p_o - \rho_o g \overline{z}\left(1 + \frac{T_o - T_1}{2T_o}\frac{\overline{z}}{h}\right). \tag{12.130}$$

Now let us define dimensional variables, denoted by a prime, that represent small perturbations from the hydrostatic limit:

$$\overline{\mathbf{v}} = \overline{\mathbf{v}}'(\overline{\mathbf{x}}, \overline{t}), \qquad \overline{T} = \tilde{T}(\overline{z}) + \overline{T}'(\overline{\mathbf{x}}, \overline{t}), \qquad \overline{p} = \tilde{p}(\overline{z}) + \overline{p}'(\overline{\mathbf{x}}, \overline{t}). \tag{12.131}$$

The governing equations become, after neglecting small terms and subtracting the hydrostatic solution:

$$\overline{\nabla}^T \cdot \overline{\mathbf{v}}' = 0, \tag{12.132}$$

$$\frac{\partial \overline{\mathbf{v}}'}{\partial \overline{t}} = \frac{\overline{T}'}{T_o}g\mathbf{k} - \frac{1}{\rho_o}\overline{\nabla}\overline{p}' + \nu \overline{\nabla}^2 \overline{\mathbf{v}}', \tag{12.133}$$

$$\frac{\partial \overline{T}'}{\partial \overline{t}} + \overline{w}'\frac{T_1 - T_o}{h} = \alpha \overline{\nabla}^2 \overline{T}'. \tag{12.134}$$

The linear temperature profile from the hydrostatic limit survives in the advection of thermal energy in the $\overline{z}$ direction. Next introduce dimensionless variables and operators:

$$\mathbf{x} = \frac{\overline{\mathbf{x}}}{h}, \ t = \frac{\alpha}{h^2}\overline{t}, \ \mathbf{v} = \frac{h}{\alpha}\overline{\mathbf{v}}', \ T = \frac{\overline{T}'}{T_o - T_1}, \ p = \frac{h^2}{\rho_o \alpha^2}\overline{p}', \ \nabla = h\overline{\nabla}, \ \nabla^2 = h^2\overline{\nabla}^2. \tag{12.135}$$

Here we anticipate the physically interesting case for which $T_0 > T_1$, and in this case a positive $\overline{T}'$ corresponds to a positive T. With such scaling, we get

$$\nabla^T \cdot \mathbf{v} = 0, \tag{12.136}$$

$$\frac{\partial \mathbf{v}}{\partial t} = (Ra)(Pr)T\mathbf{k} - \nabla p + (Pr)\nabla^2 \mathbf{v}, \tag{12.137}$$

$$\frac{\partial T}{\partial t} - w = \nabla^2 T, \tag{12.138}$$

where we have taken the *Rayleigh number*, Ra to be

$$Ra = \frac{T_o - T_1}{h} \frac{1}{T_o} \frac{gh^4}{\alpha \nu}, \qquad (12.139)$$

and recalled that $Pr = \nu/\alpha$. The Rayleigh number is a measure of the relative strength of the gravitational forces to the viscous forces. Recalling the definition of the Grashof number, Eq. (11.359), we see as an aside that

$$Ra = Gr_h Pr. \qquad (12.140)$$

It will be convenient to use Cartesian index form to better understand a series of steps in the analysis. Replacing the specific unit vector $\mathbf{k}$ with a generic unit vector n_k, the momenta equation in index form is

$$\partial_o v_k = (Ra)(Pr)T n_k - \partial_k p + (Pr)\partial_m \partial_m v_k. \qquad (12.141)$$

Take the curl to get

$$\partial_o \epsilon_{ijk} \partial_j v_k = (Ra)(Pr)\epsilon_{ijk}\partial_j T n_k - \underbrace{\epsilon_{ijk}\partial_j \partial_k p}_{0} + (Pr)\partial_m \partial_m \epsilon_{ijk}\partial_j v_k. \qquad (12.142)$$

Because ϵ_{ijk} is anti-symmetric and $\partial_j \partial_p$ is symmetric, their tensor inner product is zero, giving

$$\partial_o \epsilon_{ijk}\partial_j v_k = (Ra)(Pr)\epsilon_{ijk}\partial_j T n_k + (Pr)\partial_m \partial_m \epsilon_{ijk}\partial_j v_k. \qquad (12.143)$$

Take the curl again to give

$$\partial_o \epsilon_{ijk}\epsilon_{qni}\partial_n \partial_j v_k = (Ra)(Pr)\epsilon_{qni}\partial_n \epsilon_{ijk}\partial_j T n_k + (Pr)\partial_m \partial_m \epsilon_{qni}\partial_n \epsilon_{ijk}\partial_j v_k. \qquad (12.144)$$

Expand the term that appears twice here:

$$\epsilon_{ijk}\epsilon_{qni}\partial_n \partial_j v_k = \epsilon_{ijk}\epsilon_{iqn}\partial_n \partial_j v_k = (\delta_{jq}\delta_{kn} - \delta_{jn}\delta_{kq})\partial_n \partial_j v_k, \qquad (12.145)$$

$$= \partial_k \partial_j v_k - \partial_j \partial_j v_k = \partial_j \underbrace{\partial_k v_k}_{0} - \partial_j \partial_j v_k = -\nabla^2 \mathbf{v}, \qquad (12.146)$$

for the incompressible fluid. Now expanding the remaining term gives

$$\epsilon_{qni}\partial_n \epsilon_{ijk}\partial_j T n_k = \epsilon_{ijk}\epsilon_{iqn}\partial_n \partial_j T n_k = (\delta_{jq}\delta_{kn} - \delta_{jn}\delta_{kq})\partial_n \partial_j T n_k, \qquad (12.147)$$

$$= \partial_k \partial_j T n_k - \partial_j \partial_j T n_k = n_k \partial_k (\partial_j T) - (\partial_j \partial_j T) n_k, \qquad (12.148)$$

$$= \frac{\partial}{\partial z}\nabla T - (\nabla^2 T)\mathbf{k}. \qquad (12.149)$$

Thus, after applying the curl operator twice to the linear momenta equation, we have

$$\frac{\partial}{\partial t}\nabla^2 \mathbf{v} = (Ra)(Pr)\left((\nabla^2 T)\mathbf{k} - \nabla\frac{\partial T}{\partial z}\right) + (Pr)\nabla^4 \mathbf{v}. \qquad (12.150)$$

The z component is

$$\frac{\partial}{\partial t}\nabla^2 w = (Ra)(Pr)\left(\frac{\partial^2 T}{\partial x^2} + \frac{\partial^2 T}{\partial y^2}\right) + (Pr)\nabla^4 w. \qquad (12.151)$$

Now assume a separation of variables solution of the form

$$T(x,y,z,t) = \Theta(z)f(x,y)e^{st}, \qquad w(x,y,z,t) = W(z)f(x,y)e^{st}. \tag{12.152}$$

The energy equation, Eq. (12.138), becomes

$$s\Theta(z)f(x,y)e^{st} - W(z)f(x,y)e^{st} = \Theta(z)e^{st}\left(\frac{\partial^2 f}{\partial x^2} + \frac{\partial^2 f}{\partial y^2}\right) + f(x,y)e^{st}\frac{d^2\Theta}{dz^2}, \tag{12.153}$$

$$s\Theta(z) - \frac{\Theta(z)}{f}\left(\frac{\partial^2 f}{\partial x^2} + \frac{\partial^2 f}{\partial y^2}\right) - \frac{d^2\Theta}{dz^2} = W(z). \tag{12.154}$$

Now in order to achieve a solution via this separation of variables, we must demand that

$$-\left(\frac{\partial^2 f}{\partial x^2} + \frac{\partial^2 f}{\partial y^2}\right) = a^2 f, \tag{12.155}$$

where a^2 is some unspecified positive constant. This is a *Helmholtz equation* for f, not to be conflated with the Helmholtz vorticity transport equation. One can show from the theory of linear operators that the operator $-(\partial^2/\partial x^2 + \partial^2/\partial y^2)$ is positive definite, and it must therefore have positive eigenvalues, which we call a^2. If we guess a solution for f of the form $f(x,y) = C \sin k_x x \sin k_y y$, then the Helmholtz equation will be satisfied if $a^2 = k_x^2 + k_y^2$. Because we enforce no boundary conditions in x and y, we admit a continuum of wave numbers k_x and k_y; they need not be integers.

With this restriction of $f(x,y)$, the energy equation reduces to

$$s\Theta(z) + a^2\Theta(z) - \frac{d^2\Theta}{dz^2} = W(z). \tag{12.156}$$

Using the common notation $D = d/dz$, we recast the energy equation as

$$(D^2 - a^2 - s)\Theta = -W. \tag{12.157}$$

For the momentum equation, Eq. (12.151), a detailed analysis of the same nature yields

$$(D^2 - a^2)(D^2 - a^2 - s/Pr)W = (Ra)a^2\Theta. \tag{12.158}$$

Let us study these equations at the threshold of stability for which $s = 0$:

$$(D^2 - a^2)\Theta = -W, \qquad (D^2 - a^2)^2 W = (Ra)a^2\Theta. \tag{12.159}$$

In matrix form, this is

$$\begin{pmatrix} 1 & (D^2 - a^2) \\ (D^2 - a^2)^2 & -(Ra)a^2 \end{pmatrix} \begin{pmatrix} W \\ \Theta \end{pmatrix} = \begin{pmatrix} 0 \\ 0 \end{pmatrix}. \tag{12.160}$$

This is an eigenvalue problem, for which we seek nontrivial W and Θ that satisfy this homogeneous system. To find solutions, we must also formulate a set of homogeneous boundary conditions. For the isothermal boundaries, the temperature perturbations must be zero, giving $\Theta(0) = \Theta(1) = 0$. For no mass penetration at either boundary, we also require

$W(0) = W(1) = 0$. For the remaining conditions on W, we need to consider the physical boundary conditions. First, take $\partial/\partial z$ of the mass equation to get

$$\frac{\partial}{\partial x}\frac{\partial u}{\partial z} + \frac{\partial}{\partial y}\frac{\partial v}{\partial z} + \frac{\partial^2 w}{\partial z^2} = 0. \tag{12.161}$$

On the zero-lateral stress boundaries at $z = 0, 1$, recall from Eq. (12.126) that $\partial u/\partial z = 0$ and $\partial v/\partial z = 0$ for all x and y on the boundaries at $z = 0$ and $z = 1$. So their partial derivatives in x and y are also zero. This gives us

$$\frac{\partial^2 w}{\partial z^2} = 0, \quad z = 0, 1. \tag{12.162}$$

Thus, at either boundary we require

$$\frac{d^2 W}{dz^2} = 0, \quad z = 0, 1. \tag{12.163}$$

Let us guess a family of solutions for W and Θ. Take

$$W(z) = A \sin n\pi z, \quad \Theta(z) = B \sin n\pi z, \quad n = 1, 2, \ldots. \tag{12.164}$$

In contrast to k_x and k_y, we insist that n be an integer. This is required to satisfy all boundary conditions in the finite z domain. Substituting into Eq. (12.160) yields

$$\begin{pmatrix} 1 & -(a^2 + n^2\pi^2) \\ (a^2 + n^2\pi^2)^2 & -(Ra)a^2 \end{pmatrix} \begin{pmatrix} A \\ B \end{pmatrix} = \begin{pmatrix} 0 \\ 0 \end{pmatrix}. \tag{12.165}$$

For a nontrivial solution, the determinant of the coefficient matrix must be zero, yielding

$$Ra = \frac{(a^2 + n^2\pi^2)^3}{a^2}. \tag{12.166}$$

For a given n, this gives a relation between Ra, a controllable parameter, and a for a system that neither grows or decays in time, as $s = 0$. Such solutions are formally time-independent and thus steady. These so-called *marginal stability curves* for $n = 1$, $n = 2$, $n = 3$ are plotted in Fig. 12.11. On each curve for various n, which is proportional to the wave number of the disturbance in z, we see the values of a for which there is neutral stability. The most critical curve is that for $n = 1$, for which the initial disturbance has the longest wavelength, with half of a sine wave fitting in the region $z = [0, 1]$. For $n = 1$, we employ tools of calculus to identify the a for which Ra takes on a minimum value. We call these a_c and Ra_c and find

$$a_c = \frac{\pi}{\sqrt{2}} = 2.22, \quad Ra_c = \frac{27\pi^4}{4} = 657.5. \tag{12.167}$$

Recall that a is a parameter characterizing the form of the disturbance in $f(x, y)$, and if we took f to have a sinusoidal form, that $a^2 = k_x^2 + k_y^2$, where k_x and k_y are the wave numbers associated with f. For $Ra < Ra_c = 657.5$, all a yields a negative value of s, yielding stability. For $Ra = Ra_c$, there is exactly one value of a that is at the cusp of stability, and we could expect all such waves that had $a_c^2 = k_x^2 + k_y^2$ to persist in time. For $Ra > Ra_c$, we find a continuum of values of a to the right of the $n = 1$ marginal stability curve. All such waves would be expected to grow in time. It then is not difficult to show that for the $n = 2$ family of disturbances, the critical value is at $Ra = 108\pi^4 = 10\,520.2$ for which $a = \sqrt{2}\pi = 4.44$. For $n = 3$, we find critical

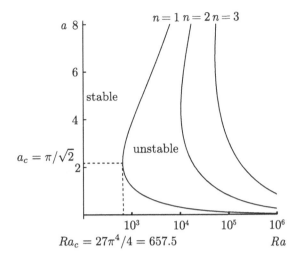

Figure 12.11 Marginal stability curves ($s = 0$) for Rayleigh–Bénard flow for $n = 1$, $n = 2$, $n = 3$.

values of $Ra = 2187\pi^4/4 = 53\,258.4$, $a = 3\pi/\sqrt{2} = 6.66$. Obviously, the threshold value of Ra increases as n increases, because as n increases, the effects of viscosity increase because the spatial gradients are steeper. For air with $T_o = 300\,\text{K}$, we take values reported by Incropera and DeWitt (1981, p. 775) of $\nu = 15.89 \times 10^{-6}\,\text{m}^2/\text{s}$, $\alpha = 22.5 \times 10^{-6}\,\text{m}^2/\text{s}$ in a gravitational field with $g = 9.81\,\text{m/s}^2$ in a gap with $h = 0.01\,\text{m}$. The temperature difference that places the system at the marginal stability threshold is $(T_o - T_1) = Ra_c T_o \alpha \nu/g/h^3 = 7.2\,\text{K}$.

We have identified some generally admissible forms of solutions, but not a unique solution, for which additional specifications would need to be provided. Let us study a few of the admissible forms to better understand the nature of the solutions obtained. First, let us consider solutions for which $n = 1$, $Ra = Ra_c = 27\pi^4/4$, and thus $a = a_c = \pi/\sqrt{2}$. We have wide latitude in selecting $f(x,y)$. Let us choose

$$f(x,y) = (\sin x)\left(\sin\sqrt{\frac{\pi^2}{2} - 1}\,y\right). \tag{12.168}$$

This has $k_x = 1$, $k_y = \sqrt{\pi^2/2 - 1}$, and clearly $k_x^2 + k_y^2 = a^2$, as required. It can be verified that this f satisfies the Helmholtz equation, Eq. (12.155). Now if we choose $A = 100$, the eigenvector nature of $(A, B)^T$ gives us $B = 200/(3\pi^2)$. We can now compose a solution for w and T, which is

$$w(x,y,z) = 100(\sin x)\sin\left(\sqrt{\frac{\pi^2}{2} - 1}\,y\right)\sin\pi z, \tag{12.169}$$

$$T(x,y,z) = \frac{200}{3\pi^2}(\sin x)\sin\left(\sqrt{\frac{\pi^2}{2} - 1}\,y\right)\sin\pi z. \tag{12.170}$$

A plot of $T(x = \pi/2, y, z)$ is given in Fig. 12.12. We see the perturbed temperature is 0 at the plate boundaries at $z = 0$ and $z = 1$, and that there is a repeating pattern of rectangular

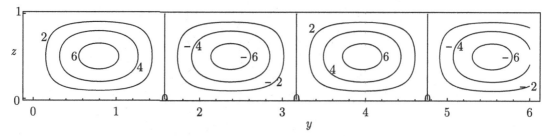

Figure 12.12 Steady temperature perturbation contours for Rayleigh–Bénard flow for $n=1$ at the marginal stability threshold, $Ra = 27\pi^4/4$, in the $x = \pi/2$ plane.

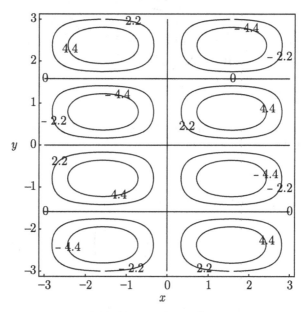

Figure 12.13 Steady temperature perturbation contours for Rayleigh–Bénard flow, $n=1$, at the stability threshold, $Ra = 27\pi^4/4$, in the $z = 1/2$ plane; rectangular cells for the chosen $f(x,y)$.

cells. One cell has positive perturbations and the next one negative. We have shown results for $y \in [0,6]$, but the results are valid for $y \in (-\infty, \infty)$. Looking down at the doubly infinite (x,y) plane, we see similar results for $T(x,y,z=1/2)$ in Fig. 12.13.

Next let us consider the same $n = 1$, $Ra = Ra_c = 27\pi^4/4$ with $a = a_c = \pi/\sqrt{2}$. However, let us choose a different f:

$$f(x,y) = \cos\left(\frac{1}{2}a\left(\sqrt{3}x - y\right)\right) + \cos\left(\frac{1}{2}a\left(\sqrt{3}x + y\right)\right) + \cos(ay). \quad (12.171)$$

This satisfies the Helmholtz equation for all a, including $a = \pi/\sqrt{2}$. We get

$$f(x,y) = \cos\left(\frac{\pi\left(\sqrt{3}x - y\right)}{2\sqrt{2}}\right) + \cos\left(\frac{\pi\left(\sqrt{3}x + y\right)}{2\sqrt{2}}\right) + \cos\left(\frac{\pi y}{\sqrt{2}}\right). \quad (12.172)$$

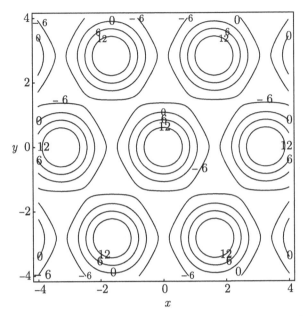

Figure 12.14 Steady temperature perturbation contours for Rayleigh–Bénard flow, $n = 1$, at the stability threshold, $Ra = 27\pi^4/4$, in the $z = 1/2$ plane; hexagonal cells for the chosen $f(x,y)$.

With this f and $A = 100$, $B = 200/(3\pi^2)$, we get

$$w(x,y,z) = 100\sin(\pi z)\left(\cos\left(\frac{\pi(\sqrt{3}x - y)}{2\sqrt{2}}\right) + \cos\left(\frac{\pi(\sqrt{3}x + y)}{2\sqrt{2}}\right) + \cos\left(\frac{\pi y}{\sqrt{2}}\right)\right),$$
(12.173)

$$T(x,y,z) = \frac{200}{3\pi^2}\sin(\pi z)\left(\cos\left(\frac{\pi(\sqrt{3}x - y)}{2\sqrt{2}}\right) + \cos\left(\frac{\pi(\sqrt{3}x + y)}{2\sqrt{2}}\right) + \cos\left(\frac{\pi y}{\sqrt{2}}\right)\right).$$
(12.174)

Looking down at the doubly infinite (x, y) plane, we see results for $T(x, y, z = 1/2)$ given in Fig. 12.14. Obviously, this choice of $f(x, y)$ induced a hexagonal pattern. And hexagonal patterns are observed in nature in special cases. However, as discussed by Leal (2007), what is observed in nature is most often the manifestation of nonlinear effects, neglected in our linear analysis. Certainly, we can make other choices for $f(x, y)$ and arrive at other patterns. Clearly our linear theory allows new structure to emerge from an unstructured state. This is the essence of the topic of *morphogenesis*, which has application in weather, material science, and other disciplines. While we will not analyze it, one can recognize that generation of structure is allowed by the second law of thermodynamics because thermal energy is admitted into the system at the isothermal boundaries. The question of which patterns are preferred and thus observed is a deep question not addressed here. All of the structures we have identified certainly satisfy the axioms of mass, momenta, energy, and entropy, and thus are admissible. In principle, one can analyze the stability of a structure such as a convection roll. And for a given set of

12.4 Thermal Convection: Rayleigh–Bénard Instability

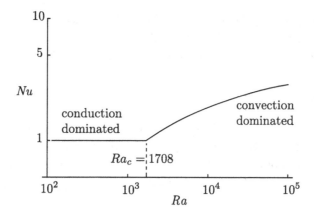

Figure 12.15 Sketch of Nusselt number versus Rayleigh number for flow between two rigid isothermal flat plates based on experimental results summarized by Drazin (2002, p. 101).

conditions, because of nonlinearity, the possibility for multiple stable structures exists. Which is achieved, if any, likely depends on the particular set of initial conditions employed. Cross and Hohenberg (1993) and Cross and Greenside (2009) give a broad discussion of spatio-temporal pattern formation and nonlinearity in many fluid systems, including for Rayleigh–Bénard flow. Kadanoff (2001) gives an incisive assessment and critique of the subject of pattern formation properties of Rayleigh–Bénard flow, concluding "...even though there is apparently no science of complexity, there is much science to be learned from studying complex systems."

A similar analysis can be performed if one replaces the zero-lateral stress condition by a rigid plate condition at $z = 0, 1$. The consequence of this is the requirement of $dW/dz = 0$ at $z = 0, 1$ instead of the condition on d^2W/dz^2. Such a condition is easier to enforce in the laboratory, but slightly more difficult to analyze. The analysis is given by Drazin (2002) and many others, and results in $Ra_c = 1708$ and $a_c = 3.117$.

This allows a more direct comparison with experiment and illustration of an important result for convective heat transfer. Let us loosely define the *Nusselt number*, Nu as

$$Nu = \frac{\text{achieved heat flux}}{\text{heat flux by conduction alone}}. \tag{12.175}$$

Now the heat flux by conduction alone in the absence of any fluid motion is for our linear temperature profile, simply

$$\text{heat flux by conduction alone} = -k\frac{\overline{T}_1 - \overline{T}_o}{h}. \tag{12.176}$$

For $Ra < Ra_c$, we expect no fluid motion and the achieved heat transfer should be entirely due to conduction, thus we expect $Nu = 1$. Because of the presence of fluid motion and convection rolls for $Ra > Ra_c$, we expect enhanced transfer of heat because the thermal energy is advected by the moving fluid. That in fact is what is observed. We sketch a plot of experimental results given in more detail by Drazin (2002, p. 101) in Fig. 12.15. The point at which Nu becomes greater than unity is well approximated by the simple theory. The system is said to have exhibited a *bifurcation* from one type of behavior to another at Ra_c. A similar plot is given by Tritton (1988, p. 361). Prediction of this curve for $Ra > Ra_c$ would require consideration of neglected nonlinearities; we provide such a prediction in Section 14.4.

SUMMARY

This chapter focused on subjecting steady solutions of fluid configurations to small time-dependent perturbations and studying whether or not the perturbations grew or decayed. Through the use of a linear theory, it was often possible to identify physical parameters that either stabilized or destabilized various flow fields.

PROBLEMS

12.1 A stationary fluid with constant density ρ, specific heat c_p, and thermal conductivity k exists within a one-dimensional domain $x \in [0, L]$. Both ends are adiabatic. Show that $T(x,t) = 0$ is a stable, but not asymptotically stable, solution. To do this, perturb a uniform state of $T(x) = 0$ by an initial perturbed temperature of $T(x,0) = \epsilon \mathsf{T}(x)$. Consider $0 < \epsilon \ll 1$, and $\mathsf{T}(x)$ to be an arbitrary function with the units of temperature. Demonstrate the total energy in the domain is constant whether or not the second law is satisfied. Find $T(x,t)$. Find the temperature distribution at long time in terms of the initial temperature perturbation. Show the long time temperature is the spatial average of the initial temperature distribution. Show the condition for stability is satisfied if the second law of thermodynamics is satisfied.

12.2 Consider a fluid with a possible Rayleigh–Taylor instability. Thus, take $U_1 = U_2 = \sigma = 0$. Take $g = 1$, $\rho_1 = 2$, $\rho_2 = 1$. With the initial amplitude of the surface displacement η' given as $A = \epsilon$, find the corresponding initial amplitudes of ϕ_1' and ϕ_2': B and C. Find expressions for the time and space evolution of $\eta'(x,t)$, $\phi_1'(x,y,t)$, $\phi_2'(x,y,t)$. Plot the real part of $\eta'(x,t)$ for $k = 1$, $\epsilon = 1$.

12.3 Consider a fluid with a possible planar inviscid instability. Take $U_1 = 1$, $U_2 = 2$, $\rho_1 = 1$, $\rho_2 = 2$, $\sigma = 1/10$, and $g = 1$. Find the range of disturbance wave numbers k that induce instability. Identify the wave number that induces the maximum growth rate α, and find the maximum growth rate. Plot the growth rate α as a function of k.

12.4 For one-dimensional waves whose oscillatory portion is described by $\exp(i(kx-\omega t))$, where k is the wave number and ω is the angular frequency, one can typically identify $\omega(k)$. Such an equation is known as a *dispersion relation*. The *phase velocity* is $c(k) = \omega(k)/k$, and the *group velocity* is $C(k) = d\omega/dk$. For surface waves with nonzero gravity and surface tension and $\rho_1 \to 0$, described by Eq. (12.81), find the dispersion relation, the phase velocity, and the group velocity. Examine the literature, for example Lighthill (1978) or Whitham (1974), and give a physical interpretation.

12.5 Consider the Orr–Sommerfeld equation for $Re \to \infty$, which induces the Rayleigh equation. Consider a dimensionless Poiseuille flow in which $\hat{u}(y) = y(1-y)$ on the domain $y \in [0,1]$. Write the Rayleigh equation for this flow. Discretize the domain into six uniformly spaced cells with $y_i = (0, 1/5, 2/5, 3/5, 4/5, 1)^T$. For $\alpha = 1$, discretize the

Rayleigh equation and identify values of c that satisfy. Ascertain the stability of each eigenfunction. Find the eigenfunction approximation for one of the eigenvalues.

12.6 Consider Rayleigh–Bénard flow for $n = 1$ at the marginal stability threshold: $Ra = 27\pi^4/4$, $a = \pi/\sqrt{2}$. With $f(x, y) = \sin k_x x \sin k_y y$, select $k_x = \pi/2$. With $A = 100$, determine the temperature and velocity profiles $T(x, y, z)$, $w(x, y, z)$. Demonstrate that the cells generated are square when viewed in the (x, y) plane.

FURTHER READING

Bodenschatz, E., Pesch, W., and Ahlers, G. (2000). Recent developments in Rayleigh–Bénard convection, *Annual Review of Fluid Mechanics*, **32**, 709–778.

Chandrasekhar, S. (1961). *Hydrodynamic and Hydromagnetic Stability*. New York: Dover.

Chang, H.-C., and Demekhin, E. A. (2002). *Complex Wave Dynamics on Thin Films*. Amsterdam: Elsevier.

Charru, F. (2011). *Hydrodynamic Instabilities*. New York: Cambridge University Press.

Criminale, W. O., Jackson, T. L., and Joslin, R. D. (2019). *Theory and Computation in Hydrodynamic Stability*, 2nd ed. Cambridge, UK: Cambridge University Press.

Currie, I. G. (2013). *Fundamental Mechanics of Fluids*, 4th ed. Boca Raton, Florida: CRC Press.

Joseph, D. D. (1976). *Stability of Fluid Motions*, 2 vols. Berlin: Springer.

Joseph, D., Funada, T., and Wang, J. (2008). *Potential Flows of Viscous and Viscoelastic Fluids*. New York: Cambridge University Press.

Kundu, P. K., Cohen, I. M., and Dowling, D. R. (2015). *Fluid Mechanics*, 6th ed. Amsterdam: Academic Press.

Lin, C. C. (1966). *The Theory of Hydrodynamic Stability*. Cambridge, UK: Cambridge University Press.

Sengupta, T. K. (2021). *Transition to Turbulence: A Dynamical Systems Approach to Receptivity*. Cambridge, UK: Cambridge University Press.

Straughan, B. (2004). *The Energy Method, Stability, and Nonlinear Convection*, 2nd ed. New York: Springer.

White, F. M. (2006). *Viscous Fluid Flow*, 3rd ed. New York: McGraw-Hill.

REFERENCES

Cross, M., and Greenside, H. (2009) *Pattern Formation and Dynamics in Nonequilibrium Systems*. Cambridge, UK: Cambridge University Press.

Cross, M. C., and Hohenberg, P. C. (1993). Pattern formation outside of equilibrium, *Reviews of Modern Physics*, **65**(3), 851–1112.

Drazin, P. G. (2002). *Introduction to Hydrodynamic Stability*. Cambridge, UK: Cambridge University Press.

Drazin, P. G., and Reid, W. H. (1981). *Hydrodynamic Stability*. Cambridge, UK: Cambridge University Press.

Incropera, F. P., and DeWitt, D. P. (1981). *Fundamentals of Heat Transfer*. New York: John Wiley.

Jordinson, R. (1970). The flat plate boundary layer. Part 1. Numerical integration of the Orr–Sommerfeld equation, *Journal of Fluid Mechanics*, **43**(4), 801–811.

Kadanoff, L. P. (2001). Turbulent heat flow: structures and scaling, *Physics Today*, **54**(8), 34–39.

Koschmieder, E. L. (1993). *Bènard Cells and Taylor Vortices*. Cambridge, UK: Cambridge University Press.

Leal, L. G. (2007). *Advanced Transport Phenomena: Fluid Mechanics and Convective Transport Processes*. New York: Cambridge University Press.

Lighthill, J. (1978). *Waves in Fluids*. Cambridge, UK: Cambridge University Press.

Osborne, M. R. (1967). Numerical methods for hydrodynamic stability problems, *SIAM Journal on Applied Mathematics*, **15**(3), 539–557.

Panton, R. L. (2013). *Incompressible Flow*, 4th ed. New York: John Wiley.

Rayleigh, J. W. S. (1916). On convection currents in a horizontal layer of fluid, when the higher temperature is on the under side, *Philosophical Magazine*, **32**(192), 529–546.

Sethian, J. A. (1996). *Level Set Methods: Evolving Interfaces in Computational Geometry, Fluid Mechanics, Computer Vision, and Materials Science*. Cambridge, UK: Cambridge University Press.

Tritton, D. J. (1988). *Physical Fluid Dynamics*, 2nd ed. Oxford: Oxford University Press.

Whitham, G. B. (1974). *Linear and Nonlinear Waves*. New York: John Wiley.

Yih, C.-S. (1977). *Fluid Mechanics*. East Hampton, Connecticut: West River Press.

13   Nonlinear Dynamics for Fluid Flow

In this chapter, we briefly introduce how to apply methods from the discipline of nonlinear dynamical systems to fluid flow. The governing equations for a fluid are a nonlinear system of partial differential equations with space and time as independent variables. Here we will adopt methods to rationally reduce the system of nonlinear partial differential equations to a system of nonlinear ordinary differential equations. Such systems are then well suited for analysis by well-established methods that transcend fluid mechanics; see Finlayson (1972), Guckenheimer and Holmes (2002), Perko (2006), Powers and Sen (2015), Robinson (2001), or Temam (1997). Holmes et al. (2012) and Perry and Chong (1987) focus on application of these methods to fluids problems and describe well the many challenges involved in simulating realistic fluid mechanics.

A typical result, adapted from the presentation of Powers and Sen (2015, Section 9.10) where a detailed discussion may be found, is given in Fig. 13.1. The figure is an unusual representation of the unsteady one-dimensional heat transfer in a stationary fluid. It describes the low-order dynamics realized when solving a dimensionless nonlinear heat equation for a fluid with a weak temperature-dependent thermal diffusivity:

$$\frac{\partial T}{\partial t} = \frac{\partial}{\partial x}\left((1 + \epsilon T)\frac{\partial T}{\partial x}\right). \tag{13.1}$$

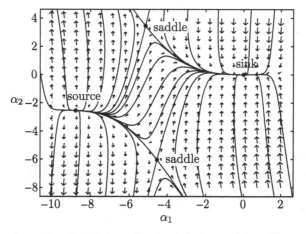

Figure 13.1 Phase plane dynamics of modal amplitude evolution, $\alpha_1(t)$, $\alpha_2(t)$, associated with unsteady heat transfer in a static fluid with temperature-dependent thermal conductivity.

Here the dimensionless thermal diffusivity is $1 + \epsilon T$. The results are achieved with $\epsilon = 1/5$ after one approximates $T(x,t)$ by the two-term expansion

$$T(x,t) \sim \alpha_1(t)\sqrt{2}\sin \pi x + \alpha_2(t)\sqrt{2}\sin 3\pi x. \tag{13.2}$$

Here the α's are time-dependent amplitudes associated with spatial basis functions and should not be interpreted as thermal diffusivity. One then uses a so-called *Galerkin*[1] *projection* to reduce the system to two autonomous, nonlinear ordinary differential equations that describe the evolution in time of the amplitudes $\alpha_1(t)$, $\alpha_2(t)$:

$$\frac{d\alpha_1}{dt} = -\pi^2 \alpha_1 + \sqrt{2}\pi\epsilon\left(-\frac{4}{3}\alpha_1^2 + \frac{8}{15}\alpha_2\alpha_1 - \frac{36}{35}\alpha_2^2\right), \qquad \alpha_1(0) = \frac{4\sqrt{2}}{\pi^3}, \tag{13.3}$$

$$\frac{d\alpha_2}{dt} = -9\pi^2 \alpha_2 + \sqrt{2}\pi\epsilon\left(\frac{12}{5}\alpha_1^2 - \frac{648}{35}\alpha_2\alpha_1 - 4\alpha_2^2\right), \qquad \alpha_2(0) = \frac{4\sqrt{2}}{27\pi^3}. \tag{13.4}$$

The method relies on numerical solution of a reduced set of ordinary differential equations such as these. In the case presented in Fig. 13.1, the dimension of the system is two, enabling efficient presentation of results in the so-called *phase plane*. Within the phase plane, one can identify equilibria of the system via a nonlinear algebraic analysis. Stability of each equilibrium may be ascertained by local linear analysis involving local eigenvalues and eigenvectors similar to that done in Chapter 12. Here four equilibria are seen: a stable *sink*, two unstable *saddles*, and an unstable *source*. There are additional equilibria at infinity that may be identified by a more sophisticated analysis, Perko (2006). Also defined within the phase plane is the vector field specified by the forcing terms of the resulting ordinary differential equations. The vectors are tangent to the solution trajectories in the phase plane. The spatial basis functions are required to be consistent with the spatial boundary conditions, and the initial conditions $\alpha_1(0)$ and $\alpha_2(0)$ are chosen to be consistent with a projection of the initial condition $T(x,0)$ onto the spatial basis functions. Knowing the functions $\alpha_1(t)$ and $\alpha_2(t)$ along with the spatial basis functions $\sqrt{2}\sin \pi x$ and $\sqrt{2}\sin 3\pi x$, one can form the two-term approximation to $T(x,t)$.

In contrast to the methods we will present that are built from deterministic mathematical equations based on axioms of physics, other approaches build low-order reduced dynamical systems directly from data. Commonly known as *dynamic mode decomposition* (DMD), which has deep connections with so-called *Koopman*[2] *operator theory*, the method draws heavily on notions from linear algebra such as the singular value decomposition (SVD) (see Section 3.8.5) to construct data-driven reduced models for the evolution in time of fluid systems. The related notions of *machine learning*, often built around *artificial neural networks* also rely on data-driven models for modeling and control of fluid systems. Brunton et al. (2019) and Schmid (2022) give full reviews with application to fluid mechanics. For more general background, see Brunton and Kutz (2022), McClarren (2021), Pollard et al. (2017), or Raissi et al. (2019). Such models might be considered to be sophisticated and useful interpolations; extrapolated prediction outside of the domain of calibration incurs risk, possibly catastrophic.

---

[1] Boris Grigoryevich Galerkin, 1871–1945, Soviet engineer and mathematician.
[2] Bernard Osgood Koopman, 1900–1981, American mathematician.

13.1 Traditional Fourier Series Expansion

The reduction we will employ for fluids problems hinges on the concept of projecting state variables onto a set of basis functions, similar to that done in traditional Fourier series analysis. Let us review that process. Consider a traditional Fourier series expansion of a function $u(x)$:

$$u(x) = \sum_{n=1}^{\infty} \alpha_n \varphi_n(x). \tag{13.5}$$

Here $\varphi_n(x)$, $n = 1, \ldots, \infty$, is a user-selected set of basis functions. If the sum runs to ∞, we have equality between the original function and its representation in terms of linear combinations of the basis functions, provided the basis is *complete*. For reduction, we truncate this series at finite N:

$$u(x) \approx \sum_{n=1}^{N} \alpha_n \varphi_n(x). \tag{13.6}$$

This approximation is known as a *projection* of $u(x)$ onto a finite-dimensional basis. It is analogous to projections of ordinary vectors in finite-dimensional spaces. For example, we can represent a three-dimensional vector as $\mathbf{u} = \alpha_1 \mathbf{e}_1 + \alpha_2 \mathbf{e}_2 + \alpha_3 \mathbf{e}_3$. Here $\mathbf{e}_i$ are orthonormal vectors. We can project $\mathbf{u}$ onto the $(1, 2)$ plane by approximating it as $\mathbf{u} \approx \alpha_1 \mathbf{e}_1 + \alpha_2 \mathbf{e}_2$. Formally, this projection is achieved by operating on $\mathbf{u}$ with a projection matrix $\mathbf{P}$, see Eq. (3.302), so the the projected vector $\mathbf{u}_p = \mathbf{P} \cdot \mathbf{u}$. Here, we have

$$\mathbf{u}_p = \mathbf{P} \cdot \mathbf{u} = \underbrace{\begin{pmatrix} 1 & 0 & 0 \\ 0 & 1 & 0 \\ 0 & 0 & 0 \end{pmatrix}}_{\mathbf{P}} \underbrace{\begin{pmatrix} \alpha_1 \\ \alpha_2 \\ \alpha_3 \end{pmatrix}}_{\mathbf{u}} = \underbrace{\begin{pmatrix} \alpha_1 \\ \alpha_2 \\ 0 \end{pmatrix}}_{\mathbf{u}_p}. \tag{13.7}$$

The usefulness of any projection relies on the neglected terms being small relative to the included terms. This is typically the case in Fourier series expansions, and standard mathematics texts can be consulted that consider the error of the approximation and its convergence properties, for example Kaplan (2003, Chapter 7).

An important problem is given $u(x)$ and $\varphi_n(x)$, find the so-called Fourier coefficients α_n. As long as the set of $\varphi_n(x)$ is linearly independent and complete, the set of α_n may be found. A set of basis functions is complete if it spans the space under consideration. For example, the basis vectors $\mathbf{i}$ and $\mathbf{j}$ span the space in the (x, y) plane, but do not span (x, y, z) space. One can consult standard texts on functional analysis for background, for example Zeidler (1995). If we choose the basis functions to be orthonormal, such that their inner product is the Kronecker delta,

$$\langle \varphi_n(x), \varphi_m(x) \rangle = \int_0^1 \varphi_n(x) \varphi_m(x) \, dx = \delta_{mn}, \tag{13.8}$$

we get significant simplification. Here we have assumed that our domain is $x \in [0, 1]$. We could extend this for more general domains.

Take the inner product of both sides of Eq. (13.5) with $\varphi_m(x)$ via

$$\varphi_m(x)u(x) = \sum_{n=1}^{\infty} \alpha_n \varphi_m(x)\varphi_n(x), \tag{13.9}$$

$$\int_0^1 \varphi_m(x)u(x)\, dx = \int_0^1 \sum_{n=1}^{\infty} \alpha_n \varphi_m(x)\varphi_n(x)\, dx = \sum_{n=1}^{\infty} \alpha_n \underbrace{\int_0^1 \varphi_m(x)\varphi_n(x)\, dx}_{\delta_{mn}}, \tag{13.10}$$

$$= \sum_{n=1}^{\infty} \alpha_n \delta_{mn} = \alpha_m, \quad \text{so} \quad \alpha_n = \int_0^1 \varphi_n(x)u(x)\, dx. \tag{13.11}$$

So given $u(x)$ and basis functions $\varphi_n(x)$, we can obtain the amplitudes α_n. For this traditional analysis, the α_n values are constants. This is precisely the method that has been used earlier in such places as Section 9.10 to recast functions in terms of Fourier series to aid in the solution of linear partial differential equations.

13.2 Galerkin Projection to a Low-order Dynamical System: Bateman–Burgers

For the nonlinear partial differential equations of fluid mechanics, we extend this method to functions of space and time, for example $u(x,t)$, and allow the amplitudes to have a time-dependency. We will thus assume

$$u(x,t) = \sum_{n=1}^{\infty} \alpha_n(t)\varphi_n(x). \tag{13.12}$$

We will again choose $\varphi_n(x)$, and our problem now will be to find the time-dependent amplitudes $\alpha_n(t)$. The method will have utility as a reduction only when we consider approximate expansions with finite N:

$$u(x,t) \approx \sum_{n=1}^{N} \alpha_n(t)\varphi_n(x). \tag{13.13}$$

This is the projection of $u(x,t)$ onto a finite set of basis functions within a function space.

We adopt a discussion given by Powers and Sen (2015, Section 9.10) so as to illustrate some important principles. (1) Solution of a partial differential equation can be cast in terms of solving an infinite set of ordinary differential equations. (2) Approximate solution of a partial differential equation can be cast in terms of solving a finite set of ordinary differential equations. (3) A linear partial differential equation induces an uncoupled system of linear ordinary differential equations. (4) A nonlinear partial differential equation induces a coupled system of nonlinear ordinary differential equations.

There are many viable methods to represent a partial differential equation as a system of ordinary differential equations. Among them are methods in which one or more dependent and independent variables are discretized; important examples are the finite difference and finite

13.2 Galerkin Projection to a Low-order Dynamical System: Bateman–Burgers

element methods, which will not be considered here. The method given here is different. We will first illustrate it by a fluids-motivated example involving a projection incorporating the method of weighted residuals.

Let us consider the nonlinear partial differential equation, initial and boundary conditions

$$\frac{\partial u}{\partial t} + u\frac{\partial u}{\partial x} = \nu\frac{\partial^2 u}{\partial x^2}, \quad u(x,0) = \sin \pi x, \quad u(0,t) = u(1,t) = 0. \quad (13.14)$$

Equation (13.14) is the viscous Bateman–Burgers equation, considered earlier in Section 9.12.1. It is not an equation that is a direct reduction of the continuum equations of the mechanics of fluids. However, it does share important key features: (a) a nonlinear term that is of the same form as acceleration with an advective nonlinearity, $\partial u/\partial t + u\partial u/\partial x$, and (b) a term that is of the same form as diffusion of momentum, $\nu\partial^2 u/\partial x^2$. It is not constrained by any mass conservation principle. Recognize that one is tempted to claim it is the linear momentum principle for an isobaric, incompressible fluid. But if it were incompressible, mass conservation would require $\partial u/\partial x = 0$, thus removing the advective nonlinearity. It thus must be viewed as a strictly mathematical equation that shares some features with physics-based models. The boundary conditions are homogeneous, and the initial condition is a simple sine wave. The equation is nonlinear due to the product $u\partial u/\partial x$. We will convert this single nonlinear partial differential equation to a system of ordinary differential equations using what is known as a Galerkin projection (see Finlayson, 1972) method that is one option within the broader method of weighted residuals.

Let us assume that $u(x,t)$ can be approximated in an N-term series by

$$u(x,t) = \sum_{n=1}^{N} \alpha_n(t)\varphi_n(x). \quad (13.15)$$

While we use an equality symbol, we recognize that it is an approximation, unless we let $N \to \infty$. This amounts to a type of separation of variables. We can consider $\alpha_n(t)$ to be a set of N time-dependent amplitudes that modulate each spatial basis function, $\varphi_n(x)$. For convenience, we insist that the spatial basis functions satisfy the spatial boundary conditions $\varphi_n(0) = \varphi_n(1) = 0$ as well as an orthonormality condition for $x \in [0,1]$:

$$\langle \varphi_n, \varphi_m \rangle = \delta_{nm}. \quad (13.16)$$

At the initial state, we have

$$u(x,0) = \sin \pi x = \sum_{n=1}^{N} \alpha_n(0)\varphi_n(x). \quad (13.17)$$

The terms $\alpha_n(0)$ are simply the constants in the Fourier series expansion of $\sin \pi x$:

$$\alpha_n(0) = \langle \varphi_n, \sin \pi x \rangle. \quad (13.18)$$

The partial differential equation expands as

$$\underbrace{\sum_{n=1}^{N} \frac{d\alpha_n}{dt}\varphi_n(x)}_{\partial u/\partial t} + \underbrace{\left(\sum_{n=1}^{N} \alpha_n(t)\varphi_n(x)\right)}_{u}\underbrace{\left(\sum_{n=1}^{N} \alpha_n(t)\frac{d\varphi_n}{dx}\right)}_{\partial u/\partial x} = \nu\underbrace{\sum_{n=1}^{N} \alpha_n(t)\frac{d^2\varphi_n}{dx^2}}_{\partial^2 u/\partial x^2}. \quad (13.19)$$

We change one of the dummy indices in each of the nonlinear terms from n to m and rearrange to find

$$\sum_{n=1}^{N} \frac{d\alpha_n}{dt}\varphi_n(x) + \sum_{n=1}^{N}\sum_{m=1}^{N} \alpha_n(t)\alpha_m(t)\varphi_n(x)\frac{d\varphi_m}{dx} = \nu \sum_{n=1}^{N} \alpha_n(t)\frac{d^2\varphi_n}{dx^2}. \quad (13.20)$$

For the method of weighted residuals, one can choose a set of weighting functions $\psi_l(x)$. Next, one takes the inner product of the equation with the weighting functions, yielding

$$\left\langle \psi_l(x), \sum_{n=1}^{N} \frac{d\alpha_n}{dt}\varphi_n(x) + \sum_{n=1}^{N}\sum_{m=1}^{N} \alpha_n(t)\alpha_m(t)\varphi_n(x)\frac{d\varphi_m}{dx} \right\rangle = \nu \left\langle \psi_l(x), \sum_{n=1}^{N} \alpha_n(t)\frac{d^2\varphi_n}{dx^2} \right\rangle. \quad (13.21)$$

For the Galerkin projection, one selects the weighting functions $\psi_l(x)$ to be the basis functions $\varphi_l(x)$, yielding

$$\left\langle \varphi_l(x), \sum_{n=1}^{N} \frac{d\alpha_n}{dt}\varphi_n(x) + \sum_{n=1}^{N}\sum_{m=1}^{N} \alpha_n(t)\alpha_m(t)\varphi_n(x)\frac{d\varphi_m}{dx} \right\rangle = \nu \left\langle \varphi_l(x), \sum_{n=1}^{N} \alpha_n(t)\frac{d^2\varphi_n}{dx^2} \right\rangle, \quad (13.22)$$

$$\sum_{n=1}^{N} \frac{d\alpha_n}{dt} \langle \phi_l(x), \varphi_n(x) \rangle + \sum_{n=1}^{N}\sum_{m=1}^{N} \alpha_n(t)\alpha_m(t) \left\langle \phi_l(x), \varphi_n(x)\frac{d\varphi_m}{dx} \right\rangle$$
$$= \nu \sum_{n=1}^{N} \alpha_n(t) \left\langle \phi_l(x), \frac{d^2\varphi_n}{dx^2} \right\rangle. \quad (13.23)$$

As discussed by Finlayson (1972) and Powers and Sen (2015, Section 6.11), one could also explore other choices for the weighting functions and obtain other approximations all within the context of the method of weighted residuals. Now let us choose a special set of basis functions. First, we shall insist they satisfy the boundary conditions. Second we require orthonormality: $\langle \varphi_l(x), \varphi_n(x) \rangle = \delta_{ln}$. Third, it will be convenient to require that the basis functions be eigenfunctions of the diffusion operator $-d^2/dx^2$:

$$-\frac{d^2\varphi_n}{dx^2} = \lambda_n^2 \varphi_n(x). \quad (13.24)$$

Here λ_n is an eigenvalue of the operator, which for the boundary conditions selected we know to be $n\pi$ with $n = 1, 2, \ldots$. This gives

$$\sum_{n=1}^{N} \frac{d\alpha_n}{dt}\delta_{ln} + \sum_{n=1}^{N}\sum_{m=1}^{N} \alpha_n(t)\alpha_m(t) \left\langle \phi_l(x), \varphi_n(x)\frac{d\varphi_m}{dx} \right\rangle = -\nu \sum_{n=1}^{N} \alpha_n(t)\lambda_n^2 \delta_{ln}, \quad (13.25)$$

$$\frac{d\alpha_l}{dt} + \sum_{n=1}^{N}\sum_{m=1}^{N} \alpha_n(t)\alpha_m(t) \underbrace{\left\langle \phi_l(x), \varphi_n(x)\frac{d\varphi_m}{dx} \right\rangle}_{C_{lnm}} = -\nu \lambda_l^2 \alpha_l(t). \quad (13.26)$$

Converting back to n, and taking $\lambda_n = n\pi$, our final form is a standard form of N autonomous, nonlinear ordinary differential equations from dynamical systems analysis:

13.2 Galerkin Projection to a Low-order Dynamical System: Bateman–Burgers

$$\frac{d\alpha_n}{dt} = -\underbrace{\nu n^2\pi^2\alpha_n(t)}_{\text{diffusion}} - \underbrace{\sum_{l=1}^{N}\sum_{m=1}^{N} C_{nlm}\alpha_l(t)\alpha_m(t)}_{\text{advection}}, \quad n=1,\ldots,N. \tag{13.27}$$

Here C_{lmn} is set of constants obtained after forming the various integrals of the basis functions and their derivatives. Note that for the limit in which nonlinear effects are negligible, $C_{nlm} = 0$, we get a set of N *uncoupled* linear ordinary differential equations for the time-dependent amplitudes. *Thus for the linear limit, the time-evolution of each mode is independent of the other modes.* In contrast, for $C_{nlm} \neq 0$, the system of N ordinary differential equations for amplitude time-evolution is fully coupled and nonlinear.

In any case, this serves to remove the explicit dependency on x, thus yielding a system of N ordinary differential equations of the classical form of an autonomous, nonlinear dynamical system:

$$\frac{d\boldsymbol{\alpha}}{dt} = \mathbf{f}(\boldsymbol{\alpha}), \tag{13.28}$$

where $\boldsymbol{\alpha}$ is a vector of length N, and $\mathbf{f}$ is in general a nonlinear function of $\boldsymbol{\alpha}$.

We summarize some important ideas for this equations of this type. For further background, one can consult Powers and Sen (2015, Sections 9.3–9.6). The system, Eq. (13.28), is in equilibrium when

$$\mathbf{f}(\boldsymbol{\alpha}) = \mathbf{0}. \tag{13.29}$$

This constitutes a system of nonlinear algebraic equations. Because the system is nonlinear, the existence and uniqueness of equilibria is not guaranteed. Thus, we could expect to find no roots, one root, or multiple roots, depending on $\mathbf{f}(\boldsymbol{\alpha})$. Equilibrium points are also known as *critical points* or *fixed points*. We define equilibrium points to be $\boldsymbol{\alpha} = \boldsymbol{\alpha}_{eq}$, and require that

$$\mathbf{f}(\boldsymbol{\alpha}_{eq}) = \mathbf{0}. \tag{13.30}$$

Stability of each equilibrium can be ascertained by a local linear analysis in the neighborhood of each equilibrium. Local Taylor series analysis of Eq. (13.28) in such a neighborhood allows it to be rewritten as

$$\frac{d\boldsymbol{\alpha}}{dt} = \underbrace{\mathbf{f}(\boldsymbol{\alpha}_{eq})}_{0} + \left.\frac{\partial \mathbf{f}}{\partial \boldsymbol{\alpha}}\right|_{\boldsymbol{\alpha}=\boldsymbol{\alpha}_{eq}} \cdot (\boldsymbol{\alpha} - \boldsymbol{\alpha}_{eq}) + \cdots. \tag{13.31}$$

Take the constant Jacobian matrix of $\mathbf{f}$ evaluated at $\boldsymbol{\alpha}_{eq}$ as $\mathbf{J}$:

$$\mathbf{J} = \left.\frac{\partial \mathbf{f}}{\partial \boldsymbol{\alpha}}\right|_{\boldsymbol{\alpha}=\boldsymbol{\alpha}_{eq}}. \tag{13.32}$$

Use this and the fact that $\boldsymbol{\alpha}_{eq}$ is a constant to rewrite Eq. (13.31) as

$$\frac{d}{dt}(\boldsymbol{\alpha} - \boldsymbol{\alpha}_{eq}) = \mathbf{J} \cdot (\boldsymbol{\alpha} - \boldsymbol{\alpha}_{eq}). \tag{13.33}$$

With $\mathbf{c}$ as an arbitrary constant, this linear system has an exact solution in terms of the matrix exponential:

$$\boldsymbol{\alpha} - \boldsymbol{\alpha}_{eq} = \mathbf{c} \cdot e^{\mathbf{J}t}. \tag{13.34}$$

With $\mathbf{S}$ as the matrix whose columns are populated by the eigenvectors of $\mathbf{J}$ and $\mathbf{\Lambda}$ as the diagonal matrix whose diagonal is populated by the eigenvalues of $\mathbf{J}$, taking care to ensure the order is such that the correct eigenvalues correspond to the correct eigenvectors, the solution can be recast as

$$\boldsymbol{\alpha} - \boldsymbol{\alpha}_{eq} = \mathbf{c} \cdot \mathbf{S} \cdot e^{\mathbf{\Lambda} t} \cdot \mathbf{S}^{-1}. \tag{13.35}$$

Obviously, the eigenvalues of $\mathbf{J}$ determine that stability of each equilibrium. For stability, the real parts of each eigenvalue cannot be positive. A *source* has all real parts positive. A *sink* has all real parts negative. A *center* has all eigenvalues purely imaginary. A *saddle* has some real parts positive and some negative.

Returning to our problem, we select our orthonormal basis functions as the eigenfunctions of $-d^2/dx^2$ that also satisfy the appropriate boundary conditions,

$$\varphi_n(x) = \sqrt{2}\sin(n\pi x), \qquad n = 1,\ldots,N. \tag{13.36}$$

We then apply Fourier expansion to find $\alpha_n(0)$, perform a detailed analysis of all of the necessary inner products, select $N = 2$, and arrive at the following nonlinear system of autonomous ordinary differential equations for the evolution of the time-dependent amplitudes:

$$\frac{d\alpha_1}{dt} = -\pi^2\nu\alpha_1 + \frac{\pi}{\sqrt{2}}\alpha_1\alpha_2, \qquad \alpha_1(0) = \frac{1}{\sqrt{2}}, \tag{13.37}$$

$$\frac{d\alpha_2}{dt} = -4\pi^2\nu\alpha_2 - \frac{\pi}{\sqrt{2}}\alpha_1^2 = 0, \qquad \alpha_2(0) = 0. \tag{13.38}$$

The system is in equilibrium at all points (α_1, α_2) where

$$f_1(\alpha_1, \alpha_2) = -\pi^2\nu\alpha_1 + \frac{\pi}{\sqrt{2}}\alpha_1\alpha_2 = 0, \tag{13.39}$$

$$f_2(\alpha_1, \alpha_2) = -4\pi^2\nu\alpha_2 - \frac{\pi}{\sqrt{2}}\alpha_1^2 = 0. \tag{13.40}$$

The general form of the Jacobian matrix $\mathbf{J}$ is

$$\mathbf{J} = \begin{pmatrix} \frac{\partial f_1}{\partial \alpha_1} & \frac{\partial f_1}{\partial \alpha_2} \\ \frac{\partial f_2}{\partial \alpha_1} & \frac{\partial f_2}{\partial \alpha_2} \end{pmatrix} = \begin{pmatrix} \frac{\pi\alpha_2}{\sqrt{2}} - \pi^2\nu & \frac{\pi\alpha_1}{\sqrt{2}} \\ -\sqrt{2}\pi\alpha_1 & -4\pi^2\nu \end{pmatrix}. \tag{13.41}$$

We find one real equilibria, $(\alpha_1, \alpha_2) = (0,0)$, and eigenvalues, (λ_1, λ_2), associated with local values of the Jacobian matrix, $\mathbf{J}$. They are

$$(\alpha_1, \alpha_2) = (0,0), \qquad (\lambda_1, \lambda_2) = (-\pi^2\nu, -4\pi^2\nu), \qquad \text{sink.} \tag{13.42}$$

Near the equilibrium, the two-term approximation is

$$u(x,t) \approx C_1 e^{-\pi^2\nu t}\sin(\pi x) + C_2 e^{-4\pi^2\nu t}\sin(2\pi x). \tag{13.43}$$

The equilibrium is locally stable as nearby initial conditions have amplitudes that are decaying exponentially with time. We see the high frequency modes decay more rapidly.

Away from the equilibrium, numerical solution is required. We do so for $\nu = 1/100$ and plot the phase plane dynamics in Fig. 13.2 for arbitrary initial conditions as well as a special trajectory representing our initial conditions. Many initial conditions lead one to the finite sink

13.2 Galerkin Projection to a Low-order Dynamical System: Bateman–Burgers

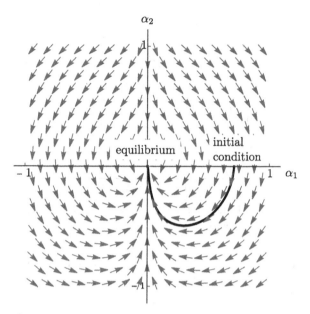

Figure 13.2 Phase plane dynamics of $N=2$ amplitudes of spatial modes of solution to viscous Bateman–Burgers equation with $\nu = 1/100$.

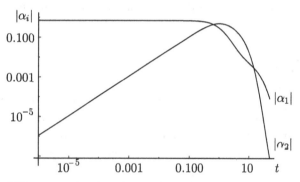

Figure 13.3 Evolution of $N=2$ amplitudes of spatial modes of solution to viscous Bateman–Burgers equation with $\nu = 1/100$.

at $(0,0)$. It is likely that the dynamics are also influenced by equilibria at infinity, not shown here. One can show that the solutions in the neighborhood of the sink are the most relevant to the underlying physical problem.

We plot results of $|\alpha_1(t)|$, $|\alpha_2(t)|$ for our initial conditions in Fig. 13.3. The absolute value is taken because the actual amplitudes could be positive or negative. In this simulation $\alpha_1 \geq 0$, $\alpha_2 \leq 0$. We see at early time the $n=2$ mode grows from zero to soon overtake the $n=1$ mode. However, as $t \to \infty$, the $n=2$ mode decays more rapidly, as expected. Both modes are decaying rapidly to the sink at $(0,0)$. The $N=2$ solution with full time and space dependency is

$$u(x,t) \approx \alpha_1(t)\sin(\pi x) + \alpha_2(t)\sin(2\pi x), \tag{13.44}$$

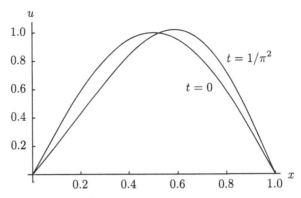

Figure 13.4 $u(x, t=0)$ and $u(x, t=1/\pi^2)$ from $N=2$ term Galerkin projection of solution to viscous Bateman–Burgers equation with $\nu = 1/100$.

and is plotted in Fig. 13.4 at two times $t = 0$ and $t = 1/\pi^2$. Similar to what was seen in independent simulations in Section 9.12.1, as time progresses, the front side of the wave steepens, and the back side flattens. It is the mode coupling due the nonlinearities in the problem that induce this asymmetric behavior, especially at early time. At late time, diffusion dominates, and all modes have their amplitudes approach zero. This illustrates some principles that carry forward to actual fluids: (1) nonlinear advective acceleration can lead to the steepening of wave forms, and (2) diffusion serves to dissipate structure.

13.3 Landau Equation

In a similar sense that the Bateman–Burgers equation is a scalar surrogate for more physically based fluid mechanical models, the so-called *Landau[3] equation* (see Landau and Lifshitz, 1959, p. 106) is a surrogate for many nonlinear phenomena predicted by fluid mechanical models. One can find an extended discussion in Drazin and Reid (1981) or Drazin (2002). Here we adapt some of the discussion from the latter, and employ the notation of this chapter. One finds many incarnations of the Landau equation in the literature. While it is possible in some cases to reduce a fluid mechanical system to a Landau equation, it is not always straightforward. Drazin's work, and citations within, discuss how the Rayleigh–Bénard flow problem of Section 12.4 may be reduced to the much simpler one-term approximation, which is one version of a Landau equation:

$$\frac{d\alpha}{dt} = k(r - r_c)\alpha - \ell\alpha^3. \tag{13.45}$$

Here α is the time-dependent amplitude of a single mode. Because we have only one mode, there is no need for a subscript on α. We have positive constants $k, \ell, r, r_c > 0$. One can consider r to be a surrogate for the Rayleigh number Ra and r_c to be a surrogate for the critical Rayleigh number Ra_c as discussed in Section 12.4. The literature uses the subscript c for various critical constants that often change meaning from problem to problem.

---

[3] Lev Davidovich Landau, 1908–1968, Soviet physicist.

13.3 Landau Equation

Equilibrium is attained when

$$k(r - r_c)\alpha - \ell\alpha^3 = 0. \tag{13.46}$$

This yields three candidates for equilibrium:

$$\alpha_{eq} = 0, \qquad \alpha_{eq} = \pm\sqrt{\frac{k(r - r_c)}{\ell}}. \tag{13.47}$$

Let us study the linear stability of each equilibria. First, in the neighborhood of $\alpha = 0$, Taylor series shows the Landau equation reduces to

$$\frac{d\alpha}{dt} = k(r - r_c)\alpha + \cdots. \tag{13.48}$$

Because all the constants are positive, we have stability if $r < r_c$. For $r > r_c$, the equilibrium $\alpha = 0$ is unstable.

Taylor series expansion near the other two equilibria show the Landau equation reduces to

$$\frac{d}{dt}\left(\alpha \pm \sqrt{\frac{k(r - r_c)}{\ell}}\right) = -2k(r - r_c)\left(\alpha \pm \sqrt{\frac{k(r - r_c)}{\ell}}\right) + \cdots. \tag{13.49}$$

For $r < r_c$, this solution is not real, and so we do not consider it. For $r > r_c$, both of these equilibria are linearly stable.

For $r = r_c$, the system has a fundamental nonlinearity. It becomes

$$\frac{d\alpha}{dt} = -\ell\alpha^3. \tag{13.50}$$

Enforcing the condition $\alpha(0) = \alpha_o$ and solving gives

$$\alpha(t) = \pm\frac{1}{\sqrt{2\ell t + \frac{1}{\alpha_o^2}}}. \tag{13.51}$$

Obviously, as $t \to \infty$, $\alpha \to 0$, so when $r = r_c$, the equilibrium at $\alpha = 0$ is *nonlinearly stable*. The variation of the equilibrium values of α with r is displayed in Fig. 13.5. The thick solid curves

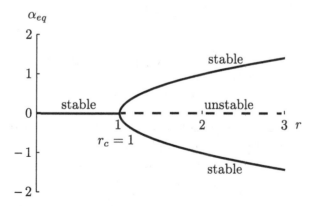

Figure 13.5 Bifurcation diagram for the Landau equation for $k = 1$, $r_c = 1$, $\ell = 1$, with stable equilibria (solid curves) and unstable equilibria (dashed curves)

represent stable equilibria and the thick dashed curve represents unstable equilibria. As r is increased, the $\alpha = 0$ equilibrium remains stable until $r = r_c = 1$, where the system undergoes a *pitchfork bifurcation*. For $r > r_c$, $\alpha = 0$ is unstable, but two stable equilibria are possible.

13.4 Lorenz Equations

We consider now one of the more well-known and important systems in nonlinear dynamics: the *Lorenz*[4] *equations*. The system has been widely discussed in the applied mathematics literature, where its foundational connection to fluid mechanics is not always obvious. One of the better mathematically oriented summaries is given by Drazin (1992). Lorenz (1963) posed and numerically solved this problem in a fashion that was fully grounded in the models of Rayleigh–Bénard flow studied in Section 12.4.

We will focus on the fluid mechanics that lead to the Lorenz equations and present some solutions that illustrate the remarkable features of nonlinear dynamics. Among the features we will see is that well defined and seemingly simple deterministic low-order nonlinear fluid dynamical systems can induce complicated solutions that, because of an extreme sensitivity to initial conditions and parameter values, have fine scale details that are effectively impossible to predict at long time. Such solutions are known as *chaotic*. This has implications for one of the most challenging unsolved problems of fluid mechanics: turbulence (see Chapter 14). They are discussed at length by Chevray and Mathieu (1993) and by Tritton (1988, p. 410), who conclude that turbulent fluid dynamics are one type of deterministic chaos in that highly disordered motion with strong sensitivity to initial conditions is predicted by the fully deterministic Navier–Stokes equations. It must of course be realized that there is a large gap between the predictions of the simple Lorenz model and most observed turbulent fluid mechanics in practical devices.

13.4.1 Derivation from Boussinesq Approximation

Consider a slightly more general version of the dimensionless Boussinesq approximation we have considered earlier, Eqs. (12.136–12.138):

$$\nabla^T \cdot \mathbf{v} = 0, \tag{13.52}$$

$$\frac{1}{Pr}\left(\frac{\partial \mathbf{v}}{\partial t} + \mathbf{v}^T \cdot \nabla \mathbf{v}\right) = (Ra)T\mathbf{k} - \frac{1}{Pr}\nabla p + \nabla^2 \mathbf{v}, \tag{13.53}$$

$$\frac{\partial T}{\partial t} + \mathbf{v}^T \cdot \nabla T - w = \nabla^2 T. \tag{13.54}$$

We have employed the same assumptions as those for the earlier analysis, except that we retained nonlinear advective terms involving the velocity vector, such as found in Eqs. (6.104, 6.105). Neglecting v as well as all variation in y, we can rewrite these as

$$\frac{\partial u}{\partial x} + \frac{\partial w}{\partial z} = 0, \tag{13.55}$$

$$\frac{1}{Pr}\left(\frac{\partial u}{\partial t} + u\frac{\partial u}{\partial x} + w\frac{\partial u}{\partial z}\right) = -\frac{1}{Pr}\frac{\partial p}{\partial x} + \frac{\partial^2 u}{\partial x^2} + \frac{\partial^2 u}{\partial z^2}, \tag{13.56}$$

---

[4] Edward Norton Lorenz, 1917–2008, American mathematician and meteorologist.

13.4 Lorenz Equations

$$\frac{1}{Pr}\left(\frac{\partial w}{\partial t} + u\frac{\partial w}{\partial x} + w\frac{\partial w}{\partial z}\right) = (Ra)T - \frac{1}{Pr}\frac{\partial p}{\partial z} + \frac{\partial^2 w}{\partial x^2} + \frac{\partial^2 w}{\partial z^2}, \quad (13.57)$$

$$\frac{\partial T}{\partial t} + u\frac{\partial T}{\partial x} + w\frac{\partial T}{\partial z} - w = \frac{\partial^2 T}{\partial x^2} + \frac{\partial^2 T}{\partial z^2}. \quad (13.58)$$

The term $-w$ in Eqs. (13.54, 13.58) appears unusual. Its origin we recall is from the advection in the z direction of the mean flow constant temperature gradient in the z direction. In our equations, T represents a perturbation of the mean flow values. In that sense the term $-w$ is analogous to the external forcing pressure gradient in the Falkner–Skan flows of Section 11.2.2. So the term $-w$ is not redundant with the term $w\partial T/\partial z$.

Following the useful exposition of Hilborn (2000), one can define a stream function ψ such that

$$u = -\frac{\partial \psi}{\partial z}, \qquad w = \frac{\partial \psi}{\partial x}. \quad (13.59)$$

Here the alternative sign convention for ψ, common in meteorology, has been employed because the problem has it origins in that discipline. With this definition, mass conservation is automatically satisfied. The remaining momenta and energy equations may be rewritten as

$$\frac{1}{Pr}\left(-\frac{\partial^2 \psi}{\partial t \partial z} + \frac{\partial \psi}{\partial z}\frac{\partial^2 \psi}{\partial x \partial z} - \frac{\partial \psi}{\partial x}\frac{\partial^2 \psi}{\partial z^2}\right) = -\frac{1}{Pr}\frac{\partial p}{\partial x} - \nabla^2\left(\frac{\partial \psi}{\partial z}\right), \quad (13.60)$$

$$\frac{1}{Pr}\left(\frac{\partial^2 \psi}{\partial t \partial x} - \frac{\partial \psi}{\partial z}\frac{\partial^2 \psi}{\partial x^2} + \frac{\partial \psi}{\partial x}\frac{\partial^2 \psi}{\partial x \partial z}\right) = (Ra)T - \frac{1}{Pr}\frac{\partial p}{\partial z} + \nabla^2\left(\frac{\partial \psi}{\partial x}\right), \quad (13.61)$$

$$\frac{\partial T}{\partial t} - \frac{\partial \psi}{\partial z}\frac{\partial T}{\partial x} + \frac{\partial \psi}{\partial x}\frac{\partial T}{\partial z} - \frac{\partial \psi}{\partial x} = \nabla^2 T. \quad (13.62)$$

Now if we take $\partial/\partial x$ of the z momentum equation and subtract from it $\partial/\partial z$ of the x momentum equation, we get, after a detailed calculation,

$$\frac{1}{Pr}\left(\frac{\partial}{\partial t}\nabla^2\psi - \frac{\partial}{\partial z}\left(\frac{\partial \psi}{\partial z}\frac{\partial^2 \psi}{\partial x \partial z} - \frac{\partial \psi}{\partial x}\frac{\partial^2 \psi}{\partial z^2}\right) - \frac{\partial}{\partial x}\left(\frac{\partial \psi}{\partial z}\frac{\partial^2 \psi}{\partial x^2} - \frac{\partial \psi}{\partial x}\frac{\partial^2 \psi}{\partial x \partial z}\right)\right)$$
$$= (Ra)\frac{\partial T}{\partial x} + \nabla^4\psi. \quad (13.63)$$

We now have two equations for two unknowns $\psi(x,z,t)$ and $T(x,z,t)$. It is at this point we could impose a formal Galerkin projection by postulating perhaps

$$\psi(x,z,t) = \sum_{n=1}^{N}\sum_{m=1}^{M} \Psi_{nm}(t)\varphi_n(x)\varphi_m(z), \quad (13.64)$$

$$T(x,z,t) = \sum_{n=1}^{N}\sum_{m=1}^{M} \mathsf{T}_{nm}(t)\varphi_n(x)\varphi_m(z). \quad (13.65)$$

and proceed as before by selecting convenient basis functions, projecting the partial differential equations onto the basis, retain only desired functions, and solving the resulting system of ordinary differential equations. This is complicated. Let us perform an equivalent procedure motivated by incisive simplifying choices made by Lorenz (1963). We take

$$\psi(x,z,t) = \Psi(t)\sin\pi z \sin ax, \quad (13.66)$$

$$T(x,z,t) = \mathsf{T}_1(t)\sin\pi z \cos ax - \mathsf{T}_2(t)\sin 2\pi z. \quad (13.67)$$

Here a is a parameter that we are free to choose, as we have not specified boundary conditions in x. We are considering one Fourier mode in x and two Fourier modes in z. The basis functions in z clearly satisfy the boundary conditions at $z = 0, 1$. Because of nonlinear mode coupling, we may expect these choices to induce higher frequency modes, which we will truncate. With these assumptions, one has

$$u(x,z,t) = -\frac{\partial \Psi}{\partial z} = -\pi \Psi(t) \cos \pi z \sin ax, \quad w(x,z,t) = \frac{\partial \Psi}{\partial x} = a\Psi(t) \sin \pi z \cos ax. \tag{13.68}$$

Clearly, with these choices, the surfaces $z = 0, 1$ have no mass flux as $w = 0$. However, u need not be zero on these surfaces, so they represent slip surfaces. However, both at $z = 0, 1$, we see that $\partial u/\partial z = 0$ and $\partial w/\partial x = 0$, so they represent surfaces where the lateral stress $\tau_{zx} = 0$, as was considered in Section 12.4. For this analysis $\tau_{zy} = 0$ because of our assumptions. The stream function at both $z = 0, 1$ is zero. At both surfaces the perturbed temperature $T(x, z = 0, 1, t) = 0$, so they are isothermal surfaces.

We make these substitutions into the momenta equation, Eq. (13.63), and recover after detailed calculation

$$\underbrace{\left(\frac{d\Psi}{dt} - \frac{a(Pr)(Ra)}{a^2 + \pi^2}T_1 + (a^2 + \pi^2)(Pr)\Psi\right)}_{0}(\sin ax)(\sin \pi z) = 0. \tag{13.69}$$

Note that the terms involving Ψ and T_1 are linear. Because of this, one sees that nonlinear advection has not influenced this limit of the momenta equation. We then make the same substitutions into the energy equation, Eq. (13.62), and get, after detailed calculation using the trigonometric identity $\sin \pi z \cos 2\pi z = (-\sin \pi z - + \sin 3\pi z)/2$ and neglecting the $\sin 3\pi z$ term,

$$\underbrace{\left(\frac{dT_1}{dt} + (a^2 + \pi^2)T_1 - a\Psi + a\pi T_2 \Psi\right)}_{0}(\cos ax)(\sin \pi z) + \underbrace{\left(\frac{dT_2}{dt} + 4\pi^2 T_2 - \frac{a\pi}{2}T_1 \Psi\right)}_{0}(\sin 2\pi z) = 0. \tag{13.70}$$

For the energy equation, we see that nonlinearity enters with a coupling involving the products $T_2 \Psi$ and $T_1 \Psi$. Now the basis functions are linearly independent, so we can guarantee the momenta and energy equations are satisfied by demanding the coefficients on the respective basis functions vanish for all time. This yields the following three autonomous ordinary differential equations in time:

$$\frac{d\Psi}{dt} = \frac{a(Pr)(Ra)}{a^2 + \pi^2}T_1 - (a^2 + \pi^2)(Pr)\Psi, \tag{13.71}$$

$$\frac{dT_1}{dt} = -(a^2 + \pi^2)T_1 + a\Psi - a\pi T_2 \Psi, \tag{13.72}$$

$$\frac{dT_2}{dt} = -4\pi^2 T_2 + \frac{a\pi}{2}T_1 \Psi. \tag{13.73}$$

Now, let us rescale so that the equations appear in a common form. We will first replace the Rayleigh number Ra by a reduced Rayleigh number r:

$$Ra = r\frac{(a^2 + \pi^2)^3}{a^2}. \tag{13.74}$$

Note that $r = Ra/Ra_{n=1}$ gives the scaled value of Rayleigh number when scaled by its neutral stability value for the $n = 1$ mode given by Eq. (12.166). We next rename the Prandtl number as σ:

$$Pr = \sigma. \tag{13.75}$$

We also remove a in favor of b via

$$a = \sqrt{\frac{4\pi^2}{b} - \pi^2}. \tag{13.76}$$

Then we scale all variables in the following fashion:

$$\hat{t} = (\pi^2 + a^2)t, \quad \hat{\Psi} = \frac{a\pi}{\sqrt{2}(\pi^2 + a^2)}\Psi, \quad \hat{\mathsf{T}}_1 = \frac{r\pi}{\sqrt{2}}\mathsf{T}_1, \quad \hat{\mathsf{T}}_2 = r\pi\mathsf{T}_2. \tag{13.77}$$

We recall from Eq. (12.167) that the critical value of a for which there is a single Rayleigh number that induces marginal stability is $a_c = \pi/\sqrt{2}$. For this value of a, we find $b = 8/3$. At this value of $a = a_c$, we have $Ra = r(27\pi^4/4) = rRa_c$. And if we study $r = 1$, we are studying the system at the margin of stability, where there is a single value of r that is at the margin. While we are tempted to identify $r = 1$ to be r_c, for the Lorenz system, r_c is typically reserved for a different value of r, and $r = 1$ is simply understood to be an important point.

Making these substitutions, then removing the "hat" notation, one arrives at the well-known Lorenz equations:

$$\frac{d\Psi}{dt} = \sigma(\mathsf{T}_1 - \Psi), \tag{13.78}$$

$$\frac{d\mathsf{T}_1}{dt} = r\Psi - \mathsf{T}_1 - \mathsf{T}_2\Psi, \tag{13.79}$$

$$\frac{d\mathsf{T}_2}{dt} = -b\mathsf{T}_2 + \mathsf{T}_1\Psi. \tag{13.80}$$

We have retained more of the traditional fluids notation that is typically found in the nonlinear dynamics literature. In that literature, following Lorenz, it is common to replace state variables $(\Psi, \mathsf{T}_1, \mathsf{T}_2)^T$ with $(X, Y, Z)^T$. Otherwise they are identical. In the remainder of this section, a discussion of some standard features of the Lorenz system is given, drawing heavily on the presentation of Powers and Sen (2015, Section 9.11), but with more appeal to the underlying fluid mechanics.

13.4.2 Equilibrium

The Lorenz system is at equilibrium at critical points in $(\Psi, \mathsf{T}_1, \mathsf{T}_2)^T$ space for which the derivative of the state variables are zero. The critical points are obtained from the nonlinear algebraic problem

$$\mathsf{T}_1 - \Psi = 0, \tag{13.81}$$
$$r\Psi - \mathsf{T}_1 - \Psi\,\mathsf{T}_2 = 0, \tag{13.82}$$
$$-b\mathsf{T}_2 + \Psi\,\mathsf{T}_1 = 0, \tag{13.83}$$

which gives, for $r \neq 1$, three distinct equilibria:

$$\begin{pmatrix} \Psi \\ \mathsf{T}_1 \\ \mathsf{T}_2 \end{pmatrix} = \underbrace{\begin{pmatrix} 0 \\ 0 \\ 0 \end{pmatrix}}_{A}, \underbrace{\begin{pmatrix} \sqrt{b(r-1)} \\ \sqrt{b(r-1)} \\ r-1 \end{pmatrix}}_{B}, \underbrace{\begin{pmatrix} -\sqrt{b(r-1)} \\ -\sqrt{b(r-1)} \\ r-1 \end{pmatrix}}_{C}. \tag{13.84}$$

If $r = 1$, there is a unique critical point at the origin. Let us call the origin Point A, the second, Point B, and the third, Point C. At Point A, we have from Eqs. (13.66, 13.67) that

$$\psi(x, z, t) = 0, \qquad T(x, z, t) = 0. \tag{13.85}$$

There is no spatial or temporal structure at this equilibrium. At Points B and C, we have from Eqs. (13.66, 13.67) that

$$\psi(x, z, t) = \pm\sqrt{b(r-1)}\,\sin \pi z \sin ax, \tag{13.86}$$
$$T(x, z, t) = \pm\sqrt{b(r-1)}\,\sin \pi z \cos ax - (r-1)\sin 2\pi x, \tag{13.87}$$

with a given from Eq. (13.76). Because this is an equilibrium, there is no temporal structure; however, there is spatial structure, corresponding to steady convection rolls.

Example 13.1 For $\sigma = 10$, $r = 10$, $b = 8/3$, study a nontrivial equilibrium structure.

Solution
This case is above the critical Rayleigh number, which exists at $r = 1$. And for the b we consider, we get $a = \pi/\sqrt{2}$ from Eq. (13.76). For these values, Eqs. (13.86, 13.87) yield the equilibrium spatial structures

$$\psi(x, z) = 2\sqrt{6}(\sin \pi z)\left(\sin \frac{\pi x}{\sqrt{2}}\right), \tag{13.88}$$

$$T(x, z) = 2\sqrt{6}(\sin \pi z)\left(\cos \frac{\pi x}{\sqrt{2}}\right) - 3\sin 2\pi z. \tag{13.89}$$

The equilibrium ψ and T for this case are plotted Fig. 13.6. Here we have considered the domain $0 < z < 1$, which is the region between the hot plate at $z = 0$ and the cold plate at $z = 1$. The domain length in x is arbitrary, and we have chosen $0 < x < 2\pi/a$ so that we have captured one wavelength in x. The patterns are periodic in x. Clearly there is a circulation pattern within the channel with cells whose circulation alternates from clockwise to counterclockwise. The temperature perturbation field is also periodic. Perturbations at $z = 0, 1$ are zero, while there is a tendency for negative perturbations to exist in the lower half and positive perturbations in the upper half, though that characterization is rough.

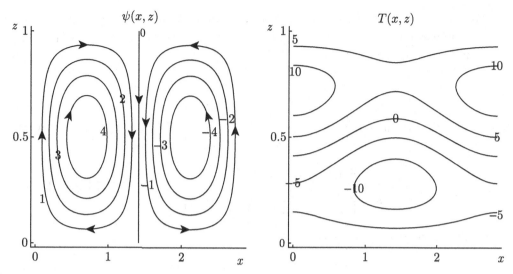

Figure 13.6 For $\sigma = 10$, $r = 10$, $b = 8/3$, equilibrium fields of $\psi(x, z)$ and $T(x, z)$ for the Lorenz system, Eqs. (13.78–13.80).

13.4.3 Linear Stability

A linear stability analysis of each of the three critical points follows.

Point A

At Point A, we have $\Psi = \mathsf{T}_1 = \mathsf{T}_2 = 0$. Small perturbations, denoted by a tilde superscript, around this point give after linearization:

$$\frac{d}{dt}\begin{pmatrix}\tilde{\Psi} \\ \tilde{\mathsf{T}}_1 \\ \tilde{\mathsf{T}}_2\end{pmatrix} = \begin{pmatrix}-\sigma & \sigma & 0 \\ r & -1 & 0 \\ 0 & 0 & -b\end{pmatrix}\begin{pmatrix}\tilde{\Psi} \\ \tilde{\mathsf{T}}_1 \\ \tilde{\mathsf{T}}_2\end{pmatrix}. \qquad (13.90)$$

We make the standard assumption that $\tilde{\Psi} = Ae^{\lambda t}$, $\tilde{\mathsf{T}}_1 = Be^{\lambda t}$, $\tilde{\mathsf{T}}_2 = Ce^{\lambda t}$ and are led to a homogeneous linear algebraic system that has nontrivial solution only if the determinant of the coefficient matrix is zero. This leads to the characteristic polynomial equation for λ, which is

$$(\lambda + b)(\lambda^2 + \lambda(\sigma + 1) - \sigma(r - 1)) = 0, \qquad (13.91)$$

from which we get the eigenvalues

$$\lambda = -b, \qquad \lambda = \frac{1}{2}\left(-(1 + \sigma) \pm \sqrt{(1 + \sigma)^2 - 4\sigma(1 - r)}\right). \qquad (13.92)$$

As we have seen before, eigenvalues with negative real parts correspond to stable modes, and any eigenvalue with a positive real part corresponds to an unstable mode. Any eigenvalue that has a nonzero imaginary part admits an oscillatory character to the local solution. For $0 < r < 1$, the eigenvalues are real and negative, because $(1 + \sigma)^2 > 4\sigma(1 - r)$, and the origin is stable to small perturbations. At $r = 1$, there is a pitchfork bifurcation with one zero eigenvalue. For $r > 1$, the origin becomes unstable. While we could name the point $r = 1$ as r_c, similar to what was done in Section 13.3, we will reserve r_c for another point of interest for the Lorenz equations.

Point B

At Point B we have $\Psi = \mathsf{T}_1 = \sqrt{b(r-1)}$, $\mathsf{T}_2 = r - 1$. We need $r \geq 1$ for a real solution. Small perturbations about this equilibrium give, after linearization:

$$\frac{d}{dt}\begin{pmatrix} \tilde{\Psi} \\ \tilde{\mathsf{T}}_1 \\ \tilde{\mathsf{T}}_2 \end{pmatrix} = \begin{pmatrix} -\sigma & \sigma & 0 \\ 1 & -1 & -\sqrt{b(r-1)} \\ \sqrt{b(r-1)} & \sqrt{b(r-1)} & -b \end{pmatrix} \begin{pmatrix} \tilde{\Psi} \\ \tilde{\mathsf{T}}_1 \\ \tilde{\mathsf{T}}_2 \end{pmatrix}. \tag{13.93}$$

The characteristic polynomial equation is

$$\lambda^3 + (\sigma + b + 1)\lambda^2 + (\sigma + r)b\lambda + 2\sigma b(r-1) = 0. \tag{13.94}$$

This system is difficult to fully analyze. Detailed analysis reveals of a critical value of r:

$$r = r_c = \frac{\sigma(\sigma + b + 3)}{\sigma - b - 1}. \tag{13.95}$$

For the Lorenz system, this is the value of r that the literature commonly labels $r = r_c$. At $r = r_c$ the characteristic polynomial, Eq. (13.94), can be factored to give the eigenvalues

$$\lambda = -(\sigma + b + 1), \qquad \lambda = \pm i\sqrt{\frac{2b\sigma(\sigma + 1)}{\sigma - b - 1}}. \tag{13.96}$$

If $\sigma > b + 1$, two of the eigenvalues are purely imaginary, and this corresponds to what is known as a *Hopf bifurcation*. The periodic solution which is created at this value of r can be shown to be unstable so that the bifurcation is what is known as *subcritical*. If $r = r_c$ and $\sigma < b + 1$, one can find all real eigenvalues, including at least one positive eigenvalue, which tells us this is unstable. We also find instability if $r > r_c$. If $r > r_c$ and $\sigma > b + 1$, we can find one negative real eigenvalue and two complex eigenvalues with positive real parts; hence, this is unstable. If $r > r_c$, and $\sigma < b + 1$, we can find three real eigenvalues, with at least one positive; this is unstable. For $1 < r < r_c$ and $\sigma < b + 1$, we find three real eigenvalues, one of which is positive; this is unstable.

For stability, we can take

$$1 < r < r_c, \quad \text{and} \quad \sigma > b + 1. \tag{13.97}$$

In this case, we can find one negative real eigenvalue and two eigenvalues (which could be real or complex) with negative real parts; hence, this is stable.

Point C

At Point C we have $\Psi = \mathsf{T}_1 = -\sqrt{b(r-1)}$, $\mathsf{T}_2 = r - 1$. Analysis of this critical point is essentially identical to that of the previous point.

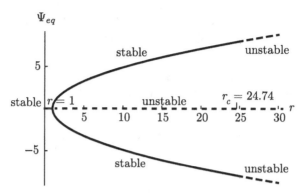

Figure 13.7 Bifurcation diagram exhibiting a pitchfork bifurcation for the Lorenz equations, Eqs. (13.78–13.80), with $\sigma = 10$, $b = 8/3$; stable equilibria (solid curves), unstable equilibria (dashed curves).

Bifurcation Diagram

For a particular case, these results are summarized in the bifurcation diagram of Fig. 13.7. It shares common features with the bifurcation diagram presented in study of the Landau equation in Section 13.3, though it is more complicated. Shown here are results when $\sigma = 10$, $b = 8/3$. For these values, Eq. (13.95) tells us that $r_c = 24.74$. Also, $\sigma > b+1$. For real equilibria, we need $r > 0$. The equilibrium at the origin is stable for $r \in [0, 1]$ and unstable for $r > 1$; an unstable equilibrium is denoted by the dashed line. At $r = 1$, there is a pitchfork bifurcation, and two new real equilibria are available. These are both linearly stable for $r \in [1, r_c]$. For $r \in [1, 1.345\,62]$, the eigenvalues are both real and negative. For $r \in [1.134\,562, r_c]$, two of the eigenvalues become complex, but all three have negative real parts, so local linear stability is maintained. For $r > r_c$, all three equilibria are unstable and are indicated by dashed lines. As an aside, because of nonlinear effects, some initial conditions yield trajectories which do not relax to a stable equilibrium for $r < r_c$. It can be shown, for example, that if $\Psi(0) = \mathsf{T}_1(0) = \mathsf{T}_2(0) = 1$, then $r = 24 < r_c$ gives rise to a trajectory that never reaches either of the linearly stable critical points.

13.4.4 Transition from Order to Chaos to Order

By varying the bifurcation parameter r, we can illustrate the bifurcation process, which includes transition to *chaos* as well as a transition back to order. We illustrate this transition for several simulations of the Lorenz equations. In each we will take as an initial condition $\Psi(0) = 0$, $\mathsf{T}_1(0) = 1$, $\mathsf{T}_2(0) = 0$. Physically, our initial condition is a fluid at rest with a low frequency temperature perturbation. The series of examples will each have $\sigma = 10$, $b = 8/3$. The reduced Rayleigh number r will vary from $r = 0, 1, 10, 28, 150$. The first case with $r = 0$ will have trajectories that relax to an asymptotically stable fixed point. The second will relax with neutral stability to the same stable fixed point. The third will exhibit damped oscillations as it relaxes to a different stable fixed point. The fourth will display a transition to chaos as trajectories relax to what is known as a *strange attractor*. However, the fifth will display a relaxation to an ordered stable *limit cycle*. A complete study of a continuum of values of r would display a wide variety of behaviors, all enabled by the nonlinearity of the system.

Example 13.2 Examine the solution to the Lorenz equations for conditions: $\sigma = 10$, $r = 0$, $b = 8/3$ with initial conditions $\Psi(0) = 0$, $\mathsf{T}_1(0) = 1$, $\mathsf{T}_2(0) = 0$.

Solution
We know that $r = 0$, there is no Rayleigh number, and thus no forcing. From Eq. (13.84), the only fixed point to examine is the origin $(\Psi, \mathsf{T}_1, \mathsf{T}_2)^T = (0, 0, 0)^T$. We find its stability by solving the characteristic equation, Eq. (13.91):

$$(\lambda + b)(\lambda^2 + \lambda(\sigma + 1) + \sigma) = 0, \quad \text{so} \quad (\lambda + 1)\left(\lambda + \frac{8}{3}\right)(\lambda + 10) = 0. \tag{13.98}$$

Solution gives $\lambda = -1$, $\lambda = -8/3$, and $\lambda = -10$. Notice that for $r = 0$, the linearization matrix of Eq. (13.90) reduces to

$$\begin{pmatrix} -\sigma & \sigma & 0 \\ 0 & -1 & 0 \\ 0 & 0 & -b \end{pmatrix}. \tag{13.99}$$

Because it is diagonal, the eigenvalues are on the diagonal and are $-\sigma$, -1, and $-b$. Because all of the eigenvalues are negative and real, the origin is stable. Figure 13.8 shows the phase space trajectories in $(\Psi, \mathsf{T}_1, \mathsf{T}_2)^T$ space and the behavior in the time domain, $\Psi(t), \mathsf{T}_1(t), \mathsf{T}_2(t)$. Here we see the nonzero T_1 induces initial growth in both Ψ and T_2. This is purely due to mode-coupling induced by nonlinearities in the system. As time advances, all state variables approach their equilibrium value of zero at the stable equilibrium point. The reciprocal of the magnitude of the least negative eigenvalue gives a good estimate of the time scale of relaxation, that being $\mathcal{O}(1)$. The final state is thus entirely at rest, with steady state heat conduction from the hot plate at $z = 0$ to the cold plate at $z = 1$.

Example 13.3 Examine the solution to the Lorenz equations for conditions: $\sigma = 10$, $r = 1$, $b = 8/3$ with initial conditions $\Psi(0) = 0$, $\mathsf{T}_1(0) = 1$, $\mathsf{T}_2(0) = 0$.

Solution
We know that $r = 1$, $b = 8/3$ places us just at the marginal stability limit for Rayleigh–Bénard flow. From Eq. (13.84), the only fixed point to examine is the origin $(\Psi, \mathsf{T}_1, \mathsf{T}_2)^T = (0, 0, 0)^T$. We find its stability by solving the characteristic equation, Eq. (13.91):

$$(\lambda + b)(\lambda^2 + \lambda(\sigma + 1) - \sigma(r - 1)) = 0, \quad \text{so} \quad \lambda\left(\lambda + \frac{8}{3}\right)(\lambda + 11) = 0. \tag{13.100}$$

Solution gives $\lambda = 0$, $\lambda = -8/3$, and $\lambda = -11$. Because none of eigenvalues has a real part greater than zero, the origin is stable. The two eigenmodes associated with the negative eigenvalues are asymptotically stable; the eigenmode associated with the zero eigenvalue is neutrally stable. Such a mode neither grows or decays. This is consistent with our parameter choice being that on the marginal stability curve.

Figure 13.9 shows the phase space trajectories in $(\Psi, \mathsf{T}_1, \mathsf{T}_2)^T$ space and the behavior in the time domain, $\Psi(t), \mathsf{T}_1(t), \mathsf{T}_2(t)$. Relative to the results of Fig. 13.8, the time scales are much longer. That is because for the eigenmode associated with neutral stability, there is no tendency to decay. Nonlinear effects are slowly decaying all the mode amplitudes to the stable sink, but an infinitesimal amplitude will persist *ad infinitum*.

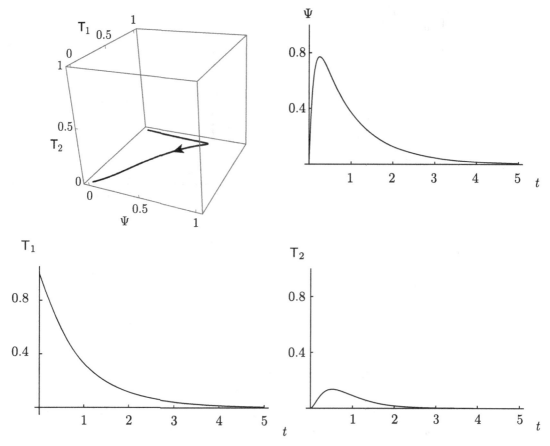

Figure 13.8 Solution to Lorenz equations, Eqs. (13.78–13.80), for $\sigma = 10$, $r = 0$, $b = 8/3$ for an asymptotically stable solution that relaxes to a state of rest. Initial conditions are $\Psi(0) = 0$, $\mathsf{T}_1(0) = 1$, $\mathsf{T}_2(0) = 0$.

Example 13.4 Examine the solution to the Lorenz equations for conditions: $\sigma = 10$, $r = 10$, $b = 8/3$ with initial conditions $\Psi(0) = 0$, $\mathsf{T}_1(0) = 1$, $\mathsf{T}_2(0) = 0$.

Solution

Note that $r > 1$, so we expect the origin to be unstable. Next note from Eq. (13.95) that

$$r_c = \frac{\sigma(\sigma + b + 3)}{\sigma - b - 1} = \frac{10(10 + \frac{8}{3} + 3)}{10 - \frac{8}{3} - 1} = \frac{470}{19} = 24.74. \quad (13.101)$$

So we have $1 < r < r_c$. We also have $\sigma > b + 1$. Thus, by Eq. (13.97), we expect the other equilibria to be stable. From Eq. (13.84), the first fixed point we examine is the origin $(\Psi, \mathsf{T}_1, \mathsf{T}_2)^T = (0, 0, 0)^T$. We find its stability by solving Eq. (13.91):

$$(\lambda + b)(\lambda^2 + \lambda(\sigma + 1) - \sigma(r - 1)) = 0, \quad \text{so} \quad \left(\lambda + \frac{8}{3}\right)(\lambda^2 + 11\lambda - 90) = 0. \quad (13.102)$$

Solution gives $\lambda = -8/3$, $\lambda = \left(-11 \pm \sqrt{481}\right)/2$. Numerically, this is $\lambda = -2.67, -16.47, 5.47$. Because one of the eigenvalues is positive, the origin is unstable. From Eq. (13.84), a second fixed point is given by

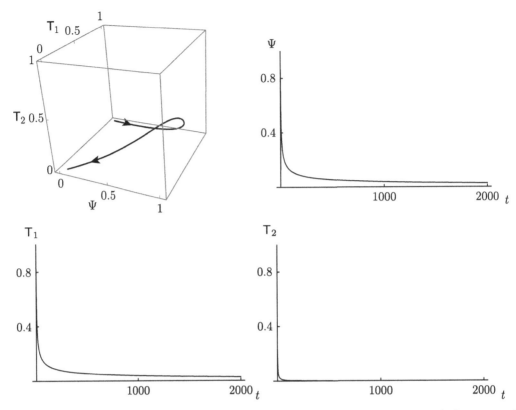

Figure 13.9 Solution to Lorenz equations, Eqs. (13.78–13.80), for $\sigma = 10$, $r = 1$, $b = 8/3$ for a neutrally stable solution. Initial conditions are $\Psi(0) = 0$, $\mathsf{T}_1(0) = 1$, $\mathsf{T}_2(0) = 0$.

$$\Psi = \sqrt{b(r-1)} = \sqrt{\frac{8}{3}(10-1)} = 2\sqrt{6} = 4.90, \tag{13.103}$$

$$\mathsf{T}_1 = \sqrt{b(r-1)} = \sqrt{\frac{8}{3}(10-1)} = 2\sqrt{6} = 4.90, \tag{13.104}$$

$$\mathsf{T}_2 = r - 1 = 10 - 1 = 9. \tag{13.105}$$

Consideration of the roots of Eq. (13.94) shows the second fixed point is stable:

$$\lambda^3 + (\sigma + b + 1)\lambda^2 + (\sigma + r)b\lambda + 2\sigma b(r-1) = 0, \tag{13.106}$$

$$\lambda^3 + \frac{41}{3}\lambda^2 + \frac{160}{3}\lambda + 480 = 0. \tag{13.107}$$

Solution gives $\lambda = -12.48$, $\lambda = -0.595 \pm 6.17i$. From Eq. (13.84), a third fixed point is

$$\Psi = -\sqrt{b(r-1)} = -\sqrt{\frac{8}{3}(10-1)} = -2\sqrt{6} = -4.90, \tag{13.108}$$

$$\mathsf{T}_1 = -\sqrt{b(r-1)} = -\sqrt{\frac{8}{3}(10-1)} = -2\sqrt{6} = -4.90, \tag{13.109}$$

$$\mathsf{T}_2 = r - 1 = 10 - 1 = 9. \tag{13.110}$$

The stability analysis for this point is essentially identical as that for the second point. The eigenvalues are identical $\lambda = -12.48, -0.595 \pm 6.17i$; thus, the root is linearly stable. Because we have two stable

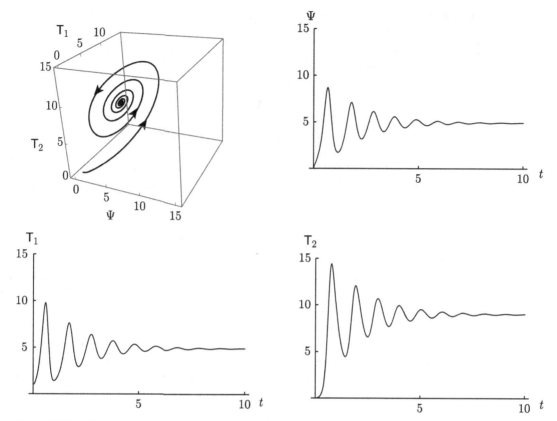

Figure 13.10 Solution to Lorenz equations, Eqs. (13.78–13.80), for $\sigma = 10$, $r = 10$, $b = 8/3$ that relaxes in a damped oscillatory fashion to an asymptotically stable solution point that represents a state of persistent convection rolls. Initial conditions are $\Psi(0) = 0$, $\mathsf{T}_1(0) = 1$, $\mathsf{T}_2(0) = 0$.

roots, we might expect some initial conditions to induce trajectories to one of the stable roots, and other initial conditions to induce trajectories to the other. Figure 13.10 shows the phase space trajectories in $(\Psi, \mathsf{T}_1, \mathsf{T}_2)^T$ space and the behavior in the time domain, $\Psi(t), \mathsf{T}_1(t), \mathsf{T}_2(t)$. Examination of the solution reveals that for this set of initial conditions, the second equilibrium is attained. The slowest time scale is $1/0.595 = 1.68$, and that is a good approximation of the time it takes to relax to the equilibrium state. The conditions of this example are precisely those that were used to generate Fig. 13.6. As such, that figure gives the equilibrium convection rolls. Physically, the enhancement of the reduced Rayleigh number r has rendered buoyancy forces to play a larger role relative to diffusion. This allows a structure to persist and not dissipate.

Example 13.5 Now consider the conditions: $\sigma = 10$, $r = 28$, $b = 8/3$. Initial conditions remain $\Psi(0) = 0$, $\mathsf{T}_1(0) = 1$, $\mathsf{T}_2(0) = 0$.

Solution

We first note that $r > 1$, so we expect the origin to be an unstable equilibrium. We next note from Eq. (13.95) that

$$r_c = \frac{\sigma(\sigma + b + 3)}{\sigma - b - 1} = \frac{10(10 + \frac{8}{3} + 3)}{10 - \frac{8}{3} - 1} = \frac{470}{19} = 24.74, \qquad (13.111)$$

remains unchanged from the previous example. So for this problem, we have $r > r_c$. Thus, we expect the other equilibria to be unstable as well.

From Eq. (13.84), the origin is again a fixed point, and again it can be shown to be unstable. From Eq. (13.84), the second fixed point is now given by

$$\Psi = \sqrt{b(r-1)} = \sqrt{\frac{8}{3}(28-1)} = 6\sqrt{2} = 8.485, \tag{13.112}$$

$$\mathsf{T}_1 = \sqrt{b(r-1)} = \sqrt{\frac{8}{3}(28-1)} = 6\sqrt{2} = 8.485, \tag{13.113}$$

$$\mathsf{T}_2 = r - 1 = 28 - 1 = 27. \tag{13.114}$$

Consideration of the roots of Eq. (13.94) shows the second fixed point is unstable:

$$\lambda^3 + (\sigma + b + 1)\lambda^2 + (\sigma + r)b\lambda + 2\sigma b(r-1) = 0, \tag{13.115}$$

$$\lambda^3 + \frac{41}{3}\lambda^2 + \frac{304}{3}\lambda + 1440 = 0. \tag{13.116}$$

Solution gives $\lambda = -13.8546$, $\lambda = 0.094 \pm 10.2i$. Moreover, the third fixed point is unstable in exactly the same fashion as the second. The consequence of this is that there is no possibility of arriving at an equilibrium point as $t \to \infty$. More importantly, numerical solution reveals the solution to approach what is known as a strange attractor. Moreover, numerical experimentation would reveal an extreme, exponential sensitivity of the solution trajectories to the initial conditions. That is, a small change in initial conditions would induce a large deviation of a trajectory in a finite time. Such systems are known as *chaotic*. Figure 13.11 shows the phase space, the strange attractor, and the behavior in the time domain of this system which has undergone a transition to a chaotic state. Importantly, the solutions are *aperiodic*. While there is some tendency to have limited windows of quasi-periodic behavior, the solution trajectory never returns to point it previously occupied. Were the solution completely random, one might expect to find the system to occupy phase space at random. Obviously, it does not, as it has approached the strange attractor. So while the position on the strange attractor is difficult to predict because of its extreme sensitivity to initial conditions, one can be sure that the solution will not approach points away from the strange attractor. Such analysis can give insights into why the general subject of fluid turbulence, of which this problem is a small subset, is so challenging. The temperature field is given by Eq. (13.67). Figure 13.12 shows the temperature field for this case at $x = 0$: $T(0, z, t) = \mathsf{T}_1(t) \sin \pi z - \mathsf{T}_2(t) \sin 2\pi z$. There is considerable variation with t and modest variation with z. The stream function also varies in a similar manner.

Example 13.6 Now consider the conditions: $\sigma = 10$, $r = 150$, $b = 8/3$. Initial conditions remain $\Psi(0) = 0$, $\mathsf{T}_1(0) = 1$, $\mathsf{T}_2(0) = 0$.

Solution

Here we have increased the reduced Rayleigh number significantly. One might expect this to further enhance the chaotic nature of the solutions as it is being driven by a much stronger temperature gradient. However, an unusual feature of nonlinear dynamics is that occasionally increased forcing can actually induce increased structure, and such will be the case here. We first note that $r > 1$, so we expect the origin to be an unstable equilibrium. And again for this problem, we have $r > r_c$. Thus, we expect the other equilibria to be unstable as well. From Eq. (13.84), the origin is again a fixed point, and again it can be shown to be unstable. From Eq. (13.84), the second fixed point is now given by

$$\Psi = \sqrt{b(r-1)} = \sqrt{\frac{8}{3}(150-1)} = 2\sqrt{\frac{298}{3}} = 19.93, \tag{13.117}$$

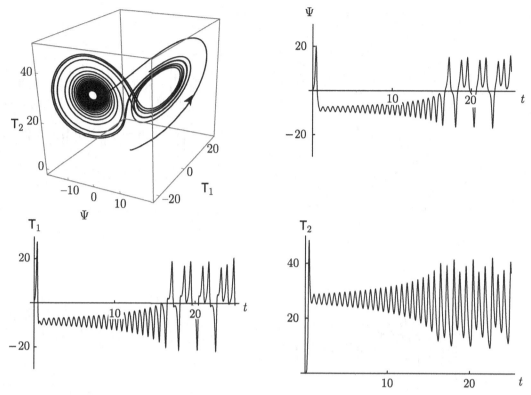

Figure 13.11 Phase space trajectory and time domain plots for solution to Lorenz equations, Eqs. (13.78–13.80), for $\sigma = 10$, $r = 28$, $b = 8/3$ that approaches an aperiodic strange attractor. Initial conditions are $\Psi(0) = 0$, $\mathsf{T}_1(0) = 1$, $\mathsf{T}_2(0) = 0$.

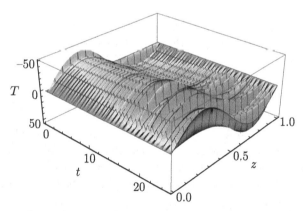

Figure 13.12 Temperature field $T(x=0, z, t)$ for solution to Lorenz equations, Eqs. (13.78–13.80), for $\sigma = 10$, $r = 28$, $b = 8/3$. Initial conditions are $\Psi(0) = 0$, $\mathsf{T}_1(0) = 1$, $\mathsf{T}_2(0) = 0$.

$$\mathsf{T}_1 = \sqrt{b(r-1)} = \sqrt{\frac{8}{3}(150-1)} = 2\sqrt{\frac{298}{3}} = 19.93, \tag{13.118}$$

$$\mathsf{T}_2 = r - 1 = 150 - 1 = 149. \tag{13.119}$$

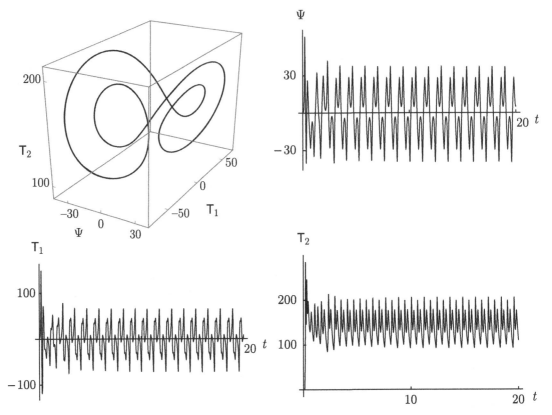

Figure 13.13 Long time phase space trajectory ($t \in [10, 20]$) and time domain plots for solution to Lorenz equations, Eqs. (13.78–13.80), for $\sigma = 10$, $r = 150$, $b = 8/3$ that approaches a stable limit cycle. Initial conditions are $\Psi(0) = 0$, $T_1(0) = 1$, $T_2(0) = 0$.

Now, consideration of the roots of the characteristic equation, Eq. (13.94), shows the second fixed point here is unstable:

$$\lambda^3 + (\sigma + b + 1)\lambda^2 + (\sigma + r)b\lambda + 2\sigma b(r - 1) = 0, \quad (13.120)$$

$$\lambda^3 + \frac{41}{3}\lambda^2 + \frac{1280}{3}\lambda + \frac{23\,840}{3} = 0. \quad (13.121)$$

Solution gives $\lambda = -16.6694$, $\lambda = 1.50 \pm 21.78i$. Once again, the third fixed point is unstable in exactly the same fashion as the second. The consequence of this is that there is no possibility of arriving at an equilibrium point as $t \to \infty$. However, in contrast to the chaotic system, this system does relax in time to a lower dimensional manifold embedded within the three-dimensional phase space.

Numerics reveals the solution to approach what is known as a limit cycle attractor. Moreover, numerical experimentation reveals a tendency for nearby initial conditions to approach the attractor. A trajectory that began at some point on the limit cycle is destined to remain on the limit cycle. In that sense, the curve describing the limit cycle is a reduced-order manifold that describes well the long time dynamics of the system. Figure 13.13 shows the limit cycle (ignoring the initial transients) embedded within the phase space, and the behavior in the time domain of this system as it relaxes to the stable limit cycle. One could use the methods of Section 3.14 to analyze the limit cycle trajectories and their attractiveness.

SUMMARY

Methods were presented to analyze spatio-temporal evolution of fluids in the presence of weakly nonlinear effects. It was shown how one could formulate such problems as a large system of nonlinear ordinary differential equations, enabling one to employ methods from dynamical system theory to better understand fluid dynamics. The problem of natural convection in the context of a nonlinear theory was studied, which led to the seminally important Lorenz equations. Their solution revealed some of the limits of deterministic science, all in the context of a problem that models warm air rising.

PROBLEMS

13.1 Consider a viscous Bateman–Burgers problem:

$$\frac{\partial u}{\partial t} + u\frac{\partial u}{\partial x} = \nu \frac{\partial^2 u}{\partial x^2}, \qquad u(x,0) = \sin \pi x, \qquad u(0,t) = u(1,t) = 0.$$

With the approximation

$$u(x,t) = \sum_{n=1}^{N} \alpha_n(t)\varphi_n(x),$$

with $N=3$, taking $\varphi_n(x) = \sqrt{2} \sin n\pi x$, do the following. With a Fourier sine expansion of the initial condition, find the coefficients $\alpha_n(0)$, $n = 1, 2, 3$. Next formulate three nonlinear ordinary differential equations for the evolution of the amplitudes $d\alpha_n/dt$, $n = 1, 2, 3$. Find a stable fixed point, and evaluate the linear stability of the system near this point. Solve for $\nu = 1/100$ with a numerical method, and plot $\alpha_n(t)$, $n = 1, 2, 3$, for $t = [0, 50]$. Plot the trajectory in the $\alpha_1, \alpha_2, \alpha_3$ phase volume. Plot the approximate solution $u(x,t)$ in a surface plot for $t = [0, 1/4]$.

13.2 A fluid system has an amplitude evolution equation for a single mode given by

$$\frac{d\alpha}{dt} = -\alpha(\alpha - (r - r_o)).$$

Here $\alpha(t)$ is the time-dependent amplitude, r_o is a fixed real constant, and r is a real parameter whose influence is to be studied. For a given simulation, both r and r_o are fixed; however, r will vary from simulation to simulation. Find all equilibria and ascertain their stability. Prepare a bifurcation diagram for what has been called a *transcritical bifurcation* showing how the amplitude at equilibrium evolves as r evolves. Ascertain the stability of the equilibria as r varies. Include an analysis of the special situation for which $r = r_o$.

13.3 A fluid system has amplitude evolution equations for two modes given by

$$\frac{d\alpha_1}{dt} = (r - r_o)\alpha_1 - \alpha_2 - \alpha_1(\alpha_1^2 + \alpha_2^2),$$

$$\frac{d\alpha_2}{dt} = \alpha_1 + (r - r_o)\alpha_2 - \alpha_2(\alpha_1^2 + \alpha_2^2).$$

Here r and r_o are both real parameters. Analyze the stability of the equilibrium at the origin $(\alpha_1, \alpha_2)^T = (0,0)^T$. Show that a Hopf bifurcation, one in which a periodic orbit appears or disappears via changes in the parameter r, occurs at $r = r_o$. For $r_o = 0$, analyze the dynamics of a system for $r = -1/5$ and for $r = 1/5$, and present relevant phase plane plots to support your analysis. By transforming to polar coordinates, analyze and determine the amplitude of the limit cycle that arises for $r = 1/5$.

13.4 Show that the dimensionless linear momenta principle in the Boussinesq limit, Eq. (13.63), can be rewritten as

$$\frac{1}{Pr}\left(\frac{\partial}{\partial t}\nabla^2 \psi + \frac{\partial(\psi, \nabla^2 \psi)}{\partial(x, z)}\right) = (Ra)\frac{\partial T}{\partial x} + \nabla^4 \psi,$$

where one has a type of Jacobian determinant for advection-based terms:

$$\frac{\partial(\psi, \nabla^2 \psi)}{\partial(x, z)} = \begin{vmatrix} \frac{\partial}{\partial x}\psi & \frac{\partial}{\partial z}\psi \\ \frac{\partial}{\partial x}\nabla^2 \psi & \frac{\partial}{\partial z}\nabla^2 \psi \end{vmatrix}.$$

13.5 Consider the Lorenz equations for $\sigma = 10$, $r = 9/10$, $b = 8/3$ with $\Psi(0) = 0$, $\mathsf{T}_1(0) = 1$, $\mathsf{T}_2(0) = 0$. Also consider the Lorenz equations linearized about the stable fixed point at $(0, 0, 0)^T$. Find an exact solution for the linearized equations. Find numerical values for the eigenvalues of the matrix that is associated with the system near $(0, 0, 0)^T$. Plot the predictions of the nonlinear and linear models of $\Psi(t)$ for $t \in [0, 50]$.

13.6 Solve the Lorenz equations for $\sigma = 10$, $r = 28$, $b = 8/3$ with $\Psi(0) = 0$, $\mathsf{T}_1(0) = 1$, $\mathsf{T}_2(0) = 0$. Solve the same system for a slightly perturbed set of initial conditions: $\Psi(0) = 0$, $\mathsf{T}_1(0) = 0.99$, $\mathsf{T}_2(0) = 0$. For $t \in [0, 50]$, plot phase space trajectories in $\Psi, \mathsf{T}_1, \mathsf{T}_2$ space for both sets of initial conditions. Plot $\Psi(t)$ and the difference in predictions of $\Psi(t)$ for the two sets of initial conditions. Assess if long time results are sensitive or insensitive to small perturbations in initial conditions, and comment on the utility of deterministic models for long term forecasting of the weather.

FURTHER READING

Bohr, T., Jensen, M. H., Paladin, G., and Vulpiani, A. (1998). *Dynamical Systems Approach to Turbulence*. Cambridge, UK: Cambridge University Press.

Doering, C. R., and Gibbon, J. D. (1995). *Applied Analysis of the Navier–Stokes Equations*. Cambridge, UK: Cambridge University Press.

Gallavotti, G. (2002). *Foundations of Fluid Dynamics*. Berlin: Springer.

Kundu, P. K., Cohen, I. M., and Dowling, D. R. (2015). *Fluid Mechanics*, 6th ed. Amsterdam: Academic Press.

Mathieu, J., and Scott, J. (2000). *An Introduction to Turbulent Flow*. Cambridge, UK: Cambridge University Press.

Sengupta, T. K. (2021). *Transition to Turbulence: A Dynamical Systems Approach to Receptivity*. Cambridge, UK: Cambridge University Press.

Strogatz, S. H. (2015). *Nonlinear Dynamics and Chaos: With Applications to Physics, Chemistry, and Engineering*, 2nd ed. Boulder, Colorado: Westview Press.

Thorne, K. S., and Blandford, R. D. (2017). *Modern Classical Physics: Optics, Fluids, Plasmas, Elasticity, and Statistical Physics*. Princeton, New Jersey: Princeton University Press.

REFERENCES

Brunton, S. L., and Kutz, J. N. (2022). *Data-Driven Science and Engineering: Machine Learning, Dynamical Systems, and Control*, 2nd ed. New York: Cambridge University Press.

Brunton, S. L., Noack, B. R. and Koumoutsakos, P. (2019). Machine learning for fluid mechanics, *Annual Review of Fluid Mechanics*, **52**, 477–508.

Chevray, R., and Mathieu, J. (1993). *Topics in Fluid Mechanics*. Cambridge, UK: Cambridge University Press.

Drazin, P. G. (1992). *Nonlinear Systems*. Cambridge, UK: Cambridge University Press.

Drazin, P. G. (2002). *Introduction to Hydrodynamic Stability*. Cambridge, UK: Cambridge University Press.

Drazin, P. G., and Reid, W. H. (1981). *Hydrodynamic Stability*. Cambridge, UK: Cambridge University Press.

Finlayson, B. A. (1972). *The Method of Weighted Residuals and Variational Principles*. New York: Academic Press.

Guckenheimer, J., and Holmes, P. H. (2002). *Nonlinear Oscillations, Dynamical Systems, and Bifurcations of Vector Fields*. New York: Springer-Verlag.

Hilborn, R. C. (2000). *Chaos and Nonlinear Dynamics*, 2nd ed. New York: Oxford University Press.

Holmes, P., Lumley, J. L., Berkooz, G., and Rowley, C. W. (2012). *Turbulence, Coherent Structures, Dynamical Systems and Symmetry*, 2nd ed. New York: Cambridge University Press.

Kaplan, W. (2003). *Advanced Calculus*, 5th ed. Boston, Massachusetts: Addison-Wesley.

Landau, L. D., and Lifshitz, E. M. (1959). *Fluid Mechanics*. Oxford: Pergamon Press.

Lorenz, E. N. (1963). Deterministic nonperiodic flow, *Journal of the Atmospheric Sciences*, **20**(2), 130–141.

McClarren, R. G. (2021). *Machine Learning for Engineers*. Cham, Switzerland: Springer.

Perko, L. (2006). *Differential Equations and Dynamical Systems*, 3rd ed. Berlin: Springer.

Perry, A. E., and Chong, M. S. (1987). A description of eddying motions and flow patterns using critical-point concepts, *Annual Review of Fluid Mechanics*, **19**, 125–155.

Pollard, A., Castillo, L., Danaila, L., and Glauser, M., eds. (2017). *Whither Turbulence and Big Data in the Twenty-first Century?* Switzerland: Springer.

Powers, J. M., and Sen, M. (2015). *Mathematical Methods in Engineering*. New York: Cambridge University Press.

Raissi, M., Perdikaris, P., and Karniadakis, G. E. (2019). Physics-informed neural networks: a deep learning framework for solving forward and inverse problems involving nonlinear partial differential equations, *Journal of Computational Physics*, **378**, 686–707.

Robinson, J. C. (2001). *Infinite-Dimensional Dynamical Systems: An Introduction to Dissipative Parabolic PDEs and the Theory of Global Attractors*. Cambridge, UK: Cambridge University Press.

Schmid, P. J. (2022). Dynamic mode decomposition and its variants, *Annual Review of Fluid Mechanics*, **54**, 225–254.

Temam, R. (1997). *Infinite-Dimensional Dynamical Systems in Mechanics and Physics*, 2nd ed. New York: Springer.

Tritton, D. J. (1988). *Physical Fluid Dynamics*, 2nd ed. Oxford: Oxford University Press.

Zeidler, E. (1995). *Applied Functional Analysis: Applications to Mathematical Physics*. New York: Springer.

14 Turbulent Flow

We close this book with a brief discussion of one of the most important and challenging unsolved problems in the mechanics of fluids: *turbulence*. As it remains as much descriptive art as predictive science, it is appropriate to call upon visual and poetic sources for inspiration to examine this daunting subject. In the visual realm, the subject has been illustrated with a well-known sketch from da Vinci, seen in Fig. 14.1. As recounted by Marusic and Broomhall (2021), da Vinci had a keen understanding of the mechanics of fluids based on observation, and his descriptions can be correlated with many modern ideas. Louis Fry Richardson (1922, p. 66), the well-known meteorologist who pioneered the use of predictive computational methods for fluid mechanics, penned poetry that provides a compact and lucid characterization of turbulence:

...big whirls have little whirls that feed on their velocity,
and little whirls have lesser whirls and so on to viscosity.

Both the sketch and the poem display one of the hallmarks of turbulence: a cascade of so-called *eddies* that exist at both large and small scales and that influence one another. We contrast this to the kinetic theory of gases. In kinetic theory, the small scale motions of molecules are well described in an average sense on a macroscale by such things as the ideal gas law

Figure 14.1 Sketch of water falling into a pool by Leonardo da Vinci; image modified with permission from an original of the Royal Collection at Windsor (RCIN 912660v). Source: Royal Collection Trust / © Her Majesty Queen Elizabeth II 2022.

and models of viscosity. Kinetic theory is successful because fluctuations on the molecular scale are far removed and largely uncoupled from those at the macroscale. This allows one to encapsulate nearly all molecular scale features within continuum constitutive models that can be used to complete the axioms of continuum mechanics. Not so for turbulence! Fluctuations such as depicted by da Vinci or described by Richardson are more tightly linked. As a result, a mathematical model as simple as the ideal gas law equivalent for turbulence is not available and likely never will be.

As summarized in two of the more persuasive discussions of turbulence, Davidson (2015) and Tritton (1988), there is little that is settled about the subject, including its definition. Tritton is reluctant to employ precise scientific criteria; one can compactly summarize his definition of turbulence as the state of a continually unstable fluid. Paraphrasing Davidson's similarly reluctantly given definition, one might consider turbulence to be a spatially and temporally complicated flow field that features chaotic advection, a random vorticity field, and a continuous distribution of features at various length and time scales. Or one might say that it is what is depicted by da Vinci. Both sources provide a comprehensive discussion of many aspects of turbulence, including the limited and standard set of topics we discuss here. Both show how nearly all of the standard ideas of turbulence remain provisional and identify numerous counter-examples to what is often treated as received wisdom. The literature of this field is vast, and the reader who needs more than a simple introduction should consult these and additional sources listed at the end of the chapter.

Our brief discussion will consider only a few aspects of turbulent fluid mechanics. We begin with a standard review of the length scales that characterize turbulent flow and compare them with those of the mean free path. We present a simple averaging procedure to arrive at a set of equations that have filtered some of the smallest scale turbulent phenomena at the expense of introducing new unknowns that require new models. Such models are often called *turbulence models*. One simple example turbulence model is presented. Serious scientific challenges remain before it can be claimed that this or any filtered model of fine scale turbulence robustly predicts observed phenomena. Even for the most basic turbulent flows – for example turbulent flow in a pipe, turbulent shear layers, turbulent boundary layers, turbulent wake flows – modeling of observed features remains difficult. For some of these phenomena, especially at a high Reynolds number, experimental observation is the better path to gain scientific insight. The cited sources give lucid descriptions coupled with measured data of these phenomena as well as detailed modeling and computational approaches; they should be consulted if basic insights on these fundament turbulent flows are required. Nevertheless, with the advent of high speed computational hardware coupled with improved numerical algorithms, *direct numerical simulation* (DNS) of the mechanics of turbulent fluids becomes feasible under moderate flow conditions. A DNS has the advantage of requiring only the deterministic set of continuum equations that employ standard evolution axioms coupled to physics-based constitutive models. When sufficient computational resources are available to capture the multiscale features of turbulence, experimental observations may be predicted by standard continuum fluid mechanics models. A DNS requires no appeal to statistics-based approaches or other turbulence models that require tuning to capture in an average sense fine-scale phenomena that have been filtered. So motivated, we close this chapter and book with an example of a DNS of an important turbulent flow field: Rayleigh–Bénard convection. This flow gives a flavor of some aspects of turbulence, was studied in less demanding conditions in earlier chapters, and incorporates many of the fully coupled multi-dimensional, unsteady, fully coupled thermal and fluid mechanics effects that this book considers. As such, it is a suitable concluding topic.

14.1 Scaling Analysis

We present here some common notions that result from scaling analysis applied to turbulent flows. Such analysis does not rely on actually solving the governing equations, but instead on simply examining them in light of some pragmatic and useful assumptions. These assumptions are fallible, but robust enough to admit insights into the nature of turbulence. Our discussion mainly follows Tennekes and Lumley (1972) and draws from Wilcox (2006).

14.1.1 Mechanical Energy Evolution

Mechanical energy of a fully compressible fluid evolves according to Eq. (4.173), rewritten here as

$$\rho \frac{d}{dt} \frac{v_j v_j}{2} = \rho v_j f_j - v_j \partial_j p + v_j \partial_i \tau_{ij}. \tag{14.1}$$

For simplicity, let us focus on incompressible, Newtonian fluids with constant μ, for which we have from Eq. (6.59),

$$\tau_{ij} = \mu(\partial_i v_j + \partial_j v_i). \tag{14.2}$$

So after scaling by ρ and using $\nu = \mu/\rho$, the mechanical energy evolution equation becomes

$$\frac{d}{dt} \frac{v_j v_j}{2} = v_j f_j - \frac{1}{\rho} v_j \partial_j p + \nu v_j \partial_i (\partial_i v_j + \partial_j v_i), \tag{14.3}$$

$$= v_j f_j - \frac{1}{\rho} v_j \partial_j p + \nu v_j (\partial_i \partial_i v_j + \partial_j \underbrace{\partial_i v_i}_{0}), \tag{14.4}$$

$$= v_j f_j - \frac{1}{\rho} v_j \partial_j p + \nu v_j \partial_i \partial_i v_j, \tag{14.5}$$

$$= v_j f_j - \frac{1}{\rho} v_j \partial_j p + \nu \left(\partial_i \partial_i \left(\frac{v_j v_j}{2} \right) - (\partial_i v_j)(\partial_i v_j) \right), \tag{14.6}$$

$$= v_j f_j - \frac{1}{\rho} v_j \partial_j p + \nu \left(\partial_i \partial_i \left(\frac{v_j v_j}{2} \right) - (\partial_{(i} v_{j)})(\partial_{(i} v_{j)}) - (\partial_{[i} v_{j]})(\partial_{[i} v_{j]}) \right). \tag{14.7}$$

We used the fact that a tensor can be decomposed into symmetric and anti-symmetric parts and that the tensor inner product of a symmetric and anti-symmetric tensor is zero. Now we recall from Eq. (3.84) that the deformation tensor $\partial_{(i} v_{j)} = \mathbf{D}$, relate $\partial_{[i} v_{j]}$ to the vorticity $\boldsymbol{\omega}$ from Eq. (3.98), and rewrite in Gibbs notation to get

$$\frac{d}{dt}\left(\frac{\mathbf{v}^T \cdot \mathbf{v}}{2}\right) = \mathbf{v}^T \cdot \mathbf{f} - \frac{1}{\rho}\mathbf{v}^T \cdot \nabla p + \nu \left(\nabla^2 \left(\frac{\mathbf{v}^T \cdot \mathbf{v}}{2}\right) - \mathbf{D}:\mathbf{D} - \frac{\boldsymbol{\omega}^T \cdot \boldsymbol{\omega}}{2} \right). \tag{14.8}$$

Other than incompressibility and constant viscosity, no approximation has been made in this mechanical energy evolution equation. Specific kinetic energy may increase or decrease due to macro-effects such as a body force or pressure gradient. However, *all viscous effects dissipate specific kinetic energy*. Obviously, viscous dissipation $\mathbf{D}:\mathbf{D} = \Phi/(2\mu)$ as well as vortical effects decrease the specific kinetic energy because $\nu, \mathbf{D}:\mathbf{D}, \omega^T \cdot \omega \geq 0$. The remaining viscous term, $\nu \nabla^2 (\mathbf{v}^T \cdot \mathbf{v}/2)$, represents diffusion of specific kinetic energy. Its presence tends to decay the amplitude of the specific kinetic energy. Focusing on the diffusion term only in the one-dimensional advection-free limit, one might say $\partial(u^2/2)/\partial t = \nu \partial^2(u^2/2)/\partial x^2$. And if one takes $u^2/2 = A(t)e^{i\alpha x}$, one gets $dA/dt = -\nu \alpha^2 A$, thus showing that viscosity decays the amplitude.

Example 14.1 Examine the evolution of mechanical energy for a general compressible flow confined in a cubical box with stationary walls. Consider the body force to be conservative and time-independent.

Solution

Let us begin with the compressible mechanical energy equation for a general compressible flow with time-independent conservative body force, Eq. (4.179):

$$\partial_o \left(\rho \left(\frac{v_j v_j}{2} + \varphi \right) \right) + \partial_i \left(\rho v_i \left(\frac{v_j v_j}{2} + \varphi \right) \right) = -v_j \partial_j p + (\partial_i \tau_{ij}) v_j, \tag{14.9}$$

$$= -(\partial_i (p \delta_{ij})) v_j + (\partial_i \tau_{ij}) v_j, \tag{14.10}$$

$$= (\partial_i (\tau_{ij} - p \delta_{ij})) v_j. \tag{14.11}$$

Rewrite to get

$$\partial_o \left(\rho \left(\frac{v_j v_j}{2} + \varphi \right) \right) + \partial_i \left(\rho v_i \left(\frac{v_j v_j}{2} + \varphi \right) \right) = \partial_i (v_j (\tau_{ij} - p \, \delta_{ij})) - (\tau_{ij} - p \, \delta_{ij}) \partial_i v_j. \tag{14.12}$$

Now we integrate over the fixed volume, use Gauss's theorem to convert some of the volume integrals to surface integrals, and apply $v_i = 0$ at the surface boundary to get

$$\int_V \partial_o \left(\rho \left(\frac{v_j v_j}{2} + \varphi \right) \right) dV + \int_V \partial_i \left(\rho v_i \left(\frac{v_j v_j}{2} + \varphi \right) \right) dV = \int_V \partial_i (v_j (\tau_{ij} - p \, \delta_{ij})) \, dV$$

$$- \int_V (\tau_{ij} - p \, \delta_{ij}) \partial_i v_j \, dV, \tag{14.13}$$

$$\int_V \partial_o \left(\rho \left(\frac{v_j v_j}{2} + \varphi \right) \right) dV + \underbrace{\int_A n_i \left(\rho v_i \left(\frac{v_j v_j}{2} + \varphi \right) \right) dA}_{0} = \underbrace{\int_A n_i (v_j (\tau_{ij} - p \, \delta_{ij})) \, dA}_{0}$$

$$- \int_V (\tau_{ij} - p \, \delta_{ij}) \partial_i v_j \, dV. \tag{14.14}$$

Thus, we get

$$\frac{d}{dt} \int_V \rho \left(\frac{v_j v_j}{2} + \varphi \right) dV = \int_V p \partial_j v_j \, dV - \int_V \underbrace{\tau_{ij} \partial_i v_j}_{\Phi} \, dV. \tag{14.15}$$

The viscous dissipation is guaranteed to decay the mechanical energy within V, while the reversible work due to the pressure force may either enhance or decay the mechanical energy.

More generally, we might expect an initially turbulent flow confined within a closed, thermally insulated, stationary volume to relax to a state of thermal and mechanical equilibrium in which the velocity is zero, and the pressure, temperature, and density distributions are consistent with hydrostatics. This is consistent with the second law of thermodynamics that insists that structures, left to themselves, tend to decay. Thus, to maintain the structures of turbulent flow, external stimuli must be continually provided, typically in the form of mass, momenta, and energy transfer into a volume of interest.

14.1.2 Approximations at the Small Scale

Here we will restrict the mechanical energy evolution equation, Eq. (14.8), so as to roughly approximate the evolution of the mechanical energy of turbulent fluctuations. Our analysis to arrive at the approximation will not be rigorous; Panton (2013, Chapter 26.6) provides a more lengthy and rigorous analysis to arrive at the same conclusion. In Section 14.2 we will formally segregate v_j into a mean quantity and fluctuating quantity. For our purposes for this section, though, we take a heuristic approach. If we temporarily take v_j to be a fluctuation about a mean, we can proceed efficiently. Now let us make the first of several plausible but nevertheless bold assumptions, all of which ultimately require both verification by detailed calculation as well as validation by comparison with experiment. It is hypothesized that at the typical small scales of turbulence, body forces and pressure imbalances are likely to be small relative to viscous effects. Returning the formulation of Eq. (14.5), we say at the small scales

$$\frac{d}{dt}\left(\frac{v_j v_j}{2}\right) = \underbrace{v_j f_j - \frac{1}{\rho} v_j \partial_j p}_{\text{small}} + \underbrace{\nu v_j \partial_i \partial_i v_j}_{\text{dissipative}} \approx \nu v_j \partial_i \partial_i v_j. \tag{14.16}$$

While body forces and pressure imbalances are likely to influence the mean flow and the largest scales of turbulence, they are less likely to influence the smallest scales of turbulence. So in Eq. (14.16), we simply assume we are at the small scale.

Using standard notation of Tennekes and Lumley (1972), we define the *dissipation rate* of specific kinetic energy as ϵ:

$$\epsilon = \nu v_j \partial_i \partial_i v_j. \tag{14.17}$$

The units of ϵ are m^2/s^3. Let us also use a standard notation to define the specific kinetic energy as k, taking

$$\mathsf{k} = \frac{1}{2} v_j v_j. \tag{14.18}$$

The units of k are m^2/s^2. And this k is not to be confused with thermal conductivity, k. Also, this k is confined to the small scale turbulent fluctuations, though it must be said that what constitutes "small" is not well defined; so here the v_j used to define k is not the macroscale fluid velocity; a more rigorous formulation is given in Section 14.2.2. So our equation for the evolution of specific kinetic energy of the small scale turbulence, Eq. (14.16), can be written as

$$\frac{d\mathsf{k}}{dt} = \epsilon, \tag{14.19}$$

and we recognize from our analysis of the viscous terms of Eq. (14.8) that ϵ will be such that it decreases the amplitude of the kinetic energy. While the definition of ϵ tells us that it can be expected to be a complicated function of space and time, we will at this stage treat it as a simple parameter that is representative of the dissipation rate.

14.1.3 Kolmogorov Microscales

We now consider a standard analysis based only on the dimensional scales involved in our model. This will be used to suggest what the relevant length, time, and velocity scales of small scale turbulence are. The only physical constant in Eq. (14.16) is the kinematic viscosity ν, which has units m^2/s. This suggests that we can define the fundamental length (η), time (t), and velocity (v) scales of small scale turbulence on purely dimensional grounds as

$$\eta = \left(\frac{\nu^3}{\epsilon}\right)^{1/4}, \quad \text{t} = \left(\frac{\nu}{\epsilon}\right)^{1/2}, \quad v = (\nu\epsilon)^{1/4}. \tag{14.20}$$

Here and throughout this discussion of scales we utilize the equals sign; truly, however, these are all only order of magnitude approximations, and one cannot expect strict equality. We can check the units:

$$[\eta] = \left(\frac{(\text{m}^2/\text{s})^3}{\text{m}^2/\text{s}^3}\right)^{1/4} = \text{m}, \quad [\text{t}] = \left(\frac{\text{m}^2/\text{s}}{\text{m}^2/\text{s}^3}\right)^{1/2} = \text{s}, \quad [v] = \left(\frac{\text{m}^2}{\text{s}}\frac{\text{m}^2}{\text{s}^3}\right)^{1/4} = \frac{\text{m}}{\text{s}}. \tag{14.21}$$

These scales are known as the *Kolmogorov*[1] *microscales*.

We now present a standard argument for linking the dissipation at the Kolmogorov microscale to the macroscale. Another bold assumption is required. Let us assume on the macroscale that we can identify the velocity and length scales of large scale motions as u and ℓ. Imagining the complexity of a variety of turbulent flows, it is not obvious how one would actually select these scales with precision, but let us imagine it can be done and proceed. Next one assumes that the kinetic energy at the large scales is proportional to u^2 and that it transfers to lower scales at a rate proportional to u/ℓ. The rate of dissipation is then taken to be the product,

$$\epsilon \sim u^2 \frac{u}{\ell} = \frac{u^3}{\ell}. \tag{14.22}$$

While this, too, is a bold assumption, it is certainly dimensionally consistent. We then use Eq. (14.22) to eliminate ϵ in Eqs. (14.20) to arrive at

$$\eta = \left(\frac{\nu^3 \ell}{u^3}\right)^{1/4}, \quad \text{t} = \left(\frac{\nu \ell}{u^3}\right)^{1/2}, \quad v = \left(\frac{\nu u^3}{\ell}\right)^{1/4}, \tag{14.23}$$

$$\frac{\eta}{\ell} = \left(\frac{\nu^3}{u^3 \ell^3}\right)^{1/4}, \quad \frac{\text{t}}{\ell/u} = \left(\frac{\nu}{\ell u}\right)^{1/2}, \quad \frac{v}{u} = \left(\frac{\nu}{u\ell}\right)^{1/4}, \tag{14.24}$$

$$\frac{\eta}{\ell} = Re^{-3/4}, \quad \frac{\text{t}}{\ell/u} = Re^{-1/2}, \quad \frac{v}{u} = Re^{-1/4}. \tag{14.25}$$

---

[1] Andrey Nikolaevich Kolmogorov, 1903–1987, Soviet mathematician.

Here we have taken the Reynolds number of the large scale motions to be

$$Re = \frac{u\ell}{\nu}. \quad (14.26)$$

14.1.4 Connection to the Mean Free Path Scale

Now from Eq. (1.17), we can infer the relationship between kinematic viscosity ν and mean free path λ:

$$\nu \sim \lambda c, \quad (14.27)$$

where $c \sim \mathcal{O}(\sqrt{RT})$ is the sound speed of a calorically perfect ideal gas. Then we find, using Eq. (14.24), that

$$\frac{\eta}{\ell} = \left(\frac{\lambda^3 c^3}{u^3 \ell^3}\right)^{1/4}. \quad (14.28)$$

Let us ignore that fact that we assumed incompressibility, and imagine that our results transfer to the compressible limit. Taking the Mach number to be $M = u/c$, we then operate to get

$$\frac{\eta}{\ell} = M^{-3/4}\left(\frac{\lambda}{\ell}\right)^{3/4}, \quad \frac{\lambda}{\ell} = M\left(\frac{\eta}{\ell}\right)^{4/3}, \quad \frac{\lambda}{\eta} = M\left(\frac{\eta}{\ell}\right)^{1/3} = M Re^{-1/4}. \quad (14.29)$$

For flows with sufficiently low Mach number and high Reynolds number, this estimate suggests the mean free path is significantly smaller than the Kolmogorov length scale. However, in a high Mach number, high Reynolds number flow, say $M = 10$, $Re = 10^7$, the ratio of mean free path length to the Kolmogorov microscale length is $10^{-3/4} \sim 0.2$, which suggests that in such a scenario, the mean free path provides a good estimate of the finest scale that needs to be resolved. Use of the mean free path as an estimator for the finest continuum scales that need to be resolved in viscous high Mach number flow is easy, unambiguous, and underutilized; moreover, it is less burdened with unclear assumptions involved in selecting u and ℓ. We have seen its value in the viscous shock calculations of Section 9.12. It typically reveals that the scales are extremely fine relative to scales of common engineering devices, rendering science-based prediction to be challenging.

The next bold assumption that is commonly made is that the dissipation rate ϵ is related to the specific kinetic energy via

$$\epsilon = \frac{\mathsf{k}^{3/2}}{\ell}. \quad (14.30)$$

We check the units:

$$\frac{\mathrm{m}^2}{\mathrm{s}^3} = \frac{(\mathrm{m}^2/\mathrm{s}^2)^{3/2}}{\mathrm{m}} = \frac{\mathrm{m}^2}{\mathrm{s}^3}. \quad (14.31)$$

Now from Eq. (14.20) and then Eq. (14.30), we can say

$$\frac{\ell}{\eta} = \frac{\ell}{\left(\frac{\nu^3}{\epsilon}\right)^{1/4}} = \frac{\ell \epsilon^{1/4}}{\nu^{3/4}} = \frac{\ell(\mathsf{k}^{3/2}/\ell)^{1/4}}{\nu^{3/4}} = \frac{\ell^{3/4}(\mathsf{k}^{1/2})^{3/4}}{\nu^{3/4}} = Re_T^{3/4}, \quad (14.32)$$

where we have defined the Reynolds number of the turbulent fluctuation kinetic energy as

$$Re_T = \frac{k^{1/2}\ell}{\nu}. \tag{14.33}$$

When $Re_T \gg 1$, the length scale of large motions ℓ is large relative to the Kolmogorov microscale length η.

In closing, we note that this standard analysis of the various quantities typically associated with turbulence such as ϵ, η, u, v, t, ℓ have no clear means of a priori determination from fundamental measurable fluid properties such as ρ, μ, and macroscale device length scales. Carefully qualified, one can imagine a posteriori determination after either an experiment or highly resolved calculation. Even then, there is some arbitrary nature to the definition of many quantities, rendering them better used for framing general discussions, rather than for detailed quantification.

14.1.5 Turbulent Kinetic Energy Cascade

Lastly, let us segregate the specific kinetic energy k into its continuous Fourier modes via

$$k = \int_0^\infty E(\kappa)\,d\kappa. \tag{14.34}$$

Here $E(\kappa)$ represents the amount of specific kinetic energy that lies between wave numbers κ and $\kappa + d\kappa$. The final bold assumption, that certainly is consistent on dimensional grounds, and importantly has experimental validation, is that

$$E(\kappa) = C\epsilon^{2/3}\kappa^{-5/3}. \tag{14.35}$$

The wave number κ has units $1/m$, and C is dimensionless Let us check the units after performing a simplistic integration ignoring the limits:

$$k = \int C\epsilon^{2/3}\kappa^{-5/3}\,d\kappa = -\frac{3}{2}C\left(\frac{\epsilon}{\kappa}\right)^{2/3}. \tag{14.36}$$

Checking the units and recalling that C is dimensionless, we get

$$\frac{m^2}{s^2} = \left(\frac{m^2/s^3}{1/m}\right)^{2/3} = \frac{m^2}{s^2}. \tag{14.37}$$

Now, Eq. (14.35) is widely considered to be one of the more important results in the study of turbulence. It describes the spectral distribution of turbulent kinetic energy. Because of nonlinear coupling, it is possible for turbulent kinetic energy at one wave number κ_1 to induce turbulent kinetic energy at another wave number κ_2. The transfer of this energy may proceed from large scales to small scales, known as a *direct energy cascade*. This is in fact the essence of Richardson's "...big whirls have little whirls that feed on their velocity...." It may also proceed from small scales to large scales, known as an *inverse energy cascade*. Given the large number of bold assumptions we made in arriving at this result, one certainly should question it. Validation with experiment is the best test, and for many flows, it appears to be remarkably accurate. For example, Sreenivasan (1995) provides a large compilation of supporting data and concludes for an unusually large number of disparate flow conditions that the Kolmogorov model is not only accurate, but has a value of $C \sim 0.5$. Verification by comparison with highly resolved numerical simulations of the Navier–Stokes equations also supports Eq. (14.35); see for example Ishihara et al. (2009) or Yeung and Zhou (1997).

14.2 Reynolds-Averaged Navier–Stokes Equations

Let us follow the development of White (2006) to obtain what is known as the Reynolds-averaged Navier–Stokes equations, often called the RANS equations.

14.2.1 Time-Averaging

Now in turbulent flows, we often observe quantities that may be effectively decomposed into the sum of a mean value and a fluctuating value. Taking q to represent a generic fluid variable, such as velocity or pressure, we say

$$q(x_i, t) = \overline{q}(x_i) + q'(x_i, t). \tag{14.38}$$

Here $\overline{q}$ represents a time-averaged value and q' represents a deviation from the time-averaged value. Here a generic time-average value is defined as

$$\overline{q} = \frac{1}{\text{t}} \int_{t_o}^{t_o + \text{t}} q \, dt. \tag{14.39}$$

Here t need not be restricted to that defined by Eq. (14.20). It is simply a time scale that will be used for averaging. One needs to ensure that t is sufficiently large relative to the time scales of turbulent fluctuation. Here, following the cautious approach of Tennekes and Lumley (1972), we restrict attention to flows for which $\overline{q}$ is time-independent. With this definition, we get the following results for generic fluid properties f, g:

$$\overline{f'} = 0, \; \overline{\overline{f}} = \overline{f}, \; \overline{\overline{f}g} = \overline{f}\,\overline{g}, \; \overline{\overline{f'}g} = 0, \; \overline{f + g} = \overline{f} + \overline{g}, \; \overline{fg} = \overline{f}\,\overline{g} + \overline{f'g'}, \; \overline{\frac{\partial f}{\partial x_i}} = \frac{\partial \overline{f}}{\partial x_i}. \tag{14.40}$$

Were we to consider flows for which the averaged variables were time-dependent, we would need to introduce a more sophisticated ensemble averaging method, described for example by Durbin and Pettersson Reif (2011) or McComb (1990).

Example 14.2 Prove $\overline{f'} = 0$.

Solution
Begin with the definition

$$f = \overline{f} + f'. \tag{14.41}$$

Now apply the operator $(1/\text{t}) \int_{t_o}^{t_o+\text{t}}$ to both sides, giving

$$\underbrace{\frac{1}{\text{t}} \int_{t_o}^{t_o+\text{t}} f \, dt}_{\overline{f}} = \frac{1}{\text{t}} \int_{t_o}^{t_o+\text{t}} \overline{f} \, dt + \frac{1}{\text{t}} \int_{t_o}^{t_o+\text{t}} f' \, dt, \tag{14.42}$$

14.2 Reynolds-Averaged Navier–Stokes Equations

$$\overline{f} = \underbrace{\frac{1}{t}\overline{f}\int_{t_o}^{t_o+t} dt}_{\overline{f}} + \frac{1}{t}\int_{t_o}^{t_o+t} f'\, dt, \tag{14.43}$$

$$0 = \frac{1}{t}\int_{t_o}^{t_o+t} f'\, dt = \overline{f'}. \tag{14.44}$$

So the time-average of a fluctuation quantity is zero.

The second identity $\overline{\overline{f}} = \overline{f}$ is obvious as the time-average of a quantity that has already been time-averaged cannot be further changed by additional time-averaging.

Example 14.3 Prove $\overline{f\overline{g}} = \overline{f}\,\overline{g}$.

Solution

$$\overline{g} = \frac{1}{t}\int_{t_o}^{t_o+t} g\, dt, \tag{14.45}$$

$$f\overline{g} = \frac{f}{t}\int_{t_o}^{t_o+t} g\, dt, \tag{14.46}$$

$$\overline{f\overline{g}} = \frac{1}{t}\int_{t_o}^{t_o+t}\left(\frac{f}{t}\int_{t_o}^{t_o+t} g\, dt\right) dt = \frac{\overline{g}}{t}\int_{t_o}^{t_o+t} f\, dt = \overline{g}\,\overline{f}. \tag{14.47}$$

So the time-average of the product of a general property and a time-averaged property is the product of the time-averaged properties.

Example 14.4 Prove $\overline{f'\overline{g}} = 0$.

Solution

$$\overline{g} = \frac{1}{t}\int_{t_o}^{t_o+t} g\, dt, \tag{14.48}$$

$$f'\overline{g} = \frac{f'}{t}\int_{t_o}^{t_o+t} g\, dt, \tag{14.49}$$

$$\overline{f'\overline{g}} = \frac{1}{t}\int_{t_o}^{t_o+t}\left(\frac{f'}{t}\int_{t_o}^{t_o+t} g\, dt\right) dt = \frac{\overline{g}}{t}\int_{t_o}^{t_o+t} f'\, dt = \overline{g}\,\underbrace{\overline{f'}}_{=0} = 0. \tag{14.50}$$

So the time-average of a the product of a fluctuation quantity and a time-averaged quantity is zero.

Example 14.5 Prove linearity of the averaging operator: $\overline{\alpha f + \beta g} = \alpha\overline{f} + \beta\overline{g}$, where α and β are scalars.

Solution

$$\overline{\alpha f + \beta g} = \frac{1}{t}\int_{t_o}^{t_o+t}(\alpha f + \beta g)\, dt = \frac{\alpha}{t}\int_{t_o}^{t_o+t} f\, dt + \frac{\beta}{t}\int_{t_o}^{t_o+t} g\, dt = \alpha\overline{f} + \beta\overline{g}. \tag{14.51}$$

The time-average of the sum is the sum of the time-averages. The averaging operator is linear.

Example 14.6 Prove $\overline{\partial f / \partial x_i} = \partial \overline{f} / \partial x_i$.

Solution

$$\overline{\frac{\partial f}{\partial x_i}} = \frac{1}{t} \int_{t_o}^{t_o+t} \frac{\partial f}{\partial x_i} \, dt = \frac{\partial}{\partial x_i} \frac{1}{t} \int_{t_o}^{t_o+t} f \, dt = \frac{\partial \overline{f}}{\partial x_i}. \tag{14.52}$$

The time-average of the gradient is the gradient of the time-average. The time-averaging operator commutes with the gradient operator.

It can also be shown that the average of the product is not the product of the average! Instead, it differs by a term related to the product of the fluctuations; $\overline{fg} = \overline{f}\overline{g} + \overline{f'g'}$. This has important implications for the study of turbulence. Lastly, the linearity of the averaging operator and the commutativity of the averaging operator and the gradient operator may not hold for weighted averages, and this has implication for some common numerical methods of turbulence modeling.

14.2.2 Averaged Incompressible Equations

Consider now the incompressible Navier–Stokes equations for a fluid with constant properties, Eqs. (0.78–0.80), written here in Cartesian index notation:

$$\partial_j v_j = 0, \tag{14.53}$$

$$\rho \left(\partial_o v_j + v_i \partial_i v_j \right) = \rho f_j - \partial_j p + \mu \partial_i \partial_i v_j. \tag{14.54}$$

These are four equations in four unknowns v_j, and p. Certainly energy could be considered, but for temperature-independent properties, the mass and momentum equations form a complete set. Now let us decompose our fluid mechanical variables:

$$v_j = \overline{v}_j + v'_j, \quad p = \overline{p} + p'. \tag{14.55}$$

With this decomposition, the mass equation transforms to

$$\partial_j (\overline{v}_j + v'_j) = 0. \tag{14.56}$$

Let us take the time-average of Eq. (14.56):

$$\frac{1}{t} \int_{t_o}^{t_o+t} \partial_j \overline{v}_j \, dt + \frac{1}{t} \int_{t_o}^{t_o+t} \partial_j v'_j \, dt = 0. \tag{14.57}$$

Because the time-average of the derivative is the derivative of the time-average, we can say

$$\partial_j \underbrace{\frac{1}{t} \int_{t_o}^{t_o+t} \overline{v}_j \, dt}_{\overline{v}_j} + \partial_j \underbrace{\frac{1}{t} \int_{t_o}^{t_o+t} v'_j \, dt}_{0} = 0. \tag{14.58}$$

And because the time-average of the fluctuations is zero and the time-average of a time-averaged quantity is that quantity, mass conservation becomes

$$\partial_j \overline{v}_j = 0. \tag{14.59}$$

Using this in the full Eq. (14.56) gives us an equivalent equation for the fluctuations:

$$\partial_j v'_j = 0. \tag{14.60}$$

Let us apply the same process to the linear momenta equation, Eq. (14.54). First, let us rewrite it in partially conservative form by adding to it the product of v_j and the mass conservation equation $\partial_i v_i = 0$ to get

$$\rho \left(\partial_o v_j + \partial_i v_j v_i \right) = \rho f_j - \partial_j p + \mu \partial_i \partial_i v_j. \tag{14.61}$$

Now in terms of time-averaged and fluctuating quantities, we get

$$\rho \left(\partial_o \left(\overline{v}_j + v'_j \right) + \partial_i \left(\left(\overline{v}_j + v'_j \right) \left(\overline{v}_i + v'_i \right) \right) \right) = \rho f_j - \partial_j (\overline{p} + p') + \mu \partial_i \partial_i \left(\overline{v}_j + v'_j \right). \tag{14.62}$$

We now time-average this equation, term by term. For the unsteady term,

$$\overline{\rho \partial_o \left(\overline{v}_j + v'_j \right)} = \rho \partial_o \left(\overline{\overline{v}}_j + \overline{v'_j} \right) = 0. \tag{14.63}$$

Because by our assumptions, $\overline{v}_j = \overline{v}_j(x_i)$, and because $\overline{f'} = 0$, the unsteady term reduces to zero. For the second term, let us use the fact that $\overline{fg} = \overline{f}\overline{g} + \overline{f'g'}$ to get

$$\overline{\rho \partial_i (\overline{v}_j + v'_j)(\overline{v}_i + v'_i)} = \rho \partial_i \overline{(\overline{v}_j + v'_j)(\overline{v}_i + v'_i)} = \rho \partial_i \left(\overline{v}_i \overline{v}_j + \overline{v'_i v'_j} \right), \tag{14.64}$$

$$= \rho \overline{v}_i \partial_i \overline{v}_j + \overline{v}_j \underbrace{\partial_i \overline{v}_i}_{0} + \rho \partial_i \overline{v'_i v'_j}, \tag{14.65}$$

$$= \rho \overline{v}_i \partial_i \overline{v}_j + \rho \partial_i \overline{v'_i v'_j}. \tag{14.66}$$

So the second term is seen to be proportional to a time-averaged advective acceleration summed to a gradient of the dyadic product of velocity fluctuations. The third term is the body force term, which has no time variation, and remains ρf_j. Fluctuations cancel on time-averaging of the fourth and fifth terms, leading to $-\partial_j \overline{p}$ and $\mu \partial_i \partial_i \overline{v}_j$. Thus, the time-averaged linear momenta equation is

$$\rho \overline{v}_i \partial_i \overline{v}_j + \rho \partial_i \overline{v'_i v'_j} = \rho f_j - \partial_j \overline{p} + \mu \partial_i \partial_i \overline{v}_j. \tag{14.67}$$

In Gibbs form, we could say

$$\rho \frac{d\overline{\mathbf{v}}}{dt} + \rho \left(\nabla^T \cdot \left(\overline{\mathbf{v'v'}^T} \right) \right)^T = \rho \mathbf{f} - \nabla \overline{p} + \mu \nabla^2 \overline{\mathbf{v}}, \tag{14.68}$$

where we recognize that for the steady time-averaged flow $d\overline{\mathbf{v}}/dt = \overline{\mathbf{v}}^T \cdot \nabla \overline{\mathbf{v}}$. Recalling that for incompressible flows $(\nabla^T \cdot \overline{\boldsymbol{\tau}})^T = \mu \nabla^2 \overline{\mathbf{v}}$ and from Eq. (6.59) that

$$\overline{\boldsymbol{\tau}} = \mu(\nabla \overline{\mathbf{v}}^T + (\nabla \overline{\mathbf{v}}^T)^T) = 2\mu \overline{\mathbf{D}}, \tag{14.69}$$

we get

$$\rho \frac{d\overline{\mathbf{v}}}{dt} = \rho \mathbf{f} - \nabla \overline{p} + \left(\nabla^T \cdot \left(\overline{\boldsymbol{\tau}} - \underbrace{\rho \overline{\mathbf{v'v'}^T}}_{\text{Reynolds stress}} \right) \right)^T. \tag{14.70}$$

The term involving the dyadic product of velocity fluctuations is known as the *Reynolds stress*, $\boldsymbol{\tau}_R$:

$$\boldsymbol{\tau}_R = -\rho\overline{\mathbf{v}'\mathbf{v}'^T}, \tag{14.71}$$

which has the units of stress and appears in the linear momenta equation in the same fashion as do the normal and viscous stresses. The presence of gradients of this turbulence-induced stress will influence the time-averaged quantities such as $\overline{\mathbf{v}}$ and $\overline{p}$.

Lastly, as an aside, following Panton (2013, p. 783), we provide a more rigorous description of the time-averaged turbulent kinetic energy than that presented in Eq. (14.18). It is better defined in terms of the fluctuations as

$$\mathsf{k} = \frac{1}{2}\overline{\mathbf{v}'^T \cdot \mathbf{v}'}. \tag{14.72}$$

Panton shows its evolution to be given by

$$\frac{d\mathsf{k}}{dt} = \overline{\mathbf{v}}^T \cdot \nabla \mathsf{k} = -\nabla^T \cdot \overline{\mathbf{v}'\mathsf{k}} - \frac{1}{\rho}\nabla^T \cdot \overline{\mathbf{v}'p} - \overline{\mathbf{v}'\mathbf{v}'^T} : \left(\nabla\overline{\mathbf{v}}^T\right) + \nu\overline{\mathbf{v}'^T \cdot \nabla^2\mathbf{v}'}, \tag{14.73}$$

which can be reformulated as

$$\frac{d\mathsf{k}}{dt} = -\nabla^T \cdot \overline{\mathbf{v}'\mathsf{k}} - \frac{1}{\rho}\nabla^T \cdot \overline{\mathbf{v}'p} + \frac{1}{\rho}\boldsymbol{\tau}_R : \left(\nabla\overline{\mathbf{v}}^T\right) + \nu\nabla^2 \mathsf{k} - \nu\overline{(\nabla\mathbf{v}'^T):(\nabla\mathbf{v}'^T)}. \tag{14.74}$$

Among other things, we see there is no unsteady term due to time-averaging. And we see the Reynolds stress $\boldsymbol{\tau}_R$ couples with the mean velocity gradient to generate turbulent kinetic energy along a particle pathline.

14.2.3 Closure Problem

The new system of mass and momenta equations are Eqs. (14.59, 14.69, 14.70). They are similar to the original incompressible flow equations, except they are now have more unknowns than equations. This is the well-known *turbulence closure problem*. The new Reynolds stress terms are new unknowns, but there are no clear choices that are able to capture all important fluid behavior in a turbulent flow environment. Finding good models for the Reynolds stress has been the subject of extensive work spanning decades. While progress has been made and is described in the cited sources, there remain serious challenges in finding a robust way to capture the physics of these fluctuating quantities.

We present here one of the simpler models for Reynolds stress, a version of the so-called *eddy viscosity* model. We follow here one of the more coherent, straightforward, and tensorially correct presentations, that of Whitaker (1968, pp. 200–202), who gives

$$\boldsymbol{\tau}_R = -\rho\overline{\mathbf{v}'\mathbf{v}'^T} = 2\mu_t \left(\frac{\nabla\overline{\mathbf{v}}^T + (\nabla\overline{\mathbf{v}}^T)^T}{2} \right) = 2\mu_t \overline{\mathbf{D}}. \tag{14.75}$$

Here μ_t is the eddy viscosity, and $\overline{\mathbf{D}}$ is the time-averaged deformation rate. The eddy viscosity coefficient μ_t is at best a crude model, and certainly is not a constant. Whitaker's presentation, based on Prandtl's mixing length theory gives

$$\mu_t = \rho\ell^2\sqrt{2\overline{\mathbf{D}}:\overline{\mathbf{D}}}, \tag{14.76}$$

where ℓ is the so-called *mixing length*, which is a parameter of the system that is not well-defined, but is often correlated with length scales of turbulent mixing. Certainly when the deformation rate $\overline{\mathbf{D}}$ has a large magnitude, there is enhanced eddy viscosity. However, Whitaker concludes the theory to be "of limited value" because in realistic flows ℓ is neither constant or a simple function of $\overline{\mathbf{D}}$. Even for simple pipe flows, the mixing length model is heuristic and requires correlation with actual data. As such it may be better classified as a *post-dictive* rather than *pre-dictive* model. Turbulence models as actually used in practice are generally complicated to state in efficient mathematical terms. A good introduction to some of the many models in use is given by Wilcox (2006) or Cummings et al. (2015, Chapter 8). All that said, with Whitaker's model of eddy viscosity, the Reynolds stress is

$$\boldsymbol{\tau}_R = 2\left(\rho\ell^2\sqrt{2\overline{\mathbf{D}}:\overline{\mathbf{D}}}\right)\overline{\mathbf{D}}. \tag{14.77}$$

Thus, the linear momentum principle, Eq. (14.70) can be written

$$\rho\frac{d\overline{\mathbf{v}}}{dt} = \rho\mathbf{f} - \nabla\overline{p} + \left(\nabla^T\cdot\left(2(\mu+\mu_t)\overline{\mathbf{D}}\right)\right)^T. \tag{14.78}$$

Written in more detail, we could say

$$\rho\frac{d\overline{\mathbf{v}}}{dt} = \rho\mathbf{f} - \nabla\overline{p} + \left(\nabla^T\cdot\left(2\left(\mu+\rho\ell^2\sqrt{2\overline{\mathbf{D}}:\overline{\mathbf{D}}}\right)\overline{\mathbf{D}}\right)\right)^T. \tag{14.79}$$

This simple model does one thing correctly: it explicitly introduces additional Reynolds stresses to the flow field in a tensorially correct fashion, and these do affect the Reynolds-averaged velocity field. To reiterate, Eq. (14.79) is one of the simplest of many eddy viscosity-based models of turbulent fluid motion. Cummings et al. (2015) discuss the strengths and weaknesses of many common eddy viscosity models. They often do well for flows for which they were designed: attached boundary layers over flat surfaces. But failure is common when they are applied to separated flows, recirculating flows, or flows with large pressure gradients.

14.3 Large Eddy Simulation

Another widely used approach in computational modeling of turbulent flow is a so-called *large eddy simulation* (LES). We will not consider it in detail. In this method, a filter of some type is applied to the governing equations in such a fashion that the smallest continuum scales are not resolved. This renders calculation of larger scale features more feasible at the expense of sacrificing some effects attributable to fine scale turbulence. These filters come in many flavors, each with their own strengths and weaknesses. The literature surrounding LES is rich and detailed; one can consult Pope (2004) or Reynolds (1976, 1990) for a review and critical assessment of its many nuances and open questions.

Compared to RANS, the ability of LES to resolve some portion of the turbulence spectrum reduces the burden placed on the closure model. The distinction due to filtering between "resolved" and "sub-filter" scales also has other beneficial features. For example, the amount of turbulent kinetic energy at the filter scale can be calculated during LES simulations. This is the basis of closure models that calculate the eddy viscosity dynamically; see Germano et al. (1991). Other models estimate the "sub-filter" portion of the kinetic energy using Taylor series expansions without appealing to the eddy viscosity hypothesis, for example Clark et al. (1979). This can be beneficial for flows in which the eddy viscosity hypothesis is invalid, for example, some flows with local heat release.

However, other nuances of LES compared to RANS must be considered. In general, the RANS averaging operation is a linear operation and so commutes with differentiation; therefore, the *only* unclosed term in the RANS momentum equation is the Reynolds stress. Conversely, LES filters in general are nonlinear functions that do not commute with differentiation. This results in additional unclosed terms in the LES-filtered Navier–Stokes equations that, in practice, are most often simply neglected. These "filter commutation" terms can become significant in certain situations, for example, if the computational grid is highly nonuniform; see Ghosal (1996).

14.4 Direct Numerical Simulation of a Rayleigh–Bénard Flow

We close this book by presenting a detailed calculation of Rayleigh–Bénard flow that advances into the turbulent flow regime. We will directly calculate the dynamics without using any RANS or LES approximations; this will come at the expense of using an extremely fine space and time-discretization. It is an appropriate final problem as it requires an appreciation of the underlying mathematics and fully couples much of the fluid physics featured throughout the book, including:

- time-dependency,
- multi-dimensional spatial dependency,
- variable density,
- fully coupled mass, momenta, energy, and thermodynamics, with advective and diffusive transport,
- linear, weakly nonlinear, and fully nonlinear behaviors,
- multiscale continuum behavior described by detailed numerical solution of partial differential equations, illuminated by consideration of various limits in which analysis tools give significant guidance, and
- physically observable phenomena.

This problem will serve to display many features of turbulence and its simulation, and it is a problem with a rich history that has been widely studied. Many of our general conclusions about the character of turbulence will extend to flows that are even more widely studied: constant density Navier–Stokes simulations at high Reynolds number, Re. For such flows, it is widely known that pipe flows often exhibit turbulence for $Re \gtrsim 2300$, and external flows for $Re \gtrsim 10^6$. For Rayleigh–Bénard flow, it is instead the Rayleigh number Ra that is the key parameter dictating whether or not a transition to turbulence has occurred.

We will extend the linear analysis given in Section 12.4 and the weakly nonlinear analysis given in studying the Lorenz model of Section 13.4. We consider the low Mach number flow of a viscous, heat conducting, calorically perfect ideal gas confined within a vertical channel.

14.4 Direct Numerical Simulation of a Rayleigh–Bénard Flow

At the base and ceiling of the channel, the fluid is in contact with hot and cold, respectively, isothermal surfaces. The thermodynamic pressure is taken to be constant in the low Mach number limit. The fluid at the hot base acquires a low density that induces it to rise in the presence of a gravitational field. The cold fluid at the top has a high density that induces it to fall. This gravity-driven fluid convection is resisted by viscous forces. As we have seen, for sufficiently low temperature differences between bottom and top, buoyancy forces cannot overcome viscous forces, and the fluid remains stationary. And for sufficiently high temperature differences, the buoyancy forces overcome the viscous forces and induce fluid acceleration. Such convection of thermal energy enhances the rate of heat transfer between the bottom and top surfaces, relative to conduction in a stationary fluid. In Section 12.4, the critical Rayleigh number for the transition from conduction- to convection-domination was given as $Ra_c = 1708$. For $Ra < Ra_c$, the stationary state is linearly stable, and is otherwise linearly unstable. In Section 13.4, we saw that as nonlinearities were introduced, the flow could become ever more complex and in fact undergo a transition to chaos. Such conclusions were confined to the highly truncated Lorenz model.

As well described by Kadanoff (2001), this configuration can be considered an "engine" for transferring heat from the hot plate to the cold plate. Motivated by a similar figure of Kadanoff, we give a cartoon sketch of this "engine" in Fig. 14.2. In Fig. 14.2(a), a stationary fluid is shown conducting energy from the hot plate to the cold plate. In Fig. 14.2(b), we depict a possible stable convection circulation pattern for Rayleigh number just above the critical value. An especially lucid description of the wide variety of stable patterns that may be realized is given by Bodenschatz et al. (2000). As the Rayleigh number increases, the patterns become more complicated, for example see Pandey et al. (2018), and subject to chaos. In Fig. 14.2(c), we depict at higher Rayleigh number a large scale clockwise circulation pattern. Though this large scale circulation is shown as clockwise, nothing prevents a dramatic transition to counterclockwise circulation due to nonlinear effects. Embedded within are mushroom-like structures that are buoyant plumes of either rising or falling fluid. Though such high Rayleigh

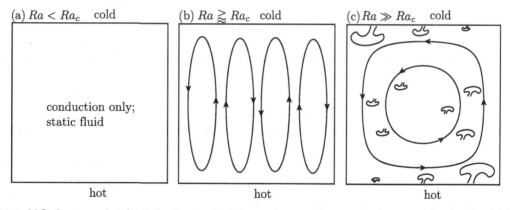

Figure 14.2 Cartoon sketch of the "engine" of thermal energy transport via convection for Rayleigh–Bénard flow: (a) stationary fluid with heat transfer via conduction only, (b) stable circulation patterns for weakly supercritical convection $Ra \gtrsim Ra_c$, (c) turbulent convection structures for $Ra \gg Ra_c$ composed of "flywheels" of large scale circulation and "mushrooms" of local thermal plumes.

number flows may exhibit structure, the structures are often highly transitory and extremely sensitive to small changes. However, the flow is not random; it is better described as complicated or perhaps simply "turbulent." The nature of the complication or turbulence does evolve with Rayleigh number, as well as Prandtl number. Such is nature. And to extend the discussion of Section 12.4, the broad spectrum of patterns, persistent or transient, is consistent with the second law of thermodynamics, as the Rayleigh–Bénard flow may receive an energy input from its surroundings. Should the energy input be withdrawn, all such structures would decay, and the ordered mechanical energy would fully dissipate into random mechanical energy that is measured by temperature.

The overall performance metric of the engine is quantified by the Nusselt number, which gives the ratio of the achieved heat transfer to that available from thermal conduction alone. The Nusselt number is a function of Rayleigh number and Prandtl number. For simplicity, we will restrict our discussion to unity or near unity Prandtl number, and focus on the Rayleigh number dependency. The goal is to be able to predict and understand how this engine works in demanding physical conditions of high Rayleigh number flow.

To enable this, we consider a more complete version of the nonlinear Navier–Stokes equations. For direct comparison with our previous results and for some computational efficiency, we restrict our model to the low Mach number limit. This allows for an effective uncoupling of the thermodynamic pressure, which we will take to be constant, from the hydrostatic pressure, which we will allow to vary. The isobaric ideal gas law applied to the thermodynamic pressure will allow for large temperature and density variation, and we will not enforce the Boussinesq approximation. We will study a series of Rayleigh numbers for which the flow will exhibit a transition from a stable stationary condition, to the onset of convection, to a weakly nonlinear regime, to a flow that exhibits many of the multiscale features of turbulence.

Our simulation will be a DNS in that no model of Reynolds stress, such as an eddy viscosity model, will be employed, and the bulk of all physical features will be resolved. We simply will consider the full equations in discrete form and employ a sufficiently fine grid to capture a broad range of the structures of turbulence. Because the continuum model is taken to be in the low Mach number limit, acoustic waves have effectively been analytically filtered, as described by Kee et al. (2018, p. 132), Paolucci (1982), and many others. This will enable the use of grids much coarser than those of the mean free path limit, such as were required in Sections 9.12 and 9.13, that considered viscous structures arising from the fully compressible Navier–Stokes equations in which acoustic phenomena could not be neglected. The grid resolution required for DNS depends on the continuum model selected and the specific set of parameters chosen to model. Ultimately, the grid requirements are set by the finest and coarsest scales that must be resolved to capture all the physics and geometric features of the problem in question in such a fashion that quantities of interest are insensitive to further grid refinement.

When possible, a DNS is a preferred modeling approach for turbulent flow simulation. It must be recognized, however, that for many practical engineering problems, DNS is not feasible because it requires simultaneous resolution of a broad spectrum of scales beyond the capacity of modern hardware. As computational hardware advances, the class of problems that are amenable to DNS expands. Ahlers et al. (2009) report that for $Pr \sim 1$, the computation time scales as $Ra^{3/2} \ln Ra$, thus rendering large Ra calculations challenging. They note at the time of writing, calculations with $Ra \sim 10^{11}$ were difficult but common. We compare this estimate for Rayleigh–Bénard to the more common simulations of constant density turbulent flows, driven by the Reynolds number, Re. From Eq. (14.25), we can conclude that the number of grid points N necessary to resolve in a single direction the Kolmogorov turbulent length scale η embedded within a larger length scale ℓ is $N \sim Re^{3/4}$. For all three directions, we then need $N^3 \sim Re^{9/4}$. For finer grids, finer time steps are needed, and it can be shown that the overall computational effort scales as Re^3. Calculation with DNS even of moderate Re flows, for example $Re = 10^6$, can be computationally challenging.

A typical goal of DNS of Rayleigh–Bénard flow is to identify a relationship between Nu, Ra, and Pr. Kadanoff (2001) comments on the challenges associated with this worthy goal; these challenges are associated with distinguishing between what amounts to curve fitting in contrast to a theory that is some asymptotic limit of a physics-based mathematical model. These dependencies are often highly qualified; a typical estimate for $Ra > 10^9$, $Pr = 0.7$ is $Nu \sim Ra^{1/3}$. One can perform DNS at any Rayleigh number. At the modestly low Ra that we will study, $Ra < 10^8$, we will predict structures of natural convection-driven turbulence.

Spatial resolution requirements for a grid-insensitive prediction of a DNS are not straightforward to determine. One reason is the flows are inherently unsteady, spatially inhomogeneous, and because of their chaotic nature, strongly sensitive to initial conditions. It is essentially impossible to converge all spatio-temporal variables of the flow. One typically compromises and inquires of the grid-sensitivity of the error of a global *quantity of interest* (QOI) such as the Nusselt number or other statistically averaged quantity. Certainly, one could expect that a spatial discretization of a few mean free paths should be sufficient and all one could expect from a continuum model. Likely this estimate is too conservative, especially in the low Mach number limit in which acoustics have been filtered. Detailed discussion is given by Grötzbach (1983) and Scheel et al. (2013). Employing extremely fine scale calculations, Scheel et al. conclude that global quantities such as Nusselt number may be well predicted even if other more fine scale structures are not well-resolved. Finally, we note for comparison purposes that typical finite difference calculations of Rayleigh–Bénard turbulence are done on grids with a few hundred cells in each dimension. For example, Bailon-Cuba et al. (2010) report a grid of $361 \times 181 \times 310$ for a DNS with $Ra = 10^9$, $Pr = 0.7$.

Let us consider a DNS of the dimensional equivalent of Eq. (6.168–6.170) taken in the limit of low Mach number and isobaric thermodynamic pressure so that viscous dissipation and material derivatives of density and pressure may be neglected. We also take a constant gravitational acceleration g pointing in the $-z$ direction. We then have the following system of equations for Rayleigh–Bénard flow in the cubical domain $x \in [0, h]$, $y \in [0, h]$, $z \in [0, h]$:

$$\partial_i v_i = 0, \tag{14.80}$$

$$\rho \frac{dv_i}{dt} = -\rho g \delta_{i3} - \partial_i p + \mu \partial_j \partial_j v_i, \tag{14.81}$$

$$\rho c_p \frac{dT}{dt} = k \partial_i \partial_i T, \tag{14.82}$$

$$p_o = \rho R T, \tag{14.83}$$

$$T(x,y,0,t) = T_o, \quad T(x,y,h,t) = T_1, \quad T(x,y,z,0) = T_o\left(1 + \frac{T_1 - T_o}{T_o}\frac{z}{h}\right), \tag{14.84}$$

$$v_i(x,y,z,0) = 0, \quad v_i(x,y,0,t) = v_i(x,y,h,t) = 0, \tag{14.85}$$

$$\rho(x,y,z,0) = \frac{p_o}{RT_o\left(1 + \frac{T_1-T_o}{T_o}\frac{z}{h}\right)}, \quad p(x,y,z,0) = p_o\left(1 - \ln\left(1 + \frac{T_1-T_o}{T_o}\frac{z}{h}\right)^{\frac{gh}{R(T_1-T_o)}}\right). \tag{14.86}$$

The initial state gives a linear temperature profile and hydrostatic distribution of density and pressure. Boundary conditions at side walls at $x = 0$, $x = h$, $y = 0$, $y = h$ are taken to be periodic.

In Table 14.1, we list a set of dimensional parameters that roughly corresponds to atmospheric pressure air at $Ra = 10^3$, $Pr = 1$. Of course actual air has slightly different values; for example, the Prandtl number is a little lower. This will not qualitatively affect our results. The hot base at $z = 0$ is denoted with the subscript "o," and the cold top at $z = h$ is denoted with the subscript 1. While parameters are associated with a particular $Ra = 10^3$, we will execute a series of calculations varying the cold plate temperature T_1, holding all other parameters constant, so that $Ra \in [10^2, 1.25 \times 10^7]$, holding $Pr = 1$.

Numerical parameters associated with this problem for $N = 512$ grid points in each dimension are found in Table 14.2. The grid is taken to be uniform and yields a grid size of about 19 μm for the values in Table 14.1. For the conditions studied, we employ the mean free path length estimate of Eq. (1.14) that gives the mean free path to be $\lambda \sim 0.066$ μm. So a given computational cell encompasses 296 mean free paths. This is consistent with the discussion of Section 14.1.4 for DNS length scale requirements for low Mach number flows; however, it not obvious how to use that discussion to provide an a priori estimate of the required length scale of a DNS, as one needs fluid particle velocity and length scale estimates to determine M and Re. We can, however, try the only estimate available. Let us take the length scale as h and the velocity scale as that estimated by a simple balance of kinetic and potential energy $v_o = \sqrt{2gh}$. Then $Re_h = v_o h/\nu_o = 305.479$ and $M = v_o/c_o = 0.00128775$. Then Eq. (14.29) gives the estimate of the ratio of the mean free path length to the Kolmogorov

Table 14.1 Parameters for Rayleigh–Bénard flow for $Ra = 10^3$, $Pr = 1$

Independent parameter	Units	Value	Dependent parameter	Units	Value
R	J/kg/K	287	$\gamma = c_p/(c_p - R)$	-	1.4
c_p	J/kg/K	1004.5	$\rho_o = p_o/R/T_o$	kg/m³	1.19897
g	m/s²	9.81	$\rho_1 = p_o/R/T_1$	kg/m³	1.22523
h	m	0.01	$\nu_o = \mu/\rho_o$	m²/s	14.5×10^{-6}
p_o	Pa	101 325	$\alpha_o = k/\rho_o/c_p$	m²/s	14.5×10^{-6}
T_o	K	294.461	$Pr = \nu_o/\alpha_o$	-	1
T_1	K	288.150	$Ra = (T_o - T_1)h^3 g/T_o/\alpha_o/\nu_o$	-	10^3
μ	kg/m/s	17.385×10^{-6}	$\lambda = (7.90864 \times 10^{-8}$ kg/m²$)/\rho_o$	m	65.9621×10^{-9}
k	W/m/K	0.0174632	$c_o = \sqrt{\gamma R T_o}$	m/s	343.969
			$v_o = \sqrt{2gh}$	m/s	0.442945
			$t_{mfp} = \lambda^2/\nu_o$	s	300.069×10^{-12}

14.4 Direct Numerical Simulation of a Rayleigh–Bénard Flow

Table 14.2 Numerical parameters for Rayleigh–Bénard flow for $N = 512$

Parameter	Units	Value
$\Delta x = \Delta y = \Delta z = h/N$	m	19.5313×10^{-6}
$\Delta x / \lambda$	-	296.098
$c_o = \sqrt{\gamma R T_o}$	m/s	343.969
$v_o = \sqrt{2gh}$	m/s	0.442945
$\Delta t_{acoustic} = \Delta x / c_o$	s	56.7821×10^{-9}
$\Delta t_{advection} = \Delta x / v_o$	s	44.0941×10^{-6}
$\Delta t_{diffusion} = \Delta x^2 / \nu_o$	s	26.3083×10^{-6}

length as $\lambda/\eta = M Re_h^{-1/4} = 0.000\,308\,025$, yielding $\eta = 214.146\,\mu$m. Our $\Delta x = 19\,\mu$m is well below this scale.

Mean free path estimates are tightly linked to acoustics, which have been filtered in the low Mach number limit. Were we to have used a spatial discretization of the same order of magnitude as the mean free path, the time scales that would have been required would have been of the order of magnitude of $\lambda^2/\nu_o = 300$ ps, considerably smaller than that based on either advection or diffusion at the finest discretization length, which are on the order of tens of microseconds. However, it remains to be seen if the chosen $N = 512$ ($\Delta x = 19\,\mu$m) yields results that are insensitive to further changes in resolution. For the chosen spatial discretization length, an estimate of the time step based on advection is similar to that based on diffusion. Both are several microseconds. Had acoustics not been filtered for the present low Mach number conditions, the time step requirement would be roughly three orders of magnitude more demanding.

There are numerous acceptable ways to discretize the equations. The key requirement is that the discretization in space and time be sufficiently fine to capture the relevant physical scales in the continuum model in such a way that quantities of interest are insensitive to further refinement. We solve the Navier Stokes equations, Eqs. (14.81–14.83), using second-order central differences, except for advective terms in the temperature equation, which, in order to suppress unphysical oscillations, are discretized using a third-order weighted essentially non-oscillatory (WENO) scheme; see Jiang and Shu (1996). While WENO schemes are most appropriate for the shock-capturing problems for which they were developed, they have become commonly used in other limits to address oscillations thought to be unphysical. Of course, many higher-order versions exist, for example Henrick et al. (2005); because of the lower order discretizations of diffusion terms in our DNS, the benefit of only extending the advection discretization to high order is difficult to quantify. And certainly, one could imagine that simpler discretizations, such as second-order central, performed on an even finer grid could achieve the same end; see Singh et al. (1999) for an example. And one should always take care with methods that advertise an "essential non-oscillatory" property that the method is not suppressing physics-based oscillations, keeping in mind the admonition of Gresho and Lee (1981): "Don't suppress the wiggles – they're telling you something!" This typically requires that length and time discretizations be sufficiently small to capture the various small wavelength and high frequency oscillations of physics-based instabilities.

The variables are arranged on a staggered mesh, Harlow and Welch (1965), with velocity components located at cell faces and scalars at cell centers; this facilitates computing second-order accurate flux terms across single mesh cells. Time integration is by a linearized trapezoidal method in an alternating-direction implicit (ADI) framework, Desjardins et al. (2008). The mass conservation constraint is enforced by projecting the velocity field to a divergence-free manifold, which is done using a fractional-step method, Kim and Moin (1985), that solves a Poisson equation for the hydrostatic portion of the pressure. This is a variable density version of Eq. (6.76). As is typical of such schemes, the Poisson equation solution constitutes the majority of the computational cost. The velocity and density fields are converged by sub-iterating over each time step; for the temperature differences we consider, four sub-iterations are typically sufficient to recover second-order accuracy, MacArt and Mueller (2016). The time step size Δt is set by the Courant–Friedrichs–Lewy (CFL) constraint, CFL = $\max(v_i \Delta t/\Delta x_i)$, where Δx is the uniform mesh spacing and v_i is the local advection velocity. To ensure accuracy, we require CFL ≤ 0.5. As seen earlier for this case, this is the same order of magnitude as a diffusion-based criterion.

The numerical solutions are accelerated by dividing the computational work between central processing units (CPUs) using the message-passing interface (MPI), a protocol for communicating information between different computing tasks, which may be located on different nodes of a cluster or supercomputer. Typically, the problem is divided into sub-domains, on which individual MPI tasks compute the numerical solution, sharing information from the borders of their sub-domains as necessary. In this manner, the problem size is limited only by the resources (CPU and memory) available on the entire machine.

Two spatial dimensions are sufficient for many laminar flows including Rayleigh–Bénard flows in the laminar convection-dominated regime. Rayleigh–Bénard flows in the conduction-dominated regime are effectively one-dimensional. For sufficiently high velocity magnitudes, the nonlinearities present in the Navier–Stokes equations will cause the initially laminar flow to become chaotic and eventually turbulent, at which point the two-dimensional approximation is no longer valid.

We first solve the equations on a square, two-dimensional spatial domain of $h = 0.01$ m, which we discretize using $N = 512^2$ mesh cells. Periodic boundary conditions are applied in the x direction, and isothermal no-slip walls are implemented at the $z = 0, h$ boundaries. Figure 14.3 gives a plot of the Nusselt number, Eq. (12.175), evaluated from DNS for unity Prandtl number and Rayleigh numbers $Ra \in [10^2, 1.25 \times 10^7]$. For these conditions, the critical Rayleigh number is approximately $Ra = 1.5 \times 10^3$, which marks the flow's transition from the conduction-dominated regime to the convection-dominated regime. This is remarkably close to the critical value of $Ra = 1708$ reported from linear stability theory as discussed in Section 12.4; see Fig. 12.15.

For this configuration, the RANS average of an arbitrary quantity ϕ is

$$\overline{\phi}(z,t) = \frac{1}{h}\int_0^h \phi(x,z,t)\,dx. \tag{14.87}$$

We note that the mean flow is time-evolving. For consistency with the form of the RANS equations typically considered for variable density flows, we also define a density-weighted average, also known as Favre average $\widetilde{\phi} = \overline{\rho\phi}/\overline{\rho}$ with fluctuations $\phi'' = \phi - \widetilde{\phi}$.

14.4 Direct Numerical Simulation of a Rayleigh–Bénard Flow

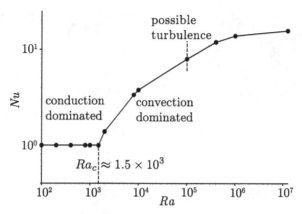

Figure 14.3 Variation of the Nusselt number with Rayleigh number for Rayleigh–Bénard flow for $Pr = 1$. Circles indicate simulations performed on two-dimensional domains with $N = 512^2$ mesh cells. Compare to Fig. 12.15.

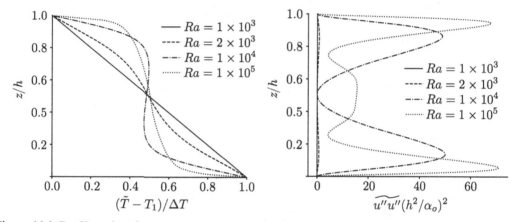

Figure 14.4 Profiles of scaled mean temperature (left) and horizontal velocity fluctuation product (right) versus scaled vertical distance for Rayleigh–Bénard flows at $Pr = 1$ and Rayleigh numbers indicated. All simulations were performed on two-dimensional domains with $N = 512^2$ mesh cells.

Figure 14.4 depicts the variation with z/h of the mean dimensionless Favre-averaged temperature $\tilde{T}$, scaled so that $(\tilde{T} - T_1)/\Delta T \in [0, 1]$ and $\widetilde{u''u''}$ for $Ra \in [1 \times 10^3, 1 \times 10^5]$, which contains the convection-dominated flow regime from Fig. 14.3. The mean temperature profile for $Ra = 1 \times 10^3$ is a straight line indicating pure conduction, and the Reynolds stress is zero. Increasing the Rayleigh number results in laminar convective heat transfer for $Ra < 1 \times 10^5$, at which point the flow becomes increasingly chaotic with nonzero Reynolds stress at the midpoint, $z/h = 0.5$. The transition from conduction heat transfer to laminar and chaotic convection is further illustrated in Fig. 14.5, which shows visualizations of the instantaneous scaled temperature field at various Rayleigh numbers once the flow has reached statistical stationarity. For $Ra = 8 \times 10^4$ and higher, increasingly chaotic vortices replace the convection rolls occurring in the laminar convection regime.

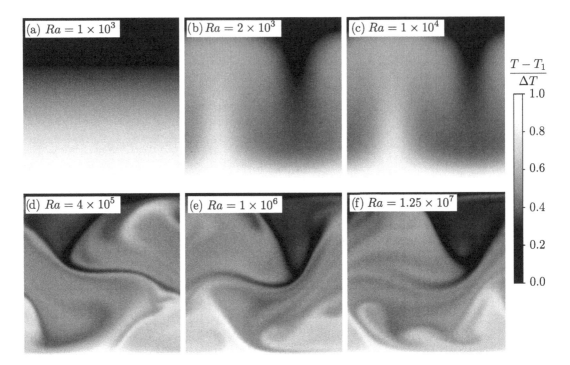

Figure 14.5 Visualizations of instantaneous scaled temperature on a scaled unit square in two-dimensional DNS Rayleigh–Bénard flow simulations at increasing Rayleigh numbers, all for $Pr = 1$: (a) $Ra = 1 \times 10^3$, $t = 40.7$ s; (b) $Ra = 2 \times 10^3$, $t = 40.0$ s; (c) $Ra = 1 \times 10^4$, $t = 38.5$ s; (d) $Ra = 4 \times 10^5$, $t = 689.6$ s; (e) $Ra = 1 \times 10^6$, $t = 344.8$ s; (f) $Ra = 1.25 \times 10^7$, $t = 34.5$ s.

Given the chaotic nature of the highest Rayleigh numbers computed in Fig. 14.5 using two-dimensional DNS, we may assume that the actual, three-dimensional flow is turbulent. To verify this, we simulate the flow on a three-dimensional domain using $N = 512^3$ mesh cells. Visualizations of the three-dimensional DNS are given in Figs. 14.6 and 14.7, in which the considerably greater range of scales and chaotic nature of the flow is apparent. These three-dimensional simulations are much more costly than the two-dimensional simulations, with each requiring approximately six days of running time on 512 CPU cores versus a few hours on several dozen cores.

Estimating the error of DNS computations is a challenging exercise. It is a critical part of the important exercise of solution verification; see Oberkampf and Roy (2010) or Roache (2009). There are three obvious sources of error. The first, known as *iterative convergence error*, is due to errors associated with solving the Poisson equation for the hydrostatic portion of pressure that must be done at every time step in order to enforce the condition $\nabla^T \cdot \mathbf{v} = 0$ associated with the low Mach number limit. This requires an iterative method, and the iteration is truncated when the error is below a user-defined tolerance. Figure 14.8 gives a scaled distribution of $\nabla^T \cdot \mathbf{v}$ for (a) a two-dimensional simulation, and (b) a two-dimensional slice of a three-dimensional simulation, both at $Pr = 1$, $Ra = 4 \times 10^5$. Clearly the scaled error is on the order of 0.002 for the two-dimensional DNS and 0.02 for the three-dimensional DNS. Both errors could be

14.4 Direct Numerical Simulation of a Rayleigh–Bénard Flow

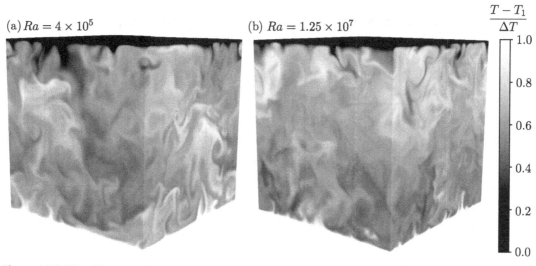

Figure 14.6 Visualization of instantaneous scaled temperature in a scaled unit cube in a three-dimensional DNS Rayleigh–Bénard flow simulation for $Pr = 1$, (a) $Ra = 4 \times 10^5$, (b) $Ra = 1.25 \times 10^7$.

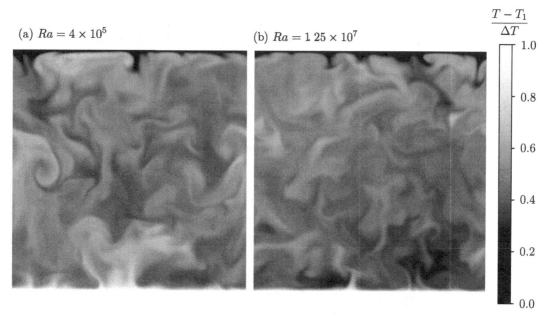

Figure 14.7 Visualization of two-dimensional slices of instantaneous scaled temperature from a scaled unit cube in a three-dimensional DNS Rayleigh–Bénard flow simulation for $Pr = 1$, (a) $Ra = 4 \times 10^5$, (b) $Ra = 1.25 \times 10^7$.

reduced by additional iteration in solving the Poisson equation. As this iteration must be done at every time step in the DNS, it may be computationally expensive. Errors of this type contribute to the global error of the DNS.

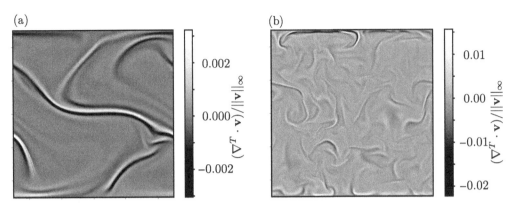

Figure 14.8 Visualization of the scaled error distribution in $\nabla^T \cdot \mathbf{v}$ that arises from solving the Poisson equation for pressure in DNS of Rayleigh–Bénard flow for $Pr = 1$, $Ra = 4 \times 10^5$: (a) two-dimensional simulation on a scaled unit square, (b) two-dimensional slice from three-dimensional simulation on a scaled unit cube.

The second source of error is due to the discrete nature of the spatial and temporal approximation to the underlying continuum partial differential equations. This is often called *truncation error*. Truncation error is a function of the number of discretizations and the nature of how the equations were discretized. To determine the effects of truncation error, it is essential to compare a prediction of some QOI against a benchmark value. In the absence of an exact solution, we take the benchmark value to be that predicted on the finest grid. While this itself introduces error that is unavoidable for problems with no exact solution, it will nonetheless illustrate the convergence of QOIs as truncation errors are reduced. In one of the strongest verifications, one predicts the error of a QOI to be converging at a rate consistent with the truncation error of the chosen numerical method. Such a solution is said to be in the desirable *asymptotic convergence regime*. For the turbulent Rayleigh–Bénard flow, there is no long time stationary solution; hence, we choose something both important and predictable: the global Nusselt number Nu, which is a time-averaged integrated effect over the conducting surfaces. We select a window in time $t \in [t_1, t_2]$ over which to estimate the average heat transfer at the top and bottom surfaces. The heat fluxes at the top and bottom surfaces are estimated by integrating over both surfaces. We estimate Nu via

$$Nu = \frac{1}{2h^2(t_2 - t_1)} \int_{t_1}^{t_1+t_2} \frac{\int_0^h \int_0^h \left| k \frac{\partial T}{\partial z} \right|_{z=0} \, dx \, dy + \int_0^h \int_0^h \left| k \frac{\partial T}{\partial z} \right|_{z=h} \, dx \, dy}{\left| \frac{k \Delta T}{h} \right|} \, dt. \tag{14.88}$$

Appropriate discrete approximations are made for all derivatives and integrals.

Figure 14.9(a) gives a plot of Nu in two-dimensional simulations for different grid resolutions for $Pr = 1$, $Ra = 4 \times 10^5$. It appears that Nu is converging to just above 16, but a more accurate prediction would require additional calculation on finer grids. We see the prediction of Nu has strong sensitivity to N_x for $N_x < 100$. Figure 14.9(b) gives an estimate of the error in Nu for the same calculations. Here the error has been estimated by subtracting from

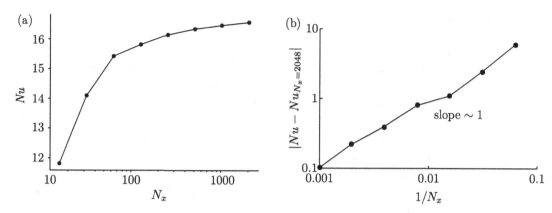

Figure 14.9 Two-dimensional DNS of Rayleigh–Bénard flow for $Pr = 1$, $Ra = 4 \times 10^5$: (a) Nu versus N_x, (b) estimated error in Nu found by subtracting the finest grid estimate ($N_x = 2048$) of Nu from that of a set of coarser grids as a function of discretization size $1/N_x$.

a local prediction of Nu the prediction of Nu on the finest grid considered, $N_x = 2048$. For the case studied, we estimate an error convergence rate of roughly order unity, as an order of magnitude finer grid results in an order of magnitude improvement in the error in Nu. Because the chosen discretization method had a greater than unity convergence rate, we are not in the asymptotic convergence regime. However, it is not clear how this relates to the convergence of statistically averaged QOIs such as Nu. The important criterion for verification of the DNS is that an error estimate is converging. Other QOIs could be examined. For example, one can expect the bifurcation point Ra_c to have a sensitivity to grid as coarse grids have the effect of adding extra pseudo-viscosity to the solution. How much grid-dependent pseudo-viscosity is added and its effects on the flow dynamics is also a function of the particulars of the discretization.

There is a third potential source of error that is typically small relative to the first two: random error due to finite precision representation of real numbers. This is often known as *roundoff error*. Typically it is negligible relative to iterative convergence and truncation errors. However, the accumulation of roundoff error occurring over many operations (about 10^{16} operations for double-precision floating-point numbers) can affect solution quality and in particular can affect sensitive flow features such as evolution near bifurcation points.

In conclusion, DNS of turbulent flows is among the most challenging of computations because of the need to capture multiscale features. At low Mach number, these scales span orders of magnitude from large device length scales to fine Kolmogorov scales. At high Mach number, the scales are more severe, and the required finest scale encroaches on the continuum limit. The demands on computational hardware are extreme, and both verification and validation present hurdles that can be difficult to overcome. That said, in contrast to common approaches that model sub-grid turbulence, typically tuned to experiment thus rendering them a posteriori post-dictions, DNS offers the hope of yielding a priori pre-diction of some of the most complicated mechanics of fluids.

SUMMARY

This chapter introduced perhaps the most challenging of problems in the mechanics of fluids: turbulence. Many standard notions were presented, including a discussion of how mechanical energy dissipates at small scales. It was shown that the small scales of turbulence are not far from the scales where the continuum assumption is invalid. A simplistic set of averaged equations was introduced, which gave rise to the well-known turbulence closure problem in which the effects of turbulence at the microscale require additional modeling; however, there is no consensus on how to formulate robust models of microscale turbulence. In closing, a calculation was presented of warm air rising turbulently that required no turbulence model; this was achieved at the expense of a highly resolved computation known as a direct numerical simulation. While difficult and often impractical, given current limitations of computational hardware, at present these offer the best method for prediction of the complicated mechanics of turbulent fluid flow.

PROBLEMS

14.1 Prove $\overline{fg} = \overline{f}\,\overline{g} + \overline{f'g'}$.

14.2 Find an expression for the generation of enstrophy $\boldsymbol{\omega}^T \cdot \boldsymbol{\omega}/2$ for a two-dimensional, incompressible flow with constant viscosity and conservative body force. Comment on the tendency of enstrophy to grow or decay for two-dimensional flows and contrast with three-dimensional flows.

14.3 Consider generation of helicity $\mathbf{v}^T \cdot \boldsymbol{\omega}$ for a two-dimensional, incompressible flow with constant viscosity and conservative body force. Contrast with equivalent three-dimensional flows.

14.4 Consider steady, incompressible, Reynolds-averaged Poiseuille flow in a channel. Neglect body forces. The channel has no-slip walls at $y = \pm h$. The fluid has viscosity μ, density ρ, and turbulent eddy length scale ℓ, each taken to be constant. Employ the simple mixing length theory for which $\mu_t = \rho \ell^2 \sqrt{2\overline{\mathbf{D}} : \overline{\mathbf{D}}}$. Find the peak velocity for the laminar limit, $\ell = 0$, and then for finite ℓ. Through use of Taylor series for small ℓ, show

$$\overline{u}_{max,turbulent} = \overline{u}_{max,laminar} - \frac{\Delta p^2 h^3}{3\mu^3 L^2}\rho\ell^2 - \cdots.$$

Plot laminar and turbulent velocity profiles for $y = [-h, h]$ for the parameters $h = 1$, $\mu = 1$, $L = 1$, $\Delta p = -1$, $\ell = 1/2$, $\rho = 1$. Evaluate the maximum velocity for laminar and turbulent Poiseuille flow.

14.5 Identify a modern calculation involving direct numerical simulation of an important problem involving turbulence. Write a short report describing the physical problem and the ability of the modern calculation to capture its physics.

14.6 Feynman et al. (1963, Vol. 1, Chapter 3) commented on the difficulty of analyzing turbulence. Read Feynman's Chapter 3, including his comments on what should be done with fluids such as wine whose properties and dynamics are not fully understood. Also consult relevant modern literature. Write a short essay ascertaining if the problems

Feynman describes remain today and if his prescription can be considered the best approach to the challenging problem of turbulence.

FURTHER READING

Batchelor, G. K. (1982). *The Theory of Homogeneous Turbulence*. Cambridge, UK: Cambridge University Press.

Bernard, P. S. (2019). *Turbulent Fluid Flow*. Hoboken, New Jersey: John Wiley.

Bohr, T., Jensen, M. H., Paladin, G., and Vulpiani, A. (1998). *Dynamical Systems Approach to Turbulence*. Cambridge, UK: Cambridge University Press.

Brodkey, R. S. (1967). *The Phenomena of Fluid Motions*. New York: Dover.

Chorin, A. J. (1993). *Vorticity and Turbulence*. New York: Springer.

Doering, C. R., and Gibbon, J. D. (1995). *Applied Analysis of the Navier–Stokes Equations*. Cambridge, UK: Cambridge University Press.

Foias, C., Manley, O., Rosa, R., and Temam, R. (2008). *Navier–Stokes Equations and Turbulence*. Cambridge, UK: Cambridge University Press.

Frisch, U. (1995). *Turbulence: the Legacy of A. N. Kolmogorov*. Cambridge, UK: Cambridge University Press.

Gallavotti, G. (2002). *Foundations of Fluid Dynamics*. Berlin: Springer.

Hinze, J. O. (1975). *Turbulence*, 2nd ed. New York: McGraw-Hill.

Holmes, P., Lumley, J. L., Berkooz, G., and Rowley, C. W. (2012). *Turbulence, Coherent Structures, Dynamical Systems and Symmetry*, 2nd ed. New York: Cambridge University Press.

Kollmann, W. (2019). *Navier–Stokes Turbulence: Theory and Analysis*. Cham, Switzerland: Springer.

Kuskin, S., and Shirikyan, A. (2012). *Mathematics of Two-Dimensional Turbulence*. Cambridge, UK: Cambridge University Press.

Landahl, M. T., and Mollo-Christensen, E. (1992). *Turbulence and Random Processes in Fluid Mechanics*, 2nd ed. New York: Cambridge University Press.

Lesieur, M. (2008). *Turbulence in Fluids*, 4th ed. Dordrecht: Springer.

Libby, P. A. (1996). *Introduction to Turbulence*. Washington, D.C.: Taylor and Francis.

Lumley, J. L., ed. (1990). *Whither Turbulence? Turbulence at the Crossroads* (*Lecture Notes in Physics*, **357**), Berlin: Springer.

Mathieu, J., and Scott, J. (2000). *An Introduction to Turbulent Flow*. Cambridge, UK: Cambridge University Press.

Pollard, A., Castillo, L., Danaila, L., and Glauser, M., eds. (2017). *Whither Turbulence and Big Data in the Twenty-first Century?* Switzerland: Springer.

Pope, S. B. (2000). *Turbulent Flows*. Cambridge, UK: Cambridge University Press.

Ruelle, D. (1989). *Chaotic Evolution and Strange Attractors*. Cambridge, UK: Cambridge University Press.

Sagaut, P., and Cambon, C. (2018). *Homogeneous Turbulence Dynamics*, 2nd ed. Cham, Switzerland: Springer.

Sengupta, T. K. (2021). *Transition to Turbulence: A Dynamical Systems Approach to Receptivity*. Cambridge, UK: Cambridge University Press.

Smits, A. J., and Dussauge, J.-P. (2006). *Turbulent Shear Layers in Supersonic Flow*. New York: Springer.

Stanišić, M. M. (1985). *The Mathematical Theory of Turbulence*. New York: Springer.

Thorne, K. S., and Blandford, R. D. (2017). *Modern Classical Physics: Optics, Fluids, Plasmas, Elasticity, and Statistical Physics*. Princeton, New Jersey: Princeton University Press.

Townsend, A. A. (1976). *The Structure of Turbulent Shear Flow*, 2nd ed. Cambridge, UK: Cambridge University Press.

Yih, C.-S. (1977). *Fluid Mechanics*. East Hampton, Connecticut: West River Press.

REFERENCES

Ahlers, G., Grossmann, S., and Lohse, D. (2009). Heat transfer and large scale dynamics in turbulent Rayleigh–Bénard convection, *Reviews of Modern Physics*, **81**(2), 503–537.

Bailon-Cuba, J., Emran, M. S., and Schumacher, J. (2010). Aspect ratio dependence of heat transfer and large-scale flow in turbulent convection, *Journal of Fluid Mechanics*, **655**, 152–173.

Berselli, L. C. (2021). *Three-Dimensional Navier–Stokes Equations for Turbulence*. London: Academic Press.

Bodenschatz, E., Pesch, W., and Ahlers, G. (2000). Recent developments in Rayleigh–Bénard convection, *Annual Review of Fluid Mechanics*, **32**, 709–778.

Clark, R. A., Ferziger, J. H., and Reynolds W. C. (1979). Evaluation of subgrid-scale models using an accurately simulated turbulent flow, *Journal of Fluid Mechanics*, **91**, 1–16.

Cummings, R. M., Mason, W. H., Morton, S. A., and McDaniel, D. R. (2015). *Applied Computational Aerodynamics: A Modern Engineering Approach*. New York: Cambridge University Press.

Davidson, P. A. (2015). *Turbulence: An Introduction for Scientists and Engineers*, 2nd ed. Oxford: Oxford University Press.

Desjardins, O., Blanquart, G., Balarac, G., and Pitsch, H. (2008). High order conservative finite difference scheme for variable density low Mach number turbulent flows, *Journal of Computational Physics*, **227**(15), 7125–7159.

Durbin, P. A., and Pettersson Reif, B. A. (2011). *Statistical Theory and Modeling for Turbulent Flows*. Chichester, UK: John Wiley.

Feynman, R. P., Leighton, R. B., and Sands, M. (1963). *The Feynman Lectures on Physics*, 3 vols., Reading, Massachusetts: Addison-Wesley.

Germano, M., Piomelli, U., Moin, P., and Cabot, W. H. (1991). Dynamic subgrid-scale eddy viscosity model, *Physics of Fluids A-Fluid Dynamics*, **3**(7), 1760–1765.

Ghosal, S. (1996). An analysis of numerical errors in large-eddy simulations of turbulence, *Journal of Computational Physics*, **125**(1), 187–206.

Gresho, P. M., and Lee, R. L. (1981). Don't suppress the wiggles – they're telling you something!, *Computers and Fluids*, **9**(2), 223–253.

Grötzbach, G. (1983). Spatial resolution requirements for direct numerical simulation of the Rayleigh–Bénard convection, *Journal of Computational Physics*, **49**(2), 241–264.

Harlow, F. H., and Welch, J. E. (1965). Numerical calculation of time-dependent viscous incompressible flow of fluid with free surface, *Physics of Fluids*, **8**(12), 2182–2189.

Henrick, A. K., Aslam, T. D., and Powers, J. M. (2005). Mapped weighted essentially non-oscillatory schemes: achieving optimal order near critical points, *Journal of Computational Physics*, **207**(2), 542–567.

Ishihara, T., Gotoh, T., and Kaneda, Y. (2009). Study of high-Reynolds number isotropic turbulence by direct numerical simulation, *Annual Review of Fluid Mechanics*, **41**, 165–180.

Jiang, G.-S., and Shu, C.-W. (1996). Efficient implementation of weighted ENO schemes, *Journal of Computational Physics*, **126**(1), 202–228.

Kadanoff, L. P. (2001). Turbulent heat flow: Structures and scaling, *Physics Today*, **54**(8), 34–39.

Kee, R. J., Coltrin, M. E., Glarborg, P., and Zhu, H. (2018). *Chemically Reacting Flow: Theory, Modeling and Simulation*, 2nd ed. Hoboken, New Jersey: John Wiley.

Kim, J., and Moin, P. (1985). Application of a fractional-step method to incompressible Navier–Stokes equations, *Journal of Computational Physics*, **59**(2), 308–323.

MacArt, J. F., and Mueller, M. E. (2016). Semi-implicit iterative methods for low Mach number turbulent reacting flows: Operator splitting versus approximate factorization, *Journal of Computational Physics*, **326**, 569–595.

Marusic I., and Broomhall, S. (2021). Leonardo da Vinci and fluid mechanics, *Annual Review of Fluid Mechanics*, **53**, 1–25.

McComb, W. D. (1990). *The Physics of Fluid Turbulence*. Oxford: Oxford University Press.

Oberkampf, W. L., and Roy, C. J. (2010). *Verification and Validation in Scientific Computing*. Cambridge, UK: Cambridge University Press.

Pandey, A., Scheel, J. D., and Schumacher, J. (2018). Turbulent superstructures in Rayleigh–Bénard convection, *Nature Communications*, **9**, 2118.

Panton, R. L. (2013). *Incompressible Flow*, 4th ed. New York: John Wiley.

Paolucci, S. (1982). On the filtering of sound from the Navier–Stokes equations, SAND82-8257, Livermore, California: Sandia National Laboratories.

Pope, S. B. (2004). Ten questions concerning the large-eddy simulation of turbulent flows, *New Journal of Physics*, **6**, 35.

Reynolds, W. C. (1976). Computation of turbulent flows, *Annual Review of Fluid Mechanics*, **8**, 183–208.

Reynolds, W. C. (1990). The potential and limitations of direct and large eddy simulations, *Whither Turbulence? Turbulence at the Crossroads*, (*Lecture Notes in Physics*, **357**), Lumley, J. L., ed. Berlin: Springer, 313–343.

Richardson, L. F. (1922). *Weather Prediction by Numerical Process*. Cambridge, UK: Cambridge University Press.

Roache, P. J. (2009). *Fundamentals of Verification and Validation*. Albuquerque, New Mexico: Hermosa.

Scheel, J. D., Emran, M. S., and Schumacher, J. (2013). Resolving the fine-scale structure in turbulent Rayleigh–Bénard convection, *New Journal of Physics*, **15**, 113063.

Singh, S., Powers, J. M., and Paolucci, S. (1999). Detonation solutions from reactive Navier–Stokes equations, AIAA-99-0966, 37th AIAA Aerospace Sciences Meeting and Exhibit, Reno, Nevada.

Sreenivasan, K. R. (1995). On the universality of the Kolmogorov constant, *Physics of Fluids*, **7**(11), 2778–2784.

Tennekes, H., and Lumley, J. L. (1972). *A First Course in Turbulence*. Cambridge, Massachusetts: MIT Press.

Tritton, D. J. (1988). *Physical Fluid Dynamics*, 2nd ed. Oxford: Oxford University Press.

White, F. M. (2006). *Viscous Fluid Flow*, 3rd ed. New York: McGraw-Hill.

Wilcox, D. C. (2006). *Turbulence Modeling for CFD*, 3rd ed. La Cañada, California: DCW Industries.

Yeung, P. K., and Zhou, Y. (1997). Universality of the Kolmogorov constant in numerical simulations of turbulence, *Physical Review E*, **56**(2), 1746–1752.

Bibliography

Abbott, I. H. (1980). *Theory of Wing Sections*. New York: Dover.

Ablowitz, M. J. (2011). *Nonlinear Dispersive Waves: Asymptotic Analysis and Solitons*. Cambridge, UK: Cambridge University Press.

Acheson, D. J. (1990). *Elementary Fluid Dynamics*. Oxford: Oxford University Press.

Ahlers, G., Grossmann, S., and Lohse, D. (2009). Heat transfer and large scale dynamics in turbulent Rayleigh–Bénard convection, *Reviews of Modern Physics*, **81**(2), 503–537.

Anderson, D. A., Tannehill, J. C., Pletcher, R. H., Munipalli, R., and Shankar, V. (2021). *Computational Fluid Mechanics and Heat Transfer*, 4th ed. Boca Raton, Florida: CRC Press.

Anderson, J. D. (2006). *Hypersonic and High Temperature Gas Dynamics*, 2nd ed. Reston, Virginia: AIAA.

Anderson, J. D. (2021). *Modern Compressible Flow with Historical Perspective*, 4th ed. New York: McGraw-Hill.

Aref, H. (1979). Motion of three vortices, *Physics of Fluids*, **22**(3), 393–400.

Aref, H., and Balachandar, S. (2018). *A First Course in Computational Fluid Dynamics*. Cambridge, UK: Cambridge University Press.

Aref, H., Blake, J. R., Budišić, M. et al. (2017). Frontiers of chaotic advection, *Reviews of Modern Physics*, **89**(2), 025007.

Aris, R. (1962). *Vectors, Tensors, and the Basic Equations of Fluid Mechanics*. New York: Dover.

Arnold, V. I., and Khesin, B. A. (2021). *Topological Methods in Hydrodynamics*, 2nd ed., Cham, Switzerland: Springer.

Arpaci, V. S., and Larsen, P. S. (1984). *Convection Heat Transfer*. Englewood Cliffs, New Jersey: Prentice-Hall.

Ashley, H., and Landahl, M. (1985). *Aerodynamics of Wings and Bodies*. New York: Dover.

Badin, G., and Crisciani, F. (2018). *Variational Formulation of Fluid and Geophysical Fluid Dynamics: Mechanics, Symmetries and Conservation Laws*. Cham, Switzerland: Springer.

Bailon-Cuba, J., Emran, M. S., and Schumacher, J. (2010). Aspect ratio dependence of heat transfer and large-scale flow in turbulent convection, *Journal of Fluid Mechanics*, **655**, 152–173.

Barenblatt, G. I. (1996). *Scaling, Self-Similarity, and Intermediate Asymptotics*. New York: Cambridge University Press.

Basset, A. B. (1888). *A Treatise on Hydrodynamics with Numerous Examples*, 2 vols. Cambridge, UK: Deighton, Bell, and Co.

Batchelor, G. K. (1982). *The Theory of Homogeneous Turbulence*. Cambridge, UK: Cambridge University Press.

Batchelor, G. K. (2000). *An Introduction to Fluid Dynamics*. Cambridge, UK: Cambridge University Press.

Bateman, H. (1915). Some recent researches in the motion of fluids, *Monthly Weather Review*, **43**(4), 163–170.

Becker, R. (1922). Stoßwelle und Detonation, *Zeitschrift für Physik*, **8**, 321–362, English translation in NACA-TM-505, 1929.

Bejan, A. (2013). *Convection Heat Transfer*, 4th ed. Hoboken, New Jersey: John Wiley.

Berkshire, F. H., Malham, S. J. A., and Stuart, J. T. (2022). *Introductory Incompressible Fluid Mechanics*. Cambridge, UK: Cambridge University Press.

Bernard, P. S. (2015). *Fluid Dynamics*. New York: Cambridge University Press.

Bernard, P. S. (2019). *Turbulent Fluid Flow*. Hoboken, New Jersey: John Wiley.

Bernoulli, D. (1738). *Hydrodynamica, Sive de Viribus et Motibus Fluidorum Commentarii*. Strasbourg: Johann Reinhold Dulsseker.

Berselli, L. C. (2021). *Three-Dimensional Navier–Stokes Equations for Turbulence*. London: Academic Press.

Bird, R. B., Armstrong, R. C., and Hassager, O. (1987). *Dynamics of Polymeric Liquids*, 2 vols. New York: John Wiley.

Bird, R. B., Stewart, W. E., and Lightfoot, E. N. (2007). *Transport Phenomena*, revised 2nd ed. New York: John Wiley.

Birkhoff, G. (1960). *Hydrodynamics: A Study in Logic, Fact and Similitude*. Princeton, New Jersey: Princeton University Press.

Bodenschatz, E., Pesch, W., and Ahlers, G. (2000). Recent developments in Rayleigh–Bénard convection, *Annual Review of Fluid Mechanics*, **32**, 709–778.

Bohr, T., Jensen, M. H., Paladin, G., and Vulpiani, A. (1998). *Dynamical Systems Approach to Turbulence*. Cambridge, UK: Cambridge University Press.

Braithwaite, J. (2017). *Essential Fluid Dynamics for Scientists*. San Rafael, California: Morgan and Claypool.

Brodkey, R. S. (1967). *The Phenomena of Fluid Motions*. New York: Dover.

Brown, J. W., and Churchill, R. V. (2014). *Complex Variables and Applications*, 9th ed. New York: McGraw-Hill.

Brunton, S. L., and Kutz, J. N. (2022). *Data-Driven Science and Engineering: Machine Learning, Dynamical Systems, and Control*, 2nd ed. New York: Cambridge University Press.

Brunton, S. L., Noack, B. R. and Koumoutsakos, P. (2019). Machine learning for fluid mechanics, *Annual Review of Fluid Mechanics*, **52**, 477–508.

Buresti, G. (2012). *Elements of Fluid Dynamics*. London: Imperial College Press.

Burgers, J. M. (1948). A mathematical model illustrating the theory of turbulence, *Advances in Applied Mathematics*. **1**, 171–199.

Burmeister, L. C. (1993). *Convective Heat Transfer*, 2nd ed. New York: John Wiley.

Callen, H. B. (1985). *Thermodynamics and an Introduction to Thermostatistics*, 2nd ed. New York: John Wiley.

Cantwell, B. J. (2002). *Introduction to Symmetry Analysis*. New York: Cambridge University Press.

Carslaw, H. S., and Jaeger, J. C. (1986). *Conduction of Heat in Solids*, 2nd ed. Oxford: Oxford University Press.

Chandrasekhar, S. (1961). *Hydrodynamic and Hydromagnetic Stability*. New York: Dover.

Chang, H.-C., and Demekhin, E. A. (2002). *Complex Wave Dynamics on Thin Films*. Amsterdam: Elsevier.

Chapman, S., and Cowling, T. G. (1991). *The Mathematical Theory of Non-Uniform Gases*, 3rd ed. Cambridge, UK: Cambridge University Press.

Charru, F. (2011). *Hydrodynamic Instabilities*. New York: Cambridge University Press.

Chemin, J.-Y. (1998). *Perfect Incompressible Fluids*. Oxford: Oxford University Press.

Chen, R.-H. (2017). *Foundations of Gas Dynamics*. New York: Cambridge University Press.

Chenoweth, D. R., and Paolucci, S. (1986). Natural convection in an enclosed vertical air layer with large horizontal temperature differences, *Journal of Fluid Mechanics*, **169**, 173–210.

Chevray, R., and Mathieu, J. (1993). *Topics in Fluid Mechanics*. Cambridge, UK: Cambridge University Press.

Childress, S. (2009). *An Introduction to Theoretical Fluid Mechanics*. Providence, Rhode Island: American Mathematical Society.

Chorin, A. J. (1993). *Vorticity and Turbulence*. New York: Springer.

Chorin, A. J., and Marsden, J. E. (2000). *A Mathematical Introduction to Fluid Mechanics*, 3rd ed. New York: Springer.

Chung, T. J. (2007). *General Continuum Mechanics*. New York: Cambridge University Press.

Chung, T. J. (2010). *Computational Fluid Dynamics*, 2nd ed. New York: Cambridge University Press.

Clark, R. A., Ferziger, J. H., and Reynolds W. C. (1979). Evaluation of subgrid-scale models using an accurately simulated turbulent flow, *Journal of Fluid Mechanics*, **91**, 1–16.

Clarke, C. J., and Carswell, R. F. (2007). *Principles of Astrophysical Fluid Dynamics*. Cambridge, UK: Cambridge University Press.

Colonna, P., and Guardone, A. (2006). Molecular interpretation of nonclassical gas dynamics of dense vapors under the van der Waals model, *Physics of Fluids*, **19**(8), 086102.

Conlisk, A. T. (2013). *Essentials of Micro- and Nanofluidics with Applications to the Biological and Chemical Sciences*, New York: Cambridge University Press.

Constantinescu, V. N. (1995). *Laminar Viscous Flow*. New York: Springer.

Cottet, G.-H., and Koumoutsakos, P. D. (2000). *Vortex Methods: Theory and Practice*, 2nd ed. Cambridge, UK: Cambridge University Press.

Courant, R., and Friedrichs, K. O. (1976). *Supersonic Flow and Shock Waves*. New York: Springer.

Criminale, W. O., Jackson, T. L., and Joslin, R. D. (2019). *Theory and Computation in Hydrodynamic Stability*, 2nd ed. Cambridge, UK: Cambridge University Press.

Crocco, L. (1937). Eine neue stromfunktion für die erforschung der bewegung der gase mit rotation, *Zeitschrift für Angewandte Mathematik und Mechanik*, **17**(1), 1–7.

Cross, M., and Greenside, H. (2009) *Pattern Formation and Dynamics in Nonequilibrium Systems*. Cambridge, UK: Cambridge University Press.

Cross, M. C., and Hohenberg, P. C. (1993). Pattern formation outside of equilibrium, *Reviews of Modern Physics*, **65**(3), 851–1112.

Cummings, R. M., Mason, W. H., Morton, S. A., and McDaniel, D. R. (2015). *Applied Computational Aerodynamics: A Modern Engineering Approach*. New York: Cambridge University Press.

Currie, I. G. (2013). *Fundamental Mechanics of Fluids*, 4th ed. Boca Raton, Florida: CRC Press.

Cushman-Roisin, B., and Beckers, J.-M. (2011). *Introduction to Geophysical Fluid Dynamics*, 2nd ed. Amsterdam: Academic Press.

Davidson, P. A. (2015). *Turbulence: An Introduction for Scientists and Engineers*, 2nd ed. Oxford: Oxford University Press.

Davidson, P. A. (2022). *Incompressible Fluid Dynamics*. Oxford: Oxford University Press.

de Groot, S. R., and Mazur, P. (1984). *Non-Equilibrium Thermodynamics*. New York: Dover.

Desjardins, O., Blanquart, G., Balarac, G., and Pitsch, H. (2008). High order conservative finite difference scheme for variable density low Mach number turbulent flows, *Journal of Computational Physics*, **227**(15), 7125–7159.

Doering, C. R. (2009). The 3D Navier–Stokes problem, *Annual Review of Fluid Mechanics*, **41**, 109–128.

Doering, C. R., and Gibbon, J. D. (1995). *Applied Analysis of the Navier–Stokes Equations*. Cambridge, UK: Cambridge University Press.

Drazin, P. G. (1992). *Nonlinear Systems*. Cambridge, UK: Cambridge University Press.

Drazin, P. G. (2002). *Introduction to Hydrodynamic Stability*. Cambridge, UK: Cambridge University Press.

Drazin, P. G., and Reid, W. H. (1981). *Hydrodynamic Stability*. Cambridge, UK: Cambridge University Press.

Drazin, P. G., and Riley, N. (2006). *The Navier–Stokes Equations: A Classification of Flows and Exact Solutions*. Cambridge, UK: Cambridge University Press.

Durbin, P. A., and Medic, G. (2007). *Fluid Dynamics with a Computational Perspective*. New York: Cambridge University Press.

Durbin, P. A., and Pettersson Reif, B. A. (2011). *Statistical Theory and Modeling for Turbulent Flows*. Chichester, UK: John Wiley.

Durst, F. (2008). *Fluid Mechanics: An Introduction to the Theory of Fluid Flows*. Berlin: Springer.

Edelen, D. G. B., and McLennan, J. A. (1973). Material indifference: a principle or a convenience, *International Journal of Engineering Science*, **11**(8), 813–817.

Eckert, E. R. G., and Gross, J. F. (1963). *Introduction to Heat and Mass Transfer.* New York: McGraw–Hill.

Emanuel, G. (1986). *Gasdynamics: Theory and Applications.* New York: AIAA.

Emanuel, G. (2016). *Analytical Fluid Dynamics*, 3rd ed. Boca Raton, Florida: CRC Press.

Eringen, A. C. (1989). *Mechanics of Continua*, 2nd ed. Malabar, Florida: Krieger.

Euler, L. (1757). Principes généraux du mouvement des fluides, *Mémoires de l'Académie des Sciences de Berlin*, **11**, 274–315.

Faber, T. E. (1995). *Fluid Dynamics for Physicists.* Cambridge, UK: Cambridge University Press.

Falkovich, G. (2018). *Fluid Mechanics*, 2nd ed. Cambridge, UK: Cambridge University Press.

Feireisl, E. (2004). *Dynamics of Viscous Compressible Fluids.* Oxford: Oxford University Press.

Feireisl, E., Karper, T. G., and Pokorný, M. (2016). *Mathematical Theory of Compressible Viscous Fluids.* Cham, Switzerland: Springer.

Feldmeier, A. (2019). *Theoretical Fluid Dynamics.* Cham, Switzerland: Springer.

Ferziger, J. H., and Peric, M. (2002). *Computational Methods for Fluid Dynamics*, revised 3rd ed. New York: Springer.

Feynman, R. P., Leighton, R. B., and Sands, M. (1963). *The Feynman Lectures on Physics*, 3 vols., Reading, Massachusetts: Addison-Wesley.

Finlayson, B. A. (1972). *The Method of Weighted Residuals and Variational Principles.* New York: Academic Press.

Finn, B. S. (1964). Laplace and the speed of sound, *Isis*, **55**(1), 7–19.

Foias, C., Manley, O., Rosa, R., and Temam, R. (2008). *Navier–Stokes Equations and Turbulence.* Cambridge, UK: Cambridge University Press.

Frisch, U. (1995). *Turbulence: the Legacy of A. N. Kolmogorov.* Cambridge, UK: Cambridge University Press.

Fung, Y. C. (1965). *Foundations of Solid Mechanics.* Englewood Cliffs, New Jersey: Prentice-Hall.

Gad-el-Hak, M. (1995). Questions in fluid mechanics: Stokes' hypothesis for a Newtonian, isotropic fluid, *Journal of Fluids Engineering-Transactions of the ASME*, **117**(1), 3–5.

Gad-el-Hak, M. (2000). *Flow Control: Passive, Active and Reactive Flow Management.* Cambridge, UK: Cambridge University Press.

Galdi, G. P. (2011). *An Introduction to the Mathematical Theory of the Navier–Stokes Equations*, 2nd ed. New York: Springer.

Gallavotti, G. (2002). *Foundations of Fluid Dynamics.* Berlin: Springer.

Garabedian, P. R. (1998). *Partial Differential Equations.* Providence, Rhode Island: American Mathematical Society.

Germano, M., Piomelli, U., Moin, P., and Cabot, W. H. (1991). Dynamic subgrid-scale eddy viscosity model, *Physics of Fluids A-Fluid Dynamics*, **3**(7), 1760–1765.

Ghia, U., Bayyuk, S., Benek, J. et al. (2021). Recommended Practice for Code Verification in Computational Fluid Dynamics, AIAA R-141-2021; portions reprinted by permission of the American Institute of Aeronautics and Astronautics, Inc

Ghosal, S. (1996). An analysis of numerical errors in large-eddy simulations of turbulence, *Journal of Computational Physics*, **125**(1), 187–206.

Gilbarg, D., and Paolucci, D. (1953). The structure of shock waves in the continuum theory of fluids, *Journal of Rational Mechanics and Analysis*, **2**(5), 617–642.

Glauert, H. (1983). *The Elements of Aerofoil and Airscrew Theory.* Cambridge, UK: Cambridge University Press.

Goldstein, H. (1950). *Classical Mechanics.* Reading, Massachusetts: Addison-Wesley.

Goldstein, S., ed. (1965). *Modern Developments in Fluid Dynamics*, 2 vols. New York: Dover.

Golub, G. H., and Van Loan, C. F. (2013). *Matrix Computations*. Baltimore: Johns Hopkins University Press.

Goodwine, B. (2010). *Engineering Differential Equations: Theory and Applications*. New York: Springer.

Gonzalez, O., and Stuart, A. M. (2008). *A First Course in Continuum Mechanics*. Cambridge, UK: Cambridge University Press.

Graebel, W. P. (2007). *Advanced Fluid Mechanics*. Amsterdam: Academic Press.

Grandy, W. T. (2008). *Entropy and the Time Evolution of Macroscopic Systems*. Oxford: Oxford University Press.

Granger, R. A. (1995). *Fluid Mechanics*. New York: Dover.

Greenspan, H. P. (1968). *The Theory of Rotating Fluids*. Cambridge, UK: Cambridge University Press.

Gresho, P. M., and Lee, R. L. (1981). Don't suppress the wiggles – they're telling you something!, *Computers and Fluids*, **9**(2), 223–253.

Grotberg, J. B. (2021). *Biofluid Mechanics: Analysis and Applications*. New York: Cambridge University Press.

Grötzbach, G. (1983). Spatial resolution requirements for direct numerical simulation of the Rayleigh–Bénard convection, *Journal of Computational Physics*, **49**(2), 241–264.

Guckenheimer, J., and Holmes, P. H. (2002). *Nonlinear Oscillations, Dynamical Systems, and Bifurcations of Vector Fields*. New York: Springer-Verlag.

Gurtin, M. E., Fried, E., and Anand, L. (2013). *The Mechanics and Thermodynamics of Continua*, New York: Cambridge University Press.

Gülçat, Ü. (2021). *Fundamentals of Modern Unsteady Aerodynamics*, 3rd ed. Cham, Switzerland: Springer.

Guyon, E., Hulin, J.-P., Petit, L., and Mitescu, C. D. (2012). *Physical Hydrodynamics*, 2nd ed. Oxford: Oxford University Press.

Gyarmati, I. (1970). *Non-equilibrium Thermodynamics. Field Theory and Variational Principles*. New York: Springer.

Han, Z., and Yin, X. (1993). *Shock Dynamics*. Dordrecht: Springer.

Harlow, F. H., and Welch, J. E. (1965). Numerical calculation of time-dependent viscous incompressible flow of fluid with free surface, *Physics of Fluids*, **8**(12), 2182–2189.

Hayes, W. D., and Probstein, R. F. (1959). *Hypersonic Flow Theory*. New York: Academic Press.

Helmholtz, H. (1954). *On the Sensations of Tone*. New York: Dover.

Henrick, A. K., Aslam, T. D., and Powers, J. M. (2005). Mapped weighted essentially non-oscillatory schemes: Achieving optimal order near critical points, *Journal of Computational Physics*, **207**(2), 542–567.

Herron, I. H., and Foster, M. R. (2008). *Partial Differential Equations in Fluid Dynamics*. New York: Cambridge University Press.

Hilborn, R. C. (2000). *Chaos and Nonlinear Dynamics*, 2nd ed. New York: Oxford University Press.

Hinch, E. J. (1991). *Perturbation Methods*. Cambridge, UK: Cambridge University Press.

Hinch, E. J. (2020). *Think Before You Compute: A Prelude to Computational Fluid Dynamics*. Cambridge, UK: Cambridge University Press.

Hinze, J. O. (1975). *Turbulence*, 2nd ed. New York: McGraw-Hill.

Hirsch, C. (1989). *Numerical Computation of Internal and External Flows*, 2 vols. New York: John Wiley.

Hirsch, M. W., Smale, S. and Devaney, R. L. (2013). *Differential Equations, Dynamical Systems, and an Introduction to Chaos*, 3rd ed. Waltham, Massachusetts: Academic Press.

Hirschfelder, J. O., Curtiss, C. F., and Bird, R. B. (1954). *Molecular Theory of Gases and Liquids*. New York: John Wiley.

Holmes, P., Lumley, J. L., Berkooz, G., and Rowley, C. W. (2012). *Turbulence, Coherent Structures, Dynamical Systems and Symmetry*, 2nd ed. New York: Cambridge University Press.

Howe, M. S. (2003). *Theory of Vortex Sound*, Cambridge, UK: Cambridge University Press.

Hubbard, J., and Burke-Hubbard, B. (2015). *Vector Calculus, Linear Algebra, and*

Differential Forms: A Unified Approach, 5th ed. Ithaca, New York: Matrix Editions.

Hughes, W. F., and Gaylord, E. W. (1964). *Basic Equations of Engineering Science*, Schaum's Outline Series. New York: McGraw-Hill.

Huilgol, R. R. (2015). *Fluid Mechanics of Viscoplasticity*. Berlin: Springer.

Hutter, K., and Wang, Y. (2016). *Fluid and Thermodynamics*, 2 vols. Switzerland: Springer.

Iannelli, J. (2013). An exact non-linear Navier–Stokes compressible-flow solution for CFD code verification, *International Journal for Numerical Methods in Fluids*, **72**(2), 157–176.

Imberger, J. (2013). *Environmental Fluid Dynamics*. Amsterdam: Academic Press.

Incropera, F. P., and DeWitt, D. P. (1981). *Fundamentals of Heat Transfer*. New York: John Wiley.

Ishihara, T., Gotoh, T., and Kaneda, Y. (2009). Study of high-Reynolds number isotropic turbulence by direct numerical simulation, *Annual Review of Fluid Mechanics*, **41**, 165–180.

Jacob, M. (1949). *Heat Transfer*. New York: John Wiley.

Jiang, G.-S., and Shu, C.-W. (1996). Efficient implementation of weighted ENO schemes, *Journal of Computational Physics*, **126**(1), 202–228.

Jog, C. S. (2015). *Fluid Mechanics: Foundations and Applications of Mechanics*, Vol. 2, 3rd ed. Delhi: Cambridge University Press.

Jones, R. T. (1990). *Wing Theory*. Princeton: Princeton University Press.

Jordinson, R. (1970). The flat plate boundary layer. Part 1. Numerical integration of the Orr–Sommerfeld equation, *Journal of Fluid Mechanics*, **43**(4), 801–811.

Joseph, D. D. (1976). *Stability of Fluid Motions*, 2 vols. Berlin: Springer.

Joseph, D. D. (1990). *Fluid Dynamics of Viscoelastic Liquids*. New York: Springer.

Joseph, D., Funada, T., and Wang, J. (2008). *Potential Flows of Viscous and Viscoelastic Fluids*. New York: Cambridge University Press.

Kadanoff, L. P. (2001). Turbulent heat flow: structures and scaling, *Physics Today*, **54**(8), 34–39.

Kaplan, W. (2003). *Advanced Calculus*, 5th ed. Boston, Massachusetts: Addison-Wesley.

Karamcheti, K. (1980). *Principles of Ideal-Fluid Aerodynamics*, 2nd ed. Malabar, Florida: Krieger.

Karniadakis, G. E., and Beskok, A. (2001). *Microflows: Fundamentals and Simulation*. Berlin: Springer.

Karniadakis, G. E., and Sherwin, S. J. (2005). *Spectral/HP Element Methods for CFD*, 2nd ed. Oxford: Oxford University Press.

Kato, S., and Fukue, J. (2020). *Fundamentals of Astrophysical Fluid Dynamics: Hydrodynamics, Magnetohydrodynamics, and Radiation Hydrodynamics*. Singapore: Springer.

Katz, J., and Plotkin, A. (2001). *Low-Speed Aerodynamics*, 2nd ed. New York: Cambridge University Press.

Kay, D. C (2011). *Tensor Calculus*, Schaum's Outline Series. New York: McGraw-Hill.

Kays, W. M., Crawford, M. E., and Weigand, B. (2004). *Convective Heat and Mass Transfer*, 4th ed. New York: McGraw-Hill.

Kee, R. J., Coltrin, M. E., Glarborg, P., and Zhu, H. (2018). *Chemically Reacting Flow: Theory, Modeling and Simulation*, 2nd ed. Hoboken, New Jersey: John Wiley.

Kim, J., and Moin, P. (1985). Application of a fractional-step method to incompressible Navier–Stokes equations, *Journal of Computational Physics*, **59**(2), 308–323.

Kirby, B. J. (2010). *Micro- and Nanoscale Fluid Mechanics: Transport in Microfluidic Devices*. New York: Cambridge University Press.

Kleinstreuer, C. (2018). *Modern Fluid Dynamics*, 2nd ed. Boca Raton, Florida: CRC Press.

Kollmann, W. (2019). *Navier–Stokes Turbulence: Theory and Analysis*. Cham, Switzerland: Springer.

Kondepudi, D., and Prigogine, I. (1998). *Modern Thermodynamics: From Heat Engines to Dissipative Structures*. New York: John Wiley.

Koschmieder, E. L. (1993). *Bènard Cells and Taylor Vortices.* Cambridge, UK: Cambridge University Press.

Kuethe, A. M., and Chow, C.-Y. (1998). *Foundations of Aerodynamics: Bases of Aerodynamic Design*, 5th ed. New York: John Wiley.

Kundu, P. K., Cohen, I. M., and Dowling, D. R. (2015). *Fluid Mechanics*, 6th ed. Amsterdam: Academic Press.

Kuskin, S., and Shirikyan, A. (2012). *Mathematics of Two-Dimensional Turbulence.* Cambridge, UK: Cambridge University Press.

Lagerstrom, P. A. (1964). *Laminar Flow Theory.* Princeton, New Jersey: Princeton University Press.

Lamb, H. (1993). *Hydrodynamics*, 6th ed. New York: Dover.

Lanchester, F. W. (1905). *Aerodynamics.* London: Constable.

Laney, C. B. (1998). *Computational Gasdynamics.* Cambridge, UK: Cambridge University Press.

Landahl, M. T. (1989). *Unsteady Transonic Flow.* Cambridge, UK: Cambridge University Press.

Landahl, M. T., and Mollo-Christensen, E. (1992). *Turbulence and Random Processes in Fluid Mechanics*, 2nd ed. New York: Cambridge University Press.

Landau, L. D., and Lifshitz, E. M. (1959). *Fluid Mechanics.* Oxford: Pergamon Press.

Langlois, W. E., and Deville, M. O. (2014). *Slow Viscous Flow*, 2nd ed. Heidelberg: Springer.

Lautrup, B. (2011). *Physics of Continuous Matter: Exotic and Everyday Phenomena in the Macroscopic World*, 2nd ed. Boca Raton, Florida: CRC Press.

Lavenda, B. H. (1978). *Thermodynamics of Irreversible Processes.* New York: John Wiley.

Leal, L. G. (1992). *Laminar Flow and Convective Transport Processes: Scaling Principles and Asymptotic Analysis.* Stoneham, Massachusetts: Butterworth-Heinemann.

Leal, L. G. (2007). *Advanced Transport Phenomena: Fluid Mechanics and Convective Transport Processes.* New York: Cambridge University Press.

Lesieur, M. (2008). *Turbulence in Fluids*, 4th ed. Dordrecht: Springer.

LeVeque, R. J. (1992). *Numerical Methods for Conservation Laws.* Basel: Birkhäuser.

LeVeque, R. J. (2002). *Finite Volume Methods for Hyperbolic Problems.* Cambridge, UK: Cambridge University Press.

Libby, P. A. (1996). *Introduction to Turbulence.* Washington, D.C.: Taylor and Francis.

Liepmann, H. W., and Roshko, A. (2002). *Elements of Gasdynamics.* New York: Dover.

Lighthill, J. (1978). *Waves in Fluids.* Cambridge, UK: Cambridge University Press.

Lighthill, J. (1986). *An Informal Introduction to Theoretical Fluid Mechanics.* Oxford: Oxford University Press.

Lin, C. C. (1966). *The Theory of Hydrodynamic Stability.* Cambridge, UK: Cambridge University Press.

Lin, C. C., and Segal, L. A. (1988). *Mathematics Applied to Deterministic Problems in the Natural Sciences.* Philadelphia: SIAM.

Lions, P.-L. (1996, 1998). *Mathematical Topics in Fluid Mechanics*, 2 vols. Oxford: Clarendon Press.

Liseikin, V. L. (2017). *Grid Generation Methods*, 3rd ed. Cham, Switzerland: Springer.

Lorenz, E. N. (1963). Deterministic nonperiodic flow, *Journal of the Atmospheric Sciences*, **20**(2), 130–141.

Lovelock, D., and Rund, H. (1975). *Tensors, Differential Forms, and Variational Principles.* New York: Dover.

Ludford, G. S. S. (1951). The classification of one-dimensional flows and the general problem of a compressible, viscous, heat-conducting fluid, *Journal of the Aeronautical Sciences*, **18**(12), 830–834.

Lumley, J. L., ed. (1990). *Whither Turbulence? Turbulence at the Crossroads*, (*Lecture Notes in Physics*, **357**), Berlin: Springer.

Ma, T., and Wang, S. (2005). *Geometric Theory of Incompressible Flows with Application to Fluid Dynamics.* Providence, Rhode Island: American Mathematical Society.

MacArt, J. F., and Mueller, M. E. (2016). Semi-implicit iterative methods for low Mach number turbulent reacting flows: Operator splitting versus approximate factorization, *Journal of Computational Physics*, **326**, 569–595.

Majda, A. J., and Bertozzi, A. L. (2002). *Vorticity and Incompressible Flow.* Cambridge, UK: Cambridge University Press.

Marchioro, C., and Pulvirenti, M. (1994). *Mathematical Theory of Incompressible Nonviscous Fluids.* New York: Springer.

Marsden, J. E., and Hughes, T. J. R. (1983). *Mathematical Foundations of Elasticity.* Mineola, New York: Dover.

Marusic I., and Broomhall, S. (2021). Leonardo da Vinci and fluid mechanics, *Annual Review of Fluid Mechanics*, **53**, 1–25.

Mathieu, J., and Scott, J. (2000). *An Introduction to Turbulent Flow.* Cambridge, UK: Cambridge University Press.

McClarren, R. G. (2021). *Machine Learning for Engineers.* Cham, Switzerland: Springer.

McComb, W. D. (1990). *The Physics of Fluid Turbulence.* Oxford: Oxford University Press.

McConnell, A. J. (1957). *Applications of Tensor Analysis.* New York: Dover.

McLeod, E. B. (2016). *Introduction to Fluid Dynamics.* Mineola, New York: Dover.

Mei, C. C. (1997). *Mathematical Analysis in Engineering.* Cambridge, UK: Cambridge University Press.

Meneveau, C. (2011). Lagrangian dynamics and models of the velocity gradient tensor in turbulent flows, *Annual Review of Fluid Mechanics*, **43**, 219–245.

Meyer, R. E. (1982). *Introduction to Mathematical Fluid Dynamics.* New York: Dover.

Mengers, J. D. (2012). Slow invariant manifolds for reaction-diffusion systems, Ph.D. Dissertation, University of Notre Dame, Notre Dame, Indiana.

Mihaljan, J. M. (1962). A rigorous exposition of the Boussinesq approximations applicable to a thin layer of fluid, *Astrophysical Journal*, **136**(3), 1126–1133.

Milne-Thompson, L. M. (1958). *Theoretical Aerodynamics*, 4th ed. New York: Dover.

Milne-Thompson, L. M. (1996). *Theoretical Hydrodynamics*, 5th ed. New York: Dover.

Moran, J. (1984). *An Introduction to Theoretical and Computational Aerodynamics*, Mineola, New York: Dover.

Morduchow, M., and Libby, P. A. (1949). On a complete solution of the one-dimensional flow equations of a viscous, heat conducting, compressible gas, *Journal of the Aeronautical Sciences*, **16**(11), 674–684.

Morita, O. (2019). *Classical Mechanics in Geophysical Fluid Dynamics.* Boca Raton, Florida: CRC Press.

Morrison, F. A. (2001). *Understanding Rheology.* New York: Oxford University Press.

Morse, P. M., and Feshbach, H. (1953). *Methods of Theoretical Physics*, 2 vols. New York: McGraw-Hill.

Müller, I. (2007). *A History of Thermodynamics: The Doctrine of Energy and Entropy.* Berlin: Springer.

Müller, I., and Ruggeri, T. (1998). *Rational Extended Thermodynamics*, 2nd ed. New York: Springer.

Nazarenko, S. (2015). *Fluid Dynamics via Examples and Solutions.* Boca Raton, Florida: CRC Press.

Newton, I. (1999). *The Principia: Mathematical Principles of Natural Philosophy.* Cohen, I. B. and Whitman, A., trans. Berkeley, California: University of California Press.

Newton, P. K. (2001). *The N-Vortex Problem: Analytical Techniques.* New York: Springer.

Novotný, A., and Straškraba, I. (2004). *Introduction to the Mathematical Theory of Compressible Flow.* Oxford: Oxford University Press.

Oberkampf, W. L., and Roy, C. J. (2010). *Verification and Validation in Scientific Computing.* Cambridge, UK: Cambridge University Press.

Ockendon, H., and Ockendon, J. R. (1995). *Viscous Flow.* Cambridge, UK: Cambridge University Press.

Osborne, M. R. (1967). Numerical methods for hydrodynamic stability problems, *SIAM Journal on Applied Mathematics*, **15**(3), 539–557.

Ottino, J. M. (1989). *The Kinematics of Mixing: Stretching, Chaos, and Transport.* Cambridge, UK: Cambridge University Press.

Pandey, A., Scheel, J. D., and Schumacher, J. (2018). Turbulent superstructures in Rayleigh–Bénard convection, *Nature Communications*, **9**, 2118.

Panton, R. L. (2013). *Incompressible Flow*, 4th ed. New York: John Wiley.

Paolucci, S. (1982). On the filtering of sound from the Navier–Stokes equations, SAND82-8257, Livermore, California: Sandia National Laboratories.

Paolucci, S. (2016). *Continuum Mechanics and Thermodynamics of Matter*. New York: Cambridge University Press.

Paterson, A. R. (1983). *A First Course in Fluid Dynamics*. Cambridge, UK: Cambridge University Press.

Pedlosky, J. (1987). *Geophysical Fluid Dynamics*, 2nd ed. New York: Springer.

Perko, L. (2006). *Differential Equations and Dynamical Systems*, 3rd ed. Berlin: Springer.

Perry, A. E., and Chong, M. S. (1987). A description of eddying motions and flow patterns using critical-point concepts, *Annual Review of Fluid Mechanics*, **19**, 125–155.

Pnueli, D., and Gutfinger, C. (1992). *Fluid Mechanics*. Cambridge, UK: Cambridge University Press.

Pollard, A., Castillo, L., Danaila, L., and Glauser, M., eds. (2017). *Whither Turbulence and Big Data in the Twenty-first Century?* Switzerland: Springer.

Pop, I., and Ingham, D. B. (2001). *Convective Heat Transfer: Mathematical and Computational Modelling of Viscous Fluids and Porous Media*. Amsterdam: Pergamon.

Pope, A. (2009). *Basic Wing and Airfoil Theory*. Mineola, New York: Dover.

Pope, S. B. (2000). *Turbulent Flows*. Cambridge, UK: Cambridge University Press.

Pope, S. B. (2004). Ten questions concerning the large-eddy simulation of turbulent flows, *New Journal of Physics*, **6**, 35.

Powers, J. M. (2004). On the necessity of positive semi-definite conductivity and Onsager reciprocity in modeling heat conduction in an anisotropic media, *Journal of Heat Transfer*, **126**(5), 670–675.

Powers, J. M. (2016). *Combustion Thermodynamics and Dynamics*. New York: Cambridge University Press.

Powers, J. M., and Sen, M. (2015). *Mathematical Methods in Engineering*. New York: Cambridge University Press.

Powers, J. M., Paolucci, S., Mengers, J. D., and Al-Khateeb, A. N. (2015). Slow attractive canonical invariant manifolds for reactive systems, *Journal of Mathematical Chemistry*, **53**(2), 737–766.

Pozrikidis, C. (2017). *Fluid Dynamics: Theory, Computation, and Numerical Simulation*, 3rd ed. New York: Springer.

Prandtl, L., and Tietjens, O. G. (1957). *Fundamentals of Hydro- and Aeromechanics*. New York: Dover.

Proudman, I., and Pearson, J. R. A. (1957). Expansions at small Reynolds numbers for the flow past a sphere and a circular cylinder, *Journal of Fluid Mechanics*, **2**(3), 237–262.

Raissi, M., Perdikaris, P., and Karniadakis, G. E. (2019). Physics-informed neural networks: a deep learning framework for solving forward and inverse problems involving nonlinear partial differential equations, *Journal of Computational Physics*, **378**, 686–707.

Ramm, H. J. (1990). *Fluid Dynamics for the Study of Transonic Flow*. Oxford: Oxford University Press.

Rayleigh, J. W. S. (1916). On convection currents in a horizontal layer of fluid, when the higher temperature is on the under side, *Philosophical Magazine*, **32**(192), 529–546.

Rayleigh, J. W. S. (1945). *The Theory of Sound*, 2 vols. New York: Dover.

Reynolds, W. C. (1968). *Thermodynamics*, 2nd ed. New York: McGraw-Hill.

Reynolds, W. C. (1976). Computation of turbulent flows, *Annual Review of Fluid Mechanics*, **8**, 183-208.

Reynolds, W. C. (1990). The potential and limitations of direct and large eddy simulations, *Whither Turbulence? Turbulence at the Crossroads*, (*Lecture Notes in Physics*, **357**), Lumley, J. L., ed. Berlin: Springer, 313–343.

Reynolds, W. C., and Colonna, P. (2018). *Thermodynamics: Fundamentals and Engineering Applications*. Cambridge, UK: Cambridge University Press.

Richardson, L. F. (1922). *Weather Prediction by Numerical Process*. Cambridge, UK: Cambridge University Press.

Rieutord, M. (2015). *Fluid Dynamics: An Introduction*. Heidelberg: Springer.

Riley, K. F., Hobson, M. P., and Bence, S. J. (2006). *Mathematical Methods for Physics and Engineering*, 3rd ed. Cambridge, UK: Cambridge University Press.

Roache, P. J. (1998). *Fundamentals of Computational Fluid Dynamics*. Albuquerque, New Mexico: Hermosa.

Roache, P. J. (2009). *Fundamentals of Verification and Validation*. Albuquerque, New Mexico: Hermosa.

Robinson, J. C. (2001). *Infinite-Dimensional Dynamical Systems: An Introduction to Dissipative Parabolic PDEs and the Theory of Global Attractors*. Cambridge, UK: Cambridge University Press.

Robinson, J. C., Rodrigo, J. L., and Sadowski, W. (2016). *The Three-Dimensional Navier-Stokes Equations: Classical Theory*. Cambridge, UK: Cambridge University Press.

Rogers, M. H., and Lance, G. N. (1960). The rotationally symmetric flow of a viscous fluid in the presence of an infinite rotating disk, *Journal of Fluid Mechanics*, **7**(4), 617–631.

Rosenhead, L. (1954). The second coefficient of viscosity: a brief review of fundamentals, *Proceedings of the Royal Society of London. Series A. Mathematical and Physical Sciences*, **226**(1164), 1–6.

Rosenhead, L., ed. (1963). *Laminar Boundary Layers*. Oxford: Oxford University Press.

Ruelle, D. (1989). *Chaotic Evolution and Strange Attractors*. Cambridge, UK: Cambridge University Press.

Sabersky, R. H., Acosta, A. J., Hauptmann, E. G., and Gates, E. M. (1999). *Fluid Flow: A First Course in Fluid Mechanics*, 4th ed. Upper Saddle River, New Jersey: Prentice-Hall.

Saffman, P. G. (1992). *Vortex Dynamics*. Cambridge, UK: Cambridge University Press.

Sagaut, P., and Cambon, C. (2018). *Homogeneous Turbulence Dynamics*, 2nd ed. Cham, Switzerland: Springer.

Samimy, M., Breuer, K. S., Leal, L. G., and Steen, P. H. (2004). *A Gallery of Fluid Motion*. Cambridge, UK: Cambridge University Press.

Samelson, R. M., and Wiggins, S. (2006). *Lagrangian Transport in Geophysical Jets and Waves: the Dynamical Systems Approach*. New York: Springer.

Sasoh, A. (2020). *Compressible Fluid Dynamics and Shock Waves*. Singapore: Springer.

Scheel, J. D., Emran, M. S., and Schumacher, J. (2013). Resolving the fine-scale structure in turbulent Rayleigh–Bénard convection, *New Journal of Physics*, **15**, 113063.

Schetz, J. A. (1993). *Boundary Layer Analysis*. Englewood Cliffs, New Jersey: Prentice-Hall.

Schey, H. M. (2004). *Div, Grad, Curl, and All That*, 4th ed. London: W. W. Norton.

Schlichting, H., and Gersten, K. (2017). *Boundary Layer Theory*, 9th ed. New York: McGraw-Hill.

Schmid, P. J. (2022). Dynamic mode decomposition and its variants, *Annual Review of Fluid Mechanics*, **54**, 225–254.

Schobeiri, M. T. (2010). *Fluid Mechanics for Engineers: A Graduate Textbook*. Berlin: Springer.

Schowalter, W. R. (1978). *Mechanics of Non-Newtonian Fluids*. Oxford: Pergamon.

Sedov, L. I. (1946). Rasprostraneniya sil'nykh vzryvnykh voln, *Prikladnaya Matematika i Mekhanika*, **10**: 241–250.

Sedov, L. I. (1959). *Similarity and Dimensional Methods in Mechanics*. New York: Academic Press.

Segel, L. A. (1987). *Mathematics Applied to Continuum Mechanics*. New York: Dover.

Sengupta, T. K. (2018). *Instabilities of Flows and Transition to Turbulence*. Boca Raton, Florida: CRC Press.

Sengupta, T. K. (2021). *Transition to Turbulence: A Dynamical Systems Approach* to Receptivity. Cambridge, UK: Cambridge University Press.

Sethian, J. A. (1996). *Level Set Methods: Evolving Interfaces in Computational Geometry, Fluid Mechanics, Computer Vision, and Materials Science*. Cambridge, UK: Cambridge University Press.

Shapiro, A. H. (1953, 1954). *The Dynamics and Thermodynamics of Compressible Fluid Flow*, Vol. 1. New York: John Wiley; Vol. 2. New York: Ronald.

Shapiro, A. H., ed. (1972). *Illustrated Experiments in Fluid Mechanics: The NCFMF Book of Film Notes*, National Committee for Fluid Mechanics Films. Cambridge, Massachusetts: MIT Press.

Sherman, F. S. (1990). *Viscous Flow*. New York: McGraw-Hill.

Singh, S., Powers, J. M., and Paolucci, S. (1999). Detonation solutions from reactive Navier–Stokes equations, AIAA-99-0966, 37th AIAA Aerospace Sciences Meeting and Exhibit, Reno, Nevada.

Slattery, J. C. (1999). *Advanced Transport Phenomena*. Cambridge, UK: Cambridge University Press.

Smits, A. J., and Dussauge, J.-P. (2006). *Turbulent Shear Layers in Supersonic Flow*. New York: Springer.

Smoller, J. (1994). *Shock Waves and Reaction-Diffusion Equations*, 2nd ed. New York: Springer.

Sohr, H. (2001). *The Navier–Stokes Equations: An Elementary Functional Analytic Approach*. Basel: Birkhäuser.

Spiegel, E. A., and Veronis, G. (1960). On the Boussinesq approximations for a compressible fluid, *Astrophysical Journal*, **131**(2), 442–447.

Spurk, J. H. (1997). *Fluid Mechanics: Problems and Solutions*. New York: Springer.

Sreenivasan, K. R. (1995). On the universality of the Kolmogorov constant, *Physics of Fluids*, **7**(11), 2778–2784.

Stanišić, M. M. (1985). *The Mathematical Theory of Turbulence*. New York: Springer.

Stokes, G. G. (1845). On the theories of internal friction of fluids in motion, *Transactions of the Cambridge Philosophical Society*, **8**, 287–305.

Stokes, G. G. (1851). On the effect of the internal friction of fluids on the motion of pendulums, *Transactions of the Cambridge Philosophical Society*, **9**(2), 8–106.

Stokes, G. G. (2009). *Mathematical and Physical Papers*, 5 vols. Cambridge, UK: Cambridge University Press.

Strang, G. (2006). *Linear Algebra and its Applications*, 4th ed. Boston, Massachusetts: Cengage.

Strang, G. (2007). *Computational Science and Engineering*. Wellesley, Massachusetts: Wellesley-Cambridge Press.

Straughan, B. (2004). *The Energy Method, Stability, and Nonlinear Convection*, 2nd ed. New York: Springer.

Strogatz, S. H. (2015). *Nonlinear Dynamics and Chaos: With Applications to Physics, Chemistry, and Engineering*, 2nd ed. Boulder, Colorado: Westview Press.

Sultanian, B. K. (2016). *Fluid Mechanics: An Intermediate Approach*. Boca Raton, Florida: CRC Press.

Swaters, G. E. (2000). *Introduction to Hamiltonian Fluid Dynamics and Stability Theory*. Boca Raton, Florida: CRC Press.

Taylor, G. I. (1910). The conditions necessary for discontinuous motion in gases, *Proceedings of the Royal Society of London Series A*, **84**(571), 271–377.

Taylor, G. I. (1950a). The formation of a blast wave by a very intense explosion. I. Theoretical discussion, *Proceedings of the Royal Society of London. Series A, Mathematical and Physical Sciences*, **201**(1065): 159–174.

Taylor, G. I. (1950b). The formation of a blast wave by a very intense explosion. II. The atomic explosion of 1945, *Proceedings of the Royal Society of London. Series A, Mathematical and Physical Sciences*, **201**(1065): 175–186.

Taylor, G. I. (1962). On scraping viscous fluid from a plane surface. In: G. K. Batchelor, ed., *The Scientific Papers of Sir Geoffrey Ingram*

Taylor. Cambridge, UK: Cambridge University Press (1971). **4**, 410–413.

Taylor, G. I. (1963). *The Scientific Papers of Sir Geoffrey Ingram Taylor*. 4 vols. Cambridge, UK: Cambridge University Press.

Telionis, D. P. (1981). *Unsteady Viscous Flows*. New York: Springer.

Temam, R. (1984). *Navier–Stokes Equations: Theory and Numerical Analysis*, 3rd ed. Amsterdam: North Holland.

Temam, R. (1997). *Infinite-Dimensional Dynamical Systems in Mechanics and Physics*, 2nd ed. New York: Springer.

Tennekes, H., and Lumley, J. L. (1972). *A First Course in Turbulence*. Cambridge, Massachusetts: MIT Press.

Thompson, P. A. (1971). A fundamental derivative in gasdynamics, *Physics of Fluids*, **14**(9), 1843–1849.

Thomson, W. (Lord Kelvin) (1911). *Mathematical and Physical Papers*. Cambridge, UK: Cambridge University Press.

Thorne, K. S., and Blandford, R. D. (2017). *Modern Classical Physics: Optics, Fluids, Plasmas, Elasticity, and Statistical Physics*. Princeton, New Jersey: Princeton University Press.

Ting, L., and Klein, R. (1991). *Viscous Vortical Flows*. Berlin: Springer.

Ting, L., Klein, R. and Knio, O. M. (2007). *Vortex Dominated Flows: Analysis and Computation for Multiple Scale Phenomena*. Berlin: Springer.

Townsend, A. A. (1976). *The Structure of Turbulent Shear Flow*, 2nd ed. Cambridge, UK: Cambridge University Press.

Trefethen, L. N., and Embree, M. (2005). *Spectra and Pseudospectra: The Behavior of Nonnormal Matrices and Operators*. Princeton, New Jersey: Princeton University Press.

Tritton, D. J. (1988). *Physical Fluid Dynamics*, 2nd ed. Oxford: Oxford University Press.

Truesdell, C. A. (1991). *A First Course in Rational Continuum Mechanics*, Vol. 1, 2nd ed. Boston, Massachusetts: Academic Press.

Truesdell, C. (2018). *The Kinematics of Vorticity*. Mineola, New York: Dover.

Truesdell, C., and Noll, W. (2004). *The Non-Linear Field Theories of Mechanics*, 3rd ed. Antman, S. S., ed. Berlin: Springer.

Tucker, P. G. (2016). *Advanced Computational Fluid and Aerodynamics*. New York: Cambridge University Press.

Turner J. S. (1973). *Buoyancy Effects in Fluids*. Cambridge, UK: Cambridge University Press.

Vallis, G. K. (2017). *Atmospheric and Oceanic Fluid Dynamics: Fundamentals and Large-Scale Circulation*, 2nd ed. Cambridge, UK: Cambridge University Press.

Van Dyke, M. (1975). *Perturbation Methods in Fluid Mechanics*. Stanford, California: Parabolic Press.

Van Dyke, M. (1982). *An Album of Fluid Motion*. Stanford, California: Parabolic Press.

Vanyo, J. P. (1993). *Rotating Fluids in Engineering and Science*. Boston, Massachusetts: Butterworth-Heinemann.

Vincenti, W. G., and Kruger, C. H. (1965). *Introduction to Physical Gas Dynamics*. New York: John Wiley.

Vinokur, M. (1974). Conservation equations of gasdynamics in curvilinear coordinate systems, *Journal of Computational Physics*, **14**(2), 105–125.

Visconti, G., and Ruggieri, P. (2020). *Fluid Dynamics: Fundamentals and Applications*. Cham, Switzerland: Springer.

von Kármán, T. (1921). Über laminare und turbulente Reibung, *Zeitschrift für Angewandte Mathematik und Mechanik*, **1**(4), 233–252.

von Mises, R. (1945). *Theory of Flight*. New York: McGraw-Hill.

von Mises, R. (1950). On the thickness of a steady shock wave, *Journal of the Aeronautical Sciences*, **17**(9), 551–554.

von Mises, R. (2004). *Mathematical Theory of Compressible Flow*. Mineola, New York: Dover.

von Neumann, J. (1961). *Collected Works*. New York: Pergamon.

Voropayev, S. I., and Afanasyev, Y. D. (1994). *Vortex Structures in a Stratified Fluid: Order from Chaos*. London: Chapman and Hall.

Wang, C. Y. (1991). Exact solutions of the steady-state Navier–Stokes equations, *Annual Review of Fluid Mechanics*, **23**, 159–177.

Weigand, B. (2015). *Analytical Methods for Heat Transfer and Fluid Flow Problems*, 2nd ed. Heidelberg: Springer.

Whitaker, S. (1968). *Introduction to Fluid Mechanics*. Malabar, Florida: Krieger.

White, F. M. (1986). *Fluid Mechanics*. New York: McGraw-Hill.

White, F. M. (2006). *Viscous Fluid Flow*, 3rd ed. New York: McGraw-Hill.

White, F. M., and Majdalani, J. (2022). *Viscous Fluid Flow*, 4th ed. New York: McGraw-Hill.

Whitham, G. B. (1974). *Linear and Nonlinear Waves*. New York: John Wiley.

Wilcox, D. C. (2006). *Turbulence Modeling for CFD*, 3rd ed. La Cañada, California: DCW Industries.

Woods, L. C. (1961). *The Theory of Subsonic Plane Flow*. Cambridge, UK: Cambridge University Press.

Woods, L. C. (1975). *The Thermodynamics of Fluid Systems*. Oxford: Clarendon Press.

Woods, L. C. (1982). Thermodynamic inequalities in continuum mechanics, *IMA Journal of Applied Mathematics*, **29**(3), 221–246.

Wu, J.-Z., Ma, H.-Y., and Zhou, M.-D. (2006). *Vorticity and Vortex Dynamics*. Berlin: Springer.

Wu, J.-Z., Ma, H.-Y., and Zhou, M.-D. (2015). *Vortical Flows*. Berlin: Springer.

Yeung, P. K., and Zhou, Y. (1997). Universality of the Kolmogorov constant in numerical simulations of turbulence, *Physical Review E*, **56**(2), 1746–1752.

Yih, C.-S. (1977). *Fluid Mechanics*. East Hampton, Connecticut: West River Press.

Zeidler, E. (1995). *Applied Functional Analysis: Applications to Mathematical Physics*. New York: Springer.

Zel'dovich, Y. B., and Raizer, Y. P. (2002). *Physics of Shock Waves and High-Temperature Hydrodynamic Phenomena*. Mineola, New York: Dover.

Zeytounian, R. Kh. (2002). *Theory and Applications of Nonviscous Fluid Flows*. Berlin: Springer.

Zeytounian, R. Kh. (2004). *Theory and Applications of Viscous Fluid Flows*. Berlin: Springer.

Zucrow, M. J., and Hoffman, J. D. (1976). *Gas Dynamics*, 2 vols. New York: John Wiley.

Index

acceleration, 3, 61–64, 82, 84, 86, 91, 123, 124, 128, 130, 132, 133, 143, 144, 151, 152, 156, 161, 190, 269–271, 273, 281, 315, 324, 335, 463, 471, 473, 480, 520, 521, 549, 568, 576, 601, 617, 657
  advective, 425, 462, 520, 521, 525, 557, 559, 568, 569, 622, 653
  centripetal, 61, 62, 64, 188, 192, 270, 273, 313, 315, 549, 581
  Coriolis, 61, 62, 64, 188, 192, 270, 576, 581, 582
  gravitational, 154, 192, 248, 516, 570, 585, 659
  vector, 62, 64, 65, 85, 86, 129, 130, 132, 143, 144, 161, 190, 259, 313, 314, 316, 335, 489, 570, 576
acoustic, 158, 201, 338, 347, 382, 385, 387, 394, 401–403, 408, 410–412, 417, 418, 437, 454–457, 496, 658, 659, 661
adiabatic, 179, 229, 348, 369, 386, 437, 476, 478, 481, 601, 610
  jump, 439
advection, 188, 339, 425, 480, 564, 601, 602, 625, 626, 661, 662
  chaotic, 643
aerodynamics, 3, 12, 297, 332, 333, 338
air, 10, 12, 13, 184, 230, 235, 240, 338, 357, 358, 362, 364, 365, 367, 368, 382, 384, 385, 390, 417, 436, 452, 453, 456, 533, 551, 569, 576, 580, 586, 606, 639, 660, 668
aircraft, 4
airfoil, 265
airplane, 364
alias transformation, 19, 78
alibi transformation, 19, 78, 81
alternating symbol, 28
amplitude, 269, 398, 400, 402, 501, 513, 515, 525, 528, 583, 591, 610, 613, 619, 622, 632, 639, 640, 645, 646
analytic, 304, 327
analytic function, 304–306
angular momenta, 4, 148, 149, 164–167, 180, 189, 232, 244
  conservation, 149
anharmonic, 314, 324
anisotropic, 200–204, 227
anisotropy, 197
anti-symmetric, 30
aperiodic, 636, 637
applied mathematics, 131, 296, 624
arbitrary volume, 50, 115–117, 182, 183, 185
arc length, 51, 97, 135–137, 178, 329
area, 50, 116, 136, 154–157, 173, 183, 207, 307–309, 317, 333, 334, 346, 349–353, 356, 357, 361, 365–367, 369, 372, 374, 376, 390, 456, 532, 534, 577
  -preserving, 59, 60, 77, 81, 191, 412
argon, 225, 372
artificial neural network, 614
astrophysics, 338
asymmetric matrix, 106, 107
asymptotic, 454, 542, 543, 552, 659
  convergence regime, 666, 667
atomic bomb explosion, 453
attractor
  strange, 631, 636, 637
autonomous, 93, 129, 130, 144, 360, 375, 614, 618–620, 626
Avogadro's number, 10, 11
axial vector, 82, 99
axiom, 4–6, 148, 149, 151, 153, 161, 164, 182, 185, 188, 189, 191, 195, 197, 207, 226, 237, 259, 260, 439, 608, 614, 643
  conservation, 150
  evolution, 115, 149, 154, 339, 349, 380
balloon, 586
Banach space, 26
baroclinic, 275, 277, 279, 299
barotropic, 251, 252, 275–278, 281, 282, 349, 352
baseball, 333
basis
  complete, 615
basis function, 614–620, 625, 626
basis vector, 9, 17, 18, 21, 53, 55, 56, 58, 62, 63, 81, 92, 615
  contravariant, 55, 56
  covariant, 55, 56, 58
  dual, 56
  reciprocal, 56, 81
Bateman–Burgers equation, 622
  inviscid, 419–421, 423, 425–427, 429, 430, 457
  viscous, 425–427, 429, 430, 617, 621, 622
Beltrami field, 52, 266, 294
  generalized, 294, 576
Bernoulli equation, 251, 255, 258–260, 271, 274, 282, 290, 297, 299, 311, 313, 323–325, 328, 333, 352, 356, 362, 364, 588
bi-modal, 399
bifurcation, 610, 630, 631, 667
  diagram, 631, 639
  Hopf, 630, 639
  pitchfork, 624, 629, 631
  subcritical, 630
  transcritical, 639
biharmonic
  equation, 523, 526, 530
  operator, 49, 526

Bingham plastic, 208
binormal, 138, 276
Biot–Savart law, 282
Blasius
  boundary layer, 520, 549–551, 556, 557, 560, 578, 598
  force theorem, 328, 332
blast wave, 441, 450, 453, 455, 456, 458
body
  couple, 166
  force, 154, 156, 160–162, 172, 174, 176, 181, 182, 190, 191, 196, 232, 234, 235, 238, 239, 241–243, 248, 254, 256, 257, 259, 260, 268, 273, 275, 276, 280, 281, 292–294, 299, 313, 324, 328, 335, 338, 352, 457, 461, 476, 507, 512, 521, 533, 576, 577, 585, 593, 601, 645, 646, 653, 668
  boundary layer, 12, 296, 297, 328, 338, 488, 491, 520, 540, 542, 544, 551, 553, 557–561, 570, 573, 576, 595, 597
    attached, 655
    Blasius, 520, 549–551, 556, 557, 560, 578, 598
    compressible, 520, 573
    equations, 576
    thermal, 552, 578
    thickness, 551, 598
    turbulent, 643
  Boussinesq approximation, 240, 241, 243, 259, 570, 578, 601, 624, 658
Boyle's law, 348
bubble, 229

caloric equation of state, 223, 225, 341, 342, 347, 354
calorically perfect, 225
calorically perfect ideal gas, 225, 234, 235, 241, 246, 249, 252, 259, 260, 277, 348, 349, 355, 356, 358, 361–365, 367, 369–372, 374–377, 381–387, 389, 390, 394, 402, 407, 408, 410, 431, 436, 441–443, 454, 456–458, 476, 569, 573, 648, 656
Cartesian
  coordinates, 20, 58, 59, 65, 68, 77, 80, 82, 145, 226, 272, 285, 286, 527, 530, 577
  index, 17–19, 29, 30, 32–35, 47, 48, 76, 160, 173, 185, 216, 268, 275, 603, 652
Cauchy integral theorem, 327, 332
Cauchy's first law of motion, 160
Cauchy's second law of motion, 166, 167
Cauchy's stress quadric, 167–170, 191, 517
Cauchy–Green tensor
  left, 109
  right, 108, 109
Cauchy–Riemann equations, 305, 306
center, 620
central processing unit (CPU), 662
centripetal acceleration, 61, 62, 64, 188, 192, 270, 273, 313, 315, 549, 581
chain rule, 90, 92, 235, 277, 340, 444, 454, 485, 504, 562, 574, 575
chaos, 624, 631, 657
chaotic, 288, 624, 636, 638, 643, 659, 662–664
  advection, 129, 643
  convection, 663
  vortices, 663
characteristic polynomial, 40, 46, 103, 202, 220, 629, 630
characteristics, 407
chemical reaction, 178
choked, 372, 374, 376, 392
  flow, 367, 371
  frictionally, 372
  thermally, 374, 456
Christoffel symbol, 61, 64
circle, 51, 102, 103, 123, 128, 190, 270, 272, 273, 309, 314, 663
  Mohr's, 40, 170
  unit, 23, 24, 77, 104, 105, 133, 301, 303, 317, 318
circulation, 265, 272, 273, 281, 282, 309, 321–323, 332, 333, 335, 502, 628, 657
Clausius–Duhem inequality, 180
  strong form, 181, 182
  weak form, 197
Clebsch decomposition, 144
codeformational derivative, 113
Colebrook equation, 369
column vector, 7, 18, 20, 21, 36, 48, 50, 105
combustion, 456

commutator, 113, 278
complex
  conjugate, 302, 599
  function, 303, 305, 306, 325, 334
  number, 299–301
  plane, 302, 304, 325
  system, 609
  variable, 299, 303, 305, 325, 499, 514
  velocity potential, 305, 318
complex lamellar vector field, 119, 145, 266, 268
compressibility, 230, 236, 270, 338, 390
compressible, 4, 124, 157, 160, 161, 178, 191, 223, 226, 230, 232–237, 241, 243, 251, 258–260, 274, 336, 338–340, 349, 350, 356, 372, 401, 441, 456–458, 476, 481, 512, 521, 573, 574, 644, 645, 648, 658
  boundary layer, 520, 573
  stream function, 574
compression, 3, 158, 176, 346, 386, 393, 439
  adiabatic, 437
  fan, 439
  isothermal, 343
  shock, 386
computational fluid dynamics (CFD), 26, 53, 171, 436
conduction, 200, 203, 490, 580, 601, 609, 632, 657, 658, 662, 663
  thermal, 376, 478
conformal mapping, 297
conservative
  body force, 161, 191, 239, 254, 259, 276, 277, 292–294, 299, 645, 668
  form, 151, 160, 173–176, 179–181, 185, 328, 352, 357, 371, 429, 431, 443, 454, 653
constitutive equation, 189, 194–198, 206, 207, 219, 232, 339, 512, 643
contact discontinuity, 230, 389
continuous spectrum, 400
continuum, 4, 6–10, 12, 13, 76, 84, 142, 148, 189, 195, 229, 431, 439, 456, 598, 604, 605, 617, 632, 643, 648, 655, 656, 658, 659, 661, 666–668
contour integral, 51, 272, 325–328, 332

contraction, 34, 48
contravariant, 57, 58
   basis vector, 55
   representation, 55
   vector, 53, 61
control volume, 52, 181, 183, 205, 329, 350, 351, 353
convected derivative, 113, 114, 278, 279
convection, 188, 580, 601, 657
   natural, 251, 338, 520, 569, 570, 572, 576, 578, 639
convergence, 142, 615, 666, 667
   rate, 667
converging nozzle, 392, 393, 456
converging-diverging nozzle, 374, 393
convex isentrope, 440, 441
coordinate stretching, 457, 503, 562
coordinate system, 16, 17, 19, 21, 24, 25, 27, 28, 44, 49, 55, 57, 61, 64, 67, 78, 88, 142, 169, 192, 195, 222, 268, 276, 297, 443, 483, 506, 522, 578
   axisymmetric spherical, 528
   Cartesian, 9, 17, 18, 24, 27, 34, 80, 443
   cylindrical, 54, 61, 64, 68, 80, 268, 271, 525
   general, 522
   intrinsic, 40
   left-handed, 35, 82
   local, 254, 268, 276
   non-Cartesian, 70, 578
   nonorthogonal, 56, 59, 63, 186
   orthogonal, 80
   polar, 62, 486, 525
   reflected, 82, 99
   right-handed, 82, 99
   rotated, 24, 25, 28, 45, 79, 82, 99, 101, 102, 167, 169, 516
   space-time, 87
   spherical, 70, 81, 336, 522
   transformed, 163, 191
coordinate transformation, 19, 59, 82, 87, 89, 90, 106–109, 226, 411, 420, 421, 423, 424, 427, 428
   general, 53
   linear, 82
   nonorthogonal, 64
   polar, 70
coordinates

Cartesian, 20, 58, 59, 65, 68, 77, 80, 82, 145, 226, 272, 285, 286, 527, 530, 577
   cylindrical, 61, 65, 66, 68, 69, 71, 190, 270, 285, 457, 461
   non-Cartesian, 20, 77, 145
   parabolic polar, 80
   spherical, 72, 79, 80, 335, 336, 441, 529
Coriolis acceleration, 61, 62, 64, 188, 192, 270, 576, 581, 582
corotational derivative, 113
Couette flow, 207, 470, 471, 473, 474, 476, 495, 496, 507, 508, 516, 577, 595
   compressible, 476, 481
   unsteady, 499
Courant–Friedrichs–Lewy (CFL), 662
covariant, 57–59
   basis vector, 55, 56, 58
   derivative, 61
   representation, 55
   vector, 53
Cramer's rule, 36, 37, 39, 63, 360
critical point, 311, 619, 627–631, 670
Crocco's equation, 251, 255, 258, 433
   extended, 255, 257, 258
cross product, 16, 28, 34, 78, 82, 92, 299
curl, 28, 48, 52, 82, 99, 119, 192, 259, 270, 271, 274, 280, 522, 603
curvature, 95, 138, 143, 313, 315, 524, 587, 588
   center of, 276, 313
   radius of, 586, 587
   streamline, 315, 317, 335, 542, 549, 564
   trajectory, 143
cylinder, 133, 321–324, 332, 333, 335, 415, 516, 565
cylindrical coordinates, 61, 65, 66, 68, 71, 190, 270, 285, 457, 461

d'Alembert solution, 394, 397, 399, 410, 512
Darcy friction, 196, 369
dashpot, 510, 513
data-driven reduced model, 614
decomposition, 32, 104, 105, 109, 110, 145, 153, 216, 336, 652

Clebsch, 144
complex lamellar, 144
diagonal, 104, 136, 168
dynamic mode (DMD), 614
Helmholtz, 145, 226, 259, 336
Jordan, 110
kinematic, 95
left polar, 109
$\mathbf{L} \cdot \mathbf{U}$, 110
polar, 107, 108
$\mathbf{Q} \cdot \mathbf{U}$, 110
right polar, 109
Schur, 110
singular value, 101, 105–107, 109, 146, 614
deformation, 3, 16, 29, 123, 124, 128, 129, 142, 179, 209, 210, 219, 220, 271, 273
   deviatoric, 176
   rate, 96, 118, 181, 654, 655
   shear, 218
   tensor, 98, 103, 129, 134, 136, 140, 143, 144, 206, 207, 217, 219, 220, 227, 313, 335, 468, 489, 510, 522, 528, 549, 644
   volumetric, 176
density, 5, 7, 8, 10–12, 19, 25, 86, 89–91, 150–152, 161, 177, 183, 190, 192, 225, 229, 230, 232, 234–242, 247, 251, 252, 259, 260, 275, 277, 281, 335, 338, 348, 363–365, 368, 372, 374, 382, 383, 385, 386, 389, 391, 392, 403, 435, 439, 444, 449, 450, 453, 456–458, 476, 480, 516–518, 569, 570, 573, 580, 592, 610, 646, 656–660, 662, 668
   spectral, 401
   stagnation, 367
derivative
   codeformational, 113
   convected, 113, 114, 278, 279
   corotational, 113
   covariant, 61
   fluid, 113
   Jaumann, 113
   Lie, 113, 278
   material, 89, 90, 112–115, 150, 160, 164, 177, 178, 196, 245, 278, 281, 442, 443, 463, 586, 659
   substantial, 89
   total, 89

upper convected, 115
determinant, 21, 29, 34, 36, 39, 92, 360, 388, 405, 464, 584, 590, 591, 605, 629
　Jacobian, 55, 88, 153, 419, 578, 640
detonation, 338
　Chapman–Jouguet, 456
diagonal, 29, 30, 79, 101, 106, 108, 136, 203, 218, 620, 632
　decomposition, 104, 136, 168
　form, 100, 102, 167, 313
　matrix, 72, 79, 101, 106, 136, 620
　tensor, 217
diffuser
　subsonic, 366
　supersonic, 366
diffusion, 188, 189, 207, 229, 243, 251, 339, 389, 395, 425, 427, 471, 473, 481, 488, 491, 505, 517, 564, 585, 602, 617, 622, 645, 661
　energy, 189
　linear momenta, 189
　mass, 488
　molecular, 585
　momentum, 207, 257, 398, 487, 561, 564, 585
　operator, 618
diffusivity
　mass, 195
　momentum, 12, 240, 482, 488, 490
　thermal, 12, 240, 241, 490, 554, 613, 614
dilatant, 208, 505, 507, 517
dilatation rate, 116
dimensionless, 11, 106, 107, 111, 190, 191, 229, 246–251, 259, 260, 280, 292, 440, 447, 457, 458, 466–470, 474, 480–482, 487, 489, 493, 507, 517, 521–523, 525, 526, 532, 536, 543, 550, 553, 560, 566, 571, 572, 576–578, 588, 596, 602, 610, 613, 614, 624, 640, 649, 663
dipole, 319, 331
　moment, 319
Dirac delta function, 494
direct energy cascade, 649
direct numerical simulation (DNS), 643, 658–662, 664–668
　three-dimensional, 664, 665

two-dimensional, 664, 667
direction cosine, 21, 24, 25, 27, 44, 45, 101, 210, 212–214
Dirichlet condition, 396
discontinuity, 229, 377, 378, 386, 387, 427, 438, 457
　contact, 230, 389
　rarefaction, 386, 438, 439, 441
　shock, 230, 443
discrete spectrum, 400
discretization, 160, 426, 439, 656, 659, 661, 666, 667
disorder, 189
dispersion relation, 610
displacement, 3, 110, 510, 610
dissipation
　rate, 646
　viscous, 180, 181, 219, 221, 227, 237, 239, 241, 464, 467, 478, 521, 552, 555, 645, 659
dissipative, 141, 182
distance, 3, 10, 13, 24, 54, 55, 64, 67, 80, 92, 96, 97, 106, 110, 111, 113, 114, 123, 165, 172, 192, 248, 255, 269, 283, 293, 314, 318, 372, 374, 444, 456, 462, 481, 510, 516, 518, 525, 528, 543, 551, 663
div, 64
divergence, 48, 49, 65, 68, 71, 80, 140–142, 152, 160–162, 176, 192, 237, 238, 273, 421, 442, 522, 524, 662
　form, 151, 160
dot product, 18, 25, 26, 32, 33, 48, 56, 59, 135, 172, 175, 253, 254, 256, 258, 292, 299
doublet, 318, 319, 321, 332
drag force, 296, 332, 533
dual vector, 32, 79, 99, 139, 146, 166
Dufour effect, 200
dummy index, 22, 30
dyadic product, 33, 48, 78, 653, 654
dynamic interface condition, 587, 589, 590
dynamic mode decomposition (DMD), 614
dynamic viscosity, 12, 431
dynamical system, 84, 129, 130, 133, 139, 141, 143, 316, 360, 613, 618, 619, 639
　low order, 614, 624

non-normal, 142
dynamics, 6, 86, 131, 133, 141, 265, 269, 288, 289, 292, 299, 312, 316, 334, 349, 419, 502, 525, 580–582, 585, 586, 588, 621, 638, 640, 656, 667, 668
　atmospheric, 192
　classical, 299
　computational fluid, 26, 53, 171, 436
　fluid, 223, 639
　gas, 402
　geophysical fluid, 576
　low order, 613
　nonlinear, 580, 624, 627, 636
　phase plane, 613, 620, 621
　rarefied gas, 9
　system, 580, 583
　turbulent, 624
　vortex, 288
　wave, 4

Eckert number, 466, 480
eddy, 642, 668
　viscosity, 654–656, 658
egg, 77
eigen-direction, 133
eigen-stretching, 106, 111
eigenvalue, 39–46, 72–75, 79, 101–107, 111, 118, 131, 133, 136–138, 140, 144, 146, 167–170, 199, 202–204, 218, 220–222, 227, 317, 396, 397, 405, 406, 410, 411, 468, 489, 497, 517, 583, 584, 590, 597, 598, 604, 611, 614, 618, 620, 629–634, 640
　left, 39
　right, 39
eigenvector, 38–40, 42, 43, 72, 74, 75, 101, 104–106, 118, 136–138, 140, 146, 168–170, 202, 313, 318, 410, 411, 468, 489, 584, 590, 606, 614, 620
　left, 39, 405, 407
　right, 39, 101
Ekman transport, 576
electro-magnetic, 178
electrodynamics, 145
ellipse, 23, 74, 77, 102–104, 296, 317, 534, 536
ellipsoid, 24, 60, 61, 77, 170

Lamé stress, 169, 170, 191
of revolution, 102
triaxial, 23, 75, 102, 169
elliptic, 294, 296, 535, 536
energy, 4, 6, 148, 149, 172–179, 185,
188, 189, 202, 204, 205, 207,
223–226, 232, 234, 236, 237,
239, 240, 245, 246, 249, 250,
259, 260, 339, 349, 353, 356,
358, 359, 361, 365, 380, 381,
390, 391, 394, 401, 403, 404,
406, 408, 431–433, 441, 443,
446, 448, 453, 454, 456, 462,
464–467, 471, 473, 476, 478,
482, 488, 491, 492, 552–555,
564, 571, 573, 576, 601, 602,
604, 608, 610, 625, 626, 646,
649, 652, 656–658
chemical, 453
diffusion, 186, 207, 240, 257, 296,
394, 425, 464, 476, 602
exothermic, 195
flux, 200, 205, 229, 230, 353
internal, 157, 171, 176–178, 191,
223, 258, 340, 451
kinetic, 61, 171, 175, 177, 188,
338, 365, 390, 441, 451, 601,
645–649, 656
mechanical, 5, 175, 176, 181, 182,
230, 353, 354, 478, 602, 644,
645, 658, 668
potential, 171, 174, 390, 660
specific, 171
thermal, 171, 173, 175, 176, 179,
181, 188, 194, 200, 239, 338,
354, 359, 365, 389, 441, 464,
471, 473, 478, 491, 580, 601,
602, 608, 609, 657
total, 171, 172, 191, 353, 441, 442,
451
turbulent kinetic, 649, 655
engineering gas constant, 225
enstrophy, 292, 668
enthalpy, 174, 177, 178, 227, 232,
234, 245, 256, 257, 381, 431, 439
stagnation, 255–258
total, 257, 258, 371, 372, 374, 574
entrance length, 462
entropy, 148, 149, 178–182, 189, 191,
202, 221, 223, 225, 244–246,
256, 257, 342, 344, 346, 356,
358, 359, 382, 386, 387, 389,
390, 395, 402–404, 406, 408,

409, 435–439, 443, 456, 457,
602, 608
inequality, 179, 181, 197, 198,
200–202, 206, 220–224, 438
jump, 387
production, 222
equation of state, 233, 244, 339, 356,
380, 403, 456
caloric, 223, 225, 341, 342, 347,
354
nonideal, 339
thermal, 223, 225, 340, 341, 346,
347
equilibrium, 3, 96, 130, 133, 134,
139, 141, 142, 144, 208, 361,
389, 433, 510, 580–584, 587,
588, 614, 619–621, 623, 624,
627–633, 635, 636, 638, 639
mechanical, 645
point, 619
thermal, 645
equipotential, 190, 298, 307, 308,
310
equipotential line, 161, 314, 319–322
equipresence, 195, 197, 198
error
roundoff, 667
truncation, 667
error function, 516
complement, 487, 518
Euclid's parallel postulate, 148
Euclidean norm, 26
Euler angle, 81
Euler equations, 243–245, 259, 260,
339, 376–378, 389, 394, 411,
414, 425, 442, 457, 458
Euler's formula, 300, 499, 500, 514
Eulerian coordinates, 87–91, 93,
128, 142, 153
evolution axiom, 115, 149, 154, 339,
349, 380
expansion, 3, 123, 126, 240
adiabatic, 386
rate, 104, 115–117, 124, 136, 140,
143–145, 153, 176, 177, 217,
219, 236
explosion, 338, 453
extension, 98, 122–124
extensional
strain, 112, 126, 489
strain rate, 102, 103, 133, 134,
136, 143, 144, 167, 169
extensive, 171, 172, 179

Falkner–Skan flow, 323, 544, 557,
559–561, 571, 578, 595, 597, 625
Fanno flow, 196, 369, 371–374, 456
fastball, 333
Favre-average, 663
field
acceleration, 190
pressure, 190–192, 238, 239, 241,
257, 271, 273, 280, 282,
290–294, 311, 313, 314, 324,
325, 328, 465, 480, 516, 520,
523, 524, 527, 578, 588
velocity, 99, 145, 151, 161, 190,
227, 286, 287, 290–294, 334,
335, 549, 550, 576, 578
figure skater, 276
finite difference, 113, 436, 597, 616,
659
finite element, 617
first law of thermodynamics, 149,
171, 175, 178, 179, 191, 205,
234, 238, 257, 258, 260, 352,
353, 404, 451
fixed point, 619
fixed volume, 53, 182, 183, 188, 191,
377, 645
fluctuation, 651
fluid, 3–5, 7, 8, 13, 16, 29, 30, 96,
104, 106, 109–116, 120, 121,
124–130, 132–134, 139,
142–146, 148, 150–154,
157–159, 161, 162, 164–166,
171, 175, 176, 179, 182, 183,
188–192, 196, 199, 206–210,
219–222, 226, 227, 229, 230,
234, 236, 237, 239–241, 243,
246, 248, 251, 253, 255–257,
259, 260, 265, 266, 268–270,
273–284, 290–292, 296, 297,
299, 313, 315, 316, 329, 332,
334, 338, 339, 349, 351, 353,
359–361, 365, 366, 372, 374,
379, 384, 389, 392, 394, 395,
402, 415, 419, 436, 437, 439,
441, 443, 454, 461, 462, 466,
470, 475, 476, 478, 481, 487,
488, 495, 499, 502, 505–508,
510–513, 516, 518, 520, 521,
526, 528, 529, 531, 533, 539,
542, 544, 549, 557, 561, 564,
565, 568–570, 572, 573,
576–578, 580, 583–585,
591–593, 600–603, 609, 610,

613, 614, 617, 622, 624, 631,
636, 639, 643, 644, 649, 650,
652, 654, 655, 657, 660, 668
derivative, 113
dilatant, 208, 505, 507, 517
friction, 196
ideal, 243
isotropic, 208
Maxwellian, 515
mechanics, 3–5, 13, 14, 17, 38, 53,
70, 87, 130, 141, 148, 189, 196,
197, 235, 305, 306, 334, 369,
376, 402, 409, 461, 511, 512,
520, 583, 584, 586, 613, 614,
616, 624, 627, 643, 644
Newtonian, 206, 208, 209, 215,
218, 219, 221–223, 226, 227,
232, 237, 243, 259, 268, 270,
273, 278, 292–294, 506, 510,
513, 515–518, 576, 577, 644
non-Newtonian, 208, 505–507, 517
nonpolar, 164, 166, 177
perfect, 243
physics, 4, 656
polar, 164, 177
rotation, 3
statics, 157
Stokesian, 227
stratified, 234
flux
energy, 200, 205, 229, 230, 353
heat, 172, 176, 179, 191, 194, 195,
199, 201, 202, 204–206, 227,
349, 357–359, 468, 470, 471,
473, 475, 601, 609, 666
mass, 151, 191, 195, 200, 349, 371,
372, 374, 474, 562, 601, 626
momentum, 562
force, 6, 100, 133, 155–157, 161, 165,
172, 207, 241, 259, 277, 280,
311, 313, 325, 329–332, 351,
389, 477, 508, 520, 521, 524,
528, 581, 585, 587
body, 154, 156, 160–162, 172, 174,
176, 181, 182, 190, 191, 196,
232, 234, 235, 238, 239,
241–243, 248, 254, 256, 257,
259, 260, 268, 273, 275, 276,
280, 281, 292–294, 299, 313,
324, 328, 335, 338, 352, 457,
461, 476, 507, 512, 521, 533,
576, 577, 585, 593, 601, 645,
646, 653, 668

drag, 296, 332, 533
field, 86, 161
friction, 157
gravity, 154, 333, 581
intermolecular, 339, 440
lift, 332, 333, 527
longitudinal, 3
normal, 3
pressure, 176, 181, 182, 245, 271,
328, 334, 351, 512, 587, 645
shear, 3, 351, 353, 521, 568
surface, 9, 153, 154, 156, 160,
161, 173, 175, 176, 185, 190,
463, 521
tension, 581
transverse, 3
viscous, 129, 173, 176, 181, 189,
226, 394, 520, 552, 558, 568,
576, 601, 657
Fourier series, 398, 538, 615–617
Fourier transformation, 494
Fourier's law, 194, 201, 227, 232,
243, 259
frame invariance, 196, 353
Galilean, 195, 226
free body diagram, 581
free energy
Gibbs, 227, 233
Helmholtz, 223, 227, 340, 342, 456
free index, 21, 30
Frenet–Serret relations, 138, 146
frequency, 401, 620, 626, 631, 661
angular, 501, 610
friction, 196, 339, 353, 356, 369, 372,
376
Froude number, 248
fully developed, 462, 463, 465, 476,
507, 508, 533, 577
functional analysis, 615
fundamental theorem of calculus, 50

Galerkin projection, 614, 617, 618,
622, 625
Galilean transformation, 6, 162–164,
196, 197, 201, 353, 369, 379,
389, 428, 431, 531
gas, 225
dynamics, 402
fluorocarbon, 457
virial, 348
gauge
invariance, 145
transformation, 145

Gauss's theorem, 50, 51, 79, 150,
159, 165, 166, 173, 181, 182,
185, 645
Gaussian elimination, 36, 37, 78
general transport theorem, 52, 53,
116
geometry, 4, 6, 9, 16, 17, 19, 23, 53,
64, 75, 84, 135, 139, 148, 156,
157, 179, 266, 297, 328, 329,
331, 349, 358, 522, 534
differential, 16, 113, 138
spherical, 192, 451
surface, 229
geostrophic flow, 192
Gibbs
equation, 178, 232, 233, 245, 257,
340, 341, 344, 349, 354, 355,
381, 386, 443
free energy, 227, 233
notation, 17, 20, 24, 26, 29, 30,
32–35, 47, 48, 50, 101, 151, 159,
160, 173, 188, 216, 239, 644
grad, 64
gradient, 47–49, 54, 66, 67, 71, 80,
160, 190, 195, 200, 239, 253,
259, 282, 285, 287, 297, 313,
427, 437, 442, 471, 477, 522,
524, 537, 625, 652, 653
density, 569
pressure, 192, 235, 240, 256, 257,
271, 314, 315, 323, 324, 462,
463, 467, 469, 471, 473–477,
482, 508, 509, 516, 518, 533,
534, 540, 541, 544, 557, 559,
561, 565, 567, 569, 571, 577,
595, 597, 598, 625, 645, 655
temperature, 195, 197, 198,
200–202, 204, 205, 425, 437,
461, 636
velocity, 95, 97, 99–101, 105–107,
109, 111, 113–115, 131, 133,
139, 142–144, 152, 177, 180,
181, 198, 207, 313, 461, 469,
488, 502, 512, 520, 553, 588
Grashof number, 572, 573, 603
gravitational field, 288, 569
group
theory, 113, 443, 484
velocity, 610
growth rate, 592–595, 610

Hagen–Poiseuille flow, 462
Hall effect, 200

Hamiltonian system, 299
harmonic, 49, 192, 522, 524
  function, 282, 298, 522
heat
  equation, 613
  flux, 172, 176, 179, 191, 194, 195, 199, 201, 202, 204–206, 227, 349, 357–359, 468, 470, 471, 473, 475, 601, 609, 666
heat transfer, 53, 149, 171, 172, 179, 201, 339, 344, 346, 349, 350, 353, 356, 359, 374, 376, 390, 395, 437, 456, 466, 472, 475, 489, 580, 600, 601, 609, 613, 657, 658, 663, 666
  convective, 609, 663
  diffusive, 260, 349
  nano-scale, 10
  reversible, 149, 178, 189
heating
  isochoric, 343
Heaviside unit step function, 457, 516
Hele-Shaw flow, 541
helicity, 292, 293, 668
helium, 14, 225, 457
Helmholtz
  decomposition, 145, 226, 259, 336
  equation, 604, 606, 607
  free energy, 223, 227, 340, 342, 456
  vorticity transport, 274, 281, 282, 289, 292–294, 299, 522, 585, 604
hemisphere, 79
heteroclinic
  connection, 141
  trajectory, 131, 139–142
Hilbert space, 26
homoentropic, 246, 258, 394, 408, 417, 433, 457, 458
homoentropy, 394, 408
homogeneous, 409, 413, 467, 496–498, 513, 538, 597, 604, 617, 629
Hopf
  bifurcation, 630, 639
  fibration, 260
Hugoniot, 381–383
hydrostatic, 217, 242, 588, 602, 646, 658, 660, 662, 664
hyperbola, 85, 102, 103, 125, 126, 190, 311, 317, 382
hyperbolic, 394, 402, 410, 419, 512

hyperboloid, 73, 102, 144, 170
hyperplane, 140
hypersonics, 338

ideal fluid, 243
ideal gas, 6, 12, 225, 226, 236, 238, 239, 241, 252, 339, 368, 382, 385, 386, 389–391, 417, 433, 438, 572, 574, 642, 643, 658
idempotent, 137
identity matrix, 29
image vortex, 284
imaginary number, 300
imaginary part, 333, 499, 514, 515, 591
impulse, 351, 488
incompressible, 116, 190, 191, 204, 205, 236–239, 241, 251, 255, 258, 259, 266, 268–271, 273, 274, 276–278, 280, 282, 288, 290–294, 296, 298, 299, 305, 317, 324, 328, 334–336, 338, 339, 362, 364, 462, 463, 465, 471, 473, 474, 476, 480, 496, 502, 505–508, 513, 516–518, 520–523, 533, 541, 542, 552, 557, 561, 564, 565, 567, 574–577, 585, 595, 599, 603, 617, 644, 652–654, 668
index, 23, 35, 56
  dummy, 18, 22, 30, 155
  free, 21, 30
  notation, 29, 185
  repeated, 22
inertia, 4, 6, 129, 156, 192, 230, 239, 271, 425, 462, 477, 480, 508, 516, 520, 521, 524, 533, 540–542, 576, 577, 587
inertial frame, 6
injection, 230
inner product, 615, 616, 618, 620
  tensor, 31–33, 99, 166, 180, 202, 238, 253, 254, 256, 603, 644
  vector, 25
instability, 580, 585, 593, 598, 599, 610
  convective, 580
  Kelvin–Helmholtz, 594
  Rayleigh–Bénard, 600
  Rayleigh–Taylor, 591, 610
integrating factor, 486, 491, 504, 511, 554, 555
intensive, 157, 171, 340, 349, 403

internal energy, 157, 171, 176–178, 191, 223, 258, 340, 451
invariance, 26, 67, 82, 164, 364
  frame, 195, 196, 226, 353
  Galilean, 164
  gauge, 145
invariant, 6, 14, 26, 28, 29, 34, 38, 40, 44–46, 55, 59, 67, 82, 114, 120, 143, 144, 162, 164, 196, 203, 217, 222, 226, 335, 411, 414, 434, 516, 517, 545, 578
  principal, 41, 118, 222, 313, 506, 524, 525, 528, 529
  Riemann, 411
  tensor, 38, 44, 45
inverse, 21, 23, 32, 61, 68, 70, 74, 75, 82, 428, 575
  additive, 29
inverse energy cascade, 649
ionized plasma, 225
irreversibility, 141, 149, 439
  production rate, 180, 189, 223, 227, 437, 439, 440
irreversible, 149, 182, 199
  adiabatic, 386
  work, 189
irrotational, 99, 128, 145, 190, 192, 252, 253, 255, 258, 271–274, 279, 282, 285, 287, 288, 290–293, 297, 299, 305, 314, 323, 336, 501–504, 585
isentrope, 440, 441
  convex, 440, 441
  non-convex, 441
  nonconvex, 440
isentropic, 191, 221, 234–236, 245, 246, 252, 260, 277, 347, 349, 352, 355, 356, 361, 362, 364, 365, 367, 368, 381, 386, 392, 393, 402, 404, 406, 408, 415, 417, 438, 443, 456, 602
isobar, 190, 274, 277, 314, 335, 524
isobaric, 192, 204, 240–242, 389, 480, 507, 513, 517, 574, 617, 658, 659
isochore, 277
isotherm, 578
isothermal, 222, 252, 343, 345, 346, 348, 457, 461, 471, 476, 481, 496, 601, 604, 608, 609, 626, 657, 662
isotropic, 28, 29, 194, 201, 207–210, 216, 223, 226, 278

tensor, 28, 29
iterative convergence error, 664

Jacobian
   determinant, 55, 88, 153, 419, 578, 640
   matrix, 54, 56, 58, 60, 66, 71, 77, 87, 88, 90, 92, 107, 153, 163, 412, 419, 483, 619, 620
Jaumann derivative, 113
jet, 576
   laminar, 86, 561, 565
   propulsion, 338

Kelvin's circulation theorem, 281, 282
Kelvin–Helmholtz instability, 594
kinematic interface condition, 586, 589, 590
kinematic viscosity, 12, 240, 241, 278, 514, 551, 576, 647, 648
kinematics, 6, 84, 85, 104, 106, 109, 122–130, 132, 139, 143–145, 265, 269, 271, 272, 276, 289, 292, 299, 311, 312, 316, 334, 585
   rigid body, 99
kinetic energy, 61, 171, 175, 188, 338, 365, 390, 441, 451, 601, 645–649, 656
   rotational, 158, 223, 226
   translational, 177
   vibrational, 158, 223, 226
kinetic theory, 158
kinetics, 6
Kolmogorov scale, 647–649, 659, 660, 667
Koopman operator theory, 614
Kronecker delta, 22, 23, 28, 29, 55, 216, 398, 615
Kutta condition, 333
Kutta–Zhukovsky lift theorem, 333
Kutta-Zhukovsky lift theorem, 334

laboratory frame, 379
Lagrange multipliers, 103, 168
Lagrangian coordinates, 87–91, 142, 153
Lamb
   surface, 254, 258, 260
   vector, 52
lamellar, 99, 119, 253
   vector field, 119

laminar, 277, 369, 520, 551, 561, 565, 566, 662, 663, 668
Lamé stress ellipsoid, 169–171, 191
Landau equation, 622, 623, 631
Laplace's equation, 49, 282, 298, 313, 394, 522, 535, 538
Laplacian, 49, 66, 68, 71, 80, 239, 269, 522, 523, 535, 541
large eddy simulation (LES), 655, 656
Laurent series, 327
law of cosines, 22
Leibniz's rule, 52, 53, 150, 181, 182, 184, 252, 378, 541
length, 4, 9, 10, 13, 26, 38, 51, 59, 60, 97, 135–137, 148, 156, 247, 248, 312, 313, 329, 357, 371, 435, 439, 462, 482, 483, 487, 489, 534, 536, 540, 543, 546, 552, 572, 619, 624, 628, 643, 647–649, 654, 655, 659–661, 667, 668
   entrance, 462
Lennard-Jones diameter, 10
level set function, 586
Levi–Civita symbol, 28, 29, 55
Lie derivative, 113, 278
limit cycle, 632, 638, 640
line, 85, 110, 113, 380, 420, 520, 525, 663
   integral, 329
   material, 110, 112
   slip, 284, 588
linear
   independence, 55
   mapping, 209
   momenta, 4, 148, 149, 153–156, 159–164, 175, 184, 185, 188, 189, 192, 196, 229, 232, 234–238, 240, 241, 243, 248, 249, 251–257, 260, 271, 274, 280, 281, 291, 292, 294, 314, 328, 330, 457, 482, 485, 521, 522, 525, 541, 576, 578, 585, 603, 640, 653, 654
   momentum, 164, 226, 351, 352, 394, 408, 431, 433, 445, 454, 456, 457, 481, 490, 499, 512, 562, 564, 617, 655
   stretching, 61
   transformation, 60, 61, 77, 81, 145, 192

linear algebra, 20, 36, 40, 78, 100, 104, 199, 202, 203, 221, 614
linear operator, 396, 497, 604
linearly
   dependent, 37, 410, 498
   independent, 54, 79, 81, 92, 101, 410, 423, 498, 615, 626
liquid, 3, 4, 158, 204, 205, 209, 225, 238–240, 461, 462
   water, 235, 551, 578
Lorenz equations, 624, 627, 630–635, 637, 638, 640
L · U decomposition, 110
lubrication theory, 539, 540

Mach number, 177, 230, 234, 237, 239–241, 249, 251, 260, 266, 352, 362–364, 367, 372, 374, 376, 384, 385, 389, 433, 435, 443, 456, 457, 480, 555, 648, 656–661, 664, 667
machine learning, 614
magnitude, 8, 9, 13, 17, 19, 24, 26, 27, 37, 39, 42, 57, 78, 79, 96, 102, 103, 111, 117, 123, 136, 139–141, 146, 167, 170, 218, 248, 249, 273, 285, 292, 301, 305, 315, 338, 352, 380, 398, 422, 426, 427, 434, 437, 440, 468–470, 489, 515, 516, 521, 573, 590, 632, 647, 655, 661, 662, 667
Magnus effect, 332
mapping, 36, 38, 60
   conformal, 297
   inverse, 21
   linear, 209
   shear, 59, 145
   squeeze, 24, 77
   stretch, 77
marginal stability curve, 605
mass, 4–7, 13, 14, 96, 115, 133, 148–151, 154, 165, 166, 172, 177, 179, 188, 189, 195, 230, 232, 234, 236–238, 243, 248, 259, 260, 265, 273, 276–279, 284, 322, 339, 349, 350, 359, 365–367, 375, 380, 391, 394, 408, 431–433, 442, 445, 447, 451, 454, 456, 457, 463, 465, 475, 476, 521, 526, 529, 540, 562, 564, 565, 567, 571, 573,

574, 576, 581, 585, 599, 604, 605, 608, 646, 652, 654, 656
conservation, 115, 149–153, 160, 177, 180, 182, 183, 191, 224, 230, 266, 270, 298, 335, 336, 375, 377, 380, 384, 463, 477, 541, 545, 562, 567, 596, 617, 625, 652, 653, 662
diffusion, 229, 296, 488, 585
diffusivity, 195
flow rate, 351, 358, 367, 376, 392, 393, 456
flux, 151, 191, 195, 200, 349, 371, 372, 374, 474, 562, 601, 626
fraction, 195
molecular, 10, 11, 225, 247
material
derivative, 89, 90, 112–115, 150, 160, 164, 177, 178, 196, 245, 278, 281, 442, 443, 463, 586, 659
indifference, 195, 210, 212
science, 608
surface, 53
volume, 53, 115–117, 136, 149, 150, 152, 154–156, 159, 165, 171, 172, 183, 191, 281
mathematics, 7, 9, 36, 39, 50, 70, 200, 201, 299, 402, 615, 656
applied, 131, 296, 624
matrix, 9, 17, 18, 20, 21, 25, 30, 34, 39, 40, 42, 44, 45, 48, 55–57, 72, 98, 101, 104–111, 137, 138, 140, 198–200, 203, 209–215, 220, 221, 360, 388, 404, 427, 464, 584, 590, 597, 604, 605, 620, 629, 632, 640
algebra, 18, 34, 47
asymmetric, 106, 107
diagonal, 72, 79, 101, 106, 136, 620
exponential, 619
identity, 29
invariants, 79
Jacobian, 54, 56, 58, 60, 66, 71, 77, 87, 88, 90, 92, 107, 153, 163, 412, 419, 483, 619, 620
multiplication, 30, 101
orthogonal, 21, 26, 72–75, 101, 104, 105
porous, 196
projection, 137, 144, 615
rectangular, 105
reflection, 21, 29, 59, 78, 79, 82

rotation, 21, 29, 30, 43–45, 59, 78, 79, 81, 82, 101, 104, 106–109, 136, 168, 206
similar, 108
square, 36, 105
symmetric, 72, 170, 198
Maxwell relation, 232, 233, 340, 341
Maxwellian
element, 510
mean free path, 10–13, 431, 434, 435, 439, 643, 648, 658–661
mean value theorem, 116, 155, 156
mechanical energy, 5, 175, 176, 181, 182, 230, 353, 354, 602, 644–646, 658, 668
mechanics, 4, 6, 9, 13, 16, 76, 201, 226, 339, 439, 617, 642, 643, 667, 668
celestial, 86
classical, 5
computational, 586
continuum, 3–6, 8, 13, 14, 19, 78, 113, 148, 189, 194, 195, 197, 200, 513, 643
fluid, 3–5, 13, 14, 17, 36, 38, 53, 70, 87, 130, 141, 148, 189, 196, 197, 235, 305, 306, 334, 369, 376, 402, 409, 461, 511, 512, 520, 583, 584, 586, 613, 614, 616, 624, 627, 642–644
molecular, 13
Newtonian, 8, 142, 164, 189
quantum, 5
rational, 4–6, 14
relativistic, 5
rigid body, 164
solid, 5, 40, 87, 153, 170, 189, 207, 216, 510
statistical, 5, 9, 12
message-passing interface (MPI), 662
meteorology, 298, 625
methane, CH_4, 225
method of characteristics, 402, 410, 419, 457
method of weighted residuals, 617, 618
metric tensor, 55, 57, 59, 67, 71, 80, 107–109
Michelson–Morley experiment, 5
mixing, 129, 564
length, 655, 668
molecular, 564, 585

mode coupling, 626
Mohr's circle, 170
momenta
angular, 4, 148, 149, 164, 165, 167, 180, 189, 232, 244
linear, 4, 148, 149, 153–156, 159–164, 175, 184, 185, 188, 189, 192, 196, 229, 232, 234–238, 240, 241, 243, 248, 249, 251–257, 260, 271, 274, 280, 281, 291, 292, 294, 314, 328, 330, 457, 482, 485, 521, 522, 525, 541, 576, 578, 585, 603, 640, 653, 654
momentum, 8, 188, 207, 230, 240, 243, 271, 339, 349, 351, 354, 359, 365, 374, 380, 433, 447, 449, 455, 463–467, 471–473, 476, 477, 481, 482, 487, 488, 491, 494, 495, 502, 503, 508, 514, 517, 533, 537, 544, 545, 547, 548, 555, 558, 561–563, 565, 567, 568, 571, 573–576, 581, 582, 584, 596, 597, 604, 625, 652, 656
diffusion, 207, 257, 398, 425, 476, 487, 561, 564, 585, 617
diffusivity, 12, 240, 482, 488, 490
flux, 562
linear, 164, 226, 351, 352, 394, 408, 431, 433, 445, 454, 456, 457, 481, 490, 499, 512, 562, 564, 617, 655
pulse, 494
mono-modal, 398
Moody chart, 369
morphogenesis, 608

natural convection, 251, 338, 520, 569, 570, 572, 576, 578, 639
Navier–Stokes equations, 230, 234, 236, 238, 241, 243, 246, 247, 259, 268, 277, 280, 339, 425, 439, 456, 462, 476, 480, 481, 516, 517, 543, 570, 577, 578, 596, 599, 624, 650, 652, 656, 658, 661, 662
negative definite, 73
Neumann condition, 396
Newton's second law of motion, 14, 132, 149, 351, 581
nitrogen N, 226
nitrogen N_2, 225, 226

no-penetration, 230, 283, 463, 530, 531
no-slip, 207, 229, 266, 268, 282, 283, 296, 310, 353, 464, 478, 481, 508, 509, 513, 517, 518, 526, 530, 531, 537, 538, 542, 543, 547, 550, 565, 599, 662, 668
non-Cartesian coordinates, 20, 145
non-Newtonian flow, 114, 507, 508, 516
nonanalytic function, 306
nonautonomous, 93, 360, 448, 449
nonconservative, 188, 352
    body force, 161, 191, 239, 275, 280, 293, 294, 299, 577
    energy equation, 178
    form, 151, 160, 175–177, 180, 181, 185, 232, 245, 252, 259, 352, 357, 403, 431, 442
nonconvex isentrope, 440, 441
nonequilibrium effects, 158
nonequilibrium thermodynamics, 197, 207, 232
nonideal gas, 225, 259, 377, 386, 402, 437, 438
nonlinear, 84, 94, 95, 129–131, 139, 287, 360, 375, 376, 391, 405, 411, 414, 427, 480, 547, 548, 560, 575, 580–584, 608, 613, 614, 616–620, 622, 624, 626, 627, 631, 639, 640, 649, 656–658
nonorthogonal, 40
nonpolar fluid, 164, 166, 177
norm, 26, 41
    p-, 26
    2-, 26
    Euclidean, 26, 41
    tensor, 41
    vector, 26
normal stretching, 140–142
nozzle
    converging, 392, 393, 456
    converging–diverging, 374, 393
    subsonic, 366
Nusselt number, 609, 658, 659, 662, 663, 666

Oberbeck–Boussinesq approximation, 240
oceanography, 298
off-diagonal, 30, 203, 218
oil, 239
Oldroyd fluid, 518

Onsager
    reciprocity, 226
    relation, 200
operator, 39, 78, 248, 274, 300, 522, 601, 602, 618, 650
    averaging, 651, 652
    biharmonic, 49, 526
    column, 48
    convected derivative, 113
    curl, 28, 48, 274, 603
    derivative, 112, 160, 432, 597
    differential, 112, 597
    diffusion, 618
    divergence, 48, 68
    ∇, 299
    E^2, 49, 336, 522
    E^4, 49, 531
    eig, 111
    finite difference, 597
    gradient, 47, 48, 67, 71, 442, 652
    group, 484
    $\Im$, 300
    integral, 478
    Koopman, 614
    Laplacian, 49, 66, 68, 71, 80, 269, 522, 523
    linear, 396, 497, 604
    material derivative, 89
    matrix, 108
    nonlinear, 95
    positive definite, 604
    $\Re$, 300
    scalar, 48, 97, 114
    self-adjoint, 497
    supremum, 41
    total derivative, 89
    transformed, 413
    transpose, 7
ordinary differential equation, 93, 129, 131, 288, 331, 360, 372, 375, 376, 379, 396, 404, 407, 411, 413, 417, 419, 436, 441, 457, 466, 497, 538, 542, 547, 562, 571, 613, 614, 616–620, 625, 626, 639
orientation
    -preserving, 21, 59, 66, 68, 71, 74, 77, 78, 81, 88, 191, 412, 422
    -reversing, 21, 59, 60, 78, 423
Orr–Sommerfeld equation, 597, 598, 610
orthogonal, 9, 18, 21, 40, 54, 56, 59, 60, 67, 71, 72, 78, 101, 108, 119, 137, 161, 168, 202, 254, 260, 266, 268, 276, 299, 315, 396, 498
    matrix, 21, 26, 72–75, 101, 104, 105
orthonormal, 9, 17, 18, 21, 43, 53, 101, 105, 137–139, 169, 170, 615, 620
oscillation, 10, 223, 401, 513, 598, 631, 661
Oseen correction, 533
outer solution, 543
oxygen, O_2, 225

pancake, 77
parabola, 271, 587
parabolic polar coordinates, 80
parallel viscous flow, 595
parallelepiped, 23, 79, 81, 156
partial differential equation, 6, 188, 229, 238, 377, 379, 394, 395, 404, 407, 409, 411, 413, 417, 419, 425, 436, 441, 455, 464, 483, 484, 494, 496, 498, 499, 503, 518, 538, 542, 562, 563, 583, 589, 613, 616, 617, 625, 656, 666
pathline, 93–96, 129, 131, 136, 142, 143, 236, 241, 242, 246, 282, 294, 315, 316, 404, 406, 408, 654
pattern, 400, 608, 628, 657, 658
    circulation, 628, 657
    formation, 609
    hexagonal, 608
    rectangular, 606
    stable, 657
    weather, 3, 265
peanut butter, 208
Peltier effect, 200
pendulum, 6, 581–583, 590
perfect fluid, 243
permutation symbol, 28
phase, 501
    plane, 614, 620, 640
    space, 632, 635, 636, 638
    speed, 501, 515
    velocity, 610
    volume, 141, 639
physics, 4, 6, 9, 14, 36, 70, 145, 148, 178, 195, 197, 216, 218, 230, 259, 323, 338, 339, 542, 614, 617, 654, 656, 658, 659, 661, 668
    mathematical, 4, 409
    Newtonian, 85

surface, 229
pitchfork bifurcation, 624, 629, 631
plane, 9, 23, 24, 27, 35, 38, 49, 78, 103, 137, 140, 156, 170, 190–192, 227, 266, 317, 325, 335, 364, 422, 423, 440, 481, 523, 531, 532, 565, 576, 577, 601, 615
poetry, 642
point vortex, 272, 282, 285, 289–292, 309, 310, 332, 335, 502
Poiseuille flow, 462, 467, 469–471, 508, 533, 534, 536, 537, 539, 595, 597, 610
   Reynolds-averaged, 668
   turbulent, 668
Poisson equation, 145, 238, 239, 314, 465, 534, 535, 537, 662, 664–666
polar
   fluid, 164, 177
   vector, 82
polar decomposition
   left, 107
   right, 107
pole, 325, 328
poly-modal, 400
polymer, 208
polytropic process, 356
position, 5, 8, 63, 82, 84, 86–88, 110, 129, 131, 138, 150, 152, 154, 172, 189, 191, 232, 266, 268, 452, 453, 591, 636
   vector, 46, 87, 88, 107, 128
positive definite, 61, 73, 75, 80, 107, 108, 396, 497, 604
positive semi-definite, 27, 55, 107, 108, 180, 199, 200, 202, 203, 222, 224, 437, 439, 507
post-diction, 667
post-dictive, 655
post-multiplication, 32, 38, 105
potential
   energy, 171, 174, 390, 660
   field, 296
   flow, 49, 282, 296–299, 318, 324, 332, 333, 335, 523, 524, 557, 558, 560
PP10, $C_{13}F_{22}$, 438
Prandtl number, 12, 431, 433, 466, 480, 490, 491, 571, 627, 658, 660, 662
Prandtl's mixing length theory, 654
pre-diction, 667
pre-dictive, 655
pre-multiplication, 32, 38, 405
pressure, 5, 8–10, 19, 25, 157, 158, 173–178, 181, 182, 191, 196, 208, 223, 225, 230, 234, 235, 239, 241, 242, 245, 247, 252, 256, 259, 260, 271, 274, 277–279, 290, 293, 311, 313, 314, 316, 323, 324, 328, 329, 333–335, 338, 342, 348, 351, 353, 363–365, 367, 368, 372, 374, 381–386, 389, 390, 392, 393, 403, 417, 418, 435, 439, 440, 443, 444, 449, 450, 452, 453, 455–458, 462, 463, 465, 470, 502, 516, 521, 522, 524, 527, 528, 532, 533, 541, 544, 568, 570, 573, 576, 577, 585–589, 645, 646, 650, 659, 660, 666
   back, 392
   coefficient, 323, 328
   dynamic, 532
   field, 190–192, 238, 239, 241, 257, 271, 273, 280, 282, 290–294, 311, 313, 314, 324, 325, 328, 465, 480, 516, 520, 523, 524, 527, 578, 588
   force, 512
   gradient, 192, 235, 240, 256, 257, 271, 314, 315, 323, 324, 462, 463, 467, 469, 471, 473–477, 482, 508, 509, 516, 518, 533, 534, 540, 541, 544, 557, 559, 561, 565, 567, 569, 571, 577, 595, 597, 598, 625, 645, 655
   hydrostatic, 217, 658, 662, 664
   mechanical, 158, 218
   stagnation, 260, 362, 364, 456
   static, 362
   thermodynamic, 157, 158, 172, 218, 657–659
principal axes, 24, 39, 41–43, 61, 77, 104, 108, 111, 129, 218, 313
   strain rate, 100
   stress, 40, 100
printer ink, 208
projection, 614–617
   Galerkin, 614, 617, 618, 622, 625
   matrix, 137, 144, 615
pseudo-
   inverse, 38
   plastic, 208, 505, 507–509, 517
   scalar, 79, 292
   tensor, 28, 29
   vector, 34, 78, 82, 99, 292
   viscosity, 667
pump, 566, 569

$Q \cdot R$ decomposition, 110
quadratic form, 72–75, 80, 81, 104, 107, 190, 191, 199, 220, 221
quadric, 74–76, 133, 134, 144, 167, 169
   strain rate, 102, 103, 133, 134, 143, 144
   stress, 167–170, 191, 517
   surface, 72–75
quadrupole, 320, 326, 332, 335
quantity of interest (QOI), 659, 666, 667
quaternion, 299

rank, 21, 27, 105, 137
Rankine half body, 320, 321
Rankine–Hugoniot jump equations, 378–380, 387, 433, 437–439, 443, 449, 457
rarefaction, 402, 414–416, 421, 425, 438, 439
   centered, 417, 418
   centered fan, 417
   discontinuity, 386, 438, 439, 441
   fan, 418
   non-centered, 418
   noncentered, 417
   nonphysical, 439
   shock, 439
ratio of specific heats, 12, 235, 252, 277, 431, 441, 457
Rayleigh
   equation, 598–600, 610, 611
   flow, 373–375, 380, 456
   line, 380–384
   number, 251, 603, 609, 622, 627, 628, 631, 632, 635, 636, 656–659, 662–664
Rayleigh–Bénard
   flow, 600, 606–609, 611, 622, 624, 632, 644, 656–667
   turbulence, 659
Rayleigh–Taylor instability, 591, 610
reacting mixing gases, 225

real function, 304
real number, 7, 107, 591, 667
real part, 131, 300, 310, 333, 512, 513, 591, 610, 629, 632
real variable, 304
reciprocal basis, 81
reciprocal basis vector, 56
recirculating flow, 655
reduced order manifold, 638
reflection, 21, 26, 29, 34, 38, 45, 61, 75, 77–79, 82, 109, 292, 323, 400, 401
reflection matrix, 21, 29, 59, 78, 79, 82
relaxation time, 511, 512, 514, 515
repeated index, 22
residue, 327, 332
reversible
    heat transfer, 149, 178
    work, 178, 189, 223, 245, 443, 645
Reynolds
    analogy, 493, 554, 556
    equation, 541, 577
    number, 249, 251, 277, 323, 369, 521, 523, 526, 529, 533, 544, 551, 553, 598, 600, 643, 648, 649, 656, 659
    stress, 654–656, 658, 663
    transport theorem, 53, 115, 150, 152, 165, 173
Reynolds-averaged Navier–Stokes (RANS), 650, 655, 656, 662
rheology, 206
rheopectic, 208
Riemann invariant, 410, 411, 414, 417
Robin condition, 396
robotics, 19, 38
rocket, 4
    propulsion, 338
rotation, 3, 19, 21, 24–30, 38, 44, 45, 61, 77, 78, 82, 107–109, 111, 122–127, 133, 139–142, 165, 169, 170, 177, 210, 212–215, 217, 265, 271, 276, 283, 291, 484
    matrix, 21, 29, 30, 43–45, 59, 78, 79, 81, 82, 101, 104, 106, 136, 168, 206
    molecular, 158, 225
    rigid body, 3, 96–99, 112, 123, 142, 177, 209, 269, 271
    rigid-body, 129

tensor, 30, 98, 115, 144, 146
rotational, 127, 259, 269, 272, 273, 280, 336, 467, 520, 523
roundoff error, 667
row vector, 18, 27, 47, 48, 50, 51
Runge–Kutta method, 331, 547

saddle, 131, 139–141, 614, 620
scalar, 8, 9, 17–19, 24–31, 33, 34, 39, 40, 47–50, 52, 79, 89, 90, 96, 97, 100, 102, 105, 114, 145, 167, 173, 180, 217, 259, 265, 266, 282, 292–294, 296, 297, 299, 336, 396, 497, 510, 522, 622, 651, 662
    function, 46, 47, 49, 53, 80, 182, 238, 285, 334
    pseudo-, 79, 292
    triple product, 79, 81
second law of thermodynamics, 149, 179, 180, 189, 195, 197, 219, 226, 227, 386, 387, 438, 439, 608, 610, 646, 658
Seeback effect, 200
self-adjoint, 396, 497
self-similar, 441, 457, 485, 503, 504, 517, 518, 562
self-similarity, 483, 560
separated flow, 323, 655
separation, 323
separation of variables, 396, 496, 530, 604, 617
series
    Fourier, 398, 538, 615–617
    Laurent, 327
    Taylor, 113, 129, 130, 236, 267, 300, 315, 327, 335, 364, 377, 387, 391, 422, 424, 425, 491, 504, 515, 583, 584, 619, 623, 656, 668
shear, 98, 123, 124, 126, 510, 533
    deformation, 125, 127, 142
    flow, 294, 585
    strain, 112
    stress, 3, 199, 207, 208, 211, 243, 369, 468–472, 474, 477, 479, 480, 488, 495, 508–510, 520, 524, 552
    thickening, 208
    viscous, 285, 462, 481, 549, 594
shear layer, 595
    inviscid, 285, 576

laminar, 564, 566
turbulent, 643
shear-thinning, 208
shock, 5, 8, 150, 230, 244, 338, 339, 376, 377, 379, 382–387, 389, 393, 410, 425, 431, 433–435, 437, 441–444, 449, 451–453, 456, 457, 648
    compression, 438
    formation, 421
    spherical, 441
    strong, 382
    structure, 435–437
    thickness, 434, 435
    tube, 439, 440
    viscous, 427, 431, 436, 437, 457
SI units, 171, 172, 175, 176, 179, 483, 571
significant digits, 11
similar matrix, 108
similarity
    solution, 328, 449, 483–485, 488, 543, 595
    variable, 441, 450, 457, 484, 485, 491, 503, 517, 518, 546, 560, 564, 573, 578
simple waves, 414
singular, 38, 66, 71, 137, 327, 360, 421, 423, 424, 495, 502, 527, 528
singular value, 41, 55, 77, 105–107, 146
singular value decomposition (SVD), 101, 105–107, 109, 112, 146, 614
singularity, 421
sink, 131, 139–142, 308, 318, 331, 332, 614, 620, 621, 632
slip surface, 230
snowflake, 484
solenoidal, 48, 80, 116, 145, 259, 336
solid, 3, 4, 87, 185, 189, 196, 208, 216, 229, 230, 238, 239, 296, 328, 329, 339, 468, 477, 478, 510–512
    elastic, 510
    flexible, 96
    mechanics, 5, 40, 87, 153, 170, 189, 207, 216, 510
sonic, 360, 361, 365–367, 371, 372, 374, 382, 386, 388, 392, 393
    singularity, 360
    throat, 367, 368
Soret effect, 200

sound speed, 158, 247, 249, 270, 338, 340, 346–348, 354, 355, 362, 385, 394, 403, 407, 416, 434, 437–440, 447, 456, 648
   isentropic, 347, 355
   isothermal, 348
source, 131, 308, 318–320, 331, 332, 336, 339, 614, 620
space, 6
space-time, 163
spaghetti noodle, 77
specific heat, 204, 225, 226, 342, 610
   constant pressure, 12, 227, 232, 246, 431
   constant volume, 12, 225, 456
specific volume, 178, 245, 356, 443
sphere, 10, 49, 61, 73, 102, 117, 118, 141, 158, 336, 451, 529–533, 576, 577
   unit, 23, 24, 60, 77, 134, 169, 191
spherical coordinates, 72, 79, 80, 335, 336, 441, 529
spheroid
   oblate, 102
   prolate, 102
spin, 165, 166, 177, 190
spring, 510, 513
square, 577, 611
   unit, 23, 24, 77, 664
square root of a matrix, 108
Squires's theorem, 599
stability, 584
   marginal, 598
stability boundary, 591
stable, 131, 142, 580, 583, 584, 591, 592, 594, 598–600, 609, 610, 620, 623, 624, 629–635, 657, 658
   asymptotic, 580, 583, 633, 635
   equilibrium, 144, 632
   fixed point, 631, 639, 640
   limit cycle, 632, 638
   neutral, 632, 634
   sink, 131, 614, 632
   solution, 426
   spiral, 131
   stagnation point, 131
stagnation, 362, 364, 392
   density, 367
   enthalpy, 255–258
   point, 130, 131, 144, 255, 313, 321, 324, 325, 333, 335, 559, 576
   pressure, 260, 364, 456
   streamline, 313
   temperature, 456
static, 362, 364, 456
static property, 361
statistics, 643
stoichiometry, 38
Stokes equations, 521, 523, 525, 528, 533
Stokes flow, 521–525, 527, 529, 531–534, 576, 577
Stokes' assumption, 158, 159, 176, 190, 218–223, 226, 227, 230, 232, 237, 243, 246, 259
Stokes' first problem, 480, 481, 484, 487, 488, 493–495, 498, 499, 503, 516–518
Stokes' second problem, 502, 513, 515
Stokes' theorem, 51, 281
Stokesian fluid, 227
strain, 96, 207, 510
   extensional, 112, 126, 489
   shear, 112
   tensor, 510
strain rate, 96, 104, 118, 206–210, 215–218, 222, 226, 239, 313, 477, 505, 506, 510–512, 524, 528
   deviatoric, 217, 219, 221, 222
   extensional, 102, 103, 133, 134, 136, 143, 144, 167, 169
   invariants, 118, 222, 525, 529
   mean, 217, 218, 221, 222
   quadric, 102, 103
   shear, 137
   tensor, 118, 181, 210, 217, 221, 222, 506, 510
straining, 96
   extensional, 96, 99–101, 103–105, 119, 121, 133, 276, 313, 468
   shear, 96, 99–101, 119, 121, 122
strange attractor, 631, 636, 637
stratified
   flow, 236, 238, 242
   fluid, 234
streakline, 94–96, 129, 132, 142, 143, 316
stream function, 297–299, 307, 311, 312, 322, 334–336, 444, 522, 523, 526, 529–532, 545, 546, 558, 562, 564, 571, 574, 576, 577, 596, 597, 625, 626, 636
   compressible, 574
streamline, 92–96, 122–129, 132, 142, 143, 238, 254, 258, 260, 266–268, 270–272, 274, 279, 284, 296, 298, 299, 307–311, 313–317, 319–322, 332, 333, 335, 524, 526, 527, 530–532, 542, 549, 550, 558, 564, 565, 576, 578
stagnation, 313
stress, 19, 40, 96, 100, 118, 154–159, 161, 162, 167–169, 175, 176, 179, 182, 190, 198, 206–209, 215–219, 221, 226, 229, 232, 237, 244, 271, 273, 280, 350, 371, 469, 470, 473, 477, 488, 495, 502, 505, 506, 510, 511, 517, 532, 541, 552, 561, 576, 577, 588, 601, 605, 609, 626, 654
   deviatoric, 158, 217, 222
   mean, 158, 222
   principal, 40
   shear, 3, 199, 207, 208, 211, 243, 369, 468–472, 474, 477, 479, 480, 488, 495, 508–510, 520, 524, 552
   shearing, 469
   tensor, 9, 17, 40, 114, 153, 156–158, 166–170, 180, 190, 191, 206, 209, 216–218, 227, 237, 292, 477, 478, 507, 517, 522, 524, 528
   thermodynamic, 157, 222
   viscous, 9, 17, 157, 158, 206, 216, 217, 227, 237, 477, 507, 517, 522
   yield, 339
stretching, 61, 107, 276
   linear, 61
   normal, 140–142
   tangential, 140
strong form, 181
strong shock limit, 385, 441
subsonic, 365, 372, 374, 384, 392, 393
substantial derivative, 89
supercomputer, 662
superposition, 95, 282, 286, 309, 318, 320, 397, 531, 538
supersonic, 230, 338, 365, 366, 368, 372, 374, 384, 386, 393, 409, 438
supremum, 41
surface
   couple, 166, 190

force, 9, 153, 154, 156, 160, 161, 173, 175, 176, 185, 190, 463, 521
   free, 601
   integral, 50, 51, 53, 79, 150, 159, 179, 185, 329, 645
   Lamb, 254, 258, 260
   material, 53
   tension, 207, 229, 585–588, 592–595, 610
   wave, 593, 610
symmetric, 29
symmetric matrix, 72, 170, 198
symmetry, 484
synthetic division, 318

tangential stretching, 140
Taylor scraper, 526, 527, 529, 577
Taylor series, 113, 129, 130, 236, 267, 300, 315, 327, 335, 364, 377, 387, 391, 422, 424, 425, 491, 504, 515, 583, 584, 619, 623, 656, 668
Taylor–Proudman theorem, 192
Taylor–Sedov, 441, 450, 458
telegraph equation, 512–514
temperature, 10, 13, 17, 19, 25, 177, 194, 200, 205, 206, 208, 223, 225–227, 229, 230, 232, 234–236, 239, 241, 242, 259, 338, 348, 355, 363–365, 367, 368, 372, 374, 386, 389, 391, 392, 417, 431, 434, 435, 437, 439, 456, 457, 462, 464, 466, 468, 469, 476, 479, 480, 489–493, 552, 555–557, 569–573, 578, 580, 600–602, 604, 606–611, 613, 625, 626, 628, 631, 636, 646, 652, 657, 658, 660–665
   absolute, 12, 149, 178, 181, 219
   field, 194, 227, 463, 468, 472, 493, 553, 554, 556, 636, 663
   gradient, 195, 197, 198, 200–202, 204, 205, 425, 437, 461, 636
   stagnation, 456
tensor, 8, 9, 17, 18, 27–35, 38–43, 45–50, 53, 75, 78, 79, 114, 118, 138, 152, 154, 157, 169, 201, 202, 209, 210, 215, 217, 238, 292, 299, 313, 524, 644
   algebra, 211
   anti-symmetric, 30, 31, 99, 202, 254, 644

   asymmetric, 44, 51
   Cartesian, 58
   deformation, 98, 103, 129, 134, 136, 140, 143, 144, 206, 207, 217, 219, 220, 227, 313, 335, 468, 489, 510, 522, 524, 528, 549, 644
   diagonal, 217
   field, 51
   inner product, 31–33, 99, 166, 180, 202, 238, 253, 254, 256, 603, 644
   invariant, 506
   invariants, 38, 44, 45
   isotropic, 28, 29
   metric, 55, 57, 59, 67, 71, 80, 107–109
   norm, 41
   product, 30, 31
   pseudo-, 28, 29
   rotation, 30, 98, 115, 144, 146
   second order, 27
   strain, 510
   strain rate, 118, 181, 210, 217, 221, 222, 506, 510
   stress, 9, 17, 40, 114, 153–158, 166–170, 180, 190, 191, 206, 209, 216–218, 227, 237, 292, 477, 478, 507, 517, 522, 524, 528
   substitution, 23
   surface couple, 190
   symmetric, 29, 31, 33, 41, 42, 99, 101, 202, 210, 254, 510, 644
   thermal conductivity, 202–204, 206
   velocity gradient, 97, 105–107, 109, 115, 143, 144, 313, 469
   viscosity, 209, 210
tetrahedron, 156
thermal
   conductivity, 12, 13, 194, 201–204, 206, 227, 230, 232, 236, 246, 349, 389, 476, 542, 569, 610
   diffusivity, 12, 240, 241, 490, 554, 613, 614
   energy, 171, 173, 175, 176, 179, 181, 188, 200, 239, 338, 354, 359, 365, 389, 441, 464, 471, 473, 478, 491, 580, 601, 602, 608, 609, 657
   equation of state, 223, 225, 340, 341, 346, 347
   expansion coefficient, 233

thermal expansion coefficient, 572
thermocouple, 200
thermodynamic, 25, 141, 157, 178, 222, 223, 225, 232, 233, 235, 243, 244, 247, 339, 342, 343, 346, 349, 355, 356, 437, 438, 456, 476, 506
   derivative, 257
   function, 403
   potential, 232
   property, 171, 223, 340, 347, 355, 362, 388, 403, 438
   state, 390
thermodynamics, 9, 145, 148, 157, 171, 174, 178, 180, 199, 232, 245, 320, 339, 340, 348, 349, 443
   first law, 149, 171, 175, 178, 179, 191, 205, 234, 238, 257, 258, 260, 352, 353, 404, 451
   nonequilibrium, 197, 207, 232
   second law, 149, 179, 180, 195, 197, 219, 226, 227, 386, 387, 438, 439, 608, 610, 646, 658
thixotropic, 208
Thomson effect, 200
three body problem, 86
throat, 367, 374, 376, 393, 456
   sonic, 367, 368
throttling device, 390
time, 3, 5–11, 35, 46, 61–63, 86, 87, 90, 93, 94, 97, 110, 113, 114, 129, 135, 138, 141–144, 148–154, 161, 165, 171, 172, 180, 182, 189, 208, 222, 223, 232, 234, 247, 248, 251, 253, 279, 281, 293, 307, 309, 311–314, 350, 351, 378, 395, 398, 400, 416, 422, 424, 426, 427, 436, 444, 453, 456, 458, 461, 487–489, 494, 495, 498, 499, 505, 511, 513, 516, 518, 542, 580, 583, 584, 589, 591, 592, 594, 596, 605, 610, 613, 614, 620–622, 624, 626, 632, 635–638, 640, 643, 647, 650, 653, 658, 659, 661, 662, 664–666
   -average, 650–654, 666
   derivative, 16, 52, 63, 64, 86, 115, 116, 135, 162, 164, 223, 224, 246, 256, 287, 455, 482, 490, 510–512
   relaxation, 511, 512, 514, 515
TNT, $C_6H_2(NO_2)_3CH_3$, 453

Index

toothpaste, 208
topology, 260
torsion, 138, 143
total
  derivative, 89
  differential, 47, 87–89, 97, 129, 298, 419
  energy, 171, 172, 191, 353, 442, 451
  enthalpy, 257, 258, 371, 372, 374, 574
  trace, 34, 158
traction, 154–156, 159, 169, 170, 191, 227, 528
trajectory, 140, 631, 636
  heteroclinic, 139–142
transcritical bifurcation, 639
transformation, 7, 14, 16, 17, 19, 25, 26, 29, 34, 54, 61, 64, 66, 68, 70, 71, 74, 78–81, 87, 88, 90–92, 99, 106, 110, 111, 153, 163, 181, 196, 205, 211, 226, 364, 379, 412, 419–424, 441, 445, 483, 484, 496, 503, 525, 538, 545, 563, 574, 578
  alias, 19, 78
  alibi, 19, 78, 81
  coordinate, 19, 53, 59, 64, 70, 82, 87, 89, 90, 106–109, 226, 411, 420, 421, 423, 424, 427, 428
  Fourier, 494
  Galilean, 6, 162–164, 196, 197, 201, 353, 369, 379, 389, 428, 431, 531
  gauge, 145
  general, 54, 87, 88, 419, 483
  inverse, 23, 68, 70, 74, 75, 428
  linear, 60, 61, 77, 81, 145, 192
  reflection, 29, 78, 79, 82, 292, 484
  rotation, 25, 28, 29, 53, 54, 78, 79, 82, 168, 211, 213–215
  rule, 64
  similarity, 417, 485, 542, 546
  stretching, 24, 191, 484
  symmetry, 484
  von Mises, 574
translation, 95, 142
transport phenomena, 188
transpose, 7, 18, 20, 21, 26, 29, 39, 47, 48, 50, 57, 58, 90, 92, 97, 101, 106, 160, 524
triangle inequality, 27

Trinity site, 453
Trkalian flow, 293
truncation error, 666, 667
turbulence, 84, 129, 624, 636, 642–644, 646, 647, 649, 652, 655, 656, 658, 659, 667, 668
  closure problem, 654, 668
  model, 643, 655
turbulent, 277, 369, 551, 624, 643–647, 649, 650, 654–656, 658, 659, 662, 664, 666–668
  convection, 657
  fluctuation, 646
  kinetic energy, 649, 654, 655
  mixing, 655

universal gas constant, 12
unstable, 131, 142, 580, 584, 591–594, 598–600, 623, 624, 629–631, 633, 636, 638
upper convected derivative, 115

vacuum, 10, 458
validation, 11, 646, 649, 667
valve, 390
van der Waals gas, 339, 342, 343, 346, 438–441, 457
vanishing viscosity solution, 426
vapor dome, 339, 437, 438, 440
vector, 3, 8, 9, 16–20, 23–28, 32, 34–39, 41, 42, 46–49, 52, 55–59, 61–64, 78, 79, 81, 82, 92, 97, 99, 101, 107, 108, 110–114, 118, 110, 120, 143, 146, 153, 154, 157, 168, 169, 172, 173, 195, 226, 252, 254, 255, 266, 267, 273–277, 279, 280, 297, 299, 313, 317, 405, 410, 423, 489, 517, 522, 524, 528, 549, 590, 597, 614, 615, 619
  acceleration, 62, 64, 65, 85, 86, 129, 130, 132, 143, 144, 161, 190, 259, 313, 314, 316, 335, 489, 570, 576
  algebra, 25
  angular momenta, 164
  axial, 82, 99
  basis, 9, 17, 18, 21, 53, 55, 56, 58, 62, 63, 81, 92, 615
  binormal, 138
  body force, 161
  calculus, 50, 145

  Cartesian, 58
  column, 7, 18, 20, 21, 36, 48, 50, 56, 105
  contravariant, 53, 61
  covariant, 53
  cross product, 34
  displacement, 510
  distance, 106, 110, 111, 165
  dual, 32, 79, 99, 139, 146, 166
  field, 48, 71, 80, 82, 119, 145, 161, 162, 190, 259, 292, 336, 549, 578, 614
  force, 172, 190
  function, 47, 80
  heat flux, 172, 179, 194, 195, 201, 202, 204, 205, 227, 470
  inner product, 25
  Lamb, 52
  linear momenta, 153
  norm, 26, 41
  normal, 140, 170, 528
  polar, 82
  position, 46, 86–88, 107, 128
  pressure gradient, 314, 315
  product, 42
  pseudo-, 34, 78, 292
  reciprocal, 56
  reciprocal basis, 56
  row, 18, 27, 47, 48, 50, 51
  tangent, 134
  traction, 154, 159, 169, 170, 191, 227
  unit, 35, 37, 39, 51, 55, 96, 100, 102, 107, 134, 136, 130, 140, 284, 317, 603
  velocity, 8, 52, 62, 70, 82, 84, 92, 97, 99, 106, 107, 110, 111, 118, 128, 134, 135, 143–145, 151, 161, 172, 175, 190–192, 227, 253, 254, 266, 268, 280, 282, 283, 286, 287, 290–294, 297, 310, 312–314, 334, 335, 510, 513, 542, 549, 550, 576, 578, 585, 624
  vorticity, 99, 129, 133, 143, 144, 265, 276, 292
velocity, 3, 5, 6, 19, 25, 53, 61, 62, 64, 70, 82, 84–86, 89, 91, 93, 95–99, 101–103, 106, 107, 110–114, 116, 117, 119, 123, 128–131, 134, 136, 138, 143–145, 148, 150, 152, 154,

156, 158, 161, 162, 165, 172,
182, 183, 190–192, 196, 197,
199, 205, 207, 208, 223, 227,
230, 234, 236, 239, 240, 247,
249, 253, 254, 258–260, 266,
269, 270, 272, 273, 281–285,
287, 290, 292–294, 297,
307–312, 315, 316, 321, 324,
333, 335, 336, 338, 352, 354,
360, 365, 367–369, 372, 374,
379–381, 384, 385, 387–390,
394, 395, 398, 406, 415, 417,
418, 425, 427, 428, 442–444,
449, 450, 452–454, 456–458,
461–463, 466–474, 476, 478,
480, 481, 484, 487, 493, 495,
498, 501–505, 507–509,
515–518, 520, 521, 523, 526,
527, 529–532, 534, 536, 539,
540, 542, 543, 545, 549, 551,
554, 559, 561, 562, 564–566,
568–573, 576–578, 588, 594,
595, 598, 599, 611, 646, 647,
650, 653–655, 660, 662, 663, 668
advection, 662
angular, 64, 192, 276, 565, 576,
582
field, 152
gradient, 95, 97, 99–101, 105–107,
109, 111, 113–115, 131, 133,
139, 142–144, 152, 177, 180,
181, 198, 207, 313, 461, 469,
488, 502, 512, 520, 553, 588, 654
group, 610
phase, 610
potential, 190, 253, 282, 285, 286,
290, 297, 307, 309, 324,
333–336, 444, 585
rotation, 99
tangential, 230
vector, 8, 52, 62, 82, 84, 92, 97,
99, 106, 107, 110, 111, 118, 128,
134, 135, 143–145, 151, 161,
172, 175, 190–192, 227, 253,
254, 266, 268, 280, 282, 283,
286, 287, 290–294, 297, 310,
312–314, 334, 335, 510, 513,
542, 549, 550, 576, 578, 585, 624
verification, 11, 297, 436, 646, 666,
667
solution, 11, 664
vibration

molecular, 158, 225
virial gas, 348
viscoelastic, 114, 208, 510, 512, 513,
516, 518
viscosity, 12–14, 207, 208, 226, 230,
232, 236, 237, 239, 243, 246,
259, 268, 278, 282, 292–294,
319, 350, 398, 425, 426, 431,
439, 456, 470, 476, 479, 480,
502, 505, 507, 516–518, 521,
522, 542, 564, 569, 574, 576,
577, 606, 642, 643, 645, 668
bulk, 216, 217
dilatational, 216
dynamic, 12, 431
eddy, 654–656, 658
first coefficient, 207, 216
kinematic, 12, 240, 241, 278, 514,
551, 576, 647, 648
matrix, 212–215
pseudo-, 667
second coefficient, 216
shear, 216
tensor, 209, 210
viscous
dissipation, 180, 181, 219, 221,
227, 237, 239, 241, 464, 467,
478, 521, 552, 555, 645, 659
flow, 49, 255, 296, 323, 338, 425,
502, 542, 576, 578, 595, 600
force, 173, 176, 181, 189, 226, 394,
520, 552, 558, 568, 576, 601,
657
shock, 431, 436, 437, 457
stress, 9, 17, 157, 158, 206, 216,
217, 227, 237, 477, 507, 517,
522
volume, 4, 7, 8, 10, 13, 21, 51, 53,
79, 81, 104, 115, 117, 118,
126–128, 135, 136, 140, 141,
149, 153, 156, 157, 165, 166,
173, 182–184, 189, 245, 260,
277, 313, 343, 350, 377, 378,
381, 451, 481, 549, 562, 645,
646
-preserving, 21, 66, 71, 78, 88
arbitrary, 50, 115–117, 182, 183,
185
control, 52, 181, 183, 205, 329,
350, 351, 353
fixed, 53, 182, 183, 188, 191, 377,
645

integral, 50, 79, 150, 154, 159,
179, 185, 645
material, 53, 115–117, 136, 149,
150, 152, 154–156, 159, 165,
171, 172, 183, 191, 281
molecular, 225
phase, 639
specific, 178, 245, 356, 443
volumetric flow rate, 308, 474, 475
von Kármán's viscous pump, 566,
569
von Mises transformation, 574
vortex, 3, 145, 269, 273, 283, 284,
286, 288–290, 293, 294, 309,
335, 501, 505
image, 284
irrotational, 128, 272, 285, 287,
501–504
line, 145, 254, 266–268
point, 272, 282, 285, 289–292,
309, 310, 332, 335, 502
rotational, 272, 273
sheet, 284–286
stretching, 277
tube, 275–277
vorticity, 82, 99, 119, 128, 129, 133,
143–145, 253, 254, 258, 260,
265–267, 272–282, 291–294,
296, 297, 299, 336, 467, 468,
480, 488, 489, 504, 505, 522,
523, 526, 527, 529, 549, 550,
576, 577, 643, 644

wake, 265, 643
water, 183, 184, 208, 235, 271
wave, 5, 8, 207, 364, 365, 376, 378,
384, 390, 398, 400, 401, 421,
422, 427, 430, 431, 437, 438,
501, 512, 515, 589, 605, 610,
617, 622, 658
acoustic, 158, 408, 412, 437
blast, 441, 450, 453, 455, 456, 458
compression, 437, 438
elastic, 512
equation, 394, 395, 399–402, 410,
512
expansion, 402, 441
isentropic, 438
nondiffusive, 398
nondispersive, 398
number, 501, 513, 591, 593, 595,
597, 598, 604, 605, 610, 649

rarefaction, 402, 438
shock, 150, 338, 339, 376, 383, 384, 387, 389, 425, 441, 442, 451, 456
simple, 414
sonic, 386
sound, 236
speed, 380–382, 384, 401, 429, 438, 589, 594
standing, 398–400, 513
surface, 593, 610
traveling, 372, 390, 589
waveform, 400, 427
wavelength, 592, 594, 605, 628, 661
weak form, 180
weak solution, 426
weather, 3, 241, 265, 338, 592, 608, 640
weighted essentially non-oscillatory (WENO), 661
weighting function, 618
wiggles, 661
wine, 668
work, 149, 171–174, 176, 178, 189, 240, 344, 346, 353, 390, 471, 473
    frictional, 239
    irreversible, 189
    rate, 172
    reversible, 178, 189, 223, 245, 443, 645
    viscous, 181, 188, 471, 473, 490

Young's modulus, 207